VIIA	VIII
							Hydrogen 1 **H** 1.0079	Helium 2 **He** 4.0026
			IIIA	IVA	VA	VIA		
			Boron 5 **B** 10.811	Carbon 6 **C** 12.011	Nitrogen 7 **N** 14.0067	Oxygen 8 **O** 15.9994	Fluorine 9 **F** 18.9984	Neon 10 **Ne** 20.1797
	IB	IIB	Aluminum 13 **Al** 26.9815	Silicon 14 **Si** 28.0855	Phosphorus 15 **P** 30.9738	Sulfur 16 **S** 32.066	Chlorine 17 **Cl** 35.4527	Argon 18 **Ar** 39.948
Nickel 28 **Ni** 58.69	Copper 29 **Cu** 63.546	Zinc 30 **Zn** 65.39	Gallium 31 **Ga** 69.723	Germanium 32 **Ge** 72.61	Arsenic 33 **As** 74.9216	Selenium 34 **Se** 78.96	Bromine 35 **Br** 79.904	Krypton 36 **Kr** 83.80
Palladium 46 **Pd** 106.42	Silver 47 **Ag** 107.8682	Cadmium 48 **Cd** 112.411	Indium 49 **In** 114.82	Tin 50 **Sn** 118.710	Antimony 51 **Sb** 121.75	Tellurium 52 **Te** 127.60	Iodine 53 **I** 126.9045	Xenon 54 **Xe** 131.29
Platinum 78 **Pt** 195.08	Gold 79 **Au** 196.9665	Mercury 80 **Hg** 200.59	Thallium 81 **Tl** 204.3833	Lead 82 **Pb** 207.2	Bismuth 83 **Bi** 208.9804	Polonium 84 **Po** (209)	Astatine 85 **At** (210)	Radon 86 **Rn** (222)

Europium 63 **Eu** 151.965	Gadolinium 64 **Gd** 157.25	Terbium 65 **Tb** 158.9253	Dysprosium 66 **Dy** 162.50	Holmium 67 **Ho** 164.9303	Erbium 68 **Er** 167.26	Thulium 69 **Tm** 168.9342	Ytterbium 70 **Yb** 173.04	Lutetium 71 **Lu** 174.967
Americium 95 **Am** (243)	Curium 96 **Cm** (247)	Berkelium 97 **Bk** (247)	Californium 98 **Cf** (251)	Einsteinium 99 **Es** (252)	Fermium 100 **Fm** (257)	Mendelevium 101 **Md** (258)	Nobelium 102 **No** (259)	Lawrencium 103 **Lr** (260)

Chemistry for Engineers and Scientists

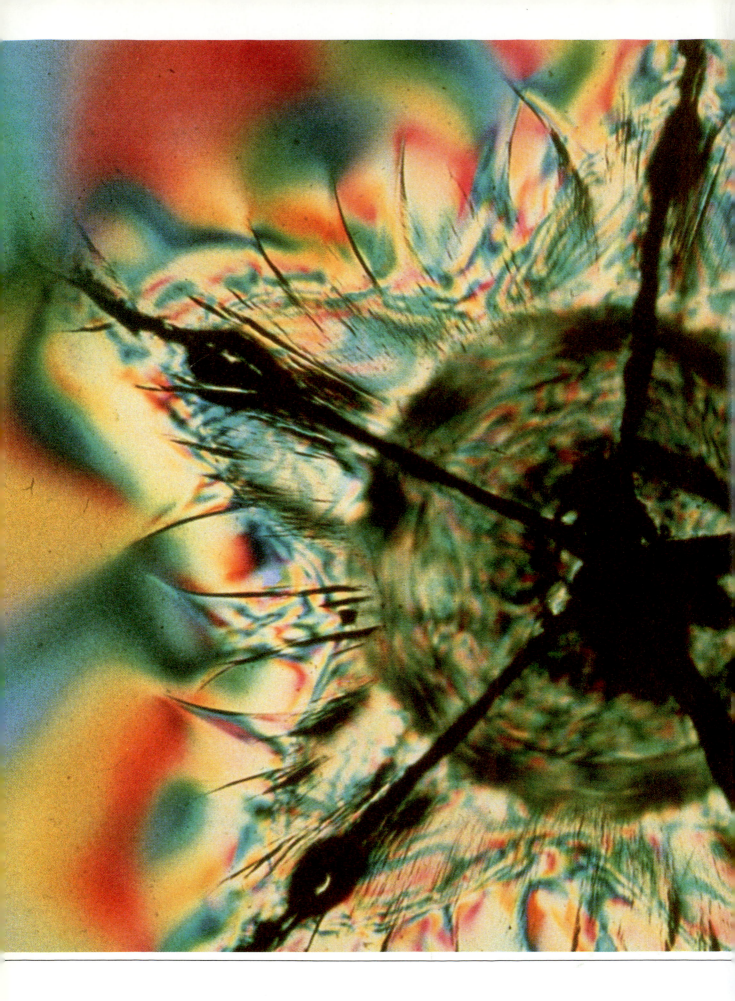

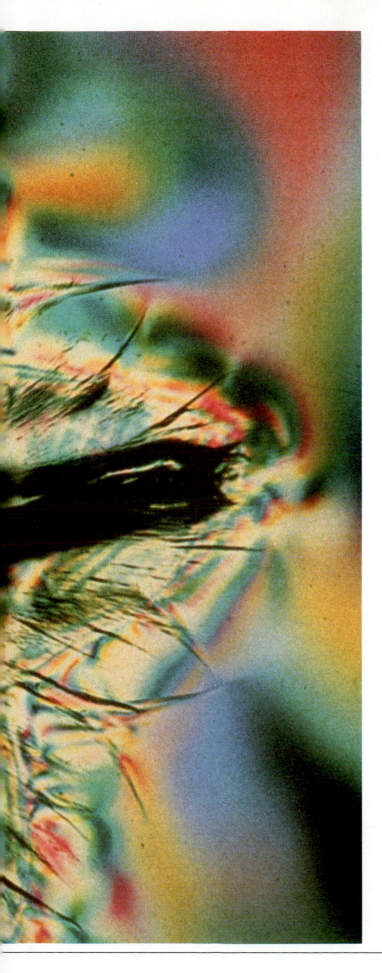

Chemistry for Engineers and Scientists

LEONARD W. FINE
Columbia University

HERBERT BEALL
Worcester Polytechnic Institute

SAUNDERS COLLEGE PUBLISHING
Philadelphia   Chicago   Fort Worth
San Francisco   Montreal   Toronto
London   Sydney   Tokyo

A LEXAN polycarbonate shield can stop a .357 magnum bullet.
Polarized light shows stresses along the fracture lines.

Text typeface: Baskerville
Compositor: General Graphic Services
Acquisitions Editor: John Vondeling
Developmental Editor: Sandi Kiselica
Managing Editor: Carol Field
Project Editor: Marc Sherman
Copy Editor: Jay Freedman
Manager of Art and Design: Carol Bleistine
Art and Design Coordinator: Doris Bruey
Text Designer and Layout Artist: Ed Butler
Cover Designer: Lawrence Didona
Text Artwork: J & R Art Services, Inc.
Director of EDP: Tim Frelick
Production Manager: Bob Butler

Cover credit: © Andrew Davidhazy, Rochester Institute of Technology

Printed in the United States of America

CHEMISTRY FOR ENGINEERS AND SCIENTISTS

ISBN 0-03-021537-4

Library of Congress Catalog Card Number: 89-043042

7 8 9 0 1 2 3 4 5 6   032   9 8 7 6 5 4 3 2

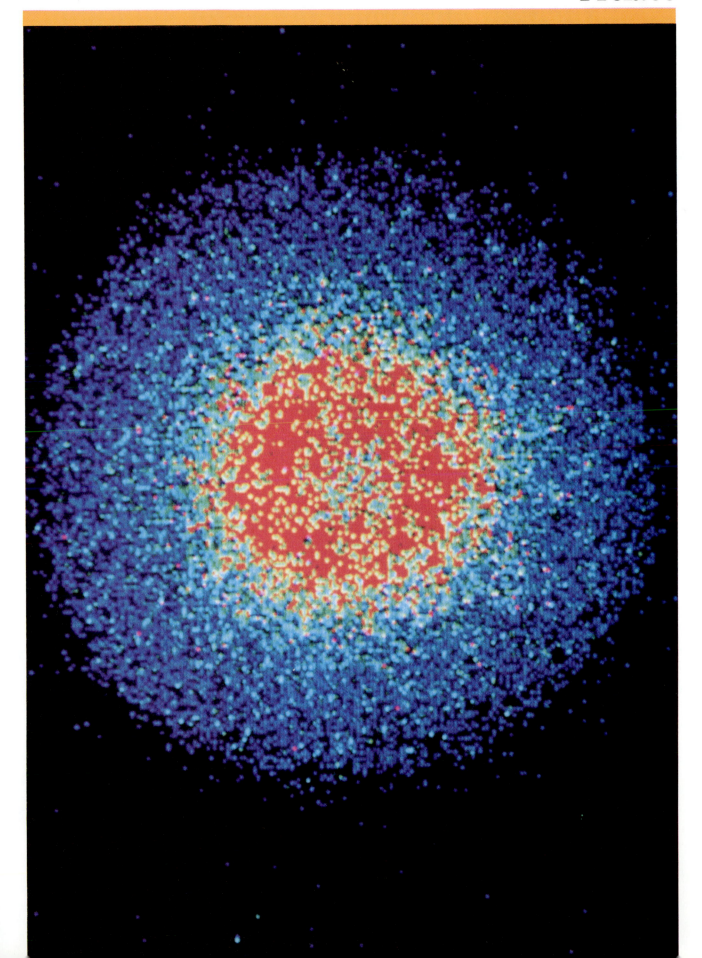

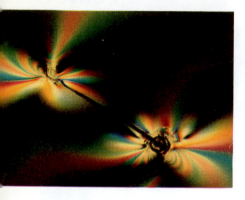

INTRODUCTION

Our textbook of general chemistry has been written for students studying engineering and science. It represents an introduction to the fundamentals of chemistry and applies those principles to relevant technological examples. We have been guided by the principle that all science students in this day and age will benefit from an exposure to engineering practices, such as chemical engineering, biomedical engineering and materials science; and unquestionably, the more science the practicing engineer learns, the better.

Toward this end, we have included cutting-edge developments in science and technology as recurring themes and have featured product and process chemistry throughout. Historical biographies of famous scientists are sprinkled throughout the book to give the text a human perspective. Nobel prizes in chemistry, in physics and in medicine and physiology have been used as benchmarks, where they are appropriate to the matter under discussion.

Our treatment of the subject is suitable for the first course in chemistry. A working knowledge of algebra is assumed, but calculus is not. Tables and graphs are used extensively. Line drawings have traditionally played an important role in illustrating how and why things work in science and technology, and we have tried to improve on what has been done before. The emphasis throughout is on simplicity and clarity. Photographs have been thoughtfully and purposefully selected to clarify or enhance ideas.

The manuscript for this textbook has been tested by more than 3000 students over three years and in five different situations at Worcester Polytechnic Institute and Columbia University. We believe that this final product will prove both interesting and challenging to the student and rewarding and effective for the teacher.

ORGANIZATION

The book is divided into two parts: **States of Matter, Energy, and Chemical Change**; and **Atomic & Molecular Structure, and Chemical Reactivity**. In the first part, you will find basic definitions, stoichiometry and calculations; descriptive chemistry of the atomic nucleus; a working knowledge of inorganic structure and reactions; gases, liquids, and solutions; equilibria and equilibrium thermodynamics. The second part includes principles of atomic structure; chemical bonding and periodic properties; applications of bonding principles to coordination chemistry, organic chemistry and biochemistry; solid state and materials science; electrochemistry and reaction kinetics. In the first half of the book we concentrate on phenomena that can be directly observed, and then we progress to the somewhat more conceptually difficult topics of atomic and molecular structure in the second half. Structure and reactivity are recurring themes throughout the text, and wherever possible, principles are reduced to practice—as early as Chapter 1, for example, you will find such topics as superconductivity introduced and discussed. On a chapter-by-chapter basis, our textbook is organized as follows:

Chapters 1 and **2** offer a classical introduction to atomic theory and the molecular hypothesis, the chemical formula and the chemical equation, and stoichiometric relationships.

Chapter 3 provides an early introduction of the properties of the nucleus, nuclear chemistry and radioactivity, but discussion of decay processes and "dating" have been deferred. We have chosen to place this chapter early in the text. It is a subject which our students normally find very interesting, and we believe it is especially important to all engineering and science students.

Chapters 4 and **6** present descriptive material on the earth and the atmospheric environment, and afford an opportunity to introduce several important substances (NaOH and H_2SO_4), classes of compounds (acids, bases, salts, oxides, metals and nonmetals), essential themes (the periodic table and Lewis structures), and industrial processes (the contact process and synthesis gas). These chapters are highly suggestive of the way we have chosen to integrate theory and practice throughout the textbook.

Chapters 5, **8**, **9**, and **19** comprise the states of matter: gases are introduced first in Chapter 5. Liquids and solutions are discussed in Chapters 8 and 9. Liquids is a more provocative topic than might be expected because students *think* the liquid state is well understood. Colligative properties make for the central theme of the chapter on solutions; a brief discussion of the colloidal state is included. Solid state (in Chapter 19) is increasingly important in this day and age, and we have treated the subject accordingly. It has been placed in the second half of the book, in conjunction with an extensive treatment of materials in Chapter 20.

The arrangement of thermodynamics and equilibrium is intended to emphasize the interdependence of these topics. **Chapter 7** introduces the concepts of energy, enthalpy, and thermochemistry necessary for the next two chapters on liquids and solutions. In these chapters the concept of phase equilibrium prepares the student for chemical equilibrium and the equilibrium constant in **Chapter 10**. Then, entropy and free energy are introduced in **Chapter 11** and the relationship between the equilibrium constant and free energy can be shown. Finally, in **Chapters 12** and **13**, thermodynamic concepts are utilized for further discussion of equilibria in water solution.

Part Two begins with **Chapters 14** and **15**, which deal with the electronic structure of the atom and modern atomic theory. Properties of electromagnetic radiation, the photoelectric effect and the quantum hypothesis, and the Bohr atom model (in Chapter 14) set the stage for the quantum mechanical atom (in Chapter 15). Important experiments are stressed. There is a philosophical undercurrent, tempered by state-of-the-art applications such as observation of quantum "jumps" and laser light.

Chapters 16 and **18** explore the nature of the chemical bond. **Chapter 17** is an introduction to the descriptive chemistry surrounding the periodic table and periodic law. Bonding theories are stressed, along with the relationships between electronic structure and observed properties. **Chapter 22** extends the discussions of bonding to coordination chemistry—a colorful chapter.

The two chapters on solid state (**Chapter 19**) and materials science (**Chapter 20**) relate structure to properties, and, we believe, constitute a unique treatment of the subjects for any introduction to college chemistry. The chapter on materials science includes metals, ceramics, and plastics in theory and practice, and includes an introduction to physical measurement. Engineering students should welcome both chapters, and science students should also be pleased as they recognize so much of what they see in the world around them, explained as part of this chemistry chapter.

Chapter 21 blends theory and practice in presenting electrochemical principles. Electron transfer at interfaces, batteries, and corrosion follow nicely on the heels of solid state and materials science in preceding chapters. Thermodynamic arguments are used to develop the main themes.

In **Chapters 23** and **24** on chemical kinetics, we have separated the discussions of mechanisms and theories (Chapter 24) from experimental rate laws, rate constants and reaction order (Chapter 23). With this two-chapter format, professors have the option of presenting a brief overview of kinetics or a more in-depth treatment of the topic. In Chapter 23, we conclude our discussions of radioactivity from Chapter 3, with a presentation of first order kinetics, decay processes and "dating."

Chapters 25 and **26** are a broad survey of topics in organic and biochemistry, concluding with an introduction to the important topic of recombinant DNA technology.

FEATURES

- Opening each of the 26 chapters is a **two-page photographic spread** designed to highlight an important contemporary theme and an element of historical perspective in keeping with the contents of the chapter. The Nobel prize is used throughout the text to further validate the importance of applications and the perspective of history by noting the work of laureates in chemistry, physics, and medicine and physiology that have proved to be especially important to the work being described in the chapter.

- As each chapter begins, there are a **few paragraphs of introduction**. This serves to put the chapter into perspective regarding its importance in chemistry and its relationship to previous chapters.

- **Technological applications** of chemistry of special interest to scientists and engineers are included in the text in appropriate places. These are not "boxed" or otherwise separated; they are integral parts of the presentation. A partial sampling of the applied topics that are covered include:

    Superconductors and superconductivity (Chapter 1)

    Nuclear reactors and nuclear medicine (Chapter 3)

    Radiocarbon dating and radiation chemistry (Chapters 3, 5, and 23)

    Chemistry of the high atmosphere, the greenhouse effect, and the chemistry leading to deterioration of the ozone layer (Chapter 6)

    Coal, oil, and energy resources (Chapter 7)

    Case study of an aluminothermic reduction (Chapter 7)

    Fractional distillation (Chapter 8)

    Colloid science (Chapter 9)

    Excited states, laser chemistry, and molecular beams (Chapter 15)

    Modern instrumental methods and diagnostic techniques, including infrared spectroscopy, magnetic resonance imaging and NMR, ESCA and Auger, mass spectrometry (Chapter 15 and several others)

    Liquid crystals, semiconductors, and solid state devices (Chapter 19)

    Industrial diamonds and synthetic gemstones (Chapter 19)

    Nylon and polycarbonate polymers—engineering plastics (Chapters 20 and 25)

    Corrosion, electrochemical manufacturing, and fuel cells (Chapter 22)

    Homogeneous and heterogeneous catalysis (Chapter 24)

    Cyclodextrins as enzyme models (Chapter 26)

    PCBs and recombinant DNA technology (Chapter 26)

- **Industrial processes** are "boxed" in order to set them apart from the body of the text. We have tried to improve upon the traditional engineering drawing with simplified process and product descriptions. Here is a sampling:

  contact process, for sulfuric acid (Chapter 4)

  synthesis gas preparation of methanol, a commodity chemical (Chapter 6)

  fertilizer ammonia—Haber process (Chapter 10)

  nitric acid—Ostwald process (Chapter 13)

  electrolytic aluminum—Hall-Heroult Process (Chapter 21)

  chloralkali—Hooker process (Chapter 22)

  paper chemistry—Kraft process (Chapter 25)

- **Examples and exercises** are vital to the flow of the text and are generally presented in pairs. First, an example is worked out in detail and then an exercise is presented to the student to affirm his or her understanding, accompanied by a validating answer.

- Since we cannot take you all on field trips to literally "see" chemistry firsthand, the next best thing is photography. To that end, **500 color photographs**, in many cases obtained from primary sources—from the people who reported the research and did the work being illustrated—have been carefully placed throughout. Historical photographs have been selected from archival materials to help portray something of the human dimension—especially scientists as young men and women, at about the age of those most likely to be reading this text.

- At the end of each chapter, there are generally 15 to 30 **QUESTIONS**, designed to help students review the material and keep them thinking about what it means.

- Following the questions are four categories of **PROBLEMS**, ranging from easy through medium difficulty to some which are quite challenging. Answers to the odd-numbered problems appear in Appendix A:

  Paired problems—one answered, followed by another that is similar, but not identical—are divided according to the main topics of the chapter.

  Additional problems are a random selection covering all the topics of the chapter.

- **Multiple Principles** problems require the student to recognize more than one essential concept to solve the problem. The principles involved often come from previous chapters.

- **Applied Principles** problems are centered around an economic or industrial element or theme—a reduction of principles to practices.

An intensive and extensive reviewing process was carried out to ensure that the information in this book is as **error-free** as possible. In addition to the many reviewers who read early drafts of this manuscript, the galleys and page proofs were read and checked by the authors and two external reviewers, John Stuehr (Case Western Reserve University) and Dorothy Swain (Oxford College).

ALTERNATE CHAPTER PRESENTATIONS

One of the important aspects of the ordering of topics in this text is the early placement of **nuclear chemistry**. Another is the concentration of observable phenomena (**macroscopic properties**) in the first half of the text and the concentration of theoretical concepts derived from observations (**microscopic properties**) in the second half. This is an order of presentation which we believe is attractive to students of engineering and science and which worked quite successfully at WPI and Columbia. However, since we realize that other valid points of view exist, this textbook has been written (and tested) in a flexible manner so that considerable rearrangement of topics is possible, if desired.

Alternate orders of topics which gave early treatment to the microscopic properties, atomic structure and bonding, were tested in full-year courses over the three years of preliminary versions of this book. The order of topics which was found to be successful for the first semester was:

Chap. 1   The Science of Chemistry
Chap. 2   The Atomic Theory and Chemical Stoichiometry
Chap. 4   Elements, Compounds, and the Earth
Chap. 5   Gases and the Pressure of the Atmosphere
Chap. 6   Chemistry in the Atmospheric Environment
Chap. 8   Liquids and Changes of State
Chap. 9   Properties of Solutions and the Colloidal State
Chap. 3   The Atomic Nucleus
Chap. 14 Atomic Structure I: Theories of the Atom
Chap. 15 Atomic Structure II: Atomic Structure and Quantum Theory
Chap. 16 Bonding I: The Properties of Bonds
Chap. 17 Main Group Elements; Periodic Properties
Chap. 18 Bonding II: Molecular Structure

For the second semester of these trial presentations, three different modifications were tested and proved to be equally satisfactory. Each emphasized different topics and used a different order of chapters. These sequences were:

Emphasis on the solid state and materials:
Chapters 7, 10, 11, 12, 13, 19, 20, 21, 22, 23, 24, 25, 26

Emphasis on organic chemistry and biochemistry:
Chapters 21, 25, 26, 7, 10, 11, 12, 13, 22, 23, 24, 19, 20

Emphasis on materials and organic chemistry:
Chapters 21, 25, 26, 7, 10, 11, 12, 13, 19, 20, 22, 23, 24

Another order of topics, which was tested, clearly separated macroscopic and microscopic properties. In this case the order of chapters was:

Macroscopic properties:
Chapters 1, 2, 4, 5, 6, 8, 9, 10, 12, 13, 7, 11
Microscopic properties:
Chapters 22, 23, 24, 3, 14, 15, 16, 17, 18, 19, 21, 20, 25, 26

If you teach a strict **one-semester course** to engineering majors, the first thirteen chapters of the textbook can be taught directly, with selected sections taken from Chapter 21 (Electrochemistry), Chapter 19 (The Solid State), Chapter 20 (Materials Science), and Chapter 23 (Kinetics I).

ACKNOWLEDGMENTS

The materials you see presented have been extensively reviewed by sympathetic but critical colleagues at colleges and universities, and by over 3000 students at WPI and Columbia during the last three years. As a result of their analyses, we have made many improvements to each draft. We thank these reviewers for their useful comments, which have been used to polish this book into its final form:

REVIEWERS

Jon Bellama, University of Maryland
George Brubaker, Illinois Institute of Technology
Robert Bryan, University of Virginia
Lawrence Conroy, University of Minnesota
Norman Eatough, California Polytechnic University, San Luis Obispo
Thomas Furtsch, Tennessee Technological University
Bruce Garetz, Brooklyn Polytechnic University
Frank Gomba, United States Naval Academy
Michael Green, City University of New York
Wilbert Hutton, University of Iowa
Michael Hampton, University of Central Florida
Harold Hunt, Georgia Institute of Technology
Larry Julien, Michigan Technological University
Zvi Kornblum, The Cooper Union School of Engineering
Walter Loveland, Oregon State University
Melvin Miller, Loyola College
Jack Powell, University of Iowa
Robert Reeves, Rensselaer Polytechnic Institute
James Richardson, Purdue University
B. Ken Robertson, University of Missouri, Rolla
Eugene Rochow, Harvard University
Mary Sohn, Florida Institute of Technology
Darel Straub, University of Pittsburgh
John Stuehr, Case Western Reserve University
Stephen Wiberley, Rensselaer Polytechnic Institute
Robert Witters, Colorado School of Mines
Gary Wnek, Rensselaer Polytechnic Institute

We would also like to acknowledge our good friends in industry who cheerfully shared with us the fruits of their research and the benefits of their wisdom. We wish to especially thank Dr. George Wise and Dr. Jeffrey Sturchio; Professor Andrew Davidhazy; Professor Francis Powell and Professor Douglas Browne; Dr. Popkin Shenian and Sam Miller. We both owe a very special vote of thanks to Professor John Stuehr, and to Dorothy Swain (Oxford College of Emory University), and would like to acknowledge the tireless assistance in the editorial process of some of our undergraduate students (who have now moved on): Dario Kunar (Johns Hopkins), John Burstein (Harvard) and AnneMarie Coffman (MIT). Special thanks to Kwanghoon Lee and Jong Park. The extraordinary assistance of Joan (JQ) Horgan can never be fully recognized but we will try by acknowledging that she may be the one truly indispensible person in the world. Thanks also to Huiling Yang, Waisum Mok, Meirav Eibschutz, Renee Fogelberg, Ted Rohm and Socorro Lugo for their special help, and to Marc Lustig and Avi Fisch.

Furthermore, we have been blessed with a thoughtful, sensitive staff of associates and assistants at Saunders College Publishing, especially our developmental editor, Sandi Kiselica, and our project editor, Marc Sherman. They calmly brought this last year through to a conclusion with patient and unfailing attention to the big issues and the small details. Our heartfelt thanks go to Associate Publisher, John Vondeling; he is a classic professional and it has been our good fortune to have him at our side throughout.

Finally, we would like to acknowledge the patience and support of our wives, Diane and Barbara, to whom this book is dedicated.

Len Fine   Herb Beall
Norwalk, Conn.   Worcester, Mass.

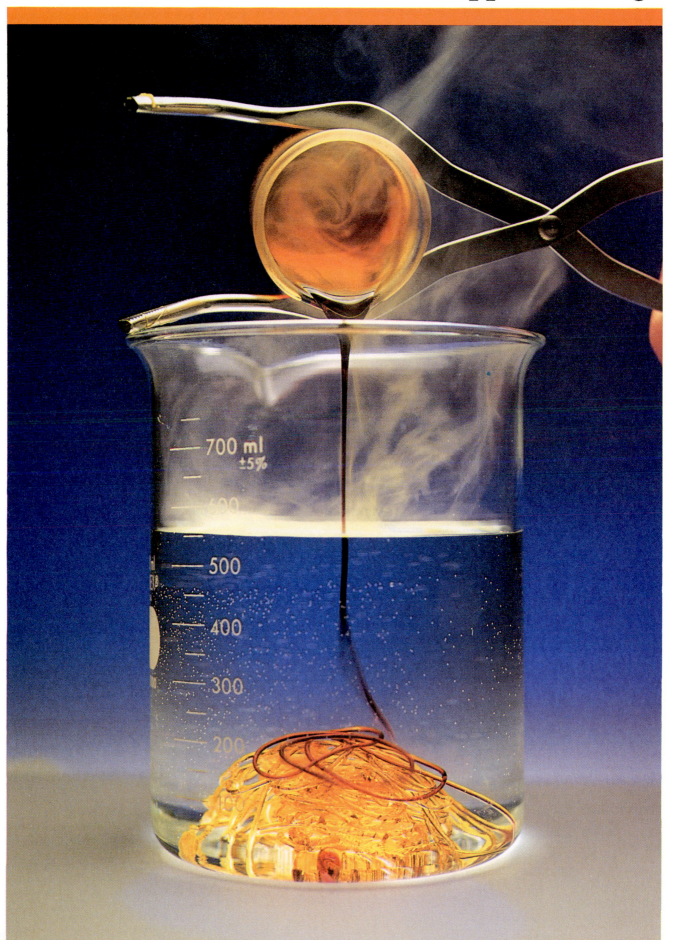

Instructor's Manual by Leonard W. Fine, Herbert Beall and Dorothy Swain. Contains chapter summaries, lecture outlines, lists of chemical demonstrations, and solutions to the even-numbered questions and problems in the text.

Student Solutions Manual by Dorothy Swain, Oxford College. Contains the complete solutions to the odd-numbered questions and problems in the text.

Student Study Guide by Robert Bryan, University of Virginia and Ken Robertson, University of Missouri, Rolla. Includes study goals, additional examples and self tests, keyed to the chapters of the text.

The Use of Estimates in Solving Chemistry Problems by Michael Green and Denise Garland. Teaches students how to solve problems in general chemistry, first intuitively and then mathematically. It contains summaries of key topics and many problems solved first by estimating the answer and then by working out the problem in detail.

Laboratory Experiments for General Chemistry by Harold Hunt and Toby Block, Georgia Institute of Technology. Provides laboratory coordinators with 42 experiments, designed to emphasize safety in the lab; includes an instructor's manual for preparing the laboratory.

Computerized Test Bank by Michael Hampton, University of Central Florida and Engineering Software Associates. Consists of more than 1000 multiple-choice questions developed around the learning objectives for each chapter. Available for IBM PC and Macintosh computers. A printed version of the test bank is also available.

Overhead Transparencies. Visual supplements, which include 125 full-color pieces of line art and photographs from the textbook.

Qualitative Analysis and the Properties of Ions in Aqueous Solution by Emil Slowinski, Macalester College and William Masterton, University of Connecticut. A qualitative analysis supplement, which encourages students to develop their own schemes of analysis.

Audio Tape Lessons and Workbook by B. Shakhashiri, R. Schreiner and P. Meyer, University of Wisconsin, Madison. Tapes to help students learn general chemistry at their own pace; students listen to the instructions on the tape and follow the illustrations and examples in the workbook.

Journal of Chemical Education: Software, Periodic Table Videodisc. A visual database of information about applications, properties, and chemical reactivity of the elements.

Tutorial Software by Charles Wilkie, Marquette University. It covers 18 major topic areas in general chemistry; available for IBM PC and Apple II computers.

Chemical Demonstration Videotapes by University of Illinois, Champaign. Schools adopting *Chemistry for Engineers and Scientists* are eligible to choose from a list of 35 available lecture/demonstration videotapes.

PART ONE: STATES OF MATTER, ENERGY, AND CHEMICAL CHANGE   1

Fundamental Concepts. Introduction to matter, stoichiometry, basic notions about the atom and descriptive nuclear chemistry. Introduction to inorganic structures, nomenclature, and reactions. The gaseous state and kinetic-molecular theory. Energetics: thermochemistry and the laws of thermodynamics. The liquid state and simple phase equilibria. Solutions and colligative properties; distillation; the colloidal state. Gas phase equilibria, ionic equilibria, and equililbria involving acids and bases:

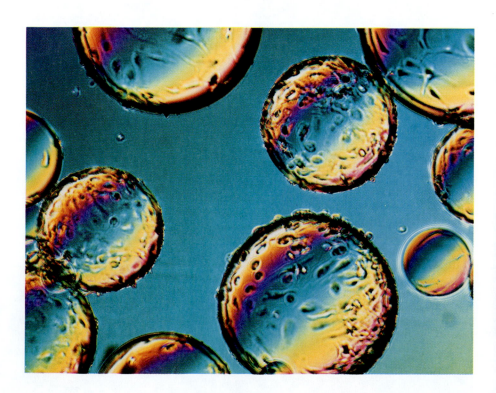

PART TWO: ATOMIC & MOLECULAR STRUCTURE, AND CHEMICAL REACTIVITY   471

Electronic structure of the atom and early quantum theory. The Bohr model. The quantum mechanical atom. Chemical bonding, bonding theories and periodic properties. Main group chemistry. Solid state and materials science. Electrochemical principles and coordination chemistry. Reaction dynamics. Organic and biochemical topics:

Contents

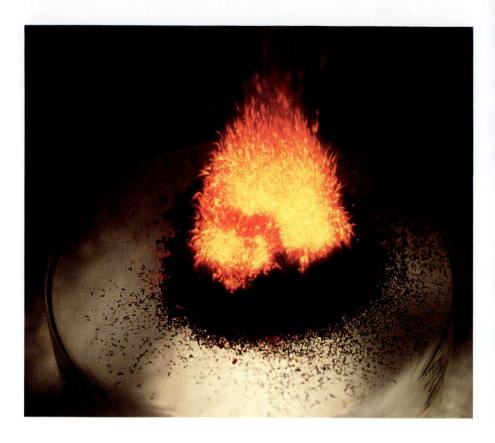

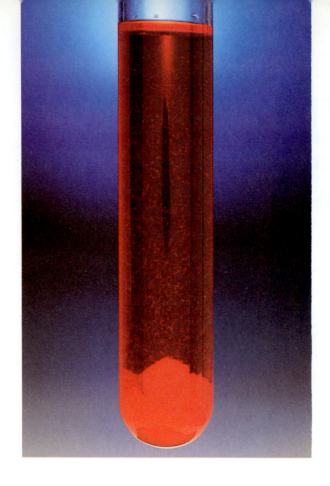

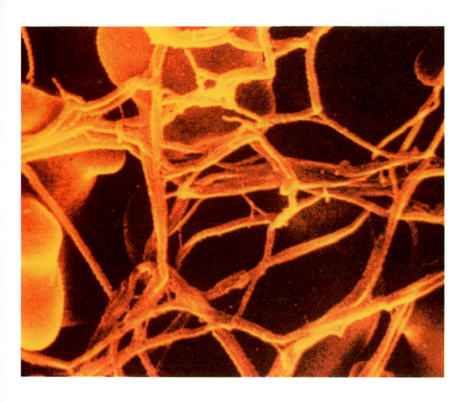

Chemistry for Engineers and Scientists

Nobel Citations

The Nobel Foundation was created under the terms of the will of Alfred Nobel (Swedish, 1833–1896), the inventor of dynamite. Each October, six Nobel Prizes may be announced, one each in chemistry, physics, medicine and physiology, economics (since 1969), literature, and peace. Part of the attraction of these Nobel Prizes lies not only in their substance and the ritual of their presentation, but also in their origin. Here was the chemist who had perfected what others turned into the embodiment of war, the explosive known as dynamite, wishing to turn men's minds toward researches for the good of all, and for world peace. The Nobel Prize in Chemistry was first awarded in 1901, and since that time, more than 100 other chemists have been so honored[†]. In each instance, the award has been for research of high principle and purpose. At the beginning of each chapter, an award citation that seems particularly appropriate, whether in chemistry, physics, or medicine and physiology, is displayed. We begin with one of the 1987 Nobel Prizes, recognizing the discovery of the first high temperature superconducting materials, the response to which has been an unprecedented advance in scientific research and development, perhaps only comparable to the discovery of X-rays, which was recognized in one of the Nobel Prize citations the first year.

[†]The Nobel Peace Prize for 1962 was awarded to Linus Pauling, the 1954 chemistry laureate.

PART ONE

States of Matter, Energy, and Chemical Change

The front side of the Nobel Prize medal awarded for chemistry and physics.

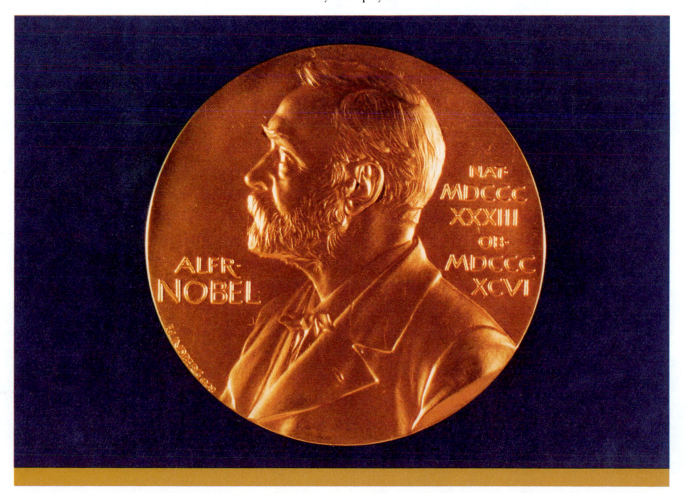

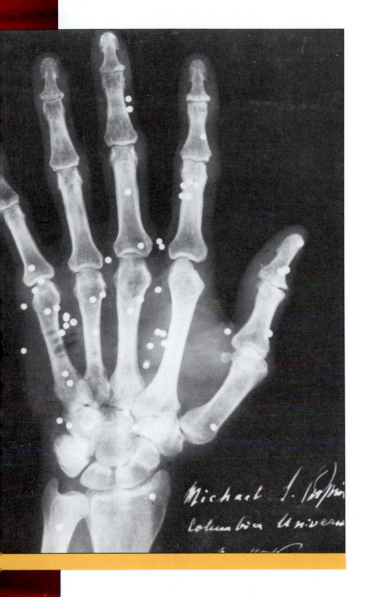

CHAPTER

1

The Science of Chemistry

Science is the great intellectual exploration of our time; engineering, the great reduction of science to practice. Together, they spread nature's mystery and magic before us. The scientific quest for a high temperature superconductor relies on experiments and applications involving deposition, production, fabrication, and characterization of superconductive materials. Plasma spray systems such as this are used to coat surfaces with yttrium-barium-copper oxides which are then specially heat-treated to produce the well-known $YBa_2Cu_3O_{-x}$ and related superconductors. In this century-old x-ray photo of a hand "pelleted" with bird shot in a hunting accident, we are witness to an early application of another monumental discovery. Brothers in discovery a century apart, Conrad Röntgen was recognized for the exciting discovery of x-rays, receiving the Nobel prize in the first year it was given (1901), while Alex Müller and Georg Bednorz are among the most recently so honored (1987), for their astonishing work in superconductivity at high temperatures.

1.1 FROM *Khemia* TO CHEMISTRY AND INDUSTRY

Chemistry is the science that describes the seemingly infinite variety of substances we encounter in the universe of our experience. Their characteristic properties provide the means by which we distinguish one substance from another. Energy in its many forms interacts with these substances, exciting them to sometimes subtle, other times dramatic, changes in both structure and reactivity. Chemists study the structure and reactivity of substances. The structures are varied and interesting (Figure 1–1): nature's proteins and nucleic acids; the synthetic polymers of modern plastics; ceramic insulators, semiconductor chips, and superconducting materials; and in some cases, the very atoms and molecules of matter themselves.

The changes are as fascinating as the structures themselves. We note the visible effects of the drugs and dyes chemists extract from nature or are able to synthesize from coal tar and petroleum; the behavior of metals, mined from the earth and extracted from seawater; the activity of plant nutrients taken from the air; the astonishing discoveries and inventions of genetic engineering; and nucleogenesis, microscopic events on the cosmological scale (Figure 1–2).

Although the science of chemistry can trace its origins back to humanity's earliest arts and crafts, it is markedly different from these practical beginnings in subject, purpose, and method. Early humans needed to know how to do things. Neolithic people had facility with agriculture and animal husbandry. They were millers, bakers, and brewmasters; makers of bricks, pots, and tools. The ancient Japanese succeeded at smelting copper more than six thousand years ago (Figure 1–3); the Egyptians, Phoenicians, and Hebrews mastered the manufacture of bronze; the Romans clearly knew how to free iron from its ore. Eventually, as how-to-do-it questions were replaced by how-does-it-work questions, crafts and trades moved toward genuine technologies. However, it was only when insight into a chemical process was first sought that chemistry began to emerge as science. Little could be gained by asking highly speculative questions such as why the world existed. But a great deal was learned by asking how the world worked. The first signs of this new way of looking at things appeared among the Greek philosophers who gave us the word **Khemia**, along with simple theories on the nature of the elements and the composition of the world. With that, the long, slow climb into the modern age of science and technology began in earnest.

The 19th century was the great period of quantitative research and systematization of information. Men like Dalton, Gay-Lussac, Dumas, Bun-

(a)

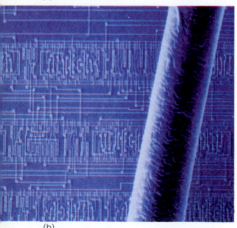

(b)

FIGURE 1–1 (a) Here, cancer cells needed for tumor research are being grown on tiny beads of oil suspended in a watery solution containing protein and other nutrients. Beads support growing cells on a special protein "skin" that forms around the droplets, which are about 0.0066 (or 1/150) in. in diameter; beads that appear transparent do not yet have cells attached to them. Commercial applications may include cultivating hybridomas for production of monoclonal antibodies. (b) Magnified 200 times, a strand of human hair dramatizes the size of the microcircuit "chip" shown in this scanning electron micrograph. The individual lines, transistors, and other elements of the experimental circuit are as small as 1 micron (10^{-6} m), about one-fiftieth the thickness of the hair. Along with increasing complexity in microelectronic circuitry has come greater "thinking" power at lower cost.

(a)   (b)

FIGURE 1–2   (a) Basic laboratory research explores the fundamental processes involved in the synthesis of pharmaceutical substances with interesting biological properties, petrochemicals as energy and materials resources, and the macromolecular structures that are the basis of plastics. (b) With new engineering designs scientists can speed the rates at which chemical intermediates are converted into commercial products.

sen, Mendeleeff, Faraday, Liebig, and Wöhler contributed significantly to this tremendous scientific expansion. By the mid-19th century, science had become less the product of the talented amateur and well-to-do hobbyist and more the fruit of dedicated professional people. This resulted from the rise of "schools of instruction," each under the direction of a highly influential individual who personally took responsibility for training a group of disciples. Many of these students in turn became great teachers, forming their own schools, guiding and inspiring still another generation of the growing ranks of scientists. Although not the first of these early laboratories, one of the most successful was Liebig's laboratory (Figure 1–4) at Giessen (Germany). Justus von Liebig's enthusiasm and the inspiration that the sheer weight of his many scientific successes gave to his students made his laboratory internationally renowned. Students from all over the world came to work and study under him. In the United States, two of the great early teaching laboratories were Remsen's laboratory at Johns Hopkins University and Chandler's laboratory (Figure 1–5) at Columbia University. Later laboratories sprang up at the University of Chicago, Harvard, and the University of Illinois. The Land Grant Act after the American Civil War led to the establishment of many large universities and experimental research stations, along with several schools of mines and technical institutes.

A burgeoning giant called industry, walking hand-in-hand with technology, pushed the new science along. Chemists had always been tinkers as well as thinkers, concerned with application and theory. As crafts left the traditional kitchen laboratory, the scene of so many of chemistry's early successes, they found new homes in pharmaceutical industries and dye manufacturing, in the gas works and the illumination industry. The manufacture of sulfuric acid, chlorine, and caustic became the basis upon which the industrialization of nations was judged. At the end of this period came the discovery of the first synthetic plastics and the invention of a curiously important object that has had a surprising impact on the growth and de-

FIGURE 1–3   Early 19th-century woodcut showing Japanese artisans freeing copper from its ore as it had been done since before the time of Christ.

FIGURE 1–4 Liebig's laboratory at Giessen (1842) became the model for the training and education of young people in chemistry. So famous did this laboratory become that it became known as "a factory for the production of professors." On the extreme right, in the top hat, is A. W. Hofmann, one of the giants of 19th-century chemistry.

velopment of chemistry and engineering—the automobile. The late 19th century was also a watershed for medicine, pharmacology, and public health. In 1901, life expectancy for males in the United States was approximately 45 years. As we near the end of the 20th century, more than 25 years have been added to life expectancy figures. Influenza and tuberculosis are no longer the number one killers in our society. Syphilis was defeated by the first chemotherapeutic agents nearly a century ago. Anesthesia and antiseptic practice became commonplace. Penicillins and tetracyclines proved

FIGURE 1–5 Chandler and students on the steps of the School of Mines in New York City (1866). His laboratory at Columbia University was in the European tradition, and the first of its kind in America.

HERALDS OF SCIENCE

Among the early Heralds of Chemistry are Sir Robert Boyle's *The Skeptical Chemist* and Antoine Lavoisier's *New System of Chemistry*. In the form of a dialog, Boyle's work presents the first modern concept of an element, based on his hypothesis that matter consisted of atoms and clusters of atoms in motion, and that every phenomenon was the result of collisions of particles in motion. Boyle appealed to chemists to do experiments and abandon their philosophical leanings, and by so doing the limiting of elements to only the classic four—earth, air, fire, and water—would be denied, and chemistry would no longer be subservient to alchemy but would rise to the status of natural science.

Lavoisier's treatise was a decisive move toward the final overthrow of alchemy and the phlogiston theory, introduced a century earlier in explanation of combustion. By the use of the balance for weight determination at every chemical change and the building of a rational system of elements, Lavoisier laid the foundation of modern chemistry. He introduced a modern definition of element and compound, explained burning and rusting as a chemical combination with oxygen, and included emission and absorption of heat in his chemical system. His concepts of the indestructibility and conservation of matter guided him in formulating his methods of analysis. Unfortunately, he was a victim of the French Revolution and the Reign of Terror, and was guillotined May 8, 1794, on the charge of "adding water to the people's tobacco."

THE SCEPTICAL CHYMIST:

OR

CHYMICO-PHYSICAL

Doubts & Paradoxes,

Touching the

SPAGYRIST'S PRINCIPLES

Commonly call'd

HYPOSTATICAL;

As they are wont to be Propos'd and Defended by the Generality of

ALCHYMISTS.

Whereunto is præmis'd Part of another Discourse relating to the same Subject.

BY

The Honourable *ROBERT BOYLE*, Esq;

LONDON,

Printed by *J. Cadwell* for *J. Crooke*, and are to be Sold at the *Ship* in St. *Paul's* Church-Yard.

M DC LXI.

High among the Heralds of Science in chemistry, the most important contributions to the field, are the works of Boyle and Lavoisier. Shown here, the title page from Robert Boyle's 17th-century treatise, "The Sceptical Chymist."

effective in treating bacterial and viral infections. Diazepines were recognized for their tranquilizing effects and amphetamines for their stimulating properties. We are beginning to understand the relationships between chemistry and nutrition. Hypertension is treatable, and the battle against the new number one killers, the chronic diseases of aging, especially heart disease and cancer, has been joined. The links between chemistry and industry are everywhere (Figure 1–6).

FIGURE 1–6 The look of modern industrial chemical synthesis, engineering and design, and manufacturing.

1.2 FUNDAMENTAL DEFINITIONS AND DESCRIPTIONS OF MATTER

A comprehensive glossary of terms can be found at the end of the text.

As of this writing, 109 different kinds of **atoms** are known. They are called **elements**. Each kind has been given a special name that often suggests discovery, origins, or some special characteristic: for example, berkelium (Bk), californium (Cf), and plutonium (Pu) (Table 1–1). Elements are distinguished from one another by their characteristic properties, which are determined by their unique combinations of **electrons**, **protons**, and **neutrons**, the three subatomic particles of which all elements are composed. Atoms of the elements chemically combine to form **molecules**. Some familiar examples of molecules containing atoms of only one kind are the major atmospheric gases N_2 (molecular nitrogen) and O_2 (molecular oxy-

---

TABLE 1–1   Names, Dates of Discovery, and Discoverers of Selected Elements

Ancient Elements—Known Before 1661

Arsenic (As)	Carbon (C)	Iron (Fe)	Tin (Sn)	Zinc (Zn)
Antimony (Sb)	Copper (Cu)	Lead (Pb)	Silver (Ag)	
Bismuth (Bi)	Gold (Au)	Mercury (Hg)	Sulfur (S)	

From 1661 to the Discovery of Radioactivity

Phosphorus (P)	Brand (1669)
Nickel (Ni)	Cronstedt (1751)
Hydrogen (H)	Cavendish (1766)
Nitrogen (N)	D. Rutherford (1772)
Oxygen (O)	Priestley (1774)/Scheele (1777)
Potassium (K)	Davy (1807–1808)
Sodium (Na)	
Barium (Ba)	
Calcium (Ca)	
Magnesium (Mg)	
Silicon (Si)/Thorium (Th)	Berzelius (1824/1828)
Cesium (Cs)/Rubidium (Rb)	Bunsen and Kirchhoff (1860/1861)
Fluorine (F)	Moissan (1886)
Helium (He)	Ramsay, Rayleigh and Travers (principally) 1894–1898
Neon (Ne)	
Argon (Ar)	
Krypton (Kr)	
Xenon (Xe)	

From Radioactivity to the Discovery of Nuclear Fission

Radium (Ra) and Polonium (Po)	M. and P. Curie (1898)
Radon (Rn)	Dorn, E. Rutherford and Owens (1900)
Hafnium (Hf)	Coster and Hevesey (1923)
Technetium (Tc)	Perrier and Segrè (1939)

The First Transuranium Elements—and Beyond

Neptunium (Np)	McMillan and Abelson (1940)
Plutonium (Pu)	Seaborg, McMillan, Kennedy and Wahl (1940)
Astatine (As)	Corson, MacKenzie and Segrè (1940)

Berkelium (Bk) and Californium (Cf) were discovered in the late 1940s and '50s by Seaborg and his colleagues at the University of California at Berkeley.

Most recently, unnamed elements 107, 108, and 109 have been synthesized by Armbruster and his colleagues at the Institute for Heavy-Ion Research at Darmstadt, West Germany.

gen); P_4 (molecular phosphorus) and S_8 (molecular sulfur), as they can exist in the free state; and the elements called halogens, which exist as diatomic molecules: F_2 (fluorine), Cl_2 (chlorine), Br_2 (bromine), and I_2 (iodine).

With the development of two special types of electron microscopes, it has become possible in recent years to produce images of individual atoms. The **field emission microscope** allows scientists to see fairly large single atoms or molecules on the tips of very fine metal needles. The **scanning electron microscope** allows scientists to form images of both atoms and molecules with resolutions typically of about 5 Å. With high voltage electron microscopes, resolutions of 2 Å or better have been achieved, and at that level images of atoms and molecules in crystals can be obtained (Figure 1–7).

Compounds are chemical combinations of atoms. A few million have been well documented to date. Compounds can be described quantitatively in terms of mass ratios and number ratios of the combined atoms. Qualitatively, compounds can be described by their characteristic properties, including the operations that transform them into other compounds or how they were originally prepared. Is the compound a liquid at room temperature? What is its color? Heat of combustion? Solubility in water? Does it taste sweet or sour, or exhibit and acrid or pungent odor? Is it elastic? Does it conduct electricity?

Elements and compounds are substances with definite composition and characteristic properties. Water is a substance. In its pure form, water always has the same properties and has never been found to be other than one part-by-weight hydrogen atoms to eight parts-by-weight oxygen atoms, regardless of the source. Similarly, common table salt is a substance, with uniform properties and definite composition. Both water and salt are *compound* substances, being composed of atoms of different elements; however, only water is considered to be molecular. Salt is composed of electrically charged particles called **ions**. Positive ions are called **cations**, whereas negative ions are called **anions**. In common salt, the cations are Na^+ and the anions are Cl^-.

Substances exist in three **phases**—solid, liquid, and gas. Whether element or compound, in any one of these phases, substances are by definition **homogeneous**, or uniform throughout. No matter where one chooses to sample and test such substances, the composition and properties prove to be the same. But homogeneity is often a function of magnification. At a microscopic level, say at a magnification of 10,000 times, sodium chloride maintains its homogeneous appearance. Under the close scrutiny of the ultramicroscope, at magnifications of 10 million times, salt reveals its separate kinds of ionic particles and no longer holds to our definition. For example, one might see a sodium ion, or perhaps a chloride ion. Only a pure element substance will be homogeneous under both macroscopic and microscopic examination, although even this concept breaks down under still higher magnification (Figure 1–8). At the subatomic level, one could even imagine distinguishing between the subatomic electrons, protons, and neutrons.

Mixtures are composed of two or more substances, which can be separated by purely physical means. Consider an example. In a **heterogeneous** mixture of table sugar and beach sand there is a distinct physical boundary between the two components. One could separate the sand from the sugar with a pair of tweezers, but that would prove to be highly impractical. A simpler, more complete separation can be accomplished by adding water to dissolve the sugar. The immediate result is to produce another heterogeneous mixture consisting of a sugar solution and water-insoluble sand.

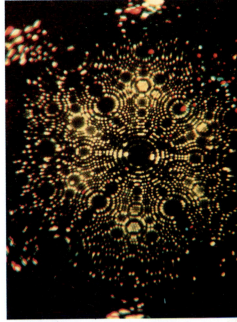

FIGURE 1–7 A clean surface of tungsten atoms photographed in a field-ion microscope.

(a)

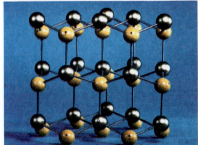

(b)

FIGURE 1–8 (a) Zinc sulfide (ZnS) in one of its naturally occurring mineral forms known as wurtzite; (b) a ball-and-stick model displaying the arrangement of the atoms of zinc (grey) and sulfur (yellow) in the crystal structure.

TABLE 1–2 Metal Alloys Are Solid Solutions		
Alloy		**Weight-Percent Composition**
Sterling silver	Ag	92%
	Cu	8%
Old pewter	Sn	80%
	Pb	20%
Common solder	Pb	66%
	Sn	34%
Type metal	Pb	80%
	Sb	15%
	Sn	5%
Bronze (gun metal)	Cu	90%
	Sn	10%
Brass	Cu	67%
	Zn	33%
Steel	Fe	99%
	C	1%

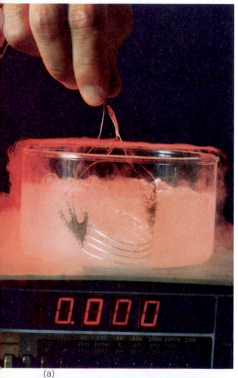

(a)

(b)

Each homogeneous component, the beach sand and the sugar solution, is called a phase. One can decant or filter the solution, separating the aqueous phase from the solid phase, and finally separate the solution into sugar and water by purely physical methods such as evaporation or distillation. Although some mixtures such as the sugar solution are homogeneous, the appearance of the mixture can be misleading. It is a mixture and not a pure substance because the relative amounts of sugar and water can vary and the solution can be separated by physical means.

Solid solutions of metals called **alloys** have special importance because of the unique properties the components lend to the composite product (Table 1–2). Classic metal alloys such as sterling silver, bronze, and pewter are well known. Many of the modern aluminum, titanium, and steel alloys have amazing engineering properties. Superconducting alloys have stretched our imagination in theory and practice to the limits of contemporary knowledge in physics, chemistry, and materials science. Alloys are homogeneous compositions of matter; nevertheless, separating these mixtures into their several component substances is not as simple as separating sugar and water from solution.

Finally, although no completely general distinction can be made among the properties typically used to characterize pure substances, there are those that are clearly **physical properties** such as freezing point and boiling point, color, and spectroscopic behavior. Observing physical properties does not alter the substance being observed. Here are some important physical properties:

FIGURE 1–9   (a) When this superconducting wire was immersed in liquid nitrogen, the wire lost all resistance to the flow of electricity, as indicated on the digital readout on the instrument below the beaker. (b) In this demonstration of the "Meissner effect," a magnet has been placed on a disc made of superconducting material and liquid nitrogen is poured into the petri dish. As the superconductor is cooled to liquid nitrogen temperatures, it creates a mirror image of the cube's magnetic field and the two objects repel each other. The cube levitates as long as the superconductor remains at liquid nitrogen temperatures.

Freezing Point or Melting Point: Both of these terms refer to the temperature at which the solid and liquid phases of a substance can coexist. This is the transition temperature over which the liquid-to-solid phase change takes place, in either direction. Water, for example, freezes at 0°C; grain alcohol at −117°C; and ethylene glycol, the common antifreeze fluid, at −15.6°C.

Boiling Point: The temperature at which the liquid-to-vapor phase change takes place, in either direction, under conditions where the pressure of the atmosphere and the pressure of the escaping vapor are the same. Boiling points are defined at a standard pressure of 1 atmosphere. The boiling point of water is 100°C; grain alcohol boils at 78.5°C; and ethylene glycol at 198°C.

Density and Specific Gravity: **Density** is mass per unit volume: for air, this is 1.29 grams per liter (g/L) at 0°C; for water 1.00 g/cm^3 at 3.98°C; for aluminum, 2.7 g/cm^3 at 25°C. **Specific gravity** is the ratio of the density of a substance to that of a common reference substance (such as air in the case of gases, or water in the case of liquids and solids); although it is not often referred to by chemists, engineers make good use of this property. Thus, the specific gravity of aluminum is 2.7, meaning that aluminum is 2.7 times as dense as water.

Solubility: The maximum quantity of a pure substance (called a solute) that will normally dissolve in a given quantity of a dissolving medium (called a solvent) at a given temperature, yielding a solution. For example, 100 grams of solvent water will dissolve 35.7 grams of solute sodium chloride at 0°C, and 39.1 grams at 100°C.

Viscosity: The internal resistance of fluids to flow. Water flows more easily than molasses. Accordingly, molasses is more viscous than water. The property can be quantified, and the usual unit of viscosity is the poise. Water at 25°C has a viscosity of 8.9×10^{-3} poise. The viscous liquid glycerol (or glycerin) has a viscosity of 9.4 poise at the same temperature.

Scientific notation for numbers is explained in Appendix C.

Electrical Conductivity: The effectiveness of a substance at carrying an electrical current by the movement of charged particles such as electrons or ions. When conducting paths are designed to offer resistance, as in the wires of a toaster or the coils of an electric stove, they slow or block the flow of electricity and heat up. Electrical energy becomes thermal energy. Copper and most other metals are good conductors; salts such as sodium chloride are good conductors of electricity in the molten state. Ceramic materials are poor conductors. **Semiconductor** materials, which have intermediate conductivities, have assumed a special importance in solid state devices.

But even the best conductors present some resistance where it isn't wanted. Thus, generators, turbines, motors of all kinds, transmission lines, and other electrical equipment always operate at less than 100% efficiency. Normally, as soon as you turn off the voltage, the moving electrons bump into vibrating atoms and into impurities, and within a few billionths of a second everything grinds to a halt. In practical terms, the current stops—the lights go out. But now imagine a loop of superconducting wire. Apply a voltage and then turn it off. This time, the current flows forever (or, if not forever, as it has been estimated, until a time somewhat longer than the age of the universe). A superconductor presents no resistance to the flow of electricity (Figure 1–9).

As of the publication date of this book, there are more than 1000 established and recognized research groups worldwide working in superconductivity.

Until recently, superconductivity had been observed only at very low temperatures, below 4 K ($-269°C$), where helium is in the liquid state, which is far too cold for any commercial applications. However, chemists have now discovered complex ceramic materials referred to as 1-2-3 compounds—yttrium–barium–copper oxide compounds—that are superconducting at liquid nitrogen temperatures. Liquid nitrogen (N_2) boils at 77 K ($-195.8°C$). This revolutionary development began in 1986 with the discovery of superconductivity at temperatures up to 30 K in compounds of lanthanum, barium, copper, and oxygen by Georg Bednorz and Alex Müller at the IBM Zurich Research Laboratory in Rüschlikon, Switzerland. (Figure 1–10). It was an important, and in many ways surprising, discovery for which the Nobel prize for physics was awarded only a year later. But now interest in the phenomenon of superconductivity has shifted to bismuth- or thallium-containing copper oxides (and, in some cases, copper-free bismuth oxides) that reach zero electrical resistance at temperatures up to 125 K. Further progress toward the goal of superconductivity in the vicinity of room temperature seems assured by the intense interest and activity.

The great advantage of these materials that offer no electrical resistance is the elimination of the heating effects (mentioned above) as electricity passes through. The energy savings and efficiencies that would be realized are substantial, to say the least. Superconducting coils of wire could store electricity (in principle) forever, since no energy is lost. Thus, the energy could be tapped as needed. Or, imagine electricity generated remotely and transmitted for hundreds, perhaps thousands of miles without losses along the way. It may well be that in some foreseeable future, clean, commercial electricity might become too cheap to bother metering—a true free lunch. Single-electron transistors based on superconductors are so sensitive that just one electron can change the current flowing between them. These experimental devices may be forerunners of all-metal, superconductor-based transistors. Similar transistors could be the basis of tomorrow's supercomputers, which would be extremely fast and use little power.

In marked contrast to physical properties, **chemical properties** such as the rusting of iron or the explosive nature of TNT result in alteration of the substance being observed. The original substance is transformed into new substances. Some properties such as taste and smell are difficult to define clearly as either physical or chemical, and are best described as simply characteristic properties.

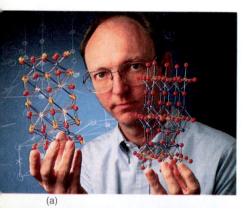

(a)

(b)

EXAMPLE 1–1

Based on your present knowledge, identify each of the following processes or properties as either chemical or physical:
(a) The effect of spreading salt on snow-covered highways.
(b) The use of antifreeze in your automobile radiator.
(c) The aging of burgundy wine and cheddar cheese.
(d) The spray from opening a warm bottle of champagne or club soda.

FIGURE 1–10 (a) Models of two early (1987) high temperature (95 K) oxide superconductors, $YBa_2Cu_3O_7$. More recently (1989), compounds of Tl/Ba/Ca/Cu/O, for example, have been discovered that have pushed the upper temperature limit for superconductivity to the vicinity of 125 K. (b) The electron-beam evaporation system deposits high temperature superconducting thin films on surfaces of silicon, the material from which nearly all microelectronic "chips" are made. Superconductors on chips might reduce time delays for electrical impulses that race within them, speeding up the rate at which devices process data.

Solution (a) The effect is the melting of ice, a phase transformation and physical change.
(b) The property is physical. A clue is provided in the marketing of "Prestone"-type radiator coolants and antifreeze solutions, which are advertised as permanent; that is, there is no change in the identity of the substances involved.
(c) The properties are chemical. In the case of wine, if it ages badly, it turns to vinegar.
(d) The escape of dissolved carbon dioxide gas is a physical property.

EXERCISE 1–1
Based on your present knowledge, identify each of the following processes or properties as either chemical or physical:
(a) Boiling an egg and baking bread.
(b) The blueness of the sky at high noon, and the red sky at sunset.
(c) Stretching a rubber band and bending a copper wire.
(d) Allowing Jell-O and concrete to set.
Answer: (a) Chemical, (b) physical, (c) physical, (d) physical (Jell-O) and chemical (concrete).

A **pure compound** is one in which there are no detectable impurities. Since at any operational level—from the macroscopic scale typical of industrial manufacture to the microscopic scale commonly encountered in laboratory research—the number of atoms and molecules one deals with is so huge, about 10^{23}, it is not possible to obtain a substance that is one hundred percent pure. It can only be described as pure within detectable limits, or it may be acceptably pure, say perhaps 99.99% or 99.9%, or possibly 99% pure.

EXAMPLE 1–2
If a liter of ethyl alcohol at 20°C contains 1.04×10^{25} molecules, how many molecules in the liter are not ethyl alcohol if the stated purity is 99.99%?

Solution If the purity of the alcohol is 99.99%, then 0.01%, or 0.0001, or one ten-thousandth of the molecules in the sample are not ethyl alcohol. This is $(0.0001)(1.04 \times 10^{25}$ alcohol molecules$) = 1.04 \times 10^{21}$ other-than-alcohol molecules.

EXERCISE 1–2
If the number of molecules in a quantity called one mole happens to be 6.022×10^{23}, and a one-mole sample of water molecules is known to be 99% pure, how many non-water molecules are present?
Answer: 6.0×10^{21}.

Physical methods can be used to separate and purify substances. **Filtration** is the process whereby one can separate two phases, usually solid from liquid, by passing the mixture through a semipermeable membrane such as a filter paper. Tea bags and modern coffee makers work this way. Cigarette filters remove solid particulates from gases. The industrial applications of filtration are virtually endless. **Extraction** is a technique for isolating a compound from a mixture by dissolving a compound in a solvent, again as in tea-making, or by the selective dissolving (or distribution) of the compound between two **immiscible** or nonmixable solvents. Very often, the two solvents are water and an organic solvent such as chloroform, gasoline-like hydrocarbons, or ethers. Slow crystallization of a solute from a solvent, a process called **recrystallization**, is a widely used technique for purification. The crucial feature of the process is the difference in solubility of the solute in hot and cold solvent. **Distillation** is a method for separating the components of a mixture based on differences in their respective boiling points.

1.3   STANDARDS FOR MEASUREMENT: MASS, LENGTH, TIME, AND TEMPERATURE

Measurement is essential to the progress of science and technology, and it depends on the existence of established standards for fundamental quantities. Four fundamental quantities can be used in beginning chemistry: mass, length, time, and temperature. The two remaining fundamental units are electrical current and luminous intensity. In Table 1–3 there is still one more fundamental unit, the quantity of substance known as the mole.

Since 1960, the accepted set of standards for the fundamental quantities has been part of a modified version of the metric system known as the **Système Internationale d'Unités**, or International System of Units (abbreviated **SI**). The unit of mass is the **kilogram**; the unit of length is the **meter**; the unit of time is the **second**; and the unit of temperature is the **kelvin**. The standards for these are given in Table 1–4. Some extreme examples of measurements are given in Table 1–5, and the standard set of prefixes is listed in Table 1–6.

The **mass** of an object is commonly measured in the metric system in kilograms (kg), grams (g), milligrams (mg), or micrograms (μg). Nanograms

The metric system was developed after the French Revolution. It is a decimal system, proceeding by tens, and was universally adopted by scientists almost at once.

TABLE 1–3   The SI Fundamental Units

Physical Quantity	Name of SI Unit	Symbol of SI Unit
Mass	kilogram	kg
Length	meter	m
Time	second	s
Thermodynamic temperature	kelvin	K
Amount of substance	mole	mol
Electric current	ampere	A
Luminous intensity	candela	cd

TABLE 1–4   Magnitudes of the Four Fundamental Quantities of Chemistry

Physical Quantity	Defined Value of SI Unit
Mass (kg)	1 kg = mass of a cylinder of a platinum-iridium alloy
Length (m)	1 m = distance traveled by light in a vacuum during the time interval of 1/299,792,458 second
Time (s)	1 s = time required for the ^{133}Cs isotope to undergo 9,192,631,770 vibrations
Temperature (K)	Defined by assigning 273.16 K to the triple point of water and 0 K to the absolute zero of temperature

TABLE 1–5 Scale of Common Usage Units

Macroscopic	Microscopic	Astronomic
Mass of 1 kg = mass of a liter of water	Mass of an electron is 9.1×10^{-31} kg	Mass of the Sun is 2×10^{30} kg
Length of 1 m = 39.37 in	Radius of an electron is 2.8×10^{-15} m	Radius of the known universe is 10^{26} m
Time between heart beats is 8×10^{-1} s	Typical lifetime of a long-lived subatomic particle is about 10^{-10} s	Estimated age of the universe is 3×10^{17} s
Temperature of 310 K is normal body temperature	The approach to the lower limit—the absolute zero of temperature—has reached 10^{-6} K	Temperature of the interior of the hottest stars is on the order of 10^{10} K

TABLE 1–6 SI Prefixes and Their Symbols

Fractions	Symbol (SI prefix)	Multiples	Symbol (SI prefix)
10^{-1}	d (deci-)	10^{1}	d (deka-)
10^{-2}	c (centi-)	10^{2}	h (hecto-)
10^{-3}	m (milli-)	10^{3}	k (kilo-)
10^{-6}	μ (micro-)	10^{6}	M (mega-)
10^{-9}	n (nano-)	10^{9}	G (giga-)
10^{-12}	p (pico-)	10^{12}	T (tera-)
10^{-15}	f (femto-)		
10^{-18}	a (atto-)		

(ng) are occasionally used, although at this low level of mass, reference is often made to parts per million (ppm) or parts per billion (ppb) instead. The fundamental quantity, the kilogram, is 1000. grams, and each succeeding unit is less by three powers of ten, or three orders-of-magnitude. The standard of mass is a platinum-iridium alloy cylinder (3.9 cm in diameter and 3.9 cm in height), kept in Sèvres, France, which is defined to have a mass of one kilogram. Since its creation in 1901, there has been no measurable change in the standard because of the unusual stability of the alloy. Duplicates can be found in London, Paris, and at the National Institute of Standards and Technology, formerly the National Bureau of Standards (in Gaithersburg, Maryland).

An object's mass is independent of the Earth's gravitational pull, as contrasted with the concept of **weight**, which is a measure of that pull. Mass is a measure of the actual quantity of matter present in an object. The gravitational effect, or weight of an object, decreases as its distance from the Earth is increased, but the object's mass remains constant. For example, in space, far from the gravitational effects of the Earth (or any other celestial body), a hammer has no weight at all, but this would not affect your ability to drive nails with it.

The metric measure of **length** is the meter, the distance light travels (through a vacuum) in 1/299,792,458 of a second. A kilometer (km) is 10^3 meters; a millimeter (mm), 10^{-3} m; a micrometer (μm), 10^{-6} m; and a nanometer (nm), 10^{-9} m. Special note should be made of the centimeter (cm), 10^{-2} m, because it is so widely used. An inch is 2.54 cm; a meter is

As a practical matter, scientists and engineers often use the terms *weight* and *mass* interchangeably.

Anders Ångström (Swedish physicist, 1814–1874) discovered hydrogen and other elements in the Sun and was the first to examine the spectrum of the aurora borealis. But he is best known for the space unit equal to one ten-billionth of a meter.

approximately 39.37 in. Because of the very small distances between atoms, and within atoms, units of length such as micrometers and nanometers are very useful in chemistry. The angstrom (Å), which is 10^{-10} m, is also used extensively.

The second (s) is the universally accepted standard of **time**. Until 1967, it was determined by a clocking device called an ammonia maser, which measured the period of vibration of the nitrogen atom in the ammonia molecule (NH_3). Such clocks lose no more than a second in several hundred years. Since then, the second has been redefined in terms of the characteristic frequency of a particular isotope of cesium, where one second is the time required for an atom of ^{133}Cs to undergo 9,192,631,770 vibrations. Such clocks are accurate to within 3 millionths of a second per year.

Temperature can be defined in operational terms by the scale of a calibrated thermometer in good thermal contact with the substance to be measured. There are many physical properties of matter that can easily be used for thermometric purposes: for example, the expansion (and contraction) of a length of metal rod or a confined liquid; changes in the pressure of a confined gas; the electrical properties of a pair of wires; or the change in color of an incandescent source. All have been used in the design and construction of thermometers for measuring temperature. The celsius scale (°C) is set with 0°C at the freezing point of water and 100°C at the boiling point. On the kelvin scale (K), 0 K is absolute zero and the freezing point of water is 273.15 K. Accordingly, the two scales are related by the equation.

$$K = °C + 273.15$$

1.4 UNITS AND CONVERSION FACTORS

The **units** tied to numbers impart physical reality to them. They are enormously useful in solving problems as well as in making measurements. Typically, problem solving involves changing from one set of units to another. To change the units tied to a number, a **conversion factor** is used. One simple illustration is the interconversion of different units representing the same quantity—for example, the interconversion of inches and centimeters. To convert 5.00 in to centimeters, multiply by the conversion factor 2.54 cm = 1 in, written as the ratio 2.54 cm per inch:

$$(5.00 \text{ in}) \left(\frac{2.54 \text{ cm}}{1 \text{ in}} \right) = 12.7 \text{ cm}$$

Note that inches in the numerator and denominator cancel, the net effect of which is to convert 5.00 in to its equivalent value, 12.7 cm.

EXAMPLE 1–3

Convert 100.0 meters to yards, using the following relationships.

$$1 \text{ m} = 100 \text{ cm} \qquad 1 \text{ in} = 2.54 \text{ cm} \qquad 1 \text{ ft} = 12 \text{ in} \qquad 1 \text{ yd} = 3 \text{ ft}$$

Solution Multiplying the original distance by the correct arrangement of conversion factors gives the numerical answer:

$$(100.0 \text{ m}) \left(\frac{100 \text{ cm}}{\text{m}} \right) \left(\frac{1 \text{ in}}{2.54 \text{ cm}} \right) \left(\frac{1 \text{ ft}}{12 \text{ in}} \right) \left(\frac{1 \text{ yd}}{3 \text{ ft}} \right) = 109 \text{ yd}$$

Note that any other arrangement of the conversion factors would have produced a number answer accompanied by some nonsensical units, rather than the desired units in yards. In a real sense, carrying the units provides a check on the logic of your calculations and solutions. ■

EXERCISE 1–3(A)

Determine the dimensions of R in the equation $PV = nRT$ if the pressure P is in atmospheres (atm), the volume V is in liters (L), n is the number of moles (mol), and T is the temperature in degrees kelvin.
Answer: L·atm/mol·K. ■

EXERCISE 1–3(B)

Check the consistency of the units in the Einstein equation $E = mc^2$, where E is the energy in ergs, m is the mass in grams, and c is speed of light in centimeters per second. An erg is one g·cm²/s². ■

Relationships between different quantities are also commonly encountered. For example, a speed of 50.0 meters per second implies that 50.0 m is equivalent to 1 s, or 1 s of time is equivalent to 50.0 m of movement at that speed. Again, expressing the conversion factor as a ratio, the distance traveled in 17 seconds at the indicated speed can be determined:

$$(17 \text{ s}) \left(\frac{50.0 \text{ m}}{1 \text{ s}} \right) = 850 \text{ m}$$

As in our earlier illustrations, the appearance of the appropriate units of distance (meters) in the solution is a strong indication that the procedure is correct.

EXAMPLE 1–4

Three identical reaction vessels in a chemical plant can together produce 1600 kg of product in 36 hours. How long would it take two of these vessels to produce 1000 kg of product?

Solution   First identify the number of vessels and the rate in kg/h at which they can produce product. Thus, 3 vessels = 1600 kg/36 h. Rearranging this relationship yields either kg/vessel·h or vessel·h/kg:

$$\frac{1600 \text{ kg}}{3 \text{ vessels} \times 36 \text{ h}} = 14.8 \frac{\text{kg}}{\text{vessel·h}}$$

or

$$\frac{3 \text{ vessels} \times 36 \text{ hr}}{1600 \text{ kg}} = 0.068 \frac{\text{vessel·h}}{\text{kg}}$$

The first conversion factor, when multiplied by hours and vessels, yields an answer in kg, which is not a sensible answer. The second conversion factor, having units of hours in the numerator, can yield the solution for this problem as follows:

$$0.068 \frac{\text{vessel·h}}{\text{kg}} \times 1000 \text{ kg} \times \frac{1}{2 \text{ vessels}} = 34 \text{ h} \quad ■$$

EXERCISE 1–4

The synthesis of phenol and acetone from cumene depends on an air oxidation process. One phenol plant has four oxidizers and produces phenol

at a rate of 400 million pounds per year, running 24 hours a day at full capacity. What is the daily rate of phenol production in lb/day? If a fifth oxidizer is added, what is the rate of production of phenol in kg/h? (1 kg = 2.2 lb)
Answer: 1.10 million lb/day; 0.026 million kg/h.   ■

1.5   ERROR IN MEASUREMENT AND ERROR ANALYSIS

Error analysis is the evaluation of the limits of uncertainty in measurement. The more careful the measurement is, the smaller the uncertainty will be. However, some uncertainty is always present in every measurement. Since the progress of science and technology depends in the most fundamental way on measurement, it is clear that in order to minimize uncertainties, their limits must be recognized and treated in a consistent fashion.

For our purposes, an **error** is the inevitable uncertainty that accompanies any measurement. This is not to be confused with **mistakes**, which can be eliminated just by being more careful. **Accuracy** answers the question: "How close am I to the truth?" **Precision** refers to the reproducibility or reliability of the experimental results, that is, the degree to which repeated determinations agree with one another. The work of the scientist and the engineer must always be done in such a way that the experimental error is as small as possible under the described conditions, and one must be able to analyze and evaluate their data on the basis of those constraints. It is worth remembering that a measurement can be precise but at the same time not be accurate.

Consider the student asked to estimate the water contained in a one-liter Erlenmeyer flask. The first response might be based on an eye-ball measurement: "It looks about 3/4 full; it contains perhaps 750 mL." But that kind of measurement is crude at best, and subject to considerable uncertainty. If challenged, our student experimenter might record the uncertainty by estimating that perhaps the volume is as little as 700 mL or as much as 800 mL. The measurement is not very precise, but is accurate to the limits of the experiment.

If greater precision is required, the liquid can be transferred to a graduated (or marked) one-liter Erlenmeyer flask, providing the basis for a much improved measurement of about 740 mL. Or was it 735 mL? Or 745 mL? Pouring the liquid into a graduated cylinder allows our student to say 739 mL. Although the measurement has improved, it is obviously still subject to uncertainties.

EXAMPLE 1–5
Excluding any mistakes, list at least three sources of error in the measurements just described, over which one could exercise some control.

Solution   One possible set of answers might be volatility of water, loss on transfer, change in temperature, and poor lighting.   ■

EXERCISE 1–5
You have been asked to estimate a quantity of sand contained in the same one-liter Erlenmeyer flask used above. Again, it appears to be 3/4 full. Fill in the details of the experiment and then suggest as many possible sources of error as you can.

Answer: Among others, include the free space between the sand particles and its effect on your estimate of the volume of the sand.   ■

Some sources of uncertainty can be minimized to the point where they are no longer significant, while others are due to the limitations of the method of measurement being used and cannot be excluded. Consider, for example, the calibration marks on the Erlenmeyer flask. If the liquid level fails to hit a **calibration** mark, then one must estimate where the liquid level rests between the marks. Or perhaps it hits the mark, in which case one still has to deal with where the level lies within the finite width of the mark. Do not forget the nature of the liquid surface, which is not perfectly flat. The **meniscus** is concave or convex, depending on the nature of the contained liquid and the container. For all these reasons (and many more), even though the liquid has an absolute volume, the experimenter cannot know the volume exactly.

But do not despair! Measurements can be made precisely, producing good data that can lead to accurate descriptions of nature. It is possible to know. Often the uncertainties in a measurement are unimportant to the questions we seek to answer. It is of no consequence to know whether the time between your classes is 15 minutes or 15.01 minutes, or whether a 100-yard dash is 100.01 yards. A good sprinter will cover that extra 0.01 yard in about a millisecond, or 0.001 second. However, where the uncertainties are important, you must not only recognize the uncertainty, but also document it correctly. For example, drug dosage at the milligram level, or 0.001 gram, can be the difference between a therapeutic dose and a lethal dose.

Consider the dilemma placed before that intrepid experimentalist, Archimedes. Anecdotal history tells us he sought to determine the value of a gold crown without defacing the crown in any way. The method he chose was to find the density of the crown, thereby determining whether it was pure 24-carat gold. If the crown was made of the less precious 18-carat gold, which is an alloy of 75% gold and 25% copper, his density measurements would be clearly different. Now, imagine Archimedes with two assistants, doing the density measurements. One exercises greater care than the other. Their data are documented below:

kind of measurement	First assistant	Second assistant
	Careful: 11%	More careful: 1%
estimate	18 ± 2	19.2 ± 0.2
range of density	16 to 20	19.0 to 19.4

The known densities for 24-carat gold and 18-carat gold alloy are:

$$d_{\text{24-carat}} = 19.3 \text{ g/cm}^3 \qquad d_{\text{18-carat}} = 17.2 \text{ g/cm}^3$$

The essential point of this discussion is that both sets of results are consistent with a crown of pure gold. Since both assistants stated the limits of their confidence in the measurements, and since the limits of confidence overlap, it is entirely reasonable to say that both statements are correct. But also note that the uncertainty in the first assistant's measurement is so large (10%) as to render the results useless, since the densities of both 24-carat gold and 18-carat gold alloy lie within that range. On the other hand, the second assistant's more careful measurement (1%) supports the conclusion that the crown could only be 24-carat gold, with the known density lying within the measured range.

1.6   SIGNIFICANT FIGURES

We have established the notion that all measurements suffer from some degree of uncertainty. Now consider the problem of expressing the limits of uncertainty in the numbers we use to represent measured physical quantities. A particular laboratory balance requires that you estimate the second decimal place. Say, for example, that the marker (Figure 1–11) registers 26.64. The calibration of the balance allows you to know the 26.6 and requires you to approximate the 0.04. Assume that you can visually divide the space between the scale marks into five parts. Accordingly, we would state the measurement as 26.64 ± 0.02. This amounts to a statement that the true value lies somewhere between 26.62 and 26.66 grams. The last figure in the original measurement (26.64 g) is the first uncertain figure. All of the digits that are known with certainty and the first uncertain (or estimated) digit are referred to as **significant figures**; they clearly bear directly on the precision of the measurement. In the case above, there are four significant figures.

When two or more measurements are combined in a calculation, the least reliable number determines the reliability of the entire group of measurements. For example, in one determination of the density of aluminum, it was found that 54.116 grams of the metal displaced (or had a volume equal to) 20.1 mL of water. The measurement of the volume of the aluminum is reliable only to the third significant figure; the mass of the piece of metal is known to five significant figures. Because the density is computed by division of the mass by the volume, the number of significant figures in the calculated density is the same as the number of significant figures in the least precise measurement—in this case, three:

$$d = \frac{54.116 \text{ g}}{20.1 \text{ mL}} = 2.69 \text{ g/mL}$$

Note that division of 54.116 by 20.1 can be carried out well beyond two decimal places (and three significant figures): for example, 2.69*23* g/mL.

The least reliable number is the one known to the least number of significant figures.

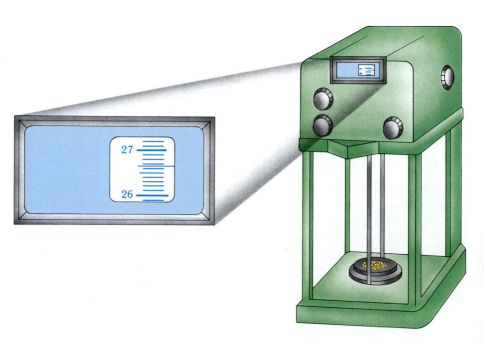

FIGURE 1–11   Estimating a value of 26.64 ± 0.01 from the scale on the face of a balance, calibrated in units of 0.1.

However, since there is no justification for that operation, the number is simply rounded off to the third significant figure.

Having experimentally determined the density of aluminum to be 2.69 g/mL, one can determine the mass of a piece of the metal with a volume of 17.35 mL:

$$m = d \times V = \left(2.69 \, \frac{g}{mL} \right) (17.35 \, mL) = 46.7 \text{ g}$$

The rule for determining the number of significant figures in the answer is the same for multiplication as for division. The multiplication process actually results in the number 46.67*15*, but only three significant figures can be considered reliable, and accordingly the number must be rounded off to the first decimal place. In this and the preceding calculation, the obvious question is: "What criteria are used for this rounding off procedure?" Usually, in **rounding off** a number, we use these rules:

- If the digit following the last significant figure is less than 5, the last significant figure remains unchanged. The following digit and all digits after it are dropped.
  For example, to round off 4.638 to two significant figures, note that the 3 is less than 5; we drop the 38 and write 4.6 as the rounded value.

- If the digit following the last significant figure is greater than 5, then 1 is added to the last significant figure. The following digit and all digits after it are dropped.
  For example, to round off 83.472 to three significant figures, note that the 7 is greater than 5; we drop the 72, add 1 to the 4, and write 83.5 as the rounded value.

- If the digit following the last significant figure is 5, then 1 is added to the last significant figure *if the last significant figure is odd*. If the last significant figure is even, it remains unchanged. The following digit and all digits after it are dropped.
  For example, to round off 27.859 to three significant figures, note that 8, the last significant digit, is even; we drop the 59 and write 27.8 as the rounded value. However, to round off 27.359 to three significant figures, we must add 1 to the 3 and write 27.4 as the rounded value.

This even-odd rule is meant to average out the rounding errors in a long series of calculations.

The question of what to do with zeros must be considered separately in any discussion of significant figures. Zeros that appear to the left of the first nonzero digit are not significant; they serve only to locate the decimal point. Those placed after—and this is a point to remember—or between other numbers, are significant. For example:

15	two significant figures	0.0015	two significant figures
15.0	three significant figures	15.00	four significant figures
10.05	four significant figures	0.01050	four significant figures
10.00005	seven significant figures	1500.	four significant figures

Although we have shown the number 1500. (with a decimal point) as having four significant figures, the number 1500 (without a decimal point) is ambiguous—it could have two or four significant figures. To remove any ambiguity on the question of zeros, we would write the mass of a large beaker that measured 1500 grams in **scientific notation**, by expressing the number as a number between 1 and 10, multiplied by 10 raised to some power:

1.5×10^3 two significant figures

1.50×10^3 three significant figures

1.500×10^3 four significant figures

EXAMPLE 1–6

State the number of significant figures in each of the following:

2001; 3.00×10^{10}; 6.626×10^{-31}; 27 billion; 10^{-6}; 0.001; 0.727.

Solution   Following the rules above, the number of significant figures are 4, 3, 4, 2, 1, 1, and 3, respectively. ∎

EXERCISE 1–6(A)

Using only the numbers 0 and 1, write numbers that have one, two three, and four significant figures.

Answer: There are many possibilities. For example, one figure: 0.0001; two figures: 0.0010; three figures: 0.0100; four figures: 0.1000 (or 1.000×10^{-1}). ∎

EXERCISE 1–6(B)

Carry out the following calculation and round off your answer to the allowable number of significant figures:

$$\frac{(6.626 \times 10^{-27})(3.00 \times 10^{10})(6.02 \times 10^{23})(1.000 \times 10^8)}{(4.184 \times 10^7)(5000.)}$$

Answer: 5.72×10^4. ∎

As a rule, when performing a series of calculations, wait until the very end to round off to the proper number of significant figures instead of rounding off each intermediate result.

Finally, when adding and subtracting measurements, the correct result contains the number of decimal places found in the number with the smallest number of decimal places. Addition of $15 + 10.05$ gives 25, not 25.05. The quality of the first number, with no significant decimal places, limits the quality of the second number when added to the first.

Many scientific procedures involve calculating the difference between two numbers. It is important to note that if the difference between the two numbers is small, the error in the difference may be relatively quite great.

An example is weighing by difference, in which the mass of substance in a crucible is determined by weighing the crucible when it is empty and when it contains the substance. Suppose that the weighing is done on a balance capable of weighing to ± 0.002 g, and the masses measured are 4.073 ± 0.002 g for the empty crucible and 4.106 ± 0.002 g for the crucible plus the substance. The question is: "What is the mass of the substance?" Here is the answer:

mass of crucible + substance	4.106 ± 0.002 g
mass of crucible	4.073 ± 0.002 g
mass of substance	0.033 ± 0.004 g

Note that the mass of the substance is known with an error of 0.004/0.033, or 12%, even though the masses determined on the balance were known with an error of 0.002/4.073 and 0.002/4.106, or 0.05% for each.

1.7 POWERS OF TEN, ORDERS-OF-MAGNITUDE CALCULATIONS, AND GOOD GUESSES

The **order of magnitude** of a quantity is the power of ten of the number that measures the quantity. Orders-of-magnitude calculations often provide approximate answers to questions and problems in the absence of sufficient data. Although such ballpark figures or guesstimates may be uncertain science, they can point the astute observer in the right direction or prevent the start of extensive, more precise studies that are clearly unwarranted. Generally speaking, the reliability of such calculations should be to one order of magnitude, or within a factor of $10^1 = 10$. A factor of 100 would be two orders of magnitude, or two powers of ten. Since atoms and molecules are tiny things and are abundant, it should not be surprising that computations in which they are typically involved require handling very small or very large numbers. A convenient way to express such number values is exponentially, using powers of ten.

FIGURE 1–12 Keep in mind that even the great eminences of science were young at one time. Here in anything but the stereotypical image is Enrico Fermi, at age 17, as he prepared to enter his first year as a university student in Italy (1918). He never lost this youthful spirit throughout his career.

EXAMPLE 1–7

Enrico Fermi (Figure 1–12), one of the great theoretical and experimental scientists of the 20th century, always considered the ability to get started on a problem to be an excellent barometer of ultimate success. He would often ask prospective students questions such as "How many piano tuners do you think live in New York City?" just to see whether they understood the essence of the good guess. He appreciated its importance in doing science. Fermi first estimated the population of New York, then the number of families, the number of families with pianos, the number of times a piano was likely to be tuned in a year, a reasonable fee that could be charged—and the cost of living in New York—and guesstimated that there probably weren't 1000 who could make a living at piano tuning, but that there probably were more than 10. So, within an order of magnitude, the number was likely to be 10^2, or 100. Not a bad guess, and probably not far wrong. ■

For an environmentalist, it is important to be able to estimate the amount of acid rain produced by the annual consumption of high-sulfur coal in the United States.

EXERCISE 1–7(A)

Estimate the number of chemistry professors in California, and the number of chemical engineers in Colorado. Estimate the number of freshman chemistry textbooks in use in the United States this academic year, and the number of hand-held calculators at Caltech, Georgia Tech, Virginia Tech, or the Tech of your choice. ■

EXERCISE 1–7(B)

Estimate the number of gallons of gasoline consumed annually by all the automobiles in the United States, and the number of grams of aspirin tablets consumed in the U.S. in a year. ■

1.8 VOLUME AND DENSITY MEASUREMENTS

The standard metric unit of **volume** (V) is the cubic meter, a derived unit of measure based on [length]³. The **milliliter** (mL) is the volume occupied by 1 gram of water at 3.98°C, the temperature at which water exhibits its maximum density at atmospheric pressure. A milliliter is equal to one cubic centimeter (cm^3); 1000 milliliters equal one liter (L).

Density (d), or mass concentration, is defined as the mass of a substance per unit volume:

$$d = m/V$$

Common units of density for solids and liquids are g/cm³ or g/mL; gas densities are often stated as g/L. For example, aluminum has a density of 2.69 g/cm³ while iron has a density of 7.86 g/cm³. On that basis, a piece of aluminum occupying a volume of 1.00 cm³ must have a mass of 2.69 g, while a piece of iron of equal volume must have a mass of 7.86 g. At 3.98°C and atmospheric pressure, 1.00 g of water occupies a volume of 1.00 mL and has a density of 1.00 g/mL. Air is a mixture of gases with a density of 1.29 g/L at 0°C and atmospheric pressure (Table 1–7).

EXAMPLE 1–8(A)

The volume of a metal plug with a mass of 50.0 g and a density of 2.69 g/cm³ can be determined directly by rearranging the basic density relationship to solve for the volume V:

$$V = m/d = \frac{50.0 \text{ g}}{2.69 \text{ g/cm}^3} = 18.6 \text{ cm}^3$$

EXAMPLE 1–8(B)

For a typical liquid whose density is 0.79 g/mL, the mass of a 20.0 mL sample can be determined by rearranging the basic density relationship to solve for the mass m:

$$m = dV = (0.79 \text{ g/mL})(20.0 \text{ mL}) = 16. \text{ g}$$

EXERCISE 1–8(A)

Calculate the density of a 453.4 g platinum bar with a measured volume of 21.13 cm³.
Answer: 21.46 g/cm³.

EXERCISE 1–8(B)

The density of water is 0.997 g/mL at 0°C; the density of ice at the same temperature is 0.915 g/mL. First determine the mass of a 10 × 10 × 10 cm block of ice, and then find the volume of the puddle of water left behind when the ice melts.
Answer: 915 g; 918 mL.

TABLE 1–7	Typical Densities at 0°C (Gases at 1 atm Pressure)				
Solids	**Density (g/cm³)**	**Liquids**	**Density (g/mL)**	**Gases**	**Density (g/L)**
Magnesium	1.75	Ether	0.714	Hydrogen	0.0899
Aluminum	2.69	Alcohol	0.789	Helium	0.178
Zinc	7.14	Water*	1.00	Ammonia	0.771
Iron	7.86	Chloroform	1.49	Carbon monoxide	1.25
Copper	8.93	Bromine	3.12	Air	1.29
Lead	11.3	Mercury	13.6	Oxygen	1.43
Uranium	18.7			Sulfur dioxide	2.93
Gold	19.3			Chlorine	3.21
				Sulfur hexafluoride	6.60

*At 3.98°C

1.9 ATOMIC WEIGHT UNITS (μ) AND THE CHEMICAL FORMULA

For the purpose of comparing the masses of the different kinds of atoms with each other, a reference standard has been chosen. The relative masses of the known elements are commonly referred to as **atomic weights**; mass is the strictly correct terminology, but weight is the common figure of speech. The unit of atomic weight is formally represented by the Greek letter μ (mu); again, one is just as likely to encounter the older **amu** (atomic **m**ass **u**nit) notation. The ^{12}C isotope of carbon, defined to have an atomic mass of 12.0000 μ, has been used as the standard of reference since 1964; the unit of atomic mass is 1.66×10^{-24} gram. On that basis, the hydrogen atom has a relative mass of 1.007825 μ. The aluminum atom is 26.98154 μ; the iron atom, 55.847 μ. Although individual atomic masses are known with considerable accuracy, in most cases it is convenient to limit the accuracy between two and four significant figures: H = 1.00; Al = 26.98 or 27.0; Fe = 55.8 or 56. The atomic weight is also the **formula weight** for the elements that occur naturally as uncombined atoms (called monatomic elements).

The term *isotope* is defined in Chapter 2 and discussed in more detail in Chapter 3.

Any combination of element symbols and numerical subscripts representing the actual number of atoms in a molecule is referred to as a **molecular formula**. The formula weight of an elemental substance with more than one atom per molecule (called a polyatomic element) is calculated by multiplying the atomic mass by the subscript in the molecular formula. For some typical elemental substances, these are:

oxygen: the formula is O_2 and the formula weight is

32.00 μ = 2 × 16.00 μ

ozone: the formula is O_3 and the formula weight is

48.00 μ = 3 × 16.00 μ

chlorine: the formula is Cl_2 and the formula weight is

71.0 μ = 2 × 35.5 μ

Most molecular formulas represent compound substances, those that contain more than one kind of atom. Consider, for example, the class of compounds known as "Freons." Generically known as **chlorofluorocarbons**, or **CFCs**, members of this family of compounds have been implicated in the critically important stratospheric chemistry resulting in depletion of the Earth's protective ozone layer. Dichlorofluoromethane, Cl_2FCH, is one example. To calculate its molecular weight, sum the relative atomic weights for the atoms indicated by the subscripts of each element. But keep in mind that the subscript 1 is always understood by the presence of the symbol itself and is never written. The molecular weight of Cl_2FCH is 103.0 μ, which is arrived at as follows:

2(Cl) + (F) + (C) + (H) =
2(35.5) + (19.0) + (12.0) + (1.00) = 103.0 μ

To calculate molecular weights from the molecular formulas of more complex chemical combinations, the procedure is the same. Cane sugar (sucrose) has the molecular formula $C_{12}H_{22}O_{11}$; the corresponding molecular weight is 342.0 μ, as determined by summing the relative atomic weights for the indicated number of atoms in the formula:

$$12(C) + 22(H) + 11(O) = 12(12.00) + 22(1.008) + 11(16.00)$$
$$= 144.0 + 22.2 + 176.0$$
$$= 342.0 \ \mu$$

For reasons that will become clear later, it is inappropriate to refer to NaCl as a molecule or to its relative weight as a molecular weight. Using the term *formula weight* overcomes this difficulty of language.

Salt-like substances such as sodium chloride are composed of huge arrays of their elements. Chemical formulas for such substances are not molecular formulas but merely represent the numbers and kinds of atoms in the formula weight. Thus:

$$\text{Sodium chloride is NaCl} = (Na) + (Cl)$$
$$= 23.0 + 35.5$$
$$= 58.5 \ \mu$$
$$\text{Potassium sulfide is } K_2S = 2(39.1) + 32.1$$
$$= 110.3 \ \mu$$
$$\text{Uranium hexafluoride is } UF_6 = 238 + 6(19)$$
$$= 352 \ \mu$$

In some chemical formulas, complex (or radical) ions such as NH_4^+, the ammonium ion, are found set apart by parentheses. The symbols within are multiplied by the subscript following the parentheses, if any. For example, high phosphate fertilizers often contain ammonium phosphate:

$$(NH_4)_3PO_4 = 3(N) + 12(H) + (P) + 4(O)$$
$$= 3(14.0) + 12(1.0) + 31.0 + 4(16.0)$$
$$= 149.0 \ \mu$$

In other cases, the formula may use parentheses or repeat groups of atoms to better indicate that the substance is a double salt, hydrate, or composite of fixed chemical composition. Again, in determining the formula weight, simply collect the atoms of each kind and add the respective sums of the atomic weights. The double salt of iron(II) sulfate and ammonium sulfate has long been known as ferrous ammonium sulfate

$$FeSO_4 \cdot (NH_4)_2SO_4 = (Fe) + 2(S) + 8(O) + 2(N) + 8(H)$$
$$= 56 + 2(32) + 8(16) + 2(14) + 8(1)$$
$$= 284 \ \mu$$

Copper sulfate can be obtained as the anhydrous salt $CuSO_4$, as the monohydrate $CuSO_4 \cdot H_2O$, or as the pentahydrate $CuSO_4 \cdot 5H_2O$. Note that the number following the *dot* in the formulas for the hydrates represents the number of H_2O's added to the formula weight—one for the monohydrate and five for the pentahydrate. Copper sulfate pentahydrate forms a brilliant, classic blue solution.

$$CuSO_4 \cdot 5H_2O = (Cu) + (S) + 9(O) + 10(H)$$
$$= 63.5 + 32.0 + 9(16.0) + 10(1.0)$$
$$= 249.5 \ \mu$$

In summary, **atomic weights** are the relative weights of the atoms of the elements, in atomic weight units. The **formula weight** is the relative weight of anything else for which a distinct chemical formula can be written. Atomic weights and formula weights can always be determined directly from the chemical formula and the table of atomic weights.

EXAMPLE 1–9

For more than a century, chemists have used the vivid yellow precipitate of ammonium phosphomolybdate as a test in phosphate analyses. Calculate the formula weight from the chemical formula, $(NH_4)_3PO_4 \cdot 12MoO_3$.

Solution

$$(NH_4)_3PO_4 \cdot 12MoO_3 = 3(N) + 12(H) + (P) + 40(O) + 12(Mo)$$
$$= 3(14) + 12(1) + (31) + 40(16) + 12(96)$$
$$= 1877 \ \mu$$

EXERCISE 1–9(A)

Calculate the formula weight for each of the following:
(a) Elemental bromine (Br_2) and phosphorus (P_4).
(b) Hydrogen peroxide (H_2O_2) and titanium dioxide (TiO_2).
Answer: (a) 160 μ and 124 μ; (b) 34 μ and 80 μ.

EXERCISE 1–9(B)

Calculate the formula weight for each of the following:
(a) Bicarbonate of soda ($NaHCO_3$) and tartar ($KC_4H_5O_6$).
(b) Aspirin ($C_9H_8O_4$) and aluminum hydroxide [$Al(OH)_3$].
Answer: (a) 84 μ and 188 μ; (b) 180 μ and 78 μ.

1.10   SPECIFIC HEAT AND THE STANDARD CALORIE

Heat is the term used to describe the transfer of energy from one place to another, as a consequence of a temperature difference. The usual symbol for heat is the letter q. Well-known units of heat are the calorie (cal) and the joule (J). The calorie is the quantity of heat needed to raise the temperature of 1 gram of water by 1 K (or 1°C) over the interval of temperature from 287.7 K to 288.7 K (or 14.5°C to 15.5°C). It is a derived unit, defined in terms of a temperature rise resulting from the application of energy to a specified mass of some standard substance. A common multiple is the kilocalorie (kcal), which equals 1000 calories. The joule is the SI unit of heat, and the conversion factor from calories to joules is:

$$1 \ cal = 4.184 \ J$$

Specific heat is defined as the quantity of heat required to raise the temperature of 1 gram of any substance by 1 K. For water, the value is 1 calorie per gram per kelvin. Other substances have their own characteristic specific heat (Table 1–8).

TABLE 1–8 Specific Heats for Some Representative Substances		
Substance	**Specific Heat**	
	cal/g·K	*J/g·K*
Water	1.000	4.184
Ice	0.538	2.25
Beryllium	0.436	1.82
Aluminum	0.215	0.900
Silicon	0.168	0.703
Iron	0.106	0.444
Nickel	0.106	0.444
Copper	0.0924	0.387
Silver	0.056	0.23
Cadmium	0.055	0.23
Mercury	0.033	0.14
Gold	0.0308	0.129
Lead	0.0305	0.128

EXAMPLE 1–10(A)

Calculate the heat required to raise the temperature of 50 grams of water from 283 K to 363 K.

Solution

Note that $\Delta T = T_{final} - T_{initial}$

$$
\begin{aligned}
\text{quantity of heat} &= (\text{mass})(\text{specific heat})(\Delta T) \\
&= (50 \text{ g})(4.184 \text{ J/g·K})[(363 - 283) \text{ K}] \\
&= 1.7 \times 10^4 \text{ J}
\end{aligned}
$$

EXAMPLE 1–10(B)

Mix 50.0 g of water at 283 K with 10.0 g of water at 323 K. Find the final temperature.

Solution
Note that the warmer water cools, and in the process gives off heat. This is an **exothermic process**. The cooler water warms, absorbing heat in the process. This is an **endothermic process**. If all the heat is transferred within the 60 g of water that are mixed together, then the following net result occurs:

$$
[\text{heat absorbed}] = -[\text{heat evolved}]
$$
$$
(50.0 \text{g})(4.184 \text{ J/g·K})([T_f - 283 \text{ K}) = -(10.0 \text{g})(4.184 \text{ J/g·K}) \times ([T_f - 323] \text{K})
$$
$$
T_f = 290. \text{ K}
$$

EXERCISE 1–10(A)

How many calories are required to raise the temperature of a liter of water from 298 K to 323 K? How many joules?
Answer: 25,000 cal; 100,000 J.

EXERCISE 1–10(B)

What final temperature results from mixing 250 mL of water at 298 K (room temperature) with 150 mL of water near the boiling point (363 K)?
Answer: 322 K.

EXERCISE 1–10(C)

A 1000.-gram ingot of a certain metal is heated to 473.0 K and then dropped into a tank containing 8.00 liters of water at 298.3 K. If the final temperature reached by the system is 300.7 K, what is the specific heat of the metal? What is the metal?
Answer: 0.111 cal/g·K; probably iron or nickel.

SUMMARY

We live in a universe of **matter** and **energy**. Our experience with the universe has proved endlessly variable. **Chemistry** is the science that describes this seemingly infinite variety of natural experience. But our universe would hardly be understandable if we were limited solely to the data, the record of physically observable experience allowed by our senses. The scientist tries to provide something more, introducing a level of understanding of that part of the universe that is not readily observable. The basis for this understanding lies cloaked in the reproducibility of events, experiments repeated (or observations made) for the *n*th time. An informal code guides the practicing scientist and engineer so that on occasion he or she is able to discover and make known nature's secret ordering principles, the **scientific laws** that govern the clockwork of the universe.

The technique of turning complex, difficult problems into simpler, more manageable ones is an important part of that process. Being able to make an educated guess, or **guesstimate**, being able to establish the magnitude of the answer when there seems to be no obvious way to proceed, is often sufficient to speed discovery (for the scientist) and invention (for the engineer)—or to speed problem-solving for the student being introduced to the theory and practice of chemistry.

Our descriptions of the compositions of matter are in terms of small particles called **atoms** and **molecules**. Because of their size, we know about these particles mostly from observing the properties of rather large collections of them. Our qualitative descriptions make use of a small dictionary of terms (such as atom, molecule, substance, compound, and solution), while our quantitative descriptions depend on our ability to measure four fundamental quantities—**mass** (in kilograms), **length** (in meters), **time** (in seconds), and **temperature** (in degrees or kelvins)—and a larger set of derived quantities such as **volume**, **density**—which is the mass of a substance per unit of volume—and **heat**, which has to do with energy transfer due to a temperature difference. **Specific heat** is the amount of heat energy needed to raise the temperature of a gram of a substance by one kelvin, and the unit of heat is the calorie (or the joule). The **SI** system of measurement commonly used today is a version of the much older **metric system**.

The atomic mass unit μ is based on the ^{12}C isotope of carbon, which is the basis for the system of relative **atomic weights**. **Formula weights** are obtained from the chemical formula and the table of relative atomic weights.

In any experiment, criteria must be established for evaluating data with regard to **accuracy** (how close to the truth) and **precision** (how reliable or reproducible). Although that is not often an apparently critical issue in this introduction to chemistry, you must not forget that all the numbers you see representing physical quantities and data from experiments are subject to the limitations of experimental procedures and their inherent uncertainties. The need to pay attention to **significant figures** is a clear illustration of those requirements.

QUESTIONS AND PROBLEMS

QUESTIONS

1. Distinguish between the following pairs of terms:
(a) atom and element
(b) physical and chemical properties
(c) atomic weight and formula weight
(d) element and compound
(e) density and specific volume
(f) accuracy and precision
(g) powers of ten and scientific notation.
2. Suggest at least two ways of proving that table salt is a compound and not an element.
3. Briefly describe some cooking or "kitchen" operation

involving the separation of a heterogeneous mixture. Repeat for a homogeneous mixture.
4. Reconcile the following facts: a piece of sulfur burns and disappears completely, while a piece of iron burns and gains weight.
5. Explain why a severe skin burn can occur if you place a wet hand on a piece of metal on a very cold day, while it is ordinarily safe to grasp a piece of wood under the same conditions.
6. Without further research, classify each of the following as an element or compound substance; as a heterogeneous

or homogeneous mixture. More than one may apply.

(a) air, water, earth, and fire.

(b) copper, iron, steel, glass, and Formica.

(c) rubber, marble, sand, and polyethylene.

(d) wood, cotton, milk, ink, paper, bread, and baking soda.

(e) Clorox bleach, Excedrin tablets, and Jell-O.

7. How many distinct phases can be identified in the following systems?

(a) A cup of tea in which a teaspoon of sugar has been dissolved by stirring.

(b) A bottle of club soda.

(c) An aerosol shaving cream.

(d) A common household thermometer.

8. Prepare a list of a dozen or so materials that you commonly "know." Which of these can be classified as pure substances? Are you able to classify any of the items in your list as elemental substances? As compound substances? As homogeneous mixtures?

9. How is a scientific law arrived at? What are the limiting considerations in establishing the validity of a scientific law?

10. Comment on the statement, "The man of science who cannot formulate an hypothesis is only an accountant of phenomena."

11. "The Earth is the center of our universe; therefore, it is natural to observe the sun's rising in the East and setting in the West." Comment on the nature of the statement as an hypothesis or as a scientific law, and on its verifiability.

12. What generalization might you conclude from the existence of two distinctly different iron oxides, three distinctly different oxides of chromium, and several oxides of manganese?

13. Comment on the statement that "underlying the complex nature of the real world exists a fundamental simplicity."

14. The mass of an object can be determined by "weighing" it; yet in an orbiting space vehicle it will have no "weight." Explain the apparent paradox.

15. What is meant by each of the following metric prefixes?

(a) deci-; centi-; milli-; micro-; nano-; pico-

(b) kilo-; mega-; giga-; tera-

16. What metric unit is represented by each of the following? ns; mg; μm; kL; K.

17. Briefly explain why it is scientifically meaningless to say that one substance is heavier (or lighter) than another.

18. Can one ever be certain that the time intervals between successive cycles of a clock's periodic motions are the same? Briefly explain.

19. Briefly explain why we may reasonably refer to density as mass concentration.

20. What is the common unit of energy and how is it defined?

21. Determine the thickness of a single penny by stacking up 50, measuring the entire pile, and taking an average value. Why would your accuracy be improved by this method, as compared to measuring a single penny? What effect would the use of a meter stick versus a yardstick have on your precision?

PROBLEMS

Measurement   [1–6]

1. The radius of a neon atom is 1.31 Å. Calculate the radius in nanometers and meters, using the correct number of significant figures.

2. The distance between hydrogen atoms in a hydrogen molecule is 0.74 Å. Calculate the distance in meters, centimeters, millimeters, and nanometers.

3. Assuming the thickness of the total number of pages of a 500-page book to be 2.54 cm, what is the thickness of one page in meters, expressed exponentially?

4. The distance from Washington to Chicago by rail is 766 miles (via Pittsburgh). Express this number in feet using exponential notation. There are 5280 feet in a mile.

5. The mass of an electron is 9.11×10^{-28} g. Calculate the number of electrons in a gram of electrons.

6. A hydrogen atom has a mass of 1.66×10^{-24} g. How many hydrogen atoms are there in a kilogram of hydrogen?

Volume and Density   [7–14]

7. The density of alcohol at 20°C is 0.79 g/mL. Calculate the weight of a liter of alcohol.

8. If a 2.5 quart tub full of butter weighs 2.27 kg, what is its density in g/L? (A quart equals 0.946 L.)

9. A 25.00 mL volumetric flask weighs 31.3 g when empty. Determine its weight when filled with

(a) water

(b) alcohol

(c) mercury.

10. A chemistry laboratory student needs 400 mL of acetic acid solution to complete an experiment. If the density of the acetic acid solution is 1.049 g/mL, will 400 g be sufficient?

11. As the story goes, Archimedes was asked to authenticate a golden crown suspected of being gold-plated rather than solid gold. He was charged with making this appraisal without breaking or scratching the crown's surface. The crown weighed 2.80×10^3 g; when submerged in a tub of water it weighed 2.65×10^3 g. Determine the density of the crown and the veracity of the crownmaker.

12. A certain irregularly shaped object weighs 200. g. It was dropped into a tank of water, became completely submerged, and displaced 60. mL of water. In a similar experiment, it displaced 50. g of an oil into which it was submerged. What is the density of the object? Determine the density of the oil.

13. The mass of a single iron atom is 9.3×10^{-23} g. Determine the number of iron atoms in a regular cube of the metal that measures exactly 10 cm on an edge ($d_{\text{iron}} = 7.86$ g/cm³).

14. A uniform cube of beryllium measures 0.254 cm

along an edge. The mass of a single beryllium atom is 1.50×10^{-23} g. Find the number of beryllium atoms contained in the block ($d_{Be} = 1.85$ g/cm^3).

Atomic Weights and Chemical Formulas   [15–22]

15. Calculate the formula weight of each of the following *second period* elements:
(a) nitrogen, N_2.
(b) oxygen, O_2.
(c) fluorine, F_2.

16. Calculate the formula weight of each of the following *halogen* elements:
(a) chlorine, Cl_2.
(b) bromine, Br_2.
(c) iodine, I_2.

17. Calculate the formula weight of each of the following compounds:
(a) water, H_2O.
(b) hydrogen peroxide, H_2O_2.
(c) the three sulfur fluorides, SF_2, SF_4, and SF_6.

18. Calculate the formula weight of each of the following compounds:
(a) the interhalogen compound ICl_3.
(b) the noble gas compound $Xe^+PtF_6^-$.
(c) the rare earth oxide Eu_2O_3.
(d) the metal carbonyl $Fe(CO)_5$.
(e) the peroxide Na_2O_2.
(f) the rhodium salt $[RhH(NH_3)_5]SO_4$.
(g) the binuclear oxorhenium complex $Re_2OCl_3[(C_6H_5)_3]_2(O_2CC_2H_5)_2$.

19. Refrigerants are used as heat transfer agents to remove heat from a low temperature region, the inside of the refrigerator, to a high temperature region, the room outside the refrigerator. The most important group are the CFCs, or chlorofluorocarbons, better known as "Freons." Freon-12 is a methane refrigerant, containing only one carbon atom per molecule. If there are four atoms attached to the carbon atom, and the formula weight is known to be 121 μ, what is the chemical formula?

20. Freon-22 has a formula weight of 86 μ. If one of the atoms attached to the lone carbon atom is H, what is the Cl/F atom ratio?

21. The chemical and engineering properties of cement and concrete are very important in the construction industries. One of the four principal chemical compounds in Portland cement is tricalcium silicate, $3CaO\cdot SiO_2$. Calculate its formula weight.

22. Another of the four principal chemical compounds in Portland cement is tetracalcium aluminoferrite, $4CaO\cdot Al_2O_3\cdot Fe_2O_3$. Calculate the formula weight.

Heating and Cooling, and Specific Heat   [23–28]

23. What quantity of heat is required to carry out each of the following?
(a) Convert a liter of water at 30°C to a liter of water at 60°C.

(b) Heat a 1 kg block of aluminum from 30° to 60°C.
(c) Heat 1 kg of mercury from 30° to 60°C.
(d) Heat a liter of mercury from 30° to 60°C.

24. Calculate the quantity of heat associated with each of the following.
(a) Converting 1 liter of water at 27°C to 82°C.
(b) Raising the temperature of a kilogram of aluminum from 27°C to 82°C.
(c) Lowering the temperature of a liter of mercury from 82°C to 27°C.

25. A sample of 45 g of water at 323 K was mixed with 15 g of water at 283 K. Determine the final temperature.

26. What is the final temperature that results after 150 g of water at 25°C is mixed with 150 g of water at 75°C? What is the result of mixing 142 g of water at 24.3°C with 156 g of water at 75.9°C?

27. A pound of lead shot is placed in a beaker, which is then placed in a pan of boiling water. After a while, the beaker is removed from the water, and the lead shot is dumped out into another beaker containing a liter of water at room temperature, which happens to be about 22°C this day. Within the limits of the experiment, what is the final temperature of the water into which the lead shot was dumped? What sources of error do you see in this experiment?

28. Exactly 453.4 g of copper shot was placed in a beaker and allowed to come to room temperature, which was recorded as 25.1°C. The copper shot was then quickly and carefully dumped into a second beaker containing 100. mL of water at a recorded temperature of 99.7°C. What is the final temperature of the water after the copper shot is dumped into it? What sources of error do you see in this experiment?

Units and Conversion Factors   [29–36]

29. Express each of the following as indicated:
(a) 100.0 miles per hour as km/h
(b) 100.0 yards as meters
(c) 440.0 meters as yards
(d) 10,000.0 km as m
(e) 186,000.0 mi/s as cm/s

30. (a) Figure out your height in meters and in centimeters.
(b) Determine your weight in kilograms and in grams.
(c) Assuming you can dash 100.0 yards in 10.0 seconds, determine your velocity in centimeters per second.
(d) How many dollars are there in a megabuck? How many cents in a millibuck?

31. (a) Consider running the 100.0-yard length of a football field. Is it a longer or shorter run to traverse a field that is 100.0 meters?
(b) Compared to the 400.0-meter freestyle swim, will it take longer for the same swimmer, swimming at the same speed, to swim 400.0 yards?

32. Which will require greater stamina and endurance on your part, an utterly boring 1-hour lecture or an equally boring microcentury lecture?

33. A light year is the distance light travels in a year at a velocity of 186,000 miles/second. Express the number of miles in a light year to the correct number of significant figures, and expressed in scientific notation.

34. The Andromeda nebula is estimated to be 2,000,000 light years from Earth. Express this distance in miles, using scientific notation and the correct number of significant figures.

35. The chemistry department purchases acetone in 5-gallon cans at a price of $17 each. What is the cost per kg? The density of acetone is 0.792 g/mL.

36. The density of alcohol at 20°C is 0.79 g/mL. The density of benzene at the same temperature is 0.88 g/mL. How many liters of alcohol would you get if you bought a pound of alcohol? How many pounds of benzene would you get if bought a liter of benzene?

Scientific Notation and Significant Figures   [37–48]

37. Write out the following numbers using scientific notation:
(a) 2.0500×10^3
(b) 0.00002050
(c) 1.0
(d) 0.727
(e) 63.5

38. Rewrite the following numbers using powers of 10 and scientific notation: 1987; 1.987; 0.01987

39. Rewrite the following numbers in non-exponential notation: 3.00×10^{-10}; 6.25×10^2; 6.25×10^{-2}; 10^4; 10^{-4}

40. Write the following numbers in standard notation:
(a) 1.54×10^{-8}
(b) 8.21×10^{-2}
(c) 1.987×10^0
(d) 1.988×10^3

41. State the number of significant figures in each of the following:
(a) 2001
(b) 7
(c) 6.62×10^{-27}

42. State the number of significant figures in each of the following:
(a) 15.9999
(b) 2.7 billion
(c) 0.000001

43. Carry out the following calculations and establish your answer to the allowed number of significant figures:
(a) A ream of paper contains 500 sheets and stands 50. mm high. What is the average thickness of a sheet of paper?
(b) Exactly 18.015 g of water was found to contain 2.016 g of hydrogen and 15.999 g of oxygen. Determine the percentage by weight of oxygen and hydrogen in the sample of water.

44. The Moon is approximately 2.37×10^5 miles distant. Light travels at 1.86×10^5 mi/s.
(a) How many seconds are required for the passage of light from the Moon to Earth?
(b) It has been estimated that the radius of the universe is 10^{28} cm. Calculate the volume of the universe.

45. Perform the following computation and report your answer to the proper number of significant figures:

$$\frac{(6.62 \times 10^{-27})(3 \times 10^{10})(6.023 \times 10^{23})}{(4.184 \times 10^7)(4785)}$$

46. Carry out the following calculation and round off the answer to the allowable number of significant figures:

$$\frac{(6.626 \times 10^{-27})(3.00 \times 10^{10})(6.02 \times 10^{23})(10^8)}{(4.184 \times 10^7)(5000)}$$

47. Paying careful attention to significant figures, calculate the density of a metallic solid, 1000. g of which occupy a volume of 0.0518 L. Can you identify the metal?

48. The element bromine is a red-brown, fuming liquid with a density of 3.119 g/mL at 20°C. Paying careful attention to significant figures, determine the volume of 3 g of bromine; of 453.4 g of bromine; and of a kg of bromine.

Additional Problems   [49–56]

49. Perchloroethylene, C_2Cl_4, an extremely stable, colorless, organic liquid with an ether-like odor, has been widely used as an industrial solvent and dry-cleaning agent, for vapor degreasing, as a vermicidal agent, and in the manufacture of fluorocarbons and CFCs.
(a) What is the formula weight of the molecule?
(b) The liquid has a density of 1.625 g/cm³. Express the density in kg/m³.
(c) What is the weight in pounds of the contents of a commercially available 5-gallon can of perchloroethylene?

50. The time of flow t for a certain liquid passing through a capillary is directly proportional to the length l of the capillary and the volume V of the flowing liquid; the time t is also found to be inversely proportional to the pressure P pushing the liquid through the capillary and to the fourth power of the radius r of the capillary tube.
(a) Using K as the constant of proportionality, write an overall algebraic expression that can be used to give t directly.
(b) What are the units of the proportionality constant K in the International System?

51. The radius of a neon atom is 1.3 Å. Express the radius in meters and in miles. Estimate the volume of a neon atom.

52. Establish the systematic relationship that exists between the variables Q and R:

Q:	0	28	56	70
R:	842	954	1066	1122

53. At one stage in the forging of a set of four 2.0-kilogram iron horseshoes, the set is taken from an annealing oven operating at 365°F and quickly dropped into a 55-gallon drum that is two-thirds full of water at 11°C. Neglecting any heat losses to the barrel or beyond, assuming no water is lost due to evaporation, and assuming the horseshoes are of equal weight, what is the final temperature of the water? The specific heat of iron is 0.444 J/g·K and °F = 9/5(°C) + 32.

54. How many grams of copper shot at 100°C must be added to a kilogram of water that is initially at 25°C in order to raise the water temperature to 28°C?

55. When the price of gold on the commodities market is quoted as $440.26 per ounce, what is the value of a gram of it?

56. (a) What is the lifting power of a spherical balloon of 100-meter diameter filled with hydrogen? *Note:* the lifting power is the difference between the weight of the hydrogen filling the balloon and the weight of an equal volume of air. See Table 1–7 for necessary density data.
(b) Helium cannot lift as much as an equal amount of hydrogen, but there are advantages to its use. By what factor is the lifting power of helium less than that of hydrogen, and what is the principal advantage to using helium?

MULTIPLE PRINCIPLES   [57–60]

57. You are asked to find the height of the liquid in a column, whose (uniform) cross-section is 2.54 cm², which contains 1000. g of alcohol at 25°C. Solve this problem by setting up a single equation with the necessary numbers and units, and express your answer (with proper units) to the correct number of significant figures.

58. James Prescott Joule, a British physicist, measured the increase in temperature of water from the top to the bottom of a falls. Assuming that all of the potential energy (PE) of the water at the top of Niagara Falls is converted into heat on falling into the water below, determine how much higher the temperature is at the base. Assume the falls to be 50. m high. The relation is PE = mgh, where m = mass in grams, g = gravitational constant of 980. cm/s², and h = height (in cm) through which the water must fall.

59. The flapability, in flips, of a sample of stomp is directly proportional to the number of flops in the sample and to its potency, expressed in pots. If the proportionality constant is 0.00849 flips per flop·pot, what is the flapability of 8.47 flops of stomp when the pot meter reads 203?

60. The common scales of temperature are related as follows:

$$°F = \frac{9}{5}(°C) + 32$$

$$K = °C + 273.15$$

The Rankine scale (°R) begins at absolute zero (0 K) but has degrees the size of Fahrenheit degrees:

$$°R = °F + 459.69$$

(a) If the melting point of ice is 273.16°K, what is the temperature on the Rankine scale?
(b) If liquid nitrogen boils at −196°C, and we have a material that is superconducting below 380°R, is liquid nitrogen a suitable coolant for carrying out superconductivity experiments with this material?

APPLIED PRINCIPLES   [61–64]

61. A mixture of table salt and cane sugar was analyzed and found to be 33.3% chlorine by weight. If pure table salt is NaCl, what percentage of the original mixture is salt?

62. The standard formulation for Clorox and similar liquid bleaches is a 5.25 weight percent water solution of sodium hypochlorite, NaClO. What is the percent by weight of chlorine?

63. The industrial process for recovering chlorine from brines of various concentrations involves passing these dilute salt solutions through a cake of sodium chloride, from which the solutions emerge near saturation levels. A near-saturated solution is found to contain 24.3% NaCl by weight and its measured density is 1.19 g/cm³ (in the vicinity of room temperature).
(a) Express the composition in units of grams of salt per kilogram of water.
(b) Express the composition in units of kilograms of salt per liter of solution.

64. Acetone is purchased for undergraduate laboratory use at a cost of $15 per 5-gallon can. After use, the acetone is collected and removed as waste at a cost of $81 per 5-gallon can. During every period, each laboratory section uses 4.0 pints of acetone. There are 20 sections, each of which meets 12 times per semester, and 20% of the acetone originally used by the students is consumed (and therefore not returned as waste). Additionally, there are two semesters and a summer session, all of which are run in an identical fashion. What is the cost per year for this one laboratory chemical?

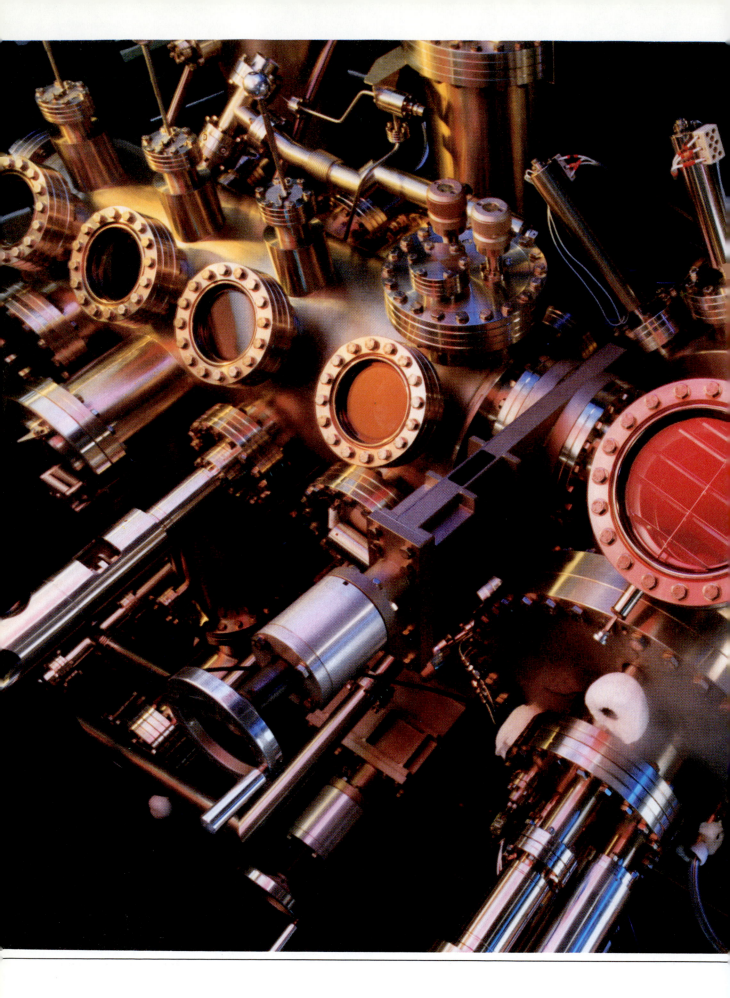

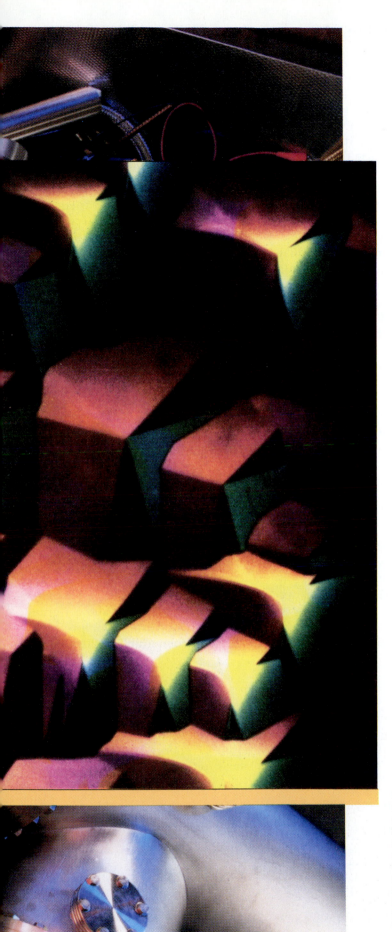

C H A P T E R
2

The Atomic Theory and Chemical Stoichiometry

Ultrahigh vacuum techniques are used to prepare single crystal thin films. The process is called molecular beam epitaxy (MBE) and can control precisely the thickness, composition, and impurity levels required to fabricate microelectronic (such as semiconductor "chips") and microoptical devices, atom-for-atom. The micrograph is of one of the earliest epitaxial layers—a thin layer of silicon atoms deposited on a germanium surface—prepared by MBE. Sixty years earlier, the Nobel Prize in chemistry was awarded to the first American to be so honored, for his equally sophisticated work with little more than an analytical balance, carefully determining and recording the relative weights—the atomic weights—of the atoms of many of the individual elements.

2.1 THE LAW OF CONSERVATION OF MASS

Long before the Christian era, Democritus noted, "Nothing can be created out of nothing, nor can it be destroyed and returned to nothing." At the beginning of the modern age of science, Lavoisier confirmed that idea experimentally, demonstrating that indeed there is no loss or gain of matter during the course of a chemical transformation. By burning phosphorus in air in a closed vessel, he was able to show that the increase in weight of the reacting phosphorus was equal to the mass of the air consumed. Lavoisier was equally successful carrying out a combustion process in reverse, where the loss in weight of red mercury (II) oxide was found to be equal to the increase in weight of the air in the reaction container. Based on these simple yet sophisticated experiments, he rightly concluded that the mass of the products of a chemical reaction must equal the mass of the reactants (Figure 2–1). Were it not for this conservation of mass principle, quantitative measurements in chemistry would be impossible and the chemical equation would not exist.

In the year 1905, Albert Einstein made a striking prediction: that mass and energy are equivalent, and that they are related by the simple equation

$$E = mc^2$$

where E is energy, m is mass, and the constant of proportionality is equal to the square of c, the speed of light, which is a very large number. That being the case, the mass associated with any particular quantity of energy must be relatively small. For example, the energy given off as light by the filament in a flashlight bulb amounts to a decrease in mass of about 10^{-10} g/s. So there is no contradiction of the old law; one must expand its scope to include energy, which is also conserved. However, in most situations only conservation of mass need be considered because the mass equivalent of the energy involved is negligible.

When the situation is reversed, the proportionality constant becomes a very large multiplication factor, and very small amounts of mass are equivalent to very large quantities of energy. For example, it has been estimated that the uranium core of the first atomic bomb lost about 1 gram of its original mass during the nuclear transformation into fission products. The energy equivalent of this gram of mass was comparable to 20,000 tons of TNT.

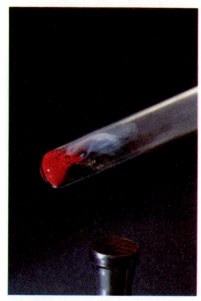

Theodore William Richards (United States, 1868-1928), Nobel Prize for Chemistry in 1914 for his accurate determination of the atomic weight of a large number of the chemical elements.

FIGURE 2–1 Decomposition of red mercury(II) oxide, a demonstration of Lavoisier's experiment, and the mass conservation principle. From his experiments on human respiration (1790), Lavoisier concluded that "respiration is merely slow combustion of carbon and hydrogen, which is similar in every respect to that which occurs in a lighted lamp or candle."

EXAMPLE 2–1

Consider a chemical transformation involving 1 pound of a substance, accompanied by a loss of 1.000 million calories of thermal energy. Compute the mass equivalent of that amount of energy using the Einstein relationship $E = mc^2$, where E is measured in ergs = g·cm²/s² and $c = 3.00 \times 10^{10}$ cm/s.

Solution Rearranging the Einstein equation to solve for mass gives

$$m = E/c^2$$

where

$$E = (1.000 \times 10^6 \text{ cal})(4.184 \times 10^7 \text{ erg/cal}) = 4.184 \times 10^{13} \text{ erg}$$
$$= 4.184 \times 10^{13} \text{ g·cm}^2/\text{s}^2$$

Substituting the appropriate values into the original equation,

$$m = \frac{4.184 \times 10^{13} \text{ g·cm}^2/\text{s}^2}{(3.00 \times 10^{10} \text{ cm/s})^2} = 4.65 \times 10^{-8} \text{ g}$$
$$= (4.65 \times 10^{-8} \text{ g}) \frac{1 \text{ lb}}{454 \text{ g}} = 1.02 \times 10^{-10} \text{ lb}$$

The loss of one ten-billionth of this pound of substance is so small as to not be detectable by any conventional means. Therefore, the simple interpretation—conservation of mass—is sufficient. ◼

EXERCISE 2–1

Calculate the energy (in ergs, kilojoules, and kilocalories) associated with conversion of 1.00 gram of mass, according to the Einstein equation. *Answer:* 9.00×10^{20} erg; 9.00×10^{10} kJ; 2.15×10^{10} kcal. ◼

2.2   THE LAW OF DEFINITE PROPORTIONS

According to the **Law of Definite Proportions**, a pure chemical compound is a product to which nature has assigned fixed proportions by weight. Sometimes referred to as the Law of Constant Composition, this law is of fundamental importance to chemistry. It stipulates that the origin or method employed in preparing a compound in no way alters or influences its composition or characteristic properties. For example, pure water has a fixed composition of 11.1% hydrogen by mass and 88.9% oxygen by mass, no matter the source; and the physical and chemical properties are the same for every sample. Fifty grams of hydrogen and 50 grams of oxygen cannot produce a 50–50 percent-by-weight combination. Because water consists of 11.1% hydrogen and 88.9% oxygen, some of one element must be left over. In this case, the element in excess is hydrogen.

EXAMPLE 2–2

In a set of experiments designed to verify the Law of Definite Proportions, very pure tin metal was quantitatively combined with elemental bromine, forming tin tetrabromide. Using the data below, confirm the law by calculating the percentage of tin in each sample of the tetrabromide:

Grams of Sn reacted	Grams of SnBr$_4$ formed
2.8445	10.4914
3.0125	11.1086
4.5236	16.6752

Solution   You will find the percentages of Sn to be 27.113, 27.119, and 27.128. The three results are all the same to 3 significant figures. The tiny fluctuations are attributable to experimental error. ◼

To find the percentage of tin in SnBr$_4$:
$$\frac{\text{Wt. of Sn reacted}}{\text{Wt. of SnBr}_4 \text{ formed}} \times 100$$

EXERCISE 2–2

Direct combination of zinc and sulfur yields zinc sulfide. In a number of experiments, the weights of zinc and sulfur reacting are as follows:

Grams of Zn reacted	Grams of S reacted
5.776	2.831
10.428	5.114
2.453	1.204

Show that these data satisfy the Law of Definite Proportions.
Answer: The Zn/S ratio is very close to 2.04 in all three cases. ∎

EXAMPLE 2–3

Calculate the percentage of each of the elements in chlorobenzene (C_6H_5Cl), an important industrial solvent and organic intermediate.

Solution   The formula weight can be found from the chemical formula and the table of atomic weights on the inside back cover.

$$C_6H_5Cl = [6(12.01) + 5(1.008) + 1(35.453)]\,\mu = 112.55\,\mu$$

Note that the six carbon atoms have a combined atomic weight of $[6(12.01)]\,\mu = 72.06\,\mu$. Thus, the percentage of carbon in the sample is

$$\%C = \frac{72.06}{112.55}\,(100) = 64.02\%$$

Percent chlorine = 100% − (64.02% + 4.48%) = 31.5%. Also note that 35.453/112.56 × 100 = 31.50%.

Similarly, the percentage of hydrogen in any sample of pure chlorobenzene is 4.48%, and the percentage of chlorine can be determined by difference to be 31.50%. ∎

EXERCISE 2–3

From the chemical formula for each of the following compounds, determine the weight-percent composition of the elements present: (a) $CuCl$; (b) ZnF_2; (c) $AlCl_3$; (d) $SiBr_4$.
Answer: (a) 64.19% Cu, 35.81% Cl; (b) 63.24% Zn, 36.76% F; (c) 20.24% Al, 79.76% Cl; (d) 8.08% Si, 91.92% Br. ∎

A relatively small number of compounds in nature display variable composition. These **nonstoichiometric compounds** might at first appear to contradict the Law of Definite Proportions. However, this variability reflects the presence of defects in the crystal structures. One example of a nonstoichiometric compound is black nickel oxide, NiO, which has an average composition ranging from $Ni_{0.97}O_{1.0}$ to very close to $Ni_{1.0}O_{1.0}$. These oxides do not violate the atomic theory; they have complex solid structures in which some atoms are missing from their "proper" places. These **defect structures** are very important to the semiconductor and microelectronics industries because they make transistors possible.

2.3   THE LAW OF MULTIPLE PROPORTIONS

When the same two elements, such as hydrogen and oxygen, combine to form more than one compound, such as water (H_2O) and hydrogen peroxide (H_2O_2), the weights of one element combining with a fixed weight of another are related to each other in a ratio of small whole numbers. This statement is called the **Law of Multiple Proportions**. Here are the facts:

• Water (H_2O) consists of 1 gram of hydrogen atoms for every 8 grams of oxygen atoms, or a 1:8 ratio of the masses of the two kinds of atoms.

- Hydrogen peroxide (H_2O_2) contains 1 gram of hydrogen for every 16 grams of oxygen or a 1:16 ratio.

Therefore, the different weights of oxygen (8 and 16 grams) combined with the same weight of hydrogen (1 gram) stand in a 1:2 ratio, a ratio of small whole numbers. The explanation in terms of the atomic theory is that twice as many oxygen atoms combine with the same number of hydrogen atoms in hydrogen peroxide (H_2O_2) as in water (H_2O), consistent with the idea of atoms as the smallest building blocks of which matter is composed.

EXAMPLE 2–4

Demonstrate the law of multiple proportions by considering three of the oxides of nitrogen: in compound 1, each 16 g of oxygen combines with 28 g of nitrogen; in compound 2, each 16 g of oxygen combines with 14 g of nitrogen; and in compound 3, each 16 g of oxygen combines with 7 g of nitrogen. What are the probable formulas of the oxides?

Solution   The weights of nitrogen, 28, 14, and 7 g, combined with the same 16 g of oxygen are in the small whole number ratios of 4:2:1, as can be seen by dividing each weight by 16 g:

$$\frac{28}{16} : \frac{14}{16} : \frac{7}{16} = 4:2:1$$

Since we know that the atomic weights of nitrogen and oxygen are 14 and 16, respectively, we could write the formulas N_2O, NO, and $N_{0.5}O$. However, the atomic theory says that a molecule can't contain half an atom, so the last formula must actually represent the compound NO_2. In three other nitrogen oxides, the weights of oxygen that combine with 28 g of nitrogen are 48 g, 64 g, and 80 g, respectively. The weights are in the ratio 3:4:5; N_2O_3, N_2O_4, N_2O_5. ∎

EXERCISE 2–4

Fluorine and oxygen combine to form a fluoride whose weight-percent composition is 70.5% fluorine and 29.5% oxygen. These same two elements combine to produce a second fluoride whose weight percent composition is 54.2% fluorine and 45.8% oxygen. Show how these data confirm the law of multiple proportions. Suggest simple formulas for the two fluorides. ∎

2.4 THE ATOMIC THEORY AND THE MOLECULAR HYPOTHESIS

At the beginning of the 19th century, John Dalton set down the following principles, which summarized his atomic theory and the molecular hypothesis. His book, published in 1808, was called *A New System of Chemical Philosophy*. Here are the essential points:

1. Matter is made up of particles called atoms.
2. Atoms are indivisible and indestructible.
3. Atoms of one kind of element are the same. Atoms of different kinds of elements are different.
4. Compounds are made up of definite numbers of atoms of the component elements.
5. The weight of a compound equals the sum of the weights of the component elements.

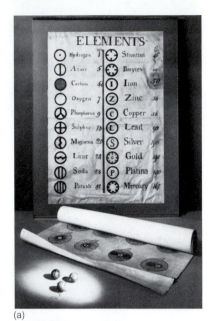

FIGURE 2–2 Dalton's lecture diagrams and atom models. John Dalton was a Quaker schoolteacher in Manchester, England. His *New System of Chemical Philosophy* was published in 1808.

An atom is the smallest unit of an element carrying the identifying properties of the element. Atoms are divisible into still smaller parts: electrons, protons, and neutrons.

By these principles, Dalton established the atomic basis of matter, correctly classifying pure substances as elements and compounds in terms of atoms and molecules (Figure 2–2). Thus, principles 1 and 2 provided the basis for understanding the indestructibility of matter in chemical reactions—Lavoisier's Law of Conservation of Mass (and later, the law of conservation of mass plus energy). Principles 3, 4, and 5 provided the basis for understanding of the Law of Definite Proportions.

The Law of Combining Volumes

In the years immediately following Dalton's statement of the atomic theory, two major observations about the properties of atoms and molecules were introduced. The first, by Joseph Gay-Lussac, concerned a simple empirical (observed) relationship involving gases, called the Law of Combining Volumes; the second, by Amadeo Avogadro, had to do with the relative numbers of particles contained in equal volumes of gases.

The **Law of Combining Volumes** states that for chemical reactions involving gases, combinations occur in simple proportions by volume. Furthermore, the volumes of the gaseous products of the reaction bear a simple relationship to the volumes of the gaseous reactants—a small, whole-number relationship. For example, under experimental conditions where temperature and pressure are kept constant, two volumes of gaseous hydrogen react with one volume of gaseous oxygen, producing two volumes of water vapor. The two reactants and the product form a 2:1:2 volume ratio:

$$2 \, H_2 \, (g) \; + \; O_2 \, (g) \longrightarrow 2 \, H_2O \, (g)$$

Note that we have added the symbol (g) following each species in the equation. It means the substance is in the gas phase. Other symbols that will be used are for liquids (liq), for solids (s), and for aqueous solutions (aq).

Hydrogen and chlorine produce hydrogen chloride in a 1:1:2 volume ratio:

$$H_2 \, (g) \; + \; Cl_2 \, (g) \longrightarrow 2 \, HCl \, (g)$$

Hydrogen and nitrogen produce ammonia in a 3:1:2 volume ratio:

$$3 \, H_2 \, (g) \; + \; N_2 \, (g) \longrightarrow 2 \, NH_3 \, (g)$$

Carbon monoxide and oxygen produce carbon dioxide in a 2:1:2 volume ratio:

$$2 \, CO \, (g) \; + \; O_2 \, (g) \longrightarrow 2 \, CO_2 \, (g)$$

Avogadro offered a rational explanation for Gay-Lussac's simple empirical statement about the behavior of gases. First, he distinguished between atoms and molecules. The term atom was to be restricted to elements. The term molecule could be applied to elements when speaking about chemical compositions of the same atoms (Figure 2–3), and it could also be applied to compounds when describing chemical combinations of different kinds of atoms. On that basis, an **atom** is considered the smallest part of an element or compound that is capable of uncombined existence. A **molecule** is a combination of atoms. Gaseous hydrogen, for example, is an element composed of pairs of H atoms, chemically combined into H_2 molecules. Nonetheless, it is very difficult to separate the pair of atoms in a molecule. Carbon dioxide (CO_2) molecules are triatomic, each molecule containing one carbon and two oxygen atoms.

(a)   (b)   (c)

FIGURE 2–3   Chlorine, bromine, and iodine, three elements composed of simple molecules. Bromine is one of only two elements that is liquid at room temperature.

Selecting Atomic Weights

The last piece of this puzzle was put in place by a 32-year-old Italian scientist working in Geneva. In 1858, Stanislao Cannizzaro was able to show how to apply Avogadro's hypothesis to the problem of selecting correct weights for the atoms of the elements. Here are his postulates:

1. The atomic theory set forth the idea that a definite weight must exist for the atoms of any element.
2. Since molecular species such as the hydrogen molecule or the water molecule must contain definite numbers of atoms, they too must have definite weights, which we refer to as formula weights.
3. These characteristic formula weights contain one atomic weight (or a whole-numbered multiple of that atomic weight) for each element present.

According to **Avogadro's Hypothesis**, at constant temperature and pressure, equal volumes of gases contain equal numbers of molecules. Therefore, the weights of these equal volumes must be in the same ratio as the weights of the molecules. In other words, the densities of these gases must be in the same ratio as their formula weights. Experimentally, formula weights for gaseous elements and compounds could then be determined by measuring their densities and comparing them to some standard of reference such as gaseous hydrogen, the lightest of the elements. Now, assigning a formula weight of 2 to molecular hydrogen, the formula of which we know to be H_2, the atomic weight for hydrogen must be 1 (Table 2–1). Therefore, the weight of H in H_2O is 2; the weight of H in NH_3 is 3; and the weight of H in CH_4 is 4. There are one, two, three, and four H atoms per formula, respectively.

Determination of the formula weights of chlorine, oxygen, nitrogen, and carbon follows directly from a percentage composition (by weight) analysis for the elements present in their compounds (Table 2–2). Finally, the **combining capacity** of an element can be determined on the basis of

TABLE 2–1   Formula Weights from Gas Density Measurements Based on H_2

Substance	Relative Weight	Weight of H from Combining Volumes*	Combining Capacity[†]
Hydrogen (H_2)	2	2	
Hydrogen chloride (HCl)	36.5	1	Cl = 1
Water (H_2O)	18	2	O = 2
Ammonia (NH_3)	17	3	N = 3
Methane (CH_4)	16	4	C = 4

* $\dfrac{\text{(volumes of hydrogen)}(2)}{\text{(volumes of product substance)}}$ = unitless weight relative to hydrogen

[†]Valence, or combining power, in HCl, H_2O, NH_3, and CH_4

TABLE 2–2   Relative Weights for Hydrogen Chloride, Water, Ammonia and Methane

Weight Percent Composition	Relative Formula Weight	Relative Atomic Weight
97.3% Cl in hydrogen chloride	36.5	Cl = (0.973)(36.5) = 35.5
89.9% O in water	18	O = (0.899)(18) = 16
82.4% N in ammonia	17	N = (0.824)(17) = 14
75.0% C in methane	16	C = (0.750)(16) = 12

how different elements combine with equivalent amounts of a common element—in this case, hydrogen. If hydrogen has a combining capacity of 1, then chlorine must also. Oxygen must have a combining capacity of 2, since one equivalent can combine with two equivalents of hydrogen. Nitrogen must have a combining capacity of 3, and carbon must have a combining capacity of 4.

2.5   ISOTOPES AND AVERAGE ATOMIC WEIGHTS

The atomic weight of an element is not identical to the actual weights of the atoms unless all of the individual atoms of the element have the same mass. There is no particle of chlorine that weighs 35.453 μ, even though 35.453 is the correct atomic weight for chlorine, according to the modern periodic table of the elements (printed on the inside front cover). Individual chlorine atoms have weights of either 35 μ or 37 μ (on the ^{12}C scale). The elemental substance called chlorine is a mixture of these two kinds of atoms in about a 3:1 ratio, with an *average* mass of 35.453 μ (Table 2–3).

Isotopes are atoms of the same element but with different atomic weights. Wherever we find chlorine in nature, the isotopic composition is the same and so is its chemistry. The mixture of isotopes behaves as if it were a single substance in every chemical process. Chlorine samples collected from widely separated sources, and from rocks that have never been in contact with the oceans, exhibit no detectable difference in atomic weight from chlorides extracted from ocean waters. Common lead, obtained from minerals not associated with radioactivity, consists of a mixture of four isotopes—masses 203.973, 205.9745, 206.9759, and 207.9766—regardless of the source. The accepted value of 207.2, reflecting the percentage com-

Differences in the relative masses of the atoms of different elements can be determined with great accuracy from the conversion of mass to energy in a nuclear reaction, according to Einstein's equation: $E = mc^2$.

TABLE 2–3   Natural Abundance of the Isotopes of Several Common Elements

Element	Relative Isotopic Weight	Natural Abundance (%)	Atomic Weight for the Average Abundance
Hydrogen	1.008	99.98	
	2.014	0.02	1.008
Carbon	12.000	98.90	
	13.003	1.10	12.011
Oxygen	15.995	99.762	
	16.999	0.038	
	17.999	0.200	15.999
Chlorine	34.97	75.77	
	36.97	24.23	35.453
Iron	53.940	5.90	55.84
	55.935	91.52	
	56.936	2.24	
	57.932	0.33	
Iodine	126.904	100	126.90
Lead	203.973	1.4	
	205.974	24.1	
	206.976	22.1	
	207.977	52.4	207.2
Uranium	234.041	0.0057	
	235.044	0.719	
	238.051	99.274	238.03

position of the four isotopes, is the proper atomic weight of lead found in the table of atomic weights and the periodic table of the elements. Note that the precision of the accepted atomic weight for lead is fairly low, since the natural abundances vary slightly.

EXAMPLE 2–5

If the average mass for Cl is 35.453, and the element is a mixture of chlorine atoms of masses 34.97 and 36.97 μ, what is the weight-percent composition of the two isotopes?

Solution   Let x = fraction of mass 34.97 and y = fraction of mass 36.97. Then $x + y$ must equal 1 exactly. We know that $34.97x + 36.97y = 35.453$. We must solve the two equations for the two unknowns.

$$x + y = 1 \tag{1}$$
$$34.97x + 36.97y = 35.453 \tag{2}$$

Solving for x in equation (1) and substituting into equation (2),

$$34.97(1 - y) + 36.97y = 35.453$$
$$y = 0.242; \quad x = 1 - 0.242 = 0.758$$

The weight-percent composition is 24.2% chlorine atoms of mass 36.97 μ and 75.8% chlorine atoms of mass 34.97 μ. ■

EXERCISE 2–5

There are two isotopic forms of lithium that occur in nature, 7.42% of mass 6.015 μ and 92.58% of mass 7.016 μ. Calculate the relative atomic weight for the natural abundance of the isotopes.
Answer: 6.94 μ. ■

FIGURE 2–4 Refractories are hard materials that are stable at high temperatures and are used for industrial polishing, grinding, and cutting. This new tungsten carbide cutting tool features a unique "ledge" shape, enabling it to sharpen itself as it wears and to operate at increased cutting speeds with metals and alloys, especially titanium.

2.6 EXPERIMENTAL DETERMINATION OF ATOMIC WEIGHTS

We can summarize the essence of the last sections in a few sentences, which are among the most important in beginning the study of chemistry. The assumption that atoms of a single isotope—every one of them—have the same mass is one of the cornerstones of modern atomic theory. The fact that the ratio of the weights of the elements that have combined to form compounds is a small, whole-number ratio confirms that assumption. Combining weight ratios are the same as the ratios of the numbers of atoms in the molecules, so that samples of any substances having masses equal to their formula weights contain the same number of molecules. In the remaining sections of this chapter we discuss applications of these ideas. But before launching into the implications and applications of the atomic theory, it is useful to illustrate how one obtains atomic weights experimentally. Since atomic weights are a central issue to all of chemistry, it should not be surprising that there are many ways of obtaining them, which we will point out as we go. To begin with, here are two variations on the same theme—atomic weights by direct combination.

Atomic Weights by Direct Combination

Accurate values for the relative atomic weights of many elements can be obtained by direct chemical combination. The reaction of an element of unknown atomic weight with an element of known atomic weight establishes a combining ratio for the two elements. Since chemical reactions bring atoms together in small, whole-number combinations, the ratio in which two such elements combine will be equal to, or some simple multiple of, the ratio of their respective atomic weights. Experimentally, two kinds of data are needed: (1) accurate chemical analysis and (2) information about the chemical formula.

EXAMPLE 2–6

Determine the atomic weight of copper from the following facts: 63.5 g of copper combine with 16.0 g of oxygen, forming a 1:1 oxide of copper.

Solution The combining weight of copper, the amount that combines with 16.0 g of oxygen, is 63.5 g. If the chemical combination of Cu and O is known to be CuO, then the relative atomic weight of Cu must be 63.5 μ. ∎

EXERCISE 2–6

Refractories are materials specifically designed to withstand the high temperatures encountered in furnaces (Figure 2–4). Magnesia refractories such as forsterite are composed of magnesia and silica. (a) Without consulting the periodic table of the elements, if the Mg/O combining weight ratio in magnesia (MgO) is 1.52:1.00, what is the relative atomic weight of Mg? Compare your answer to the value in the periodic table. (b) If the Si/O combining weight ratio in silica (SiO_2) is 0.878, what is the combining weight of silicon and what is its relative atomic weight?
Answer: (a) 24.3 μ; (b) 14.0 grams, 28.0 μ. ∎

In order to understand the importance of the chemical formula in establishing atomic weights, consider the following example. In this case,

50.77 g of iodine combine with 16.00 g of oxygen, forming a stable oxide. If this oxide turned out to be a 1:1 combination, then you might rightly conclude that the atomic weight of iodine in grams is 50.77. But the chemical combination is not 1:1, it is 2:5, and thus the correct formula is I_2O_5. Therefore, the atomic weight of iodine is 2.5 times 50.77, or 126.9 μ. Think of it this way: You need 2.5 atomic weights of oxygen to combine with 1 atomic weight of iodine, or $I_1O_{2.5}$. But since our theory of atoms cannot accommodate fractions of atoms—only integer values—multiply through by 2, giving 5 atomic weights of oxygen for every 2 atomic weights of iodine. Hence, the correct formula is I_2O_5.

> The combining weight is either the atomic weight or some multiple or fraction thereof.

EXAMPLE 2–7

Kaolinite refractories contain alumina, Al_2O_3. Making use of the atomic weight of aluminum from the periodic table of the elements, calculate the Al/O ratio and the combining weight of aluminum.

Solution   The Al/O ratio is obtained from the chemical formula and the atomic weights: $(2)(27.0\ \mu)/(3)(16.0\ \mu) = 1.125$. On that basis, the combining weight of aluminum is $(1.125)(16.0\ \mu) = 18.0\ \mu$. ■

EXERCISE 2–7

Chromite is a mixed iron oxide and chromium oxide refractory. This particular chromium oxide is known to have a combining weight of 34.667 g for the metal. Taking the values for the atomic weights from the periodic table of the elements, determine the chemical formula for the oxide. *Answer:* Cr_2O_3. ■

Atomic Weights from Specific Heats and Chemical Analysis

Depending on the chemical formula, the weight of a metal M combined with 16 g of oxygen can be the atomic weight or some simple fraction or multiple thereof. That being the case, we need to be able to select the correct formula from among the possibilities: MO if the combining weight is the atomic weight; M_xO if the combining weight is a fraction of the atomic weight; or MO_x if the combining weight is multiple of the atomic weight. The **Law of Dulong and Petit** is an empirical relationship that furnishes just that kind of information. In 1819, these two French physicists discovered that the product of the specific heat and the atomic weight of most solid elements is approximately 25 J/mol·K. This number is called the molar specific heat, the quantity of heat necessary to raise the temperature of a formula weight in grams, or one mole (mol), by one kelvin, and it is about the same for many solid elements (Table 2–4):

> Nearly a century passed before Einstein explained the sweeping, fundamental implications of the existence of Dulong and Petit's empirical relationship between specific heat and atomic weight.

$$(\text{specific heat})(\text{atomic weight}) \cong 25\ \text{J/mol·K}$$

EXAMPLE 2–8

Consider an iron oxide sample and the problem of finding its exact atomic weight based on the following information: Chemical analysis indicates the weight-percent iron to be 69.956. The weight x of iron combined with 16 g of oxygen is

			Specific Heat ×
	Atomic Weight	Specific Heat	Atomic Weight
Element	(g/mol)	(J/g·K)	(J/mol·K)
Calcium	40.08	0.6552	26.3
Copper	63.55	0.3866	24.6
Molybdenum	95.94	0.2498	24.0
Gold	197.0	0.1298	25.4

TABLE 2–4   Specific Heat Data for Some Solid Elements at 25°C

$$\frac{x \text{ g Fe}}{16 \text{ g O}} = \frac{69.956 \text{ g Fe}}{[100.00 - 69.956] \text{ g O}}$$

$$x = 37.255 \text{ g Fe}$$

which is the combining weight of iron. Turning to the law of Dulong and Petit, the known specific heat of iron metal is 0.444 J/g·K, so the approximate atomic weight is

$$\frac{25 \text{ J/mol·K}}{0.444 \text{ J/g·K}} = 56.3 \text{ g/mol}$$

Comparing 56.3 g/mol, the approximate value for the atomic weight, with the combining weight of 37.255 g/16 g O suggests that the actual value is 1.5 times the value obtained by direct combination of iron and oxygen. On that basis, the atomic weight of iron is

$$(37.255)(1.5) = 55.88 \text{ g/mol}$$ ■

EXERCISE 2–8

$$\frac{56}{37.255} = 1.5$$

The combining weight of molybdenum metal in one of its several oxides is known to be 31.98 g/16 g O. The specific heat of molybdenum is known to be 0.250 J/g·K. What is the atomic weight of molybdenum?
Answer: 95.94 g/mol. ■

There are many more ways of experimentally determining atomic weights, ranging from classical methods based on gas densities (measured at low pressures) to modern instrumental techniques such as mass spectroscopic analysis (Figure 2–5) and nuclear reaction energy measurements. Results from these last two methods, which are the most accurate known, are combined to give the best current values as shown in Table 2–4. There are also many ways to double check the atomic weights of the elements, such as specific heats, periodic relationships between properties of its compounds and other compounds with similar properties, and characteristic X-ray spectra. More will be said about many of these important ideas and practices as we go.

2.7   THE MOLE CONCEPT

The methods we described for determining atomic and molecular weights constitute the beginnings of a general approach to chemical calculations and quantitative methods called chemical **stoichiometry**. Stoichiometry is that part of chemistry and chemical engineering on which the determination of the quantities of substances entering into a chemical reaction, or

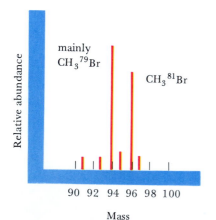

FIGURE 2–5 The mass spectrum of methyl bromide (CH_3Br) is in the region of molecular weight 95 μ, showing the resolution of the two bromine isotopes, $CH_3\,^{79}Br$ (94 μ) and $CH_3\,^{81}Br$ (96 μ).

produced as products of a chemical reaction, is based. Although the quantities of substances range from laboratory scale—generally grams and milligrams—to plant production scale that could be pounds or tons, it is the number of particles contained in those quantities that is essential. But since the particles of elements and compounds are atoms and molecules and very small indeed, chemists and engineers of necessity resort to weighing procedures as a convenient means for counting out particles.

To state the case, recall that samples of any substances having weights equal to their formula weights contain the same number of molecules. Weighing out a quantity in grams equal to the formula weight of a substance is the same as counting out 6.022×10^{23} particles of that substance. The formula weight in grams is called a **mole**, abbreviated mol, and the number 6.022×10^{23} particles per mole is called N_A, Avogadro's number, where

$$N_A = N/n$$

Here N is the number of particles present in n moles of a substance. The mole is defined as the quantity of a substance that contains as many particles as there are atoms of the ^{12}C isotope of carbon in 12.0000 grams of the isotope: 6.022×10^{23} particles. Simply put, an atomic or molecular weight (or a formula weight) contains Avogadro's number of atoms or molecules (or particles). The units of atomic weight and formula weight can be stated as μ or g/mol. Thus, the atomic weight of H can be given as $1.0079\ \mu$ or as 1.0079 g/mol.

Some simple illustrations can help us to understand this relationship between weight and number. The elements carbon, potassium, silver, and uranium have characteristically different properties, and their relative atomic weights are also different. However, one mole of each kind of atom contains the same number of particles, Avogadro's number. In a similar way, the count in a dozen grapes, oranges, grapefruit, or watermelons is 12. They are all different particles, and the separate collections of particles have different (collective) weights, which will be very clear to anyone carrying a dozen grapes or a dozen watermelons. But the particle count in each set is the same—12. For substances at the atomic and molecular level, the particle count is in multiples or fractions of Avogadro's number—6.022×10^{23}. Figure 2–6 shows a mole of each of several elements and compounds.

Avogadro's number N_A gives us the interesting and useful relationship between the mass of a mole of particles and the mass of a single particle. For example, the atomic weight of the H atom is 1.0079 g/mol, and therefore the mass of a single H atom can be calculated directly:

$$m_H = (1.0079 \text{ g/mol}) \left(\frac{1 \text{ mol}}{6.022 \times 10^{23} \text{ H atoms}} \right)$$
$$= 1.674 \times 10^{-24} \text{ g/atom}$$

The mass of a single iron atom can be calculated in the same way. From the table of atomic weights, we know that the atomic weight of iron is 55.847 g/mol. Thus:

$$m_{Fe} = (55.847 \text{ g/mol}) \left(\frac{1 \text{ mol}}{6.022 \times 10^{23} \text{ Fe atoms}} \right)$$
$$= 9.274 \times 10^{-23} \text{ g/atom}$$

One mole of H_2O molecules weighs 18.0 g, and therefore one water molecule weighs

Avogadro's number has been counted many times—and in many different ways—since it was first determined. The current best value is about 6.02252×10^{23} particles/mol.

FIGURE 2–6   (a) One mole of some common elements. Back row (left to right) bromine, aluminum, mercury, and copper. Front row (left to right) sulfur, zinc, and iron. (b) One-mole quantities of a range of compounds. The white compound is NaCl (58.44 g/mol); the blue compound is $CuSO_4 \cdot 5H_2O$ (249.68 g/mol); the deep red compound is $CoCl_2 \cdot 6H_2O$ (165.87 g/mol); the green compound is $NiCl_2 \cdot 6H_2O$ (237.70 g/mol); and the orange compound is $K_2Cr_2O_7$ (294.19 g/mol).

(a)

(b)

$$(18.0 \text{ g/mol}) \left(\frac{1 \text{ mol}}{6.022 \times 10^{23} \text{ H}_2\text{O molecules}} \right)$$

$$= 2.99 \times 10^{-23} \text{ g/molecule}$$

A formula weight in grams is a mole of particles. 18.0 g of H_2O contains 6.022×10^{23} H_2O molecules.

To count one mole of molecules, it is only necessary to weigh out the molecular weight in grams. If instead you are in need of some fraction of that number of molecules, or some multiple, then simply weigh out that fraction or multiple of the molecular weight in grams. For example, weighing out 180 grams of H_2O molecules is the same as counting out 10 times Avogadro's number of water molecules; weighing out 1.80 grams is the same as counting out 1/10 of Avogadro's number of water molecules.

The point of this discussion should be kept clearly in mind. Chemical processes involve interactions between atoms and molecules. Chemical formulas contain small, whole-number ratios of atoms. Consequently, it is the relative numbers of atoms and molecules, not their relative weights, that are of primary concern. Formula weights only provide the means for determining the numbers of atoms.

EXAMPLE 2–9

Calculate the mass of one P_4O_{10} molecule.

Solution   The formula weight in grams of P_4O_{10} is 284 g. Therefore,

$$\left(\frac{284 \text{ g } P_4O_{10}}{1 \text{ mol}}\right)\left(\frac{1 \text{ mol}}{6.022 \times 10^{23} \text{ } P_4O_{10} \text{ molecules}}\right)$$
$$= 4.72 \times 10^{-22} \text{ g/molecule} \quad \blacksquare$$

(a)

EXERCISE 2–9(A)

Give the formula weight in g/mol for each of the following workhorse industrial and laboratory reagent chemicals:
(a) Ammonia (NH_3) and sodium hydroxide (NaOH)
(b) Nitric acid (HNO_3) and acetic acid (CH_3COOH)
Answer: (a) 17.0 g/mol, 40.0 g/mol; (b) 63.0 g/mol, 60.0 g/mol. $\blacksquare$

EXERCISE 2–9(B)

For a 150-gram sample of phosphorus trichloride (PCl_3), calculate each of the following:
(a) The mass of a single PCl_3 molecule
(b) The number of moles of PCl_3 in the sample
(c) The number of grams of Cl atoms in the sample
(d) The number of molecules of PCl_3 in the sample
Answer: (a) 2.283×10^{-22} g; (b) 1.09 mol; (c) 116 g of Cl; (d) 6.56×10^{23} molecules. $\blacksquare$

2.8   CHEMICAL EQUATIONS AND REACTIONS

Conventions for Writing Chemical Equations

A **chemical equation** is a shorthand notation for a chemical reaction (Figure 2–7). Bonds break and bonds form, rearrangements take place at the atomic level, and new substances, often with radically different properties, form (Table 2–5). Mass and energy are conserved. Chemical equations state:

1. the formula of each reactant and product, thus identifying them;
2. the quantitative relationships between reactants and products, thus emphasizing conservation of mass and atoms (that is, the equation is balanced). A more complete discussion of the nomenclature or naming of inorganic compounds will be presented in Chapter 4. For the moment, keep in mind that many names are historical rather than systematic.

Consider the combustion of potassium in an oxygen atmosphere (Figure 2–8):

$$4 \text{ K (s)} + O_2 \text{ (g)} \longrightarrow 2 \text{ K}_2O \text{ (s)}$$

A great deal of information is conveyed by such a formalized chemical statement:

(b)

(c)

FIGURE 2–7   (a) Sodium chloride crystals. (b) Sodium metal is soft and can be cut easily. Chlorine gas is shown in Figure 2–3. (c) When sodium metal reacts with chlorine gas it produces solid sodium chloride, seen as a white smoke rising out of the test tube.

FIGURE 2–8 As potassium metal burns in an oxygen atmosphere, it displays its characteristic purple flame.

TABLE 2–5   Some Properties of the Substances Involved in the Cycle Starting with Electrolytic Decomposition of NaCl*

Substance	Appearance	Melting Point/ Boiling Point (°C)	Reactivity
Salt (NaCl)	white, crystalline solid	801/1413	unreactive; soluble in water
Sodium (Na)	soft, lustrous metallic solid	97.5/883	highly reactive; generates hydrogen gas in water
Chlorine (Cl_2)	greenish-yellow gas	−103/−33	slightly soluble in water; oxidizing agent
Hydrogen (H_2)	colorless gas	−259/−253	highly reactive; explosive in O_2
Caustic soda (aqueous NaOH)	colorless solution		corrosive

*These chemicals rank near the top, close to sulfuric acid and ammonia, in dollar value of use. Their applications are so diverse that it would be fair to say that hardly a consumer product is sold that does not depend in some way upon them.

- Potassium reacts with oxygen, to produce potassium oxide, in a 4:1:2 particle ratio, or a weight ratio of 39:8:47

- Four moles of potassium atoms react with one mole of oxygen molecules to produce two moles of potassium oxide.

- 156 g of potassium react with 32 g of oxygen to produce 188 g of potassium oxide.

- 4 moles of K atoms, originally present as reactant on the left side, are now incorporated into the potassium oxide product on the right side.

- 2 moles of O atoms end up chemically incorporated into 2 moles of potassium oxide on the product side.

- Reactants and products are solids or gases under the reaction conditions, as indicated by the symbol (s) or (g) following each formula.

In summary, here are the standard conventions that chemists and chemical engineers worldwide generally observe in writing chemical equations for reactions and processes:

- Reactants are written on the left side of the arrow; products appear on the right.

- Only those atoms or molecules experiencing a net change during the reaction are written into the equation.

- The equation must be **balanced** (the number of atoms of each element must be the same on both sides of the equation), demonstrating the conservation of mass principle.

- The state of matter should be shown in parentheses, following the chemical formula; **(s)** refers to the solid state and **(g)** to the gaseous state. We will use **(liq)** for the liquid state. Finally, **(aq)** is used for a substance dissolved in water—a substance in the aqueous phase.

Here are four examples of chemical reactions and balanced chemical equations, and a sentence on how each should be read.

1. Solid calcium carbonate can be decomposed to produce solid calcium oxide and gaseous carbon dioxide in a 1:1:1 ratio:

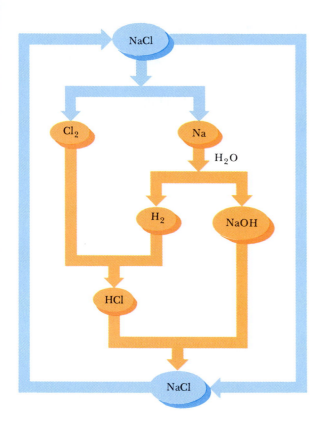

FIGURE 2–9 Schematic representation of a series of chemical reactions involving sodium chloride, sodium metal, and chlorine gas. Beginning with white, crystalline NaCl (Table 2–5 and Figure 2–7), electrolytic decomposition produces Na and Cl_2. Reaction of Na with H_2O vigorously produces NaOH and liberates H_2, which can react with Cl_2, producing HCl. The cycle is coupled with the reaction of HCl and NaOH, regenerating NaCl.

$$CaCO_3 \text{ (s)} \longrightarrow CaO \text{ (s)} + CO_2 \text{ (g)}$$

2. In a reaction occurring entirely in the gas phase, sulfur dioxide can interact with oxygen to yield gaseous sulfur trioxide in a 2:1:2 ratio:

$$2 \, SO_2 \text{ (g)} + O_2 \text{ (g)} \longrightarrow 2 \, SO_3 \text{ (g)}$$

3. Silver nitrate reacts with sodium chloride in aqueous solution, and the products of the reaction are aqueous (soluble) sodium nitrate and solid (insoluble) silver chloride. The silver chloride precipitates from the medium.

$$AgNO_3 \text{ (aq)} + NaCl \text{ (aq)} \longrightarrow AgCl \text{ (s)} + NaNO_3 \text{ (aq)}$$

4. In a reaction taken from the reaction flow chart in Figure 2–9, solid sodium reacts with liquid water, producing an aqueous solution of caustic soda and hydrogen gas:

$$2 \, Na \text{ (s)} + 2 \, H_2O \text{ (liq)} \longrightarrow 2 \, NaOH \text{ (aq)} + H_2 \text{ (g)}$$

Balancing Chemical Equations

Chemical equations can be balanced by one of several methods. For complex chemical processes (such as the class of reactions known collectively as oxidation-reduction or redox reactions), a systematic procedure is used to avoid a tedious trial-and-error ritual. On the other hand, many simple chemical equations can be successfully balanced by trial and error or, as it has come to be known, by inspection. Consider the oxidation of iso-octane, a major constituent of automotive and aviation fuels. The *unbalanced* equation is:

$$C_8H_{18} \text{ (liq)} + O_2 \text{ (g)} \longrightarrow CO_2 \text{ (g)} + H_2O \text{ (g)}$$

To begin the balancing process, select an atom present in only one molecule on each side of the arrow—in this case, C in C_8H_{18} (on the left) and in CO_2 (on the right), and H in C_8H_{18} (left) and in H_2O (right)—and balance it:

$$C_8H_{18} \text{ (liq)} + O_2 \text{ (g)} \longrightarrow \mathbf{8}\ CO_2 \text{ (g)} + \mathbf{9}\ H_2O \text{ (g)}$$

Then, complete the balancing process for these same atoms on the left side:

$$C_8H_{18} \text{ (liq)} + \frac{25}{2}\ O_2 \text{ (g)} \longrightarrow \mathbf{8}\ CO_2 \text{ (g)} + \mathbf{9}\ H_2O \text{ (g)}$$

The fraction can be eliminated by multiplying through by a factor of 2:

$$\mathbf{2}\ C_8H_{18} \text{ (liq)} + \mathbf{25}\ O_2 \text{ (g)} \longrightarrow \mathbf{16}\ CO_2 \text{ (g)} + \mathbf{18}\ H_2O \text{ (g)}$$

EXAMPLE 2–10
Balance the equation

$$Fe_2(SO_4)_3 \text{ (aq)} + BaCl_2 \text{ (aq)} \longrightarrow BaSO_4 \text{ (s)} + FeCl_3 \text{ (aq)}$$

Solution   Begin the balancing process by comparing Fe in $Fe_2(SO_4)_3$ on the left of the arrow and Fe in $FeCl_3$ on the right of the arrow. You'll need **2** $FeCl_3$ and **3** $BaSO_4$ on the right to balance the S, O, and F atoms:

$$Fe_2(SO_4)_3 \text{ (aq)} + BaCl_2 \text{ (aq)} \longrightarrow 3\ BaSO_4 \text{ (s)} + 2\ FeCl_3 \text{ (aq)}$$

Then balance the Ba and Cl atoms with 3 $BaCl_2$ on the left:

$$Fe_2(SO_4)_3 \text{ (aq)} + 3\ BaCl_2 \text{ (aq)} \longrightarrow 3\ BaSO_4 \text{ (s)} + 2\ FeCl_3 \text{ (aq)} \quad \blacksquare$$

EXERCISE 2–10
Balance the following equations by inspection. They all represent real processes of considerable economic value and historic significance.

1. The **LeBlanc process** for the manufacture of caustic soda by the batch reaction of caustic ash with lime. The success of this process depends on the nearly total insolubility of calcium carbonate in aqueous solutions.

$$Na_2CO_3 \text{ (aq)} + Ca(OH)_2 \text{ (aq)} \longrightarrow NaOH \text{ (aq)} + CaCO_3 \text{ (s)}$$

2. The **Deacon process** is widely used because most commercial chlorination processes (that is, addition of Cl_2 to other compounds) produce by-product HCl in huge quantities, and therefore there is an economic incentive to find something to do with it. Regeneration of chlorine was the best possible solution:

$$HCl \text{ (g)} + O_2 \text{ (g)} \longrightarrow Cl_2 \text{ (g)} + H_2O \text{ (liq)}$$

3. Sodium pyrophosphate is widely used as a chemical leavening agent in producing donuts, cakes, and biscuit doughs. It is manufactured by partially dehydrating sodium dihydrogen phosphate at elevated temperatures:

$$NaH_2PO_4 \text{ (s)} \longrightarrow Na_2H_2P_2O_7 \text{ (s)} + H_2O \text{ (g)}$$

4. Synthesis of boron carbide (B_4C), a refractory material approaching diamond in hardness. It is used in applications in which wear-resistance is important.

$$B_2O_3 \text{ (s)} + C \text{ (s)} \xrightarrow{2500°C} B_4C \text{ (s)} + CO \text{ (g)} \quad \blacksquare$$

2.9 STOICHIOMETRIC COMPUTATIONS

We have said that quantitative relationships in chemistry depend upon the number of particles participating in a process, upon the chemical formula, and upon the balanced chemical equation. Let's illustrate these statements by example. Consider the reaction of 100 g of hydrogen gas with sufficient oxygen gas to produce the **stoichiometric quantity** of steam; (that is, the amount predicted by the balanced equation):

$$2 \ H_2 \ (g) + O_2 \ (g) \longrightarrow 2 \ H_2O \ (g)$$

What is the weight of the water produced? First find the number of moles of H_2 involved:

$$\text{moles of } H_2 = \frac{\text{weight of } H_2}{\text{formula weight of } H_2}$$

$$= (100. \ \text{g of } H_2) \left(\frac{1 \ \text{mol}}{2.0 \ \text{g of } H_2} \right) = 50 \ \text{mol } H_2$$

According to the logic of the balanced chemical equation, 2 moles of H_2 produce 2 moles of H_2O, so we can write:

$$\text{moles of } H_2O \text{ produced} = \text{moles of } H_2 \text{ consumed} = 50 \text{ moles}$$

$$\text{weight of } H_2O \text{ produced} = (\text{moles of } H_2O)(\text{formula weight of } H_2O)$$

$$= (50 \ \text{mol } H_2O) \left(\frac{18 \ \text{g}}{\text{mol } H_2O} \right)$$

$$= 900 \ \text{g } H_2O$$

An important feature of this problem is the use of conversion factors. As we pointed out in Chapter 1, correct units are a part of every measurement and cannot be separated from that measurement. Therefore, in all mathematical operations, remember to do unto the units as you would do unto the numbers, and follow the units to make sure the answer comes out in the units desired.

EXAMPLE 2–11

Hydrogen reacts with nitrogen to produce ammonia:

$$3 \ H_2 \ (g) + N_2 \ (g) \longrightarrow 2 \ NH_3 \ (g)$$

Determine how much ammonia would be produced if 100 g of hydrogen reacts.

Solution   First find the number of moles of H_2 involved:

$$\text{moles of } H_2 = \frac{\text{weight of } H_2}{\text{formula weight of } H_2} = \frac{100. \ \text{g}}{2.0 \ \text{g/mol}} = 50. \ \text{mol}$$

From the balanced equation,

$$\text{mol of } NH_3 \text{ produced} = \tfrac{2}{3}(\text{mol of } H_2 \text{ consumed})$$

$$= \tfrac{2}{3}(50 \ \text{mol}) = 33.3 \ \text{mol}$$

$$\text{g } NH_3 \text{ produced} = (33.3 \ \text{mol } NH_3) \left(\frac{17 \ \text{g}}{\text{mol } NH_3} \right)$$

$$= 566 \ \text{g } NH_3$$

EXERCISE 2–11

Methanation involves passing synthesis gas, a mixture of carbon monoxide and hydrogen, over a catalyst of nickel supported on alumina:

$$CO \ (g) \ + \ H_2 \ (g) \ \xrightarrow{\text{Ni/Al}_2\text{O}_3, \ 180°C} \ CH_4 \ (g) \ + \ H_2O \ (g)$$

Balance the equation. Then, consider the reaction of 100 g of H_2 with a *stoichiometric* quantity of carbon monoxide and determine (a) the number of moles of H_2 and CO that react and (b) the number of grams of CH_4 and H_2O produced. (c) Finally, check the material balance by comparing total grams of reactants to total grams of products.

Answer: The coefficients in the balanced chemical equation are 1:3:1:1. (a) 50 mol H_2 and 50/3 mol CO; (b) 267 g CH_4 and 300 g H_2O; (c) 567 g = total mass of reactants = total mass of products. ∎

The Empirical Formula

An important application of the mole concept is the calculation of empirical formulas for particular compounds. The **empirical formula** is the simplest expression of the ratio of the numbers of atoms of the elements present in a compound. The usual procedure involves first determining all the elements present (by analysis), along with the fraction or weight-percent composition. Consider a compound of hydrogen and oxygen that is found to be 5.88% H and 94.1% O by weight. In order to establish the formula for this compound, divide the weight-percentage of each by its relative atomic weight. That establishes the mole ratio of the elements present in the compound. In every 100.0 g of this compound there is 5.88% or 5.88 g of H, and 94.1% or 94.1 g of O, according to the data above:

$$\text{Hydrogen} \qquad \frac{5.88 \ \text{g H}}{1.01 \ \text{g H/mol H}} = 5.82 \ \text{mol H in every 100 g}$$

$$\text{Oxygen} \qquad \frac{94.1 \ \text{g O}}{16.00 \ \text{g O/mol O}} = 5.88 \ \text{mol O in every 100 g}$$

$$\frac{\text{mol H}}{\text{mol O}} = \frac{5.82}{5.88} \cong 1/1$$

We know the atoms combine in small whole-number ratios and that some error in the experiment is unavoidable. Therefore, the empirical formula for this compound of hydrogen and oxygen must be HO, since only those two elements are present and they are found in a 1:1 atom ratio.

Note that the empirical formula for a compound provides no information about structure; it contains only composition data. For example, the empirical formula for carbon dioxide is CO_2, which might lead you to conclude erroneously that the atoms are connected that way, namely C—O—O. In fact, the atoms are connected O—C—O.

The Molecular Formula

The empirical formula tells you no more than the ratio of the different atoms in a substance. In order to arrive at the **molecular formula**, which states the numbers of atoms in each molecule, you must also know the true formula weight. The formula weight must be some integral multiple of the empirical formula weight. In our previous example, the empirical formula weight for HO is 17. However, since the true formula weight is known (from other experiments) to be 34, the simple species (HO) must actually be $(HO)_2$. The compound in question is hydrogen peroxide:

$$(HO)_2 = H_2O_2 = 34 \text{ g/mol}$$

For glucose, the true formula weight is 180 g/mol. The empirical formula is CH_2O, for which the formula weight would be 30 g/mol. Since the molecular formula must be some integral multiple of the empirical formula, $(CH_2O)_x$ is the molecular formula, where

$$x = \text{an integer} = \frac{\text{molecular formula weight}}{\text{empirical formula weight}}$$

Therefore, $x = 180/30 = 6$, and

$$(CH_2O)_x = (CH_2O)_6 = C_6H_{12}O_6 = 180 \text{ g/mol}$$

Problems in Stoichiometry

Solving problems in stoichiometry requires a firm grasp of the mole concept, familiarity with balanced chemical equations, and care in the consistent use of units. Work along as we go through the following examples.

Problem 1. Moles, Formula Weight, and Percent Composition: For the Compound Fe_3O_4:
(a) The empirical formula indicates 3 moles of Fe atoms and 4 moles of O atoms per mole of Fe_3O_4.
(b) Formula weight of $Fe_3O_4 = 3(55.85 \text{ g/mol}) + 4(16.00 \text{ g/mol})$
$$= (167.55 + 64.00) \text{ g/mol}$$
$$= 231.55 \text{ g/mol}$$

(c) Percent Fe $= \dfrac{167.55}{231.55} (100) = 72.36\%$

Percent O $= \dfrac{64.00}{231.55} (100) = 27.64\%$

Problem 2. A Product Formed in a Chemical Reaction: When 12.00 g of carbon was completely burned in an atmosphere of pure oxygen, carbon dioxide was the only product. To determine the number of grams of carbon dioxide formed, proceed as follows:

$$C \text{ (s)} + O_2 \text{ (g)} \longrightarrow CO_2 \text{ (g)}$$

The balanced equation states that for every mole of carbon consumed, an equal number of moles of carbon dioxide must be formed:

The coefficients in front of C and CO_2 in the equations are both 1.

$$\text{moles of C} = \frac{12.00 \text{ g C}}{12.01 \text{ g/mol C}} = 0.999 \text{ mole C}$$

Therefore 0.999 mole of CO_2 must be produced:

$$\text{g } CO_2 = (\text{mole of } CO_2)(\text{formula weight of } CO_2)$$

$$= (0.999 \text{ mol } CO_2) \left(\frac{44.0 \text{ g}}{\text{mol } CO_2} \right)$$

$$= 44.0 \text{ g } CO_2$$

All the features of this problem could have been set into a single statement, rather than solving the problem in pieces as we did above:

$$(12.00 \text{ g } \cancel{C}) \frac{(1 \text{ mol } \cancel{C})}{(12.01 \text{ g } \cancel{C})} \frac{(1 \text{ mol } \cancel{CO_2})}{(1 \text{ mol } \cancel{C})} \frac{(44.0 \text{ g } CO_2)}{(1 \text{ mol } \cancel{CO_2})} = 44.0 \text{ g } CO_2$$

Problem 3. Reaction Products and Stoichiometric Ratios:

A standard laboratory procedure for generating oxygen gas involves the catalytic decomposition of potassium chlorate ($KClO_3$). Potassium chloride (KCl) and oxygen (O_2) are the only products of the reaction. If we need 5 mol of oxygen for an experiment, how many grams of potassium chlorate must we decompose? The first step in the solution of this problem, before any calculations are made, is to write down the balanced equation. According to the statement of this problem, the following equation can be written:

$$KClO_3 \text{ (s) } \longrightarrow KCl \text{ (s) } + O_2 \text{ (g)}$$

Balancing this equation by inspection:

$$2 \ KClO_3 \text{ (s) } \longrightarrow 2 \ KCl \text{ (s) } + 3 \ O_2 \text{ (g)}$$

Coefficients in balanced chemical equations are generally reduced to their lowest common denominator, in this case 2:2:3. Although it is sometimes awkward, it is not incorrect when dealing with moles rather than molecules to use fractions and write the balanced equation as follows:

$$KClO_3 \text{ (s) } \longrightarrow KCl \text{ (s) } + \tfrac{3}{2} O_2 \text{ (g)}$$

Continuing with the problem, we need to produce 5 mol of O_2. The equation states that for every 3 moles of O_2 produced, 2 moles of $KClO_3$ are required. In other words, for each mole of O_2 we need $\tfrac{2}{3}$ mol of $KClO_3$. So, for 5 mol of oxygen, we need

$$5(\tfrac{2}{3}) = \tfrac{10}{3} = 3.33 \text{ moles of } KClO_3$$

$$\text{g } KClO_3 = (3.33 \text{ mol } \cancel{KClO_3}) \left(\frac{122.5 \text{ g}}{\text{mol } \cancel{KClO_3}} \right) = 408 \text{ g } KClO_3$$

Again, setting the problem up as a single statement:

$$(5 \text{ mol } \cancel{O_2}) \frac{(2 \text{ mol } \cancel{KClO_3})}{(3 \text{ mol } \cancel{O_2})} \frac{(122.5 \text{ g } KClO_3)}{(\text{mol } \cancel{KClO_3})} = 408 \text{ g } KClO_3$$

Problem 4. Reaction Product in a Multi-Step Synthesis:

Silver sulfide (Ag_2S) can be converted into silver chloride (AgCl) by following the sequence of chemical reactions shown below. Starting with 100 g of silver sulfide, find the maximum quantity (in grams) of silver chloride that could be produced.

$$2 \ Ag_2S \text{ (s) } + 10 \ NaCN \text{ (aq) } + O_2 \text{ (g) } + 2 \ H_2O \text{ (liq) } \longrightarrow$$
$$4 \ NaAg(CN)_2 \text{ (aq) } + 4 \ NaOH \text{ (aq) } + 2 \ NaCNS \text{ (aq)}$$

$$2 \ NaAg(CN)_2 \text{ (aq) } + Zn \text{ (s) } \longrightarrow$$
$$2 \ NaCN \text{ (aq) } + Zn(CN)_2 \text{ (aq) } + 2 \ Ag \text{ (s)}$$

$$3 \ Ag \text{ (s) } + 4 \ HNO_3 \text{ (aq) } \longrightarrow 3 \ AgNO_3 \text{ (aq) } + NO \text{ (g) } + 2 \ H_2O \text{ (liq)}$$

$$AgNO_3 \text{ (aq) } + NaCl \text{ (aq) } \longrightarrow AgCl \text{ (s) } + NaNO_3 \text{ (aq)}$$

To solve this problem, you need not go through all four reactions or even be concerned with any particular one of them. The key lies in the conservation of silver atoms. All the Ag originally present in 100 g of Ag_2S

ends up as Ag in AgCl. Determination of the Ag originally present will fix the upper limit or maximum possible in the final product:

$$\text{formula weight of } Ag_2S = 2(107.9 \text{ g/mol}) + (32.1 \text{ g/mol})$$

$$= 247.9 \text{ g/mol}$$

$$\text{mol of } Ag_2S = \frac{100 \text{ g } Ag_2S}{248 \text{ g/mol } Ag_2S} = 0.403 \text{ mol } Ag_2S$$

From the chemical formula, 0.403 mol of Ag_2S contains 0.806 mol of Ag. Thus, the maximum AgCl possible in the final product is 0.806 mole. Since 1 mol of AgCl contains 1 mol of Ag, 0.806 mol of Ag will be present in 0.806 mol of AgCl:

2 mol Ag:1 mol Ag₂S.

$$\text{g of AgCl} = (0.806 \text{ mol AgCl}) \left(\frac{143.3 \text{ g AgCl}}{\text{mol AgCl}} \right) = 116 \text{ g AgCl}$$

Problem 5. Reaction Product and Percent Purity of Starting Material: When the iron oxide known as hematite reacts with the form of carbon known as coke under certain conditions, the iron can be freed from its combination with oxygen. The process is generally referred to as reduction of the metal oxide, or "freeing" the metal:

$$Fe_2O_3 \text{ (s)} + 3 \text{ C (s)} \longrightarrow 2 \text{ Fe (s)} + 3 \text{ CO (g)}$$

Given the balanced equation and the known atomic weights, the combining weights can be determined. First, identify the proper proportions by weight for all the substances involved and check the mass balance:

hematite		coke		iron		carbon monoxide
Fe_2O_3	+	3 C	$\longrightarrow$	2 Fe	+	3 CO
2(55.85) + 3(16.00)		3(12.01)	=	2(55.85)		3(12.01 + 16.00)
159.7	+	36.03	=	111.7	+	84.03
		195.7	=	195.7		

The unit factor method can be used to check the results:

$$(111.7 \text{ g Fe}) \left(\frac{1 \text{ mol Fe}}{55.85 \text{ g Fe}} \right) \left(\frac{1 \text{ mol } Fe_2O_3}{2 \text{ mol Fe}} \right) \left(\frac{159.70 \text{ g}}{1 \text{ mol } Fe_2O_3} \right) = 159.7 \text{ g } Fe_2O_3$$

$$(111.7 \text{ g Fe}) \left(\frac{1 \text{ mol Fe}}{55.85 \text{ g Fe}} \right) \left(\frac{3 \text{ mol coke}}{2 \text{ mol Fe}} \right) \left(\frac{12.01 \text{ g}}{1 \text{ mol coke}} \right) = 36.03 \text{ g C}$$

This kind of analysis provides information concerning the limitations of the reaction process. For example, if you want 112 g of iron from this process, the furnace will have to be charged with 160. g of hematite and 36 g of coke, assuming hematite and coke are pure Fe_2O_3 and C, respectively. At the same time, you should be alert to the need for taking proper precautions because 84 g of poisonous carbon monoxide will be released to the atmosphere. On the other hand, if only 100. g of iron was produced, in a situation in which a large excess of carbon was used to ensure complete reaction with the 160. g of hematite originally charged into the furnace, then you can say that the original oxide was only 89.3% pure:

$$\frac{\text{grams Fe produced}}{\text{grams Fe possible}} \times 100 = \% \text{ purity of original oxide charge}$$

$$\frac{100.}{112} \times 100 = 89.3\% \text{ pure}$$

Problem 6. The Limiting Reactant: Reaction stoichiometry depends on the balanced chemical equation. The amounts of products formed from given amounts of reactants depend on the reactant present in the smaller quantity, according to the stoichiometric ratio established by the coefficients in the balanced chemical equation. In other words, the reactant present in shortest supply is the limiting factor, or **limiting reactant**. To illustrate what we mean by the limiting reactant consider the following situation. It is well known that limestone ($CaCO_3$) is decomposed by acid, with generation of carbon dioxide and water:

$$CaCO_3 \text{ (s)} + 2 \text{ HCl (aq)} \longrightarrow CaCl_2 \text{ (aq)} + H_2O \text{ (liq)} + CO_2 \text{ (g)}$$

If 10.0 g of $CaCO_3$ is treated with 10.0 g of HCl, how many grams of CO_2 can be generated? The first step is to find which of the two reactants is the limiting reactant. To resolve that issue, you need only ask how much of each is needed to react with the other, according to the chemical equation. Arbitrarily choosing $CaCO_3$,

$$\text{g } CaCO_3 = (10.0 \text{ g HCl}) \left(\frac{1 \text{ mol HCl}}{36.5 \text{ g HCl}} \right) \left(\frac{1 \text{ mol } CaCO_3}{2 \text{ mol HCl}} \right) \left(\frac{100 \text{ g } CaCO_3}{1 \text{ mol } CaCO_3} \right)$$

$$= 13.7 \text{ g } CaCO_3$$

The answer says that 10.0 g HCl requires 13.7 g $CaCO_3$. Since there are only 10.0 g $CaCO_3$, calcium carbonate must be the limiting reactant, and the HCl is present in excess of the stoichiometric ratio established by the balanced chemical equation. Assuming that the reaction goes to completion, we can turn back to the original problem and ask: "If all the limiting reactant reacts, how much CO_2 forms?"

$$\text{g } CO_2 = 10.0 \text{ g } CaCO_3 \left(\frac{1 \text{ mol } CaCO_3}{100 \text{ g } CaCO_3} \right) \left(\frac{1 \text{ mol } CO_2}{1 \text{ mol } CaCO_3} \right) \left(\frac{44.0 \text{ g } CO_2}{1 \text{ mol } CO_2} \right)$$

$$= 4.40 \text{ g } CO_2$$

Problem 7. Product Yields: The principles established in these sample problems are generally applicable to calculating the **theoretical yield**, the maximum amount(s) of product(s) possible, based on complete reaction of the limiting reactant in a chemical reaction. In turn, knowledge of the theoretical yield makes it possible to state the actual product yield for a reaction as the **percent yield**, a percentage of the theoretical yield. For example, consider the laboratory synthesis of benzaldehyde (C_6H_5CHO) from toluene ($C_6H_5CH_3$), using manganese dioxide (MnO_2):

$$C_6H_5CH_3 \text{ (liq)} + 2 \text{ } MnO_2 \text{ (s)} + 2 \text{ } H_2SO_4 \text{ (aq)} \longrightarrow$$
$$C_6H_5CHO \text{ (liq)} + 3 \text{ } H_2O \text{ (liq)} + 2 \text{ } MnSO_4 \text{ (aq)}$$

In order to calculate the theoretical yield of benzaldehyde, starting with 18.4 g of toluene in the presence of excess MnO_2 and H_2SO_4, note that benzaldehyde is formed from toluene on a 1:1 mole basis. The balanced chemical equation tells us that. Therefore,

$$(18.4 \text{ g } C_6H_5CH_3) \left(\frac{1 \text{ mol } C_6H_5CH_3}{92.0 \text{ g } C_6H_5CH_3} \right) = 0.200 \text{ mol } C_6H_5CH_3$$

$$0.200 \text{ mol } C_6H_5CH_3 = 0.200 \text{ mol } C_6H_5CHO$$

$$(0.200 \text{ mol } C_6H_5CHO) \left(\frac{106 \text{ g } C_6H_5CHO}{1 \text{ mol } C_6H_5CHO} \right) = 21.2 \text{ g } C_6H_5CHO$$

The theoretical yield of C_6H_5CHO is 21.2 g, the maximum amount of product to expect from pure reactants in 100% yield.

What if, as frequently happens, the actual yield is less than the theoretical yield? This may happen because the starting materials are not pure, because of the presence of competing reactions that lead to side-products, because the reaction does not go to completion, or perhaps because of loss of product in handling. Let's say 19.4 grams of pure C_6H_5CHO were recovered. Then

$$\% \text{ yield} = \frac{\text{actual}}{\text{theoretical}} \times 100 = \frac{19.4}{21.2} \times 100 = 91.5\%$$

Problem 8. Composition of a Reacting Mixture of Iron Oxides: You are given a sample of mixed iron oxides, FeO and Fe_2O_3, and the following information: When 1.000 g of the mixture was reduced to the pure metal, 0.738 g of Fe was obtained. The composition of the original mixture can be found, based on the conservation principle that atoms (and mass) are conserved. We write this conservation principle as:

moles Fe in mixed oxides = moles Fe in pure metal

The molecular formulas tell us that

moles Fe in mixed oxides = moles FeO + 2(moles Fe_2O_3)

Each mole of Fe_2O_3 gives twice that many moles of Fe.

Therefore,

moles FeO + 2(moles Fe_2O_3) = moles Fe in pure metal

To relate the given weights to numbers of moles, let x represent the mass of FeO present in the mixture. Then $1.000 - x$ is the mass of Fe_2O_3. The formula weights show that

$$\text{mol FeO} = \frac{x \text{ g FeO}}{71.85 \text{ g/mol}}$$

$$\text{mol Fe}_2\text{O}_3 = \frac{(1.000 - x) \text{ g Fe}_2\text{O}_3}{159.7 \text{ g/mol}}$$

$$\text{mol Fe} = \frac{0.738 \text{ g Fe}}{55.85 \text{ g/mol}}$$

Substituting these expressions into the conservation equation gives:

$$\frac{x \text{ g}}{71.85 \text{ g/mol}} + 2 \frac{(1.000 - x) \text{ g}}{159.7 \text{ g/mol}} = \frac{0.738 \text{ g}}{55.85 \text{ g/mol}}$$

Solving this equation for x, we find that the composition of the original mixture is

mass FeO = x = 0.495 g

mass Fe_2O_3 = $1.000 - x$ = 0.505 g

SUMMARY

Mass is conserved in all chemical reactions, and atoms maintain their discrete identities although they may enter into new combinations, thereby displaying new identifying properties. Mass and energy are conserved in chemical reactions but may be interconverted according to a principle first stated by Einstein in his equation $E = mc^2$. But that does not alter the usual meaning of conservation of mass for most chemical reactions and processes.

The **Law of Definite Proportions** (or constant composition) recognizes the importance of the chemical formula and the discreteness that exists in nature at the atomic and molecular level: a pure chemical compound always contains the same elements in the same proportions (by weight), no matter what the source. H_2O is H_2O, whatever and wherever.

The **Law of Multiple Proportions** provides further affirmation of the idea of atoms and the discrete nature of chemical combination: when more than one combination of different kinds of atoms is possible, the combinations must reflect a small whole-number ratio in their combining weights. H_2O and H_2O_2 contain H and O atoms in 2:1 and 1:1 ratios. But the more insightful comparison is the 1:2 ratio of oxygen combined with the same weight of hydrogen in H_2O and H_2O_2.

Chemical transformations involving gaseous reactants and products are governed by **Gay-Lussac's Law of Combining Volumes**, which says that they combine in simple integer relationships by volume: 2 volumes of hydrogen react with 1 volume of oxygen to yield 2 volumes of water vapor. That's because equal volumes of gases under the same conditions of temperature and pressure contain the same number of particles, which is a statement of **Avogadro's Hypothesis**.

As a result of these early ideas, the hierarchy among atoms and molecules that begins with elemental substances is well understood. **Elemental substances** can be atomic or molecular; **compounds** result from combinations of one element with definite numbers of atoms of another. Elemental substances can be atomic or molecular.

The system of **relative atomic weights**—actually atomic masses—is currently based on the ^{12}C isotope of carbon as the standard of reference. The isotopic masses of all the carbon isotopes combine to produce an average atomic weight of 12.011 μ. **Exact atomic weights** can be obtained experimentally in a number of ways including direct chemical combination and the empirical relationship known as the **Law of Dulong and Petit**.

The **mole concept** provides the basis for relating the average mass of an element or compound substance to the number of particles present in a formula weight in grams. In an engineering sense, since atoms (and molecules) are too small to be seen by any common means, let alone be counted out in the fixed numbers required for chemical reactions, the mole concept makes it possible to measure out the right numbers of particles into a laboratory-sized beaker, flask, or reactor, or an industrial-scale hopper. By weighing out formula weights in grams (or any other convenient units), we in effect count out moles of particles. **Avogadro's number** is $N_A = N/n$, where N is the number of particles present in n moles. Its value is $N_A = 6.022 \times 10^{23}$.

The **empirical formula** gives us the simplest ratio of atoms in any discrete chemical combination. Such formulas are obtained with the aid of two kinds of experimental data: analysis for the elements present, and percentage composition (by weight). By adding a third piece of information, the true formula weight, the **molecular formula** can be obtained. The **chemical formula** and the **balanced chemical equation** are the statements used by chemists to represent the shuffling of atoms that occurs in chemistry.

The **theoretical yield** of product(s) in a reaction is the maximum quantity possible, based on the stoichiometric ratios of reactants to products established by the balanced chemical equation. The **actual yield** is the realized quantity of product(s). The **percentage yield** is the ratio of the **actual** to the **theoretical** yields (multiplied by 100). When reactants are not present in the stoichiometric quantities demanded by the balanced chemical equation, the **limiting reactant**, the one present in shortest supply, determines the maximum quantity of product(s) possible.

QUESTIONS AND PROBLEMS

QUESTIONS

1. Give several examples of conservation of mass during chemical transformations, from your daily experience, that could be proved by a simple before-and-after weighing procedure. Examples include boiling an egg and using the somewhat outdated flash bulb.

2. Give several illustrations of apparent contradictions to the law of conservation of mass, with appropriate explanation. Examples that come to mind are most combustion processes.

3. What is the basis for the truth of the laws of definite and multiple proportions?

4. Is the existence of three distinct chromium oxides consistent with, or a contradiction of, the law of definite proportions? The law of multiple proportions?

5. In the upper atmosphere, oxygen is converted to ozone, forming a layer that is very important to the maintenance of our environment. The volume relationship for the gaseous reaction is given by the equation:

$$\text{oxygen (3 vol)} \longrightarrow \text{ozone (2 vol)}$$

What is the chemical formula for ozone?

6. What are isotopes? How are they alike and how do they differ from each other?

7. Of what value is a knowledge of the specific heat of an element in determining its atomic weight?

8. In what way does the Dulong and Petit relationship suggest something more than a simple empirical law?

9. Distinguish between the following:
(a) The atomic weight of an atom and the weight of an atom.
(b) The ^{35}Cl and ^{37}Cl isotopes.
(c) The three forms of oxygen: O, O_2, and O_3.

10. Why must the weight of an element in a mole of a compound always be an integral multiple of the atomic weight of the element?

11. If Avogadro's number were on the order of 10^{30} instead of 10^{23}, what effect would that have on the molecular weight of oxygen?

12. What do formaldehyde (CH_2O), acetic acid ($C_2H_4O_2$), and grape sugar ($C_6H_{12}O_6$) all have in common besides being organic compounds of C, H, and O?

13. Why is the formula for hydrazine written as N_2H_4, rather than as the simpler ratio NH_2?

14. Phosphorus forms two oxides, a phosphorus(III) oxide that may be written P_4O_6 or $2(P_2O_3)$, and a phosphorus(V) oxide that may be written P_4O_{10} or $2(P_2O_5)$.
(a) What is the difference between P_4O_6 and $2(P_2O_3)$?
(b) Could you chemically detect the difference between P_4O_{10} and $2(P_2O_5)$ by elemental analysis?
(c) Both P_4O_6 and P_4O_{10} react with water to give H_3PO_3 and H_3PO_4 respectively. Would the two different ways of writing the formulas, as noted in (a) and (b), require that different weights of the oxides be employed to react with a fixed amount of water? Briefly explain.

15. Dulong and Petit's relationship is applicable to solid elementary substances, in particular the metallic elements at ordinary temperatures. For the metals lithium, aluminum, copper, tin, and lead, verify to what extent the relationship is valid. Why has this empirical relationship become so important? (Specific heats for these metals are, respectively, 3.6, 0.890, 0.38, 0.21, and 0.13 J/g·K.)

16. Arsenic has an effective atomic mass of 74.92 μ; the effective atomic mass of molybdenum is 95.94 μ. Is it necessarily true that they are predominantly single isotopic forms, ^{75}As and ^{96}Mo? Check yourself by looking up the percentage natural abundances of the isotopes of both in the *Handbook of Chemistry and Physics*.

PROBLEMS

Conservation of Mass   [1–4]

1. How many ergs are associated with complete conversion of exactly 4.0 g of mass into energy?

2. Calculate the mass increase associated with the addition of one unit of energy (1 joule) to a material object.

3. A 10-watt light bulb loses about 10^{-10} g/s by radiating the equivalent amount of energy into the space that surrounds it. For how many hours will this bulb have to operate at that rate for a measurable amount of mass to have radiated away, if the limit of your ability to measure mass is 10^{-6} g?

4. The Sun loses about 10^6 tons of mass per second by radiating the equivalent energy into space. How much energy is the Sun radiating into space per second? Assume the ton to be a metric ton, or 10^3 kg, and state the answer in both kilocalories and kilojoules.

Definite/Multiple Proportions   [5–8]

5. Sodium enters into two distinctly different chemical combinations with oxygen. Here are the products of these two reactions:

Sodium oxide: 74.2% sodium, 25.8% oxygen
Sodium peroxide: 59.0% sodium, 41.0% oxygen

How do these data demonstrate the law of multiple proportions?

6. Here are some data on the analysis of a set of compounds composed of only two elements, carbon and hydrogen (hydrocarbons). Show how these data can be used to demonstrate the law of multiple proportions.

ethane	C_2H_6	79.89% C
ethylene	C_2H_4	85.63% C
acetylene	C_2H_2	92.26% C

7. On the basis of the information below, determine the relative atomic weights for the atoms in each compound. Assume the formulas are correct and that H = 1.0.

HCl	2.74% H, 97.26% Cl
NaCl	39.3% Na, 60.7% Cl
Na$_2$O	74.15% Na, 25.85% O
CO$_2$	27.3% C, 72.7% O
CS$_2$	15.80% C, 84.20% S
H$_2$S	5.9% H, 94.1% S

8. Here are three weight ratios, obtained from combining weight data, that have been used in the experimental determination of atomic weights:

I:O $= 50.768:16.000$

I$_2$O$_5$:2Ag $= 100:64.630$

Ag:Cl $= 100:32.867$

Accepting the formula for the combination of iodine and oxygen as I$_2$O$_5$ and the atomic weight of oxygen as 16.000 g/mol, perform each of the following operations:

(a) Establish the atomic weight of iodine.

(b) Then establish the atomic weight of silver.

(c) Finally, establish the atomic weight of chlorine.

Combining Volumes   [9–10]

9. It was found that the ratio of the weights of equal volumes of chlorine and oxygen was 2.22 (32 g/mol). What is the apparent molecular weight of chlorine, assuming oxygen to have an atomic weight of 16.0 g/mol?

10. At room temperature and atmospheric pressure, a gram of oxygen occupies a volume of 0.764 L whereas a gram of an oxide of nitrogen under the same conditions occupies a volume of 0.266 L. What is the formula weight of this oxide of nitrogen?

Average Atomic Weights   [11–14]

11. The isotopic composition of ordinary lead was found by one researcher to be the following:

Isotopic Mass	Percent
203.973 μ	1.3
205.974 μ	27.3
206.976 μ	20.0
207.977 μ	51.4
	100.0

What is the average atomic mass for the natural abundance of lead according to these data?

12. There are three isotopes of magnesium, mass 23.985 μ (78.70%), mass 24.986 μ (10.13%), and mass 25.982 μ (11.17%). Calculate the effective atomic mass of natural magnesium.

13. The natural abundance of the four isotopes of iron is as follows:

Mass of Fe isotopes	53.940	55.935	56.936	57.933
Percent composition	5.82	91.66	2.19	0.33

Calculate the atomic weight for the natural abundance of iron.

14. The atomic weight for the natural abundance of bromine is 79.909 μ. If there are two bromine isotopes of masses 78.918 μ and 80.916 μ, what is the percent composition of the two isotopes in nature?

Assigning Atomic Weights   [15–22]

15. The specific heat of lead is 0.13 J/g·K; its combining weight, in lead chloride, has been found to be exactly 103.605 g/35.453 g Cl.

(a) Using the law of Dulong and Petit, determine the approximate atomic weight of Pb.

(b) Since the atomic weight must be an integral multiple of the combining weight, determine the correct atomic weight of Pb.

(c) What is the empirical formula for this particular chloride?

16. Gold forms a chloride in which the metal accounts for 65.0% of the mass. First determine the combining weight by finding how much gold will react with 35.453 g of Cl; then find the approximate atomic weight by employing the Dulong-Petit relationship; finally, determine the correct atomic weight.

17. The specific heat of lead is 0.13 J/g·K; its combining weight in a certain lead oxide is found to be exactly 138.133 g. What is the empricial formula for this particular oxide?

18. The combining weight of elemental chromium in one of its common oxides was found to be 17.332; its specific heat is 0.510 J/g·K. What is the correct atomic weight of chromium?

19. The atomic weight of hydrogen is known to be exactly 1.008 mass units. Nitrogen (N$_2$) can be combined with hydrogen (H$_2$) to produce ammonia (NH$_3$), and data (obtained experimentally) have shown the weight-percentage of nitrogen in ammonia to be 82.25%. Calculate the atomic weight of nitrogen.

20. When a carefully weighed sample of an unknown metal M reacted completely with oxygen, it was found that the resulting oxide was exactly 10.30% oxygen (by weight). If the empirical formula is known to be M$_2$O$_3$, what are the atomic weight and specific heat of the metal?

21. The fluoride of a particular element contains 32.4% fluorine. An oxide of the same element contains 11.8% oxygen. Suggest what the atomic weight of the metal might be, and determine the formulas for both the fluoride and the oxide. Molecular weight determinations gave results of 270 g/mol for the oxide and 352 g/mol for the fluoride.

22. The chloride of a particular element contains 46.5% chlorine. An oxide of the same element contains 24.8% oxygen. Determine the formulas for both the chloride and the oxide, and suggest what the atomic weight of the element might be. Molecular weight determinations gave results of 324 g/mol for the oxide and 228 g/mol for the chloride.

Moles and Molecular Weights   [23–38]

23. Determine each of the following for the chemical formula UO$_2$(NO$_3$)$_2$:

(a) There are _____ U atoms.

(b) There are _____ O atoms.

(c) There are _____ N atoms.

(d) The formula weight is _____ .

24. Consider the chemical formula $Al_2(SO_4)_3$:

(a) There are _____ Al atoms.

(b) There are _____ S atoms.

(c) There are _____ N atoms.

(d) The formula weight is _____ .

25. Carry out the following conversions:

(a) g KOH to mol KOH

(b) mol K_2SO_4 to kg K_2SO_4

(c) g $KClO_3$ to mg $KClO_3$

26. Carry out the following conversions.

(a) mg NH_3 to molecules NH_3

(b) molecules H_2O to mmol H_2O

(c) mol HF to ng HF

27. A uniform block of beryllium measures 0.254 cm along an edge. Find the number of beryllium atoms contained in the block. ($d_{Be} = 1.85$ g/cm^3)

28. A regular cube of nickel was found to have a distance along an edge of 100 Å. Calculate the number of nickel atoms in the cube. ($d_{Ni} = 8.90$ g/cm^3)

29. A solid sphere of silicon weighs 1 kg. What is the radius of the sphere? ($d_{Si} = 2.33$ g/cm^3)

30. What is the volume occupied by a mole of zinc atoms in the solid state? By a mole of copper atoms? (Refer to Table 1–7).

31. Calculate the weight of one chlorine atom and that of one hydrogen atom. What is the ratio of the weight of 1000. atoms of chlorine to the weight of 1000. atoms of hydrogen? Compare this ratio to the ratio of their respective atomic weights.

32. A mole of sodium atoms weighs 23.0 g; a mole of chlorine atoms weighs 35.5 g. What weight of sodium must you buy in order to get the same number of sodium atoms as there are in a mole of chlorine atoms?

33. In a 120. g sample of NaH_2PO_4 determine the number of

(a) moles of the compound substance present

(b) moles of each element present

(c) grams of each element

(d) sodium atoms present

34. Consider a 100. g sample of potassium hydrogen carbonate ($KHCO_3$). Indicate the total number of

(a) moles of the compound substance present

(b) moles of each element present

(c) grams of each element

(d) potassium atoms present

35. Calculate the number of H_2O molecules in a 180. g snowball.

36. A flask contains 28 g each of carbon monoxide (CO), ethylene (C_2H_4), and nitrogen (N_2). How many molecules are present?

37. A commercial grade of natural gas was found to be 95% methane (CH_4) and 5% propane (C_3H_8). Calculate the average molecular weight of the gas.

38. Calculate the average molecular weight of a fuel gas mixture reported to be 26% methane (CH_4), 5% ethane (C_2H_6), and 3% propane (C_3H_8). The remainder is carbon dioxide (CO_2).

Balancing Chemical Equations   [39–40]

39. Balance each of the following chemical equations by inspection:

(a) $Al + Cl_2 \longrightarrow Al_2Cl_6$

(b) $N_2 + H_2 \longrightarrow NH_3$

(c) $C_3H_8 + O_2 \longrightarrow CO_2 + H_2O$

(d) $Fe_2O_3 + CO \longrightarrow Fe + CO_2$

(e) $Mg_3N_2 + H_2O \longrightarrow NH_3 + Mg(OH)_2$

(f) $Ca_3(PO_4)_2 + H_2SO_4 \longrightarrow Ca(H_2PO_4)_2 + Ca(HSO_4)_2$

(g) $K_2CO_3 + Al_2Cl_6 \longrightarrow Al_2(CO_3)_3 + KCl$

(h) $KClO_3 + C_{12}H_{22}O_{11} \longrightarrow KCl + CO_2 + H_2O$

(i) $KOH + H_3PO_4 \longrightarrow KH_2PO_4 + H_2O$

40. Balance each of the following equations by inspection:

(a) $UO_2 + HF \longrightarrow UF_4 + H_2O$

(b) $NaCl + H_2O + SiO_2 \longrightarrow HCl + Na_2SiO_3$

(c) $Ca(HCO_3)_2 + Na_2CO_3 \longrightarrow CaCO_3 + NaHCO_3$

(d) $NH_3 + O_2 \longrightarrow NO + H_2O$

(e) $PCl_3 + O_2 \longrightarrow POCl_3$

(f) $NaH_2PO_4 + Na_2HPO_4 \longrightarrow Na_5P_3O_{10} + H_2O$

(g) $NaVO_3 + H_2S \longrightarrow Na_2V_4O_9 + S + NaOH + H_2O$

(h) $PbO_2 + SO_2 \longrightarrow PbSO_4$

(i) $K_2O + (NH_4)_2SO_4 \longrightarrow K_2SO_4 + H_2O + NH_3$

Percent Composition/Chemical Formulas   [41–54]

41. State the percent composition for each of the following:

(a) Li in LiOH

(b) C in $SrCO_3$

(c) O in Mn_2O_7

(d) H_2O in $CuSO_4 \cdot 5H_2O$

(e) SO_3 in H_2SO_4

42. Calculate the percent composition of the elements for each of the following chlorine oxide formulas:

$$Cl_2O, \quad ClO, \quad ClO_2$$

43. A compound is composed of atoms of only two elements, carbon and oxygen. If the compound contains 53.1% carbon, what is its empirical formula?

44. A compound of chromium and sulfur was found to be 67.6% chromium. What is the empirical formula for the compound?

45. A compound is composed of only three elements: calcium, carbon, and oxygen. Given the following analysis for an 2.200-g sample, what is the empirical formula? Ca = 0.880 g; C = 0.264 g.

46. A compound is composed of only three elements, sodium, nitrogen, and oxygen. Given the following analysis for a 0.834-g sample, what is the empirical formula? Na = 0.279 g; N = 0.169 g.

47. Find the empirical formula for each of the following oxides of chromium:

(a) Red oxide: 47.99% oxygen

(b) Green oxide: 31.57% oxygen

(c) Black oxide: 23.52% oxygen

48. When silicon was burned in air, 1.00 g was found to combine with 1.14 g of oxygen. Determine the empirical formula of the oxide.

49. A yellowish-orange crystalline solid contains only the elements Fe, C, and H, in the proportions 30.0:64.5:5.5 by weight.
(a) Calculate the empirical formula.
(b) Determination of the formula weight gives a value of 186 g/mol. What is the true formula?
50. The formula weight of a certain hydrocarbon was found to be 100.2 g/mol. It contained 84.0% C and 16.0% H by weight. Determine the empirical formula and the true formula.
51. A compound containing only carbon and hydrogen is burned in air to give 3.210 g of carbon dioxide (CO_2) gas and 1.751 g of water, and no other products. Calculate the empirical formula of the compound.
52. A compound of only sulfur and carbon is burned to give 12.60 g of sulfur dioxide (SO_2) and 4.33 g of carbon dioxide. What is the empirical formula (C_xS_y) of the original compound?
53. A sample of 10.0 g of a certain sugar, known to contain only carbon, hydrogen, and oxygen, is completely burned, yielding 14.7 g of carbon dioxide and 6.0 g of water. What is the empirical formula ($C_xH_yO_z$) of the sugar?
54. A 7.842. g sample of a compound is burned to give 22.96 g of carbon dioxide and 14.09 g of water vapor as the only products. What is the empirical formula of the unknown compound?

Stoichiometry: Amounts of Reactants and Products   [55–66]

55. In the process of changing ammonia to nitrates for agriculture or armaments, the following sequence of reactions takes place:

$$NH_3(g) + O_2(g) \longrightarrow NO(g) + H_2O\ (g)$$
$$NO\ (g) + O_2\ (g) \longrightarrow NO_2\ (g)$$
$$NO_2\ (g) + H_2O\ (liq) \longrightarrow HNO_3\ (aq) + NO\ (g)$$
$$HNO_3\ (aq) + NH_3\ (g) \longrightarrow NH_4NO_3\ (aq)$$

(a) Balance each equation.
(b) How many moles of nitrogen atoms are required for every mole of ammonium nitrate (NH_4NO_3)?
(c) How much ammonia is needed to prepare a kilogram of ammonium nitrate (NH_4NO_3)?
56. One chemical equation depicting the rusting of iron can be written as follows:

$$Fe\ (s) + H_2O\ (liq) \longrightarrow Fe_3O_4\ (s) + H_2\ (g)$$

(a) Balance the equation by inspection.
(b) Determine the number of moles of hydrogen liberated per gram of iron rusted.
(c) Calculate the number of grams of iron oxide that could be generated by the complete corrosion (rusting) of a 60.0-g beer can that is 17.0% iron by weight.
57. When pure iso-octane is burned in an oxygen atmosphere, the chemical reaction is:

$$C_8H_{18}\ (liq) + O_2\ (g) \longrightarrow CO_2\ (g) + H_2O\ (g)$$

(a) Balance the equation.
(b) For every gram of iso-octane consumed, how many grams of carbon dioxide form?

(c) If 124 g of C_8H_{18} reacts with 32 g of O_2, what materials will be present after reaction is complete?
58. How many pounds of $FeCl_3$ can be prepared from a ton of iron ore that is 43% Fe_2O_3? The unbalanced reaction is

$$Fe_2O_3\ (s) + HCl\ (aq) \longrightarrow FeCl_3\ (aq) + H_2O\ (liq)$$

59. Consider the reduction of iron oxide by carbon according to the equation

$$Fe_2O_3\ (s) + 3\ C\ (s) \longrightarrow 2\ Fe\ (s) + 3\ CO\ (g)$$

(a) How much carbon is required per kilogram of pure iron oxide?
(b) If the carbon monoxide is removed by catalytic oxidation according to the equation

$$2\ CO\ (g) + O_2\ (g) \longrightarrow 2\ CO_2\ (g)$$

how much oxygen is required?
60. Oxygen can be produced by thermal decomposition of potassium chlorate or potassium nitrate, according to the equations that follow:

$$KClO_3\ (s) \longrightarrow KCl\ (s) + O_2\ (g)$$
$$KNO_3\ (s) \longrightarrow KNO_2\ (s) + O_2\ (g)$$

(a) Balance the two equations by inspection.
(b) If potassium chlorate costs half as much by weight as potassium nitrate, which is the more cost-effective process for producing oxygen?
61. For a long time now, alkali metal oxides and hydroxides have been successfully used to scavenge carbon dioxide (CO_2) from the breathing space in closed human environments in everything from the first submersible vehicles to modern spacecraft:

$$Na_2O\ (s) + CO_2\ (g) \longrightarrow Na_2CO_3\ (s)$$
$$2\ NaOH\ (s) + CO_2\ (g) \longrightarrow$$
$$Na_2CO_3\ (s) + H_2O\ (liq)$$

Calculate the theoretical removal of CO_2 as grams of CO_2 per kilogram of the reagent (NaOH or Na_2O). Write similar equations for Li_2O and LiOH and determine whether on a weight basis LiOH is more effective than NaOH? Is LiOH more effective than Li_2O?
62. Aluminum metal can be prepared by reducing aluminum chloride with sodium metal, producing sodium chloride at the same time. When a charge of 34.5 g of sodium (atomic weight = 23.0 g/mol) is used, 13.5 g of aluminum (atomic weight = 27.0 g/mol) is produced from 66.8 g of aluminum chloride. Determine the simplest chemical equation that describes the process and agrees with these data.
63. A certain metallic element has an atomic weight of 24. g/mol; a certain nonmetallic element has an atomic weight of 80. g/mol. When this metal and nonmetal are combined chemically, they do so in a ratio of 1 atom to 2 atoms, respectively.
(a) Determine the number of grams of metal that would react with 5.00 g of the nonmetal.
(b) How many grams of product will form?
64. The following sequence of reactions is used to prepare sodium sulfate (Na_2SO_4):

$$S \text{ (s)} + O_2 \text{ (g)} \longrightarrow SO_2 \text{ (g)}$$
$$2 SO_2 \text{ (g)} + O_2 \text{ (g)} \longrightarrow 2 SO_3 \text{ (g)}$$
$$SO_3 \text{ (g)} + H_2O \text{ (liq)} \longrightarrow H_2SO_4 \text{ (liq)}$$
$$2 NaOH \text{ (aq)} + H_2SO_4 \text{ (liq)} \longrightarrow$$
$$Na_2SO_4 \text{ (aq)} + 2 H_2O \text{ (liq)}$$

(a) How many moles of sodium sulfate are possible, starting with 1 mole of sulfur?
(b) With 1 mole of molecular oxygen?
(c) With 4.5 g of water?

65. The synthesis of potassium hypochlorite (KClO) is carried out by passing chlorine gas (Cl_2) into a hot solution of potassium hydroxide (KOH):

$$Cl_2 + 2 KOH \longrightarrow KCl + KClO + H_2O$$

The chlorine gas used in this reaction can be produced by manganese dioxide (MnO_2) reacting with hydrochloric acid (HCl):

$$MnO_2 + 4 HCl \longrightarrow MnCl_2 + Cl_2 + 2 H_2O$$

What weight in grams of manganese dioxide is required to prepare enough chlorine gas to produce 25.0 g of potassium hypochlorite?

66. The synthesis of potassium chlorate ($KClO_3$) is carried out by passing chlorine gas (Cl_2) into a hot solution of potassium hydroxide (KOH):

$$3 Cl_2 + 6 KOH \longrightarrow 5 KCl + KClO_3 + 3 H_2O$$

The resulting solution is evaporated to dryness, leaving behind a mixture of potassium chlorate ($KClO_3$) and potassium chloride (KCl). The chlorate decomposes on heating:

$$2 KClO_3 \longrightarrow 2 KCl + 3 O_2$$

If 25.0 g of potassium chloride remains after complete decomposition of any potassium chlorate present, what weight of potassium hydroxide was initially needed?

Stoichiometry and Yields [67–72]

67. For each of the following reactions:

$$2 HCl \text{ (g)} \longrightarrow H_2 \text{ (g)} + Cl_2 \text{ (g)}$$
$$2 H_2O \text{ (liq)} \longrightarrow 2 H_2 \text{ (g)} + O_2 \text{ (g)}$$
$$2 NH_3 \text{ (g)} \longrightarrow 3 H_2 \text{ (g)} + N_2 \text{ (g)}$$

(a) Determine the maximum possible number of moles of each reaction product from the decomposition of 1 mole of reactant.
(b) Consider each of the reactions as taking place in the reverse direction. Then calculate the number of grams of hydrogen required to fully react with 1 mole of chlorine, oxygen, and nitrogen, respectively.
(c) Again, consider the above reactions as taking place in the reverse direction. If 10 g of hydrogen is available for reaction with 10 g of chlorine, oxygen, or nitrogen, what substances will be present, and in what amounts, at the completion of the reaction?

68. Magnesium oxide (MgO) contains 60.3% magnesium.
(a) What size sample would contain 1.00 g of magnesium?

(b) If MgO were prepared from 50. g of magnesium atoms and 50. g of oxygen atoms, how many grams of MgO would you have left?

69. Given the following chemical equation for the reaction of lithium aluminum hydride ($LiAlH_4$) and boron trifluoride (BF_3):

$$\_\_ LiAlH_4 \text{ (s)} + \_\_ BF_3 \text{ (g)} \longrightarrow$$
$$\_\_ LiF \text{ (s)} + \_\_ AlF_3 \text{ (s)} + \_\_ B_2H_6 \text{ (g)}$$

(a) Balance the equation.
(b) Exactly 100. g of $LiAlH_4$ was allowed to react with 225 g of BF_3. What is the theoretical yield of B_2H_6? Identify the limiting reactant.

70. How many grams of AgCl can be produced from 17 g of silver nitrate ($AgNO_3$) and 20. g of barium chloride ($BaCl_2$)? From 20. g of $AgNO_3$ and 17 g of $BaCl_2$? Be certain to balance the equation before proceeding!

$$AgNO_3 \text{ (aq)} + BaCl_2 \text{ (aq)} \longrightarrow$$
$$AgCl \text{ (s)} + Ba(NO_3)_2 \text{ (aq)}$$

71. Balance the equation and calculate each of the following:

$$HCl \text{ (g)} + O_2 \text{ (g)} \longrightarrow H_2O \text{ (g)} + Cl_2 \text{ (g)}$$

(a) Moles of HCl needed to produce 10. moles of Cl_2.
(b) Moles of Cl_2 produced from 10. moles HCl and excess oxygen.
(c) Moles of Cl_2 produced from 10. moles HCl and 10 moles O_2.

72. The unbalanced equation for the reaction of ammonia and oxygen is

$$NH_3 \text{ (g)} + O_2 \text{ (g)} \longrightarrow H_2O \text{ (g)} + N_2 \text{ (g)}$$

Balance the equation and calculate:
(a) The number of moles of oxygen needed to produce 15 moles of water vapor.
(b) The number of moles of water vapor produced from 15 moles of ammonia and 15 moles of oxygen.

Stoichiometry: Chemical Analysis [73–78]

73. A 2.000-g sample of a silver alloy was dissolved in nitric acid and then precipitated as AgBr. The dried sample of silver bromide weighed 2.000 g. Calculate the percentage of silver in the alloy.

74. A 0.4273-g sample of sodium chloride was analyzed by precipitation of all the chloride present as silver chloride (AgCl). The weight of pure silver chloride so obtained was 0.8094 g. What was the percent purity of the original sample?

75. A mixture of K_2SO_4 and Na_2SO_4 weighed 1.000 g. It was treated with an aqueous solution of silver nitrate, which resulted in complete conversion of all the SO_4 present to Ag_2SO_4, and 1.992 g of Ag_2SO_4 were obtained.

$$K_2SO_4 \text{ (aq)} + 2 AgNO_3 \text{ (aq)} \longrightarrow$$
$$2 KNO_3 \text{ (aq)} + Ag_2SO_4 \text{ (s)}$$

$$Na_2SO_4 \text{ (aq)} + 2 AgNO_3 \text{ (aq)} \longrightarrow$$
$$2 NaNO_3 \text{ (aq)} + Ag_2SO_4 \text{ (s)}$$

What was the percent composition of the original mixture?

76. A mixture consisting of SO_2 and SO_3 weighed exactly 1.78 g. By a special technique, the SO_2 was completely oxidized to SO_3, and the mixture then weighed 2.08 g. Determine the mole ratio of SO_2 to SO_3 in the original mixture.

77. When silver nitrate reacts with sodium chloride (NaCl) or potassium chloride (KCl), insoluble silver chloride (AgCl) is precipitated from the solution. If a 1.00-g sample reacts with silver nitrate and forms 2.15 g of silver chloride, was the original sample pure KCl, pure NaCl, or some mixture of the two?

78. A mixture was believed to contain potassium chlorate ($KClO_3$), potassium bicarbonate ($KHCO_3$), potassium carbonate (K_2CO_3), and potassium chloride (KCl). Exactly 1000. g of this mixture was heated and the following gases evolved, according to the equations below (assume complete decomposition): 18 g of water (H_2O), 132 g of carbon dioxide (CO_2), and 40. g of oxygen (O_2).

$$2\ KClO_3\ (s) \longrightarrow 2\ KCl\ (s) + 3\ O_2\ (g)$$
$$2\ KHCO_3\ (s) \longrightarrow K_2O\ (s) + H_2O\ (g) + 2\ CO_2\ (g)$$
$$K_2CO_3\ (s) \longrightarrow K_2O\ (s) + CO_2\ (g)$$

KCl is inert under the conditions of the reaction. Determine the composition of the original mixture.

Additional Problems [79–83]

79. A flask contains 30. g of nitrogen oxide (NO) and 30. g of diimide (N_2H_2). How many molecules are present in the flask?

80. A 1.00-g mixture of copper(I) oxide (Cu_2O) and copper(II) oxide (CuO) is completely reduced to 0.839 g of the pure metal. Determine the percentage composition (by weight) of the original mixture of oxides.

81. Calcium sulfate, the essential ingredient in two important materials of construction, plaster and sheet rock, is produced as an unusable by-product of many industrial processes. One way of recycling it is to treat it with carbon at elevated temperatures, producing carbon monoxide, quicklime, and sulfur dioxide. In the vicinity of 1200 K, the reaction can be thought of as proceeding in two stages:

$$CaSO_4\ (s) + 4\ C\ (s) \longrightarrow CaS\ (liq) + 4\ CO\ (g)$$
$$CaS\ (liq) + 3\ CaSO_4\ (s) \longrightarrow 4\ CaO\ (s) + 4\ SO_2\ (g)$$

What weight of sulfur dioxide (in grams) could be obtained from 1.000 kg of calcium sulfate?

82. Equal weights of iron and sulfur are heated together, reacting to form iron(II) sulfide:

$$Fe + S \longrightarrow FeS$$

Which reactant is in excess and what fraction of the original sample will be left unchanged?

83. The industrial synthesis of carbon monoxide can be carried out by roasting carbon (in the form of coke) in an environment of carbon dioxide and oxygen.
(a) Assuming the only reaction product to be carbon monoxide, write a complete, balanced equation for the process.
(b) Carbon monoxide can also be prepared by reaction of steam with methane, according to the following reaction:

$$CH_4 + H_2O\ (steam) \longrightarrow CO + 3H_2$$

If 10.0 g of methane is allowed to react with 10.0 g of steam, which reactant would be described as *limiting*?
(c) If 10.0 g of carbon monoxide is actually produced, what is the percentage yield (the actual yield as a percentage of the theoretical yield) for this process?
(d) Again, if 10.0 g of CO is actually produced, how many grams of hydrogen would be produced?

MULTIPLE PRINCIPLES [84–89]

84. When potassium chlorate is heated, it decomposes according to the following equation by loss of oxygen gas:

$$2\ KClO_3\ (s) \longrightarrow 2\ KCl\ (s) + 3\ O_2\ (g)$$

Determine the expected loss in weight for a 100. g sample that is 50.0% decomposed.

85. How many grams of HCl can be obtained from 85 g of 85% pure H_3PO_4 if the maximum yield of reaction products is 85%? Remember to balance the equation.

$$CaCl_2\ (s) + H_3PO_4\ (liq) \longrightarrow Ca_3(PO_4)_2\ (s) + HCl\ (g)$$

86. The commercial synthesis of chlorine dioxide, ClO_2, a bleaching agent widely used by the pulp and paper industry, is complicated by a secondary process in which the essential reagent, $NaClO_3$ (sodium chlorate), is consumed:

$$2\ NaClO_3\ (s) + 4\ HCl\ (g) \longrightarrow$$
$$2\ ClO_2\ (g) + Cl_2\ (g) + 2\ NaCl\ (s) + 2\ H_2O\ (liq)$$
$$NaClO_3\ (s) + 6\ HCl\ (g) \longrightarrow$$
$$3\ Cl_2\ (g) + NaCl\ (s) + 3\ H_2O\ (liq)$$

(a) If equal numbers of moles of chlorine dioxide and elemental chlorine are produced, what percentage of the sodium chlorate starting material is undergoing the secondary reaction?
(b) If 1 kg of sodium chlorate reacts, how much chlorine dioxide and chlorine will be produced?

87. Nitrogen and hydrogen are mixed under the proper conditions and ammonia is formed. But the ammonia product decomposes as well, in a reverse reaction, reforming nitrogen and hydrogen. When all reaction has ceased, there are exactly 2.0 mol of hydrogen, 2.0 mol of nitrogen, and 2.0 mol of ammonia present. How many moles of hydrogen and nitrogen were originally present? What is the maximum number of moles of ammonia possible with the substances at hand?

88. Manganese dioxide has interesting catalytic properties, which some scientists believe are due to its nonstoichiometric nature. Writing the structure in the general form Mn_aO_b, the oxide typically displays compositions that range from $a = 0.81$ and $b = 2.00$ to $a = 1.00$ and $b = 1.60$. Determine the *maximum* and the *minimum* weight-percentage of oxygen in this nonstoichiometric compound.

89. A nonstoichiometric nickel oxide sample, Ni_aO_b, was found to be 76.1% of the metal (by weight). If the value of b is set as 1.00, what is the chemical formula for the oxide?

APPLIED PRINCIPLES [90–96]

90. In the entire universe, it has been estimated that approximately 93% of all atoms are hydrogen and almost all the rest are helium. On that basis, determine the percentage composition (by weight) of the universe.

91. Peroxydisulfuric acid is composed of only H, S, and O in the following proportions: 1.04% H; 33.04% S; 65.92% O. First calculate the simple empirical formula. Then, based on the apparent molecular weight of 194 g/mol, determine the molecular formula.

92. Depending on the amount of carbon black (a form of elemental carbon) present, potassium perchlorate ($KClO_4$) decomposes to potassium chloride (KCl) and either carbon monoxide (CO) or carbon dioxide (CO_2).

(a) Write a separate balanced chemical equation for each reaction.

(b) How much carbon black should be added to 1 kg of potassium perchlorate to produce complete decomposition to KCl and CO_2?

93. A sample of Indiana dolomite is analyzed and found to contain 89.29% $CaCO_3$, 10.51% $MgCO_3$, and 0.20% inert material. Calculate the number of pounds of calcium oxide (CaO) and magnesium hydroxide [$Mg(OH)_2$] that can be produced from a 10.-ton load of the crushed stone.

94. Potassium hydrogen carbonate ($KHCO_3$) can be purchased from large-scale suppliers in shipping containers called gaylords that hold 1000. pounds of the chemical. How many moles of the formula substance are present in such a container?

95. Benzoic acid (C_6H_5COOH) is a particularly effective antimicrobial agent that is widely used against molds. The recommended dose in foods such as breads is 2 mg/kg. How many moles of benzoic acid would you expect to find in a 1-lb loaf that had been so preserved?

96. Carbon tetrachloride (CCl_4) is manufactured on a commercial scale from carbon disulfide and chlorine:

$$CS_2 \text{ (g)} + Cl_2 \text{ (g)} \xrightarrow{FeCl_3,\ 30°C} CCl_4 \text{ (liq) (at 30°)} + S_2Cl_2 \text{ (g)}$$

By comparing the formula weights of the reactants to those of the products, show that the equation is not balanced as written. Then balance the equation and show that material balance follows directly.

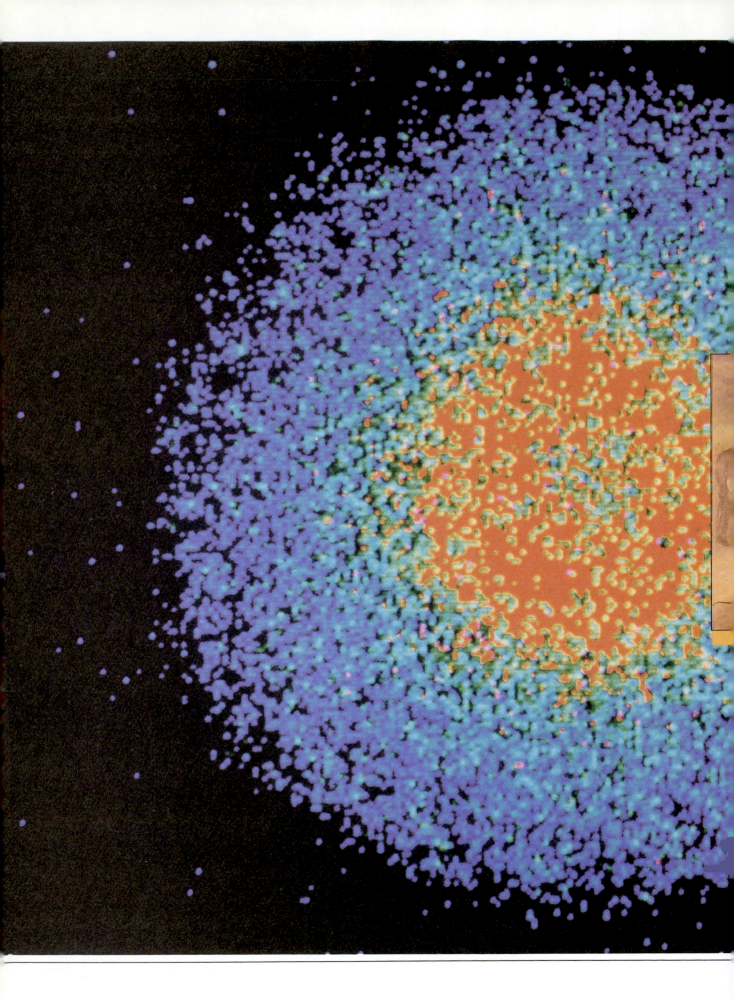

CHAPTER
3

The Atomic Nucleus

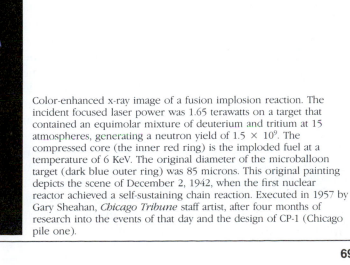

Color-enhanced x-ray image of a fusion implosion reaction. The incident focused laser power was 1.65 terawatts on a target that contained an equimolar mixture of deuterium and tritium at 15 atmospheres, generating a neutron yield of 1.5×10^9. The compressed core (the inner red ring) is the imploded fuel at a temperature of 6 KeV. The original diameter of the microballoon target (dark blue outer ring) was 85 microns. This original painting depicts the scene of December 2, 1942, when the first nuclear reactor achieved a self-sustaining chain reaction. Executed in 1957 by Gary Sheahan, *Chicago Tribune* staff artist, after four months of research into the events of that day and the design of CP-1 (Chicago pile one).

3.1   THE STRUCTURE AND COMPOSITION OF THE NUCLEAR ATOM

Considering the range of contemporary human experience, from nuclear medicine to nuclear weapons, no thoughtful person can doubt the importance of understanding the nature of the nuclear atom. At the beginning of the 20th century, the idea that atoms have structure simplified our understanding of chemistry. Nature's foundation stones were not Dalton's indivisible atomic bricks after all.

The first conclusive information about the nucleus within the atom appeared as a direct result of **radioactivity**. Strange and unexpected radiations were being observed, and they appeared to be spontaneously emitted from certain elements. These emanations were given now famous names: alpha-particles, beta-particles, and gamma-rays; and they were characterized according to their ability to penetrate matter and ionize air. Alpha-particles penetrated matter the least, but produced the greatest degree of ionization; gamma-rays were most penetrating. As it turned out, alpha-particles could be used to probe the atom. Scattering experiments with alpha-particles led to the first successful models for a nuclear atom

Protons were produced when nitrogen atoms were bombarded with these same alpha-particles, as scientists recorded the first artificial nuclear disintegrations. The long-suspected neutron was discovered in 1932, followed by a veritable flood of important new discoveries and inventions, including giant devices called particle accelerators. These instruments could accelerate particles to tremendous energies and hurl them into other particles, making nuclear scientists into modern day alchemists. Nucleosynthesis, the creation of new nuclei, had become a reality. We could indeed transmute base metals into gold. Unfortunately, the economics has not proved to be very favorable. The very largest and most costly devices known to science are required to bring about these nuclear reactions between the very smallest particles of matter.

Electrons and X-rays

Electrons were the first subatomic particles to be discovered. They are the fundamental particles responsible for chemical bonds and the passage of electric current. Electrons are the cathode rays in highly evacuated discharge tubes and the beta-particles ejected from nuclei during natural radioactivity. To adequately characterize the essential properties of electrons we need to describe their charge and mass.

The discovery of the electron emerged from practical experiments designed to study the nature of electricity and the observation of sometimes mysterious phenomena broadly described as luminescence. When metal electrodes were sealed in the ends of a glass tube, the pressure within the tube was reduced, and the electrodes were connected to a source of high potential, a luminous glow was observed. As the internal pressure was further reduced, the luminous glow within this device began to break down and was eventually replaced at very low pressures by a green fluorescence on the walls of the tube farthest from the cathode (Figure 3–1). The fluorescence appeared to be caused by some kind of rays emanating from the cathode. Called cathode rays by the scientists who first observed them, they were eventually shown to be streams of negatively charged particles. These

Light emission not due to the temperature of the radiating object or substance is referred to as luminescence—for example, chemiluminescence, bioluminescence, phosphorescence, and fluorescence.

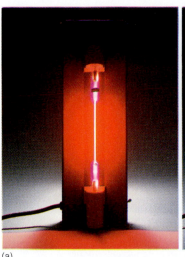

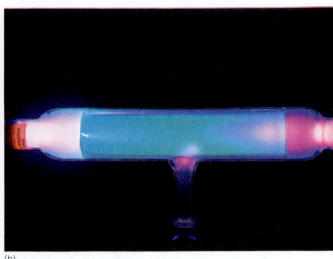

(a)   (b)

FIGURE 3–1 (a) Movement of positively and negatively charged particles carries the electric current through neon (left) and helium (right) at low pressure, producing characteristic colors; (b) fluorescent lighting is also produced by the passage of a high voltage discharge through a rarified gas (Chapter 15). Here, green light is emitted in direct response to a stimulating source—fluorescence.

highly evacuated discharge tubes became known as **cathode ray tubes**; the particles were electrons.

By deflecting a narrow beam of cathode rays in a modified discharge tube (with the aid of a magnetic field), and observing the deflection of the beam on a fluorescent screen, J. J. Thomson experimentally obtained an average value for the charge-to-mass ratio (e/m) (Figure 3–2). Assume you have an electron of mass m and charge e, moving at a velocity v in the presence of an external magnetic field of strength B. According to classical physics, the force acting on the electron is Bev. Now, that force must be exactly enough to make the electron follow the observed circular path of radius r, which is mv^2/r. Set the expressions equal to each other and rearrange the equation to give e/m, the charge-to-mass ratio:

$$Bev = \frac{mv^2}{r}$$

$$\frac{e}{m} = \frac{v}{Br} \qquad (3–1)$$

To solve this equation, the velocity v of the moving particle had to be known. But since it wasn't known at the time, the term had to be cleverly removed from the equation. That was accomplished by an ingenious experiment. Thomson imposed an electrostatic field E with its force Ee upon the moving particle, simultaneously with the magnetic field but in a direction that caused the deflected beam to return to its original (straight) path. Then the force from the electric field had to be equal in magnitude but opposite in sign to the force from the magnetic field. Thus:

$$Ee = Bev$$

which gives

$$\frac{E}{B} = v$$

J. J. Thomson (1856–1940) was the Master of Trinity College, Cambridge University (England). In 1906 he won the Nobel Prize in physics. Six of his students subsequently were so honored in physics and chemistry, the most famous being Ernest Rutherford.

(a)

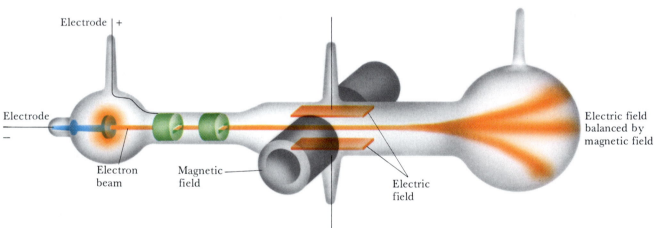

Electrode | +

Electrode –

Electron beam

Magnetic field

Electric field

Electric field balanced by magnetic field

(b)

FIGURE 3–2 (a) J. J. Thomson's original apparatus for measuring the charge-to-mass (e/m) ratio for the electron. Beam of rays from the cathode (C) narrow on passing through slit in anode (A), and (B), falling on screen at end of tube, producing sharply defined, phosphorescent lines. Upward deflections were brought about by connecting the upper plate (D) to the (+) pole of a battery; downward deflections by connecting lower plate (E) to (+) pole. When placed between poles of an electromagnet (foreground), the effect of a magnetic field on the rays could be tested; (b) schematic diagram of the apparatus and experiment.

and substituting back into equation (3–1):

$$\frac{e}{m} = \frac{E}{B^2 r} \tag{3–2}$$

Thomson's experiments were brilliantly conceived (Figure 3-3), yielding the proper order of magnitude for e/m, giving values for v of about one-tenth the velocity of light, and showing these particles to be fundamental, universal particles of matter that are independent of the composition of the cathode or the chemical nature of the residual gas in the tube. The currently accepted value of e/m is 1.76×10^8 coulombs/gram (C/g).

The tiny charge on a single electron was determined by Robert Millikan in an equally ingenious experiment. His apparatus consisted of a pair of metal plates within a chamber containing air (Figure 3–4). These plates could be charged to produce an electrostatic field between them. Droplets of oil produced by an atomizer were allowed to fall through a small hole in the top plate. As they did so, they became charged by collisions with residual gaseous ions, or x-rays. With the aid of a microscope eyepiece, an observer could watch an individual oil droplet passing between the plates. Downward motion was due to the gravitational field (mg). The mass of a particular droplet could be determined from its rate of fall when the electrostatic field was turned off. The electric field was then adjusted so that the downward force mg was exactly balanced by an upward electrostatic force neE, where E is the strength of the electrostatic field, e is the charge on the electron, and n is an integer representing the number of electronic charges on the droplet.

On the basis of a very large number of measurements, Millikan was able to conclude that ne was an integral multiple of a number very close to 4.8×10^{-10} electrostatic units (esu) or 1.6×10^{-19} coulomb (C), the currently accepted value for the charge on the electron. For the mass of the electron (m_e), the current value is $0.00055\ \mu$, or 9.1×10^{-28} g.

Experiments with cathode rays accidentally led to the discovery of **x-rays**. Wilhelm Röntgen found certain barium salts that were **fluorescent**; that is, they could be stimulated to emit visible light by unknown radiations coming from a cathode ray tube. This occurred even when the barium salt was placed some distance away and the cathode ray tube was specially modified by a cover of heavy black paper to mask fluorescence. Additional experiments showed that the stimulating rays from the cathode ray tube, with their considerable range and penetrating power, were being generated at the end of the tube where the rays themselves fell upon the anode or upon the glass. When these radiations passed through solids of differing densities onto photographic plates, the shadows produced strange and important results—namely, the first x-ray photographs (Figure 3–5).

FIGURE 3–3 (a) At the age of 27, Joseph John (J. J.) Thomson became director of the famed Cavendish Laboratories at Cambridge University where he remained for more than a generation.

The compound was $BaPt(CN)_6$, which was then known by the name barium platinocyanide. Its modern name is barium hexacyanoplatinate.

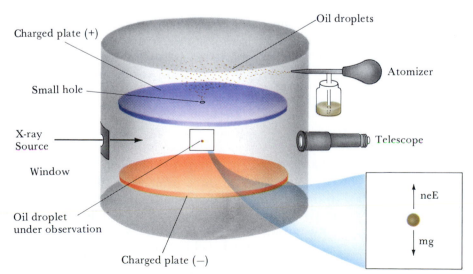

Charged plate (+)

Oil droplets

Atomizer

Small hole

X-ray Source

Telescope

Window

Oil droplet under observation

Charged plate (−)

neE

mg

FIGURE 3–4 Tiny oildrops produced by an atomizer fall between the plates where they are charged by x-rays. By controlling the voltage between the plates, it is possible to balance the electric (neE) and gravitational (mg) forces and determine the charge on the electron.

When the net velocity is zero, the droplet stands suspended in space because the downward force acting on the droplet of mass m equals the upward force due to the strength of the electric field E, and $mg = neE$.

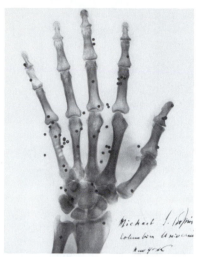

FIGURE 3–5   Just three months after their discovery in Germany, x-rays were already being used here to examine the hand of a New York attorney who had been accidently shot with a shotgun.

Phosphorescence differs from fluorescence in that visible light continues to be emitted after excitation ceases.

FIGURE 3–6   Pierre and Marie Curie, shown here in their Paris laboratory, shared the Nobel Prize for Physics in 1903 with their countryman, Henri Becquerel.

Radioactivity

X-rays were discovered emanating from the glow on the glass of the cathode ray tube where it was struck by the beam of high energy electrons. However, when the electron beam was shut down, the glow disappeared and the x-rays stopped. That would be an expected phenomenon; fluorescence should stop when the excitation source is shut off or removed. The discovery of radioactivity is yet another example of an open mind responding to apparently strange results. In this case, it was investigations into fluorescence that lead to an understanding of **radioactivity**, the spontaneous disintegration of certain nuclei because of their inherent instability.

Fluorescence and phosphorescence are two phenomena related to radioactivity. **Fluorescence** is emission of visible light while a sample is being irradiated by a nonvisible light source. **Phosphorescence** differs in that the emission continues after the irradiation of the sample has been stopped. While studying these phenomena using a particular uranium sulfate salt, the French physicist Henri Becquerel found that invisible radiations were affecting his photographic plates. Although he had carefully protected them by a double thickness of heavy black paper, they were exposed when he placed them in a darkened drawer along side a crust of potassium uranyl sulfate crystals. Furthermore, it was clear that the extent to which they were exposed was several times greater than could reasonably be attributed to the phosphorescence of the uranium salt in the drawer or to any possible light source.

What Becquerel had discovered was typical of neither fluorescence nor phosphorescence. The uranium salt sealed away inside the drawer was not being excited by cathode rays or sunlight. Yet it succeeded in giving off sufficient radiations to expose the protected photographic plates. Furthermore, all uranium salts exhibited this property. Even the uncombined metal produced the effect, with the extent of exposure of the photographic plates being solely a function of how much uranium was present. In addition, these strange radiations persisted, undiminished by time or temperature.

Did nuclei other than uranium exhibit this phenomenon? How widespread was radioactivity? What was the nuclear driving force of these curious uranium emanations? Marie and Pierre Curie (Figure 3–6), a husband-and-wife team undertaking a systematic study of the new phenomenon in Bequerel's laboratory at the time, began to examine these questions. Polonium and radium were discovered and isolated by the Curies in weighable quantities from pitchblende, a uranium ore. Later, thorium was isolated from monazite sands; and radon, a radioactive gas, was discovered. Polonium proved to be the most intensely radioactive of these natural elements. Although eventually an international army of scientists contributed to this robust new science, it was the Curies who were justifiably recognized for their pioneering work.

Radioactivity's Radiations

Radioactivity involves the spontaneous disintegration of nuclei, accompanied by the emission of one or more of three principal radiations, which can be distinguished from each other experimentally (1) by observing their range and penetrating power in passing through metal foils of various thicknesses and (2) by measuring their charge and mass by means of deflections caused by magnetic and electric fields to which the rays are sub-

TABLE 3–1   Relative Penetrating Power, Charge, and Mass Data for Radium Radiation

Type	Foil Thickness	Relative Penetrating Power	Charge	Rest Mass
α	0.005 mm	1	2 +	4 amu
β	0.50 mm	100	1 −	1/1837 amu
γ	5.0 mm	1000	0	0

jected. The three radiations were arbitrarily labeled alpha- (α) and beta- (β) particles, and gamma- (γ) rays. The **γ-rays** turned out to be similar to x-rays. They are highly penetrating and are not deflected by electric or magnetic fields. The **β-particles** were identified as high energy electrons, and the **α-particles** were shown to be energetic helium nuclei (Table 3–1).

In another ingenious experiment, conducted by Ernest Rutherford and his students, a sample of the newly discovered radioactive gaseous element called radon was sealed in a thin-walled glass capillary tube contained within a larger tube. The capillary glass wall was about 0.01 mm thick, thin enough to allow fast-moving α-particles, but not radon atoms, to pass through and accumulate over a week's time in the space between the two tubes (Figure 3–7). The trapped gas was then identified (spectroscopically) as helium.

FIGURE 3–7 Rutherford and Royds successfully designed a "mouse trap" for α-particles. (a) During week-long experiments, alpha particles were emitted, passed through the thin-walled glass tube, and (b) collected in the outer tube. (c) Mercury was pumped into the enclosure to compress the collected gas into a discharge tube where spectral lines, produced on excitation, identified the gas as helium. (d) Characteristic 6-line spectral "fingerprint" for helium.

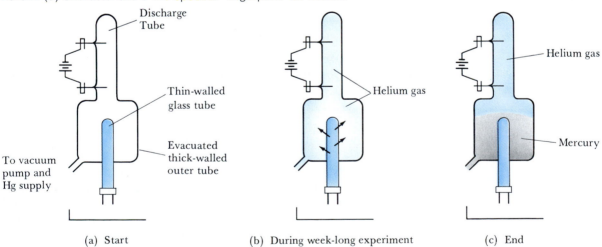

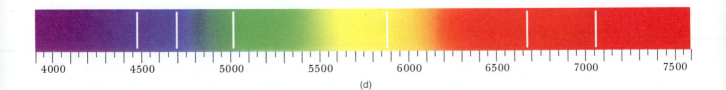

FIGURE 3–8 Separating α-particles, β-particles, and γ-rays, the three kinds of natural radiation, or "radioactivity." The central beam of gamma rays is undeflected; the beta-particles are deflected by a large amount, being negatively charged and having little mass; the alpha particles being more massive and oppositely charged, are deflected less and in the opposite direction.

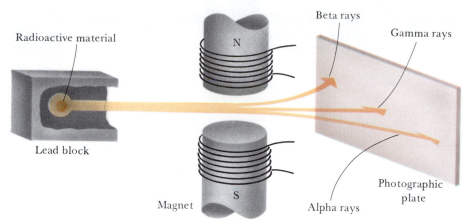

The fact that α-particles and β-particles are charged particles emerged largely as a result of experiments on magnetic and electrostatic deflection. These rays were directed through a magnetic field to see whether they deviated from the incident path as a result. J. J. Thomson had successfully used that method when he orignally showed that cathode rays were beams of high-speed electrons and determined their charge-to-mass ratio. The technique is based on fundamental principles of physics and is still widely used in studying the behavior of nuclear events. When a charged particle moves across a magnetic field, the force of the field acts at right angles to the direction of motion of the charged particle. The particle experiences a continual deflection and, if directed into a uniform field at a right angle, will move along the arc of a circle.

Begin the experiment by placing a radium sample at the base of a narrow passage through a lead block. A pencil-thin beam of α-particles, β-particles, and γ-rays is then directed along the passage toward the only escape path. The beam emerges into a strong, uniform magnetic field, which will separate the three types of radiation (Figure 3–8). The γ-rays continue along the original line-of-flight, while the β-particles are deflected to one side and the α-particles to the other, in circular arcs of different radii. Now the charge-to-mass ratio (e/m) for each type of particle can be determined. For β-particles it coincides with known results for electrons from Thomson's earlier work. Very strong magnetic fields must be used for the α-particles, since e/m is about 8000 times smaller than for beta-particles.

In real terms, α-particles can be stopped by little more than a sheet of tissue paper or a few centimeters of air. The β-particles, though more penetrating, will be stopped by traveling several meters through an air space or by passing through a thin sheet of aluminum foil. The γ-rays, however, will pass unhindered through several inches of lead or several feet of concrete; when γ-rays are involved in experiments, heavy shielding is required for protection of the experimenters. In the early days of radiation research, many workers were exposed to dangerous levels of radiation, which led to serious health problems later in life.

The Nuclear Atom

Electrons, protons, and neutrons were quickly accepted as fundamental subatomic particles. Yet, questions were raised about what the new, composite atom must look like inside, and how the fundamental particles were

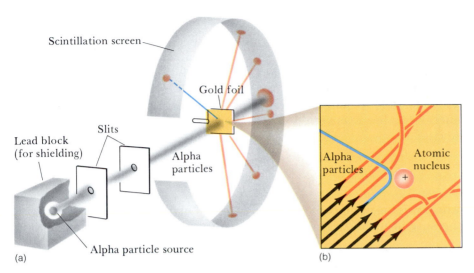

Scintillation screen

Gold foil

Slits

Lead block
(for shielding)

Alpha
particles

Alpha
particles

Atomic
nucleus

Alpha particle source

(a)

(b)

FIGURE 3–9 (a) In Rutherford's "scattering" experiment, a gold foil was bombarded with alpha particles. Most passed through undeflected, as though the atoms in the foil were so much empty space. Surprisingly, some were deflected through moderate angles from the line-of-flight of incident beam. (b) Astonishingly, some were deflected at acute angles, and on occasion, bounced directly backward, as though they had hit a stone wall.

held together. Close to a century later, precise answers are still not available, although practical solutions have been worked out.

After discovering the electron, J. J. Thomson suggested an arrangement of positive and negative charges uniformly distributed throughout a neutral atom. But no test of the hypothesis could be suggested. It remained for Ernest Rutherford to propose one of the most insightful experiments of the 20th century. The results suggested in no uncertain terms that the microcosmic atom—like our solar system—contains a well-defined center of force. That, in turn, required a central concentration of charge to provide the force, and a central mass to provide a firm inertial base. Rutherford called this center of charge and mass a nucleus. His experiment was definitive in leading directly to our present picture of atomic structure.

The basic idea of Rutherford's scattering experiment (which was carried out by his students, Hans Geiger and Ernest Marsden) was to use α-particles to probe the atom. When atoms in a metal foil were bombarded with α-particles from a radium source, most passed directly through. Therefore, metal foils were not constructed of closely packed, impenetrable atoms, but rather were mostly empty space. Other α-particles were deflected, a few through very great angles with respect to their original path, leading to the conclusion that a localized, strongly positive electric charge existed within the atom. A careful evaluation of the angular distributions for the deflected α-particles suggested that the scattering was due to a variety of close encounters with an atomic nucleus (Figure 3–9a). Rutherford's data also made it possible to estimate the dimensions of both the atom and its nucleus to a reasonable approximation. The diameter of this dense, positively charged nucleus was about 10^{-15} m, and electrons apparently revolved around it through otherwise empty space in orbits with a radius close to 10^{-10} m.

The underlying principle can be simply illustrated by considering the special case of closest approach (Figure 3–9b). An α-particle, on a collision course with a nucleus, is deflected backwards along its path. From classical physics we know that the repulsive force between the positively charged nucleus and the positively charged α-particle is $(2e)(Ze)/r^2$, where e is the unit electric charge, Z is the total positive charge within the nucleus of the target metal atoms, and r is the distance between the two particles. The potential energy PE, which is a measure of the work done by the repulsive force to bring the probing particle to rest, is

$$\frac{2Ze^2}{r} \tag{3-3}$$

Now the kinetic energy KE of any mass m traveling at a velocity v is $\frac{1}{2}mv^2$. At the instant at which the α-particle has been brought to rest and is just about to start its journey back along its original track, all of its kinetic energy has been converted to potential energy, so $KE = PE$, which we rewrite as

$$\tfrac{1}{2}mv^2 = \frac{2Ze^2}{r_0} \tag{3-4}$$

where r_0 is the distance of closest approach. This distance is an approximate measure of the nuclear radius.

EXAMPLE 3-1

Consider a 6 million electron-volt (6-MeV) α-particle approaching an atom in an aluminum foil target. Calculate the distance r_0 of closest approach.

Solution   The α-particle is characterized by its kinetic energy, 6 MeV. An electron-volt is a unit of energy, where 1 eV = 1.6×10^{-19} J = 1.6×10^{-12} erg, so the kinetic energy can be expressed as

$$\tfrac{1}{2}mv^2 = 6 \text{ MeV}$$
$$= (6 \text{ MeV})(1.6 \times 10^{-6} \text{ erg/MeV})$$
$$= 9.6 \times 10^{-6} \text{ erg}$$

For an aluminum atom, $Z = 13$. Rearranging equation (3-4), the radius of the aluminum nucleus is

The electrostatic unit (esu), the erg, and the centimeter are consistent units in the centimeter-gram-second (CGS) system.

$$r_0 = \frac{2Ze^2}{\tfrac{1}{2}mv^2} = \frac{2(13)(4.8 \times 10^{-10} \text{ esu})^2}{9.6 \times 10^{-6} \text{ erg}} = 6.2 \times 10^{-13} \text{ cm} \quad \blacksquare$$

EXERCISE 3-1

The same experiment is done using a gold foil. What is the radius of the gold nucleus?
Answer: 4×10^{-12} cm. $\blacksquare$

Nuclear Protons and Neutrons

When an electric discharge is passed through a gas at low pressure, cathode rays aren't the only particles produced. Almost from the outset, investigators were aware of the presence of positively charged particles as well. The lightest such particle is the proton, the electron's positively charged counterpart—equal in magnitude of charge, but opposite in sign, with a mass 1836 times larger.

James Chadwick received the Nobel prize in Physics for his discovery of the neutron.

Rutherford was the first to predict the existence of a neutral nuclear particle. But it was one of his students who identified the particle (in 1932) as a product of bombarding beryllium atoms with α-particles. The mass of the neutron was estimated to be slightly less than, but very close to, the sum of the masses of a proton and an electron.

EXAMPLE 3-2

Making use of the Einstein equation, calculate the energy equivalent of an atomic mass unit.

Solution   Begin with $E = mc^2$, the atomic mass unit μ of 1.66×10^{-24} g, and the square of the velocity of light:

$$E = mc^2$$
$$= (1.66 \times 10^{-24} \text{ g}) (3.00 \times 10^{10} \text{ cm/s})^2$$
$$= 1.49 \times 10^{-3} \text{ g·cm}^2/\text{s}^2$$
$$= 1.49 \times 10^{-3} \text{ erg}$$

In MeV units,

$$E = \frac{1.49 \times 10^{-3} \text{ erg}}{1.60 \times 10^{-6} \text{ erg/MeV}}$$
$$= 930 \text{ MeV}$$

1.60×10^{-6} erg = 1 MeV

EXERCISE 3–2(A)
Calculate the energy equivalent of the mass of an electron in MeV.
Answer: 0.51 MeV.

EXERCISE 3–2(B)
Calculate the mass equivalent of a 10^{12} J of energy.
Answer: 0.01 g.

3.2   THE PROPERTIES OF ATOMIC NUCLEI

Standard Terminology for Atoms and Nuclei

The basic building blocks of the nucleus, neutrons and protons, are referred to as **nucleons**. The **neutron number**, N, represents the number of neutrons present in a nucleus; Z is the **proton number**—perhaps better known as the **atomic number**; A is the **mass number**, the sum of the numbers of neutrons and protons:

> N = neutron number
>
> Z = proton number = atomic number
>
> $A = N + Z$ = mass number

The mass number is an integer, very nearly equal to the atomic weight of a particular isotope. A **nuclide** is a unique atomic species, with particular A and Z values. Each nuclide can be specified by its atomic symbol (which identifies the Z values), with the A value written as a superscript on the left. The N values can be found by subtracting Z from A. For example, in ^{7}Li there are 3 protons and $7 - 3 = 4$ neutrons.

Nuclides with the same chemical identity (same Z value), such as ^{1}H, ^{2}H and ^{3}H, or ^{35}Cl and ^{37}Cl, are **isotopes**. Those with the same N values, such as ^{14}N and ^{15}O, are called **isotones**. Nuclides with the same A values, such as ^{40}Ca and ^{40}K, are called **isobars**.

As the name implies, **antiparticles** are nuclear opposites. **Positrons** (β^+) are antiparticles of electrons (β^-), which they resemble except that they are opposite in charge. One of the most important properties of antiparticles is that when one comes together with its particle counterpart, they annihilate each other and energy is liberated in an amount equal to their combined mass, producing two γ-rays:

$$\beta^- + \beta^+ \longrightarrow \gamma + \gamma$$

Keep in mind that although the occurrence of isotopes among the 83 natural elements is widespread (only 21 do not have isotopic forms), their separation is difficult, complicated, and costly. Approximately 1900 nuclides have been observed altogether. Do not refer to elements that have only one atomic form as "being an isotope." The term requires at least two elemental forms (just as "twins" refers to the existence of a pair).

Positrons were predicted by Cambridge theoretician Paul Dirac in 1930, and discovered by the American experimentalist Carl Anderson. They are very rare in the universe, being produced only in violent, unusual nuclear events.

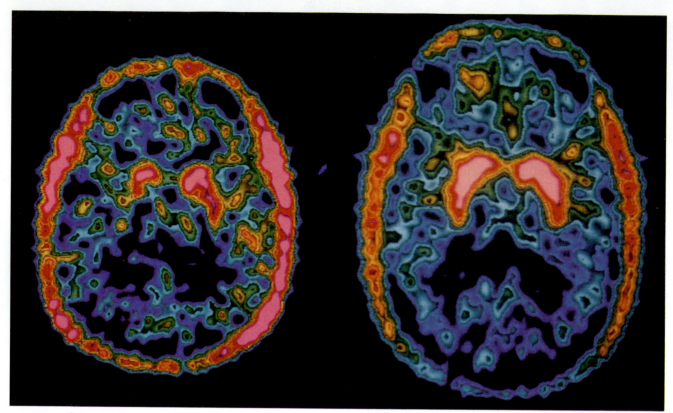

FIGURE 3–10 PET image of a brain slice in a normal human being (right) shows high uptake of [^{18}F]-fluorodopamine in the putamen (pink areas in center). In a patient with moderately severe Parkinson's disease, the uptake is markedly reduced (left).

Neutrinos were predicted by Wolfgang Pauli (Swiss) and Enrico Fermi (Italian). Working together in the 1930s, they first suggested the existence of the neutrino, or "little neutral one," which was confirmed experimentally in 1956.

Most nuclei are baseball-shaped (or spherical). There are some exceptions among the rare earth elements ($Z = 57$ through 71), which are football-shaped (or ellipsoidal).

Electron/positron annihilation produces two 0.51-MeV γ-rays, representing complete conversion of mass to energy, in agreement with the predictions of the Einstein equation $E = mc^2$.

Organic molecules tagged with ^{18}F have been used to explore the recesses of the brain by positron emission tomography (PET). As the ^{18}F isotope decays, positrons are emitted which almost immediately encounter electrons, resulting in annihilation, producing a pair of gamma-ray photons. These distinctive photon pairs are picked up, computer-enhanced, and displayed (Figure 3-10). The technique has been particularly successful in studying and treating Parkinson's disease.

Neutrinos (ν) are unusual particles, to say the least. They are described as having zero charge and near-zero mass. Their existence was predicted from theory, and they were demonstrated experimentally only after **antineutrinos** ($\bar{\nu}$) were found. Faced with the unexpected experimental fact that β-particles ejected from the *same* atomic species during radioactive decay displayed a *continuous* distribution of energies, Wolfgang Pauli suggested that a second particle must be simultaneously emitted, carrying off some of the kinetic energy with it. As a result, the total energy for the radioactive decay process was being *shared*, resulting in a distribution of energies. The neutrino is associated with positron (β^+) decay; the antineutrino, with electron (β^-) emission.

In the ground state, most nuclei are just about spherical, so their sizes can be estimated from their respective radii. Good values for nuclear radii can be calculated using an empirical formula obtained by many techniques:

$$R = R_0 A^{1/3}$$

where $R_0 \cong 1 \times 10^{-13}$ cm. That relationship is also important for its suggestion that nuclear forces are the same for most (if not all) nuclei. The

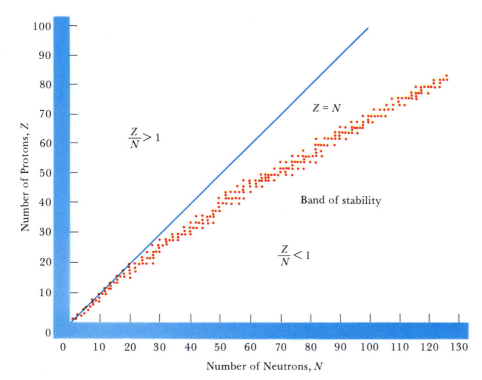

FIGURE 3–11   Plot of proton (Z) versus neutron (N) number and the line of maximum stability along which the ratio $N/Z = 1$. All isotopes of a given element are found in a vertical column of red dots, centered on the element's atomic number (Z).

fact that the radius is proportional to $A^{1/3}$ means that densities are the same for all nuclei, on the order of 10^{14} g/cm^3. Compare that to densities in the macroscopic world of everyday experience, which are much smaller, on the order of 10^0 to 10^1.

Charting Nuclides

Since nuclides are defined by their numbers of protons and neutrons, we can draw a graph (Figure 3–11) consisting of vertical columns of the former and horizontal rows of the latter. In the resulting grid, each square represents a single nuclide, one possible collection of protons and neutrons. Such a chart turns out to be very useful, not only as an organizing scheme but also for predicting pathways for radioactive decay and nuclear transformations. To the nuclear chemist, this systematic display of the nuclides is as important as the periodic table of the elements itself. The line of maximum nuclear stability has a slope corresponding to a 1:1 proton-to-neutron ratio for the lightest nuclei. Typical examples are ^{4}He, ^{14}N, and ^{40}Ca. For heavier nuclei, the slope gradually falls away toward a value of about 1:1.6. The position of a given unstable nuclide with respect to this line determines the type of process by which it is likely to decay. Nuclides to the right of the line mostly decay by β^- emission, whereas those to the left decay by β^+ emission or electron capture. The distance from the line to the nuclide is a qualitative measure of relative nuclear stability—the closer to the line, the more stable the nuclide.

Nuclear Instability

There are certain stable proton/neutron configurations. Unstable nuclear arrangements arise from three general sets of circumstances, and the instability tends to be self-correcting through a number of decay processes.

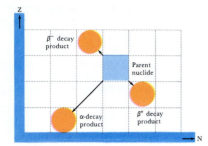

Parent nuclides and decay products.

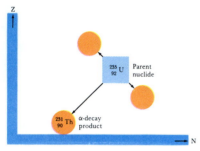

(1) alpha-decay.

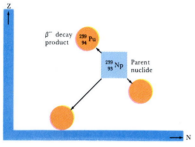

(2) beta-decay.

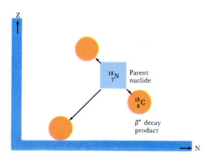

(3) positron-decay.

1. Excessively Large Number of Protons and Neutrons: In this case there is just too much mass. Nature's solution to the problem is to have the unstable nuclide lose an α-particle. The driving force for this process is the high stability of the alpha-particle. With some exceptions, α-decay only occurs among nuclides with mass 200 or greater, such as ^{235}U:

$$\underset{\text{(parent)}}{^{235}_{92}\text{U}} \longrightarrow \underset{\text{(daughter)}}{^{231}_{90}\text{Th}} + \underset{(\alpha\text{-particle})}{^{4}_{2}\text{He}} \qquad (\alpha\text{-decay})$$

The mass number is reduced by four units, and the atomic number is reduced by two units, producing a lighter daughter nuclide. (Here we show Z for each nuclide as a subscript, although it is implied by the atomic symbol.)

2. Excess of Neutrons Over Protons (to the right of the line of stability): Unstable nuclides tend to rectify this situation by converting a neutron to a proton, an electron, and an antineutrino:

$$\underset{\text{(parent)}}{^{239}_{93}\text{Np}} \longrightarrow \underset{\text{(daughter)}}{^{239}_{94}\text{Pu}} + \underset{\text{(electron)}}{\beta^-} + \underset{\text{(antineutrino)}}{\bar{\nu}} \qquad (\beta^-\text{-decay})$$

By this process, the atomic number increases to the next element and at the same time an electron is shot out, preserving electrical neutrality. The nuclear mass number is unchanged. Since many artificial isotopes are produced in nuclear reactions by bombarding target nuclides with neutrons, one would expect isotopic products that are neutron-rich and decay by beta-emission. $^{239}_{94}$Pu (above) is also unstable, eventually undergoing α-decay.

3. Excess of Protons Over Neutrons (to the left of the line of stability): Unstable nuclides correct this situation by two different mechanisms:

a. Conversion of a proton into a neutron, a positron, and a neutrino:

$$\underset{\text{(parent)}}{^{13}_{7}\text{N}} \longrightarrow \underset{\text{(daugher)}}{^{13}_{6}\text{C}} + \underset{\text{(positron)}}{\beta^+} + \underset{\text{(neutrino)}}{\nu} \qquad (\beta^+\text{-decay})$$

b. Capture of one of the atom's electrons by the nucleus:

$$\underset{\text{(parent)}}{^{79}_{36}\text{Kr}} + \underset{\text{(electron)}}{\beta^-} \longrightarrow \underset{\text{(daughter)}}{^{79}_{35}\text{Br}} + \underset{\text{(neutrino)}}{\nu}$$

Either route leads to daughters with the atomic number reduced by one, but with the same mass number as the parent. Note that a neutron is produced from the excess proton and the captured electron. **Electron capture** is especially interesting to chemistry because this decay process, unlike the others, is known to be influenced by chemical composition.

Mass Defect, Binding Energy, and Nuclear Stability

Largely because of the discovery of the neutron and the development of the high resolution mass spectrograph, which allowed scientists to measure atomic masses with remarkable accuracy, it became possible to explain why atomic masses are not exact integers. Each mass is slightly less than the sum of the masses of the nucleons present. According to Einstein's special theory of relativity, this mass deficiency or **mass defect** has an energy equivalent, the **nuclear binding energy**. To illustrate what this means, consider the following balance sheet for the α-particle:

Measured masses for the isolated (or free) nucleons involved:

$$2 \text{ neutrons} = 2(1.008665 \ \mu) = 2.017330 \ \mu$$
$$\underline{2 \text{ protons} \ \ = 2(1.007276 \ \mu) = 2.014552 \ \mu}$$
$$\text{Total mass} \qquad\qquad\quad = 4.031882 \ \mu$$

This is what one would expect for the mass of one α-particle. However, its actual mass is 4.00151 μ. The mass defect is the difference between what the α-particle's mass "should" be and what it is experimentally known to be, 0.03037 μ. The binding energy is the equivalent of the difference. From the Einstein relationship expressing the equivalence of mass and energy, $E = mc^2$, the energy equivalent of 1 μ is 931.5 MeV. Therefore,

$$\text{(4.031882} - 4.00151) \ \mu = 0.03037 \ \mu$$

$$(931.5 \text{ MeV}/\mu)(0.03037 \ \mu) = 28.29 \text{ MeV}$$

and the binding energy per nucleon is

$$\frac{28.29 \text{ MeV}}{4 \text{ nucleons}} = 7.072 \text{ MeV/nucleon}$$

One way of looking at nuclear binding energy is as something the nucleus lacks. Consider a thought experiment in which we bring together two protons and two neutrons to form an α-particle. To do that, energy equivalent to the mass defect (or 28.29 MeV) would have to be released. The nucleons would stay together as an α-particle because each nucleon lacks the mass equivalent of about 7 MeV that would be required in order to sustain an independent existence. Furthermore, the nucleus must stay together (as an α-particle) until the requisite energy is available. So the binding energy is the energy equivalent of the missing mass, the energy with which the nucleus is held together.

The orders-of-magnitude energy differences between the usual chemical reactions, which involve the extranuclear electrons, and the unusual chemical reactions involving nuclear processes can be demonstrated by the next example.

EXAMPLE 3–3

Take the mass of the ^{16}O atom as 15.994915 μ with its complement of eight electrons. For eight neutrons and eight hydrogen atoms, each with its nucleus and lone electron, find the total mass. Then, find the mass defect for the oxygen atom.

Solution

$$8 \text{ neutrons} \qquad\quad = 8(1.008665) = \quad 8.069320 \ \mu$$
$$\underline{8 \text{ hydrogen atoms} = 8(1.007825) = \quad 8.062600 \ \mu}$$
$$\text{Total mass} \qquad\qquad\qquad\qquad\quad\ = 16.131920 \ \mu$$

mass of neutron = 1.008665 μ
mass of proton = 1.007276 μ
mass of electron = 0.000549 μ

The mass of the proton is equivalent to the mass of a hydrogen atom minus the mass of an electron.

Therefore the atom of oxygen is lighter. It has less mass than its nucleons and electrons—by the difference:

$$\text{sum of the parts} = \ 16.131920 \ \mu$$
$$\underline{\text{known mass} \ \ \ \ = \ 15.994915 \ \mu}$$
$$\text{difference} \qquad = \ \ \ 0.137005 \ \mu$$

That mass difference is equivalent to 128 MeV/atom or 8 MeV/nucleon.

By comparison, the energies involved in the usual chemical reactions are on the order of electron-volts, eV, not MeV. For example, when an oxygen molecule is formed from a pair of oxygen atoms, the energy released is about 5.1 eV. ■

EXERCISE 3–3
Calculate the energy released (in kJ/mol and in kcal/mol) when 2 moles of H atoms combine with 2 moles of neutrons to produce 1 mole of helium atoms. $\frac{4}{2}$He has an atomic weight of 4.00260 g/mol.
Answer: 2.73×10^9 kJ/mol He; 6.53×10^8 kcal/mol He. ■

To summarize this discusson of mass defect and binding energy, here is a set of simple formulas where Δm is the missing mass, or mass defect:

$$\Delta m = [Z(m_p + m_e) + (A - Z)m_n] - m_{atom}$$

As a reasonable approximation, considering the H atom to be a mass of one proton and one electron, we can rewrite this equation in a slightly modified form:

$$\Delta m = [Z(m_{H\ atom}) + (A - Z)m_n] - m_{atom}$$

The energy associated with the mass defect, which we call the binding energy (*BE*), is written as $BE = \Delta mc^2$.

EXAMPLE 3–4
Calculate the mass defect (Δm) and binding energy (*BE*) for a deuteron, $\frac{2}{1}$H, with an atomic weight of 2.0140 g/mol.

Solution $\Delta m = [1(1.0078) + (2 - 1)(1.0087)] - (2.0140) = 0.0025\ \mu$
$BE = (0.0025\ \mu)(931.5\ \text{MeV}/\mu) = 2.3\ \text{MeV}$ ■

1 μ = 931.5 MeV

EXERCISE 3–4
Calculate the mass defect (Δm) and binding energy (*BE*) for a tritium atom, $\frac{3}{1}$H, with an atomic weight of 3.01605 g/mol.
Answer: $\Delta m = 0.00910\ \mu$, $BE = 8.48\ \text{MeV}$. ■

The binding energy per nucleon falls generally within the range of 8 MeV or less, as our calculations for the α-particle and the oxygen atom indicate. A plot of binding energy per nucleon versus mass number illustrates the point (Figure 3–12); the maximum occurs at a mass number of about the vicinity of iron and nickel. In other words, according to the curve of binding energy, iron and nickel are the most stable nuclei in the periodic table of the elements. Therefore, it should not be at all surprising that these elements are common components of meteorites and account for a significant fraction of the Earth's core.

The existence of a maximum in the middle of the binding energy curve is the basis for understanding how we can generate energy from both fission of the heaviest elements and fusion of the lightest elements. There is no difference in principle between these energy-releasing reactions and normal chemical reactions, except, of course, the magnitude of the energy released:

nuclear fission
$$^1n + ^{235}_{92}U \longrightarrow ^{140}_{54}Xe + ^{94}_{38}Sr + 2^1n + \text{energy}$$

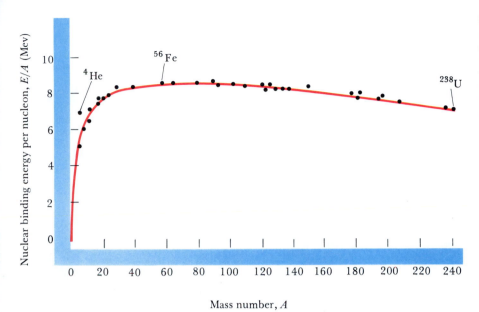

FIGURE 3–12   The curve of binding energy, a plot of binding energy per nucleon (E_b/A) versus mass number (A). The maximum in the curve corresponds to the most stable nuclei, those of the elements in the middle of the periodic table.

nuclear fusion

$$^2_1\text{H} + ^3_1\text{H} \longrightarrow ^4_2\text{He} + ^1n + \text{energy}$$

chemical reaction

$$2\,\text{H}_2 + \text{O}_2 \longrightarrow 2\,\text{H}_2\text{O} + \text{energy}$$

Synthetic Nuclear Reactions

Most of our information about nuclei has been obtained by hurling particles at them and watching what happens. Early experiments were limited. One nuclide could be converted to another by bombarding the parent with an α-particle or a γ-ray photon emitted from a naturally radioactive source. Rutherford and his students in 1911 carried out the first such transformations, successfully injecting α-particles from a radium or polonium source into nitrogen nuclei:

$$^{14}_7\text{N} + ^4_2\text{He} \longrightarrow ^{18}_9\text{F}^* \longrightarrow ^{17}_8\text{O} + ^1_1\text{H}$$

The written reaction includes an intermediate fluorine nucleus in an excited (or unstable) state, as indicated by the asterisk. This is typical of nuclear bombardment reactions. The target nucleus, or parent, absorbs the α-particle and forms a compound nucleus in an excited state, which subsequently decays by one or more routes to more stable daughters.

The first artificial transmutations were carried out in the early 1930s. Protons were accelerated through an evacuated tube by application of a very high potential. Particles carrying an electric charge e will be accelerated by the application of a potential V to some kinetic energy eV:

$$KE = \tfrac{1}{2}mv^2 = eV$$

The units used to express the kinetic energy (KE) of the particle are electron-volts (eV) and kiloelectron-volts (keV) but mostly megaelectron-volts (MeV). These correspond respectively to the energies acquired by a charged particle carrying a unit electric charge when it is accelerated through potential differences of 10^0 volt, 10^3 volts, and 10^6 volts. Keep in mind (as a reference point) that 1 MeV = 1.60×10^{-6} erg. This is approximately

In 1932, two Rutherford students (Crockcroft and Walton) produced a successful transmutation using artificially accelerated protons.

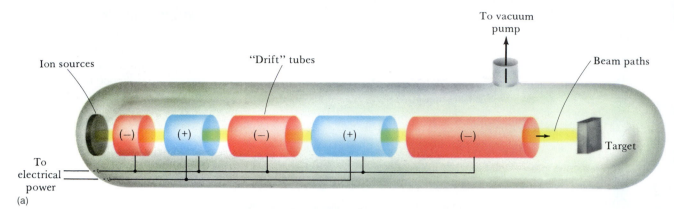

Ion sources

"Drift" tubes

To vacuum pump

Beam paths

(−) (+) (−) (+) (−)

To electrical power

Target

(a)

FIGURE 3–13 (a) Schematic drawing, illustrating the basic principles of a linear accelerator; (b) bending magnets in the interior of the main accelerator at the National Accelerator Laboratory (Fermi Lab in Illinois). The main accelerator is four miles in circumference.

(b)

the potential barrier a proton experiences when it collides with light nuclei such as lithium and beryllium. For heavier nuclei, which often carry much larger nuclear charges, the proton kinetic energy required to cause a nuclear reaction is correspondingly higher.

At about the same time as these transmutation experiments, van de Graaff generators, linear accelerators, and the first cyclotrons appeared, greatly improving the experimental potential for such reactions. The **van de Graaff generator** was an electrostatic generator capable of producing carefully controlled potentials of several million volts. Among other things, the device served as an early source of neutrons by making it possible to bombard beryllium nuclei with deuterons, the nuclei of deuterium atoms:

$$^{9}_{4}\text{Be} + ^{2}_{1}\text{H} \longrightarrow ^{1}_{0}n + ^{10}_{5}\text{B}$$

In a **linear accelerator** (Figure 3–13), an ion is accelerated through a tube set in a magnetic field, which serves to hold the particle in a straight line. As the ion emerges from the tube, the polarity is reversed, accelerating the particle into a second tube. By repeating the process, the charged particle is subjected to a series of electric kicks. Each tube must be longer than the one before it, so that the alternating polarity of all the tubes can be synchronized. Today, GeV (10^9 eV) energies are available.

Cyclotron designs (Figure 3–14) also use multiple acceleration of ions, and in many ways they can be thought of as linear accelerators wrapped into a spiral configuration. The tubes are replaced by a pair of flat, hollow, semicircular electrodes called **dees** (because they resemble the letter D).

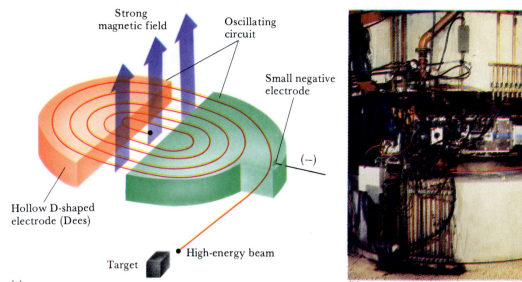

FIGURE 3–14 (a) Cyclotron: schematic drawing, illustrating the basic principles; (b) modern cyclotron, showing proton beam (in blue).

Every time the particle crosses from one dee to the other, it gets an energy boost. During the last 50 years, largely because of the development of this kind of technology for accelerating and detecting particles, **nucleosynthesis** has become a distinct subfield of chemistry.

As with all chemical reactions, nucleosynthesis may either release energy or absorb it. Consider, for example, the Crockcroft-Walton experiment. Using a high voltage generator, 0.4-MeV protons were hurled into lithium nuclei, producing 8.8-MeV α-particles. The energy accounting is interesting:

$$^{7}_{3}\text{Li} + ^{1}_{1}\text{H} + \text{energy} \longrightarrow ^{8}_{4}\text{Be}^* \longrightarrow 2\,^{4}_{2}\text{He} + \text{energy}$$

energy in $= 0.4$ MeV

energy out $= 2(8.8\text{ MeV}) = 17.6$ MeV

energy net $= (17.6 - 0.4)\text{MeV} = 17.2$ Mev

Hence, the process is net energy-releasing, and there should be an equivalent amount of mass lost in going from reactant nuclides to product nuclides:

reactant nuclides	$7.016003\ \mu + 1.007825\ \mu$	$= 8.023828\ \mu$
product nuclides	$2\ (4.00260)\ \mu$	$= 8.00520\ \mu$
mass difference		$= 0.01863\ \mu$

$$0.01863\ \mu \left(\frac{931.5\text{ MeV}}{\mu}\right) = 17.4\text{ MeV}$$

The worldwide synthetic radioisotope industry has expanded rapidly due in large part to the booming radiopharmaceuticals industry, which produces a steadily increasing number of radiolabeled drugs for both diagnostic and therapeutic use in medicine. All synthetic isotopes, with the exception of those derived from uranium decay, must be produced by neutron irradiation or ion bombardment. Both procedures require cyclotrons and other costly equipment. For 50 of the elements and their 240 stable isotopes, there is only one reliable source in the world, a bank of 30 electromagnetic separators owned by the U.S. Department of Energy (DOE)

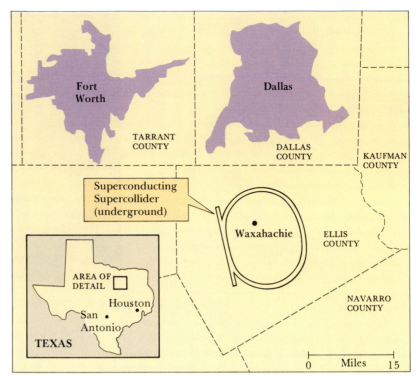

FIGURE 3–15 Site of a proposed superconducting supercollider.

at the Oak Ridge National Laboratories in Tennessee. Preventing shortages is of vital interest to U.S. research scientists, U.S. pharmaceuticals companies, and foreign customers of all types. For example, in 1982, Oak Ridge ran out of 60 of the 240 isotopes it produces. Unfortunately, a single pharmaceutical company cannot afford to build its own facilities (a multi-million-dollar investment). Plans to regulate production or ration products

---

THE SUPERCONDUCTING SUPERCOLLIDER

It is interesting that the largest scientific machines are required to carry out many of the most important experiments involving nature's smallest particles. In November of 1988, the Department of Energy awarded the State of Texas the federal project to build a $4.4 billion atom smasher, the superconducting supercollider, which will be the largest and most costly scientific instrument in the world and is likely to be so for decades to come. The proposed supercollider, an underground tunnel 53 miles in circumference, would encircle Waxahachie and an area nearly as big as Dallas (Figure 3–15).

The key feature of the supercollider design is its series of very powerful superconducting magnets, which can generate a virtually perpetual magnetic field (see Chapter 1). Narrow beams of protons collide at almost the speed of light, creating even smaller particles that exist at only very high energies. The protons are propelled by thousands of superconducting magnets, which achieve great power because they suffer no energy losses to electrical resistance (Figure 3–16). Protons would first be boosted to high energy levels by a linear accelerator and three progressively larger circular accelerators. In the main ring, some protons would be directed clockwise, others counterclockwise. At certain sites, special magnets will force the protons to collide while detectors record the results.

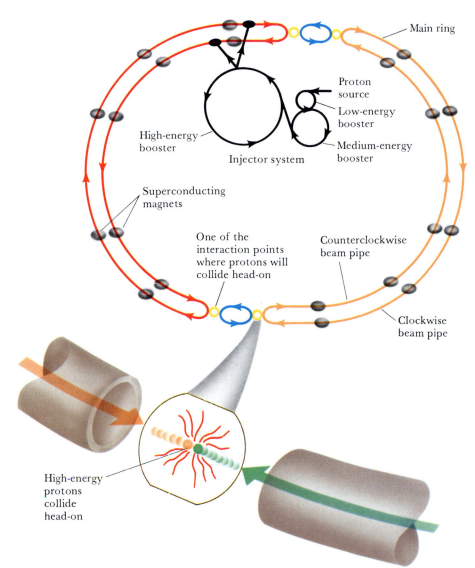

FIGURE 3–16   Operational design of the proposed supercollider.

The largest scientific instrument yet built came to life in August of 1989 as opposing beams of matter and antimatter began surging through Europe's Large Electron-Positron Collider, known as the LEP. It is built into a circular tunnel 16.6 miles in circumference that passes under villages, farms and the Jura Mountains along the French-Swiss border. LEP cost nearly a billion dollars and took seven years to build. It is larger than Fermilab's Tevatron (in Illinois) and the Stanford Linear Collider (in California). All of these large machines are designed to produce particle collisions that may provide keys to the nature of matter and the formation of the universe.

on the world market would be politically unacceptable solutions. The problem remains unresolved.

3.3   SPONTANEOUS NUCLEAR DECAY IN NATURE

As nuclides of radioactive elements decay, they change into other elements, so the number of atoms of the original element decreases. The rate of this process is usually measured by the **half-life**, the time it takes for half the nuclei of the original element to decay. With the notable exception of $^{14}_{6}C$ formed in the upper atmosphere, and perhaps tritium ($^{3}_{1}H$), natural radioactive nuclides fall into two main categories: those with half-lives about as long as the age of the Earth, estimated to be about 4.5×10^9 years, and the nuclear daughters that are continually produced from one of three disintegration (or decay) series, beginning with particularly long-lived parent nuclides. Representatives of the first group, with half-lives on the order of geologic time, can be found both early in the periodic table and among the heavy elements:

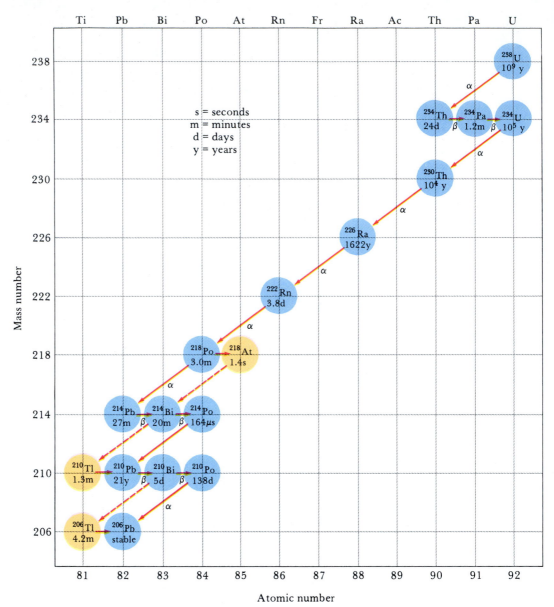

FIGURE 3–17 The uranium decay series, from U-238 to the stable lead isotope, Pb-206.

Lighter Nuclides		**Half-life, years**
potassium-40	$^{40}_{19}K$	1.3×10^9 y
rubidium-87	$^{87}_{37}Rb$	5.2×10^{10} y
Heavier Nuclides		**Half-life, years**
thorium-232	$^{232}_{90}Th$	1.4×10^{10} y
uranium-235	$^{235}_{92}U$	7.1×10^8 y
uranium-238	$^{238}_{92}U$	4.5×10^9 y

The three principal natural radioactive disintegration series originate with the heavier nuclies, eventually terminating in stable lead isotopes (Figure 3–17):

Thorium Series: Ten steps, beginning with thorium-232, during which six α- and four β-particles are emitted:

$$^{232}_{90}\text{Th} \longrightarrow ^{208}_{82}\text{Pb} + 6\,^4_2\text{He} + 4\,\beta^-$$

$$Z = 90 = 82 + 12 - 4$$

$$A = 232 = 208 + 24$$

Actinium Series: Eleven steps, beginning with uranium-235, during which seven α- and four β-particles are emitted:

$$^{235}_{92}\text{U} \longrightarrow ^{207}_{82}\text{Pb} + 7\,^4_2\text{He} + 4\,\beta^-$$

$$Z = 92 = 82 + 14 - 4$$

$$A = 235 = 207 + 28$$

Uranium Series: Fourteen steps, beginning with uranium-238, during which eight α- and six β-particles are emitted:

$$^{238}_{92}\text{U} \longrightarrow ^{206}_{82}\text{Pb} + 8\,^4_2\text{He} + 6\,\beta^-$$

$$Z = 92 = 82 + 16 - 6$$

$$A = 238 = 206 + 32$$

EXAMPLE 3–5

The actinium series, beginning with ^{235}U, ends up eleven steps later at the stable ^{207}Pb isotope. The first three steps are α-decay, followed by β-decay, and then α-decay:

$$^{235}_{92}\text{U} \longrightarrow ^{231}_{90}\text{Th} + ^4_2\text{He}$$

$$^{231}_{90}\text{Th} \longrightarrow ^{231}_{91}\text{Pa} + \beta^-$$

$$^{231}_{91}\text{Pa} \longrightarrow ^{227}_{89}\text{Ac} + ^4_2\text{He}$$

Write the reactions for the next three steps of the actinium series. They follow the same decay pattern: α-decay, β-decay, and α-decay.

EXERCISE 3–5

One important fissionable nuclide, the ^{233}U isotope, does not exist in nature. However, it has been successfully produced by nuclear reactions beginning with ^{232}Th. The process begins with neutron capture, followed by successive β-emissions. Write the nuclear reactions. ■

3.4   NUCLEAR FISSION AND FUSION

Fission

In the Fall of 1938, chemists working in Germany performed one of the most profoundly important experiments in modern times. Otto Hahn and Fritz Strassmann showed that when uranium nuclei were bombarded with neutrons, they became unstable, split apart in roughly equal fragments, and formed more stable nuclei with intermediate values of A. The fission process that followed absorption of the neutron is analogous to the behavior of a liquid drop as it contracts, elongates, and eventually comes apart (Figure 3–18).

In addition to the release of enormous amounts of energy, the most significant feature of nuclear fission is that more neutrons are produced than consumed. As a result, a **chain reaction** is feasible, with neutrons released in one fission reaction bringing about a second fission reaction,

Otto Hahn, Fritz Strassmann, and Lise Meitner conducted experiments in Germany that first demonstrated nuclear fission.

FIGURE 3–18 (a) Schematic diagram representing the uranium fission process and the Bohr-Wheeler liquid drop model for heavy nuclei. (b) The beginning of a chain reaction; the center nucleus has undergone fission, coming apart into two pieces, releasing gamma rays and more neutrons, some of which are captured by other nuclei, propagating the growing chain reaction.

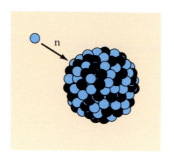

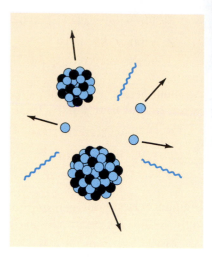

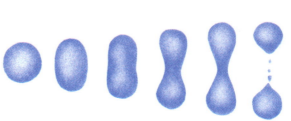

FIGURE 3–18a

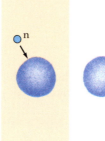

First atom bomb being test-fired in the New Mexico desert at Alamagordo, July 1945.

and so on. To guarantee a self-sustaining chain reaction, such as that in a nuclear reactor, each fission must produce excess neutrons that can initiate successive fission reactions. It is necessary to achieve maximum neutron production along with minimum loss due to nonfission absorption or escape from the surface of the fuel. The number of neutrons produced is proportional to the volume of reactor fuel, which is a function of r^3 (where r is a characteristic dimension of the fuel). However, neutron losses from the surface must be a function of r^2. For example, if we double the size of the fuel, we increase neutron production eightfold while neutron loss increases only fourfold. As a consequence, there must be a certain minimum or critical size of fuel mass at which the chain reaction produces enough neutrons to become self-sustaining. Applications for weapons and nuclear electric power should be immediately evident. Table 3–2 lists three nuclides commonly used as fission reactor fuels.

Uranium ore is mined by both open pit (surface) and underground operations, mostly in the western United States, especially in New Mexico. Major worldwide resources are found in Australia, Canada, South Africa, and the USSR. Typical U.S. deposits contain about 0.25 percent uranium metal. Despite the low content of metal, uranium ore is 30 to 50 times more efficient than coal on the basis of available energy per ton. Since environmental impacts are proportional to the amount of ore mined, this is one of the advantages of nuclear electric power over coal-fired power plants.

Natural uranium is composed of two isotopes; one (^{235}U, 0.711%) is **fissile**, or supports a self-sustaining chain reaction, whereas the other (^{238}U,

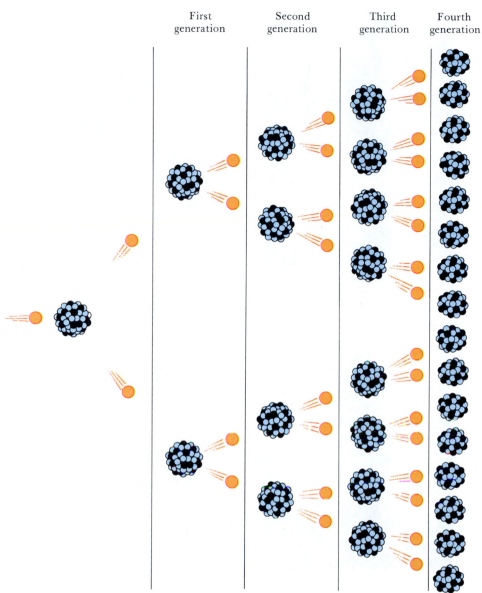

First
generation

Second
generation

Third
generation

Fourth
generation

FIGURE 3–18b

99.3%) is fissionable by other means. Therefore, methods for separating, converting, and enriching the isotopes are important. They cannot be separated by chemical means, but they can be separated physically, and fuels that are even slightly enriched have proved satisfactory for power reactors. Weapons-grade fuels need to be enriched considerably more.

TABLE 3–2 Fission Fuels	
Uranium-235	(^{235}U)
Occurs in nature and constitutes 0.7% of natural uranium.	
Uranium-233	(^{233}U)
Formed by neutron capture in thorium.	
Plutonium-239	(^{239}Pu)
Produced by neutron capture in ^{238}U.	

Uranium has been deposited on an electrode in the electrorefining phase of fuel processing. The crystalline mass is mostly lithium and potassium chlorides; the amethyst color is due to about 3% uranium chloride.

The enrichment process begins with purification of **yellow cake**, or U_3O_8, so named because of its color. Reaction with fluorine produces uranium hexafluoride (UF_6). This compound is a gas at temperatures above 56°C (at atmospheric pressure), and it has been widely used by the United States in a number of nuclear enrichment schemes. The most famous is the **gaseous diffusion** method, in which UF_6 is forced against a porous barrier. The lighter isotope ($^{235}UF_6$) molecules penetrate the barrier in greater numbers than do the heavier ($^{238}UF_6$) isotope molecules. By passing the gas successively through many barrier stages, any desired enrichment can be achieved.

Only slightly enriched ^{235}U (at 2 to 4%) is necessary for a light-water reactor (LWR) power plant. Because the amount of enrichment per stage is very small, thousands of individual stages are required, and the power consumption of the interstage pumps is tremendous. Yet, the power produced by such a plant is on average 20 times greater than the expended energy.

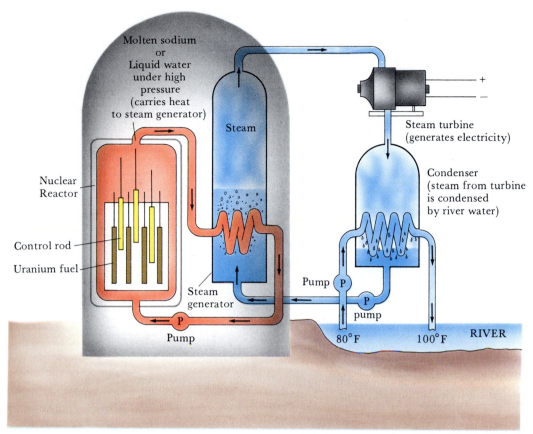

FIGURE 3–19 Basic design for the light water (LWR) fission electric power plant reactor that is typical of the American nuclear industry. Heat produced in a reactor by flying fission fragments does not turn directly to steam. As this simplified diagram indicates, the water is heated in an "exchanger" by a fluid that circulates through the reactor core.

Newer technologies of separation and enrichment are under development, including laser separation techniques that can be expected to compete favorably with the energy-intensive gaseous diffusion process.

The nuclear reactor that has become the most common commercial source of nuclear electric power and isotope production is known as the **thermal reactor** (Figure 3–19). It consists of assemblies of fuel rods, all surrounded by a medium containing atoms of low atomic number. This medium is called a **moderator**. The role of the moderator is to make propagation of the chain reaction possible by slowing the neutrons to the point where they can be absorbed by the heavy nuclei. (The probability that a neutron will be absorbed in a collision with a nucleus increases as the neutron's energy decreases.) When the high energy neutrons collide with light nuclei, some of their energy is transferred, and the neutrons are sufficiently slowed to bring them into thermal equilibrium with the fuel rods so that they can be captured. Graphite and heavy water are commonly used as the moderators in uranium reactors. Ordinary water can be used with enriched uranium reactors; beryllium can also be used, but the cost is prohibitive.

EXAMPLE 3–6

The following diagram shows the two different decay chains that occur when ^{235}U undergoes neutron irradiation, with the resulting radiations noted. In the initial step, a neutron is absorbed, after which the resulting nuclide becomes unstable and comes apart into elements in the middle of the periodic table. If the upper track begins with ^{90}Br, what nuclide does the lower track begin with?

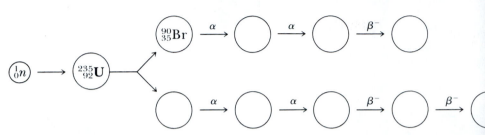

Solution   The starting point is the $^{236}_{92}U$ nuclide, produced from $^{235}_{92}U$ and $^{1}_{0}n$, which comes apart into $^{90}_{35}Br$. The other isotope must be $^{(236-90)}_{(92-35)}M$ or $^{146}_{57}M$. Thus, M is a lanthanum isotope.   ■

EXERCISE 3–6

What are the two nuclides at the ends of the tracks?
Answer: ^{82}Ge; ^{138}Cs.   ■

The estimated 1988 total use of energy by the United States for industry, transportation, heating, and lighting is about 75 quads, where 1 quad = 10^{15} Btu = 10^{18} J, the energy used by a metropolitan area such as greater San Francisco and Oakland in a year.

The first central-station nuclear electric power plant operated by a public utility in the United States was built in Shippingport, Pennsylvania and commissioned in 1957. Its capacity was about 100 megawatts. Some thirty years later, there are about 100 nuclear electric power stations in the United States, many with capacities of about 1000 megawatts, which is about the size of the largest fossil-fuel-fired electric power stations. Together, they provide about 15 percent of the electrical needs of the United States, and place the U.S. 19th on the world list of nuclear power users. Most reactors are located east of the Mississippi and are concentrated along the Northeast Corridor and near the Great Lakes, while smaller concentrations exist on the West Coast. It normally takes ten or more years to build a 1000-MW reactor station, at costs approaching 5 billion dollars. Nuclear plants are more expensive to construct than coal- or oil-fired stations, and the cost of nuclear power has risen from the lowest among all power sources to become not much different from those of the other sources.

There are several types of nuclear power station reactors. The first commercial station was a **pressurized water reactor** (PWR) or **light water reactor** (LWR), and that design is still popular today, along with the **boiling water reactor** (BWR). France leads the way in **liquid metal fast breeder reactors** (LMFBR), a fundamental design that has never passed the development stage in the U.S.

Various types of breeder reactors have been proposed in anticipation of the time when the world's limited supplies of ^{235}U have been depleted. Natural uranium contains only 0.72 (atom) percent ^{235}U, the rest being ^{238}U and 0.006% ^{234}U. But that large remainder of ^{238}U can undergo several kinds of nuclear reactions with neutrons, one of which is neutron capture. That is the first step in the production of plutonium, which is itself fissionable by thermal neutrons:

The failed nuclear power plant at Chernobyl, site of the world nuclear industry's most catastrophic event.

$$\begin{array}{cccc} (1) & (2) & (3) & (4) \\ {}^{238}_{92}\text{U} \longrightarrow & {}^{239}_{92}\text{U*} \longrightarrow & {}^{239}_{93}\text{Np} \longrightarrow & {}^{239}_{94}\text{Pu} \end{array}$$

(1) n,γ neutron capture
(2) β^- half-life = 23.5 months
(3) β^- half-life = 2.35 days
(4) ^{239}Pu α-emitter; half-life = 24,390 years

The troubled graphite-moderated Soviet reactor at Chernobyi was of a generally old-fashioned design not commonly found in the West. One of the most attractive developments in nuclear power technology today is the modular high-temperature gas-cooled reactor. Because it uses helium (instead of water) as the coolant, the reactor can operate at higher temperatures—hence the description "high-temperature" reactor. Because it can operate at higher temperatures, it can operate at efficiencies closer to 50% rather than 30%. But the key safety feature of the new design is that the fuel is more dilute than usual, made by forming uranium into billions of sand-sized grains, each covered with a tough ceramic shell that can withstand temperatures higher than any reactor temperatures. Therefore, the reactor cannot have a "meltdown." In the conventional reactor design, the dense, reactive fuel is concentrated in rods that begin to fail at temperatures of 1500–1800°C. Temperatures that high are reached in seconds or minutes if the complex cooling system of a conventional plant should fail, as it did at Three Mile Island, Pennsylvania, in 1979. In the high-temperature reactor, these little glassy spheres are packed into billiard-ball-sized containers that transfer heat and trap fission waste products but remain intact up to temperatures of 3300°C (Figure 3–20). Because the fuel is more dilute, the reactor would have to be much larger to achieve the same power output, typically 1000 MW of electricity. It is more reasonable to go to a design that includes several small reactors, say 150 MW each, strung together into one power plant—hence the term "modular."

A catastrophic near-meltdown and explosion in April of 1986 contaminated thousands of square miles of Northern Europe and the Soviet Union.

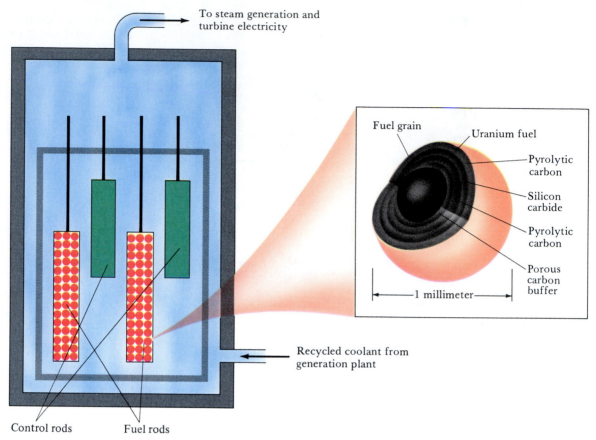

FIGURE 3–20 Schematic representation of the high-temperature gas-cooled reactor, a prototype of which is being built in Idaho. Note the modular design and the novel fuel capsules.

Fusion

Production of electric power from the energy produced by fusing light nuclei, in effect duplicating the Sun's thermonuclear furnace here on Earth, holds great promise for inexpensive, safe, clean nuclear electric power. The hydrogen isotopes that will likely be used as fuels are available everywhere; the reactions produce little in the way of hazardous wastes (compared with fission reactors); and there is no possibility of producing nuclear fuels for weapons in any direct way. Attempts have been underway for more than 30 years to develop such stations, but unfortunately only the science is reasonably known at present. It will likely be well into the first decades of the next century before the technology is equally developed and we can look forward to tapping fusion energy for electric power.

There are several possible fusion reactions. Consider the experimental prototype, the so-called deuterium-tritium or DT reaction. This is still the best bet for use in the first generation of fusion power plants, perhaps about the middle of the 21st century.

$$\mathrm{^{2}_{1}H} + \mathrm{^{3}_{1}H} \longrightarrow \mathrm{^{4}_{2}He} + \mathrm{^{1}_{0}}n + 17.6 \text{ MeV}$$

Because of the electrical repulsion between nuclei, the hydrogen nuclei must have kinetic energies on the order of 10 MeV, corresponding to a temperature of about 100,000,000 degrees. At such temperatures, which are indeed found in the interior of stars, atoms separate into nuclei and

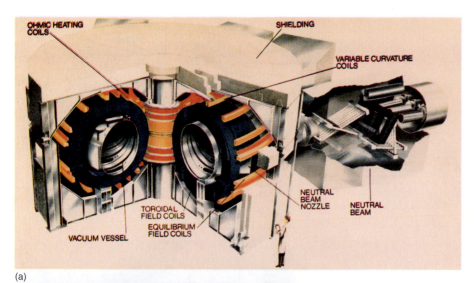

(a)

FIGURE 3–21   (a) A model of the Princeton Tokamak Fusion Test reactor (TFTR) that was built in 1983. Note the characteristic "toroidal" or doughnut-shaped chamber for generating and containing the plasma. (b) Photograph of a test plasma inside a reactor chamber.

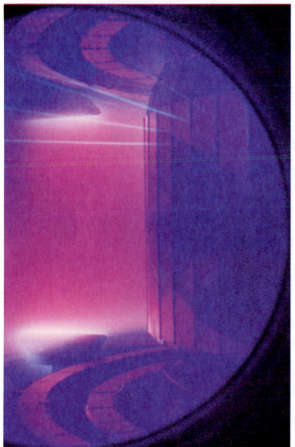

(b)

The hydrogen-3 isotope, better known as tritium, is an essential material in the design of both fusion reactors (as a fuel) and fission and fusion weapons (to boost fission yields and trigger thermonuclear warheads). Because tritium decays rapidly, declining by half every 12.3 years, supplies must be freshly prepared and there must be regular production. Arms negotiators, recognizing that control of tritium production could therefore be used to drive arms reduction and verification, have introduced this idea into East-West nuclear arms negotiations as the "Tritium Factor."

electrons, a state of matter called a **plasma** (Figure 3–21). The main problem that scientists face in achieving controlled thermonuclear fusion is confining the plasma long enough and at sufficient density to get the nuclei to fuse. At present, magnetic fields are used to confine the hot plasma in a doughnut-shaped device called a tokamak. Scientists and engineers have reached the break-even condition at which they are able to extract at least as much energy as is required to generate the fusion conditions in the first place.

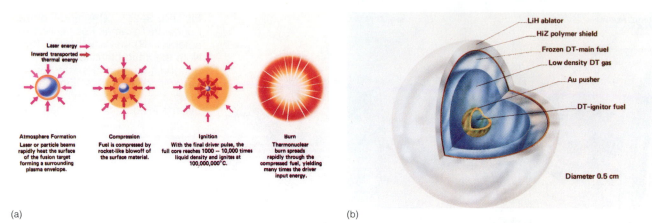

(a)   (b)

FIGURE 3–22 (a) Chemistry of the energy-producing implosion sequence (see also chapter opener photograph). (b) The fuel pellet is no larger than a tiny grain of sand, yet the design is very intricate. (c) NOVA, the giant lasers at Lawrence Livermore National Laboratories used to test inertial confinement fusion.

(c)

In a second scheme, which may seem as though it was taken from *Buck Rogers in the 25th Century*, solid DT fuel pellets are struck by many intense pulsed laser beams, which compress the pellet to a density four orders of magnitude greater than normal (Figure 3–22). The sudden compression heats the fuel pellet to fusion temperatures. This approach, called **inertial confinement fusion** (ICF), is in a much earlier stage of development than tokamak fusion. Nevertheless, it has been demonstrated experimentally.

3.5   RADIATION CHEMISTRY AND THE ENVIRONMENT

Ionizing radiations have been present in the environment since the Earth's formation approximately 4.5 billion years ago. Atoms of primordial, long-lived radioisotopes and their short-lived daughters have made the Earth a complex environment of ionizing radiation since that ancient time. In addition, the Earth is continuously assaulted from space by high energy cosmic radiations. Beginning late in the 19th century, mankind began to add significantly to the radiation in the environment, especially in the period since 1939 with the discovery of nuclear fission.

When nuclear radiations pass through matter, atoms and molecules are *ionized*. That is, the high energy of the radiation causes electrons to be shifted in a way that creates pairs of oppositely charged ions. A single α-particle or β-particle or a γ-ray is capable of producing about 10^5 to 10^6 ion pairs. Radiation damage is due primarily to this displacement of electrons, and the release of much energy in a small place. It can affect biological systems as well as material systems. At a chemical level, radiation damage is largely due to the ability of ionizing radiation to disrupt chemical bonds. Unfortunately, the chemical bonds most sensitive to radiation often turn out to be the very bonds responsible for fine-tuning the biological systems. At the cellular level, damage is found to be greatest in cells that multiply most rapidly, especially in the lymphatic system (Table 3–3).

Two distinct types of hazards are generally encountered from radioactive materials. **Contamination hazards** involve ingestion of radioactive materials, usually through food cycles. The α-emitters and β-emitters are the worst risks. Because of their chemical properties, radioisotopes may be deposited in critical tissues in the body. Isotopes of elements in the same group as calcium, for example, accumulate in the bone where they can irradiate blood-forming cells; for that reason, ^{90}Sr and ^{226}Ra are highly radiotoxic. Except for workers in the nuclear industry, most people are not subject to radiation hazards other than background radiation, the topical effects of a sunny day at the beach, television viewing, and exposure to occasional medical or dental x-rays. However, contamination of water supplies by radon from minerals and exposure of the pilots of high-altitude flights to radiation are exceptions. Radon also tends to collect in well-insulated homes built over rocks that contain its parent nuclides.

The effects of ionizing radiation on the human body can be divided into two categories. **Somatic effects** are those produced in the body after exposure to radiation. At 25 rems or less (the rem is a unit of radiation dose equivalent), effects are not generally detectable, although the number of leukocytes (white blood cells) may decrease. At 200 rems, nausea, fatigue, and increased susceptibility to infectious disease occur. The **LD-50**, a radiation dose lethal to 50% of those receiving it, is 400 rems; others recover over a period from weeks to months. At 600 rems, only about 20% of

TABLE 3–3	Effects of Ionizing Radiation at the Cellular Level
lymph blood bone nerve brain muscle	↑ direction of increasing susceptibility

GEIGER COUNTER

From the very beginning of radiation chemistry, the ionizing effects of radiations have been used as detectors. Early researchers used low-power optical magnifiers, actually counting individual hits, as radiations fell onto a fluorescent screen. Geiger and Marsden did just that, making the tedious measurements that provided Rutherford with the raw data for the scattering experiments that led to the planetary atom model. A **Geiger counter** (see figure next page) consists of a wire electrode within a tube filled with neon or argon. The central wire electrode is positive at a potential of 1000 to 5000 volts with respect to the walls of the cylinder, which serve as the negative electrode. When a high energy particle or a γ-ray photon enters the tube, the gas becomes ionized at that point, producing positive ions and negative electrons. They accelerate toward the opposite electrodes, ionizing gas particles along the way. The result is a cascade of particles that form an electrical pulse, which is capable of activating a counting device. An audible click is heard as each event is recorded, provided the pulse is amplified and fed through a speaker.

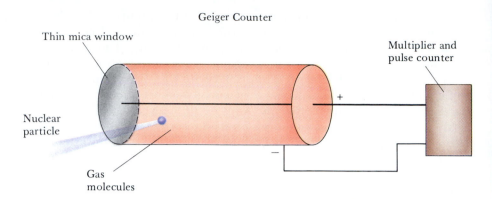

Geiger Counter

Thin mica window

Multiplier and
pulse counter

Nuclear
particle

Gas
molecules

exposed populations can be expected to survive, recovering from the effects only after many months. Chronic exposure to radiation can result in skin cancer, sterility, cataracts, and blood disorders. **Genetic effects** are those transferred to the next generation by alterations in the reproductive cells. It should be noted that ionizing radiations are mutagenic, capable of causing gene mutations and chromosome aberrations at any exposure.

3.6   ISOTOPES IN INDUSTRY, AGRICULTURE, AND MEDICINE

In addition to the best known applications of isotopes, such as the use of ^{235}U as a fission fuel for power reactors and weapons, separated isotopes serve many practical purposes in basic and applied research, in industry, and in medicine. A few of the many interesting and important ones are mentioned here. These applications can be roughly divided into three categories: **tracer applications**, where the nuclear properties of the isotope provide a label for studying the properties or behavior of another substance or material; **radiation applications**, where radioactivity is used to affect a substance, or allow some measurement to be made on the substance; and **dating techniques**, where quantitative measurement of an isotope present in a substance can be used to determine its age.

Tracer application: The Army Corps of Engineers has used water-insoluble $^{140}BaSO_4$ to study the movement of silt in rivers—the mouth of the Mississippi, for example. After the compound is released to the river bed, the movement of the material under various tidal conditions can be followed with a suitable radiation detector hung over the side of a boat.

Tracer application: Hydrologists sometimes add a water-soluble radioisotope to underground water supplies through a bore hole, and then sample the water supply from adjacent bore holes. In this way they can study the direction and velocity of large water tables. Iodine-131 has been used for this purpose.

Radiation application: "Go-devils" have been used for many years for detecting leaks in long water and oil pipelines. The go-devil is a packaged radiation detector that is inserted into the pipeline and is pushed along by the fluid. Cobalt-60 sources are attached along the pipeline at fixed intervals, and as the go-devil passes each marker a signal is recorded. If there is a leak in the pipeline, however, a second signal is received at a different location.

Radiation application: Thickness gauging, based on the reduction of the intensity of radiation as it passes through a medium, is widely used in industry to control the uniformity of films, sheets, and laminates. **Trans-**

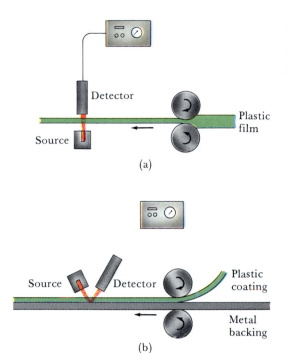

FIGURE 3–23 An industrial transmission thickness gauge for tracking the uniformity of plastic film and paper sheet.

mission thickness gauges use β-emitters for paper sheets, plastic films, linoleum, and thin metal foils, or γ-emitters for heavy gauge metal sheet. The source is placed on one side of the material to be measured, and the detector on the other (Figure 3–23). **Backscatter gauges**, having both the source and the detector on the same side of the material to be measured, allow one to control the thickness of backing materials and laminates.

Tracer application: Suitably labeled ^{32}P phosphates have been used to study the uptake rates and mechanisms whereby plants absorb nutrients, leading to the development of improved fertilizers and crops and a better understanding of root systems and plant foliage. Plant respiration and photosynthetic studies have been carried out using $^{14}CO_2$.

Dating application: Radioactive isotopes provide information to earth scientists and geologists, especially in dating old rocks. A favorite and highly reliable isotope system is uranium-lead preserved in zircons. These radioactive "clocks" tell us a great deal about the history of the Earth and will be discussed in some detail in Chapter 23 (on Chemical Kinetics).

Tracer application: ^{59}Fe-labeled hemoglobin has been used in studies of red blood cells and bone marrow; casein labeled with ^{32}P has been used to study metabolism and nutrition in laboratory animals as well as in human beings; and important drug absorption and metabolite studies of penicillin have been carried out with ^{35}S-labeled penicillin-G. In applications such as these, often painstaking synthetic procedures are first required in order to prepare the properly labeled compounds. Sometimes biosynthetic techniques are used, as in the preparation of the labeled penicillin by adding $^{32}SO_4^{2-}$ to the medium in which the mold is growing.

Tracer application: ^{131}I has been used in studies of thyroid activity and brain tumors; and ^{58}Co isotopes are used in the detection of pernicious anemia and the ability to absorb vitamin B_{12}.

Tracer application: ^{32}P and ^{35}S were used in experiments that demonstrated that DNA is indeed the carrier of genetic information. ^{15}N was used in studies that showed DNA is passed from generation to generation in a semiconservative fashion. Radioisotopes have been used to induce mutations and for the purpose of carrying out genetic studies.

Radiation application: ^{60}Co has been used for sterilization of foods, pharmaceuticals, packaging materials and containers; it is widely used as a chemotherapeutic agent.

SUMMARY

In a period spanning a professional lifetime, our knowledge of nuclear properties expanded from the basics of structure and properties of the atom to enormously significant applications. These range from electric power and medicine in the benefit of mankind to weapons of imponderable destruction. Thomson described the electron. Becquerel and the Curies discovered the phenomenon of radioactivity. Einstein told us about the equivalence of mass and energy. Rutherford and his students, colleagues, and co-workers produced transmutations of elements, discovered neutrons and split the atom, and along the way established the nuclear atom model.

In summarizing this chapter, remember that all nuclei are made up of N **neutrons** and Z **protons**. The nuclear radius is proportional to $A^{1/3}$, which means that the nuclear volume is proportional to the number of nucleons (A) and the density of the nucleus is relatively constant (or independent of A), being about 10^{15} g/cm^3. The nuclear **binding energy** averages about 8 MeV per nucleon, with a maximum in the curve of binding energy at iron and nickel. Both **fission** of heavier elements and **fusion** of lighter elements can be energy-releasing.

Nuclear fission begins with capture of a neutron by heavy nuclides such as ^{235}U, leading to an unstable state of the nuclide. The nucleus then comes apart into more stable nuclides of intermediate elements, nearer the middle of the periodic table, while emitting two or three neutrons. A chain reaction can result if the emitted neutrons are captured by other fissionable nuclei nearby, causing them to come apart in a similar fashion. **Nuclear fusion** has not yet been reduced to practice in any controlled fashion, although it has been demonstrated in tokamak and inertial confinement experiments with deuterium and tritium.

Detection of nuclear events is based on the fact that these high energy particles often interact with matter by ionizing atoms. With the aid of a **Geiger counter** we can detect individual nuclear events as they happen. Finally, the role of radioactivity and radioisotopes in medicine and industry is still surprising if not astonishing, nearly a century after their discovery. The impact of fission and fusion, half a century later, goes without saying.

QUESTIONS AND PROBLEMS

QUESTIONS

1. What is deuterium, and what is a deuteron? What is heavy water? What is "protium," and what is a proton?

2. What is the proper order of (increasing) penetrating power for α-, β-, and γ-rays? Why is penetrating power an inverse function of ability to cause ionization?

3. What was the evidence in favor of β-particles being electrons?

4. A Rutherford experiment established the identity of the α-particle as a helium nucleus. What was the confirming experiment?

5. Why is it assumed that nuclear forces are short-ranging?

6. Why is it necessary to propose the existence of special nuclear forces to hold the nucleus together?

7. How are radioactive emissions from an element affected by chemical combination?

8. Why is it so difficult to separate radioisotopes and decay products?

9. Starting with isotopically pure ^{238}U, what is the first decay product found, and how does the decay process change over time?

10. By how many units of mass does a radioisotope change after undergoing α-decay? β-decay?

11. If nuclei do not contain electrons, how is β-emission possible?

12. Why was it necesary to invent the existence of a strange particle like the neutrino in order to explain β-decay?

13. Why do protons generally make better bullets for nuclear reactions than α-particles? What makes neutrons particularly useful in that role?

14. Chain reactions are made possible by what fission product?

15. Briefly discuss the role of the moderator in a nuclear reactor.

16. Why is water a good moderator? What is its principal disadvantage?

17. How is the rate of reaction controlled in a nuclear reactor?

18. Especially high temperatures are required to bring about fusion reactions. Why?

19. What are plasmas and under what conditions do they exist?

20. What is the essential containment "trick" used to hold a 12,000-KeV plasma (which has a temperature of about 10^6 degrees) within a fusion reactor?

21. Answer the following questions true or false. If they are false, briefly explain why:

(a) When water is used as a moderator, surrounding the fissionable material in the core of a nuclear reactor, the purpose is to maintain temperature control.

(b) Extreme changes in temperature won't affect radioactivity.

(c) The penetrating radiation emitted from beryllium nuclei when bombarded with α-particles consists of x-ray photons.

22. (a) There are three radioactive decay processes in which the mass remains unchanged while the proton number and the neutron number change by ± 1. Name them.

(b) When uranium-235 nuclei undergo fission, they are excited by capture of (α-particles/β-particles/neutrons/γ-rays), splitting into nuclei nearer the middle of the periodic table.

(c) The (most/least) penetrating particles among the principal emanations observed in radioactive decay are also the most ionizing.

23. In studying the movement of silt in river estuaries and harbors using isotopic tracers, why must the tracer have a moderate half-life?

24. In studying the movement of underground waters with radioisotopes, the tracer should be water-soluble, with a relatively short half-life. Why?

25. Why do you think nuclei of low atomic number are necessary to ensure that neutrons are properly slowed for capture by the heavy nuclei in a reactor fuel?

PROBLEMS

Repulsion Forces   [1–4]

1. Calculate the force of repulsion between two protons at their point of closest approach, say 2.0×10^{-13} cm, the approximate diameter of a helium nucleus. What does your answer suggest about nuclear forces of attraction?

2. Calculate the force of repulsion between two electrons at a distance of 8.0×10^{-13} cm, the approximate diameter of a lead nucleus.

3. Consider a 6-MeV α-particle approaching an atom in a silver foil.

(a) Calculate the distance of closest approach (r_0) of the α-particle.

(b) By what factor would the energy of an α-particle have to change in order to approach this close to an atom in a gold foil?

4. Calculate the distance of the closest approach of a 10.-MeV α-particle to a lead atom.

Mass/Energy Relationships; Binding Energies   [5–12]

5. How much energy would be released in a proton/antiproton annihilation event?

6. What is the minimum energy released during an elec-

7. Compute the total binding energy and the binding energy per nucleon for each of the following nuclides: $^2_1$H, $^{12}_6$C, and $^{56}_{26}$Fe, given their isotopic masses of 2.0140 amu, 12.0000 amu, and 55.9349 amu, respectively.

8. Calculate the total binding energy and binding energy per nucleon for $^7_3$Li, $^{60}_{28}$Ni, and $^{235}_{92}$U, given their isotopic masses of 7.0160 amu, 59.9308 amu, and 235.0439 amu, respectively.

9. Calculate the mass defect and binding energy per nucleon for ^{16}O, given the following information:

proton mass	$= 1.007276 \ \mu$
neutron mass	$= 1.008665 \ \mu$

electron mass   = 0.000549 μ

isotopic mass for ^{16}O = 15.9994 μ

binding energy   = 931.5 MeV/μ

10. Compute the mass defect, binding energy, and binding energy per nucleon for the ^{12}C isotope of carbon, given the following data:

mass of neutron   = 1.008665 μ

mass of a hydrogen atom = 1.007825 μ

The atomic mass unit is defined to be exactly 1/12 the mass of a ^{12}C atom (and 1 μ is equivalent to 931.5 MeV).

11. Calculate the energy released during the deuterium/tritium reaction:

$$^2_1H + {}^3_1H \longrightarrow {}^4_2He + {}^1_0n + \text{energy}$$

12. Calculate the energy released in the following reaction:

$$^1_1H + {}^1_0n \longrightarrow {}^2_1H + \text{energy},$$ if the isotopic mass of ^2_1H is 2.0140 amu

Atomic Structure   [13–16]

13. State the numbers of electrons, protons, and neutrons in an atom of the $^{10}_5$B isotope of the element boron.

14. Given the following symbolic representation for an isotope of an unknown element, provide the information requested: $^{37}_{17}$M.

(a) the element's mass number

(b) the element's atomic number

(c) the proton number

(d) the number and identity of the extranuclear particles

(e) the number and identity of the nucleons (nuclear particles)

(f) the identity of the element M

(g) the atomic weight for the natural abundance of all the isotopes of the element

15. For each of the following, write the symbol of an appropriate nuclide:

(a) an isotope and an isotone of ^{16}O; of ^{208}Pb; of ^{120}Sn.

(b) an isotope and an isobar of ^{14}N; of ^{63}Cu; of ^{238}U.

16. Write the symbol of an isotope, an isotone, and an isobar of each of the following nuclides:

$$^{115}_{49}\text{In}, \quad ^{40}_{20}\text{Ca}, \quad \text{and} \quad ^{107}_{47}\text{Ag}$$

Radioactivity; Unstable Nuclides   [17–22]

17. Given that the stable isotope of sodium is ^{23}Na, what kind of radioactivity would you expect for ^{22}Na and ^{24}Na respectively, and why?

18. Predict the modes of decay for $^{26}_{13}$Al and for $^{28}_{13}$Al, and write the nuclear equations for the processes. Naturally occurring aluminum is 100% $^{27}_{13}$Al.

19. Rutherford's source of α-particles was the decay of radium-226. Write the nuclear equation.

20. The nuclide $^{201}_{84}$Po decays by α-emission. Write the nuclear equation for the process.

21. Carbon-14 decays by β-emission. Write the nuclear equation for that process.

22. Arsenic-81 is an unstable nuclide that decays by β-emission. Write the nuclear equation for this decay process.

Nuclear Reactions   [23–34]

23. Complete the following nuclear equations:

(a) $^{10}_5$B + $^4_2$He $\longrightarrow$ __ + $^1_1$H

(b) $^9_4$Be + $^1_1$H $\longrightarrow$ __ + $^2_1$H

(c) $^{27}_{13}$Al + $^1_0$n $\longrightarrow$ __ + $^{28}_{13}$Al

24. Complete the following nuclear transformations by listing the missing nuclide:

(a) $^2_1$D + α $\longrightarrow$ __ + $^1_0$n

(b) $^9_4$Be + $^4_2$He $\longrightarrow$ __ + $^1_0$n

(c) The ^{197}Au isotope of gold, decaying by β,γ-emission.

25. Fission of $^{235}_{92}$U is induced by neutron capture. The fission fragments are elements in the middle of the periodic table, along with neutrons that provide the means for generating a self-sustaining chain reaction. One mode of disintegration produces ^{90}Sr and ^{143}Xe. Write the nuclear equation.

26. Write a possible nuclear reaction for a fission of $^{239}_{94}$Pu that produces two nuclides of smaller mass and two neutrons.

27. The first artificial transmutation of an element occurred when Rutherford bombarded a sample of ^{14}N with α-particles, producing protons and an isotope of oxygen. Write the nuclear equation for artificial transmutation.

28. Write the nuclear equation for each of the following processes:

(a) A deuterium atom absorbs gamma radiation and then emits a neutron.

(b) The $^{40}_{20}$Ca isotope of calcium absorbs a neutron and emits an α-particle.

(c) The $^{238}_{92}$U isotope is bombarded with a ^{12}C isotope, yielding four neutrons and an isotope of californium.

29. The nucleus $^{235}_{92}$U is unstable and decays by sequentially emitting alpha (α) and beta (β) particles in the following order:

$$\alpha \; \beta \; \alpha \; \beta \; \alpha \; \alpha \; \alpha \; \beta \; \beta \; \alpha$$

Write the series of nuclei (the complete nuclear symbols) that are produced by the disintegration process.

30. One important fissionable nuclide, the $^{239}_{94}$Pu isotope, does not exist in nature, but it has been successfully produced by nuclear reactions beginning with $^{238}_{92}$U. The process begins with neutron capture, followed by successive β-emissions. Write the nuclear reactions.

31. It is useful to make some comparisons between coal and uranium as fuels for power generation, considering typical coal to have a heat content of 19 to 28 GJ/ton of mined material. Uranium as employed in the most widely used commercial reactor design, the LWR reactor, has a heat content of 460. GJ/kg of uranium metal. On that basis, calculate the heat content of a ton of uranium and its ratio to the coal value.

32. It is also useful to make some comparisons between megawatt (MW) electrical and thermal energy. Considering that electrical conversion efficiencies (MW$_{\text{electrical}}$/MW$_{\text{thermal}}$) are about 32 percent for an LWR nuclear electric power plant and 38 percent for a modern coal-fired electric power plant, calculate the electric energy ratio using data from the previous problem.

33. Here is a set of nuclear reactor neutron irradiation chains. The horizontal arrows indicate neutron capture, and the vertical arrows indicate β-emission. Fit the appropriate nuclides in the boxes.

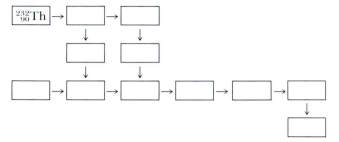

34. Here is a set of nuclear reactor neutron irradiation chains. The horizontal arrows indicate neutron capture, and the vertical arrows indicate β-emission. Fit the appropriate nuclides in the boxes.

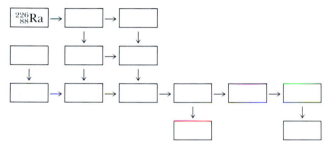

Additional Problems   [35–44]

35. Estimate the nuclear radius and the density of an aluminum atom and a gold atom. Are the assumptions of this chapter about nuclear densities (all being about the same) valid?

36. What fraction of the mass of an atom of ^{238}U is due to the electrons? Are the assumptions of this chapter about the mass of an electron (being negligible) valid?

37. From the curve of binding energy, the mass defect (or loss) for tin (Sn) is known to be about 0.00915 μ per nucleon. Using values for the neutron mass and the mass of the hydrogen atom (or the proton and the electron) listed in this chapter, estimate the mass of an atom of the ^{120}Sn isotope of tin.

38. Find the binding energy for the lithium-7 isotope, whose mass is known to be 7.01601 μ.

39. A hypothetical nucleus with mass 150 undergoes a fission reaction, and two fission fragments of mass 80 and 70 result. Using the curve of binding energy (Figure 3–12), estimate the total binding energy of the original nuclide and the two fission fragments. Is energy released or must it be supplied? In either case, estimate how much.

40. Repeat Problem 39 for a nuclide of mass 80 dividing into two equal fission fragments of mass 40.

41. (a) Yttrium-90 and copper-66 are both β-emitters. Identify both daughter isotopes.
(b) Plutonium-236 and protactinium-226 are α-emitters. Identify both daughter isotopes.

42. Plutonium-246 successively emits two β-particles, two α-particles and several γ-rays. What is the isotope that results?

43. Actinium-231 emits two β-particles, four α-particles, one β-particle, one α-particle, and several γ-rays as it decays to what isotope of which element?

44. Chlorine-36 is an unstable artificial isotope of chlorine. It decays by either β-decay or electron capture. Write nuclear equations for both possible processes.

MULTIPLE PRINCIPLES   [45–48]

45. (a) Three neon isotopes occur naturally: neon-20 (90.9%), neon-21 (0.30%), and neon-22 (8.8%). Find the average mass for the natural abundance of the isotopes of neon. Compare your value to that given in the periodic table.
(b) Two principal isotopes of uranium occur naturally. Find the approximate percent composition of the natural abundance of these two uranium isotopes from the fact that the masses of the two isotopes are 235.044 and 238.051, respectively, using the value for the average mass given in the Table of Atomic Weights.

46. The radius of the Earth is approximately 6×10^6 m and its density is 6 g/cm^3. If the Earth were shrunk so that its density was of the same order of magnitude as nuclear densities, what would its radius have to be?

47. In a particle accelerator experiment, the common isotope of magnesium, $^{24}_{12}$Mg, is bombarded by two different nuclear particles, producing an uncommon isotope of sodium, $^{22}_{11}$Na. Write likely/possible nuclear equations for the two-step nucleosynthesis of sodium-22.

48. Recall that fusion energy is being considered as an alternative for the 21st century. Determine whether it is a viable alternative on an energy per unit weight basis by comparing the energy release per kilogram for deuterium-tritium fusion with that for nuclear fission.

APPLIED PRINCIPLES   [49–51]

49. Uranium-238 is not a useful fuel for nuclear electric power reactors but it is widely available, making up 99.3% of natural uranium. On the other hand, plutonium-239 is fissionable and therefore a very useful fuel, but it does not occur naturally. However, the plutonium isotope can be synthesized from the uranium isotope by neutron capture, followed by decay of the unstable uranium-239 isotope. Write nuclear equations for the nucleosynthesis.

50. Calculate the energy released, in kilowatt-hours per kilogram, for the fission of uranium-235, assuming the energy release to be 200 MeV per fission.

51. Calculate the masses of ^{235}U and ^{238}U per metric ton of uranium ore, assuming ^{235}U is 0.25 weight-percent of the total uranium content.

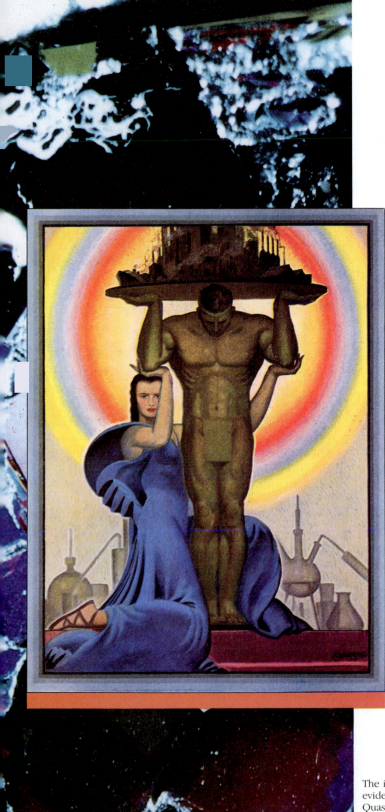

CHAPTER
4

Elements, Compounds, and the Earth

The interface between chemistry and industry is never more in evidence than when theory and practice come together. These large Quasicrystals consisting of aluminum, lithium, and copper are a new form of solid state matter. Recent experiments have shown that Quasicrystals might be used for a variety of applications from avionics to microelectronics. The painting depicts "Chemical Industry, upheld by Pure Science, sustaining the production of Man's necessities." Created during the Great Depression as part of the 300th anniversary celebration of the founding of chemical industries in America (1935), it reflects an idealized world on the verge of dramatic change.

4.1   SOME ASPECTS OF THE EARTH AND THE ELEMENTS

Elements represent the basic building blocks of matter, including planet Earth and all the other pieces of the universe, from the largest to the smallest. This chapter begins with a few words about the structure of the Earth and the elements and compounds of which it is composed. To say that the Earth is an extraordinarily complex and fragile chemical entity would be a gross trivialization. While it is relatively easy to visualize the Earth's surface or regions near the surface, evidence of its internal structure and composition is not as easily obtained. We have to work with indirect evidence based on studies by geologists, geophysicists, and geochemists. **Geochemistry** is the study of the Earth and its six-mile-high envelope of surrounding air (Figure 4–1). Solid samples for study are obtained by scratching at that part of the Earth called the crust. The gases of the atmosphere are often sampled at a distance, by spectroscopic examination.

In simplified terms, the Earth's structure can be likened to that of a golf ball with a fluid center (Figure 4–2). There is a thin outer crust, from 5 to 65 kilometers in thickness. The average thickness is about 15 km, and all of our currently available supply of mineral resources is contained in it. For our animal and vegetable resources, we depend on the thin cover of soil immediately at the surface. The deepest penetrations of the crust have been about 3.5 km (or 2 miles) in a South African mine shaft, and an 8-km Texas oil well. Neither comes close to penetrating the second layer.

Underlying the crust, at depths extending to about 2900 km below the surface, lies the mantle. There is a boundary region at the base of the crust called the Mohorovicic (often shortened to Moho) discontinuity. Our knowledge of this area is primarily from the observation that a sudden change takes place in the velocity of earthquake waves in that region.

The mantle extends down to where the core begins, and it constitutes 83% of the Earth's total volume and 68% of its mass. Geologists speculate that during the early history of the Earth there was only one layer, a primitive mantle, and that the crust developed slowly through volcanic events. The mantle today is the source of most of the Earth's internal energy; it is responsible for the forces that cause spreading of the ocean floors, continental drift, and major earthquakes. The mantle is composed of dense magnesium and iron silicates (which are complex compounds of these metals with silicon and oxygen) under very high pressures, and it is likely that it also contains aluminum, sodium, potassium, and calcium. This is known from seismic density measurements and chemical examination of meteorites, which are presumed to have similar compositions.

The Earth's core extends downward for more than 3000 km, from the mantle to the geometric center, and appears to be divided into a liquid outer core and a solid inner core of extremely high density. It is assumed to be metallic, mostly iron, but with some nickel. Again, what we know has come from seismic data and the analysis of meteorite materials. The data are qualitative at best. Therefore, geochemists and physicists continue to study the exact composition of the Earth's core.

The envelope of the atmosphere is composed primarily of oxygen and nitrogen, along with several other elements in much smaller concentrations. Table 4–1 lists the elements in the continental crust, and Table 4–2 lists the principal components of the atmosphere. There is no obvious scheme that explains which elements are present or why they have their abundances. There is a considerable mix of the inert and the reactive, the rare

FIGURE 4–1 The geochemical laboratory called Planet Earth as viewed from space.

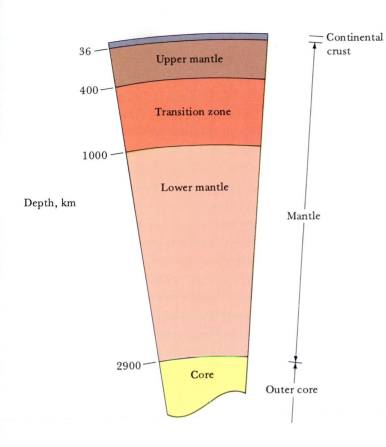

FIGURE 4-2 Section through the layered structure of planet Earth.

TABLE 4-1	Average Composition of the Earth's Continental Crust	
Element	**Symbol**	**Percent by Mass**
oxygen	O	46.59
silicon	Si	27.72
aluminum	Al	8.13
iron	Fe	5.01
copper	Cu	3.63
sodium	Na	2.85
potassium	K	2.60
magnesium	Mg	2.09
titanium	Ti	0.63
hydrogen	H	0.13
phosphorus	P	0.13
manganese	Mn	0.10
sulfur	S	0.052
chlorine	Cl	0.048
chromium	Cr	0.037
nickel	Ni	0.020
cobalt	Co	0.0011
others	—	0.23

TABLE 4-2	Average Composition of the Earth's Atmosphere Near the Surface
Components	**Fraction by Volume**
Major Components (in dry air)	
nitrogen (N_2)	0.78084
oxygen (O_2)	0.20946
argon (Ar)	0.00934
carbon dioxide (CO_2)	0.00033
water vapor (H_2O)	highly variable
ozone (O_3)	variable
Lesser Components	
neon (Ne)	1.1818×10^{-5}
helium (He)	5.24×10^{-6}
krypton (Kr)	1.14×10^{-6}
methane (CH_4)	1.6×10^{-6}
hydrogen (H_2)	5×10^{-7}
others including NO, CO, and NO_2	small amounts

and the commonplace, those that have proved useful and widely applicable and those that have little demonstrated utility.

There are more than a hundred elements in the whole Earth catalog, and although that number can be expected to grow, it will grow slowly and the new elements will almost surely be produced only in minute quantities.

Of the known elements, 89 can be concentrated, weighed, and made to undergo reactions in order to study their compounds. The rest of the elements are the rare, unstable, radioactive kinds described in Chapter 3. Of the 89 elements, less than half occur to a significant extent in the world around us and in the things we use. However, the story changes when we describe the compounds that these elements form. The elements combine in an almost endless number of ways to form literally millions of different compounds.

The great mass of information on chemical compounds can be condensed into a learnable amount of material. In order to do this, chemists have developed a number of schemes. Three of the most important ones are the systematic arrangement of the elements into a periodic table, the Lewis system of symbols and structures that represent the electronic structures of atoms and provide the basis for some simple models for bonding, and the modern methods of chemical naming or nomenclature. In this chapter we will use these devices to establish a working knowledge of the common events of chemistry and the more important chemical compounds. Later in this book we will look into the reasons why such tools work for us in simplifying and organizing the theory and practice of modern chemistry.

If we wrote just one short paragraph for each of the 18,000 or so most important chemical compounds, this text would consist of not one but perhaps ten volumes, each about 1000 pages!

4.2 THE PERIODIC TABLE

If two elements are very similar, we can expect that they will probably form similar compounds. Knowing this, in turn, will make classifying their compounds easier. On the other hand, if two elements are very different, they will probably form very different compounds. For example, sodium is similar to potassium. Both are soft metals with low densities; both react quickly and violently when dropped onto water. Fluorine is similar to chlorine. Both are yellowish, poisonous gases, and both quickly corrode the surfaces of most metals. On the other hand, sodium and fluorine have very little in common. Since few similarities exist among their characteristic properties, we can reasonably conclude that (1) a compound formed of sodium and fluorine would be very similar to a compound formed of sodium and chlorine, and (2) both of these compounds would be very different from a compound formed of fluorine and chlorine. To see how good these predictions are, look at the data in Table 4–3.

The most useful tool for classifying the elements is the **periodic table** (Figure 4–3). If the chemical elements are listed in the order of increasing atomic number, elements with similar properties repeat in a regular way. This periodic occurrence of repeating properties of the elements is called the **periodic law**. Elements with similar properties line up under each other to form vertical columns, which are known as **groups** or **families**. Hydrogen, the first element, is unique, and does not fall nicely into any of the

TABLE 4–3 Some Characteristic Properties of the Simple Combinations of Na, Cl, and F

Compound	Physical State at 25°C	Melting Point °C	Boiling Point °C
NaF	solid	993	1700
NaCl	solid	801	1413
ClF	gas	−154	−101

Group IA																	VIII
1 H 1.0079	IIA											IIIA	IVA	VA	VIA	VIIA	2 He 4.0026
3 Li 6.941	4 Be 9.0122											5 B 10.811	6 C 12.011	7 N 14.0067	8 O 15.9994	9 F 18.9984	10 Ne 20.1797
11 Na 22.9898	12 Mg 24.3050	IIIB	IVB	VB	VIB	VIIB	VIIIB			IB	IIB	13 Al 26.9815	14 Si 28.0855	15 P 30.9738	16 S 32.066	17 Cl 35.4527	18 Ar 39.948
19 K 39.0983	20 Ca 40.078	21 Sc 44.9559	22 Ti 47.88	23 V 50.9415	24 Cr 51.9961	25 Mn 54.9380	26 Fe 55.847	27 Co 58.9332	28 Ni 58.69	29 Cu 63.546	30 Zn 65.39	31 Ga 69.723	32 Ge 72.61	33 As 74.9216	34 Se 78.96	35 Br 79.904	36 Kr 83.80
37 Rb 85.4678	38 Sr 87.62	39 Y 88.9059	40 Zr 91.224	41 Nb 92.9064	42 Mo 95.94	43 Tc (98)	44 Ru 101.07	45 Rh 102.9055	46 Pd 106.42	47 Ag 107.8682	48 Cd 112.411	49 In 114.82	50 Sn 118.710	51 Sb 121.75	52 Te 127.60	53 I 126.9045	54 Xe 131.29
55 Cs 132.9054	56 Ba 137.327	57 *La 138.9055	72 Hf 178.49	73 Ta 180.9479	74 W 183.85	75 Re 186.207	76 Os 190.2	77 Ir 192.22	78 Pt 195.08	79 Au 196.9665	80 Hg 200.59	81 Tl 204.3833	82 Pb 207.2	83 Bi 208.9804	84 Po (209)	85 At (210)	86 Rn (222)
87 Fr (223)	88 Ra (226)	89 **Ac (227)	104 Unq (261)	105 Unp (262)	106 Unh (263)	107 Uns (262)	108 Uno (265)	109 Une (266)									

*Lanthanide Series

58 Ce 140.115	59 Pr 140.9076	60 Nd 144.24	61 Pm (145)	62 Sm 150.36	63 Eu 151.965	64 Gd 157.25	65 Tb 158.9253	66 Dy 162.50	67 Ho 164.9303	68 Er 167.26	69 Tm 168.9342	70 Yb 173.04	71 Lu 174.967

**Actinide Series

90 Th 232.0381	91 Pa 231.0359	92 U 238.0289	93 Np (237)	94 Pu (244)	95 Am (243)	96 Cm (247)	97 Bk (247)	98 Cf (251)	99 Es (252)	100 Fm (257)	101 Md (258)	102 No (259)	103 Lr (260)

Metals

Metalloids

Nonmetals

Note: Atomic masses are 1987 IUPAC values (up to four decimal places).

FIGURE 4–3 The Periodic Table of the Elements.

The correlation between the number of electrons and chemical properties will be carefully explained in Chapters 14 and 15, when we discuss the electronic structure of atoms in some detail.

groups; rather, it has properties similar to those of several groups. Each of the other elements can be assigned to a specific group in the periodic table.

The atomic number of a nuclide, used to order the elements in the periodic table, is equal to the number of protons in the nucleus of the atom. It is also equal to the number of electrons in the neutral atom. The electrons determine the chemical properties of an element because chemical reactions involve removing, adding, and sharing electrons. Therefore, it should not be surprising that the number of electrons is related to the chemical properties of the atoms of an element. For the present, the periodic table will serve as a very useful method of classifying the elements without worrying about why the method works.

The groups in the periodic table are commonly designated by Roman numerals from I to VIII, accompanied by either A or B. Groups IA through VIIIA are called **main groups**, and the properties of the elements of these groups run between extremes—from solid, metallic elements to gaseous, nonmetallic elements. The other elements, including those in Groups IIIB to IIB, elements 58 to 71 (which are called lanthanides), and elements 90 to 103 (which are known as actinides) are all generally classified as **transition elements**. The transition elements are all metals and, except for mercury, all are solids at room temperature. More importantly, the transition elements are far less diverse in their chemical properties than the main group elements. In recent years, some changes in the naming of groups in the periodic table have been proposed. One involves trading the main group element designation IIIA through VIIIA for IIIB to VIIIB, and giving the transition metal groups headed by Sc through Ni the IIIA to VIIIA designation, which is in full agreement with European usage. Another suggestion is simply numbering the eighteen columns consecutively from 1 through 18.

We will concentrate almost exclusively on the main group elements in this chapter on descriptive chemistry. The transition metal elements will be discussed in Chapter 22.

Another useful distinction can be made by observing the zig-zag line stepping down the right side of the periodic table. Atoms of elements that lie below and to the left of this line are classified as **metals**. Here are some of their general properties: (1) Metals are almost all solids. (2) They are generally good conductors of heat and electricity. (3) Most are lustrous, malleable, and ductile, at least to some extent. Elements above and to the right of this line, along with H, are classified as **nonmetals**, and their properties are radically different from those of the metallic elements. Since properties change continuously across the periodic table, those elements adjacent to the zig-zag line tend to bridge the two classifications, combining both metallic and nonmetallic properties. They are often called **metalloids**.

4.3 LEWIS SYMBOLS FOR THE ELEMENTS

One whole group, the VIIIA elements (called the **noble gases**), were discovered within the span of a few years at the end of the last century. These elements were found to be extraordinarily unreactive, and this inertness led to the realization that there is something special about the number of electrons possessed by each of these elements. According to the periodic table, the atomic numbers of these elements are 2, 10, 18, 36, 54, and 86. Their atoms have little or no tendency to gain, lose, or share electrons. Thus, these atomic numbers represent very stable configurations of electrons, and only electrons beyond those in a complete noble gas configuration can be expected to participate in chemical reactions. Electrons in

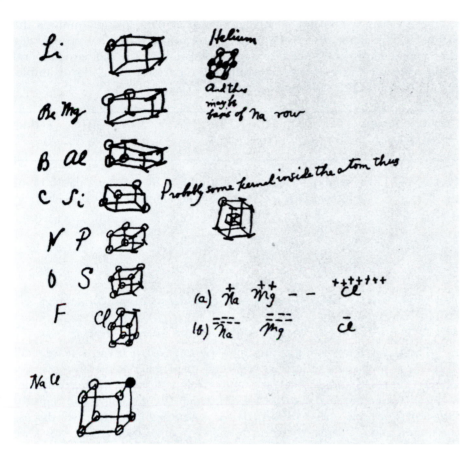

FIGURE 4–4 The first Lewis symbols. The cube was not significant except that it had eight corners. The photo is from a memorandum Lewis wrote in 1902.

excess of the noble gas configurations are called **valence electrons** (and the remaining electrons are referred to as core electrons). Hydrogen has one valence electron, as do all of the elements in Group IA of the periodic table; oxygen has six valence electrons, and fluorine has seven.

In 1916, Gilbert Newton Lewis (Figure 4–4) proposed a scheme for describing the electron configurations of the elements from lithium (Li) through neon (Ne)—the second period elements—and from sodium (Na) through argon (Ar) in the third period. He imagined the atom as occupying a cubic space. Starting with the one valence electron in lithium, he added successive electrons to the corners of the cube until all eight corners were occupied (the neon configuration). The same thing could be done starting with the one valence electron of sodium and proceeding to the configuration of argon. The only significance of the cube, however, is that it has the right number of corners, eight. **Lewis symbols** for atoms are just as useful (and more simply drawn), with the electrons shown as dots arranged around the symbol for the element. The Lewis symbols for the elements from Li through Ne are as follows:

$$\text{Li} \cdot \quad \text{Be} : \quad \cdot \text{B} : \quad \cdot \dot{\text{C}} : \quad \cdot \ddot{\text{N}} : \quad \cdot \ddot{\text{O}} : \quad \cdot \ddot{\text{F}} : \quad : \ddot{\text{Ne}} :$$

Note that electrons are customarily paired for numbers of valence electrons greater than four. Also, since the octet of electrons in any noble gas (other than helium) corresponds to a filled valence shell, its symbol can be shown either with all eight electron dots or without any. For helium (only), the core electrons and the valence shell are the same—two—and we write it as a pair (He:) or as He without the electron dots.

The Lewis symbol for each member of a main group family is the symbol of the element with the appropriate number of electron dots for the valence electrons. That number is the same as the Roman numeral designation above the element's group in the periodic table. For example, the fourth-period Lewis symbols are:

Families/Groups	IA	IIA	IIIA	IVA	VA	VIA	VIIA	VIIIA
4th row/period	K·	Ca:	·Ga·	·Ge·	·As:	·Se:	·Br:	Kr

Finally, the Lewis symbol of hydrogen is traditionally written, like those of the Group IA elements, as the symbol for the element with one electron dot:

IA
H·
Li·
Na·
K·
Rb·
Cs·

The Lewis symbols for atoms, and the Lewis structures that we shall draw for compounds, should be regarded principally as useful tools, not as exact representations. They are quite effective for the main group elements in predicting correct formulas and structures of possible compounds. However, they are not as useful for transition elements. We will discuss the underlying principles that support Lewis symbols and structures as we get deeper into our study of chemistry.

4.4   FORMATION OF MONATOMIC IONS

Because the noble gas configurations are so stable, Lewis argued reasonably that other atoms gain or lose electrons to achieve such configurations. Once an atom gains or loses an electron, there is an imbalance between the total positive charge of the protons in the nucleus and the total negative charge of the extranuclear electrons. Removal of electrons thus creates a positive ion or **cation**, and addition of electrons creates a negative ion or **anion**. Such ions involving only a single atom are called **monatomic ions**. Particularly stable ions have the noble gas electron configuration, and a positive or negative charge of 1 or 2. Thus Group IA elements can be expected to lose one electron, forming ions with an electronic charge of $+1$:

$$Li^+, Na^+, K^+, Rb^+, Cs^+$$

The formation of these ions can be represented by equations of the type

$$Li \longrightarrow Li^+ + e^-$$

Similarly, IIA atoms will form ions with an electronic charge of $+2$:

$$Be^{2+}, Mg^{2+}, Ca^{2+}, Sr^{2+}, Ba^{2+}, Ra^{2+}$$

A typical equation representing the process would be

$$Ca \longrightarrow Ca^{2+} + 2\,e^-$$

Stable negative ions are typically formed by the VIIA elements by adding an electron to their valence shell, giving F^-, Cl^-, Br^- and I^-. The reactions forming these ions are of the type

Large flat salt plates of sodium chloride (Na^+ and Cl^-) such as these can be used for infrared spectral analyses, a laboratory tool for determining chemical structures (see Chapters 14, 16, and 25).

The unit of ionic charge has the same magnitude as e, the electronic charge, or 1.6×10^{-19} C.

$$F + e^- \longrightarrow F^-$$

Ions with -2 charges arise from VIA elements, particularly

$$O^{2-}, S^{2-}, Se^{2-}, Te^{2-}$$

Ions with charges $+3$ or -3 electronic units are less common than the lower charges; nevertheless, they do exist. Al^{3+} is the most common of the IIIA ions, and N^{3-} is the most common of the ions resulting from atoms of the VA elements. The P^{3-} ion also exists.

Formation of ions with noble gas configurations from Group IVA elements would result in ions with charges of either $+4$ or -4. These can occur only under most unusual circumstances. Some compounds with C^{4-} ions do exist; and Sn^{4+} and Pb^{4+} ions appear to exist in some compounds.

Main group metals also form a few monatomic cations that do not have noble gas configurations. Important examples of this state of affairs are found in the bottom two members of main groups IIIA, IVA, and VA. These elements form cations with charges that are two less than would be needed to reach the noble gas core. Thus, the ions In^+, Tl^+, Sn^{2+}, Pb^{2+}, Sb^{3+}, and Bi^{3+} all exist.

Finally, we must consider hydrogen to be a unique element. A hydrogen atom can add one electron to form the hydride ion (H^-), with the He electron-dot configuration,

$$H\cdot + e^- \longrightarrow H\!:^-$$

However, the hydrogen atom more commonly forms a hydrogen cation (a proton) by losing its electron:

$$H\cdot \longrightarrow H^+ + e^-$$

4.5 SALTS—NOMENCLATURE AND SIMPLE CHEMISTRY

Naming Simple Salts and Acids

Many of the important monatomic cations and anions can be derived from main group elements, using the periodic table as a guide. None of these ions normally exist separately, but are found instead combined with other ions. Compounds of one monatomic cation and one monatomic anion are called **binary** combinations, since they contain only two different components. These compounds are classified as **salts** unless the cation is H^+, in which case the compound is most likely to be classified either as an **acid** or as a covalent compound such as ammonia, NH_3. Combination of the 12 cations and 12 anions listed in Table 4–4 leads to the formation of more than one hundred salts and a dozen hydrogen compounds; most are known and many have proved to be very important.

The only rule for combining cations and anions to form compounds is that the resulting product must be electrically neutral. If the two ions have opposite signs of equal magnitude, the ions in the resulting compounds are present in a simple one-to-one ratio. Here are some typical examples:

Lithium fluoride	LiF	$+1$ and -1
Sodium bromide	NaBr	$+1$ and -1
Magnesium sulfide	MgS	$+2$ and -2
Calcium oxide	CaO	$+2$ and -2
Aluminum nitride	AlN	$+3$ and -3

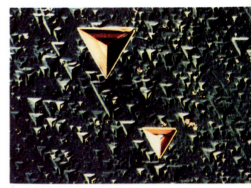

The triangular depressions in this single crystal of arsenic formed during studies of the evaporation properties of the element. Arsenic, along with gallium and aluminum, is a key constituent of semiconductor devices such as injection laser and light-emitting diodes (see chapters 19 and 20).

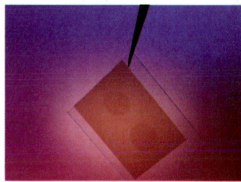

Indium has been used in developing an immunoassay for diagnosing schistosomiasis (see chapter 25). An indium-coated substrate is coated with antigens of the parasite and then spotted with two drops of patient's blood. The presence of antibodies is positively indicated by a darkening of the indium coating over the location of the blood spots due to reaction with antigens, as antibodies fight the disease.

TABLE 4–4 Monatomic Cations and Anions of the Main Group Elements Which Form One Significant Monatomic Ion

Cations		Anions	
H^+	hydrogen	H^-	hydride
Li^+	lithium	F^-	fluoride
Na^+	sodium	Cl^-	chloride
K^+	potassium	Br^-	bromide
Rb^+	rubidium	I^-	iodide
Cs^+	cesium	O^{2-}	oxide
Be^{2+}	beryllium	S^{2-}	sulfide
Mg^{2+}	magnesium	Se^{2-}	selenide
Ca^{2+}	calcium	Te^{2-}	telluride
Sr^{2+}	strontium	N^{3-}	nitride
Ba^{2+}	barium	P^{3-}	phosphide
Al^{3+}	aluminum	C^{4-}	carbide

Note that the charge that is formally assigned to the ions in each case—anion and cation—does not appear in the formulas. This is a historical precedent, because the formulas were established before the existence of ions was known. Subscripts representing the lowest factors that will balance the charges must be added to the formulas if the constituent ions have different magnitudes of charge. Typical examples are:

$$2\,K^+ + S^{2-} \longrightarrow K_2S$$
$$Ba^{2+} + 2\,Cl^- \longrightarrow BaCl_2$$

Potassium sulfide	K_2S	$2(+1)$ and -2
Barium chloride	$BaCl_2$	$+2$ and $2(-1)$
Aluminum oxide	Al_2O_3	$2(+3)$ and $3(-2)$
Lithium nitride	Li_3N	$3(+1)$ and -3
Germanium tetrafluoride	GeF_4	$+4$ and $4(-1)$

In order to discuss such compounds it is necessary to give names to ions that can be combined to form salts. After you learn a relatively small number of names for the ions, you will be able to provide a large number of names for their salts. The rules for monatomic ions are simple. The cation simply has the same name as the element: Na^+ is sodium ion; Ca^{2+} is calcium ion; Al^{3+} is aluminum ion. The anions have the suffix *-ide* substituted for the last one or two syllables of the name of the element. Accordingly, F^- is fluoride ion, O^{2-} is oxide ion, and N^{3-} is nitride ion.

The binary salts are named simply by combining the names of the ions, cation first, without mentioning the subscripts in the formula. Thus, RbF is rubidium fluoride, and Mg_3N_2 is magnesium nitride. If the metal forms ions with more than one charge, the charge of the metal ion is given in Roman numerals enclosed in parentheses. Thus, $SnCl_2$ is tin(II) chloride and PbO_2 is lead(IV) oxide. Some main group elements that form more than one kind of ion are listed in Table 4–5.

Quite a few cations, especially those of the transition elements, do not have noble gas configurations. Consequently, you cannot predict the charges on these ions by relying only on the periodic table. Most of the transition elements form ions with more than one charge, and charges of $+1$, $+2$, and $+3$ are common. Since ions of different charges are formed, the Roman numeral designation is necessary to specify the charge on a particular transition-element ion. Thus Fe^{3+} is iron(III), Ni^{2+} is nickel(II), and Cu^+ is copper(I). The charge on the transition-element cation in the formula of a salt can be deduced if the charge on the anion is known.

TABLE 4–5 Some Main Group Elements Forming Ions with More Than One Charge

In^+	indium(I)
In^{3+}	indium(III)
Tl^+	thallium(I)
Tl^{3+}	thallium(III)
Sn^{2+}	tin(II)
Sn^{4+}	tin(IV)
Pb^{2+}	lead(II)
Pb^{4+}	lead(IV)
Sb^{3+}	antimony(III)
Sb^{5+}	antimony(V)
Bi^{3+}	bismuth(III)
Bi^{5+}	bismuth(V)

EXAMPLE 4–1

In order to name the compounds Fe_2O_3 and FeO, proceed as follows. We know that the oxide anion is normally O^{2-}. Therefore, to maintain electrical neutrality, the cation in Fe_2O_3 must be Fe^{3+}. The compound is unambiguously titled iron(III) oxide. Likewise for FeO, the cation must be Fe^{2+} and the compound is iron(II) oxide.

EXERCISE 4–1(A)

Name the following compounds: CrS and Cr_2S_3; SnO and SnO_2.
Answer: Chromium(II) sulfide and chromium(III) sulfide; tin(II) oxide and tin(IV) oxide.

EXERCISE 4–1(B)

Write the formulas for the following salts: lithium nitride, magnesium oxide, aluminum iodide, sodium hydride, lead(II) sulfide, antimony(III) chloride, manganese(II) bromide, nickel(II) oxide.
Answer: Li_3N, MgO, AlI_3, NaH, PbS, $SbCl_3$, $MnBr_2$, NiO.

The compounds HF, HCl, HBr, HI, H_2S, H_2Se, and H_2Te are gases at room temperature. For the compounds in the pure state, the names are taken from the anion name coupled to the word "hydrogen":

HCl is hydrogen chloride

H_2S is hydrogen sulfide

When they are dissolved in water, these binary hydrogen compounds are called acids, and their naming follows a different rule. The anion is named first, with the prefix *hydro-* and the suffix *-ic*, followed by the word "acid." Important examples are:

Formula	Pure Compound	Aqueous Solution
HF	hydrogen fluoride	hydrofluoric acid
HCl	hydrogen chloride	hydrochloric acid
HBr	hydrogen bromide	hydrobromic acid
HI	hydrogen iodide	hydriodic acid

The following compounds fall outside our rules:

1. The hydrogen compounds of O^{2-}, N^{3-}, and C^{4-} (that is, H_2O, NH_3, and CH_4) are not considered acids and will be discussed later with covalent compounds.
2. New ions are created if you add to an anion fewer than the number of H^+ ions necessary to form the neutral acid. In particular, this would occur if we added just one H^+ ion to the O^{2-}, S^{2-}, or Se^{2-} ions. That results in OH^-, HS^-, and HSe^-. The OH^- ion is especially important and is called **hydroxide**. The other two are the hydrogen sulfide ion and the hydrogen selenide ion.
3. In passing, we note that the compound formed of H^+ and H^- is simply H_2, hydrogen gas.

EXAMPLE 4–2

To name the compounds $NaHS$ and $Ca(HS)_2$, note that the constituent ions are Na^+, Ca^{2+}, and HS^-. The names of the salts are sodium hydrogen sulfide and calcium hydrogen sulfide. Note that parentheses are used to

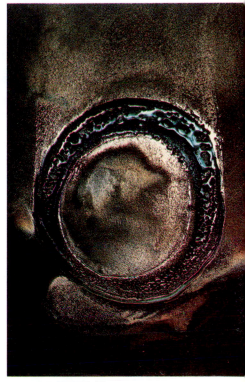

In this application, a drop of solder flux was placed on a copper film to test for corrosion. The ring of copper(II) sulfate (bluish-green) indicates that the copper was corroded significantly. The "copper ring" test thus indicates that this particular flux is not suitable for use in assembling electronic equipment.

group the atoms of an ion. Unfortunately, historical precedents die slowly (if ever), so you are likely to encounter the prefix *bi-* instead of "hydrogen." Thus, NaHS is also called sodium bisulfide, and $Ca(HS)_2$ is also called calcium bisulfide.

EXERCISE 4–2
Give the formulas for calcium hydrogen selenide, barium hydrogen selenide, and lithium hydrogen selenide.
Answer: $Ca(HSe)_2$; $Ba(HSe)_2$; LiHSe.

Some Chemistry of Binary Salts and Acids

Most of the binary salts described in Table 4–6 are solids with high melting and boiling points, resulting from structures held together by strong coulombic (electrical) interactions between ions of opposite charge. Many of these ionic structures dissolve freely in water, but a few display only limited water solubility. When an ionic salt dissolves in water, the ions separate and become mobile, and the solution conducts electricity. We say that an ionic salt "dissociates" in solution. The straightforward method of **direct reaction of the elements** is often successful for preparing salts of monatomic ions. For example, KBr can be prepared from potassium metal and bromine gas in a reaction which proceeds with explosive ferocity. Direct reaction between sodium and bromine is more manageable. We could write this as an equation involving the atoms themselves:

$$Na\cdot \; + \; \cdot \overset{..}{\underset{..}{Br}}: \; \longrightarrow \; Na^+ \; + \; : \overset{..}{\underset{..}{Br}}:^-$$

where the sodium atom has lost an electron to the bromine atom, resulting in the formation of the two ions. But the facts are different. Under standard conditions, bromine exists in the liquid state as diatomic molecules,

TABLE 4–6   Melting and Boiling Points for Some Binary Salts

Salt	Melting Point, °C	Boiling Point, °C
Al_4C_3	1400 (dec.)*	—
Al_2Br_6	98	264
AlF_3	1040	—
$CaCl_2$	772	>1600
CaO	2580	2850
$CsBr_2$	636	1300
TlCl	430	720
$SnCl_2$	246	623
$BiCl_3$	230	447
$MgCl_2$	714	1412
NaCl	801	1413
Na_2S	1180	—
Na_2Se	>875	—
$PbCl_2$	501	950

*(dec.) means that the salt decomposes before melting. Compounds listed without boiling points decompose (as the temperature is raised) before boiling is observed.

(a)

FIGURE 4–5 Direct combination of the elements: (a) reaction of potassium and chlorine; (b) reaction of magnesium and oxygen; and (c) fireworks. The elegant colors are due to finely divided metallic elements.

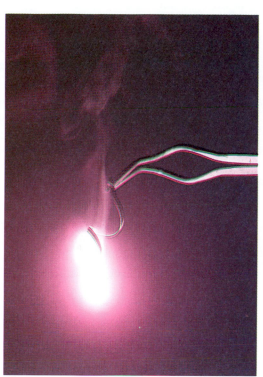

(b)

(c)

Br_2 (liq). The Na^+ and Br^- ions form the solid salt, NaBr(s). Thus, the equation for the reaction as it occurs under standard conditions is:

$$2\ Na\ (s)\ +\ Br_2\ (liq)\ \longrightarrow\ 2\ NaBr\ (s)$$

Other examples of the synthesis of binary salts by direct reactions (Figure 4–5) between elements are

$$2\ Ca\ (s)\ +\ O_2\ (g)\ \longrightarrow\ 2\ CaO\ (s)$$

$$2\ K\ (s)\ +\ S\ (s)\ \longrightarrow\ K_2S\ (s)$$

$$2\ Al\ (s)\ +\ 3\ F_2\ (g)\ \longrightarrow\ 2\ AlF_3\ (s)$$

$$6\ Li\ (s)\ +\ N_2\ (g)\ \longrightarrow\ 2\ Li_3N\ (s)$$

Another method commonly used to prepare salts of monatomic ions is the reaction of a metal with a binary acid. For example, magnesium metal dissolves in an aqueous hydrochloric acid solution to form a solution of magnesium chloride. In this case, the ions formed remain in solution, and the hydrogen ion of the acid is lost as hydrogen gas:

$$Mg\ (s)\ +\ 2\ HCl\ (aq)\ \longrightarrow\ MgCl_2\ (aq)\ +\ H_2\ (g)$$

The designation (aq), for aqueous, following HCl means that this compound is dissolved in water, the medium for this reaction. Similarly, aluminum and lithium react with hydrofluoric and hydriodic acids, respectively:

$$2\ Al\ (s)\ +\ 6\ HF\ (aq)\ \longrightarrow\ 2\ AlF_3\ (aq)\ +\ 3\ H_2\ (g)$$

$$2\ Li(s)\ +\ 2\ HI\ (aq)\ \longrightarrow\ 2\ LiI\ (aq)\ +\ H_2\ (g)$$

The binary acids included in Table 4–7 are all gases at room temperature and atmospheric pressure (except hydrogen fluoride, which has a

TABLE 4–7	Melting and Boiling Points for Some Binary Acids	
Binary acid	**Melting Point, °C**	**Boiling Point, °C**
HF	−92	19
HCl	−112	−84
HBr	−88	−67
HI	−51	—
H_2S	−83	−62
H_2Se	−64	−42

boiling point in the vicinity of room temperature). They dissolve freely in water and their water solutions have typical *acidic properties*, which we will describe in considerable detail in later chapters. All should be handled with more than routine care; hydrofluoric acid is extraordinarily corrosive, causing terrible skin burns and etching glass and metal surfaces. Hydrogen fluoride (HF) and hydrogen chloride (HCl) can be prepared by direct combination of the elements, and the reactions are violent:

$$H_2 \text{ (g)} + F_2 \text{ (g)} \longrightarrow 2 \text{ HF (g)}$$

$$H_2 \text{ (g)} + Cl_2 \text{ (g)} \longrightarrow 2 \text{ HCl (g)}$$

Another useful general preparation of the gaseous binary acids is the reaction of a salt of the desired acid with an acid of low volatility, usually sulfuric acid (H_2SO_4) or phosphoric acid (H_3PO_4). This is a standard procedure for the preparation of hydrochloric acid:

$$NaCl \text{ (s)} + H_2SO_4 \text{ (liq)} \longrightarrow HCl \text{ (g)} + NaHSO_4 \text{ (s)}.$$

HCl (g) escapes from the reaction mixture as it is produced and can be either liquefied or dissolved in water to produce hydrochloric acid. Hydrogen sulfide and hydrogen selenide can be produced in the same way:

$$Na_2Se \text{ (s)} + 2 H_2SO_4 \text{ (liq)} \longrightarrow H_2Se \text{ (g)} + 2 NaHSO_4 \text{ (s)}$$

Hydrogen bromide and hydrogen iodide are normally prepared by the following reactions:

$$H_3PO_4 \text{ (liq)} + KBr \text{ (s)} \longrightarrow HBr \text{ (g)} + KH_2PO_4 \text{ (s)}$$

$$H_3PO_4 \text{ (liq)} + KI \text{ (s)} \longrightarrow HI \text{ (g)} + KH_2PO_4 \text{ (s)}$$

Two common reactions of acids are dissolving metals and neutralizing bases. The dissolving of metals by acids to give the metal cation and H_2 (g) was mentioned before as a method of preparing salts. This reaction can be used to dissolve a number of familiar metals, including main group metals and transition elements. Examples are:

$$Sn \text{ (s)} + 2 HCl \text{ (aq)} \longrightarrow SnCl_2 \text{ (aq)} + H_2 \text{ (g)}$$

$$Fe \text{ (s)} + 2 HBr \text{ (aq)} \longrightarrow FeBr_2 \text{ (aq)} + H_2 \text{ (g)}$$

The special class of reactions involved in dissolving such metals will be discussed later in this chapter.

A few well-known metals, however, are not attacked by aqueous solutions of binary acids such as HCl or HBr. These include copper, silver, and mercury. Dissolving these metals requires a so-called oxidizing acid (Figure 4–6).

Acids also react with compounds known as bases in a process that neutralizes the characteristic properties of both the acid and the base by

FIGURE 4–6 Copper will not dissolve in hydrochloric acid, but as shown, it will dissolve in nitric acid, releasing Cu^{2+} ions (which color the solution) and liberating oxides of nitrogen (notably brown NO_2 gas).

converting them into salts and water. (Such a reaction is called **neutralization**.) The most common base is sodium hydroxide, NaOH:

$$\text{HCl (aq)} + \text{NaOH (aq)} \longrightarrow \text{NaCl (aq)} + \text{H}_2\text{O (liq)}$$

For acids with more than one H atom, such as H_2S or H_2Se, neutralization can produce two different salts, depending on the relative amounts of reacting materials:

$$\text{NaOH (aq)} + \text{H}_2\text{S (aq)} \longrightarrow \text{NaHS (aq)} + \text{H}_2\text{O (liq)}$$
$$2\text{ NaOH (aq)} + \text{H}_2\text{S (aq)} \longrightarrow \text{Na}_2\text{S (aq)} + 2\text{ H}_2\text{O (aq)}$$

4.6 COVALENT COMPOUNDS

Lewis Structures for Compounds

The binary salts described in Section 4.5 are called ionic compounds because they contain only ions: metal cations and nonmetal anions. There are many more compounds that include elements bonded to each other by electrons shared between them. Such bonds are called **covalent bonds** and occur particularly between nonmetallic atoms. In fact, hydrogen is a nonmetal, and the binary acids we have described are essentially covalently bonded compounds. We can describe compounds with covalent bonds by notations called **Lewis structures**, which are analogous to our earlier descriptions of Lewis symbols for atoms of the elements.

EXAMPLE 4–3

As an example of covalent bonding in a molecule, consider two atoms of chlorine coming together to form a compound with the formula Cl_2. From the characteristic properties of molecular chlorine, we know that this compound is not composed of monatomic Cl^+ and Cl^- ions. The Lewis symbols for the Cl atoms are

$$\cdot \ddot{\underset{\cdot\cdot}{\text{Cl}}} : \quad \text{and} \quad \cdot \ddot{\underset{\cdot\cdot}{\text{Cl}}} :$$

Each has one electron less than the stable electron configuration of the nearest noble gas, argon (Ar). We already know that each chlorine atom could achieve noble gas configuration by transfer of an electron in a reaction with a metal atom during salt formation, say from sodium in Group IA or

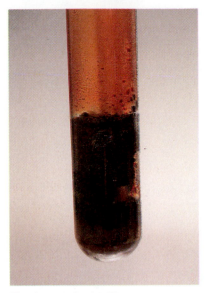

The brown liquid ICl can be prepared by passing chlorine gas into solid iodine. Drops of the ICl can be seen above the iodine in the test tube.

magnesium in Group IIA. However, each could also accomplish the same configuration by sharing a pair of electrons between them:

$$: \ddot{C}l : \ddot{C}l :$$

This drawing is called a Lewis structure. Note that the same number of valence electrons are present here as in the original Lewis symbols for the atoms. However, now each atom is surrounded by eight electrons and has the octet configuration. This strong tendency toward acquiring eight valence electrons is often called the **octet rule**. The shared pair of electrons between the Cl atoms represents a bond holding them together, since it allows each to have a stable electron configuration. Such a shared electron-pair bond is called a covalent **single bond**. It is often depicted as a line connecting the two atoms:

$$: \ddot{C}l - \ddot{C}l :$$

■

EXERCISE 4–3

Draw the Lewis structure for iodine monochloride, a covalently bonded molecule whose chemical formula is ICl.

Answer: $: \ddot{I} : \ddot{C}l :$

■

 EXAMPLE 4–4

To draw the Lewis structure for the molecule resulting from combination of four hydrogen atoms and a carbon atom in methane (CH_4), begin with the Lewis symbols for the atoms:

$$4 \text{ H} \cdot \text{ and } \cdot \dot{\underset{.}{C}} \cdot$$

In order to achieve the necessary noble gas configurations, each of the four hydrogen atoms needs to share one additional electron—remember that hydrogen seeks the two-electron helium core—whereas the carbon atom needs to share four more. This has been achieved in the following two equivalent representations of the Lewis structure:

$$
\begin{array}{ccc}
& & \text{H} \\
\text{H} & & | \\
\text{H} : \overset{\cdot\cdot}{C} : \text{H} & & \text{H} - \text{C} - \text{H} \\
\text{H} & & | \\
& & \text{H}
\end{array}
$$

■

EXERCISE 4–4

Form a covalent molecule by combining six hydrogen and two carbon atoms; the compound is called ethane. Do the same for the compound called propane, C_3H_8. Remember that an octet of electrons is needed for carbon, and a pair for hydrogen. Both compounds are important components of petroleum gas.

Answer:

$$
\begin{array}{cc}
\quad\text{H}\quad\text{H} & \quad\text{H}\quad\text{H}\quad\text{H} \\
\quad|\quad\quad| & \quad|\quad\quad|\quad\quad| \\
\text{H}-\text{C}-\text{C}-\text{H} & \text{H}-\text{C}-\text{C}-\text{C}-\text{H} \\
\quad|\quad\quad| & \quad|\quad\quad|\quad\quad| \\
\quad\text{H}\quad\text{H} & \quad\text{H}\quad\text{H}\quad\text{H} \\
\text{ethane} & \text{propane}
\end{array}
$$

■

EXAMPLE 4–5

To draw the Lewis structure for the covalently bonded ammonia molecule, NH_3, note that two of nitrogen's five valence electrons are not involved in covalent bonds. Nevertheless, they are counted in the octet rule in the same way. The only way the atoms can be combined to give each atom a noble gas configuration is

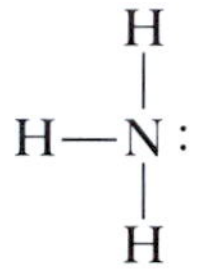

$$
\begin{array}{c}
\text{H} \\
| \\
\text{H}\!-\!\text{N}\!:\\
| \\
\text{H}
\end{array}
$$

■

EXERCISE 4–5

Draw Lewis structures for water and hydrogen sulfide, H_2O and H_2S.

Answer: H—Ö—H   H—S̈—H   ■

Pairs of electrons that are "left over" after drawing all the covalent bonds are called **lone electron pairs** or **nonbonding electron pairs**. Ammonia has one lone pair and water has two. The nonbonding or lone pair electrons have considerable effect on the structure and reactions of the compound.

For some molecules, the Lewis structure contains more than one shared pair of electrons bonding the same two atoms together.

EXAMPLE 4–6

Keeping in mind the octet rule, the Lewis structure for ethylene, C_2H_4, can be conceived as follows. The two carbon atoms are first bonded together by a single covalent bond; then the hydrogen atoms are attached, two to each carbon. That leaves us with the following picture:

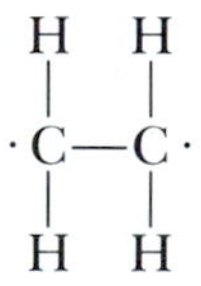

$$
\begin{array}{cc}
\text{H} & \text{H} \\
| & | \\
\cdot\text{C}\!-\!\text{C}\cdot \\
| & | \\
\text{H} & \text{H}
\end{array}
$$

If the unpaired electron on each carbon atom is shared between the two carbon atoms, an octet has been achieved for each

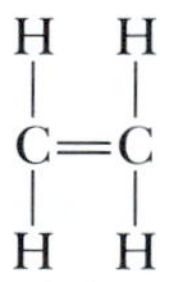

$$
\begin{array}{cc}
\text{H} & \text{H} \\
| & | \\
\text{C}\!=\!\text{C} \\
| & | \\
\text{H} & \text{H}
\end{array}
$$
ethylene

The pair of bonds is called a **double bond**. The presence of a double bond confers special structural and reactive properties on a compound. Double bonds and **triple bonds** (consisting of three pairs of electrons shared between the same two atoms) are classified as multiple bonds. Many compounds contain lone pairs and multiple bonds.   ■

EXERCISE 4–6

Draw the Lewis structures for acetylene, C_2H_2, and for diimide, N_2H_2.

Answer: H—C≡C—H   H—N̈=N̈—H
            acetylene   diimide

Note that the two carbon atoms in acetylene are triple-bonded, and the nitrogen atoms in diimide are double-bonded with each having a non-bonding or lone electron pair. Both structures are consistent with the octet rule.   ■

For more complicated compounds, a systematic method of drawing Lewis structures is helpful. Here is an operating procedure:

1. First determine the sum of the valence electrons of all the atoms in the compound. Going back to a couple of our earlier examples, ammonia (NH_3) has eight valence electrons, one each from the three hydrogen atoms and five from the central nitrogen atom; ethane (C_2H_6) has 14 valence electrons, one each from the six hydrogen atoms and eight from the two carbon atoms.

2. Then divide this number by 2 in order to arrive at the number of bonding and nonbonding electron pairs in the structure.

3. Arrange the atoms into a reasonable structure, based on prior knowledge (and occasionally a bit of intuition). For example, by now you should realize that hydrogen atoms are always connected by single bonds—shared electron pairs—and that carbon atoms always have four bonds attached to neighboring atoms.

4. Connect the atoms with single bonds first, and complete the octet of electrons for each atom with nonbonding electron pairs as necessary. This gives you a trial structure.

5. Now, compare the total number of shared and nonbonding electron pairs in your calculation of step 2 and those of your trial structure in step 4. If the numbers agree, as they do for NH_3 and C_2H_6, then your trial structure is a proper model.

6. However, if your trial structure has more electron pairs than required by your valence shell calculation, remove the correct number of nonbonding electron pairs and reconstruct the octets by shifting an additional nonbonding pair into a bonding position.

For example, ethylene (Example 4–6) needs six pairs—12 valence shell electrons in all—to satisfy the octet rule for two carbon and four hydrogen atoms. But the trial structure contains seven pairs of valence electrons, one pair too many.

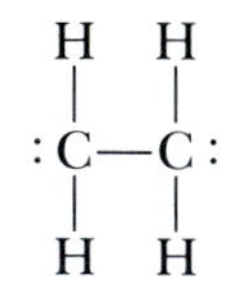

```
    H   H
    |   |
  : C—C :
    |   |
    H   H
```

So we remove one lone pair and shift the other lone between the carbon atoms to reconstruct the required octets on the carbon atoms:

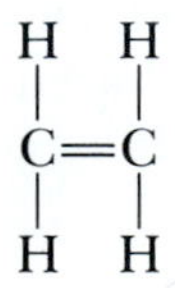

```
    H   H
    |   |
    C=C
    |   |
    H   H
```

Keep in mind that this is only a useful device and is not related to the way that compounds actually form.

Continuing this discussion, in order to draw the Lewis structure for nitric acid, HNO_3, you need to know that the one hydrogen atom is bonded to one of the three oxygen atoms, and the oxygen atoms are all bonded to a central nitrogen atom. We arrange the five constituent atoms as described, connect them by single bonds between atoms, and add lone pairs to complete the required octets. The resulting **trial** structure looks like this:

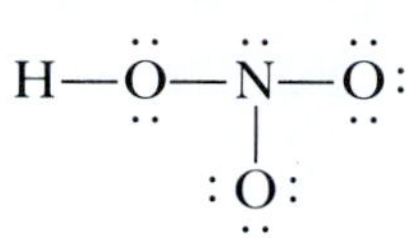

$$H-\overset{..}{\underset{..}{O}}-\overset{..}{N}-\overset{..}{\underset{..}{O}}:$$
$$\mid$$
$$:\overset{}{\underset{..}{O}}:$$

There are 26 electrons appearing as shared pairs in bonds or as nonbonding pairs, completing the octets on the three O atoms and the N atom. However, the total number of valence electrons available for these five atoms is only 24, including 1 for H, (3×6) for O, and 5 for N. Therefore, two electrons must be removed, requiring that one lone pair be converted into a bonding pair to maintain the octets. Thus the trial structure becomes

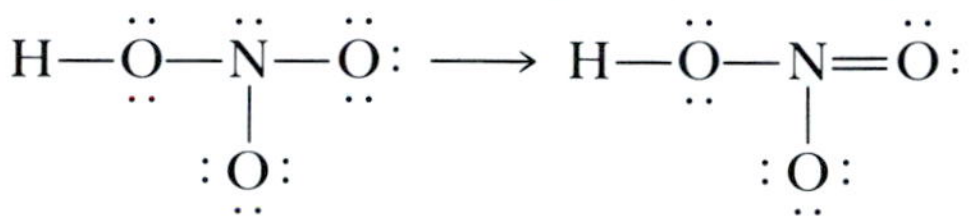

$$H-\overset{..}{\underset{..}{O}}-\overset{..}{N}-\overset{..}{\underset{..}{O}}: \longrightarrow H-\overset{..}{\underset{..}{O}}-N=\overset{..}{\underset{..}{O}}:$$
$$\mid \qquad\qquad\qquad \mid$$
$$:\overset{}{\underset{..}{O}}: \qquad\qquad\quad :\overset{}{\underset{..}{O}}:$$

Formal Charges

Chemists sometimes use a set of bookkeeping rules to assign **formal charges** to the atoms in a molecule or ion. These formal charges do not necessarily correspond to the real charges on the atoms. Instead, they are computed under the assumption that all the bonding electrons are *equally* shared between the atoms that they connect. This assumption is not always accurate, but it can be used to determine which of two or more possible molecular structures is more likely.

The formal charge on each atom in the molecule or ion is found as follows:

1. Note the number of valence electrons in the free atom, according to the element's group number in the periodic table.
2. Form the sum of the number of lone *electrons* (that is, 2 times the number of lone *pairs*) plus half the number of bonding electrons that the atom shares.
3. Subtract the number found in step 2 from the number found in step 1. The result is the formal charge on the atom.

In equation form, for each atom in a molecule or ion,

formal charge = group number

$$- \left[\left(\begin{array}{c} \text{number of} \\ \text{lone electrons} \end{array} \right) + \frac{1}{2} \left(\begin{array}{c} \text{number of} \\ \text{bonding electrons} \end{array} \right) \right]$$

As an example, consider the ammonia molecule:

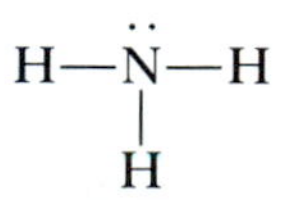

$$H-\overset{..}{N}-H$$
$$\mid$$
$$H$$

First let us calculate the formal charge of the nitrogen atom. Its group

Nitric acid reacts with protein-containing materials such as skin, or in this case, the tip of a feather, staining them yellow.

number, and the number of valence electrons, is 5. The N atom "owns" both of the lone electrons, and half of the six bonding electrons. Therefore,

$$\text{formal charge of N} = 5 - \left[2 + \frac{1}{2}(6) \right] = 0$$

For each of the hydrogen atoms, the calculation gives

$$\text{formal charge of H} = 1 - \left[0 + \frac{1}{2}(2) \right] = 0$$

The sum of the formal charges of all atoms in a molecule or ion must equal the total charge of the assembly. Thus, the sum of formal charges for a neutral molecule such as ammonia must be zero.

EXAMPLE 4–7

For the nitric acid molecule, the numbers of valence electrons for the neutral atoms are 1 for H, 5 for N, and 6 for O. The number of lone electrons plus half the number of bonding electrons for each atom is as follows:

$$
\begin{array}{cccc}
1 & 6 & 4 & 6 \\
& \ddot{} & & \ddot{} \\
\mathrm{H} - \ddot{\mathrm{O}} - \mathrm{N} = \ddot{\mathrm{O}} : \\
& \ddot{} & | & \\
& & : \underset{\ddot{}}{\mathrm{O}} : & \\
& & 7 &
\end{array}
$$

Performing the subtraction, we find the following formal charges:

$$
\begin{array}{cccc}
0 & 0 & +1 & 0 \\
& \ddot{} & & \ddot{} \\
\mathrm{H} - \ddot{\mathrm{O}} - \mathrm{N} = \ddot{\mathrm{O}} : \\
& \ddot{} & | & \\
& & : \underset{\ddot{}}{\mathrm{O}} : & \\
& & -1 &
\end{array}
$$

As expected, the sum of the formal charges for the neutral HNO_3 molecule is zero.

EXERCISE 4–7

Sulfuric acid, H_2SO_4, has a central sulfur atom surrounded by four oxygen atoms, two of which are also bonded to hydrogen atoms. Draw the Lewis structure for sulfuric acid and assign the formal charges to the atoms. *Answer:* The formal charge on S is $+2$. The formal charge on each H is 0. The formal charge on each O bonded to H is 0. The formal charge on each O not bonded to H is -1.

$$
\begin{array}{c}
-1 \\
\ddot{\textrm{O}}\textrm{:} \\
| \\
\textrm{H}-\ddot{\textrm{O}}-\textrm{S}-\ddot{\textrm{O}}-\textrm{H} \\
|+2 \\
\textrm{:}\ddot{\textrm{O}}\textrm{:} \\
-1
\end{array}
$$

The formal charge can sometimes be used as an aid in deciding which of two possible structures of a molecule or ion is more reasonable. This decision is guided by two principles: (1) A molecule is more stable if the formal charges of its atoms are as close to zero as possible. (2) If a negative formal charge is necessary, the molecule is more stable when that charge is on the atom nearest to fluorine in the periodic table. (That is, the atom with the negative charge should have the largest *electronegativity;* this property of atoms will be discussed in Chapter 16.)

Consider these two possible structures for carbon dioxide:

$$
\textrm{:}\ddot{\textrm{O}}\textrm{=}\textrm{C}\textrm{=}\ddot{\textrm{O}}\textrm{:} \qquad {}^{+1}\textrm{O}\textrm{≡}\textrm{C}-\ddot{\textrm{O}}\textrm{:}^{-1}
$$

In the double-bonded structure, all three atoms have a formal charge of zero. In the other structure, in contrast, the triply bonded oxygen atom has a formal charge of $+1$ while the singly bonded O has a formal charge of -1. Thus, we can expect that the double-bonded structure is more likely to occur.

Naming Binary Covalent Compounds

There are many covalent compounds, and unfortunately more than one system for naming them. In one common system, the naming of binary covalent compounds is similar to the naming of binary salts except that charges do not enter into their names. Recall that in binary salts the cation charge is either indicated in parentheses or understood (in cases where elements form ions of only one charge). For covalent compounds, we use prefixes before the names of the elements to indicate the number of atoms of the element in the molecule. The prefixes for 1 through 10 are *mono-, di-, tri-, tetra-, penta-, hexa-, hepta-, octa-, nona-,* and *deca-*. For many elements (particularly oxygen, which begins with a vowel) the final *o* or *a* of the prefix is dropped. For example, we say tetroxide, not tetraoxide.

It is common practice to omit the prefix mono-, indicating one atom per molecule, when it refers to the first element named. Thus, NF_3 is commonly called nitrogen trifluoride, not mononitrogen trifluoride. In binary covalent compounds, the second element in the formula has the suffix *-ide* as in binary salts. However, since there are no ions, we cannot go by the cation/anion order used for salts. For covalent compounds the formulas are usually written with the atoms in the order of position in the periodic table. That is, the element more to the left and to the bottom of the periodic table is placed first and the element more to the right and to the top is second. ICl is iodine monochloride; it is not written as ClI, chlorine monoiodide.

EXAMPLE 4–8

If asked to give the formula and name for the covalent compound with one atom of oxygen and two atoms of fluorine (per molecule), you should respond as follows. Since fluorine is to the right on the periodic table, the formula should be given as OF_2. The name would be oxygen difluoride, not monoxygen difluoride.

EXERCISE 4–8(A)

Give the name and formula for the compound containing one atom of phosphorus per molecule and three atoms of chlorine. Do the same for the covalent molecule containing one P and five Cl atoms.
Answer: PCl_3, phosphorus trichloride; PCl_5, phosphorus pentachloride.

EXERCISE 4–8(B)

Name the compound and give the formula for the molecule containing one iodine and seven fluorine atoms; the molecule with two chlorine and seven oxygen atoms; and the molecule with one chlorine and three fluorine atoms.
Answer: IF_7, iodine heptafluoride; Cl_2O_7, dichlorine heptoxide; ClF_3, chlorine trifluoride.

Hydrogen is a nonmetal, and its compounds with other nonmetals are mostly covalent. Because many of these compounds are familiar, their naming tends to be irregular. Some important hydrogen compounds, including those classified as acids, and their names are listed in Table 4–8.

Some Binary Covalent Compounds and Their Chemistry

Good evidence for the range and variability of the chemical and physical properties of binary covalent compounds can be found in Table 4–9. In contrast to the binary salts (refer back to Table 4–6), they generally exist

Excess phosphorus reacts vigorously with chlorine to produce mainly the trichloride, PCl_3.

In the absence of a catalyst, sulfur burns in oxygen with a blue flame, forming the dioxide, SO_2.

TABLE 4–8	Covalent Compounds of Hydrogen with Elements from the Second and Third Rows of the Periodic Table
Name	**Formula**
hydrogen fluoride	HF
hydrogen chloride	HCl
hydrogen sulfide	H_2S
diborane	B_2H_6
methane	CH_4
silane	SiH_4
ammonia	NH_3
phosphine	PH_3
water	H_2O

TABLE 4–9   Selected Properties of Several Binary Covalent Compounds

Compound	Melting Point, °C	Boiling Point, °C	Reaction on Exposure to Air or Moisture
CH_4	-184	-161	stable—no reaction
SiH_4	-185	-112	$SiH_4 \ (g) + 2 \ O_2 \ (g) \longrightarrow SiO_2 \ (s) + 2 \ H_2O \ (liq)$
P_4O_{10}	563	—	$P_4O_{10} \ (s) + 6 \ H_2O \ (liq) \longrightarrow 4 \ H_3PO_4 \ (aq)$
SF_4	-124	-40	$SF_4 \ (g) + 2 \ H_2O \ (liq) \longrightarrow SO_2 \ (g) + 4 \ HF \ (aq)$
SO_3	17	45	$SO_3 \ (g) + H_2O \ (liq) \longrightarrow H_2SO_4 \ (aq)$ stable
CS_2	-109	46	stable—no reaction

as gases, as liquids (with relatively low boiling points and melting points), or in a few cases as low-melting solids. Binary covalent compounds can often be prepared by direct combination of the elements. However, the actual reactions are not always so simple, often requiring special conditions. For example,

$$P_4 \ (s) + 5 \ O_2 \ (g) \xrightarrow{\text{room temperature}} P_4O_{10} \ (s)$$

$$S \ (s) + 3 \ F_2 \ (g) \longrightarrow \text{produces mainly } SF_6 \ (g)$$

$$2 \ S \ (s) + 3 \ O_2 \ (g) \xrightarrow{\text{a catalyst such as Pt}} 2 \ SO_3 \ (liq)$$

$$C \ (s) + 2 \ S \ (g) \xrightarrow{\text{elevated temperature}} CS_2 \ (g)$$

Methane (CH_4) cannot be conveniently prepared directly because of the nonselectivity of the reaction, which produces many stable products, including $CH_4 \ (g)$, $C_2H_6 \ (g)$, $C_2H_4 \ (g)$, $C_3H_8 \ (g)$, and so on:

$$x \ C \ (s) + y \ H_2(g) \longrightarrow C_xH_{2y}$$

Sulfur hexafluoride (SF_6) is particularly interesting. It is manufactured by burning sulfur in fluorine and washing the crude product to remove the easily hydrolyzed sulfur tetrafluoride (SF_4), which also happens to be highly toxic. A stable and essentially nontoxic product is produced. It has excellent dielectric properties (a dielectric is a nonconducting material like rubber or glass) and is used in high voltage transformers and current breakers. An unusual and important application for sulfur hexafluoride is as a tracer gas in studying ventilation and air flow characteristics of buildings. The first step in the synthesis is direct combination:

$$2 \ S \ (s) + 5 \ F_2 \ (s) \longrightarrow SF_4 \ (g) + SF_6 \ (g)$$

This is followed by hydrolysis (reaction with water) to remove SF_4:

$$SF_4 \ (g) + 2 \ H_2O \ (liq) \longrightarrow SO_2 \ (aq) + 4 \ HF \ (aq)$$

4.7   POLYATOMIC IONS AND THEIR SALTS

A large number of compounds are made up of complex ions, which themselves contain covalent bonds. Sodium nitrate ($NaNO_3$) is such an ionic compound, and another is ammonium chloride (NH_4Cl); both dissociate in aqueous solutions into their respective ions:

$$NaNO_3 \ (aq) \longrightarrow Na^+ \ (aq) + NO_3^- \ (aq)$$

$$NH_4Cl \ (aq) \longrightarrow NH_4^+ \ (aq) + Cl^- \ (aq)$$

The **nitrate ion** is composed of a central nitrogen atom bonded to three oxygen atoms. There is a net charge of −1 for the entire ion (there is an additional electron beyond the number of valence electrons on the four atoms):

$$\left[\begin{array}{c} \overset{-1}{\underset{\cdot\cdot}{:\overset{\cdot\cdot}{O}}}-\overset{+1}{N}=\overset{}{\underset{\cdot\cdot}{O}}: \\ | \\ \underset{\cdot\cdot}{:\overset{\cdot\cdot}{O}:} \\ -1 \end{array}\right]^{-}$$

Note that the net charge of −1 for the complex ion is the sum of the formal charges on the atoms of the ion.

The **ammonium ion** is composed of a central nitrogen atom bonded to four hydrogen atoms. There is a net charge of +1 for the ion, which means there is one electron less than the number of valence electrons on the five atoms:

$$\left[\begin{array}{c} H \\ \overset{+1}{|} \\ H-N-H \\ | \\ H \end{array}\right]^{+}$$

Again, the charge on the ion is equal to the sum of the formal charges for all the atoms of the ion.

EXAMPLE 4–9

To draw the Lewis structure of the chlorite ion, ClO_2^{-}, you must know that both oxygen atoms are bonded directly to the central chlorine atom. Then, add on a complete octet of electrons for each atom and compare the result to the total number of valence shell electrons:

$$[:\overset{\cdot\cdot}{\underset{\cdot\cdot}{O}}-\overset{\cdot\cdot}{\underset{\cdot\cdot}{Cl}}-\overset{\cdot\cdot}{\underset{\cdot\cdot}{O}}:]^{-}$$

Seven electrons are available from the chlorine atom and $2 \times 6 = 12$ from the pair of oxygen atoms. Do not forget to add the extra electron which gives the ion its net negative charge. The total is 20, which is the number required for the Lewis structure. ■

EXERCISE 4–9

Write out the Lewis structures for the sulfate (SO_4^{2-}) and hydrogen sulfate (HSO_4^{-}) ions.

Answer: The former is a divalent anion, SO_4^{2-}; the latter is a monovalent anion, HSO_4^{-}:

$$SO_4^{2-} \qquad\qquad HSO_4^{-}$$

$$\left[\begin{array}{c} \overset{-1}{\underset{\cdot\cdot}{:O:}} \\ | \\ \overset{-1}{:\underset{\cdot\cdot}{O}}-S-\overset{-1}{\underset{\cdot\cdot}{O}:} \\ \underset{+2}{|} \\ \underset{\cdot\cdot}{:O:} \\ -1 \end{array}\right]^{2-} \qquad \left[\begin{array}{c} \overset{-1}{\underset{\cdot\cdot}{:O:}} \\ | \\ H-\underset{\cdot\cdot}{O}-S-\overset{-1}{\underset{\cdot\cdot}{O}:} \\ \underset{+2}{|} \\ \underset{\cdot\cdot}{:O:} \\ -1 \end{array}\right]^{-}$$

The formal charges add up to -2 for SO_4^{2-} and -1 for HSO_4^-.   ■

These complex ions are examples of **polyatomic ions**, each containing several atoms. Polyatomic cations are not common and NH_4^+, the ammonium ion, is the one you are most likely to encounter. However, there are a number of important polyatomic anions. Those containing oxygen and one other element are called **oxo anions**, and they are quite common. Table 4–10 lists several, along with their Lewis structures and names.

TABLE 4–10 Lewis Structures and Names for Some Oxo Anions

Formula	Lewis Structure	Name
CO_3^{2-}		carbonate
NO_2^-		nitrite
NO_3^-		nitrate
SO_3^{2-}		sulfite
SO_4^{2-}		sulfate
ClO^-		hypochlorite
ClO_2^-		chlorite
ClO_3^-		chlorate
ClO_4^-		perchlorate
PO_4^{3-}		phosphate

The name of an oxo anion is formed from the name of the (non-oxygen) central atom with the suffix *-ate* or *-ite* according to these rules:

1. If a particular central atom forms only one oxo anion, as is the case for C, the suffix *-ate* is used (carbonate ion, for example).
2. If two common oxo anions are formed, as is the case for N and S, the one with fewer oxygen atoms has the suffix *-ite* added; the other, with more oxygen atoms, bears the suffix *-ate*.
3. When an element forms more than two oxo anions with a single central atom, as is the case for Cl, the prefixes *hypo-* and *per-* are added. *Hypo-* is used with the ion having fewest oxygen atoms; *per-* is added to the ion having the most.

There are several more polyatomic cations and anions, which will be discussed as needed.

EXAMPLE 4–10
Write down the empirical formulas for ammonium sulfate and calcium carbonate.

Solution   Note that the ammonium ion is a complex cation with a $+1$ charge and the sulfate ion is a complex anion with a -2 charge. Therefore, the combination that yields a neutral compound is $2:1$, or $(NH_4)_2SO_4$. That same kind of logic produces $CaCO_3$ for calcium carbonate, since the cation carries a double positive charge and the complex anion has a charge of -2, leading to a $1:1$ combination.   ■

EXERCISE 4–10
Write down the empirical formulas for sodium bisulfite (also called sodium hydrogen sulfite) and aluminum nitrate.
Answer: $NaHSO_3$; $Al(NO_3)_3$.   ■

Oxy Acids

If a neutral compound is created by adding one or more protons to the formula for an oxo anion, the result is called an **oxy acid**. To write Lewis structures for oxy acids, add the necessary H^+ ions to oxygen atoms of the anion that do not already carry hydrogen atoms. Table 4–11 lists the Lewis structures for the oxy acids formed from the ions in Table 4–10, along with their names. The prefix and suffix changes on converting from oxo anions to oxy acids (using chlorine as an example) are as follows:

Name of the oxo anion	Name of the oxy acid
hypo**chlor**ite	hypo**chlor**ous acid
chlorite	**chlor**ous acid
chlorate	**chlor**ic acid
per**chlor**ate	per**chlor**ic acid

Several ions containing hydrogen can be obtained for oxo anions with -2 or -3 charges. Thus, addition of one proton to the carbonate ion gives the hydrogen carbonate (or bicarbonate ion); for sulfate, the hydrogen sulfate ion. The phosphate ion, PO_4^{3-}, with a charge of -3, gives rise to two hydrogen ions: HPO_4^{2-} (hydrogen phosphate) and $H_2PO_4^-$ (dihydrogen phosphate).

TABLE 4–11   Lewis Structures and Names for Some Oxy Acids

Formula	Lewis Structure	Name
H_2CO_3		carbonic acid
HNO_2	$H—\ddot{O}—\ddot{N}=\ddot{O}$	nitrous acid
HNO_3		nitric acid
H_2SO_3		sulfurous acid
H_2SO_4		sulfuric acid
$HClO$	$H—\ddot{O}—\ddot{Cl}:$	hypochlorous acid
$HClO_2$	$H—\ddot{O}—\overset{+}{Cl}—\overset{-}{\ddot{O}}:$	chlorous acid
$HClO_3$		chloric acid
$HClO_4$		perchloric acid
H_3PO_4		phosphoric acid

HCO_3^-	hydrogen carbonate from CO_3^{2-}, the carbonate ion
HSO_4^-	hydrogen sulfate from SO_4^{2-}, the sulfate ion
HPO_4^{2-}	hydrogen phosphate from PO_4^{3-}, the phosphate ion
$H_2PO_4^-$	dihydrogen phosphate from PO_4^{3-}, the phosphate ion
HSO_3^-	hydrogen sulfite (or bisulfite) from SO_3^{2-}, the sulfite ion

(a)

(b)

FIGURE 4–7   (a) Dendritic growth of copper crystals; (b) a white hot wire touches off explosive material on the back of a thin gold foil, bonding the foil to a contact finger on a printed circuit board. The "explosive bonding" technique is used to repair gold coatings that are imperfect or otherwise too thin for microelectronics applications.

EXAMPLE 4–11

The formulas for calcium hydrogen carbonate and for calcium hydrogen phosphate should follow directly from the preceding list. We determine the net charge on the complex ion and the charge on the metal ion. The calcium ion is $+2$ and the hydrogen carbonate and hydrogen phosphate ions are -1 and -2, respectively. Therefore, the empirical formulas are $Ca(HCO_3)_2$ and $CaHPO_4$. ■

EXERCISE 4–11

Give the empirical formulas for calcium dihydrogen phosphate and for calcium bisulfate.
Answer: $Ca(H_2PO_4)_2$; $Ca(HSO_4)_2$. ■

The Salts and Acids of Some Polyatomic Ions

Reactions of Polyatomic Acids: Direct reaction of the elements is not a useful method for preparing polyatomic acids, anions, and cations, mostly because too many elements are involved and the chances of getting a decent yield of the right product are not very good. Instead, it is better to proceed in steps that lead from available materials to the desired products. Salts are normally prepared from aqueous solutions of the respective acids by reaction with either the metal or a base (generally a hydroxide) containing the desired metal cation. For example,

$$4\ Al\ (s)\ +\ 6\ H_2SO_4\ (aq)\ \longrightarrow\ 2\ Al_2(SO_4)_3\ (aq)\ +\ 3\ H_2\ (g)$$

$$6\ Mg\ (s)\ +\ 4\ H_3PO_4\ (aq)\ \longrightarrow\ 2\ Mg_3(PO_4)_2\ (aq)\ +\ 3\ H_2\ (g)$$

Some metals such as copper, silver, and gold are notably unreactive. That makes them suitable for some interesting applications such as fabrication into the coin of the realm. Because these metals have been used for that purpose for so many centuries, they are often called **coinage metals** (Figure 4–7). They will not decompose water, as will more reactive metals such as Na and K; nor will they decompose the simple proton acids such as hydrochloric acid, as will Fe and Zn. However, Cu and Ag will decompose oxidizing acids such as HNO_3 and Au will react with a strangely powerful $3:1$ mixture of nitric and hydrochloric acids known as *aqua regia:*

They are hardly "coinage" metals any more, having long since become too valuable to industry and commerce to use for that purpose. Gold has not been used since 1933; silver has been out of circulation since 1963.

$$3\ Cu\ (s)\ +\ 8\ HNO_3\ (aq)\ \longrightarrow$$
$$3\ Cu(NO_3)_2\ (aq)\ +\ 2\ NO\ (g)\ +\ 4\ H_2O\ (liq)$$

$$3\ Ag\ (s)\ +\ 4\ HNO_3\ (aq)\ \longrightarrow$$
$$3\ AgNO_3\ (aq)\ +\ NO\ (g)\ +\ 2\ H_2O\ (liq)$$

In marked contrast to the reactions of more active metals with aqueous acids, coinage metals do not generate hydrogen gas. Furthermore, the chemical equations are not easily balanced by inspection.

Oxy acids typically react with hydroxides to yield salts and water. Examples include the following:

$$NaOH\ (aq)\ +\ HClO_2\ (aq)\ \longrightarrow\ NaClO_2\ (aq)\ +\ H_2O\ (liq)$$

$$Ca(OH)_2\ (aq)\ +\ 2\ HNO_2\ (aq)\ \longrightarrow\ Ca(NO_2)_2\ (aq)\ +\ 2\ H_2O\ (liq)$$

For acids with two or three protons (also called **diprotic** and **triprotic** acids), different products are possible depending on the ratios of starting materials used. Thus, sodium hydroxide and sulfuric acid can react to give sodium bisulfate, sodium sulfate, or mixtures of the two:

$$NaOH\ (aq)\ +\ H_2SO_4\ (aq)\ \longrightarrow\ NaHSO_4\ (aq)\ +\ H_2O\ (liq)$$

$$2\ NaOH\ (aq)\ +\ H_2SO_4\ (aq)\ \longrightarrow\ Na_2SO_4\ (aq)\ +\ 2\ H_2O\ (liq)$$

With phosphoric acid, three different salts are possible from potassium hydroxide, namely potassium dihydrogen phosphate, potassium hydrogen phosphate, and potassium phosphate:

$$KOH\ (aq)\ +\ H_3PO_4\ (aq)\ \longrightarrow\ KH_2PO_4\ (aq)\ +\ H_2O\ (liq)$$

$$2\ KOH\ (aq)\ +\ H_3PO_4\ (aq)\ \longrightarrow\ K_2HPO_4\ (aq)\ +\ 2\ H_2O\ (liq)$$

$$3\ KOH\ (aq)\ +\ H_3PO_4\ (aq)\ \longrightarrow\ K_3PO_4\ (aq)\ +\ 3\ H_2O\ (liq)$$

Neutralization reactions such as these can also be used to prepare ammonium salts. Aqueous ammonia is the base, and it reacts with aqueous acids as follows:

$$NH_3\ (aq)\ +\ HCl\ (aq)\ \longrightarrow\ NH_4Cl\ (aq)$$

$$NH_3\ (aq)\ +\ HNO_3\ (aq)\ \longrightarrow\ NH_4NO_3\ (aq)$$

Preparation of Oxy Acids: Oxy acids can be prepared by adding the oxide of a nonmetallic element to water. For example:

$$CO_2\ (g)\ +\ H_2O\ (liq)\ \longrightarrow\ H_2CO_3\ (aq) \qquad \text{carbonic acid}$$

$$SO_2\ (g)\ +\ H_2O\ (liq)\ \longrightarrow\ H_2SO_3\ (aq) \qquad \text{sulfurous acid}$$

$$SO_3\ (g)\ +\ H_2O\ (liq)\ \longrightarrow\ H_2SO_4\ (aq) \qquad \text{sulfuric acid}$$

$$P_4O_6\ (s)\ +\ 6\ H_2O\ (liq)\ \longrightarrow\ 4\ H_3PO_3\ (aq) \qquad \text{phosphorous acid}$$

$$P_4O_{10}\ (s)\ +\ 6\ H_2O\ (liq)\ \longrightarrow\ 4\ H_3PO_4\ (aq) \qquad \text{phosphoric acid}$$

These equations can all be readily balanced by inspection. Nitric acid is made commercially by the addition of NO_2 (g) to water, but the reaction is more complex than the preceding ones:

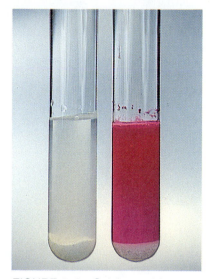

FIGURE 4–8 Calcium oxide, the white solid that has settled to the bottom (left) reacts with water to produce calcium hydroxide (right), as **indicated** by the pink color of phenolphthalein in the now alkaline (basic) solution.

$$3 \ NO_2 \ (g) + H_2O \ (liq) \longrightarrow 2 \ HNO_3 \ (aq) + NO \ (g)$$

Oxides of nonmetallic elements such as sulfur, nitrogen, carbon, chlorine, and fluorine can generally be prepared by direct combination. In some cases, adding the oxide of a metal to water results in formation of hydroxides (Figure 4–8):

$$K_2O \ (s) + H_2O \ (aq) \longrightarrow 2 \ KOH \ (aq) \qquad \text{potassium hydroxide}$$

$$CaO \ (s) + H_2O \ (liq) \longrightarrow Ca(OH)_2 \ (aq) \qquad \text{calcium hydroxide}$$

Oxy acids that are gases at the reaction temperature can be prepared by reaction of an acid of low volatility with the appropriate salt. Sulfuric acid is particularly useful for this. Thus, nitric acid is prepared at temperatures above its normal boiling point of 83°C, from sulfuric acid and potassium nitrate, and the product is collected as vapor above the reaction mixture:

$$H_2SO_4 \ (liq) + KNO_3 \ (s) + heat \longrightarrow HNO_3 \ (g) + KHSO_4 \ (s)$$

4.8   CHEMICAL PROCESSING AND ENGINEERING

An Introduction to the Four Workhorses of the Chemical Industry: H_2SO_4, H_3PO_4, NaOH, and NaCl

Many of the compounds and reactions described in this chapter are of great commercial and industrial importance, but few stand higher in value and significance than sulfuric acid, phosphoric acid, sodium hydroxide, and sodium chloride. They are the workhorses of chemistry and industry. It has been said that the wealth of industrialized nations rests on their backs.

Sulfuric Acid

The raw material for sulfuric acid (other than air and water) is elemental sulfur. Close to 100 billion pounds of it were produced in the United States in 1987, more than 40% of it by the Frasch process. The rest is obtained from petroleum refining, desulfurization of natural gas, and sulfur recovered as a byproduct of the smelting of copper, zinc, and lead from the ores $CuFeS_2$, ZnS, and PbS. More than 90% of all sulfur produced is used for the preparation of sulfuric acid.

Domestic sulfur is extracted by the **Frasch process** from the great salt domes of Texas and Louisiana. A sink hole is drilled down into the sulfur, and a pipe well consisting of three concentric pipes is built within it. One pipe conducts superheated water to melt the sulfur (m.p. 119°C); a second provides compressed air, which is used to force the melted sulfur up through the third pipe. At the surface, the sulfur is allowed to solidify in huge lagoons and then is broken up by blasting for shipment.

Manufacture of Sulfuric Acid: Sulfur burns readily in air to produce sulfur dioxide:

$$S \ (s) + O_2 \ (g) \longrightarrow SO_2 \ (g)$$

However, dissolving sulfur dioxide in water would produce sulfurous acid rather than sulfuric acid. The sulfur dioxide will react with more oxygen gas if a suitable catalyst is present to produce sulfur trioxide,

Sulfur used in the manufacture of agricultural fertilizer in Australia.

$$2 \ SO_2 \ (g) \ + \ O_2 \ (g) \xrightarrow{\text{platinum catalyst}} 2 \ SO_3 \ (g)$$

which dissolves in water to produce sulfuric acid, H_2SO_4:

$$SO_3 \ (g) \ + \ H_2O \ (liq) \longrightarrow H_2SO_4 \ (aq)$$

Increasingly, hydrogen sulfide is recovered from sour natural gas, coking gas, and petroleum refinery gas by "scrubbers." In these devices, the H_2S is dissolved in aqueous potassium carbonate solution, from which it can be regenerated by simple heating:

$$H_2S \ (g) \ + \ K_2CO_3 \ (aq) \rightleftharpoons KHS \ (aq) \ + \ KHCO_3 \ (aq)$$

The hydrogen sulfide collected in this fashion can then be burned to give sulfur dioxide for sulfuric acid manufacture. Alternatively, it is converted to elemental sulfur by allowing it to react with the sulfur dioxide formed from combustion of part of the hydrogen sulfide in the presence of an iron oxide (Fe_2O_3) catalyst:

$$2 \ H_2S \ (g) \ + \ 3 \ O_2 \ (g) \longrightarrow 2 \ SO_2 \ (g) \ + \ 2 \ H_2O \ (g)$$
$$2 \ H_2S \ (g) \ + \ SO_2 \ (g) \longrightarrow 3 \ S \ (liq) \ + \ 2 \ H_2O \ (g)$$

The **contact process** for manufacturing sulfuric acid begins with the catalytic oxidation of sulfur dioxide to sulfur trioxide:

$$2 \ SO_2 \ (g) \ + \ O_2 \ (g) \longrightarrow 2 \ SO_3 \ (g)$$

Conditions favoring SO_3 formation are high pressure, an excess of oxygen, and low temperature. In order to operate at a comparatively low temperature, however, a catalyst must be used to speed the reaction. Vanadium pentoxide, V_2O_5, is commonly used. The yield of SO_3 is about 99.7% at 700 to 725°C in a special four-pass converter. Converted gas is then absorbed into a concentrated H_2SO_4 solution, producing a substance called fuming sulfuric acid or oleum:

SULFURIC ACID, SULFUR TRIOXIDE, AND OLEUM—
THE CONTACT PROCESS

Sulfuric acid (H_2SO_4) is a colorless, viscous liquid with a boiling point of 270°C. Sulfur trioxide (SO_3), needed for sulfuric acid production, is also a liquid at atmospheric pressure, with a boiling point of 45°C. Oleum is fuming sulfuric acid, $H_2SO_4{\cdot}SO_3$ (or $H_2S_2O_7$). Collectively, these three are by far the largest volume commodity chemicals produced in the world, mostly by a 150-year old process known as the contact process. The basic raw material feed is sulfur dioxide produced from liquid sulfur or hydrogen sulfide (H_2S):

$$S\ (liq) + O_2\ (g) \longrightarrow SO_2\ (g)$$

$$H_2S\ (g) + O_2\ (g) \longrightarrow H_2O\ (g) + SO_2\ (g)$$

The heart of the contact process is the converter, in which sulfur dioxide is catalytically converted to sulfur trioxide. Today, vanadium catalysts are used in several different forms, such as a diatomaceous earth (clay) impregnated with 7% V_2O_5.

$$2\ SO_2\ (g) + O_2\ (g) \rightleftharpoons 2\ SO_3\ (g)$$

The reaction can run in either direction. The direction written is favored if the temperature is low and the pressure is high.

The reaction between sulfur trioxide and water, the next step in the contact process, is so vigorous that if any SO_3 escapes into the atmosphere, it immediately produces an acid mist with the moisture usually present in air.

$$SO_3\ (g) + H_2O\ (g) \longrightarrow H_2SO_4\ (aq)$$

Finally, absorption of sulfur trioxide by concentrated sulfuric acid produces what is known as oleum or fuming sulfuric acid:

$$H_2SO_4\ (liq) + SO_3\ (g) \longrightarrow H_2SO_4{\cdot}SO_3\ (liq)$$

As the name implies, it "fumes" in moist air. The word oleum comes from the Latin, meaning oil.

$$SO_3\ (g) + H_2SO_4\ (liq) \longrightarrow H_2S_2O_7\ (liq)$$

By adding the stoichiometric amount of water to the fuming sulfuric acid, nearly pure sulfuric acid is formed under carefully controlled conditions:

$$H_2S_2O_7\ (liq) + H_2O\ (liq) \longrightarrow 2\ H_2SO_4\ (liq)$$

Oleums are marketed on the basis of the percentage of SO_3 present. For example, 20% oleum contains 20 kg of SO_3 and 80 kg of H_2SO_4 per 100 kg. If 100 kg of this 20% oleum were very carefully diluted with water, 104.5 kg of 100% H_2SO_4 would be obtained.

EXAMPLE 4–12

To calculate the number of moles of SO_3 and H_2SO_4 in the 20% oleum just described, use the figures of 20 kg of SO_3 and 80 kg of H_2SO_4 per 100 kg and the respective formula weights of 80 g/mol and 98 g/mol:

$$2.00 \times 10^4\ g\ SO_3 \left(\frac{1\ mol}{80.0\ g}\right) = 250\ mol\ SO_3$$

$$8.00 \times 10^4\ g\ H_2SO_4 \left(\frac{1\ mol}{98.0\ g}\right) = 816\ mol\ H_2SO_4$$

EXERCISE 4–12

Calculate the number of moles of water required to produce 100% sulfuric acid from the 20% oleum described in the preceding example. Show that these figures do indeed add up to 104.5 kg.

If 654 pounds of sulfur are input to a contact sulfuric acid process and 2030 pounds of 98.5 percent sulfuric acid is the output, what is the percent yield? To how many pounds of 20% oleum is this equivalent?
Answer: 99.83%; 1914 pounds.

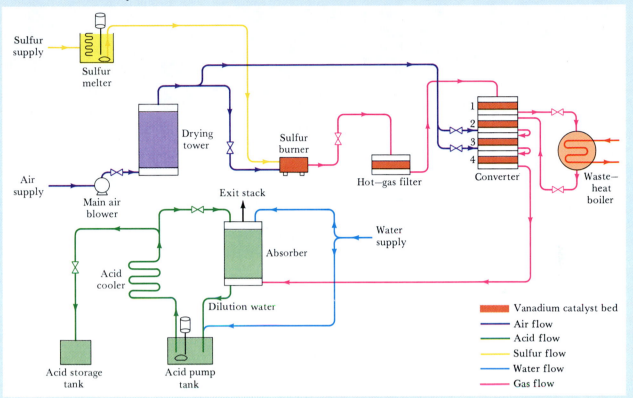

Schematic diagram for the industrial scale synthesis of sulfuric acid by the Contact process. The Environ-chem four-pass converter.

Answer: 250 moles SO_3 react with 250 moles H_2O, which weigh 4.5 kg; the total is 104.5 kg. ∎

Properties of Sulfuric Acid: The sulfuric acid of the chemistry laboratory is a colorless, odorless, syrupy liquid. It was formerly called oil of vitriol because of its consistency. It is 98% H_2SO_4 and is soluble in water in all proportions with the evolution of considerable amounts of heat. Commercial sulfuric acid ranges from colorless to pink, yellow, brown, or black and may be milky or opaque due to impurities. It is highly corrosive in dilute solutions, attacking nearly all common metals. Lead, however, is quite resistant. Sulfuric acid rapidly decomposes rubber, wood, and most organic substances.

Sulfuric acid is injurious to the skin, mucous membranes, and eyes. Dilute solutions of the acid in the presence of metals may liberate dangerous quantities of flammable hydrogen gas. **Concentrated solutions of sulfuric acid must never be diluted by adding water to the acid**, because the heat released by the reaction would boil the water drops and spatter acid out of the container. Dilution can be carried out safely by carefully adding the acid to water, with adequate stirring. Sulfuric acid is a powerful dehydrating agent and will strongly remove water from other sources with which it is in contact. For example, sulfuric acid removes water from cane (table)

This concentration is 18 moles of H_2SO_4 per liter of solution, or 18 M (see Chapter 12).

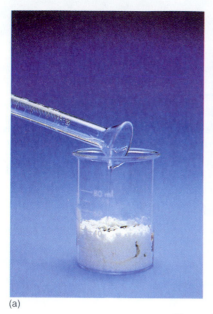
(a)

(b)

(c)

Carmelizing of sugar by the powerful dehydrating action of sulfuric acid.

sugar, leaving behind a carbonaceous residue that has a characteristic "caramel" scent to it.

Uses of Sulfuric Acid: The fertilizer industry is the largest single consumer of sulfuric acid, mostly for the production of superphosphates. The chemical industry accounts for the second largest fraction, for production of phosphoric acid and aluminum sulfate for water purification, for paper manufacture, and for petrochemicals. The petroleum industry is the third largest consumer for the alkylation process that produces high-octane motor fuels and for the refining of petroleum distillates. Other uses include production of titanium pigments, steel pickling, the manufacture of rayon, dyes, and various chemical intermediates—especially nitrocellulose, nitroglycerin, and TNT—and the industrial scale synthesis of detergents (Table 4–12).

Misuses of Sulfuric Acid: Wherever acids are manufactured or used, acid wastes accumulate. Sulfuric acid is by far the most commonly encountered. The effects of acid wastes, especially sulfuric acid wastes, take several forms. The operators of a chemical plant that successfully recovers 99.99% of the sulfuric acid used might feel that there is no waste at all and that the process

Many of the features of this discussion of sulfuric acid chemistry and applications are continued in our discussion of the oxides of sulfur in Chapter 6.

TABLE 4–12   World Consumption of Sulfuric Acid in 1987

Use	Percent	Use	Percent
fertilizer/agriculture	26.4	chemicals	
phosphate fertilizer	20.6	plastics	2.8
ammonium sulfate	15.4	metal sulfates	2.2
paints and pigments	11.7	HF	1.2
natural and manmade		HCl	1.1
fiber and films	10.2	dyes and intermediates	2.3
soaps and detergents	7.1	petroleum refining	1.4
steel manufacture	3.1	all others	4.5

is quite good. However, large quantities of effluent containing 0.01% acid can cause the death of all the aquatic life in the region. Rivers and streams in general are incapable of absorbing large amounts of acid.

Acid discharges can disrupt food cycles and interfere with natural purification processes. Growth of molds, wild yeasts, and other organisms that are unsuitable for the normal botanical inhabitants may succeed in acid environments. The hardness (mineral content) of the waters may be affected by conversion of carbonates and bicarbonates to sulfates, which in turn directly affects the economics of water treatment plants and sewage systems. Limestone and cement erode in acid. Secondary bacterial corrosion and the release of noxious gases such as hydrogen sulfide also occur.

Phosphoric Acid

Phosphoric acid can be prepared by reaction of calcium phosphate (from an ore called phosphate rock) with sulfuric acid:

$$Ca_3(PO_4)_2 \text{ (s)} + 3 H_2SO_4 \text{ (aq)} \longrightarrow 2 H_3PO_4 \text{ (aq)} + 3 CaSO_4 \text{ (s)}$$

Calcium sulfate occurs in nature in a hydrated form ($CaSO_4 \cdot 2H_2O$) that often goes by the names gypsum and plaster of paris.

The calcium sulfate that is formed is an insoluble solid precipitate, which can be separated from the desired phosphoric acid by filtration.

The element phosphorus is one of the three major nutrients needed for plant growth—the others are potassium and nitrogen—and it is therefore extremely important in the fertilizer industry. Bones were once a common source of phosphorus for agriculture, but now phosphate rock, which is mined in Florida, Africa, and some Pacific islands, has taken that role. Phosphate rock is not particularly soluble in water, however, and so it cannot be easily absorbed by plants. Treating phosphate rock with sulfuric acid produces the desired water solubility, and this is now the most important use for this widely used chemical. When the amount of sulfuric acid is limited, two moles of sulfuric acid react with one mole of calcium phosphate:

$$Ca_3(PO_4)_2 \text{ (s)} + 2 H_2SO_4 \text{ (aq)} \longrightarrow Ca(H_2PO_4)_2 \text{ (aq)} + 2 CaSO_4 \text{ (s)}$$

The calcium dihydrogen phosphate produced in this reaction is somewhat soluble in water; this salt, mixed with the calcium sulfate, is sold as a fertilizer called superphosphate of lime. A fertilizer with a higher phosphorus content can be produced by treating calcium phosphate with phosphoric acid as prepared previously from $Ca_3(PO_4)_2 + 3 H_2SO_4$:

$$Ca_3(PO_4)_2 \text{ (s)} + 4 H_3PO_4 \text{ (aq)} \longrightarrow 3 Ca(H_2PO_4)_2 \text{ (aq)}$$

The product of this reaction is sold as triple superphosphate.

Sodium Hydroxide

Sodium hydroxide is a white solid that is soluble in water and corrosive to the skin. Its melting point is 318.4°C and it boils at 1390°C. Solid NaOH tends to form a crust of sodium carbonate on its surface by reaction with CO_2 in the atmosphere. The principal means of industrial production is the electrolytic **chlor-alkali process**. This process involves applying electric currents to saturated salt solutions in one of a variety of cells, the most common being mercury cathode or diaphragm cells. The reaction is as follows:

$$2 \text{ H}_2\text{O} + 2 \text{ NaCl (aq)} \xrightarrow{\text{electrolysis}} 2 \text{ NaOH (aq)} + \text{Cl}_2 \text{ (g)} + \text{H}_2 \text{ (g)}$$

The different types of cells are designed to prevent the sodium hydroxide, hydrogen, and chlorine reaction products from reacting with each other.

Traditionally, NaOH has been used in producing soap, processing textiles, and refining petroleum. It is also widely used in the dye, detergent, and paper industries. NaOH is rarely used for industrial neutralization because there are cheaper chemicals, such as lime (CaO), for such purposes.

Sodium Chloride

Solid NaCl or salt is a crystalline, transparent substance that melts at 803°C and boils at 1430°C. It is widely distributed in nature. Sea water is 2.68% NaCl, and rock salt is found in tremendous deposits all across North America. There are significant deposits in Europe as well. It is usually recovered by one of three principal methods. **Shaft mining** uses techniques similar to those of coal mining, involving cutting, shaving, drilling, blasting, and transportation. In **solution mining**, equipment pumps water into a rock salt deposit, dissolving the salt, and brings the brine (salt water) to the surface. In **solar evaporation**, brine is placed in enormous outdoor basins, and the water evaporates and leaves solid salt behind. Although this is the least used technique in the United States, it is the oldest known and still acounts for 50% of the world production. The uses of salt run into the thousands, but most important are those in which it is the raw material for the production of other chemicals, as we will discuss in later sections.

Austria salt mine.

SUMMARY

In order to help organize and make digestible the huge amount of information that exists about the chemical elements and their compounds, chemists have developed a number of useful tools. Three of special importance are the **periodic table**, the system of **Lewis symbols** for atoms and **Lewis structures** for molecules, and the system of chemical naming or **nomenclature**. The whole earth catalog is, as a practical matter, a limited set of nature's 89 elements. Of the millions of known chemical combinations, only a few thousand need concern us on a day-to-day basis. Still, even that would present us with an insurmountable task were it not for the periodic table. The periodic table places the elements in different **groups** or **families**. Thus, we need not learn the properties of each element independently. The periodic table also distinguishes between **main group** and **transition** elements, and between metals and nonmetals.

Using the periodic table we can construct **Lewis symbols** for the elements. These serve two useful purposes. First, they are a guide for determining which elements form ions and what the charges on these ions will be. Second, by extending the idea to combinations of atoms, **Lewis structures** of covalent compounds help us identify the presence of special features such as lone pairs, multiple bonds, and ionic charges.

The **monatomic ions** formed by adding or removing electrons from atoms can be combined to form **binary salts**. The combination of protons with some monatomic ions, particularly F^-, Cl^-, Br^-, I^-, S^{2-}, and Se^{2-}, yields gaseous covalent compounds that behave as **binary acids** when dissolved in water.

Polyatomic ions are charged species that include covalent bonds. The **ammonium** ion, NH_4^+, is the most important polyatomic cation. The most important polyatomic anions are the **oxo anions**. All of these species can form salts, and the oxo anions can be combined with protons to form **oxy acids**.

A number of chemical reactions were discussed, particularly interconversions between acids and salts.

QUESTIONS AND PROBLEMS

QUESTIONS

1. Explain what is meant by the periodic law.
2. Listing as many ways as you can, show that the properties of sulfur are characteristically nonmetallic.
3. In what way(s) did the discovery of the noble gas elements influence our present theory of atomic structure and chemical reactivity?
4. Why do we not draw Lewis symbols as Lewis originally did, placing electrons at the corners of a cube?
5. Under what circumstances is it necessary to write a Roman numeral after the symbol of an element in a salt? Give an example where such a Roman numeral is necessary and one where it is not.
6. What is the likely reason that the charges of ions are not given in the formula of a salt?

7. The ions Na^+, Ne, and F^- are said to be isoelectronic with respect to their valence electrons. What do you think that means? Compare the reactivities of the three species. Name two species that are isoelectronic with Ar.
8. What properties distinguish salts from covalent compounds?
9. Why is direct combination of the elements rarely used as a method of preparation of compounds of polyatomic ions?
10. Why can't phosphate rock be used directly for fertilizer?
11. The Group IA and VIIA elements are never found in the free state. Why? Oxygen is highly reactive, yet it is found in the free state, whereas chlorine is not.

PROBLEMS

Lewis Symbols [1–2]

1. (a) Draw Lewis symbols for Ga, Te, Ba, Bi, and Rb. (b) Do the same for Be, Mg, Si, and Kr.

2. (a) Draw Lewis symbols for Al, Se, Ca, and Ge. (b) Draw Lewis symbols for Cs, Ba, Pb, and Rn.

Monatomic Ions; Binary Salts and Acids [3–20]

3. What monatomic ions would you expect to be formed by the following elements: Sr, At, Fr, Se, Sn, N, Ne, H?

4. Predict what monatomic ions are formed by the following elements: Cs, Br, Ra, Ar, Tl.

5. List the binary salts formed by the following pairs of elements:
(a) Ca and Br
(b) K and Se
(c) Al and Cl
(d) Na and P

6. Write the formulas for the binary salts that you would predict are formed by direct reaction of the following pairs of elements:
(a) Tl and Br
(b) Mg and N
(c) Ba and H
(d) H and S

7. Write the names of the following salts:
(a) SrO
(b) AlF_3
(c) Be_2C
(d) RbOH
(e) Li_3N
(f) NaH
(g) $SnCl_2$
(h) TlI
(i) KHSe

8. Name the following salts:
(a) CaH_2
(b) $BaCl_2$
(c) Mg_3N_2
(d) $AlCl_3$
(e) KOH
(f) LiHS
(g) $PbBr_2$
(h) $TlCl_3$

9. State the names of the following compounds when pure, and when dissolved in water:
(a) HF
(b) HBr
(c) H_2S
(d) H_2Se

10. Each of the following is a gas when pure but an acid when dissolved in water. Give its name in each form:
(a) HBr and HI
(b) H_2Se and H_2Te

11. Give the formulas for the following salts:
(a) potassium iodide
(b) calcium hydroxide

(c) strontium hydride
(d) bismuth(III) chloride
(e) barium hydrogen sulfide

12. Write the formula for each of the following binary salts:
(a) tin(II) oxide
(b) magnesium nitride
(c) lead(IV) bromide
(d) barium hydroxide

13. Write balanced equations for formation of the following compounds by direct combination of the elements:
(a) NaCl
(b) BaO
(c) $AlBr_3$
(d) HCl

14. Write a balanced equation for the process in which each of the following compounds is prepared from its elements:
(a) Mg_3N_2
(b) HF
(c) Al_2O_3
(d) H_2S

15. Write a balanced equation for the preparation of each of the following salts from the reaction of an acid with a metal:
(a) $AlCl_3$
(b) $MgBr_2$
(c) LiI

16. Show the preparation of each of the following from the reaction of an acid with a metal. Be certain your equations are balanced:
(a) CaF_2
(b) $SnCl_2$
(c) MgI_2

17. Write a balanced equation for the preparation of each of the following compounds from the reaction of an acid with a salt:
(a) H_2S
(b) HF
(c) HBr

18. Write the balanced equation for the formation of each of the following compounds from the reaction of an acid and a salt:
(a) H_2Te
(b) HCl
(c) HI

19. Give the balanced equation for the preparation of each of the following compounds from the reaction of an acid with a base:
(a) KCl
(b) CaF_2
(c) NaHSe

20. Use balanced equations to show the preparation of each of the following compounds from the reaction of an acid and a base:
(a) LiBr
(b) $Ca(HS)_2$
(c) $BaCl_2$

Lewis Structures Covalent Compounds [21–28]

21. Draw the Lewis structures of the following covalent compounds and give the formal charges on the atoms:
(a) HF
(b) H_2S
(c) ClF
(d) NF_3
(e) CO

22. Write the Lewis structure for each of the following compounds, including formal charges. Except for H_2O_2, none of these compounds have oxygen atoms bonded together.
(a) H_2O_2
(b) CO_2
(c) SO_2
(d) SO_3
(e) HNO_3

23. Name the following covalent compounds:
(a) $AsCl_3$
(b) IF_7
(c) P_4O_6
(d) N_2O_4
(e) NH_3
(f) PH_3
(g) N_4S_4

24. Give the names of each of the following compounds:
(a) $AsCl_5$
(b) BrF_3
(c) P_4O_{10}
(d) N_2O
(e) CH_4
(f) $SiCl_4$

25. Write the formulas of the following compounds:
(a) carbon disulfide
(b) phosphorus pentachloride
(c) dichlorine heptoxide
(d) methane
(e) ditellurium decoxide
(f) silane

26. Give the formula for each of the following:
(a) ammonia
(b) dinitrogen pentoxide
(c) water
(d) xenon trioxide

27. Write balanced equations for the preparation of the following covalent compounds from their elements.
(a) ClF_3
(b) P_4O_6
(c) SF_6

28. Use a balanced equation to show how each of the following compounds can be prepared from its elements:
(a) SeF_2
(b) BrF_3
(c) SiO_2

Polyatomic Ions; Oxy Acids [29–40]

29. Draw the Lewis structure and give the formal charges on the atoms for each of the following polyatomic ions,
without referring to Table 4–10. Note that none of these ions have oxygen atoms bonded together.
(a) ClO_3^-
(b) NH_4^+
(c) CO_3^{2-}
(d) PO_4^{3-}
(e) SO_3^{2-}

30. Give the Lewis structure and formal charges on the atoms for each of the following ions. Try not to refer back to the text except to check your work. None of these ions have oxygen atoms bonded together.
(a) SO_4^{2-}
(b) ClO_4^-
(c) NO_2^-

31. Name the following compounds:
(a) $NaClO_3$
(b) NH_4Cl
(c) H_2SO_3
(d) $Pb(NO_3)_2$
(e) NH_4NO_3
(f) HNO_3

32. Name the following compounds:
(a) $NaHCO_3$
(b) KClO
(c) $Ca(H_2PO_4)_2$
(d) $Fe(ClO_4)_3$
(e) $Ba(HSO_4)_2$
(f) Na_3PO_4

33. Write the formulas of the following compounds.
(a) aluminum nitrate
(b) nitrous acid
(c) ammonium chlorate
(d) magnesium nitrate
(e) perchloric acid

34. Write the formulas of the following compounds.
(a) calcium hydrogen carbonate
(b) tin(II) sulfate
(c) sodium monohydrogen phosphate
(d) copper(II) sulfate
(e) barium nitrate

35. Write balanced equations for reactions between an acid and a base to produce the following salts:
(a) $CaCO_3$
(b) $(NH_4)_2SO_4$
(c) $NaClO_2$
(d) K_3PO_4

36. Write balanced equations for reactions between an acid and a base to produce the following salts:
(a) NH_4HSO_4
(b) $Sr(ClO_2)_2$
(c) $Mg_3(PO_4)_2$
(d) $KHCO_3$

37. Write a balanced equation for the reaction of an acid with a metal to produce the following:
(a) $LiClO_2$
(b) $SnSO_4$
(c) $Ca(NO_3)_2$

38. Write a balanced equation for the reaction of an acid with a metal to produce the following:

(a) $Mg_3(PO_4)_2$
(b) Li_2SO_4
(c) $Ba(NO_3)_2$

39. Write the balanced equations for a set of reactions, starting with only pure elements and water, that would lead to each of the following products.
(a) H_2SO_3
(b) $NaHCO_3$
(c) NH_4Cl

40. Write the balanced equations for a set of reactions, starting with only pure elements and water, that would lead to each of the following products.
(a) $(NH_4)_2SO_4$
(b) K_3PO_4
(c) HNO_2

Stoichiometry   [41–48]

41. How many grams of H_2S would be needed to react with 10.0 g of $Ca(OH)_2$ to produce CaS? to produce $Ca(HS)_2$?

42. If 15.0 g of sodium hydroxide are treated with sulfuric acid to produce sodium sulfate, how many grams of H_2SO_4 are required? If the desired product is sodium hydrogen sulfate, how many grams of sulfuric acid are required?

43. How much sulfur is needed to produce 1000 kg of H_2SO_4?

44. How much chlorine is needed to produce 1000 kg of HCl?

45. How much H_3PO_4 can be prepared from 1.00 metric ton of phosphate rock? Consider phosphate rock to be $Ca_3(PO_4)_2$, although it actually also contains apatite, $Ca_5(PO_4)_3F$.

46. Calculate the percentages of phosphorus in superphosphate of lime and triple superphosphate.

47. By calculation, explain why 40% oleum is said to be 109% sulfuric acid.

48. A commercial sulfuric acid sample falls into the range of 60 to 65% oleum. If it turns out to be 114% acid, what was the actual percentage oleum?

Additional Problems   [49–55]

49. Give names for the following salts: K_2Se, MgI_2, $AlBr_3$, SnO, $BiCl_3$, $NiBr_3$, CuCl, NaH.

50. A special formulation for an industrial strength drain cleaner is a 50% by weight aqueous barium hydroxide solution.
(a) Express this concentration as kg barium hydroxide per kg water.
(b) If potassium hydroxide is available at half the price per kilogram for this same special application, what is the economic advantage (or disadvantage) in switching to KOH?

51. Consider the following process for the synthesis of sodium thiosulfate:

$$Na_2CO_3\ (aq) + 2\ Na_2S\ (aq) + 4\ SO_2\ (g) \longrightarrow$$
$$3\ Na_2S_2O_3\ (aq) + CO_2\ (g)$$

(a) How many grams of sodium carbonate are required to react with 10.0 g of sodium sulfide?

(b) How many grams of sodium carbonate are needed to produce 10.0 g of sodium thiosulfate?
(c) What is the maximum number of grams of sodium thiosulfate produced along with 10.0 g of carbon dioxide?
(d) If 10.0 g of sodium carbonate are combined with 10.0 g of sodium sulfide and 10.0 g of sulfur dioxide, how many grams of sodium thiosulfate is it possible to obtain? How many grams of each of the reagent(s) in excess remain?
(e) Write the equation for the synthesis of potassium thiosulfate.

52. Alkali metal hydroxides have been used to remove carbon dioxide from the air in enclosed spaces, first in submarines and more recently in spacecraft.
(a) Write the equation for the reaction of lithium hydroxide with carbon dioxide.
(b) Determine the capacity of lithium hydroxide for removing carbon dioxide, and express your answer as kg CO_2 per kg LiOH.
(c) Write the equation for the reaction of sodium hydroxide with carbon dioxide and determine which would be more cost-effective if sodium hydroxide is half the price of lithium hydroxide on a weight basis.

53. A century after it was first developed, the Hooker process based on the electrolysis of salt solutions is still the most economic method of recovering chlorine:

$$2\ NaCl + 2\ H_2O \longrightarrow 2\ NaOH + H_2 + Cl_2$$

In one commercial operation, it is possible to process 1600 tons per day of the needed 23% salt solution. How many tons of chlorine are produced? How many tons of hydrogen?

54. By direct reaction of the mineral stibnite with scrap iron, antimony can be obtained in the free state:

$$Sb_2S_3 + 3\ Fe \longrightarrow 2\ Sb + 3\ FeS$$

On heating 400 pounds of the ore with 167 pounds of scrap iron, 133 pounds of antimony were obtained.
(a) What is the limiting reagent?
(b) How many grams of the other reagent are present in excess?
(c) What is the percent yield for the process?

55. The production of aluminum sulfate from bauxite depends on the following reaction:

$$Al_2O_3 + 3\ H_2SO_4 \longrightarrow Al_2(SO_4)_3 + 3\ H_2O$$

A bauxite sample is 44.4 percent aluminum oxide, with the remainder being inert material. One kg of bauxite is treated with one kg of an industrial grade of sulfuric acid known to be a 66.7% aqueous solution. What fraction of the excess reagent reacted?

MULTIPLE PRINCIPLES   [56–59]

56. Table 4–1 gives the percent composition of the Earth's continental crust by mass. Calculate the mole percentages in the continental crust for the eight most abundant elements.

57. We wish to prepare the following compounds by the reaction of an acid and a metal. Tell which acid you would choose for each preparation and what mass of the pure acid would be required to produce 1.00 g of the desired compound:

(a) NaCl

(b) AlF_3

(c) KBr

(d) CaI_2

58. Industrial grade concentrated sulfuric acid is 98% acid (the rest being water), and it has a density of 1.834 g/mL at 15.6°C. How many grams of sulfur are required to produce 1.55 L of this sulfuric acid?

59. Osmium forms three ions that combine with the oxide ion: Os^{2+}, Os^{3+}, and Os^{4+}. The specific product formed in the reaction between two elements can usually be controlled by the relative amounts of the reactants used, that is, using the stoichiometry of the reaction. Write the bal-anced equation for the reaction leading to each oxide from osmium metal and oxygen gas. How many grams of oxygen would react with 1.00 g of osmium metal to produce each of the three oxides?

APPLIED PRINCIPLES   [60–61]

60. How many tons of superphosphate of lime would be necessary to deliver the same amount of phosphorus plant nutrient as 1.00 metric ton of triple superphosphate?

61. The normal strengths of commercial oleums fall into three categories, expressed as percent free sulfur trioxide: 10–35%, 40%, and 60–65%. Calculate the weight limits for a 100 kg sample of the 10–35% oleum, after it has been diluted with the stoichiometric quantity of water for preparing 100% sulfuric acid.

CHAPTER

5

Gases and the Pressure of the Atmosphere

5.1 COMPOSITION OF THE ATMOSPHERE

5.2 PRESSURE AND ITS MEASUREMENT

5.3 AVOGADRO'S HYPOTHESIS: THE MASS AND VOLUME OF A GAS

5.4 CHARLES'S LAW: TEMPERATURE AND ITS EFFECT ON THE VOLUME OF A GAS

5.5 BOYLE'S LAW: PRESSURE AND ITS EFFECT ON THE VOLUME OF A GAS

5.6 THE IDEAL GAS LAW: COMBINED P, V, and T RELATIONSHIPS

5.7 DALTON'S LAW: MIXTURES OF GASES AND PARTIAL PRESSURES

5.8 REAL GASES AND IDEAL BEHAVIOR

5.9 AVOGADRO'S HYPOTHESIS AND REAL GASES

5.10 THE KINETIC THEORY OF GASEOUS BEHAVIOR

5.11 GRAHAM'S LAW: EFFUSION AND DIFFUSION

5.12 COMPRESSIBILITY, COMPRESSED GASES, AND COMPRESSORS

Although pneumatic tires and blimps have been around for a long time, the idea of large air-supported all-tension structures first took shape in the late 1960s. The atmospheric pressure within the 61,000-seat Hoosier Dome (Indianapolis) supports only the fabric roof covering this sports stadium. The dome is cheaper to build and maintain than a conventional roof structure; fans are needed to maintain the proper tension of the air. Dirigibles such as the *Hindenberg* were not all-tension structures, having a steel frame over which the fabric shell was stretched. These lighter-than-air ships were popular between the two world wars, until the tragic destruction of the hydrogen-filled *Hindenberg* in May, 1937, recorded here in early color photos. Since then, helium has been used to fill non-commercial lighter-than-air craft.

5.1   COMPOSITION OF THE ATMOSPHERE

Above the Earth's surface lie the successive layers of the atmosphere (Figure 5–1). The composition changes because higher layers are subjected to smaller gravitational forces and are bombarded by unfiltered and powerful solar radiations. Our lives are spent in the bottom layer, the one we call the troposphere, surrounded by a mixture of gases composed mostly of nitrogen and oxygen. Respiration in humans and animals depends upon the presence of oxygen, and modern civilization requires oxygen for the combustion of coal, oil, wood, and other fuels to provide warmth and motive power. Without nitrogen to dilute the oxygen in the air, however, the needed combustion reactions would proceed uncontrolled. In an atmosphere of pure oxygen, the surface of the Earth would long ago have become a charred ruin.

Although we now take for granted the knowledge that air is a mixture of gases, the idea persisted for centuries that air was a simpler substance. Ancient peoples considered it an element (one of four, along with fire, earth, and water). However, by the seventeenth century, chemical experiments began to be interpreted in terms of two components in air. One component supported combustion and respiration (oxygen); the other (nitrogen) did not. By 1785, Henry Cavendish had shown that air consists of not only these two major components, but also small amounts of other gases. Much later, these were identified as the noble gases, particularly argon (Table 5–1).

5.2   PRESSURE AND ITS MEASUREMENT

The atmosphere is held in place by the force of the Earth's gravitational field acting against the mass of the air. The resulting force is spread out over the surface of the Earth. We commonly measure the force acting upon a unit area of the Earth's surface, say one square meter, as the response

FIGURE 5–1   In the early 1990s, the space shuttle will deliver the NASA Hubble Space Telescope into the Earth's orbit. Lying beyond the distorting haze of the Earth's atmosphere (in this artist's rendering), the space telescope will be able to peer to the edge of the universe, expanding our view to a volume 350 times larger than ever seen before.

TABLE 5–1   Composition of Clean Dry Air Near Sea Level

Component	Formula	Percent by Volume (Mole %)
nitrogen	N_2	78.084
oxygen	O_2	20.9476
argon	Ar	0.934
carbon dioxide	CO_2	0.0314
neon	Ne	0.001818
helium	He	0.000524
methane	CH_4	0.0002
krypton	Kr	0.000114
dinitrogen monoxide	N_2O	0.00005
hydrogen	H_2	0.00005
xenon	Xe	0.0000087
ozone	O_3	0 to 0.000007 (summer)
		0 to 0.000002 (winter)
ammonia	NH_3	trace
carbon monoxide	CO	trace
iodine	I_2	0 to 0.000001
nitrogen oxide	NO	0 to 0.000002
sulfur dioxide	SO_2	0 to 0.001

TABLE 5–2   Equivalent Units of Pressure

1 atmosphere = 760 mm of mercury = 760 torr = 33.7 feet of water
= 1,013,000 dynes per square centimeter
= 101,300 newtons per square meter
= 101,300 pascals = 101.3 kilopascals
= 14.7 pounds per square inch

to gravity of the total mass of the column of air above the area. This force divided by the area (F/A) is called the pressure or **atmospheric pressure**:

$$P = \frac{F}{A}$$

If the force is measured in newtons (N) and the area in square meters (m^2), the pressure will be in units of newtons per square meter (N/m^2). A pressure of $1\ N/m^2$ has been defined as 1 pascal (Pa), the standard SI unit of pressure. However, pressure is a commonly measured quantity, and many other units besides pascals are still used, including millimeters of mercury, torr, and atmospheres (Table 5–2).

The pressure of the atmosphere acting on a liquid surface is responsible for such phenomena as drawing soda up a straw or pumping water out of a well. Figure 5–2 depicts water being pumped the old-fashioned way. Movement of the piston reduces the pressure of the air above the water in the pipe; pressure of the atmosphere pressing on the surface of the water in the well drives the water up the pipe. The same process drives soda up through a straw as you suck the air out, reducing the pressure above the liquid surface within the straw, relative to the pressure acting on the surface of the liquid in the bottle.

The height of a column of liquid that can be supported in a tube by the pressure of the atmosphere is limited. If the part of the column above the liquid is completely evacuated, then the downward force of the liquid in the column can be determined. The density d of the liquid is

FIGURE 5–2 The pressure of the atmosphere forcing water up the pipe of a pump.

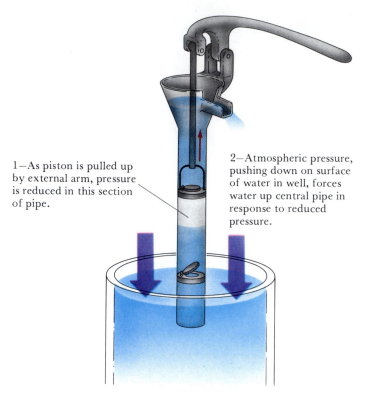

1—As piston is pulled up by external arm, pressure is reduced in this section of pipe.

2—Atmospheric pressure, pushing down on surface of water in well, forces water up central pipe in response to reduced pressure.

$$d = \frac{\text{mass}}{\text{volume}} = \frac{m}{V}$$

Thus, the mass m of the liquid in the column is $m = dV$. In turn, the volume V of the liquid in the column is

$$V = (\text{height of column})(\text{area of column cross-section}) = hA$$

Since force is measured as the product of the mass m and the acceleration of gravity g, the downward force of the liquid in the column is

$$F = dhAg$$

Division of both sides of this equation by A gives

$$\frac{F}{A} = dhg$$

which is the pressure resulting from the mass of a column of liquid that is just balanced by the pressure of the atmosphere. Thus,

$$P_{\text{atm}} = dhg \tag{5–1}$$

The standard value for atmospheric presure is 1.01×10^5 newtons per square meter. Since one newton is $1\ \text{kg·m/s}^2$, this pressure is

$$1.01 \times 10^5\ \frac{(\text{kg})(\text{m/s}^2)}{(\text{m}^2)} = 1.01 \times 10^5\ \frac{\text{kg}}{(\text{m})(\text{s}^2)}$$

We could choose water for the barometer fluid to fill the column. The density of water is $1.00 \times 10^3\ \text{kg/m}^3$, and the acceleration due to gravity is $9.807\ \text{m/s}^2$, so the maximum height h to which the column of water could be supported is

$$h = \frac{1.01 \times 10^5\ \text{kg/m·s}^2}{(1.00 \times 10^3\ \text{kg/m}^3)(9.807\ \text{m/s}^2)} = 10.3\ \text{m}$$

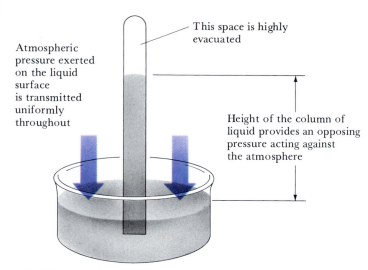

Atmospheric pressure exerted on the liquid surface is transmitted uniformly throughout

This space is highly evacuated

Height of the column of liquid provides an opposing pressure acting against the atmosphere

FIGURE 5–3 A barometer device for measuring the pressure of the atmosphere.

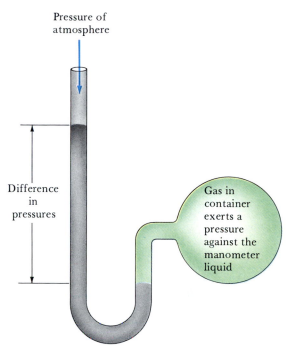

Pressure of atmosphere

Difference in pressures

Gas in container exerts a pressure against the manometer liquid

FIGURE 5–4 An open-ended manometer for determining the difference between the pressure of the gas in a container and the pressure of the atmosphere.

Thus, the greatest height to which water can be pumped by reducing the pressure in the tube above the water column is 10.3 m or 33.8 ft, at standard atmospheric pressure (which is the average pressure at sea level).

One of the earliest scientific instruments used for measuring the pressure of the atmosphere is the **barometer** (Figure 5–3). The operating principle is based on the fact that changes in atmospheric pressure alter the height of a supported column of liquid. The height of the liquid column is directly proportional to the atmospheric pressure, as shown by Eq. (5–1). Unfortunately, water has two serious shortcomings for use as a barometer liquid. First, a column that is several meters in height would be cumbersome. Second, vaporization of water into the column space above the liquid surface creates an additional force, acting downward, and a significant correction must be made for this effect. Mercury is much better in both respects. The density of mercury is 13.5 times that of water, requiring a column less than 1 m high to balance standard atmospheric pressure. In addition, pressure in the column due to vaporized mercury in the space above the liquid is negligible.

Because mercury is commonly used in barometers for determining atmospheric pressure, the height of the mercury column—normally measured in mm Hg—is used as a unit of pressure. One mm Hg is defined as one torr, named after the Italian physicist and mathematician Evangelista Torricelli. A comparison of different units of pressure is given in Table 5–2.

A gas in a container also exerts pressure. If the container is open to the atmosphere, then the pressure within the container must be equal to the atmospheric pressure. If the container is closed, however, the pressure within the container can be different than the pressure of the atmosphere. The pressure of a gas within a closed container can also be measured with a column of mercury, but the configuration used is slightly different from the barometer and is called a manometer. An **open-ended manometer** is illustrated in Figure 5–4. The difference between the heights of the mercury columns on either side of the U-shaped tube represents the difference

Evangelista Torricelli (1608–1647) is credited with the invention of the barometer. To his eternal credit, one of his students was Galileo.

between the pressure of the gas in the container and that of the atmosphere. Use of this apparatus requires a second measurement, using a barometer to determine the pressure of atmosphere, before the pressure of the gas within the container can be precisely determined. The difference between the heights of the mercury columns on either side of the U-tube changes as the atmospheric pressure changes.

EXAMPLE 5–1

Consider an open-ended manometer, attached to a cylinder that contains a gas. The mercury in the manometer is 121 mm higher on the open side. In a separate reading using a barometer, the pressure of the atmosphere is found to be 743 mm of mercury. What then is the pressure of the gas in the cylinder in mm of Hg, in torr, in atm, in Pa, and kPa?

Solution   The pressure of the gas in the container equals the atmospheric pressure plus the pressure resulting from the extra height of the column of mercury on the open side. Thus:

$$P_{gas} = P_{atm} + P_{extra}$$
$$= 743 \text{ mm Hg} + 121 \text{ mm Hg}$$
$$= 864 \text{ mm Hg} = 864 \text{ torr} \left(\frac{1 \text{ atm}}{760 \text{ torr}}\right) = 1.14 \text{ atm}$$

Since 1 mm Hg = 1 torr = 133.3 Pa,

$$1 \text{ kPa} = 7.50 \text{ torr}$$

Therefore, either of the following calculations will lead you to the answers in pascals and kilopascals:

$$P_{gas} = 864 \text{ torr} \left(\frac{133.3 \text{ Pa}}{\text{torr}}\right) = 115,000 \text{ Pa} = 115 \text{ kPa}$$

$$P_{gas} = 864 \text{ torr} \left(\frac{1 \text{ kPa}}{7.50 \text{ torr}}\right) = 115 \text{ kPa}$$

EXERCISE 5–1(A)

If you were to construct a barometer for measuring the pressure of the atmosphere in the vicinity of sea level using ethyl alcohol (whose density is 0.789 g/mL) instead of mercury as the barometer fluid, what length of glass tubing with an area of 1 cm² is needed? What length of glass tubing with an area of 2 cm²?
Answer: >13 m; same for any area.

EXERCISE 5–1(B)

Consider the open-ended manometer pictured in Figure 5–4. What is the pressure of the gas trapped in the container if the pressure of the atmosphere is 746 torr and the mercury levels are 50 mm in the closed arm and 200 mm in the open arm? What is the pressure if atmospheric pressure is 0.950 atm?
Answer: 896 mm; 872 mm.

An alternative form of the manometer, which allows measurement of the pressure within a container but does not require measuring the atmospheric pressure, is the **closed-ended manometer** shown in Figure 5–5. In this case, there is no gas above the mercury on the closed side of the

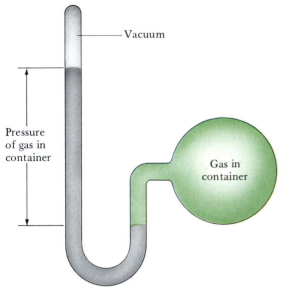

Vacuum

Pressure
of gas in
container

Gas in
container

U-shaped tube except for a very small amount of mercury vapor. The near-vacuum on the closed side exerts essentially no downward force on the mercury, and the difference in heights between the columns can be taken as the pressure of the gas within the container in torr or mm Hg.

5.3 AVOGADRO'S HYPOTHESIS: THE MASS AND VOLUME OF A GAS

The essential features of a trapped gas can be seen in the relationships between mass, volume, temperature, and pressure. **Avogadro's Hypothesis** states that equal volumes of gases at the same temperature and pressure contain the same numbers of molecules. Because this simple statement holds, regardless of the identity of the gas, we are led to important generalizations about gases.

If equal numbers of molecules of all gases have the same volume at the same temperature and pressure, then (as long as pressure and temperature are held constant) the volume of a gas will be proportional to the number of molecules it contains. Since the number of molecules contained in even the smallest measurable amount of gas is enormous, we choose a convenient standard, the mole—a mass equal in grams to the formula weight, or 6.022×10^{23} molecules. Thus Avogadro's Hypothesis can be stated

$$V_{T,P} = k'n \qquad (5\text{–}2)$$

V is the volume of the gas and n is the number of moles. The subscripts T and P specify that the temperature and pressure are held constant. The quantity k' is a proportionality constant; this value depends on the temperature and pressure and has units of volume/mole. Now, the number of moles n of a gas can be calculated from

$$n = \frac{m}{M}$$

where m is the mass of the gas and M is its formula weight. Substituting into Eq. (5–2) gives

$$V_{T,P} = \frac{k'm}{M}$$

which can be rearranged to

$$\frac{1}{k'} M = \frac{m}{V_{T,P}} = d$$

where mass divided by volume is equal to the density d of the gas. Now, since $1/k'$ is a constant, we can conclude that **the density of a gas is proportional to its formula weight**, another important consequence of Avogadro's Hypothesis.

EXAMPLE 5–2

A mass of 15.0 g of oxygen gas (O_2) at a pressure of 1.00 atm and a temperature of 298 K occupies a volume of 11.5 L. At the same pressure and temperature, a sample of 10.0 g of an unknown gas is found to occupy 5.55 L. Calculate the formula weight of the unknown gas.

Solution　Because pressure and temperature are held constant, Eq. (5–2) can be used and the value of k' for this particular combination of T and P can be evaluated from the data for the oxygen sample:

$$k' = \frac{V}{n} = \frac{11.5 \text{ L}}{(15.0 \text{ g})/(32.0 \text{ g/mol})} = 24.5 \text{ L/mol}$$

Using the same equation, the number of moles of the unknown gas can now be determined:

$$n = \frac{V}{k'} = \frac{5.55 \text{ L}}{24.5 \text{ L/mol}} = 0.226 \text{ mol}$$

Finally, dividing of the number of grams of the unknown gas by the number of moles gives the formula weight:

$$M = \frac{10.0 \text{ g}}{0.226 \text{ mol}} = 44.2 \text{ g/mol}$$

EXERCISE 5–2

The density of CO gas at 1.00 atm and 323 K is 1.06 g/L. What is the density of NO_2 gas at the same temperature and pressure?
Answer: 1.74 g/L.

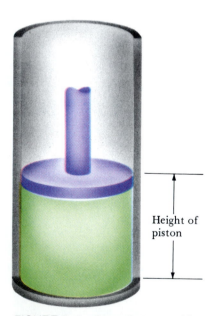

Height of piston

FIGURE 5–6　A gas is trapped in a vessel by a freely moving piston so the pressure remains constant. As the temperature of the gas rises, the piston rises because the volume increases. Thus, the volume of the gas (as determined by the height of the piston) is a measure of the temperature. The device serves as a gas thermometer.

5.4　CHARLES'S LAW: TEMPERATURE AND ITS EFFECT ON THE VOLUME OF A GAS

Suppose that a fixed amount of a gas is contained in a vessel, such as that in Figure 5–6, fitted with a movable piston so that the volume of the gas can change but the pressure remains constant. If the temperature of the gas is increased, the piston is observed to rise as the volume of the gas increases. In fact, the increase in the volume of a gas at constant pressure can be used as an operational definition of temperature, and the device pictured in the figure can be used as a practical thermometer. The rela-

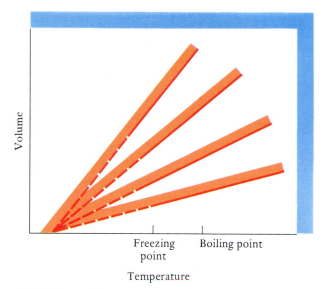

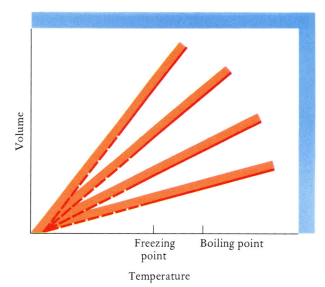

FIGURE 5–7 Plots of *V* as a function of *T* in °C at constant *P*. The different lines arise from experiments with different pressures or numbers of moles of gas. In each experiment, the pressure must be held constant. Extrapolating all the lines to zero gives the absolute zero of temperature.

FIGURE 5–8 The *V* versus *T* lines drawn so the origin is at *V* = 0 and *T* = 0 on the Absolute (Kelvin) scale. V_o for each line is the volume at 0°C (273 K).

tionship between the temperature and the volume of a gas is called **Charles's Law**, after the French experimenter Jacques Charles. One statement of the law is that **at constant pressure, the volume of a fixed mass of gas is directly proportional to its temperature**. Another way of stating the relationship is to plot volume against temperature; a linear (or directly proportional) relationship is shown by such a graph.

A change in temperature can be defined as that property which causes the volume of a quantity of gas (at constant pressure) to change, and the magnitude of the volume change gives the magnitude of the temperature change. Each line in Figure 5–7 corresponds to one combination of pressure and number of moles of gas. All of these lines can be extrapolated to the origin, corresponding to a common minimum of temperature at zero volume. As the graph indicates, any lower temperature would require an impossible negative volume. We refer to this minimum temperature as the **absolute zero** of temperature. It is a very important reference point and appears in many calculations.

In Figure 5–8 the origin has the coordinates *V* = 0 and *T* = 0. Such a temperature scale (with zero set at absolute zero) is called an **absolute temperature scale**. Remember from geometry that the equation for a straight line is

$$y = mx + b$$

where *m* is the slope and *b* is the intercept on the *y*-axis. In this case, *V* = *y* and *T* = *x*. In Figure 5–8 we have redefined the temperature scale and set the origin so that *b* = 0, and the equation reduces to

$$y = mx$$

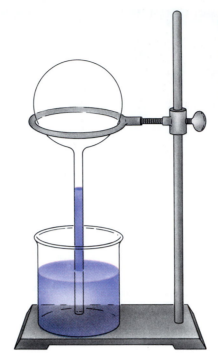

FIGURE 5–9 The operation of the thermoscope was straightforward—the bulb was warmed until some air was forced out. On cooling, the column of water rose part of the way back. Then, as the temperature around the bulb changed, the level of the liquid in the column rose and fell.

or in other words,

$$V_{P,n} = k''T \tag{5–3}$$

The subscripts (P and n) specify the conditions of constant pressure and moles, and k'' is another constant with units of volume per degree of temperature. Its value depends on the magnitudes of P and n.

Devices for making such measurements, called gas thermometers, can give very accurate temperature measurements (Figure 5–9). Early temperature scales such as those of Fahrenheit and Celsius had their zero points set at the temperatures of commonly observed phenomena. Zero on the **Celsius scale** is the normal freezing point of water; the normal boiling point of water is marked at 100. Accordingly, one Celsius degree represents $\frac{1}{100}$ of the temperature range between the melting and boiling points of water. The **Kelvin scale** uses the same magnitude for a degree as the Celsius scale but places zero at the absolute zero of temperature, so it is alternately referred to as the **absolute scale of temperature**. Accurate determinations of absolute zero show that it is 273.15 degrees below zero Celsius. Thus, the freezing point and the boiling point of water are 273.15 K and 373.15 K, respectively. The conversion from Celsius to Kelvin is

$$K = °C + 273.15$$

The **Fahrenheit scale** of temperature is still widely used, especially in North America. Curiously, the fixed points on this scale were originally set as the temperature of a mixture of snow and salt (thought to be the lowest temperature that could be reproducibly obtained) as 0°F and body temperature as 100°F. Slight modification of the Fahrenheit scale and more precise determinations (made subsequently) have placed the normal oral body temperature at only 98.6°F. On the Fahrenheit scale, the fixed points are now set as 32°F, the normal freezing point of water, and 212°F, its normal boiling point. Conversions between scales can be done with the simple linear relationship.

$$°F = \frac{9}{5}°C + 32$$

Temperatures can be measured from near absolute zero to well beyond 3000 K, and some temperatures of important phenomena within this range are given in Table 5–3.

___

EXAMPLE 5–3

At what temperature is the temperature on the Fahrenheit scale numerically equal to that on the Celsius scale?

Solution To answer this question, let x equal the temperature in question. Then

$$°F = \frac{9}{5}°C + 32$$

$$x = \frac{9}{5}x + 32 = -40$$

and

$$-40°C = -40°F$$

___

EXERCISE 5–3

At what temperature is the Fahrenheit scale numerically equal but opposite in sign to the Celsius scale?
Answer: $11.4°F = -11.4°C$.

___

TABLE 5–3	Some Benchmark Temperatures in Fahrenheit, Celsius, and Kelvin		
	°F	°C	K
iron boils	5400	3000	3273
gold boils	4700	2600	2873
silver boils	3542	1950	2223
lead boils	2948	1620	1893
iron melts	2795	1535	1808
NaCl boils	2575	1413	1686
gold melts	1945	1063	1336
silver melts	1762	961	1234
NaCl melts	1474	801	1074
sulfur boils	833	445	718
mercury boils	675	357	630
lead melts	621	327	600
water boils	212	100	373
ice melts	32	0	273
mercury melts	−38	−39	234
oxygen boils	−297	−183	90
nitrogen boils	−320	−196	77
nitrogen melts	−346	−210	63
oxygen melts	−360	−218	55
hydrogen boils	−423	−253	20
hydrogen melts	−434	−259	14
helium boils	−452	−269	4
absolute zero	−459.67	−273.15	0

ROBERT BOYLE

Robert Boyle discovered that air is compressible and that compressibility varies inversely with the compressing force (the pressure). He trapped a gas (with mercury) in the closed end of a very tall J-shaped glass tube and observed that adding more mercury to the long, open end caused the volume of the trapped gas to diminish. Doubling the pressure halved the volume; tripling the pressure reduced the volume to a third. This "spring of the air," as Boyle referred to compressibility, led to the conclusion that gases must be composed of discrete, widely separated particles (in a void). The ability to compress a gas was understandable—nothing more than squeezing the particles of the gas closer together.

5.5 BOYLE'S LAW: PRESSURE AND ITS EFFECT ON THE VOLUME OF A GAS

Here is a simple experiment that can be performed on a gas. Using the apparatus shown in Figure 5–10, we can measure its volume for each of a number of different pressures. In this experiment, the temperature remains constant. The pressure exerted on the gas in the cylinder can be calculated from the atmospheric pressure plus the pressure exerted by the mass of the piston. This is equal to the acceleration of gravity times the total mass of the piston plus any weights placed upon it, divided by the area of the piston. When the pressure exerted on the gas is the same as the pressure of the gas in the cylinder, the position of the piston will be fixed. At that point, the volume of the gas can be calculated after measuring the height of the piston above the base of the cylinder. Changing the masses on the cylinder in Figure 5–10 results in a new fixed position for the piston and new pressure-volume data for the gas. The relationship between pressure and volume can be determined from a series of such measurements.

FIGURE 5–10 A simple apparatus for determining V as a function of P for a gas. The piston comes to rest when P for the gas equals the external pressure. The volume of the trapped gas decreases as the external pressure increases.

TABLE 5–4   The Results of a Gas Law Experiment*

Pressure, atm	Volume, L	$1/V$, L^{-1}	k''', L·atm
0.10	17.5	0.0571	1.75
0.30	5.83	0.172	1.75
0.50	3.50	0.286	1.75
0.80	2.19	0.457	1.75
1.10	1.59	0.629	1.75
1.30	1.35	0.741	1.76
1.50	1.17	0.854	1.76
1.70	1.04	0.961	1.77
1.90	0.92	1.09	1.75

*The pressure on n moles of a gas is varied at constant T and the volume is measured. Note the nearly constant value for the PV product.

Since gases are compressible substances, increasing the pressure causes the volume to decrease accordingly, whereas reducing the pressure results in a larger volume. Table 5–4 gives the results of a typical experiment of this kind. Figure 5–11 shows the curves resulting from plotting the results of several such experiments at different temperatures. As the curves in Figure 5–11 move away from the origin, each represents a higher temperature. Furthermore, each curve takes the form of a rectangular hyperbola with the equation

$$x \cdot y = c$$

where c is a constant. Robert Boyle is credited with discovering the mathematical nature of this relationship. He found that as long as the pressure was not too high, the pressure-volume product is nearly constant. The algebraic statement of **Boyle's Law** is

$$(PV)_{n,T} = k''' \tag{5–4}$$

where the subscripts note that the number of moles n of gas and the temperature T remain constant. The value for the constant k''' depends on the amount of gas and the temperature. It should also be noted that this constant has units of volume times pressure: for example, liter·atmospheres. It is a distinctly different constant than the one for Avogadro's Hypothesis or the Charles's Law relationship.

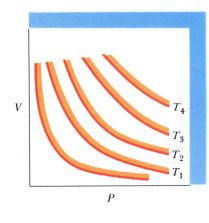

FIGURE 5–11   A Boyle's law plot of V versus P for a fixed number of moles of an ideal gas. Each line represents a different constant temperature; the higher the temperature, the farther from the origin.

FIGURE 5–12 Divers carry tanks containing compressed air in order to expose the lungs to high alveolar gas pressures. This prevents lung collapse at depths where the surrounding pressure is very high.

Gases that are easily liquefied, such as ammonia and chlorine, usually deviate more from predictions of ideal behavior than do very low-boiling gases such as hydrogen and oxygen.

For the scuba diver, the proper application of Boyle's Law is of vital interest (Figure 5–12). With every 10 meters of depth, divers experience an additional pressure of one atmosphere, which in turn affects gases trapped anywhere in the body cavities—the middle ear, sinuses, and most importantly the lungs—or dissolved in body fluids such as the blood. If a diver descends without scuba gear, the amount of gas contained in the body cavities is constant and the volume of these cavities decreases as the surrounding water pressure increases. However, a diver using scuba gear does not experience the crush of the great external pressure, because the regulator on the tank delivers air to the lungs at the same pressure as the surroundings. This means that at a depth of 30 meters, the air in the diver's lungs is at a pressure of 4 atm. If the diver is going to ascend quickly from such depth, he or she must remember to breath out regularly while returning to the surface. If not, the pressure of the air in the lungs will cause the lungs to expand. The extreme distortion of the lungs can cause the small air sacs called alveoli to burst; this could allow air bubbles to enter the bloodstream and cause a dangerous blockage, followed by loss of consciousness, heart attack, or brain damage.

Boyle's Law can be used to show the relationship between the pressure of a gas and its density d. For a given mass m of a gas, the density is given by

$$d = \frac{m}{V}$$

According to Boyle's Law,

$$V = \frac{k'''}{P}$$

Substituting into the previous equation gives the following:

$$d = \left(\frac{m}{k'''}\right) P$$

This shows that for a given quantity of a gas at constant temperature, the density is directly proportional to the pressure.

It is important to note that Avogadro's Hypothesis, Charles's Law, and Boyle's Law do not perfectly describe any real gas. They accurately describe **ideal gases**. In fact, an ideal gas can be defined as a gas that obeys Boyle's Law, Charles's Law, and Avogadro's Hypothesis. Consider why a real gas cannot strictly follow Boyle's Law under all conditions: First, if a gas is compressed by a sufficiently high pressure at a low temperature, it will liquefy and no longer obey Boyle's Law. Liquefaction is the result of attractions between the molecules, which tend to hold the substance in the liquid state. A second factor that causes real gases to deviate from Boyle's Law at high pressures is the effect of the actual volume of each molecule. Figure 5–13 is a schematic representation of the compression of a gas to the point where the volume of the gas molecules themselves has an effect on the pressure-volume relationship. At high pressures and low volumes, as the molecules become tightly packed together like so many spheres in a box, there is no longer any empty space and therefore the volume can no longer decrease according to Boyle's Law.

An ideal gas would have no tendency to liquefy because there are no forces of attraction between molecules. There would also be no repulsive

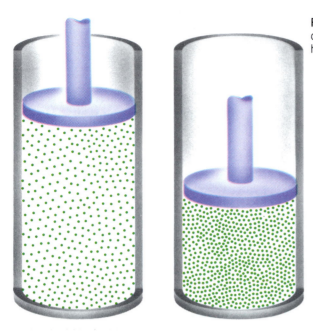

FIGURE 5–13 In a highly compressed state, the molecules occupy a significant fraction of the total volume, and the general behavior of the gas is no longer ideal.

forces between the molecules of such a gas. In fact, for an ideal gas, there are no intermolecular forces whatsoever. Furthermore, the molecules of an ideal gas have zero volume. Of course, perfect Boyle's Law behavior must be considered as the limiting case. Real gases can only approach these criteria to varying degrees. However, if the molecules are kept far apart, the effects of intermolecular forces and those of the actual volume occupied by individual molecules are minimized. Thus, under conditions of low pressure and high temperature, ideal behavior is approached and Boyle's Law holds.

EXAMPLE 5–4

When 4.0 mol of an ideal gas at 35°C occupies a volume of 67.4 L, the pressure must be 1.50 atm. Find the volume of the gas when the pressure is raised to 10.0 atm.

Solution First, note that the number of moles of gas and the temperature remain constant. The initial volume (67.4 L) and pressure (1.50 atm) can be substituted into Eq. (5–4) to obtain a value of k''' under these conditions:

$$(67.4 \text{ L})(1.50 \text{ atm}) = k''' = 101 \text{ L·atm}$$

The new volume (at the new pressure) can then be determined by reapplication of Eq. (5–4):

$$V = \frac{101 \text{ L·atm}}{10.0 \text{ atm}} = 10.1 \text{ L}$$

Alternatively, Eq. (5–4) can be used to create a relationship between two sets of conditions. For convenience, assume that n and T are constant and omit the subscripts from the equation. Specify the initial pressure and volume (1.50 atm and 67.4 L) with the subscript i and the final pressure and volume (10.0 atm and volume to be calculated) with the subscript f. Since Boyle's Law holds for an ideal gas for all combinations of P and V as long as n and T remain constant, then

$$P_i V_i = P_f V_f = k'''$$

and

$$V_f = \frac{P_i V_i}{P_f}$$

Substitution into this expression gives the earlier result, namely

$$V_f = \frac{(1.50 \text{ atm})(67.4 \text{ L})}{10.0 \text{ atm}} = 10.1 \text{ L}$$

EXERCISE 5–4

6.50 mol of an ideal gas at 100°C occupies a volume of 99.5 L at a pressure of 2.00 atm. What is the pressure of this quantity of gas if the volume expands to 150 L at constant temperature?

Answer: 1.33 atm.

5.6 THE IDEAL GAS LAW: COMBINED *P*, *V*, AND *T* RELATIONSHIPS

The term *equation of state* refers to any expression of the relationship among the set of variables that fix the properties of a particular composition of matter. There are others besides the Ideal Gas Law.

Avogadro's Hypothesis, Boyle's Law, and Charles's Law can be combined into a single relationship, an equation of state called the **Ideal Gas Law**. For an ideal gas:

$$PV = nRT \tag{5-5}$$

where P is the pressure of the gas, V is its volume, n is the number of moles, and T is the absolute temperature in kelvins. The constant of proportionality R in the equation is the **universal gas constant**, and its value depends on the units of the variables P, V, and T. This relationship is subject to the same limitations as Boyle's law. Only an ideal gas would obey this law strictly. However, if the gas molecules are kept far apart by avoiding high pressures and low temperatures, agreement between experiment and theory is reasonable.

It has been established by experiment that one mole of an ideal gas at the standard temperature of 0°C or 273.15 K and the standard pressure of 1.0000 atm—a special set of conditions called **standard temperature and pressure** or **STP**—occupies a volume of 22.414 L. Thus, R can be calculated by rearranging Eq. 5–5 and solving:

$$R = \frac{PV}{nT} = \frac{(1.0000 \text{ atm})(22.414 \text{ L})}{(1.0000 \text{ mol})(273.15 \text{ K})} = 0.08206 \text{ L·atm/mol·K}$$

In this form, R is useful for the pressure-volume-temperature calculations that we will perform in this chapter. Values of R in other units will be presented later for other types of calculations.

The separate relationships—Boyle's Law, Charles's Law, and Avogadro's Hypothesis—are all contained in the Ideal Gas Law. This can be seen most easily by solving it for V:

$$V = \frac{RT}{P} n$$

If temperature and pressure are held constant,

$$V_{T,P} = k'n \qquad \text{(Avogadro's Hypothesis)} \tag{5-2}$$

where

$$k' = \frac{RT}{P}$$

Likewise, if the pressure and number of moles are held constant,

$$V_{P,n} = k''T \qquad \text{(Charles's Law)} \qquad\qquad (5\text{–}3)$$

where

$$k'' = \frac{nR}{P}$$

Finally, if the temperature and the number of moles are held constant,

$$(PV)_{n,T} = k''' \qquad \text{(Boyle's Law)} \qquad\qquad (5\text{–}4)$$

where

$$k''' = nRT$$

An alternative way of expressing the effect of temperature on the properties of a gas is through pressure rather than volume. For example, at constant volume and for a fixed number of moles:

$$P = \frac{nR}{V} T = k''''T \qquad\qquad (5\text{–}6)$$

where

$$k'''' = \frac{nR}{V}$$

This expression suggests a different form for the gas thermometer (Figure 5–14), which contains a gas confined in a vessel of fixed volume whose pressure can be determined by a manometer. In this case, the observed pressure is related to the absolute temperature through a simple proportionality constant.

Many different kinds of problems can be solved by using the Ideal Gas Law. The following examples illustrate two of them.

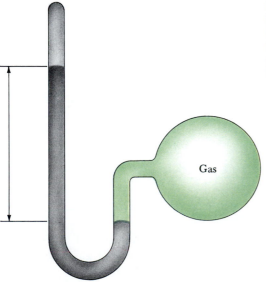

Gas

FIGURE 5–14   A gas thermometer. The difference between the heights of the two columns is a measure of the pressure of the gas and is proportional to the temperature.

EXAMPLE 5–5

What pressure would be exerted by 50.0 g of oxygen gas (O_2) at 40.0°C in a volume of 100.0 L?

Solution   The quantity of oxygen gas is given in grams and must be converted to number of moles, using the molecular weight as the conversion factor:

$$n = \frac{m}{M} = \frac{50.0 \text{ g}}{32.0 \text{ g/mol}} = 1.56 \text{ mol}$$

Also, the temperature is given in degrees Celsius and must be converted to kelvins:

$$T = 273.15° + 40.0° = 313.2 \text{ K}$$

Now we can simply substitute values in the Ideal Gas Law and solve for P:

$$P = \frac{nRT}{V} = \frac{(1.56 \text{ mol})(0.08206 \text{ L·atm/mol·K})(313.2 \text{ K})}{100.0 \text{ L}}$$
$$= 0.401 \text{ atm}$$

In any gas law problem, you should always check the units of the given quantities and make any required conversions before you substitute into the Ideal Gas Law equation. The most common error is forgetting to convert the temperature to kelvins. ■

EXERCISE 5–5

A 12.5 g sample of CO_2 gas occupies a volume of 8.00 L at a pressure of 1.50 atm. What must the temperature of the gas be?
Answer: 515 K or 242°C. ■

EXAMPLE 5–6

A sample of gas occupies 4.5 L at 300 K and a pressure of 0.67 atm. At what temperature would it occupy a volume of 12.0 L and exert a pressure of 0.55 atm?

Solution   One method of solving this problem would be to calculate n from the first set of conditions and then use this value of n to calculate T for the second set of conditions. However, a one-step calculation is possible, since the number of moles of gas n is unchanged:

$$\frac{P_iV_i}{T_i} = nR = \frac{P_fV_f}{T_f}$$

Again, the subscript i is used to denote the initial conditions and the subscript f, the final conditions. The quantity nR can be dispensed with, and the expression involving P's, V's, and T's can be rearranged to give T_f as the answer:

$$T_f = \frac{P_fV_fT_i}{P_iV_i} = \frac{(0.55 \text{ atm})(12.0 \text{ L})(300 \text{ K})}{(0.67 \text{ atm})(4.5 \text{ L})} = 660 \text{ K} \quad ■$$

EXERCISE 5–6

A gas occupies a volume of 10.0 L and exerts a pressure of 1.25 atm at 55.0°C. What volume will it occupy if this same amount of gas is heated to 155°C and its pressure is allowed to increase to 2.00 atm?
Answer: 8.16 L. ■

Molecular Weights from Gas Density Data

The molecular weight M of a substance is equal to its mass m divided by the number of moles n. Substituting $n = m/M$ and solving Eq. (5–5) for M gives the following result:

$$M = \frac{mRT}{PV}$$

But m/V is equal to density. Therefore,

$$M = \frac{dRT}{P} \qquad\qquad (5\text{–}7)$$

Thus, if the temperature and the pressure are unchanged, the density of a gas is proportional to its molecular weight. This is one of the consequences of Avogadro's Hypothesis.

EXAMPLE 5–7

The molecular weight of an unknown, volatile liquid can be obtained by determining the density of its vapor by what is called the Dumas method. Consider, for example, a liquid contained in a 250.0 mL flask and heated at 100.°C until all of the liquid has been vaporized, filling the container completely. The container is open to the atmosphere by a small hole so that the pressure of the vapor inside the flask equals the atmospheric pressure, which in this experiment is 1.05 atm. When condensed, the vapor filling the container weighs 0.720 g. What is the molecular weight of the unknown liquid?

Solution The density of the vapor is

$$\frac{m}{V} = \frac{0.720 \text{ g}}{0.250 \text{ L}} = 2.88 \text{ g/L}$$

The molecular weight of the unknown liquid is given by the ideal gas relationship:

$$M = \frac{dRT}{P} = \frac{(2.88 \text{ g/L})(0.0821 \text{ L·atm/mol·K})(373 \text{ K})}{1.05 \text{ atm}} = 84.0 \text{ g/mol}$$

EXERCISE 5–7

The Dumas method is used to determine the molecular weight of a volatile liquid as in the previous example. The volume of the container is 400.0 mL; the atmospheric pressure registers 0.995 atm (which is the pressure in the container). The vapor completely fills the flask at a temperature of 100.°C, and the condensed vapor weighs 1.274 g. What is the molecular weight of the liquid?
Answer: 98.0 g/mol. ■

Because of their especially low densities, some gases can be used in lighter-than-air craft. According to Avogadro's Hypothesis, the density of an ideal gas is proportional to its molecular weight. Thus, any gas with a molecular weight less than the average molecular weight of air (28.8 μ) has potentially useful lifting power. This includes CH_4 (16.0 μ) and NH_3 (17.0 μ) as well as the more buoyant He (4.0 μ) and H_2 (2.0 μ). The limit of the ability of a volume of a buoyant gas to lift in an air atmosphere is equal to the difference in mass between the gas and an equal volume of air. At this limit, the volume of gas and its load will float in air without either rising or falling. Helium is commonly used in balloons and blimps instead of hydrogen because the greater lifting power of hydrogen is more than offset by its extreme flammability. Hot air balloons (Figure 5–15) take advantage of the lower density of heated air resulting from its increase in volume according to Charles's Law.

EXAMPLE 5–8

Consider one mole of each of H_2, He, CH_4, and air. Calculate their relative lifting powers.

Solution   First recognize that at the same temperature and pressure, one mole of each of these gases would occupy the same volume. The difference in mass between a mole of air and a mole of any of the other gases represents the load that a mole of each gas could carry while balancing the same volume of air (at the same temperature and pressure). These values are

$$28.8 \text{ g} - 2.0 \text{ g} = 26.8 \text{ g} \quad \text{for } H_2$$
$$28.8 \text{ g} - 4.0 \text{ g} = 24.8 \text{ g} \quad \text{for He}$$
$$28.8 \text{ g} - 16.0 \text{ g} = 12.8 \text{ g} \quad \text{for } CH_4$$

Thus H_2 can lift 26.8 g/24.8 g = 1.08 times as much as He, and 26.8/12.8 = 2.08 times as much as CH_4. ■

EXERCISE 5–8

Calculate the relative lifting capacities of Ne and NH_3 gases.
Answer: NH_3 will lift 1.4 times as much as the equal number of moles of Ne. ■

5.7   DALTON'S LAW: MIXTURES OF GASES AND PARTIAL PRESSURES

There are no forces operating between the molecules of an ideal gas—neither attractive forces nor repulsive forces. Each molecule acts independently of every other in the gas sample. Therefore, it does not matter whether the molecules are the same or different; that is, the Ideal Gas Law holds for a mixture of gases as it does for a single substance. Thus,

$$P_t V = n_t RT \tag{5–8}$$

where P_t is the total pressure resulting from a total of n_t moles of gas in a volume V at a temperature T.

Since the molecules of a mixture of ideal gases act independently, we can define a **partial pressure** for each kind of molecule from the number of moles of each that are present. Thus, for a mixture of gases A, B, and C, the partial pressures of the three gases are:

$$p_A = \frac{n_A RT}{V}; \qquad p_B = \frac{n_B RT}{V}; \qquad p_C = \frac{n_C RT}{V} \tag{5–9}$$

where n_A, n_B, and n_C are the corresponding numbers of moles of the components; V is the volume of the vessel in which the mixture is contained; and T is the temperature. If the mixture has more components, there will be a similar expression for each component.

FIGURE 5–15   Hot air ballooning.

The total number of moles of gas n_t must be the sum of the numbers of moles of the individual gaseous components in the mixture; for the three-component mixture,

$$n_t = n_A + n_B + n_C \qquad (5\text{--}10)$$

Because ideal gas behavior is assumed, each term in this expression can be replaced by its equivalent from Eq. (5–9):

$$\frac{P_t V}{RT} = \frac{p_A V}{RT} + \frac{p_B V}{RT} + \frac{p_C V}{RT}$$

Finally, dividing each term by V/RT:

$$P_t = p_A + p_B + p_C$$

This means that the total pressure P_t of a mixture of gases is equal to the sum of the partial pressures $(p_A + p_B + p_C)$ of the gases in the mixture. This is a statement of **Dalton's law of partial pressures**.

A useful form of this law can be obtained from the ratio of p_A to P_t:

$$\frac{p_A}{P_t} = \frac{n_A RT/V}{n_t RT/V} = \frac{n_A}{n_t}$$

The ratio n_A/n_t is the **mole fraction** of A, the fraction of the total number of moles contributed by component A in the mixture. The symbol used to represent mole fraction is X_A:

$$p_A = X_A P_t \qquad (5\text{--}11)$$

Similarly,

$$p_B = X_B P_t; \qquad p_C = X_C P_t$$

This simple statement, that the partial pressure of a gas in a mixture is equal to the product of its mole fraction and the total pressure, is a very useful idea that appears in a variety of calculations.

EXAMPLE 5–9

A gaseous mixture of oxygen, nitrogen, and hydrogen has a total pressure of 1.50 atm and contains 5.00 g of each gas. Find the partial pressure of oxygen in this mixture.

Solution   First find the total number of moles:

$$n_{O_2} = \frac{5.00 \text{ g}}{32.00 \text{ g/mol}} = 0.156 \text{ mol}$$

$$n_{N_2} = \frac{5.00 \text{ g}}{28.01 \text{ g/mol}} = 0.179 \text{ mol}$$

$$n_{H_2} = \frac{5.00 \text{ g}}{2.02 \text{ g/mol}} = 2.48 \text{ mol}$$

$$n_t = n_{O_2} + n_{N_2} + n_{H_2}$$

$$= 0.156 \text{ mol} + 0.179 \text{ mol} + 2.48 \text{ mol}$$

$$= 2.82 \text{ mol}$$

Consequently, the mole fraction of oxygen is

$$X_{O_2} = \frac{n_{O_2}}{n_t} = \frac{0.156 \text{ mol}}{2.82 \text{ mol}} = 0.0553$$

Using Eq. (5–11):

$$p_{O_2} = P_t X_{O_2} = (1.50 \text{ atm})(0.0553) = 0.0830 \text{ atm}$$

EXERCISE 5–9

A gaseous mixture with a total pressure of 0.750 atm contains 1.00 g of CO and 2.00 g of CO_2. What is the partial pressure of CO_2 in this mixture? *Answer:* 0.420 atm.   ∎

5.8   REAL GASES AND IDEAL BEHAVIOR

The Ideal Gas Law, $PV = nRT$, is also called the **equation of state for an ideal gas**. This equation provides us with limiting *P-V-T* values for real gases. Reasonable agreement is observed if the pressure is low—generally about one atmosphere or less—and the temperature is high—usually well above the condensation point for the gas. The Ideal Gas Law may be rearranged as follows:

$$\frac{PV}{nRT} = 1$$

For a gas that does not follow the Ideal Gas Law,

$$\frac{PV}{nRT} \neq 1$$

The quantity PV/nRT is called Z, the **compressibility factor**, and it has been used in many instances to show deviations from ideal behavior among real gases. Figure 5–16 is a plot of the compressibility factor Z against P for a number of gases. Clearly, deviations from ideal behavior are considerable at high pressures.

Real gases fail to satisfy two essential criteria for ideal gases: **(1) The molecules of a real gas occupy space**. They have an inherent volume, although it is usually small compared to the total volume occupied by the gas. **(2) Attractive forces exist between molecules**. That is shown by the tendency for gases to change to liquids (and solids) under suitable conditions.

The **van der Waals equation**, probably the most often used equation of state for real gases, corrects the Ideal Gas Law by taking both of these factors into consideration. P_{CORR} and V_{CORR} are the corrected values of pressure and volume.

$$P_{CORR} V_{CORR} = nRT \tag{5–12}$$

The new value for V takes into account the actual volumes of the individual gas molecules. Thus, the volume of space available for any gas molecule will be the measured volume of the container minus the volume occupied by all the other gas molecules. If the volume of the molecules in one mole of gas is b L/mol, then for n moles of gas

$$V_{CORR} = V - nb \tag{5–13}$$

where V is the measured volume of the container in liters. The constant b depends on the nature of the gas and is approximately equal to the volume of one mole of gas molecules when condensed to a liquid.

The pressure P for a real gas will be lower than predicted for an ideal gas as a result of attractions between molecules. The attractive forces be-

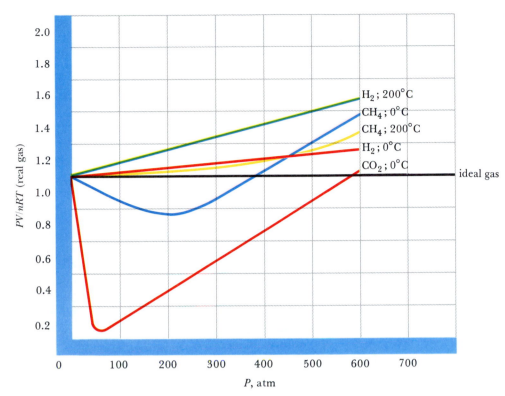

FIGURE 5-16 A plot of
PV/nRT versus *P* for a number of real gases, showing
deviations from ideal behavior at high pressures.

tween molecules tend to cause deviations in their otherwise straight-line
paths as they move about. This reduces the frequency of impacts with the
walls of the container, resulting in diminished pressure. The effect on a
single gas molecule is proportional to the number of gas molecules N in a
unit volume V, or N/V. The total effect is N/V times the number of gas
molecules per unit volume, or $(N/V)^2$. Now for n moles of gas there are
nN_A molecules, where $N_A = 6.022 \times 10^{23}$ molecules per mole, and so the
attractive forces between molecules that effectively reduce the pressure are
proportional to

$$\frac{n^2 N_A{}^2}{V^2}$$

If the proportionality constant is combined with the constant $N_A{}^2$ to give
a new constant a, then the corrected value for the pressure P is

$$P_{\text{CORR}} = \left(P + \frac{n^2 a}{V^2} \right) \tag{5-14}$$

Thus, a correction has been added to the observed pressure P to account
for the reduction in pressure caused by attractions between gas molecules.
Substitution of V_{CORR} and P_{CORR} into Eq. (5-12) gives the van der
Waals equation:

$$\left(P + \frac{n^2 a}{V^2} \right)(V - nb) = nRT \tag{5-15}$$

Note what happens as V becomes larger:

P_{CORR} approaches P

V_{CORR} approaches V

and the van der Waals equation of state approaches and eventually becomes
the same as the equation of the state for an ideal gas. Because V will be

A number of equations of state have been
devised that more accurately describe the
behavior of real gases and their deviations from ideal behavior. But the price
that must be paid for a better fit between
theory and practice is a loss of simplicity.

TABLE 5–5   van der Waals Constants for Some Common Gases

Gas	$a, \dfrac{L^2 \cdot atm}{mol^2}$	$b, \dfrac{L}{mol}$
Ar	1.345	0.03219
CO_2	3.592	0.04267
Cl_2	6.493	0.05622
He	0.03412	0.02370
H_2	0.2444	0.02661
N_2	1.390	0.03913
O_2	1.360	0.03183

relatively large at high values of T and low values of P, this result is equivalent to our earlier statement that real gases approach ideal behavior at high temperatures and low pressures. Values for the van der Waals constants a and b for some common gases are listed in Table 5–5. They provide useful information about the nature of gases themselves. For example, the constant b (in liters per mole) is related to the actual sizes of gas molecules, allowing us to infer that Cl_2 molecules are the largest (in the list) while He atoms and H_2 molecules are the smallest.

Because the van der Waals equation of state is an approximation, it is only as reliable as its underlying assumption. Is it reasonable to estimate the volume of a gas molecule based on one characteristic constant? You might ask as well whether it is reasonable to consider a molecule as though it were a hard sphere with a fixed volume in the first place. It is generally accepted that the magnitude of the van der Waals constant a reflects the tendency of molecules in a gas to attract one another. Wide variations are clearly evident in Table 5–5. The constant b, which estimates the volume of a gas excluded because of the finite size of the molecules themselves, is roughly four times the volume of a mole of molecules as measured by other techniques.

EXAMPLE 5–10

Determine the pressure of 1.00 mol of helium and chlorine, stored in separate 1.00-L containers at 323 K.

Solution   For an ideal gas:

$$P = \frac{nRT}{V} = \frac{(1.00 \text{ mol})(0.0821 \text{ L·atm/mol·K})(323 \text{ K})}{1.00 \text{ L}} = 26.5 \text{ atm}$$

For a real gas described by the van der Waals equation:

$$\left(P + \frac{n^2 a}{V^2}\right)(V - nb) = nRT$$

$$P = \frac{nRT}{V - nb} - \frac{n^2 a}{V^2}$$

For one mole of helium,

$$P = \frac{(1.00 \text{ mol})(0.0821 \text{ L·atm/mol·K})(323 \text{ K})}{1.00 \text{ L} - (1.00 \text{ mol})(0.02370 \text{ L/mol})}$$

$$- \frac{(1 \text{ mol})^2 (0.03412 \text{ L}^2 \cdot atm/mol^2)}{(1.00 \text{ L})^2}$$

$$= 27.1 \text{ atm}$$

REFRIGERATION AND THE JOULE-THOMSON EFFECT

One of the spin-offs of van der Waals' work was the discovery that the Joule-Thomson effect (that gases cool when allowed to expand) holds true only below a certain characteristic temperature for every gas. The Joule-Thomson effect is an essential factor in modern refrigeration and air-conditioning. For most gases, the characteristic "Joule-Thomson temperature" is high enough to allow cooling by expansion. However, this is not the case for hydrogen and helium. For a long time, these two were considered to be permanent gases because they could not be conveniently liquefied by carrying them through a series of expansions. Their temperatures had to be first lowered to the Joule-Thomson temperature by other methods. That was finally accomplished in the period between 1900 and 1910 by James Dewar and Heike Kamerlingh-Onnes.

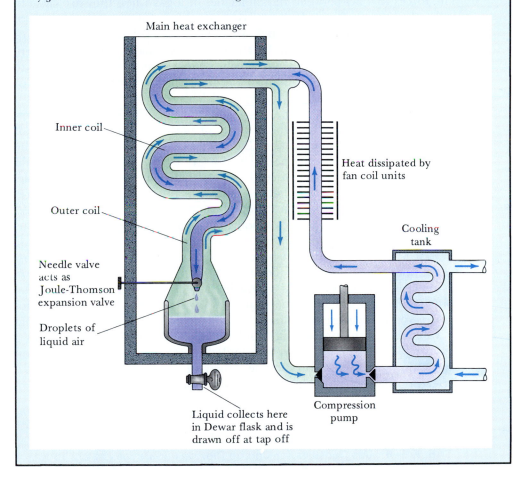

Main heat exchanger

Inner coil

Outer coil

Needle valve acts as Joule-Thomson expansion valve

Droplets of liquid air

Liquid collects here in Dewar flask and is drawn off at tap off

Heat dissipated by fan coil units

Cooling tank

Compression pump

For one mole of chlorine,

$$P = \frac{(1.00 \text{ mol})(0.0821 \text{ L·atm/mol·K})(323 \text{ K})}{1.00 \text{ L} - (1.00 \text{ mol})(0.05622 \text{ L/mol})}$$
$$- \frac{(1 \text{ mol})^2(6.493 \text{ L}^2\text{·atm/mol}^2)}{(1.00 \text{ L})^2}$$

$$= 21.6 \text{ atm}$$

As shown by this example, depending on the values of the constants a and b, the van der Waals equation may predict a pressure that is greater or less than that predicted by the Ideal Gas Law. Note also that these calculations agree with our intuitive expectation that He is more nearly ideal than Cl_2. ■

EXERCISE 5–10

Using the van der Waals equation of state, calculate the pressure exerted by 100 moles of Cl_2 gas in a 20.0-L tank at 25°C.
Answer: 7.84 atm. ■

5.9 AVOGADRO'S HYPOTHESIS AND REAL GASES

The forces that act between the molecules of a gas when they are close to each other are different for various gases. Therefore, equal volumes of gases do not contain exactly equal numbers of molecules, and the name of Avogadro's Hypothesis should be changed to Avogadro's approximation. However, that should not be surprising. Nor should we be surprised to learn that as gas pressures diminish and the space between the particles of a gas becomes increasingly large, Avogadro's approximation becomes a very good approximation. Finally, at limiting values of pressure near zero, gas densities (and therefore the number of particles in a given volume of a gas) are indeed proportional to the atomic or molecular weights of the gaseous particles. When the particles of a gas are far apart, as they would be near zero pressure, the forces that exist between them are negligible. Therefore, the density of the gas should be proportional to the pressure.

If the ratio of the gas density d and the pressure P is a linear function of pressure, then linear extrapolation can be used to extend the d/P versus P plot to the limiting value of zero pressure. This gives the value for the gas density under conditions where it can be considered to be an ideal gas. Under those conditions Avogadro's approximation becomes Avogadro's Hypothesis, and one can say with a high degree of certainty that equal volumes of gases contain equal numbers of particles. Very accurate molecular weight measurements have been made on that basis. From the data in Table 5–6, a plot of d/P versus P for CO_2 (Figure 5–17) results in a nearly straight line which, when extended to zero pressure, gives the limiting ratio (d/P) for carbon dioxide. Since $d/P = M/RT$, the value of d/P for carbon dioxide gives the limiting value for the molecular weight:

TABLE 5–6 Gas Densities for Carbon Dioxide and Oxygen at 273.16 K				
	Carbon dioxide		**Oxygen**	
P, atm	**Density, g/L**	**Ratio *d/P***	**Density, g/L**	**Ratio *d/P***
1	1.97676	1.97676	1.42896	1.42896
$\frac{3}{4}$	—	—	1.07149	1.42865
$\frac{2}{3}$	1.31485	1.97228	—	—
$\frac{1}{2}$	0.98504	1.97008	0.71415	1.42830
$\frac{1}{3}$	0.65596	1.96788	—	—
$\frac{1}{4}$	0.49169	1.96676	0.35699	1.42796
$\frac{1}{6}$	0.32761	1.96566	—	—
—	—	—	—	—
0	—	1.96346	—	—

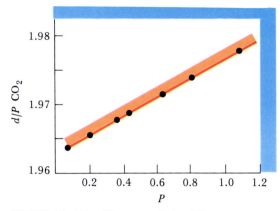

FIGURE 5–17 d/P versus P for CO_2 extrapolated to $P = 0$

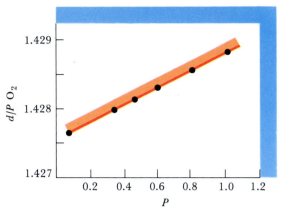

FIGURE 5–18 d/P versus P for O_2 extrapolated to $P = 0$.

$$1.96346 \text{ g/L·atm} = \frac{M}{(0.08206 \text{ L·atm/mol·K})(273.15 \text{ K})}$$

$$M = 44.01 \text{ g/mol}$$

Similar data for O_2 (Figure 5–18) show its limiting ratio (d/p) to be 1.4277 g/L·atm. Since the ratio of the limiting d/p values would be the ratio of their molecular weights, the molecular weight for CO_2 can be computed directly from the known value for O_2:

$$\left(\frac{1.9635 \text{ g/L·atm}}{1.4277 \text{ g/L·atm}}\right) (31.999 \text{ } \mu) = 44.008 \text{ } \mu$$

This method is particularly suited to determining atomic weights for the noble gas elements, since they do not usually combine with other elements to form compounds.

EXAMPLE 5–11
For neon, the limiting ratio of d/P is 0.90043 g/L·atm at 273.1 K. To calculate the atomic weight of neon, proceed as follows:

$$\left(\frac{0.90043 \text{ g/L·atm}}{1.4277 \text{ g/L·atm}}\right) (31.999 \text{ } \mu) = 20.181 \text{ } \mu$$

EXERCISE 5–11(A)
Using data for oxygen from Example 5–11, calculate the atomic weight of argon from the limiting ratio d/P at 273.1 K, which is known to be 1.78204 g/L·atm.
Answer: 39.94 μ.

EXERCISE 5–11(B)
Using the following data (to plot d/P versus P), find the limiting ratio of d/P for phosphine (PH_3) at 273.15 K.

P, atm	d/P, g/L·atm
1.0000	1.5307
0.7500	1.5272
0.5000	1.5238
0.2500	1.5205

Then, using the limiting ratio of d/P for oxygen, calculate the molecular weight of PH_3. Finally, taking the atomic weight of hydrogen to be 1.008, calculate the atomic weight of phosphorus.
Answer: 34.00 μ; 31.0 μ. ■

5.10   THE KINETIC THEORY OF GASEOUS BEHAVIOR

To this point in our discussion, we have been looking at the macroscopic properties of gases—pressure, volume, and temperature. Our microscopic model for the behavior of ideal gases is called the **kinetic theory**. Here are the essential features of the model:

1. An ideal gas consists of infinitely small particles. The particles of a gas have mass but no volume. We know they have mass, because there is ample evidence that the bulk sample of the gas has mass. Therefore, each particle of an ideal gas is assumed to have its mass concentrated at a single point in space.

2. Particles of an ideal gas are separated from each other by relatively long distances, so that the volume containing an ideal gas can be considered empty space. Support for this assumption lies in the great compressibility of gases. If the particles of a gas were close together, the compressibility would be limited by contacts among the gas particles.

3. Ideal gas particles are in constant motion. The basis for this assumption is evident in the very short time required for a gas to fill an evacuated vessel into which it is admitted. Another property consistent with this assumption is their ability to diffuse and effuse. **Diffusion**, defined as the spreading of molecules of one substance through another substance, is a matter of common experience; for example, a strong scent liberated in one part of a room is quickly evident throughout. **Effusion** is also a matter of experience; it is the process by which molecules of a gas escape a small opening in a container. Further evidence for this continual movement of the gas particles is the rapid mixing of gases in homogeneous mixtures and the failure of gas mixtures to separate completely into their lighter and heavier components.

4. The ceaseless, random motion that is characteristic of the behavior of the particles of a gas leads to collisions between particles and with the walls of the container. In these collisions, the energy of motion (the kinetic energy of the particles) is conserved. That is, the average velocity of the gas particles does not increase or decrease as a result of the collisions. In the terminology of physics, the collisions between the particles of an ideal gas are perfectly elastic.

> If collisions did alter the average kinetic energy, then the motion of the particles would eventually stop and the substance would cease to be a gas. This, of course, is not observed.

5. Pressure is the result of collisions of the particles with the walls of the container. Pressure is force per unit area, and force is mass times acceleration. Acceleration is change in speed with time, and thus

$$f = ma = m\frac{\Delta v}{t}$$

where v is speed and t is time. We can rewrite the equation to read

$$f = \frac{\Delta(mv)}{t} \tag{5-16}$$

and interpret it as saying that force is change of momentum (mv) with time. The force per unit area, or pressure experienced by the walls of the container, is the result of the transfer of momentum from the gas particles that strike it.

The pressure of the gas is proportional to the number of collisions occurring per unit time—the frequency of collisions—and the momentum transferred per collision. The frequency of collisions is proportional to an average speed of the gas particles, which we call v, and the number of particles per unit volume. For one mole of gas per unit volume, the number of particles is N_A, the Avogadro number, and P is proportional to

$$(v)\left(\frac{N_A}{V}\right)(mv) = \frac{N_A mv^2}{V}$$

where m is the mass of an individual molecule. The proportionality constant can be shown to be $\frac{1}{3}$, and

$$P = \frac{N_A mv^2}{3V}$$

which can be rearranged to

$$PV = \frac{N_A mv^2}{3} \tag{5-17}$$

The kinetic energy of a body moving from one point to another (translation) is $\frac{1}{2}mv^2$, and this term can be factored out of Eq. (5-17):

$$PV = \tfrac{2}{3}(N_A)\left(\frac{mv^2}{2}\right) = \tfrac{2}{3}N_A(KE)$$

where KE is the average kinetic energy of the gas molecules. Substitution into the Ideal Gas Law gives

$$PV = \tfrac{2}{3}N_A(KE) = RT$$

and

$$KE = \frac{3RT}{2N_A} \tag{5-18}$$

The quantity R/N_A is called the **Boltzmann constant** k, the gas constant per molecule. Making this substitution gives the important result—the average kinetic energy of a single particle of an ideal gas is proportional to the absolute temperature:

$$KE = \frac{3kT}{2} \tag{5-19}$$

The second major assumption made for an ideal gas is that the particles neither attract nor repel each other. No intermolecular forces exist! Certain conclusions about the internal energy E of an ideal gas follow:

1. Attractive or repulsive forces between objects can result in potential energy (PE). For example, a brick held above the ground has potential energy due to its position with respect to the Earth and the Earth's gravitational field. When the brick is released, the potential energy is converted to kinetic energy as the brick starts to fall. The **internal energy** E of an object is the sum of the potential energy and the kinetic energy:

$$E = PE + KE \tag{5-20}$$

2. Since the particles of an ideal gas have no attractions or repulsions for each other, they can have no potential energy—no energy related to position. Thus, the internal energy for an ideal gas molecule is equal to its kinetic energy, and for the average molecule this is

$$E = KE = \tfrac{3}{2}kT$$

3. For a mole of gas, the total internal energy is the average internal energy per particle times N_A, the number of particles per mole, or

$$E = \tfrac{3}{2}N_A kT$$

which simplifies to

$$E = \tfrac{3}{2}RT$$

4. For n moles of gas this becomes

$$E = \tfrac{3}{2}nRT \tag{5-21}$$

Thus, for a fixed amount of an ideal monatomic gas, the internal energy depends only on the temperature.

EXAMPLE 5–12

Calculate the internal energy of 100.0 g of neon at 50°C (assuming ideal gas behavior).

Solution   Begin with Eq. (5–21) and proceed as follows. First, determine the number of moles of neon and convert the temperature to kelvins. For Ne, $M = 20.18$ g/mol, so

$$n = \frac{100.0 \text{ g}}{20.18 \text{ g/mol}} = 4.955 \text{ mol}$$

$$E = \frac{3nRT}{2} = \frac{(3)(4.955 \text{ mol})(0.0821 \text{ L·atm/mol·K})(323 \text{ K})}{2}$$

$$= 197 \text{ L·atm}$$

Note that the energy units are L·atm, or pressure times volume. The standard SI unit of energy is the joule, and the customary unit of energy is the calorie. Proper selection of R, the universal gas constant, gives the answer with the desired units. The conversions are straightforward:

$$
\begin{aligned}
1 \text{ L·atm} &= (1 \text{ L})(1.01325 \times 10^5 \text{ Pa}) \\
&= (10^{-3} \text{ m}^3)(1.01325 \times 10^5 \text{ kg/m·s}^2) \\
&= 1.01325 \times 10^2 \text{ kg·m}^2\text{/s}^2 \\
&= 101.325 \text{ J} \\
&= 24.22 \text{ cal}
\end{aligned}
$$

Therefore,

$$R = (0.08206 \text{ L·atm/mol·K})(24.22 \text{ cal/L·atm}) = 1.987 \text{ cal/mol·K}$$
$$R = (0.08206 \text{ L·atm/mol·K})(101.325 \text{ J/L·atm}) = 8.314 \text{ J/mol·K}$$

Returning to our original calculation for E, we can show that

$$E = \frac{3nRT}{2}$$
$$= \frac{3(4.955 \text{ mol})(8.314 \text{ J/mol·K})(323 \text{ K})}{2}$$
$$= 2.00 \times 10^4 \text{ J}$$

Some values of R with various units are given in Table 5–8.

TABLE 5–7	Values of R, the Ideal Gas Constant
Value of R	**Units**
0.08206	L·atm/mol·K
8.314	J/mol·K
1.987	cal/mol·K
8.314×10^7	erg/mol·K

EXERCISE 5–12(A)

What is the internal energy of 150.0 g of helium gas at 400. K, assuming ideal behavior? Express your answer in L·atm, cal, and J.
Answer: 1.83×10^3 L·atm; 4.44×10^4 cal; 1.86×10^5 J.   ■

EXERCISE 5–12(B)

Calculate the temperature of a mole of oxygen molecules if the internal energy is known to be 1.16×10^4 J; if the energy is known to be 2.77×10^3 cal; if the energy is 114 L·atm. Assume ideal gas behavior.
Answer: 930 K; 929 K; 926 K.   ■

The Average Kinetic Energy and the Distribution of Molecular Speeds

The arguments of kinetic theory in the preceding section have their place, but it is a matter of common experience that there is a relationship between temperature and motion. Not only does a piece of metal become warm, and then hot, on being bent back and forth, but simple friction created by two sticks can start a fire, and warmth is created by just rubbing your hands together.

In a gas, some particles travel faster than others. As particles collide with each other and with the walls of the container, some speed up and others slow down. The relationship between temperature and the speeds of the particles can be determined from Eq. (5–19) and the expression for the average kinetic energy of the ideal gas particles:

$$\overline{KE} = \frac{m\overline{v}^2}{2} = \frac{3kT}{2}$$

where the overbar signifies an average quantity. Solving for $\overline{v}^2$ gives

$$\overline{v}^2 = \frac{3kT}{m}$$

Taking the square root of both sides of the equation gives

$$\sqrt{\overline{v}^2} = \sqrt{\frac{3kT}{m}} = \sqrt{\frac{3RT}{M}} \qquad (5-22)$$

This equation determines $\sqrt{\overline{v}^2}$, the square root of the average of the squares of the speeds of the particles, which is called the **root mean square speed** or v_{rms}. It is not identical to the average speed $\overline{v}$, although the two are close in value and often considered to be identical for practical purposes. The m in Eq. (5–22) is the mass of an individual gas particle, or M/N_A, where M is the molecular weight (or the atomic weight) of the gas. For an ideal gas, v_{rms} depends only on its molecular weight and the temperature.

EXAMPLE 5–13

Determine the root mean square speeds for hydrogen and chlorine gases at 25°C, assuming ideal gas behavior.

Solution Begin by making the necessary temperature conversion and then apply Eq. (5–22) with suitable units for R.

$$T = (25 + 273) \text{ K} = 298 \text{ K}$$

Use of R in units of erg/mol·K gives speeds in cm/s. Note the conversion factor:

$$1 \text{ erg} = 1 \text{ g·cm}^2/\text{s}^2$$

For hydrogen gas, $M = 2.02$ g/mol. Substituting into the equation:

$$v_{rms} = \sqrt{\frac{(3)(8.314 \times 10^7 \text{ g·cm}^2/\text{s}^2\text{·mol·K})(298 \text{ K})}{2.02 \text{ g/mol}}}$$

$$= \sqrt{3.68 \times 10^{10} \text{ cm}^2/\text{s}^2}$$

$$= 1.92 \times 10^5 \text{ cm/s}$$

Similarly, for chlorine ($M = 70.91$ g/mole),

$$v_{rms} = \sqrt{\frac{(3)(8.314 \times 10^7 \text{ g·cm}^2/\text{s}^2\text{·mol·K})(298 \text{ K})}{70.91 \text{ g/mol}}}$$

$$= 3.24 \times 10^4 \text{ cm/s} \qquad \blacksquare$$

EXERCISE 5–13
Determine v_{rms} for Ar gas at 200 K.
Answer: 3.53×10^4 cm/s. ■

To show the difference between the average of several numbers and the root mean square of the numbers, consider the following set:

$$2.2, \quad 2.5, \quad 2.8, \quad 3.0, \quad 3.2, \quad 3.5, \quad 3.8$$

• The sum of the seven members of the set is 21.0, and the average is 3.00.
• The sum of the squares is 64.86, and the average of the squares is 9.266.
• The square root of this last number—the root mean square—is 3.044.

Note that the root mean square accentuates the higher values and is slightly higher than the average.

Calculations performed using Eq. (5–22) give the root mean square speed v_{rms} of the gas particles at a particular temperature, and this value will be close to the average speed of the particles. However, all particles are not traveling at the same speed, and what we see is a distribution of speeds—some greater than v_{rms} and some less. The distribution of speeds can be expressed graphically as in Figure 5–19, which gives the fraction of the total number of gas particles as a function of their speed. Note that the function is zero at a speed of zero, which in effect means that all of the particles are in motion and that no measurable number of them are standing still for any finite period of time. At somewhat greater kinetic energy and speed, the fraction (of particles) increases and reaches a peak at the most probable speed or v_{mp}. The fraction then decreases at still higher speeds, approaching but never quite reaching zero. Thus, among the huge number of particles in a macroscopic sample, a few can be found having very high speeds. Because there are small numbers at very high speeds but none at zero speed, the average speed $\bar{v}$ is slightly higher than v_{mp}. Furthermore, since v_{rms} gives more weight to higher speeds, it is slightly higher than $\bar{v}$. All of these speeds are shown in Figure 5–19, which is referred to as a **Maxwell-Boltzmann distribution**.

Eq. (5–22) tells us that if the temperature is raised, v_{rms} increases. This will be accompanied by similar increases in v_{mp} and $\bar{v}$, and by a general

James Clerk Maxwell (British, 1831–1879) wrote down sets of equations, which together with Ludwig Boltzmann's kinetic theory form the basis for modern understanding of the behavior of solids and fluids.

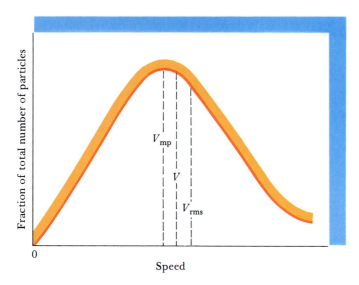

FIGURE 5–19 The distribution of speeds for a collection of gas particles at a given temperature. The most probable speed is V_{mp}, the average speed is V and the root mean square speed is V_{rms}.

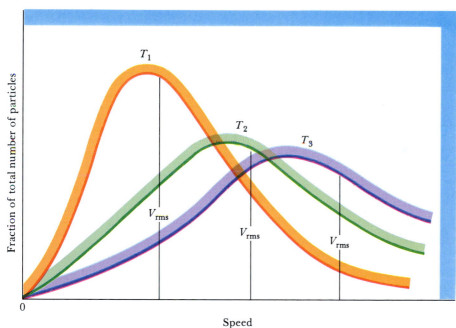

FIGURE 5–20 How the distribution of velocities changes with increasing temperature. $T_3 > T_2 > T_1$.

stretching of the function in Figure 5–19 toward higher speeds. Finally, since the fraction of particles under the curve must be constant, totaling 1.00 for the entire sample, the value of the fraction at v_{mp} must drop as the function is stretched out. Figure 5–20 shows the Maxwell-Boltzmann distribution function at three successively higher temperatures. Eq. (5–22) also tells us that v_{rms} increases with decreasing molecular weight M of the gas. The distribution of speeds for several gases at the same temperature takes the same form as Figure 5–20, except with T replaced by $1/M$.

5.11 GRAHAM'S LAW: EFFUSION AND DIFFUSION

One of the simplest experiments for demonstrating the speeds of particles in a gas is the process of effusion. **Effusion** is said to occur when a gas passes through a hole so tiny that the particles cannot simply pour through.

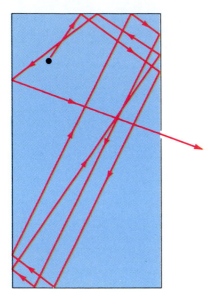

FIGURE 5–21   In effusion, gas particles escape from a small opening in the wall of the container.

The mechanism for effusion, shown in Figure 5–21, amounts to little more than repeated rebounds off the container walls and other particles until the tiny opening is finally found. Every particle will trace its own unique course to the opening, but there is an average path length $\bar{a}$ for all of the particles that depends upon the shape of the container and the size of the opening. The average time for a gas particle to effuse through the hole will be $\bar{t} = \bar{a}/\bar{v}$, and the rate of effusion will be proportional to $\bar{v}$ and v_{rms}. For two different gases A and B at the same partial pressure in a container:

$$\text{rate of effusion of A} \propto v_{rms} = \sqrt{\frac{3RT}{M_A}}$$

$$\text{rate of effusion of B} \propto v_{rms} = \sqrt{\frac{3RT}{M_B}}$$

Taking the ratio of the rates gives the following result:

$$\frac{\text{rate of effusion of A}}{\text{rate of effusion of B}} = \frac{v_{rms}(A)}{v_{rms}(B)} = \frac{\sqrt{3RT/M_A}}{\sqrt{3RT/M_B}}$$

$$\frac{\text{rate of effusion of A}}{\text{rate of effusion of B}} = \frac{\sqrt{M_B}}{\sqrt{M_A}} \qquad (5\text{–}23)$$

This is called **Graham's Law** of effusion.

EXAMPLE 5–14

To calculate the relative rate of effusion of hydrogen gas compared to nitrogen gas at 25°C, use the relationship in Eq. (5–23).

$$\frac{\text{rate}_{H_2}}{\text{rate}_{N_2}} = \sqrt{\frac{M_{N_2}}{M_{H_2}}} = \sqrt{\frac{28.0 \text{ g/mol}}{2.02 \text{ g/mol}}} = 3.72 \qquad \blacksquare$$

EXERCISE 5–14

He and NO_2 are allowed to effuse through a porous barrier. Calculate their relative rates of effusion.

Answer: He effuses 3.4 times as fast as NO_2. Note that the temperature is irrelevant to the relative rates. ∎

Diffusion is similar to effusion. Gases diffuse into each other when they mix. If a sample of an easily detectable gas such as chlorine is released in one part of a large room, its presence will soon be detectable throughout as the chlorine diffuses through the air. If the chlorine could spread directly throughout the room at the speed calculated in Example 5–13, it would take only about one-tenth of a second to traverse an auditorium 40 meters long! It does not diffuse that fast because each Cl_2 molecule collides with many other molecules along its path. The path is irregular and very much longer than the straight-line distance across the room. However, the diffusion process is still rapid. As for effusion, rates of diffusion are directly proportional to the average speeds of the particles and inversely proportional to the square roots of their masses.

The high rates of effusion and diffusion of gases result in very rapid mixing and often very fast chemical reactions between gases. The gasoline-powered internal combustion engine depends on many reactions taking place in the cylinders every second. The reactants, oxygen from the air and vaporized gasoline, are both gases so they mix and react very rapidly.

Gaseous insecticides and fumigants such as methyl bromide (CH_3Br) rapidly permeate all of an enclosed space.

We mentioned in Chapter 3 that the relative rates of diffusion of different gases are used in the enrichment of uranium for fission processes. Naturally occurring uranium consists of 99.28% ^{238}U and only 0.72% ^{235}U. The ^{235}U content must be raised by about an order of magnitude to make a useful reactor fuel. Fortunately, uranium forms a volatile fluoride, UF_6, and the slight mass of difference between $^{235}UF_6$ and $^{238}UF_6$ means that the former will diffuse through a porous barrier slightly more quickly. Successive diffusion stages eventually result in enriched uranium.

5.12   COMPRESSIBILITY, COMPRESSED GASES, AND COMPRESSORS

The behavior of gases depends on their unique characteristics in general: compressibility, low density, high rates of effusion and diffusion. Of these, compressibility may be the most obviously important. Compressed air is an especially interesting illustration. It is used as an inexpensive, convenient, and spark-free form of power transmission in every imaginable industry:

- For driving pneumatic tools that produce torque or rotary motion for drills and drivers for screws and nuts
- For devices producing linear motion by means of diaphragms, bellows, and pistons for operating clamps, presses, and automatic feed devices
- For moving or breaking up solid masses with jackhammers or pile drivers.

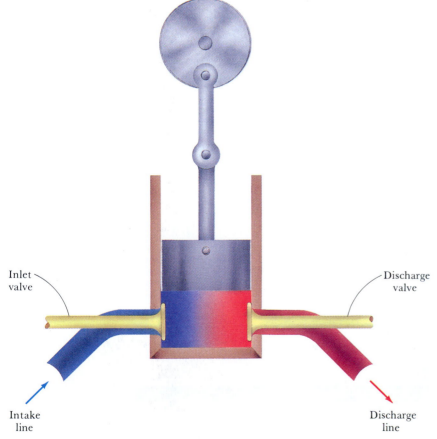

Inlet valve

Discharge valve

Intake line

Discharge line

FIGURE 5–22   A one-stage reciprocating compressor.

Compressed air is used in chemical processes for mixing and agitating liquids. It is also used to prevent icing of dams and channels by moving subsurface waters, which are above the freezing point, to the surface.

Many kinds of air compressors have been constructed. Figure 5–22 is a schematic diagram of a simple one-stage reciprocating compressor. It consists of an intake line, inlet valve, piston and cylinder, discharge valve and discharge line, and the necessary reciprocating linkages. When the gas in the cylinder is compressed, its temperature rises; this lowers the efficiency of the process, since the air pressure will drop when the air cools again. Therefore, effective cooling of the cylinder is important.

Boyle's Law: $PV = k$

The pressure versus volume cycle for this compressor is shown in Figure 5–23. Let us assume that cooling is totally effective, that the temperature is constant, and that Boyle's Law holds. At point 1, the piston is positioned so that the cylinder volume is at a minimum, both valves are closed, and the intake stroke begins. The piston then begins to move, increasing the volume in the cylinder, and the pressure of the air decreases according to Boyle's Law until point 2 is reached. At this point the pressure in the cylinder is equal to the pressure of the intake line, and the inlet valve opens. The piston continues to increase the volume of the cylinder, but now the pressure remains constant at the intake line pressure until point 3 is reached and the inlet valve closes. The piston then begins its compression and discharge stroke. The volume is decreased and the pressure increases according to Boyle's Law until point 4 is reached, at which the pressure in the cylinder is equal to that in the discharge line. The discharge valve then opens and the piston continues decreasing the volume in the cylinder at the discharge line pressure. At the end of this stroke, point 1 is reached again and the discharge valve closes.

From a graph such as Figure 5–23 we can make two useful calculations. First, the pressure at the discharge line 4-1 divided by the pressure at the

FIGURE 5–23 *P* versus *V* curve for the operation of a one-stage reciprocating compressor. A constant temperature resulting from ideally effective cooling is assumed.

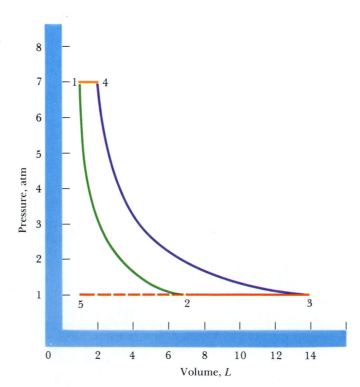

intake line 2-3 is the **compression ratio** of the machine. This is fixed by the compressor's geometry. Second, the efficiency of the compressor—often called volumetric efficiency—is the fraction of the intake stroke in which air is actually being drawn into the cylinder (part 2–3 of the cycle). Since the entire intake stroke of the piston can be represented by a line from point 5 to point 3, then the compressor efficiency can be given as the length of line 2–3 divided by the length of the total line 5–3.

Because the compression ratio of a compressor depends on its design, the pressure of intake air is significant. Suppose we are operating an air compressor at a high altitude, say in Cheyenne, Wyoming, which is 6062 ft above sea level and where the atmospheric pressure is only 609 torr. If the inlet air is at this pressure and the compression ratio of our compressor is 5.0, then it can deliver air at 5×609 torr. This means that our compressor can deliver only $609/760 = 0.801$ as much air in Cheyenne as it can at sea level.

The relationship between standard atmospheric pressure (760 torr) and altitude is approximated by the equation

$$H = 17000 \left(\frac{760 - P}{760 + P} \right) \tag{5–24}$$

where H is the altitude in meters and P is the pressure at that altitude in torr. The performance factor F at altitude H is the ratio of the pressure P divided by the standard sea level pressure of 760 torr if the temperature at both places is 25°C. That is,

$$F = \frac{P}{760} \tag{5–25}$$

SUMMARY

Atmospheric pressure—measured as force per unit area—is due to the mass of the atmosphere being acted upon by the Earth's gravity, and is commonly measured with a barometer. A gas in a closed container can have a pressure different from atmospheric pressure, and this pressure is measured with a manometer.

The **equation of state for an ideal gas** (also called the **Ideal Gas Law**) is $PV = nRT$, where R is the ideal gas constant. It contains the **Boyle's Law** relationship, that the PV product is a constant (at constant temperature), and the **Charles's Law** relationship, that V is directly proportional to T (at constant pressure). The laws described by this equation apply perfectly only for a hypothetical ideal gas, one in which the individual molecules occupy no space and do not attract or repel each other. These requirements are never fully realized, but ideal behavior is approximated at relatively low pressures and relatively high temperatures. Based on the equation of state for an ideal gas and the **Dumas method**, it is possible to obtain molecular weights by making density measurements.

For mixtures of gases, the Ideal Gas Law applies just as it does for pure gases. The **partial pressure** of a gas in a mixture can be calculated from the mole fraction and the total pressure:

$$p_A = P_t X_A$$

where p_A is the partial pressure of gas A, X_A is the mole fraction of gas A, and P_t is the total pressure of the gas mixture. **Dalton's law of partial**

pressures states that the total pressure of a gas mixture is equal to the sum of the partial pressures of all the constituent gases.

The **van der Waals** equation of state includes corrections to the Ideal Gas Law in order to obtain more accurate results for real gases. The van der Waals equation takes the form

$$\left(P + \frac{n^2 a}{V^2} \right) (V - nb) = nRT$$

where a and b are constants specific for each gas. The constant a makes a correction for the forces of attraction between gas molecules, and the constant b makes a correction for the actual volume of the individual particles in a real gas. The van der Waals equation provides reasonably accurate results under moderate conditions.

A fact that has proved very useful in obtaining accurate values for atomic and molecular weights by the method of limiting densities is that real gases behave ideally at very low pressures.

The Ideal Gas Law and the van der Waals equation deal with easily observable macroscopic properties of gases, whereas the kinetic molecular theory is a model for the behavior of gases at the microscopic level. The model begins with a collection of particles, each with its mass concentrated at a point in space. These particles are at relatively large distances from each other (in what is essentially empty space) and are in constant motion. No attractive or repulsive forces exist between particles, and collisions between particles or with the container walls are perfectly elastic (there is total conservation of kinetic energy as well as momentum). The phenomenon of pressure results from the collision of particles with the walls of the container. Using these premises and a few simplifying assumptions, it is possible to show that the PV product is a constant (Boyle's Law) and that for n moles of an ideal gas

$$E = KE = \frac{3nRT}{2}$$

where E is the total internal energy of the ideal gas, which is equal to its kinetic energy. It is also possible to determine the root mean square velocity of the particles in a gas from the equation

$$V_{\text{rms}} = \sqrt{\frac{3RT}{M}}$$

where M is the molecular (or atomic) weight of the particles. The **Maxwell-Boltzmann distribution function** describes molecular speeds among the particles in a collection.

QUESTIONS AND PROBLEMS

QUESTIONS

1. What is the advantage of a closed-ended manometer over an open-ended manometer?

2. Making use of the properties of gases, explain how you would establish that oxygen gas is a collection of diatomic molecules while neon gas is a collection of atoms.

3. Consider the proportionality constant in the Boyle's Law statement given by Eq. (5–4). What is the effect on the constant if the temperature is increased?

4. What advantages would a gas thermometer have over an ordinary mercury thermometer? What disadvantages?

5. Why must T be kelvins in the Ideal Gas Law?

6. Suppose a gas thermometer is constructed according

to Figure 5–6 with a closed-ended mercury manometer attached to a container of a gas. What would be a source of error in these temperature measurements? How could you correct for this error or eliminate it?

7. Describe a gas that would perfectly fit the van der Waals equation. In what respects is this model unrealistic for a real gas?

8. The quantity PV/nRT is defined as the compressibility factor for a gas. For an ideal gas this quantity is, of course, equal to 1. For a real gas the compressibility factor can range in value from less than 1 to greater than 1. Describe conditions under which the compressibility of a real gas would be (a) less than 1; (b) greater than 1.

9. Why is the method of limiting densities especially useful in determining the atomic weight of elements such as He, Ne, and Ar?

10. Briefly state each of the assumptions made in the kinetic theory of gases, and the experimental observations that justify each assumption.

11. Why is the total internal energy of an ideal monatomic gas equal to its kinetic energy?

12. What happens to the speed of gas particles at the absolute zero of temperature?

13. Explain why v_{mp} is less than $\bar{v}$, which in turn is less than v_{rms}.

14. How can effusion be used to separate a mixture of gases?

15. Briefly explain why the rate of diffusion of hydrogen through air is only a few meters per second, even though v_{rms} for H_2 is on the order of 10^5 m/s.

16. Gases do not settle; in other words, if the temperature and volume of a given quantity of an ideal gas are kept constant, the pressure also remains constant. Briefly explain.

17. Which would you expect to display the greater difference from ideal behavior: nitrogen (N_2, b.p. 77 K), or nitrous oxide (N_2O, b.p. 183 K)? Offer a brief explanation.

18. Barometric pressure tends to fall before a rainstorm. Why do you think that happens?

PROBLEMS

Pressure and Temperature   [1–8]

1. The height of the column of liquid bromine pentafluoride (BrF_5) that could be supported by a pressure of 1.33 atm is 5.55 meters at 25°C. What is the density of bromine pentafluoride at that temperature?

2. How high a column of ethyl alcohol (density = 0.789 g/cm³) would be supported by an atmospheric pressure of 785 torr?

3. An open-ended manometer (Figure 5–4) is used to determine the pressure of a gas in a container. The height of the mercury column is 33.1 mm higher in the leg of the U-tube opening into the container than the level in the open leg. Atmospheric pressure, determined in a separate experiment, is 748.4 mm Hg. What is the pressure of the gas in the container?

4. A closed-ended manometer (Figure 5–5) is used to determine the pressure of a gas in a container. The height

of the mercury column is 48.7 mm lower in the leg of the U-tube opening into the container than the level of the closed leg. What is the pressure of the gas in the container?

5. Mercury boils at 630. K. What is its boiling point in °C and °F?

6. In a yet-to-be-discovered world, the principal liquid present is chlorine, not water as on Earth. Therefore, the temperature scale in this different world in °D is based on the melting and boiling points of chlorine: 0°D is −103°C, and 100°D is −35°C. Derive equations relating °D to °C and °F. Also calculate absolute zero in °D.

7. There is a gas trapped in a bulb by the mercury in a manometer as shown in the following sketch. What is the pressure of the gas if atmospheric pressure is 1.00 atm? If atmospheric pressure is 1.00 kPa?

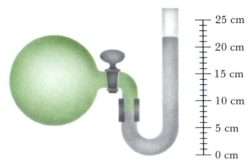

8. There is a gas trapped in a bulb by the mercury in a manometer of the same design used in Problem 7. This time the difference between the heights of the two mercury levels is 47 mm Hg. If barometric pressure at that time is 756 torr, what is the pressure of the gas in the bulb, in kPa?

Equation of State for an Ideal Gas   [9–26]

9. A sample of oxygen gas (O_2) occupies 3.12 L at 1.55 atm pressure and 25°C. Calculate the mass of oxygen (in grams), assuming ideal behavior.

10. At what temperature would 5.00 g of hydrogen gas (H_2) occupy 50.0 L at 0.755 atm pressure, assuming ideal behavior?

11. A sample of 10.0 L of an ideal gas at 350 K exerts a pressure of 0.785 atm. At what temperature would the same amount of gas occupy 15.0 L at 0.656 atm?

12. A spherical balloon has a diameter of 1.15 meters at a temperature of 10.°C and an atmospheric pressure of 775 mm of mercury. What will be the change in the diameter of the balloon if the temperature rises to 30.°C and the atmospheric pressure drops to 755 mm of mercury? (Assume ideal gas behavior and $V_{sphere} = \frac{4}{3}\pi r^3$.)

13. Calculate the missing quantities in the following table of data, assuming constant temperature:

	P_i	V_i	P_f	V_f
1.	1.00 atm	22.4 L	_____ atm	2.24 L
2.	_____ kPa	100. mL	101.3 kPa	1.00 L
3.	100. Pa	_____ m³	100. kPa	1.00 L
4.	2.5 atm	2.50 L	_____ atm	250. mL

14. Calculate the missing quantities in the following table of data, assuming constant pressure:

	V_i	T_i	V_f	T_f
1.	27.4 L	0°C	____ L	100.°C
2.	10.0 L	____ K	2.25 L	298 K
3.	250 mL	373 K	240 mL	____ K
4.	____ m³	273 °C	10.0 L	273 K

15. A vacuum pump exhausts a heavy-walled 1.00-L flask to a pressure of 10^{-6} torr. How many molecules of gas are present if the temperature is 273 K?

16. A heavy-walled flask equipped with an escape valve initially contains half a mole of O_2 at 1 atm and 0°C. The escape valve is left in the open position and the temperature of the flask is raised to 373 K so the pressure in the flask remains at 1 atm. What fraction of the molecules originally present escaped?

17. Limestone ($CaCO_3$) decomposes when heated according to

$$CaCO_3 \ (s) \longrightarrow CaO \ (s) + CO_2 \ (g)$$

If 200. g of limestone is heated to 1000.°C, how many liters of CO_2 (g) will be produced at that temperature and a pressure of 750 torr?

18. How many kilograms of CS_2 must be burned with the stoichiometric quantity of O_2 in order to produce a total of 200. m³ of gas at 300.°C and 100. kPa? The reaction is

$$CS_2 \ (g) + 3 \ O_2 \ (g) \longrightarrow CO_2 \ (g) + 2 \ SO_2 \ (g)$$

19. Determine the density of CCl_4 vapor at 1.00 atm pressure and 100.°C.

20. What is the density of N_2 gas at 27°C and 100. kPa? Express your answer in SI units of kg/m³ as well as g/cm³.

21. An unknown gas has a density of 8.06 g/L at a pressure of 1.50 atm and 298 K. Calculate the molecular weight of the gas.

22. In a gas density experiment, the following data were obtained in order to determine the molecular weight of a sample at its normal boiling point of −33°C. Pressure = 756 torr; gas density = 7.05 g/L. What is the molecular weight of the vapor?

23. An organic compound has the following composition:

C = 24.26%   H = 4.08%   Cl = 71.66%

Further, it was found that 140.2 mL of the vapor of this liquid at 100.°C and 740 torr weighed 441.6 mg. What is the molecular (or true) formula of this compound?

24. Calculate the molecular weight for each of the following:
(a) An unknown gas whose density is 1.79 g/L at 100. kPa and 300. K
(b) An unknown gas whose density is 3.80 g/L at 100.°C and 0.973 atm.

25. You are not sure whether to fill a balloon with He or hot air. To what temperature would the air have to be heated to have the same lifting power as helium at 25°C?

26. A particular balloon when uninflated weighs 175 kg, and when inflated with He at 25°C has a gas volume of 552 m³. What is the maximum mass of the cargo that this balloon can carry at an inside pressure of 1.05 atm and an outside pressure of 0.99 atm?

Mixtures of Ideal Gases   [27–36]

27. Calculate the partial pressures of nitrogen and argon in the air if the atmospheric pressure is 772 mm of mercury.

28. A mixture of O_2, CO_2, and Ar has a total pressure of 1.20 atm and contains 7.75 g of each gas. Calculate the partial pressure of each gas in the mixture.

29. Air is a mixture of a number of gases. However, it is principally nitrogen and oxygen, containing 79 mole percent N_2 and 21 mole percent O_2. Calculate the weight percent of each of these components and the average molecular weight of air on that basis.

30. Producer gas is prepared by the reaction of hot coal with air and steam. It is useful as gaseous fuel and as a feedstock for producing other fuels and chemicals. A typical producer gas has the following mole percent composition:

CO_2 (g)	9.15%
H_2 (g)	19.65%
N_2 (g)	46.10%
CO (g)	21.70%
CH_4 (g)	3.40%

Calculate the average molecular weight of this producer gas.

31. A gas-tight reactor vessel has a volume of 1000. m³ and is filled with air at 27°C and 1.00 atm. Assuming air to be 21% O_2 and 79% N_2 (by volume), calculate the following:
(a) Partial pressure of O_2
(b) Partial pressure of N_2
(c) If all the O_2 were removed chemically, what would the pressure in the vessel be?

32. A weight-percent analysis of a flue gas is 15% CO_2, 6% O_2, and 79% N_2. The recorded temperature and pressure were 400.°F and 765 mm of Hg. Calculate the partial pressures of all three gases.

33. A fuel gas is 20.0 mole percent CH_4, 5.0 mole percent C_2H_6, and the remainder CO_2. Calculate the weight-percent analysis of this gas.

34. If the composition of air is roughly 78 mol% N_2, 21 mol% O_2, and 1 mol% Ar, calculate the partial pressure of each gas in a tank containing air at 10,000. kPa. What is the pressure in the tank after the N_2 is removed?

35. A hydrocarbon (C_nH_m) is burned completely in oxygen to produce a mixture of CO_2 (g) and H_2O (g). The total pressure of the mixture is 1.200 atm and the partial pressure of H_2O (g) is 0.686 atm. What is the empirical formula of the hydrocarbon?

36. A compound composed only of carbon and hydrogen is burned in oxygen to CO_2 (g) and H_2O (g). This mixture of product gases is found to occupy a volume of 2.000 L at a pressure of 1.387 atm and a temperature of 150°C. The H_2O is then condensed and found to weigh 0.800 g. What is the empirical formula of the compound?

Real Gases: Limiting Densities [37–44]

37. Use the van der Waals equation to calculate the pressure needed to contain 10.0 mol of He gas in a volume of 15.0 L and 100. K. Calculate the pressure needed to contain the same amount of chlorine gas, Cl_2 (g), under the same conditions.

38. Use the van der Waals equation to calculate the temperatures at which one mole of each of the gases He, O_2, and Cl_2 would occupy 1.00 L at a pressure of 10.0 atm.

39. Given the following information, and assuming gas molecules are rigid spheres, determine the fraction of the molar volume of argon atoms at STP that represents the molecular volume, and calculate the diameter of an argon atom. The value of the van der Waals constant b for argon is 0.03219 L/mol where b is the excluded volume of one mole of gas, which is four times the actual volume of the molecules: $b = 4 N_A V$ where N_A = Avogadro's number and V = actual volume of a molecule.

40. Although the compressibility factor is frequently greater than 1 for real gases—called positive deviation from ideal behavior—there are circumstances that favor negative deviations (values less than 1). Using the general expression for the compressibility factor for a van der Waals gas, determine the conditions that favor negative deviations from ideal behavior based on the following:

- Excluded volume due to molecular size: Small or large?
- Intermolecular forces of attraction: Weak or strong?
- Temperature effects: High or low?
- Pressure effects: High or low?

41. Calculate the molecular weight of the oxide of nitrogen described by the following density data, measured at 273.16 K:

P, atm	d/P, g/L·atm
1	1.9804
$\frac{2}{3}$	1.9746
$\frac{1}{2}$	1.9722
$\frac{1}{3}$	1.9694

42. Determine the molecular weight of a gas that has the following densities at 298.0 K:

P, atm	1	$\frac{3}{4}$	$\frac{1}{2}$	$\frac{1}{4}$
d/P, g/L·atm	1.1357	1.1338	1.1319	1.1300

43. One variation on the limiting ratio (d/P) procedure for determining atomic and molecular weights compares the pressures at which two gases have identical densities. When the ratio of pressures for equal density has been measured at a series of pressures, the limiting value for the ratio of pressures can be obtained by (linear) extrapolation. For example, at 293 K when the pressures for oxygen are $\frac{3}{4}$, $\frac{1}{2}$, and $\frac{1}{4}$ atm, the ratio of these pressures to those for carbon monoxide *at the same density* are 0.87574, 0.87523, and 0.87500, respectively. Plot the values, obtain the limiting

value for the ratio of pressures, calculate the molecular weight of CO, and determine the atomic weight of carbon.

44. Similar data to those described in Problem 43 for oxygen and N_2O lead to a limiting value (from linear extrapolation for equal densities) for P_{O_2}/P_{N_2O} of 1.37543. Calculate the molecular weight for N_2O and the atomic weight of nitrogen.

Kinetic Theory: Effusion and Diffusion [45–54]

45. Calculate the total kinetic energy for one mole of an ideal gas at 298 K.

46. A sample of 20.0 g of O_2 gas occupies 18.0 L at 0.750 atm and an unknown temperature. Calculate the total kinetic energy of this gas sample, assuming ideal behavior.

47. Calculate the root mean square speed of He gas molecules at 300. K.

48. Calculate v_{rms} for Rn gas at 25°C.

49. Two gases are trapped on opposite sides of a movable piston. The piston is allowed to move until the gas pressures on both sides are equal, at which point the entire apparatus is immersed in a constant temperature bath at 273 K.

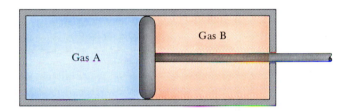

(a) What conclusions can be drawn concerning the average *KE* of Gas A and Gas B?

(b) When *P-T* equilibrium is reached between the two gases, the volume of Gas A is 5 L while the volume of Gas B is 10. L. What conclusions can be drawn concerning the relative number of moles of A molecules versus B molecules?

(c) If Gas A is acetaldehyde (CH_3CHO) and Gas B is nitrous oxide (N_2O), what is the speed of acetaldehyde molecules relative to that of nitrous oxide molecules?

50. An equimolar mixture of methane (CH_4) and propane (C_3H_8) is trapped in a container of fixed volume at constant temperature.

(a) Is the average kinetic energy of methane greater than, the same as, or less than that of propane?

(b) Is the partial pressure of methane greater than, the same as, or less than that of propane?

(c) If the container develops a leak, allowing the trapped gases to effuse through, what is the relative rate of escape of the two gases, initially?

51. Calculate the relative rates at which Ar and Ne at equal partial pressures effuse through a tiny opening in a container.

52. An unknown gas effuses through a very small opening at a rate 0.63 times as fast as O_2 gas at the same partial pressure. What is the molecular weight of the unknown gas?

53. Calculate the rate of effusion of phosphine (PH_3) molecules through a small opening if ammonia (NH_3) molecules pass through the same opening at a rate of 8.02 cm³/s. Assume the same temperature and equal partial pressures of the two gases.

54. A glass tube 100. cm long is fitted with cotton plugs at either end. One plug is soaked in aqueous ammonia, the other in hydrochloric acid. Assume that vapors of the two gases escape simultaneously from the two plugs and diffuse through the glass tube toward each other. At what distance from the hydrochloric acid plug will the white cloud of solid ammonium chloride form?

Compressors and Air Pressure   [55–58]

55. Calculate the compression ratio and volumetric efficiency of the compressor described in Figure 5–23.

56. Draw the P versus V diagram for an air compressor with an inlet pressure of 1.00 atm and discharge pressure of 5.55 atm. The cylinder volume varies from a minimum of 1.0 L to a maximum of 10.0 L. Calculate the volumetric efficiency of this compressor.

57. Write an equation based on Eq. 5–24 that can be used to determine the barometric pressure at sea level, given the measured barometric pressure at altitude H.

58. Calculate the performance factor for an air compressor operating at an altitude of 6000. m and a temperature of 25°C.

Additional Problems   [59–71]

59. A quantity of a gas has a volume of 3.9 cubic feet at 273 K. Find its volume at 400. K if the pressure has not changed.

60. A gas is contained in a balloon whose volume is 2.0 L at 0°C. The temperature is lowered to −196°C by pouring liquid nitrogen over the balloon, but the pressure is not allowed to change. What is the new volume of the balloon?

61. Imagine a bubble of swamp gas that has formed at the bottom of a primordial swamp. It is 1.0 cm in diameter, the temperature is 10.°C, and the pressure is 2.5 atm. The bubble rises to the surface, where the temperature is 30°C and the pressure is 1.0 atm. Based on this information alone, what is the diameter of the bubble at the surface?

62. The molecular weight of ozone was determined by a modified version of the vapor density method by weighing a known volume of the gas and measuring its temperature and pressure. Given the following data, calculate the molecular weight: the temperature is 28.2°C; the volume of the flask is 235.67 mL; the weight of the evacuated flask is 6.5998 g; the weight of the flask, filled with ozone to a pressure of 274.4 torr, is 6.7624 g.

63. A 10.0-L flask contains 2.00 moles of gasoline vapor at 25°C. Calculate the pressure of the vapor in atmospheres, using first the equation of state for an ideal gas and then the van der Waals equation of state. Assume this gasoline to be pure iso-octane, C_8H_{18}, for which the van der Waals constants are known to be $a = 25.43$ L²·atm/mol² and $b = 0.1453$ L/mol. The term "vapor" is usually reserved for the gaseous state of compounds that would normally be considered liquids.

64. A certain mixture of neon and krypton weighs 10.0 g and occupies a volume of 20.0 L at 25°C and 1 atm. What is the percent composition (by weight) of the mixture of noble gases?

65. You have a container of gas X and a container of gas Y, and the following information: the two gases are ideal; they are at the same temperature; the molecular weight of Y is twice that of X; the density of Y is half that of X. What is the ratio of the pressures of the gases in the two containers?

66. At 2300 K and 15 atm, hydrogen is a collection of molecules and atoms, due to the following dissociation reaction:

$$H_2 \rightleftharpoons 2\,H$$

Assuming the behavior of the atoms and the molecules to be ideal, if a mole of hydrogen molecules is one-third dissociated into atoms, what is its density?

67. A large glass bulb is evacuated and weighed, and then filled with an unknown gas to a pressure of 1.00 atm at a temperature of 25.0°C. The bulb weighed 82.3 grams when empty and 89.6 grams when filled. In a second experiment, the same (evacuated) bulb was filled with oxygen gas, again to a pressure of 1.00 atm at a temperature of 25.0°C. When filled, the bulb weighed 84.7 grams.
(a) What is the molecular weight of the unknown gas?
(b) What is the density of the unknown gas under these conditions?

68. A number of interesting B_xH_y compounds (known as boranes) have been characterized. A certain gaseous compound of B and H has a density of 0.002334 g/cm³ at STP, and a determination of its empirical formula produced a B:H ratio of 0.445. Calculate the molecular weight and the molecular formula of the gaseous borane.

69. Consider two separate gas samples, one of helium (He) and the other of sulfur dioxide (SO_2), under the following conditions:

helium sample: 1.00 L at 0°C and 2.00 atm

sulfur dioxide sample: 2.00 L at 273 K and 4.00 atm

Supply the information that is missing, using quantitative phrases such as *half*, *twice*, and *equal to*. Show briefly how you arrived at each answer.
(a) The number of moles of sulfur dioxide is _____ the number of moles of helium.
(b) The average kinetic energy of sulfur dioxide molecules is _____ the average kinetic energy of helium atoms.
(c) The average speed of sulfur dioxide molecules is _____ the average speed of helium atoms.
(d) If the separate containers are connected by a small orifice, helium atoms will diffuse into sulfur dioxide molecules at a rate _____ as great as that at which sulfur dioxide molecules will diffuse into helium atoms.

70. Ten grams of A were placed in an evacuated flask and the pressure was found to be 400. torr. After adding fifteen grams of B the pressure was observed to be 600. torr. If the temperature was maintained at 25°C throughout, and ideal gas behavior is assumed, what is the ratio of molecular weights for the two gaseous substances, M_A/M_B?

71. At what temperature will hydrogen molecules have the same velocities that oxygen molecules have at 25°C?

MULTIPLE PRINCIPLES   [72–79]

72. A certain metallic element forms a chloride that boils without decomposition at 346°C under 1 atm pressure, and under these conditions the density of the vapor was found to be about 8.0 g/L. By other methods, the chloride was found to contain 53.6% chlorine, while the specific heat of the metal was established as 0.138 J/g·k.
(a) Calculate the apparent formula weight of the gaseous chloride using the vapor density data provided.
(b) Determine the combining weight of the metal (relative to chlorine).
(c) Determine the approximate atomic weight of the metal.
(d) Determine the exact atomic weight of the metal.
(e) What is the empirical formula for the chloride?

73. Calculate the missing quantities from the table of data below:

	Pi	Vi	Ti	Pf	Vf	Tf
1.	101 kPa	22.4 L	273 K	50.5 kPa	____ L	273°C
2.	500. torr	500. ml	500.°C	560. torr	430. ml	____ °C
3.	1.25 atm	____ L	300. K	2.25 atm	2.25 L	323 K
4.	83.3 kPa	100. L	373 K	____ kPa	110. L	363 K

74. Consider the air around us (at STP) to be composed of three kinds of particles: 78.1% of them are nitrogen molecules, 20.9% are oxygen molecules, and 1.0% are argon atoms.
(a) What is the average molecular mass of the air?
(b) What is the average density of the air (at STP)?

75. In the first step of ^{235}U decay, the products are ^{231}Th and an α-particle. The α-particle is very reactive and will pick up the necessary two electrons to form a helium atom. What volume of helium, measured at 400.°C and 0.450 atm, would result from decay of 0.500 g of ^{235}U?

76. One of the raw materials needed for the manufacture of sulfuric acid is the oxygen in the air, which is used to convert sulfur to sulfur trioxide. What volume of air, measured at 1.10 atm and 500.°C, would be needed to produce 1000. kg of 98% sulfuric acid?

77. A gas mixture is known to contain equal numbers of moles of two gases. The mixture has a density of 1.47 g/L at 1.00 atm and 298 K. In a diffusion experiment, one of the gases was found to diffuse 1.25 times faster than the other under the same conditions. What are the respective molecular weights of the two gases?

78. A mixture of CO and acetone is trapped in a 1.0 L flask at 298 K. The pressure in the flask registers 100 torr initially, but after the acetone in the flask is caused to (catalytically) decompose according to the following reaction, the pressure registers 114 torr.

$$CH_3COCH_3 \text{ (g)} \longrightarrow C_2H_4 \text{ (g)} + CO \text{ (g)} + H_2 \text{ (g)}$$

If everything that happens and all the materials present are in the gas phase, and CO is inert to any process, what were the initial and final pressures of CO, assuming complete reaction of the acetone?

79. For xenon atoms, the van der Waals constants are

$$a = 4.194 \text{ L}^2\cdot\text{atm/mol}^2 \qquad b = 0.05105 \text{ L/mol}$$

(a) Calculate the pressure due to a mole of Xe atoms in a 10,000 liter tank at 273 K.
(b) How much larger is the atomic volume of Xe atoms than that of Ar atoms (as calculated in Problem 39)?

APPLIED PRINCIPLES   [80–82]

80. The gas meter of a home determines the volume of natural gas that passes through it. If the meter is located outside the house, the temperature of the gas can vary considerably. If the gas company makes no compensation for the temperature change, is the price of the gas per unit mass greater or less in winter than in summer? What is the ratio of the price of gas per unit mass at 70°F to the price at 10°F?

81. A gas company is storing methane in a tank with a volume of 40,000 m³. The pressure of the methane in the tank is kept at 1.04 atm. How much more methane (in kg) can be stored in the tank when the temperature is 0.0°F than when the temperature is 80°F?

82. A bituminous coal is found to have the following composition: C, 83.5%; H, 5.1%; S, 0.9%; O, 9.7%; ash, 0.8%. Calculate the volume of air at 35°C and 0.920 atm necessary to burn 1.00 kg of this coal completely. Assume the ash is inert to combustion, and that the combustion products are CO_2, SO_2, and H_2O.

CHAPTER
6

Chemistry in the Atmospheric Environment

A cross section through the Earth's atmosphere over the Pacific Ocean, from the Space Shuttle *Challenger*. The stratosphere is in blue, the troposphere in white and pink. This low sun angle photograph was taken by Robert Crippen, STS-17 Mission Commander, on October 10, 1984, from an altitude of 123 nautical miles. The inset photo shows Arno Penzias checking the inside of the horn reflector antenna used to measure the 3K background noise radiation in the galaxy, which according to modern cosmological theory may be radiation left over from the universe's beginnings as a collapsed "fireball."

Steven Weinberg, *The First Three Minutes,* Basic Books (New York, 1977).

The Nobel Prize for Physics in 1981 was awarded to Arno Penzias and Robert Wilson, for their discovery of the background radiation.

6.1   THE PRINCIPAL GASES OF THE UNIVERSE

The universe is made up of mostly hydrogen, some helium, and trace amounts of all the other elements (Table 6–1). Far from being evenly distributed, this cosmic composition consists of an irregular display of galaxies with an enormous range of density. Our galaxy, like its closest neighbor Andromeda (approximately 3 million light years away), is a spiraling, disklike structure, about 100,000 light years in diameter and 15,000 light years thick at the center.

When and where the universe originated, and how the galaxies formed, are questions that have interested scientists and philosophers (and theologians) since antiquity. Opinion falls into two categories—either the universe had a beginning in time, an instant "Genesis" referred to as the "big bang," or the universe did not have a beginning in time and is in a state of continuous creation. The continuous creation theory was popular some years ago, but has recently fallen from favor because of the growing scientific evidence supporting the big bang hypothesis.

According to "big bang" theorists, there was an explosion of truly astronomical proportions within a highly dense, undifferentiated mix of primordial mass and energy, and all of this "stuff" of the universe is still hurtling outward as a result. It has been suggested by Penzias and Wilson that the universal background temperature of 3 K that exists today is the energy residue of the big bang.

At the instant of creation and in the seconds and minutes that followed, neutrons decayed into protons and electrons. Then, as the temperature of the universe dropped, the protons and neutrons in the primordial plasma began to combine. The first combinations might have led to stable helium nuclei by the following sequence of reactions, for example:

$$^1_1H + {}^1n \longrightarrow {}^2_1H$$

$$^2_1H + {}^1_1H \longrightarrow {}^3_2He$$

$$^3_2He + {}^1n \longrightarrow {}^4_2He$$

Production of ^4He nuclei could continue until the neutrons were used up. Note that the production of one ^4He nucleus requires two protons and two neutrons, and leads directly to the observed abundance of helium in the universe. Further cooling allowed available electrons to combine with the nuclei to form hydrogen and helium atoms.

Condensation of some of the gaseous material into stars produced reactions between helium nuclei:

$$^4_2He + {}^4_2He \longrightarrow {}^8_4Be$$

The ^8_4Be nucleus itself is unstable, but when it combined with another ^4_2He nucleus perhaps the stable $^{12}_6C$ nuclide formed:

$$^8_4Be + {}^4_2He \longrightarrow {}^{12}_6C$$

Further fusion of helium nuclei could then account for nuclei of even greater atomic number, multiples of ^4_2He such as $^{16}_8O$ and $^{20}_{10}Ne$. Continuing the process, other even-numbered nuclei are produced:

$$^{12}_6C + {}^{12}_6C \longrightarrow {}^{24}_{12}Mg$$

$$^{16}_8O + {}^{16}_8O \longrightarrow {}^{32}_{16}S$$

TABLE 6–1   The Universal Abundances of the First 36 Elements

Atomic Number	Symbol	Abundance, ppm	Notes
1	H	739,000	
2	He	240,000	
3	Li	0.006	Notably lacking.
4	Be	0.001	
5	B	0.001	
6	C	4,600	Examine the relative abundances of the
7	N	970	even-numbered elements between 6 and
8	O	10,700	20, and compare them to the odd-num-
9	F	0.4	bered elements. Note their much greater
10	Ne	1,300	abundance.
11	Na	22	
12	Mg	580	
13	Al	55	
14	Si	650	
15	P	7	
16	S	440	
17	Cl	1	
18	Ar	220	
19	K	3	
20	Ca	67	
21	Sc	0.03	
22	Ti	3	
23	V	0.7	
24	Cr	14	
25	Mn	8	
26	Fe	1,900	Note the peak abundance of iron and nickel.
27	Co	3	
28	Ni	60	
29	Cu	0.06	The rapid fall-off in abundance with in-
30	Zn	0.3	creasing mass number begins with Cu.
31	Ga	0.01	
32	Ge	0.2	
33	As	0.008	
34	Se	0.03	
35	Br	0.007	
36	Kr	0.04	

Reactions such as those shown so far may be the reason that nuclei with even atomic numbers predominate over those with odd numbers. However, some combinations that produce odd atomic numbers also exist. For example, the following reactions are likely:

$$^{12}_{6}C + ^{12}_{6}C \longrightarrow ^{23}_{11}Na + ^{1}_{1}H$$

$$^{16}_{8}O + ^{16}_{8}O \longrightarrow ^{31}_{15}P + ^{1}_{1}H$$

Fusion reactions producing atoms with atomic numbers much above that of iron are energetically unfavorable and are therefore unexpected. Still, such elements do exist and their presence is significant, although their existence on Earth is not well understood. The most likely mechanism is rapid accretion of neutrons and protons during the explosion of a supernova.

6.2 CHEMISTRY IN THE SUN AND THE OTHER STARS

As you might expect from the apparent difficulties in obtaining data and setting up model experiments, no one theory of stellar synthesis for the elements has proved entirely satisfactory. One theory of stellar synthesis—**nucleogenesis**, as it is called—has received a great deal of attention. The process begins with a star formed by condensation of hydrogen nuclei. As gravity causes the mass of gas to contract, it heats up until the nuclei fuse, forming helium and releasing a great deal of energy:

$$^1_1H + {}^1_1H \longrightarrow {}^2_1H + \beta^+ + \gamma$$

$$^1_1H + {}^2_1H \longrightarrow {}^3_2He + \gamma$$

$$^3_2He + {}^3_2He \longrightarrow {}^4_2He + {}^1_1H + {}^1_1H$$

If a small number of carbon and nitrogen nuclei happen to be present, having been injected into the star from some external cosmic source or process, they can act as catalysts for further helium synthesis:

$$^{12}_6C + {}^1_1H \longrightarrow {}^{13}_7N + \gamma$$

$$^{13}_7N \longrightarrow {}^{13}_6C + \beta^+ + \gamma$$

$$^{13}_6C + {}^1_1H \longrightarrow {}^{14}_7N + \gamma$$

$$^{14}_7N + {}^1_1H \longrightarrow {}^{15}_8O + \gamma$$

$$^{15}_8O \longrightarrow {}^{15}_7N + \beta^+ + \gamma$$

$$^{15}_7N + {}^1_1H \longrightarrow {}^{12}_6C + {}^4_2He$$

These two processes convert hydrogen to helium. They are referred to as hydrogen burning reactions, and they are the processes that fuel the stars, including our own Sun.

When the star's hydrogen supply is exhausted, the diminished energy release from fusion can no longer keep the nuclei separated against the force of gravity. The resulting gravitational collapse raises the temperature to the point at which helium fuses. In this stage, which happens much faster than hydrogen burning, carbon and oxygen are formed in abundance. At this point we begin to see the emergence of heavier elements:

$$^4_2He + {}^4_2He + {}^4_2He \longrightarrow {}^{12}_6C$$

$$^{12}_6C + {}^4_2He \longrightarrow {}^{16}_8O$$

As the last of the helium in very massive stars is consumed, the temperature rises again, and in a still faster phase, carbon nuclei fuse:

$$^{12}_6C + {}^{12}_6C \longrightarrow {}^{24}_{12}Mg + \gamma$$

Gravitational collapse continues, and the pace of events quickens:

$$^{16}_8O + {}^4_2He \longrightarrow {}^{20}_{10}Ne$$

$$^{24}_{12}Mg + {}^4_2He \longrightarrow {}^{28}_{14}Si$$

These processes continue toward production of the most stable nuclides in the highest yields and in a state of thermal equilibrium. Iron-56 is the nuclide with the greatest binding energy per nuclide of all.

The ultimate fate of a star depends on how much mass it started with. Small stars, up to about four times the mass of our Sun, pass through a "red giant" phase and eventually become stable "white dwarf" stars. Much more massive stars, however, become "red supergiants" that eventually form substantial amounts of iron in the core. When the iron nuclei begin to fuse, they absorb rather than release energy (because iron is the most

As an example of the magnitude of the energy released by fusing together these smallest nuclei, recall the deuterium-tritium fusion reaction from our discussion in Chapter 3:

$$D + T \longrightarrow {}^4_2He + {}^1n + 17.6 \text{ MeV}$$

Remember, that's MeV, not eV.

(a)

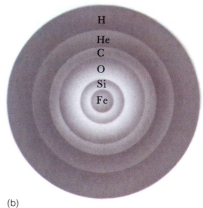

(b)

FIGURE 6–1 (a) Supernova 1987A, an exploding star just 160,000 light-years from Earth, shines forth shortly after its discovery on February 23, 1987. On the left, at the point of the arrow, is the location of a blue supergiant star about four times the size of our sun, hours before it collapsed and exploded into the picture on the right. This was the first such observation since Kepler in 1604 (before the invention of the telescope). (b) By the time the star reached the threshold of collapse, thermonuclear fusion was taking place in the many concentric layers of this cosmic onion. The "ash" from reactions in each layer rained down and ignited the layer below. At the center—hottest and densest of all—was the iron core of the collapsing star.

stable nucleus). This cools the core of the star, leading to gravitational collapse and sudden reheating. The outer layers of the star are blown off in a tremendous **supernova** explosion (Figure 6–1). Most of the star's matter is flung into space, where the atoms may eventually become part of another star or its planets.

Although our theory of stellar synthesis is incomplete, it does successfully account for the natural abundance of the elements throughout the universe as we know it today.

6.3 PLANET EARTH AND ITS ATMOSPHERE

The Layers of the Atmosphere

Far from being a simple distribution of gases whose collective density trails off into the apparent void of deep space, the Earth's atmosphere is highly structured. It is dynamic and rapidly changing in response to changes in geography, the position of the Sun, and stellar activity. The temperature profile displayed in Figure 6–2 provides a great deal of useful information about the structure of the atmosphere. The **exosphere** approaches the composition of outer space, and few reactions occur there. The **thermosphere** is a region of increasing temperatures resulting from the absorption of high energy ultraviolet radiation by both oxygen atoms and molecules and nitrogen molecules. At an altitude of about 200 km one might find temperatures on the order of 1000°C. Proceeding down through this layer, the intensity of the radiation decreases as a result of this absorption. Hence, the temperature drops with decreasing altitude.

A minimum temperature of about 160 K is reached on descending to about 90 km. This is the region containing ozone, O_3. The concentration of ozone increases with decreasing altitudes until it reaches a maximum at about 25 km. Because ozone absorbs lower energy ultraviolet radiation that is not absorbed by oxygen and hydrogen in the thermosphere, this is an especially important layer. A maximum temperature of about 280 K is reached at about 50 km at the boundary with the stratosphere, when suf-

In our galaxy, supernovae occur about once every 200 years. The Crab Nebula is the remains of one such stellar explosion that was recorded by Chinese astronomers in the year 1054 A.D. In 1987, astronomers photographed the first signal of a stellar explosion that took place more than 100,000 years ago (Figure 6–1).

This is the *kinetic* temperature determined from the molecular speed distribution, as described in Chapter 5. However, there are so few particles in this region that you would not feel warm.

FIGURE 6–2 Temperature profile of the Earth's atmosphere. Some of the radiation falling to Earth is reflected by the surface, the cloud cover in the lower atmosphere, or by the atmosphere itself. The rest is absorbed at the Earth's surface (radiational heating) only to be returned as thermal radiation (radiational cooling). Weather and climate result directly.

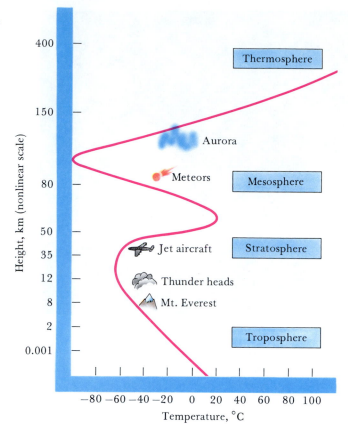

ficient radiation has been absorbed to decrease the solar heating. The temperature falls with decreasing altitude in the stratosphere until the boundary with the **troposphere** is reached at about 12 km, reaching a minimum value of about 210 K.

Temperatures again increase with decreasing altitude in the troposphere. Heating in this region comes from the surface of the Earth, which absorbs visible radiation from the Sun, along with any ultraviolet radiation not absorbed by higher layers of the atmosphere. The warm surface of the Earth re-radiates this heat mainly in the form of infrared radiation, which is absorbed by the atmospheric gases. This absorption results in a decreasing intensity of radiation from the Earth at increasing altitudes, and thus the temperatures are highest in the region nearest the Earth's surface.

Two other divisions of the atmosphere are sometimes made. Below 115 km the principal mixing of the atmosphere is by winds and air turbulence. In this region, called the **homosphere**, the main gases of the atmosphere are always found in the same proportions (essentially 78% N_2, 21% O_2, and 1% Ar). Above 115 km, particle diffusion is the dominant mode of transport of the gases. In this region, the **heterosphere**, gravitational separation takes place as the lightest and fastest atoms and molecules move farther from the Earth.

The region called the **ionosphere** extends from about 60 km. In the ionosphere, high energy solar rays are sufficiently intense to cause the formation of ions. Absorption of this high energy radiation results in the expulsion of electrons, leaving molecular ions behind. For example,

$$O_2 \text{ (g)} \longrightarrow O_2^+ \text{ (g)} + e^-$$

Carbon dioxide is currently present in the atmosphere at a level of 0.0314% by volume.

These reactive species are capable of initiating chain reactions with other atoms and molecules present in the atmosphere.

The Greenhouse Effect

Not all of the infrared radiation from the Earth is absorbed by the normal gases of the troposphere. Carbon dioxide, a product of the burning of fossil fuels such as coal and petroleum products, is a very good infrared absorber. For example, complete combustion of a typical fuel oil component would be

$$23\ O_2\ (g)\ +\ C_{15}H_{32}\ (liq)\ \longrightarrow\ 15\ CO_2(g)\ +\ 16\ H_2O\ (g)$$

There is good evidence that the increasing use of fossil fuels is raising carbon dioxide concentrations in the atmosphere. If this is allowed to continue, the result will be an increasing absorption of the infrared rays emitted from the Earth's surface, with a corresponding increase in the temperature of the troposphere. Some researchers claim to have detected this atmospheric warming. They warn that one result could be melting of the polar ice caps, thus raising ocean levels. Coastal cities and low-lying areas such as New York and Los Angeles, southern Florida, and Louisiana would be inundated. Even though the data are inconclusive at present, a prudent person would consider the matter seriously. The debate over the "greenhouse effect," as it has come to be known, continues.

6.4 CHEMISTRY IN THE HIGH ATMOSPHERE

Chemistry of Atomic Oxygen

Below 160 km the principal chemical reactions occur because the Sun's radiation is absorbed by molecules of oxygen (O_2) and ozone (O_3). In each case, very reactive oxygen atoms are generated:

$$O_2\ (g)\ \longrightarrow\ 2\ O\ (g)$$
$$O_3\ (g)\ \longrightarrow\ O_2\ (g)\ +\ O\ (g)$$

Oxygen atoms react with many molecules. Of particular importance are the reactions with ozone and dinitrogen monoxide (nitrous oxide):

$$O\ (g)\ +\ O_3\ (g)\ \longrightarrow\ 2\ O_2\ (g)$$
$$O\ (g)\ +\ N_2O\ (g)\ \longrightarrow\ 2\ NO\ (g)$$

The reaction with O_3 provides a pathway for reducing the ozone concentration in the **mesosphere**. This reaction is relatively slow unless NO or chlorofluorocarbons, compounds called Freons, are present. Dinitrogen monoxide (N_2O) does not arise from reactions in the atmosphere but is mainly a product of decomposing animal wastes on the Earth's surface. A balance between production of N_2O on the Earth and removal by reaction with oxygen atoms maintains the concentration of this gas.

The Ozone Layer

Possible destruction of the ozone layer around the Earth has raised considerable concern. Without a sufficient quantity of this gas to absorb the intense ultraviolet radiation, damage to plants and animals would increase

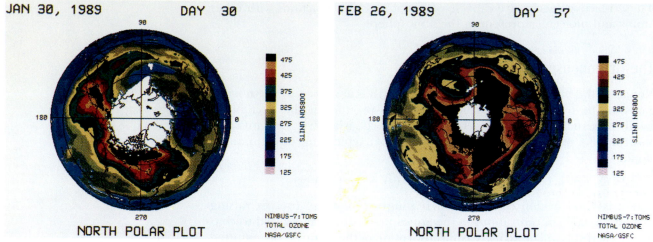

FIGURE 6–3 As active chlorine increases, it is now generally accepted that Antarctic ozone decreases. The effect seems to be seasonal, being most pronounced in winter. More recently, Arctic ozone destruction has also been observed. In these striking NASA photos, an ozone minimum (blue) has been observed across Iceland, Scotland, and Scandanavia, a maximum (green, red, and black) in an arch across the Soviet Union and North America. By late February, the diminished ozone effect (blue), though still evident, has lessened considerably. Dobson Units measure total ozone concentration.

significantly, such as skin cancer in humans. Recently, an alarming seasonal disappearance of the ozone layer has been found over Antarctica (Figure 6–3). At least two possible pathways for ozone depletion resulting from human activities have been proposed. One is an increase in NO concentration resulting from N_2O from animal wastes or high flying jet aircraft. Nitrogen monoxide then can react with ozone

$$NO \text{ (g)} + O_3 \text{ (g)} \longrightarrow NO_2 \text{ (g)} + O_2 \text{ (g)}$$

and then be regenerated in a reaction with oxygen atoms

$$NO_2 \text{ (g)} + O \text{ (g)} \longrightarrow NO \text{ (g)} + O_2 \text{ (g)}$$

The net reaction for this cycle is

$$O_3 \text{ (g)} + O \text{ (g)} \longrightarrow 2 O_2 \text{ (g)}$$

but the NO provides a pathway for rapid reactions since it can be used over and over in a chain reaction process. A substance such as NO (in this case) that provides a pathway without itself being consumed is called a **catalyst**. It is said to catalyze the reaction.

> The chlorofluorocarbons are the well-known "Freons," now often referred to in the popular press as CFCs.

A cause of depletion of the ozone layer—perhaps the major cause—is worldwide industrial and commercial use of chlorofluorocarbons (CFCs) such as dichlorodifluoromethane (CF_2Cl_2). These compounds have been widely used since the end of World War II, and especially in the last 25 years, as refrigerants, foaming agents for plastics in disposable coffee cups and packing materials, and (in countries other than the United States) as propellants in aerosol sprays. Eventually, the CFCs find their way into the upper atmosphere where they absorb radiation, breaking chlorine atoms off the central carbon atom:

$$CF_2Cl_2 \text{ (g)} \longrightarrow CF_2Cl \text{ (g)} + Cl \text{ (g)} \qquad \text{(chain initiation)}$$

The chlorine atom then enters a cycle that converts ozone to molecular oxygen:

$$Cl\ (g)\ +\ O_3\ (g)\ \longrightarrow\ ClO\ (g)\ +\ O_2\ (g) \qquad \text{(chain propagation)}$$
$$ClO\ (g)\ +\ O\ (g)\ \longrightarrow\ Cl\ (g)\ +\ O_2\ (g)$$

Here the Cl atom is the catalyst for the overall reaction:

$$O_3\ (g)\ +\ O\ (g)\ \longrightarrow\ 2\ O_2\ (g) \qquad \text{(overall reaction)}$$

Recent studies (1988) have now provided seemingly irrefutable evidence that CFCs do indeed destroy stratospheric ozone. The mechanisms that lead to this catastrophic loss of ozone are a complex combination of both chemistry and the dynamic behavior of the atmosphere. What is clear at this point, based on convincing aircraft experiments, is that chlorine monoxide is the "smoking gun" of ozone depletion. The chain reaction mechanism proposed in the preceding paragraph depends on ClO in a propagation step that returns Cl atoms to destroy large numbers of ozone molecules. Unfortunately, CFCs (and Halons, CFCs containing bromine) have been extraordinarily beneficial chemicals, and weaning ourselves away from them won't be easy. In the same context, loss of ozone above the Arctic has also been recently measured over Greenland, and although it is very small compared to the Antarctic depletion, there is evidence of diminished levels.

6.5 HYDROGEN

The natural abundance of the elements on the Earth, including the Earth's crust, the oceans, and the atmosphere, are given in Table 6–2 (and in Table 4–1). The most notable difference between this table and the abundances of the elements in the universe (in Table 6–1) is the much lower abundance on Earth of hydrogen, neon, and particularly helium. Presumably the Earth's gravitational field is too weak to hold the lightest atoms and molecules, allowing them to escape to outer space. Since helium and neon do not occur in any known compounds, their abundances on Earth have dropped to very low values. However, considerable quantities of bound hydrogen—atoms bonded into compounds—are present; water is by far the most prevalent compound of hydrogen. In fact, the number of compounds containing hydrogen here on Earth is enormous.

There are three isotopes of hydrogen and the two less common ones have their own names:

hydrogen	^1H (1.008 μ)
deuterium	^2H or simply D (2.014 μ)
tritium	^3H or simply T (3.016 μ)

Because their respective masses are so radically different, deuterium being twice and tritium being three times the mass of hydrogen, the three isotopes tend to display markedly different physical (but not chemical) properties. For example, the boiling point of H_2 is 20.4 K, whereas D_2 boils at 23.5 K. The natural abundance of hydrogen contains 0.0156% deuterium; tritium is produced in trace quantities during high energy nuclear reactions in the upper atmosphere and in nuclear reactors on Earth. Heavy water (D_2O) is used as a moderator in nuclear reactors and is prepared in ton quantities. Tritium is a radioactive β-particle emitter with a half life of 12.5 years.

Elemental hydrogen is a colorless, odorless, tasteless gas of diatomic molecules. It is insoluble in water, chemically neutral, and highly flammable

Deuterium oxide (D_2O) is commonly referred to as "heavy" water.

TABLE 6–2	The Abundances in Parts per Million of Elements in the Earth's Crust, the Oceans, and the Atmosphere That Occur Over 0.003 ppm				
Atomic Number	**Symbol**	**Abundance**	**Atomic Number**	**Atomic Symbol**	**Abundance**
1	H	8,700	40	Zr	220
2	He	0.003	41	Nb	24
3	Li	65	42	Mo	8
4	Be	6	46	Pd	0.010
5	B	3	47	Ag	0.10
6	C	800	48	Cd	0.15
7	N	300	49	In	0.1
8	O	495,000	50	Sn	0
9	F	270	51	Sb	1
10	Na	26,000	53	I	0.3
12	Mg	19,000	55	Cs	46.1
13	Al	75,000	56	Ba	5.53
14	Si	257,000	57	La	18.3
15	P	1,200	58	Ce	46.1
16	S	600	59	Pr	5.53
17	Cl	1,900	60	Nd	23.6
18	Ar	400	62	Sm	6.47
19	K	24,000	63	Eu	1.06
20	Ca	34,000	64	Gd	6.36
21	Sc	5	65	Tb	0.91
22	Ti	5,800	66	Dy	4.47
23	V	150	67	Ho	1.15
24	Cr	200	68	Er	2.47
25	Mn	1,000	69	Tm	0.20
26	Fe	47,000	70	Yb	2.66
27	Co	23	71	Lu	0.75
28	Ni	80	72	Hf	4.5
29	Cu	70	73	Ta	2.1
30	Zn	132	74	W	34
31	Ga	15	78	Pt	0.005
32	Ge	7	79	Au	0.005
33	As	5	80	Hg	0.30
34	Se	0.09	81	Tl	1.8
35	Br	1.62	82	Pb	16
37	Rb	310	83	Bi	0.2
38	Sr	300	90	Th	11.5
39	Y	28.1	92	U	4

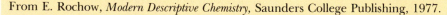

From E. Rochow, *Modern Descriptive Chemistry*, Saunders College Publishing, 1977.

FIGURE 6–4 Laboratory preparation of hydrogen gas by the reaction of zinc, an "active" metal, and hydrochloric acid.

First gaseous element isolated, by Henry Cavendish, after whom the world-renowned physics laboratories at Cambridge University were named. In 1766, Cavendish obtained H_2 by the action of dilute sulfuric acid on Zn, Fe, or Sn, a method still used to prepare H_2 in the laboratory.

(see Chap. 5 photo p.151). In the laboratory, hydrogen is usually prepared by the reaction of certain metals with a dilute acid. Zinc and dilute HCl are frequently used (Figure 6–4):

$$Zn \ (s) \ + \ 2 \ HCl \ (aq) \longrightarrow H_2 \ (g) \ + \ ZnCl_2 \ (aq)$$

Industrially, hydrogen is made by the action of steam on red-hot carbon:

$$H_2O \ (steam) \ + \ C \ (s, \ coke) \longrightarrow CO \ (g) \ + \ H_2 \ (g)$$

Hydrogen is also prepared by a catalytic re-forming process involving steam and gasoline-range hydrocarbons from petroleum at temperatures of about 1000°C in the presence of a nickel catalyst. For example, using octane:

$$C_8H_{18} \text{ (g)} + 16 \text{ H}_2O \text{ (g)} \longrightarrow 25 \text{ H}_2 \text{ (g)} + 8 \text{ CO}_2 \text{ (g)}$$

The carbon dioxide is removed by scrubbing the gaseous mixture of CO_2 and H_2 with water (under pressure). Hydrogen is also recovered as a by-product of the industrial preparation of chlorine by electrolysis of brine.

Molecules containing covalently bonded hydrogen are the most important known, the simplest being the diatomic H_2 molecule itself. The energy required to break the single shared pair of electrons is quite high, about 434 kJ/mol, and therefore at low temperatures it should not be surprising to find that hydrogen is not very reactive.

Compounds with hydrogen and the nonmetallic elements (excluding the noble gases He, Ne, and Ar) are many and well-documented. The simplest covalent compounds of hydrogen with the elements of the second period—from beryllium through fluorine—have chemical compositions related to their position in the periodic table. Beryllium, for example, with two electrons to share, bonds with two hydrogen atoms to form the covalent hydride BeH_2. Boron, having three electrons to share, might be expected to form BH_3, an unstable species. Instead, a dimeric species called diborane, B_2H_6, exists. Carbon compounds such as CH_4 display a full octet of electrons. Nitrogen, oxygen, and fluorine complete their octets to form NH_3, H_2O, and HF. Analogous compounds are formed by the elements in the other periods, along the right side of the periodic table. For example, the third-period compounds are AlH_3, SiH_4, PH_3, H_2S, and HCl (Figure 6–5).

A number of these elements combine with hydrogen to form much larger covalent molecules than any of the compounds just mentioned. The most notable example is the huge family of organic compounds called hydrocarbons and its derivatives. Methane (CH_4) is the simplest. Others include C_2H_6, C_3H_8, and C_4H_{10}. Each is characterized by a chain of carbon atoms linked consecutively, a phenomenon referred to as **catenation**. Some chains are straight; others are branched or form five- or six-membered rings; and some of the chains are of great length, with the number of carbon atoms numbering in the hundreds and thousands.

methane	CH_4
ethane	CH_3-CH_3
propane	$CH_3-CH_2-CH_3$
n-butane	$CH_3-CH_2-CH_2-CH_3$
n-pentane	$CH_3-CH_2-CH_2-CH_2-CH_3$
candle wax	$CH_3-[CH_2]_{25}-CH_3$
iso-octane	

$$CH_3-\overset{\overset{\displaystyle CH_3}{|}}{C}-CH_2-\overset{\overset{\displaystyle CH_3}{|}}{CH}-CH_3$$
$$\underset{\underset{\displaystyle CH_3}{|}}{}$$

FIGURE 6–5 Hydrogen bromide (HBr), a colorless gas, is formed on combustion of colorless hydrogen gas in an atmosphere of reddish brown bromine vapor.

TABLE 6–3 Some Important Covalent Compounds of Hydrogen

Name	Formula	Melting Point, °C	Boiling Point, °C
hydrogen fluoride	HF	−92.3	19.4
hydrogen chloride	HCl	−112	−83.7
hydrogen sulfide	H_2S	−82.9	−61.8
diborane	B_2H_6	−165	−93
methane	CH_4	−184	−161.5
silane	SiH_4	−185	−111.8
ammonia	NH_3	−77.7	−34.5
phosphine	PH_3	−133.3	−87.4
water	H_2O	0	100

Silane (SiH_4) is the first member of a much smaller series of compounds called silicon hydrides—or **silanes**—containing chains of up to eight or so silicon atoms. Diborane (B_2H_6) is the first member of the group of boron hydrides—or **boranes**—that includes B_4H_{10}, B_5H_9, B_6H_{10}, and $B_{10}H_{14}$. In these compounds, the boron atoms are arranged in clusters rather than in chains. Nitrogen and oxygen form two-atom chains in H_2NNH_2 (hydrazine) and HOOH (hydrogen peroxide). Sulfur forms a series of compounds, H—$[S]_n$—H, in which n ranges from 1 to 6.

Covalent compounds of hydrogen tend to have low melting and boiling points. Table 6–3 lists some selected data. Note the striking difference between HF and H_2O, which is due to the higher degree of association between water molecules. Much more will be said in a later chapter about this important phenomenon, known as hydrogen bonding.

Hydrogen reacts with the metallic elements of Group IA and Ca, Sr, and Ba of Group IIA. In these reactions, the hydrogen atom gains an electron to form H:$^-$, the hydride ion:

$$2\ Na\ (s)\ +\ H_2\ (g)\ \longrightarrow\ 2\ NaH\ (s)$$

The compound NaH is composed of sodium ions and hydride ions, Na^+ and H^-. Calcium and hydrogen yield calcium hydride:

$$Ca\ (s)\ +\ H_2\ (g)\ \longrightarrow\ CaH_2\ (s)$$

Metallic elements of the first two families of the periodic table form such hydride salts with essentially ionic properties. These metal hydrides are white solids with high melting points, which rapidly decompose in the presence of water or moisture to produce a metal hydroxide. We say they **hydrolyze**:

$$CaH_2\ (s)\ +\ 2\ H_2O\ (liq)\ \longrightarrow\ Ca(OH)_2\ (s)\ +\ 2\ H_2\ (g)$$

In this example, the metal hydroxide is sparingly soluble calcium hydroxide (Figure 6–6).

Hydrogen reacts with many transition metal elements. The products generally do not exhibit stoichiometric relationships between the elements. Nonstoichiometric compounds such as $LaH_{2.87}$ and $TiH_{1.7}$ are typical. What initially appears to be a violation of the law of definite proportions is the result of incomplete filling by hydrogen atoms of available sites between metal atoms in the solid. Such hydrogen compounds are known only in the solid state. In general, hydrogen gas can be re-formed (in the reverse reaction), leading to the suggestion to use these compounds to store hydrogen for hydrogen-powered vehicles.

FIGURE 6–6 Calcium hydride reacts vigorously with water, yielding an aqueous calcium hydroxide and liberating hydrogen gas.

Applications of hydrogen include production of very pure molybdenum metal by reaction of the oxide with hydrogen gas.

$$MoO_3 \text{ (s)} + 3\ H_2 \text{ (g)} \longrightarrow 3\ H_2O \text{ (g)} + Mo \text{ (s)}$$

A spongy form of iron metal that has proved useful as a catalyst—because of its especially large surface area—can be prepared by the reaction of iron(II) oxide with hydrogen:

$$FeO \text{ (s)} + H_2 \text{ (g)} \longrightarrow H_2O \text{ (g)} + Fe \text{ (s)}$$

The reaction of oxygen with hydrogen gives a high temperature flame, which makes it useful in the oxy-hydrogen welding torch:

$$2\ H_2 \text{ (g)} + O_2 \text{ (g)} \longrightarrow 2\ H_2O \text{ (g)}$$

However, because some of the available energy must first be used to break the H-to-H bond, an even higher temperature flame can be produced by the atomic hydrogen torch. In this torch the H_2 gas is passed between tungsten electrodes just before it reaches the burner. An electric discharge between these electrodes breaks molecules into atoms:

$$H_2 \text{ (g)} \longrightarrow 2\ H \text{ (g)}$$

The energy given off on recombination of these atoms plus combustion with O_2 gas gives a flame temperature close to 4000°C.

The atomic hydrogen torch was invented by Irving Langmuir (1881–1957). Among the achievements of his long and distinguished career at General Electric, Langmuir introduced use of argon as a filler in incandescent bulbs. He won the Nobel Prize for chemistry in 1932.

6.6 NITROGEN

The most abundant of the gases that fill the troposphere, the layer of the atmosphere in which we spend our lives, is nitrogen—78% by volume. It exists as N_2 molecules in which the atoms are triple-bonded:

$$:N:::N:$$

To rupture a triple bond requires a great deal of energy. Consequently, molecular nitrogen is relatively unreactive. Furthermore, although nitrogen compounds are important, they are relatively hard to synthesize. By contrast, oxygen is extremely reactive, but it is effectively diluted by the fourfold excess of nitrogen present in the air. Thus, the chemical reactivity of the air can best be understood as that of a dilute solution of oxygen in nitrogen.

The usual source of gaseous nitrogen is the air, which can be liquefied and then distilled to separate N_2 (boiling point, −195.8°C) from O_2 (boiling point, −182.96°C) (Figure 6–7). Liquid nitrogen is a very effective and relatively low-cost refrigerant. The recent development of superconducting materials that operate at temperatures above −196°C will likely result in a massive demand for liquid nitrogen.

Fairly pure nitrogen can be obtained by a simple laboratory procedure that makes use of some interesting chemical reactions that do not actually involve nitrogen. Instead, the other major components are removed, leaving nitrogen behind. Air is pushed slowly through a potassium hydroxide (KOH) solution to remove the small amount of carbon dioxide that is present:

$$KOH \text{ (aq)} + CO_2 \text{ (g)} \longrightarrow KHCO_3 \text{ (aq)}$$

Then, fuming sulfuric acid—which acts as a powerful dehydrating agent—removes the water vapor that is present.

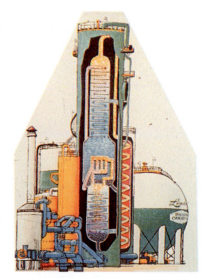

FIGURE 6–7 In the center of the cut-away central tower of the distillation unit is the Linde main condenser, which separates liquid air into the more volatile nitrogen (bp 77.4 K) and the less volatile oxygen (bp 90.2 K). The bottom kettle is rich in oxygen (35%); condensed oxygen (100%) is tapped off from the wide center section where it has accumulated; cold nitrogen vapors lead off to the main nitrogen condenser, the adjacent column.

FIGURE 6–8 Rhizobium in the root systems of nitrogen-fixing leguminous plants.

Fritz Haber (German, 1868–1934) successfully designed a practical ammonia process. The Haber process freed Germany from reliance on Chilean saltpeter, $NaNO_3$, as a source of nitrogen compounds for explosives in World War I. See Chapter 10.

$$H_2S_2O_7 \text{ (liq)} + H_2O \text{ (liq)} \longrightarrow 2\ H_2SO_4 \text{ (liq)}$$

Finally, when raised to red heat, copper turnings remove the oxygen from the air by forming black copper(II) oxide (CuO), leaving only nitrogen behind:

$$2\ Cu \text{ (s)} + O_2 \text{ (g)} \longrightarrow 2\ CuO \text{ (s)}$$

The nitrogen separated from air in this way contains some other gases as impurities—about 1% noble gases, mostly argon.

Pure N_2 for laboratory use can be prepared by heating sodium azide (NaN_3) or ammonium nitrite (NH_4NO_2) solutions:

$$2\ NaN_3 \text{ (s)} \longrightarrow 2\ Na \text{ (s)} + 3\ N_2 \text{ (g)}$$

$$NH_4NO_2 \text{ (aq)} \longrightarrow N_2 \text{ (g)} + 2\ H_2O \text{ (liq)}$$

One of the few direct reactions of nitrogen at room temperature is the reaction with lithium metal to form lithium nitride:

$$6\ Li \text{ (s)} + N_2 \text{ (g)} \longrightarrow 2\ Li_3N \text{ (s)}$$

Another occurs biologically in the root nodules of leguminous plants such as beans, peas, clover, and alfalfa, where nitrogen-fixing bacteria in the root nodules convert atmospheric nitrogen to compounds suitable for the plant's nutritional needs (Figure 6–8). Using light as their only energy source, these bacteria convert the nitrogen into ammonia, which is then absorbed by the plant and incorporated into amino acids for building proteins. Biological nitrogen fixation accounts for some 70% of all the nitrogen taken up in the environmental nitrogen cycle. It is far more efficient than any industrial nitrogen fixation process. Plants that lack nitrogen-fixing bacteria absorb necessary nitrogen from the soil in the form of naturally occurring nitrates or synthetic fertilizers. It has long been realized that if bacterial nitrogen fixation could be understood and translated into practical fertilization of the soil, a major step toward an economic solution of world food production would be taken. Unfortunately, nitrogen isn't easily fixed.

Nitrogen enters into many other reactions at high temperatures. The most important of these reactions is the **Haber process**, in which nitrogen and hydrogen react to produce ammonia:

$$N_2 \text{ (g)} + 3\ H_2 \text{(g)} \longrightarrow 2\ NH_3 \text{ (g)}$$

Pressures of 100 to 1000 atmospheres and a mixed iron/iron oxide catalyst are required for a suitable conversion. Ammonia produced by the Haber process is one of the two starting materials for nearly all commercially produced and industrially important nitrogen compounds. Reaction of ammonia with sulfuric acid yields ammonium sulfate, used as a water-soluble nitrogen-containing plant nutrient:

$$2\ NH_3 \text{ (aq)} + H_2SO_4 \text{ (aq)} \longrightarrow (NH_4)_2SO_4 \text{ (aq)}$$

The principal source of naturally occurring inorganic nitrogen is sodium nitrate, or Chilean saltpeter. Nitric acid can be produced by treating sodium nitrate with sulfuric acid. The reaction is run at elevated temperatures to drive off the volatile nitric acid (b.p. 86°C):

$$NaNO_3 \text{ (s)} + H_2SO_4 \text{ (liq)} \longrightarrow NaHSO_4 \text{ (s)} + HNO_3 \text{ (g)}$$

Nitric acid produced this way then becomes the starting material for the synthesis of a host of important nitrogen compounds, especially fertilizers and explosives.

In the classic laboratory synthesis of nitric acid, the materials of choice are potassium nitrate and concentrated sulfuric acid. The equipment must be all glass—no corks or rubber stoppers—because of the corrosive nature of the product. The flask fills with brown fumes due to formation of NO_2 (as a side-product), and the distilled acid has a yellowish color because of dissolved NO_2. The acid product is almost 100% HNO_3 and it is referred to as "fuming" nitric acid because it literally fumes. The concentrated nitric acid in the laboratory reagent bottle is a 69% aqueous solution of HNO_3. Commercial grades of nitric acid are synthesized by the **Ostwald process**, which is based on the catalytic oxidation of ammonia and will be described in detail in a later discussion.

An interesting reaction of molecular nitrogen occurs with calcium carbide in an electric furnace at 1100°C.

$$CaC_2 \text{ (s)} + N_2 \text{ (g)} \longrightarrow CaCN_2 \text{ (s)} + C \text{ (s)}$$

The reaction product $CaCN_2$ is known as calcium cyanamide. It has a number of potentially important industrial uses, and is also a plant nutrient and a source of ammonia:

$$CaCN_2 \text{ (s)} + 3 H_2O \text{ (g)} \longrightarrow CaCO_3 \text{ (s)} + 2 NH_3 \text{ (g)}$$

However, this process is more expensive than the Haber process for ammonia synthesis. The reaction of calcium cyanamide with sodium carbonate and carbon yields sodium cyanide, a very useful although highly toxic salt:

$$Na_2CO_3 \text{ (s)} + CaCN_2 \text{ (s)} + C \text{ (s)} \longrightarrow 2 NaCN \text{ (s)} + CaCO_3 \text{ (s)}$$

Sodium cyanide is used to extract gold and silver from their ores and in electroplating these metals. The reaction of sodium cyanide with an aqueous acid such as H_2SO_4 produces a highly toxic gas, hydrogen cyanide:

$$H_2SO_4 \text{ (aq)} + 2NaCN \text{ (s)} \longrightarrow 2HCN \text{ (g)} + Na_2SO_4 \text{ (s)}$$

Therefore, all cyanides must be handled with caution. Reaction of HCN with acetylene yields acrylonitrile, which is used to prepare acrylic plastics and synthetic textile fibers:

$$C_2H_2 \text{ (g)} + HCN \text{ (g)} \longrightarrow H_2C{=}CHCN \text{ (liq)}$$

For many years, HCN was an important intermediate in the synthesis of nylon.

Life processes require nitrogen for construction of nitrogen-containing organic molecules called proteins. Green plants use nitrates for this purpose. They are usually obtained in aqueous solution through the roots or through nitrogen-fixing bacteria. Animals obtain the nitrogen required for protein synthesis either by eating plants or by eating animals that have eaten plants. Metabolism of nitrogen-containing animal proteins ends with excretion of nitrogen as urea and the return of the nitrogen to nature, completing a complex cycle called the nitrogen cycle. Modern agriculture mainly uses synthetic fertilizers to replenish the nitrogen in the soil.

6.7 OXYGEN AND OZONE

Elemental oxygen exists in two different forms: as diatomic O_2, the **oxygen** molecule of the air, and as triatomic O_3, the **ozone** molecule of the upper atmosphere. Oxygen (O_2; m.p. $= -218$°C; b.p. $= -183$°C) is a colorless, odorless, and tasteless gas that is soluble in water to the extent of

Joseph Priestley (1733–1804), an English chemist and Unitarian clergyman, discovered oxygen (1774) by allowing the heat of the Sun's rays, when concentrated by a large lens, to decompose red mercury(II) oxide. The new gas supported combustion and metabolism: a glowing splint burst into flame and a mouse survived in the presence of the gas.

0.007 g/100 mL H_2O at 0°C. This limited quantity of dissolved oxygen supports the underwater existence of fish and other aquatic species. The condensed states of oxygen—liquid and solid—are characteristically pale blue.

Ozone (O_3; m.p. = −193°C; b.p. = −112°C) can be prepared by continuously passing an electric discharge through an oxygen atmosphere. Concentrations of up to 10% can be obtained in this way, following which the ozone can be separated by fractional liquefaction. The odor of ozone can be detected near an electric discharge, such as a sparking motor or a summer thunderstorm. It is a powerful bleaching agent and germicide, and must be handled with care because of its considerable toxicity. The gas is perceptibly blue, the liquid is deep blue and very explosive, and the solid has a black-violet appearance. Ozone plays a critical role in the Earth's upper atmosphere, acting as a shield against dangerous ultraviolet radiation.

Ozone reacts with silver at room temperature to produce black silver oxide or with potassium iodide to produce iodine:

$$O_3 \text{ (g)} + 2 \text{ Ag (s)} \longrightarrow Ag_2O \text{ (s)} + O_2 \text{ (g)}$$

$$H_2O \text{ (liq)} + 2 \text{ KI (s)} + O_3 \text{ (g)} \longrightarrow I_2 \text{ (s)} + 2 \text{ KOH (s)} + O_2 \text{ (g)}$$

Oxygen is produced commercially, along with nitrogen, by the fractional distillation of liquid air. In Priestley's original laboratory preparation, mercury(II) oxide was thermally decomposed:

$$2 \text{ HgO (s)} \longrightarrow 2 \text{ Hg (liq)} + O_2 \text{ (g)}$$

Quite a few substances are known to give off oxygen on heating, including manganese(IV) oxide and lead(II) nitrate. The latter decomposes when heated in a test tube over a Bunsen flame, with a crackling sound (as the crystals shatter). The reaction products are lead(II) oxide, dinitrogen tetroxide, and oxygen:

$$2 \text{ MnO}_2 \text{ (s)} \longrightarrow 2 \text{ MnO (s)} + O_2 \text{ (g)}$$

$$2 \text{ Pb(NO}_3)_2 \text{ (s)} \longrightarrow 2 \text{ PbO (s)} + 2 \text{ N}_2O_4 \text{ (g)} + O_2 \text{ (g)}$$

Heating potassium chlorate with catalytic quantities of manganese(IV) oxide is the traditional laboratory preparation:

$$2 \text{ KClO}_3 \text{ (s)} \longrightarrow 2 \text{ KCl (s)} + 3 O_2 \text{ (g)}$$

Oxygen is far less reactive than ozone, normally requiring elevated temperatures for reaction. However, under appropriate conditions, oxygen will combine with almost every other element in the periodic table. One of the greatest difficulties encountered by firefighters is a well-established fire that races through a building or across a forest, due to the presence of combustible materials at elevated temperatures. Devastating fire storms destroyed Tokyo and Dresden in the aftermath of incendiary air raids during World War II. Perhaps because nitrogen serves to dilute the oxygen in the air, we don't commonly recognize oxygen as being particularly reactive, but it is! The Apollo 1 crew died in 1967 in a matter of seconds in the fire that erupted in the pure oxygen atmosphere of their cabin.

6.8   OXYGEN COMPOUNDS IN THE ATMOSPHERE

The most familiar compounds of oxygen are the oxides. Those of non-metallic elements are covalent compounds that tend to be gases, liquids, or low-melting solids. In marked contrast, most metallic oxides are high-melting ionic solids. Here, we will consider water—the oxide of hydrogen—and the oxides of carbon, nitrogen, and sulfur.

Water

Most of the hydrogen present during the formation of the Earth has long since dissipated, and the percentage abundance of this element on Earth is much lower than in the universe as a whole. However, the one principal compound of hydrogen on Earth may well qualify as the most important compound of them all. Water is everywhere, and it is essential to all forms of life. It is present in all plants and animals, and is the solvent commonly used for many chemical reactions in both the laboratory and the world around us. We note in advance of more detailed discussions in Chapters 8, 9, 12, and 18 its unusually high boiling point because of intermolecular attractions between water molecules.

Water readily reacts with active metals, giving gaseous hydrogen and metal hydroxides:

$$Ca\ (s)\ +\ 2\ H_2O\ (liq)\ \longrightarrow\ Ca(OH)_2\ (s)\ +\ H_2\ (g)$$

Less reactive metals are attacked only by steam at high pressures, again generating hydrogen gas, along with the corresponding metal oxides:

$$3\ Fe\ (s)\ +\ 4\ H_2O\ (g)\ \longrightarrow\ Fe_3O_4\ (s)\ +\ 4\ H_2\ (g)$$

$$Ni\ (s)\ +\ H_2O\ (g)\ \longrightarrow\ NiO\ (s)\ +\ H_2\ (g)$$

Steam corrodes steel, an alloy of carbon and iron, at 350°C and will corrode stainless steel, an alloy of iron, chromium, and nickel, at 500°C. The high reactivity of steam at elevated temperatures limits the temperature of steam used in power plants.

Of all of the water on the planet Earth, 98% is seawater, which covers 71% of the Earth's surface. The remaining 2% of water on the Earth is found in freshwater bodies such as lakes, ponds, rivers, and streams, in the ground, in the tissues of plants and animals, frozen into the polar ice caps, and contained in the atmosphere as water vapor, clouds, and falling rain and snow. Surface water on the Earth participates in a huge cycle, in which liquid water evaporates to enter the atmosphere and then returns by condensation as either rain or snow. Most of the evaporation occurs from the surface of the oceans because of their overwhelming size, and the water is returned to the oceans either directly or by the rivers and streams that empty into them. Given sufficient time, every water molecule will make this round trip. However, the vast capacity of the oceans and the small fraction of the water molecules that evaporate at any time result in an estimated average time per cycle of 40,000 years.

Carbon Oxides

Carbon forms two important oxides, a monoxide (CO) and a dioxide (CO_2). Both can be prepared from the elements by proper adjustment of the stoichiometry:

$$C\ (s)\ +\ excess\ O_2\ (g)\ \longrightarrow\ CO_2\ (g)$$

$$C\ (s)\ +\ limited\ O_2\ (g)\ \longrightarrow\ CO\ (g)$$

compound	MW, μ	b.p., °C
CH_4	16	−184
NH_3	17	−34
H_2O	18	+100
C_7H_{16}	100	+98

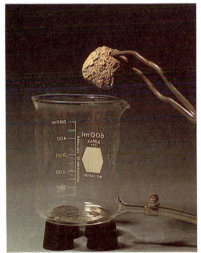

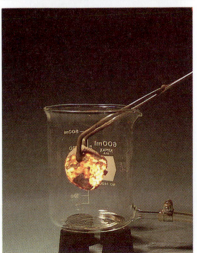

A carbon briquette burns white hot in an oxygen atmosphere, forming carbon dioxide.

METHYL ALCOHOL—ALIAS METHANOL—ALIAS WOOD ALCOHOL

Before 1926, all methyl alcohol produced in North America was obtained from the destructive distillation of wood (hence the name wood alcohol). In that year, synthetic alcohol from petroleum was introduced in Germany; today, only negligible quantities are made from wood. The modern industrial synthesis of this commodity chemical is almost exclusively based on the reaction of pressurized mixtures of hydrogen and carbon monoxide (synthesis gas) and carbon dioxide. This reaction takes place on the surface of metallic catalysts, typically zinc-chromium (for high pressure processes) or copper (for low pressure processes). Choices of catalysts and reaction conditions are critical, since the reactants are capable of combining into so many different organic products:

$$CH_4 \text{ (g)} + H_2O \text{ (steam)} \rightleftharpoons CO \text{ (g)} + 3 H_2 \text{ (g)}$$

$$CO_2 \text{ (g)} + H_2 \text{ (g)} \rightleftharpoons CO \text{ (g)} + H_2O \text{ (g)}$$

$$CO \text{ (g)} + 2 H_2 \text{ (g)} \rightleftharpoons CH_3OH \text{ (g)}$$

The catalysts are easily poisoned by sulfur compounds, so sulfur-free CO/H_2 mixtures are required.

The primary uses for methanol are the syntheses of formaldehyde and dimethyl terephthalate, used respectively in the manufacture of plastics used to laminate kitchen countertops (Formica) and in the ubiquitous transparent soda bottle (polyethylene terephthalate). At the height of the international petroleum crisis of 1970's, when cars were lined up at the pumps and the price of motor fuels more than quadrupled, it was recognized that methanol is an excellent carburated motor fuel, either alone (neat) or mixed with gasoline. It has better "knocking" behavior because of the high octane number (>100), and fewer pollutants are exhausted into the atmosphere. Methanol is also becoming increasingly important as a source of carbon for single cell protein production (SCP) via microorganisms, particularly yeasts and bacteria, in the presence of aqueous nutrient salt solutions containing the essential N, P, and S compounds.

It is worth noting that, unlike its two-carbon sister alcohol (ethanol, which is selectively metabolized), methanol is highly toxic. The most generally known hazard of drinking methanol is blindness.

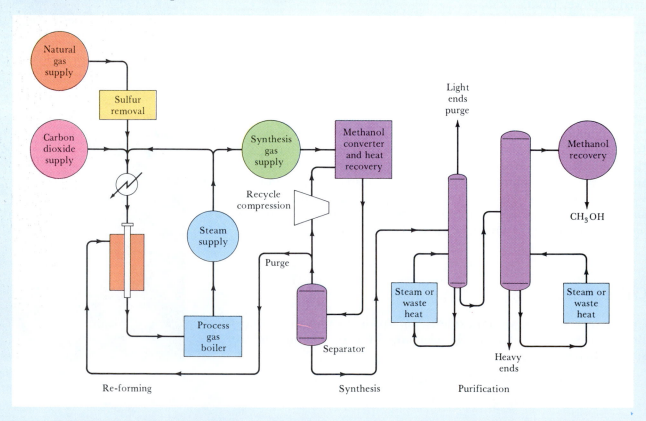

Furthermore, CO will burn in air to give CO_2, and hot carbon will convert CO_2 to CO:

$$2\ CO\ (g)\ +\ O_2\ (g)\ \longrightarrow\ 2\ CO_2\ (g)$$

$$CO_2\ (g)\ +\ C\ (s)\ \longrightarrow\ 2\ CO\ (g)$$

All four reactions take place in different parts of a furnace when coal, coke, or charcoal is burned. Coal burned in a shortage of air yields a product mixture rich in the flammable gas carbon monoxide; the mixture is called **producer gas**. One common industrial method for the large scale industrial preparation of carbon monoxide is the action of steam on methane (natural gas):

$$CH_4\ (g)\ +\ H_2O\ (steam)\ \longrightarrow\ CO\ (g)\ +\ 3\ H_2\ (g)$$

The product is sometimes called **water gas**, for obvious reasons; after adjusting the hydrogen content, it is called **synthesis gas** because it is so widely used for producing industrial scale quantities of important molecules such as methanol:

$$CO\ (g)\ +\ 2\ H_2\ (g)\ +\ catalyst\ \longrightarrow\ CH_3OH\ (g)$$

Carbon monoxide chemistry and synthesis gas are subjects of intense scientific interest because they could be sources of many important carbon compounds currently obtained from petroleum, when petroleum supplies diminish to the point that they are no longer economical. For example, gasoline-range hydrocarbons—which have the general formula C_nH_{2n+2}—can be obtained from synthesis gas by the **Fischer-Tropsch process**:

$$n\ CO\ (g)\ +\ (2n\ +\ 1)\ H_2\ (g)\ \longrightarrow\ C_nH_{2n+2}\ (g)\ +\ n\ H_2O\ (g)$$

Carbon monoxide melts at $-205°C$ and boils at $-190°C$. It is a colorless, odorless, highly toxic gas with about the same density as air. The carbon monoxide produced in the incomplete combustion of hydrocarbons such as motor fuels is a familiar hazard. The toxic action of carbon monoxide is well understood. It is able to bind blood hemoglobin more strongly than can oxygen, which is thereby excluded from normal respiration. This results in suffocation. The bright red characteristic color of oxygenated blood is due to the presence of oxyhemoglobin, which readily decomposes to make the bound oxygen available for metabolism. However, scarlet-colored carboxyhemoglobin is much more stable. In some cases, CO poisoning can be reversed by emergency administration of hyperbaric (high-pressure) oxygen.

The chemistry of carbon monoxide is not particularly extensive. Among its commercially important reactions are those with metals such as nickel, yielding the surprisingly volatile and often highly toxic metal carbonyls such as nickel tetracarbonyl (m.p. = $-19°C$; b.p. = $43°C$):

$$Ni\ (s)\ +\ 4\ CO\ (g)\ \longrightarrow\ Ni(CO)_4\ (g)$$

Ultrapure nickel metal can be obtained by first forming nickel tetracarbonyl, distilling this product to high purity, and then decomposing it again to Ni and CO at high temperature. This process is known as the **Mond process**.

Carbon monoxide combines directly with anhydrous chlorine gas in the presence of ultraviolet light or a suitable catalyst to form carbonyl chloride, better known as phosgene:

$$CO\ (g)\ +\ Cl_2\ (g)\ \longrightarrow\ COCl_2\ (g)$$

In practice, a mixture of the CO and Cl_2 is passed through a bed of activated charcoal at about 200°C. Used ineffectively (but dramatically) for gas warfare during World War I by both sides, phosgene today is widely used in the manufacture of dyes and drugs. More recently, it has been employed in the synthesis of plastics called polycarbonates that are fabricated into everything from automobile bumpers to McDonald's golden arches.

Carbon monoxide is considered to be a serious atmospheric pollutant. The quantities released by the engine of an automobile can be reduced through the use of a catalytic converter installed into the exhaust system. This introduces additional air to the exhaust and then passes the mixture over a catalyst—usually platinum—that has been heated by passage of previous exhaust. The oxygen in the air converts CO to the much less harmful CO_2:

$$2\ CO\ (g)\ +\ O_2\ (g)\ \xrightarrow{\text{platinum catalyst}}\ 2\ CO_2\ (g)$$

Carbon dioxide is a colorless, odorless, essentially nontoxic gas. It **sublimes**, changing directly from a solid to a gas, at atmospheric pressure and any temperature above −78°C, and the liquid state cannot be obtained without increasing the pressure to more than 5 atm. In CO_2 fire extinguishers, the internal pressure is sufficient (about 60 atm at 25°C) for the liquid state to exist. Solid CO_2 is commonly used as a refrigerant, often under the trade name Dry Ice. As a waste product of animal metabolism, carbon dioxide is carried to the lungs by hemoglobin in the blood and exhaled during normal respiration. The overall process reflects a delicately tuned balance—if the CO_2 concentration in the air is too high, suffocation can result.

When CO_2 dissolves in water, it reacts to some extent to form an unstable solution called carbonic acid:

$$CO_2\ (g)\ +\ H_2O\ (liq)\ \rightleftharpoons\ H_2CO_3\ (aq)$$

In the laboratory, carbon dioxide is usually produced by the reaction of a mineral acid such as aqueous HCl on calcium carbonate in the form of marble chips, chalk, or crushed oyster shells (Figure 6–9):

$$CaCO_3\ (s)\ +\ 2\ HCl\ (aq)\ \longrightarrow\ CaCl_2\ (aq)\ +\ H_2O\ (liq)\ +\ CO_2\ (g)$$

Carbon monoxide can be prepared in the laboratory by passing CO_2 through a hardened glass tube packed with charcoal at elevated temperatures:

$$C\ (charcoal)\ +\ CO_2\ (g)\ \longrightarrow\ 2\ CO\ (g)$$

A second method for preparing carbon monoxide in the laboratory is the removal of water from formic acid, either by heating or by the dehydrating action of sulfuric acid:

$$HCOOH\ (liq)\ \longrightarrow\ CO\ (g)\ +\ H_2O\ (liq)$$

In addition to the combustion processes, carbon dioxide is the product of many other industrial reactions. Examples are heating of calcium carbonate and fermentation of sugars:

$$CaCO_3\ (s)\ \longrightarrow\ CaO\ (s)\ +\ CO_2\ (g)$$

$$\underset{\text{(glucose or fructose)}}{C_6H_{12}O_6\ (aq)}\ \xrightarrow{\text{yeast}}\ 2\ \underset{\text{(ethanol)}}{C_2H_5OH\ (aq)}\ +\ 2\ CO_2\ (g) \qquad \text{(fermentation)}$$

Because carbon dioxide does not support combustion, it is widely used in fire extinguishers. However, oxygen can be removed from the CO_2 by

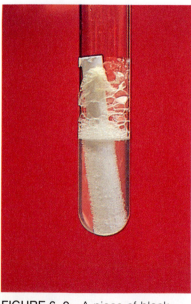

FIGURE 6–9 A piece of blackboard chalk, which is mostly calcium carbonate ($CaCO_3$), reacts rapidly with hydrochloric acid (HCl), producing an aqueous solution of calcium chloride ($CaCl_2$) and liberating bubbles of carbon dioxide (CO_2).

vigorously burning substances that generate very high temperatures, supporting further combustion. Thus, other types of extinguishers (Figure 6–10) must be used. The problem is critical when construction materials include magnesium or aluminum, which is typically the case in jet aircraft and office towers, because of reactions like this one:

$$CO_2 \text{ (g)} + 2 \text{ Mg (s)} \longrightarrow 2 \text{ MgO (s)} + C \text{ (s)}$$

Nitrogen Oxides

There are many oxides of nitrogen. The most important are N_2O, NO, NO_2, and N_2O_4. Dinitrogen monoxide, commonly referred to as nitrous oxide or "laughing gas," has been used as a local and general anesthetic (Figure 6–11) and as a propellant for aerosol cans containing everything from shaving cream and hair sprays to whipped cream and cheese spreads. N_2O has also been implicated in chemical processes in the upper atmosphere that could contribute to depletion of the ozone layer. It can be prepared by heating ammonium nitrate at 250°C:

$$NH_4NO_3 \text{ (s)} \longrightarrow N_2O \text{ (g)} + 2 \text{ H}_2O \text{ (g)}$$

Dinitrogen monoxide decomposes to nitrogen and oxygen at elevated temperatures. For example, the heat of a candle flame is sufficient to produce the reaction:

$$2 \text{ N}_2O \text{ (g)} \longrightarrow 2 \text{ N}_2 \text{ (g)} + O_2 \text{ (g)}$$

The O_2 produced supports the continued combustion of the candle.

Nitrogen and oxygen react at the high temperatures of an electric arc, or an internal combustion or jet aircraft engine, to form nitrogen monoxide (nitric oxide):

$$N_2 \text{ (g)} + O_2 \text{ (g)} \longrightarrow 2 \text{ NO (g)}$$

Nitrogen monoxide is a colorless gas that reacts instantly with more oxygen

FIGURE 6–10 Fire extinguisher.

FIGURE 6–11 The humorous effects of "laughing gas" are shown in this 1802 cartoon by James Gilray. The caption reads ". . . an experimental lecture on the powers of air." It is a parody of the public science lectures given at the Royal Institution (London) and still popular today. Sir Humphrey Davy (the director, with bellows) assists.

FIGURE 6–12 Effect of temperature on nitrogen dioxide. At 0°C (right) N_2O_4, which is colorless, predominates. At 50°C (left), some of the N_2O_4 has dissociated to give the deep brown color of NO_2.

There was a time when this simple experiment could be done with a penny, but pennies haven't been "coppers" for more than 25 years now.

FIGURE 6–13 Copper reacts vigorously with nitric acid, yielding copper(II) nitrate, $Cu(NO_3)_2$, and brown nitrogen dioxide (NO_2).

to form a red-brown gas, nitrogen dioxide:

$$2 \text{ NO (g)} + \text{O}_2 \text{ (g)} \longrightarrow 2 \text{ NO}_2 \text{ (g)}$$

It is not possible to draw Lewis structures for NO or NO_2 with an octet of electrons around each atom, because each molecule has an odd number of valence electrons—13 for NO and 19 for NO_2. For NO_2 the best structure is

$$:\ddot{O}—\overset{\textstyle .}{N}=\ddot{O}:$$

with only seven electrons around N. Combination of two NO_2 molecules produces a new "dimeric" molecule that does satisfy the octet rule:

The product is N_2O_4, dinitrogen tetroxide. At room temperature and pressure, reactant molecules (NO_2) and product molecules (N_2O_4) are both present:

$$\underset{\text{(colorless)}}{\text{N}_2\text{O}_4 \text{ (g)}} \rightleftharpoons \underset{\text{(deep red-brown)}}{2 \text{ NO}_2 \text{ (g)}}$$

At very low temperatures (in the solid state) this chemical system is colorless, since only colorless dinitrogen tetroxide molecules are present. At the normal boiling point of N_2O_4 (21°C), the liquid has a deep brown color because of the presence of a small amount (0.1%) of monomeric nitrogen dioxide (NO_2) molecules. As the temperature increases, the color of the vapor intensifies greatly as the dimer dissociates into the monomer; at 140°C the dissociation is complete (Figure 6–12).

The laboratory preparation of NO can be accomplished by the direct reaction of nitric acid with copper metal:

$$3 \text{ Cu (s)} + 8 \text{ HNO}_3 \text{ (aq)} \longrightarrow$$
$$2 \text{ NO (g)} + 4 \text{ H}_2\text{O (liq)} + 3 \text{ Cu(NO}_3)_2 \text{ (aq)}$$

The reaction is easily demonstrated by placing a strip of copper sheet in concentrated nitric acid (in a fume hood). Copious quantities of brown fumes of NO_2 are produced from the reaction of NO and oxygen in the air (Figure 6–13). The direct reaction of N_2 and O_2 in an electric arc has been used for the industrial production of NO. Because the electrical energy required is considerable, this method has been restricted to places such as Canada, Norway, and upstate New York, which have access to cheap electricity. In general, NO is produced commercially by the catalytic oxidation of ammonia, an intermediate step in the **Ostwald process** for nitric acid manufacture:

$$4 \text{ NH}_3 \text{ (g)} + 5 \text{ O}_2 \text{ (g)} \xrightarrow{\text{Pt catalyst}} 4 \text{ NO (g)} + 6 \text{ H}_2\text{O (g)}$$

In this process, a mixture of Haber process ammonia and air—which serves as the source of oxygen—is passed over a gauze of platinum/rhodium alloy at red heat. The initial reaction is highly exothermic, and the heat generated is used to sustain the temperature of about 900°C at the catalyst surface. Ammonia reacts with oxygen in the absence of a catalyst, but N_2 (not NO) is the product:

$$4 \text{ NH}_3 \text{ (g)} + 3 \text{ O}_2 \text{ (g)} \longrightarrow 2 \text{ N}_2 \text{ (g)} + 6 \text{ H}_2\text{O (g)}$$

The nitrogen monoxide produced in the Ostwald process reacts with more O_2 to form NO_2, which reacts with water to form nitric acid:

$$2 \text{ NO (g)} + O_2 \text{ (g)} \longrightarrow 2 \text{ NO}_2 \text{ (g)}$$

$$3 \text{ NO}_2 \text{ (g)} + H_2O \text{ (liq)} \longrightarrow 2 \text{ HNO}_3 \text{ (aq)} + \text{NO (g)}$$

Byproduct NO, produced in the last reaction, is recycled in order to improve the overall yield. Nitric acid, a valuable industrial product, has many varied uses, but principally it serves as a nitrogen source in fertilizers. Because of its especially high nitrogen content and easy transportability, ammonium nitrate is used heavily for fertilizer applications:

$$\text{HNO}_3 \text{ (aq)} + \text{NH}_3 \text{ (g)} \longrightarrow \text{NH}_4\text{NO}_3 \text{ (aq)}$$

Nitrogen monoxide is a serious atmospheric pollutant. It is produced in significant quantities in the internal combustion engine, and it forms NO_2 in the air. Nitrogen dioxide (NO_2) is an acidic oxide, reacting in turn with airborne water droplets to form nitric acid. This reaction is one of the sources of both acid rain and plant nutrient nitrate.

The catalytic converter of a modern automobile brings about the decomposition of NO to the normal atmospheric gases, nitrogen and oxygen:

$$2 \text{ NO (g)} \xrightarrow{\text{Pt catalyst}} N_2 \text{ (g)} + O_2 \text{ (g)}$$

Most commercial nitrogen-containing compounds (except those derived from plant or animal protein) have their origin in ammonia produced from atmospheric nitrogen. Beyond the essential few already described, many useful and sometimes problematic compounds exist. Some are highly energetic and relatively unstable, and are used as fuels and explosives. Reaction of nitric acid with a number of compounds derived from plants or animals produces explosive substances. For example, glycerine (a byproduct of the rendering of plant or animal fats for the manufacture of soaps) reacts with nitric acid to give glyceryl trinitrate, better known by its common name, nitroglycerine:

$$\underset{\text{(glycerine)}}{C_3H_8O_3 \text{ (liq)}} + 3 \text{ HNO}_3 \text{ (aq)} \longrightarrow \underset{\text{(glyceryl trinitrate)}}{C_3H_5N_3O_9 \text{ (s)}} + 3H_2O \text{ (liq)}$$

When glyceryl trinitrate detonates, it undergoes an abrupt and enormous volume change:

$$4 \text{ C}_3\text{H}_5\text{N}_3\text{O}_9 \text{ (liq)} \longrightarrow 10 \text{ H}_2\text{O (g)} + 6 \text{ N}_2 \text{ (g)} + 12 \text{ CO}_2 \text{ (g)} + O_2 \text{ (g)}$$

Glyceryl trinitrate is dangerous to use because it is sensitive to shock. Nobel discovered that the sensitivity is greatly diminished if the compound is absorbed by a clay known as diatomaceous earth. The resulting composite material is called **dynamite**.

The reaction of cotton with nitric acid leads to the formation of nitrocellulose, or "gun cotton." It is used in firearms (and firecrackers) as smokeless powder. Another important nitrogen-containing explosive is trinitrotoluene (TNT), $C_7H_5N_3O_6$. Lead azide, $Pb(N_3)_2$, explodes when struck sharply and is used in percussion caps.

Although they are not nitrogen oxides, hydrazines are so readily oxidized that they are notable in the general context of these compounds. Hydrazine, N_2H_4, has been used as a rocket fuel. In fact, gasolines laced with hydrazine are widely used by drag-racing enthusiasts in autos equipped

The combustion of ammonia vapor at the tip of a jet delivering oxygen gas. The solution is concentrated aqueous ammonia. Note that the flame is quite independent of the presence of water.

Alfred Nobel (1833–1896), a Swedish chemist and industrialist, was shocked by the results of the military use of dynamite. He used much of the profits from its production to found the Nobel Foundation, dedicated to recognition of science in the benefit of humanity and the pursuit of peace.

Propulsion system for the Lunar Excursion Module (LEM).

with special engines. The Lewis structure of hydrazine is

$$\begin{array}{ccc} H & & H \\ | & & | \\ :N & \!\!-\!\! & N: \\ | & & | \\ H & & H \end{array}$$

and it can be made by the reaction

$$2\ NH_3\ (aq)\ +\ NaOCl\ (aq)\ \longrightarrow\ N_2H_4\ (aq)\ +\ NaCl\ (aq)\ +\ H_2O\ (liq)$$

Hydrazine burns easily and cleanly, and gives off considerable heat:

$$N_2H_4\ (liq)\ +\ O_2\ (g)\ \longrightarrow\ N_2\ (g)\ +\ 2\ H_2O\ (liq)$$

Note the formation of N_2 as a reaction product! Again, that contributes significantly to the heat evolved. Several hydrazine derivatives have also been tested as high energy explosives and fuels. One important example is dimethyl hydrazine, $(CH_3)_2N_2H_2$, and its reaction with dinitrogen tetroxide:

$$(CH_3)_2N_2H_2\ (liq)\ +\ 2\ N_2O_4\ (liq)\ \longrightarrow$$
$$3\ N_2\ (g)\ +\ 4\ H_2O\ (g)\ +\ 2\ CO_2\ (g)$$

Dimethyl hydrazine ignites spontaneously on mixing with dinitrogen tetroxide, so the reaction was used in the Lunar Excursion Modules (LEMs) of the Apollo program.

Sulfur Oxides

There are two important sulfur oxides, the dioxide (SO_2) and the trioxide (SO_3). Sulfur burns readily in air to produce the dioxide, a colorless gas with the choking, suffocating odor of burning sulfur:

$$S\ (s)\ +\ O_2\ (g)\ \longrightarrow\ SO_2\ (g)$$

With a relatively high boiling point of $-10°C$ at atmospheric pressure, SO_2 is easily liquefied. It can be produced in the laboratory either by warming a metal (usually copper) in concentrated sulfuric acid or by warming a sulfite salt in dilute acid:

$$Cu\ (s)\ +\ 2\ H_2SO_4\ (liq)\ \longrightarrow\ CuSO_4\ (aq)\ +\ 2\ H_2O\ (liq)\ +\ SO_2\ (g)$$
$$Na_2SO_3\ (aq)\ +\ 2\ HCl\ (aq)\ \longrightarrow\ 2\ NaCl\ (aq)\ +\ H_2O\ (liq)\ +\ SO_2\ (g)$$

Industrially, in addition to direct combination of the elements, sulfur dioxide is produced by roasting iron pyrites (FeS_2) in air:

$$4\ FeS_2\ (s)\ +\ 11\ O_2\ (g)\ \longrightarrow\ 2\ Fe_2O_3\ (s)\ +\ 8\ SO_2\ (g)$$

Sulfur dioxide (SO_2) is an effective yet moderate bleaching agent that is less destructive to fabrics than the common chlorine bleaches such as "Clorox," which bleach by oxidation.

The reaction of SO_2 with oxygen to produce SO_3 is impractically slow unless a catalyst such as platinum metal, an oxide of vanadium (V_2O_5), or NO is used:

$$2\ SO_2\ (g)\ +\ O_2\ (g)\ \xrightarrow{\text{catalyst}}\ 2SO_3\ (g)$$

Both sulfur oxides are water-soluble, reacting with water to form aqueous acidic solutions. **Sulfurous acid** is obtained from SO_2. It is a very weak acid and cannot be concentrated without loss of SO_2:

$$SO_2\ (g)\ +\ H_2O\ (liq)\ \longrightarrow\ H_2SO_3\ (aq)$$

Sulfuric acid is obtained from SO_3:

$$SO_3 \text{ (g)} + H_2O \text{ (liq)} \longrightarrow H_2SO_4 \text{ (liq)}$$

Bubbling the sulfur trioxide gas stream into concentrated sulfuric acid is an important part of the industrial process, generating the "polynuclear" sulfuric acid

$$SO_3 \text{ (g)} + H_2SO_4 \text{ (liq)} \longrightarrow H_2S_2O_7 \text{ (liq)}$$

which is carefully decomposed with the stoichiometric quantity of water to give the industrial strength, concentrated form of sulfuric acid:

$$H_2S_2O_7 \text{ (liq)} + H_2O \text{ (liq)} \longrightarrow 2 H_2SO_4 \text{ (liq)}$$

Sulfuric acid is a very strong acid. In its 98% concentrated form (b.p. = 338°C) it is of enormous commercial importance in the processing and manufacture of everything from drugs and fertilizers to plastics and steels.

Sulfur oxides in the atmosphere are dangerous pollutants, resulting primarily from the combustion of sulfur-containing coal and petroleum fuels and from the smelting of sulfur-containing metal ores. Gasolines often contain small amounts of sulfur that exit with the exhaust gases, mainly as SO_2, from the combustion processes in an engine. This sulfur dioxide reacts with moisture in the atmosphere to form sulfurous acid, which then oxidizes to sulfuric acid. Such "smogs"—which have become all too common in industrialized nations—cause everything from lung problems for individuals to acid rains capable of destroying forests and farm lands. One way of removing sulfur oxides from stack gases is by reaction with lime:

$$CaO \text{ (s)} + SO_2 \text{ (g)} \longrightarrow CaSO_3 \text{ (s)}$$

$$CaO \text{ (s)} + SO_3 \text{ (g)} \longrightarrow CaSO_4 \text{ (s)}$$

A promising new method for burning high-sulfur coal stocks without atmospheric contamination by sulfur oxides is to form the coal into a fluidized (finely powdered) bed. The dustlike coal particles are kept mobile as a liquid by air or oxygen that is forced through a perforated plate beneath the bed. Powdered lime is mixed with the coal to react with and remove the sulfur oxides as they are formed. It has also been reported recently that certain microorganisms hold promise in economical desulfurization of coal for use in electricity-generating plants and other large industrial applications.

> When sulfur cannot be removed from coal before it is burned, expensive postcombustion desulfurization equipment must be used to prevent great quantities of sulfurous wastes from entering the environment. Up to 90% of inorganic (pyritic) sulfur can be removed by chemical means. Now the more difficult task of removing sulfur bonded to carbon (organic sulfur) is being tackled. Certain microbial agents—dubbed "IGT" organisms after the Institute of Gas Technology, where they were developed by genetic engineering—hold promise for precombustion desulfurization.

Air Pollution

Contamination of the Earth's atmosphere is not a new phenomenon, nor are humans responsible for all of it. In a sense, air pollution is part of the normal life cycle. Problems arise because our contribution, perhaps no more than 0.05% of the total emissions from all sources entering the 12 miles or so of the troposphere blanketing the Earth, happens to be particularly obnoxious. Air pollution began to develop epic proportions with the onset of the industrial revolution and urbanization in the 18th and 19th centuries. Lethal fogs and smogs of historic proportions have been recorded. In 1909 in Glasgow, for example, there were 1063 deaths; in 1952, more than 4000 deaths due to a killer smog were reported in London. Donora, Pennsylvania, a coal-mining town of 15,000 inhabitants, found half its population ill and 20 dead due to an extreme concentration of atmospheric pollutants in 1948.

In 1951, Haagen-Smit pinpointed a key mechanism involving an atmospheric reaction between unburned gasoline-range hydrocarbons—presumably unburned automobile emissions—and nitrogen dioxide in the presence of bright sunlight, a photochemical reaction. The products were peroxyacyl nitrates (PANs) and ozone. In poorly ventilated, well-irradiated urban areas such as the Los Angeles basin, photochemical smog has been superimposed upon our earlier air pollution problems of smoke and particulates.

Substances exhausted into the atmosphere are subject to a number of physical and chemical processes that pollute the air. Photochemical smog has three main causes:

1. Nitrogen oxides enter the air chiefly by fixation of nitrogen from the air during combustion in the internal combustion engine. This is followed by a rapid, light-induced oxidation to nitrogen dioxide.
2. Hydrocarbons such as unburned gasoline and fuel oil are a second factor. Their droplets are sites upon which other reactions can occur.
3. The third factor is bright sunlight. In the sequence of reactions leading to photochemical smog, nitrogen dioxide decomposes to nitrogen monoxide and atomic oxygen:

$$NO_2 \text{ (g)} \longrightarrow NO \text{ (g)} + O \text{ (g)}$$

Oxygen atoms then react with molecular oxygen in the presence of an inert third body (presumably nitrogen) to yield ozone:

$$O \text{ (g)} + O_2 \text{ (g)} \longrightarrow O_3 \text{ (g)}$$

Ozone molecules are then free to react with nitrogen monoxide, forming more nitrogen dioxide:

$$O_3 \text{ (g)} + NO \text{ (g)} \longrightarrow NO_2 \text{ (g)} + O_2 \text{ (g)}$$

The unburned hydrocarbons are consumed by this set of compounds, showing up as a complex mixture of oxidized products and organic nitrates such as peroxyacetyl nitrate (PAN). Nitrogen oxide is consumed and nitrogen dioxide is regenerated.

The primary effects of air pollution generally include the following:

1. Direct and significant economic loss due to increased maintenance and replacement costs for equipment as well as diminished property values.
2. Restricted visibility.
3. Acid rain.
4. Varying degrees of physiological distress for plants and animals and, in the extreme, significant kills of local plant and animal populations and destruction of arable lands. In humans, the effects are mainly chronic or acute respiratory and cardiac problems.

Acid Rain

Acid precipitation in the form of rain, snow, or fog was first reported late in the 19th century. However, it has been recognized as a serious environmental problem only since the 1960s as a result of Rachel Carson's revealing book, *Silent Spring*. The damaging effects are now known to be due to the release of sulfur oxides and nitrogen oxides into the atmosphere, mainly at the sites of industrial power plants, petroleum refineries, and iron and steel mills. The oxides eventually mix with moisture and form sulfuric and nitric acid rains. In the vicinity of smelters in which high-sulfur coals or sulfide ores are oxidized, the concentrations of sulfur dioxide, sulfur trioxide, and the sulfur acids can be so great as to create local deserts. Unfortunately, industries attempted to dilute these effects by building high stacks to carry the acidic oxides and acids away from the local environment, and the compounds have been spread across whole continents.

These acids are extremely destructive to aquatic and terrestrial ecosystems, making lakes, ponds, streams, and soils too acidic to support life. A major change in soil acidity also makes it possible for important minerals to be leached from the soil, and produces highly toxic run-offs that affect farm lands and waters at a distance from the original source of the contamination. As a striking and well-understood illustration of the complexity of these environmental problems, fish kills are sometimes caused not by the increased acidity of the aquatic environment, but by aluminum ions leached out of neighboring soils that irritate their gills, causing them to produce an extra mucus coating that eventually suffocates them.

In a second dramatic illustration, Figure 6–14 shows a mountain stream near Leadville, Colorado that receives acidic, metal-enriched drainage from an abandoned mine. Photochemical reduction of Fe(III) oxides results in dissolved Fe(II) ions during the day, imparting the bright orange color to the stream. The Fe(II) is formed at a rate some four times as great as the nighttime oxidation that reverses the process.

Rachel L. Carson, *Silent Spring*, Houghton Mifflin (Boston, 1962).

FIGURE 6–14   A mountain stream near Leadville, Colorado which receives acidic, metal-enriched drainage from an abandoned mine.

6.9   THE NOBLE GASES

In 1784, Henry Cavendish performed a landmark experiment with air. Using an electric spark, he planned to convert oxygen and nitrogen to nitrogen monoxide. Then he would add extra oxygen so that not only would all the nitrogen react, but also all the nitrogen oxide formed would be converted to nitrogen dioxide, which would then react with water to form nitric acid. Finally, Cavendish planned to convert the nitric acid to potassium nitrate:

step 1: $N_2 (g) + O_2 (g) \longrightarrow 2 NO (g)$

step 2: $2 NO (g) + O_2 (g) \longrightarrow 2 NO_2 (g)$

step 3: $3 NO_2 (g) + H_2O (liq) \longrightarrow 2 HNO_3 (aq) + NO (g)$

step 4: $KOH (aq) + HNO_3 (aq) \longrightarrow KNO_3 (aq) + H_2O (liq)$

Considering the apparatus available at the time, Cavendish must indeed have been an excellent experimentalist. He observed that not all of the original air (oxygen and nitrogen, he thought) was used up. In fact, about 0.8% of the original volume of the air remained. In this incredibly careful work, he had unknowingly isolated argon, which we now know occupies 0.934% by volume of the atmosphere. However, because argon undergoes no ordinary chemical reactions, the significance of his finding could not be appreciated for about a hundred years.

John William Strutt, Lord Rayleigh (1842–1919) won the Nobel Prize for physics in 1904 for his discovery of argon and the properties of gases, especially gas densities and gaseous diffusion.

Sir William Ramsay (1852–1916) received the Nobel Prize for chemistry in 1904, the year Lord Rayleigh received the physics award. Ramsay was discoverer or codiscoverer of five elements in as many years, and was deeply involved in discovering the nature of the atom.

Sir William Ramsay, shown in this Spy cartoon, is pointing to the group of inert gases of the atmosphere that he helped discover. Notice the symbols and atomic weights of neon, argon, krypton and xenon; helium presumably sits on top of the list but is not shown. Also note the composition of air written on the blackboard.

In 1882, Lord Rayleigh noted that at 0.00°C and 1.00 atm, a liter of nitrogen separated from the air weighed 1.2572 g, whereas a liter of nitrogen prepared by heating pure ammonium nitrite weighed only 1.2506 g. Finding no reasonable explanation for this result, he repeated the experiment Cavendish had performed a century earlier and prepared a small amount of the residual gas. On passing an electric spark through this gas he noted that it emitted light of different frequencies than any known gas. He concluded that the difference in mass between nitrogen prepared synthetically and that isolated from the air in his own experiment was due to the presence of a new element, and that the new element was the same as the residual in Cavendish's experiment. He called the new element "argon," meaning inert. Larger quantities of argon were prepared by Rayleigh and others, and attempts to cause it to react with different elements and compounds failed. The new element seemed to have no place in the periodic table.

Ramsay, who had worked with Rayleigh on the discovery of argon, proposed that a whole new column of elements must exist in the periodic table. He then set out to identify them. The other members of Group VIIIA, the **noble gases**, were discovered in short order. Helium was found trapped in radioactive rocks, particularly uranium minerals, as a result of alpha particle emission. Interestingly, it had already been identified much earlier by spectroscopists analyzing the solar spectrum. Krypton, neon, and xenon were discovered as minor components of air, which is the commercial source of these gases and argon. Radon results from the loss of an alpha particle from radium:

$$^{226}Ra \longrightarrow {}^{222}Rn + {}^{4}He$$

The noble gases show no tendency to react with other species under ordinary conditions, and they exhibit little attraction for each other. This results in the low melting and boiling points listed in Table 6–4. The boiling points decrease with decreasing atomic weight, and helium has the lowest boiling point of any known substance. Furthermore, helium has so little tendency to form condensed states that this cannot be accomplished at atmospheric pressure. To produce the solid state, pressures greater than 25 atm and temperatures near absolute zero are required. At just below 2.2 K, liquid helium undergoes a remarkable transformation into a form known as helium II. Characterized by an immeasurably low viscosity, helium II spreads out in extremely thin films and wets any surface it contacts. It will even climb up the sides of a suspended beaker, climb down the outside, and drip off the bottom.

Helium is used in a mixture with oxygen for underwater divers (instead of ordinary air that contains nitrogen). At high underwater pressures, considerable nitrogen dissolves in the blood. Upon decompression on coming to the surface, bubbles of gaseous nitrogen form, producing a painful

TABLE 6–4 The Noble Gases		
Element	**Melting Point, °C**	**Boiling Point, °C**
He	−272*	−269
Ne	−249	−246
Ar	−189	−186
Kr	−157	−153
Xe	−112	−107
Rn	−71	−65

*At 26 atm.

and sometimes fatal condition known as decompression sickness or "the bends." This condition does not occur in an atmosphere of 20% O_2 and 80% He, since helium is much less soluble in the blood than nitrogen. However, because helium atoms are so much lighter than nitrogen molecules, they have a much higher average speed and transport heat much more effectively, creating another problem—the occupants of an underwater vessel in an oxygen-helium atmosphere complain of the cold.

The extreme lack of chemical reactivity of the noble gases has been useful in providing a model for bonding and chemical behavior based on stable electron configurations. Unfortunately, that model also reinforced the belief that all of the noble gases were totally incapable of forming ordinary chemical compounds. This view was widely held until 1962, the year that Bartlett and Lohman prepared the first compounds of xenon. Bartlett had previously reported the reaction of platinum hexafluoride with oxygen to yield a substance believed to be $O_2{}^+PtF_6{}^-$. Noting (from the scientific literature) that it was easier to remove an electron from xenon than from oxygen, he felt that PtF_6 should be able to oxidize xenon. Direct reaction between PtF_6 and Xe gas at room temperature did in fact produce a red solid they believed to have the formula $XePtF_6$—the first noble gas compound. More syntheses of noble gas compounds followed almost immediately, based on work at the National Bureau of Standards. The most easily prepared are the three xenon fluorides, formed by reaction of the elements in appropriate proportions:

$$Xe\ (g)\ +\ F_2\ (g) \longrightarrow XeF_2\ (s)$$
$$Xe\ (g)\ +\ 2\ F_2\ (g) \longrightarrow XeF_4\ (s)$$
$$Xe\ (g)\ +\ 3\ F_2\ (g) \longrightarrow XeF_6\ (s)$$

The reactions are run in nickel containers at about 400°C and moderate pressure. Xenon hexafluoride reacts with water to form xenon trioxide:

$$XeF_6\ (s)\ +\ 3\ H_2O\ (liq) \longrightarrow 6\ HF\ (aq)\ +\ XeO_3\ (s)$$

This is acidic, and forms perxenates such as K_2XeO_4.

A krypton-nitrogen bond has recently (1988) been reported that is stable below $-50°C$:

$$HC{\equiv}CN{-}KrF^+,AsF_6{}^-$$

As you might expect, it is unstable at higher temperatures; in fact, it decomposes explosively when allowed to warm slightly.

Argon: the idle one
Neon: the new one
Krypton: the secret one
Xenon: the strange one
Radon: the radiant one
Helium: derived from *helios*, for its discovery in the Sun's spectrum.

NEIL BARTLETT

Neil Bartlett (now at the University of California at Berkeley) was a professor at the University of British Columbia at the time of the discovery. As he described the experiment: "The predicted interaction of xenon and platinum hexafluoride was confirmed in a simple and visually dramatic experiment. The deep red platinum hexafluoride vapor of known pressure was mixed by breaking a glass diaphragm, with the same volume of xenon the pressure of which was greater than that of the hexafluoride. Combination, to produce a yellow solid, was immediate at room temperature, and the quantity of xenon which remained was commensurate with a combining ratio of 1:1."

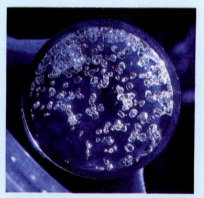

Crystals of xenon tetrafluoride, XeF_4.

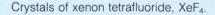

SUMMARY

Hydrogen and helium make up most of the atoms of the universe. Scientists have speculated about possible series of nuclear reactions that would lead to the observed cosmic abundances of the other elements. The possible nuclear reactions occurring in the Sun have also challenged chemists. Although we have some good ideas about **nucleogenesis** and stellar chemistry, it is a speculative business at best.

The atmosphere of the Earth can be divided into different regions according to its temperature profile. Reactions in the high atmosphere are initiated by absorption of high energy solar rays that break molecules into very reactive atoms or ions.

Hydrogen is the simplest of all of the chemical elements. Its chemistry places it in no single group of the periodic table, but suggests instead that it has aspects in common with several groups. Because the relative weight differences among the three isotopes of hydrogen are considerable—deuterium is twice the mass of hydrogen—there are significant differences in physical properties. Thus they alone among the isotopes of the elements are given distinguishing names and symbols: hydrogen (H), deuterium (D), and tritium (T). Hydrogen forms compounds with almost all other elements. With the principal nonmetals such as carbon and silicon, nitrogen and phosphorus, oxygen and sulfur, and the halogens—fluorine, chlorine, bromine, and iodine—hydrogen forms a wide range of covalent compounds. Included in this list are the halogen acids and sulfuric acid, which characteristically produce H^+ ions in aqueous solutions. With the Group IA metals and Ca, Sr, and Ba of Group IIA, ionic hydrides containing the H^- ion are formed. Hydrogen also reacts with transition metals to form **nonstoichiometric compounds**.

Nitrogen is generally considered to be an unreactive gas, but some of its compounds are very reactive, and at elevated temperatures and in the presence of certain catalysts, N_2 is reactive, too. The most important source of nitrogen compounds for fertilizers or explosives is the Haber process, which forms ammonia from nitrogen and hydrogen gases. Nitric acid can be synthesized from ammonia by the Ostwald process, and in turn reacts with ammonia to produce ammonium nitrate. The Ostwald process begins with the catalytic oxidation of nitrogen, producing nitric oxide; nitric oxide is easily oxidized to the dioxide, even in air, after which reaction with water yields HNO_3, an important article of commerce.

Elemental oxygen exists in two forms, **ozone** (O_3) and oxygen (O_2). Ozone can be prepared from oxygen by an electric discharge and is far more reactive than O_2. Our discussions in this chapter included an introduction to the oxides of carbon, nitrogen, and sulfur, notably CO and CO_2; N_2O, NO, and NO_2; and SO_2 and SO_3. The principal oxide of hydrogen is water. Of special significance are (1) the natural greenhouse effect, and the intensification of this effect by human activity, and (2) the atmospheric chemistry leading to depletion of the ozone layer.

Surprisingly, the inert noble gases aren't as unreactive as once thought.

QUESTIONS AND PROBLEMS

QUESTIONS

1. Why do elements with even atomic numbers predominate over those with odd atomic numbers?

2. Under what conditions do regular atoms with a nucleus and electrons not exist?

3. Why is the abundance of neon on the earth so low, whereas the abundances of carbon and oxygen are much higher?

4. Why are the different isotopes of hydrogen given different chemical symbols?

5. Explain why helium is found trapped in some rocks. How would you determine which rocks would likely contain helium?

6. The "big bang" theory and "cosmogenesis" suggest the ultimate end of the Sun's solar activity and the heat death of the universe. Why should those events not be of practical concern?

7. How many different forms of molecular hydrogen are there? Remember, there are three isotopes. Write all the forms in order of increasing molecular weight.

8. Explain why the temperature of the atmosphere rises and falls, then rises again with increasing elevation above the Earth's surface.

9. The increase of concentration of CO_2, with a resulting warming of the troposphere, is often called the "greenhouse effect." What is the reason for this effect?

10. What is the ionosphere and how does it arise?

11. What is the difference between the homosphere and the heterosphere?

12. In air pollution emergencies, usually caused by temperature inversions in the atmosphere, the air takes on a sickly yellowish cast and one can sometimes detect the early symptoms of "monoxide" poisoning such as drowsiness and nausea. Briefly explain the chemical basis for these observations.

13. Why can hydrogen not be assigned to any specific family in the periodic table?

14. How does helium protect divers from "the bends"? Are there any disadvantages in using helium to replace nitrogen?

15. Compare low versus high ozone levels in the atmosphere in terms of pollution versus protection.

16. Which elements form covalent compounds of hydrogen of more than one formula?

17. List several examples of the operation of the law of multiple proportions from the reactions mentioned in this chapter.

18. Why does ozone react more readily than oxygen?

19. Suggest one factor that restricts the temperature of the steam that may be used in a power plant.

20. Why is "Dry Ice" called that?

21. Why does NO_2 tend to form the dimer N_2O_4?

22. Why are many nitrogen compounds useful as explosives or propellants?

23. Briefly describe the chemistry of the Haber process for ammonia and the Ostwald process for nitric acid.

24. What is synthesis gas? Producer gas?

25. Describe laboratory preparations of nitrogen, oxygen, and hydrogen. Describe their industrial preparations.

26. Describe the laboratory synthesis of carbon monoxide and carbon dioxide. How are they prepared industrially?

27. What characteristic properties of carbon dioxide make it a good extinguisher for some classes of fires, but not others?

28. Distinguish between carboxy- and oxyhemoglobin, and briefly comment on the nature of the toxicity of the former.

29. In what way was the periodic law and classification system for the elements important in the discovery of the noble gas elements?

30. If the noble gases are inert and unreactive, how can they be characterized and definitively identified?

31. Compare the problems of CFC-induced depletion of the ozone layer and the greenhouse effect.

PROBLEMS

Nuclear Reactions [1–4]

1. A nuclear reaction that occurs in the Sun is

$$^{16}O + {}^{16}O \longrightarrow {}^{32}S$$

Calculate the energy released when one mole of ^{32}S is formed in this reaction. (Atomic weights of isotopes: ^{16}O, 15.9994; ^{32}S, 31.9706)

2. It is estimated that 1.6675×10^{16} kJ of energy from the Sun enters the Earth's atmosphere every day. Calculate the mass of protons used up per day to provide this energy.

3. Nuclei may disintegrate in cosmogenic processes by emission of high energy α-particles, leading eventually to construction of higher elements by capturing surviving nuclei. Thus, the reactions in the following sequence. Balance the nuclear processes, illustrating that possibility:
(a) $^{28}Si \longrightarrow {}^4He$
(b) $^{28}Si + {}^4He \longrightarrow {}^{56}Ni$
(c) $^{28}Si \longrightarrow {}^{56}Ni$

4. Nickel-56 is an unstable iostope that decays through cobalt-56 to the stable iron-56 nuclide. Write equations for the nuclear transformations.

High Atmosphere [5–6]

5. Write the high altitude reactions by which nitrogen monoxide introduced into the atmosphere could destroy the ozone layer.

6. Write chemical reactions leading to destruction of the ozone layer by the introduction of "Freons" into the atmosphere.

Stoichiometry and the Gas Laws [7–18]

7. You have iron, cobalt, and nickel metal turnings at your disposal for the purpose of generating hydrogen gas in the laboratory. If you needed to use the smallest weight of metal possible to generate 10.0 mL of the gas, measured at 0°C and 1.00 atm, by reaction with dilute sulfuric acid, which metal would you choose and how many grams would be required?

8. If, in Problem 7, you had only cobalt and nickel, how would you answer the question?

9. What volume of nitrogen, initially at 755 torr and 30°C, would be required to prepare 1 kg of ammonia gas by the Haber process, assuming 10% conversion?

10. What is the maximum volume of hydrogen produced by the action of 10 g of zinc metal turnings and an aqueous solution of hydrochloric acid in a laboratory preparation, where the temperature is 20°C and the pressure of the atmosphere is 750 torr?

11. How many kilograms of Chilean saltpeter are required in order to prepare 100 kg of constant boiling (69%) nitric acid solution?

12. You are out to produce oxygen as cost-effectively as possible in the laboratory. If potassium chlorate, mercury(II) oxide, and lead nitrate all cost a dollar a pound, which should you buy?

13. Suppose that, in the explosion of glyceryl trinitrate, the temperature of the gases produced is 1000°C. What

would be the total volume of the gases produced from the detonation of 1.0 g of glyceryl trinitrate at 1.00 atm pressure?

14. What quantity of glyceryl trinitrate could be prepared in theory from the nitrogen in a cubic meter of air at 1 atm and 0°C?

15. Both carbon dioxide and carbon monoxide are by-products from the isolation of pure manganese metal from a $MnCO_3$ ore. In the first step, the ore is roasted, producing the oxide and CO_2, after which it is reduced to the free metal and CO by reaction with elemental carbon. Determine the number of kilograms of Mn that can be won from a metric ton of the crushed carbonate ore known to be 83.7% pure manganese carbonate.

16. In the laboratory analysis of a sample of a manganese carbonate ore, a 1.500-g sample of finely crushed ore yielded 0.410 g of carbon dioxide. How many grams of carbon monoxide could be recovered on conversion to the free metal, and what was the percentage of manganese carbonate present in the original ore sample? Assume that the process is as described in Problem 15.

17. An electric discharge was passed through a quantity of oxygen until eventually it was 10% (by weight) ozone. If the temperature and pressure were kept constant throughout the process, what (if anything) happened to the weight of the gas? To the volume of the gas?

18. You are given 40.0 mL of a mixture of gases of the atmosphere. After the sample is shaken with a concentrated solution of potassium hydroxide, the volume of the residual gas measures only 20.0 mL. That gas is then passed over a copper metal surface at red heat, which serves to reduce the volume still further, to 10.0 mL. Finally, that gas is passed over magnesium metal at elevated temperatures, reducing the volume once again, this time to 5.0 mL. If this residual gas is spectroscopically identified as argon, what was the original weight-percent composition of the mixture? Assume the temperature and pressure are the same through all volume measurements.

Laboratory and Industrial Reactions   [19–32]

19. Write balanced equations for chemical methods of preparation for each of the following species:
(a) H_2 (g)   (d) CO_2 (g)
(b) synthesis gas   (e) SO_3 (g)
(c) NO (g)

20. Use balanced equations to show how the following elements and compounds can be prepared:
(a) N_2 (g)   (d) CO (g)
(b) O_2 (g)   (e) SO_2 (g)
(c) producer gas

21. Write balanced equations for each of the following:
(a) the Haber process and the Ostwald process
(b) heating of sodium azide, NaN_3
(c) the preparation of calcium cyanamide, $CaCN_2$
(d) the reaction of sodium cyanide with sulfuric acid

22. Give a balanced equation for each of the following processes:
(a) thermal decomposition of mercury(II) oxide
(b) thermal decomposition of lead(II) nitrate
(c) catalytic decomposition of potassium chlorate

23. Write balanced equations showing that an automobile's catalytic converter helps reduce air pollution due to carbon monoxide and nitrogen oxides (called NO_x in the environmental business) but aggravates pollution due to sulfur oxides.

24. Write the balanced chemical equations involved in the production of ultrapure nickel metal using carbon monoxide.

25. Write balanced equations for the syntheses of NO, NO_2, and N_2O_4 from the elements.

26. Write balanced equations for the preparation of NH_4NO_3 from the elements.

27. Sodium can be used to "fix" nitrogen by forming the nitride, followed by decomposition with water to ammonia. Write the equation(s) for the fixation of nitrogen by lithium and magnesium.

28. Write the reactions showing how calcium oxide can be used to remove sulfur oxides from stack gases.

29. Complete and balance the following reactions:
(a) FeS (s) $+ O_2$ (g) $\longrightarrow$
(b) ZnS (s) $+ O_2$ (g) $\longrightarrow$
(c) $NaNO_3$ (s) $+ H_2SO_4$ (liq) $+$ heat $\longrightarrow$

30. Complete and balance the following reactions:
(a) KNO_3 (s) $+ H_2SO_4$ (liq) $+$ heat $\longrightarrow$
(b) NH_3 (g) $+ O_2$ (g) $+$ heat $+$ catalyst $\longrightarrow$
(c) CO (g) $+ Cl_2$ (g) $+$ catalyst $\longrightarrow$

31. Write the balanced equation for the reaction of an aqueous solution of hydrochloric acid with each of the following:
(a) NaCN (s)   (b) NH_3 (aq)   (c) $CaCO_3$ (s)
(d) Zn (s)   (e) Na_2SO_3 (aq)

32. Write the balanced equation for the reaction of water with each of the following substances:
(a) NO_2 (g)   (b) Ca (s)   (c) CO_2 (g)
(d) CaH_2 (s)

Additional Problems   [33–41]

33. The Sun's energy can be explained in terms of a "carbon cycle." Complete the steps in the carbon cycle without consulting the discussion in the text:

$$^{12}C + {}^1H \longrightarrow (\quad) + \gamma$$
$$(\quad) \longrightarrow {}^{13}C + \beta^+ + \nu$$
$$^{13}C + {}^1H \longrightarrow (\quad) + \gamma$$
$$(\quad) + {}^1H \longrightarrow {}^{15}O + \gamma$$
$$^{15}O \longrightarrow (\quad) + \beta^+ + \nu$$
$$(\quad) + {}^1H \longrightarrow {}^{12}C + {}^4He$$

34. Write balanced equations for the main reactions by which the sulfur in a high-sulfur coal is converted into a sulfuric acid rain.

35. (a) Sulfur dioxide can be "scrubbed" (removed) from industrial process stack gases by reaction with magnesium oxide. Write the chemical reaction for the process.
(b) Acid rain is corroding many historic monuments and statues, yet lakes that are in contact with limestone deposits are protected from the environmental effects usually imposed on aquatic ecosystems by acid rains. Briefly explain the fundamental chemistry common to both effects.

36. Use the reaction in Problem 35 to determine how

many grams of magnesium oxide would be required to "scrub" the stack gases from the burning of a kilogram of high-sulfur coal (#6 industrial grade) that is 3.0% elemental sulfur.

37. Write chemical reactions describing the processes that lead to air pollution problems arising from the introduction of oxides of carbon and nitrogen into the atmosphere.

38. Write chemical reactions for correcting or compensating for the pollutants formed in the processes described in Problem 37.

39. The structures of many covalent oxides of the nonmetallic elements are dimers, trimers, tetramers and larger combinations sometimes collectively referred to as "oligomers." For example, phosphorus(III) oxide is correctly the dimer, P_4O_6, not the simple monomer, P_2O_3.
(a) Write Lewis structures for phosphorus(III) oxide, assuming that phosphorus atoms bond only to oxygen, not to phosphorus.
(b) Would you suppose that arsenic(III) oxide would also be dimeric? Why? Write a Lewis structure, assuming only arsenic-to-oxygen bonds.
(c) Dinitrogen tetroxide is the "dimer" of nitrogen dioxide. Write a Lewis structure for it.
(d) There is good evidence that sulfur trioxide is an oligomeric "trimer" under certain atmospheric conditions. Write a Lewis structure for it.

40. In the lower atmosphere, nitrogen dioxide is involved in a complex series of "photochemical" reactions in polluted air (containing unburned gasoline hydrocarbons) in the presence of bright sunlight. These reactions produce organic nitrates such as peroxyacetyl nitrate (PAN). Briefly explain the chemistry, using reactions where possible, and briefly discuss the short-term insult to plants and people. How might you chemically destroy some of these pollutants or inhibit their activity?

41. Briefly describe the properties of (yet-to-be-discovered) element 118. What kinds of chemical compounds can you predict (or expect), if any?

MULTIPLE PRINCIPLES [42–49]

42. How many kilograms of Zn metal are necessary to produce the H_2 necessary to lift a mass of 1.00 kg at STP?

43. Molybdenum metal is produced by the reaction of hydrogen gas with MoO_3. If sufficient hydrogen is contained in an 0.500-L container to produce 1.00 g of molybdenum metal, what is the pressure in the container at a temperature of 25°C?

44. A sample of 10.0 g of oxygen gas reacts with sulfur to produce a mixture of SO_2 and SO_3. All of the O_2 gas is consumed in the process. The gaseous mixture produced has a volume of 6.2 L at a temperature of 350 K and a pressure of 1.15 atm. What is the mole percent of each component in the gaseous mixture?

45. Lithium metal reacts with air to produce lithium oxide and lithium nitride. What mass of lithium is necessary to react completely with 1.00 m³ of air at 20.°C and 1.10 atm? What is the volume of the gas remaining, at the same temperature and pressure, after the reaction?

46. Ammonia burns in a pure oxygen atmosphere according to the following equation:

$$4\ NH_3\ (g)\ +\ 3\ O_2\ (g)\ \longrightarrow\ 2\ N_2\ (g)\ +\ 6\ H_2O\ (liq)$$

Calculate the volume of oxygen required for complete combustion of 10,000. L of ammonia, both measured at STP. Determine the volume of nitrogen produced, also measured at STP.

47. Ammonia burns in a pure chlorine atmosphere according to the following equation:

$$2\ NH_3\ (g)\ +\ 3\ Cl_2\ (g)\ \longrightarrow\ N_2\ (g)\ +\ 6\ HCl\ (g)$$

If 10,000. L of chlorine and 5000. L of ammonia are brought together and allowed to react, what volumes of what gases are present at the end of the reaction? Assume that STP exists before the reaction begins and after the reaction is complete.

48. Consider the combustion of ammonia with oxygen and chlorine as described in Problems 46 and 47. Beginning with equimolar quantities of ammonia, chlorine, and oxygen, first carry out the chlorine oxidation of ammonia to the extent that it is possible, followed by the oxygen oxidation to the extent that it is possible. What substances do you have and in what mole ratio?

49. Stable nuclei near ^{56}Fe can be produced by the same high temperature nuclear disintegration processes described in Problems 3 and 4. Write sequences of nuclear reaction that lead to ^{46}Ti and ^{52}Ni, both of which are stable isotopes.

APPLIED PRINCIPLES [50–53]

50. Here are some approximate costs of several purified gases, in dollars per kilogram:

gas	N_2	O_2	Ar	CO_2	Ne	He	Kr	CH_4	H_2
$/kg	4.9	8.8	3.8	2.8	3000	60	4000	34	122

Using Table 5–1, calculate the value of the air in a room that measures 8.0 m by 12 m by 3.1 m at STP, if it were separated into its component parts.

51. Typical coals range from 0.2 to 2.0% sulfur by mass. When the coal is burned, sulfur oxides form; they can be trapped by conversion to the respective sulfites and sulfates by reaction with lime (CaO). Write the reactions and determine the upper and lower limits of the masses of lime necessary to remove the sulfur oxides (from the combustion of coal) completely.

52. A limestone deposit is 85.5% calcium carbonate. A charge of 5.00 kg of this particular limestone is heated in a kiln and decomposed cleanly to calcium oxide and carbon dioxide. The carbon dioxide is then compressed and placed in tanks with a total volume of 165 L. Use the van der Waals equation to calculate the pressure of the carbon dioxide in these tanks at 30.°C. Compare this result to the calculated pressure according to the Ideal Gas Law.

53. Assume that a typical family living space requires 1×10^6 m³ of natural gas to provide adequate heating for a year, and that natural gas is essentially pure methane, which burns cleanly to carbon dioxide and water. Calculate the volume of air required at the same temperature and pressure as the natural gas.

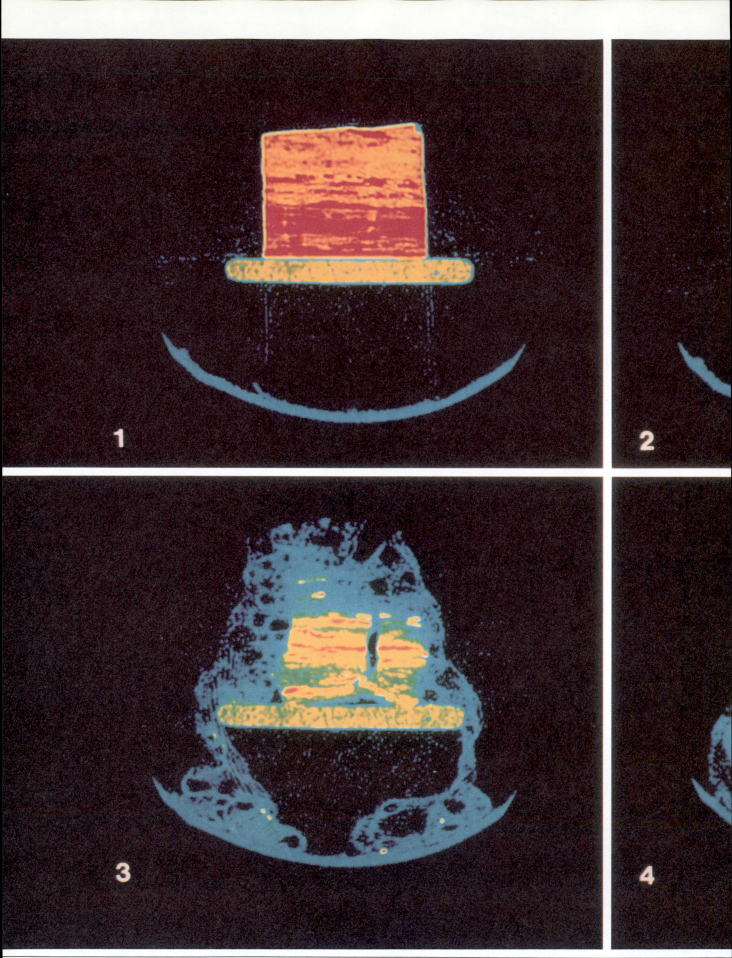

C H A P T E R
7

Chemical Thermodynamics I: Heat, Work, and Energy

Using a computed tomography medical scanner, researchers obtained these unique X-ray images of the interior of a lump of coal as it was heated and gasified. The sequence shows clearly how the coal expands and "melts" as it is heated, going from the relatively dense solid visible in view 1 (at 212°F) to the frothy foam in view 4 (at 750°F). Data such as these aid in the design of improved coal gasification, combustion equipment, and more efficient utilization of energy resources. It was not until the late 18th century that the combustion of gases was used for practical illumination, giving rise to the development of gasworks and commercial lighting—the gaslight era. The gaslight shown here was originally used in one of the early chemistry laboratories in America (at Union College).

Walther Nernst (German), Nobel Prize for Chemistry in 1920 for his studies on combustion and heat changes during chemical reactions.

7.1   CHEMISTRY AND ENERGY

One of the great lessons learned by humans was that energy could be usefully converted into work but that you needed a good machine to do it, and good fuel to run that machine. A strong back, bulging biceps, and a meal were fine for doing a day's work; a good horse and a bag of oats were better; a steam engine and a cord of wood, much better; and a coal-fired power plant, incomparably better at delivering energy (Table 7–1). Today, there is no question that energy is basic to life, and that is understood very early in everyone's experience. At present, the world's energy resources are largely chemical and almost all based on combustion—the burning of coal, oil, and wood, and to some extent organic waste. Wind, water, geothermal, solar, and especially nuclear power are real energy resources, but they account for perhaps no more than a 5% piece of the annual energy pie.

Predicting energy resources, especially petroleum reserves, continues to prove difficult. What does seem clear is that within the short span of perhaps less than 200 years from the first discovery of petroleum in the United States (in 1859), our estimated total recoverable resources may well be more or less exhausted. The noted geologist M. King Hubbert's 1956 and 1972 studies of resources and consumption (Figure 7–1) have been largely confirmed and we expect to witness his predicted peak in production during the 1990's. The consequences of the loss of these invaluable energy and materials resources is uncertain but could be enormous economically and politically. Hubbert's graph shows production in billions of barrels per year. Note that roughly 80% of the total resources has been exhausted in only 30% of the 200-year time span, suggesting the importance of petroleum to our industrial and economic base.

Fuels are the vital connection between energy and work. It is important for each of us to know something about how energy is stored, released, transferred, and used; also, we need to understand the nature and consequences of the transfer of energy. This is the subject called **thermodynamics**. The name suggests something to do with heat, or thermal energy, and motion or movement. The laws of thermodynamics are based on more than 200 years of practical experience drawn from basic science and engineering practice. The great triumph of thermodynamics lies in the explanation of the macroscopic (or bulk) properties of matter—especially the thermal properties of matter—in terms that are consistent with our microscopic view of a material world made up of atoms and molecules.

A barrel (bbl) of oil is 42 gallons.

TABLE 7–1 Manpower, Horsepower, and Machine-power*		
	Rate of Energy Delivery	
Energy Source	**Horsepower**	**Watts**
man	1/30	25
horse	1	750
waterwheel	5	3750
windmill	8	6000
early steam engine	100	75,000
modern electric power station	10^6	1000 megawatts

*The power of a machine is a statement of how fast it can work.

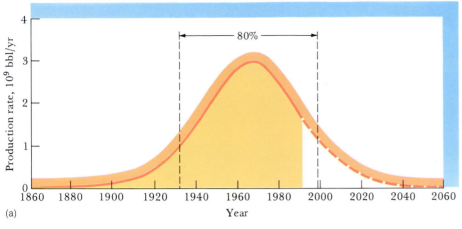

(a)

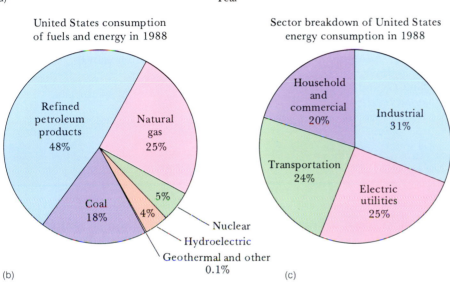

United States consumption
of fuels and energy in 1988

Sector breakdown of United States
energy consumption in 1988

(b)

(c)

FIGURE 7-1 (a) Complete cycle of crude oil production in the United States. (b) U.S. consumption of fuels and energy for the year 1988. (c) Sector breakdown of U.S. energy consumption, 1988. (d) Borrowing one of the medical profession's most advanced diagnostic tools, with the aid of a computed tomography X-ray scanner and a specially built furnace that slides into it, just as a human patient normally would, scientists have obtained a first ever look inside a piece of coal that is being heated and gasified. (See Chapter opening photo.)

(d)

To begin, we need to establish some basic definitions and a few fundamental ideas in order to better understand two important rules that are the backbone of this first chapter on thermodynamics: (1) energy is conserved, or to put it another way, the amount of energy in the universe is constant; and (2) that energy can be converted from one form to another—from gasoline to motive power, for example.

Unfortunately, all the energy of the universe is not available, and all too often energy conversions are woefully inefficient (see Chapter 11).

7.2 THERMODYNAMIC TERMS AND CONCEPTS

Thermal Equilibrium and the Thermodynamic Scale of Temperature

It is common experience that objects can be arranged according to how hot they are. Accordingly, we should be able to assign numbers to these objects that correspond to their relative hotness, in effect creating a hotness

TABLE 7–2   Fixed Points of Temperature, Based on the International Practical Temperature Scale of 1960* (Based on Equilibrium States)

	°C	°F
• oxygen point liquid oxygen and its vapor	−182.970	−297.346
• ice point ice and air saturated water	0.0	32.0
• steam point liquid water and its vapor	100.0	212.0
• sulfur point liquid sulfur and its vapor	444.600	832.28
• silver point solid and liquid silver	960.8	1761.44
• gold point solid and liquid gold	1063.0	1945.40
Additional Fixed Points (Based on Equilibrium States)		
• carbon dioxide solid CO_2 and its vapor	−78.5	−109.3
• mercury freezing mercury	−38.87	−37.97
• triple point for H_2O ice, water, and vapor	+0.0100	32.018

*At standard atmospheric pressure. Data revised in 1968, 1971, and 1986.

Temperature enters into descriptions of nature in our daily lives in so many ways that we tend to lose sight of (or miss) the underlying significance of the concept.

scale. All we need to translate this idea into a recognizable temperature scale (the Celsius scale, for example) are the familiar reference points, the ice point (0°C) and the steam point (100°C) for water, with the interval between them divided into 100 parts called degrees (Table 7–2). The Kelvin scale of temperature was universally adopted in 1954 by the International Committee on Weights and Measures. It is based on the absolute zero of temperature and the **triple point of water**, which corresponds to the single temperature and pressure (namely 4.58 torr and 0.01°C, which has been set as 273.16 K) at which ice, water, and water vapor all exist in equilibrium. The basic SI unit (the kelvin) is 1/273.16 of the triple point temperature for water. The new scale is called the **thermodynamic temperature scale**. It is singularly important because it is an absolute scale (having as its zero point the lowest possible temperature), whereas the Celsius scale is a relative scale whose zero point is the arbitrary creation of its inventor.

The choice of 273.16 was made so that the older centigrade scale would correspond closely to the newer single fixed point scale. The word "centigrade" was dropped in favor of "Celsius" to avoid confusion that might be caused by the 100-degree implication of the centi- prefix.

The Fahrenheit scale, along with its absolute scale counterpart, the Rankine scale (named after a Swedish engineer), is still widely used in engineering applications. Zero on the Rankine scale is −459.7°F. The relationship between the Fahrenheit and Rankine scales is

$$°R = °F + 459.7$$

Therefore, the normal boiling point of water (at atmospheric pressure) should be very close to 672°R.

EXAMPLE 7–1
Calculate the ice point of water on the Rankine scale.

Solution   °R = 32 + 460 = 492°R

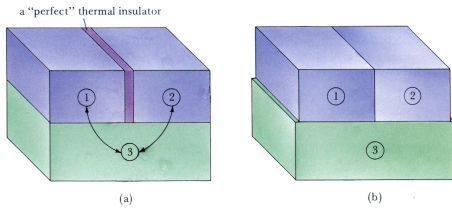

a "perfect" thermal insulator

(a)   (b)

FIGURE 7–2   The Zeroth law of thermodynamics. (**a**) After 1 and 2 have come to thermal equilibrium with 3, they are (**b**) in thermal equilibrium with each other.

EXERCISE 7–1(A)
If your body temperature is reported to be 558.3°R, should you be concerned about your health?
Answer: No, 98.6°F is normal, although recording it on the Rankine scale isn't. ∎

EXERCISE 7–1(B)
Science fiction buffs may recall the title of Ray Bradbury's novel, "Fahrenheit 451." What is the source of the number 451? ∎

It is also a fact that two objects in thermal contact between which there is a temperature difference will exchange energy with each other until they reach a state of thermal equilibrium, meaning that energy is no longer exchanged between them. At this point, the temperatures of the two objects are the same. Now, if one of those two objects happens to be a thermometer, the reading on the thermometer should become constant and the calibration mark should give the temperature of the object. These ideas are summarized by the **Law of Temperature Equilibrium**, sometimes called **the zeroth law of thermodynamics**:

• Two objects in thermal contact with each other tend to move toward the same temperature and reach equilibrium at the same temperature (Figure 7–2).

• Two objects in thermal equilibrium with a third are in thermal equilibrium with each other.

This entire discussion is of fundamental importance to thermodynamics because it is the definition of temperature—the property that determines whether objects are in thermal equilibrium (Table 7–3).

Heat and Thermal Energy

A substance has a measurable amount of energy. The same thing cannot be said about the amount of heat in a substance. The terms "heat" and

TABLE 7–3	Commonly Used Thermometers and Their Thermometric Properties
Thermometer	**Property**
A liquid contained in a bulb at the base of an enclosed capillary column (mercury and alcohol are typical liquids)	Length of the liquid in the capillary column as a function of temperature
Constant volume (and constant pressure) gas thermometers	Pressure (and volume) as a function of temperature
Electrical resistance	Electrical resistance as a function of temperature
Thermocouple	Thermal emf (voltage) as a function of temperature
Optical pyrometer	Color of a heated filament as a function of temperature

"energy" are often used interchangeably, but in fact they are not synonymous. **Heat** refers to the transfer of energy from one substance to another. Imagine heat flowing from one place to another as a direct consequence of a difference in temperature—energy on the move. If energy is exchanged between a system and its surroundings by thermal conduction when a process takes place within a system, we call that exchange heat.

By contrast, **internal energy** is the energy a substance has at a particular temperature. We know, for example, that the particles (atoms or molecules) contained in a given quantity of an ideal gas are ceaselessly in motion due to their kinetic energy. In other words, the internal energy of a gas is essentially its kinetic energy on a microscopic scale, and we may say: "The higher the temperature, the greater the internal energy." In fact, our definition of energy follows directly from our earlier discussion of the kinetic theory of gases. For one mole of a monatomic ideal gas, the remarkable result was

See Chapter 5 for a detailed discussion of thermometry, the expansion of an ideal gas, and the absolute scale of temperature.

$$E = \tfrac{3}{2} RT \qquad (7–1)$$

The internal energy normally changes when heat is transferred.

Heat is involved in almost every form of chemical change, not just the familiar burning of hydrocarbon fuels such as gasoline, home heating oil, or coal. If heat q is released by the system to the surroundings, we describe the process as **exothermic** and by convention assign a negative value to q. The conversion of a metal to a salt is an exothermic process:

$$2 \text{ Li (s)} + \text{F}_2 \text{ (g)} \longrightarrow 2 \text{ LiF (s)} + \text{heat}$$

The combustion of any hydrocarbon is an exothermic reaction:

$$\text{C}_8\text{H}_{18} \text{ (g)} + \tfrac{25}{2} \text{ O}_2 \text{ (g)} \longrightarrow 8 \text{ CO}_2 \text{ (g)} + 9 \text{ H}_2\text{O (g)} + \text{heat}$$

The displacement of certain metals by others can be exothermic:

$$2 \text{ K (s)} + \text{MgCl}_2 \text{ (s)} \longrightarrow 2 \text{ KCl (s)} + \text{Mg (s)} + \text{heat}$$

In each of these cases, the net internal energy of the products is less than that of the reactants.

On the other hand, if the system absorbs heat from the surroundings, we describe the process as **endothermic** and by convention assign a positive value to q. The decomposition of limestone is an endothermic process:

$$CaCO_3 \text{ (s)} + \text{heat} \longrightarrow CaO \text{ (s)} + CO_2 \text{ (g)}$$

Likewise, the electrolysis of water and the decomposition of red mercury(II) oxide are endothermic processes:

$$2 H_2O \text{ (liq)} + \text{heat} \longrightarrow 2 H_2 \text{ (g)} + O_2 \text{ (g)}$$

$$2 HgO \text{ (liq)} + \text{heat} \longrightarrow 2 Hg \text{ (liq)} + O_2 \text{ (g)}$$

Here, the net internal energy of the reactants is less than that of the products. Exothermic and endothermic processes can be plotted on a scale of internal energy as shown in Figure 7–3.

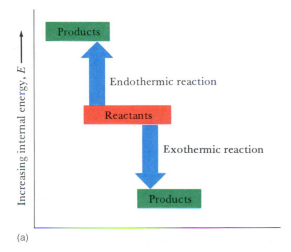

(a)

(b)

FIGURE 7–3 Exothermic and endothermic reactions: (a) An exothermic reaction is accompanied by a release of energy, while an endothermic reaction is accompanied by an increase in internal energy; (b) The aluminothermic reduction of iron(III) oxide is clearly exothermic.

FIGURE 7–4   A gas-filled cylinder with a movable piston acting against an external force.

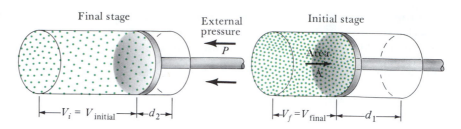

Heat and Work

The concept of work w comes from classical mechanics and can be defined operationally as the product of a force F on an object that moves through a distance d (in the direction of the force):

$$w = Fd \tag{7-2}$$

Work done by the system on the surroundings, as when the system expands, is conventionally given a negative sign. If work is done on the system by the surroundings, as when the system is made to contract, then the sign of w is positive. Therefore, d is given a positive value in the direction in which the system is made to contract.

Much of the work to be discussed here involves the change in volume of a gas working against some constant external pressure. This is referred to as pressure-volume (PV) work. For the purposes of this discussion, consider the system shown in Figure 7–4. The cylinder contains one mole of an ideal gas in a state of thermal equilibrium. In order to characterize the system completely, it requires that the pressure and the volume be fixed. Now push on the piston and displace it a distance d, increasing the pressure on the gas within the cylinder as a direct result of work done. By convention, this work has a positive sign since it is being done on the system. By pushing on the piston, a force has acted through a distance. Since the volume of the gas in the cylinder is decreasing as it is being compressed, the pressure of the gas must be increasing, according to the Ideal Gas Law.

Here is another way of looking at this same compression. Before anything happens, consider the piston to be locked in place so it can't move. Call the pressure outside the cylinder P_{ex} and set it equal to the pressure of the trapped gas after compression is complete. Before the compression, the pressure of the trapped gas P is less than P_{ex}. The lock is released, and the piston is pushed to the proper point, where $P = P_{ex}$. The gas in the cylinder was compressed from an initial volume V_i to a final volume V_f, in response to the external (compressing) force F. The external force is related to the external pressure P_{ex} and the cross-sectional area A of the piston as follows:

$$P_{ex} = \frac{F}{A}$$

Rearranging and substituting for F, the work done on the system by the surroundings can be restated:

$$w = P_{ex}Ad$$

The Ad product corresponds to the volume change that takes place as the piston of surface area A sweeps through the distance d in moving from its initial position V_i (before the push) to its final position V_f (after the push):

$$w = -P_{ex}(V_f - V_i)$$

$$w = -P_{ex}\Delta V \tag{7-3}$$

The negative sign in this expression is necessary to give the conventional sign for w. The net result is that the surroundings do work on the system, compressing it; and since V_f is less than V_i (for compression), the sign of w is positive. On the other hand, if the external pressure is decreased, the gas within the cylinder expands, and work is done by the system on the surroundings; the sign of w will be negative, since the final volume (V_f) is greater than the initial volume (V_i).

If it is not clear why the external pressure (not the pressure of the gas within the cylinder) determines the work done on (or by) the system, consider this example. A gas-filled cylinder is placed in deep space where, for all practical purposes, there is no external pressure:

$$P_{ext} = 0$$

No useful work can be done, no matter how great the internal pressure, because the expanding gas is pushing against nothing. We call this a free expansion, and for a free expansion

$$w = 0$$

It should also be evident that the work done by the system during a constant volume process must equal zero, since $\Delta V = 0$.

EXAMPLE 7–2

(a) When 2 mol of an ideal gas is compressed by a constant external pressure of 1.50 atm from an initial volume of 40.0 L to a final volume of 30.0 L, what is the work?
(b) Determine the work done when a sample of an ideal gas expands from 1.00 L to 1.33 L against an external pressure of 2.00 atm.

Solution

(a) $V_i = 40.0$ L; $V_f = 30.0$ L; $P_{ex} = 1.50$ atm

$$w = -P_{ex}\Delta V = -P_{ex}(V_f - V_i)$$
$$= -1.50 \text{ atm } [(30.0 - 40.0) \text{ L}] = +15.0 \text{ L·atm}$$

The value of w is positive since work is being done on the system.
(b) To determine w, proceed as above:

$$w = -P_{ex}\Delta V = -P_{ex}(V_f - V_i) = -2.00 \text{ atm } [(1.33 - 1.00) \text{ L}]$$
$$w = -0.66 \text{ L·atm}$$

In this case, the system is doing work, pushing back the surrounding atmosphere, and the sign of w is negative. ■

EXERCISE 7–2

Calculate w when a gas expands from 1.0 L to 3.0 L against a constant external pressure of 2.0 atm.
Answer: -4.0 L·atm. ■

FIGURE 7–5 Joule's experiment for determining the mechanical equivalent of heat gave surprisingly good (accurate) results.

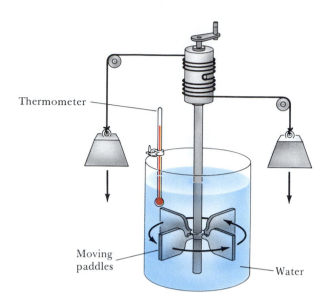

Thermometer

Moving paddles

Water

The Equivalence of Heat and Energy

In a series of experiments carried out about 150 years ago, James Joule demonstrated that energy can be transferred in the form of heat. Joule succeeded in raising the temperature of a well-insulated quantity of water by the motion of a paddle wheel driven by a falling weight (Figure 7–5). He found a proportional relationship between the amount of work done on the liquid by the paddle wheel and the increase in temperature of the water. Within the limits of accuracy of the experiment, he arrived at a value of 4.16 J/g·K for the specific heat of water, which is very close the modern value of 4.184 J/g·K. In other words, 4.184 J of mechanical energy will raise the temperature of one gram of water one degree Celsius (from 14.5°C to 15.5°C).

Systems and States

The term **thermodynamic system** refers to that part of the universe being studied at the moment. A system may be as simple or complex as you wish, just as long as it is carefully defined. Everything else that can interact with the system in any way is considered part of the **thermodynamic surroundings**. The surroundings can affect the system by the transfer of heat or energy:

- An *open system* can exchange or transfer matter and energy between system and surroundings. A cup of coffee is an open system—open to the atmosphere and allowing rapid transfer of heat and water vapor.

- An *isolated system* cannot interact with its surroundings under any circumstances. A properly stoppered Thermos bottle containing coffee is very nearly an isolated system—the stopper prevents water vapor from escaping, while the vacuum construction keeps heat from being lost to the surroundings.

- A *closed system* can exchange energy with its surroundings, but no matter can be transferred. To turn an isolated system into a closed system, simply

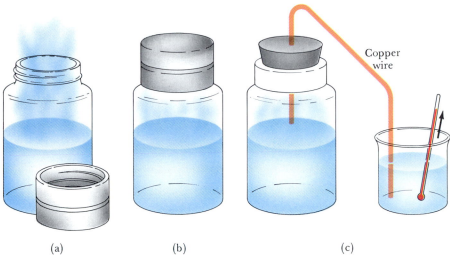

(a)   (b)   (c)

(d)

FIGURE 7–6 (a) An open system can exchange matter and energy with its surroundings; (b) an isolated system can exchange neither matter nor energy; (c) a closed system can exchange energy, but not matter; (d) hot springs—at Strokkur geyser in Iceland—an open system.

pierce the stopper with a piece of heavy gauge copper wire. No material escapes, but heat leaves the system by conduction along the copper wire (Fig. 7–6).

The state of a system is defined when a certain required number of variables have been fixed. These variables are macroscopic properties such as pressure, volume, temperature, mass, and chemical composition—properties that do not depend on the single atom (or microscopic) composition but rather on the aggregate properties. A system in a state of **thermodynamic equilibrium** is one in which the macroscopic properties do not change with time. It is a system at rest.

The **initial state** of a system refers to the starting materials in a state of equilibrium, prior to the onset of some kind of chemical interaction or physical transformation. The **final state** of a system refers to the products of the chemical interaction or physical process after everything that is going to happen has happened, and a new thermodynamic equilibrium has been established. The **process** is the path by which the change from the initial equilibrium state to the final equilibrium state takes place.

7.3 THE FIRST LAW OF THERMODYNAMICS

The **first law of thermodynamics** simply says, "Energy is conserved." It is a demonstration of the equivalence of heat and work, a connection between macro and micro states of matter, and most importantly a law of experience. Although it isn't always obvious that energy is conserved, and that heat and work are equivalent manifestations of the same thing, 300 years of experience have produced no contradictions to these ideas. Automobiles, for example, return little more than 15 to 20% on our energy investment, and it is not at all clear that the motive power and the heat taken away by the car's cooling system are in any semblance of balance. Electric power stations return only 30 to 40% of our energy investment. Biological systems operate at higher efficiencies (80 to 98%), but even they are not perfect

Gottfried Wilhelm von Leibniz, German mathematician (1646–1716) wrote in 1693 that kinetic energy is not lost due to inelastic collisions of macroscopic masses but is transferred to the microscopic particles of which larger bodies are composed.

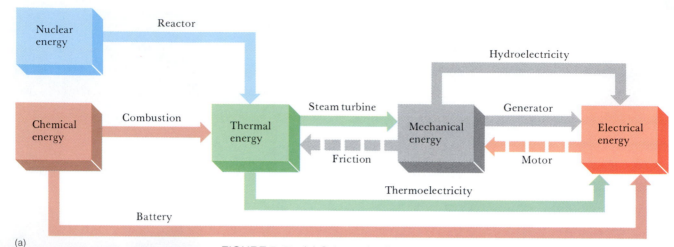

(a)

(b)

FIGURE 7–7   (a) Schematic diagram of the common energy conversion processes. (b) A tomato being pierced by a 30-caliber bullet traveling at a supersonic speed of 900 m/s. This collision was photographed with a microflash stroboscope, using a microsecond exposure. Shortly after the photograph was recorded, the tomato disintegrated completely. Note that the entry and exit points of the bullet are both "exploded."

demonstrations of energy conversion and conservation. Still, our confidence in the first law of thermodynamics will stand up to careful inspection (Figure 7–7).

Consider the expansion of an ideal gas caused by adding heat. The added heat could be used to do the work needed to push back the atmosphere, equal to force multiplied by distance. That's $-P_{ex}\Delta V$, where ΔV is the change in volume of the expanded gas. The added heat could also be used to separate the particles of the gas further from each other and to increase their speeds, or in other words, to increase their internal energy. So heat can be converted to internal energy and external work, and the overall effect can be stated as an equality:

$$\Delta E = q + w$$

(7–4)

where ΔE is the change in internal energy, q represents the heat added, and w is the work done. This is the mathematical form of the first law of thermodynamics, a statement that energy is conserved. Two important points need to be made about this statement:

1. A change ΔE in the internal energy stored in a system occurs when net energy is transferred to or from the system as heat q and work w.
2. The change ΔE in the internal energy of the system (the sum of q and w) is a characteristic property of the thermodynamic process by which it occurs. However, the individual contributions q and w are the equivalent of energy in transit, and are not energy in the same sense as E. Both q and w are significant during the change from the initial state E_i to the final state E_f:

$$\Delta E = E_f - E_i = q + w$$

We say that the internal energy E is a **state function** because it depends only on variables such as n, T, P, and V. As values of a state function, E_i and E_f represent thermodynamic properties, energy stored initially and energy retained at the end. On the other hand, since q and w depend on the manner of change—being energy transfers during a process—they are not considered thermodynamic properties. Upper- and lowercase letters

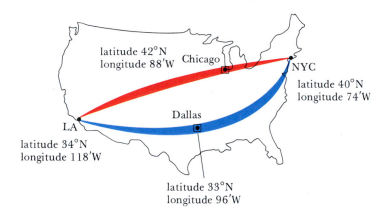

FIGURE 7–8 The coordinates on the map are analogous to state functions (or thermodynamic properties), and it does not really matter, with respect to the coordinates of latitude and longitude for New York (40°N by 74'W) and Los Angeles (34°N by 118'W), whether one is navigating from one to the other via Chicago (42°N by 88'W) or Dallas (33°N by 96'W).

are used to distinguish between those properties that are state functions and those that are not. The initial state E_i and the final state E_f are often specified as E_1 and E_2, respectively.

Keep in mind that the first law of thermodynamics does not tell you how the difference in the internal energy (ΔE) between two states (E_1 and E_2) is divided up between q and w, any more than two pairs of coordinates of longitude and latitude on a map tell you how you might get from the first point to the second. As you see from Figure 7–8, the coordinates on the map are analogous to state functions (or thermodynamic properties), and it does not really matter which of the many possible pathways was followed in navigating from the initial state to the final state. But do not underestimate the importance of features of a thermodynamic system such as q and w. Consider the social, economic, and political implications of the efficient use of our energy resources—how critically important wasted q and useful w are for a given change of state from E_1 to E_2.

It is a more difficult matter to illustrate that E is a state function than that q and w are not. For example, suppose we convert the internal energy in a gallon of oil to heat and work, using a water heater or the engine of a truck. In the truck, the energy stored in the fuel is used for locomotion but some clearly ends up as heat, dissipated by the engine's cooling system. In fact, approximately 80% of the stored energy radiates away uselessly as the truck moves along. In either case,

$$\Delta E = q + w$$

and ΔE is the same, as long as the initial and final states are the same. The change in the internal energy ΔE is the difference between E of the oil and E of the products of combustion. But q and w are different in each case. An oil-fired water heating system is designed to use the internal energy locked in the chemical composition of the fuel to heat water. The internal combustion engine in a truck is capable of using the same store of energy to provide locomotion and to heat water in the radiator at the same time.

Imagine work being done by a system that is perfectly insulated from its surroundings so heat cannot be transferred. If the initial state of the system is a gas trapped in a perfectly insulated cylinder (as in Figure 7–4), the equilibrium state is determined by P, V, T, and n (according to the Ideal Gas Law). Furthermore, if the only interaction between system and surroundings involves displacement of the piston (which is itself perfectly insulated and friction-free), the equilibrium state at the end of the process must be determined solely by the new values for P, V, T, and n. A process such as this in which heat cannot be transferred is called an **adiabatic process** (Figure 7–9a). Adiabatic processes are performed either with suf-

In Chapter 21 (Electrochemistry), we will discuss energy converters that can change the internal energy of similar hydrocarbon fuels into useful work without producing much heat. Batteries and fuel cells do that as they produce electrical work.

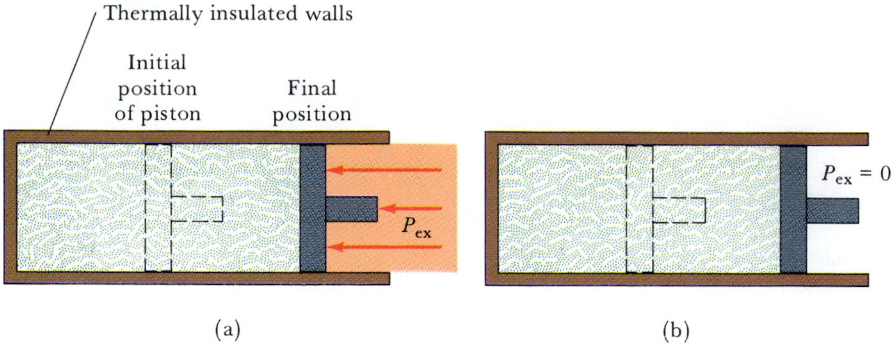

FIGURE 7–9   (a) For an adiabatic expansion against an external pressure, $q = 0$ and $\Delta E = w$. (b) For the special case of an adiabatic free expansion, note that both q and w equal zero, and as a consequence, $\Delta E = 0$.

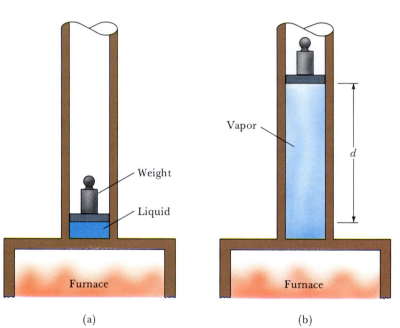

FIGURE 7–10   Vaporization of a liquid at constant pressure. (a) The liquid at its boiling point, trapped in a cylinder by a movable piston and a weight. Vaporization is about to begin. (b) After the vaporization process has been completed.

ficient insulation to prevent the flow of heat between the system and the surroundings, or by performing the process so quickly that there is not sufficient time for any significant quantity of heat to flow. Under these circumstances, since

$$q = 0$$

we can say that for an adiabatic process

$$\Delta E = w$$

For the special case of an **adiabatic free expansion**, both q and w equal zero, and as a consequence, ΔE must also equal zero (Figure 7–9b).

To summarize the terms and ideas embodied in the first law of thermodynamics, consider an experiment in which we add heat to 1.00 mol of water, causing it to vaporize at constant pressure. It is known that 40.67 kJ of heat is necessary for this process. We wish to answer the following questions: how much work is done and what is the change in the internal energy of the water? In our experiment (Figure 7–10a), the water is trapped in a cylinder fitted with a friction-free piston of negligible mass, and the base of the cylinder is in contact with a furnace capable of supplying heat. A weight is placed on top of the piston to provide a pressure P, and A is the cross-sectional area of the piston, so that

$$P = \frac{mg}{A} = \frac{F}{A}$$

Assume the liquid is at the same temperature as the furnace. Now select a weight of mass m such that the pressure produced by the piston is a little less than the pressure produced by the vapor of the liquid in the cylinder at the boiling point. In this example, assume that mass produces a pressure of 1.00 atm. Finally, set the temperature of the furnace just slightly higher than the boiling point.

In what amounts to a single stroke of a steam engine, heat flows from the furnace to the liquid until vaporization is complete (Figure 7–10b) and the vapor forces the piston upward. Since the process is carried out under the constant pressure of the weight,

$$w = PAd$$

where d is the distance moved by the piston (and the weight), and w has a negative value. But since Ad is equal to $V_2 - V_1$,

$$w = -P(V_2 - V_1) = -P\Delta V$$

To evaluate the work w, we use

$$V_2 - V_1 = \Delta V$$

where V_2 equals the volume of 1.0 mol of water vapor at 100°C and V_1 is the volume of 1.0 mol of liquid water at the same temperature:

$$V_2 = \frac{nRT}{P} = \frac{(1.00 \text{ mol})(0.0821 \text{ L·atm/mol·K})(373 \text{ K})}{1.00 \text{ atm}} = 30.6 \text{ L}$$

$$V_1 = \frac{\text{mass}}{\text{density}} = \frac{18.0 \text{ g}}{0.96 \text{ g/mL}} = 19 \text{ mL} = 0.019 \text{ L}$$

$$V_2 - V_1 = \Delta V = (30.6 - 0.019) \text{ L} = 30.6 \text{ L}$$

$$w = -P\Delta V = -(1.00 \text{ atm})(30.6 \text{ L})(0.1013 \text{ kJ/L·atm}) = -3.10 \text{ kJ}$$

To calculate the change in internal energy ΔE, you need to know that it takes 40.7 kJ to vaporize 1.00 mol of water. Then

$$\Delta E = q + w = (40.7 - 3.10) \text{ kJ}$$
$$\Delta E = 37.6 \text{ kJ}$$

Note the consequences of the result:

- Less than 10% of the energy added to the system from the furnace was used to do useful work, the mechanical work of expansion that can lift or move things.
- More than 90% of the energy input to the system was used for changing the liquid into the vapor state. Clearly the work done produced a very large increase in the average distance between H_2O molecules.

EXAMPLE 7–3
Calculate q, w, and ΔE for the vaporization of 2.00 mol of ammonia at its normal boiling point (-33.4°C) and a pressure of 1.00 atm.

Solution You need to know (1) that 23.3 kJ are required to vaporize 1.00 mol of ammonia under these conditions and (2) that the density of liquid ammonia at its boiling point is 0.80 g/mL. Assume that the volume of the liquid is negligible. Then the calculation proceeds as follows:

$$q = (2.00 \text{ mol})(23.3 \text{ kJ/mol}) = 46.6 \text{ kJ}$$

$$\Delta V = (2.00 \text{ mol})(0.0821 \text{ L·atm/mol·K})(240 \text{ K})/1.00 \text{ atm}$$
$$- \frac{(2.00 \text{ mol})(17.0 \text{ g/mol})}{(800 \text{ g/L})} = 39.4 \text{ L}$$

$$w = -(1.00 \text{ atm})(39.4 \text{ L})(0.1013 \text{ kJ/L·atm}) = -4.00 \text{ kJ}$$

$$\Delta E = q + w = (46.6 - 4.00) \text{ kJ} = 42.6 \text{ kJ}$$

EXERCISE 7–3

A gas expands against a constant pressure of 5 atm, from 10 to 20 L, while absorbing 2 kJ of heat. Calculate the work done and the change in the internal energy of the gas.
Answer: $w = -5 \text{ kJ}$; $\Delta E = -3 \text{ kJ}$.

An important consequence of Eq. (7–1) is that $\Delta E = 0$ for isothermal (constant temperature) processes of ideal gases. This is useful in heat and work calculations using the first law of thermodynamics.

EXAMPLE 7–4

(a) Consider a system containing 1.00 mol O_2 (g) that is allowed to expand from 12.2 L to 24.4 L against an external pressure of 1.00 atm at a constant temperature of 298 K. Calculate ΔE, q, and w.
(b) Calculate ΔE, q, and w for the free expansion of an ideal gas at constant temperature.

Solution

(a) $\quad w = -P_{ex}\Delta V = -1.00 \text{ atm } (24.4 - 12.2) \text{ L} = -12.2 \text{ L·atm}$

$\qquad \Delta E = 0$

$\qquad q = \Delta E - w = 0 - (-12.2 \text{ L·atm}) = +12.2 \text{ L·atm}$

(b) For a free expansion,

$\qquad w = 0 \quad$ and $\quad \Delta E = 0$

Therefore, $q = 0$.

EXERCISE 7–4(A)

Early in the American space program, the Apollo 13 mission had to be aborted (some 200,000 miles from Earth) due to the rupture of an oxygen tank used for powering certain fuel cells. Considering the oxygen to be an ideal gas, freely escaping into the vacuum of deep space, calculate q, w, and ΔE for the process, and briefly explain your reasoning. Assume constant T.
Answer: Since T is constant, $\Delta E = 0$. Since it is a free expansion, $w = 0$; therefore, $q = 0$.

EXERCISE 7–4(B)

A kilogram of water at 100°C is allowed to vaporize against an external pressure of 1.00 atm (according to the description in Figure 7–9). Calculate q, w, and ΔE for the process in kJ.
Answer: $\Delta E = 2.09 \times 10^3 \text{ kJ}$; $q = 2.26 \times 10^3 \text{ kJ}$; $w = -172 \text{ kJ}$.

7.4 ENTHALPY AND ΔH

Chemical systems are not typically designed to do useful work. Under ordinary circumstances, the only work they can do is the PV work of expansion or compression. The best-known exceptions are batteries and other electrochemical cells. Since our definition of work is the product of a force moved through a distance,

$$w = -P\Delta V$$

if there is no volume change (no distance) then no useful work can be done. Under these circumstances, say for a reaction run in a closed container,

$$\Delta V = 0$$
$$w = -P\Delta V = 0$$

and

$$\Delta E = q_v \tag{7-5}$$

where the subscript v indicates a constant volume process. For a chemical system, then, an increase in the internal energy ΔE is due to the heat absorbed at constant volume, q_v. Such a system can do no useful work, and therefore q_v must be a state property just as ΔE is.

Most chemical experiments, however, are carried out at constant pressure, not constant volume. Under these conditions, pressure-volume work is not zero, and the heat of reaction q_p is ΔE plus the $P\Delta V$ term due to the work of expansion:

$$q_p = \Delta E + P\Delta V$$

The heat of reaction at constant pressure, q_p, is also a state function, having a definite value for each state (since E, P, and V have definite values). This new state function is called the **enthalpy** H, and the change in enthalpy ΔH is defined as follows:

$$\Delta H = \Delta E + P\Delta V \tag{7-6}$$
$$\Delta H = q_p \tag{7-7}$$

That is, ΔH for a process is numerically equal to the heat exchanged with the surroundings at constant pressure. Assuming that the work done is of the $P\Delta V$ (pressure-volume) kind, the increase in the enthalpy of a system ΔH equals the heat absorbed at constant pressure, q_p.

The importance of the enthalpy function will become apparent in the study of thermochemistry. For example, consider the following **thermochemical equation**:

$$CS_2\,(g) + 3\,O_2\,(g) \longrightarrow CO_2\,(g) + 2\,SO_2\,(g)$$
$$\Delta H = -1110 \text{ kJ/mol } CS_2$$

What does this mean? When one mole of CS_2 gas reacts, the enthalpy of the system decreases by 1110 kJ, because that amount of energy is liberated from the system in the form of heat at constant pressure.

For processes involving solids and liquids, ΔH and ΔE are essentially the same because $P\Delta V$ is negligible. That is not the case, however, for gas phase reactions or for heterogeneous reactions involving gaseous components. ΔH and ΔE can be significantly different because of the work term $P\Delta V$. By definition, at constant pressure

$$P\Delta V = PV_2 - PV_1$$

Substituting nRT for PV from the Ideal Gas Law,

$$P\Delta V = n_2 R T_2 - n_1 R T_1$$

For an **isothermal** (constant temperature) expansion,

$$P\Delta V = (n_2 - n_1)RT = \Delta nRT$$

Finally, substituting for $P\Delta V$ in our original expression, the heat of reaction at constant pressure (the enthalpy of reaction) is

$$\Delta H = \Delta E + \Delta nRT \tag{7–8}$$

It is important to remember when using this equation that Δn refers only to substances in the gaseous state. Note that in the vicinity of room temperature, the value for RT is about 2.5 kJ/mol, which is not a negligible quantity:

$$RT = (8.314 \text{ J/mol·K})(298 \text{ K}) = 2.5 \text{ kJ/mol}$$

Consider the combustion of liquid benzene in a constant volume "bomb" calorimeter at 298 K. The heat of reaction is 3264 kJ/mol, and the reaction is exothermic:

$$C_6H_6 \text{ (liq)} + \tfrac{15}{2} \, O_2 \text{ (g)} \longrightarrow 6 \, CO_2 \text{ (g)} + 3 \, H_2O \text{ (liq)} + \text{heat}$$

Since this is stated to be a constant volume process, the heat of reaction is ΔE, the change in the internal energy:

$$q = q_v = \Delta E = -3264 \text{ kJ/mol } C_6H_6$$

We can calculate ΔH for this reaction. Since a gas has a volume so much larger than that of an equal mass of liquid, it is only necessary to consider the volumes of the gases. Note that the reactants include $\tfrac{15}{2}$ moles of gas while the products include only 6 moles of gas. Then

$$
\begin{aligned}
\Delta H &= \Delta E + \Delta nRT \\
&= -3264 \text{ kJ/mol} + (n_2 - n_1)RT \\
&= -3264 \text{ kJ/mol} + [(6 - \tfrac{15}{2})]RT \\
&= -3264 \text{ kJ/mol} + [-\tfrac{3}{2}][(8.314 \text{ J/mol·K})(0.001 \text{ kJ/J})(298 \text{ K})] \\
&= -3264 \text{ kJ/mol} + (-4 \text{ kJ/mol}) = -3268 \text{ kJ/mol } C_6H_6
\end{aligned}
$$

Note that Δn is the difference between the number of moles of gaseous products and the number of moles of gaseous reactants, and that the difference between ΔH and ΔE reflects the pressure-volume work the system would have to do in pushing back the atmosphere.

EXAMPLE 7–5

A mole of methane is oxidized to carbon dioxide and water at 25°C:

$$CH_4 \text{ (g)} + 2 \, O_2 \text{ (g)} \longrightarrow CO_2 \text{ (g)} + 2 \, H_2O \text{ (?)}$$

For this particular reaction,

$$q_v = -887 \text{ kJ/mol } CH_4$$

$$q_p = -891 \text{ kJ/mol } CH_4$$

Determine whether the H_2O formed in the reaction is a gas or a liquid.

Solution Determine Δn. Remember that the difference between ΔE and ΔH is negligible except for reactions that include gases. If H_2O is a gas, then $\Delta n = 0$ and ΔE equals ΔH, which is not the case in our example.

Therefore, the H_2O reaction product is in the liquid state, and $\Delta n = 1 - 3 = -2$. To validate that conclusion, calculate ΔnRT and ΔH as follows:

$$\Delta nRT = [(1 - 3)][8.314 \text{ J/K·mol})(0.001 \text{ kJ/J})(298 \text{ K})$$
$$= -5 \text{ kJ/mol}$$
$$\Delta H = \Delta E + \Delta nRT = q_v + \Delta nRT = -887 \text{ kJ/mol} + (-5 \text{ kJ/mol})$$
$$= -892 \text{ kJ/mol CH}_4 \qquad ■$$

EXERCISE 7–5

The constant volume combustion of toluene in a bomb calorimeter takes place according to the following reaction at 298 K:

$$C_6H_5CH_3 \text{ (liq)} + 9 O_2 \text{ (g)} \longrightarrow 7 CO_2 \text{ (g)} + 4 H_2O \text{ (liq)}$$

The heat of the reaction is -3912 kJ/mol $C_6H_5CH_3$. Determine the values of q_v, q_p, ΔE, and ΔH. By what factor would the work term ΔnRT change if the H_2O were formed in the vapor state?
Answer: $q_v = -3912$ kJ; $\Delta E = -3912$ kJ; $q_p = -3917$ kJ; $\Delta H = -3917$ kJ. By -1.   ■

One of the most useful methods for determining q, ΔE, and ΔH for a chemical reaction is by **calorimetry**, which consists of running a reaction in a specially designed and well-insulated vessel called a **calorimeter**, at either constant volume or constant pressure, and recording the temperature change that takes place (Fig. 7–11). For most processes, it is desirable to carry out chemical reactions under isothermal (constant temperature) or nearly isothermal conditions, since that simplifies many thermochemical equations. In order for this to happen, the calorimeter must be able to absorb the heat released by the reaction with an increase in temperature of no more than a few degrees or so. Since the recorded temperature change of the calorimeter and its contents is small, the temperature measurements must be precisely made, and especially accurate thermometers have been developed for use in calorimetry.

The chemical process in the calorimeter is assumed to be isothermal at the initial temperature, a simplifying assumption that usually does not impair the results.

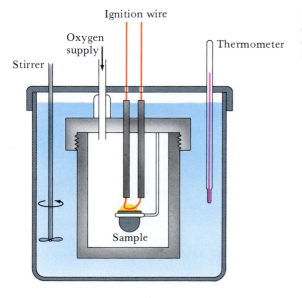

Ignition wire

Oxygen supply

Thermometer

Stirrer

Sample

FIGURE 7–11   A constant volume bomb calorimeter. Sample in cup is ignited by a hot wire in an oxygen atmosphere. The heat generated is measured by the rise in temperature of the surrounding water bath.

The fundamental calorimeter equation is

$$q = -C\Delta T \tag{7-9}$$

It is convenient to describe the energy-storing capacities of systems and substances in terms of heat capacity C (the heat required to increase the temperature by one degree), specific heat capacity (per gram of substance), or molar heat capacity (per mole of substance).

where C is the heat capacity of the calorimeter and its contents—the amount of heat it takes to raise the temperature of the calorimeter by one kelvin. The negative sign in the equation is consistent with the definition that an exothermic process (one with a negative value of q) causes the temperature of the calorimeter to rise. (ΔT has a positive value). The heat capacity of the calorimeter and its contents is sometimes called the **calorimeter constant**. For a constant volume calorimeter,

$$q = q_v = \Delta E$$

and for a constant pressure calorimeter,

$$q = q_p = \Delta H$$

EXAMPLE 7–6

Benzoic acid ($C_7H_6O_2$) is commonly used to calibrate calorimeters (determine their heat capacities) because it is an easily weighed solid that burns readily and completely. Its heat of combustion is known to be -3227 kJ/mol. A particular calorimeter is calibrated by burning 1.000 g of benzoic acid at constant pressure, and the temperature is observed to rise from 23.00°C to 25.12°C. Next, a 3.100 g sample of solid citric acid ($C_6H_8O_7$) is completely oxidized in the same calorimeter, and a temperature change from 23.00°C to 25.57°C is observed. Calculate ΔH and ΔE for the combustion of citric acid.

Solution   Because the reaction is run at constant pressure, $q = \Delta H$. The heat released by the benzoic acid calibration reaction is

$$(-3227 \text{ kJ/mol}) \left(\frac{1.000 \text{ g}}{122.1 \text{ g/mol}} \right) = -26.43 \text{ kJ}$$

Therefore, the heat capacity of the calorimeter is

$$C = -\frac{q}{\Delta T} = -\frac{-26.43 \text{ kJ}}{2.12 \text{ K}} = 12.47 \text{ kJ/K}$$

Since the citric acid is burned in the same calorimeter, the value of C is the same. The heat released by combustion of the citric acid sample is

$$q = -C\Delta T = -(12.47 \text{ kJ/K})(2.57 \text{ K}) = -32.0 \text{ kJ}$$

The heat of combustion per mole of citric acid is

$$q_p = \Delta H = \frac{-32.0 \text{ kJ}}{3.100 \text{ g}} (192.1 \text{ g/mol}) = -1.98 \times 10^3 \text{ kJ/mol}$$

To determine ΔE we need to find Δn, which we get from the balanced reaction:

$$C_6H_8O_7 \text{ (s)} + \tfrac{9}{2} O_2 \text{ (g)} \longrightarrow 6 \text{ CO}_2 \text{ (g)} + 4 \text{ H}_2O \text{ (liq)}$$

$$\Delta n = (6 - \tfrac{9}{2}) = \tfrac{3}{2} = 1.5$$

$$\begin{aligned} \Delta E &= \Delta H - \Delta n R T \\ &= (-1980 \text{ kJ/mol}) - (1.5)(8.314 \text{ J/mol·K})(0.001 \text{ kJ/J})(296 \text{ K}) \\ &= -1.98 \times 10^3 \text{ kJ/mol} \end{aligned}$$

EXERCISE 7–6

In order to calibrate a constant pressure calorimeter, a 2.000-g sample of naphthalene ($C_{10}H_8$) is burned in it and the temperature is observed to change from 26.50 to 27.68°C. The heat of combustion of naphthalene is -5153.9 kJ/mol. A 3.210 g sample of liquid pentene (C_5H_{10}) is then burned to liquid water and carbon dioxide in the same calorimeter, and the temperature rises from 25.50 to 26.69°C. Determine ΔE and ΔH for the combustion of pentene.

Answer: $\Delta H = -1772$ kJ/mol; $\Delta E = -1766$ kJ/mol. ■

7.5 THERMODYNAMICS AND THERMOCHEMICAL METHODS; HESS'S LAWS

Thermochemistry describes the heat transfer that takes place between a chemical system and its surroundings when a chemical reaction or phase change takes place. The quantities of heat transferred are related to the energy and enthalpy changes occurring in the system.

- Because both ΔE and ΔH are state functions, the sum of the changes in each of these quantities must be zero over a complete cycle, whether two or twenty-two steps are involved in getting you around the cycle and back to the beginning. In the same way, the total change in altitude for a race around a closed track is zero. Thus, altitude is a state function, and as long as you finish the race at the starting point, you must return to the same altitude.

- Because ΔH is a state function, any difference between two states is constant and independent of the path taken between those states (Figure 7–12).

These two principles were first stated by G. H. Hess 150 years ago. Here are several illustrations of Hess's laws and their implications.

1. The heat associated with a chemical reaction depends on the amounts of substances present as well as on the pressure-volume conditions and the physical states of the reactants and products. For example, 1 g of hydrogen is oxidized by 35.5 g of chlorine according to the equation

$$\tfrac{1}{2} H_2 \,(g) + \tfrac{1}{2} Cl_2 \,(g) \longrightarrow HCl \,(g) \qquad \Delta H = -92.5 \text{ kJ}$$

The change in enthalpy is -92.5 kJ for each mole of HCl produced, and the sign of ΔH is negative because the reaction is exothermic. But if 2 g of hydrogen is similarly oxidized, twice the amount of heat is released by the system. The heat associated with a chemical reaction is directly proportional to the amounts of substances involved:

$$H_2 \,(g) + Cl_2 \,(g) \longrightarrow 2 \, HCl \,(g) \qquad \Delta H = -185 \text{ kJ}$$

These equations are called **thermochemical equations** because each includes the heat of reaction in addition to the stoichiometric information. In this case, it is the heat of reaction at constant pressure, $q_p = \Delta H$.

2. In dealing with thermochemical equations, it is important to identify clearly the physical state of each reactant and product at the stated pressure and temperature. Consider the oxidation of hydrogen by oxygen:

$$H_2 \,(g) + \tfrac{1}{2} O_2 \,(g) \longrightarrow H_2O \,(g) \qquad \Delta H = -242 \text{ kJ}$$

$$H_2 \,(g) + \tfrac{1}{2} O_2 \,(g) \longrightarrow H_2O \,(liq) \qquad \Delta H = -286 \text{ kJ}$$

Because pressure is not a variable for enthalpy, ΔH is a very convenient measurement for chemists and engineers who are working under atmospheric conditions.

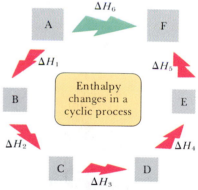

FIGURE 7–12 Hess's Law and the conservation of energy. Enthalpy changes are shown for a cyclic process. Hess's Law requires that $\Delta H_6 = \Delta H_1 + \Delta H_2 + \Delta H_3 + \Delta H_4 + \Delta H_5$. Hess's Law also requires that $\Delta H_1 + \Delta H_2 + \Delta H_3 + \Delta H_4 + \Delta H_5 + (-\Delta H_6) = 0$.

At constant pressure, $q_p = \Delta H$.

The heat of reaction associated with formation of 1 mol of H_2O in the gaseous state differs from that for the reaction where liquid water is formed. The difference equals the heat required by the system to separate liquid H_2O molecules and create the vapor state—a difference of 44 kJ/mol H_2O.

3. The enthalpy change ΔH for a chemical reaction has a definite value—the net difference between the enthalpy H of the products and the enthalpy H of the reactants. When it is stated that

$$\tfrac{1}{2} H_2 \text{ (g)} + \tfrac{1}{2} Cl_2 \text{ (g)} \longrightarrow HCl \text{ (g)} \qquad \Delta H = -92.5 \text{ kJ}$$

what is really being said is that

$$\Delta H = H_{products} - H_{reactants}$$
$$\Delta H = H \text{ for HCl} - [\tfrac{1}{2} H \text{ for } H_2 + \tfrac{1}{2} H \text{ for } Cl_2] = -92.5 \text{ kJ}$$

When 1 mol of HCl is formed from $\tfrac{1}{2}$ mol of hydrogen and $\tfrac{1}{2}$ mol chlorine, the enthalpy of the reactants exceeds that of the products by 92.5 kJ, the heat given off by the system. Similarly, for the formation of liquid H_2O, the enthalpy of the reactants is greater than the enthalpy of the products by 286 kJ, which is the amount of heat given off by the system.

4. When reactions are run in reverse, ΔH has the same numerical value but with the opposite sign:

$$\Delta H_{forward} = -\Delta H_{reverse}$$

In the following pair of reactions, note that the second is $-\Delta H$ for the reverse of the reaction for the formation of H_2O liquid in (2) above. Adding the two reactions gives ΔH for the phase change from liquid to gas, and adding the ΔH values gives ΔH for the phase change:

$$H_2 \text{ (g)} + \tfrac{1}{2} O_2 \text{ (g)} \longrightarrow H_2O \text{ (g)} \qquad \Delta H = -242 \text{ kJ}$$
$$H_2O \text{ (liq)} \longrightarrow H_2 \text{ (g)} + \tfrac{1}{2} O_2 \text{ (g)} \qquad \Delta H = +286 \text{ kJ}$$

$$H_2 \text{ (g)} + \tfrac{1}{2} O_2 \text{ (g)} + H_2O \text{ (liq)} \longrightarrow$$
$$H_2O \text{ (g)} + H_2 \text{ (g)} + \tfrac{1}{2} O_2 \text{ (g)}$$

Canceling like species on opposite sides of the arrow gives the net result:

$$H_2O \text{ (liq)} \longrightarrow H_2O \text{ (g)} \qquad \Delta H = +44.4 \text{ kJ}$$

Thus, when a mole of water is vaporized at 298 K, 44.4 kJ of energy is absorbed (note that 298 K is not the normal boiling point). Since ΔH is positive, the reaction is endothermic.

Thermochemical reactions are dependent upon the states of the reactants and products. As a convenience, the **standard state** of a substance is defined as the most stable state at 1 atm and a specified temperature, usually 298 K. For example, oxygen can exist in a number of forms, including O atoms, O_2 molecules, and O_3 molecules. Oxygen can also be solid, liquid, or gas, depending on conditions. However, at 1 atm and 298 K its most stable form is O_2 (g) molecules. This is therefore its standard state. On the other hand, aluminum metal is Al (s) in its standard state. For clarity, ΔH for a process with all reactants and products in their standard states is given the symbol $\Delta H°$. Standard states for other common elements are H_2 (g), N_2 (g), O_2 (g), F_2 (g), Cl_2 (g), Br_2 (liq), I_2 (s). All metals are solids in their standard states except mercury, which is Hg (liq) (Figure 7–13).

FIGURE 7–13 Standard states for three of the halogen elements: chlorine (left) is a greenish-yellow gas; bromine (center) is a reddish, volatile liquid and only one of two liquid elements at room temperature; iodine (right) crystals.

EXAMPLE 7–7

Find the enthalpy change accompanying the formation of benzene from its elements, both reactants and products being in their standard states. This is referred to as the standard enthalpy of formation, $\Delta H_f°$.

Solution

(1) $6 \text{ C (s, graphite)} + 3 \text{ H}_2 \text{ (g)} \longrightarrow \text{C}_6\text{H}_6 \text{ (liq)}$ $\Delta H_f° = ?$

The standard state of carbon at 298 K is graphite. It is highly unlikely that very much benzene could ever be succesfully prepared by the reaction of graphite with hydrogen regardless of the pressure and temperature. But it is relatively easy to measure the heat of combustion ($\Delta H_c°$) of all three substances in this equation in a suitable calorimeter. Then we can combine that information by applying Hess's Law.

Benzene is easily oxidized and $\Delta H_c°$ can be measured:

(2) $\text{C}_6\text{H}_6 \text{ (liq)} + \frac{15}{2} \text{ O}_2 \text{ (g)} \longrightarrow 6 \text{ CO}_2 \text{ (g)} + 3 \text{ H}_2\text{O} \text{ (liq)}$
$$\Delta H_c° = -3268 \text{ kJ/mol}$$

Carbon dioxide can be formed from its elements, and the heat of this combustion reaction is stated as the standard heat of formation, $\Delta H_f°$, of CO_2 (g):

(3) $\text{C (s)} + \text{O}_2 \text{ (g)} \longrightarrow \text{CO}_2 \text{ (g)}$ $\Delta H_f° = -393 \text{ kJ/mol}$

Water can be formed from its elements, and its $\Delta H_f°$ value can be measured:

(4) $\text{H}_2 \text{ (g)} + \frac{1}{2} \text{ O}_2 \text{ (g)} \longrightarrow \text{H}_2\text{O} \text{ (liq)}$ $\Delta H_f° = -286 \text{ kJ/mol}$

Properly combining the standard enthalpy of combustion $\Delta H_c°$ for benzene with the standard enthalpies of formation $\Delta H_f°$ for carbon dioxide and water—equations 2, 3, and 4—gives the enthalpy of formation $\Delta H_f°$ for benzene in equation 1. Remember that you must include the necessary coefficients—six times $\Delta H_f°$ for carbon dioxide and three times $\Delta H_f°$ for water—and watch the signs, reversing $\Delta H_c°$ for benzene:

$6 \text{ [C (s)} + \text{O}_2 \text{ (g)} \longrightarrow \text{CO}_2 \text{ (g)]}$ $\Delta H = (6)(-393 \text{ kJ/mol})$
$3 \text{[H}_2 \text{ (g)} + \frac{1}{2} \text{ O}_2 \text{ (g)} \longrightarrow \text{H}_2\text{O} \text{ (liq)]}$ $\Delta H = (3)(-286 \text{ kJ/mol})$
$6 \text{ CO}_2 \text{ (g)} + 3 \text{ H}_2\text{O} \text{ (liq)} \longrightarrow \text{C}_6\text{H}_6 \text{ (liq)} + \frac{15}{2} \text{ O}_2 \text{ (g)}$
$$\Delta H = -(1)(-3268 \text{ kJ/mol})$$

The net result is

$6 \text{ C (s)} + 3 \text{ H}_2 \text{ (g)} \longrightarrow \text{C}_6\text{H}_6 \text{ (liq)}$ $\Delta H = +52 \text{ kJ/mol}$

which is obtained from

$\Delta H_f° \text{ [C}_6\text{H}_6 \text{ (liq)]} = 6 \Delta H_f° \text{ [CO}_2 \text{ (g)]} + 3 \Delta H_f° \text{[H}_2\text{O (liq)]}$
$- \Delta H_c° \text{[C}_6\text{H}_6 \text{ (liq)]}$
$= [6(-393) + 3(-286) - (-3268)] \text{ kJ/mol}$
$= [(-2358) + (-858) - (-3268)] \text{ kJ/mol}$
$= +52 \text{ kJ/mol}$

EXERCISE 7–7(A)

Using data from Tables 7–4 and 7–5 and following the logic of Example 7–7, calculate $\Delta H_f° \text{ [C}_2\text{H}_2 \text{ (g)]}$.
Answer: $\Delta H_f° = 228 \text{ kJ/mol C}_2\text{H}_2 \text{ (g)}$.

TABLE 7–4   Standard Enthalpies of Formation (ΔH_f°) at 298 K

	Substance and State	ΔH_f°, kJ/mol
	H_2O (liq)	−286
	H_2O (g)	−242
	H_2O_2 (liq)	−187
	HCl (g)	−92.5
	HBr (g)	−36.4
	HI (g)	+25.9
	H_2S (g)	−20.1
	SO_2 (g)	−297
	SO_3 (g)	−395
	H_2SO_4 (liq)	−811
	NH_3 (g)	−46.0
	NO (g)	+90.4
	NO_2 (g)	+33.9
	HNO_3 (liq)	−173
	CO (g)	−110
	CO_2 (g)	−393
	Ag_2O (s)	−30.5
	AgCl (s)	−127
	$AgNO_3$ (s)	−123
	Al_2Cl_6 (s)	−1291
	Al_2O_3 (s)	−1676
	FeO (s)	−269
	Fe_2O_3 (s)	−822
	Fe_3O_4 (s)	−1121
	FeS_2 (s)	−178
	NaCl (s)	−411
	Na_2SO_4 (s)	−1384
	Na_2CO_3 (s)	−1131
	PbO (s)	−217
	$CaCl_2$ (s)	−795
	$CaCO_3$ (s)	−1207
	CuO (s)	−155
	CuS (s)	−53.5
methane	CH_4 (g)	−74.9
ethane	C_2H_6 (g)	−84.5
propane	C_3H_8 (g)	−103.8
n-butane	C_4H_{10} (g)	−125
n-pentane	C_5H_{12} (liq)	−146
n-octane	C_8H_{18} (liq)	−208
ethylene	C_2H_4 (g)	+52.3
acetylene	C_2H_2 (g)	+227
benzene	C_6H_6 (liq)	+49.0
methyl alcohol	CH_3OH (liq)	−238
ethyl alcohol	C_2H_5OH (liq)	−278
urea	$CO(NH_2)_2$ (s)	−330
glucose	$C_6H_{12}O_6$ (s)	−1274

EXERCISE 7–7(B)

Given the data in Tables 7–4 and 7–5, calculate the enthalpy of formation of sucrose:

$$12 \text{ C (s)} + 11 \text{ H}_2 \text{ (g)} + \tfrac{11}{2} \text{ O}_2 \text{ (g)} \longrightarrow C_{12}H_{22}O_{11} \text{ (s)}$$

Answer: $\Delta H_f^\circ = -2221$ kJ/mol.

TABLE 7–5   Heats of Combustion ($\Delta H_c°$) at 298 K

	Substance and State	$\Delta H_c°$ (kJ/mol)
methane	CH_4 (g)	−890
ethane	C_2H_6 (g)	−1560
propane	C_3H_8 (g)	−2220
n-butane	C_4H_{10} (g)	−2879
n-octane	C_8H_{18} (liq)	−5450
ethylene	C_2H_4 (g)	−1411
acetylene	C_2H_2 (g)	−1300
benzene	C_6H_6 (liq)	−3268
benzene	C_6H_6 (g)	−3293
naphthalene	$C_{10}H_8$ (s)	−5153.9
benzoic acid	C_6H_5COOH (s)	−3227
acetic acid	CH_3COOH (liq)	−875
methyl alcohol	CH_3OH (liq)	−727
ethyl alcohol	C_2H_5OH (liq)	−1367
urea	$CO(NH_2)_2$ (s)	−632
glucose	$C_6H_{12}O_6$ (s)	−2803
sucrose	$C_{12}H_{22}O_{11}$ (s)	−5641

EXERCISE 7–7(C)

Given the following information and the data in Tables 7–4 and 7–5, calculate the standard enthalpy of formation ($\Delta H_f°$) of ethyl alcohol (C_2H_5OH) vapor:

$$C_2H_5OH \text{ (liq)} \longrightarrow C_2H_5OH \text{ (g)} \qquad \Delta H° = +38.5 \text{ kJ}$$

Answer: $\Delta H_f° = -239.5$ kJ/mol.   ■

Values for $\Delta H_f°$ and $\Delta H_c°$ for complete combustion have been obtained and tabulated for large numbers of organic compounds. Combustion data are available for most of the hydrocarbons isolated from coal and crude oil and for many other organic substances such as alcohols, sugars, and organic acids (Tables 7–4 and 7–5). Note that where heats of combustion happen to be heats of formation (as is the case for the burning of carbon and hydrogen to give carbon dioxide and water), it is common practice to list the data as $\Delta H_f°$ values.

The enthalpies of formation of the elements in their standard states are defined as zero. This fact is very useful for calculating $\Delta H°$ values for reactions in general, using the equation

$$\Delta H° = \sum \Delta H_f°(\text{products}) - \sum \Delta H_f°(\text{reactants})$$

and the thermodynamic data in Table 7–4.

Since thermochemical quantities are stated as differences in heat content, the reference point is immaterial. Setting the heat content of all elements in their standard states equal to zero at all temperatures is simple and convenient, making the heat content of a compound equal to the heat of formation. It does not imply that the heat content of an element is actually zero.

EXAMPLE 7–8

Calculate $\Delta H_f°$ for Fe_2O_3 (s).

Solution The measured heat of reaction for the oxidation of iron is the heat of formation of iron(III) oxide. If the reaction is carried out under standard state conditions, then $\Delta H = \Delta H_f°$, and such data are commonly listed in thermodynamic tables such as Table 7–4.

$$2 \text{ Fe (s)} + \tfrac{3}{2} O_2 \text{ (g)} \longrightarrow Fe_2O_3 \text{ (s)}$$
$$\Delta H_f^\circ = -822 \text{ kJ/mol } Fe_2O_3 \text{ (s)}$$

$$\Delta H^\circ = \sum \Delta H_f^\circ (\text{products}) - \sum \Delta H_f^\circ (\text{reactants})$$
$$= \Delta H^\circ \text{ for } Fe_2O_3 \text{ (s)} - [2 \ \Delta H^\circ \text{ for Fe (s)} + \tfrac{3}{2} \ \Delta H^\circ \text{ for } O_2 \text{ (g)}]$$
$$= \Delta H^\circ \text{ for } Fe_2O_3 \text{ (s)} - [0 + 0]$$
$$= \Delta H^\circ \text{ for } Fe_2O_3 \text{ (s)} = \Delta H_f^\circ = -822 \text{ kJ/mol } Fe_2O_3 \text{ (s)} \quad \blacksquare$$

EXERCISE 7–8

Given the following reaction at 298 K:

$$C \text{ (s)} + O_2 \text{ (g)} + 2 NH_3 \text{ (g)} \longrightarrow CO(NH_2)_2 \text{ (s)} + H_2O \text{ (liq)}$$
$$\Delta H = -1207 \text{ kJ}$$

and any other necessary thermochemical data (Tables 7–4 and 7–5), calculate the heat of reaction for the synthesis of urea from ammonia and carbon dioxide:

$$2 NH_3 \text{ (g)} + CO_2 \text{ (g)} \longrightarrow CO(NH_2)_2 \text{ (s)} + H_2O \text{ (liq)} \qquad \Delta H = \ ?$$

Answer: $\Delta H = -814 \text{ kJ/mol}$. $\quad \blacksquare$

EXAMPLE 7–9

Consider the special case of **allotropic modifications** (Figure 7–14), in which an elemental substance exists in more than one form. Calculate ΔH° for the conversion of graphite to diamond.

Solution Heats of combustion allow direct determination of energy differences between allotropes of an element. For example, for diamond and graphite, the two allotropic forms of carbon:

$$C \text{ (diamond)} + O_2 \text{ (g)} \longrightarrow CO_2 \text{ (g)}$$
$$\Delta H_c^\circ = -395.388 \text{ kJ/mol (diamond)}$$
$$C \text{ (graphite)} + O_2 \text{ (g)} \longrightarrow CO_2 \text{ (g)}$$
$$\Delta H_c^\circ = -393.505 \text{ kJ/mol (graphite)}$$

Graphite is the lower energy, more stable form (at room temperature), and the enthalpy change associated with the transition from one form to the other can be calculated (based on Hess's laws of thermochemistry):

$$C \text{ (graphite)} \longrightarrow C \text{ (diamond)} \qquad \Delta H^\circ = +1.883 \text{ kJ/mol C} \quad \blacksquare$$

EXERCISE 7–9

Given the following thermochemical data, calculate the enthalpy change (in joules per mole) accompanying the interconversion of the two allotropic forms of sulfur:

$$S \text{ (rhombic)} + O_2 \text{ (g)} \longrightarrow SO_2 \text{ (g)} \qquad \Delta H_f^\circ = -296.980 \text{ kJ}$$
$$S \text{ (monoclinic)} + O_2 \text{ (g)} \longrightarrow SO_2 \text{ (g)} \qquad \Delta H_f^\circ = -297.148 \text{ kJ}$$

Which is the more stable allotrope under standard conditions?
Answer: $\Delta H^\circ = 168 \text{ J/mol}$; S (rhombic). $\quad \blacksquare$

FIGURE 7–14 Finely divided graphite crystals and single crystals of gemstone quality synthetic diamonds, grown in a special apparatus for converting graphite into diamond (see Chapter 19). Both are allotropic modifications of elemental carbon. Graphite is also commonly used in "lead" pencils.

Thermochemical Treatment of a Metallurgical Problem

A classic metallurgical technique for producing a pure metal from a suitable compound, often an oxide of the metal, involves reduction by a second reactive metal. In this process, it is often desirable to have the chemical reaction produce sufficient heat to raise all the products to the temperature of the liquid state, thereby allowing for easier separation (of the fluid reaction products). In the **aluminothermic reduction** (Figure 7–3b) used to prepare free manganese, a number of different oxides can be used. The naturally occurring ore called pyrolusite contains manganese in an oxidation state of $+4$ as the dioxide (MnO_2). However, MnO, Mn_2O_3, and Mn_3O_4 are also naturally occurring oxides of definite chemical composition. Enthalpies of formation are, respectively:

$$Mn\ (s) + \tfrac{1}{2}\ O_2\ (g) \longrightarrow MnO\ (s) \qquad \Delta H_f^\circ = -385\ kJ$$

$$Mn\ (s) + O_2\ (g) \longrightarrow MnO_2\ (s) \qquad \Delta H_f^\circ = -520\ kJ$$

$$2\ Mn\ (s) + \tfrac{3}{2}\ O_2\ (g) \longrightarrow Mn_2O_3\ (s) \qquad \Delta H_f^\circ = -957\ kJ$$

$$3\ Mn\ (s) + 2\ O_2\ (g) \longrightarrow Mn_3O_4\ (s) \qquad \Delta H_f^\circ = -1387\ kJ$$

For the formation of aluminum oxide,

$$2\ Al\ (s) + \tfrac{3}{2}\ O_2\ (g) \longrightarrow Al_2O_3\ (s) \qquad \Delta H_f^\circ = -1674\ kJ$$

During the reduction, the temperature of the products reaches about 2300 K. The heat required to raise manganese metal and aluminum oxide to this temperature has been calculated. For manganese metal:

$$H^\circ_{2300} - H^\circ_{298} = 100\ kJ/mol$$

For aluminum oxide:

$$H^\circ_{2300} - H^\circ_{298} = 356\ kJ/mol$$

The heat required and the heat available to bring the products to 2300 K for the different possible manganese oxide starting materials can now be calculated and tabulated. For MnO:

$$3\ MnO\ (s) \longrightarrow 3\ Mn\ (s) + \tfrac{3}{2}\ O_2\ (g) \qquad \Delta H^\circ = 3(+385)\ kJ$$
$$2\ Al\ (s) + \tfrac{3}{2}\ O_2\ (g) \longrightarrow Al_2O_3\ (s) \qquad \Delta H^\circ = -1674\ kJ$$

$$\overline{3\ MnO\ (s) + 2\ Al\ (s) \longrightarrow 3\ Mn\ (s) + Al_2O_3\ (s) \qquad \Delta H^\circ = -519\ kJ}$$

The reduction produces 519 kJ for each mole of aluminum oxide produced. The heat required for the desired liquefaction is:

$$
\begin{array}{ll}
1\ mol\ Al_2O_3\ (s) & 356\ kJ \\
3\ mol\ Mn\ (s) & 3(100) = 300\ kJ \\
\hline
 & 656\ kJ\ required
\end{array}
$$

Tabulating these data for all the oxides, calculated in the same fashion:

Manganese Oxide	Heat Required to Raise Products to 2300 K, kJ/mol	Heat Evolved kJ/mol Al_2O_3
MnO (s)	656	519
Mn_3O_4 (s)	582	636
Mn_2O_3 (s)	556	717
MnO_2 (s)	506	895

Analysis of these data gives the following results:

- The process is self-sustaining if an oxide mixture somewhere between MnO and Mn_3O_4 is used.

- If pure MnO_2 is used, the reaction might be explosive once initiated because of the large excess of available heat.

7.6   HEAT CAPACITIES

Thermal energy can be added to a substance in various ways. Some goes to the **bulk energy** of the material, the mechanical energy a body has because of its positon and momentum, such as a bullet on its way. Part goes to **internal energy**, the randomly distributed energy that has to do with the motion of the atoms and molecules within the substance as they collide and move about. The internal energy includes **translational kinetic energy**, the energy of particles moving along straight lines, which serves to raise the temperature of a substance. For most substances, part of the internal energy goes to **rotational** and **vibrational energy**, which serve to excite the atoms and molecules; if the temperatures are high enough, some energy may go to **electronic excitations** that serve to ionize and dissociate atoms and molecules.

For a given substance, the experimentally measured quantities that summarize the relationship between the temperature change and the energy change are the **specific heat** (the quantity of heat needed to raise the temperature of a gram of the substance by one kelvin) and the **heat capacity** C (the molar specific heat). Since heat capacities depend on external conditions (as well as chemical composition), these must be specified. Remember that heat and work are not state functions, so neither can be used to describe the state of a system. There are two standard conditions for measuring heat capacities—at constant volume (C_v) and at constant pressure (C_p). For solids and liquids, C_v and C_p are not very different; however, this is not the case for gases, where both values are important and the difference between them is significant.

Recall our discussion of calorimetry in Section 7.4.

EXAMPLE 7–10

By definition, the specific heat capacity (or simply the specific heat) is the quantity of heat needed per degree temperature change per unit mass of substance. Magnesium, one of the space-age materials so important in the engineering design of aircraft surfaces, has a specific heat of 1.0 J/g·K. What is its heat capacity, or molar specific heat?

Solution   The heat capacity is simply the product of the specific heat and the molar weight:

$$\left(\frac{24.3 \text{ g}}{\text{mol}}\right)\left(\frac{1.0 \text{ J}}{\text{g·K}}\right) = \frac{24.3 \text{ J}}{\text{mol·K}}$$

EXERCISE 7–10(A)

The heat capacity of silver is 25.48 J/mol·K. What is the specific heat of metallic silver?
Answer: 0.2362 J/g·K.

EXERCISE 7–10(B)

A sphere of fixed volume is filled with 1.00 mol of helium gas at room temperature (25°C). Calculate the quantity of heat that must be transferred to the gas in order to double its absolute temperature. Use C_v for He from Table 7–7.
Answer: 3.81 kJ.

Thermodynamically, the simplest substances to treat are those in which all the internal energy is of the translational (kinetic) variety. The closest approximation to this is an ideal monatomic gas. He, Ne, and the other noble gases are good examples. For an ideal monatomic gas,

$$C_v = \frac{q_v}{\Delta T} = \frac{\Delta E}{\Delta T} \tag{7-10}$$

$$C_p = \frac{q_p}{\Delta T} = \frac{\Delta H}{\Delta T} \tag{7-11}$$

Although these definitions are exact only if ΔT is an infinitesimally small interval, they are still very good approximations for the reasonable temperature differences (of up to several hundred degrees) commonly encountered.

For the heating of a gas, because $q_p > q_v$, we can make the following statement:

$$C_p > C_v$$

Given that

$$C_p = \frac{\Delta H}{\Delta T}$$

and making the appropriate substitution for ΔH,

$$C_p = \frac{\Delta E + P\Delta V}{\Delta T}$$

Separating the terms gives

$$C_p = \frac{\Delta E}{\Delta T} + \frac{P\Delta V}{\Delta T}$$

But since $\Delta E/\Delta T = C_v$ and $P\Delta V/\Delta T = R$ for one mole of an ideal gas, then

$$C_p = C_v + R \tag{7-12}$$

The difference between C_p and C_v is R, the universal gas constant, the work done by a gas pushing back against a constant external pressure.

The physical interpretation of this difference between C_p and C_v lies in the fact that for C_v, all the energy imparted to the gas goes to increase the kinetic energy of the molecules. For C_p, because there is a volume increase, additional heat must be supplied to do work against the external pressure to permit the expansion (Tables 7–6 and 7–7). One of the conclusions of kinetic-molecular theory was that for one mole of an ideal monatomic gas

$$E = \tfrac{3}{2} RT \tag{7-13}$$

If we now increase the temperature, the energy of the gas particles will increase proportionately:

$$\Delta E = \tfrac{3}{2} R\Delta T$$

$$\frac{\Delta E}{\Delta T} = \tfrac{3}{2} R$$

$$C_v = \tfrac{3}{2} R$$

Substituting this value for C_v into Eq. (7–12),

It is not necessary to restrict this conclusion to 1 mole:

$$\frac{P\,\Delta V}{\Delta T} = nR$$

Then $C_p = C_v + nR$. But for the sake of simplicity, we have defined the molar heat capacity and Eq. (7–12) instead.

TABLE 7–6   Specific Heats and Heat Capacities for Some Substances, Recorded Under Constant Pressure at 298 K

Substance	Specific Heat (J/g·K)	Heat Capacity (J/mol·K)
air	0.720	20.8
water	4.184	75.4
ammonia	2.06	35.1
hydrogen chloride	0.797	29.1
hydrogen bromide	0.360	29.1
ozone	0.817	39.2
neon	1.03	20.7
chlorine	0.477	33.8
bromine	0.473	75.6
iron	0.460	25.1
copper	0.385	24.7

TABLE 7–7   Heat Capacities for Some Gases (J/mol·K)

Gas	C_p	C_v	$C_p - C_v$	C_p/C_v
Monatomic*				
helium	20.9	12.8	8.28	1.63
argon	20.8	12.5	8.33	1.66
iodine	20.9	12.6	8.37	1.66
mercury	20.8	12.5	8.33	1.66
Diatomic†				
hydrogen	28.6	20.2	8.33	1.41
oxygen	29.1	20.8	8.33	1.39
nitrogen	29.0	20.7	8.30	1.40
hydrogen chloride	29.6	21.0	8.60	1.41
carbon monoxide	29.0	21.0	8.00	1.39
Triatomic†				
nitrous oxide	39.0	30.5	8.50	1.28
carbon dioxide	37.5	29.0	8.50	1.29
Polyatomic†				
ethane	53.2	44.6	8.60	1.19

*Translational kinetic energy only.
†Translational, vibrational, and rotational energy.

$$C_p = \tfrac{3}{2} R + R$$

$$C_p = \tfrac{5}{2} R$$

For diatomic and polyatomic molecules, other energy terms must be included and different results are obtained. For example, the diatomic oxygen molecule has values of

$$C_v = \tfrac{5}{2} R$$
$$C_p = \tfrac{7}{2} R$$

It is particularly intriguing (and has proved especially valuable) to observe the C_p/C_v ratio. For a monatomic gas, the ratio is 1.67. For a diatomic gas it is 1.40. Consequently, one should be able to use the ratio to determine whether a gas is diatomic. Early studies on iodine vapor did just that.

EXAMPLE 7–11

Is iodine monatomic or diatomic in the vapor state?

Solution   Original data indicated that iodine vapor was diatomic. But when heat capacity data became available (about 100 years ago), the ratio turned out to be very close to 1.67. Today, it is well known that in the vapor state the behavior of iodine is better understood if it is assumed to be monatomic. Early in the 20th century, similar measurements proved to be valuable in establishing the monatomic nature of the newly discovered noble gas elements.

EXERCISE 7–11(A)

Given that C_p and C_v for chlorine gas at 298 K are very close to 33.8 and 24.0 J/mol·K, respectively, determine whether chlorine gas is best described as monatomic or diatomic.
Answer: diatomic: 33.8/24.0 = 1.41.

EXERCISE 7–11(B)

Given that C_p and C_v for bromine in the vapor state at 298 K are very close to 75.6 and 50.0 J/mol·K, respectively, determine whether bromine gas is best described as monatomic or diatomic.
Answer: diatomic: 75.6/50.0 = 1.52. However, there is some evidence that bromine may dissociate in the vapor state to a small degree, perhaps accounting for the somewhat higher number.

SUMMARY

If energy is exchanged by simple mechanical means, we call it **work**. When energy is exchanged between system and surroundings by thermal conduction, we call it **heat**. Heat flows as a consequence of differences in temperature. Internal energy is also a function of temperature, increasing with increasing temperature. The mechanical equivalent of heat is 1 cal = 4.184 J. The **zeroth law** of thermodynamics is an intuitive statement that bodies in good thermal contact with each other will eventually arrive at an equilibrium temperature, which is the basis for thermometry and the whole idea of measuring temperature.

The **first law** of thermodynamics is a generalization of the principle that energy is conserved: $\Delta E = q + w$. As a direct consequence of the first law, when a system changes from one state to another, the change in its internal energy ΔE is the sum of the heat transferred into (or out of) the system and the work done on (or by) the system. Keep in mind that although q and w depend on the path followed from the initial to the final state, the change in the internal energy does not. ΔE is independent of how you get from one state to another. Furthermore, for a cyclic process, the change in the internal energy must be zero. Thus, heat transferred must be equal to work done.

An **isothermal process** takes place at constant temperature. For an isothermal volume change of an ideal gas, $\Delta E = 0$ and $q = -w$. An **adiabatic process** displays no transfer of heat between system and surroundings, so $q = 0$ and $\Delta E = w$.

Hess's laws govern thermochemistry: (1) The quantity of heat necessary for the decomposition of a compound into its elements is equal to the heat given off when the same compound is formed from its elements. (2) The net heat change resulting from a particular chemical reaction is the same, no matter whether one step or several are involved in the transformation. Thermochemical equations are chemical reactions that document the accompanying enthalpy change, the heat of reaction at constant pressure, ΔH. Two special cases of importance to chemistry are heats of formation ($\Delta H_f°$) and heats of combustion ($\Delta H_c°$).

The **heat capacity** of an ideal monatomic gas at constant volume C_v is $\frac{3}{2} R$; the heat capacity at constant pressure C_p is $\frac{5}{2} R$. They are related by $C_p - C_v = R$, and $C_p/C_v = 1.67$. For diatomic and polyatomic molecules where there are rotational and vibration energy terms as well as translational energies, the heat capacities are different, yet characteristic. Two useful heat capacity relationships are $\Delta H = nC_p\Delta T$ and $\Delta E = nC_v\Delta T$.

Table 7–8 summarizes the special terms used in this chapter, and Table 7–9 lists the most useful equations and the conditions under which they are true.

QUESTIONS AND PROBLEMS

QUESTIONS

1. Two objects in thermal equilibrium with each other must be at the same temperature; two objects in thermal equilibrium with a third object must be in thermal equilibrium with each other. Briefly explain these two statements (which are collectively referred to as the zeroth law of thermodynamics) by making use of a simple algebraic argument.

2. Noting that the zeroth law of thermodynamics says nothing about the time required to establish equivalence of temperature between two systems in thermal contact, give an example of each of the following:
(a) Equilibrium is established instantly (in less than a second).
(b) Equilibrium is established in a few hours.
(c) Millions of years pass, and equilibrium still hasn't been established.

3. What are the sign conventions regarding the transfer of heat and doing work between a thermodynamic system and its surroundings?

4. Illustrate each of the following thermodynamic concepts:
(a) An open system.
(b) A closed system.
(c) An isolated system.

5. Can you identify some other thermodynamic quantities besides temperature that tend toward equivalence when two systems are in direct and intimate contact?

6. Out of your common experience, illustrate what is meant by:
(a) An exothermic process.
(b) An endothermic process.
(c) An isothermal process.
(d) An adiabatic process.
(e) A cyclic process.

7. Distinguish between the following pairs of terms:
(a) Energy and enthalpy.
(b) Heat and work.
(c) Heat and temperature.
(d) Work and energy.

8. Fakirs (and fakers) have walked on burning coals barefooted without apparent injury. How is that possible?

9. Why is it necessary to keep liquid nitrogen in a Thermos bottle?

10. Briefly explain the role of the double-silvered glass walls in a "Dewar" flask and the role of the vacuum jacket between the walls.

11. Briefly explain the small (negligible) difference between C_p and C_v for solids and liquids but the significant difference for gases.

12. Why is there a difference between the molar heat capacities for hydrogen and helium?

13. An iron horseshoe is taken from a furnace at red heat—about 900°F—and plunged into a bathtub-sized trough of water at room temperature. What is the approximate final thermal equilibrium temperature after this "quenching" operation?

14. Ethyl alcohol has about half the molar heat capacity of water. If equal masses of alcohol and water in separate containers are supplied with the same amount of heat, how will their respective temperature changes compare?

PROBLEMS

Temperature Measurements   [1–4]

1. Write linear algebraic equations relating:
(a) K to °C   (b) °C to °F   (c) °R to K
2. The boiling temperature for acetic acid on the Celsius scale is 118.1°C at atmospheric pressure. What is the corresponding temperature on the two absolute scales, K and °R?
3. For all the primary fixed points of temperature listed in Table 7–2, convert °F to °C, °R, and K.
4. For all the primary fixed points of temperature listed in Table 7–2, convert °C to °F, °R, and K.

First Law Calculations: w, q, ΔE, ΔH   [5–20]

5. A mole of liquid potassium is allowed to vaporize at its normal boiling point at atmospheric pressure in an apparatus such as is described in Figure 7–10. How much heat is required? How much work is done? By what quantity is the internal energy of the potassium vapor in-

TABLE 7–8 Summary of Definitions from Thermodynamics I

System The part of the universe in which we are studying a particular process. The system is separated from the rest of the universe by boundaries that we define.

Surroundings Technically, the surroundings are all of the universe except the system. In practical terms, the surroundings are that part of the universe that can be affected in any way by the system. The way a system can affect its surroundings is determined by the nature of its boundaries.

Isolated System A system with boundaries such that it cannot interact with its surroundings in any way.

Closed System A system with boundaries that allow flow of energy but not matter between it and its surroundings.

Open System A system that can interchange matter and energy with its surroundings.

Internal Energy The total of the kinetic energy and the potential energy for a system or part of a system.

Kinetic Energy Energy related to movement of atoms and molecules. Kinetic energy increases with increase of temperature.

Potential Energy Energy resulting from attractions and repulsions of an object as a result of its position.

State A situation within a system that can be defined by properties such as pressure, temperature, and chemical composition. The important states in thermodynamics are equilibrium states in which none of the properties is changing.

Process A change in a system from one equilibrium state to another.

Isothermal Process A process conducted at a constant temperature.

Adiabatic Process A process conducted so that no heat flows between the system and the surroundings.

Path The route taken during a process from one equilibrium state to another. The path is described by a series of very small steps (microsteps).

State Function A function for a process for which the value depends only on the initial and final states of the system. Internal energy is an example.

Path Function A function for a process for which the value depends on the path of the process. Heat and work are examples.

TABLE 7–9 Key Thermodynamics I Equations and the Conditions Under Which They Apply

Equation	Condition
$E = \frac{3}{2}RT$	1 mol of ideal monatomic gas
$w = -P\Delta V$	Constant external pressure, P
$\Delta E = q + w$	**All conditions**
$\Delta H = \Delta E + P\Delta V$	Constant pressure
$\Delta E = q_v$	Constant volume
$\Delta H = q_p$	Constant pressure
$\Delta H = \Delta E + \Delta nRT$	Constant temperature, solid and liquid volumes negligible
$C_v = \dfrac{q_v}{\Delta T} = \dfrac{\Delta E}{\Delta T}$	Constant volume
$C_p = \dfrac{q_p}{\Delta T} = \dfrac{\Delta H}{\Delta T}$	Constant pressure
$C_p = C_v + R$	Ideal gas
$C_v = \frac{3}{2}R$	Ideal monatomic gas

creased? The heat of vaporization for potassium is known to be 77.08 kJ/mol.

6. A kilogram of liquid potassium at 1400°F, contained in a cylinder similar to that described in Figure 7–10, was allowed to vaporize at constant temperature and pressure. During the process, 217 Btu of work were done and the change in the internal energy was +1597 Btu. (The British thermal unit or Btu is the amount of energy required to raise the temperature of one pound of water by 1°F, so 1 Btu = 1055 J.) Calculate the quantity of heat added from the reservoir. Record your answer in both Btu and joules.

7. If the internal energy of a thermodynamic system is decreased by 500 J because 150 J of work is done on the system, how much heat was transferred, and in what direction—to or from the system?

8. A system undergoes a process in which its internal energy drops by 3660 kJ. The system gives off 2950 kJ of heat during the process. Give the values of w, q, and ΔE for the process. Does the system do work on the surroundings or do the surroundings do work on the system as the process proceeds?

9. Given 5 g of nitrogen gas at 530°R and a pressure of 1 atm, slowly being compressed in a cylinder such that the temperature remains constant and 65 cal of heat is rejected:
(a) What are the corresponding Fahrenheit, Celsius, and Kelvin temperatures?
(b) Why is ΔE zero for the process?
(c) How much work was done on the nitrogen?

10. A gas is compressed from 10.0 L to 5.00 L at a constant pressure of 0.500 atm. During the course of the compression the gas rejects 400. J. Calculate the work done on the gas and the change in the internal energy of the gas.

11. As a gas expands against a constant pressure of 5 atm, its volume changes from 10.0 L to 20.0 L while it absorbs 2.00 kJ of heat. Calculate the change in internal energy experienced by the gas.

12. An ideal gas expanded against a constant external pressure of 700. torr, and as it did so, its volume changed from 50.0 L to 150. L. During the process, 6485 J of heat was absorbed. Calculate the change in the internal energy of the gas.

13. A kilogram of liquid potassium is allowed to vaporize at a constant temperature of 760°C and an unknown but constant pressure. During the vaporization process 216.2 kJ of work was done and the internal energy increased by 1530 kJ. The volume of potassium vapor produced is 2134 L.
(a) Calculate the heat of vaporization for potassium.
(b) Find the constant pressure against which the system did work.

14. Calculate the work done in vaporizing a mole of water under the following conditions, assuming ideal behavior.
(a) At its normal boiling point (100°C).
(b) By letting it evaporate at room temperature (25°C).

15. One liter of an ideal gas at 0°C and 10.0 atm was allowed to expand against a constant external pressure of 1 atm. The temperature of the gas remained constant throughout.
(a) Calculate q, w, ΔE, and ΔH.

(b) If the expansion were instead a free expansion, how would these values be changed?

16. Calculate w, q, ΔE, and ΔH for the adiabatic expansion of 2.00 L of an ideal gas to a final volume of 5.00 L against a constant external pressure of 2.50 atm.

17. During the course of a certain constant volume process, 200. J of heat is transferred to 2.0 mol of an ideal monatomic gas initially at room temperature (25°C). Calculate:
(a) The change in the internal energy of the gas.
(b) The work done by the gas.
(c) The final temperature of the gas.

18. During the course of a certain constant pressure process, the temperature of 2 mol of nitrogen, initially at 25°C, was observed to double. Calculate:
(a) The amount of heat (in joules) transferred to the gas.
(b) The change in the internal energy of the gas.
(c) The work done by the gas.

19. Calculate the change in the internal energy per kelvin for a mole of helium.

20. Calculate the difference between the quantity of heat transferred to a mole of argon when its Celsius temperature is allowed to double at constant volume and the heat transferred during the same change of state at constant pressure.

Calorimetry   [21–22]

21. A very pure sample of benzoic acid (C_6H_5COOH), a combustible organic solid, weighed 1.221 g. It was placed in a bomb calorimeter and ignited in a pure oxygen atmosphere, and a temperature rise from 25.240°C to 31.668°C was noted. The heat capacity of the calorimeter was 5.020 kJ/°C. The combustion products were carbon dioxide and water.
(a) Write the stoichiometric equation for the combustion reaction.
(b) Calculate ΔE, q, and w for the calorimeter process.
(c) Calculate ΔH for the reaction.

22. A constant volume calorimeter has a heat capacity (calorimeter constant) of 32.606 kJ/K. Combustion of a 2.316-g sample of menthol, $C_{10}H_{20}O$ (s), causes the temperature to increase from 25.010°C to 27.897°C. Calculate ΔE and ΔH for the combustion of 1 mol of menthol.

Thermochemistry; Hess's Laws   [23–36]

23. Using the data in Table 7–4, calculate the enthalpy change for the hydration of ethylene according to the following equation. The product is ethyl alcohol.

$$C_2H_4 \text{ (g)} + H_2O \text{ (liq)} \longrightarrow C_2H_5OH \text{ (liq)}$$

24. Using the appropriate thermochemical data from Tables 7–4 and 7–5, calculate ΔH and ΔE for the following reaction at 298 K:

$$C \text{ (graphite)} + 2 H_2O \text{ (g)} \longrightarrow CO_2 \text{ (g)} + 2 H_2 \text{ (g)}$$

25. From the thermochemical data in Tables 7–4 and 7–5, calculate the heat of reaction for each of the following transformations carried out at 298 K:
(a) $CO \text{ (g)} + \frac{1}{2} O_2 \text{ (g)} \longrightarrow CO_2 \text{ (g)}$
(b) $H_2 \text{ (g)} + \frac{1}{2} O_2 \text{ (g)} \longrightarrow H_2O \text{ (liq)}$

(c) H_2O (liq) $\longrightarrow$ H_2O (g)

(d) Then, using these data, calculate the heat of reaction at 298 K for the following reaction:

$$CO\ (g)\ +\ H_2O\ (g)\ \longrightarrow\ CO_2\ (g)\ +\ H_2\ (g)$$

26. You are given the following thermochemical data:

$$2\ P\ (s)\ +\ 3\ Cl_2\ (g)\ \longrightarrow\ 2\ PCl_3\ (liq)$$
$$\Delta H° = -636\ kJ$$

$$PCl_3\ (liq)\ +\ Cl_2\ (g)\ \longrightarrow\ PCl_5\ (s)$$
$$\Delta H° = -138\ kJ$$

(a) Write the equation for $\Delta H_f°$ for PCl_5 (s) and then calculate the value.

(b) Then, given that $\Delta H_f°$ for H_3PO_4 (s) is 1280 kJ, calculate $\Delta H°$ for the following reaction:

$$PCl_5\ (s)\ +\ 4\ H_2O\ (liq)\ \longrightarrow\ H_3PO_4\ (s)\ +\ 5\ HCl\ (g)$$

27. The Goldschmidt process is used industrially to free metallic iron by aluminothermic reduction:

$$2\ Al\ (s)\ +\ Fe_2O_3\ (s)\ \longrightarrow\ Al_2O_3\ (s)\ +\ 2\ Fe\ (s)$$

Calculate the heat liberated per gram of metal produced.

28. Methane can be used to convert iron(II) oxide to iron metal. The other products of the reaction are carbon monoxide gas and water vapor:

$$CH_4\ (g)\ +\ 3\ FeO\ (s)\ \longrightarrow$$
$$3\ Fe\ (s)\ +\ CO\ (g)\ +\ 2\ H_2O\ (g)$$

How much heat is required to produce one kg of iron by this process?

29. Taking the necessary thermochemical data from Table 7–4, calculate the heat of the following reaction at 298 K:

$$SO_2\ (g)\ +\ 2\ H_2S\ (g)\ \longrightarrow$$
$$3\ S\ (rhombic)\ +\ 2\ H_2O\ (liq)\qquad \Delta H° = ?\ kJ$$

30. Taking the additional thermochemical data necessary from Table 7–4, calculate $\Delta H_f°$ for AgBr (s):

$$AgBr\ (s)\ +\ \tfrac{1}{2}\ Cl_2\ (g)\ \longrightarrow\ AgCl\ (s)\ +\ \tfrac{1}{2}\ Br_2\ (g)$$
$$\Delta H° = -27.5\ kJ$$

31. Write the equation and then calculate $\Delta H_f°$ for $Ca(OH)_2$ (s), given the following thermochemical data:

$$H_2\ (g)\ +\ \tfrac{1}{2}\ O_2\ (g)\ \longrightarrow\ H_2O\ (liq)$$
$$\Delta H° = -285\ kJ$$

$$CaO\ (s)\ +\ H_2O\ (liq)\ \longrightarrow\ Ca(OH)_2\ (s)$$
$$\Delta H° = -64\ kJ$$

$$Ca\ (s)\ +\ \tfrac{1}{2}\ O_2\ (g)\ \longrightarrow\ CaO\ (s)$$
$$\Delta H° = -635\ kJ$$

32. Given the following thermochemical data, calculate $\Delta H_f°$ for ICl (g):

$\tfrac{1}{2}\ I_2\ (s)\ +\ \tfrac{1}{2}\ Cl_2\ (g)\ \longrightarrow\ ICl\ (g)$	$\Delta H_f° = ?\ kJ$	
$Cl_2\ (g)\ \longrightarrow\ 2\ Cl\ (g)$	$\Delta H_1° = +242\ kJ$	
$I_2\ (g)\ \longrightarrow\ 2\ I\ (g)$	$\Delta H_2° = +151\ kJ$	
$I_2\ (s)\ \longrightarrow\ I_2\ (g)$	$\Delta H_3° = +63\ kJ$	
$ICl\ (g)\ \longrightarrow\ I\ (g)\ +\ Cl\ (g)$	$\Delta H° = +211\ kJ$	

33. Calculate the heat of combustion of *n*-pentane (C_5H_{12}), a gasoline fraction obtained from petroleum, on carrying out the reaction in a bomb calorimeter at room temperature (25°C). Write the balanced equation first.

34. Strictly on a weight basis, determine which is the preferred rocket fuel, assuming that they cost the same, dimethyl hydrazine or hydrogen.

$$(CH_3)_2NNH_2\ (liq)\ +\ 4\ O_2\ (g)\ \longrightarrow$$
$$N_2\ (g)\ +\ 4\ H_2O\ (liq)\ +\ 2\ CO_2\ (g)$$
$$\Delta H° = -1694\ kJ$$

$$H_2\ (g)\ +\ \tfrac{1}{2}\ O_2\ (g)\ \longrightarrow\ H_2O\ (liq)$$
$$\Delta H_f° = -286\ kJ$$

35. Sufficient heat is liberated on ignition and burning of 1.00 g of rhombic sulfur to raise the temperature of 100. g of water by 22.2°C. What is $\Delta H_c°$ for rhombic sulfur?

36. On burning 1 kg of low-carbon coal in a water-jacketed furnace, the 47 liters of water in the jacket increased in temperature from 25.0°C to 38.3°C If the coal is 90% carbon (and 10% inert ingredients) by weight and we need not consider any other heating effects, what is $\Delta H_c°$ for carbon?

Heat Capacity   [37–40]

37. Calculate the heat capacity and the specific heat for water from the following data:

$$H_2O\ (liq,\ 0°C)\ \longrightarrow\ H_2O\ (liq,\ 25°C)$$
$$\Delta H = 1883\ kJ$$

38. Calculate the molar heat capacity and the specific heat for carbon tetrachloride from the following data:

$$CCl_4\ (liq,\ 0°C)\ \longrightarrow\ CCl_4\ (liq,\ 50°C)$$
$$\Delta H = 6468\ kJ$$

39. Based on the heat capacity measured at constant pressure, $C_p = 20.9\ J/mol\cdot K$, determine whether iodine vapor behaves as a monatomic or diatomic species.

40. The heat capacity of carbon monoxide at constant pressure is 29.0 J/mol·K. Does carbon monoxide behave as a monatomic or diatomic gas?

Additional Problems   [41–50]

41. (a) Calculate the heat of formation of $PbCl_2$ (s) from the following thermochemical reaction:

$$2\ Ag\ (s)\ +\ PbCl_2\ (s)\ \longrightarrow\ 2\ AgCl\ (s)\ +\ Pb\ (s)$$
$$\Delta H°_{298} = +105\ kJ$$

(b) Calculate $\Delta H_f°$ for naphthalene ($C_{10}H_8$), using the data in Tables 7–4 and 7–5.

42. A gas expands against a constant external pressure of 1.50 atm from an initial volume of 1.50 L to a final volume of 6.25 L. The container in which this process is taking place is well insulated, and you may assume no heat enters or leaves the system.

(a) Determine the work w done by the system (in joules).

(b) Calculate the energy change ΔE for the system (in joules).

(c) What happens to the temperature of the gas—does it rise, fall, or stay the same? Briefly explain.

43. Both acetylene (C_2H_2) and ethylene (C_2H_4) yield ethane (C_2H_6) on hydrogenation, according to the following thermochemical reactions:

$$C_2H_2 \text{ (g)} + 2\ H_2 \text{ (g)} \longrightarrow C_2H_6 \text{ (g)}$$
$$\Delta H = -311 \text{ kJ}$$

$$C_2H_4 \text{ (g)} + H_2 \text{ (g)} \longrightarrow C_2H_6 \text{ (g)}$$
$$\Delta H = -136 \text{ kJ}$$

Calculate ΔH (in kJ) for the formation of ethylene from acetylene.

44. (a) Calculate the work done when 1.00 mol of ethyl ether ($C_4H_{10}O$) is allowed to vaporize completely at its normal boiling point (34.5°C) and atmospheric pressure (1 atm). Assume ideal behavior on the part of the gas.
(b) Based on your answer to part (a), and the known heat of vaporization for ethyl ether (20.0 kJ/mol), find ΔE for the process.

45. Here is a laboratory experiment on calorimetry, designed to determine (1) the heat capacity of the calorimeter and (2) the heat of combustion of an unknown compound. A 605-mg sample of naphthalene ($C_{10}H_8$) was burned, the calorimeter was a constant volume bomb calorimeter, and the recorded change in temperature was 2.255°C. A 1.67-g sample of the unknown compound, burned in the same way and in the same calorimeter, resulted in a temperature change of 2.030°C. Make the calculations for the heat capacity and the heat of combustion.

46. The volume of a sample of methane was measured at 23.4°C and 0.987 atm and found to be 0.129 L. The sample was then burned in a constant volume bomb calorimeter that has a heat capacity (or calorimeter constant) of 5432.10 J/K. On burning the methane sample completely, the temperature in the calorimeter was observed to rise from 23.456°C to 24.321°C. Calculate ΔH_c for methane in kJ/mol.

47. The following problems apply to 1 mol of an ideal monatomic gas:
(a) The value of w for a cyclic process is 500 J per cycle. What is the value of q for the cycle, and why?
(b) The volume of the gas, initially at a pressure of 2 atm, is doubled in an isothermal process. What is the new pressure, and why?
(c) In either case, what are C_p and C_v in J/mol·K? Find the C_p/C_v ratio.

48. The following problems apply to 1 mol of an ideal monatomic gas:
(a) Calculate ΔE, q, and w on heating the gas from 300. K to 600. K, at a constant volume of 1 L.
(b) Calculate ΔE, q, and w on heating the gas from 300. K to 600. K, at a constant pressure of 1 atm.

49. Assume that neon is an ideal monatomic gas and hydrogen chloride is diatomic. Starting with the C_p values in Table 7–7, what are the corresponding C_v values?

50. In a classic experiment, the heat of combustion of benzoic acid was determined by burning it in a constant volume bomb calorimeter. Initially, 1.870 g of benzoic acid was placed in the calorimeter, whose temperature was 25.010°C. After combustion, the calorimeter's temperature was observed to have risen to 29.315°C.

(a) Neglecting any other effects, if the calorimeter constant is 11.485 kJ/K, what is the standard heat of combustion at 298 K in kJ/mol?
(b) Compare this result to that obtained from Table 7–5.
(c) Calculate the heat of combustion of benzoic acid if H_2O (g) is a product.

MULTIPLE PRINCIPLES   [51–56]

51. The industrial process for the synthesis of acetylene is based on the partial combustion of liquefied petroleum gas (LPG), which is a mixture of 95% propane (C_3H_8) and 5% methane (CH_4) by weight. Consider only the acetylene synthesis, shown by the following reaction:

$$C_3H_8 \text{ (liq)} + 2\ O_2 \text{ (g)} \longrightarrow$$
$$C_2H_2 \text{ (g)} + CO \text{ (g)} + 3\ H_2O \text{ (liq)}$$

(a) Calculate the standard heat of reaction at 298 K for the acetylene synthesis.
(b) Starting with a kilogram of LPG, what is the maximum yield in grams of acetylene?
(c) Calculate the heat for the 5% methane side-reaction:

$$2\ CH_4 \text{ (liq)} + 3\ O_2 \text{ (g)} \longrightarrow 2\ CO \text{ (g)} + 4\ H_2O \text{ (liq)}$$

52. Beginning with a kilogram of the LPG mixture described in Problem 51 and assuming that only the two indicated reactions occur, what fraction of the total heat from the reaction is due to the acetylene synthesis? What is the overall standard heat of reaction at 298 K as measured in kJ/kg?

53. If a kilogram of water is produced in the synthesis of acetylene from LPG (as described in Problem 51), how many kilograms of LPG were consumed?

54. Determine the difference between ΔE and ΔH for the following reaction at 298 K:

$$2\ C \text{ (s, graphite)} + O_2 \text{ (g)} \longrightarrow 2\ CO \text{ (g)}$$

Before the reaction begins, there is an excess of oxygen gas and the pressure in the reactor is 1.00 atm; after reaction, the pressure is 1.05 atm. Both pressure readings are taken at the same temperature.
(a) What fraction of the oxygen present has reacted?
(b) If the volume of the reactor is 1.5 L, how much heat has the reactor absorbed or given off? Be sure to state which it is—absorbed or given off.

55. Consider the industrial synthesis of methanol from synthesis gas:

$$CO \text{ (g)} + 2\ H_2 \text{ (g)} \longrightarrow CH_3OH \text{ (liq)}$$

A 10.0-L reactor is initially charged with CO (g) and H_2 (g) in a 1:2 ratio. The initial and final reaction temperatures are 298 K, and the initial pressure is 22.0 atm. If the reaction is complete, what is the final pressure and how much heat was transferred into or out of the reactor? Be sure to state which it is—into or out of the reactor.

56. A sample of iron pyrite ore, known to be 87.3% FeS_2 with the remainder being inert materials (mainly dirt and gravel), was roasted in an excess of oxygen. The reaction is as follows:

$$4 \text{ FeS}_2 \text{ (s)} + 11 \text{ O}_2 \text{ (g)} \longrightarrow 2 \text{ Fe}_2\text{O}_3 \text{ (s)} + 8 \text{ SO}_2 \text{ (g)}$$

If 93.6% of the FeS_2 (s) reacts, what is the standard heat of reaction at 298 K per kilogram of crude ore?

APPLIED PRINCIPLES　　[57–67]

57. An important industrial application of Hess's Law is the determination of heats of hydrogenation from heats of combustion, such as the heat of hydrogenation of ethylene. Using the data in Tables 7–4 and 7–5, determine $\Delta H°$ at 298 K and express your answer in kJ/mole and Btu/lb.

$$\text{C}_2\text{H}_4 \text{ (g)} + \text{H}_2 \text{ (g)} \longrightarrow \text{C}_2\text{H}_6 \text{ (g)}$$

58. The heat change that occurs during the following reaction is − 1172 kJ.

$$\text{Al}_2\text{Cl}_6 \text{ (s)} + 6 \text{ Na (s)} \longrightarrow 2 \text{ Al (s)} + 6 \text{ NaCl (s)}$$

Given that the heat of formation of solid sodium chloride is − 411 kJ/mol, calculate the heat of formation of Al_2Cl_6 (s) in kJ/mole and Btu/lb.

59. Tungsten carbide (WC) is a refractory material of considerable industrial interest. As a result, it has been studied extensively. On the basis of the following data and any other thermodynamic data you might need from Tables 7–4 and 7–5, calculate the standard enthalpy of formation ($\Delta H_f°$) of tungsten carbide at 298 K:

$$2 \text{ WC (s)} + 5 \text{ O}_2 \text{ (g)} \longrightarrow 2 \text{ WO}_3 \text{ (s)} + 2 \text{ CO}_2 \text{ (g)}$$
$$\Delta H°_{298} = -2392 \text{ kJ}$$
$$2 \text{ W (s)} + 3 \text{ O}_2 \text{ (g)} \longrightarrow 2 \text{ WO}_3 \text{ (s)}$$
$$\Delta H°_{298} = -1675 \text{ kJ}$$

60. In making the necessary calculations to answer Problem 59, you needed to obtain the standard heat of formation of carbon dioxide:

$$\text{C (graphite)} + \text{O}_2 \text{ (g)} \longrightarrow \text{CO}_2 \text{ (g)}$$
$$\Delta H°_{298} = 393.505 \text{ kJ/mol}$$

Would your answer be the same or different if you had chosen diamond instead of graphite? If your answer is the same, give reasons to support your answer. If your answer is different, give reasons and calculations to support your answer. Necessary information and data can be found in this chapter.

61. The standard enthalpy of combustion of animal fats per gram is an important value that is commonly referred to by nutritionists as the "calorific value." It can be determined by measuring the heat produced on burning a known

sample in a bomb calorimeter. If a 1.00-g sample of a C_{20} component of human fat known as arachidonic acid released 42.0 kJ at 98.6°F, what is its calorific value in kJ/g? Assume the unbalanced combustion reaction to be the following:

$$\text{C}_{20}\text{H}_{32}\text{O}_2 \text{ (s)} + \text{O}_2 \text{ (g)} \longrightarrow \text{CO}_2 \text{ (g)} + \text{H}_2\text{O} \text{ (liq)}$$

62. The heat required to sustain animals that hibernate—polar bears and hamsters alike—arises from the combustion of fatty acids such as arachidonic acid (see Problem 61). Calculate the weight of fatty acid needed to warm a 500-kg bear from 5°C to 25°C, assuming the average specific heat of bear tissue is 4.18 J/g·K.

63. If a ton of TNT (trinitrotoluene) can deliver 10^6 kJ of energy, how many Btu are released in a 20-kiloton nuclear detonation, and what is the explosive yield? How many quads of energy is that equivalent to? It is convenient to speak of explosive yields as "energy released by a standard kiloton of TNT," or 10^9 kJ, and a quad is 10^{15} Btu.

64. It has been estimated that (on average) 180 MeV are released every time an atom of uranium-235 "fissions." How many uranium-235 atoms must split in a megaton nuclear blast? How many quads of energy is that equivalent to?

65. Liquefied propane (C_3H_8) is widely used as a portable heating fuel. Given the heats of formation of water, carbon dioxide, and liquid propane to be − 286 kJ/mol, − 393 kJ/mol, and − 119 kJ/mol, respectively:
(a) Calculate the heat of combustion of liquefied propane.
(b) Calculate the heat of combustion per cubic foot of propane gas at 298 K and 101.3 kPa. One cubic foot is 28.3 L, or about equal to the molar volume at STP.
(c) Based on the assumption that the hot water needs of an average home, housing a family of four, are about 100 gallons per day, estimate the volume of propane gas needed per day.

66. Estimate the heat of combustion of the kerosine-range diesel fuel hydrocarbon known as dodecane ($\text{C}_{12}\text{H}_{26}$), given only the data in Table 7–5. What assumptions have you made? Estimate the heat of combustion that would be determined experimentally in a constant volume bomb calorimeter.

67. The CO content in an enclosed (or poorly ventilated) space can be determined by converting the CO to CO_2 by oxidizing a sample (of polluted air) and measuring the rise in the temperature that results from the oxidation, using a particularly sensitive thermometer. Estimate how sensitive a thermometer would have to be in order to detect parts per million (ppm) by volume of CO in air samples taken at tunnel check points such as the Brooklyn Battery tunnel, the Houston Baytown tunnel, the Baltimore Harbor tunnel, or the tunnel of your choice. What assumptions have you made?

CHAPTER 8

Liquids and Changes of State

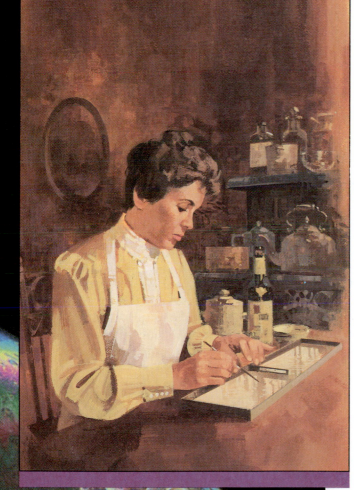

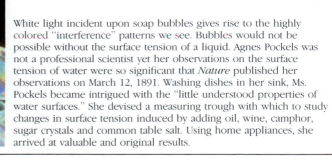

White light incident upon soap bubbles gives rise to the highly colored "interference" patterns we see. Bubbles would not be possible without the surface tension of a liquid. Agnes Pockels was not a professional scientist yet her observations on the surface tension of water were so significant that *Nature* published her observations on March 12, 1891. Washing dishes in her sink, Ms. Pockels became intrigued with the "little understood properties of water surfaces." She devised a measuring trough with which to study changes in surface tension induced by adding oil, wine, camphor, sugar crystals and common table salt. Using home appliances, she arrived at valuable and original results.

Johannes Diderik van der Waals (Dutch) and Jean Perrin (French) shared the Nobel Prize for Physics in 1904, for their work concerning the equation of state of gases and liquids, and the discontinuous structure of matter, respectively.

8.1 CHARACTERIZING THE LIQUID STATE

The study of gases is simplified by the fact that, except for brief collisions, molecules in the gaseous state are sufficiently far apart that molecular interactions are negligible. The study of solids is similarly simplified by the fact that their atoms and molecules are close together, arranged in regular geometric patterns, and interact strongly with their neighbors. The study of liquids, however, has few simplifying factors; liquids are neither well understood nor easily defined.

Molecular interactions are quite strong in liquids, and liquids occupy definite volumes. That would suggest that the liquid state is more like the solid state and less like the gaseous state. However, liquids flow easily and assume the shapes of their containers, suggesting that the liquid state is at least somewhat like the gaseous state.

It is reasonable, therefore, to consider the liquid state as intermediate between the solid and gaseous states. Liquids might be thought of as dense gases, or perhaps as extremely disordered solids. Liquids and gases share the property of fluidity—when either is subjected to a shear stress, it simply yields and flows without fracturing. Liquids and solids share the property of cohesion—both are capable of keeping their particles from rapidly flying apart, and therefore are capable of maintaining a definite volume.

8.2 GENERAL PROPERTIES OF LIQUIDS

Density, Compressibility, Diffusion, and Evaporation

Molecules in the liquid and solid states are closely packed, so that the densities of the liquid and solid forms of the same substance usually differ by no more than 10%. Gas densities, in contrast, may be 100 to 1000 times less than those of the liquid and solid densities of the substance (Table 8–1). Gas densities depend strongly on temperature, and solid densities vary very little with temperature. Liquid densities display intermediate behavior; their variation with temperature is large enough that it cannot be ignored. For example, the density of ethyl alcohol changes from 0.806 g/mL at 0°C to 0.772 g/mL at 40°C, a difference of more than 4%. Although that may not seem impressive compared to the effect of temperature on gas densities, it is sufficient to permit the use of alcohol as a thermometer fluid.

Large pressures are required to compress solids or liquids by small amounts, again pointing up the close packing of the particles in those states. At ordinary temperatures, approximately 100 atm of pressure is needed to reduce the volume of a sample of water by only 5%. A similar 5% reduction in the volume of a gas at 1 atm requires only an additional 0.05 atm. The **compressibility** of a substance is defined as the fractional change in volume per unit of applied pressure; it is much larger for gases than for liquids or solids.

The relative speeds with which gases diffuse or mix is apparent even to the most casual observer. An aromatic pipe tobacco or a perfume quickly announces its presence, over considerable distances and in short times. In liquids, the mean free path (the average distance that the particles travel between collisions) is much shorter and the space through which the particles move is much more congested. Therefore, many more collisions and longer times are required before the diffusion process brings about the same degree of mixing.

TABLE 8–1 Densities of Different States for Several Pure Substances and Solid Materials

Pure Substances	Density (g/cm³)
mercury	
liquid (−38.8°C)	13.090
solid (−38.8°C)	14.193
sodium	
liquid (97.6°C)	0.9287
solid (97.6°C)	0.9519
ammonia	
gas (0°C)	0.000771
liquid (−79°C)	0.817
nitrogen	
gas (0°C)	0.001251
liquid (−195.8°C)	0.80811

Solid Materials (at ordinary temperatures)	Density (g/cm³)
asphalt	1.1–1.5
bone	1.7–2.0
brick	1.4–2.2
butter	0.86–0.87
diamond	3.01–3.52
ice	0.917
marble	2.6–2.84
paper	0.7–1.15
rock salt	2.18
rubber	0.91–1.19
sugar	1.19
wax	1.8
wood (ash, hickory, mahogany, walnut)	0.66–0.93

Basis for Comparison	
Water, for solids and liquids:	1.00000 g/cm³ at 3.98°C
Air, for gases:	0.001204 g/cm³ at 20°C

The atoms and molecules of any liquid at a given temperature are in constant motion. Some move faster, some slower, but the overall sample can best be described by the average speed. Those particles that have enough kinetic energy—the small fraction of faster moving particles—can escape from the liquid at the surface, evaporating into the space above (Figure 8–1). They have succeeded in overcoming the attractive forces of the liquid. As a direct result of this loss of the faster moving molecules, the molecules remaining in the liquid are increasingly the slower moving ones. Because their kinetic energy is lower (on the average), the temperature

FIGURE 8–1 Diagram of molecular events leading to evaporation. Those particles possessing sufficient kinetic energy can escape from the liquid at the surface.

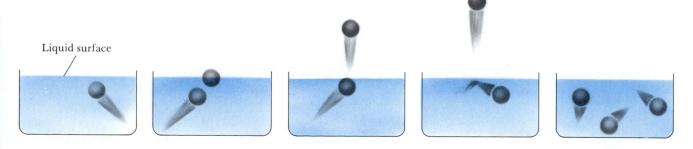

Liquid surface

FIGURE 8–2 The evaporation process continues to its ultimate conclusion as escaping molecules are swept away.

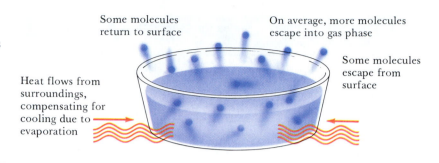

Some molecules return to surface

On average, more molecules escape into gas phase

Some molecules escape from surface

Heat flows from surroundings, compensating for cooling due to evaporation

FIGURE 8–3 Establishing (vapor ⇌ liquid) equilibrium under an evacuated bell jar.

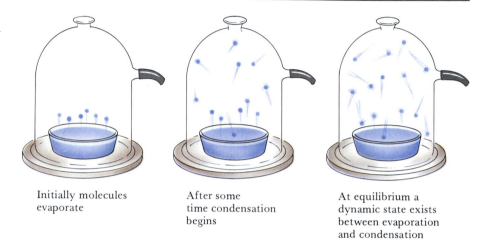

Initially molecules evaporate

After some time condensation begins

At equilibrium a dynamic state exists between evaporation and condensation

must fall accordingly, which is entirely consistent with experience—evaporation is a cooling process. If the liquid is open to the atmosphere, the evaporation process continues until no more liquid remains; escaping molecules are swept away and heat flows in from the surroundings, replacing the energy lost to the evaporating molecules and maintaining the rate of evaporation (Figure 8–2).

Equilibrium Vapor Pressure

By placing enough liquid in an evacuated chamber, such as the bell jar in Figure 8–3, the equilibrium vapor pressure can be measured. At the point where no more liquid appears to be evaporating, the pressure on the walls of the bell jar is entirely due to the vapor of the liquid and is called the **equilibrium vapor pressure** of the liquid. For a given temperature, the equilibrium vapor pressure is a characteristic property of the liquid, and is a measure of the forces of attraction between the molecules. For a given liquid, the vapor pressure increases with temperature, along with the escaping tendency of the molecules of the liquid.

The temperature at which the vapor pressure of the liquid equals the external pressure is the **boiling point**; the **normal boiling point** of a liquid is the temperature at which its vapor pressure equals 760 torr (Figure 8–4). If the external pressure is reduced, the boiling point is lower. For example, water typically boils at 97°C to 98°C in Denver, where atmospheric pressure is near 700 torr. Increased pressure produces the opposite effect; in a pressure reactor operating at 1500 torr, the boiling point of water is about 120°C. Some liquids, especially in very clean, smooth-surfaced containers, may boil in a bumpy and uneven fashion, heating several degrees beyond their normal boiling point. This phenomenon is known as **superheating**.

The temperature rises past the normal boiling point and continues to rise until the vapor pressure is finally large enough to produce small bubbles, which then expand violently due to rapid evaporation of superheated liquid.

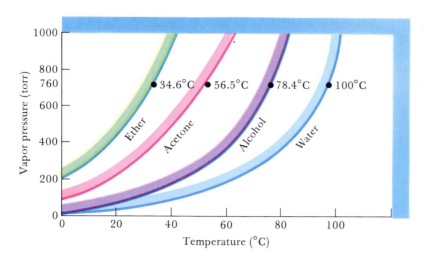

FIGURE 8–4 Comparing equilibrium vapor pressures for several different liquids, as a function of temperature. The normal boiling point for each liquid is recorded at 760 torr at sea level.

It is due in part to very large pressures, which sometimes build in small bubbles. Under these circumstances the normal vapor pressure is inadequate for expanding the bubbles, thus inhibiting the boiling process.

The vapor pressures of solids are generally much smaller than those of liquids at the same temperature; yet many solids such as camphor are still easily sensed by their characteristic odor. This difference in vapor pressures occurs because the attractive forces between atoms in solids are so much greater. For a few solids, there exists a temperature below the normal melting point at which the vapor pressure is equal to the pressure of the atmosphere. Under these conditions, the solid is said to sublime, passing directly from the solid to the vapor state. The process is called **sublimation.**

Experimental Determination of Vapor Pressures

There are several methods for experimentally determining vapor pressures of liquids. We will describe two that have been commonly used. In the **static method** (Figure 8–5), a pair of barometer tubes are used, one of which serves as reference standard. The liquid in question is introduced

When solid iodine is heated at temperatures below its triple point, 114°C, it sublimes; the vapor condenses to solid on the cool upper surface of the tube.

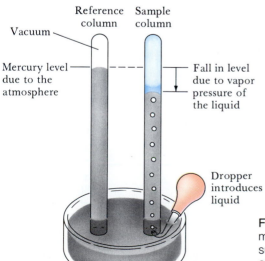

FIGURE 8–5 An experimental method for determining vapor pressures. The difference in the heights of the two columns measures the vapor pressure.

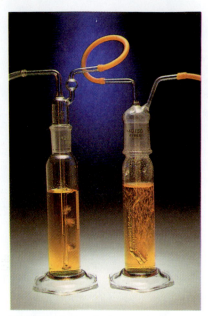

FIGURE 8–6   A dry gas being saturated by passing it successively through two columns filled with the saturating liquid.

into the other. Since mercury is much denser than most liquids, the sample floats to the surface, where liquid-vapor equilibrium is established. The difference between the levels of the two columns (in mm Hg) equals the vapor pressure (in torr) of the liquid at the temperature of the apparatus.

In the **gas saturation method,** a dry inert gas such as nitrogen is slowly bubbled through a carefully weighed amount of liquid, maintained at constant temperature (Figure 8–6). The bubbling rate must be slow enough to guarantee that the equilibrium vapor pressure is established in the bubbles. The carrier gas—the gas now mixed with vapor—is collected. The loss in weight of the liquid is measured, and the number of grams (g) of vapor is determined by difference. The volume of the gas-vapor mixture is V, the molecular weight of the liquid is M, and p is the pressure of the vapor in equilibrium with the liquid at temperature T. To calculate the vapor pressure p for the liquid, we use the equation of state for an ideal gas:

$$pV = nRT$$

Substituting $n = g/M$ into the equation gives

$$p = \frac{gRT}{MV} \tag{8–1}$$

EXAMPLE 8–1

Dry nitrogen at 1 atm pressure is passed through a sample of carbon tetrabromide (CBr_4) at 25°C, and 30.00 L of combined CBr_4 and N_2 is collected. As a result, the original 50.000 g sample of CBr_4 now weighs only 48.175 g. Assuming equilibrium conditions were established between the N_2 and CBr_4, determine the vapor pressure of carbon tetrabromide at 25°C.

Solution   The quantities needed to use Eq. (8–1) are:

$$g(CBr_4) = (50.000 - 48.175) \text{ g} = 1.825 \text{ g}$$

$$M(CBr_4) = [12 + 4(80)] \text{ g/mol} = 332 \text{ g/mol}$$

$$V(CBr_4) = 30.00 \text{ liters} = \text{volume of } N_2 \text{ carrier gas} + CBr_4 \text{ vapor}$$

$$T(CBr_4) = (273 + 25) \text{ K} = 298 \text{ K}$$

Therefore, the vapor pressure of CBr_4 at 25°C is

$$p = \frac{(1.825 \text{ g})(0.0821 \text{ L·atm/mol·K})(298 \text{ K})(760 \text{ torr/atm})}{(332 \text{ g/mol})(30.00 \text{ L})}$$

$$= 3.41 \text{ torr}$$

EXERCISE 8–1(A)

Graph the following vapor pressure data for carbon tetrachloride, and determine the normal boiling point of the liquid:

t, °C	20	40	60	70	80	90
Pressure, torr	91	213	444	670	836	1110

Answer: 76.5°C at 760 torr.

EXERCISE 8–1(B)

The vapor pressure of water was determined by the gas saturation method, collecting 100 L of water-vapor-saturated helium carrier gas. After passage

of the helium through a saturator originally containing 65.44 g of water, 63.13 g of water remained. The temperature of the water was 25°C. Calculate the vapor pressure of water at that temperature.
Answer: 23.9 torr. ■

Humidity

Dalton's Law of Partial Pressures is directly concerned with personal comfort because it is the basis for our understanding of **humidity**, the concentration of water vapor in the air. Ordinarily, the partial pressure of water vapor is not more than a few torr. The partial pressure due to water vapor at any air temperature cannot be greater than the vapor pressure for water at that temperature.

• On a relatively cold day when the thermometer registers 5°C, the partial pressure of water vapor cannot exceed 6.54 torr.

• At a comfortable room temperature of 20°C, the partial pressure of water vapor cannot exceed 17.54 torr.

• On a sizzling hot day, when the thermometer registers 35°C, the partial pressure due to water vapor cannot exceed 42.18 torr (Table 8–2).

The ratio of the partial pressure of water to its vapor pressure is the **relative humidity**, which is commonly stated as a percentage:

$$\text{relative humidity} = \frac{\text{partial pressure of water vapor}}{\text{vapor pressure at that temperature}} \times 100\%$$

If the partial pressure of water vapor in the air is the same as the vapor pressure at that temperature, the air is said to be **saturated** and the relative humidity is 100%. The saturated condition happens to be a useful limit. Such a state can be established by humidifying the air, adding more water vapor and increasing the partial pressure; the relative humidity can also be increased by cooling the air until the existing partial pressure of water equals the vapor pressure for water at that temperature.

Consider the situation in which the partial pressure is found to be 20.0 torr and the thermometer registers 20°C, so that the calculated relative humidity is 114%. The partial pressure is greater than the vapor pressure,

TABLE 8–2 Vapor Pressures of Water	
Temperature (°C)	**Vapor Pressure (torr)**
0	4.58
5	6.54
10	9.21
15	12.79
20	17.54
25	23.76
30	31.82
35	42.18
40	55.32
50	92.51
60	149.38
75	289.1
100	760.0

and this should suggest an unstable condition. That is the case, and condensation takes place until the partial pressure falls to at least 17.5 torr. Clouds and fog develop out of these circumstances. The condensation called dew forms in the same way as the Earth's surface cools at night. If the partial pressure is very low, night temperatures may fall to 0°C before reaching the saturation level for water vapor in the air, in which case condensed water vapor takes the form of tiny ice crystals called frost.

EXAMPLE 8–2

A comfortable summer day may register thermometer readings of 86°F (30°C) and a pleasantly dry 35% relative humidity:

$$\text{relative humidity} = \frac{11.14 \text{ torr}}{31.82 \text{ torr}} \times 100 = 35\%$$

Calculate the partial pressure of water vapor for a much less comfortable daytime relative humidity of 86%, with the same thermometer reading.

Solution   Recognizing that the relative humidity is the ratio of the partial pressure of water vapor to the vapor pressure at any given temperature,

$$31.82 \text{ torr} \times \tfrac{86}{100} = 27.36 \text{ torr}$$

EXERCISE 8–2(A)

The weather report for today includes the following information: noontime temperatures of about 84°F and relative humidity near 53%. Estimate the vapor pressure of water at that temperature and the partial pressure of water vapor.
Answer: 30 torr; 16 torr. ■

EXERCISE 8–2(B)

The evening temperature on the same day as in Exercise 8–2(A) is expected to fall to 55°F. Is it likely that fog will develop that evening?
Answer: Yes. ■

The three phases of H_2O are clearly evident in this scene. Note especially the ground fog.

8.3   INTERMOLECULAR FORCES IN LIQUIDS

Although much more will be said in later chapters about the nature of **intermolecular forces**, the attractive and repulsive interactions that occur between atoms and molecules, some introduction is necessary at this point. These forces are weaker than the usual forces that bond atoms together in molecules or the electrostatic interactions between the charged particles called ions, yet they are responsible for many of the bulk properties of matter in the liquid state. Intermolecular forces account for the surprisingly high boiling point of water compared to compounds of nearly identical molecular weight, such as methane. They can also account for the fact that "permanent gases" such as helium can be condensed to the liquid state at all.

Many molecules containing different atoms bonded to each other can be considered as having positive and negative charges that are separated by some distances; that is, they are electrostatic dipoles. For example, the covalent molecule Br—Cl has negative charge concentrated on the Cl atom and positive charge concentrated on the Br atom. The dipole is usually drawn as an arrow with the head pointing toward the negative end and a "plus" on the positive end:

Br $+\!\!\longrightarrow$ Cl

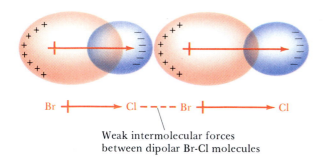

FIGURE 8–7 Dipole-dipole attractions among Br—Cl molecules. The Br atom at the more positive end of each molecule is shown as the larger circle; molecules are arranged so the positive end of one dipole is near the negative end of another, creating the overall dipole-dipole attraction.

Weak intermolecular forces
between dipolar Br-Cl molecules

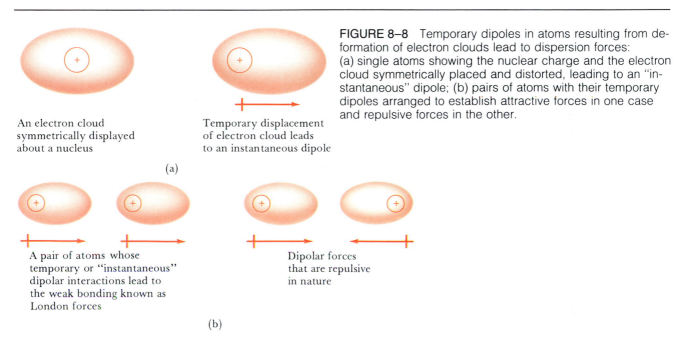

FIGURE 8–8 Temporary dipoles in atoms resulting from deformation of electron clouds lead to dispersion forces: (a) single atoms showing the nuclear charge and the electron cloud symmetrically placed and distorted, leading to an "instantaneous" dipole; (b) pairs of atoms with their temporary dipoles arranged to establish attractive forces in one case and repulsive forces in the other.

An electron cloud
symmetrically displayed
about a nucleus

Temporary displacement
of electron cloud leads
to an instantaneous dipole

(a)

A pair of atoms whose
temporary or "instantaneous"
dipolar interactions lead to
the weak bonding known as
London forces

Dipolar forces
that are repulsive
in nature

(b)

In the liquid phase the Br—Cl dipoles line up, with the negative end of one molecule near the positive end of another (Figure 8–7). The electrostatic attractions among the molecules, called **dipole-dipole attractions**, constitute one of the principal intermolecular forces holding these molecules in the liquid phase.

Dipole-dipole attractions are particularly strong for compounds containing hydrogen atoms bonded to oxygen, nitrogen, or fluorine. The attractions between such compounds are often called **hydrogen bonds**. They lead to the abnormally high boiling points observed for compounds such as H_2O, NH_3, and HF.

Intermolecular and interatomic attractions also exist for substances in which dipole-dipole attractions are impossible, such as diatomic elements (O_2, N_2, F_2, Cl_2, and Br_2) and even the noble gases. In these cases, the attractions are explained as being the result of *induced* dipole attractions, also known as **London dispersion forces**. This interaction occurs when the diffuse electron cloud surrounding the nucleus of an atom (or the nuclei within a molecule) becomes "momentarily" distorted in an unsymmetrical way to create an instantaneous dipole (Figure 8–8). One of these temporary dipoles can induce similar distortions in neighboring atoms or molecules, thus creating a system of dipole-dipole attractions. Induced dipoles are most easily achieved among heavier atoms having large and diffuse electron clouds. This accounts for the observation that boiling points increase from F_2 to I_2 and from He to Rn, for example. Now let's look at some of the consequences of intermolecular forces.

TABLE 8–3	Temperature Dependence of the Viscosity of Water				
T (°C)	0.0	20.	25.	55.	100.
η (centipoise)	1.8	1.0	0.9	0.5	0.3

TABLE 8–4	Representative Viscosities at 25°C		
Gases	**Viscosity η (centipoise)**	**Liquids**	**Viscosity η (centipoise)**
methane (CH_4)	0.010	benzene (C_6H_6)	0.60
nitrogen (N_2)	0.019	water (H_2O)	0.90
oxygen (O_2)	0.020	sulfuric acid (H_2SO_4)	19
		olive oil	80
		glycerin	954

1 poise = 0.1 kg/m·s
As a point of interest, the viscosity of whole blood at 98.6°F is 2.7 centipoise; for molten glass at 800°C, $\eta = 10^9$ centipoise

Viscosity

For many liquids, the temperature dependence of η can be described by the exponential function $\eta = Ae^{b/T}$ where A and b are constants, positive numbers characteristic of the liquid.

An important property of the liquid state is its viscosity. Molecules tend to push past neighboring molecules with varying degrees of ease. Maple syrup flows slowly, indicating a high degree of friction between adjacent layers of molecules. Water, on the other hand, exhibits a low level of friction and flows quite freely. **Viscosity** (η) is simply a measure of the internal liquid friction opposing any change in the liquid's movement. As shown in Table 8–3, the phenomenon is temperature-dependent, with increasing temperature bringing about a decrease in viscosity. The unit of measure for viscosity is the **poise** (after Jean Louis Poiseuille) (Table 8–4).

Experimentally, the viscosity of a liquid is determined relative to that of some reference liquid such as water in a device called a viscometer (Figure 8–9). A definite quantity of the liquid in question is introduced into the viscometer and drawn by suction into the bulb until its upper surface, the meniscus, lies above the upper mark. The liquid is then allowed to drain, and the time required for its meniscus to fall from the upper to the lower mark is noted. The procedure is repeated for the reference liquid. The driving force propelling the liquid through the capillary depends upon the difference in the liquid levels, the liquid density, and the acceleration due to gravity.

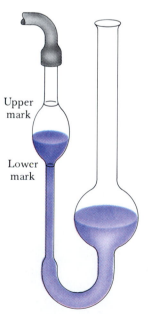

Upper mark

Lower mark

FIGURE 8–9 The Ostwald viscometer for measuring the viscosity of a liquid, relative to that of a reference liquid.

The Liquid Surface

The **gas-liquid interface** is the very thin but exceedingly important boundary layer between a liquid and its vapor. Think of it as a third phase, with properties intermediate between those of the liquid and those of the gas. The gas-liquid interface is only a few molecules thick, but in the space of this thin film the properties undergo drastic changes from those of a bulk liquid to those of a gas. A large change in density occurs, along with a marked alteration in molecular energies.

Other than at the surface or in the vicinity of the walls of the container, molecules in a liquid are subject to attractive forces that are essentially

(a)   (b)

FIGURE 8–10   A molecule depicted (a) within the body of the liquid where the forces (arrows) are symmetric, and (b) at the liquid surface where the acting forces are radically different.

(b)

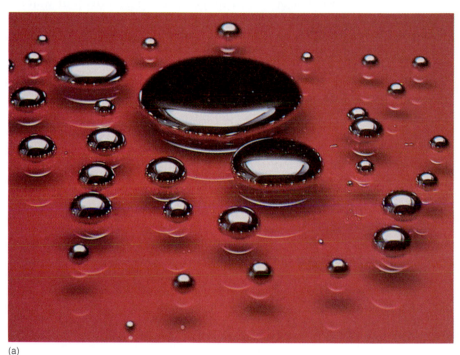

(a)

(c)

symmetric (Figure 8–10). However, the resultant forces acting on a molecule at the surface are unequal and are on average directed toward the body of the liquid. Because of this tendency to draw inward from the surface, there is a **surface tension** that results in (1) a membrane-like character to the surface, which has an unusual ability to resist penetration, and (2) a tendency for drops of liquid to become spherical in shape, contracting to the smallest possible area (Figure 8–11). As you might anticipate from the relationship between temperature and kinetic energy, surface tension is a temperature-dependent property of the liquid surface, decreasing with increasing temperature (Table 8–5). The tautness or tension in the liquid surface is measured in energy per unit area (Table 8–6).

Adhesion and Cohesion

The molecules of a liquid exhibit cohesive and adhesive forces. Attractive forces between the molecules of a liquid are **cohesive forces**. Forces existing between the molecules of the liquid and those of the container are referred

FIGURE 8–11   (a) Droplets of mercury lying on a glass surface. Mercury is the only metal that is liquid at room temperature. Note that the smaller droplets are almost spherical, while the large droplets are flattened, showing the effect of surface tension, which has greater influence on the shape of the smaller droplets. (b) Razor blade "floating" on oil; (c) In this microsecond flash photograph, a water droplet descends into a still pool of water and the impact produces a momentary liquid "crown." Note the almost perfect spherical shape of the exploded water droplets at the crown points and beyond.

TABLE 8–5	Surface Tension at the Water/Air Interface at Several Temperatures				
T (°C)	20.	40.	60.	80.	100.
Surface Tension (dynes/cm)	72.8	69.6	66.2	62.6	58.9

TABLE 8–6	Surface Tension of Several Fluids at the Liquid/Air Interface at 298 K
Fluids	Surface Tension (dynes/cm)
diethyl ether	17.1
n-hexane	18.4
ethylene glycol	47.7
sulfuric acid	55.1
glycerol (glycerin)	63.4
water	72.0

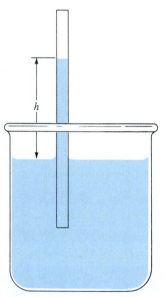

FIGURE 8–12 Liquid rise in a capillary column. Liquid will climb to the point where the downward force (due to the height of the column) equals the upward force due to adhesive forces.

to as **adhesive forces**. The spherical shape assumed by a drop of water in free fall is due to the cohesive forces of attraction between the like molecules of the liquid. On the other hand, the pages of a wet newspaper are difficult to separate because of adhesive forces between the liquid molecules and the material of the paper.

Adhesive and cohesive forces are the cause of the commonly observed curvature of the boundary surface between a liquid and the air above it at the point of contact with the container. It is especially evident in narrow containers or tubes. The meniscus is concave upward if the liquid wets or is strongly attracted to the walls of the container. The meniscus is concave downward if the molecules of the liquid are attracted to each other more strongly than to the container walls, as can be seen in the liquid mercury in a thermometer or barometer tube.

Capillary rise in narrow tubes is a measure of the pressure difference across the meniscus of the liquid. The curvature of the liquid surface is a function of the radius of the tube and the contact angle between the liquid and the tube. Depending on the viscosity of the liquid, the contact angle, and the resultant pressure difference, liquid may rise up the tube above the surface of the liquid outside the tube. The liquid will climb to the point at which the downward force due to the height of the column of liquid equals the upward force due to adhesive forces (Figure 8–12).

8.4 PHASE TRANSFORMATIONS

Fusion and Vaporization

When heat is transferred to a substance from its surroundings, there is usually a change in temperature. One situation in which heat flows but no change in temperature is observed is during a change of state—a **phase transformation**. There are three common transformations, and a fourth that we need to consider:

ΔH values for all transitions are written with a positive sign for the endothermic, energy-absorbing direction of the process.

Fusion	solid → liquid	ΔH_{fusion}
Vaporization	liquid → vapor	$\Delta H_{\text{vaporization}}$
Sublimation	solid → vapor	$\Delta H_{\text{sublimation}}$
Allotropic modifications	the transformation from one form of an element, with its unique properties, to another	$\Delta H_{\text{allotropy}}$

Table 8-7 lists heats of fusion and vaporization for a number of substances at their normal melting and boiling points.

TABLE 8–7 **Selected Heats of Fusion and Vaporization at the Normal Phase Transition Temperatures**

Substance	Melting Point °C	ΔH_{fusion} (J/g)	Boiling Point °C	$\Delta H_{vaporization}$ (J/g)
aluminum	660.	395.	2450.	10,500.
bromine	−7.2	67.8	58.8	97.
copper	1083.	134.	1187.	5065.
ethanol	−114.	104.	78.	854.
gallium	29.8	79.9	1983.	3710.
gold	1063.	64.0	2660.	1580.
helium	−269.7	5.23	−268.9	20.9
iron	1535.	267.		
lead	327.	24.5	1750.	858.
mercury	−38.9	11.3	356.	290.
nickel	1455.	299.		
oxygen	−218.8	13.8	−183.0	213.
potassium	63.2	59.7	760.	1995.
silver	961.	88.3	2193.	2350.
sodium	97.5	115.	880.	4260.
sulfur	119.	38.1	445.	326.
tin	182.	60.2		

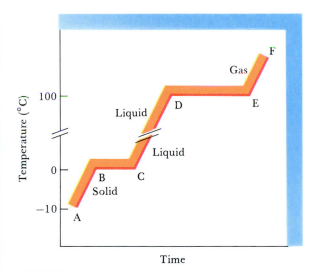

FIGURE 8–13 Heating curve for H_2O, plotted as temperature versus time.

If these is no temperature change, what drives transformations such as these? All four involve a change in internal energy. This is the hidden or **latent heat** of a substance, or simply the heat of the transformation—the heat of fusion ΔH_{fusion}, for example. To illustrate what we mean, consider the following experiment, which is described by the graph of temperature versus time in Figure 8–13. The experiment begins with 10.0 g of ice at −10.0°C; the plan is to transform it into steam at +110.0°C by adding heat at a steady rate, and all at atmospheric pressure. Since we are beginning at a temperature below the normal melting point of ice and ending above the normal boiling point, we'll need to know the specific heats of ice, water, and steam, all of which are different. Finally, keep in mind that our (linear) time scale—the abscissa—really represents a heat transfer scale, since heat is being added at a constant rate.

Step One—Ice: To begin with, as heat is added, the temperature of the ice rises from $-10.0°C$ to $0.0°C$, the normal melting point. Increasing the temperature of the solid increases the kinetic energy due to the increased vibrational motion of the particles. Since the specific heat of ice in this range is known to be 2.09 J/g·K, we can determine the quantity of heat required:

$$q_P \text{ (step 1)} = (10.0 \text{ g})(2.09 \text{ J/g·K})(10.0°C) = 209 \text{ J}$$

Step Two—Ice and Water: At the normal melting point, from the first appearance of liquid until the last bit of solid disappears into the puddle, the temperature is well-defined and constant. For all practical purposes, all the heat transferred during the time it takes to complete the phase change is used to overcome the attractive forces between the particles, increasing only the potential energy. Since there is no change in the kinetic energy, the temperature does not rise. The quantity of heat required can be calculated, since we know the heat of fusion to be 335 J/g:

$$q_P \text{ (step 2)} = (10.0 \text{ g})(335 \text{ J/g}) = 3.35 \times 10^3 \text{ J}$$

Step Three—Water: Between 0.0°C and 100.0°C, events are similar to those of step one except that it is the kinetic energy of the liquid that is increasing. From the known specific heat of water, 4.18 J/g·K, the heat required can be calculated:

$$q_P \text{ (step 3)} = (10.0 \text{ g})(4.18 \text{ J/g·K})(100.0°C) = 4.18 \times 10^3 \text{ J}$$

Step Four—Water and Steam: In a manner analogous to step two, from the time the normal boiling point (the point where the vapor pressure is 1 atm) is reached to the disappearance of the last trace of liquid, the temperature remains constant. All the heat transferred is used to increase the potential energy of the substance, overcoming the attractive forces that existed between the particles in the liquid state in order to separate the particles from each other. Since there is no change in kinetic energy, there is no change in temperature. We know the heat of vaporization to be 2259 J/g, so the heat required can be calculated:

$$q_P \text{ (step 4)} = (10.0 \text{ g})(2259 \text{ J/g}) = 2.26 \times 10^4 \text{ J}$$

Step Five—Steam: Finally, the heat transferred is again used to raise the kinetic energy of the particles, now in the vapor phase. The specific heat of steam (water vapor) in this temperature range is 2.01 J/g·K, so

$$q_P \text{ (step 5)} = (10.0 \text{ g})(2.01 \text{ J/g·K})(10.0°C) = 201 \text{ J}$$

The quantity of heat that must be transferred to 10.0 g of ice initially at $-10°C$ in order to transform it into steam at $+110.0°C$ is the sum of the five steps:

$$q_P = 3.03 \times 10^4 \text{ J} = 30.3 \text{ kJ}$$

Because the process is reversible, the heating curve in Figure 8–13 could just as well be a cooling curve, and we could recover the stored potential energy. Farmers sometimes spray vegetables or citrus fruits with water before freezing weather hits; the latent heat of fusion of the surface layer of water is released as it freezes, protecting the fruit from frost damage. The tremendous destructive power of hurricanes and typhoons comes from the heat liberated when billions of gallons of water, picked up over warm tropical oceans, condense into rain.

EXAMPLE 8–3

Calculate q_p for each of the steps and for the entire process required to transform 2.00 g of solid bromine (Br_2) at $-20.0°C$ to vapor at $70.0°C$ (at 1 atm pressure). Use the following data:

melting point $-7.2°C$ boiling point $58.8°C$

ΔH_{fusion} 67.8 J/g $\Delta H_{vaporization}$ 183 J/g

Specific heats for Br_2:

solid	0.37 J/g·°C
liquid	0.448 J/g·°C
vapor	0.23 J/g·°C

Solution   In a stepwise manner, calculate each q and then sum them:

$$q_1 = (2.00 \text{ g})(0.37 \text{ J/g·°C})[(-7.2 + 20)°C] \quad = \quad 9.5 \text{ J}$$
$$q_2 = (2.00 \text{ g})(67.8 \text{ J/g}) \quad = 136 \text{ J}$$
$$q_3 = (2.00 \text{ g})(0.448 \text{ J/g·°C})[(58.8 + 7.2)°C] \quad = \quad 59.1 \text{ J}$$
$$q_4 = (2.00 \text{ g})(183 \text{ J/g}) \quad = 366 \text{ J}$$
$$q_5 = (2.00 \text{ g})(0.23 \text{ J/g·°C})[(70.0 - 58.8)°C] \quad = \quad 5.2 \text{ J}$$
$$\text{Total } q_p \quad = 576 \text{ J}$$

EXERCISE 8–3

Calculate q_p for each of the five steps required to transform 1000 g of steam at 500°F into a block of ice at 5°F. Then calculate the total q_p.
Answer: -3.37×10^3 kJ (total).

Pressure Dependence

Phase transformations depend on pressure as well as temperature. A plot of temperature versus pressure for the phase transformations of a substance is called a **phase diagram**. As an example, we can construct such a diagram for H_2O, starting with the solid-to-liquid phase change. Ice is an unusual substance in that its volume decreases on melting; melting should therefore be favored by an increase in pressure. It should be possible to melt ice at temperatures lower than the normal melting point by applying an increased pressure. It also follows that any decrease in the applied pressure is accompanied by an increase in the melting point, as illustrated in Figure 8–14. The more commonly encountered, opposite situation is illustrated in Figure 8–15.

Consider, for the sake of illustration, the ease with which one can skate over a frozen lake in winter on a narrow skate blade. In the area immediately under the blade, the considerable pressure causes local melting, and one skates along smoothly on an invisible film of water. As you skate along, the pressure is removed and the water refreezes. However, if the temperature is too low to begin with, the pressure increase is likely to be insufficient to melt the ice, and skating becomes more difficult.

A demonstration of the pressure-dependent melting of ice is the passing of a piano wire, with two suitable weights suspended from its ends, through a block of ice without apparently cutting the ice. The trick has to do with the pressure-dependence of the freezing point. Extreme local pressure under the wire causes the ice to melt, freeing water molecules that

The skating actually improves a bit due to additional melting caused by heat generated by friction between the blade and the ice.

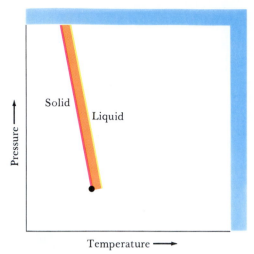

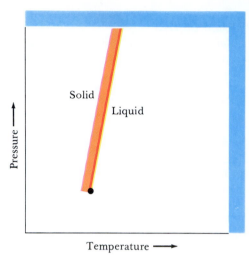

FIGURE 8–14 The [solid ⇌ liquid] phase change for H_2O, a solid whose volume contracts on melting.

FIGURE 8–15 The [solid ⇌ liquid] phase change for a solid whose volume expands on melting.

(a)

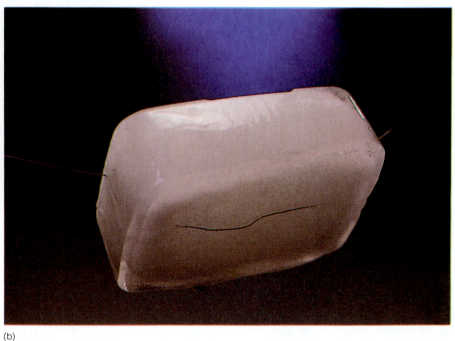

(b)

FIGURE 8–16 A weighted wire cutting through a block of ice, without cutting the ice in half, as the water molecules are squeezed out from underneath the wire and quickly refreeze.

are then squeezed out to the other side of the wire. Since the pressure of the wire and the weights has now passed, the water refreezes (Figure 8–16).

The pressure-dependence of the vaporization process is just as easily demonstrated. Water boils at a lower temperature in Mexico City and in Denver than in New York or any other coastal city, due to the diminished pressure of the atmosphere. That needs to be included on our *P-T* plot, too (Figure 8–17).

Sublimation

Although most solids melt long before their vapor pressures reach the pressure of the atmosphere, there are some that do not. The change of

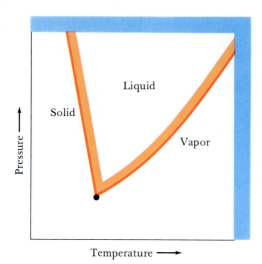

FIGURE 8-17 Pressure-temperature equilibria for the (solid $\rightleftharpoons$ liquid) and (liquid $\rightleftharpoons$ vapor) phase transitions for H_2O.

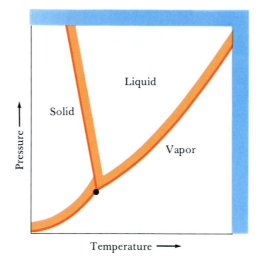

FIGURE 8-18 Adding the [solid $\rightleftharpoons$ vapor] sublimation equilibrium to our pressure-temperature phase diagram for H_2O.

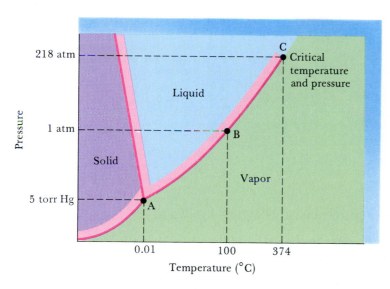

FIGURE 8-19 The complete pressure-temperature phase diagram for H_2O, showing the triple point (A), the normal boiling point (B), and the critical point (C).

state that occurs when a solid substance is transformed directly into the vapor state is known as **sublimation**. Carbon dioxide and iodine are two common examples. At about 1 atm and at temperatures above $-78°C$, solid carbon dioxide (sold under the trade name Dry Ice) sublimes rather than melts. Iodine crystals in a sealed capillary tube melt at 184°C, but crystals left in an open dish at room temperature readily evaporate. As is the case for the solid-liquid and liquid-vapor transformations, the solid-vapor phase transformation is pressure-dependent. In Figure 8–18, our *P-T* diagram for H_2O, we have now included the sublimation line along which the solid-to-vapor transformation takes place. Even though H_2O is a normal substance in the sense that solid commonly melts to liquid and then vaporizes, there are conditions of pressure and temperature under which ice will sublime. Figure 8–19 is the complete phase diagram for H_2O.

On cold days, or especially over long stretches of weather in the temperature range of 0 to 15°F, ice gradually disappears from city streets by sublimation or direct evaporation. Actually, all solids are evaporating all the time, some more than others.

FIGURE 8–20 At the triple point, solid-liquid-vapor are all in equilibrium.

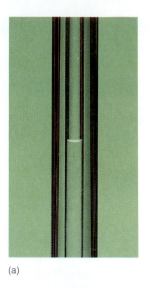
(a)

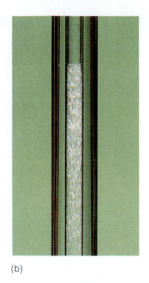
(b)

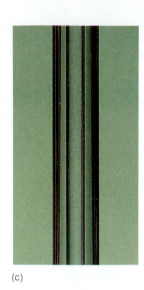
(c)

FIGURE 8–21 The pressure-temperature phase diagram for CO_2.

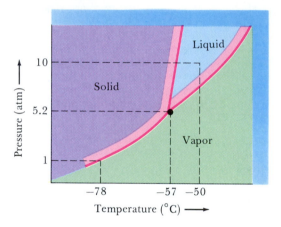

Triple Points and the Phase Diagram for Water

The combined **fusion-vaporization-sublimation curves** for a substance intersect at one point, a set of coordinates known as the **triple point**. Under those conditions of temperature and pressure, all three phases—solid, liquid, and vapor—are simultaneously in equilibrium; that is, they coexist indefinitely as long as the conditions do not change (Figure 8–20). Consider the triple point in the phase diagram for CO_2 in Figure 8–21. The triple point conditions are −57°C and 5.2 atm. Therefore, the solid-to-liquid phase transformation for CO_2 cannot take place unless the pressure is at least 5.2 atm. It is important to note that these phase diagrams refer only to closed systems (e.g., a substance contained in a sealed jar) so that there is no loss of material to the surroundings.

EXAMPLE 8–4

At −50°C and 10. atm, is CO_2 solid, liquid, or gas? What happens if there is a shift in temperature by 10°C in either direction? Read the data from the phase diagram for CO_2. What do you conclude?

Solution At −40°C and 10 atm, vapor state; at −60°C and 10 atm, solid state.

EXERCISE 8–4

Compare the behavior of CO_2 and H_2O with respect to their phase diagrams. At the triple point, describe the net effect of a temperature increase and of a pressure increase. For CO_2, what will be the effect of a slight temperature increase at $-78°C$ and 1 atm? For H_2O, what will be the effect of a slight pressure increase at $101°C$ and 1 atm?

Answer: For CO_2, solid sublimes to vapor; for H_2O, vapor condenses to liquid. ∎

8.5 CRITICAL TEMPERATURE AND PRESSURE

For each substance there exists a temperature above which the gas cannot be liquefied no matter how great the pressure becomes. At any temperature below this **critical temperature**, a vapor can be condensed if the applied pressure exceeds the equilibrium vapor pressure for that temperature. If the critical temperature is very low, then the gas must be cooled to very low temperatures before it can exist in the liquid state.

Faraday was the first to liquefy chlorine. In the very simple experiment shown schematically in Figure 8–22, he decomposed a chlorine compound in a sealed tube. A few drops of liquid chlorine condensed in the cold end of the tube. By immersing the cold end in a cooling mixture, Faraday was able to isolate and liquefy a number of gases that had previously resisted efforts to condense them. However, all attempts at condensing hydrogen, nitrogen, and oxygen were unsuccessful, and many scientists of the day believed them to be "permanent gases." With advances in low-temperature technology, all of the uncondensable gases of Faraday's generation have now been condensed to the liquid state. James Dewar liquefied hydrogen in 1898, and Heike Kamerlingh-Onnes successfully condensed the last and most difficult, helium, in 1908.

The critical temperature is at the end of the liquid-vapor line in the phase diagram in Figure 8–19. This point is called the **critical point**, and its corresponding pressure is the **critical pressure**. The critical points of some common substances are listed in Table 8–8. Note the very low critical temperatures for oxygen, nitrogen, and hydrogen, the so-called permanent gases.

EXAMPLE 8–5

Using critical data available in Table 8–8, explain why chlorine is easier (or more difficult) to liquefy than ammonia.

Solution   Ammonia is more difficult because its critical pressure is higher. ∎

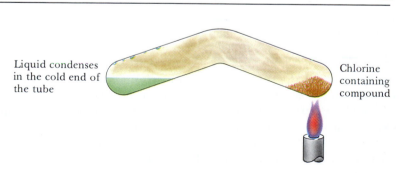

Liquid condenses in the cold end of the tube

Chlorine containing compound

FIGURE 8–22 Faraday's liquefaction experiment (1823). At the time, Michael Faraday was an assistant to Sir Humphrey Davy (at the Royal Institution, Great Britain) who was studying the properties of chlorine compounds. On heating one of these compounds in a sealed, heavy-walled glass tube, he noted some droplets of an oily liquid in the cold end and correctly assumed they were chlorine liquid.

TABLE 8–8 Critical Data for Selected Substances

Substance	Critical Temperature (°C)	Critical Pressure (atm)
helium (He)	−267.9	2.3
hydrogen (H_2)	−239.9	12.8
neon (Ne)	−228.6	27.2
nitrogen (N_2)	−147.1	33.5
oxygen (O_2)	−118.8	49.7
carbon dioxide (CO_2)	31.1	73.0
radon (Rn)	104.	62.0
ammonia (NH_3)	132.4	111.5
chlorine (Cl_2)	144.0	76.1
water (H_2O)	374.1	218.3
mercury (Hg)	1485	1490.

EXERCISE 8–5

Make the same comparison for nitrogen and oxygen.
Answer: Oxygen is easier to liquefy.   ■

The phase diagram for water, including the critical constants, is shown in Figure 8–19. The triple point is the lower terminus of both the fusion and vaporization curves; but whereas the former extends up indefinitely, the latter reaches only to the critical point. At any point along the vaporization curve, a molecule escaping from the liquid surface into the vapor state must have acquired an energy much higher than the average energy of the remaining molecules in the liquid. If the temperature of the liquid is now increased, the average excess energy needed to escape the liquid will decrease. This excess energy is the heat of vaporization at that temperature, and it becomes lower with increasingly higher temperatures. Finally, approaching the critical temperature, the average energy of the molecules in the liquid state approaches that of the molecules in the vapor state, and the heat of vaporization approaches zero. Note how the heat of vaporization of water varies with temperature:

Temperature	ΔH_{vap}
0°C	44.88 kJ/mol
100°C	40.67 kJ/mol
200°C	34.87 kJ/mol

In the case of water as well as most other substances, the heat of fusion is considerably less than the heat of vaporization. For water:

0°C	$\Delta H_{fus} = 6.0$ kJ/mol
0°C	$\Delta H_{vap} = 45$ kJ/mol

To bring about the solid-to-liquid transformation, sufficient energy must be put into the substance to disrupt the structure of the solid, to the point where the particles can slide by each other—to flow—in response to an external stress of some kind. In the case of the liquid-to-vapor transformation, however, one must additionally expand the distance between the particles by a factor of a thousand (or more), against the external pressure, to the volume of the gaseous state. It is estimated that approximately 10%,

JAMES DEWAR

James Dewar (Scottish; 1842–1923) was Fullerian Professor of Chemistry at the Royal Institution (London). His most important work in a long and successful career was in the field of low temperatures. In 1892, he constructed double-walled evacuated glass flasks generally shaped like large test tubes. Because the space between the walls was evacuated, heat could not be transmitted by thermal conduction or convection due to air currents. Heat could be transferred only by radiation, and even that could be cut down by "mirroring" the walls with silver so that radiated heat would be reflected rather than absorbed. Today, these "cryostats" are called Dewar flasks and are little changed from Dewar's original designs. The household variety is sold under the trade name of Thermos bottle. Dewar was able to liquefy hydrogen in 1898. He was subsequently able to solidify it in 1899 at a temperature only 14 degrees above absolute zero, a temperature at which all substances except helium are solids.

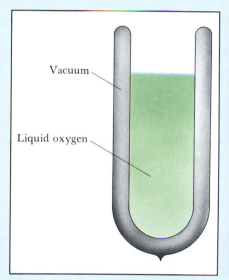
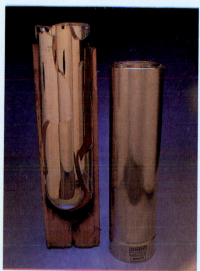

Vacuum

Liquid oxygen

First used on a regular basis a century ago, and named "Dewar" flasks after their inventor, these cryostats, or low-temperature containment vessels, are double-walled and evacuated to exclude air and water vapor and the inner walls are silvered, reducing heat exchange by radiation. The schematic diagram is accompanied by a photo showing an external view of a new Dewar flask and one that was mounted in a wooden protective frame and broken open for illustration.

or about 3 kJ/mol, of the heat of vaporization of water at 100°C is required to accomplish the expansion while the remaining 37 kJ/mol is used to separate the molecules from each other in the first place

EXAMPLE 8–6

What is the heat of vaporization of chlorine (or ammonia) at their respective critical temperatures?

Solution The heat of vaporization of any liquid at its critical temperature is zero.

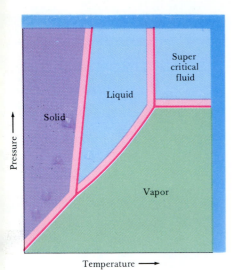

Pressure →

Temperature →

FIGURE 8–23 But, like a liquid, a supercritical fluid can dissolve substances. Shown here is the supercritical fluid state.

EXERCISE 8–6

What is the heat of vaporization of water at the critical point?
Answer: Zero. ∎

Supercritical Fluids

When the temperature and pressure are both above the critical point, a substance becomes a supercritical fluid. Like a gas, a supercritical fluid expands to fill its container completely. But like a liquid, a supercritical fluid can dissolve substances (Figure 8–23). By carefully controlling the pressure, temperature, and composition of a supercritical fluid it is even possible to dissolve solids selectively, and that of course has great commercial as well as purely scientific possibilities:

- For the pulp and paper industry in the manufacture of cellulosic materials
- For the plastics industry in the important area of recycling polymeric materials
- In the food processing industries to improve process chemistry and solve engineering problems such as the decaffeination of coffee

This new field combines physics, chemistry, and engineering. Its promise and limits are just beginning to be explored.

Erdogan Kiran and his group of chemical engineering researchers at the University of Maine at Orono have designed a special high-pressure chamber that allows one to view the strange behavior of supercritical fluids directly—at pressures of 1000 atmospheres—through half-inch-thick, colorless sapphire windows. With this pressure cell, experiments with polymers and plastics have been observed directly. For example, as one increases the temperature of a supercritical fluid, the visible outline of a chunk of polystyrene gradually dissolves and disappears in a display of brilliant colors. When the temperature and the pressure are again lowered, the supercritical fluid returns to its original state and the polystyrene precipitates out of solution in finely divided particles. One might imagine selective separation of cellulose, hemicellulose, and lignins, wood products that are normally very difficult to separate.

Many substances become supercritical fluids. For water, the conditions are temperatures above 374°C and pressures above 218 atm. Keep in mind, however, that under such severe conditions many organic molecules are irreparably damaged. Paper products, for example would be ignited by water under these conditions. A more practical supercritical fluid might be carbon dioxide. At a temperature of about 30°C and at pressures of around 73 atm—relatively modest conditions—CO_2 displays supercritical behavior and has proved capable in limited, commercial scale processes of penetrating coffee beans and extracting the caffeine, leaving a product that is free of any solvent residue. "Sanka" brand coffee is decaffeinated by such a process. Everything is left in the bean, including the smell and the flavor. Only the caffeine is removed by the selective dissolving action of the supercritical fluid. A supercritical phase of liquid helium called helium II displays some astonishing behavior (Figure 8–24).

In research and development laboratories, chromatographic techniques for separating substances have begun to take advantage of supercritical fluid behavior, replacing conventional gas and liquid methods. There are recent reports in the scientific literature of good results in the analysis of certain antibiotics, pesticides, narcotics, and drugs of plant origin.

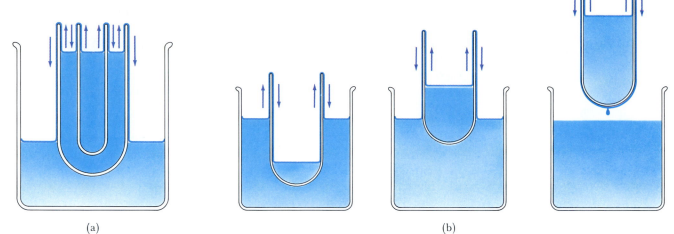

(a)   (b)

FIGURE 8–24 Demonstrating superfluidity by the "creeping film" experiment. A liquid that wets a surface with which it is in contact will form a film: (a) Helium(II) forms a film so rapidly that a siphon effect is produced; (b) Helium(II) flows (left) into a flask partially submerged in a reservoir of the liquid, or (center) out of a flask, or (right) falls in drops from the bottom of a flask.

SUMMARY

All gases may be changed to liquids by sufficiently increasing the pressure and decreasing the temperature. The liquid state is best understood as an intermediate state of matter having some of the properties of gases (an extremely dense gas) and some of the properties of solids (a highly disordered solid).

Equilibrium vapor pressure is the pressure due to vapor in equilibrium with the liquid in a closed container, at a given temperature. It is a characteristic property of the liquid state that can be measured in a straightforward manner, based on manometric measurements and the equation of state for an ideal gas. Besides temperature, equilibrium vapor pressure depends on molecular character, but it does not depend on the quantity of liquid present as long as some liquid is present at equilibrium. The temperature at which the equilibrium vapor pressure of a liquid equals the pressure of the atmosphere is the **boiling point**.

Molecules in the liquid state have the same average kinetic energies as those that characterize the gaseous state for a given temperature, but they travel much shorter **mean free paths** between collisions and have much less freedom of motion.

Perhaps the most characteristic properties of liquids have to do with phenomena that occur at the **liquid surface**, such as **evaporation** and **surface tension**. Both have to do with the intermolecular forces of cohesion and adhesion. The **cohesive forces** exist between the molecules of the liquid, and the **adhesive forces** exist between the liquid molecules and the surface of the container.

The rate of evaporation and the surface tension are both temperature-dependent. **Viscosity**, or the resistance of a liquid to flow in response to a stress, is also a characteristic (and temperature-dependent) phenomenon. The surface tension and the viscosity increase with decreasing temperature, while the rate of evaporation increases with temperature.

Heating and cooling curves can be extracted from P versus T plots called **phase diagrams**. Heating and cooling curves show relationships between kinetic and potential energy and the states of matter for a given substance. Two-dimensional phase diagrams illustrate under what conditions of pressure and temperature the solid, liquid, and vapor states can exist (areas on the plot); under what conditions two states of matter can coexist indefinitely (along the curves or lines of the P versus T plot); and under what conditions all three states of matter can coexist indefinitely (the triple point). Phase diagrams are based on having a substance contained in a closed system so that there is no loss of material to the surroundings.

Based on the hidden or **latent heat** (the heat required to bring about a phase change) and the specific heats for the different states of matter, it is possible to make simple thermochemical calculations of the heat necessary to bring about seemingly complex changes (such as conversion of ice to steam).

There is a certain **critical temperature** for each substance above which no condensation of a gas to a liquid occurs, no matter how great the pressure. The pressure necessary to liquefy a gas at the critical temperature is the **critical pressure**. Substances with low critical temperatures are difficult to liquefy.

Supercritical fluids can exist at high temperatures and pressures. They have exhibited some remarkable and commercially useful physical properties.

QUESTIONS AND PROBLEMS

QUESTIONS

1. Briefly define, discuss, identify, or otherwise distinguish between each of the following pairs of terms:
(a) liquids and fluids
(b) sublimation and vaporization
(c) cohesion and adhesion
(d) surface tension and meniscus
(e) density and compressibility
(f) boiling and evaporation
(g) critical temperature and critical pressure
2. Briefly explain each of the following:
(a) In order for a liquid to evaporate at constant temperature, heat must be absorbed.
(b) A person emerging from the ocean onto a warm beach, especially on a breezy day, experiences a cooling effect.
(c) Boiling points are more pressure-dependent than melting points.
(d) A three-minute egg boiled on top of Pike's Peak isn't very palatable.
(e) The insect called a water strider can walk on water.
3. Consider the theory of liquid behavior and briefly explain:
(a) the processes of fusion and vaporization.
(b) the processes of diffusion and evaporation.
(c) why the heat of vaporization is greater than the heat of fusion.
4. Briefly explain why:
(a) the heat of vaporization approaches zero at temperatures approaching the critical temperature.
(b) the surface tension is zero at the critical temperature.
5. Account for each of the following:
(a) Lakes freeze in winter from the top down.
(b) Liquids can be understood as "dense" gases.
(c) Liquids can be understood as "disordered" solids.
(d) Gasoline producers blend lower-boiling fuels into their mix in winter.
(e) The pages of a wet newspaper are hard to separate.

6. Offer a brief explanation of each of the following:
(a) Faraday was able to liquefy chlorine without difficulty by simply cooling it under modest pressure in a sealed tube.
(b) The upper limit of the vapor-liquid equilibrium curve is the critical temperature.
(c) A car easily skids on ice when the temperature is just below freezing.
(d) A freshly made cube of ice loses its sharp edges after standing for a long while in a freezer, even though the temperature is well below 0°C.
(e) A winter golfer standing on a frozen pond finds his spikes have sunk into the ice.
(f) A summer golfer standing on a tarred road finds his spikes have sunk into the tar.
7. Why can wet clothing be freeze-dried on an outdoor clothes line in winter?
8. A fire extinguisher containing liquid carbon dioxide sprays snow-like solid carbon dioxide through the nozzle when discharged. How does this happen?
9. Why have hydrogen, nitrogen, and oxygen been referred to as permanent gases?
10. Briefly explain each of the following:
(a) the effect of a pressure change on the critical point of water.
(b) the relative ease of liquefaction of two gases whose critical temperatures are, respectively, −40°C and +40°C.
(c) ΔH of vaporization at the critical temperature.
(d) the surface tension of a liquid at the critical point.
(e) the effect of lowering the temperature on the viscosity of a liquid.
(f) the change in kinetic energy as pure liquid vaporizes at the phase transformation temperature.
(g) the change in potential energy as pure liquid freezes at its phase transformation temperature.
(h) the time required to cook boiled eggs at elevations below sea level.

11. Explain why rain drops are roughly spherical.

12. Why are the shapes of the liquid surface in a narrow column—the meniscus—different for mercury and water?

13. Molten copper has a larger volume than the same weight of solid copper. How is the melting point of copper affected by a pressure change?

14. It is well known that solid benzene sinks in its liquid. What happens to an equilibrium mixture of liquid and solid benzene if the pressure on the mixture is lowered and the temperature is unchanged?

15. Briefly explain why the vapor pressure of a liquid in a closed container is independent of the surface of the liquid, the volume of the container, and the mass of the liquid as long as some is present.

16. Heat is liberated when steam at 100°C is converted to water at the same temperature. Briefly explain.

17. When oxygen is drawn off from a cylinder in which it is contained, the pressure falls. However, that is not the case with chlorine cylinders, in which the pressure stays at approximately 6.6 atm until nearly all the chlorine has been removed. Why?

18. Offer a brief explanation of why viscosities of gases might reasonably be expected to increase with temperature while liquid viscosities commonly decrease with rising temperature.

19. In the liquid and vapor states, hydrogen fluoride molecules are associated due to the existence of intermolecular forces of attraction. Briefly explain.

20. A room in which the humidity has been lowered feels relatively cooler. Yet the dehumidifier has not cooled the room. Briefly explain.

21. Surface tension of a liquid diminishes with rising temperature. What should the value of the surface tension be at the critical temperature? Briefly explain.

22. Liquids are much denser than their vapors. Compare liquid and vapor densities near the critical temperature.

23. What experimental evidence can you use to suggest that intermolecular forces exist in liquids to a significant degree?

PROBLEMS

Calculations of Vapor Pressure [1–12]

1. Graph the following vapor pressure data for CCl_4 and determine the normal boiling point of the liquid:

Temp. (°C)	20.0	40.0	60.0	70.0	80.0	90.0	100.
V.P. (torr)	91.0	213.3	444.3	617.4	836.0	1110	1459

2. Graph the following vapor pressure data for ethanol (ethyl alcohol) and determine the normal boiling point of the liquid:

Temp. (°C)	22.7	34.9	48.3	56.9	68.3	76.4
V.P. (torr)	50.0	100.	200.	300.	500.	700.

3. Vapor pressures for water at several temperatures are recorded below:

Temp. (°C)	0.0	20.	40.	60.	80.
V.P. (torr)	4.58	17.54	55.32	149.38	355.10

Make a graph of vapor pressure (as ordinate) versus temperature (as abscissa). Then make a second graph, by plotting the logarithm of vapor pressure versus the reciprocal of the absolute temperature (multiplied by 10^3). Now, smoothly extend both curves to 1 atm pressure to obtain two estimates of the normal boiling point. Compare the accuracy of the two ways of handling the experimental data.

4. Plot the vapor pressure data for ethanol in Problem 2 as log vapor pressure versus $1/T$, where T is temperature in K. Determine the normal boiling point of ethanol using this graph, and compare this result with that obtained in Problem 2.

5. At Denver, Colorado the pressure of the atmosphere is 697 torr, while on top of one of the peaks in nearby Rocky Mountain National Park it is only 550 torr. Determine the boiling points of water at these two locations.

6. Determine the boiling points of ethanol in Denver, Colorado and in Rocky Mountain National Park. Use the data of Problems 2 and 5.

7. The vapor pressure of water was determined by the gas saturation method, by passing dry helium through a saturator containing 65.44 g of water. When 100. L of He was collected after passage through the saturator, 63.13 g of water remained. The temperature of the water was 25°C. Calculate the vapor pressure of water at that temperature.

8. After 10.0 L of air at 20.°C and atmospheric pressure was bubbled through several consecutive tubes of water at 20.°C, the moisture in the saturated air was absorbed in sulfuric acid and weighed. In this experiment it was found that 0.178 g of water saturated the air. Determine the vapor pressure of water at that temperature.

9. The liquid in a molecular weight experiment using the gas saturation method is shown to be ethyl alcohol, and the measured vapor pressure is 59.0 torr at 25.°C. What must have been the weight loss in the original sample if 25.00 L of dry nitrogen passed through?

10. The gas saturation method is used to determine the vapor pressure of acetone at 21.8°C, and this is found to be 200. torr. Calculate the weight loss of the original sample if 40.0 L of N_2 (g) passed through.

11. Consider the experimental determination of the vapor pressure of benzene, an organic liquid. Only two measurements are made: at 20°C, the pressure is 76.7 torr; and at 60.°C, it is 384.6 torr. How would you best handle the data in order to determine the boiling point of benzene? What is the result? Compare your answer to the values in the *CRC Handbook of Chemistry and Physics*.

12. Determine the normal boiling point of acetic acid from the following two data points:

vapor pressure = 100 torr at 63.2°C

vapor pressure = 500 torr at 106.0°C

Phase Transformations [13–16]

13. Calculate the quantity of heat required to do each of the following:

(a) melt 1.0 kg of ice at its melting point.

(b) vaporize 1.0 kg of water at its boiling point.

14. Calculate the heat required to melt 1.0 mol of solid gallium and the heat required to convert the mole of molten gallium to gallium vapor.

15. Calculate the final temperature reached by this system at equilibrium:
(a) 10. g of ice at 0°C in contact with 10. g of water at 50°C.
(b) 10. g of ice at 0°C in contact with 50. g of water at 10°C.

16. Calculate the final temperature reached by a system containing 20. g of ice at 0°C, 50. g of water at 80°C, and 10. g of aluminum at 90°C when it reaches equilibrium.

Phase Diagrams   [17–24]

17. Draw heating curves for CO_2 at the following pressures, beginning at some temperature below $-78°C$. See Figure 8–21:
(a) 1.0 atm   (b) 5.2 atm   (c) 30. atm

18. Draw heating curves for CO_2 at the following pressures. See Figure 8–21:
(a) 10. atm   (b) 30. atm

19. Consider the accompanying diagram and answer the following:
(a) How does the melting point change with pressure?
(b) Describe what happens at 375 K and 1.25 atm when the pressure is decreased at constant temperature. What happens when the temperature is decreased at constant pressure?
(c) Give approximate values for the melting point, the boiling point, the vapor pressure at room temperature, and the sublimation pressure at room temperature.

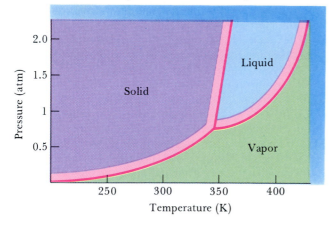

20. The triple point (*P-T*) phase diagram for an element Z is shown below:
(a) In what phases does Z exist at 263 K and 660. torr?
(b) In what phases does Z exist at 303 K and 300. torr?
(c) At what temperature does vapor at a pressure of 400 torr condense?
(d) What is the sublimation pressure of Z at 268 K?
(e) At what temperature and pressure can solid, liquid, and vapor all exist in equilibrium in the same container?

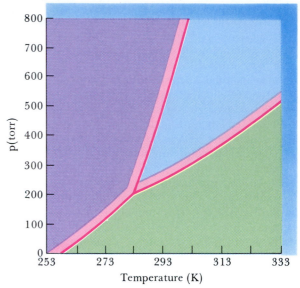

21. Here is the phase diagram for a substance called R:
(a) What is the critical pressure of R?
(b) Will solid R sink in its own liquid?
(c) At what temperature and pressure can solid, liquid, and vapor exist together at equilibrium?
(d) Does this substance have a normal boiling point? Briefly explain.

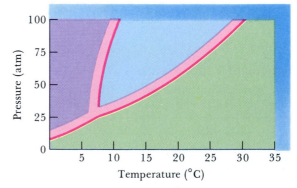

22. Here is the phase diagram for a fictitious one-component system:

(a) What is the state of the system at 100.°C and 4.0 atm; at 100.°C and 2. atm; at 0.°C and 4. atm; at 0.°C and 2. atm?

(b) What is the pressure if $T = -10°C$ and two phases are present?

(c) Does liquid of this substance expand or contract on freezing? What is the effect of increased pressure on the freezing point?

(d) Sketch a heating curve at a constant pressure of 1.75 atm, starting at point A, passing through B, to point C. How many phases are present at each point, and what are they?

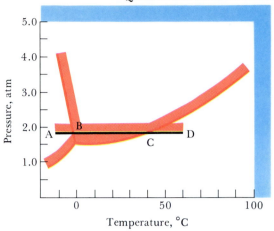

23. The figure that follows is the phase diagram for sulfur:

(a) There are three triple points. Identify them and describe the phase equilibria.

(b) Discuss the possible meaning of the dotted lines and the meta-stable, fourth triple point.

(c) What is the normal state of sulfur at room temperature and atmospheric pressure?

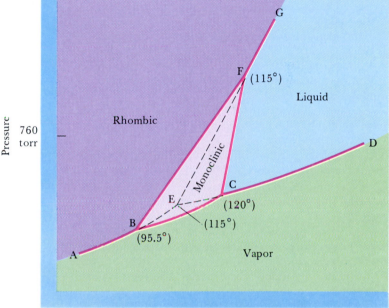

24. The phase diagram for carbon over a wide range of conditions of temperature and pressure is illustrated:

(a) Are diamonds really "forever"?

(b) At what pressure will diamond melt in a furnace that registers 1250. K?

(c) What is the history of events when graphite at 2500 K is compressed in a super press? Note especially points A and B.

(d) What is point C, and what phases are present?

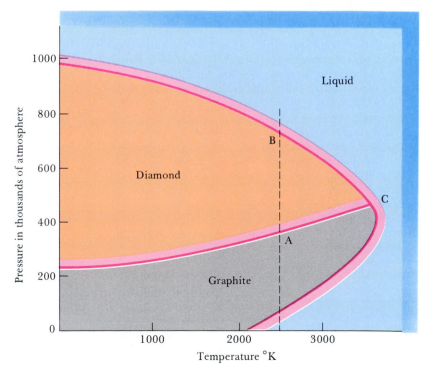

Additional Problems   [25–37]

25. A kilogram of steam initially at 373 K is cooled by extracting 3000. kJ. Given data available in this chapter on heats of fusion and vaporization, and the specific heat of water, determine which phases are present after the heat is extracted, and in what quantities.

26. The vapor pressure of decane ($C_{10}H_{22}$) is determined at 55°C using the gas saturation method with helium. The mass of the liquid in question was 10.00 g before the experiment and 6.70 g after 200. L of helium had passed through the saturator. What is the equilibrium vapor pressure of decane at this temperature?

27. How much heat is required to melt 1.0 kg of sodium solid at its normal melting point? How much must be extracted to condense 1.0 kg of sodium vapor at its normal boiling point?

28. Calculate the final temperature if 20.0 g of ice at 0.0°C is placed in 50.0 g of water at 25.0°C. If 50.0 g of ice at 0.0°C is placed in 20.0 g of water at 25.0°C, what phases will be present and in what quantities when temperature equilibrium is reached?

29. A molar volume of dry air, measured at STP, is placed over water at 298 K and 753 torr. What is the percent change in the volume?

30. The partial pressure of water vapor in an air sample was found to be 10.0 torr at 20.°C.

(a) Determine the relative humidity.

(b) Find the approximate dew point for such an air sample.

(c) Calculate the absolute humidity, the mass of water vapor present in a liter of moist air at the relative humidity found in part a.

31. It is uncomfortably warm today. The thermometer reads 30°C. A smoothly polished metal can is cooled by adding cold water, and at a water temperature of 5.0°C, the metal surface becomes fogged over. What is the relative humidity in the room?

32. In Example 8–1, the equilibrium vapor pressure of carbon tetrabromide was determined. If the student doing the experiment measured the volume of the carrier gas *before* it passed through the carbon tetrabromide liquid, and proceeded to use that value for the volume of CBr_4 vapor, his or her results would be in error. Why? If the recorded value for V of the carrier gas was 30.00 L, what was the percent error?

33. In Exercise 8–1(B), if the experiment was incorrectly performed, and what was actually done was to pass 100 L of dry He *into* the saturator and use that value for the volume in the calculations of the vapor pressure for water, what is the source of the error in the experiment, and what is the percent error?

34. Graph the following viscosity measurements made on ethylene glycol, the liquid commonly mixed with water and used as an automotive radiator fluid:

Temp. (°C)	20.	40.	60.	80.	100.
η (centipoise)	19.9	9.13	4.95	3.02	1.99

Find the viscosity at 85°C, note the general temperature dependence of viscosity, and comment on the shape of the curve.

35. Given the following data, construct a phase diagram for hydrogen, labeling all areas, points, and lines.

  Solid hydrogen sinks in its own liquid.
  The normal boiling point is 20.4 K.
  The critical temperature is 32 K.
  The critical pressure is 20 atm.
  The triple point is 14 K at 140 torr.

Using the diagram, predict the state of hydrogen under the following conditions:

(a) 25 K and 760. torr

(b) 25 K and 100. torr

(c) 75 K and 30. atm

36. Draw and completely label a cooling curve for any substance listed in Table 8–7. Start at a temperature above the normal boiling point and end at a temperature below the normal freezing point. Indicate all the phase transformations, and distinguish those processes that result in a change in kinetic energy from those that involve a change in potential energy.

37. A rectangular ingot of silver measures 6.0 × 3.0 × 2.0 inches at room temperature (25°C). It is placed in a crucible, and heat is added until it all melts and eventually vaporizes at its normal melting point. What are the heat requirements for the individual steps? If you had a gold ingot instead, would the same quantity of heat vaporize all of it? If not, how much would vaporize? Use the following data.

Silver
density of solid = 10.5 g/cm³
C_p (s) = 0.235 J/g·K C_p (liq) = 0.28 J/g·K
T_{mp} = 961°C = 1234 K T_{bp} = 2212°C = 2485 K
Gold
density of solid = 19.3 g/cm³
C_p (s) = 0.128 J/g·K C_p (liq) = 0.15 J/g·K
T_{mp} = 1063°C = 1336 K T_{bp} = 2660°C = 2933 K

MULTIPLE PRINCIPLES   [38–40]

38. How many grams of sodium peroxide must be used to produce 100. L of oxygen gas at 298 K and 1.0 atm? Assume the oxygen gas that is collected is saturated with water vapor.

$$2 \ Na_2O_2 \ (s) \ + \ 2 \ H_2O \ (liq) \longrightarrow$$
$$4 \ NaOH \ (aq) \ + \ O_2 \ (g)$$

If instead you wanted 100. L of dry oxygen gas, how much sodium peroxide would be required?

39. A 300.-mg sample of an unknown liquid is vaporized in an apparatus that allows the volume of the vapor to be measured by displacement of air, which is collected over water at 20.°C and 750. torr. In this particular experiment, 40.0 cm³ of air was collected. What is the formula weight of the original sample?

40. A 300.-mg sample of CCl₄ was placed in a 2-L bulb-shaped flask fitted with a capillary tip that was open to the air. The flask (except for the capillary tip) was immersed in a water bath at 98°C. The barometric pressure was 748 torr. After all the liquid in the flask had vaporized, the flask was removed from the bath, sealed, and cooled to 25°C. What was the weight of the remaining CCl₄ in the flask?

APPLIED PRINCIPLES   [41–44]

41. Calculate the minimum number of liters of dry air at 19°C and 750. torr needed to strip off 10,000. grams of ethyl alcohol, the solvent used in an industrial synthetic process. Assume that the pressure remains constant and that the process is isothermal, that the air is blown through the solvent in a standing column that is tall enough to saturate the air, and that as the air escapes at the top of the evaporator column it has a pressure of 750. torr. The partial pressure of ethyl alcohol at 19.°C is 40. torr. *Hint:* The minimum volume is for the saturated mixture since anything less than saturation would require more air. The ratio of the number of moles of alcohol to the number of moles of air in the final gaseous mixture is the ratio of their partial pressures. Since we know the number of moles of alcohol (from the problem) and the partial pressures (from the problem and from Dalton's law), this is really a gas law problem.

42. Consider the accompanying diagram for a one-component system and give approximate values for each of the following:
(a) critical temperature
(b) normal boiling point
(c) melting point
(d) equilibrium vapor pressure at room temperature

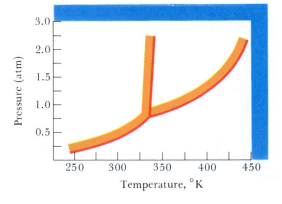

43. Referring to the one-component system described in Problem 42, answer each of the following:
(a) How does the melting point change with changing pressure?
(b) What happens to the substance at 1.5 atm and 350 K when the pressure is increased under isothermal conditions?
(c) What happens to the substance at 0.5 atm and 350 K when the temperature is lowered under isobaric conditions?

44. From the following data, determine the vapor pressure of ether and alcohol at the normal freezing point of water:

Ether						
Temp. (°C)	10	15	20	25	30	35
V.P. (torr)	292	361	442	537	647	776

Alcohol							
Temp. (°C)	20	40	60	70	80	90	100
V.P. (torr)	91.0	213.3	444.3	617.4	836.0	1110	1459

Then, using the data in Fig. 8–4, determine the temperature of a water bath that could supply enough heat to cause ether to boil. What qualifications must be placed on both answers?

CHAPTER
9

Properties of Solutions and the Colloidal State

This plant consists of six distillation process trains designed to permit the recovery of commercial-grade propane and butane. Supporting the plant are facilities for power generation, product storage, seawater desalination to provide potable water, and fire protection. The unique solvent properties of water make it easy to contaminate and difficult to purify. To this point, water pollution has become one of chemistry's benchmark problems. In Faraday's 19th-century London the fabled Thames River was heavily polluted. Here the learned professor presents his card to Father Thames.

297

Jacobus Henricus van't Hoff (Dutch), Nobel Prize for Chemistry in 1901 for the discovery of the laws of chemical dynamics and of osmotic pressure.

9.1   SOLUTES AND SOLUTIONS

Our discussions so far have been limited to the properties of pure substances. The next step is to consider the behavior of simple mixtures of pure substances, especially homogeneous mixtures (solutions). In particular, we are interested in the relationships between the observed properties of solutions and the relative concentrations of their components. It is an important subject for chemists and engineers, and in everyday life.

Carrying out chemical reactions between reactants dissolved in each other or in a common solvent can have the advantages of (1) *accelerating* the reaction by improving contact between molecules, because the process is now homogeneous (rather than heterogeneous), and (2) *moderating* the reaction because the solvent medium acts as a diluent. There can also be negative effects due to the presence of the dissolving medium:

- Its own potential for reactivity.
- The possible introduction of contaminants.
- The difficulty and cost of removing the dissolving medium, and perhaps recovering or disposing of it.

Finally, solutions have a profound effect in our daily lives. Their properties are important in such widely different phenomena as blood dialysis and the salting of ice-covered highways in winter.

On occasion, what appears to be a solution turns out to be a stable suspension of finely divided particles. Such suspensions are as important and ubiquitous as are solutions. Grape jelly and blood are about as different a pair of examples as easily come to mind. In this chapter, we will also introduce some aspects of the properties of such colloidal systems.

Terms and Concepts Defining Solutions

Three situations may result when two substances are mixed that do not chemically react with each other.

1. When sand and salt are mixed, the result is a **heterogeneous mixture**. The individual particles of such a gross dispersion are easily seen and separated by purely mechanical means. Colloids are also heterogeneous mixtures, though not as easily seen or simply separated.
2. **Colloidal dispersions** such as jelly or fog are created by dispersing a liquid in a gas or a solid, or vice versa. The particles tend to be finely divided and not obviously distinguishable, but are still part of a heterogeneous system that can be mechanically separated; the spinning action of a centrifuge will separate milk into clearly defined solid and liquid phases.
3. On dissolving sugar in water, a true **solution** results. The components of a solution cannot be partitioned by mechanical means. The properties of solutions are continuous throughout; they are analytically identical at every point.

When we refer to the components of a solution, the substance that dissolves is the **solute** and the dissolving substance is the **solvent**. In cases where the identification of the solute and the solvent is ambiguous, the choice is usually arbitrary. For example, there is no question which is the solute or the solvent when a teaspoon of sugar is dumped into a glass of

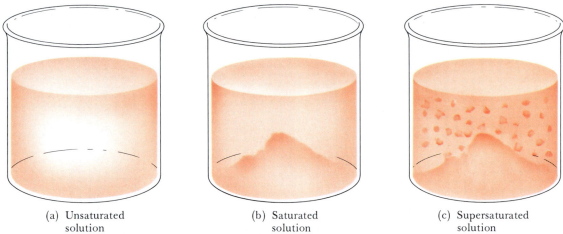

(a) Unsaturated
    solution

(b) Saturated
    solution

(c) Supersaturated
    solution

FIGURE 9–1 An unsaturated solution (a) dissolves added solute until it reaches the equilibrium saturation concentration (b), at the limit of solubility for that solute in the given solvent at a particular temperature. If excess solute is present, dissolving and precipitation occur at the same rate and an equilibrium condition is established between the two processes. A supersaturated solution (c) is unstable and precipitation of excess solute can begin at any time, and once begun, continues until the concentration of the equilibrium saturated solution is reached.

water; sugar is the solute and water is the solvent. But when 2 gallons of ethylene glycol (antifreeze) are mixed into 2 gallons of water in an automobile radiator, it is not as clear-cut. Each dissolves completely in the other.

The Solubility of Substances in Solution

The **concentration** of a solution is the amount of solute per unit volume (or per unit mass) of solvent. At any particular temperature, a solution that has dissolved as much solute as it is capable of dissolving is said to be **saturated**. In a saturated solution, dissolved and undissolved solute are in equilibrium. When the amount of solute contained in a solution is less than the saturation level, the solution is said to be **unsaturated**. If the solution contains more than the saturation concentration, it is **supersaturated**. Supersaturated solutions are unstable; stirring, introduction of more solute, or some other disturbance will cause precipitation of the excess solute, resulting in a saturated solution. To determine which of the three types of solutions you are confronted with, add more solute to the existing solution and observe what happens (Figure 9–1). If the added solute dissolves, the solution was unsaturated; if the added solute will not dissolve, the solution is saturated; if precipitation occurs upon addition of solute, the original solution must have been supersaturated (Figure 9–2).

The **solubility** of any solute in a given solvent is the concentration of the saturated solution at equilibrium (Figure 9–3). To determine the solubility of a substance, all that is required is an analysis of the saturated solution. The simplest way would be to allow the solution to evaporate to dryness, followed by weighing of the residue. For solid solutes, solubility is most often expressed as grams of solute per 100 grams of solvent, at a given temperature. Predicting solubility in advance (or explaining it after the fact) with any degree of certainty is tricky business because the process depends on the properties of the solvent and the solute and on their interaction. Solubility differences make it possible to purify substances by fractional crystalization (Figure 9–4).

FIGURE 9–2 Precipitating the excess of dissolved sodium acetate by pouring a supersaturated solution onto seed crystals.

FIGURE 9–3 A homogeneous solution prepared by dissolving crystals of nickel nitrate, the solute, in solvent water.

Qualitative Consideration of Temperature Effects

As some substances dissolve in water, there is a pronounced cooling effect—the process is endothermic. Many more substances dissolve in water with a pronounced heating effect—the process is exothermic. The source of the cooling and heating can be thought of as the summation of two processes: (1) the energy required to disrupt the **lattice structure**, the arrangement

(a)

(b)

(c)

(d)

(e)

FIGURE 9–4 Potassium dichromate crystals can be purified by fractional crystallization. Here, the solid salt is dissolved in a minimum amount of near-boiling water and then cooled in an ice bath. At temperatures near the freezing point, most of the potassium dichromate "crystallizes" from solution, leaving impurities behind in the solution. Filtration is then used to separate precipitated solid from solution.

of the atoms (or ions) in a crystalline solid; and (2) the **hydration energy**, the energy involved in the association of the resulting fragments with surrounding water molecules.

Separation of the lattice takes place by absorption of sufficient energy to overcome the forces that normally hold the solid together. Hydration is an exothermic process. Qualitatively, if the first step (disruption of the lattice) is dominant, then the overall effect of dissolving a solute in a solvent will be cooling, with heat flowing into the system from the surroundings. Warming the mixture responds to that demand, aiding solution of the solute. On the other hand, if the second step (hydration) is the dominant process, the net effect is the transfer of heat from the system to the surroundings, releasing energy. Warming results in decreased solubility (Figure 9–5).

Many salts absorb heat as they dissolve, and accordingly, their solubilities in water increase with increasing temperature. Other salts show little change in solubility with temperature and exhibit little or no heat of solution. Still others give off considerable heat upon dissolution and show decreased solubility with increased temperatures. We will return to question the basis for these differences in solubility when we discuss the quantitative aspects of ionic equilibrium and sparingly soluble salts in Chapter 12.

Molecular Aspects of Solubility

Our model for explaining gaseous behavior is straightforward. The particles are far apart, to the point where one can assume no intermolecular forces operate between any of them. That is not the case for liquids, however, because as soon as molecules are as close as they must be in the condensed phase, molecular interactions become quite important. The theory of liquids is a complicated business, but qualitatively it can be said that similar substances tend to be soluble in each other ("like dissolves like"). Methyl alcohol (CH_3OH) has a significant degree of water-like OH character. This allows for intermolecular attraction through hydrogen bonding between H_2O and CH_3—OH:

$$CH_3-O-H\text{---}O\begin{matrix} \diagup H \\ \diagdown H \end{matrix}$$

Because of this hydrogen bonding, methanol and water are soluble in each other. Similarly, it is not surprising that ethyl alcohol (CH_3—CH_2—OH) and water are soluble in each other in all proportions. Both methyl alcohol and ethyl alcohol are said to be **miscible** with water; this means that they are mutually soluble in all proportions. By contrast, the oil-like petroleum hydrocarbon known as octane

$$CH_3-CH_2-CH_2-CH_2-CH_2-CH_2-CH_2-CH_3$$

has no OH character and consequently there are no intermolecular, hydrogen bonding forces of attraction to water. As a result, oil and water are predictably insoluble in each other—we say they are **immiscible**. Immiscible liquids form distinct liquid phases, with the denser of the two on the bottom. Should the densities be similar, globules of one liquid may be found suspended throughout the other (Figure 9–6).

Between the two limiting examples (miscibility and immiscibility) lie the majority of molecular solutions. Limited solubility is often the case for solutions of liquids. Octyl alcohol, for example,

"Solvation" is the general term. "Hydration" is used when the solvent is water.

FIGURE 9–5 Calcium acetate is more soluble in cold water than in hot water. On warming a saturated solution of calcium acetate, crystals of the substance precipitate.

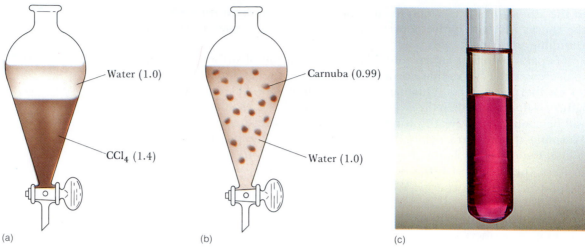

FIGURE 9–6 Immiscible liquids of differing densities. (a) The much denser CCl_4 drawn out of the separatory funnel, effecting a clean physical separation. (b) If the densities happen to be similar, globules of one may be found suspended in the other; it may be physically difficult to separate the two phases. (c) Water and carbon tetrachloride are not miscible and the less dense water layer sits on top of the carbon tetrachloride layer (which has been colored purple by dissolved iodine, which is only sparingly soluble in water).

$$CH_3-CH_2-CH_2-CH_2-CH_2-CH_2-CH_2-CH_2-OH$$

has limited water solubility, reflecting both its oil-like and OH character.

To set this qualitative argument for understanding solubility in molecular solutions on a firmer technical footing, consider dielectric behavior. Water-like liquids have high dielectric constants. The **dielectric constant** (ϵ) is a general property of matter that measures the ability of a substance to decrease the force of attraction (F) between two oppositely charged particles (q_1 and q_2) separated by a distance r along the line joining them. In the inverse-square law describing the electrostatic force of attraction (or repulsion) between charged particles, the dielectric constant shows up in the denominator:

$$F = \frac{q_1 q_2}{\epsilon r^2}$$

Since an ionic crystal (such as sodium chloride and most other inorganic salts) is a regular arrangement of oppositely charged particles, solvents with high dielectric constants effectively lower the forces of attraction that would otherwise exist between these ions. That means easier disintegration of the lattice structure and dissolution of the solid into the solvent. For example, a great deal of sodium chloride can be dissolved by a small amount of a high-dielectric medium such as water ($\epsilon \cong 80$ C^2/N·m^2); its solubility is 36 g NaCl/100 g H$_2$O. Methyl alcohol, a solvent medium with lower dielectric properties ($\epsilon \cong 33$ C^2/N·m^2), dissolves very little sodium chloride (1 g/100 g). On the other hand, sodium chloride is almost completely insoluble in low-dielectric hydrocarbon solvents such as gasoline ($\epsilon \cong 2$ C^2/N·m^2).

Pressure and Temperature Effects on Solubility

Unless gases are involved, the effect of pressure on solubility is of little importance. Beverages such as soft drinks and sodas are "carbonated" by dissolving carbon dioxide gas in water and sealing the container under pressure, or by generating CO_2 in place by fermentation or reaction of carbonates with acid. The carbon dioxide concentration in water depends directly upon the partial pressure of CO_2 above the water surface. When the cap or cork is removed from a bottle of soda or champagne (releasing the pressure), there is an accompanying drop in the solubility of the carbon dioxide in solution. The bubbles of gas flee to the surface of the liquid, where they escape. Aerosol containers of all kinds have been pressured by dinitrogen monoxide (N_2O) and low-boiling Freons, although their general use has lately been restricted and in some cases eliminated because of environmental issues.

It is now forbidden by law to use the "Freons" known as CFCs, the chlorofluorocarbons, as aerosol propellents, because of the adverse effect on the ozone layer. Recall our discussion in Chapter 6.

At a fixed temperature, the solubility of gases in dilute solutions is predicted by **Henry's Law**, which states that the amount of gas dissolved in a given quantity of solvent is proportional to the partial pressure of the gas above the solution:

$$p = KX$$

In this relationship, p is the partial pressure of the gas above the liquid, X is the mole fraction of the dissolved gas in the solution, and K is the constant of proportionality, the Henry's Law constant. In water, the Henry's Law constants for oxygen and carbon dioxide are 3.00×10^7 torr and 1.25×10^6 torr, respectively. The practical consequences of Henry's Law include the concentration of dissolved oxygen available to aquatic species and biodegradation, and corrosion problems. The magnitude of the effect of temperature on gas solubilities is quite pronounced and often a matter of common experience. Open a warm can of soda and note the larger quantity of gas evolved; compare that to opening a cold can of soda.

Illustration of Henry's Law. The greater the partial pressure of CO_2 gas over the liquid in a can of soda, the greater the mole fraction of CO_2 that is dissolved. There is more CO_2 dissolved in the closed can than in the can open to the atmosphere.

EXAMPLE 9–1

The solubility of oxygen in water follows Henry's Law. Under ordinary atmospheric conditions, say 1 atm pressure and 20°C, what is the solubility of oxygen in water?

Solution To begin with, since the air is about 21% oxygen,

$$p(O_2) = (0.21)(1 \text{ atm})(760 \text{ torr/atm}) = 160 \text{ torr}$$

Since the Henry's Law constant is known, the mole fraction of oxygen in solution can be found:

$$X(O_2) = \frac{p(O_2)}{K} = \frac{160 \text{ torr}}{3.00 \times 10^7 \text{ torr}} = 5.3 \times 10^{-6}$$

$$\frac{n(O_2)}{n(O_2) + n(H_2O)} = 5.3 \times 10^{-6}$$

Since solubility is traditionally expressed as g solute per 100 g solvent, choose 100 g H_2O as a convenient quantity:

$$n(H_2O) = 100 \text{ g } (1 \text{ mol}/18.0 \text{ g}) = 5.56 \text{ mol}$$

Making the further assumption that for a dilute solution, the number of moles of oxygen present is small compared to 5.56 moles of water, we may write

$$\frac{n(O_2)}{n(O_2) + n(H_2O)} = \frac{n(O_2)}{5.56} = 5.3 \times 10^{-6}$$

$$n(O_2) = (3.0 \times 10^{-5} \text{ mol})(32.0 \text{ g/mol}) = 9.5 \times 10^{-4} \text{ g}/100 \text{ g } H_2O$$

or approximately 0.001 g O_2/100 g H_2O. ■

EXERCISE 9–1

A gas in contact with water contains 1 mole percent carbon dioxide. The temperature is 20°C and the total gas pressure is 2.0 atm. What are the mole fraction and solubility of carbon dioxide in water under these conditions?

Answer: $X = 1.2 \times 10^{-5}$ atm; 0.003 g/100 g H_2O. ∎

THREE LANDMARK DISCOVERIES ON THE BEHAVIOR OF SOLUTIONS

François-Marie Raoult established the existence of a relationship between the formula weight of a dissolved substance and the vapor pressure of a dilute solution of the dissolved substance. A century ago, the accurate determination of formula weights by simple experimental measurements was a major concern for chemists, and to be able to provide such data with little more than a barometer and a good thermometer was significant. The essential thread in this fabric was the idea that certain characteristic physical properties of solutions depend on the choice of solvent but not on the choice of solute. Raoult was professor at the French university of Grenoble, and although a number of scientists of the day had studied the effects of solutes on solutions, his was the definitive work.

Jacobus Henricus van't Hoff noted that electrolytes—solutions that conduct an electric current—exhibited anomalous results, consistently deviating from the behavior predicted by Raoult and de Coppett's relationship for solutions. Salt solutions, for example, produced nearly twice as large a colligative effect as sugar solutions, behaving as though there were twice as many moles of particles present in solution. This Dutch physical chemist was the first Nobel laureate in chemistry.

Svante Arrhenius suggested that the marked deviation van't Hoff observed in the behavior of electrolyte solutions such as sodium chloride could be explained by assuming that the salt split into ions—sodium ions and chloride ions. Thus, there would be twice as many particles and twice the effect in dilute solutions. Arrhenius was the first to postulate that electrolyte solutions result from a separation of neutral species into positively and negatively charged atoms called ions. First presented as part of his Ph.D. thesis, these ideas were in such conflict with prevailing theories that he received the lowest possible passing grade. Years later, so much hostility remained that Arrhenius was almost prevented from being appointed professor at Stockholm. But it was for this very work that he was awarded the Nobel prize for chemistry in 1903.

van't Hoff in 1901, the year he won the Nobel Prize.

9.2 COLLIGATIVE PROPERTIES OF SOLUTIONS

An ideal solution is a homogeneous mixture of two or more components that exhibit the same forces of attraction for each other as for themselves, and that mix without change in volume or heat content. Our discussions of liquid vapor pressures in Chapter 8 were based on a simple line of reasoning derived from such an ideal model for behavior. We continue this line of reasoning based on ideal behavior as we extend our discussion of solutions to include their colligative properties—the properties of solutions that acccount for such widely different phenomena as melting of ice on highways in winter by salt and the passage of nutrients through cell membranes in plants. **Colligative properties** depend only on the number of solute particles present in solution—the number of moles of particles—without regard for the nature of the particles.

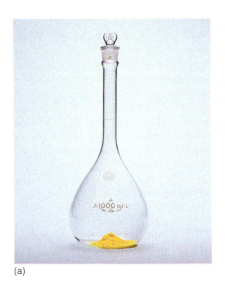
(a)

(b)

(c)

FIGURE 9–7 To prepare a liter of 0.1000 M potassium chromate solution, add 19.4 g of $K_2Cr_2O_4$ to a 1000-mL volumetric flask, add enough water with shaking to facilitate dissolution, and then dilute with solvent water to the mark on the neck of the flask. Note the important difference if you were making up molal, rather than molar, solutions—you would dissolve the solute in 1000 g of solvent, not just dilute to a final volume of 1000 mL.

Although there are solutions made up of many components, in the discussions that follow we will deal mostly with those containing only two components. The properties of such *binary* solutions depend upon the **concentration** of the solution, the relative amounts of solute and solvent present. There are many ways of expressing concentration, but the three most common that will be used in the rest of this book are:

- **mole fraction** (defined in Chapter 5), the ratio of the number of moles of solute to the total number of moles present. Mole fraction is a unitless quantity, represented as **X**.

- **molality**, the number of moles of solute per kilogram of solvent. The units of mol/kg are also represented as **m**.

- **molarity** (which will be discussed further in Chapter 12), the number of moles of solute per liter of solution. The units of mol/L are also represented as **M** (Figure 9–7).

To calculate the molality of a solution prepared by dissolving 46.0 g of ethyl alcohol (C_2H_5OH) in 250 g of water, we first determine the number of moles of solute (alcohol):

$$\frac{46.0 \text{ g } C_2H_5OH}{46.0 \text{ g } C_2H_5OH/\text{mol } C_2H_5OH} = 1.00 \text{ mol } C_2H_5OH$$

The number of kilograms of solvent (water) is:

$$(250 \text{ g } H_2O)(1 \text{ kg}/1000 \text{ g}) = 0.250 \text{ kg } H_2O$$

Thus the molality is

$$m = \frac{1.00 \text{ mol } C_2H_5OH}{0.250 \text{ kg } H_2O} = 4.0 \text{ mol } C_2H_5OH/\text{kg } H_2O$$

EXAMPLE 9–2

Calculate the number of grams of ethyl alcohol that must be added to 50.0 g of water to prepare a 0.15 m solution.

Solution By definition, such a solution contains 0.15 mol of C_2H_5OH per kg of H_2O. For only 50.0 g of H_2O, or

$$(50.0 \text{ g } H_2O) \left(\frac{1 \text{ kg}}{1000 \text{ g}} \right) = 0.050 \text{ kg } H_2O$$

the number of moles of alcohol required is

$$\left(\frac{0.15 \text{ mol } C_2H_5OH}{\text{kg } H_2O} \right) (0.050 \text{ kg } H_2O) = 0.0075 \text{ mol } C_2H_5OH$$

Since the formula weight of ethyl alcohol is 46.0 g/mol, the required number of grams of alcohol is

$$(0.0075 \text{ mol } C_2H_5OH) \left(\frac{46.0 \text{ g } C_2H_5OH}{\text{mol } C_2H_5OH} \right) = 0.34 \text{ g } C_2H_5OH \quad \blacksquare$$

EXERCISE 9–2
Calculate the mole fraction of acetone (C_3H_6O) in a solution composed of 112.0 g of water and 32.2 g of acetone. What is the molality of the solution? *Answer:* $X_{\text{acetone}} = 0.0818$; molality = 4.95 m. $\blacksquare$

9.3   LOWERING OF THE VAPOR PRESSURE IN DILUTE SOLUTIONS

It has been known for a long time that the vapor pressure of a solution is less than that of the pure solvent because of the presence of the dissolved solute particles. The vapor pressure lowering is usually small, and may go unnoticed. However, if we place solvent and solution in separate beakers in a closed container as shown in Figure 9–8, eventually all the solvent will evaporate from its beaker and move into the solution. In the process, the solution becomes more dilute.

Although a number of scientists of the day had studied the effects of solutes on the properties of solutions, François-Marie Raoult did the definitive work in 1887.

The quantitative statement expressing the relationship between vapor pressure lowering and the concentration of a solution is known as **Raoult's Law**. It is analogous to the behavior of an ideal mixture of gases, in which total pressure is equal to the sum of the partial pressures. In an ideal solution, the escaping tendency of the molecules is the same, whether they are surrounded by similar or dissimilar molecules. Thus, Raoult's Law, in its usual form, is given by Eq. (9–1):

$$p = p°X \tag{9–1}$$

where $p°$ is the vapor pressure of the pure solvent and X is the mole fraction of the solvent in the solution. The total pressure P_T is given by

$$P_T = p_A + p_B = p_A° X_A + p_B° X_B$$

FIGURE 9–8 Demonstration of the effect of lowering of the vapor pressure. With solvent and solution in separate beakers in a closed container, solvent distills into the solution, which becomes more dilute.

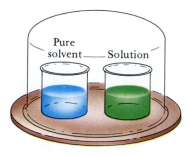
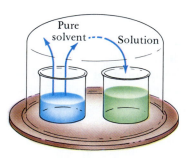

Now consider the special case of a typical automotive antifreeze solution composed of a volatile solvent A such as water and a nonvolatile solute B such as ethylene glycol [$(CH_2OH)_2$]. The vapor pressure of the solution is proportional to the mole fraction of the solvent in the solution, and the constant of proportionality is the vapor pressure of the pure solvent itself:

$$p_A = p_A^\circ X_A \tag{9-2}$$

Equation (9–2), which is a statement of Raoult's Law, is plotted on the graph in Figure 9–9. Note that for pure solvent,

$$X_A = \frac{n_A}{n_A + n_B} = \frac{n_A}{n_A} = 1$$

and the vapor pressure is, as it must be, equal to the vapor pressure of the pure solvent:

$$p \text{ (solvent)} = p_A^\circ X_A = p_A^\circ(1) = p_A^\circ$$

For the pure solute, p_B is zero, since B is nonvolatile. The curve in Figure 9–9 is linear between these two extremes.

The lowering of the vapor pressure, Δp, is the difference between the vapor pressure of the pure solvent p° and p(solution), referred to simply as p. Since p° is a larger number than p, let's define Δp as $p^\circ - p$ and obtain that value from Raoult's Law by subtracting both sides of Eq. (9–2) from p_A°:

$$p_A = p_A^\circ X_A$$
$$p_A^\circ - p_A = p_A^\circ - p_A^\circ X_A = p_A^\circ(1 - X_A)$$

Since $X_A + X_B = 1$, we know that $1 - X_A = X_B$ and therefore

$$p_A^\circ - p_A = p_A^\circ X_B$$
$$\Delta p = p_A^\circ X_B \tag{9-3}$$

which is a consequence of Raoult's Law. This law states that the lowering of the vapor pressure of the solvent in a solution is proportional to the mole fraction of the solute. Therefore, if the solvent is water and the solute is ethylene glycol, it is the mole fraction of ethylene glycol present that determines the vapor pressure of the ethylene glycol solution.

By modifying the mathematical statement of Raoult's Law in Eq. (9–3), it is possible to determine the molecular weight of the solute. If the solution is composed of g_A grams of solvent of molecular weight M_A and g_B grams of solute of molecular weight M_B, then the number of moles of each component can be given as follows:

$$n_A = \frac{g_A}{M_A} \qquad n_B = \frac{g_B}{M_B}$$

Since the molecular weight of the solvent will generally be known (water, for example), the molecular weight of a soluble, nonvolatile solute (ethylene glycol in water, for example) can be determined in a straightforward manner: simply measure Δp, the lowering of the vapor pressure, accurately.

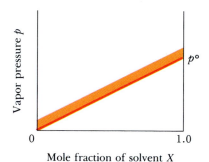

FIGURE 9–9 Raoult's law states that $p = p^\circ X$. Pure solvent is found where $X = 1.0$.

EXAMPLE 9–3

The vapor pressure of water at 25°C is 23.76 torr. After adding 25.0 g of an unknown compound to 200.0 g of water, the vapor pressure was measured to be 23.42 torr.

Find the formula weight of the unknown compound, assuming it is nonvolatile.

Solution

A = solvent water, H_2O; B = unknown, solute

$$\Delta p = p_A^\circ - p_A = (23.76 - 23.42) \text{ torr} = 0.34 \text{ torr}$$

$$\frac{\Delta p}{p_A^\circ} = X_B = \frac{0.34 \text{ torr}}{23.76 \text{ torr}} = 0.014$$

$$n_A = \frac{200.0 \text{ g}}{18.0 \text{ g/mol}} = 11.1 \text{ mol}$$

$$X_B = \frac{n_B}{n_A + n_B}$$

$$n_B = \frac{X_B n_A}{1 - X_B} = \frac{(0.014)(11.1 \text{ mol})}{1 - 0.014} = 0.16 \text{ mol}$$

$$M_B = \frac{25.0 \text{ g}}{0.16 \text{ mol}} = 1.6 \times 10^2 \text{ g/mol}$$

■

EXERCISE 9–3

When 18.0 g of a certain unknown sugar is dissolved in 100 g of water, the known vapor pressure of pure water at 20°C (17.54 torr) is lowered to 17.23 torr. What is the molecular weight of the unknown sugar?
Answer: 1.8×10^2 g/mol. ■

9.4   BOILING POINT ELEVATION

One of the consequences of lowering the vapor pressure of a solvent by adding a nonvolatile solute is that the boiling point of the solution is higher than that of the pure solvent. In other words, the temperature at which the vapor pressure of a solution equals the pressure of the atmosphere will be affected by the presence of solute molecules in the solvent. Consider the graph in Figure 9–10, showing the effect of temperature on the vapor pressure of a pure solvent and a solution. The boiling point of the pure solvent is T_0, the boiling point of the solution is T, and the elevation of the boiling point is $T - T_0$, the higher temperature minus the lower:

FIGURE 9–10   The elevation of the boiling point. A plot of temperature versus vapor pressure for pure solvent and solution.

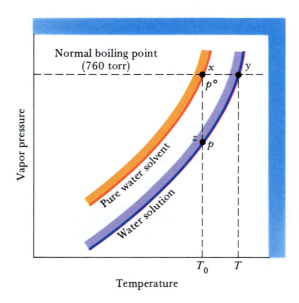

$$T - T_0 = \Delta T_b$$

The vapor pressure of the pure solvent at temperature T_0 is p_A° at point x; the vapor pressure for the solution at the same temperature T_0 is p_A at point z. As a result, the lowering of the vapor pressure $p_A^\circ - p_A$ is the line segment xz.

In dilute solutions, it can be assumed that the two vapor pressure curves parallel each other and are essentially linear. Therefore, xz is proportional to the distance xy. To state it another way, the lowering of the vapor pressure is proportional to the elevation of the boiling point:

$$p_A^\circ - p_A \propto T - T_0 \propto \Delta T_b$$

According to Raoult's Law, the relative lowering of the vapor pressure is proportional to the mole fraction of solute, and therefore

$$p_A^\circ - p_A \propto T - T_0 \propto \frac{n_B}{n_A + n_B}$$

Converting the proportionality into an equality by introducing a proportionality constant k:

$$\Delta T = k \frac{n_B}{n_A + n_B}$$

In dilute solutions, the number of moles of solute n_B is small compared to the number of moles of solvent n_A. Therefore $n_A + n_B$ in the denominator will be approximately equal to n_A. On that assumption,

$$\frac{n_B}{n_A + n_B} = \frac{n_B}{n_A}$$

and

$$\Delta T_b = k \frac{n_B}{n_A} \tag{9-4}$$

It is convenient to relate ΔT_b to the molality m of the solution, the number of moles of solute per kilogram of solvent:

$$m = \frac{n_B}{kg_A} = 1000 \frac{n_B}{g_A} \tag{9-5}$$

For the solvent, we can write

$$n_A = \frac{g_A}{M_A} \tag{9-6}$$

Solving Eq. (9-6) for g_A and substituting into Eq. (9-5) gives

$$m = 1000 \frac{n_B}{n_A M_A}$$

which can be solved for the mole ratio n_B/n_A. Substituting back into Eq. (9-4) finally gives

$$\Delta T_b = k \frac{n_B}{n_A} = k \frac{m M_A}{1000}$$
$$\Delta T_b = K_b m \tag{9-7}$$

where K_b is the **molal boiling point elevation constant**:

$$K_b = k \frac{M_A}{1000}$$

TABLE 9–1 Molal Boiling Point Elevation Constants	
Solvent	K_b(°C/m)
water	0.52
ethyl alcohol	1.20
benzene	2.67
acetic acid	2.93
chloroform	3.85
carbon tetrachloride	5.02

The constant K_b is a characteristic physical property of the solvent (Table 9–1) that is independent of the solute. In the case of water, for example, the molal boiling point elevation constant is 0.52°C/m. That means that dissolution of 1 mol of any substance (as long as it remains undissociated) in 1000 g of water will bring about an 0.52°C elevation of the boiling point; 0.10 mol would produce one-tenth the effect. Actually, 1 m solutions often prove to be too concentrated for our dilute solution approximation to hold, and poor results are obtained on comparing experiment with theory; 0.01 m gives better results, and 0.001 m, better still.

In the following discussion, the boiling point elevation constant K_b for benzene, which is 2.67°C/m, will be used to determine the formula weight of sulfur. One mole of a nonvolatile and undissociated solute dissolved in 1000 g of a benzene will result in an elevation of the boiling point from the normal value of 80.10°C to 82.77°C. When 1.2 g of elemental sulfur is dissolved in 50.0 g of benzene, the boiling point is elevated to 80.36°C. To calculate the apparent molecular weight of the dissolved sulfur, proceed as follows. On a molal basis, by straight proportion, 1.2 g solute in 50 g solvent is equivalent to 24 g solute in 1000 g solvent. That amount of solute raises the boiling point by 0.26°C, or 0.097 times the molal boiling point elevation constant for the solvent. Therefore, 24 g solute must be about one-tenth mole, or about one-tenth the formula weight. The apparent formula weight of sulfur must be

$$\frac{24 \text{ g}}{0.097 \text{ mol}} = 2.5 \times 10^2 \text{ g/mol}$$

Rhombic sulfur, the stable (up to 95°C) crystalline modification, forms puckered, eight-atom rings. Hence, the experimental result is a molecular weight of 2.5×10^2 g/mol, or about eight times the atomic weight of 32.

 EXAMPLE 9–4

Calculate the molecular weight of an unknown compound that causes an elevation in the boiling point of carbon tetrachloride of 0.055°C when 0.100 g is dissolved in 50.0 g of solvent.

Solution   We begin by finding the molality of the resulting solution, using the proper K_b value from Table 9–1:

$$m = \frac{\Delta T_b}{K_b} = \frac{0.055°C}{5.02°C/m} = 0.0110 \text{ mol/kg}$$

To find the number of moles,

$$(0.0110 \text{ mol/kg})(0.0500 \text{ kg}) = 5.5 \times 10^{-4} \text{ mol}$$

Finally, to find the molecular weight,

$$\frac{0.100 \text{ g}}{5.50 \times 10^{-4} \text{ mol}} = 182 \text{ g/mol}$$

EXERCISE 9–4

Calculate the experimental K_b value for water from the following data. A urea solution is prepared by mixing 1.800 g of urea [$(NH_2)_2CO$] and 90.0 g of water. Its boiling point was recorded as 100.170°C (at 1 atm). *Answer:* 0.510°C/m. ■

9.5 FREEZING POINT DEPRESSION

A further consequence of the lowering of the vapor pressure of the solvent in a solution is that the freezing point of the solution will be lower than that of the pure solvent. Sprinkling salt on ice-covered roads in winter is a typical illustration; ethylene glycol is widely used to prevent auto radiators from freezing. The freezing point of pure water is depressed by the presence of a dissolved substance. For an ideal solution, the presence of 1 mol of a nonvolatile solute such as urea in 1 kg of water causes the resulting solution to freeze at $-1.86°C$, rather than at the normal freezing point of water, 0°C. Using a similar line of reasoning (Figure 9–11) to that leading to Eq. (9–7) for the elevation of the boiling point leads to a parallel expression for ΔT_f, the depression of the freezing point:

$$\Delta T_f = K_f m \tag{9-8}$$

where K_f is the **molal freezing point depression constant**. This constant (Table 9–2) is characteristic of the solvent and independent of the nature of the solute present—it is a typical colligative property, depending only on the number of particles present. Ideally, the freezing point of pure solvent water is depressed by 1.86°C when 1 mol of an undissociated solute is dissolved in 1000 g of it. One-tenth mole of solute lowers the normal freezing point by one-tenth that value, or 0.186°C.

The effect of a solute on the freezing point of a solvent can be used for the determination of molecular weights. The principle is analogous to

Adding "Prestone," a commercial-grade ethylene glycol, to an automobile radiator. The otherwise colorless glycol is made green and translucent by small amounts of corrosion inhibitors and antioxidants.

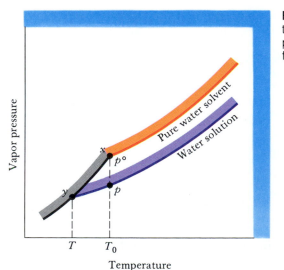

FIGURE 9–11 The lowering of the freezing point. A plot of temperature versus vapor pressure for pure solvent and solution.

TABLE 9–2	**Molal Freezing Point Depression Constants**
Solvent	K_f(°C/m)
water	-1.86
acetic acid	-3.90
chloroform	-4.68
benzene	-5.12
naphthalene	-7.00
camphor	-40.0

the use of boiling point elevation for that purpose. As each solvent has its own characteristic molal freezing point depression constant, the amount that a given number of grams of solute depresses the freezing point will be some percentage of the molal freezing point constant and consequently, some percentage of the molecular weight of the solute.

EXAMPLE 9–5

The molal freezing point depression constant for naphthalene is $-7.0°C/m$. In other words, 1.0 mol of a solute dissolved in 1000 g of naphthalene produces a change in the freezing point from the expected value of 80.2°C to 73.2°C. Determine the apparent molecular weight of elemental sulfur if 6.0 g was dissolved in 250 g of naphthalene and the freezing point was observed to be 79.5°C.

Solution   Changing the concentrations to a molal basis, 6.0 g in 250 g is equivalent to 24 g sulfur in 1000 g naphthalene. That amount lowers the freezing point by 0.7°C, or 10% of the molal freezing point depression. Therefore, 24 g must be 10% of the formula weight, and the apparent formula weight must be 2.4×10^2 g/mol. This result compares favorably with the value calculated earlier from the elevation of the boiling point.

EXERCISE 9–5

A sample of 1.820 g of carbon tetrachloride is mixed with 100.0 g of benzene, and the freezing point of the resulting solution is measured. As a result, the normal freezing point of benzene is depressed by 0.605°C. Calculate the experimental value for the molecular weight of carbon tetrachloride.
Answer: 154 g/mol. ∎

EXAMPLE 9–6

When 0.0250 g of an unknown compound with empirical formula CH_2 is dissolved in 5.00 g of benzene, the freezing point of solvent benzene is depressed from 5.510°C to 5.280°C. Determine the molecular formula of the compound.

Solution   From Eq. (9–8) for the lowering of the freezing point, both the molality of the solution and the molecular weight can be determined:

$$\Delta T_f = (5.280 - 5.510)°C = -0.230°C$$

$$m = \frac{\Delta T_f}{K_f} = \frac{-0.230°C}{-5.12°C/m} = 0.0449 \text{ mol/kg}$$

$$(0.0449 \text{ mol/kg})(0.00500 \text{ kg}) = 2.24 \times 10^{-4} \text{ mol}$$

$$\frac{0.0250 \text{ g}}{2.24 \times 10^{-4} \text{ mol}} = 112 \text{ g/mol}$$

Since the formula weight of CH_2 is 14.0 g/mol,

$$\frac{112 \text{ g/mol}}{14.0 \text{ g/mol}} = 8$$

Thus, the molecular formula must contain 8 empirical formulas, or

$$(CH_2)_8 = C_8H_{16}$$

EXERCISE 9–6(A)
When 1.00 g of elemental phosphorus was carefully dissolved in 100 g of benzene, the freezing point of the solution was observed to be 5.09°C. If K_f for benzene is -5.12°C/m and the freezing point (at 1 atm) is 5.50°C, what is the apparent molecular weight of elemental phosphorus? Based on the atomic weight (obtained from the table of atomic weights), what is the apparent molecular formula for phosphorus?
Answer: 1.2×10^2 g/mol; P_4.   ■

EXERCISE 9–6(B)
For the solution resulting from dissolving 0.32 g of naphthalene ($C_{10}H_8$) in 25 g of benzene (C_6H_6) at a temperature of 26.1°C, calculate the vapor pressure lowering, the boiling point elevation, and the freezing point depression. The vapor pressure of benzene at the temperature of the experiment is 100 torr.
Answer: 0.78 torr; 0.27°C; 0.51°C.   ■

9.6   OSMOSIS AND OSMOTIC PRESSURE

Osmotic pressure commands special attention because of its importance in fluid transport in all living systems throughout the plant and animal kingdom. In 1748, the Abbe Nollet described some experiments in which pure water and aqueous solutions were separated from each other by animal membranes. He observed that the pure solvent moved through the membrane and entered the solution, which became progressively more dilute, but never the other way around. The Abbe described one case in which wine was placed in the open end of a porous cylinder sealed with an animal bladder and submerged in pure water. The animal bladder selectively allowed water molecules to move into the alcohol solution in the cylinder, but did not allow alcohol molecules to move in the opposite direction. We say that such a membrane is **semipermeable**. As the process continued, the bladder gradually swelled and eventually burst. The increased pressure in the cylinder, caused by water diffusing through the bladder into the solution, we now call **osmotic pressure**. The process is called **osmosis**.

If a glass tube, sealed at one end by a semipermeable membrane and partially filled with a salt solution (Figure 9–12), is immersed (membrane

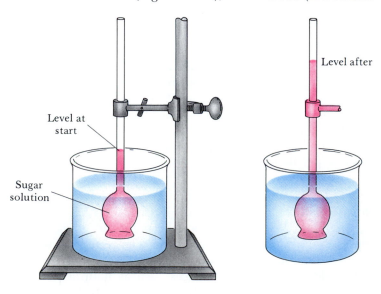

Level at start

Sugar solution

Level after

FIGURE 9–12 The process of osmosis brings the level of the solution above the level of the surrounding solvent.

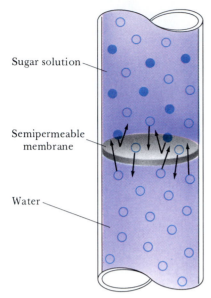

FIGURE 9–13 The cylinder is separated into two chambers by a semipermeable membrane which allows free passage of water molecules in both directions. The rate of transfer of solvent differs between solvent on one side and solution on the other, or between solutions of differing concentrations.

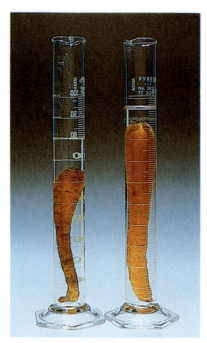

After some hours in a concentrated NaCl solution, the carrot on the left shows the effects of osmosis. Water flowed out of the carrot into the NaCl solution, leaving the carrot limp. The slightly swollen carrot on the right is in pure water.

down) in pure water so the levels of the salt solution in the tube and the water outside are equal, the level of the solution will eventually rise above the level of the surrounding solvent. The difference between the levels represents a hydrostatic pressure, brought about by the process of osmosis causing water to pass from solvent to solution. The hydrostatic pressure is just balanced by the osmotic pressure of the solution when the levels come to rest.

As a model, consider a uniform cylinder, open at the top and bottom and separated into two chambers by a semipermeable membrane. It is placed in a beaker of water as in Figure 9–13, and a glucose solution fills the top chamber. Because it is semipermeable, the membrane does not allow glucose molecules to pass through. However, the membrane permits water molecules to penetrate it freely in both directions, at a speed determined by the number of water molecules at its surface. Since there are more water molecules at a given surface in pure water than in an aqueous solution, there will be a greater tendency for water to flow from the pure solvent to the solution than in the reverse diection. Thus, the same mechanism that causes the lowering of the vapor pressure due to the presence of the solute is the driving force for the transfer of solvent in osmosis.

The net result is that water passes through the membrane, from solvent to solution, and the level of the glucose solution rises until the hydrostatic pressure equals the osmotic pressure, the force due to the difference in concentration of solvent molecules between the two sides of the membrane barrier. At that point, a state of equilibrium exists, with water moving equally rapidly in both directions. The osmotic pressure π is the excess pressure (the hydrostatic pressure in our example) necessary to equalize the rates of passage of solvent molecules in both directions through the barrier. It is the pressure required to establish osmotic equilibrium.

In 1885, van't Hoff noted that for dilute solutions the osmotic pressure π obeys the relationship

$$\pi = \frac{nRT}{V}$$

where n = moles of solute, R = universal gas constant, T = absolute temperature, and V = volume in liters of solution. Recognizing that n/V represents the molar concentration of the solute, the previous equation can be rewritten as

$$\pi = cRT \tag{9–9}$$

where $c = n/V$. This leads to an interesting and useful conclusion, namely that molecular weights can be calculated from yet another colligative property of solutions—osmotic pressure. Note that the units of concentration used here are moles of solute per *liter* of solution. This is called **molarity**, or molar concentration, and it is clearly and importantly different from molality, which is the number of moles of solute per *kilogram* of solvent. The symbol for molarity is M. Thus, a 0.15 M solution contains 0.15 mol of solute per liter of solution:

0.15 M = 0.15 mol/L

Molarity will be discussed further in Chapter 12.

In marked contrast to boiling point elevation and freezing point depression, osmotic pressures are very large for dilute solutions. Consequently, osmotic pressure measurements are useful for determining very high molecular weight substances such as natural and synthetic macromolecules—proteins and plastics, for example. From osmotic pressure measurements, the number average of the particles per unit volume of solution can be determined as well as actual molecular weights.

EXAMPLE 9–7

When 5.00 g of a compound is dissolved in sufficient water to make 2.00 L of solution at 298 K, the measured osmotic pressure (for the solution) was 0.220 atm. Determine the formula weight of the compound.

Solution   Find the concentration c in mol/L and proceed from there:

$$\pi = cRT$$

$$c = \frac{\pi}{RT} = \frac{0.220 \text{ atm}}{(0.0821 \text{ L·atm/mol·K})(298 \text{ K})} = 8.99 \times 10^{-3} \text{ mol/L}$$

$$(8.99 \times 10^{-3} \text{ mol/L})(2.00 \text{ L}) = 0.0180 \text{ mol}$$

$$\frac{5.00 \text{ g}}{0.0180 \text{ mol}} = 278 \text{ g/mol} \qquad ■$$

EXERCISE 9–7(A)

Calculate the osmotic pressure of a 0.0734 M glucose solution ($C_6H_{12}O_6$) at 27°C.
Answer: 1.81 atm.   ■

EXERCISE 9–7(B)

When 4.0 g of a PVC (polyvinyl chloride) polymer was dissolved in 1 L of dioxane, the osmotic pressure at 25°C was measured to be 0.494 torr. Calculate the molecular weight of the polymer.
Answer: 1.5×10^5 g/mol.   ■

9.7   MEMBRANES AND PERMEABILITY

The physical phenomenon of osmosis depends on the presence of an imperfect barrier between a solvent and a solution or between two solutions containing different amounts of the same solute. Membranes may be of animal or vegetable matter or synthetic material, and they can vary in their permeability toward a particular molecule. Different membranes allow rapid transport, or slow transport, or do not allow transport of a particular solute molecule at all. For example, cellophane and polyethylene are excellent commercial packaging materials because of their particularly low permeability toward water and oxygen. Parchment and collodion are also commonly used membrane materials.

The phenomenon that we refer to as **permeation** is a two-step process. First, the molecules present in the fluid (gas or liquid) dissolve in the membrane until an equilibrium saturation of the membrane has been attained. Long before the actual achievement of that stage, however, transport through the membrane begins. This movement through the membrane is due to the pressure difference between the fluids on the two sides of the membrane (Figure 9–14).

The filtration and penetration phenomena based on the variable permeability of membranes is the basis for some of the most important industrial processes and biomedical applications. These phenomena show up in food, chemical, and metallurgical applications; in the processing of pulp and paper, in water desalination, in the solving of pollution problems,

FIGURE 9–14 Section across a semipermeable membrane.

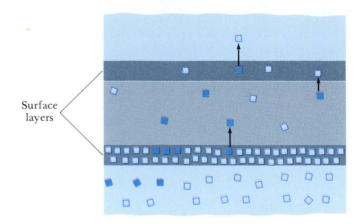

Surface layers

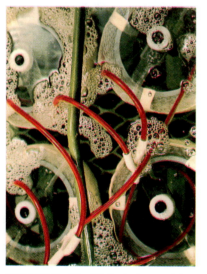

FIGURE 9–15 Dialysis is used to treat patients suffering from kidney failure. In this artificial kidney, blood from the patient is circulated through cellophane tubes that act as semipermeable membranes. Waste products dialyze through the membrane, thus purifying the blood.

and in sterilization, pasteurization, purification, and a host of other fields. In the medical field, blood dialysis and artificial kidney machines (Figure 9–15) depend upon membrane technology, as do heart-lung machines and some prosthetic devices such as heart valves. Although much improved by recent engineering advances, machine dialysis is expensive, clumsy, and inconvenient, and can lead to serious side effects such as osteoporosis.

9.8 COLLIGATIVE PROPERTIES OF DILUTE ELECTROLYTES

Van't Hoff's studies revealed that for dilute solutions containing 1 mol of a nonelectrolyte in V liters of solution, osmotic pressure is given by the equation $\pi = RT/V$. It was van't Hoff's further experience that the osmotic pressure of an electrically conducting (electrolyte) solution was always higher than his equation predicted. He suggested a more general equation:

$$\pi = \frac{iRT}{V}$$

The factor i was always greater than 1 for electrolytes, and changed with concentration.

The basic premise from the beginning of this discussion has been that colligative properties of solutions depend on the number and not the kind of particles present. If a dissolved substance dissociates into separated ions, and if each ion acts in the same manner as an un-ionized molecule with respect to the colligative properties, then the increased colligative effects observed by van't Hoff could be explained. An electrolyte in water ($K_f = -1.86°C/m$) could be expected to yield twice the freezing point depression of a nonelectrolyte if every molecule dissociated into two ions. A mole of molecules that dissociated into three moles of completely separate ions should depress the freezing point by 5.58°C, a factor of three compared to a nonelectrolyte:

$$(-1.86°C/m)(3\ m) = -5.58°C$$

Van't Hoff's experimental i factor represents colligative properties for electrolytes in terms of the behavior of nonelectrolytes. The i factor is defined as the ratio of the colligative effect produced by a concentration m of an electrolyte to that effect due to the same concentration of a nonelectrolyte:

TABLE 9–3 Dissociation of Some Typical Electrolytes in Water*
hydrochloric acid: $\quad$ HCl (aq) + H_2O (liq) $\longrightarrow$ H_3O^+ (aq) + Cl^- (aq)
sulfuric acid: $\qquad$ H_2SO_4 (aq) + 2 H_2O (liq) $\longrightarrow$ 2 H_3O^+ (aq) + SO_4^{2-} (aq)
sodium chloride: $\quad$ NaCl (aq) $\longrightarrow$ Na^+ (aq) + Cl^- (aq)
potassium sulfate: $\;$ K_2SO_4 (aq) $\longrightarrow$ 2 K^+ (aq) + SO_4^{2-} (aq)
*The protons are written as hydrated protons: $H^+(H_2O)$ = H_3O^+ = hydronium ions. More will be said about that in our discussions of ionic equilibrium in Chapters 12 and 13.

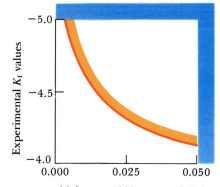

FIGURE 9–16 K_f values for dilute sulfuric acid solutions approach the limiting value of $-5.58°C/m$.

$$i = \frac{\Delta T_f(\text{electrolyte})}{\Delta T_f(\text{nonelectrolyte})}$$

$$i = \frac{\Delta T_f(\text{electrolyte})}{K_f m(\text{nonelectrolyte of same concentration})}$$

For example, the complete dissociation of sulfuric acid in water generates three fragments, the two (hydrated) protons and the sulfate ion (Table 9–3). Consequently, the observed freezing point depression should be three times the expected value (for an ideal solution). Note the graph in Figure 9–16. As sulfuric acid solutions become more dilute, the degree of dissociation approaches 100% and K_f approaches $-5.58°C/m$.

Figure 9–17 shows i values for different concentrations of several electrolytes and nonelectrolytes. As solutions become more dilute, i values for nonelectrolytes approach the limiting value of 1. For solutions of NaCl and $CaCl_2$, i values approach 2 and 3, respectively, indicating the presence of 2 and 3 times as many particles as in nonelectrolyte solutions at the same concentrations.

Colligative properties of solutions can be used to determine the degree of dissociation. Aqueous solutions of most strong acids, strong bases, and salts are fully dissociated. Weak acids and weak bases, and a few salts are not fully dissociated and are classified as weak electrolytes (see Chapter 12).

EXAMPLE 9–8

Predict the limiting i value (in dilute solutions) for $FeCl_3$.

Solution $FeCl_3$ would form an Fe^{3+} ion plus three Cl^- ions. The total of four ions would lead to a limiting value of 4. ■

EXERCISE 9–8

Predict the limiting i values (in dilute solutions) for $MnSO_4$, for $K_2S_2O_8$, and for NH_3.
Answer: 2; 3; 1. ■

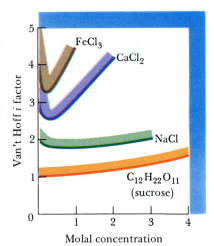

FIGURE 9–17 Experimental i-values for several solutes as a function of molal concentration.

9.9 VOLATILE COMPONENTS, IDEALITY, AND DISTILLATION

Raoult's Law and Dalton's Law

For two liquids that are volatile and mutually soluble in all proportions, one need not make the distinction between solute and solvent. As an example, consider the case of alcohol and water. In principle, if alcohol and water behave ideally, the presence of alcohol molecules should have no effect on the forces among water molecules, and likewise, water molecules should in no way affect alcohol molecules. One would therefore expect no heat effect or volume change when alcohol and water are mixed. That is not strictly the case, but let's assume for the sake of illustration that both

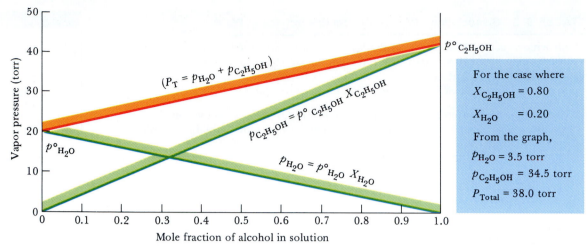

FIGURE 9–18 Vapor pressure diagram for an ideal binary solution of two miscible liquids.

constituents obey Raoult's Law at all temperatures and over the entire range of concentrations from pure alcohol to pure water. Let A represent ethyl alcohol (C_2H_5OH) and B represent water (H_2O):

For pure alcohol $X_A = 1$ and $X_B = 0$

For pure water $X_A = 0$ and $X_B = 1$

According to Dalton's Law of Partial Pressures, the total pressure exerted by the system equals the sum of the partial pressures due to each component:

$$p_T = p_A + p_B$$

According to Raoult's Law

$$p_A = p_A^{\circ} X_A$$
$$p_B = p_B^{\circ} X_B$$

Substituting these results into the equation for total pressure gives

$$P_T = p_A^{\circ} X_A + p_B^{\circ} X_B$$

Figure 9–18 shows how vapor pressure varies with the composition of a solution. For ideal solutions of alcohol and water, the partial pressure of each component can be read along a straight-line Raoult's Law plot from $X = 0$ to $X = 1$ for that component, as the other component changes from $X = 1$ to $X = 0$. The total pressure of the system is the sum of the partial pressures at any combination of mole fractions for the two components of the binary mixture.

Ideal and Real Solutions

As you might expect, real solutions differ from the ideal behavior predicted by Dalton's Law of Partial Pressures. The best fit between theory and practice is found among mixtures of similar liquids. Liquid pairs that tend toward nearly ideal behavior include the following:

1. Carbon tetrachloride and silicon tetrachloride. Although the central atom differs, C and Si are members of the same family of elements in the periodic table, and the basic structure and geometry are the same:

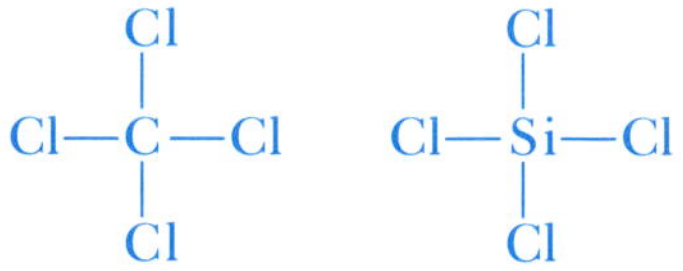

$$
\begin{array}{ccc}
 & \text{Cl} & & & \text{Cl} \\
 & | & & & | \\
\text{Cl}-\text{C}-\text{Cl} & & \text{Cl}-\text{Si}-\text{Cl} \\
 & | & & & | \\
 & \text{Cl} & & & \text{Cl}
\end{array}
$$

2. Pentane and heptane. This pair of gasoline-range hydrocarbons are as close to an ideal pair as can be found.

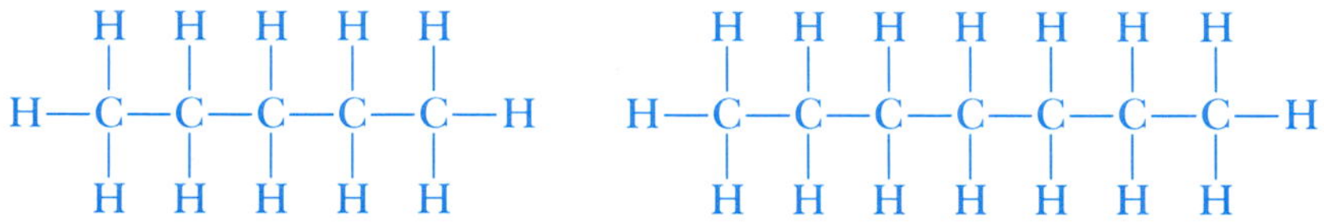

$$
\begin{array}{ccccc}
\text{H} & \text{H} & \text{H} & \text{H} & \text{H} \\
| & | & | & | & | \\
\text{H}-\text{C}-\text{C}-\text{C}-\text{C}-\text{C}-\text{H} \\
| & | & | & | & | \\
\text{H} & \text{H} & \text{H} & \text{H} & \text{H}
\end{array}
\qquad
\begin{array}{ccccccc}
\text{H} & \text{H} & \text{H} & \text{H} & \text{H} & \text{H} & \text{H} \\
| & | & | & | & | & | & | \\
\text{H}-\text{C}-\text{C}-\text{C}-\text{C}-\text{C}-\text{C}-\text{H} \\
| & | & | & | & | & | & | \\
\text{H} & \text{H} & \text{H} & \text{H} & \text{H} & \text{H} & \text{H}
\end{array}
$$

Methyl alcohol and ethyl alcohol solutions show some evidence of variation from ideal behavior, and there is actually a marked variation from ideal behavior when either alcohol is paired with water.

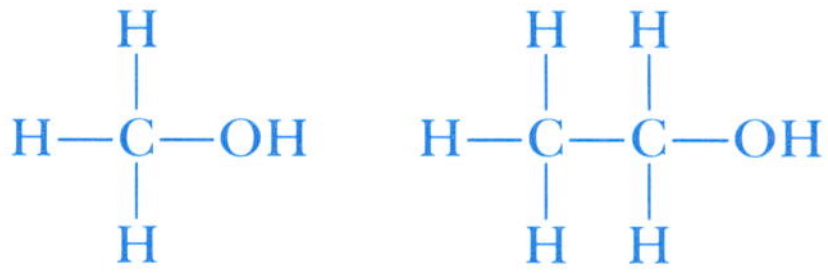

$$
\begin{array}{cc}
\text{H} & & \text{H} \quad \text{H} \\
| & & | \quad\; | \\
\text{H}-\text{C}-\text{OH} & \quad \text{H}-\text{C}-\text{C}-\text{OH} \\
| & & | \quad\; | \\
\text{H} & & \text{H} \quad \text{H}
\end{array}
$$

When the components of a pair of liquids in solution differ markedly from each other, Raoult's Law does not hold and the resulting solutions are nonideal in one of two major ways:

1. If the forces of attraction among like molecules in solution are greater, **positive deviation** from ideal behavior is observed. That is to say, the A molecules bind more strongly to each other than to the B molecules, and vice versa: A-A and B-B, rather than A-B. The result is that both A and B escape from solution more easily. Each exhibits a higher vapor pressure (based on their behavior in dilute solutions, according to Raoult's Law) than would be expected ideally.
2. If the reverse is true, and A molecules attract B molecules in preference to attracting their own kind, the vapor pressure of each one in solution is less than that indicated by Raoult's Law, and **negative deviation** from ideal behavior is observed.

Both types of behavior are shown by the dashed lines in Figure 9–19. Two dissimilar liquids such as acetone and carbon disulfide exhibit positive deviations, whereas two dissimilar liquids that strongly interact with each other such as chloroform and acetone exhibit negative deviations.

carbon disulfide and acetone

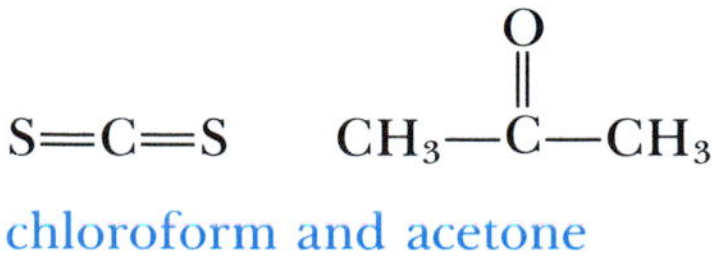

$$
\begin{array}{cc}
 & & & \text{O} \\
 & & & \| \\
\text{S}{=}\text{C}{=}\text{S} & \quad \text{CH}_3-\text{C}-\text{CH}_3
\end{array}
$$

chloroform and acetone

$$
\begin{array}{cc}
\text{H} & & \text{O} \\
| & & \| \\
\text{Cl}-\text{C}-\text{Cl} & \quad \text{CH}_3-\text{C}-\text{CH}_3 \\
| & \\
\text{Cl} &
\end{array}
$$

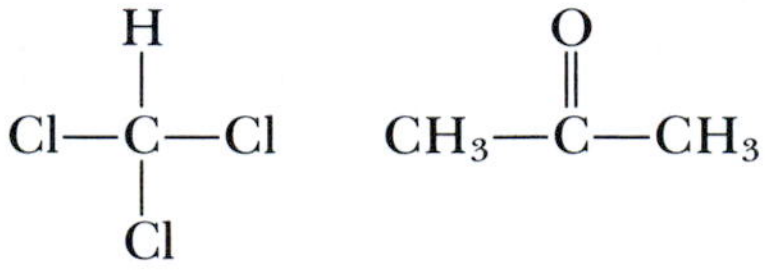

If heat is evolved upon mixing A and B, the resulting solution must

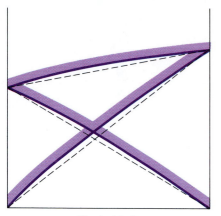

Nearly ideal

Minor differences between intra- and intermolecular forces between molecules

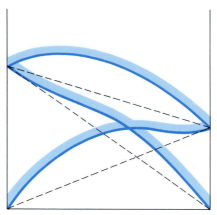

Positive deviation

Intramolecular forces favored: A–A and B–B types

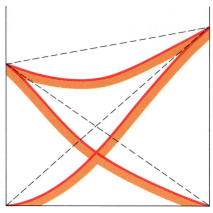

Negative deviation

Intermolecular forces favored: A–B and B–A types

FIGURE 9–19 Vapor pressure diagrams showing deviations from ideal behavior for three pairs of liquids in solution.

be in a lower energy state than the separated components. This would generally be the case if intermolecular interactions are established on mixing—strong forces between A and B molecules, in preference to A-to-A or B-to-B interactions. Since the molecules in the resulting solution are in a lower energy state, one would expect a lower vapor pressure and a negative deviation from Raoult's Law. If heat is absorbed on mixing A and B, the resulting solution must be in a higher energy state than the separated components, indicating a preference for intramolecular attractions of the A-A and B-B type. A higher energy state is consistent with a higher vapor pressure and a resulting positive deviation from Raoult's Law.

EXAMPLE 9–9

Use Raoult's Law to calculate the partial pressures and total pressure of a water-alcohol solution at 20°C containing an equal number of moles of each. At this temperature, $p°(H_2O) = 17.5$ torr and $p°(C_2H_5OH) = 44.5$ torr. Check your results against the data in Figure 9–18.

Solution

$$X(H_2O) = X(C_2H_5OH) = 0.50$$

$$p°(H_2O) = 17.5 \text{ torr}$$

$$p°(C_2H_5OH) = 44.5 \text{ torr}$$

$$p(H_2O) = (0.50)(17.5 \text{ torr}) = 8.75 \text{ torr}$$

$$p(C_2H_5OH) = (0.50)(44.5 \text{ torr}) = 22.3 \text{ torr}$$

$$P_T = p(H_2O) + p(C_2H_5OH) = 31.0 \text{ torr}$$

EXERCISE 9–9

Using Raoult's Law, calculate the vapor pressure of a 1:1 (by weight) alcohol-water solution at 20°C using the $p°$ values from Example 9–9. Check your answer with the numerical value that can be read directly from the graph for the components and for the solution.

Answer: $P_T = 25$ torr.

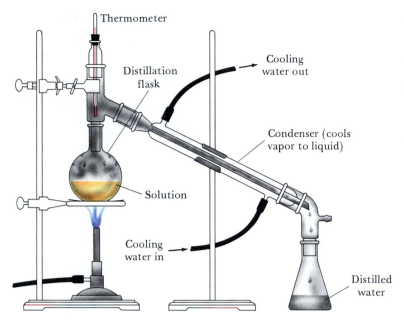

Thermometer

Distillation flask

Cooling water out

Condenser (cools vapor to liquid)

Solution

Cooling water in

Distilled water

FIGURE 9–20 Laboratory scale equipment and experimental set-up for simple distillation. Here, the components of a water solution are separated. During the distillation process, more volatile components "distill" and are eventually condensed and collected in the receiving flask. Separation is largely a function of differences in boiling points, and if the differences are significant, or if one or more components are nonvolatile, simple distillation can effectively separate the components of a solution.

Distillation

The composition of the vapor above the surface of a solution is not the same as the composition of the solution with which it is in equilibrium, because the vapor contains more of the component with the higher vapor pressure. That is the basis for transferring and separating liquids and liquid solutions by **distillation**. With an appropriate arrangement of laboratory or industrial equipment (Figure 9–20), the processes of vaporization and condensation can be used to transfer a volatile liquid or to separate and transfer the components of mixtures of volatile liquids. If all that is done is to heat a liquid in a reactor to the boiling point and condense the vapors produced, mass is transferred. It is an effective means of moving liquids from one place to another and for separating liquids from the nonvolatile solids and solutes often used as catalysts for chemical reactions (see also Chapter 25). If the vapors are condensed in several fractions and each is separately heated, vaporized, and condensed again, a considerable separation can be achieved.

EXAMPLE 9–10

Consider a mixture of two similar liquids, benzene (C_6H_6) and toluene ($C_6H_5CH_3$). Let B represent benzene molecules and A represent toluene. The vapor pressures for the two pure liquid solvents are

$$p_A^\circ = 22 \text{ torr} \qquad p_B^\circ = 75 \text{ torr}$$

Assume that the solution's composition happens to be

$$X_A = 0.67 \qquad X_B = 0.33$$

According to Raoult's Law:

$$p_A = p_A^\circ X_A(\text{liq}) = (22 \text{ torr})(0.67) = 15 \text{ torr}$$
$$p_B = p_B^\circ X_B(\text{liq}) = (75 \text{ torr})(0.33) = 25 \text{ torr}$$
$$P_T = p_A + p_B = 40. \text{ torr}$$

Calculate the composition of the vapor in equilibrium with the solution.

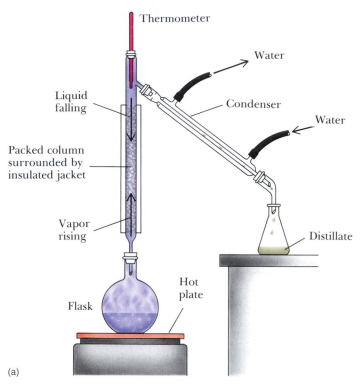

(a)

(b)

FIGURE 9–21 (a) The tall, insulated, and heat-controlled central column is packed with inert materials, often ceramic chips or glass beads, affording a large surface area over which distilling vapors pass, condense, and then drip back down into the hotter vapors rising from the boiling solution in the flask below. Repeating the process many times gradually takes the more volatile components through the column to the top and then to the condenser. (b) On an industrial scale, the sophistication of the fractional distillation process is lost in the massive scale of the two towers.

Solution   Use Dalton's Law of Partial Pressures, where X_A(vap) and X_B(vap) now refer to mole fractions in the vapor phase:

$$p_A = X_A(\text{vap})P_T$$

$$\frac{p_A}{P_T} = X_A(\text{vap}) = \frac{15\text{ torr}}{40\text{ torr}} = 0.38$$

$$\frac{p_B}{P_T} = X_B(\text{vap}) = \frac{25\text{ torr}}{40\text{ torr}} = 0.62$$

Note that the vapor is nearly twice as rich in the more volatile liquid—benzene, in this case—than in the solution with which it is in equilibrium. Now, if one were to repeat the entire process, condensing the vapor phase to liquid and then establishing a new liquid-vapor equilibrium, the vapor will become still richer in the more volatile component. Repeating the process of condensation and vaporization, it should be possible to reach the point where the vapor is essentially pure benzene.   ■

EXERCISE 9–10

Calculate the composition of the vapor in equilibrium with the condensed liquid in Example 9–10.
Answer: $X(C_6H_6) = 0.85$; $X(C_6H_5CH_3) = 1 - X(C_6H_6) = 0.15$.   ■

Fractional distillation (Figure 9–21) is a distillation process carried out in a special apparatus designed for more efficient separations than can be obtained with an ordinary distillation setup. In fractional distillation multiple, successive distillations occur in a continuous operation. Only a small fraction of the distillate is condensed and collected, while the rest

FIGURE 9–22 The industrial-scale low-temperature distillation and liquefaction equipment shown here is capable of separating, purifying and recovering methane (LNG), propane (LPG), liquid oxygen and hydrogen, and helium. Built near natural gas supplies, plants such as this in Algeria and Malaysia are capable of processing as much as a billion cubic feet of natural gas per day.

flows back into the main distillation column, bringing about a significantly greater degree of separation. A **fractionation column** is generally a tall column (or tower) through which vaporized liquid rises and condensed liquid descends. A dynamic equilibrium is established in the column so that there is a uniform temperature gradient—high temperature nearer the column/reactor connection, and lower temperature nearer the column/take-off connection. The process is called **refluxing**, and it leads to a gradual enrichment of the lower-boiling, more volatile components in the rising vapors and an enrichment of the higher-boiling, less volatile components in the liquid as it descends in the column. But even with the most efficient columns, it is impossible to separate two volatile liquids completely by distillation. For example, ethyl alcohol (b.p. 78.3°C) is often used in the preparation of an orange-yellow dye known as azobenzene (b.p. 293°C). It would seem that because of the very large difference in their boiling points (more than 200°), they could be cleanly separated by distillation. Yet when an ordinary laboratory distillation flask, column, and head are employed, even the very few first drops of distillate—which should be pure alcohol—are tinted by the presence of azobenzene molecules carried over with the alcohol. However, for most practical purposes—in the laboratory and on an industrial scale—it is possible to separate volatile liquids by fractionation adequately when there is as little difference in boiling points as two or three degrees. (See Figure 9–22.)

Vacuum distillation, distillation under reduced pressure, is often used to distill high-boiling substances. Because substances boil when their vapor pressures exceed the opposing pressure at the surface of the liquid, reducing the pressure allows distillation at lower temperatures. Vacuum distillation allows one to reduce the higher temperatures normally required to distill at atmospheric pressure and, at the same time, the degree of thermal decomposition (which can be considerable, especially for organic compounds). Azobenzene, which distills with a boiling point of 293°C at 760 torr and with considerable decomposition, distills smoothly at 179°C at 35 torr with little decomposition.

TABLE 9–4 Some Industrial Water Requirements

Industrial Products and Processes	Estimated Minimum Water Use
gunpowder	200,000 gal/ton
beer	470 gal/barrel
meatpacking	55,000 gal/100 hog unit
milk	450 gal/100 gal
sugar	20,000 gal/ton
rayon	200,000 gal/ton yarn
wool	40,000 gal/ton goods
leather	18,000 gal/ton hides
oxygen	2,000 gal/1000 ft^3 liq. O_2
soda ash (Na_2CO_3)	15,000 gal/ton
sulfuric acid (contact)	4,000 gal/ton
coke	1,500 gal/ton
oil refining	77,000 gal/100 barrel crude
steel	20,000 gal/ton

9.10 WATER POLLUTION

In the world of living things, water is essential. Animal and plant life cannot function without it. The water content of the protoplasm of most living cells approaches 80%, and almost all biochemical processes that occur during cell metabolism and growth take place in an aqueous environment (Table 9–4). In a scientific sense, water comes closest to fulfilling the definition of the proverbial "universal solvent." This unique natural resource, taken for granted where plentiful but coveted when scarce, has become a major factor in social, economic, and political decision making. Its nature allows it to be easily and cheaply transported and stored but, unfortunately, easily polluted—and once it is polluted, reclamation is difficult and expensive.

Pollution of fresh water occurs by both natural and man-made processes. The former can lead to some serious industrial problems such as boiler scale due to "hardness" or aesthetic problems due to high concentrations of certain minerals. Human contributions (Table 9–5) to the pollution of fresh water can be traced directly to the Industrial Revolution and the invention of the water-based sewage system. A century ago, the problem reached crisis proportions, leading the British Parliament to pass the landmark River Pollution Prevention Act of 1868. It became internationally accepted by industrialized nations. Today, direct pollution by toxic substances has been virtually eliminated in those nations, as has pollution by disease-bearing bacteria. Both still crop up as local problems, but waterborne bacteria still pose a problem in developing nations.

The pollution problem, however, rather than abating, has intensified. There are two essential problems: (1) the simple mathematics of a world population that is increasing geometrically in the face of a constant water supply, and (2) the ecological problems resulting from tampering with the biosphere. The United States Public Health Service (USPHS) classifies water pollutants into eight separate categories:

1. Oxygen-demanding wastes include organic wastes of both plant and animal origin, from domestic and industrial sources. Their aerobic (oxygen-using) decomposition depletes the available oxygen in the water at the expense of the oxygen demands of natural biological processes.

TABLE 9–5   General Nature of Industrial Wastes and Pollutants

Industry	Contributing Processes	Effects
agriculture	fertilizing, irrigation	nitrate and phosphate runoff
brewing	malting, fermenting liquors	organic overload
dairy	milk processing, butter/cheese manufacturing	acids, organic overload (manure-contaminated runoff)
dyeing	spent dyes, sizing, bleaching	acids, bases, bleaches
food processing	canning, freezing	organic overload
meat packing	slaughterhouse preparation and processing	organic overload
paper	pulp and paper manufacturing	organic overload, bleaches, waste, wood fibers and chips, sulfites
steel	pickling, plating, processing and manufacturing	acid wastes, dissolved iron
tanning	leather cleaning and manufacturing	organic overload, acids and bases
textiles	wood scouring, dyeing, bleaching, cotton treatment	organic overload, bleaches, acids

2. Infectious agents include bacteria and viruses from sewage and other types of municipal waste. Many of the viruses are particularly insidious because they are resistant to disinfectants and they can grow under either aerobic or anaerobic conditions.

3. Plant nutrients such as nitrates and phosphates are washed into streams and lakes, where they encourage the growth of algae and aquatic plants. The subsequent death and decomposition of these plants further increases the biological oxygen demand.

4. Industrial organic chemicals often present unusual problems. In addition, the cumulative effects of large-scale use of chemicals such as PCBs and DDT have caused much concern. PCBs (polychlorinated biphenyls; see Chapter 25) were widely used as dielectric fluids in transformers and in related industrial applications before 1976. Their use has been severely restricted because they may be carcinogenic and persist in the environment.

The pesticide DDT is another chlorinated hydrocarbon that is perhaps best known for its effect on the reproductive cycles of several species of birds. Among other things, it causes the birds to lay eggs with thin shells that break prematurely. The general use of DDT as an insecticide was prohibited in the United States in 1972, but it is still used in developing countries and for special applications in the U.S.

5. Other minerals and chemicals include a variety of substances from various mining, agricultural, industrial, and domestic operations. A continuing problem is the widespread use of detergents.

6. Sediment from land erosion reduces the ability of natural streams to assimilate oxygen-demanding wastes and affects the normal habitat of fish and shellfish, as well as aquatic plants.

7. Radioactive substances have rapidly turned into a major problem because of their increased use and disposal requirements.

8. Heat from industry or thermal pollution results from the use of water as a coolant because of its availability and unusually high heat capacity. Heated water is subsequently returned to the biosphere, raising stream temperatures markedly, thus decreasing oxygen solubility. This affects the local ecological balance by lowering the capacity to assimilate waste through depletion of available oxygen.

Many environmental changes occur slowly and many are reversible with time if the ecosphere can be left alone. In view of the impossibility of

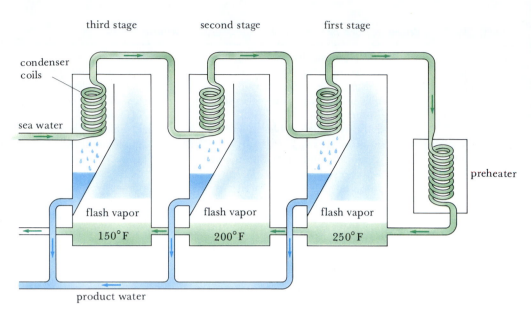

FIGURE 9–23 Multi-stage flash distillation process. Seawater that has been previously heated is rapidly brought to the boiling point by carrying it into an evaporator where the pressure has been reduced below the equilibrium vapor pressure (at the seawater temperature). The remaining seawater (at a lowered temperature) is pumped into a second evaporator where it is flash-distilled (at a still lower pressure), and then into a third. The seawater coils also function as condenser coils.

"Water, water, everywhere/Nor any drop to drink." The salt in seawater is not completely sodium and chloride ions. There is much Mg^{2+}, K^+, and Ca^{2+} in solution; also Br^-, SO_4^{2-}, and HCO_3^-.

"leaving nature alone" to allow natural restoring measures to work, post-treatment of fouled waters prior to their return into the natural habitat should be considered imperative no matter what the cost. Sedimentation, flocculation, aeration, filtration, and even sterilization are all used in a variety of applications, singly or in combination, to process fresh water successfully.

Saline seawater is as much a paradox to modern people as it was to the Ancient Mariner. The 3.5% NaCl that is present prevents drinking it (it is unpalatable and biologically incompatible) or using it industrially (because it is highly corrosive). Yet, it covers three-fourths the surface of the earth. The demand for fresh water (which has less than 0.05% salt) has led to massive efforts to develop economical means for recovering it from sea water. Currently, there are four major technical approaches to desalination: distillation, reverse osmosis, electrodialysis, and freezing.

1. The multi-stage flash **distillation** process is the most commonly used desalination technique (Figure 9–23). Preheated seawater is rapidly brought to its boiling point, or "flashed," by introducing it into a chamber where the pressure is below the equilibrium vapor pressure (at the seawater temperature). Unevaporated seawater passes on to a second-stage evaporator where it is "flashed" at still lower pressure. The seawater intake coils serve as condensation surfaces for the pure water product.

2. Reverse osmosis depends upon exertion of a hydrostatic pressure greater than the osmotic pressure to cause water to flow through a semipermeable membrane from the more concentrated salt solution to the pure water side. The direction of flow is opposite the usual direction, in which water flows through a membrane from a solution of lower concentration to one of higher concentration—hence the name "reverse" osmosis. The special membranes needed are of two types, either bundles of very fine capillaries or spiral-wound sheets made of cellulose acetate (for brackish water) or polymeric polyamide materials (for more concentrated seawater salt solutions) (Figure 9–24).

3. Electrodialysis is also membrane-based, making use of an ion-transfer process for the separation of salt from water. It uses the fact that the salt in solution exists completely as negatively and positively charged ions. Prop-

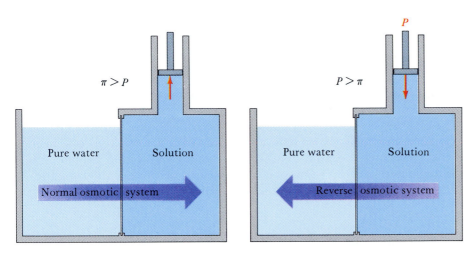

FIGURE 9–24 The principles of reverse osmosis, which is widely used to obtain fresh water from seawater, can be illustrated by considering how "normal" osmosis can be prevented—by applying a pressure P to a solution in a normal osmotic system that just balances the osmotic pressure π. If P is less than π, osmosis occurs in the normal fashion; but if P is greater than π, water flows in the opposite direction, reversing the normal osmotic process.

erly positioned electrodes of opposite charge attract the ions through a series of semipermeable membranes that are permeable to either anions or cations but not both. Areas of high salt concentration and low salt concentration result. When the level of desalination (or deionization) becomes high enough, the water is drawn off from that area of the system.

4. Vacuum-freeze desalination is based on the fact that pure water—not an ice-salt mixture—crystallizes from a freezing salt water solution. Ice crystals are removed, washed free of any remaining brine solution, and finally melted to yield pure water. The latent heat of fusion, recovered when the water crystallized, is used to melt the crystals in recovering water. Critical stages of the entire process are run under pressures of approximately 3 torr, where the boiling point and the melting point are virtually the same, separated in energy only by the latent heats of fusion and vaporization. One of the intriguing features of this process is that ice crystals form as water vapor boils away.

At 4.6 torr, the freezing point and melting point are 0°C. Salt water freezes at somewhat lower temperatures, so the process operates at 3 torr.

9.11 SYSTEMS OF COLLOID-SIZED PARTICLES

True solutions are homogeneous mixtures of solid, liquid or gaseous solutes dissolved in solvents. Common experience also reveals many examples of nonhomogeneous systems in which solids or liquids are dispersed in liquids. For example, sparingly soluble barium sulfate ($BaSO_4$) dispersed in water is an opaque medium for diagnostic x-rays. Some air and water pollutants and certain body fluids are dispersions; so are stained glass windows. A **dispersion** consists of a homogeneous, continuous phase in which a second, discontinuous phase is scattered. There are three distinct types, distinguished by the degree of subdivision of the discontinuous phase. Consider a flowing system of coarsely ground calcium carbonate ($CaCO_3$) in water (Figure 9–25a); such a system is characterized by the bulk properties of each separate phase and the laws of hydromechanics. If the size of the limestone particles is reduced to a few nanometers (Figure 9–25b), the characteristics of the system become quite different, possibly like those of a semisolid, free-flowing mass. A further reduction in the size of the limestone particles, down to molecular dimensions (Figure 9–25c), results in the behavior characteristic of liquids. The coarse limestone particles constitute one phase of a **gross** dispersion; the finely divided particles of the second case make up one phase of a **colloidal** dispersion; and the third case, the true solution, may be looked upon as a **molecular** (or ionic) dispersion.

Dust and smog are two examples of dispersed air pollutants. Blood, turbid water, and milk are examples of aqueous dispersions; urine and blood plasma are not.

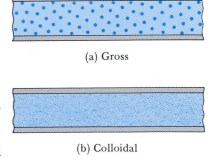

(a) Gross

(b) Colloidal

(c) Molecular

FIGURE 9–25 Gross, colloidal, and molecular dispersions.

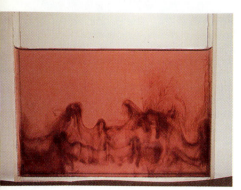

FIGURE 9–26 Colloidal gold particles dispersed and suspended in an aqueous medium.

Many questions concerning the nature of colloidal-sized particles were answered by Richard Zsigmondy with the aid of the ultramicroscope, a microscope capable of seeing particles ranging in size from 500 nm all the way down to 10 nm.

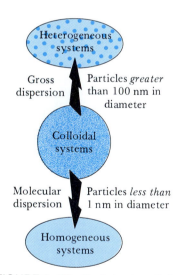

FIGURE 9–27 Particle size distinguishes true solutions, colloidal suspensions and dispersions, and heterogeneous mixtures.

In optical experiments carried out 150 years ago, Michael Faraday observed that colloidal-sized particles of gold, suspended in a liquid, produced vivid red dispersions (Figure 9–26). Faraday commented that these fluids:

"... when in their finest state, often remain unchanged for many months and have all the appearance of solutions. But they never are such, containing in fact no dissolved but only diffused gold. The particles are easily rendered evident by gathering the rays of the Sun (or a lamp) into a cone by a lens, and sending the part of the cone near the focus into the fluid. The cone becomes visible and though the illuminated particles cannot be distinguished because of their minuteness, yet the light they reflect is golden in character."

Characteristics of the Colloidal State

Colloidal dispersions sometimes appear to be true solutions, but close inspection reveals major differences in particle size and behavior. They are best considered to be intermediate states between true solutions and coarse mixtures (Figure 9–27). As an order-of-magnitude classification, particles whose diameters fall into the 10-nm range are typically colloidal; with respect to behavior of colloidal dispersions, we can state the following:

- Colloidal dispersions can generally be separated by physical means.
- Colligative properties such as osmotic pressure are notably different than for true solutions.
- Turbidity is often visible.
- Rates of diffusion are negligible.
- Unusually high electrostatic charges are common (Figure 9–28).

Foams, Emulsions, and Gels

When a relatively large volume of a gas is dispersed into a relatively small volume of a liquid, the resulting colloidal dispersion is referred to as a **foam** (Table 9–6). Foams can be produced by both condensation and disintegration methods. In the condensation method, the gas phase (which is initially present as separate molecules) is liberated within the liquid to form bubbles. Many of the solid foams used for construction and insulation are manufactured in this way. The source of the bubbles is a "blowing agent," usually a liquid with a very high vapor pressure that produces many times its own volume of foam.

Emulsions result when one liquid is colloidally dispersed into another by some form of agitation or homogenization. Some emulsions are very stable—milk, for example. Others are not stable and settle upon standing, such as oil-and-water based salad dressings. Emulsifying agents (usually soaps and detergents) can be added to help stabilize the emulsion. Stable emulsions can be purposely broken by freezing, heating, adding salts, or centrifuging. The latter has been successfully used in the laboratory, and on an industrial scale in separating cream from milk.

Under proper conditions, colloidal dispersions can be precipitated in a gelatinous form in which all the liquid medium is trapped. Such a product is a **gel**; the process is called gelation. A familiar example is the gelatin dessert, which is generally prepared simply by cooling the aqueous colloidal dispersion.

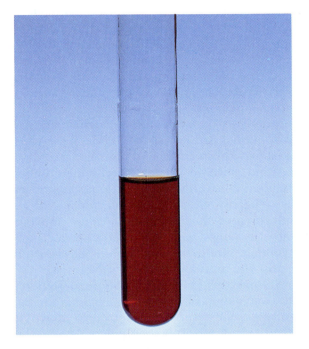

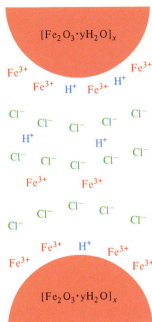

FIGURE 9–28 Stabilization of an iron(III) oxide sol by electrostatic forces. Each colloidal Fe_2O_3 particle of this red colloidal dispersion is a cluster of many "hydrated" formula units. Each attracts positively charged Fe^{3+} ions to its surface. Since Fe^{3+} ions fit readily into the crystal (lattice) structure, they are preferentially adsorbed rather than chloride (Cl^-) ions. Because each particle is surrounded by a shell of positively charged ions, the particles repel each other and cannot agglomerate (combine) to the extent necessary to cause actual precipitation.

TABLE 9–6   Nine Possible Colloidal Dispersions

Phase	Medium	Classification	Example
1. gas	gas	(Unknown)	(Unknown)
2. gas	liquid	foam	aerosol shaving creams
3. gas	solid	solid foam	volcanic ash
4. liquid	gas	aerosol	fogs and clouds
5. liquid	liquid	emulsion	oil in water
6. liquid	solid	gel	gelatin, jellies
7. solid	gas	aerosol	smoke
8. solid	liquid	sol	Faraday's "gold" solution
9. solid	solid	solid sol	ruby glass*

*"A ruby glass . . . is best obtained when gold is added to the molten glass either in the form of purple of Cassius . . . or in the form of some gold salt or other."

SUMMARY

The **theory of solutions** that has developed over the last century has both qualitative and quantitative features. It has proved useful in dealing with many practical scientific and engineering problems. In a general way, we can say that **solubility** depends on the nature of the solute and solvent, on the temperature, and, for gases, on the pressure. Solute cannot be separated from solvent by any mechanical means because the particle sizes of both are of the same order of magnitude, of molecular dimensions.

Lattice energies are endothermic; **hydration energies** are exothermic; what we see when a salt dissolves is the net effect. In most cases heat aids dissolution, but in some cases dissolving is favored by cooling. The latter is especially true for gases. A high dielectric solvent such as water will dissolve ionic crystal lattices such as sodium chloride by lowering the force of attraction between the oppositely charged ions. On the other hand, a low dielectric medium such as a petroleum hydrocarbon will not dissolve much salt at all. We would not expect oil and water to mix.

Colligative properties of solutions are those that depend on the number, not the kind, of particles present. Ideal solutions are governed by **Raoult's Law**, a consequence of which is the lowering of the vapor pressure of a pure solvent due to the presence of solute particles is proportional to the mole fraction of solute present. This results in some easily measured physical quantities: elevation of the boiling point, ΔT_b; lowering of the freezing point, ΔT_f; and *osmotic pressure*, π. Colligative phenomena such as these permit measurements of formula weights. Osmotic pressure is especially useful for such measurements on substances of high molecular weight, such as the polymers of plastics and proteins.

Raoult's Law was developed for undissociated and nonvolatile solutes. It is important to keep that in mind, since we are dealing with the number of particles in solution. Subsequently, it was shown that ionic substances dissociate in solution, and the colligative effects are greater than if dissociation had not occurred. Sulfuric acid, for example, in dilute aqueous solution shows three times the effect expected for an undissociated molecule. The **van't Hoff** i factor compares the expected colligative effect based on dissociation with the effect for the undissociated electrolyte.

Distillation is a means of transferring mass and separating a solution of volatile liquids based on a combination of ideas: Raoult's Law for the solution, and Dalton's Law of Partial Pressures for the vapor above it; equilibrium vapor pressures and volatility; and boiling points.

Water pollution and the recovery of potable water from seawater are two of the important solubility problems of our times. Water is such a good solvent that it is difficult and expensive to remove (or recover) dissolved substances from it—it is easily polluted, accidentally or on purpose.

Although the particles are too small to be distinguished by the naked eye, dispersions of **colloid**-sized particles (whose dimensions are in the nanometer range) are not solutions in the usual sense. Their unique properties make them important in understanding natural phenomena and industrial processes.

Formulas for this chapter:

$$m = 1000 \, \frac{n_B}{g_A} \qquad \text{molality}$$

$$p_A = p_A^\circ X_A \text{ (liq)} \qquad \text{Raoult's Law}$$

$$\Delta p = p_A^\circ X_B \text{ (liq)} \qquad \text{Raoult's Law}$$

$$P_T = p_A^\circ X_A + p_B^\circ X_B + \cdots \qquad \text{Dalton's Law}$$

$$X_A \text{ (vap)} = \frac{p_A}{p_T} = \frac{p_A^\circ X_A \text{ (liq)}}{p_T}$$

$$\Delta T_b = K_b m \qquad \text{Boiling point elevation}$$

$$\Delta T_f = K_f m \qquad \text{Freezing point depression}$$

$$\pi = \frac{nRT}{V} \qquad \text{Osmotic pressure}$$

$$\pi = cRT \qquad \text{Osmotic pressure}$$

QUESTIONS AND PROBLEMS

QUESTIONS

1. A warm saturated solution containing a small amount of undissolved solute is cooled, causing the solute to dissolve completely. Is the dissolving process exothermic or endothermic? Compare the hydration and lattice energies for the solute.

2. Is a saturated solution necessarily a concentrated solution?

3. How might one go about establishing whether a solution of sodium sulfate in water is saturated, unsaturated, or supersaturated?

4. Based on the discussions of this chapter, how many ways can you come up with for determining the molecular weight of a substance?

5. Here's an experiment you can do at home. Slice a raw potato across the bottom so it will stand flat. Dig a 2-inch hole in the top side and fill it two-thirds full of sugar. After a while, the sugar will be moist, perhaps even watery. Explain.

6. Why do wilted flowers revive when placed in a pitcher of water? Explain what would happen (and why) if you inadvertently placed the wilted flowers in a pitcher of salt water.

7. State your understanding of each of the following terms:
(a) colligative properties
(b) mole fraction
(c) molality
(d) Raoult's Law
(e) osmosis and osmotic pressure
(f) osmotic pressure and hydrostatic pressure

8. Briefly identify, explain, or otherwise indicate what is meant by each of the following terms:
(a) van't Hoff i factor
(b) the Arrhenius theory (of electrolytic dissociation)
(c) K_f and K_b
(d) electrolyte solutions
(e) cRT
(f) Dalton's Law of Partial Pressures

9. Why does the addition of a solute alter the escaping tendency—the vapor pressure—of the solvent?

10. Why would you expect greater deviation from ideal behavior in the liquid state than the gaseous state?

11. Explain the reasons for the vertical and horizontal trends in the van't Hoff i factors listed in the following table.

Molar Concentration	KCl	K_2SO_4	$K_3Fe(CN)_6$
0.005	1.96	2.76	3.51
0.010	1.94	2.71	3.48
0.050	1.88	2.67	3.46
0.100	1.86	2.64	
0.200	1.83		
0.500	1.80		

12. Briefly explain why the rising moon is often redder in the fall of the year than in the winter.

13. Grape jelly is a colloidal suspension; alcohol and water is a true solution. How would you demonstrate that each statement is true?

14. Ordinary methods for molecular weight determinations based on colligative behavior are unsuitable for colloidal-sized particles. Why would that be the case?

15. Explain the physical basis for dialysis as a means of purifying colloidal suspensions such as human blood.

16. At 4.6 torr, water both boils and melts at 0°C. Briefly explain why this seemingly absurd statement is not absurd.

17. Briefly explain how "reverse osmosis" causes water to flow spontaneously out of solution, apparently in the wrong direction, forming the basis for a method of recovering potable water from seawater.

18. Distinguish between Raoult's Law and Henry's Law.

19. Explain why the vapor pressure of the solvent in a dilute solution is given by Raoult's Law:

$$\frac{p_A}{p_A^\circ} = X_A$$

20. How would you show that the total vapor pressure of a solution of alcohol and water (or any other pair of liquids ideally miscible in all proportions) is a straight-line function of the mole fraction of each component?

21. Briefly explain the decreased effectiveness of highway lighting under conditions of heavy fog or smog.

22. Which of the following are true, and which are false? How might you demonstrate that to be the case?
(a) Grape jelly is colloidal.
(b) Fog is colloidal.
(c) Smoke is colloidal.
(d) Glucose forms a true solution in water.
(e) Jell-O dessert is a true solution.
(f) Water-based latex paints are colloidal.
(g) Milk is colloidal.

23. If you are going to try to melt snow off city streets, should your trucks spread NaCl, $Ca(NO_3)_2$, or $CaCl_2$? Assume that each salt dissociates completely and that they all cost the same per ton.

PROBLEMS
Expressing Concentrations [1–4]

1. Calculate the mole fraction of each component in a mixture composed of equal masses of ethanol (C_2H_5OH), water (H_2O), and ethylene glycol [$(CH_2OH)_2$].

2. Calculate the mole fraction of each component of a solution that contains 10.0 g of carbon tetrachloride (CCl_4), 20.0 g of benzene (C_6H_6), and 30.0 g of decane ($C_{10}H_{22}$).

3. Determine the molal concentration of each of the following solutions:
(a) 40. g of NaOH added to 400. g of H_2O
(b) 0.032 g of CH_3OH dissolved in 10. ml of H_2O
(c) 1.0 g of naphthalene ($C_{10}H_8$) dissolved in 1.0 mol of benzene (C_6H_6)

4. Calculate the molality of each of the following solutions:
(a) 15.0 g of NaCl dissolved in 220. mL of water
(b) 2.50 g of acetone (C_3H_6O) dissolved in 1.0 mol of water
(c) 0.010 mol of acetic acid (CH_3COOH) dissolved in 100. g of H_2SO_4

Raoult's Law   [5–8]

5. A solution is prepared by adding 1.00 g of solid naphthalene ($C_{10}H_8$) to 10.0 g of benzene (C_6H_6). Determine each of the following:
(a) mole fraction of the solute
(b) molality of the resulting solution
(c) vapor pressure of the resulting solution at 26.1°C, where that of the pure solvent is found to be 100 torr.
6. The vapor pressure of pure water at 298 K is 23.76 torr. A dilute solution was prepared by dissolving 1.70 g of a nonvolatile nonelectrolyte in 50.0 g of water. The vapor pressure of the solution (at the same temperature) was found to be 23.60 torr. Calculate:
(a) the mole fraction of the solute
(b) the molecular weight of the solute
(c) the freezing point of the solution
7. As part of a microscale chemistry experiment, 13.10 mg of a nonvolatile substance was dissolved in 111.4 mg of camphor. The substance is known to have a molecular weight of 178.24 mg/mmol and is not known to dissociate in solution. By how many degrees is the normal freezing point of the solvent depressed in solution? The K_f value for camphor can be found in Table 9–2.
8. When 5 g of an unknown sugar is dissolved in a kilogram of water, the resulting solution freezes at −0.050°C. What is the molecular weight of the unknown sugar? The K_f value for water can be found in Table 9–2.

Vapor Pressure, Boiling Point Elevation, and Freezing Point Depression   [9–20]

9. When 10. g of $CaCl_2 \cdot 2H_2O$ is dissolved in 150. g of water, at what temperature will the solution freeze?
10. When 10. g of sucrose ($C_{12}H_{22}O_{11}$) is dissolved in 150. g of ethyl alcohol, at what temperature will the solution boil?
11. A certain undissociated and nonvolatile solute raises the boiling point of water by 0.52°C when 3.0 g is added to 30. mL of water. Calculate the molecular weight of the solute.
12. A sample of 5.0 g of an unknown organic substance is dissolved in 80. g of acetic acid, resulting in an increase in the boiling point of the solvent by 0.97°C. Calculate the molecular weight of the unknown organic substance from the known value of K_b for acetic acid (from Table 9–1).
13. An automobile has a cooling system with a 4-gal capacity. If you were to fill it with a solution that contained equal parts by volume of ethylene glycol ($CH_2OH)_2$ and water, to what lowest temperature could the car be exposed before freezing becomes a problem? The density of ethylene glycol is 1.115 g/mL.
14. A substance that has been commonly used as an antifreeze is methanol (CH_3OH). Its current cost is about 8

cents/lb, while the cost of ethylene glycol is about 12 cents/lb.
(a) How much money would you save if you used methanol in Problem 13 instead of ethylene glycol? (The density of methanol is 0.79 g/mL.)
(b) Based on your knowledge of liquids and solutions, why would ethylene glycol be a more desirable antifreeze in spite of its much higher cost?
15. Consider 300. g of urea [$(NH_2)_2CO$] dissolved in 1000. g of water in order to prevent it from freezing at its normal freezing point. Assuming that the resulting solution does not deviate from ideal behavior, determine the following:
(a) the vapor pressure of the solvent at 0°C
(b) the vapor pressure of the solvent at 100°C
(c) the boiling point of the solution
(d) the freezing point of the solution
16. Repeat the calculations in Problem 15 for 300. g of sodium chloride dissolved in 1000. g of water. Assume ideal behavior.
17. On dissolving 0.412 g of naphthalene ($C_{10}H_8$) in 10.0 g of camphor, the normal melting point of camphor was lowered by 13°C. When 1.00 g of a second substance was dissolved in 8.55 grams of camphor, the observed depression of the freezing point was 9.5°C. Calculate:
(a) the molal freezing point depression constant (K_f) for camphor
(b) the apparent molecular weight of the second substance
18. A solution of 86.52 g of solvent and 1.4020 g of an undissociated, nonvolatile solute has a boiling point 0.220°C higher than that of the pure solvent. In a second experiment, it was found that 1 mol of ethylene glycol, when added to 100. g of the same solvent, resulted in a boiling point elevation of 23.7°C.
(a) Calculate the boiling point elevation constant for the solvent.
(b) Calculate the molecular weight of the solute.
19. The known freezing point depression constant for camphor is 40.3°C/mol. In one particular experimental determination of the molecular weight of the organic compound called phenanthrene, 0.0113 g resulted in a lowering of the usual freezing point of 0.0961 g of camphor by 27.3°C.
(a) What is the *apparent* molecular weight of phenanthrene in camphor?
(b) Briefly state why the word "apparent" should be emphasized in an experiment such as this.
20. The normal boiling point of toluene ($C_6H_5CH_3$) is 110.6°C. When 10.0 g of biphenyl [$(C_6H_5)_2$] is dissolved in enough toluene to make 100. g of solution, the resulting mixture boils at 112.8°C. Based on that information, determine the molecular weight of an unknown solute, 10.0 g of which causes 333 g of a toluene solution to boil at 112.0°C.

Osmotic Pressure   [21–24]

21. An aqueous glucose solution contains 20. g of glucose per liter. Calculate the expected osmotic pressure at 27°C.
22. An aqueous sucrose solution registers an osmotic pressure of 5.06 atm at 20°C. Determine its concentration.
23. Of all the colligative properties of a solution, only

osmotic pressure offers an approach to the molecular weights of proteins, the polymers of plastics, and other macromolecular structures.

(a) If 10. g of an unknown protein is present per liter of water at 25°C and the osmotic pressure is carefully determined to be 9.25 torr, what must its molecular weight be?

(b) Based on the number of moles in 10. g of the protein, what would the freezing point depression and boiling point elevation be? (Assume 1 liter = 1000. g.)

(c) Which of the three techniques—osmotic pressure, freezing point depression, or boiling point elevation—would be most accurate in determining the molecular weight of particles such as these?

24. The osmotic pressure of a solution of 4.00 g of sucrose in 100. mL of solution was found to be 2.75 atm at 15°C.

(a) Calculate the apparent molecular weight for sucrose using $\pi = cRT$.

(b) If the density of the solution is 1.04 g/mL, what would the molecular weight be if you used molal concentration ($\pi = mRT$) in your calculation?

(c) Compare your answers to parts (a) and (b).

Electrolyte Solutions; van't Hoff *i* Factor   [25–26]

25. Calculate the van't Hoff *i* factor for a 0.50 m aqueous solution whose freezing point is −3.27°C.

26. A 0.020 M aqueous solution gives an osmotic pressure of 0.489 atm at 298 K. What is the van't Hoff *i* factor for the solute?

Vapor Pressure and Distillation   [27–32]

27. You have a solution made up of 1.0 mol of liquid A and 4.0 mol of liquid B. The vapor pressure of pure A is 80. torr, while that of liquid B is 20. torr. Calculate the vapor pressures of A and B above the solution and calculate the composition of the vapor. (Assume the solution is ideal.)

28. The vapor pressure of diethyl ether ($C_4H_{10}O$) at 29.5°C is 645 torr. For acetone (C_3H_6O), the value is 280. torr at the same temperature. Using Raoult's Law and Dalton's Law, calculate the mole fractions of diethyl ether and acetone in the vapor state for a mixture of equal weights of the two.

29. Ethyl alcohol ($p° = 44.5$ torr) and methyl alcohol ($p° = 88.7$ torr) are miscible in all proportions, forming solutions that are essentially ideal.

(a) Calculate the composition of the solution of these two substances whose vapor pressure is 66.1 torr.

(b) What is the composition of the vapor in equilibrium with this solution?

30. A solution of benzene (C_6H_6) and toluene ($C_6H_5CH_3$) is prepared by thoroughly mixing together equal volumes of the pure liquids at 25°C. At this temperature, the densities of both liquids are the same. Given the following information:

$$p°(C_6H_6) = 100.\ torr$$

$$p°(C_6H_5CH_3) = 30.\ torr$$

(a) Determine the vapor pressure of the resulting solution.

(b) Determine the composition of the vapor (as mole fraction) in equilibrium with the solution.

(c) If the vapor in equilibrium with the solution were condensed to liquid, removed to another container, and allowed to stand at least long enough for equilibrium to be reestablished, what would be the composition of the (new) vapor?

31. You have at your disposal two separate solutions of a pair of substances (A and B). Solution 1 contains 1.0 mol of A and 3.0 mol of B, and the vapor pressure of the solution is 1.0 atm. Solution 2 contains 2.0 mol of A and 2.0 mol of B, and the vapor pressure is greater than 1 atm; however, by adding 6.0 mol of C (for which $p° = 0.80$ atm) the vapor pressure of solution 2 can be reduced to 1.0 atm. Determine the original vapor pressures of substances A and B.

32. Two organic solvents, X and Y, form an ideal solution at 25°C. At that temperature, the following pressure data are known:

$$p°(X) = 100.\ torr$$

$$p°(Y) = 0.0\ torr$$

The vapor pressure of a solution of 10.0 g of X and 1.00 g of Y was found to be 95 torr. Find the ratio of the molecular weights of X and Y.

Additional Problems   [33–46]

33. A solution is prepared by dissolving 50.0 g of CH_3COOH (acetic acid) in 500. g of water. Calculate X_B and m for the solution.

34. A liter of methyl alcohol (CH_3OH, $d = 0.796$ g/cm^3) and a liter of ethyl alcohol (CH_3CH_2OH, $d = 0.789$ g/cm^3) are mixed together. What is the mole fraction of methyl alcohol in the resulting solution? What is the molal concentration of the ethyl alcohol?

35. Hydrogen is soluble in water at 20°C only to the extent of 0.000164 g/100. g at atmospheric pressure. Find the mole fraction of hydrogen and the Henry's Law constant for hydrogen.

36. A solution consisting of 86.52 g of CS_2 and 1.4020 g of anthracene has a boiling point 0.220°C higher than that of pure CS_2 under the conditions of the experiment. On adding 1.0 mol of solute to 100. g of CS_2, the boiling point of the resulting solution was found to be 23.7°C higher than that of pure CS_2. What is the molecular weight of anthracene, assuming it is undissociated in CS_2?

37. An aqueous sugar solution freezes at −0.200°C. Find the vapor pressure of the solution at 25°C, given that the vapor pressure of pure water at that temperature is 23.76 torr and K_f for water is −1.86°C/m.

38. A solution was prepared by saturating 10. cm^3 of water with 650. mg of a particular protein. If the osmotic pressure of the resulting solution was 91.2 torr, what is the molecular weight of the protein in question?

39. On dissolving 9.0 g of a sugar in 500. g of water at 27°C, the osmotic pressure was found to be 2.46 atm. Determine the molecular weight of this particular sugar.

40. A 0.010 m solution of $TlCl_3$ was found to freeze at −0.0744°C. Assuming ideal behavior,

(a) what is the van't Hoff *i* factor?

(b) what is the boiling point of the solution?

(c) what species are likely to be present in solution?

41. Pure carbon tetrachloride (CCl_4) has a vapor pressure of 223 torr at 40°C. Pure dichloroethane ($C_2H_4Cl_2$) has a vapor pressure of 159 torr at the same temperature. Calculate the partial pressure of each vapor and the total vapor pressure over a solution made by combining 50.0 g of each liquid. Assume ideal behavior.

42. Two liquids (A and B) combine, forming an ideal solution. At 50°C, the total vapor pressure for a solution of 1.0 mol of A and 2.0 mol of B is 250. torr. On adding a second mol of A, the vapor pressure of the solution rises to 300. torr. Calculate the vapor pressures of pure A and B.

43. Benzene (C_6H_6) and toluene ($C_6H_5CH_3$) are known to form nearly ideal solutions. A solution of 2.0 mol of benzene and 3.0 mol of toluene has a vapor pressure of 280. torr at 60°C. On adding a third mole of benzene to the solution, the vapor pressure of the solution rises to 300. torr. Calculate the vapor pressure of pure benzene and toluene.

44. Using the data in the following table, construct a graph of the vapor pressure curves for solutions of acetone and chloroform at 35°C.

(a) Plot mole fraction along the ordinate versus vapor pressure along the abscissa.

(b) Draw a dashed line extending from the vapor pressure of each pure component (100% concentration) to the zero vapor pressure point (0% concentration). These will be the Raoult's Law curves for ideal behavior.

(c) Plot the Raoult's Law curves for real behavior (from the data).

X(chloroform)	p(chloroform)	p(acetone)
0.00	—	358
0.10	17	306
0.20	33	270
0.30	56	230
0.40	82	185
0.50	112	145
0.60	157	103
0.70	185	70
0.80	225	41
0.90	263	18
1.00	328	—

45. Discuss the deviation observed from ideal behavior (as predicted by Raoult's Law) in Problem 44. Given the following structures for the two molecules, develop a simple explanation of the observed behavior.

$$CH_3\text{—}C=O\text{—}CH_3 \qquad H\text{—}C\text{—}Cl$$

acetone   chloroform

46. Since $\pi = nRT/V$, it ought to be possible to design a graphical technique for calculating molecular weights. What should you plot, and how would you get the molecular weight value?

MULTIPLE PRINCIPLES [47–54]

47. Calculate the molality, mole fraction, and molarity of a solution that is 50% water and 50% ethanol by volume. Should you worry about this alcohol solution freezing if it is left outdoors and the temperature drops to −10°C? Assume the density of the ethyl alcohol to be 0.79 g/mL and that of the solution to be 0.94 g/mL.

48. When 1.00 g of arsenic is dissolved in 86.0 g of benzene, the freezing point of the resulting solution is found to be 5.31°C. If the normal freezing point of benzene is 5.50°C and K_f is known to be −4.90°C/m, what is the true molecular formula for the element?

49. It is well known that nitrous acid (HNO_2) dissociates to a limited extent in dilute aqueous solutions by means of a proton-transfer reaction with the solvent, producing nitrite ions and hydronium ions:

$$HNO_2\ (aq) \rightleftharpoons H^+\ (aq) + NO_2^-\ (aq)$$

If a 0.100 m solution of aqueous nitrous acid freezes at −0.200°C, what fraction of the nitrous acid molecules have undergone this proton-transfer reaction with solvent water? $K_f(H_2O) = -1.86°C/m$.

50. When 1.00 g of p-dichlorobenzene ($C_6H_4Cl_2$) is dissolved in 10.0 g of benzene (C_6H_6), the freezing point of the resulting solution is 2.17°C. In a separate experiment, 1.00 g of acetic acid (CH_3COOH) is dissolved in 25.0 g of benzene and the freezing point of the solution is observed to be 2.48°C.

(a) Calculate K_f for solvent benzene.

(b) Calculate the molality of the acetic acid solution, based on the usual formula, CH_3COOH.

(c) It is well known that acetic acid molecules dimerize, according to the following:

$$2\ CH_3COOH \rightleftharpoons (CH_3COOH)_2$$

Determine the fraction of acetic acid molecules that exist as dimers.

51. (a) Assuming KI (potassium iodide) is completely dissociated in dilute aqueous solutions, what is the normal freezing point of a 1.00 m solution and what is the value of the van't Hoff i factor?

(b) On adding mercury(II) iodide to KI solutions, the following reaction occurs:

$$HgI_2\ (s) + 2\ I^-\ (aq) \longrightarrow HgI_4^{2-}\ (aq)$$

If sufficient HgI_2 is added to a 1.00 m KI solution to react with all the iodide ions present, what is the freezing point of the resulting solution? *Hint:* In solving this part of the problem, keep in mind that there are cations present, and they also contribute to the colligative effect and the freezing point.

52. Because of the marked sensitivity of osmotic pressure measurements (which produce relatively large values for very dilute solutions), molecular weights for very large molecules such as proteins can be obtained. Now suppose that a typical synthetic nylon or natural protein was soluble to the extent of 0.1 g/L, and that the limit of your ability to measure osmotic pressure is 10^{-3} torr at 298 K. Estimate the highest molecular weight you can measure by this method.

53. A sample of 10. g of an unknown nonvolatile molecular solute is dissolved in 100. g of benzene (C_6H_6). A stream of dry air is then bubbled through the resulting solution, and the loss in weight of the solution due to saturation of the air by benzene vapor is found to be 1.205 g. In a separate experiment using pure benzene, the loss in weight (on passing the same volume of dry air through) was 1.273 g. What is the formula weight of the unknown solute?

54. If 0.0093 L of helium at 20°C and 0.96 atm is required to saturate 100. g of H_2O, what is the Henry's Law constant for the gas?

APPLIED PRINCIPLES [55–62]

55. An industrial grade of "concentrated sulfuric acid" is 98.0% by weight H_2SO_4 and has a density of 1.836 g/cm³. Calculate the molality of the solution.

56. Calculate the vapor pressure, freezing point, boiling point, and osmotic pressure for the following solutions. Assume both sodium chloride and calcium chloride are completely dissociated. The vapor pressure of water at 25°C is 23.8 torr.
(a) 1 M NaCl solution, d = 1.040 g/cm³
(b) 1 M $CaCl_2$ solution, d = 1.084 g/cm³

57. Assuming the price of sodium chloride to be half that of calcium chloride, by what factor is which salt in Problem 56 more cost-effective for the purpose of salting the streets to melt ice and snow in winter?

58. A saturated solution of sodium hydrogen carbonate (sodium bicarbonate) contains 16.4 g $NaHCO_3$/100. g H_2O at 60°C. If the solution is cooled to 20°C and is then found to contain 9.6 g $NaHCO_3$/100. g H_2O, is the solution still saturated? Briefly explain. If it is still saturated, then what percentage of the dissolved salt must have crystallized from the solution?

59. The pressure of CO_2 in a 1-L bottle of soda is 1.5 atm. On opening the bottle and letting the gas escape, the pres-

sure falls to 1.0 atm. If the temperature is 20°C, what volume of gas escaped into the atmosphere?

60. If the osmotic pressure of blood is 7.6 atm at normal body temperature (98.6°F), what is its freezing point?

61. Assuming seawater to have a salt concentration of 0.52 M (moles/liter) and a density of 1.024 g/cm³, determine what pressure (in atmospheres) is needed in order to cause seawater at 25°C to move through a semipermeable membrane in a reverse osmosis desalination plant. Disregarding the presence of calcium, magnesium, and sulfate ions will not affect an order-of-magnitude calculation. At what temperature will seawater freeze?

62. The vapor pressure equilibrium curve on the phase diagram for water and for many other liquid substances can be represented by the following equation:

$$\ln p = -\left(\frac{\Delta H}{R}\right)\left(\frac{1}{T}\right) + \text{constant}$$

where p = vapor pressure
ΔH = heat of vaporization
R = universal gas constant
T = kelvin temperature
and ln denotes the natural logarithm. This is a form of an equation known as the Clausius-Clapeyron equation. Consider the following data.

Temperature, °C	p, torr
10	9.21
20	17.54
30	31.82
40	55.32
50	92.51
60	149.38

(a) Using a graphical procedure, find the heat of vaporization for water in kJ/mol.
(b) Then explain how the given equation can be used to find vapor pressures in general.

CHAPTER
10

Equilibrium I: Equilibrium Processes and Gas Phase Equilibria

Large-scale plants such as this one in The Netherlands are designed to produce upwards of 1000 to as much as 7000 metric tons of ammonia per day. There are now close to a hundred such plants around the world, especially in developing and third world countries. In the years just before and after World War I in Germany, Fritz Haber and Karl Bosch explored the equilibrium conditions of temperature and pressure, and reactor and catalyst design. The reactor shown here dates back to the 1920s and was used to prepare catalysts for ammonia synthesis.

10.1   EQUILIBRIUM IN CHEMICAL PROCESSES

An interesting model process for studying chemical equilibrium is the reversible dissociation of colorless dinitrogen tetroxide molecules into a reddish-brown gas of nitrogen dioxide molecules (Figure 10–1):

$$N_2O_4 \text{ (g)} \longrightarrow 2 \text{ NO}_2 \text{ (g)} \qquad \text{dissociation}$$

$$2 \text{ NO}_2 \text{ (g)} \longrightarrow N_2O_4 \text{ (g)} \qquad \text{dimerization}$$

At equilibrium, both gases are present and the two reactions—dissociation and dimerization—are occurring at the same rate. When equilibrium has been established, the partial pressures of nitrogen dioxide and dinitrogen tetroxide are constant. Yet the turnover between the two gaseous species continues ceaselessly. To state the equilibrium condition, one equation and double half-headed arrows are used:

$$N_2O_4 \text{ (g)} \rightleftharpoons 2 \text{ NO}_2 \text{ (g)}$$

Another interesting example is the thermal decomposition of ammonium chloride where a solid phase is present in equilibrium with a gas phase in a closed system. This is different from the NO_2/N_2O_4 system, in which the process was homogeneous with all components in the gas phase. Here, a heterogeneous equilibrium between the solid salt and the two gaseous reaction products results (Figure 10–2):

$$NH_4Cl \text{ (s)} \rightleftharpoons NH_3 \text{ (g)} + HCl \text{ (g)}$$

At room temperature, this reaction shows very little tendency to proceed from left to right, and almost none of the gaseous products can be detected over a sample of solid NH_4Cl. However, if some solid is placed in a spoon and heated over a burner, ammonia and hydrogen chloride are produced. Away from the heat source, if the gases are brought together they recombine by the reverse reaction:

$$NH_3 \text{ (g)} + HCl \text{ (g)} \rightleftharpoons NH_4Cl \text{ (s)}$$

FIGURE 10–1 The dinitrogen tetroxide/nitrogen dioxide equilibrium: (a) colorless N_2O_4; (b) red-brown NO_2; (c) the concentration of NO_2 in an equilibrium mixture of the two gases is directly related to the depth of color.

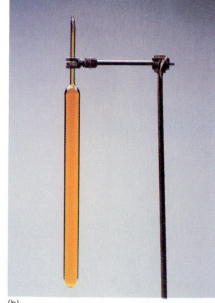

(a)

(b)

(c)

A dense white "smoke" of very small NH_4Cl crystals forms (Figure 10–3). In fact, this process has been used on a large scale in military combat to produce "smoke screens."

We can do a "labeling" experiment to show that the chemical reaction at equilibrium is dynamic. By tracking a single, isotopically tagged N atom, it is possible to demonstrate that over a sufficiently long period it spends part of its time in ammonium chloride crystals and part of its time in ammonia molecules in the gas phase. Ammonia enriched in ^{15}N, which is normally present as only 0.36% of naturally occurring nitrogen, has been used for this purpose. When introduced as ammonia into a vessel containing NH_4Cl crystals in equilibrium with NH_3 (g) and HCl (g), the labeled ^{15}N is quickly detectable in the solid.

Another important feature of chemical equilibrium is that the same equilibrium results regardless of whether the reaction starts from reactants or from products. This happens because equilibrium is a dynamic process going in both directions. If NH_3 and HCl are injected into a closed container (Figure 10–4), solid ammonium chloride forms; the resulting equilibrium

FIGURE 10–2 Heating solid ammonium chloride produces ammonia and hydrogen chloride vapors with the expected result, a white "smoke" of ammonium chloride crystals.

FIGURE 10–3 When concentrated aqueous ammonia and hydrogen chloride solutions come together, solid ammonium chloride is produced (the white smoke).

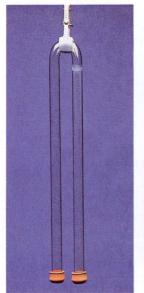

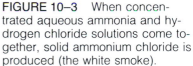

FIGURE 10–4 Diffusion brings ammonia (left) and hydrogen chloride (right), on injection into the arms of the U-tube, into contact at a point closer to the hydrogen chloride end. Eventually, equilibrium will be established. The same equilibrium can be achieved by the reverse reaction, decomposing ammonium chloride into ammonia and hydrogen chloride.

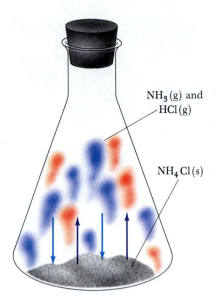

NH$_3$(g) and HCl(g)

NH$_4$Cl(s)

$$NH_3(g) + HCl(g) \rightleftharpoons NH_4Cl(s)$$

FIGURE 10–5 Solid ammonium chloride in equilibrium with gaseous ammonia and gaseous hydrogen chloride in a closed container. The rates of both processes are the same, and the partial pressures of the gases are constant.

is the same as that produced by allowing solid ammonium chloride to generate partial pressures of NH$_3$ and HCl.

The tendency toward equilibrium in a chemical system is a spontaneous process. **Spontaneous** in a chemical sense means that the process occurs at a finite rate without having to change any outside variable such as temperature or pressure. Sometimes there is no question about spontaneity, as in reactions that occur explosively, in an instant. But many spontaneous processes occur only very slowly, and there are cases in which equilibrium takes millions of years to be reached. Even if it is very slow, though, the spontaneous process will continue in its relentless pursuit of equilibrium. Once equilibrium is reached there will be no tendency for spontaneous changes unless the controlling conditions are changed. The question that will concern us throughout this chapter is, "What happens to a system at equilibrium if conditions such as temperature, pressure, and concentration change?"

10.2   LE CHATELIER'S PRINCIPLE

Stressing a System at Equilibrium by Pressure/Volume Changes

According to **Le Chatelier's principle,** when the conditions that govern a system at equilibrium change, the system changes in such a way as to restore the equilibrium condition. Put another way, if a stress is applied to a system at equilibrium, the system will try to change in order to relieve the stress. Let's look at the decomposition of ammonium chloride (Figure 10–5):

$$NH_4Cl\ (s) \rightleftharpoons NH_3\ (g) + HCl\ (g)$$

Place a stress on the system (initally at equilibrium) by increasing the volume, which has the effect of reducing the partial pressures of each of the gases. According to Le Chatelier's principle, the chemical process moves in the direction that relieves this stress or change. In this case the reaction proceeds to the right, increasing the partial pressures of the gases and reestablishing the equilibrium condition as more ammonium chloride dissociates.

If the volume is reduced, the partial pressures of the gases increase accordingly. The reaction then proceeds to the left, returning some of the gases back to solid NH$_4$Cl. The net result is to reduce the partial pressures of the gases, and again establish the equilibrium conditions.

EXAMPLE 10–1

The following chemical process is at equilibrium:

$$2\ H_2\ (g) + O_2\ (g) \rightleftharpoons 2\ H_2O\ (g)$$

How would the process respond if the pressure increased at constant temperature?

Solution   Remember the Boyle's Law relationship. Pressure and volume are inversely proportional. Increasing the pressure should favor a decreased volume, and therefore fewer moles of particles. In other words, the reaction should move left (3 moles) to right (2 moles). ■

EXERCISE 10–1

Given the following reaction at equilibrium:

$$3 \; F_2 \; (g) \; + \; Cl_2 \; (g) \rightleftharpoons 2 \; ClF_3 \; (g)$$

(a) Suppose the volume is reduced by increasing the pressure at constant temperature. In which direction does the reaction proceed in response to the change? (b) Suppose the pressure is reduced at constant temperature. In what way does the system compensate to restore the equilibrium? *Answer:* (a) To the right, forming more product molecules at the expense of the reactants, decreasing the total number of particles. (b) To the left, forming more reactant molecules at the expense of the products. ■

Stressing a System at Equilibrium by Temperature Changes

Temperature is another factor that affects the equilibrium condition. In order to determine the effect of temperature, however, we need to know whether the reaction is endothermic and absorbs heat or is exothermic and gives off heat. Experiments show, for example, that the formation of NO_2 from N_2O_4 equilibrium is endothermic. That is to say,

$$N_2O_4 \; (g) \rightleftharpoons 2 \; NO_2 \; (g) \qquad \Delta H > 0$$

Therefore, if this system is at equilibrium and the temperature is raised, the reaction will proceed from left to right because that process, and not the reverse process, is the energy-demanding process.

What will happen if the N_2O_4/NO_2 reaction is at equilibrium and the temperature is lowered? A reaction that is endothermic in one direction is always exothermic to the same degree in the opposite direction. If the temperature is lowered, the system reacts by generating heat. Thus, the system moves from right to left, favoring the exothermic direction.

Stressing A System at Equilibrium by Composition Changes

Suppose a reaction is at equilibrium and some amount of either reactants or products are added to the equilibrium mixture. For example, imagine adding more N_2O_4 (g) or NO_2 (g) to the reaction at equilibrium:

$$N_2O_4 \; (g) \rightleftharpoons 2 \; NO_2 \; (g)$$

If NO_2 is added, the reaction will proceed to the left in order to remove some of the excess of NO_2. If N_2O_4 is added the reaction will proceed to the right, removing some of the excess N_2O_4. By the same token, it is often possible to remove products or reactants from a reaction. Removal of NO_2 drives the reaction to the right, replacing the NO_2 removed, whereas removal of N_2O_4 drives the reaction to the left, producing more N_2O_4.

EXAMPLE 10–2(A)

Ethane can be produced by hydrogenation of ethylene, according to the following reaction:

$$C_2H_4 \; (g) \; + \; H_2 \; (g) \rightleftharpoons C_2H_6 \; (g) \qquad \Delta H° = -149 \; kJ/mol \; C_2H_4$$

What conditions of pressure and temperature would be most favorable for promoting the reaction in the direction written, producing ethane at the expense of ethylene and hydrogen?

Solution *Pressure:* Since the number of moles of molecules is reduced (by 2:1) as the reaction proceeds to the right, higher pressure should favor a shift in that direction. Increasing the pressure favors a smaller volume, which in turn is favored by the presence of fewer particles.

Temperature: Since the reaction is exothermic, lower temperatures would be preferred. The system spontaneously gives off heat; lowering the temperature aids removal of the heat, favoring ethane synthesis. ■

EXAMPLE 10–2(B)

In the decomposition of ammonium chloride, what is the effect on the equilibrium of increasing the partial pressure of ammonia?

$$NH_4Cl \ (s) \rightleftharpoons NH_3 \ (g) + HCl \ (g)$$

Solution   Adding NH_3 (g) drives the reaction to the left, reducing excess NH_3 (g). Note that when the reaction runs to the left, some HCl (g) must also react, reducing its partial pressure. Consequently, when the new equilibrium is established, the partial pressure of HCl (g) will be less than before the extra NH_3 was added. ■

EXERCISE 10–2(A)

The preparation of HCl (g) from the elements in their standard states can be written as follows:

$$H_2 \ (g) + Cl_2 \ (g) \rightleftharpoons 2 \ HCl \ (g) \qquad \Delta H = -185 \ kJ/mol \ H_2$$

How will the equilibrium expressed by this reaction respond to changes in temperature and pressure?

Answer: Raising the temperature will favor the reverse reaction; the process is independent of pressure. ■

EXERCISE 10–2(B)

In the decomposition of ammonium hydrogen sulfide, what is the effect on the equilibrium of increasing the partial pressure of hydrogen sulfide?

$$NH_4HS \ (s) \rightleftharpoons NH_3 \ (g) + H_2S \ (g)$$

Answer: Favors formation of NH_4HS. ■

Le Chatelier's principle is of enormous practical value as a guide for regulating conditions in order to promote the spontaneity of desired chemical reactions and to limit the spontaneity of undesired ones. In summary, reactions can be promoted by adding reactants or removing products. Endothermic reactions can be promoted by raising the temperature; exothermic reactions can be promoted by lowering the temperature. Reducing the volume (or increasing the pressure) favors reactions in which the total number of moles of gas is reduced, whereas increasing the volume (or decreasing the pressure) favors reactions in which the total number of moles of gas is increased. However, as we have suggested, these rules give no clues about how rapidly the system approaches equilibrium.

The Industrial Synthesis of Ammonia

Le Chatelier's principle and an understanding of gas phase equilibria have been useful in the industrial preparation of many important substances. For example, the successful production of ammonia on an industrial scale was the result of a careful study of the factors affecting the equilibrium. Ammonia is an exceptionally important chemical intermediate in the production of fertilizers, explosives, and other products. Before World War

HABER AMMONIA SYNTHESIS

The Haber ammonia process is the most important chemical synthesis devised in the 20th century. By 1913, Fritz Haber had worked out the essential chemistry and Karl Bosch had done the engineering needed to obtain the product inexpensively. Although this energy-intensive process has been subject to considerable modification over the years, the main theme is the same—namely, trade-offs between pressure and temperature that favor the equilibrium in the direction of ammonia synthesis, along with use of efficient catalyst materials.

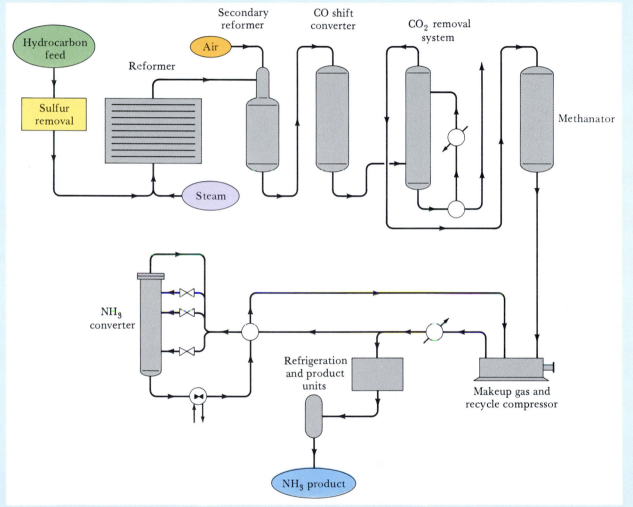

Raw materials feedstocks can be LNG (liquefied natural gas, which is largely methane), LPG (liquefied petroleum gas, which is largely propane), light naphthas (such as pentanes, hexanes, heptanes, and octanes), coal, or coke. The overall process begins with the high-pressure catalytic re-forming of the feedstock, freed of sulfur, in the presence of steam over a nickel catalyst in the primary reformer, producing **synthesis gas**. If the feedstock is methane, this reaction is

$$CH_4 \text{ (g)} + H_2O \text{ (g)} \rightleftharpoons CO \text{ (g)} + 3 H_2 \text{ (g)}$$

This is followed by a catalyzed carbon monoxide **shift reaction** in the secondary reformer to eliminate the CO and produce more H_2:

$$CO \text{ (g)} + H_2O \text{ (g)} \rightleftharpoons CO_2 \text{ (g)} + H_2 \text{ (g)}$$

(continued on next page)

If the process begins with partial combustion of coal or naphthas, then carbon and ash must be removed, followed by scrubbing to remove hydrogen sulfide:

$$2\ H_2S\ (g)\ +\ K_2CO_3\ (aq)\ \longrightarrow\ 2\ KHS\ (aq)\ +\ H_2O\ (liq)\ +\ CO_2\ (aq)$$

Then a high-pressure reaction of the hydrogen with nitrogen over a promoted (or activated) iron catalyst produces product ammonia, which is removed as liquid by refrigeration:

$$N_2\ (g)\ +\ 3H_2\ (g)\ \rightleftharpoons\ 2\ NH_3\ (g)$$

With methane or propane as raw materials, steam refining is used to produce hydrogen. Air is bled into a secondary tower to supply nitrogen, and residual carbon monoxide and carbon dioxide are removed by **methanation**:

$$CO\ (g)\ +\ 3\ H_2\ (g)\ \rightleftharpoons\ CH_4\ (g)\ +\ H_2O\ (g)$$

Compressed nitrogen and hydrogen are then fed into the ammonia synthesis unit, which uses promoted iron catalyst.

As Le Chatelier's Principle would predict, the ammonia synthesis reaction depends on both temperature and pressure. The percentage ammonia at equilibrium, beginning with a stoichiometric 3:1 mixture of hydrogen and nitrogen, at various temperatures and pressures are shown.

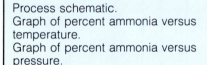

Process schematic.
Graph of percent ammonia versus temperature.
Graph of percent ammonia versus pressure.

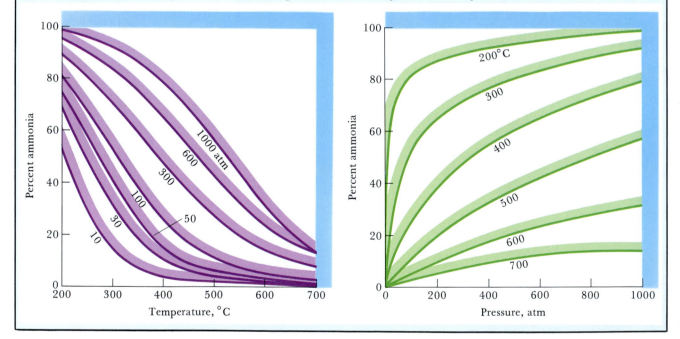

I, the traditional precursor for these products had been the Chilean nitrates, particularly sodium nitrate ($NaNO_3$). By the beginning of the 20th century, Chilean nitrates supplied two-thirds of the world's industrial and agricultural needs. At the same time, there were obvious concerns about depletion of these resources and about being dependent on one source of supply.

The production of ammonia from inexpensive gases, hydrogen and nitrogen, presented an attractive alternative if the synthesis could be carried out economically in reasonable yields. The exothermic reaction converts four moles of gaseous reactants to two moles of gaseous products:

$$N_2\ (g)\ +\ 3\ H_2\ (g)\ \rightleftharpoons\ 2\ NH_3\ (g)\qquad \Delta H\ =\ -92\ kJ/mol\ N_2$$

Applying Le Chatelier's principle leads to two conclusions: (1) raising the pressure increases the yield of NH_3; (2) raising the temperature decreases the yield of NH_3. Raising the pressure helps move the process in the right direction, shifting the equilibrium in favor of ammonia synthesis. However, the rate of the reaction is very slow, and since raising the temperature increases the reaction rate of most chemical reactions, the temperature effect seems to be in conflict with any hope of obtaining a successful process.

Fritz Haber studied this reaction carefully in the early years of the 20th century as a Professor of physical chemistry at the University of Karlsruhe in Germany. Haber was well aware of Le Chatelier's scientific work on the control of chemical equilibrium. One of the most tantalizing pieces of information was a little-known report Le Chatelier published in 1901, in which he reported that increased pressure resulted in definite improvements of the yield of ammonia from the elements. Haber found that high pressure (about 500 atm) and moderately high temperature (about 450°C) speeded the reaction without lowering the yield too much, and an iron catalyst speeded the reaction further. These conditions resulted in a workable, economical process.

SYNTHETIC AMMONIA

It took Haber, his students, and scientific associates nearly a decade to work out the details of the process to produce ammonia. It took another five years to produce ammonia on a commercial scale. The Haber process was an immediate success. By 1913, the first major plant based on the process went "on-stream." Seventy-five years later, there are some 335 active synthetic ammonia plants worldwide, based on the essential chemistry and engineering of Le Chatelier and Haber.

10.3   THE EQUILIBRIUM CONSTANT

Reactants and Products at Equilibrium

Having now discussed the effects of changing conditions on a chemical system at equilibrium in general terms, let's consider some actual data from experiments. We start with only NH_4Cl (s) in a closed container at some temperature T. When equilibrium is reached, we would expect to find at least a trace of NH_4Cl (s) left, along with equal numbers of moles, and therefore equal partial pressures, of NH_3 (g) and HCl (g):

$$NH_4Cl \text{ (s)} \rightleftharpoons NH_3 \text{ (g)} + HCl \text{ (g)}$$

Once the equilibrium condition has been established, add a quantity of either gas into the container. When equilibrium conditions have been reestablished, the partial pressures of NH_3 (g) and HCl (g) will be different, reflecting the effects of the added gas and Le Chatelier's principle. If the reaction vessel were initially charged with different partial pressures of the two gases, solid ammonium chloride would form as the system moved to establish equilibrium conditions. Table 10–1 gives the partial pressures resulting from different starting conditions for each of the gases at 300°C. The partial pressures of HCl (g) are plotted as a function of the partial pressure of NH_3 in Figure 10–6. The curve is a rectangular hyperbola that is asymptotic to the X and Y axes. The equation of such a curve is $X \cdot Y = C$, where C is a constant. In this case,

TABLE 10–1   Partial Pressures of NH₃ (g) and HCl (g) in Equilibrium with NH₄Cl (s) at 300°C, in Atmospheres

Exp. #	p_{NH_3}	p_{HCl}	$K = p_{NH_3} \cdot p_{HCl}$
1	2.3×10^{-1}	2.3×10^{-1}	5.3×10^{-2}
2	3.1×10^{-1}	1.8×10^{-1}	5.6×10^{-2}
3	4.0×10^{-1}	1.4×10^{-1}	5.6×10^{-2}
4	4.9×10^{-1}	1.1×10^{-1}	5.4×10^{-2}
5	6.0×10^{-1}	9.2×10^{-2}	5.5×10^{-2}
6	1.7×10^{-1}	3.2×10^{-1}	5.4×10^{-2}
7	1.2×10^{-1}	4.5×10^{-1}	5.4×10^{-2}
8	8.9×10^{-2}	6.2×10^{-1}	5.5×10^{-2}

FIGURE 10–6 Plot of the partial pressure data in Table 10–1.

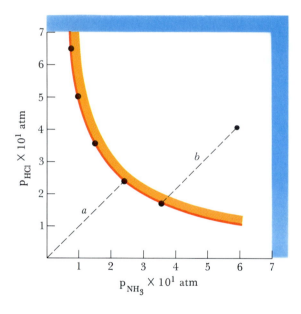

$$p(NH_3) \cdot p(HCl) = K_p$$

The constant K is called the **equilibrium constant**. An equilibrium constant calculated from partial pressures is often subscripted K_p. It has a unique value for a particular reaction at a particular temperature. K_p for this reaction can be determined by multiplying any pair of equilibrium partial pressures.

The equilibrium constant is a useful concept that can be generally applied to chemical reactions that have achieved thermodynamic equilibrium. However, when it is applied to partial pressures of reactants and products, its precision is limited by the same kinds of restraints as the Ideal Gas Law. Thus, the equilibrium law expression for the ammonium chloride decomposition and its characteristic constant K_p reasonably represent the equilibrium condition when the gas molecules are well separated from each other, under conditions of relatively low pressure and relatively high temperature. Under other circumstances, K_p isn't constant, and other methods of calculation have been developed for such situations.

Figure 10–6 can be used to describe the reaction pathway that starts with a combination of partial pressures of NH_3 and HCl moving toward equilibrium. First, note that only those partial pressure combinations that lie on the line in Figure 10–6 are at equilibrium. All others are not at equilibrium and spontaneously move toward a combination on the equi-

librium line. The balanced chemical equation demands that when the reaction runs to the right, equal partial pressures of NH_3 and HCl are produced.
That is, the partial pressures change along a diagonal line on Figure 10–6,
moving upward and to the right. If the reaction runs to the left, the partial
pressures of NH_3 and HCl are reduced by equal amounts and change along
a diagonal line on Figure 10–6, moving downward and to the left.

Two paths for non-equilibrium situations are shown on Figure 10–6.
Path *a* starts with just NH_4Cl (s) before any NH_3 (g) and HCl (g) have been
produced. Production of the gases proceeds on the line moving upward
and to the right from the origin, until the equilibrium line is reached at a
partial pressure of 2.3×10^{-1} atm for each gas. Path *b* refers to the
following example.

EXAMPLE 10–3

Consider an evacuated vessel at 300°C. NH_3 (g) and HCl (g) are quickly
injected into it so that the initial partial pressures are 6.0×10^{-1} atm for
NH_3 and 4.0×10^{-1} atm for HCl. Use Figure 10–6 to predict the spontaneous direction of the reaction and the equilibrium partial pressures of
the two gases.

Solution   From Figure 10–6 we can see that proceeding from the initial
partial pressures to the equilibrium line requires a movement downward
and to the left. Therefore, the reaction proceeds in the right-to-left (reverse)
direction, lowering the partial pressures of the NH_3 and HCl and producing
an equivalent quantity of NH_4Cl (s).

$$NH_4Cl \ (s) \rightleftharpoons NH_3 \ (g) \ + \ HCl \ (g)$$

The point of intersection for the reaction line and the equilibrium line
represents the composition of the equilibrium mixture of gases. This can
be estimated from the graph as $p(NH_3) = 3.60 \times 10^{-1}$ atm and $p(HCl) =
1.55 \times 10^{-1}$ atm.

EXERCISE 10–3

Use Figure 10–6 to predict the final composition of a gas mixture at 300°C
starting with $p(NH_3) = 1.0 \times 10^{-1}$ atm and $p(HCl) = 2.0 \times 10^{-1}$ atm.
Answer: At equilibrium, $p(NH_3) = 1.9 \times 10^{-1}$ atm, $p(HCl) = 2.8 \times 10^{-1}$
atm.   ∎

The dimerization of nitrogen dioxide results in measurable amounts
of reactant and product molecules at 25°C.

$$2 \ NO_2 \ (g) \rightleftharpoons N_2O_4 \ (g)$$

According to Le Chatelier's principle, this reaction proceeds more to the
right if the total pressure in the reaction vessel is increased. If the temperature is held constant at 25°C and the partial pressures of NO_2 and
N_2O_4 are determined for a number of different total pressures, a collection
of data can be compiled (Table 10–2).

When these data are plotted for $p(N_2O_4)$ as a function of $p(NO_2)$, the
curve in Figure 10–7 results. This curve takes the form of a parabola with
the equation $y = Kx^2$ or $K = y/x^2$. Thus, in our case

$$K_P = \frac{p(N_2O_4)}{p(NO_2)^2}$$

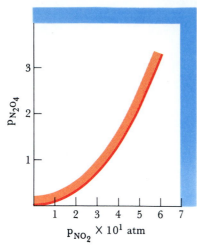

FIGURE 10–7 Plot of the partial pressure data in Table 10–2.

p_{NO_2}	$p_{N_2O_4}$	P_{total}	$K = \dfrac{p_{N_2O_4}}{p_{NO_2}^2}$
1.0×10^{-2}	8.8×10^{-4}	1.1×10^{-2}	8.8
2.0×10^{-2}	3.5×10^{-3}	2.4×10^{-2}	8.8
4.0×10^{-2}	1.4×10^{-2}	5.4×10^{-2}	8.8
8.0×10^{-2}	5.6×10^{-2}	1.36×10^{-1}	8.8
1.6×10^{-1}	2.3×10^{-1}	3.9×10^{-1}	9.0
3.2×10^{-1}	9.0×10^{-1}	1.22×10^{0}	8.8
6.4×10^{-1}	3.6×10^{0}	4.2×10^{0}	8.8

TABLE 10–2 Equilibrium Mixtures of NO_2 (g) and N_2O_4 (g) at 25°C

where K_p is the equilibrium constant for the reaction at 25°C. Every set of equilibrium partial pressures (for 25°C) lies on this curve, and there is a similar curve for each temperature.

The Generalized Equilibrium Expression

The two equilibria we have just described, the heterogeneous NH_4Cl (s)/ NH_3 (g)/HCl (g) reaction and the homogeneous NO_2 (g)/N_2O_4 (g) reaction, follow a general rule. We can express the general case by the "reaction"

$$aA + bB + \cdots \rightleftharpoons cC + dD + \cdots$$

where the uppercase letters represent reactants and products and the lowercase letters represent the coefficients in the balanced chemical equation. The equilibrium law expression then takes the following form:

$$K_p = \frac{p(C)^c p(D)^d \cdots}{p(A)^a p(B)^b \cdots} \tag{10–1}$$

In this case, reactants and products appear in the equilibrium law expression as partial pressures for the gaseous components of the system. However, if the reaction takes place in a solution, the concentrations (mol/L) of dissolved reactants and products (represented by enclosing the species in square brackets) can be used in the same way as partial pressures for gases to give K_c, the equilibrium constant based on concentrations:

$$K_c = \frac{[C]^c[D]^d \cdots}{[A]^a[B]^b \cdots} \tag{10–2}$$

K_c can also be computed for gas-phase reactions. The relationship between K_p and K_c is straightforward, assuming ideal behavior and the equation of state for an ideal gas:

$$PV = nRT$$

Given that the concentration $c = n/V$, then

$$P = cRT$$

We show the calculation for two reactants and two products, but it is easily extended to any number of species as in Eqs. (10–1) and (10–2). From the definition of K_p,

$$K_p = \frac{p(C)^c p(D)^d}{p(A)^a p(B)^b}$$

$$= \frac{(c_C RT)^c (c_D RT)^d}{(c_A RT)^a (c_B RT)^b}$$

$$= \left[\frac{(c_C)^c (c_D)^d}{(c_A)^a (c_B)^b}\right]\left[\frac{(RT)^c (RT)^d}{(RT)^a (RT)^b}\right]$$

$$= K_c (RT)^{(c+d)-(a+b)}$$

$$K_p = K_c(RT)^{\Delta n} \qquad\qquad (10\text{--}3)$$

where Δn is the difference between the numbers of moles of gaseous products and reactants.

EXAMPLE 10–4

At 523 K and 1 atm, the equilibrium constant K_p for the thermal dissociation of phosphorus pentachloride in the gas phase is 1.78:

$$PCl_5\ (g) \rightleftharpoons PCl_3\ (g) + Cl_2\ (g) \qquad K_p = 1.78$$

Calculate Δn and K_c for the reaction.

Solution For the reaction as written, there are two moles of gaseous products and one mole of gaseous reactant. Therefore,

$$\Delta n = n_{products} - n_{reactants} = [2 - 1]\ mol = 1\ mol$$

$$K_c = \frac{K_p}{(RT)^1} = \frac{1.78}{(0.0821\ L\cdot atm/mol\cdot K)(523\ K)} = 0.0415$$

Note the absence of units in expressing the equilibrium constant—a common (though not universal) convention that we adopt for simplicity. ■

EXERCISE 10–4

Calculate K_c for the following reaction at 298 K:

$$2\ NO\ (g) + O_2\ (g) \rightleftharpoons N_2O_4\ (g) \qquad K_p = 1.4 \times 10^{13}$$

Answer: 8.38×10^{15}. ■

Five Features of the Equilibrium Expression

1. Pure solids or liquids do not appear in equilibrium expressions. For example, in the expression for the ammonium chloride decomposition,

$$NH_4Cl\ (s) \rightleftharpoons NH_3\ (g) + HCl\ (g)$$

NH_4Cl (s) does not appear:

$$K_p = p(NH_3)\cdot p(HCl)$$

A large amount of experimental evidence has shown that the amount of pure solid or pure liquid present in an equilibrium reaction does not affect the partial pressures (or concentrations) of other reactants or products. It is of course a requirement that some of the pure solid or pure liquid be present to satisfy the equilibrium conditions.

EXAMPLE 10–5
We can write equilibrium expressions for the following processes:
(1) Because $CaCO_3$ and CaO are both pure solids, only the CO_2 appears in the equilibrium expression for the decomposition of limestone:

$$CaCO_3 \text{ (s)} \rightleftharpoons CaO \text{ (s)} + CO_2 \text{ (g)} \qquad K_p = p_{CO_2}$$

(2) For ammonia synthesis, all species are present in the gas phase:

$$N_2 \text{ (g)} + 3 \, H_2 \text{ (g)} \rightleftharpoons 2 \, NH_3 \text{ (g)} \qquad K_p = \frac{p_{NH_3}^2}{p_{N_2} \cdot p_{H_2}^3}$$

(3) For the phase transformation that describes the vaporization of water, assuming we are dealing with pure liquid in equilibrium with its vapor,

$$H_2O \text{ (liq)} \rightleftharpoons H_2O \text{ (g)} \qquad K_p = p(H_2O)$$

EXERCISE 10–5
Write equilibrium expressions for the following:
(a) $Cl_2 \text{ (g)} + F_2 \text{ (g)} \rightleftharpoons 2 \, ClF \text{ (g)}$
(b) $CO_2 \text{ (s)}$ in equilibrium with $CO_2 \text{ (g)}$

Answer: (a) $K_p = \dfrac{p_{ClF}^2}{p_{Cl_2} \cdot p_{F_2}}$; (b) $K_p = p_{CO_2}$.

2. If concentrations of dissolved substances appear in the equilibrium expression, the conventional shorthand for the concentration (usually the molarity, mol/L) of a dissolved substance is to enclose the formula in brackets. Thus,

$$AgCl \text{ (s)} \rightleftharpoons Ag^+ \text{ (aq)} + Cl^- \text{ (aq)} \qquad K = [Ag^+][Cl^-]$$

Partial pressures and concentrations can appear in the same expression. For example,

$$2 \, HCN \text{ (aq)} + Zn \text{ (s)} \rightleftharpoons H_2 \text{ (g)} + 2 \, CN^- \text{ (aq)} + Zn^{2+} \text{ (aq)}$$

$$K = \frac{p_{H_2}[CN^-]^2[Zn^{2+}]}{[HCN]^2}$$

EXAMPLE 10–6
We write equilibrium expressions for the following reactions:
(1) Dissociation of calcium fluoride, a salt that is sparingly soluble in water, is straightforward; the equilibrium expression states only the concentrations of the ions in solution:

$$CaF_2 \text{ (s)} \rightleftharpoons Ca^{2+} \text{ (aq)} + 2 \, F^- \text{ (aq)} \qquad K = [Ca^{2+}][F^-]^2$$

(2) The reaction of iron and hydrochloric acid has the added wrinkle of producing a gaseous product, which is included in the equilibrium expression in terms of its partial pressure:

$$2 \, Fe \text{ (s)} + 6 \, HCl \text{ (aq)} \rightleftharpoons 2 \, Fe^{3+} \text{ (aq)} + 6 \, Cl^- \text{ (aq)} + 3 \, H_2 \text{ (g)}$$

$$K = \frac{[Fe^{3+}]^2[Cl^-]^6 p_{H_2}^3}{[HCl]^6}$$

EXERCISE 10–6
Write equilibrium expressions for the following reactions:

(a) Formation of iodine monochloride, a representative of an important group of compounds known as interhalogens:

$$I_2 \text{ (s)} + Cl_2 \text{ (g)} \rightleftharpoons 2 \text{ ICl (g)}$$

(b) The equilibrium between a "hydrate" and water vapor:

$$CuSO_4 \cdot 5H_2O \text{ (s)} \rightleftharpoons CuSO_4 \cdot 3H_2O \text{ (s)} + 2 \text{ } H_2O \text{ (g)}$$

Answer: (a) $K = \dfrac{p_{ICl}^2}{p_{Cl_2}}$; (b) $K = p_{H_2O}^2$.   ∎

---

3. The form of the equilibrium expression depends on the way the chemical equation is written. For example, consider the two equations

$$2 \text{ CO (g)} + O_2 \text{ (g)} \rightleftharpoons 2 \text{ CO}_2 \text{ (g)}$$

$$CO \text{ (g)} + \tfrac{1}{2} O_2 \text{ (g)} \rightleftharpoons CO_2 \text{ (g)}$$

Both describe the same process. The difference lies in the coefficients used to balance the equations. The first equation is the same as the second, multiplied by two. The equilibrium expressions corresponding to these equations are

$$K_1 = \frac{p_{CO_2}^2}{p_{CO}^2 \cdot p_{O_2}}$$

for the first equation and

$$K_2 = \frac{p_{CO_2}}{p_{CO} \cdot p_{O_2}^{1/2}}$$

for the second, from which one can conclude that

$$K_1 = K_2^2$$

The general principle is that if two reactions differ from one another only by a simple multiplication factor, then the equilibrium constant for the first reaction equals the equilibrium constant for the other reaction raised to a power equal to the multiplication factor.

4. If a pair of equations describe the same equilibrium in the forward and reverse directions, the equilibrium constants are reciprocals. For example, the following pair of reactions represent the same equilibrium written in opposite directions:

$$C \text{ (s)} + CO_2 \text{ (g)} \rightleftharpoons 2 \text{ CO (g)}$$
$$2 \text{ CO (g)} \rightleftharpoons C \text{ (s)} + CO_2 \text{ (g)}$$

The equilibrium constant for the first reaction (we'll call it the forward reaction) is

$$K_f = \frac{p_{CO}^2}{p_{CO_2}}$$

For the second reaction (which is called the reverse reaction),

$$K_r = \frac{p_{CO_2}}{p_{CO}^2}$$

Note that K_f is the reciprocal of K_r:

$$K_f = \frac{1}{K_r}$$

EXAMPLE 10–7

Show the relationship between the equilibrium constants for the following reactions:

(a) $NO\ (g) + \frac{1}{2}O_2\ (g) \rightleftharpoons NO_2\ (g)$

(b) $2\ NO_2 \rightleftharpoons 2\ NO\ (g) + O_2\ (g)$

Solution   Equation (b) is the reverse of equation (a), multiplied by a factor of 2. Therefore,

$$K_{(b)} = \frac{1}{K_{(a)}^2}$$

■

EXERCISE 10–7

State the relationship between the equilibrium constants for the following two processes:

(a) $2\ B\ (s) + 3\ F_2\ (g) \rightleftharpoons 2\ BF_3\ (g)$

(b) $BF_3\ (g) \rightleftharpoons B\ (s) + \frac{3}{2}F_2\ (g)$

Answer: $K_{(a)} = \dfrac{1}{K_{(b)}^2}$

■

---

5. The magnitude of the equilibrium constant is a measure of the tendency of a chemical reaction to proceed along the corresponding path. Possible values for K range from very large to very small. Spontaneous reactions can have K values equal to many powers of ten. Reactions with K near 10^0, for example the NO_2/N_2O_4 reaction, have very substantial amounts of both reactants and products present at equilibrium. Reactions with very small values of K are spontaneous in the reverse direction.

10.4   CHANGING PARTIAL PRESSURES AND CHEMICAL EQUILIBRIA

The Law of Mass Action

If a reaction product is added to an equilibrium mixture, the reaction proceeds toward the left in order to restore the equilibrium condition; if additional reactant molecules are added to a chemical system at equilibrium, the reaction proceeds toward the right as the system moves to restore equilibrium. One of the guiding principles of equilibrium thermodynamics is that the amounts of reacting substances influence the extent of a chemical reaction. This idea is important enough to have a name of its own, the **Law of Mass Action**. It is an essential feature of Le Chatelier's principle. Our purpose in this section is to examine the impact of such effects on the equilibrium in terms of the equilibrium constant.

There are many ways to demonstrate that a reversible reaction has arrived at a state of thermodynamic equilibrium. For example, reconsider an earlier illustration, the time-dependency of concentration. When the concentrations of reactants and products are no longer changing with time, we say that equilibrium has been established, and we can represent that equilibrium graphically as in Figure 10–8. The same demonstration can be accomplished algebraically. Simply substitute existing concentrations (in mol/L) or partial pressures (in atm) of reactants and products into the

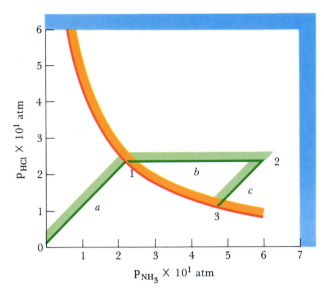

FIGURE 10–8 Three processes involving the ammonium chloride/ammonia/hydrogen chloride equilibrium at 300°C: (a) Equilibrium is reached starting with pure ammonium chloride; (b) extra ammonia is injected into the system; (c) system returns to a new equilibrium.

equilibrium expression and determine at what time successive calculations give the same unique constant—the equilibrium constant K.

The Reaction Quotient Q and the Equilibrium Constant K

The number calculated by substituting the existing values for the amounts of reactants and products is frequently called the **reaction quotient Q**. When $Q = K$, the reaction is at equilibrium. If $Q > K$, then the amounts of the products are too great and the reaction must proceed to the left to reach equilibrium. When $Q < K$, the amounts of the products are too small and the reaction must proceed to the right to reach equilibrium.

EXAMPLE 10–8

Consider once again the nitrogen dioxide/dinitrogen tetroxide equilibrium,

$$2 NO_2 (g) \rightleftharpoons N_2O_4 (g)$$

for which $K = 8.8$ at 25°C. A mixture of the two gases is sampled and the partial pressures are found to be 9.8×10^{-2} atm for NO_2 (g) and 2.1×10^{-1} atm for N_2O_4 (g). Is this system at equilibrium? If not, what will be the spontaneous direction of change?

Solution   To answer these questions, begin by calculating the reaction quotient Q:

$$Q = \frac{(2.1 \times 10^{-1})}{(9.8 \times 10^{-2})^2} = 22$$

Since $Q > K$, we conclude that the reaction is not at equilibrium. The reaction must proceed spontaneously to the left, reducing $p(N_2O_4)$ and increasing $p(NO_2)$, in order to reduce Q to the value of K. ■

EXERCISE 10–8

A closed vessel contains SO_2Cl (g) at 0.150 atm, SO_2 (g) at 0.250 atm, and Cl_2 (g) at 0.350 atm. The equilibrium constant for the following reaction is 2.4 and the temperature of the system is 100.°C:

$$2 \ SO_2Cl \ (g) \rightleftharpoons 2 \ SO_2 \ (g) + Cl_2 \ (g)$$

Is the system at equilibrium, and how do you know? If it is not, what is the spontaneous direction of the reaction?

Answer: No, since Q_p is less than K_p; to the right. ■

We know that a reversible reaction at equilibrium shifts (that is, proceeds in one direction or the other) in response to the addition of reactants or products. For example, once again allow NH_4Cl (s) to come to equilibrium with its decomposition products:

$$NH_4Cl \ (s) \rightleftharpoons NH_3 \ (g) + HCl \ (g)$$

Figure 10–8 shows the path of this reaction at 300.°C with increasing partial pressure of each gas, from 0 to 2.3×10^{-1} atm, as line *a*. Now instantly inject sufficient extra NH_3 (g) to raise p_{NH_3} to 6.0×10^{-1} while p_{HCl} remains at 2.3×10^{-1} atm; the result is shown as line *b*. Calculating Q gives the following results:

$$Q = (6.0 \times 10^{-1})(2.3 \times 10^{-1})$$
$$= 1.38 \times 10^{-1}$$

If we know $K = 5.5 \times 10^{-2}$ at 300.°C, then $Q > K$ and the reaction must proceed spontaneously to the left, reducing p_{NH_3} and p_{HCl} from their instantaneous nonequilibrium values. Note that at the new equilibrium, p_{HCl} is now lower than its original equilibrium value while p_{NH_3} is higher. That is expected, since some HCl (g) was converted to NH_4Cl (s) in reestablishing the equilibrium after extra NH_3 was added. If p_{HCl} is lower at the new equilibrium than at the original equilibrium, then p_{NH_3} will have to be higher than p_{HCl} to satisfy the equilibrium expression.

Changing the Total Pressure

According to Le Chatelier's principle, changing the total pressure displaces a system at equilibrium when the number of moles of gaseous reactants differs from the number of moles of gaseous products.

EXAMPLE 10–9

For the dimerization of dinitrogen dioxide at 25°C,

$$2 \ NO_2 \ (g) \rightleftharpoons N_2O_4 \ (g) \qquad K = \frac{p_{N_2O_4}}{p^2_{NO_2}} = 8.8$$

Suppose this system is at equilibrium in a closed vessel. What is the effect of increasing the total pressure suddenly, by reducing its volume?

Solution According to the Ideal Gas Law, $P = nRT/V$, so the partial pressures of N_2O_4 (g) and NO_2 (g) rise. The relative effect on the partial pressure will be the same for both gases. However, the value of p_{NO_2} in the denominator of the equilibrium expression is raised to the second power. Therefore, the value of Q drops below K for this reaction when the total pressure is raised, and the reaction proceeds spontaneously toward the right. This is in agreement with Le Chatelier's principle, because the system reacts so as to reduce the total amount of gas when the pressure is increased. If, on the other hand, the total pressure is reduced, we would witness the reverse effect. ■

EXERCISE 10–9

Consider the following reversible chemical reaction at equilibrium at 298 K:

$$NH_4HCO_3 \text{ (s)} \rightleftharpoons NH_3 \text{ (g)} + H_2O \text{ (liq)} + CO_2 \text{ (g)}$$

Write the equilibrium expression; state the effect of injecting CO_2 (g) on the partial pressures of both gases after equilibrium has been reestablished; and compare the new product of p_{NH_3} and p_{CO_2} to K_p.
Answer: $K_p = p_{NH_3} \cdot p_{CO_2}$; p_{PH_3} is reduced; p_{CO_2} is increased; same. ■

An interesting and unusual heterogeneous gas phase equilibrium can be found in the industrial process for the production and purification of potassium. The basic chemistry is the reaction of molten KCl with sodium vapor at 1150 K. At that temperature, sodium and potassium are in the vapor state while the two salts are liquids:

$$Na \text{ (g)} + KCl \text{ (liq)} \rightleftharpoons NaCl \text{ (liq)} + K \text{(g)}$$

The equilibrium constant for this reaction is less than 1, which reflects the greater reactivity of potassium than of sodium and favors the reactant side of the equation. However, potassium is more volatile than sodium. Therefore, potassium can be removed from the reaction vessel by distillation during the process, causing the reaction to proceed to the product side, producing potassium at the expense of sodium. In this way a practical yield of potassium metal can be obtained.

10.5 CHANGING TEMPERATURE AND CHEMICAL EQUILIBRIA

Le Chatelier's principle predicts that if the temperature of a chemical system at equilibrium is raised, the ratio of products to reactants will shift in such a way as to favor the energy-demanding, or endothermic, direction. This is evident in plots of the equilibrium constant versus temperature, since the magnitude of K is a measure of the spontaneity of the reaction. Figure 10–9 shows K as a function of temperature for the NH_4Cl (s) decomposition reaction, which is energy-demanding:

$$NH_4Cl \text{ (s)} \rightleftharpoons NH_3 \text{ (g)} + HCl \text{ (g)}$$

Figure 10–10 shows K as a function of temperature for the NO_2 dimerization reaction,

$$2 \, NO_2 \text{ (g)} \rightleftharpoons N_2O_4 \text{ (g)}$$

which is exothermic, or energy-demanding in the reverse direction. Thus, increasing temperature favors dissociation of dinitrogen tetroxide.

Two heterogeneous gas phase equilibria that are highly temperature-dependent are the decomposition reactions of $CaCO_3$ (s) and $NaHCO_3$ (s):

$$CaCO_3 \text{ (s)} \rightleftharpoons CaO \text{ (s)} + CO_2 \text{ (g)}$$
$$2 \, NaHCO_3 \text{ (g)} \rightleftharpoons Na_2CO_3 \text{ (s)} + H_2O \text{ (g)} + CO_2 \text{ (g)}$$

At room temperature the equilibrium constants for both of these reactions are very low and no significant yield of products can be obtained. However, each of these reactions is endothermic. Therefore, by raising the temperature and removing the product gases as they are produced, complete conversion to products can be achieved. The decompositon of calcium carbonate in this manner is the important industrial process by which lime-

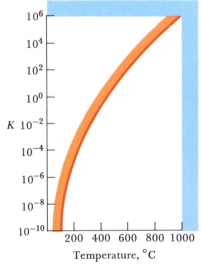

FIGURE 10–9 The equilibrium constant as a function of temperature for the ammonium chloride thermolysis:

NH_4Cl (s) $\rightleftharpoons$ NH_3 (g) + HCl (g)
$\Delta H°(298 \text{ K}) = 176.9 \text{ kJ/mol}$

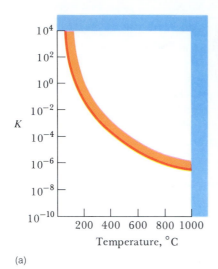

(a)

(b)

FIGURE 10–10   (a) The equilibrium constant as a function of temperature for the dimerization of nitrogen dioxide:

$$2NO_2 \text{ (g)} \rightleftharpoons N_2O_4\text{(g)} \qquad \Delta H°(298 \text{ K}) = -14.5 \text{ kJ/mol}$$

(b) The concentration of the highly colored NO_2 gas is greater in the right-hand photograph where the temperature is 50°C than in the left-hand photograph where the temperature is 0°C. Lowering the temperature favors the exothermic reaction, formation of colorless N_2O_4 gas.

stone is converted to lime for construction purposes. Sodium hydrogen carbonate ($NaHCO_3$) decomposes at a much lower temperature and is widely used to generate effervescence in beverages and "leavening" in baking.

Baking powders—there are several types—are composed of carbonates, weakly acidic substances such as tartaric acid or the monobasic phosphate of calcium ($CaHPO_4$), and a filler (perhaps corn starch). The mechanism for "double acting baking powders," which cause cookies and other baked goods to rise quickly and retain the carbon dioxide being generated as a stable solid foam, provides an interesting bit of "kitchen" chemistry. During the first few minutes of baking, the following reaction occurs:

$$2 \text{ NaHCO}_3 \text{ (s)} + 2 \text{ CaHPO}_4 \text{ (s)} \rightleftharpoons$$
$$2 \text{ H}_2\text{O (g)} + 2 \text{ CO}_2 \text{ (g)} + 2 \text{ CaNaPO}_4 \text{ (s)}$$

Substances such as wheat flour gluten are used because they are sufficiently elastic (when dampened) to hold bubbles of CO_2 long enough to give structure to the baking bread or cookie.

10.6   COMBINING CHEMICAL REACTIONS

Chemical reactions often appear to be straightforward statements of reactants and products, when in fact they are the net result of combinations of sets of reactions. It is important to know how combining chemical reactions affects the equilibrium constant for the net reaction. Consider the following three equations:

(1) $C(s) + CO_2(g) \rightleftharpoons 2\,CO(g)$ $\qquad K_1 = \dfrac{p_{CO}^2}{p_{CO_2}}$

(2) $2\,CO(g) + O_2(g) \rightleftharpoons 2\,CO_2(g)$ $\qquad K_2 = \dfrac{p_{CO_2}^2}{p_{CO}^2 \cdot p_{O_2}}$

(3) $C(s) + O_2(g) \rightleftharpoons CO_2(g)$ $\qquad K_3 = \dfrac{p_{CO_2}}{p_{O_2}}$

The third equation is the sum of the first and the second, and
$K_3 = K_1 \cdot K_2$

$$K_1 \cdot K_2 = \left[\frac{p_{CO}^2}{p_{CO_2}}\right]\left[\frac{p_{CO_2}^2}{p_{CO}^2 \cdot p_{O_2}}\right] = \frac{p_{CO_2}}{p_{O_2}} = K_3$$

The general result is that whenever two or more reactions are added together, the equilibrium constant for the net reaction is the product of the equilibrium constants for the individual steps.

EXAMPLE 10–10

Using the equilibrium constants for the following reactions at 25°C,

$N_2(g) + 3\,H_2(g) \rightleftharpoons 2\,NH_3(g)$ $\qquad K_1 = 6.80 \times 10^5$

$NH_4Cl(s) \rightleftharpoons NH_3(g) + HCl(g)$ $\qquad K_2 = 7.31 \times 10^{-17}$

$H_2(g) + Cl_2(g) \rightleftharpoons 2\,HCl(g)$ $\qquad K_3 = 2.52 \times 10^{33}$

calculate the equilibrium constant for the following reaction:

$\frac{1}{2}N_2(g) + 2\,H_2(g) + \frac{1}{2}Cl_2(g) \rightleftharpoons NH_4Cl(s)$ $\qquad K_4 = ?$

Solution   To find K_4, combine the equilibrium constants as follows:

$\frac{1}{2}N_2(g) + \frac{3}{2}H_2(g) \rightleftharpoons NH_3(g)$ $\qquad\qquad K_5 = \sqrt{K_1}$

$NH_3(g) + HCl(g) \rightleftharpoons NH_4Cl(g)$ $\qquad\qquad K_6 = \dfrac{1}{K_2}$

$\frac{1}{2}H_2(g) + \frac{1}{2}Cl_2(g) \rightleftharpoons HCl(g)$ $\qquad\qquad K_7 = \sqrt{K_3}$

$\overline{\frac{1}{2}N_2(g) + 2\,H_2(g) + \frac{1}{2}Cl_2(g) \rightleftharpoons NH_4Cl(s)}$ $\qquad \overline{K_4 = ?}$

$$K_4 = K_5 \cdot K_6 \cdot K_7 = \frac{\sqrt{K_1 \cdot K_3}}{K_2}$$

$$= \frac{\sqrt{(6.80 \times 10^5)(2.52 \times 10^{33})}}{7.31 \times 10^{-17}} = 5.66 \times 10^{35}$$

EXERCISE 10--10

Use the equilibrium constants for the following reactions to calculate the equilibrium constant for the formation of phosgene ($COCl_2$). All reactions are at 1100. K.

$2\,CO(g) \rightleftharpoons C(s) + CO_2(g)$ $\qquad\qquad K_1 = 7.7 \times 10^{-15}$

$COCl_2(g) \rightleftharpoons CO(g) + Cl_2(g)$ $\qquad\qquad K_2 = 1.7 \times 10^2$

$C(s) + CO_2(g) + 2\,Cl_2(g) \rightleftharpoons 2\,COCl_2(g)$ $\qquad K_3 = ?$

Answer: 4.5×10^9.

10.7 EQUILIBRIUM CALCULATIONS

In the simplest equilibrium calculations, you are asked to find one unknown quantity in an equilibrium expression when the others are known. The unknown quantity could be the amount of one of the reactants or products at equilibrium, or it could be the equilibrium constant itself.

EXAMPLE 10–11

The equilibrium constant for the reversible formation of nitrosyl chloride (NOCl) in the gas phase at 25°C is known:

$$Cl_2 \text{ (g)} + 2 \text{ NO (g)} \rightleftharpoons 2 \text{ NOCl (g)} \qquad K_p = 1.35 \times 10^7$$

Given that the partial pressures of Cl_2 (g) and NOCl (g) in a particular equilibrium mixture are $p_{Cl_2} = 2.7 \times 10^{-3}$ atm and $p_{NOCl} = 1.2 \times 10^{-1}$ atm, what must be the partial pressure of NO (g)?

Solution For the reaction as written, the equilibrium expression is

$$K = \frac{p_{NOCl}^2}{p_{Cl_2} \cdot p_{NO}^2}$$

Substitution and rearrangement of the terms in the equation yield the desired result, the partial pressure of NO (g):

$$p_{NO} = \frac{p_{NOCl}}{[p_{Cl_2} \cdot K]^{1/2}}$$

$$= \frac{1.2 \times 10^{-1}}{[(2.7 \times 10^{-3})(1.35 \times 10^7)]^{1/2}} = 6.3 \times 10^{-4} \text{ atm}$$

EXERCISE 10–11(A)

The equilibrium constant for the following reaction has been measured at 1000 K:

$$2 \text{ NO}_2 \text{ (g)} \rightleftharpoons 2 \text{ NO (g)} + O_2 \text{ (g)} \qquad K_p = 160.$$

An equilibrium mixture of these gases is analyzed, and the partial pressures of NO (g) and O_2 (g) are found to be 10.5 atm and 5.80 atm, respectively. What is the equilibrium partial pressure of NO_2?
Answer: 2.00 atm.

EXERCISE 10–11(B)

At 700 K, the measured values for the partial pressures of nitrogen, hydrogen, and ammonia are 2.40 atm, 7.20 atm, and 0.400 atm, respectively. What are the values for K_p and K_c at 700 K for the ammonia synthesis:

$$N_2 \text{ (g)} + 3 H_2 \text{ (g)} \rightleftharpoons 2 \text{ NH}_3 \text{ (g)}$$

Answer: $K_p = 1.79 \times 10^{-4}$; $K_c = 5.89 \times 10^{-1}$.

Starting from an initial set of conditions, equilibrium quantities of reactants and products can be found, but the calculations are somewhat more complex. In examples of this kind, it is useful to imagine the establishment of equilibrium in two stages: (1) at the instant the starting mixture is prepared, just before the spontaneous move toward equilibrium begins, and (2) after equilibrium has been reached.

EXAMPLE 10–12

The equilibrium constant for the following reaction has been measured at 250°C:

$$PCl_5 \text{ (g)} \rightleftharpoons PCl_3 \text{ (g)} + Cl_2 \text{ (g)} \qquad K_p = 2.15$$

What are the equilibrium partial pressures of all three gases in a closed container, initially charged with PCl_5 (g) at 0.100 atm and held at 250°C?

Solution The stoichiometry of the reaction is 1:1:1, which means that one mole of PCl_3 (g) and one mole of Cl_2 (g) are produced for every mole of PCl_5 (g) that reacts. Since the data are in terms of pressures rather than moles, it is convenient to use pressures (instead of concentrations in mol/L) in our calculations. This presents no problem (assuming ideal behavior), since partial pressures and moles are directly proportional, according to the Ideal Gas Law and Dalton's Law of Partial Pressures. As the reaction proceeds to the right, the increase in the partial pressure of PCl_3 equals the increase in the partial pressure of Cl_2. At the same time, it is also true that the increase in the partial pressure of PCl_3 equals the decrease in the partial pressure of PCl_5.

Set up the problem in two stages: (1) the initial conditions, before the move toward equilibrium begins; and (2) the conditions at equilibrium. Let x, the unknown in the problem, equal the increase in partial pressure of PCl_3 (g). Set up a table for the initial and equilibrium conditions:

	PCl_5 (g) $\rightleftharpoons$	PCl_3 (g) +	Cl_2 (g)
Initial pressures	0.100 atm	0 atm	0 atm
Equilibrium pressures	$(0.100 - x)$ atm	x atm	x atm

Substituting the equilibrium values for the respective partial pressures into the equilibrium expression:

$$K = \frac{p_{PCl_3} \cdot p_{Cl_2}}{p_{PCl_5}}$$

$$= \frac{(x)(x)}{0.100 - x} = \frac{x^2}{0.100 - x} = 2.15$$

Combining and rearranging the terms leaves us with a second-order equation,

$$x^2 + 2.15x - 0.215 = 0$$

which can be solved for x in a straightforward fashion, using the quadratic formula:

$$x = \frac{-b \pm \sqrt{(b^2 - 4ac)}}{2a}$$

where $a = 1$, $b = 2.15$, and $c = -0.215$. The two roots, or values of x, for this equation are

$$x = 0.0957 \text{ atm} \qquad \text{and} \qquad x = -2.25 \text{ atm.}$$

Although both roots satisfy the quadratic formula, only one has physical meaning. The other can be eliminated—an equilibrium partial pressure of negative 2.25 atm for PCl_3 (g) can be rejected, since any pressure less than zero makes no sense. Finally, the equilibrium partial pressures are:

$$p_{PCl_3} = p_{Cl_2} = x = 0.0957 \text{ atm}$$

$$p_{PCl_5} = (0.100 - x) \text{ atm} = (0.100 - 0.0957) \text{ atm} = 0.004 \text{ atm}$$

In performing these calculations, it is often useful to check the results by inserting the calculated partial pressures back into the equilibrium expression and determining whether the equilibrium constant is correct within the limits of round-off error:

$$K_p = \frac{p_{PCl_3} \cdot p_{Cl_2}}{p_{PCl_5}} = \frac{(0.0957)^2}{0.004} = 2.29$$

which is not far from the given value of 2.15.   ■

EXERCISE 10–12

Calculate the equilibrium partial pressures for the reaction in Example 10–12, starting with $p_{PCl_5} = 0.750$ atm.

Answer: $p_{PCl_3} = p_{Cl_2} = 0.588$ atm; $p_{PCl_5} = 0.162$ atm.   ■

Matters become more complicated when more than one of the gases is present in the initial mixture. Consider the reaction in Example 10–12 again.

EXAMPLE 10–13

Given initial partial pressures of $p_{PCl_5} = 0.0500$ atm, $p_{PCl_3} = 0.150$ atm, and $p_{Cl_2} = 0.250$ atm at 250°C for the following reaction, what must the equilibrium partial pressures be?

$$PCl_5 \text{ (g)} \rightleftharpoons PCl_3 \text{ (g)} + Cl_2 \text{ (g)} \qquad K_p = 2.15$$

Solution   First, calculate the reaction quotient Q to determine whether the reaction is at equilibrium:

$$Q = \frac{(0.150)(0.250)}{0.0500} = 0.750$$

Since we know $K_p = 2.15$, the reaction is clearly not at equilibrium. Because $Q < K$, the reaction must proceed to the right as it moves toward equilibrium. Letting x be the increase in partial pressure of PCl_3 (g), set up a table of initial and equilibrium conditions:

	PCl₅ (g)	⇌ PCl₃ (g)	+ Cl₂ (g)
Initial pressures	0.0500	0.150	0.250
Equilibrium pressures	$(0.0500 - x)$	$(0.150 + x)$	$(0.250 + x)$

Substitute into the equilibrium expression and solve for x:

$$2.15 = \frac{(0.150 + x)(0.250 + x)}{0.0500 - x}$$

which can be rearranged into the second-order equation

$$x^2 + 2.55x - 0.0700 = 0$$

The two solutions to the quadratic equation for x are 0.0272 atm and -2.58 atm. Reject the negative value for x and calculate the equilibrium partial pressures as you defined them in your table. At equilibrium:

$p_{PCl_5} = 0.0500 - x = 0.0500 - 0.0272 = 0.023$ atm

$p_{PCl_3} = 0.150 + x = 0.150 + 0.0272 = 0.177$ atm

$p_{Cl_2} = 0.250 + x = 0.250 + 0.0272 = 0.277$ atm

As a check on the correctness of this solution,

$$K_p = \frac{(0.177)(0.277)}{0.023} = 2.13$$

EXERCISE 10–13

Calculate the equilibrium partial pressures for the reaction in Example 10–13 if the initial partial pressures are $p_{PCl_5} = 0.850$ atm, $p_{PCl_3} = 0.440$ atm, and $p_{Cl_2} = 0.935$ atm.

Answer: $p_{PCl_5} = 0.485$; $p_{PCl_3} = 0.805$ atm; $p_{Cl_2} = 1.30$ atm. ∎

Another variation on this theme occurs when only the total pressure at equilibrium is known. Since the starting conditions are generally not specified, the method of treatment is slightly different.

EXAMPLE 10–14

Consider the dimerization of nitrogen dioxide, for which the equilibrium constant is known at 298 K:

$$2 \ NO_2 \ (g) \rightleftharpoons N_2O_4 \ (g) \qquad K_p = 8.8$$

What are the equilibrium partial pressures if the total pressure is 0.220 atm?

Solution To begin with, write the equilibrium expression for K_p:

$$K_p = \frac{p_{N_2O_4}}{p_{NO_2}^2}$$

Assign x to be the partial pressure of N_2O_4 (g). Since the total pressure P_T is the sum of the partial pressures, we can define the partial pressure of NO_2 (g):

$$P_T = p_{N_2O_4} + p_{NO_2}$$

$$p_{NO_2} = (0.220 - x) \ \text{atm}$$

Substitution into the equilibrium expression gives

$$\frac{x}{(0.220 - x)^2} = 8.8$$

which we rearrange to

$$8.8x^2 - 4.87x + 0.426 = 0$$

Solving for x using the quadratic equation gives roots of 0.444 and 0.109. The root 0.444 can be discarded, since it gives a negative value of p_{NO_2}. The partial pressures are thus:

$$p_{N_2O_4} = 0.109 \ \text{atm}$$

$$p_{NO_2} = 0.220 - 0.109 \ \text{atm} = 0.111 \ \text{atm}$$

Check

$$\frac{p_{N_2O_4}}{p_{NO_2}^2} = \frac{0.109}{(0.111)^2} = 8.8$$

EXERCISE 10–14

For the following process at 700°C, what is the partial pressure of each gas at equilibrium if the total pressure is 0.750 atm?

$$C \text{ (s)} + CO_2 \text{ (g)} \rightleftharpoons 2 \text{ CO (g)} \qquad K_p = 1.50$$

Answer: $p(CO_2) = 0.203$ atm; $p(CO) = 0.547$ atm. ■

SUMMARY

Equilibrium exists when the forward and reverse reactions of a reversible process proceed at equal rates. Equilibrium vapor pressures and the solubility of substances in saturated solutions are good examples. For a reversible chemical reaction, the observable result is that the concentrations and partial pressures of reactants and products change with time to the point where equilibrium is established. At that point, the amounts of reactants and products are no longer changing with time. **Chemical equilibrium** is a dynamic process involving continual interconversion between reactants and products. The tendency for a reversible process to move to the equilibrium condition is spontaneous in either direction, starting with reactants or products.

Le Chatelier's principle summarizes the impact of changing conditions on a reversible process or system at equilibrium. A system or process responds to relieve the stress caused by changes in pressure or volume, changes in temperature, and removal or addition of reactants or products. For the general case,

$$aA + bB \rightleftharpoons cC + dD$$

When all the components are in the gas phase and measured by their partial pressures at equilibrium,

$$K_p = \frac{P_C^c P_D^d}{P_A^a P_B^b}$$

where the equilibrium constant K_p is unique for a particular temperature. When concentrations (in moles per liter) are used in place of partial pressures (for reactions in solution), the equilibrium constant is specified as K_c; the values of K_p and K_c are related by a simple equation:

$$K_p = K_c(RT)^{\Delta n}$$

Pure solids and liquids do not appear in equilibrium expressions because the concentration of a solid in moles per liter of itself is constant. Equilibrium expressions are accurate at low partial pressures or in dilute solutions.

The reaction quotient Q is a useful measure of whether a system is at equilibrium or on the way and in which direction it is proceeding, forward or reverse. Q is calculated from existing partial pressures or concentrations. If Q is equal to K, the system is at equilibrium. If $Q < K$, then the reaction will proceed spontaneously to the right; if $Q > K$, then the reaction proceeds spontaneously in the opposite direction, to the left.

Temperature-dependence of the equilibrium constant is a function of the direction of energy demand. Le Chatelier's principle predicts that if the temperature of a chemical system at equilibrium is raised, the equilibrium will shift in such a way as to favor the energy-demanding (endothermic) direction; lowering the temperature will produce the opposite effect.

When a chemical equilibrium is the net result of two or more reactions, the equilibrium constant for the net reaction is the product of the equilibrium constants for the individual steps whose sum is that net result. If an equilibrium reaction is written in reverse order, the equilibrium constant is the reciprocal of that for the original expression.

Although it is difficult to summarize "problem-solving," note that practice at solving equilibrium problems is essential to understanding these ideas and to success in future chapters, many of which depend on principles established here. Representative examples are homogeneous reactions such as the ammonia synthesis,

$$N_2 \ (g) \ + \ 3H_2 \ (g) \longrightarrow 2 \ NH_3 \ (g)$$

the thermal decomposition of phosphorus pentachloride,

$$PCl_3 \ (g) \ + \ Cl_2 \ (g) \longrightarrow PCl_5 \ (g)$$

the dissociation of dinitrogen tetroxide,

$$N_2O_4 \ (g) \longrightarrow 2 \ NO_2 \ (g)$$

and the heterogeneous decomposition of solid ammonium chloride,

$$NH_4Cl \ (s) \longrightarrow NH_3 \ (g) \ + \ HCl \ (g)$$

These reactions illustrate pressure and temperature dependence of the equilibrium, and the effects of the law of mass action. All four reactions are pressure dependent because Δn, the difference between the numbers of moles of reactants and products, is not zero.

QUESTIONS AND PROBLEMS

QUESTIONS

1. What tests would you perform to determine whether a chemical reaction is really at equilibrium or is just proceeding very slowly?

2. Define the word "spontaneous" as used to describe chemical processes. Will a spontaneous process always occur immediately?

3. By some simple experiments, describe methods for demonstrating the reversibility of the following reactions:

$$H_2 \ (g) + I_2 \ (g) \rightleftharpoons 2 \ HI \ (g)$$

$$C \ (s) + O_2 \ (g) \rightleftharpoons CO_2 \ (g)$$

$$CaCl_2 \ (s) + Na_2CO_3 \ (s) \rightleftharpoons 2 \ NaCl \ (s) + CaCO_3 \ (s)$$

4. The mixture of the two flammable gases produced in the following endothermic reaction is very useful for synthetic chemistry, and as an energy-rich fuel:

$$C \ (s) + H_2O \ (g) \rightleftharpoons CO \ (g) + H_2 \ (g)$$

What conditions would you use to achieve the best possible yield of products?

5. Briefly explain Le Chatelier's principle and discuss the several aspects of a chemical reaction that affect the equilibrium condition.

6. Iodine in the vapor state is known to dissociate into a gas of atoms according to the following equation:

$$I_2 \ (g) \rightleftharpoons 2 \ I \ (g) \qquad \Delta H \ = \ +150 \ kJ/mol$$

What happens to this reversible process at equilibrium in each of the following cases?
(a) T increases
(b) P increases
(c) V increases
(d) $I_2 \ (g)$ concentration increases
(e) a catalyst is added

7. In the production of ammonia from nitrogen and hydrogen, the percentage ammonia at equilibrium changes from 2% at 10. atm to 97% at 3500. atm. Why then do industrial plants synthesizing ammonia on a commercial scale settle for only 50 to 70% yields and operate at 600. to 1000. atm?

8. Can you compare the magnitudes of the equilibrium constants of two reactions as measures of their relative spontaneity? Suggest a pair of reactions for which such a direct comparison is possible, and a pair for which it is not.

9. What is an equilibrium constant? Of what value is it in describing the equilibrium thermodynamics of reversible reactions? What are the limitations of the equilibrium constant expressions using concentrations and/or partial pressures?

10. At 400°C and at a nitrogen-to-hydrogen ratio of 1:3, the equilibrium constant for ammonia synthesis by the Haber process is 1.64×10^{-3}, and ΔH is -25.14 kcal. Qualitatively assess the effect on the equilibrium of the following changes:

(a) raising the temperature
(b) running the reaction in a smaller vessel
(c) adding more nitrogen

11. Distinguish between K_p, K_c, and Q.

12. Consider each of the following reversible reactions at equilibrium and qualitatively determine what would happen if

• the temperature decreases while pressure is held constant.

• the pressure decreases by increasing the volume at constant temperature.

• the partial pressure of the substance shown in **bold** type increases if it is a reactant, or is removed if it is a product.

(a) $\mathbf{N_2}$ (g) + O_2 (g) $\rightleftharpoons$ 2 NO (g) $\Delta H = 45.2$ kJ/mol
(b) $2 H_2$ (g) + O_2 (g) $\rightleftharpoons$ 2 $\mathbf{H_2O}$ (g) $\Delta H = -122$ kJ/mol
(c) $3 O_2$ (g) $\rightleftharpoons$ 2 $\mathbf{O_3}$ (g) $\Delta H = 67.8$ kJ/mol
(d) $\mathbf{C_2H_2}$ (g) + H_2 (g) $\rightleftharpoons$ C_2H_4 (g) $\Delta H = 172$ kJ/mol
(e) $\mathbf{CO_2}$ (g) + H_2 (g) $\rightleftharpoons$ CO (g) + H_2O (g)
                                    $\Delta H = -1.88$ kJ/mol

13. For each reaction in Question 12, is K_p qualitatively larger, smaller, or the same as K_c? Briefly explain.

14. Why is there no difference between K_p and K_c for the following reaction?

$$H_2 \text{ (g)} + I_2 \text{ (g)} \rightleftharpoons 2 \text{ HI (g)}$$

PROBLEMS

Le Chatelier's Principle; Equilibrium Expressions   [1–14]

1. Given the following reversible chemical reactions at equilibrium, write the equilibrium expression and state the effect of increasing the total pressure on production of the indicated products.

(a) $2 \text{ NO (g)} + O_2 \text{ (g)} \rightleftharpoons 2 \text{ NO}_2 \text{ (g)}$
(b) $3 \text{ NO (g)} \rightleftharpoons N_2O \text{ (g)} + NO_2 \text{ (g)}$

2. Write the equilibrium expressions for the following reactions and state the effect on the direction of the reaction that would result from decreasing the total pressure.

(a) $NH_4HS \text{ (s)} \rightleftharpoons NH_3 \text{ (g)} + H_2S \text{ (g)}$
(b) $C \text{ (s)} + CO_2 \text{ (g)} \rightleftharpoons 2 \text{ CO (g)}$

3. Given the following reversible chemical reactions at equilibrium, write the equilibrium expression and state the effect of increasing the partial pressure of the substance shown in **bold** type.

(a) $CaC_2O_4 \text{ (s)} \rightleftharpoons CaCO_3 \text{ (s)} + \mathbf{CO} \text{ (g)}$
(b) $PCl_5 \text{ (g)} \rightleftharpoons \mathbf{PCl_3} \text{ (g)} + Cl_2 \text{ (g)}$

4. Write the equilibrium expression for each of the fol-

lowing reversible reactions and state the effect of increasing the partial pressure of the substance shown in **bold** type.

(a) CH_4 (g) + $\mathbf{H_2O}$ (g) $\rightleftharpoons$ CO (g) + $3 H_2$ (g)
(b) $2 SO_2$ (g) + O_2 (g) $\rightleftharpoons$ 2 $\mathbf{SO_3}$ (g)

5. Write the equilibrium expression for each of the following reactions:

(a) $2 H_3O^+ + Zn \text{ (s)} \rightleftharpoons$
                    $H_2 \text{ (g)} + Zn^{2+} \text{ (aq)} + 2 H_2O \text{ (liq)}$
(b) $CaF_2 \text{ (s)} \rightleftharpoons Ca^{2+} \text{ (aq)} + 2 F^- \text{ (aq)}$
(c) $Cu \text{ (s)} + Cu^{2+} \text{ (aq)} + 2 Cl^- \text{ (aq)} \rightleftharpoons 2 \text{ CuCl (s)}$
(d) $3 V \text{ (s)} + 2 Cr^{3+} \text{ (aq)} \rightleftharpoons 3 V^{2+} \text{ (aq)} + 2 \text{ Cr (s)}$

6. Write the equilibrium expression for each of the following:

(a) $CuO \text{ (s)} + H_2 \text{ (g)} \rightleftharpoons Cu \text{ (s)} + H_2O \text{ (g)}$
(b) $4 CuO \text{ (s)} \rightleftharpoons 2 Cu_2O \text{ (s)} + O_2 \text{ (g)}$
(c) $CuSO_4 \cdot 5H_2O \text{ (s)} \rightleftharpoons CuSO_4 \text{ (s)} + 5 H_2O \text{ (g)}$

7. State the effect upon the position of the equilibrium that would be caused by each of the following changes on this reaction.

$$4 \text{ HCl (g)} + O_2 \text{ (g)} \rightleftharpoons 2 H_2O \text{ (g)} + 2 Cl_2 \text{ (g)}$$
$$\Delta H = +30 \text{ kJ/mol HCl}$$

(a) adding of 1 mol of Cl_2 (g)
(b) increasing the partial pressure of O_2 (g)
(c) decreasing the temperature
(d) increasing the total pressure
(e) using a smaller reaction flask
(f) adding a catalyst

8. Consider the dissociation of phosgene in the gas phase to carbon monoxide and chlorine, according to the following equation:

$$COCl_2 \text{ (g)} \rightleftharpoons CO \text{ (g)} + Cl_2 \text{ (g)}$$

(a) Write the equilibrium expression for the process as written.

(b) State the effect of each of the following on the equilibrium (proceeds left, proceeds right, or remains unchanged):

• raising the partial pressure of $COCl_2$ (g)
• raising the partial pressure of Cl_2 (g)
• raising the partial pressure of CO (g)

(c) If the process is exothermic, is K_p larger, smaller, or unchanged at higher temperatures? Why?

9. State the relationship that exists between the equilibrium constants K_1 and K_2 for the following pairs of reaction by expressing the second in terms of the first:

(a) $2 H_2$ (g) + O_2 (g) $\rightleftharpoons$ 2 H_2O (g) K_1
     H_2 (g) + $\frac{1}{2} O_2$ (g) $\rightleftharpoons$ H_2O (g) K_2
(b) CO (g) + Cl_2 (g) $\rightleftharpoons$ $COCl_2$ (g) K_1
     $COCl_2$ (g) $\rightleftharpoons$ CO (g) + Cl_2 (g) K_2

10. By expressing the equilibrium constant for the second reaction in terms of the first, state the relationship that exists between the equilibrium constants K_1 and K_2 for the following pairs of reactions:

(a) Cl_2 (g) + 2 NO (g) $\rightleftharpoons$ 2 NOCl (g) K_1
     NOCl (g) $\rightleftharpoons$ $\frac{1}{2} Cl_2$ (g) + NO (g) K_2
(b) NO (g) $\rightleftharpoons$ $\frac{1}{3} N_2O$ (g) + $\frac{1}{3} NO_2$ (g) K_1
     3 NO (g) $\rightleftharpoons$ N_2O (g) + NO_2 (g) K_2

11. Use the equilibrium constants for the first two reactions to calculate the equilibrium constant for the third:

$$2 \text{ NO (g)} + \text{O}_2 \text{ (g)} \rightleftharpoons \text{N}_2\text{O}_4 \text{ (g)}$$
$$K = 1.49 \times 10^{13}$$

$$2 \text{ NO (g)} + \text{O}_2 \text{ (g)} \rightleftharpoons 2 \text{ NO}_2 \text{ (g)}$$
$$K = 1.66 \times 10^{12}$$

$$\text{N}_2\text{O}_4 \text{ (g)} \rightleftharpoons 2 \text{ NO}_2 \text{ (g)} \qquad K = ?$$

12. Use the equilibrium constants for the first two reactions to calculate the equilibrium constant for the third reaction:

$$\text{CoO (s)} + \text{H}_2 \text{ (g)} \rightleftharpoons \text{Co (s)} + \text{H}_2\text{O (g)}$$
$$K = 67$$

$$\text{CO (g)} + \text{H}_2\text{O (g)} \rightleftharpoons \text{CO}_2 \text{ (g)} + \text{H}_2 \text{ (g)}$$
$$K = 7.3$$

$$\text{Co (s)} + \text{CO}_2 \text{ (g)} \rightleftharpoons \text{CoO (s)} + \text{CO (g)}$$
$$K = ?$$

13. Given the following information about the dissociation of sulfuryl chloride at 373 K, calculate K_c for the reaction:

$$\text{SO}_2\text{Cl}_2 \text{ (g)} \rightleftharpoons \text{SO}_2 \text{ (g)} + \text{Cl}_2 \text{ (g)} \qquad K_p = 2.4$$

14. Given the following information about the ammonia synthesis at 298 K, calculate K_c for the reaction:

$$\text{N}_2 \text{ (g)} + 3 \text{ H}_2 \text{ (g)} \rightleftharpoons 2 \text{ NH}_3 \text{ (g)}$$
$$K_p = 6.80 \times 10^5$$

Reaction Quotients [15–18]

15. The equilibrium constant at 600°C for the reaction

$$\text{CO (g)} + \text{Cl}_2 \text{ (g)} \rightleftharpoons \text{COCl}_2 \text{ (g)}$$

is 0.20. A mixture of these three gases in a container at this temperature is sampled, and the partial pressures are found to be $p(\text{CO}) = 0.35$ atm, $p(\text{Cl}_2) = 0.52$ atm, and $p(\text{COCl}_2) = 0.12$ atm. Is the system at equilibrium? If not, what will be the spontaneous direction of reaction?

16. An experiment involved the reversible dissociation of phosphorus pentachloride according to the following equation:

$$\text{PCl}_5 \text{ (g)} \rightleftharpoons \text{PCl}_3 \text{ (g)} + \text{Cl}_2 \text{ (g)}$$

It was found that at a certain temperature, K_p was 1.15. If 0.250 atm of PCl_5, 0.250 atm of Cl_2, and 0.500 atm of PCl_5 are mixed at that same temperature, does any net reaction occur? If so, will Cl_2 be formed or consumed? Show the calculation necessary to arrive at your answers.

17. Given the equilibrium constants at 1300 K for the following reversible reactions, find the equilibrium constant for the third:

$$\text{C (s)} + 2 \text{ H}_2\text{O (g)} \rightleftharpoons \text{CO}_2 \text{ (g)} + 2 \text{ H}_2 \text{ (g)}$$
$$K_1 = 3.8$$

$$\text{CO}_2 \text{ (g)} + \text{H}_2 \text{ (g)} \rightleftharpoons \text{CO (g)} + \text{H}_2\text{O (g)}$$
$$K_2 = 0.70$$

$$\text{C (s)} + \text{CO}_2 \text{ (g)} \rightleftharpoons 2 \text{ CO (g)} \qquad K_3 = ?$$

In which direction will the reaction proceed if the initial partial pressures of carbon monoxide and carbon dioxide are 1.50 atm and 1.40 atm, respectively?

18. Consider the following set of equilibrium reactions:

$$\text{Fe}_2\text{O}_3 \text{ (s)} + 3 \text{ H}_2 \text{ (g)} \rightleftharpoons 2 \text{ Fe (s)} + 3 \text{ H}_2\text{O (g)}$$
$$K_1 = 100.$$

$$\text{Fe}_2\text{O}_3 \text{ (s)} + 3 \text{ CO (g)} \rightleftharpoons 2 \text{ Fe (s)} + 3 \text{ CO}_2 \text{ (g)}$$
$$K_2 = ?$$

$$\text{CO}_2 \text{ (g)} + \text{H}_2 \text{ (g)} \rightleftharpoons \text{CO (g)} + \text{H}_2\text{O (g)}$$
$$K_3 = 0.100$$

(a) Write the equilibrium expressions for the three reversible reactions, and calculate the missing equilibrium constant K_2.
(b) Given 5 mol of CO_2, 4 mol of H_2, 3 mol of CO, and 2 mol of H_2O, determine whether the third reaction is at equilibrium. If it is not at equilibrium, indicate whether the reaction is forming more CO_2 or more CO as it proceeds.
(c) The third reaction is known to be endothermic. What will be the effect of raising the temperature on the extent of the reaction?
(d) How would a catalyst effect the equilibrium position?

Equilibrium Calculations [19–28]

19. At 250°C the equilibrium constant for the reaction

$$\text{PCl}_5 \text{ (g)} \rightleftharpoons \text{PCl}_3 \text{ (g)} + \text{Cl}_2 \text{ (g)}$$

is 2.15. Calculate the partial pressures of all these gases after equilibrium has been reached in a reaction vessel in which initially only PCl_5 was present at 0.012500 atm.

20. At 600 K the equilibrium constant is 38.6 for the reaction

$$2 \text{ HI (g)} \rightleftharpoons \text{H}_2 \text{ (g)} + \text{I}_2 \text{ (g)}$$

Start with a 1.000-L closed vessel containing 1.500 g of HI (g). Calculate the equilibrium partial pressures of HI, H_2, and I_2 at equilibrium.

21. At 900 K the equilibrium constant is 0.587 for the reaction

$$2 \text{ SO}_2 \text{ (g)} + \text{O}_2 \text{ (g)} \rightleftharpoons 2 \text{ SO}_3 \text{ (g)} \qquad K_p = 0.587$$

Calculate the partial pressures of all three gases at 900 K after equilibrium has been reached, starting with $p(\text{O}_2) = 0.100$ atm and $p(\text{SO}_2) = 0.100$ atm. Assume $p(\text{SO}_3) = 0$ initially.

22. At 100°C the equilibrium constant is 2.4 for the reaction

$$\text{SO}_2\text{Cl}_2 \text{ (g)} \rightleftharpoons \text{SO}_2 \text{ (g)} + \text{Cl}_2 \text{ (g)}$$

Calculate the equilibrium partial pressures of all three gases if 1.000 g of SO_2 and 1.000 g Cl_2 are injected into a 3.000-L vessel at 100°C.

23. CO, H_2O, CO_2, and H_2 are injected into a reaction vessel at 500°C so that the initial partial pressure of each gas is 0.100 atm. The equilibrium constant for the reaction is 3.9 at this temperature:

$$\text{CO (g)} + \text{H}_2\text{O (g)} \rightleftharpoons \text{CO}_2 \text{ (g)} + \text{H}_2 \text{ (g)}$$

Calculate the equilibrium partial pressure of each of the four gases.

24. When 5.000 g of phosgene ($COCl_2$) is placed in a 5.000-L container at 600 K and allowed to come to equilibrium, $p(COCl_2)$ is 0.0420 atm. Calculate the equilibrium constant for the dissociation of phosgene:

$$COCl_2 \text{ (g)} \rightleftharpoons CO \text{ (g)} + Cl_2 \text{ (g)}$$

25. The equilibrium constant for the following reaction at 690 K is known:

$$CO \text{ (g)} + H_2O \text{ (g)} \rightleftharpoons CO_2 \text{ (g)} + H_2 \text{ (g)}$$
$$K_p = 0.10$$

(a) Write the equilibrium expression.
(b) Starting with one mole of CO (g) and one mole of H_2O (g), determine the number of moles at equilibrium for all compounds present.
(c) If equimolar concentrations of all reactants and products are present initially, does the reaction favor carbon monoxide or carbon dioxide formation?

26. Consider the thermal decomposition of iron(II) sulfate, according to the following reaction:

$$2 FeSO_4 \text{ (s)} \rightleftharpoons Fe_2O_3 \text{ (s)} + SO_2 \text{ (g)} + SO_3 \text{ (g)}$$

At equilibrium at 373 K, both solids are present and the total gas pressure is 1.0 atm.
(a) Write the equilibrium expression.
(b) Calculate K_p and K_c.
(c) Calculate the total pressure at equilibrium, starting with both solids and sulfur dioxide at an initial partial pressure of 1.0 atm.

27. In a 5.00-L container, 7.50 mol of carbon dioxide was heated until 20.0% had dissociated at 3400 K, according to the following reaction:

$$2 CO_2 \text{ (g)} \rightleftharpoons 2 CO \text{ (g)} + O_2 \text{ (g)}$$
$$\Delta H = 47.5 \text{ kJ/mol } CO_2 \text{ (g)}$$

(a) Calculate K_p and K_c for the reaction as written, and for the reverse reaction.
(b) What effect would a pressure increase have on the original equilibrium?
(c) What effect would a decrease in temperature have?
(d) What effect would the presence of an inhibitor (negative catalyst) have on the position of the equilibrium?

28. Synthesis gas (or "Syngas") is a mixture of CO and hydrogen. It is extremely useful for synthesizing a variety of important commercial chemicals and articles of commerce. Methanol (methyl alcohol) can be produced as follows:

$$CO \text{ (g)} + 2 H_2 \text{ (g)} \rightleftharpoons CH_3OH \text{ (g)}$$

Along with the catalyst needed to make the reaction happen at a reasonable rate, 0.10 mol of carbon monoxide was placed in a 2.0-L flask that was carefully maintained at 427°C. Hydrogen was then added until the total pressure reached 7.0 atm. Assume that equilibrium has been established at that point and that 0.060 mol of methanol has formed in the reactor.
(a) What is K_p for the reaction?
(b) Determine the total pressure if the same quantities of carbon monoxide and hydrogen were introduced into the

reaction but the technician failed to add the catalyst, so that no measurable reaction took place.

Additional Problems   [29–39]

29. Write the equilibrium expression for each of the following reversible reactions:
(a) Synthesis of the volatile compound uranium hexafluoride:

$$UF_4 \text{ (s)} + F_2 \text{ (g)} \rightleftharpoons UF_6 \text{ (g)} \qquad K_p = ?$$

(b) Synthesis of the interhalogen compound bromine monochloride:

$$Br_2 \text{ (liq)} + Cl_2 \text{ (g)} \rightleftharpoons 2 BrCl \text{ (g)} \qquad K_p = ?$$

(c) Xenon hexafluoride is known to react with silica to form tetrafluoroxenon(VI) oxide:

$$SiO_2 \text{ (s)} + 2 XeF_6 \text{ (liq)} \rightleftharpoons$$
$$SiF_4 \text{ (g)} + 2 XeF_4O \text{ (liq)} \qquad K_p = ?$$

30. The chemistry of the oxyacetylene torch can be written as a direct combustion reaction:

$$2 C_2H_2 \text{ (g)} + 5 O_2 \text{ (g)} \rightleftharpoons$$
$$4 CO_2 \text{ (g)} + 2 H_2O \text{ (g)} \qquad \Delta H = -2600 \text{ kJ}$$

(a) What happens when the temperature is raised at constant pressure?
(b) What happens when the pressure is raised at constant temperature?
(c) What is the K_p/K_c ratio?

31. The equilibrium constant for the reaction

$$NH_4OCONH_2 \text{ (s)} \rightleftharpoons 2 NH_3 \text{ (g)} + CO_2 \text{ (g)}$$

is 5.78×10^{-5} at a certain constant temperature. Calculate the equilibrium partial pressures of NH_3 and CO_2 starting with pure NH_4OCONH_2.

32. The interhalogen compound iodine monobromide is added to a 10.0-L reaction vessel until its pressure is 380. torr. The reaction that then begins is allowed to proceed to equilibrium, at which point the partial pressure of bromine is 129 torr. Calculate K_p for the reaction:

$$2 IBr \text{ (g)} \rightleftharpoons I_2 \text{ (s)} + Br_2 \text{ (g)}$$

33. The equilibrium constant for the reaction

$$PCl_5 \text{ (g)} \rightleftharpoons PCl_3 \text{ (g)} + Cl_2 \text{ (g)}$$

is 2.15 at 250°C. A container at this temperature is charged with PCl_5 (g) to a pressure of 0.30 atm. Later the total pressure of the gas in there is found to be 0.54 atm. Has equilibrium been reached? Why, or why not?

34. Consider the equilibrium stated by the following equation:

$$2 SO_2 \text{ (g)} + O_2 \text{ (g)} \rightleftharpoons 2 SO_3 \text{ (g)}$$
$$\Delta H = 394 \text{ kJ/mol } SO_2$$

In a 10.-L container at 250°C, analysis of the equilibrium mixture showed 6.0 mol of SO_3, 2.0 mol of SO_2, and 3.0 mol of O_2.
(a) Calculate K_c.
(b) Calculate K_p.

(c) When Q_p is greater than K_p, is SO_2 or SO_3 being formed? Why?
(d) What is the effect of raising the pressure at constant temperature? Why?
(e) What is the effect of raising the temperature at constant pressure? Why?

35. A mixture of NO_2 (g) and N_2O_4 (g) at equilibrium has a total pressure of 0.135 atm. The equilibrium constant is 8.8 for the reaction

$$2\ NO_2\ (g) \rightleftharpoons N_2O_4\ (g)$$

Calculate the equilibrium partial pressures of NO_2 and N_2O_4.

36. The reaction between nitrogen oxide, bromine, and nitrosyl bromide can be studied in the gas phase by following the change in the total pressure, under conditions of constant volume:

$$2\ NO\ (g)\ +\ Br_2\ (g) \rightleftharpoons 2\ NOBr\ (g)$$

In one experiment, 0.0103 mol of NO (g) and 0.0044 mol of Br_2 (g) were sealed into a heavy-walled glass tube of volume 1055 cm³. At equilibrium at 350 K, the total pressure was 0.345 atm.
(a) Quantitatively demonstrate that this reaction does not proceed directly and completely to the indicated product, NOBr (g), according to the reaction as written.
(b) Determine the equilibrium constant for the forward reaction at 350. K.
(c) Determine the equilibrium constant for the reverse reaction at 350. K.
(d) Determine the equilibrium constant for the reaction when writen as follows:

$$NOBr\ (g) \rightleftharpoons NO\ (g)\ +\ \tfrac{1}{2}\ Br_2\ (g)$$

37. The equilibrium constant for the dissociation of phosgene at 373 K is known:

$$COCl_2\ (g) \rightleftharpoons CO\ (g)\ +\ Cl_2\ (g)$$
$$K_p = 8.0 \times 10^{-19}$$

Calculate the percentage of the phosgene that is dissociated when the total pressure of the system is 2.0 atm.

38. Phosphorus pentachloride dissociates into phosphorus trichloride and chlorine according to the following:

$$PCl_5\ (g) \rightleftharpoons PCl_3\ (g)\ +\ Cl_2\ (g)$$

At 503 K and 1.0 atm pressure, the density of the gaseous mixture is 3.03 g/L. Calculate the following:
(a) the fraction of PCl_5 that has dissociated
(b) the partial pressure of each component present
(c) the value of K_p

39. Consider a system composed of a 100.-L sphere filled with ozone and oxygen (and nothing else) at equilibrium at 1700°C. If the total pressure is 10.0 atm, estimate the partial pressure of ozone in the sphere:

$$3\ O_2\ (g) \rightleftharpoons 2\ O_3\ (g) \qquad K_p = 2.40 \times 10^{-15}$$

MULTIPLE PRINCIPLES [40–46]

40. A liquid and its vapor are in equilibrium in a container in which the volume can be changed by a movable piston.

The volume of the vapor is suddenly decreased and equilibrium is reestablished at constant temperature. Use Le Chatelier's principle to predict how the pressure of the vapor changes during the process. Prepare a sample plot of the partial pressure of the vapor versus time.

41. For the dimerization of nitrogen dioxide at 298 K,

$$2\ NO_2\ (g) \rightleftharpoons N_2O_4\ (g) \qquad K_p = 8.8$$

The total pressure of one particular equilibrium mixture of the gases is 1.00 atm. If the volume of the container is increased to 3.00 times its original value, what is the equilibrium pressure of the gases in the container at 298 K?

42. For the dissociation of sulfuryl chloride at 373 K,

$$SO_2Cl_2\ (g) \rightleftharpoons SO_2\ (g)\ +\ Cl_2\ (g) \qquad K_p = 2.4$$

Beginning with a mixture of sulfur dioxide and chlorine in which the mole fraction of sulfur dioxide is 0.40 and the total pressure of the initial mixture is 1.00 atm, what is the percentage yield of sulfuryl chloride at equilibrium?

43. Crystals of $CuSO_4 \cdot 3H_2O$ are placed on a watch glass open to the atmosphere at 25°C and relative humidity 20%. Given the following additional information, what changes do you expect?

$$CuSO_4 \cdot 3H_2O\ (s) \rightleftharpoons CuSO_4 \cdot H_2O\ (s)\ +\ 2\ H_2O\ (g)$$

The experimentally determined vapor pressure for the system is 5.60 torr and the saturation vapor pressure for water at the same temperature is 23.8 torr. (The process taking place is called **efflorescence**, the spontaneous loss of water by a crystal hydrate.)

44. Crystals of $CuSO_4 \cdot 3H_2O$ are placed on a watch glass open to the atmosphere at 25°C and relative humidity 50%. Under these conditions, the following reaction proceeds to completion:

$$CuSO_4 \cdot 3H_2O\ (s)\ +\ 2\ H_2O\ (g) \rightleftharpoons CuSO_4 \cdot 5H_2O\ (s)$$

Given the following additional information, explain why that occurs. The experimentally determined vapor pressure for this system is 7.80 torr and the saturation vapor pressure for water at the same temperature is 23.8 torr. ($CuSO_4 \cdot 3H_2O$ (s) is said to be **hygroscopic** because it will spontaneously pick up water. **Deliquescence** is an extreme condition in which the quantity of water picked up by the crystal hydrate is sufficient to cause it to dissolve, forming a solution.)

45. The fraction of iodine molecules dissociated into atoms in the gas phase at 901 K and 9.2 atm is 0.799. Find K_p for the dissociation process and estimate the C_p/C_v ratio under these conditions.

46. Iron(III) oxide can be successfully reduced in a hydrogen atmosphere to the pure metal, according to the following equation:

$$Fe_2O_3\ (s)\ +\ 3\ H_2\ (g) \rightleftharpoons 2\ Fe\ (s)\ +\ 3\ H_2O\ (g)$$

(a) If the equilibrium mixture of steam and hydrogen at 1000 K contains 60% steam by volume, what is K_p for the reaction?
(b) A second mixture of steam and hydrogen from the same reaction at 1000 K was found to be 50% steam. Was iron(III) oxide being reduced or formed? Explain.

APPLIED PRINCIPLES   [47–49]

47. For the reaction of carbon monoxide with super-heated steam at 800 K, the equilibrium constant is known to be 0.160:

$$CO\ (g) + H_2O\ (g) \rightleftharpoons CO_2\ (g) + H_2\ (g)$$

What fraction of the carbon monoxide is converted to the dioxide if the initial partial pressure for the steam is 4.00 times greater than that of the monoxide? The total pressure is kept at 1.00 atm and the temperature is kept at 800 K.

48. A burner gas is 8.0% sulfur dioxide and 12.0% oxygen by volume. The burner gas is passed through a catalytic converter at a total pressure of 1.00 atm:

$$2\ SO_2 + O_2\ (g) \rightleftharpoons 2\ SO_3\ (g) \qquad K_p = 83.1$$

Calculate the percent conversion, assuming that equilibrium is established in the converter.

49. The preparation of potassium metal takes place at 1200 K according to the following reaction:

$$Na\ (g) + KCl\ (liq) \rightleftharpoons K\ (g) + NaCl\ (liq)$$

(a) Write the equilibrium expression.

(b) Sketch a plot of the equilibrium partial pressure of potassium as a function of the sodium partial pressure, and state how the value for the equilibrium constant is reflected in this plot.

(c) Sketch a plot of the course of the partial pressures of the two gases, starting with pure sodium vapor in contact with liquid potassium chloride.

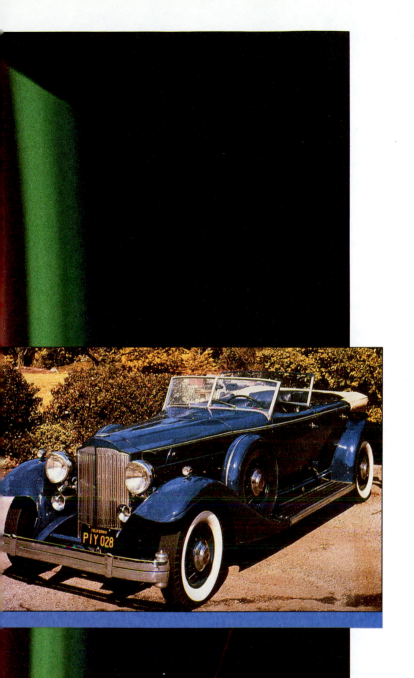

CHAPTER
11

Chemical Thermodynamics II: Reversibility, Entropy, and Free Energy

Thermodynamics has been concerned with the study of the burning of fuels and the release of energy in combustion engineering, heat transfer processes and the efficiencies of heat engines. Because Schlieren optics and photography can make visible density gradients in transparent media, they have been useful in such thermodynamics studies. Here, a collimated beam of light has been made to pass through the density gradient in air created above a combustion heat source—in this case, no engine at all, but a set of four birthday candles. The colors are due to different filters intercepting the light as it traverses the density gradients. Still, it is combustion engineering, engines, and motor fuels that we are most interested in studying; and of course, the engine of engines is the internal combustion automobile engine, like the one inside this classic touring car of half a century ago.

William Francis Giauque and Lars Onsager (United States, 1949 and 1968) and Ilya Prigogine (Belgium, 1977) won Nobel Prizes in Chemistry for their separate researches on nonequilibrium thermodynamics and the behavior of substances at extremely low temperatures.

11.1 SPONTANEOUS CHANGE AND DISORDER

To be able to answer the questions "Why do chemical reactions occur?" and "Why do chemical reactions proceed in a given direction?" is essential to our understanding of chemistry. They are not simple questions. Just look at the world around you for a moment. The leaves on the trees and the trees themselves flourish on the photosynthetic conversion of CO_2 and H_2O to the cellulosic materials of which they are composed. Yet right alongside them, we are just as likely to find dead leaves, branches of trees, and tree trunks decaying, returning CO_2 and H_2O to the atmosphere. Both processes happen without help—spontaneously, simultaneously, and apparently in reverse of each other:

Photosynthesis: $n\ CO_2\ (g)\ +\ n\ H_2O\ (liq)\ \longrightarrow\ (C_nH_{2n}O_n)\ +\ n\ O_2\ (g)$

Decay: $(C_nH_{2n}O_n)\ +\ n\ O_2\ (g)\ \longrightarrow\ n\ CO_2\ (g)\ +\ n\ H_2O\ (liq)$

The blast furnace at the steel plant frees iron from the oxide, while in the parking lot a steelworker's auto rusts away, returning iron and oxygen to iron oxide. If rusting is clearly spontaneous, at least it is mercifully slow so you don't have to replace your car each year.

It would seem that a closer look at the ideas behind the terms "spontaneity" and "reversibility" is in order if we are to come away with some idea of what drives and directs chemical reactions and processes. To begin with, consider the combustion of "coked" coal:

$$C\ (s,\ coke)\ +\ O_2\ (g)\ \longrightarrow\ CO_2\ (g)$$

The process occurs as written, and large quantities of energy are released, reflecting the enthalpy difference between the products and the reactants:

$$\Delta H°\ (\text{reaction})\ =\ \Delta H_f°\ (\text{products})\ -\ \Delta H_f°\ (\text{reactants})\ <\ 0$$

We would expect the reverse reaction

$$CO_2\ (g)\ \longrightarrow\ C\ (s,\ coke)\ +\ O_2\ (g)$$

to be endothermic, and it is. Furthermore, we know from common experience that CO_2 shows little tendency to decompose into its elements. Thus, one might conclude that exothermic reactions have something to do with **spontaneous change**, the tendency of a reaction to follow a particular pathway. At 298 K, all the following are spontaneous:

$HCl\ (g)\ +\ NH_3\ (g)\ \longrightarrow\ NH_4Cl\ (s)$ $\qquad\qquad$ $\Delta H°\ =\ -177$ kJ/mol

$Na\ (s)\ +\ H_2O\ (liq)\ \longrightarrow\ NaOH\ (aq)\ +\ \frac{1}{2}\ H_2\ (g)$

$\qquad\qquad\qquad\qquad\qquad\qquad\qquad\qquad\qquad\quad$ $\Delta H°\ =\ -184$ kJ/mol

$H_2\ (g)\ +\ \frac{1}{2}\ O_2\ (g)\ \longrightarrow\ H_2O\ (liq)$ $\qquad\qquad$ $\Delta H°\ =\ -286$ kJ/mol

$2\ Fe\ (s)\ +\ \frac{3}{2}\ O_2\ (g)\ \longrightarrow\ Fe_2O_3\ (s)$ $\qquad\qquad$ $\Delta H°\ =\ -822$ kJ/mol

$C_6H_{12}O_6\ (s)\ +\ 6\ O_2\ (g)\ \longrightarrow\ 6\ CO_2\ (g)\ +\ 6\ H_2O\ (liq)$

$\qquad\qquad\qquad\qquad\qquad\qquad\qquad\qquad\qquad\quad$ $\Delta H°\ =\ -1275$ kJ/mol

There are, however, spontaneous **endothermic** reactions. For example, if barium hydroxide and ammonium thiocyanate are mixed in a flask, and the flask and contents are then set down on a wet surface, the flask will soon be frozen to the surface (Figure 11–1). As the reaction in the flask proceeds, spontaneously, heat is extracted from the surroundings and there is an obvious liberation of ammonia gas:

(a)

(b)

FIGURE 11–1   An endothermic process. (a) Equal masses of hydrated barium hydroxide and ammonium nitrate are mixed in the flask and the endothermic reaction takes place, liberating ammonia and sufficient water to dissolve the excess of ammonium nitrate in a second endothermic process. The net result (b) is enough heat is absorbed so that when the flask is placed on a wet wooden block, the flask quickly ''freezes'' to the surface.

$$Ba(OH)_2 \cdot 8H_2O \ (s) \ + \ 2 \ NH_4SCN \ (s) \ \longrightarrow$$
$$Ba(SCN)_2 \ (s) \ + \ 2 \ NH_3 \ (g) \ + \ 10 \ H_2O \ (liq)$$

If the fact that a reaction is exothermic is not a sufficient criterion of spontaneity, then what is the driving force for chemical reactions? How is it that some reactions run uphill, spontaneously (Figure 11–2)? How can a chemical system go from a state of lower enthalpy to one of higher enthalpy on its own? One important clue lies in the change of state.

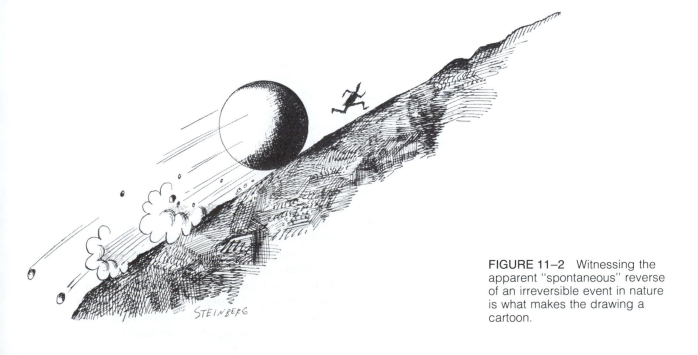

FIGURE 11–2   Witnessing the apparent ''spontaneous'' reverse of an irreversible event in nature is what makes the drawing a cartoon.

Entropy on a small scale, as potassium permanganate crystals diffuse through water. About twelve seconds after the first photo was taken, the second was snapped, showing clearly that entropy had visibly increased.

NH_3 (g) is produced. The spontaneous change from the solid state to the liquid and gaseous states is accompanied by increasing **disorder**. The degree of disorder in a system is related to the number of different states in which its parts can be arranged. Since the particles of a gas are free to move throughout the volume of their container without restriction, the gas phase must be the most disordered state of matter. Hence, it appears that spontaneous change in our reaction is in the direction producing gaseous products, even though the reaction is endothermic. We say, "The reaction proceeds in the direction leading to gaseous products, toward the state that is less ordered and has the highest **entropy**." Entropy is the thermodynamic term used to describe the spontaneous tendency of a system to move toward a less ordered state. The concept is very important and has far-reaching consequences. But as we shall see, this is not the whole story.

11.2   IRREVERSIBLE AND REVERSIBLE PROCESSES

Before discussing the factors driving spontaneous processes, it is useful to consider the so-called reversible process. You may ask, "What do we mean by **thermodynamic reversibility**?" Thermodynamic reversibility does not simply mean that a process can somehow be stopped, placed in reverse, and forced to go in the opposite direction—running the motion picture of an event backwards. For a process to be thermodynamically reversible, it must be so delicately balanced between forward and reverse that an infinitely small change in some single feature of the process is enough to change the direction of the process. Are there reversible processes that meet such stringent conditions?

As an example of a thermodynamically reversible process, recall the phase transformation taking place between a liquid and its vapor in equilibrium at a given temperature and pressure. The temperature is the boiling point of the liquid at the existing pressure. Since this is a dynamic process, molecules in the liquid state are constantly entering the vapor phase as molecules of vapor return to the liquid state. Because this is an equilibrium situation, the forward and reverse processes are occurring at equal rates and there is no net change in the amount of either liquid or vapor.

Running a motion picture of an event in reverse is not the same as the reverse of the event.

An infinitesimal change in an outside variable, however, changes the situation. For example, if the temperature is raised ever so slightly above the boiling point, more liquid boils away. It does not matter how slightly above the boiling point the temperature is raised—even an infinitesimal rise. Now lower the temperature slightly, dropping it back down to the boiling point and stopping the process. Condensation will occur until the original equilibrium vapor pressure is re-established. Lowering the temperature infinitesimally below the boiling point results in steady condensation of the vapor. Likewise, at the boiling point temperature, an infinitesimal increase in pressure results in the formation of liquid from vapor whereas an infinitesimal decrease in pressure will have the opposite effect. Thus, the change in state taking place at the boiling point is a reversible phase change. The same thing is true for fusion, the phase change between liquid and solid at the melting point, or between gas and solid at the sublimation point.

The change in state from liquid to vapor at temperatures other than at the boiling point is not a reversible process. For the process at atmospheric pressure and 120°C

$$H_2O \text{ (liq)} \rightleftharpoons H_2O \text{ (g)}$$

vapor forms from liquid spontaneously, and no infinitesimal change in an outside variable can reverse this. It would take a change of temperature of 20°C to reverse the process—far from infinitesimal!

Chemical reactions such as nitrogen dioxide dimerization are also reversible; at equilibrium, infinitesimal changes in conditions can cause the reaction to proceed in one direction or the other:

$$2\ NO_2 \text{ (g)} \rightleftharpoons N_2O_4 \text{ (g)}$$

For this particular reaction, these changes include temperature, total pressure, and the partial pressures of NO_2 (g) and N_2O_4 (g).

Irreversible processes are spontaneous. For example, at atmospheric pressure and 120°C, H_2O (liq) $\rightleftharpoons$ H_2O (g) is a spontaneous process.

11.3 REVERSIBILITY AND WORK

In general terms, a process is spontaneous if it is not at equilibrium. A spontaneous process proceeds in a direction that takes it closer to the equilibrium condition. When equilibrium is finally reached, the process is reversible. This is as it must be, according to Le Chatelier's principle. The task at hand is to consider how the work w for a particular process is affected by whether the process is reversible or spontaneous.

Having discussed reversible and spontaneous processes in terms of phase changes and chemical reactions, let's now consider the simple compression or expansion of a gas in which no phase change occurs. Take 1.00 L of an ideal gas at 25°C and 1.00 atm pressure and compress it isothermally to 0.50 L at 2.00 atm. A one-step compression is not a reversible compression because the change in external pressure to 1.00 atm from 0.50 atm is much more than an infinitesimal change in conditions. In order for this process to be reversible, the change from 0.50 atm to 1.00 atm must take place by a large number of small steps—not in one step, or even a small number of steps. The reversible compression of a gas must take place very slowly (in principle, infinitely slowly). The more steps used to bring about the total volume change in the gas, the closer to truly reversible the process becomes.

If we compress our 1.00 L of gas to 0.50 L in one step we have performed the process in the least reversible way. This can be accomplished

For quick reference, one L·atm is approximately 100 joules.

by suddenly changing the external pressure to 2.00 atm, for which

$$w = -P\Delta V = (-2.00 \text{ atm})(0.50 - 1.00 \text{ L}) = 1.00 \text{ L·atm}$$

Next, carry out the same compression in two steps, first to 1.50 atm and then to 2.00 atm. Call the initial pressure and volume P_1 and V_1 and increase the values of the subscripts by 1 for each change. The volume after the first change of pressure is as follows:

$$V_2 = \frac{P_1 V_1}{P_2} = \frac{(1.00 \text{ L})(1.00 \text{ atm})}{1.50 \text{ atm}} = 0.67 \text{ L}$$

For the first step,

$$w = (-1.50 \text{ atm})(0.67 - 1.00 \text{ L}) = 0.50 \text{ L·atm}$$

For the second step,

$$w = (-2.00 \text{ atm})(0.50 - 0.67 \text{ L}) = 0.34 \text{ L·atm}$$

The total work for the two-step process is 0.84 L·atm. Note that less work is done during the two-step process than during the one-step process.

EXAMPLE 11–1

Calculate the total work for compressing the ideal gas described above from 1.00 L to 0.050 L in three steps, using equal-sized pressure changes.

Solution   The pressures are $P_1 = 1.00$ atm, $P_2 = 1.33$ atm, $P_3 = 1.67$ atm, and $P_4 = 2.00$ atm. The volumes V_2, V_3, and V_4 after the compression steps are

$$V_2 = \frac{P_1 V_1}{P_2} = \frac{(1.00 \text{ atm})(1.00 \text{ L})}{1.33 \text{ atm}} = 0.75 \text{ L}$$

$$V_3 = \frac{P_1 V_1}{P_3} = \frac{(1.00 \text{ atm})(1.00 \text{ L})}{1.67 \text{ atm}} = 0.60 \text{ L}$$

$$V_4 = 0.50 \text{ L}$$

The work for each step is

$$w_1 = (-1.33 \text{ atm})(0.75 - 1.00 \text{ L}) = 0.33 \text{ L·atm}$$

$$w_2 = (-1.67 \text{ atm})(0.60 - 0.75 \text{ L}) = 0.25 \text{ L·atm}$$

$$w_3 = (-2.00 \text{ atm})(0.50 - 0.60 \text{ L}) = 0.20 \text{ L·atm}$$

and the total work is 0.78 L·atm.   ■

EXERCISE 11–1

Calculate the work for the same compression as above but performed in four steps, each of which takes place because of an identical incremental pressure change.
Answer: 0.77 L·atm.   ■

The following table summarizes our results so far:

Number of steps	1	2	3	4
Total work, L·atm	1.00	0.84	0.78	0.77

We observe that the work decreases as the process is performed in a more and more reversible manner. These results are stated graphically in Figure

FIGURE 11–3 The stepwise isothermal change in volume of an ideal gas in (a) one step, (b) two steps, (c) three steps, and (d) four steps. The pressure is in atmospheres and the volume is in liters. The work $-P\Delta V$ is equal to the total area of the shaded boxes.

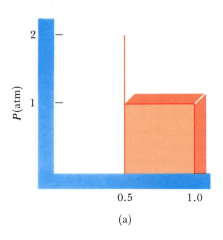

(a)

11–3. If the compression is performed completely reversibly, the graph of pressure versus volume is that given in Figure 11–4. The equation stating the work done is no longer simply $-P\Delta V$; it has become

$$w_{rev} = -nRT \ln\left(\frac{V_f}{V_i}\right) \tag{11–1}$$

where ln denotes the natural logarithm. The value of n calculated from the Ideal Gas Law for our sample is 0.041 mol, and Eq. (11–1) says that $w_{rev} = 0.69$ L·atm.

The observed result, that the work done on the system along a reversible path is less than the work along an irreversible path, is always true and can be stated as

$$w_{rev} < w_{irrev} \tag{11–2}$$

From the first law of thermodynamics,

$$\Delta E = q + w$$

By definition, ΔE is a state function, independent of the path. If an irreversible path is followed, then

$$\Delta E = q_{irrev} + w_{irrev} \tag{11–3}$$

If a reversible path is taken, then

$$\Delta E = q_{rev} + w_{rev} \tag{11–4}$$

Subtraction of Eq. (11–3) from Eq. (11–4) and rearranging terms gives

$$q_{rev} - q_{irrev} = w_{irrev} - w_{rev}$$

As a result of Eq. (11–2), the right side of this equation must be greater than zero and

$$q_{rev} > q_{irrev} \tag{11–5}$$

The concept that q_{rev}, the maximum heat absorbed by a process, is the heat realized along the reversible path allows us to calculate a value for the change in the disorder or entropy for the process.

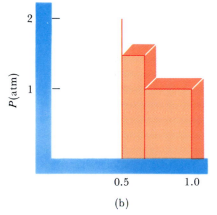

(b)

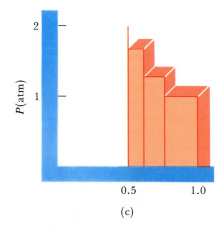

(c)

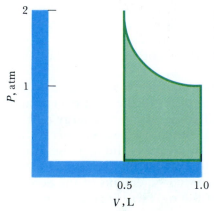

FIGURE 11–4 The isothermal reversible compression of an ideal gas. The work is equal to the area within the darkened lines.

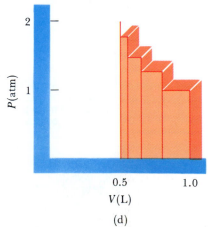

(d)

11.4 ENTROPY AND THE SECOND LAW

The fundamental aim of chemical thermodynamics is to predict the degree of spontaneity of a chemical process. There is a tendency in chemical processes to proceed toward lower enthalpy and energy; that is, there is a tendency for exothermic reactions to be spontaneous. But this can be only part of the story since there are also well-known examples of endothermic processes that occur spontaneously. A glass of water left open to the air will spontaneously evaporate until the glass is empty. The process absorbs energy but is driven by the greater disorder of the water molecules in the vapor state compared to the liquid state. This disorder is quantified in terms of the **entropy**, S.

In order to make exact calculations of the spontaneity of a reaction, we need to do two things: first, calculate a numerical value for the entropy; second, compare the relative effects that the energy and the entropy have on the spontaneity of a chemical process. Consider the first of these problems. For the entropy S to be a function of the same usefulness as energy E or enthalpy H, it is essential that it be a state function. Thus, the change in entropy ΔS must be independent of the path taken. For an isothermal process, it is defined as

$$\Delta S = \frac{q_{rev}}{T} \qquad (11-6)$$

where q_{rev} is the heat absorbed when the process is carried out reversibly and T is the constant temperature of the system. Since q_{rev} is the maximum amount of heat transferred between any two thermodynamic states, it is independent of the actual path between states. Thus, ΔS, as defined by Eq. (11-6), is also independent of path and therefore S is a state function. The selection of q_{rev}/T as the ultimate measure of disorder can be justified on the grounds that the absorption of heat into a system is the fundamental factor that increases the disorder of the system. That is, the more heat the system absorbs, the more disordered it becomes. Furthermore, the lower the temperature of the system (when the heat is absorbed), the more disordered the system becomes for a given amount of heat, accounting for T in the denominator of Eq. (11-6).

Now, having an equation that allows you to calculate entropy changes, it is possible to formulate a rule for determining the spontaneity of processes. This rule, called the **Second Law of Thermodynamics**, is not something that can be derived from first principles. The Second Law is a matter of experience. Furthermore, processes forbidden by this law—for example, the spontaneous flow of heat from a colder body to a warmer body, or the spontaneous compression of an ideal gas without any change in the external conditions—have never been observed. The actual observance of either of these forbidden processes (or a host of others) would require a rethinking of not only the Second Law of Thermodynamics but the nature of the universe. Until then, the Second Law stands as one of the basic building blocks of our scientific understanding.

There are a number of different but equivalent statements of the Second Law. Our statement of choice is in terms of the total energy and entropy of the universe. The principle of conservation of energy tells us that the energy content of the universe is constant. However, there is no conservation of entropy in the universe. In fact, it is the increase in entropy—or disorder—that provides the only possible driving force for the universe as a whole. This leads to the following statements of the Second Law of Thermodynamics: **the entropy of the universe increases for any**

Keep in mind that mass and energy are equivalent manifestations of the same thing. It is actually the total of mass plus energy in the universe that is constant.

spontaneous process, and that the entropy of the universe is constant for any reversible process.

11.5 ENTROPY CALCULATIONS

Isothermal Changes in the Volume of an Ideal Gas

Consider the isothermal (constant-temperature) reversible expansion or compression of an ideal gas. For such a process, $\Delta E = 0$ and therefore $q = -w$. Furthermore, from Eq. (11–1),

$$q_{rev} = -w_{rev} = nRT \ln \left(\frac{V_f}{V_i} \right) = nRT \ln \left(\frac{P_i}{P_f} \right)$$

Substitution into Eq. (11–6) gives

$$\Delta S = \frac{q_{rev}}{T} = nR \ln \left(\frac{V_f}{V_i} \right) = nR \ln \left(\frac{P_i}{P_f} \right) \tag{11–7}$$

Note that ΔS refers to the system in which the gas is expanding. We call this ΔS_{sys}. The Second Law deals with ΔS of the universe, which we call ΔS_{univ}:

$$\Delta S_{univ} = \Delta S_{sys} + \Delta S_{surr} \tag{11–8}$$

ΔS_{surr} is ΔS for the surroundings, which nominally includes all of the universe except for the system itself. The surroundings are taken to be an infinitely vast energy sink, which can reversibly absorb or give up the heat of any thermodynamic process. Thus,

$$\Delta S_{surr} = \frac{q_{surr}}{T} \tag{11–9}$$

where T is the temperature of the process and q_{surr} is the heat absorbed by the surroundings. Since the surroundings will simply absorb whatever heat is given off by the system or give up whatever heat is absorbed by the system,

$$q_{surr} = -q_{sys} \tag{11–10}$$

Since q_{sys} is simply what we have been calling q,

$$\Delta S_{surr} = -\frac{q}{T} \tag{11–11}$$

where T is taken to be the temperature of the surroundings and also of the system.

EXAMPLE 11–2

Calculate ΔS_{sys} and ΔS_{univ} for the reversible expansion of 1.00 L of an ideal gas at 1.00 atm to a final volume of 2.00 L.

Solution The process is isothermal and is carried out at 298 K. The number of moles of gas in the system is

$$n = \frac{(1.00 \text{ L})(1.00 \text{ atm})}{(0.08206 \text{ L·atm/mol·K})(298 \text{ K})} = 0.0409 \text{ mol}$$

ΔS_{sys} for this process can be calculated using Eq. (11–7),

$$\Delta S_{sys} = nR \ln \left(\frac{P_i}{P_f} \right)$$

$$= (0.0409 \text{ mol})(0.08206 \text{ L·atm/mol·K}) \left[\ln \left(\frac{2.00 \text{ L}}{1.00 \text{ L}} \right) \right]$$

$$= 0.00233 \text{ L·atm/K}$$

For an ideal gas in a reversible isothermal process,

$$q_{rev} = -w_{rev} = nRT \ln \left(\frac{P_i}{P_f} \right)$$

$$= (0.0409 \text{ mol})(0.08206 \text{ L·atm/mol·K})(298 \text{ K}) \left[\ln \left(\frac{2.00 \text{ L}}{1.00 \text{ L}} \right) \right]$$

$$= 0.693 \text{ L·atm}$$

Using Eq. (11–11),

$$\Delta S_{surr} = -\frac{q}{T} = -\frac{0.693 \text{ L·atm}}{298 \text{ K}} = -0.00233 \text{ L·atm/K}$$

Using Eq. (11–8),

$$\Delta S_{univ} = \Delta S_{sys} + \Delta S_{surr} = (0.00233 - 0.00233 \text{ L·atm/K}) = 0 \text{ L·atm/K}$$

For the reversible process, ΔS_{univ} is zero, in agreement with the Second Law. Notice that expansion of the gas resulted in an increase in ΔS_{sys}, which is entirely consistent, since the disorder of the gas is greater if it has a greater volume in which to move. ■

EXERCISE 11–2

Calculate ΔS_{sys} and ΔS_{surr} for the reversible, isothermal compression of 30.0 L of an ideal gas at 2.00 atm to a volume of 8.00 L. The process is carried out isothermally at 298 K.
Answer: $\Delta S_{sys} = -0.266 \text{ L·atm/K}$; $\Delta S_{surr} = 0.266 \text{ L·atm/K}$. ■

EXAMPLE 11–3

The same kinds of calculations can be used to treat irreversible changes, such as the free expansion of an ideal gas. Calculate the values of ΔS_{sys} and ΔS_{univ} for the free expansion of an ideal gas from 1.00 L at 1.00 atm to 2.00 L, at a constant temperature of 298 K.

A free expansion takes place, for example, under conditions where $P_{ext} = 0$.

Solution   First note that although this is not a reversible process, the value of ΔS_{sys} is the same as for the reversible expansion between the same volumes at the same constant temperature because S is a state function. As before, $n = 0.0409$ mol and

$$\Delta S_{sys} = nR \ln (V_f/V_i) = 0.00233 \text{ L·atm/K}$$

For the free expansion of an ideal gas,

$$w = 0 \qquad \Delta E = 0 \qquad \text{and} \qquad q = 0$$

Therefore,

$$\Delta S_{surr} = -\frac{q}{T} = 0$$

$$\Delta S_{univ} = \Delta S_{sys} + \Delta S_{surr} = (0.00233 + 0) \text{ L·atm/K} = 0.00233 \text{ L·atm/K}$$

Because ΔS_{univ} is positive, the process is spontaneous, which is consistent with experience. ■

EXERCISE 11–3

Calculate ΔS_{sys} and ΔS_{univ} for the free expansion of 3.00 L of an ideal gas at 1.00 atm to 11.0 L at a constant temperature of 298 K.
Answer: $\Delta S_{sys} = 0.0131$ L·atm/K; $\Delta S_{univ} = 0.0131$ L·atm/K. ∎

To test the Second Law, let us calculate the change of entropy of the universe for the spontaneous contraction of 0.041 mol of gas from 2.00 L to 1.00 L at 298 K in such a way that no energy is exchanged between the system and the surroundings. In other words, let's see whether a gas could spontaneously contract in the absence of a pressure or temperature change.

$$\Delta S_{sys} = (0.041 \text{ mol})(0.08206 \text{ L·atm/mol·K}) \left[\ln \left(\frac{1.00 \text{ L}}{2.00 \text{ L}} \right) \right]$$

$$= -0.0023 \text{ L·atm/K}$$

Because no heat flows, $q = 0$ and $\Delta S_{surr} = 0$, so

$$\Delta S_{univ} = -0.0023 \text{ L·atm/K}$$

Because ΔS_{univ} is negative, the process of spontaneous contraction is forbidden by the Second Law. Note that this does not mean that it is impossible to cause a gas to contract, only that a gas will not spontaneously contract without any change in external conditions.

Changing the Temperature of a Substance

Fundamental to our discussion is the notion that adding heat to a system disturbs the organization of the system, increasing the degree of disorder and entropy. Thus, it should not be surprising that heating a substance results in an increase in ΔS whereas cooling results in a decrease in ΔS. The heat q involved in either heating or cooling a substance can be obtained from its heat capacity C and the relationship

$$q = C\Delta T$$

Recall (from Chapter 7) that C is the heat capacity, a characteristic constant for a given substance over the temperature range ΔT. However, we cannot simply substitute this equation into Eq. (11–6) because that equation requires T be a fixed value, not a range. What can be done is to calculate ΔS for a reversible process, in which T changes, from the equation

$$\Delta S = nC \ln \left(\frac{T_f}{T_i} \right) \tag{11–12}$$

But the overall temperature change (from T_i to T_f) must be small enough that the heat capacity C can be assumed to be constant. C can be either C_p or C_v depending on whether P or V (respectively) is constant. Because this heating process proceeds reversibly, during each step the temperature changes by an infinitesimally small amount from the previous temperature.

EXAMPLE 11–4

Calculate ΔS_{sys} for reversibly heating 1.00 mol of water from 303 K to 313 K at constant P. The molar heat capacity C_p of liquid water is 75.4 J/mol·K.

Solution From Eq. (11–12),

$$\Delta S = C \ln \left(\frac{T_f}{T_i}\right)$$

$$= (1.00 \text{ mol})(75.4 \text{ J/mol·K}) \left[\ln \left(\frac{313 \text{ K}}{303 \text{ K}}\right)\right] = 2.45 \text{ J/K}$$

EXERCISE 11–4

Calculate ΔS for the process of cooling 2.00 mol of ice from $-5°C$ to $-18°C$. C_p for H_2O (s) is 37.6 J/mol·K.

Answer: $\Delta S = -3.74$ J/K.

Phase Transformations

Phase transformations reflect the relative change in organization and disorder of a system in the accompanying entropy change. The quantity ΔS_{sys} is positive for transformations from solid to liquid to gas, and is negative for those processes when carried out in the reverse direction. At the normal melting, boiling, and sublimation points, ΔS_{univ} is zero for the phase change, since that is a reversible transformation.

EXAMPLE 11–5

Find ΔS_{sys} and ΔS_{univ} for the transformation of 1.0 mol of liquid benzene to benzene vapor at 1.00 atm pressure and its normal boiling point (which is 353 K).

Solution ΔH_{vap} for benzene must be known.

$$\Delta H_{vap} = 3.08 \times 10^4 \text{ J/mol}$$

Since this is a reversible process at constant pressure,

$$q_{rev} = \Delta H_{vap} = (1.00 \text{ mol})(3.08 \times 10^4 \text{ J/mol}) = 3.08 \times 10^4 \text{ J}$$

$$S_{sys} = \frac{q_{rev}}{T} = \frac{(3.08 \times 10^4 \text{ J})}{353 \text{ K}} = 87.2 \text{ J/K}$$

$$\Delta S_{surr} = \frac{-q}{T} = -87.2 \text{ J/K}$$

$$\Delta S_{univ} = \Delta S_{sys} + \Delta S_{surr} = (87.2 - 87.2) \text{ J/K} = 0$$

Note that ΔS_{univ} is zero, as must be the case for a reversible phase change such as a phase transformation at the normal boiling point, and ΔS_{sys} is positive in this same process since the resulting gas phase is more disordered than the initial liquid phase.

EXERCISE 11–5

Calculate ΔS for the vaporization of 1.00 mol of diethyl ether at its normal boiling point, 308 K. The enthalpy of vaporization is 27.2 kJ/mol.

Answer: $\Delta S = 88.3$ J/K.

It is worth mentioning that many liquids have a molar entropy of vaporization very close to 88 J/mol·K, a fact that has come to be known as **Trouton's rule**. The implication is that the gain in disorder accompanying the change from the liquid to the vapor state is generally the same for many liquids.

All phase changes at the normal phase transformation temperatures are reversible, and $\Delta S_{univ} = 0$. For example, that would be true for

$$H_2O \text{ (s)} \rightleftharpoons H_2O \text{ (liq)}$$

at 0°C and 1 atm or for

$$H_2O \text{ (liq)} \rightleftharpoons H_2O \text{ (g)}$$

at 100°C and 1 atm. Calculations at other temperatures are more complex, as the next example and exercise illustrate.

In summary, ΔS_{sys} is positive for (1) an increase in the volume of a gas, (2) an increase in temperature, and (3) a phase change to a more disordered state.

EXAMPLE 11–6

Calculate ΔS_{sys} for the freezing of 1.00 mol of water that has been super-cooled to $-15°C$.

Solution In order to make this calculation, you need to know ΔH_{fus} of water (6025 J/mol) and the heat capacities of liquid water (75.4 J/mol·K) and ice (37.6 J/mol·K).

The difficulty in solving this problem is that freezing is not a reversible process except at the normal freezing point. However, we can imagine performing three separate reversible processes whose sum brings the system to the same final state:

1. Reversibly warm the supercooled water to 0°C, the normal freezing point.
2. Reversibly freeze the liquid at the normal freezing point.
3. Reversibly cool the ice to $-15°C$.

Schematically, the **thermodynamic cycle** we have described looks like this:

$$\begin{array}{ccc}
1.00 \text{ mol } H_2O \text{ (liq), 273 K} & \xrightarrow{\Delta S_2} & 1.00 \text{ mol } H_2O \text{ (s), 273 K} \\
\Delta S_1 \uparrow & & \downarrow \Delta S_3 \\
1.00 \text{ mol } H_2O \text{ (liq), 258 K} & \xrightarrow{\Delta S_4} & 1.00 \text{ mol } H_2O \text{ (s), 258 K}
\end{array}$$

Because entropy is a state function, the change ΔS_4 for the irreversible process from the initial state to the final state must be the sum of the entropy changes for the three reversible steps:

$$\Delta S_1 = (1.00 \text{ mol})(75.4 \text{ J/mol·K})\left[\ln\left(\frac{273 \text{ K}}{258 \text{ K}}\right)\right] = 4.26 \text{ J/K}$$

$$\Delta S_2 = \frac{(1.00 \text{ mol})(-6025 \text{ J/mol})}{273 \text{ K}} = -22.1 \text{ J/K}$$

$$\Delta S_3 = (1.00 \text{ mol})(37.6 \text{ J/mol·K})\left[\ln\left(\frac{258 \text{ K}}{273 \text{ K}}\right)\right] = -2.12 \text{ J/K}$$

$$\Delta S_4 = \Delta S_1 + \Delta S_2 + \Delta S_3 = -20.0 \text{ J/K}$$

Note that the quantity ΔS_4 is negative, which is consistent since ice has more order and lower entropy than water. ∎

EXERCISE 11–6

Calculate the entropy change for the system when 2.00 mol of water su-percooled to $-7°C$ freezes to ice.
Answer: $\Delta S = -42.1$ J/K. ∎

11.6   STATISTICAL INTERPRETATION OF ENTROPY

It is a remarkable feature of the principles of thermodynamics that it is possible to make predictions about the spontaneity of chemical processes based on little more than properties such as heat and work. It requires no knowledge of the actual workings of the process. Thermodynamics looks only at the **macroscopic properties** we measure on the entire system, such as temperature, pressure, and volume, without concern for the **microscopic properties**—what the atoms and molecules inside the system are actually doing. Although we have associated entropy with disorder and defined it as a quantity that will increase when heat is added to a system, it is useful to provide a more physical interpretation of what we actually mean by entropy and disorder.

In the preceding section, we were able to show that the entropy of the system increases as the volume of the gas expands. But why is that? The reason is that disorder in a thermodynamic sense refers to the number of different orientations that a system can have. Consider a thought experiment in which we construct a system analogous to the gas with its volume expanded. Suppose we have a checker board and eight identical checkers (Figure 11–5). We make a rule that only one checker can occupy a square. To begin with, the eight checkers are confined to one row of the board. How many different ways can we draw this picture? There is only one way, with one checker per square. If the space that the checkers can occupy is now expanded to two rows, then the eight checkers can occupy sixteen possible squares and there are many more ways that the picture can be drawn. A couple of examples are given in Figure 11–6. In fact, there are 12,870 ways of arranging the eight checkers onto the sixteen squares:

$$\frac{16 \times 15 \times 14 \times 13 \times 12 \times 11 \times 10 \times 9}{1 \times 2 \times 3 \times 4 \times 5 \times 6 \times 7 \times 8} = 12{,}870$$

Thus, the situation in Figure 11–6 is more disordered than the situation in Figure 11–5. The actual calculation of the disorder depends on the number of ways the checkers can be arranged.

Where only one arrangement is possible, the disorder or entropy is set at zero. In statistical thermodynamics the entropy is proportional to the natural logarithm of the number of possible arrangements, and the logarithm of one is zero. The disorder of the system in Figure 11–6 would be proportional to

$$\ln 12{,}870 = 9.46$$

The equation relating number of arrangements and entropy is

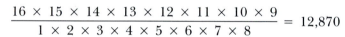

$$S = k \ln \Omega \qquad \qquad (11\text{–}13)$$

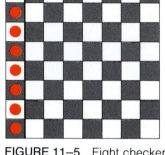

FIGURE 11–5   Eight checkers confined to one row of squares on a checker board.

FIGURE 11–6   Two of the 12,870 possible arrangements for eight checkers confined to two lines of a checker board.

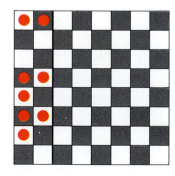
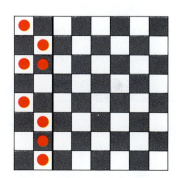

where S is the absolute entropy of the system, k is the Boltzmann constant (R/N_A), and Ω is the number of ways the system can be arranged. Clearly the disorder of our system would increase drastically upon expanding the number of squares available. Allowing a third line of squares increases the number of possibilities to 735,471 with a relative disorder proportional to

$$\ln 735{,}471 = 13.51$$

This example is analogous to the particles of a gas in a container. The larger the volume available to them, the more possible ways they can arrange themselves, and the greater the entropy of the system.

Now, go back to the checkers and allow them to roam the entire board. There are millions of possibilities available to them. What is the chance that if we moved the checkers at random we would find all of the checkers lined up in one row, just as they were confined in Figure 11–5? There is a chance, but it is only one chance among millions of other possible arrangements. Therefore we must conclude that although the arrangement is possible, it is improbable.

It is just the same with gases. The particles of a gas move in a random manner. A spontaneous contraction of a gas would require all the particles move in the same direction at the same time. Possible? Yes, but highly improbable. Thermodynamics does not say that any processes are impossible on a microscopic basis. Remember, thermodynamics is not concerned with the microscopic description of the system. Rather, thermodynamics points out processes that are so statistically improbable that we may safely assume they have never and will never be observed.

Using Eq. (11–13), we can derive the entropy change accompanying an isothermal expansion (or compression) of an ideal gas. For a volume change from V_i to V_f there is a corresponding change in the number of arrangements or microstates from Ω_i to Ω_f. The change in entropy for this process is

$$\Delta S = \Delta S_f - \Delta S_i = k \ln \Omega_f - k \ln \Omega_i = k \ln \left(\frac{\Omega_f}{\Omega_i} \right)$$

Our job is to determine the ratio Ω_f/Ω_i. If our gas were composed of only one particle, the number of possible arrangements would simply be proportional to the volume. For example, doubling the volume would allow twice as many locations. For n particles, the number of arrangements is proportional to V^n, that is, the number of arrangements available to particle one, times the number available to particle two, and so forth. Therefore

$$\frac{\Omega_f}{\Omega_i} = \frac{V_f^n}{V_i^n} = \left(\frac{V_f}{V_i} \right)^n$$

For one mole of gas, consisting of N_A particles,

$$\frac{\Omega_f}{\Omega_i} = \left(\frac{V_f}{V_i} \right)^{N_A} \quad \text{and} \quad \Delta S = k \ln \left(\frac{V_f}{V_i} \right)^{N_A} = k N_A \ln \left(\frac{V_f}{V_i} \right) = R \ln \left(\frac{V_f}{V_i} \right)$$

for one mole of gas since $R = k N_A$. For n moles of gas the result is

$$\Delta S = nR \ln \left(\frac{V_f}{V_i} \right)$$

which is identical to Eq. (11–7).

The benchmark of thermodynamics in 19th-century technology, the steam locomotive is depicted in a classic portrait.

In theory, the operation of a heat engine depends on the existence of a temperature differential between two regions.

HEAT ENGINES

The dynamics of heat and the power of the steam engine were essential themes in 19th-century science and technology. It has been said of the century before nuclear power that the steam engine was its most impressive symbol, using heat continuously to create power on command. Could anyone doubt the majesty of the steam locomotive? But the problem then, as today, is a very practical one, namely the efficiency of a heat engine. The production of motion—moving force—is due to the transfer, not the consumption, of heat. If you plan to get work out of heat, a temperature difference is necessary. A classic steam engine has two heat reservoirs, one at a higher temperature T_2 and one at a lower temperature T_1, and a working fluid (steam) that is in contact with both. The ratio of the net work done by the system and the heat put into the working fluid is the efficiency of the engine:

$$\% \text{ efficiency} = \frac{T_2 - T_1}{T_2} (100)$$

The heat delivered to the low-temperature reservoir represents wasted heat that is not converted to work. Thus, the efficiency of a heat engine is increased by having as great a difference as possible between the high-temperature reservoir and the low-temperature reservoir. Since the low-temperature reservoir cannot usually be varied by much, operating temperatures are made as high as possible. However, the materials available limit the higher temperatures. Much research is being performed in order to produce materials for turbines that can withstand increasingly high temperatures. Efficiencies of about 50% are realized in commercial steam turbine electrical plants, where the low-temperature reservoir can be maintained at a fairly low temperature. Where it is necessary to discharge the waste heat at a relatively high temperature, as in an automobile engine, the efficiency drops to about 15 to 20%. Consider that in terms of gasoline consumption; 80 to 85% of your fuel dollar is waste heat, taken away through the car's radiator.

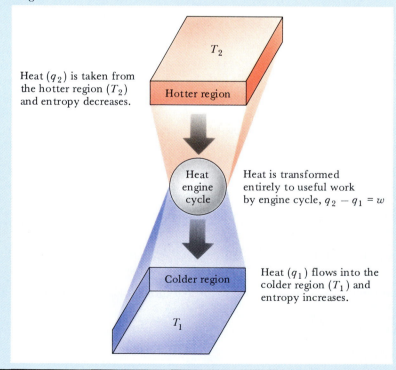

Heat (q_2) is taken from the hotter region (T_2) and entropy decreases.

T_2

Hotter region

Heat engine cycle

Heat is transformed entirely to useful work by engine cycle, $q_2 - q_1 = w$

Colder region

Heat (q_1) flows into the colder region (T_1) and entropy increases.

T_1

11.7 ABSOLUTE ENTROPIES AND THE THIRD LAW

It is not possible to establish any absolute zero for H the enthalpy function. With entropy, however, it is possible to set an absolute zero. In Section 11.6 we stated that a system with only one possible arrangement would have an entropy of zero. To establish a real chemical situation, think in terms of atoms. As we proceed from gases to liquids to solids, the states have less and less disorder. In the solid or crystalline state each atom has its own location in the structure. However, this is not a zero entropy situation since the atoms are in thermal vibrational motion about some average position and the crystal also has some defects. Even the presence of impurities of the smallest imaginable quantities serves to introduce elements of disorder; the impurity atoms could be situated in different locations, which would allow for alternate arrangements. If the crystal had no impurities, and if there were no defects, and if each atom was at its regular site and at minimum energy, we would have a zero entropy situation. For a perfect crystal, this would occur at the absolute zero of temperature, the temperature at which all vibrational motion ceases.

This concept leads to the **Third Law of Thermodynamics**: At a temperature of absolute zero, perfect crystals of all pure elements and compounds have an entropy of zero. The fact that we have an absolute zero of entropy allows the calculation of an **absolute entropy**, $S°$, for a compound or element for any given set of conditions. It is customary to calculate the absolute entropies of compounds and elements at standard conditions, and these are given the symbol $S°$. The method of this calculation follows methods that we have already discussed.

Suppose we wish to calculate $S°$ for one mole of oxygen gas at 298 K. Start with one mole of perfect pure crystalline solid O_2 at absolute zero. Heat is added reversibly. Applying Eq. (11–12),

$$\Delta S = C_s \ln \left(\frac{T_f}{T_i} \right)$$

Unfortunately, things are actually not so straightforward. The value of the heat capacity of the solid oxygen, C_s, isn't constant over the long ranges of temperature involved, and so additional details must be included. Continue reversibly raising the temperature of the solid until its melting point is reached. At this point,

$$\Delta S = \frac{q_{rev}}{T} = \frac{\Delta H_{fus}}{T_{mp}}$$

where the heat of fusion is q_{rev} and T is the melting point. The temperature is then raised reversibly, making use of the heat capacity of liquid oxygen, until the boiling point is reached. At this point,

$$\Delta S = \frac{q_{rev}}{T} = \frac{\Delta H_{vap}}{T_{bp}}$$

is again employed, this time for the vaporization process, after which the gas is warmed reversibly to 298 K.

TABLE 11–1 Absolute Entropies (J/mol K) of Compounds and Elements at 298 K

Solid Elements		Solid Compounds		Liquids	
Ag	42.68	BaO	70.3	Br_2	152
B	7.1	$BaCO_3$	112	H_2O	70.00
Ba	63.2	$BaSO_4$	132	Hg	76.02
C (graphite)	5.73	CaO	40	CH_3OH	127
C (diamond)	2.5	$Ca(OH)_2$	72.8	C_2H_5OH	161
Ca	41.6	$CaCO_3$	92.9	CH_3COOH	160
Cu	33.3	CuO	43.5	C_6H_6	203
Fe	27.2	Fe_2O_3	90.0		
S (rhombic)	31.9	ZnO	43.9		
Zn	41.6	ZnS	57.7		
Monatomic Gases		**Diatomic Gases**		**Polyatomic Gases**	
He	126.1	H_2	130.6	H_2O	189
Ne	146.2	D_2	145	CO_2	214
Ar	154.7	F_2	203	SO_2	248
Kr	164.0	Cl_2	223	H_2S	205
Xe	169.6	Br_2	245	NO_2	240
H	114.6	CO	198	N_2O	219.7
F	158.6	NO	210	NH_3	192
Cl	165.1	N_2	191	O_3	238
Br	174.9	O_2	205	CH_4	186
I	180.7	HF	174	C_2H_6	229
N	153.2	HCl	187	C_3H_8	270
C	158.0	HBr	198	n-C_4H_{10}	310
O	161	HI	206	i-C_4H_{10}	294.6
				C_2H_2	167
				C_2H_4	219.4
				C_3H_6	266.9
				C_4H_8 (1-butene)	307.4
				C_4H_8 (*cis*-2-butene)	301
				C_4H_8 (*trans*-2-butene)	297
				C_4H_8 (*iso*-butene)	294

The absolute entropies of a number of compounds and elements at 298 K are given in Table 11–1. From the recorded values, it is evident that large atoms and molecules generally have higher standard absolute entropies than small atoms and molecules. For molecules of about the same size, the expected result that gases have the highest absolute entropies and solids the lowest is observed.

Using tabulated standard absolute entropies, it is possible to calculate $\Delta S°$, the standard entropy change of a chemical reaction, in exactly the same way that standard enthalpies of formation were used to calculate $\Delta H°$ values. Thus, for a reaction

$$aA + bB \longrightarrow cC + dD$$

we can say

$$\Delta S° = (cS_C° + dS_D°) - (aS_A° + bS_B°)$$

and in general,

$$\Delta S° = \Sigma S°_{products} - \Sigma S°_{reactants} \tag{11–14}$$

EXAMPLE 11–7

Calculate the standard entropy change for the reaction

$$CaCO_3 (s) \longrightarrow CaO (s) + CO_2 (g)$$

Solution   Using the values of $S°$ from Table 11–1 and Eq. (11–14),

$$\Delta S = (40 + 214 - 92.9 \text{ J/mol·K}) = 161 \text{ J/mol·K}$$

It is reasonable that this process should show a gain in entropy, since we started with a solid reactant, and ended with a gas as one of the reaction products. ■

EXERCISE 11–7

Calculate $\Delta S°$ for the reaction

$$2 H_2 (g) + O_2 (g) \longrightarrow 2 H_2O (liq)$$

Answer: $\Delta S° = -326$ J/mol·K.   ■

11.8   ENTROPY AND FREE ENERGY OF CHEMICAL CHANGE

Although ΔS_{univ} is a useful thermodynamic term for predicting the spontaneous direction of a process undergoing change, it is not a simple matter to obtain the necessary data. That is because knowledge of the system *and* its surroundings is required:

$$\Delta S_{univ} = \Delta S_{sys} + \Delta S_{surr}$$

A far more convenient measure of spontaneity is available that depends only on knowledge of the system, a feature that makes it particularly attractive for chemistry. That measure is the Gibbs free energy, or simply the **free energy** G, defined as

$$G = H - TS$$

The free energy can be used to predict spontaneity under the kinds of conditions typically encountered in chemical reactions. Under conditions of constant temperature, it follows that

$$\Delta G = \Delta H - T\Delta S \qquad (11–15)$$

This may well be the most important equation in all of chemical thermodynamics. Under standard state conditions, at temperature T,

$$\Delta G° = \Delta H° - T\Delta S° \qquad (11–16)$$

We need to know how this new function ΔG is related to spontaneity. For an isothermal process at constant pressure,

$$\Delta G = \Delta H - T\Delta S = q - T\Delta S$$

In this equation, ΔS is the entropy change of the system. Dividing each side of the equation by T gives

$$\frac{\Delta G}{T} = \frac{q}{T} - \Delta S_{sys} = -\Delta S_{surr} - \Delta S_{sys} = -\Delta S_{univ}$$

Thus, we can write

$$\Delta G = -T\Delta S_{univ}$$

For reversible processes, ΔS_{univ} is zero and $\Delta G = 0$. For spontaneous processes, in which ΔS_{univ} is positive, ΔG is negative. A process with a positive value for ΔG is not spontaneous in the direction written, but is spontaneous in the opposite direction.

The change in free energy ΔG thus allows us to assess the relative importance of the forces that drive chemical processes—energy (or enthalpy) and entropy. Processes that have a negative value of ΔH are said to be energy- (or enthalpy-) driven processes, and those that have positive values of ΔS are said to be entropy-driven. From Eq. (11–15) it is clear that processes that are driven by both energy and entropy will have negative values of ΔG and thus be spontaneous at all temperatures. Processes driven by enthalpy but not entropy are spontaneous if the temperature is low enough, and those driven by entropy but not enthalpy are spontaneous if the temperature is sufficiently high. Processes driven by neither energy nor enthalpy have positive values of ΔG and are not spontaneous under any conditions. Here is a summary:

- If ΔH is negative and ΔS is positive, ΔG is less than zero for all values of T.

- If ΔH and ΔS are both negative, ΔG can be less than zero at low T.

- If ΔH and ΔS are both positive, ΔG can be less than zero at high T.

- If ΔH is positive and ΔS is negative, ΔG is never less than zero, no matter what the value of T.

Calculations Using Free Energy Changes

We can use Eq. (11–16), along with the data in the tables of standard enthalpies of formation $\Delta H°$ and absolute entropies $S°$, to calculate $\Delta G°$ values for chemical reactions.

EXAMPLE 11–8

Consider the oxidation of nitrogen monoxide at 298 K:

$$2 \text{ NO (g)} + \text{O}_2 \text{ (g)} \longrightarrow 2 \text{ NO}_2 \text{ (g)}$$

Find $\Delta G°$ for the reaction at 298 K and determine whether it is spontaneous.

Solution   Calculations using values from the tables give $\Delta H° = -113.0$ kJ and $\Delta S° = -145$ J/K per mole of reaction as written. Thus,

$$\Delta G° = -113.0 \text{ kJ/mol} - (298 \text{ K})(-145 \text{ J/mol·K})(1 \text{ kJ/1000 J})$$
$$= -69.8 \text{ kJ/mol}$$

Since $\Delta G° < 0$, the reaction is spontaneous under standard conditions.

EXERCISE 11–8

Determine $\Delta G°$ for the ammonia synthesis at 298 K, and evaluate the spontaneous direction of this reaction:

$$3 \text{ H}_2 \text{ (g)} + \text{N}_2 \text{ (g)} \rightleftharpoons 2 \text{ NH}_3 \text{ (g)}$$

Answer: $\Delta G° = -33.4$ kJ/mol; spontaneous as written.

In much the same way as standard enthalpies of formation $\Delta H_f°$ simplified $\Delta H°$ calculations (in an earlier discussion), a considerable simplifi-

TABLE 11–2 Standard Free Energies of Formation, ΔG_f°, kJ/mol at 298 K

Solid Compounds		Simple Gaseous Molecules	
BaO	−528.4	HCl	−95.27
BaCO$_3$	−1139	H$_2$O	−228.6
BaSO$_4$	−1465	H$_2$O$_2$	−103.3
CaO	−604.2	H$_2$S	−33.1
Ca(OH)$_2$	−896.6	CO	−137.3
CaCO$_3$	−1129	CO$_2$	−394.4
CuO	−127	SO$_2$	−300.4
Cu$_2$O	−146	SO$_3$	−370.4
Fe$_2$O$_3$	−741	NO$_2$	51.84
ZnO	−318.2	N$_2$O	104
SiO$_2$	−805	NH$_3$	−16.6
PbO$_2$	−219	O$_3$	163.4
		NO	86.69
Liquids		CH$_4$ (methane)	−50.79
CH$_3$OH	−166.2	C$_2$H$_6$ (ethane)	−32.8
C$_2$H$_5$OH	−174.8	C$_3$H$_8$ (propane)	−23.6
CH$_3$COOH	−392	n-C$_4$H$_{10}$ (n-butane)	−17.2
C$_6$H$_6$	−124.5	i-C$_4$H$_{10}$ (iso-butane)	−20.9
		C$_2$H$_2$ (acetylene)	209.2
Monatomic Gases		C$_2$H$_4$ (ethylene)	68.24
H	203.2	C$_3$H$_6$ (propylene)	62.84
F	59.4	C$_4$H$_8$ (1-butene)	71.30
Cl	105.4	C$_4$H$_8$ (cis-2-butene)	65.86
Br	82.38	C$_4$H$_8$ ($trans$-2-butene)	62.97
C	672.95	C$_4$H$_8$ (iso-butene)	58.07
I	70.16		
N	340.9		
O	161		

cation is introduced into ΔG° calculations for chemical processes by making use of ΔG_f° data. The standard free energy of formation ΔG_f° is the free energy of reaction for the formation of the compound in its standard state from its elements in their standard states. Taking ΔG_f° for the elements in their standard states to be zero, we may write the following:

$$\Delta G^\circ = \Sigma \Delta G_{f(products)}^\circ - \Sigma \Delta G_{f(reactants)}^\circ \qquad (11\text{–}17)$$

Table 11–2 lists standard free energies of formation for some compounds and for the monatomic forms of common diatomic elements. Using Eq. (11–17) and ΔG_f° values from Table 11–2, the ΔG° value for the oxidation of carbon monoxide can be calculated as a simple example:

$$2 \text{ CO (g)} + \text{O}_2 \text{ (g)} \longrightarrow 2 \text{ CO}_2 \text{ (g)}$$
$$\begin{aligned} \Delta G^\circ &= (2 \text{ mol})(-394.4 \text{ kJ/mol}) - (2 \text{ mol})(-137.3 \text{ kJ/mol}) \\ &= -514.2 \text{ kJ/(2 mol CO)} \\ &= -257.1 \text{ kJ/mol CO} \end{aligned}$$

EXAMPLE 11–9

The cleavage of inorganic phosphate from the energy-carrying ATP (adenosine triphosphate) molecule can be written as follows, where P$_i$ represents inorganic phosphate:

$$\text{ATP (aq)} \rightleftharpoons \text{ADP (aq)} + \text{P}_i \text{ (aq)}$$

Calculate ΔG° and determine the spontaneous direction of the process.

Solution   Unfortunately, for a variety of reasons, $\Delta G°$ for this reaction is difficult to measure directly. However, by looking at two related biochemical reactions involving these same three components (ATP, ADP, and P_i) and glucose $\Delta G°$ can be calculated:

ATP (aq) + glucose (aq) $\rightleftharpoons$ ADP (aq) + glucose-6-phosphate (aq)

$\Delta G_1° = -16.7$ kJ/mol

glucose-6-phosphate (aq) $\rightleftharpoons$ glucose (aq) + P_i (aq)

$\Delta G_2° = -13.8$ kJ/mol

By adding the separate reactions, we obtain the net result, which is our original equation:

ATP (aq) $\rightleftharpoons$ ADP (aq) + P_i (aq)

$\Delta G_3° = \Delta G_1° + \Delta G_2°$

$\quad\quad = -30.5$ kJ/mol

Since the sign is negative, we may conclude that cleavage of inorganic phosphate is spontaneous.   ■

EXERCISE 11–9(A)

Use $\Delta G_f°$ values to calculate the standard free energy change for the following reaction; $\Delta G_f°$ (H_2S) = -33.0 kJ/mol:

$$2\ H_2S\ (g) + 3\ O_2\ (g) \longrightarrow 2\ H_2O\ (g) + 2\ SO_2\ (g)$$

Answer: $\Delta G° = -992.0$ kJ/mol.   ■

EXERCISE 11–9(B)

You are planning a chemical plant for the manufacture of hydrogen peroxide, an important industrial reagent. There are two possible routes to the synthesis; $\Delta G_f°$ (H_2O)$_{liq}$ = -243 kJ/mol:

(1) Reduction of oxygen with hydrogen:

$$H_2\ (g) + O_2\ (g) \longrightarrow H_2O_2\ (liq)$$

(2) Direct oxidation of water:

$$2\ H_2O\ (liq) + O_2\ (g) \longrightarrow 2\ H_2O_2\ (liq)$$

Using the appropriate data from the tables, predict which process might be possible.

Answer: $\Delta G° < 0$ for the first process, which makes it possible. That is, in fact, the basis for a commercial process.   ■

11.9   FREE ENERGY AND THE EQUILIBRIUM CONSTANT

The standard free energy change $\Delta G°$ refers to reactants and products at 1 atm for gases, or for solutes in 1 M solutions. For example, $\Delta G°$ is positive for the process

$$2\ N_2\ (g) + O_2\ (g) \rightleftharpoons 2\ N_2O\ (g) \quad\quad \Delta G° = 103.5\ \text{kJ}$$

Because of this, we know that if nitrogen gas, oxygen gas, and dinitrogen monoxide gas are mixed together so that each is at a partial pressure of 1 atm, the reaction proceeds from right to left. The partial pressures of nitrogen and oxygen increase at the expense of the partial pressure of the dinitrogen monoxide gas. To state it another way, this is a reaction with an equilibrium constant less than one. What we would like to know is how to convert $\Delta G°$ values to equilibrium constants.

For a chemical reaction, the free energy change is the difference between the free energy of the products and that of the reactants. The free energy of a gas depends on the pressure and temperature of the gas; at one atmosphere of pressure, G is the standard free energy $G°$ at temperature T, or $G°(T)$. It can be shown that for a gas, G and $G°$ are related by the equation

In Chapter 7, the same approach was applied to heats of reactions and enthalpy changes.

$$G(T,P) = G°(T) + nRT \ln p \qquad (11\text{–}18)$$

where p is the pressure of the gas.

For the general reaction

$$aA + bB \longrightarrow cC + dD$$

where the pressures are not necessarily standard, ΔG can be calculated using the following relationship:

$$\Delta G = \Sigma G_{\text{products}} - \Sigma G_{\text{reactants}} = [cG_C + dG_D] - [aG_A + bG_B]$$

Substituting Eq. (11–18) into this equation gives

$$\Delta G = [cG_C° + dG_D°] - [aG_A° + bG_B°] + [cRT \ln p_C + dRT \ln p_D] - [aRT \ln p_A + bRT \ln p_B]$$

where $p_A, p_B, p_C,$ and p_D are the partial pressures of the reactant and product gases. The first four terms on the right side of this equation equal $\Delta G°$. Therefore,

$$\Delta G = \Delta G° + RT \ln \left[\frac{(p_C)^c(p_D)^d}{(p_A)^a(p_B)^b} \right] \qquad (11\text{–}19)$$

The quantity in the brackets is the reaction quotient, Q. If the reaction is at equilibrium, ΔG equals zero and the reaction quotient equals the equilibrium constant K, leading to the equation

$$\Delta G° = -RT \ln K \qquad (11\text{–}20)$$

This is the essential equation relating $\Delta G°$ and the equilibrium constant. When the reactants and products are in their standard states, spontaneous reactions have equilibrium constants greater than one, whereas reactions that move spontaneously in the reverse direction have equilibrium constants less than one. For the special case where $\Delta G°$ is zero, the equilibrium constant is one.

Just as the standard state for gases is a pressure of 1 atm, that for solutes in solution is a 1 M concentration. Therefore, Eq. (11–19) or Eq. (11–20) work equally well if the particles of the reaction are in solution. In these cases, the concentration of the dissolved species is used in the same way as the partial pressure of a gas. As before, pure solids and liquids do not appear in equilibrium expressions.

EXAMPLE 11–10

Calculate the equilibrium constant at 298 K for the reaction

$$2 SO_2 (g) + O_2 (g) \longrightarrow 2 SO_3 (g)$$

Solution From the table of free energies of formation, $\Delta G°$ is -140.0 kJ. Rearrangement of Eq. (11–20) gives

$$K = e^{-\Delta G°/RT} = 10^{-\Delta G°/2.303RT}$$
$$= 10^{(140.0 \text{ kJ})(1000 \text{ J/kJ})/(2.303)(8.314 \text{ J/mol·K})(298 \text{ K})}$$
$$= 10^{24.5} = 3.16 \times 10^{24}$$

EXERCISE 11–10

Determine the equilibrium constant for the following reaction at 298 K.

$$H_2O \text{ (g)} + O_3 \text{ (g)} \longrightarrow H_2O_2 \text{ (g)} + O_2 \text{ (g)}$$

Answer: $K = 4.79 \times 10^6$.

11.10   EFFECT OF TEMPERATURE ON EQUILIBRIUM

Two major tasks have been accomplished to this point. First, we showed that the change in free energy $\Delta G°$ for a process under standard state conditions and at 298 K can be obtained from the relationship

$$\Delta G° = \Delta H° - T\Delta S°$$

Then, from $\Delta G°$ values, equilibrium constants can be obtained. However, since not everything happens at 298 K, we need to be able to determine $\Delta G°$ at other temperatures. In other words, we need to know how to calculate equilibrium constants at different temperatures. Can this be done? In order to answer that question, we need to know how much $\Delta H°$ and $\Delta S°$ of a reaction vary with temperature.

We can test the effect of T on $\Delta H°$ and $\Delta S°$ by considering a prototype reaction at some temperature T other than 298 K and setting up a thermodynamic cycle. That is possible because both enthalpy and entropy are state functions.

Define ΔT as $(T - 298 \text{ K})$, C_i as the total heat capacity of the reactants, and C_f as the total heat capacity of the products. The standard enthalpy and entropy changes at T, $\Delta H(T)$ and $\Delta S(T)$, can then be calculated as follows:

$$\Delta H° (T) = \Delta H_1 + \Delta H° (298 \text{ K}) + \Delta H_2$$
$$= \Delta H° (298 \text{ K}) + C_i\Delta T - C_f\Delta T$$
$$= \Delta H° (298 \text{ K}) + \Delta T(C_i - C_f)$$
$$\Delta S° (T) = \Delta S_1 + \Delta S° (298 \text{ K}) + \Delta S_2$$
$$= \Delta S° (298 \text{ K}) + C_i \ln (T/298) - C_f \ln (T/298)$$
$$= \Delta S° (298 \text{ K}) + \ln (T/298)(C_i - C_f)$$

In each case the difference between the value of the function at temperature T and the value at 298 K is related to the magnitude of ΔT. However, in those cases where the heat capacities of the reactants and the products are relatively close, the changes in ΔH and ΔS are reasonably small. Since that is often the case, $\Delta H°$ and $\Delta S°$ can be assumed to be constant over reasonable ranges in temperature and Eq. (11–16) can be used to calculate $\Delta G°$ values for temperatures other than 298 K.

$\Delta H_f°$ for H_2O (liq) is 286 kJ/mol at 298 K. At 398 K, $\Delta H_f°$ is 284 kJ/mol, a difference of less than 1%.

EXAMPLE 11–11(A)

Based on available thermochemical data for the following reaction at 298 K and 1 atm, calculate the enthalpy change accompanying the same reaction at 373 K.

$$H_2 \text{ (g)} + \frac{1}{2} O_2 \text{ (g)} \longrightarrow H_2O \text{ (g)} \qquad \Delta H_f^\circ = -242 \text{ kJ/mol}$$

Solution

$$C_p \text{ (H}_2\text{)} = 28.6 \text{ J/mol·K}; \qquad C_p \text{ (O}_2\text{)} = 29.1 \text{ J/mol·K};$$
$$C_p \text{ (H}_2\text{O)} = 33.6 \text{ J/mol·K}$$

$$\Delta C_p = C_p \text{ (H}_2\text{O)} - [C_p \text{ (H}_2\text{)} + \frac{1}{2} C_p \text{ (O}_2\text{)}]$$

$$= 33.6 - [28.6 + \frac{1}{2} (29.1)]$$

$$= -9.55 \text{ J/mol·K} = -0.00955 \text{ kJ/mol·K}$$

$$\Delta C_p = \frac{\Delta H_2 - \Delta H_1}{\Delta T}$$

$$-0.00955 = \frac{\Delta H_2 - (-242)}{373 - 298}$$

$$\Delta H_2 = -243 \text{ kJ/mol}$$

The difference is very small, only about a kilojoule, which is generally the case. If the temperature range is not very large, to a good approximation ΔH is independent of temperature.

EXAMPLE 11–11(B)

Calculate ΔG° at 400 K for the reaction

$$2 \text{ NO (g)} + O_2 \text{ (g)} \longrightarrow 2 \text{ NO}_2 \text{ (g)}$$

Solution First obtain ΔH° and ΔS° values from the tables of ΔH_f° and S°:

$$\Delta H^\circ = -113.0 \text{ kJ}$$
$$\Delta S^\circ = -145 \text{ J/K}$$
$$\Delta G^\circ = \Delta H^\circ - T\Delta S^\circ$$
$$= -113.0 \text{ kJ} - (400 \text{ K})(-145 \text{ J/K})(1 \text{ kJ/1000 J})$$
$$= -54.2 \text{ kJ}$$

EXERCISE 11–11(A)

The mean molar heat capacities for nitrogen, hydrogen, and ammonia are 29.0, 28.6, and 35.1 J/mol·K, respectively. Find ΔH_{450} for this reaction:

$$N_2 \text{ (g)} + 3 H_2 \text{ (g)} \longrightarrow 2 \text{ NH}_3 \text{ (g)} \qquad \Delta H_{298} = -92.5 \text{ kJ}$$

Answer: -93.2 kJ.

EXERCISE 11–11(B)

Calculate ΔG° for the following reaction at 500 K; $\Delta G_f^\circ \text{ (BaCO}_3\text{)} = -1218$ kJ/mol:

$$\text{BaCO}_3 \text{ (s)} \longrightarrow \text{BaO (s)} + \text{CO}_2 \text{ (g)}$$

Answer: $\Delta G^\circ = 181$ kJ/mol.

To determine the temperature for a reaction at which $\Delta G°$ is zero, solve Eq. (11–16) for T. At this temperature the reaction is at equilibrium with all participating species in their standard states.

EXAMPLE 11–12

Calculate the temperature at which $\Delta G°$ is zero for the reaction

$$2 \text{ NO (g)} + O_2 \text{ (g)} \longrightarrow 2 \text{ NO}_2 \text{ (g)}$$

Solution   If $\Delta G° = 0$, then

$$0 = \Delta H° - T\Delta S°$$

$$T = \frac{\Delta H°}{\Delta S°}$$

$$= \frac{(-113.0 \text{ kJ})(1000 \text{ J/kJ})}{-145 \text{ J/K}} = 779 \text{ K}$$

EXERCISE 11–12

Calculate the temperature at which the reaction in Exercise 11–11(B) would be at equilibrium with CO_2 (g) at a pressure of 1 atm.
Answer: $T = 1.55 \times 10^3$ K.

Note that these calculations of temperature will give a meaningful result—a positive value of the temperature—only for reactions where both $\Delta H°$ and $\Delta S°$ have the same sign. These are reactions driven by either enthalpy or entropy, but not both. Reactions driven by both will be spontaneous at all temperatures, while reactions driven by neither are not spontaneous at any temperature.

The temperature dependence of the equilibrium constant can be obtained by combining Eqs. (11–16) and (11–20):

$$-RT \ln K = \Delta H° - T\Delta S°$$

This can be arranged to give

$$\ln K = \left(-\frac{\Delta H°}{RT}\right) + \left(\frac{\Delta S°}{R}\right) \tag{11–21}$$

To the extent that $\Delta H°$ and $\Delta S°$ are independent of temperature, this is a useful equation. If that assumption is valid, then a plot of $\ln K$ as a function of $1/T$ yields a straight line. From Eq. (11–21) it can be seen that if $\Delta H°$ is negative, then K will decrease with increasing temperature. If $\Delta H°$ is positive, then an increase in temperature will increase K. This is in agreement with Le Chatelier's principle. Heating an endothermic reaction drives it forward, whereas heating an exothermic reaction has the opposite effect.

If the equilibrium constant for a reaction is determined at two temperatures, T_1 and T_2, Eq. (11–21) can be used to determine the average $\Delta H°$ value for the reaction. At these two temperatures

$$\ln K_1 = -\frac{\Delta H°}{RT_1} + \frac{\Delta S°}{R}$$

and

$$\ln K_2 = -\frac{\Delta H°}{RT_2} + \frac{\Delta S°}{R}$$

If the first of these equations is subtracted from the second, the result is

$$\ln K_2 - \ln K_1 = \ln\left(\frac{K_2}{K_1}\right) = -\frac{\Delta H^\circ}{R}\left(\frac{1}{T_2} - \frac{1}{T_1}\right) \qquad (11\text{-}22)$$

EXAMPLE 11–13

Equilibrium constants for the combustion of sulfur dioxide are 3.98×10^{24} at 298 K and 1.56×10^7 at 600 K.

$$2\ SO_2\ (g) + O_2\ (g) \longrightarrow 2\ SO_3\ (g)$$

Calculate the average ΔH° for this reaction.

Solution Substitute appropriate values for K_1, T_1, K_2, and T_2 into Eq. (11-22). The result is a ΔH° value of -197 kJ/mol. ■

EXERCISE 11–13

For the reaction

$$H_2O\ (g) + O_3\ (g) \longrightarrow H_2O_2\ (g) + O_2\ (g)$$

the equilibrium constant is 4.89×10^6 at 298 K and 1.28×10^4 at 500 K. What is the average ΔH° for this reaction over the given temperature range? *Answer:* -36.5 kJ/mol. ■

For the equilibrium between a liquid and its vapor, the equilibrium constant is the vapor pressure of the liquid:

$$\text{liquid} \rightleftharpoons \text{vapor} \qquad K = p_{vap}$$

If this is substituted into Eq. (11–22), the result is

$$\ln\left(\frac{p_2}{p_1}\right) = -\frac{\Delta H_{vap}}{R}\left(\frac{1}{T_2} - \frac{1}{T_1}\right) \qquad (11\text{-}23)$$

where ΔH_{vap} is the enthalpy of vaporization of the liquid. Using this equation, known as the **Clausius-Clapeyron equation**, it is possible to calculate ΔH_{vap} if the vapor pressures p_1 and p_2 of the liquid are known at the corresponding temperatures T_1 and T_2. With the same equation, given the heat of vaporization and the vapor pressure at a certain temperature, the vapor pressure at another temperature can be determined. Note that since the Clausius-Clapeyron equation is written as a ratio of pressures, measurements can be made in any convenient units.

EXAMPLE 11–14

The vapor pressure of water at 363 K is 526 torr, and the average heat of vaporization across the temperature range from 363 K to 373 K is 40.8 kJ/mol. Calculate the vapor pressure of water at 373 K.

Solution

$$\ln\left(\frac{p_2}{526\ \text{torr}}\right) = -\frac{(40.8\ \text{kJ/mol})(1000\ \text{J/kJ})}{8.314\ \text{J/mol·K}}\left(\frac{1}{373\ \text{K}} - \frac{1}{363\ \text{K}}\right)$$
$$= 0.362$$

$$\left(\frac{p_2}{526\ \text{torr}}\right) = e^{0.362} = 1.44$$

$$p_2 = (1.44)(526\ \text{torr}) = 757\ \text{torr}$$ ■

EXERCISE 11-14

The vapor pressure of ethyl alcohol (ethanol) is 100.0 torr at 34.9°C and 400.0 torr at 63.5°C. Calculate the average heat of vaporization across this range.

Answer: 4.18×10^4 J/mol.   ∎

Substitute Natural Gas—A Synthesis Problem

The thermodynamics of chemical reactions provides the essential guide for planning the processes of chemical production. At first, thermodynamics tells us whether a particular reaction can be performed at all; and if it is possible, thermodynamics is necessary to help us find the best conditions for carrying out the reaction. Thus, to the extent that it is possible, the chemical reactions of any industrial process have been subjected to careful thermodynamic scrutiny. As an example, consider the production of substitute natural gas, SNG, from coal.

Production of a natural gas substitute from coal is attractive because of uncertainties in the supplies of petroleum and natural gas, and because of the huge existing coal reserves. In particular, let's examine the synthesis of methane (CH_4), the principal constituent of natural gas. **Coal** is a complicated material containing carbon, hydrogen, oxygen, nitrogen, sulfur, and other elements in a complex structure, as well as interspersed mineral matter. On heating, coal is converted to **coke**, which consists principally of carbon in the form of microscopic graphite crystals and mineral matter ash. Treatment of coke with steam produces carbon monoxide and hydrogen, a mixture known as **synthesis gas**:

(1) $C (s) + H_2O (g) \longrightarrow CO (g) + H_2 (g)$

Methane can be manufactured directly from synthesis gas. However, the synthesis gas is first enriched with hydrogen for the obvious reason that in order to produce CH_4, four hydrogen atoms are needed per carbon atom. This is done by taking some of the carbon monoxide from reaction (1) and letting it react with steam in what is called the **shift reaction**:

(2) $CO (g) + H_2O (g) \longrightarrow CO_2 (g) + H_2 (g)$

The carbon dioxide produced in reaction (2) is absorbed on lime (CaO) and disposed of as calcium carbonate ($CaCO_3$). In the final step of the process, carbon monoxide and hydrogen react to form methane:

(3) $CO (g) + 3 H_2 (g) \longrightarrow CH_4 (g) + H_2O (g)$

That's the chemistry! Now, consider the thermodynamics of these reactions.

The values of $\Delta H°$, $\Delta S°$, and $\Delta G°$ can be calculated for all of these reactions from tables in this and the preceding chapter. For reaction (1), the preparation of synthesis gas, $\Delta H_1° = 131.3$ kJ, $\Delta S_1° = 134.0$ J/K, and $\Delta G_1° = 91.35$ kJ. This is a strongly endothermic reaction. The value of $\Delta G°$ states that the reaction is not spontaneous at 298 K. In fact, the equilibrium constant at that temperature is 9.54×10^{-17}. However, $\Delta S°$ is positive because two moles of gas are produced starting with one mole of gas and one of solid. Since the reaction is endothermic, raising the temperature drives the reaction. Calculation of the temperature at which $\Delta G°$ is zero (that is, at which the reaction is at equilibrium with all partial pressures at 1 atm) gives a value of 980 K. Therefore, at temperatures above 980 K, conversions should start to become significant. In fact, this reaction is normally run at temperatures in excess of 1090 K.

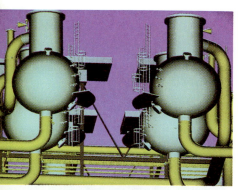

One of the primary commercial products prepared from synthesis gas is methanol (Chapter 6). Shown here is a computer-aided design for a commercial methanol reactor that was used in establishing the final design for a working reactor.

Note our earlier discussion and schematic diagram of the methanol synthesis, which is based on synthesis gas, in Chapter 4.

TWO 19TH-CENTURY THERMODYNAMICISTS: CARNOT AND GIBBS

Sadi Carnot (French, 1796–1832) devoted himself to the study of physics, engineering, and economics. He anticipated much of this applied science in his 118-page booklet, published in 1824 under the title "Reflections on the Motive Power of Heat and on the Machines Equipped to Develop This Power." Josiah Willard Gibbs (American, 1839–1903) entered Yale College in 1854. He received the Doctor of Philosophy degree in 1863 in mathematics and physics, only the fifth to be granted in the United States. He studied in Paris, Berlin, and Heidelberg before returning to become Professor of mathematical physics at Yale, a position he held until his death. Gibbs' work is essentially a unified theory of physical-chemical behavior; he proposed that the conditions for equilibrium derived from the first two laws of thermodynamics were universally applicable. The sesquicentennial of his birth was celebrated in New Haven in May, 1989.

Sadi Carnot in the year 1813, about the time he entered the Ecole Polytechnique as a first-year student.

Willard Gibbs in his yearbook portrait as a graduating senior at Yale, class of 1858.

For reaction (2), the shift reaction, $\Delta H_2^\circ = -41.2$ kJ, $\Delta S_2^\circ = -42.2$ J/K, and $\Delta G_2^\circ = -28.6$ kJ. This is a spontaneous reaction, and it is exothermic. Therefore, raising the temperature will not increase the equilibrium constant and the conversion of the reaction. In fact, raising the temperature drives the reaction in the reverse direction. To get the reaction to proceed at a reasonable rate, a catalyst of chromium and iron and a somewhat raised temperature of about 400°C are employed. It is often the case in chemical processes that temperatures are raised somewhat to increase the speed of an exothermic reaction, even though this makes the equilibrium constant less favorable.

In reaction (3), the methane synthesis reaction, $\Delta H_3^\circ = -206.1$ kJ, $\Delta S_3^\circ = -215$ J/K and $\Delta G_3^\circ = -142.0$ kJ. Again, this is an exothermic and spontaneous reaction. In order to speed it along, a nickel-iron catalyst is used and the temperature is raised to about 575 K.

The overall process is the sum of reaction (1) taken twice, reaction (2), and reaction (3):

$$2\ C\ (s) + 2\ H_2O\ (g) \longrightarrow CH_4\ (g) + CO_2\ (g)$$

The total $\Delta H°$ for the entire sequence is 15.3 kJ. This means that the entire process would require the input of very little energy *if* the energy evolved in reactions (2) and (3) could be put into reaction (1). Unfortunately, reactions (2) and (3) are run at reasonably low temperatures in order to keep their equilibrium constants (and the ratio of reactants to products) favorable to the desired product. Therefore, waste heat from these reactions is not at nearly the high temperature needed to sustain reaction (1). Consequently, a considerable amount of coal must also be burned to provide the energy requirements of the first reaction.

SUMMARY

In this chapter, the Second Law and the Third Law of Thermodynamics have been developed with the aim of predicting whether a process proceeds spontaneously. Reversible processes are at equilibrium and a spontaneous process is any other process with a tendency to proceed in the direction as written.

The relationship between reversibility, work, and heat can be demonstrated by examining the change in volume of an ideal gas. There are many possible paths for a change between two different thermodynamic states. For an isothermal process, the reversible path has the lowest value of w and the highest value of q.

Entropy is a thermodynamic state function associated with disorder. For a change between two thermodynamic states, the entropy is q_{rev} for the reversible path between these states, divided by the temperature T. According to the Second Law of Thermodynamics, the entropy of the universe is constant for a reversible process and increases for a spontaneous process. The statistical basis of entropy is the number of possible arrangements, or microstates, a system can assume. On this basis, it is possible to explain why the entropy of a gas increases when it expands and why the spontaneous contraction of a gas has never been observed.

Making use of the Third Law of Thermodynamics, it is possible to calculate the absolute entropies of elements and compounds. This is the basis for the calculation of entropy changes for chemical reactions.

The relationship between the two essential driving forces for chemical processes, entropy and enthalpy, is contained in the free energy relationship:

$$\Delta G = \Delta H - T\Delta S$$

ΔG is called the Gibbs free energy change. For processes at constant temperature and pressure—the usual conditions of practice in chemical research and development—ΔG is negative for spontaneous processes, zero for reversible processes, and positive for processes that would be spontaneous in the reverse direction (to that written). The free energy change at standard conditions is $\Delta G°$. In many cases $\Delta H°$ and $\Delta S°$ are essentially independent of temperature, so $\Delta G°$ values can be approximated at other temperatures. The standard change in free energy is related to the equilibrium constant by

$$\Delta G°(T) = -RT \ln K$$

In summary form, here are the most important equations of the chapter, and the conditions under which they apply.

$w_{rev} < w_{irrev}$	All conditions
$q_{rev} > q_{irrev}$	All conditions
$\Delta S = q_{rev}/T$	Constant temperature
$\Delta S_{univ} = \Delta S_{sys} + \Delta S_{surr}$	All conditions
$\Delta S_{surr} = -q/T$	Constant temperature
$\Delta S = nR \ln(V_f/V_i)$	Ideal gas at constant temperature
$\Delta S = nC \ln(T_f/T_i)$	Reasonably small range of temperature
$S = k \ln \Omega$	All conditions
$\Delta G = \Delta H - T\Delta S$	Constant temperature
$\Delta G = \Delta G° + RT \ln Q$	Constant temperature
$\Delta G° = -RT \ln K$	Constant pressure and temperature
$\ln K = (-\Delta H°/RT) + (\Delta S°/R)$	Constant pressure

QUESTIONS AND PROBLEMS

QUESTIONS

1. Give five examples of processes that are spontaneous but proceed very slowly.

2. Why is the following phase transformation referred to as a reversible process? H_2O (s, 0°C) $\rightleftharpoons$ H_2O (liq, 0°C)

3. Why is the following process not considered to be a reversible process? H_2O (liq, -5°C) $\rightleftharpoons$ H_2O (s, -5°C)

4. Why is entropy defined as a state function?

5. With the passage of time, what happens to the total energy of the universe? What happens to the total entropy of the universe?

6. We often read in the newspapers of an energy crisis. Can there really be an energy crisis? What might be a better name for the crisis that they have in mind?

7. What assumptions do we make about the nature of "the surroundings" in our discussions of the Second Law?

8. What is the basis of our belief in the Second Law of Thermodynamics? Is this the same as our basis for believing the Ideal Gas Law? Briefly explain.

9. When solid ammonium chloride is dissolved in water, the temperature of the solution falls. Try to explain this spontaneous change in temperature in terms of the Second Law of Thermodynamics.

10. When sodium hydroxide pellets dissolve, the temperature of the solution rises spontaneously. Briefly explain, in terms of the Second Law of Thermodynamics.

11. Comment on each of the following.
(a) If you think things are mixed up now, just wait.
(b) Entropy is not disorder, but a measure of the degree of disorder.
(c) Philosophical: If the entropy of the universe is increasing, does this imply that there must have been an initially ordered state from which everything has come?

12. Give the best reason you can, in terms of the disorder that exists in liquids and gases, that Trouton's rule usually holds.

13. Tell which of the following situations result in an increase in the entropy of the universe. Briefly state why.
(a) A tree is blown over by the wind.
(b) The water in your swimming pool leaks away into the ground.

(c) A building burns down.
(d) A balloon bursts.
(e) One hundred cars are driven from scattered locations to a central parking lot. Think carefully about this one. Consider what is going on in the engines of all these cars.

14. Tell which of the following situations result in an increase in the entropy of the system. Briefly state why.
(a) H_2O (liq) $\longrightarrow$ H_2O (g)
(b) $2\ NO_2$ (g) $\longrightarrow$ N_2O_4 (g)
(c) $CaCO_3$ (s) $\longrightarrow$ CaO (s) + CO_2 (g)
(d) $3\ H_2$ (g) + N_2 (g) $\longrightarrow$ $2\ NH_3$ (g)
(e) HCl (g) + NH_3 (g) $\longrightarrow$ NH_4Cl (s)
(f) C (s) + H_2O (g) $\longrightarrow$ CO (g) + H_2 (g)

15. Seawater is a vast storehouse of thermal energy. That is, the ambient temperature is well above 0 K. Why, then, hasn't anyone designed a ship that uses this source of energy for propulsion, or built a power plant at Newport Beach to light Los Angeles?

16. Why is Eq. (11–12) insufficient for calculating the standard absolute entropy of a solid compound or element?

17. What is the basis for the assumption that for many reactions, $\Delta H°$ and $\Delta S°$ are relatively independent of temperature?

18. If you stretch a rubber band suddenly and press it quickly against your lips, you will notice an increase in the temperature of the band. Offer a thermodynamic explanation of this effect.

19. Which of the following are state functions?
(a) $PV + q$
(b) $-w_{rev}$
(c) $\Delta H - \Delta E$

20. Using the sign conventions and ideas established in this chapter, state the sign of the heat q for each of the following:
(a) The vaporization of liquid butane in a lighter:

$$C_4H_{10} \text{ (liq, 298 K)} \longrightarrow C_4H_{10} \text{ (g, 298 K)}$$

(b) The burning of liquid iso-octane in a high-compression internal combustion engine cylinder at normal operating temperatures:

$$C_8H_{18} \text{ (g)} + \tfrac{25}{2} O_2 \text{ (g)} \longrightarrow 8 CO_2 \text{ (g)} + 9 H_2O \text{ (g)}$$

21. Using the sign conventions and ideas established in this chapter, state the sign of the heat q for each of the following:
(a) water freezes
(b) water vaporizes
(c) ice sublimes

22. Addition of thionyl chloride to hexaaquocobalt(II) chloride is an example of an endothermic but spontaneous chemical reaction. There is an immediate drop in temperature, accompanied by liberation of gaseous HCl and SO_2:

$$Co(H_2O)_6Cl_2 \text{ (s)} + 6 SOCl_2 \text{ (liq)} \longrightarrow$$
$$CoCl_2 \text{ (s)} + 12 HCl \text{ (g)} + 6 SO_2 \text{ (g)}$$

Briefly explain the spontaneous drive from reactants to products.

PROBLEMS

Reversible and Irreversible Processes   [1–4]

1. Calculate the total work, w, for the expansion of an ideal gas from 2.00 L to 3.00 L in one, two, and three irreversible steps. The initial gas pressure is 1.00 atm and the temperature of 298 K is kept constant throughout the process. The expansion in one step is performed against a constant external pressure equal to the final pressure of the gas in the cylinder. The two- and three-step expansions are done by equal stepwise decreases in the external pressure, reaching the same final pressure. Calculate the value for the work for the three expansions and compare the results.

2. You are given 10.0 L of an ideal gas contained in a cylinder with a movable piston at 298 K at an initial pressure of 2.00 atm. The gas is then compressed to 1.00 L in one, two, and three irreversible steps. The compression in one step is performed with a constant external pressure equal to the final pressure of the gas in the cylinder. The two- and three-step compressions are done by equal stepwise increases of the external pressure, reaching the same final pressure as the contained gas. Calculate the total value of the work for each process.

3. Calculate the work done during the reversible compression of 2.0 mol of an ideal gas from 10.0 L to 1.00 L, at 298 K.

4. Calculate the value of w for the reversible isothermal expansion of 1.0 mol of an ideal gas at 298 K, from 12.0 L to 30.0 L.

Entropy Calculations   [5–22]

5. A sample of 1.50 mol of carbon dioxide at 298 K and initially at 2.50 atm is allowed to expand isothermally and reversibly to a pressure of 1.50 atm. Assume ideal gas behavior. Calculate the entropy change of the system and the universe for this process.

6. A sample of 4.50 L of oxygen gas at 350 K and 0.500 atm is compressed isothermally and reversibly to a pressure of 7.50 atm. Calculate ΔS for the system, assuming ideal gas behavior.

7. Start with 1.50 mol of hydrogen gas confined in a cylinder with a movable piston at a pressure of 2.00 atm and a constant temperature of 298 K. While the temperature is kept constant, the pressure on the gas is changed to 3.00 atm and the gas is compressed. Calculate the change of entropy for the system.

8. A 50.-g sample of oxygen gas is contained in a cylinder at a pressure of 1.00 atm and a temperature of 298 K. The external pressure is then suddenly changed to 10.0 atm and the gas is compressed isothermally. Calculate the change in the entropy of the system.

9. We have a system consisting of two gas cylinders at 298 K connected by copper tubing with a valve in the middle. Each of the cylinders has a volume of 10.0 L. One cylinder contains N_2 (g) at 1.00 atm pressure. The other cylinder is empty (under vacuum). The valve is then opened, and the entire system is held at 298 K during the process. Make a sketch of the system before and after the valve is opened. Indicate the pressure and volume of the gas in all parts of the system. Assume that the nitrogen gas shows ideal behavior and that the volume of the tubing is negligible. Calculate the entropy change of the system and the universe resulting from this process.

10. Suppose that in Problem 9 the second cylinder contained O_2 (g) at 1.00 atm instead of being empty. The gases would diffuse into each other until each cylinder contained the same homogeneous mixture. Under ideal behavior, the molecules of the gas do not interact with each other. Therefore, effectively, each gas would expand to fill the entire apparatus. The *entropy of mixing* for this process would be the sum of the entropies of change of volume for the two gases. Calculate the entropy of mixing for this process.

11. A 5.00 L container is divided so that 1.00 L of O_2 (g) at 1.00 atm and 4.00 L of N_2 (g) at 1.00 atm are on either side of a thin membrane.

1 lit	4 L
1 atm	1 atm
oxygen	nitrogen

The membrane is then broken, allowing the gases to mix while the temperature is maintained at 298 K. Calculate the entropy of mixing for this isothermal process.

12. Show by a thermodynamic calculation why we would not expect a 1.00 L sample of air to separate itself into its components, say 20% O_2 (g) and 80% N_2 (g). Assume standard temperature and pressure.

13. In each of the following cases, one mole of the substance undergoes the indicated transformation. Calculate the average heat capacity.

Process	Heat (kJ/mol)
(a) H_2O (liq, 273 K) $\longrightarrow$ H_2O (liq, 298 K)	1.8
(b) H_2O (g, 373 K) $\longrightarrow$ H_2O (g, 473 K)	3.6

14. In each of the following cases, one mole of the substance undergoes the indicated transformation. Calculate the average heat capacity.

Process	Heat (kJ/mol)
(a) Cu (s, 323 K) $\longrightarrow$	-0.75
Cu (s, 298 K)	
(b) CCl$_4$ (liq, 298 K) $\longrightarrow$	-3.1
CCl$_4$ (liq, 273 K)	

15. The heat capacity of silver metal is 25.3 J/mol·K. Calculate the changes in entropy for the system and the surroundings if a 10.0 g block of silver metal at 90.0°C is allowed to cool very slowly to 20.0°C.

16. A sample of 10.0 g of oxygen gas at 1.00 atm and 100.0°C is heated reversibly under constant pressure to 155.0°C. The value of C_p for O$_2$ gas in this temperature range is 29.2 J/mol·K. Assume ideal gas behavior and calculate ΔS_{sys} for this process.

17. Calculate ΔS_{sys} and ΔS_{univ} for the melting of 50.0 g of ice at 0°C. The heat of fusion of water is 6025 J/mol.

18. Calculate the entropy change for the system and the universe that accompanies the melting of 1 kg of lead at its normal melting point of 327°C. The heat of fusion of lead is 21.3 kJ/mol.

19. The heat of vaporization of chloroform (CHCl$_3$) is 247 J/g. Estimate the boiling point of chloroform from Trouton's rule.

20. Methane boils at -159°C and its heat of vaporization is 9240 J/mol. Calculate the entropy of vaporization of methane at its boiling point. Is Trouton's rule obeyed?

21. Determine the entropy change for the process that takes place when 10.0 g of water, supercooled to -5.0°C, freezes. The heat of fusion is 6025 J/mol and the heat capacities for liquid and solid H$_2$O are 75.3 and 37.6 J/mol·K, respectively.

22. Bromine liquid freezes at -7.3°C and its heat of fusion is 10.8 kJ/mol. The heat capacity of solid bromine is 58.9 J/mol·K and that of liquid bromine is 71.6 J/mol·K. Calculate the entropy change for the bromine and for the universe if 200. g of the liquid is supercooled to -15°C and then freezes. Assume the temperature of the surroundings is -15°C.

Statistical Entropy Calculations   [23–24]

23. Calculate the entropy of a single crystal of iron at absolute zero. The crystal has a mass of 0.100 g, and it is perfect in every respect except for one single location in the crystal that should contain an iron atom but is vacant.

24. Calculate the entropy of a 0.050-g crystal of nickel at absolute zero. The crystal is perfect except for two vacant locations.

Free Energy and Equilibrium Constant Calculations   [25–34]

25. Use Tables 11–1 and 11–2, and Eq. (11–16) to calculate $\Delta S°$, $\Delta G°$ and $\Delta H°$ for each of the following reactions at 298 K, assuming all substances present to be in their standard states:

(a) 2 Ca (s) + O$_2$ (g) $\longrightarrow$ 2 CaO (s)
(b) 3 C (graphite) + 4 H$_2$ (g) $\longrightarrow$ C$_3$H$_8$ (g)
(c) HCl (g) $\longrightarrow$ H (g) + Cl (g)
(d) CH$_4$ (g) + $\frac{1}{2}$ O$_2$ (g) $\longrightarrow$ CH$_3$OH (liq)
(e) H$_2$ (g) + CO$_2$ (g) $\longrightarrow$ H$_2$O (g) + CO (g)
(f) 2 H$_2$S (g) + SO$_2$ (g) $\longrightarrow$
$\qquad\qquad$ 2 H$_2$O (g) + 3 S (rhombic)

26. Use Tables 11–1 and 11–2, and Eq. (11–16) to calculate $\Delta S°$, $\Delta G°$ and $\Delta H°$ for each of the following reactions at 298 K, assuming all substances present to be in their standard states:

(a) 2 F$_2$ (g) + 2 H$_2$O (liq) $\longrightarrow$ 4 HF (g) + O$_2$ (g)
(b) 3 Zn (s) + Fe$_2$O$_3$ (s) $\longrightarrow$ 3 ZnO (s) + 2 Fe (s)
(c) CuO (s) + H$_2$ (g) $\longrightarrow$ Cu (s) + H$_2$O (liq)
(d) Ca(OH)$_2$ (s) $\longrightarrow$ CaO (s) + H$_2$O (g)
(e) 2 C$_6$H$_6$ (liq) + 15 O$_2$ (g) $\longrightarrow$
$\qquad\qquad$ 12 CO$_2$ (g) + 6 H$_2$O (g)
(f) 2 HBr (g) + Cl$_2$ (g) $\longrightarrow$ 2 HCl (g) + Br$_2$ (liq)

27. Assuming all substances to be present in their standard states, determine which of the reactions in Problem 25 are spontaneous in the indicated direction.

28. Assuming all substances to be present in their standard states, determine which of the reactions in Problem 26 are spontaneous in the indicated direction.

29. Use Table 11–2 to calculate $\Delta G°$ for each of the following reactions at 298 K:

(a) 2 Cu$_2$O (s) + O$_2$ (g) $\longrightarrow$ 4 CuO (s)
(b) C$_2$H$_5$OH (liq) $\longrightarrow$ C$_2$H$_4$ (g) + H$_2$O (g)
(c) C$_3$H$_6$ (g) + H$_2$ (g) $\longrightarrow$ C$_3$H$_8$ (g)

30. Use Table 11–2 to calculate $\Delta G°$ for each of the following reactions at 298 K:

(a) BaO (s) + SO$_3$ (g) $\longrightarrow$ BaSO$_4$ (s)
(b) 2 Al (s) + Fe$_2$O$_3$ (s) $\longrightarrow$ Al$_2$O$_3$ (s) + 2 Fe (s)
(c) O$_3$ (g) $\longrightarrow$ O$_2$ (g) + O (g)

31. Calculate the equilibrium constants for the three reactions in Problem 29.

32. Calculate the equilibrium constants for the three reactions in Problem 30.

33. Calculate $\Delta G°$ and the equilibrium constant for the following reaction performed at 410°C:

$$2 \text{ N}_2\text{O (g)} \longrightarrow 2 \text{ N}_2 \text{ (g)} + \text{O}_2 \text{ (g)}$$

34. Determine the standard free energy change and the equilibrium constant at -100°C for the reaction (using data from the appropriate table):

$$\text{H}_2\text{S (g)} + \tfrac{3}{2} \text{O}_2 \text{ (g)} \longrightarrow \text{H}_2\text{O (g)} + \text{SO}_2 \text{ (g)}$$

Free Energy, Equilibrium, and Temperature Effects   [35–46]

35. You carry out the following reaction at 25°C. Use $\Delta H_f°$ and C_p data from Tables 7–4, 7–6 and 7–7.

$$4 \text{ NH}_3 \text{ (g)} + 3 \text{ O}_2 \text{ (g)} \longrightarrow 2 \text{ N}_2 \text{ (g)} + 6 \text{ H}_2\text{O (liq)}$$

(a) Calculate $\Delta E°$ and $\Delta H°$ for the reaction at 25°C.
(b) Calculate $\Delta H°$ for the reaction at 50°C.

36. Given the thermochemical equation for the synthesis of ammonium chloride from ammonia and hydrogen chloride at 298K,

$$NH_3 (g) + HCl (g) \longrightarrow NH_4Cl (s)$$
$$\Delta H_f^\circ = -177 \text{ kJ}$$

(a) Derive a relationship between heat capacity and enthalpy change that can be used to calculate the enthalpy change at 323 K.

(b) Then, calculate the value. Note any simplifying assumptions or approximations required.

37. Determine the heat of formation of HBr (g) at 546 K from the value of 298 K and the heat capacities (at constant pressure) of H_2 (g), Br_2 (liq), and HBr (g). Data are available in Tables 7–4, 7–6 and 7–7.

38. Find the heat of formation of yellow PbO (s) at 500 K, given the heat of formation at 298 K in Table 7–4. The mean specific heats (at constant pressure) for lead, oxygen, and the yellow form of lead(II) oxide are 0.155, 0.909, and 0.238 J/g·K, respectively.

39. Determine the temperature at which each of the following reactions would be at equilibrium with all participating species in their standard states:

(a) C_3H_6 (g) + H_2 (g) $\longrightarrow$ C_3H_8 (g)
(b) CH_3COOH (liq) $\longrightarrow$ CH_4 (g) + CO_2 (g)

40. Determine T at which $K = 1$ for the following reactions:

(a) C_2H_5OH (liq) $\longrightarrow$ C_2H_4 (g) + H_2O (g)
(b) Br_2 (liq) $\longrightarrow$ 2 Br (g)

41. The equilibrium constant for the reaction

$$I_2 (s) + Br_2 (liq) \longrightarrow 2 IBr (g)$$

was found to be 4.52×10^{-2} at 25°C and 3.05×10^3 at 177°C. What is ΔH° for this reaction?

42. The equilibrium constant for the reaction

$$2 NO_2 (g) \longrightarrow N_2O_4 (g)$$

at several different temperatures gives the following results:

Temperature (K)	Equilibrium Constant
300	7.49
350	2.70×10^{-1}
400	2.23×10^{-2}
450	3.21×10^{-3}

Use a plot of ln K versus $1/T$ to determine ΔH° and ΔS° of the reaction.

43. Slaked lime for mortar is made from limestone by the following two reactions:

$$CaCO_3 (s) \longrightarrow CaO (s) + CO_2 (g)$$
$$CaO (s) + H_2O (liq) \longrightarrow Ca(OH)_2 (s)$$

Use thermodynamic data to suggest the best conditions for performing these reactions.

44. Consider the thermodynamics for the possible preparation of benzene (C_6H_6) from carbon monoxide and hydrogen gas. Would this process be possible and, if so, what do you think would be the best conditions?

45. The Haber process for the preparation of ammonia from hydrogen and nitrogen gases was formulated after a careful thermodynamic study. Look at the thermodynamics of this process and tell what you think would be the best reaction conditions.

46. (a) Using the following thermodynamic data, determine the temperature at which the decomposition of magnesium carbonate first becomes feasible:

Substance	ΔH_f° (kJ/mol)	S° (J/mol·K)
$MgCO_3$ (s)	−1113	66
MgO (s)	−602	27
CO_2 (g)	−394	214
Mg (s)	0	33
O_2 (g)	0	206

(b) By considering the data further, indicate why the reaction stops at MgO (s) and does not proceed further to Mg (s) and $\frac{1}{2}$ O_2 (g).

The Clausius-Clapeyron Equation   [47–48]

47. The vapor pressure of hexane (C_6H_{14}) is 40.0 torr at −2.3°C and 100.0 torr at 15.8°C. Calculate the average enthalpy of vaporization for hexane within this range.

48. The vapor pressure of ethyl acetate ($C_4H_8O_2$) is 400 torr at 59.3°C. Calculate the normal boiling point of ethyl acetate. The average enthalpy of vaporization in that range is 34,900 J/mol.

Additional Problems   [49–57]

49. Calculate the work for the expansion of 1.0 L of an ideal gas to a final volume of 5.0 L, against a constant external pressure of 1.5 atm.

50. Calculate the change of entropy in the system and the universe for the spontaneous compression of 4.00 mol of hydrogen gas at a pressure of 5.00 atm to 15.00 atm without any changes in external conditions. The temperature is kept constant at 350 K during this process. Assume ideal gas behavior. How does this result agree with the Second Law?

51. Calculate q, w, ΔE, ΔH, ΔS, and ΔG for the reversible vaporization of one mole of water at 100°C and 1.00 atm.

52. For the reversible expansion of one mole of an ideal gas at 298 K from 10. L to 20. L, calculate q, w, ΔE, ΔH, ΔS, and ΔG.

53. We never observe the spontaneous flow of heat from a cooler to a warmer body. Define a system consisting of a cooler body at a temperature T_1, and a warmer body at a temperature T_2. Allow a quantity of heat to flow from the cooler to the warmer body in a way such that no heat is transferred between the system and the surroundings. Show that this process is forbidden by the Second Law.

54. (a) If 1000. J of heat is added reversibly to a body at a constant temperature, increasing its entropy by 10. J/K, what is the temperature of the substance? (b) If the same amount of heat was added to the same body at 298 K, what increase in entropy would result?

55. For the following process at 373 K,

$$H_2O (liq) \longrightarrow H_2O (g)$$

$\Delta S^\circ = 109$ J/mol K. How do you explain this unexpectedly high value in light of the fact that for a great many liquids $\Delta S°$ is approximately 88 kJ/mol·K?

56. Calculate the theoretical maximum efficiency of a heat engine that uses steam superheated to 175°C and a condenser operating at 20°C.

57. What is the theoretical maximum work that can be obtained from a heat engine if 850 J of heat is supplied at 250°C and the heat sink is maintained at 50°C?

MULTIPLE PRINCIPLES [58–62]

58. Calculate the equilibrium constant at 400°C for the reaction

$$CO \ (g) \ + \ H_2O \ (g) \longrightarrow CO_2 \ (g) \ + \ H_2 \ (g)$$

assuming that $\Delta H°$ and $\Delta S°$ do not change significantly with temperature. Calculate the equilibrium partial pressures of all components at this temperature, starting with CO (g) and H_2O (g) each at a partial pressure of 0.100 atm.

59. Determine the equilibrium constant at 25°C for the reaction

$$2 \ N_2O \ (g) \ + \ 3 \ O_2 \ (g) \longrightarrow 4 \ NO_2 \ (g)$$

In an equilibrium mixture of these gases at 25°C, the partial pressures of both N_2O and NO_2 were 0.500 atm. What was the partial pressure of oxygen?

60. The vapor pressures in torr for water at room temperatures are as follows:

T (K)	273	283	293	303	313	323	333
p_{vap} (torr)	4.58	9.21	17.54	31.82	55.32	92.51	149.4

Create a table containing values of T, $1/T$, and $\ln p$. Then graph the data as $\ln p$ versus $1/T$, and find $\Delta H°$ and $\Delta S°$ for the vaporization process, recording your answers in kJ/mol and J/K, respectively.

61. Here are some data for the following gas phase equilibrium at several different temperatures:

$$SO_2 \ (g) \ + \ \tfrac{1}{2} \ O_2 \ (g) \rightleftharpoons SO_3 \ (g)$$

T (K)	800	850	900	950	1000	1100
K_p	31.3	13.8	6.55	3.24	1.85	0.628

(a) Write the equilibrium expression for K_p.
(b) Algebraically, calculate $\Delta H°$ and $\Delta S°$ over the range of T values.
(c) Graphically, calculate $\Delta H°$ and $\Delta S°$ over the range of T values.

62. The average specific heat of ice is 0.492 cal/g·K; $\Delta H_f° = 1434$ cal/mol, and $\Delta H_{vap} = 9712$ cal/mol at 1 atm and 373 K.

(a) Calculate the energy accompanying the transformation of 1.00 kg of H_2O as follows:

$$H_2O \ (s, \ 263 \ K) \longrightarrow H_2O \ (liq, \ 298 \ K)$$

(b) Calculate ΔH for the transformation of 1.00 kg of H_2O as follows:

$$H_2O \ (g, \ 373 \ K) \longrightarrow H_2O \ (liq, \ 373 \ K)$$

(c) Calculate the entropy change accompanying the fusion and condensation of 1.00 kg of H_2O.
(d) On the basis of your results in (a), (b), and (c), what would you conclude about the degree of disorder associated with the solid, liquid, and gas states of H_2O?

APPLIED PRINCIPLES [63–67]

63. Carbon monoxide and steam can be used as a source of hydrogen, according to the following reaction:

$$CO \ (g) \ + \ H_2O \ (g) \longrightarrow CO_2 \ (g) \ + \ H_2 \ (g)$$

Taking $\Delta H°$ and $\Delta S°$ to be constants over the temperature range under consideration, calculate the fraction of CO converted to CO_2 at 100° intervals from 300°C to 800°C. In each case the equilibrium partial pressure of H_2O is 0.900 atm and that of CO is 0.100 atm.

64. Calculate the values of the equilibrium constant at 0°C, 200°C and 400°C for ammonia synthesis, assuming that $\Delta H°$ and $\Delta S°$ do not change significantly with temperature:

$$N_2 \ (g) \ + \ 3 \ H_2 \ (g) \longrightarrow 2 \ NH_3 \ (g)$$

Comment on the effect of increasing the temperature on the extent of conversion of reactants into products for this reaction.

65. Propane is a clean-burning fuel that has been widely used for such things as hand-held torches and home heating; it is used in Europe for internal combustion heat engines. What is the maximum work that an engine can deliver from burning 1.00 kg of propane if the heat is utilized at 600.°C and exhausted at 400.°C?

66. It has been estimated that in every cubic kilometer of seawater there are 6 kg of gold (in combined form). At current gold prices of about $450/ounce, that would suggest a sizeable fortune for anyone clever enough to mine it successfully. Should you try to exploit this gold mine or not? Use thermodynamic arguments to make your case.

67. In Chapter 8 we described an important water purification system called reverse osmosis, in which salt water is forced through tubes that are made of a semipermeable material, which allows pure water to pass through the walls. Explain why this highly effective and commercially successful process is not in conflict with the Second Law of Thermodynamics.

CHAPTER
12

Equilibrium II: Ionic Equilibria in Aqueous Solutions

In contrast to instructional laboratories, it is not uncommon for industrial analytical laboratories to receive thousands of samples each year. Bar codes are now used to improve data integrity while productivity is enhanced through greater accuracy and precision. Here, acid-base titrations are being carried out by an automatic titrator coupled to a 10-sample changer and a laserscan box code reader. Missing from the photo are an automatic recording analytical balance and PC data station. An "indicator" solution has been added to each sample to enhance the illustration—red before neutralizing the acid solution, yellow after neutralization. Of the strong acids used on an industrial scale, nitric acid is among the most important, shown here in a late 18th-century French painting depicting its early commercial preparation.

Svante Arrhenius (Swedish), Nobel Prize for Chemistry in 1903 for his theory of electrolytic dissociation and ionic equilibria in aqueous solution.

The five oceans and the seven seas are more a figure of speech than a definitive list. But for the record (according to Webster): Atlantic, Pacific, Indian, Arctic, and Antarctic Oceans; Mediterranean, China, North, Japan, Red, Black, and Baltic Seas.

12.1   IONS IN SOLUTION

Water is the very essence of our environment and our existence, so aqueous solutions and reactions involving dissolved species are of enormous importance in theory and practice. More than 97% of the hydrosphere is in the five great oceans and the seven seas. This vast blanket of water covers more than 70% of the Earth's surface to an average depth of nearly 4 km; its mass is an almost unfathomable quantity, estimated to be 1.4×10^{18} kg. This enormous and spectacular environment is a 0.5 M aqueous electrolyte solution of mostly sodium chloride, with a bit of magnesium sulfate, and trace quantities of a laundry list of other elements and compounds. Except for the marked enrichment of potassium ions over sodium ions in cellular fluid, the composition of the human body fluids is surprisingly similar to that of seawater. The unique solvating properties of water lead to one of its most characteristic qualities, the ability to support the reaction of ions in solution. These hydrated, electrically charged particles are the principal species involved in nearly all the reactions in aqueous solutions. This world of ions in solution and aqueous reactions is the subject of this chapter and the one that follows. But before we begin, recall a few of the important ideas from the preceding two chapters that will prove useful here.

Chapter 10 introduced the equilibrium expression and the general features of the equilibrium state. From experiment, we know that the equilibrium condition is dynamic; we also know that reversible systems move spontaneously toward thermodynamic equilibrium; and we know that the nature and properties of the equilibrium state are the same, regardless of the direction from which the equilibrium condition is approached. Chapter 11 introduced the relationship between the equilibrium constant and the change in free energy ΔG. We found that the tendency for a chemical system to change depends upon a competition between the drive to lower the enthalpy of the system and the drive to increase its entropy or disorder; and further, that the drive to increase the entropy becomes more important at elevated temperatures. Because Chapter 10 dealt almost entirely with reversible processes in the gas phase, it was convenient to state the amounts of reactants and products as partial pressures. Thus, equilibrium constants in the preceding two chapters were expressed mostly as K_p; but since partial pressures are proportional to moles, the examples and illustrations could just as well have been expressed in terms of K_c.

A more detailed discussion of electrodes and electrode processes is coming in Chapter 21, Electrochemistry.

12.2   ELECTROLYTES, SOLUBILITY, AND SPARINGLY SOLUBLE SALTS

For nearly two centuries, we have known experimentally that water solutions of some salts and other compounds are conductors of electricity. We also know that these solutions differ from metallic conductors; that electrical conductivity increases with increasing temperature; and that chemical changes occur at surfaces called electrodes. For nearly a century, we have understood that current is carried in solution by positively charged species called **cations** and negatively charges species called **anions**. The mobility of ions within solutions increases at higher temperatures, with a resulting increase in conductivity. Chemical reactions take place when the ions reach the surface of each electrode.

Electrolytes and Nonelectrolytes

Substances that show significant conductivity because they furnish or otherwise produce ions when they dissolve in water are called **electrolytes**. Substances that dissolve without furnishing ions are called **nonelectrolytes**. We were introduced briefly to both as part of an earlier discussion of colligative properties of solutions in Chapter 9. Most organic molecules are nonelectrolytes; typical examples we have encountered before include alcohols, glycols, and sugars. Electrolytes include classes of compounds known as acids, bases, and salts; typical examples are hydrochloric and sulfuric acids, and sodium hydroxide and sodium chloride.

The conductivity of aqueous solutions of electrolytes has many important implications and applications, some helpful and some detrimental. Because of the presence of traces of electrolytes, the ground is a conductor when moisture is present. Under most conditions, driving a metal rod just a few feet below the surface will encounter permanently conducting conditions, and this is used for grounding lightning rods, antennae, telephone lines, TV cables, and exposed metal structures. The ground provides the return path for the electric circuit for electric fences and, in the 19th century, for telephone wires.

Strong and Weak Electrolytes

Suppose the conductivity experiment in Figure 12–1 is performed with various familiar samples in the beaker: distilled water; acetic acid (vinegar solution), or ammonia (household cleaner); possibly table salt dissolved in water. With distilled water, there is no visual evidence of current flowing—

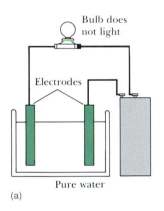

(a) Bulb does not light — Electrodes — Pure water

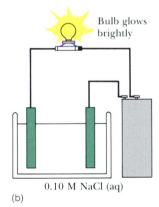

(b) Bulb glows brightly — 0.10 M NaCl (aq)

FIGURE 12–1 (a) Being only slightly dissociated, the concentration of ions in water is very low—on the order of 10^{-7} M in H_3O^+ or OH^-—and there is no evidence of electrical conductivity. The light bulb in the external circuit does not glow; (b) a 0.10 M NaCl solution conducts electricity very well because of the presence of a high concentration of Na^+ and Cl^- ions in solution and the light bulb glows intensely; (c) the dim glow of the light bulb connected into the circuit of a 0.10 M acetic acid solution indicates low conductivity due to the presence of a low concentration of ions in solution; (d) finally, sugar is un-ionized in aqueous solution, there is no electrical conductivity and the bulb does not glow.

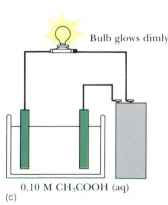

(c) Bulb glows dimly — 0.10 M CH_3COOH (aq)

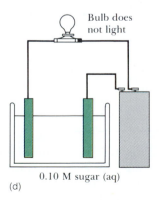

(d) Bulb does not light — 0.10 M sugar (aq)

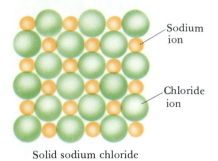

Sodium ion

Chloride ion

Solid sodium chloride

Step 1

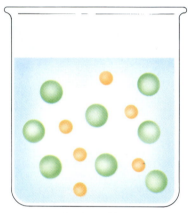

Separated ions

Step 2

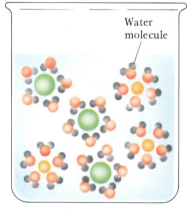

Water molecule

Hydrated ions

FIGURE 12–2 The ionization of sodium chloride in water can be pictured as a stepwise process, beginning with the coming apart of the crystal structure into a gas of ions which then become "hydrated" ions as the salt dissolves into solution.

no glow in the light bulb. The bulb does glow faintly with acetic acid solution and aqueous ammonia, but shines brightly when the electrodes are immersed in sodium chloride solution. What we observed with acetic acid and ammonia is typical of **weak electrolytes**, whereas the behavior of sodium chloride is typical of **strong electrolytes**. In general, strong electrolytes include soluble salts, which dissolve freely in water, aqueous solutions of strong acids, and aqueous solutions of strong bases. Acetic acid is weakly acidic, and a typical weak electrolyte. Ammonia is weakly basic, and also a typical weak electrolyte.

The fundamental distinction between strong and weak electrolytes can be examined in terms of equilibrium constants. In general, we can write the following:

$$\text{electrolyte} \rightleftharpoons \text{aqueous cation} + \text{aqueous anion}$$

For weak electrolytes, the equilibrium constants are small numbers, much less than 1. These equilibria lie far to the left, representing substances that produce relatively small numbers of ions in solution. For strong electrolytes, the hydration reaction proceeds essentially to completion, and the equilibrium constant is a very large number, approaching infinity as the dissociation into hydrated ions approaches completion. Consider once again the dissociation of sodium chloride into hydrated ions in solution:

$$\text{NaCl (s)} \rightleftharpoons \text{Na}^+ \text{ (aq)} + \text{Cl}^- \text{ (aq)}$$

It should not be surprising that earlier ideas on energy (from Chapter 7) and entropy (from Chapter 11) are important in understanding the processes involved.

It is useful to reduce the reaction to simple steps, remembering that enthalpy and entropy (which are state functions) depend only on the initial and final states and not on the nature of intermediate steps. (Refer to discussion of dissolving and dissolution in Chapter 9.) First, we imagine that the solid structure comes apart into a gas of ions (Figure 12–2). Then these separated ions react with solvent water, forming hydrated ions. The water molecule acts as an electrical dipole, with the negative end at the oxygen atom and the positive end between the hydrogen atoms. Thus, oxygen atoms in water molecules are drawn toward positive sodium ions, while hydrogen atoms in water molecules are attracted to negative chloride ions.

The first step is endothermic since energy must be added to the system to separate the oppositely charged ions; and ΔS is positive because a solid is being converted into a gas of ions. The second step in Figure 12–2 is an exothermic process since interactions between the ions and water molecules are being established; and ΔS is negative because the disorganization of the gas of ions is being lost, and the interactions between the ions and the solvent result in some ordering of the otherwise more randomly oriented water molecules. In summary,

Step 1	$\Delta H > 0$	$\Delta S > 0$
Step 2	$\Delta H < 0$	$\Delta S < 0$
Overall	?	?

Note that this analysis fails to predict the sign of either ΔH or ΔS for the overall process, because for each function the signs of the first and second steps are opposite. Thus, the signs of ΔH and ΔS depend upon the relative magnitudes of the values for the two steps. Table 12–1 gives $\Delta H°$ and $\Delta S°$ values for dissolution of some salts in water. We have already observed (in

TABLE 12–1	Thermodynamic Data for Dissolving Some Common Salts in Water		
	$\Delta H°$ (kJ/mol)	$\Delta S°$ (J/mol·K)	$\Delta G°$ (kJ/mol)
$Ca(NO_3)_2$	− 18.9	42.5	− 32.1
$CaSO_4$	− 16.7	− 144.3	26.3
NaCl	3.89	43.1	− 8.95
NH_4NO_3	14.8	75.3	− 7.66

Chapter 11) that some salts dissolve in water exothermically whereas others dissolve endothermically. For example, dissolving NH_4NO_3 absorbs enough energy that it is the active ingredient for the instant cold packs used by medical laboratories, hospitals, and injured athletes (Figure 12–3).

The standard free energy change for solution has also been tabulated in Table 12–1, based upon calculations using $\Delta G° = \Delta H° - T\Delta S°$ at 298 K. Three of the four salts have $\Delta G < 0$, indicating a spontaneous process. Such salts are considered to be freely soluble in water. However, $CaSO_4$ is one of a large group of salts for which the equilibrium is far to the reactant side, with the spontaneous reaction being in the reverse direction. Thus, if $CaSO_4$ is added to distilled water in the conductivity experiment in Figure 12–1, most of the salt would remain in the solid phase, lying at the bottom of the beaker. Only a relatively small number of ions would be present in solution and in our conductivity (bulb) experiment, the light would glow dimly at best. Calcium sulfate and salts like it are referred to as **sparingly soluble salts**.

FIGURE 12–3 The "cold pack" shown here is used to preserve blood samples for short periods of time. The inner pouch is broken by squeezing, allowing the ammonium nitrate to mix with and dissolve into an aqueous solution with the absorption of considerable heat from the surroundings. A blood vial inserted into the pack will be kept cold for up to half an hour.

Sparingly Soluble Salts

The equilibrium expression for the dissolution of the sparingly soluble salt calcium sulfate is

$$CaSO_4 \text{ (s)} \rightleftharpoons Ca^{2+} \text{ (aq)} + SO_4^{2-} \text{ (aq)} \qquad K_{sp} = [Ca^{2+}][SO_4^{2-}]$$

Solid $CaSO_4$ does not appear in the equilibrium expression because the concentration of calcium sulfate in the pure solid is constant. The subscript "sp" stands for **solubility product**, the product of the concentrations of the ions that result from the dissolving and dissociation of the salt. The exponents in the equilibrium expression are the same as the coefficients in the dissociation process. For $CaSO_4$ both coefficients are 1; for $Ca_3(PO_4)_2$ they are 3 and 2, respectively:

$$Ca_3(PO_4)_2 \text{ (s)} \rightleftharpoons 3 Ca^{2+} \text{ (aq)} + 2 PO_4^{3-} \text{ (aq)} \qquad K_{sp} = [Ca^{2+}]^3[PO_4^{3-}]^2$$

The value of K_{sp} can be calculated from the $\Delta G°$ of solution. For $CaSO_4$ (s) at 298 K, for which $\Delta G°$ of solution is 26.3 kJ/mol, we use

$$K_{sp} = e^{-\Delta G/RT}$$

where

$$\frac{\Delta G}{RT} = \frac{26.3 \text{ kJ/mol}}{(8.314 \text{ J/mol·K})(1 \text{ kJ/1000 J})(298 \text{ K})}$$

$$= 10.62$$

$$K_{sp} = e^{-10.62} = 2.4 \times 10^{-5}$$

Solubility products are normally determined experimentally, although calculation from direct analysis of the ion concentrations is difficult because

TABLE 12–2   Solubility Product Constants at 25°C

Compound	K_{sp}	Compound	K_{sp}
AgBr	5.2×10^{-13}	MgCO$_3$	4.0×10^{-5}
AgCl	1.6×10^{-10}	Mg(OH)$_2$	1.2×10^{-11}
Ag$_2$CrO$_4$	1.9×10^{-12}	Mn(OH)$_2$	2.0×10^{-13}
AgI	8.5×10^{-17}	MnS	1.4×10^{-15}
Ag$_2$S	1.6×10^{-49}	Ni(OH)$_2$	1.6×10^{-16}
Al(OH)$_3$	1.8×10^{-33}	NiS	1.4×10^{-24}
BaCO$_3$	1.6×10^{-9}	PbCO$_3$	1.5×10^{-13}
BaCrO$_4$	8.5×10^{-11}	PbCrO$_4$	1.8×10^{-14}
BaF$_2$	1.7×10^{-6}	Pb(OH)$_2$	1.8×10^{-16}
BaSO$_4$	1.1×10^{-10}	PbS	3.4×10^{-28}
CaCrO$_4$	7.1×10^{-4}	PbSO$_4$	1.3×10^{-8}
CaF$_2$	1.7×10^{-10}	Sn(OH)$_2$	5×10^{-26}
Ca$_3$(PO$_4$)$_2$	1.3×10^{-32}	SnS	8×10^{-29}
Cu(OH)$_2$	1.6×10^{-19}	SrCO$_3$	1.6×10^{-9}
CuS	8.5×10^{-45}	SrF$_2$	2.8×10^{-9}
Fe(OH)$_2$	1.6×10^{-15}	ZnCO$_3$	2×10^{-10}
FeS	3.7×10^{-19}	Zn(OH)$_2$	4.5×10^{-24}
HgS	3×10^{-53}	ZnS	4.5×10^{-24}

the concentrations are so low. Electrochemical methods (Chapter 21) have been particularly useful for this. Solubility products of some salts are given in Table 12–2.

Solubility Products and Solubility

The relationships that exist between the solubility product constant and the solubility of the salt in solution are illustrated by the following examples and exercises. Note that moles of solute per liter of solution is the unit typically used to express concentrations of ions in solution.

EXAMPLE 12–1

Determine the solubility of BaSO$_4$ (s) in pure water at 298 K in moles per liter and grams per liter.

Solution   In the dissolution reaction, the number of moles of BaSO$_4$ (s) dissolved is equal to the number of moles of Ba^{2+} (aq) or SO$_4^{2-}$ (aq) ions formed:

$$\text{BaSO}_4 \text{ (s)} \rightleftharpoons \text{Ba}^{2+} \text{ (aq)} + \text{SO}_4^{2-} \text{ (aq)}$$

We assign the variable x to be the solubility at equilibrium. Remember that brackets around the formula of a species denotes its concentration in moles/liter. Thus,

$$x = \text{solubility in moles/liter} = \text{moles of BaSO}_4 \text{ (s)/liter of solution}$$
$$= [\text{Ba}^{2+}] = [\text{SO}_4^{2-}]$$

Using the K_{sp} value from Table 12–2 and substituting into the solubility product expression,

$$K_{sp} = 1.1 \times 10^{-10} = (x)(x) = x^2$$
$$x = \sqrt{1.1 \times 10^{-10}} = 1.0 \times 10^{-5}$$

Thus, the solubility of $BaSO_4$ is 1.0×10^{-5} mol/L, or

$$(1.0 \times 10^{-5} \text{ mol/L})(233 \text{ g/mol}) = 2.3 \times 10^{-3} \text{ g/L} \qquad \blacksquare$$

EXERCISE 12–1

Calculate the solubility in moles per liter and grams per liter of PbS in pure water at 25°C.
Answer: 1.8×10^{-14} mol/L; 4.3×10^{-12} g/L.   ■

EXAMPLE 12–2

Determine the solubility in moles per liter of $Ca_3(PO_4)_2$ in pure water at 25°C. The reaction is

$$Ca_3(PO_4)_2 \text{ (s)} \rightleftharpoons 3 \text{ Ca}^{2+} \text{ (aq)} + 2 \text{ PO}_4^{3-} \text{ (aq)}$$

Solution   If we let the solubility of $Ca_3(PO_4)_2$ be x mol/L, then the stoichiometric coefficients require that $[Ca^{2+}] = 3x$ and $[PO_4^{3-}] = 2x$. Substituting into the equilibrium expression,

$$K_{sp} = 1.3 \times 10^{-32} = (3x)^3(2x)^2 = 108x^5$$

$$x^5 = 1.2 \times 10^{-34}$$

$$x = \sqrt[5]{1.2 \times 10^{-34}} = 1.6 \times 10^{-7}$$

The solubility of $Ca_3(PO_4)_2$ in pure water is 1.6×10^{-7} mol/L.   ■

EXERCISE 12–2

Determine the solubility of $Al(OH)_3$ in pure water at 25°C in moles per liter.
Answer: 2.9×10^{-9} mol/L.   ■

EXAMPLE 12–3

The solubility of PbF_2 in pure water at 25°C is 2.1×10^{-3} mol/L. What is the solubility product constant?

Solution   The reaction is

$$PbF_2 \text{ (s)} \rightleftharpoons Pb^{2+} \text{ (aq)} + 2 \text{ F}^- \text{ (aq)}$$

$$[Pb^{2+}] = \text{solubility of } PbF_2$$

$$[F^-] = \text{twice the solubility of } PbF_2$$

Therefore,

$$K_{sp} = [Pb^{2+}][F^-]^2 = (2.1 \times 10^{-3})(2 \times 2.1 \times 10^{-3})^2 = 3.7 \times 10^{-8}$$

■

EXERCISE 12–3

At 15°C, the solubility of PbI_2 in pure water is 1.23×10^{-3} mol/L. What is the solubility product constant for PbI_2 at this temperature?
Answer: 7.44×10^{-9}.   ■

As we described in Chapter 9, the **molarity** (M) of a solution is the number of moles of solute per liter of solution. A 1 M solution contains one formula weight of solute in grams dissolved in sufficient solvent to produce a final volume of one liter of solution. Thus, a 1.00 M hydrochloric

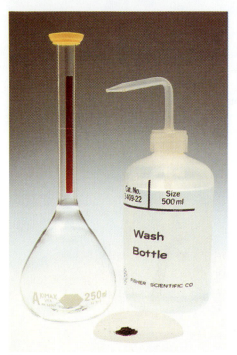

(a)

(b)

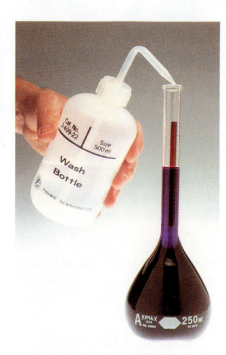

(c)

FIGURE 12–4 Preparation of an 0.0100 M solution of potassium permanganate ($KMnO_4$) solution: 250. mL of 0.0100 M $KMnO_4$ solution contains 0.395 g (0.0025 mol). The solution is prepared by (a) dissolving the solid in enough water to make a solution (b) and then (c) diluting the solution to a final volume of 250 mL in a volumetric flask especially designed and calibrated to contain just that volume, when filled to the mark on the neck. (See also Figure 9–7.)

acid solution contains 36.5 g of HCl dissolved in a liter of solution; and 0.150 M ammonia solution contains 2.55 g of NH_3 in a liter of solution (Figure 12–4).

EXAMPLE 12–4

A solution is prepared by dissolving 125 g of acetic acid (CH_3COOH) in enough water to make 1.25 L of solution. Determine the molarity of the solution.

Solution First we determine how many moles of acetic acid are present:

$$(125 \text{ g } CH_3COOH) \left(\frac{1 \text{ mol}}{60.0 \text{ g}} \right) = 2.08 \text{ mol } CH_3COOH$$

Then we find the molarity:

$$\frac{2.08 \text{ mol } CH_3COOH}{1.25 \text{ L}} = 1.67 \text{ mol/L} = 1.67 \text{ M}$$

EXERCISE 12–4

Calculate the molarity of the solution that results from dissolving 10.0 g of NaCl (s) in sufficient water to produce 450.0 mL of solution.
Answer: 0.380 M.

EXAMPLE 12–5

In another situation, 0.0250 mol of sodium hydroxide (NaOH) is required for a certain chemical reaction. The available laboratory reagent is a sodium hydroxide solution known to be 0.525 M. Find how much solution to take.

Solution   Note the generally useful relationship that the product of volume and molarity equals moles:

$$V \times M = (\text{L})(\text{mol/L}) = \text{mol}$$

Therefore, in this case

$$\frac{\text{mol}}{\text{mol/L}} = \text{L} = \frac{0.0250 \text{ mol NaOH}}{0.525 \text{ mol/L}} = 0.0476 \text{ L} = 47.6 \text{ mL solution} \quad \blacksquare$$

EXERCISE 12–5
How much 0.430 M $CaCl_2$ solution is required to deliver 1.55 mole of $CaCl_2$?
Answer: 3.60 L.   ◼

The Common Ion Effect

If a sparingly soluble salt is in equilibrium with its ions, and then the concentration of one of these ions is increased, Le Chatelier's principle tells us that the reaction moves toward the solid salt, decreasing the solubility of the salt. For example, if the reaction

$$BaSO_4 \text{ (s)} \rightleftharpoons Ba^{2+} \text{ (aq)} + SO_4^{2-} \text{ (aq)}$$

is at equilibrium, the concentration of either of the ions can be increased by adding a freely soluble salt of one of them. Thus, adding $BaCl_2$ increases $[Ba^{2+}]$, driving the reaction to the left and decreasing $[SO_4^{2-}]$. At equilibrium, $[Ba^{2+}]$ must be higher, and $[SO_4^{2-}]$ lower, in order to satisfy the equilibrium expression and K_{sp}. The effect of the presence of a common ion is to diminish the solubility of a sparingly soluble salt; this is called the **common ion effect**.

In order to demonstrate the common ion effect, consider the amount of $BaSO_4$ (s) that dissociates into hydrated ions in water and in an aqueous barium chloride solution by working through the following example. It is a classic demonstration of the principle.

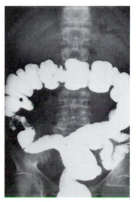

Barium sulfate is the insoluble "opaque-ing" substance used to create a background image for gastrointestinal X-rays. Even though Ba^{2+} is highly toxic to the central nervous system, because of the extremely low concentration of these ions when present as the sparingly soluble sulfate salt, it is perfectly safe to ingest orally for this purpose.

EXAMPLE 12–6
Compare the solubility of $BaSO_4$ (s) in pure water and in a 0.10 M $BaCl_2$ solution.

Solution $BaCl_2$ dissociates completely, according to the following equation:

$$BaCl_2 \text{ (s)} \longrightarrow Ba^{2+} \text{ (aq)} + 2 \text{ Cl}^- \text{ (aq)}$$

For a 0.10 M $BaCl_2$ solution,

$$[Ba^{2+}] = 0.10 \text{ M} \quad \text{and} \quad [Cl^-] = 0.20 \text{ M}$$

Write down the concentrations of the ions in the dissolution of $BaSO_4$ (s) in tabular form, first before ionization begins, and then after equilibrium is reached. Let x represent the number of moles per liter of $BaSO_4$ (s) dissolved:

	$BaSO_4$ (s) $\rightleftharpoons$	Ba^{2+} (aq) +	SO_4^{2-} (aq)
before ionization		0.10	0
at equilibrium		$0.10 + x$	x

Substitution into the K_{sp} expression gives

$$K_{sp} = 1.1 \times 10^{-10} = (0.10 + x)(x)$$

Multiplication and rearrangement yields a quadratic equation, which could be solved by the customary formula. However, we can save some work by recognizing that the value of x will be very small due to the small value of K_{sp}; thus, $0.10 + x$ is not significantly different from 0.10 once the significant figures have been accounted for. Therefore, we can make the simplifying assumption that since $x \ll 0.10$, the quantity x can be dropped from the $0.10 + x$ term. This gives

$$1.1 \times 10^{-10} = (0.10)(x)$$
$$x = 1.1 \times 10^{-9}$$

Note that x is indeed much, much less than 0.10, validating our assumption. In 0.10 M $BaCl_2$, the solubility of $BaSO_4$ is 1.1×10^{-9} M, so $[Ba^{2+}] = 0.10$ M and $[SO_4{}^{2-}] = 1.1 \times 10^{-9}$ M. We can compare these results with those in Example 12–1, where $BaSO_4$ was dissolved in pure water:

	In Pure Water	**In 0.10 M BaCl₂ Solution**
solubility of $BaSO_4$	1.0×10^{-5} M	1.1×10^{-9} M
equilibrium $[Ba^{2+}]$	1.0×10^{-5} M	0.10 M
equilibrium $[SO_4{}^{2-}]$	1.0×10^{-5} M	1.1×10^{-9} M

The solubility of $BaSO_4$ in 0.10 M $BaCl_2$ is reduced significantly; the equilibrium concentration of Ba^{2+} is much higher than in pure water; and the equilibrium concentration of $SO_4{}^{2-}$ is much lower. ■

EXERCISE 12–6

Compare the solubility and equilibrium ion concentrations of $PbCO_3$ in pure water and in 0.010 M $Pb(NO_3)_2$.
Answer: In pure water, solubility $= [Pb^{2+}] = [CO_3{}^{2-}] = 3.9 \times 10^{-7}$ mol/L. In 0.010 M $Pb(NO_3)_2$, solubility $= [CO_3{}^{2-}] = 1.5 \times 10^{-11}$ mol/L and $[Pb^{2+}] = 0.010$ M. ■

Unfortunately, the very convenient simplifying assumption used in Example 12–6 is not always justified. We see that in the following illustration.

EXAMPLE 12–7

Determine the solubility and equilibrium ion concentrations of $CaCrO_4$ in a 0.080 M solution of $CaCl_2$.

Solution Let x = solubility of $CaCrO_4$ (s).

$$CaCrO_4 \text{ (s)} \rightleftharpoons Ca^{2+} \text{ (aq)} + CrO_4{}^{2-} \text{ (aq)}$$

before ionization	0.080	0
at equilibrium	0.080 + x	x

$$K_{sp} = 7.1 \times 10^{-4} = (0.080 + x)(x)$$

Let us assume (for the moment) that $x \ll 0.080$, so that x can be dropped from the $0.080 + x$ term. Thus,

$$7.1 \times 10^{-4} = (0.080)(x)$$
$$x = 8.9 \times 10^{-3} = 0.0089$$

Our original term was $0.080 + x$:

$$0.080 + x = 0.080 + 0.0089 = 0.089$$

So x is not insignificant compared to 0.080. In fact, it is about 10% of the value and cannot be ignored, leaving us with a computational problem. Various approaches can be used for solving this problem. One could simply solve the quadratic equation, using the quadratic formula:

$$7.1 \times 10^{-4} = 0.080x + x^2$$
$$x^2 + 0.080x - (7.1 \times 10^{-4}) = 0$$

The roots of this equation are

$$x = 8.1 \times 10^{-3} \quad \text{and} \quad x = -8.8 \times 10^{-2}$$

Ignoring the negative root,

$$\text{solubility of } CaCrO_4 = 8.1 \times 10^{-3} \text{ M}$$
$$\text{equilibrium } [Ca^{2+}] = 0.088 \text{ M}$$
$$\text{equilibrium } [CrO_4^{2-}] = 8.1 \times 10^{-3} \text{ M}$$

EXERCISE 12–7

K_{sp} for $CaSO_4$ is 2.4×10^{-5}. Calculate the solubility and equilibrium ion concentrations of $CaSO_4$ in 0.0050 M $CaCl_2$.
Answer: Solubility $= [SO_4^{2-}] = 0.0030$ M; $[Ca^{2+}] = 0.0080$ M.   ■

The presence of coefficients in the chemical equation complicates equilibrium problems involving sparingly soluble salts to some extent. Consider the following example and exercise.

EXAMPLE 12–8

Determine the solubility and the equilibrium ion concentrations of CaF_2 in 0.10 M BeF_2.

Solution   A 0.10 M BeF_2 solution produces $[F^-] = 0.20$ M, according to the following equation:

$$BeF_2 \text{ (s)} \longrightarrow Be^{2+} \text{ (aq)} + 2 \text{ F}^- \text{ (aq)}$$

Let $x =$ solubility of CaF_2 in the equilibrium reaction. Then

	CaF_2 (s) $\rightleftharpoons$	Ca^{2+} (aq) +	$2 F^-$ (aq)
before ionization		0	0.20
at equilibrium		x	$0.20 + 2x$

$$K_{sp} = 1.7 \times 10^{-10} = (x)(0.20 + 2x)^2$$

Assuming $2x \ll 0.20$,

$$1.7 \times 10^{-10} = (0.20)^2(x)$$
$$x = 4.2 \times 10^{-9}$$

The simplifying assumption is justified, and

$$\text{solubility of } CaF_2 = 4.2 \times 10^{-9} \text{ M}$$
$$\text{equilibrium } [Ca^{2+}] = 4.2 \times 10^{-9} \text{ M}$$
$$\text{equilibrium } [F^-] = 0.20 \text{ M}$$

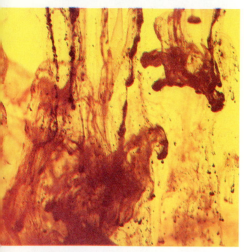

In-situ precipitation of the insoluble red pigment, silver chromate, on addition of aqueous silver ions to the yellow solution of sodium chromate:

$$2\,Ag^+\,(aq) + CrO_4^{2-}\,(aq) \longrightarrow Ag_2CrO_4\,(s)$$

For purposes of comparison, here are some heats of solution, obtained by dissolving a mole of solute in 200 mol of water at 25°C:

$\Delta H_{solution}(NH_4NO_3) = 14.8$ kJ/mol

$\Delta H_{solution}(NaCl) = 3.89$ kJ/mol

$\Delta H_{solution}(CaSO_4) = -16.7$ kJ/mol

$\Delta H_{solution}(Na_2SO_4 \cdot 10H_2O)$
$\qquad\qquad\qquad = 79.1$ kJ/mol

$\Delta H_{solution}(Na_2SO_4) = -23.0$ kJ/mol

EXERCISE 12–8

How many moles of Ag_2CrO_4 will dissolve in 1.0 L of 0.010 M K_2CrO_4 solution, and what will be the equilibrium ion concentrations?
Answer: Solubility $= 6.9 \times 10^{-6}$ mol/L; $[Ag^+] = 1.5 \times 10^{-5}$ M; $[CrO_4^{2-}] = 0.010$ M. ∎

The Effect of Temperature on Solubility and Solubility Rules

Early in our discussions of liquids and solutions (in Chapter 9), we noted the general effect of temperature on solubility. We need to look at that a little more closely. When the temperature changes, the solubility product for a salt changes, and therefore its solubility must change accordingly. That follows from Le Chatelier's principle. For example, a reversible reaction at equilibrium that is endothermic forms more product at higher temperatures, while an exothermic process produces less product if the temperature is raised. Because the dissolution of ammonium nitrate is accompanied by the absorption of heat—the reaction is endothermic—it is not surprising that the solubility increases from 118 to 871 g/100 g of water in the temperature range between 0°C and 100°C. Sodium chloride also dissolves with the absorption of heat, but with a more modest ΔH of solution there is a more modest gain in solubility in the same temperature range: from 35.7 to 39.1 g/100 g of water. The solubility of calcium sulfate drops from 0.209 to 0.162 g/100 g of water in the temperature range from 30°C to 100°C, and as you might expect, dissolution is exothermic—accompanied by liberation of heat. Solubilities of several salts as a function of temperature are displayed in Figure 12–5. The curious break in the solubility curve for $Na_2SO_4 \cdot 10H_2O$ is due to a chemical change that occurs at 32°C, where ten chemically bound water molecules are lost to the solvent.

In general it is possible to differentiate between **freely soluble salts**, of which at least several grams dissolve in 100 g of water, and **sparingly**

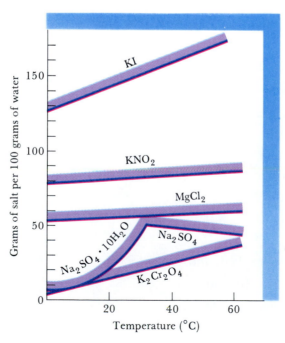

FIGURE 12–5 Solubility versus temperature plot for some selected salts in water. Note the behavior of potassium iodide, which is very soluble, and sodium sulfate, with the marked break at about 32°C.

soluble salts, for which the mass that dissolves in 100 g of water is considerably less than one gram. Only a few salts fall in between: silver acetate (1.02 g/100 g of water at 20°C), silver sulfate (0.57 g/100 g of water at 0°C), and calcium hydroxide (0.185 g/100 g of water at 0°C), for example. A fairly straightforward set of rules can be used to predict the relative solubilities of a large number of salts in water. For the more commonly encountered salts:

1. Practically all sodium, potassium, and ammonium salts are freely soluble.
2. All chlorides, bromides, and iodides are freely soluble except those of Ag^+, Pb^{2+}, and Hg_2^{2+}.
3. All nitrates, chlorates, perchlorates, and acetates ($C_2H_3O_2^-$) are freely soluble except for two borderline cases, silver acetate and mercury(I) acetate.
4. All sulfates are freely soluble except those of Sr^{2+}, Ba^{2+}, and Pb^{2+}. Borderline solubility is observed for calcium sulfate and silver sulfate.
5. All carbonates, phosphates, oxalates ($C_2O_4^{2-}$), and chromates (CrO_4^{2-}) are sparingly soluble except those of sodium, potassium, and ammonium.
6. All sulfides are sparingly soluble except those of Group IA and IIA metal ions and the ammonium ion.
7. All oxides and hydroxides are sparingly soluble except those of sodium and potassium. The oxides and hydroxides of Ca^{2+}, Sr^{2+}, and Ba^{2+} have borderline solubility. Note that soluble oxides, when dissolved in water, are converted to the hydroxides.

Precipitation Reactions

The precipitation reaction of a sparingly soluble salt represents the reverse of the dissolution reaction. For example,

$$Ba^{2+} (aq) + SO_4^{2-} (aq) \rightleftharpoons BaSO_4 (s)$$

is the net ionic equation for the reaction of a freely soluble barium salt such as barium chloride with a freely soluble sulfate salt such as sodium sulfate. The equilibrium constant for this reaction is

$$\frac{1}{K_{sp}} = \frac{1}{1.1 \times 10^{-10}} = 9.1 \times 10^9$$

The very large equilibrium constant for the precipitation of a sparingly soluble salt indicates that the reaction can be viewed as having gone to completion. Therefore, precipitation reactions are very useful in analytical chemistry, where "quantitative" removal of a species is often required. For example, suppose we want to determine the concentration of SO_4^{2-} in a solution. A classical method of analysis makes use of the insolubility of barium sulfate formed on addition of a $BaCl_2$ solution, followed by removal and weighing of the insoluble sulfate formed. By adding Ba^{2+} (aq) until its concentration is 0.10 M, the equilibrium concentration of SO_4^{2-} is reduced:

$$[SO_4^{2-}] = \frac{K_{sp}}{[Ba^{2+}]} = \frac{1.1 \times 10^{-10}}{0.10} = 1.1 \times 10^{-9} \text{ M}$$

Thus, if the initial $[SO_4^{2-}]$ was 0.10 M, the fraction remaining unprecipitated and in solution has been reduced to a very small percentage of the original concentration:

$$\left(\frac{1.1 \times 10^{-9}}{0.10}\right)(100) = 1.1 \times 10^{-6}\%$$

A particularly useful application of precipitation reactions among sparingly soluble salts is **selective precipitation**. Two ions of the same charge can be separated from each other by using an ion of opposite charge that forms a sparingly soluble salt with each. In the following example, consider how F^- ion is used to separate Ca^{2+} from Ba^{2+} by selectively precipitating the sparingly soluble $CaF_2(K_{sp} = 1.7 \times 10^{-10})$ and separating it from solution before any $BaF_2(K_{sp} = 1.7 \times 10^{-6})$ precipitates.

EXAMPLE 12–9

A solution is 0.10 M in both Ca^{2+} and Ba^{2+} ions. NaF is slowly added to precipitate the CaF_2. (a) How high can the $[F^-]$ be allowed to rise before BaF_2 begins to precipitate? (b) What fraction of the Ca^{2+} remains unprecipitated at that point?

Solution   The concentration of F^- can continue to rise until K_{sp} for BaF_2 is reached. In this example,

$$K_{sp}(BaF_2) = [Ba^{2+}][F^-]^2$$
$$1.7 \times 10^{-6} = [0.10][F^-]^2$$
$$[F^-] = \sqrt{\frac{1.7 \times 10^{-6}}{0.10}} = 4.1 \times 10^{-3} \text{ M}$$

The concentration of Ca^{2+} at this point can be calculated from the known K_{sp} for CaF_2 and the known concentration of 4.1×10^{-3} M for $[F^-]$:

$$K_{sp}(CaF_2) = 1.7 \times 10^{-10} = [Ca^{2+}][4.1 \times 10^{-3}]^2$$
$$[Ca^{2+}] = \frac{1.7 \times 10^{-10}}{1.7 \times 10^{-5}} = 1.0 \times 10^{-5} \text{ M}$$

The fraction of the original Ca^{2+} ion concentration remaining in solution when Ba^{2+} ions are about to precipitate is

$$\frac{1.0 \times 10^{-5}}{0.10} = 1.0 \times 10^{-4}$$

or 0.01 percent, from which one can conclude that this would be an effective means of quantitatively separating barium and calcium ions in solution.

EXERCISE 12–9

A solution contains Sr^{2+} and Pb^{2+}, each at 0.050 M. A solution of Na_2CO_3 will be added to precipitate the Pb^{2+} as $PbCO_3$. What will be the concentration of Pb^{2+} just as $SrCO_3$ starts to precipitate?
Answer: 4.7×10^{-6} M.

Precipitation reactions have many practical uses. For example, mortar when wet includes some $Ca(OH)_2$ in solution. Carbon dioxide in the air dissolves in the wet mortar to produce carbonic acid, which precipitates $CaCO_3$ or limestone:

$$Ca^{2+} \text{ (aq)} + 2 OH^- \text{ (aq)} + H_2CO_3 \text{ (aq)} \rightleftharpoons$$
$$CaCO_3 \text{ (s)} + 2 H_2O \text{ (liq)}$$

Stalactites and stalagmites are formed as insoluble carbonates crystallize from groundwater.

The resulting rock-like material has held temples and aqueducts together since Roman times. Stalactites and stalagmites are formed by the crystallization of carbonates dissolved in groundwater.

In a more recent example of a precipitation reaction, drinking water is quickly assayed for the presence of salt by adding Ag^+ from $AgNO_3$ and looking for the evidence of precipitated $AgCl$:

$$Ag^+ \text{ (aq)} + Cl^- \text{ (aq)} \rightleftharpoons AgCl \text{ (s)}$$

Another application of precipitation reactions is the softening of water by Na_2CO_3 or washing soda. Most of the hardness in home water supplies is the result of Ca^{2+} ions, which precipitate as a scum when mixed with soaps. That is not the problem it was 40 years ago, before the widespread use of detergents, but it is still a local problem and especially so in Third World countries. Na_2CO_3 removes these Ca^{2+} ions as the insoluble carbonate:

$$Ca^{2+} \text{ (aq)} + CO_3^{2-} \text{ (aq)} \rightleftharpoons CaCO_3 \text{ (s)}$$

Magnesium is removed from seawater by a process in which precipitation is a key step. The magnesium ions in solution precipitate, forming $Mg(OH)_2$, when OH^- ions are added. The most inexpensive source of OH^- ions is $Ca(OH)_2$ from sea shells, which are normally available near these plants. $Ca(OH)_2$ has limited solubility in water, only 0.021 mol/L at 25°C:

$$Ca(OH)_2 \text{ (s)} \rightleftharpoons Ca^{2+} \text{ (aq)} + 2 OH^- \text{ (aq)}$$

This gives $[OH^-]$ of only 0.042 M. However, since K_{sp} for $Mg(OH)_2$ is 1.2×10^{-11}, this means that the concentration of Mg^{2+} has been reduced to

$$[Mg^{2+}] = \frac{1.2 \times 10^{-11}}{(0.042)^2} = 6.8 \times 10^{-9} \text{ M}$$

after precipitation with OH^- from $Ca(OH)_2$.

Silver chloride precipitates when aqueous solutions of silver ions are added to solutions containing chloride ions.

12.3 AUTOIONIZATION OF WATER

If the conductivity experiment in Figure 12–1 is performed with pure water and a sensitive meter instead of a light bulb, a very small but definite conductivity can be observed. Water itself is the source of a small concentration of ions, and on that basis it is reasonable to conclude that an ionic equilibrium exists:

$$H_2O \text{ (liq)} \rightleftharpoons H^+ \text{ (aq)} + OH^- \text{ (aq)} \qquad (12\text{–}1)$$

However, the H^+ ion (or proton) does not have an independent existence in water. Because of its positive charge and small ionic radius (about one-hundred-thousandth of the radius of other ions), the proton is able to approach any atom with a partial negative charge and establish a strong bonding interaction. For a proton surrounded by a large excess of water, there is an immediate bonding of the proton with oxygen atoms. Although the nature of this interaction is not necessarily straightforward and the product almost certainly involves more than one H_2O molecule, a reasonable representation of the interaction of a proton with water is

$$H_2O \text{ (liq)} + H^+ \text{ (aq)} \longrightarrow H_3O^+ \text{ (aq)} \qquad (12\text{–}2)$$

This reaction can be viewed as going to completion to produce H_3O^+, the **hydronium ion**. Combining Eqs. (12–1) and (12–2) gives

$$2 \; H_2O \; (liq) \rightleftharpoons H_3O^+ \; (aq) + OH^- \; (aq) \qquad (12\text{–}3)$$

for which the following equilibrium expression can be written:

$$K = \frac{[H_3O^+][OH^-]}{[H_2O]^2}$$

Now, assume $[H_2O]$ is constant and combine the denominator with K to give the equilibrium expression for the autoionization of water:

$$K_w = [H_3O^+][OH^-] \qquad (12\text{–}4)$$

K_w is the autoionization constant, or simply the **ionization constant of water**. At 25°C,

$$K_w = [H_3O^+][OH^-] = 1.0 \times 10^{-14}$$

and since there must be equimolar concentrations of H_3O^+ and OH^-,

$$[H_3O^+] = [OH^-] = 1.0 \times 10^{-7} \; M$$

for each ion in pure water. The practical range for this equilibrium expression spans $[H_3O^+]$ from 10^{-14} M to 10^0 M. If a reagent is added to water or to an aqueous solution and $[H_3O^+]$ increases, $[OH^-]$ must decrease accordingly. Such a reagent is called an **acid**. Likewise, a reagent that increases $[OH^-]$, causing a reduction in $[H_3O^+]$, is called a **base**.

> Since the concentrations of H_3O^+ and OH^- are on the order of parts per million, the following is a reasonable approximation for $[H_2O]$:
>
> (1000 mL/L)(0.997 g/mL)
> $\qquad = 997$ g/L
> (997 g/L)/(18.0 g/mol)
> $\qquad = 55.4$ mol/L $= 55.4$ M
>
> Therefore,
>
> $K(55.4)^2 = K_w = [H_3O^+][OH^-]$
> $\qquad = 1.0 \times 10^{-14}$

12.4 THE pH SCALE

The **pH** scale is used extensively to describe the acidity or basicity of aqueous solutions and to simplify descriptions of $[H_3O^+]$ and $[OH^-]$ concentrations in water solutions. It is defined as the negative log of the hydrogen ion concentration:

$$pH = -\log[H_3O^+]$$

The practical lower limit for $[H_3O^+]$ is represented by a pH of 14,

$$pH = -\log[H_3O^+] = -\log(1 \times 10^{-14}) = 14$$

whereas the practical higher limit for $[H_3O^+]$ is represented by a pH of 0:

$$pH = -\log[H_3O^+] = -\log(1 \times 10^0) = 0$$

Pure (neutral) water has a pH of 7; pH values lower than 7 represent acidic solutions; and pH values higher than 7 represent basic solutions (Figure 12–6). Table 12–3 lists pH values for some familiar solutions.

> Since the scale is logarithmic, a solution with a pH of 6 has a hydronium ion concentration 10 times greater than a solution with a pH of 7.

The rule for significant figures in pH measurements (and in other calculations involving logarithms) is that the number of digits to the right of the decimal point equals the number of significant figures in the original expression of concentration. Thus, if $[H_3O^+] = 1.3 \times 10^{-2}$, then the pH = 1.89 with two digits following the decimal point because $[H_3O^+]$ has two significant figures.

EXAMPLE 12–10

What is the pH when the concentration of H_3O^+ is 0.010 M?

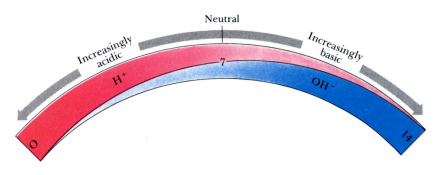

FIGURE 12–6 The pH scale. A "neutral" solution has a pH of 7, which means that the H_3O^+ ion concentration is equal to the OH^- ion concentration. As the pH falls into the acid range, below 7, the H_3O^+ and OH^- ion concentrations are increased and decreased, accordingly. The reverse is true as the pH rises into the alkaline (or basic) range, above 7.

TABLE 12–3 Approximate pH Values for Some Familiar Solutions	
Solution	**pH**
1 M NaOH (lye)	14
1 M NH_3 (household ammonia)	11.6
saturated $Mg(OH)_2$ (milk of magnesia)	10.5
blood	7.3–7.5
saliva	6.5–7.5
urine	5.5–7.5
coffee	4.5–5.5
beer	4.0–5.0
tomato juice	4.0–4.4
wine	2.8–3.8
vinegar	2.4–3.4
lemon juice	2.2–2.4
gastric juice	1.0–3.0
battery acid	0.5
1 M HCl	0

Solution $[H_3O^+] = 0.010$ M

$$pH = -\log[H_3O^+] = -\log(0.010) = 2.00$$ ■

EXERCISE 12–10

Determine the pH of the acidic solution formed when 0.050 mol of HCl is added to enough water to make the final volume 2.0 L. $[H_3O^+] = 0.025$ M.

Answer: pH = 1.60. ■

EXAMPLE 12–11

When 0.20 mol of NH_3 is dissolved in sufficient water to produce 1.0 L of solution, the $[H_3O^+]$ is 5.3×10^{-12} M. What is the pH of this alkaline solution?

Solution $pH = -\log[H_3O^+] = -\log(5.3 \times 10^{-12}) = 11.28$ ■

EXERCISE 12–11

The mid-range pH value for human blood is just slightly on the alkaline side of neutral at 7.44. Calculate the hydronium ion concentration.

Answer: $[H_3O^+] = 3.6 \times 10^{-8}$ M. ■

12.5   INTERACTION OF ACIDS WITH WATER

Our notion of an acid at this point is a compound that ionizes to produce a proton, which then combines with water to produce a hydronium ion. For example,

$$HCN\ (aq) + H_2O\ (liq) \rightleftharpoons H_3O^+\ (aq) + CN^-\ (aq)$$

$$HCl\ (aq) + H_2O\ (liq) \rightleftharpoons H_3O^+\ (aq) + Cl^-\ (aq) \tag{12–5}$$

By our definition, HCN and HCl are both acids. However, there is a significant difference between them. With HCN, the concentration of ions produced is very small and the equilibrium lies very far to the left, on the reactant side. Since the increase of the H_3O^+ concentration is very small, HCN is a weak acid. However, with HCl the conversion to ions is essentially complete so long as the concentration of HCl is kept moderately low. Accordingly, the increase in concentration of H_3O^+ is considerable when HCl is added to water and so HCl is a strong acid.

EXAMPLE 12–12

Calculate the concentrations of H_3O^+ and OH^- and the pH of the solution when 0.10 mol of HCl is added to sufficient water to produce 1.0 L of solution.

Solution   According to Eq. (12–5), 0.10 mol/L HCl yields 0.10 M H_3O^+. Substituting this value into the rearranged equation for K_w gives

$$[OH^-] = \frac{K_w}{[H_3O^+]} = \frac{1.0 \times 10^{-14}}{0.10} = 1.0 \times 10^{-13}\ M$$

$$pH = -\log(0.10) = 1$$

■

EXERCISE 12–12

Calculate $[H_3O^+]$ and $[OH^-]$ and the pH in one liter of a solution prepared from water and 3.5×10^{-6} mol of HCl.
Answer: $[H_3O^+] = 3.5 \times 10^{-6}\ M$; $[OH^-] = 2.9 \times 10^{-9}\ M$; pH = 5.46.

■

Commonly encountered strong acids include nitric (HNO_3), sulfuric (H_2SO_4), perchloric ($HClO_4$), chloric ($HClO_3$), hydrochloric (HCl), hydrobromic (HBr), and hydriodic (HI) acids. But most acids are weak acids, including aqueous solutions of inorganic acids such as H_2S, HClO, and HF and organic acids such as formic acid (HCOOH) and acetic acid (CH_3COOH).

Ionization Constants of Weak Acids

Equilibrium expressions for the ionization of weak acids can be applied to provide useful information about ionization and acidity. Let us consider a model for a weak acid, which we will call HB. It ionizes in water according to the equation

$$HB\ (aq) + H_2O\ (liq) \rightleftharpoons H_3O^+\ (aq) + B^-\ (aq)$$

Such an acid, capable of releasing one proton, is called a **monoprotic acid**. Here are some examples:

TABLE 12–4 K_a **Values for Some Monoprotic Weak Acids**

Name	Formula	K_a
acetic	CH_3COOH	1.8×10^{-5}
benzoic	C_6H_5COOH	6.3×10^{-5}
formic	$HCOOH$	1.8×10^{-4}
hyrdazoic	HN_3	1.9×10^{-5}
hydrocyanic	HCN	7.2×10^{-10}
hydrofluoric	HF	6.8×10^{-4}
hypochlorous	$HOCl$	3.7×10^{-8}
nitrous	HNO_2	4.5×10^{-4}

$$HF\ (aq) + H_2O\ (liq) \rightleftharpoons H_3O^+\ (aq) + F^-\ (aq)$$

$$HOCl\ (aq) + H_2O\ (liq) \rightleftharpoons H_3O^+\ (aq) + OCl^-\ (aq)$$

$$CH_3COOH\ (aq) + H_2O\ (liq) \rightleftharpoons H_3O^+\ (aq) + CH_3COO^-\ (aq)$$

As in our treatment of the autoionization of water, the concentration of water is essentially constant and does not appear in the equilibrium expression:

$$K_a = \frac{[H_3O^+][B^-]}{[HB]}$$

K_a is the **acid ionization constant**. Values of K_a for a number of monoprotic weak acids are given in Table 12–4.

The equilibrium expression can be used to calculate concentrations of important species present in weakly acidic aqueous solutions.

EXAMPLE 12–13

Determine the concentrations of H_3O^+, OH^-, $HOCl$, and OCl^- and the pH when 0.50 mol of $HOCl$ is added to sufficient water to make 1.0 L of solution.

Solution For the ionization reaction, let x be the concentration of H_3O^+ produced, and construct a table of the initial concentrations and the equilibrium concentrations:

$$H_2O\ (liq) + HOCl\ (aq) \rightleftharpoons H_3O^+\ (aq) + OCl^-\ (aq)$$

before ionization	0.50	0	0
at equilibrium	$0.50 - x$	x	x

Then substitute into the expression for K_a:

$$K_a = 3.7 \times 10^{-8} = \frac{x^2}{0.50 - x}$$

Because K_a is small, assume that $x \ll 0.50$. Then

$$x = \sqrt{1.8 \times 10^{-8}} = 1.3 \times 10^{-4}$$

so $[OCl^-] = [H_3O^+] = 1.3 \times 10^{-4}$ M. The autoionization of water yields

$$[OH^-] = \frac{K_w}{[H_3O^+]} = \frac{1.0 \times 10^{-14}}{1.3 \times 10^{-4}} = 7.7 \times 10^{-11}\ M$$

TABLE 12–5 K_a Values for Some Diprotic Acids

Name	Formula	$K_a(1)$	$K_a(2)$
carbonic	H_2CO_3	4.3×10^{-7}	4.4×10^{-11}
hydrosulfuric	H_2S	9.1×10^{-8}	1.2×10^{-15}
oxalic	$H_2C_2O_4$	6.5×10^{-2}	6.1×10^{-5}
selenious	H_2SeO_3	3×10^{-3}	5×10^{-8}
sulfuric	H_2SO_4	very large	2×10^{-2}
sulfurous	H_2SO_3	1.7×10^{-2}	6.2×10^{-8}

Since $x \ll 0.50$, [HOCl] is essentially unchanged at 0.50 M. Finally,

$$pH = -\log (1.3 \times 10^{-4}) = 3.9 \qquad \blacksquare$$

EXERCISE 12–13

Add 0.0150 mol of HCN to enough water to make 1.00 L of solution and determine $[CN^-]$, $[H_3O^+]$, and $[OH^-]$ and the pH of the solution. *Answer:* $[CN^-] = [H_3O^+] = 3.3 \times 10^{-6}$ M; $[OH^-] = 3.0 \times 10^{-9}$ M; $pH = -\log(1.3 \times 10^{-4}) = 3.87$. $\qquad \blacksquare$

Some acids are capable of producing more than one proton per molecule. For these acids, the first ionization produces an anion that is itself an acid. For example, the following sequential reactions occur when H_2S is added to water:

$$H_2S \text{ (aq)} + H_2O \text{ (liq)} \rightleftharpoons H_3O^+ \text{ (aq)} + HS^- \text{ (aq)}$$

$$K_a (1) = \frac{[H_3O^+][HS^-]}{[H_2S]}$$

$$HS^- \text{ (aq)} + H_2O \text{ (liq)} \rightleftharpoons H_3O^+ \text{ (aq)} + S^{2-} \text{ (aq)}$$

$$K_a (2) = \frac{[H_3O^+][S^{2-}]}{[HS^-]}$$

Acids that produce two H_3O^+ ions per molecule of acid are called **diprotic acids**. The K_a values for a few such acids are given in Table 12–5. Note that the second ionization constant is always much smaller than the first, reflecting the greater difficulty of separating the second proton from an anion, compared with separation of the first proton from the neutral acid molecule. For example, the first proton in sulfuric acid is almost completely dissociated, so H_2SO_4 is a strong acid; but the ionization constant for the second proton is much smaller, so HSO_4^- is not nearly as strong an acid:

$$H_2SO_4 \text{ (liq)} + H_2O \text{ (liq)} \longrightarrow H_3O^+ \text{ (aq)} + HSO_4^- \text{ (aq)}$$

$$K_a(1) = \frac{[H_3O^+][HSO_4^-]}{[H_2SO_4]}$$

$$HSO_4^- \text{ (aq)} + H_2O \text{ (liq)} \rightleftharpoons H_3O^+ \text{ (aq)} + SO_4^{2-} \text{ (aq)}$$

$$K_a(2) = \frac{[H_3O^+][SO_4^{2-}]}{[HSO_4^-]}$$

Triprotic acids are capable of transferring three protons per molecule

TABLE 12–6 K_a **Values for Some Triprotic Acids**

Name	Formula	$K_a(1)$	$K_a(2)$	$K_a(3)$
citric	$C_6H_8O_7$	8.4×10^{-4}	1.8×10^{-5}	4×10^{-6}
phosphoric	H_3PO_4	7.5×10^{-3}	6.2×10^{-8}	1.7×10^{-12}
pyrophosphoric	$H_4P_2O_7$	1.4×10^{-1}	1.1×10^{-2}	2.9×10^{-7}

to water molecules in the solvent. Phosphoric acid is an example:

$$H_3PO_4 \text{ (aq)} + H_2O \text{ (liq)} \rightleftharpoons H_3O^+ \text{ (aq)} + H_2PO_4^- \text{ (aq)}$$

$$K_a(1) = \frac{[H_3O^+][H_2PO_4^-]}{[H_3PO_4]}$$

$$H_2PO_4^- \text{ (aq)} + H_2O \text{ (liq)} \rightleftharpoons H_3O^+ \text{ (aq)} + HPO_4^{2-} \text{ (aq)}$$

$$K_a(2) = \frac{[H_3O^+][HPO_4^{2-}]}{[H_2PO_4^-]}$$

$$HPO_4^{2-} \text{ (aq)} + H_2O \text{ (liq)} \rightleftharpoons H_3O^+ \text{ (aq)} + PO_4^{3-} \text{ (aq)}$$

$$K_a(3) = \frac{[H_3O^+][PO_4^{3-}]}{[HPO_4^{2-}]}$$

Ionization constants for some triprotic acids are given in Table 12–6. Note that for any acid the first constant is larger than the second, which is larger than the third. Diprotic and triprotic acids are collectively referred to as **polyprotic acids**. Because successive K_a's of di- and triprotic acids such as sulfuric and phosphoric acids are separated by such large factors, each successive step has very little effect on the step preceding it. Therefore, determination of equilibrium concentrations can be made by stepwise calculation of each separate stage in the ionization process, starting with the first (largest) equilibrium constant, $K_a(1)$. Such an approach is not satisfactory for an acid such as citric acid, for which the successive K_a values are relatively close to each other.

EXAMPLE 12–14

Calculate the concentrations of all species present at equilibrium when 0.10 mol of H_2S is dissolved in water to produce 1.00 L of solution.

Solution Let $x = [H_3O^+]$ produced in the $K_a(1)$ equilibrium:

	$H_2O \text{ (liq)} + H_2S \text{ (aq)} \rightleftharpoons H_3O^+ \text{ (aq)} + HS^- \text{ (aq)}$		
before ionization	0.10	0	0
at equilibrium	$0.10 - x$	x	x

$$K_a(1) = 9.1 \times 10^{-8} = \frac{x^2}{0.10 - x}$$

Assuming $x \ll 0.10$,

$$x^2 = 9.1 \times 10^{-9}$$

$$x = [H_3O^+] = [HS^-] = \sqrt{9.1 \times 10^{-9}} = 9.5 \times 10^{-5} \text{ M}$$

$[H_2S] = 0.10$ M, and x is insignificant compared to 0.10.

Now we can use the concentrations from $K_a(1)$ to calculate the concentrations produced in the $K_a(2)$ equilibrium. This time let $y = [S^{2-}]$ produced.

$$HS^- \text{ (aq)} + H_2O \text{ (liq)} \rightleftharpoons H_3O^+ \text{ (aq)} \quad + S^{2-} \text{ (aq)}$$

before ionization 9.5×10^{-5} $\qquad\qquad 9.5 \times 10^{-5} \qquad\quad 0$

equilibrium $\qquad 9.5 \times 10^{-5} - y \qquad\qquad 9.5 \times 10^{-5} + y \qquad y$

$$K_a(2) = 1.2 \times 10^{-15} = \frac{(9.5 \times 10^{-5} + y)(y)}{9.5 \times 10^{-5} - y}$$

Assuming $y \ll 9.5 \times 10^{-5}$, we find $y = [S^{2-}] = 1.2 \times 10^{-15}$ M. The other concentrations are essentially unchanged, and

$$[H_3O^+] = [HS^-] = 9.5 \times 10^{-5} \text{ M}$$

$$[H_2S] = 0.10 \text{ M}$$

$$[OH^-] = \frac{K_w}{[H_3O^+]} = \frac{1.0 \times 10^{-14}}{9.5 \times 10^{-5}} = 1.0 \times 10^{-10} \text{ M}$$

$$pH = -\log(9.5 \times 10^{-5}) = 4.0 \qquad\qquad \blacksquare$$

EXERCISE 12–14

Determine the concentrations of all species present and the pH at equilibrium in a solution that is initially 0.010 M in H_2CO_3.
Answer: $[H_2CO_3] = 0.010$ M; $[H_3O^+] = [HCO_3^-] = 6.6 \times 10^{-5}$ M; $[CO_3^{2-}] = 4.4 \times 10^{-11}$ M; $[OH^-] = 1.5 \times 10^{-10}$ M; pH $= -\log$ $(9.5 \times 10^{-5}) = 4.02$. $\blacksquare$

The common ion effect applies to ionization of weak acids in water; the presence of a freely soluble salt containing the acid anion suppresses the ionization of the acid. The phenomenon is analogous to the common ion effect with sparingly soluble salts that we discussed earlier.

EXAMPLE 12–15

Determine the equilibrium concentration of $[H_3O^+]$ in a solution containing 0.50 mol/L of HOCl that is additionally 0.010 M in OCl^- (from NaOCl). Compare the $[H_3O^+]$ in this case with that in which the common ion is not present.

Solution Let $x = [H_3O^+]$ produced by the ionization of HOCl. As before, set up a table of concentrations:

$$HOCl \text{ (aq)} + H_2O \text{ (aq)} \rightleftharpoons H_3O^+ \text{ (aq)} + OCl^- \text{ (aq)}$$

before ionization 0.50 $\qquad\qquad\qquad\qquad 0 \qquad\quad 0.010$

at equilibrium $\quad 0.50 - x \qquad\qquad\qquad\quad x \qquad\quad 0.010 + x$

$$K_a = 3.7 \times 10^{-8} = \frac{(x)(0.010 + x)}{0.50 - x}$$

Assume that $x \ll 0.010$, which also implies $x \ll 0.50$, so that

$$x = 1.8 \times 10^{-6}$$

Thus, $[H_3O^+] = 1.8 \times 10^{-6}$ M, compared with 1.3×10^{-4} M when the additional common ion is not present (see Example 12–13). The addition of only 0.010 M OCl^- reduces $[H_3O^+]$ to 1.4% of its original value. $\blacksquare$

EXERCISE 12–15

Calculate the equilibrium value of $[H_3O^+]$ in a solution prepared from 0.150 mol of CH_3COOH, 0.0100 mol of CH_3COO^- (from CH_3COONa),

and enough water to make one liter of solution.
Answer: $[H_3O^+] = 2.7 \times 10^{-4}$ M. ■

The acidity of an aqueous solution is one of its most important properties. For example, the pH of rainwater is normally 5.5. Note that this is lower than the value of 7.0 that we would expect if rainwater were neutral. The slightly acid pH of normal rain results from CO_2 in the atmosphere, which dissolves to form carbonic acid,

$$H_2O \text{ (liq)} + CO_2 \text{ (g)} \rightleftharpoons H_2CO_3 \text{ (aq)}$$

Carbonic acid in turn ionizes to form H_3O^+ ions,

$$H_2CO_3 \text{ (aq)} + H_2O \text{ (liq)} \rightleftharpoons H_3O^+ \text{ (aq)} + HCO_3^- \text{ (aq)}$$

However, the low $K_a(1)$ for H_2CO_3 (4.3×10^{-7}) and its low concentration combine to give only a very weakly acidic pH.

Burning of sulfur-containing fuels leads to contamination of the atmosphere by large amounts of the sulfur oxides, SO_2 and SO_3 (Figure 12–7). Sulfur trioxide is particularly harmful because of its reaction with water to form sulfuric acid:

$$SO_3 \text{ (g)} + H_2O \text{ (liq)} \rightleftharpoons H_2SO_4 \text{ (aq)}$$

This results in rainwater containing a strong acid, having a low pH, and resulting in a devastating effect on pH-sensitive plant and animal life in freshwater lakes. Some lakes in northern New York State and as far south as Georgia are essentially lifeless as a result of their high acidity and low pH. Acid rain is especially prevalent at higher elevations (along the crest of the Appalachians), where clouds literally dip down into mountain lakes. In addition to polluting bodies of water, acid rain appears to dissolve clay minerals, releasing soluble aluminum salts that are harmful to trees (Figure 12–8). Furthermore, statues and buildings constructed of limestone are very susceptible to acid rain. Limestone is composed of the salt $CaCO_3$:

$$CaCO_3 \text{ (s)} + H_3O^+ \text{ (aq)} \rightleftharpoons Ca^{2+} \text{ (aq)} + HCO_3^- \text{ (aq)} + H_2O \text{ (liq)}$$

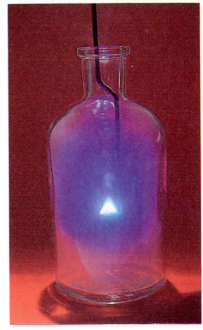

FIGURE 12–7 When the sulfur in high sulfur coals is burned, harmful sulfur dioxide and sulfur trioxide gases are formed just as in this air oxidation of sulfur, carried out in a spoon lowered into a heavy-walled, air-filled bottle.

FIGURE 12–8 The effects of acid rain on the evergreen population atop Camel's Hump in Vermont's Green Mountain range. The left photo was taken 15 years earlier than the right photo.

FIGURE 12–9   The skills of an early sculptor are still clearly evident in this 600-year-old limestone relief, photographed in 1903. But in less than 100 years since, the acid environment of the heavily industrialized European Rhein-Ruhr valley has taken its toll.

The HCO_3^- ion will react with another H_3O^+ ion,

$$HCO_3^- \ (aq) + H_3O^+ \ (aq) \rightleftharpoons H_2CO_3 \ (aq) + H_2O \ (liq)$$

The H_2CO_3 (aq) is an unstable solution of carbon dioxide and water:

$$H_2CO_3 \ (aq) \rightleftharpoons H_2O \ (liq) + CO_2 \ (g)$$

These acid rain reactions have inflicted irreversible damage on priceless art treasures (Figure 12–9).

12.6   INTERACTIONS OF BASES WITH WATER

Metal hydroxides that dissolve freely in water also ionize completely to metal ions and hydroxide ions. Although this is a limited group, it does include the very important alkali metal hydroxides—the Li group—and the alkaline earth metal hydroxides, Sr, Ba, and Ra. $Ca(OH)_2$ is only somewhat soluble in water.

EXAMPLE 12–16

Calculate the concentrations of H_3O^+ and OH^- when 0.15 mol of $Ba(OH)_2$ is added to sufficient water to make 1.0 L of solution.

Solution $Ba(OH)_2$ ionizes in water according to the equation

$$Ba(OH)_2 \ (s) \longrightarrow Ba^{2+} \ (aq) + 2 \ OH^- \ (aq)$$

so that 0.15 mol $Ba(OH)_2$ yields 0.30 mol of OH^-. Therefore,

$$[OH^-] = 0.30 \ M$$

$$[H_3O^+] = \frac{K_w}{[OH^-]} = \frac{1.0 \times 10^{-14}}{0.30} = 3.3 \times 10^{-14} \ M$$

EXERCISE 12–16

Calculate $[H_3O^+]$ and $[OH^-]$ in a solution of 0.00250 mol of KOH made up to one liter with water.
Answer: $[OH^-] = 0.00250 \ M$; $[H_3O^+] = 4.00 \times 10^{-12} \ M$. ∎

Sparingly soluble metal hydroxides are sometimes used as sources of low concentrations of OH^-. However, the presence of large amounts of the undissolved compound generally makes this a less than satisfactory approach. Other weak bases produce modest concentrations of OH^- by removing protons from water. Ammonia is one example; more will be described in the next chapter:

$$NH_3 \ (aq) + H_2O \ (liq) \rightleftharpoons NH_4^+ \ (aq) + OH^- \ (aq) \qquad (12\text{–}6)$$

Ionization Constants of Weak Bases

The equilibrium constant for the ammonia equilibrium in Eq. (12–6) is:

$$K_b = \frac{[NH_4^+][OH^-]}{[NH_3]}$$

K_b is called the **base ionization constant**. Values of K_b for a few bases in aqueous solution are given in Table 12–7.

TABLE 12–7 K_b Values for Some Weak Bases

Name	Formula	K_b
ammonia	NH_3	1.8×10^{-5}
methylamine	$(CH_3)NH_2$	4.4×10^{-4}
dimethylamine	$(CH_3)_2NH$	5.0×10^{-4}
trimethylamine	$(CH_3)_3N$	6.3×10^{-5}

EXAMPLE 12–17

Determine the OH^- and H_3O^+ concentrations for a solution prepared from 0.20 mol of NH_3 and sufficient water to make a final volume of 1.0 L.

Solution   Let x be the concentration of OH^- from ionization of NH_3 in H_2O. We make the usual table:

$$NH_3 \text{ (aq)} + H_2O \text{ (liq)} \rightleftharpoons NH_4^+ \text{ (aq)} + OH^- \text{ (aq)}$$

before ionization	0.20	0 0
at equilibrium	$0.20 - x$	x x

$$K_b = 1.8 \times 10^{-5} = \frac{x^2}{0.20 - x}$$

Assume $x \ll 0.20$, which leads to

$$x = \sqrt{3.6 \times 10^{-6}} = 1.9 \times 10^{-3}$$

The assumption is valid and

$$[OH^-] = 1.9 \times 10^{-3} \text{ M}$$

$$[H_3O^+] = \frac{K_w}{[OH^-]} = \frac{1.0 \times 10^{-14}}{1.9 \times 10^{-3}} = 5.3 \times 10^{-12} \text{ M}$$

EXERCISE 12–17

Calculate $[H_3O^+]$ and $[OH^-]$ at equilibrium when 0.350 mol of $(CH_3)_3N$ is dissolved in enough water to make 1.00 L of solution.

$$(CH_3)_3N \text{ (aq)} + H_2O \text{ (liq)} \rightleftharpoons (CH_3)_3NH^+ + OH^- \text{ (aq)}$$

Answer: $[OH^-] = 4.7 \times 10^{-3}$ M; $[H_3O^+] = 2.1 \times 10^{-12}$ M.

The following example and exercise demonstrate the common ion effect in the ionization of a weak base.

EXAMPLE 12–18

Compare $[OH^-]$ in the following two solutions:
Solution A: 1.0 L of a solution containing 0.20 mol of NH_3 in pure water.
Solution B: 1.0 L of a solution containing 0.20 mol of NH_3 and 0.010 mol of the freely soluble salt NH_4Cl dissolved in water.

Solution   The result for solution A is already known (from Example 12–17) to be $[OH^-] = 1.9 \times 10^{-3}$ M. For solution B, NH_4Cl is a strong electrolyte and dissociates completely:

$$NH_4Cl \text{ (s)} \longrightarrow NH_4^+ \text{ (aq)} + Cl^- \text{ (aq)}$$

So 0.010 mol/L of NH_4Cl is 0.010 M in NH_4^+ ions.

Now let x be the concentration of OH^- from ionization of aqueous NH_3. Setting up our table,

$$NH_3 \text{ (aq)} + H_2O \text{ (liq)} \rightleftharpoons NH_4^+ \text{ (aq)} + OH^- \text{ (aq)}$$

before ionization 0.20   0.010   0
at equilibrium $0.20 - x$ $0.010 + x$ x

$$K_b = 1.8 \times 10^{-5} = \frac{(0.010 + x)(x)}{0.20 - x}$$

Assuming $x \ll 0.010$ M,

$$x = \left(\frac{0.20}{0.010}\right)(1.8 \times 10^{-5}) = 3.6 \times 10^{-4}$$

The assumption is valid and

$$[OH^-] = 3.6 \times 10^{-4} \text{ M}$$

The reduction in the $[OH^-]$ as a result of the presence of 0.010 M NH_4^+ is

$$\frac{[OH^-]_B}{[OH^-]_A} = \frac{3.6 \times 10^{-4}}{1.9 \times 10^{-3}} = 0.19$$

Note that $[OH^-]$ has been reduced to 19% of its original value. ■

EXERCISE 12–18

Calculate $[OH^-]$ in a solution containing 0.350 mol of $(CH_3)_3N$, 0.050 mol of $(CH_3)_3NH^+$, and sufficient water to make up one liter of solution. *Answer:* $[OH^-] = 4.4 \times 10^{-4}$ M. ■

As in other examples, the addition of a small amount of a common ion results in a shift in the equilibrium away from the common ion.

12.7 DISSOLVING PRECIPITATES

The problem of dissolving sparingly soluble salts can often be solved by adding acid or base. This can be explained by combining the solubility product equilibrium with the equilibrium for the ionization of a weak acid or base. Consider a hypothetical sparingly soluble salt MB, which dissolves according to the equation

$$MB \text{ (s)} \rightleftharpoons M^+ \text{ (aq)} + B^- \text{ (aq)}$$

Le Chatelier's principle tells us that removal of either M^+ or B^- causes the reaction to move to the right. This, in turn, causes more of the sparingly soluble salt MB to dissolve. Either M^+ or B^- can be removed if either one happens to be the reactant in a subsequent reaction that has a large equilibrium constant.

For example, consider sparingly soluble calcium fluoride:

$$CaF_2 \text{ (s)} \rightleftharpoons Ca^{2+} \text{ (aq)} + 2 F^- \text{ (aq)} \tag{12-7}$$
$$K_{sp} = [Ca^{2+}][F^-]^2 = 1.7 \times 10^{-10}$$

F^- is also a product of the ionization of HF in water.

$$HF \text{ (aq)} + H_2O \text{ (liq)} \rightleftharpoons H_3O^+ \text{ (aq)} + F^- \text{ (aq)} \tag{12-8}$$
$$K_a = \frac{[H_3O^+][F^-]}{[HF]} = 6.8 \times 10^{-4}$$

If we reverse the direction of Eq. (12–8), F^- becomes a reactant:

$$H_3O^+ \text{ (aq)} + F^- \text{ (aq)} \rightleftharpoons HF \text{ (aq)} + H_2O \text{ (liq)} \qquad (12–9)$$

$$K = \frac{1}{K_a} = 1.5 \times 10^3$$

Thus, we have a reaction with a large equilibrium constant for which F^- is a reactant. Now, combining Eq. (12–7) and twice Eq. (12–9) yields

$$CaF_2 \text{ (s)} + 2\, H_3O^+ \text{ (aq)} \rightleftharpoons Ca^{2+} \text{ (aq)} + 2\, HF \text{ (aq)} + 2\, H_2O \text{ (liq)}$$

$$K = (K_{sp})\left(\frac{1}{K_a}\right)^2 = (1.7 \times 10^{-10})(1.5 \times 10^3)^2$$
$$= 3.8 \times 10^{-4}$$

We now have a reaction with an equilibrium constant much larger than K_{sp} for CaF_2 in pure water. The H_3O^+ needed to help dissolve CaF_2 can be produced by dissolving any of the strong acids in water.

EXAMPLE 12–19

Compare the solubility of CaF_2 in pure water and in 1.0 M HNO_3.

Solution   Let x be the solubility of CaF_2 in pure water, and construct the usual table:

$$CaF_2 \text{ (s)} \rightleftharpoons Ca^{2+} \text{ (aq)} + 2F^- \text{ (aq)}$$

before ionization	0	0
at equilibrium	x	$2x$

$$K_{sp} = 1.7 \times 10^{-10} = (x)(2x)^2 = 4x^3$$
$$x = 3.5 \times 10^{-4} \text{ M}$$

Thus, the solubility of CaF_2 in pure water is 3.5×10^{-4} mol/L.
Now let x be the solubility of CaF_2 in 1.0 M HNO_3:

$$CaF_2 \text{ (s)} + 2\, H_3O^+ \text{ (aq)} \rightleftharpoons Ca^{2+} \text{ (aq)} + 2\, HF \text{ (aq)}$$

before ionization	1.0	0	0
at equilibrium	$1.0 - 2x$	x	$2x$

$$K = 3.7 \times 10^{-4} = \frac{(x)(2x)^2}{1.0 - 2x}$$

Assume $2x \ll 1.0$, so that

$$4x^3 = 3.7 \times 10^{-4}$$
$$x = 4.5 \times 10^{-2}$$

The solubility of CaF_2 in 1.0 M HNO_3 is, without assumption, 4.4×10^{-2} M.
The increase in solubility resulting from addition of 1.0 M HNO_3 is

$$\frac{4.5 \times 10^{-2}}{3.5 \times 10^{-4}} = 130$$

Note that the solubility has increased 130-fold, or 13,000%. ■

EXERCISE 12–19

Compare the solubility of BaF_2 in water and in 1.00 M HCl.
Answer: Due to the fact that the method of solution to this problem involves a third degree equation, students may leave the answer expressed in terms of the polynomial $4x^3 - 14.72x^2 + 14.72x - 3.68 = 0$. ■

One important general observation from all these examples is this: If a sparingly soluble salt ionizes to form the anion of a weak acid, then addition of a strong acid will help dissolve the salt. Often the addition of a high concentration of a strong acid will completely dissolve such a salt. A strong acid will also help dissolve a hydroxide such as $Mg(OH)_2$, which can be considered the Mg^{2+} salt of H_2O. The reaction is

$$Mg(OH)_2 \text{ (s)} + 2\,H_3O^+ \text{ (aq)} \longrightarrow Mg^{2+} \text{ (aq)} + 4\,H_2O \text{ (liq)}$$

$$K = \frac{K_{sp}}{K_w{}^2} = \frac{1.2 \times 10^{-11}}{(1.0 \times 10^{-14})^2} = 1.2 \times 10^{17}$$

However, addition of a strong acid will not help dissolve a salt that ionizes to form the anion of a strong acid. For example,

$$AgCl \text{ (s)} \rightleftharpoons Ag^+ \text{ (aq)} + Cl^- \text{ (aq)} \qquad K_{sp} = 2.8 \times 10^{-10}$$

The H_3O^+ will not remove Cl^-, since HCl is a strong acid itself and is ionized by the reaction

$$HCl \text{ (aq)} + H_2O \text{ (liq)} \rightleftharpoons H_3O^+ \text{ (aq)} + Cl^- \text{ (aq)}$$

A precipitate may also be dissolved by adding a weak acid, such as NH_4^+ ion to help dissolve $Mg(OH)_2$:

$$Mg(OH)_2 \text{ (s)} \rightleftharpoons Mg^{2+} \text{ (aq)} + 2\,OH^- \text{ (aq)} \qquad K_{sp} = 1.2 \times 10^{-11}$$
$$NH_3 \text{ (aq)} + H_2O \text{ (liq)} \rightleftharpoons NH_4^+ \text{ (aq)} + OH^- \text{ (aq)}$$
$$K_b = 1.8 \times 10^{-5}$$

Reversing the second reaction, multiplying it by 2, and adding it to the first gives

$$Mg(OH)_2 \text{ (s)} + 2\,NH_4^+ \text{ (aq)} \rightleftharpoons$$
$$Mg^{2+} \text{ (aq)} + 2\,H_2O \text{ (liq)} + 2\,NH_3 \text{ (aq)}$$

$$K = \frac{K_{sp}}{K_b{}^2} = 3.7 \times 10^{-2}$$

This is a much more favorable equilibrium constant for the dissolution reaction.

Boiling hard water causes deposits of $CaCO_3$ to accumulate in kettles and boilers. Since $CaCO_3$ contains the anion of a weak acid, it can be dissolved by treatment with a solution of stronger acid in water:

$$CaCO_3 \text{ (aq)} + 2\,H_3O^+ \text{ (aq)} \rightleftharpoons Ca^{2+} \text{ (aq)} + H_2CO_3 \text{ (aq)}$$

Aqueous acetic acid solutions such as vinegar can be used to remove $CaCO_3$ deposits from tea kettles.

The dissolving of precipitates can cause environmental problems. For example, FeS (which can exist in coal) contains the anion of a weak acid, the sulfide ion S^{2-}. Thus acid rain can leach FeS out of coal piles, causing serious environmental damage to streams into which the leached material runs.

SUMMARY

Compounds that dissolve in water to produce ionic and therefore conductive solutions are called **electrolytes**. Compounds that are completely converted into hydrated ions when dissolved in water are called **strong electrolytes**; those that produce relatively low concentrations of hydrated ions are called **weak electrolytes**. Salts that are freely soluble in water, along with aqueous solutions of strong acids and bases, are typically strong electrolytes.

The equilibrium constant for the ionization of a sparingly soluble salt in water is called the **solubility product constant**, K_{sp}, the value of which is related to the solubility of the salt in water. Introducing anions or cations that are common to a sparingly soluble salt supresses its solubility; this is called the **common ion effect**. Based on differences in solubility, selective precipitation can be used to quantitatively separate ions of the same charge from solution.

Water is itself a weak electrolyte because of its self-ionization, or **autoionization**. The equilibrium constant for the autoionization of water is K_w:

$$2 \, H_2O \, (liq) \rightleftharpoons H_3O^+ \, (aq) + OH^- \, (aq) \qquad K_w = 1.0 \times 10^{-14}$$

Substances that produce either H_3O^+ or OH^- and alter this equilibrium are called **acids** and **bases**, respectively. The **pH** scale is a way of expressing the acidity of a solution as the negative logarithm of $[H_3O^+]$.

A few acids are termed **strong acids** because they ionize extensively, if not completely, in water. Most acids are **weak acids**, and their equilibria in water are described by their **acid ionization constants**, K_a. Addition of the anion of an acid to a solution of the acid lowers the concentration of H_3O^+.

A few metal hydroxides dissolve freely in water to produce high concentrations of OH^-. Such compounds are **strong bases**. The most commonly encountered **weak bases** are compounds such as ammonia and its derivatives; their combination with protons in aqueous solutions leads to increased concentrations of OH^-.

Knowledge of K_{sp}, K_a, and K_b can be used to enhance the solubilities of sparingly soluble salts.

QUESTIONS AND PROBLEMS

QUESTIONS

1. State the approximate degree of conductivity of water solutions of each of the following substances. Give a brief reason for each answer.
(a) KCl
(b) H_2SO_4
(c) NaOH
(d) $CaSO_4$
(e) $Mg(OH)_2$
(f) NH_3
(g) H_2S
(h) HCOOH

2. Consider a solution of NaCl in water. What happens to the conductivity of this solution as the temperature diminishes?

3. When added to water, some salts cause the temperature of the solution to increase. Some cause it to decrease. Why is this differing behavior observed for what appears in each case to be the same kind of reaction?

4. Why are solubility products normally not used for studying freely soluble salts?

5. Ag_2SO_4 and $BaCl_2$ are both reasonably soluble in water. Predict the results observed in the conductivity experiment (Figure 12–1) if a solution of Ag_2SO_4 is placed in the beaker and a solution of $BaCl_2$ is slowly added to it.

6. What is the basis for the separation technique referred to as "selective precipitation," and how is it carried out? Is it possible to precipitate AgCl selectively from a solution containing Cl^- and Br^-? Why or why not?

7. Suppose you wish to precipitate Pb^{2+} selectively from a solution containing Pb^{2+} and Ba^{2+}. What anion would you choose and why?

8. Why are there no "free" H^+ ions in aqueous solutions?

9. The $[H_2O]$ concentration term does not usually appear in equilibrium expressions involving acids and bases in water. Briefly explain why.

10. Why is $K_a(2)$ always less than $K_a(1)$ for diprotic acids?

11. Where possible, suggest reagents for dissolving each of the following compounds.
(a) MnS
(b) $MgCO_3$

(c) AgCl

(d) Na_2SO_4

(e) $Mg(OH)_2$

12. Will mortar that has hardened be affected by acid rain? Support your answer with chemical equations where possible.

PROBLEMS

Solubility Products and Solubility   [1–14]

1. Determine the solubility of AgI in pure water in moles per liter.

2. Determine the solubility of Ag_2CrO_4 in pure water in grams per liter.

3. If 0.062 g of CuCl will dissolve in 1.0 L of water at 25°C, what is the solubility product of CuCl at 25°C?

4. Ag_2CO_3 will dissolve in water to give a concentration of Ag^+ equal to 2.3×10^{-4} M. What is the solubility product of Ag_2CO_3?

5. For $MnCO_3$ the heat of solution, $\Delta H°$, is -12.1 kJ/mol and the entropy of solution, $\Delta S°$, is 66.1 J/mol·K. Calculate the solubility product of $MnCO_3$.

6. The solubility product of $PbBr_2$ at 18°C is 5.0×10^{-5} and the standard enthalpy of solution at this temperature is -42.0 kJ/mol. Calculate the standard entropy of solution of $PbBr_2$ at 18°C.

7. Calculate the solubility and equilibrium ion concentrations of NiS in a 0.010 M Na_2S solution. Express your answer as molarity, in units of mol/L.

8. Calculate the equilibrium ion concentrations if CaF_2 (s) is added to a 0.10 M solution of $CaCl_2$.

9. A solution contains 0.10 M Mg^{2+} and 0.10 M Pb^{2+}. If a sodium carbonate solution is slowly added, how high can the CO_3^{2-} concentration be allowed to rise before $MgCO_3$ starts to precipitate? What will be the concentration of Pb^{2+} at this point?

10. A classical method for separating zinc ions (Zn^{2+}) from ferrous ions (Fe^{2+}) is by selective precipitation of their insoluble sulfides. Consider a solution that is 0.10 M in each metal ion, and determine the concentration of sulfide ions (S^{2-}) when precipitation is first evident. Which metal sulfide is it that is precipitated at that point, and what is the ion concentration of that species remaining in solution when the second sulfide just begins to precipitate?

$$K_{sp} \text{ (ZnS)} = 4.5 \times 10^{-24} \qquad K_{sp} \text{ (FeS)} = 3.7 \times 10^{-19}$$

11. We wish to precipitate $Mn(OH)_2$ selectively from a solution containing 0.10 M Mn^{2+} and 0.10 M Mg^{2+}. What fraction of the Mn^{2+} present can be precipitated in this fashion?

12. Could you successfully separate Pb^{2+} ions from Sr^{2+} ions in solution by precipitation of the insoluble fluorides? Briefly explain why or why not.

$$K_{sp} \text{ (PbF}_2) = 8.0 \times 10^{-8}$$

$$K_{sp} \text{ (SrF}_2) = 2.8 \times 10^{-9}$$

13. Write net ionic equations for the reactions that occur when the following solutions are mixed:

(a) solutions of NaOH and $MgCl_2$

(b) solutions of $CaCl_2$ and Na_3PO_4

14. Write net ionic equations for the reactions that occur when the following solutions are mixed:

(a) solutions of $AgNO_3$ and Na_2S

(b) solutions of $MgCl_2$ and Na_2CO_3

Molarity Calculations   [15–20]

15. A sugar solution is prepared by dissolving 15.5 g of glucose, $C_6H_{12}O_6$, in enough water to make 335 mL of solution. What is the molarity of the solution?

16. You have at your disposal an HCl solution that is 0.050 M. How much of this solution do you take if you need 100.0 g HCl for a particular reaction?

17. Concentrated hydrochloric acid is 36% HCl by weight and its density is 1.189 g/mL at 25°C. How many milliliters of acid and water must be mixed to make a liter of 0.50 M HCl solution? Assume volumes are additive.

18. What is the molarity of an H_2SO_4 solution prepared by diluting 100. mL of the concentrated reagent to a volume of 300. mL? The density of the concentrated acid is 1.84 g/mL and it is 95.6% by weight.

19. Commercial concentrated sulfuric acid is 95.6% H_2SO_4 (by weight); its density is 1.84 g/mL.

(a) Calculate the molarity of the concentrated acid.

(b) Calculate the volume necesary to make up one liter of 0.100 M sulfuric acid solution.

20. Concentrated hydrochloric acid (HCl) is available commercially as a 36% aqueous solution with a density of 1.189 g/mL at 25°C.

(a) Calculate the molarity of the solution.

(b) Determine the number of milliliters of concentrated acid required to prepare a liter of 0.100 M HCl solution.

pH Calculations   [21–24]

21. Convert the following H_3O^+ concentrations to pH:

(a) 1.3×10^{-2} M

(b) 0.0231 M

(c) 7.9×10^{-8} M

22. Convert the following OH^- concentrations to pH:

(a) 0.133 M

(b) 6.2×10^{-9} M

23. Calculate $[H_3O^+]$ and $[OH^-]$ for each of the following pH values:

(a) 1.2

(b) 6.7

(c) 13.4

24. Calculate the pH and the pOH (that is, $-\log[OH^-]$) for each of the following hydronium ion solutions:

(a) 1.0×10^{-5} M

(b) 8.5×10^{-11} M

(c) 0.847 M

Acid-Base Equilibria   [25–36]

25. Calculate the $[H_3O^+]$ and $[OH^-]$ and the pH that result when 0.10 mol of each of the following acids is added to sufficient water to produce 1.00 L of solution.

(a) HOCl

(b) HNO_2

26. Calculate the $[H_3O^+]$ and $[OH^-]$ and the pH that result when 0.010 mol of each of the following acids is added to sufficient water to produce 1.00 L of solution.

(a) HCN

(b) HNO_3

27. Calculate $[H_3O^+]$ and $[OH^-]$ and the pH at equilibrium when 0.15 mol of each of the following bases is added to sufficient water to make one liter of solution.

(a) KOH

(b) methylamine

28. Calculate $[H_3O^+]$ and $[OH^-]$ and the pH at equilibrium when 0.15 mol of each of the following bases is added to sufficient water to make one liter of solution.

(a) ammonia

(b) trimethylamine

29. Calculate $[H_3O^+]$ and $[OH^-]$ at equilibrium when 1.50 g of HCN and 0.300 g of NaCN are added to sufficient water to produce 0.750 L of solution.

30. Calculate the equilibrium values of $[H_3O^+]$ and $[OH^-]$ when 0.150 mol of NH_3 and 0.020 mol of NH_4Cl are dissolved in sufficient water to produce 1.00 L of solution.

31. The dissociation constant for formic acid (HCOOH) at 25°C is 1.8×10^{-4}. For an 0.10 M solution, calculate the percentage of the acid that is dissociated into ions.

32. Calculate the dissociation constant at 25°C for nitrous acid (HNO_2) in a 0.10 M solution if it is known to be 6.3% dissociated.

33. Calculate the change in $[H_3O^+]$ when 0.100 g of KCN is added to one liter of a 0.100 M solution of HCN.

34. Calculate the pH change that takes place when 10.0 g of sodium acetate is added to 1.000 L of a 1.00 M solution of acetic acid.

35. Calculate the concentration of all species present at equilibrium in a solution that is initially 0.50 M in selenious acid, H_2SeO_3.

36. Calculate the equilibrium concentrations of all species present, starting with a 0.150 M solution of carbonic acid, H_2CO_3.

Dissolving Precipitates   [37–38]

37. Calculate the solubility of $Mg(OH)_2$ in 0.50 M NH_4Cl.

38. Determine the solubility of CaF_2 in a 0.50 M HCl solution.

General Problems   [39–48]

39. How many liters of water are needed to dissolve one molecule of HgS?

40. If 0.77 g of $Cu(IO_3)_2$ will dissolve in 1.0 L of 0.010 M $CuSO_4$ solution, what is the solubility product of $Cu(IO_3)$?

41. Lanthanum iodate, $La(IO_3)_3$, is sparingly soluble in water and has a K_{sp} value equal to 6.10×10^{-12}.

(a) Calculate its solubility in water, and determine the molar concentration of iodate ion, IO_3^-.

(b) What would be the effect on the solubility of introducing lanthanum nitrate, a soluble salt? Briefly explain.

42. Calculate the hydronium ion concentration of each of the following solutions:

(a) pH = 2.97

(b) pH = 8.31

(c) pH = 0.04

43. (a) Calculate the solubility of silver chromate (Ag_2CrO_4) in a 0.10 M sodium chromate (Na_2CrO_4) solution. K_{sp} (Ag_2CrO_4) = 1.9×10^{-12}.

(b) Determine the solubility of silver chloride in water. $K_{sp} = 1.6 \times 10^{-10}$.

(c) Determine the solubility product for silver bromide (AgBr) if the solubility in water is 0.000125 g/L at 298 K.

(d) Determine how many grams of $Ba(OH)_2$ can be dissolved in a liter of water at 298 K. $K_{sp} = 4.9 \times 10^{-3}$.

44. The dilute solution of HCl on the laboratory shelf happens to be 16.8% by weight. Its density is 1.085 g/mL. Calculate the molarity of the solution.

45. Calculate the pH of each of the following:

(a) pure water

(b) a 0.10 M sodium chloride solution

(c) a 0.10 M ammonia solution ($K_b = 1.8 \times 10^{-5}$)

46. Given K_a for acetic acid (HOAc) = 1.8×10^{-5}, determine the hydronium ion concentrations for the following solutions:

(a) 1.0 M HOAc solution

(b) 0.01 M HOAc solution

47. Calculate the equilibrium constant for the reaction of $Al(OH)_3$ with NH_4^+. Give the balanced equation for the reaction.

48. A solution containing 0.100 mol of lactic acid and 0.0100 mol of the soluble salt sodium lactate has a H_3O^+ concentration of 1.24×10^{-3} M. What is K_a for lactic acid?

MULTIPLE PRINCIPLES   [49–55]

49. Use the thermodynamic data in Table 12–1 to calculate the solubility of $CaSO_4$ in water at 25°C, 60°C, and 95°C. Assume that $\Delta H°$ and $\Delta S°$ are constant over the temperature range under consideration.

50. Write balanced equations for the reactions that occur when the following aqueous reagents are mixed:

(a) ammonia and hydrogen chloride

(b) trimethylamine and nitric acid

(c) dimethylamine and hydrogen fluoride

(d) sulfuric acid and excess dimethylamine

51. Calculate the freezing points of the following aqueous solutions:

(a) 0.100 M HCl

(b) 0.100 M HF

(c) 0.100 M H_2SO_4

52. Calculate the boiling point of a saturated aqueous solution of $CaCrO_4$.

53. Because there is a greater tendency to ionize at higher temperatures, the pH of neutral water varies with temperature. The value of K_w for water rises from 1.1×10^{-15} at 0°C to 9.6×10^{-14} at 60°C. Calculate the pH of neutral water at 0°C.

54. Using the data in Problem 53, calculate the pH of neutral water at 60°C.

55. From the following data, first determine the apparent heats of solution for the three salts as the difference between the lattice energies and the hydration energies. Then, predict the effect of a temperature increase on their respective solubilities in water. What other conclusions might be drawn from the systematic way in which the data change in going from CaF_2 to CaI_2? Note that lattice energy is energy of step 1 in Figure 12–2 and hydration energy is energy of step 2.

	Lattice energy	**Hydration energy**
CaF_2	$+2611$ kJ/mol	-251 kJ/mol
$CaCl_2$	$+2247$ kJ/mol	-2293 kJ/mol
CaI_2	$+2059$ kJ/mol	-2163 kJ/mol

APPLIED PRINCIPLES   [56–58]

56. The solubility product constant K_{sp} for $CaCO_3$ is 8.7×10^{-9} at 25°C. What would be the maximum mass of $CaCO_3$ that could be dissolved in a 10,000-L tank of water at 25°C?

57. Calcium ions, the most common cause of hardness in water, are often introduced into water supplies by the freely soluble salt $CaHCO_3$. Suppose you have a 10,000-L tank of water at 25°C containing $CaHCO_3$ in which you measure a Ca^{2+} concentration of 0.0145 M. You then add 50. kg of Na_2CO_3 to precipitate the Ca^{2+} ions as $CaCO_3$. What is the concentration of Ca^{2+} ions remaining after equilibrium has been reached? K_{sp} for $CaCO_3$ is 8.7×10^{-9} at 25°C.

58. Seawater contains 0.13% magnesium by mass. What fraction of this can be removed by addition of a stoichiometric quantity of $Ca(OH)_2$? Use a density for seawater of 1.025 g/mL.

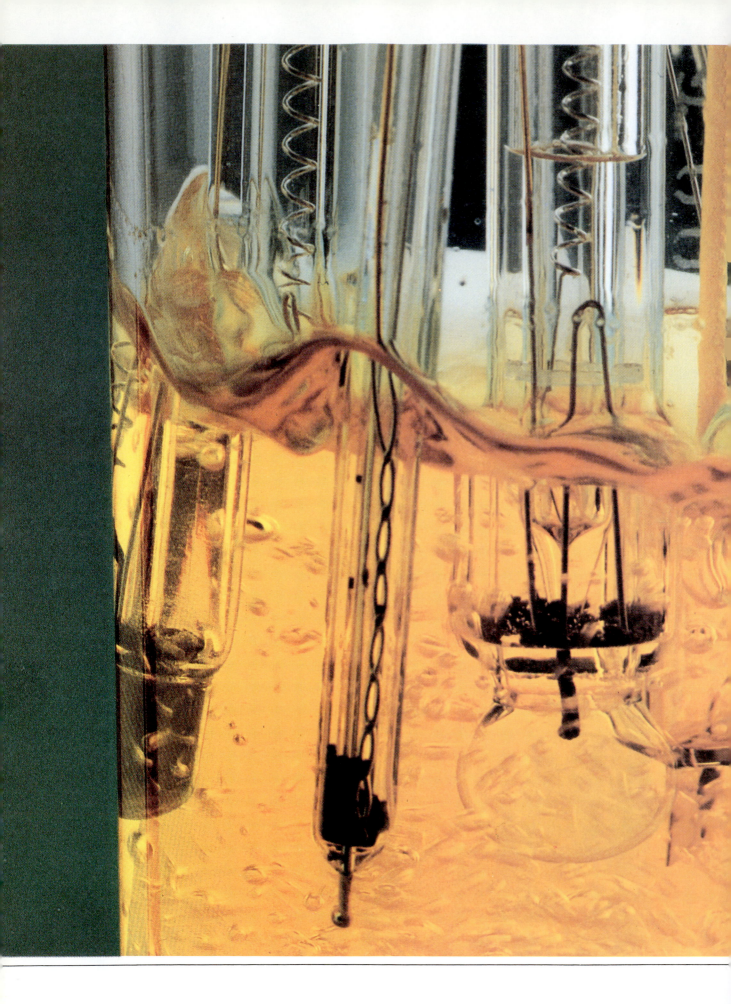

CHAPTER 13

Equilibrium III: Acids and Bases

The Beckman Acidimeter was the first of a generation of modern electronic instruments for chemical analysis. One of the first of these instruments, now so common we know them simply as pH meters, is shown here in an original promotional pamphlet (1936) against a larger display of modern glass electrodes routinely used to measure pH across a wide range of temperatures in continuous operation from pH 0 to 14. Other electrodes serve to analyze or monitor specific ions or carry out unique electroanalytical assays for atoms, molecules, and processes. And although the Nobel Prize was never awarded for this contribution to chemical methods and analytical instrumentation, others have been so recognized, beginning with X-ray crystallography (1914, 1915, and 1917 prizes) and more recently, magnetic resonance and chromatography (1952) and microscopy (1986).

Fritz Pregl (Austrian), Nobel Prize for Chemistry in 1923 for establishing reliable methods and microanalytical procedures for chemical substances.

Acid: from the Latin, meaning sour.

13.1 DEFINING ACIDS AND BASES

Acids and bases are among the most important chemical reagents. Their structure and reactivity have been topics of scientific study by chemists since the time of Lavoisier. They have been a source of manufacturing and environmental problems for engineers for about as long. There are few substances produced on a larger scale than sulfuric acid and sodium hydroxide, and few substances need to be handled with more care and attention. At first, acids were noted for their characteristic sour taste; bases were articles of commerce, used in the early manufacture of soaps, for rendering animal fats, and in tanning hides. Some properties of acids were probably discovered accidentally: their reactivity in water with certain metals; the effect they have on the color of certain vegetables (red cabbage) and flowers (rose petals); causing chalk to effervesce. It was quickly discovered that bases—alkalies, as they came to be known—could be distinguished by their ability to react with acids, neutralizing their characteristic acidic properties.

Many famous scientists have contributed to our understanding of acid-base chemistry. Antoine Lavoisier, who set so much of chemistry straight in the early years of the modern era, stated erroneously that oxygen was required in all acids (1787). In fact, he gave oxygen its name–it means "acid producer." But it remained for Humphrey Davy (1811) and others to recognize that acids such as hydrochloric acid contain no oxygen at all, leading to the general conclusion that hydrogen was the important element in acids. Then, Michael Faraday (1834) showed acids and bases to be substances he called "electrolytes," setting the stage for modern understanding of acidic substances. Finally, through the innovative research of Svante Arrhenius (1887), a young graduate student working in Sweden, even the presence of hydrogen was shown to be insuffecnt for distinguishing acids from the host of hydrogen-containing compounds that were clearly not acidic.

The Arrhenius concept of an acid followed from his theory of electrolytic dissociation in aqueous solution. Acids produce hydrogen ions; bases produce hydroxide ions; and acids "neutralize" bases, producing salts and water:

$$HCl \ (aq) \longrightarrow H^+ \ (aq) \ + \ Cl^- \ (aq) \qquad \text{acid}$$

$$NaOH \ (aq) \longrightarrow Na^+ \ + \ OH^- \ (aq) \qquad \text{base}$$

$$HCl \ (aq) \ + \ NaOH \ (aq) \longrightarrow NaCl \ (aq) \ + \ H_2O \ (liq) \qquad \text{neutralization}$$

But it remained for Bronsted and Lowry (forty years later) to recognize that isolated protons cannot exist in aqueous solutions because of the high charge density of a bare nucleus. An acid, according to Bronsted-Lowry theory, was any hydrogen-containing substance capable of producing hydronium ions in aqueous solution, by donating protons to solvent molecules. Thus, HCl was still an acid in aqueous solution, but the reaction could now be better stated:

$$HCl \ (aq) \ + \ H_2O \ (liq) \longrightarrow H_3O^+ \ (aq) \ + \ Cl^- \ (aq)$$

A tetrahydrate, $H(H_2O)_4{}^+$, or $H_9O_4{}^+$, is probably a better representation for the hydrated proton than simply H_3O^+, and recent estimates have put the number of hydrating water molecules at more than a dozen, perhaps as many as sixteen.

Although it is a considerable oversimplification, it is traditional to show the hydronium ion as H_3O^+, the product of a proton and only one "hydrating" water molecule. To complete the picture, a Bronsted base is a proton acceptor—hydroxide ion, for example—and here is the neutralization reaction:

$$H_3O^+ \ (aq) \ + \ OH^- \ (aq) \longrightarrow H_2O \ (liq) \ + \ H_2O \ (liq)$$

There is a still more general definition, the Lewis acid and base. In this definition, an acid is an electron-pair acceptor and a base is an electron-pair donor. Thus,

In the next section, we will examine these ideas in more detail.

13.2 THE BRONSTED-LOWRY DEFINITION OF ACIDS AND BASES

Proton-Transfer Reactions

A **Bronsted-Lowry acid** is a proton donor and a **Bronsted-Lowry base** is a proton acceptor, regardless of the solvent. According to the Bronsted-Lowry definition, HCN is an acid:

$$HCN \rightleftharpoons H^+ + CN^-$$

and NH_3 is a base:

$$NH_3 + H^+ \rightleftharpoons NH_4^+$$

Because H^+ cannot exist independently in any commonly encountered chemical environment, these two chemical reactions are incomplete and should be regarded only as half-reactions. On adding the acid half-reaction to the base half-reaction, the complete reaction is generated:

$$HCN + NH_3 \rightleftharpoons NH_4^+ + CN^-$$

The proton produced in one half-reaction is used up in the other half-reaction and does not appear in the overall reaction.

When a Bronsted-Lowry acid loses a proton, it becomes a species that can gain a proton in the reverse reaction; that is, every acid has a **conjugate base**, differing from the original acid only in the missing proton. Of course, every Bronsted-Lowry base has its **conjugate acid**, differing from the original base only in the added proton. For example, HCN is converted into its conjugate base CN^-, and the base NH_3 is converted into its conjugate acid NH_4^+. For the acid-base reaction

$$HF + NH_3 \rightleftharpoons NH_4^+ + F^-$$

the acid HF is converted to its conjugate base F^-, and the base NH_3 is simultaneously converted to its conjugate acid NH_4^+.

The spontaneous direction of any Bronsted-Lowry acid-base reaction is determined by the following rule: the acid-base reaction proceeds spontaneously toward the weaker acid and the weaker base. Applying this rule to the reaction

$$\underset{\text{acid}}{NH_4^+} + \underset{\text{base}}{OH^-} \rightleftharpoons \underset{\text{acid}}{H_2O} + \underset{\text{base}}{NH_3}$$

and comparing the strengths (K_b) of the two bases, OH^- and NH_3, clearly demonstrates that OH^- is a much stronger base than NH_3. Likewise, H_2O is a much weaker acid than NH_4^+. Therefore, with the weaker acid and base on the right side of the equation, the reaction can be expected to proceed spontaneously to the right. Note that in the reverse reaction, H_2O is an acid since reaction in this direction involves transfer of a proton from H_2O to NH_3.

The autoionization of water can be explained in Bronsted acid-base terms. Because two species are needed on each side of the equation, write H_2O twice:

$$\underset{\text{acid}}{H_2O} + \underset{\text{base}}{H_2O} \rightleftharpoons \underset{\text{acid}}{H_3O^+} + \underset{\text{base}}{OH^-}$$

One H_2O molecule donates a proton to the other in the forward reaction, so we designate one H_2O molecule as an acid and call the other a base.

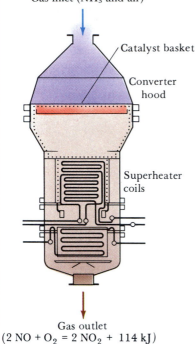

Ammonia Combustion
Gas inlet (NH₃ and air)

Catalyst basket

Converter
hood

Superheater
coils

Gas outlet
$(2 NO + O_2 = 2 NO_2 + 114 kJ)$

OXIDATIVE CONVERSION OF AMMONIA INTO NITRIC ACID

Nitric acid (HNO_3) has been known since Glauber's synthesis from concentrated sulfuric acid and sodium nitrate in the 13th century. However, it was Lavoisier, late in the 18th century, who showed that nitric acid contains oxygen, and it was Cavendish who first synthesized it in a modern way, by passing an electric spark through moist air. The oldest preparative-scale technique was the reaction of nitrate salts with concentrated sulfuric acid in heated cast iron retorts; as the nitric acid vapors condensed, they were collected in stoneware containers.

The importance of nitric acid as an industrial reactant emerged along with the explosives and dyestuffs industries in the 19th century. Today, virtually all industrial grades of nitric acid are manufactured by three-stage ammonia oxidation over a precious metal catalyst, typically a gauze of a platinum-rhodium alloy containing 10% rhodium (see Figures). Here is the basic chemistry:

1. Catalytic oxidation of ammonia with excess oxygen is rapid, and goes virtually to completion:

$$4 NH_3 (g) + 5 O_2 (g) \longrightarrow 4 NO (g) + 6 H_2O (liq)$$
$$\Delta H = -226 \text{ kJ/mol}$$

2. Oxidation of nitric oxide takes place in a slow, gas phase process, yielding nitrogen dioxide:

$$2 NO (g) + O_2 (g) \rightleftharpoons 2 NO_2 (g) \qquad \Delta H = -114 \text{ kJ/mol}$$

The equilibrium favors nitrogen dioxide formation at lower temperatures; below 150°C, if there is sufficient residence time in the reactor, almost all the nitrogen oxide present combines with any oxygen present. Nitrogen dioxide in turn establishes an equilibrium mixture with the dimer, dinitrogen tetroxide. According to Le Chatelier's Principle, that equilibrium should be sensitive to temperature and pressure:

$$2 NO_2 (g) \rightleftharpoons N_2O_4 (g) \qquad \Delta H = -57.4 \text{ kJ/mol}$$

3. Finally, in a process that is known to be much more complex than the reaction suggests, the nitrogen oxides are absorbed in water, leading to a constant-boiling solution that is 68% HNO_3 by weight:

$$3 NO_2 (g) + H_2O (liq) \rightleftharpoons 2 HNO_3 (aq) + NO (g)$$
$$\Delta H = -135.6 \text{ kJ/mol}$$

and the NO is cycled back into step (2)

Now H_3O^+ is clearly a stronger acid than H_2O, and OH^- is clearly a stronger base than H_2O. So the spontaneous direction of reaction is to the left. Thus, the autoionization of H_2O has a small equilibrium constant ($K_w = 1.0 \times 10^{-14}$). On the other hand, when writing the reaction in the reverse direction,

$$H_3O^+ + OH^- \rightleftharpoons H_2O + H_2O$$
$$\text{acid} \quad\quad \text{base} \quad\quad\quad \text{acid} \quad\quad \text{base}$$

the stronger acid and base appear on the left side, and the reaction is spontaneous to the right:

$$K = \frac{1}{K_w} = 1.0 \times 10^{14}$$

Regarding the relative strengths of the acid and the base in a conjugate acid-base pair, the stronger the acid, the weaker the base. The corollary is also true: the stronger the base, the weaker the acid. To see the reasoning behind this rule, consider the proton transfer between members of a general conjugate acid-base pair:

$$BH \rightleftharpoons B^- + H^+$$
$$\text{acid} \quad\quad \text{base}$$

If BH is a strong acid, there is a strong tendency for the B—H bond to break, and a correspondingly weak tendency to form the bond; therefore, B^- is a weak base. Should B^- be a strong base, there would be a strong tendency to form the B-H bond and a weak tendency for it to come apart; in this case, BH would be a weak acid.

Relative Strengths of Acids and Bases

A great deal of information on the relative strengths of acids and bases in aqueous solutions can be derived from the K_a and K_b values listed in Tables 12–4 to 12–7. For example, consider the following reaction:

$$HNO_2 \text{ (aq)} + H_2O \text{ (liq)} \rightleftharpoons H_3O^+ \text{ (aq)} + NO_2^- \text{ (aq)}$$
$$K_a = \frac{[H_3O^+][NO_2^-]}{[HNO_2]} = 4.5 \times 10^{-4}$$

Since $K_a < 1$, the spontaneous direction of the reaction is to the left. This means that in aqueous solution, H_3O^+ is a stronger acid than HNO_2, and NO_2^- is a stronger base than H_2O.

Now consider the acid-base equilibrium in an aqueous solution of hydrogen cyanide, also called hydrocyanic acid, HCN (aq):

$$HCN \text{ (aq)} + H_2O \text{ (liq)} \rightleftharpoons H_3O^+ \text{ (aq)} + CN^- \text{ (aq)}$$
$$K_a = \frac{[H_3O^+][CN^-]}{[HCN]} = 7.2 \times 10^{-10}$$

Since $K_a < 1$, H_3O^+ is a stronger acid than HCN and CN^- is a stronger base than H_2O in aqueous solution. In addition, comparing K_a values for HNO_2 and HCN in aqueous solution shows that HNO_2 is a stronger acid than HCN, because its K_a is larger. Furthermore, since HNO_2 is a stronger acid than HCN, the reverse order of strength will be observed for their conjugate bases; thus, CN^- is a stronger base than NO_2^-. That brings us to the point where we can construct a new reaction and predict its spontaneous direction:

Using this principle, we can classify the Bronsted-Lowry acid-base pairs already mentioned:

strong acid/weak base H_3O^+/H_2O
moderately weak acid and base
                                        NH_4^+/NH_3
                                        HCN/CN^-
weak acid/strong base H_2O/OH^-

The historical name is "prussic" acid.

$$HNO_2 \text{ (aq)} + CN^- \text{ (aq)} \rightleftharpoons HCN \text{ (aq)} + NO_2^-$$

acid   base   acid   base

As written, the reaction spontaneously moves to the right, forming the weaker acid and base. Since this reaction is the net result of combining the ionization equations representing the two acids in aqueous solution, the equilibrium constant can be found directly. Because we are adding the first reaction to the reverse of the second,

$$K = \frac{K_a(HNO_2)}{K_a(HCN)} = \frac{4.5 \times 10^{-4}}{7.2 \times 10^{-10}} = 6.3 \times 10^5$$

The very large value of the equilibrium constant is consistent with the very strong driving force to the right.

EXAMPLE 13–1

Predict the spontaneous direction of the reaction between hydrazoic acid, HN_3 (aq), and fluoride ion in aqueous solution:

$$HN_3 \text{ (aq)} + F^- \text{ (aq)} \rightleftharpoons HF \text{ (aq)} + N_3^- \text{ (aq)}$$

Solution   Since HF is a stronger acid than HN_3 in aqueous solution (Table 12–4), the spontaneous direction of the reaction is to the left. The equilibrium constant for this reaction can be calculated as the ratio

$$K = \frac{K_a(HN_3)}{K_a(HF)} = \frac{1.9 \times 10^{-5}}{6.8 \times 10^{-4}} = 2.8 \times 10^{-2}$$

EXERCISE 13–1

Predict the direction of the reactions

$$HCOOH \text{ (aq)} + OCl^- \text{ (aq)} \rightleftharpoons HCOO^- \text{ (aq)} + HOCl \text{ (aq)}$$

$$HCN \text{ (aq)} + F^- \text{ (aq)} \rightleftharpoons HF \text{ (aq)} + CN^- \text{ (aq)}$$

Answer: To the right; to the left.

13.3   ANIONS AS WEAK BASES

Hydrolysis

As we have stated, anions of weak acids such as cyanide ion (CN^-) are Bronsted bases because of their ability to accept protons. This basic quality is also seen in their effect on the pH of water. Such reactions are referred to as **hydrolysis reactions**. For example,

$$H_2O \text{ (liq)} + CN^- \text{ (aq)} \rightleftharpoons HCN \text{ (aq)} + OH^- \text{ (aq)}$$

$$K_b = \frac{[HCN][OH^-]}{[CN^-]}$$

Production of OH^- ions due to hydrolysis of CN^- ions makes the solution basic. Multiplying the numerator and the denominator by $[H_3O^+]$,

$$K_b = \frac{[HCN][OH^-][\mathbf{H_3O^+}]}{[CN^-][\mathbf{H_3O^+}]} = \frac{K_w}{K_a(HCN)}$$

So the equilibrium constant K_b for hydrolysis is made up of K_a and K_w.

This is a general result for the anions of all weak acids in water, and the equilibrium constants for these reactions are referred to as the hydrolysis constants.

$$K_b = \frac{K_w}{K_a}$$

where K_a is the ionization constant of the conjugate acid. It is important to note that although the hydrolysis constant is often given its own unique subscript identification (K_h) because of its practical importance and ubiquitous nature in aqueous systems, it is only a special case of K_b (which is what we will call these equilibrium constants). Hydrolysis reactions are acid-base reactions in which the base is the anion of a weak acid.

EXAMPLE 13–2

Calculate $[OH^-]$, $[H_3O^+]$, and the pH when 0.10 mol of $NaNO_2$ is dissolved in sufficient water to produce 1.0 L of solution.

Solution Begin by assuming that the freely soluble $NaNO_2$ ionizes completely, generating a solution that is 0.10 M in Na^+ and NO_2^-. The K_h for NO_2^- is

$$K_b = \frac{K_w}{K_a} = \frac{1.0 \times 10^{-14}}{4.5 \times 10^{-4}} = 2.2 \times 10^{-11} = K_h$$

Let x be the concentration of OH^- produced in the hydrolysis reaction, and make a table of concentrations:

$$NO_2^- \text{ (aq)} + H_2O \text{ (liq)} \rightleftharpoons HNO_2 \text{ (aq)} + OH^- \text{ (aq)}$$

before hydrolysis	0.10	0	0
at equilibrium	$0.10 - x$	x	x

$$K_b = 2.2 \times 10^{-11} = \frac{x^2}{0.10 - x}$$

Assume $x \ll 0.10$ M. Then

$$x^2 = 2.2 \times 10^{-12}$$

$$x = [OH^-] = 1.5 \times 10^{-6} \text{ M}$$

$$\frac{1.0 \times 10^{-14}}{1.5 \times 10^{-6}} = [H_3O^+] = 6.7 \times 10^{-9} \text{ M}$$

$$pH = -\log [H_3O^+] = -\log [6.0 \times 10^{-9}] = 8.17$$

EXERCISE 13–2(A)

Determine $[OH^-]$, $[H_3O^+]$, and pH when 15.0 g of sodium acetate (CH_3COONa) is added to sufficient water to make up one liter of solution. $K_a = 1.8 \times 10^{-5}$
Answer: $[OH^-] = 1.0 \times 10^{-5}$ M; $[H_3O^+] = 1.0 \times 10^{-9}$ M; pH = 9.00.

EXERCISE 13–2(B)

Determine $[OH^-]$, $[H_3O^+]$, and pH when 15.0 g of sodium fluoride (NaF) is added to sufficient water to make up one liter of solution. $K_a = 6.8 \times 10^{-4}$
Answer: $[H_3O^+] = 4.3 \times 10^{-9}$ M; pH = 8.37; $[OH^-] = 2.3 \times 10^{-6}$ M.

Note that the value of K_b and the degree of hydrolysis are greater for bases with conjugate acids having small K_a values. This is consistent with our previous statement about relative acid and base strengths for conjugate acid-base pairs. Consequently, weak acids have strong conjugate bases.

The hydrolysis reaction causes the conjugate base of a weak acid to increase $[OH^-]$ in aqueous solutions. It is also true that the conjugate acid of a weak base will increase $[H_3O^+]$ in water solution. For example, NH_4^+, the conjugate acid of the weak base NH_3, will react with solvent water:

$$NH_4^+ \text{ (aq)} + H_2O \text{ (liq)} \rightleftharpoons H_3O^+ \text{ (aq)} + NH_3 \text{ (aq)}$$

$$K_a = \frac{[H_3O^+][NH_3]}{[NH_4^+]}$$

Here, multiplying both numerator and denominator by $[OH^-]$ gives

$$K_a = \frac{[H_3O^+][NH_3][OH^-]}{[NH_4^+][OH^-]} = \frac{K_w}{K_b}$$

In fact, we often find that K_b's for weak bases are not listed in tables of equilibrium data; K_a values of their conjugate acids are often given instead.

EXAMPLE 13–3

Calculate $[OH^-]$, $[H_3O^+]$, and the pH when 0.150 mol of NH_4Cl (s) is dissolved in sufficient water to produce one liter of solution.

Solution   Given $K_b(NH_3) = 1.8 \times 10^{-5}$ and that the NH_4Cl dissolves and dissociates completely, producing a solution that is 0.150 M in both NH_4^+ and Cl^-:

$$K_a = \frac{K_w}{K_b} = \frac{1.0 \times 10^{-14}}{1.8 \times 10^{-5}} = 5.6 \times 10^{-10}$$

Let x be the $[H_3O^+]$ produced by the NH_4^+ ions generated on adding ammonium chloride to water:

$$NH_4^+ \text{ (aq)} + H_2O \text{ (liq)} \rightleftharpoons H_3O^+ \text{ (aq)} + NH_3 \text{ (aq)}$$

before ionization	0.150	0	0
at equilibrium	$0.150 - x$	x	x

$$K_a = 5.6 \times 10^{-10} = \frac{x^2}{0.150 - x}$$

Assume $x \ll 0.150$ M. Then

$$x^2 = 8.4 \times 10^{-11}$$

$$x = [H_3O^+] = 9.2 \times 10^{-6} \text{ M}$$

$$\frac{1.0 \times 10^{-14}}{9.2 \times 10^{-6}} = [OH^-] = 1.1 \times 10^{-9} \text{ M}$$

$$pH = -\log [H_3O^+] = -\log [9.2 \times 10^{-6}] = 5.04$$ ■

EXERCISE 13–3(A)

Determine the pH when 0.0550 mol of $[CH_3NH_3^+, Cl^-]$ (s) is added to sufficient water to make up 1.00 L of solution.
Answer: pH = 5.96.   ■

EXERCISE 13–3(B)

Determine $[OH^-]$, $[H_3O^+]$, and pH when 15.0 g of ammonium chloride

(NH₄Cl) is added to sufficient water to make up one liter of solution.
Answer: $[H_3O^+] = 1.3 \times 10^{-5}$; pH = 4.89; $[OH^-] = 7.7 \times 10^{-10}$. ∎

Buffers

When an acid or a base, even in relatively small amounts, is added to water, there is a large change in the pH of the solution. For example, addition of as litle as 0.010 mol of HCl lowers the pH of solvent water from 7 to 2, a change of 5 pH units. A **buffer** is a solution that is capable of absorbing excess acid or base without major changes in pH. Because such a solution must be able to react with a large fraction of added acid or base, the buffer solution must contain both an acid and a base. However, neither can be strong, since they would simply neutralize each other:

$$H_3O^+ \text{ (aq)} + OH^- \text{ (aq)} \longrightarrow 2\,H_2O \text{ (liq)}$$

Instead, a buffer must contain a weak acid and a weak base. More specifically, a buffer must contain a weak acid and its conjugate base.

For example, a liter of solution containing 0.50 mol of formic acid (HCOOH) and 0.50 mol of sodium formate (HCOONa) can provide effective buffering action. Before any dissociation occurs, the resulting solution is 0.50 M in formic acid and sodium formate. To determine the $[H_3O^+]$ and pH of this solution, keep in mind that sodium formate is a freely soluble salt, which dissociates essentially completely:

$$HCOONa \text{ (s)} \longrightarrow Na^+ \text{ (aq)} + HCOO^- \text{ (aq)}$$

Thus, the solution is 0.50 M in formate ion. HCOOH ionizes according to the following equation:

$$HCOOH \text{ (aq)} + H_2O \text{ (aq)} \rightleftharpoons H_3O^+ \text{ (aq)} + HCOO^- \text{ (aq)}$$

$$K_a = \frac{[H_3O^+][HCOO^-]}{[HCOOH]} = 1.8 \times 10^{-4}$$

$$[H_3O^+] = (1.8 \times 10^{-4}) \frac{[HCOOH]}{[HCOO^-]}$$

If we let $[H_3O^+]$ formed as a result of this ionization be x, then, assuming $x \ll 0.500$

$$x = [H_3O^+] = (1.8 \times 10^{-4}) \left(\frac{0.500}{0.500}\right) = 1.8 \times 10^{-4} \text{ M}$$

$$pH = -\log [H_3O^+] = -\log (1.8 \times 10^{-4}) = 3.74$$

Now suppose we dilute this buffer solution by adding enough water to make 2.0 L. The concentrations of HCOOH and HCOO⁻ drop accordingly, to 0.25 M, and we may write the equilibrium expression as follows:

$$[H_3O^+] = (1.8 \times 10^{-4}) \left(\frac{0.25}{0.25}\right) = 1.8 \times 10^{-4} \text{ M}$$

Note that the $[H_3O^+]$ and pH are unchanged, demonstrating one very important feature of buffer solutions: moderate dilution of a buffer solution does not significantly change its pH.

Let's now go back to our original, undiluted 1.0 L of buffer solution and add 0.010 mol of strong acid. What effect does this have on the $[H_3O^+]$ and pH of the resulting solution? Remember what happened when we added the same small amount of acid to pure water—the pH fell from 7 to 2. In this case, however, the added hydronium ions react with the weakly basic formate anions ($HCOO^-$) according to the following equation:

$$H_3O^+ \text{ (aq)} + HCOO^- \text{ (aq)} \rightleftharpoons HCOOH \text{ (aq)} + H_2O \text{ (aq)}$$

The equilibrium constant for this reaction is

$$K = \frac{[HCOOH]}{[H_3O^+][HCOO^-]} = \frac{1}{K_a} = \frac{1}{1.8 \times 10^{-4}} = 5.6 \times 10^3$$

With an equilibrium constant this large, we can reasonably assume that the reaction is complete. Therefore, the new concentration of HCOOH consists of the formic acid originally present plus the formic acid formed by reaction of formate ions with hydronium ions:

$$[HCOOH] = 0.50 + 0.01 = 0.51 \text{ M}$$

and the new concentration of $HCOO^-$ consists of the formate ions originally present minus the formate ions that reacted:

$$[HCOO^-] = 0.50 - 0.01 = 0.49 \text{ M}$$

The hydronium concentration is obtained from the equilibrium relationship and K_a for formic acid:

$$[H_3O^+] = 1.8 \times 10^{-4} \frac{[HCOOH]}{[HCOO^-]} = 1.9 \times 10^{-4} \text{ M}$$

$$pH = -\log [H_3O^+] = -\log 1.9 \times 10^{-4} = 3.71$$

Recall that the pH of the buffer solution was 3.74 before we added the strong acid. Thus, there is no significant change in the pH. Compare this result with a change of 5 full pH units when 0.010 mol of strong acid is added to 1.0 L of pure water.

Our buffer should also show a minimal pH change on adding 0.010 mol of a strong base such as NaOH to it. The NaOH will ionize completely in water to give 0.010 M OH^-, which will react with HCOOH:

$$HCOOH \text{ (aq)} + OH^- \text{ (aq)} \rightleftharpoons H_2O \text{ (liq)} + HCOO^- \text{ (aq)}$$

The equilibrium constant for this reaction is

$$\frac{K_a}{K_w} = \frac{1.8 \times 10^{-4}}{1.0 \times 10^{-14}} = 1.8 \times 10^{10}$$

Again, this reaction can be considered to go to completion, giving the following new concentrations:

$$[HCOOH] = 0.50 - 0.01 = 0.49 \text{ M}$$
$$[HCOO^-] = 0.50 + 0.01 = 0.51 \text{ M}$$

The hydronium ion concentration and the pH are

$$[H_3O^+] = 1.8 \times 10^{-4} \frac{[HCOOH]}{[HCOO^-]} = 1.8 \times 10^{-4} \left(\frac{0.49}{0.51}\right)$$

$$= 1.7 \times 10^{-4} \text{ M}$$

$$pH = 3.77$$

Note that the pH remained virtually unchanged.

Buffers are remarkable in their ability to maintain pH in the face of significant additions of strong acids or bases. The preceding calculations are good evidence of that. However, here is one cautionary note. The concentrations of the components of the buffer must be considerably greater than the concentration of added H_3O^+ or OH^-. If a relatively large amount of acid was added, on the order of the concentration of the buffer, the buffer would be overpowered. Consider adding 0.45 mol of strong acid to 1.0 L of our 0.50 M buffer solution. The new concentrations and the final pH are

$$[HCOOH] = 0.50 + 0.45 = 0.95 \text{ M}$$
$$[HCOO^-] = 0.50 - 0.445 = 0.05 \text{ M}$$
$$[H_3O^+] = 1.8 \times 10^{-4} \left(\frac{0.95}{0.05}\right) = 3.4 \times 10^{-3} \text{ M}$$
$$pH = 2.47$$

Now a pH of about 2.5 is a change approaching 1.3 pH units. That magnitude of change is sufficient to drastically alter most biological and ecological systems by changing the course of the chemical processes involved. Changes in pH of that magnitude could be expected to alter the speed of acid-catalyzed industrial processes, affecting their economics as well as the chemistry. Our very first example in this discussion of buffers illustrated the point that diluting a buffer does not change its pH. But lowering the concentration of the buffer reduces the concentrations of strong acid or base that the buffer can absorb without a drastic pH change.

We can develop a simple and straightforward equation relating the pH, the equilibrium constant K_a for the weak acid, and the ratio of the concentrations of the weak acid and its conjugate base. For the general case where the weak acid HA ionizes in water,

$$HA + H_2O \rightleftharpoons H_3O^+ + A^-$$

for which we can write the equilibrium expression

$$K_a = \frac{[H_3O^+][A^-]}{[HA]}$$

Rearranging the expression,

$$[H_3O^+] = K_a \frac{[HA]}{[A^-]}$$

Taking the negative logarithms of both sides and using the p notation (from our definition of pH),

$$-\log [H_3O^+] = -\log \left(K_a \frac{[HA]}{[A^-]}\right)$$
$$pH = -\log K_a - \log \frac{[HA]}{[A^-]}$$
$$pH = pK_a + \log \frac{[A^-]}{[HA]}$$

The last expression is known as the **Henderson-Hasselbalch equation**. This equation immediately allows you to determine the following:

1. The effective range of buffer activity for a given weak acid whose K_a is known and its conjugate base. It is a good idea to start at the midrange, where $[A^-] = [HA]$ and therefore since $\log 1 = 0$, $pH = pK_a$.

2. The relative quantities of weak acid and conjugate base required to construct a buffer of given pH range. As a rule of thumb, the limits of buffering action are a pH unit on either side of the midrange pH, corresponding to values of $[A^-]$ from 0.1 [HA] to 10 [HA].

A similar analysis of a buffer made with a weak base and its conjugate acid leads to the analogous result:

$$pOH = pK_b + \log \frac{[BH^+]}{[B]}$$

EXAMPLE 13–4

Calculate the midrange of buffer activity for a formic acid/formate ion buffer.

Solution   The pH at the midrange equals the pK_a for formic acid:

$$pH = pK_a + \log \frac{[HCOO^-]}{[HCOOH]} = pK_a + \log 1$$

Knowing the equilibrium constant for formic acid, we find

$$pH = pK_a = -\log (1.8 \times 10^{-4}) = 3.74$$

This is the midrange of buffering activity. The limits of the range for effective buffering action are typically one order of magnitude of concentration, which is one pH unit, above and below the midrange concentration:

$$range = pK_a \pm 1 \text{ unit}$$

In other words, for a formic acid/formate ion buffer, the useful range is approximately between pH 2.7 and 4.7. ■

EXERCISE 13–4

What is the useful range of a buffer composed of CN^- and HCN?
Answer: 8.1 to 10.1. ■

EXAMPLE 13–5

Prepare a buffer that is effective within the pH range from 3.00 to 5.00.

Solution   Since the midrange is pH 4.00, pick an acid with a pK_a close to 4.00. Formic acid would be an appropriate choice:

$$K_a = 1.8 \times 10^{-4}$$

$$pK_a = 3.74$$

Use the Henderson-Hasselbalch equation to determine the concentration ratio for a pH of 4.00:

$$pH = pK_a + \log \frac{[HCOO^-]}{[HCOOH]}$$

$$4.00 = 3.74 + \log \frac{[HCOO^-]}{[HCOOH]}$$

$$0.26 = \log \frac{[HCOO^-]}{[HCOOH]}$$

$$\frac{[HCOO^-]}{[HCOOH]} = 10^{0.26} = 1.8$$

A buffer solution in which the ratio of formate ion to formic aid is $1.8:1$ has a pH of 4.00. ■

EXERCISE 13–5

What would be the ratio of $[HCN]/[CN^-]$ in a buffer with a pH of 8.50?
Answer: $[HCN]/[CN^-] = 4.4$. ■

Such a relationship between the buffer concentrations and $[H_3O^+]$, however, is possible only if the concentrations of conjugate acid and conjugate base are relatively close. For an example in which these simple calculations fail, consider a buffer prepared from 0.5000 mol of formic acid and 0.0050 mol of sodium formate in one liter of solution. The sodium formate ionizes completely in solution to Na^+ and $HCOO^-$ ions. If we let x be the $[H_3O^+]$ produced in the ionization of HCOOH, then we can write the following:

$$HCOOH\,(aq) + H_2O\,(liq) \rightleftharpoons H_3O + (aq) + HCOO^-\,(aq)$$

before ionization	0.5000	0	0.0050
at equilibrium	$0.5000 - x$	x	$0.0050 + x$

leading to the expression

$$K_a = 1.8 \times 10^{-4} = \frac{(x)(0.0050 + x)}{(0.5000 - x)}$$

Our usual assumption that $x \ll 0.0050$ does not work in this case. To see the problem, we can try this assumption and then solve:

$$1.8 \times 10^{-4} = \frac{(x)(0.0050)}{(0.5000)}$$

$$x = 1.8 \times 10^{-2} = 0.018$$

Clearly, our assumption fails. The reason is that the high concentration of HCOOH compared to that of $HCOO^-$ allows considerable additional $HCOO^-$ to be produced as a result of ionization of HCOOH. A higher concentration of $HCOO^-$ would have suppressed this ionization considerably by means of the common ion effect. Easy calculation of $[H_3O^+]$ in buffers is possible only if the concentrations of conjugate acid and base are comparable.

Solving for $[H_3O^+]$ by the quadratic equation without making the simplifying assumption gives the result that

$$[H_3O^+] = 0.0072 \text{ M}$$

$$pH = 2.14$$

Thus, the $HCOO^-$ produced by the ionization of HCOOH (0.0072 M) is actually greater than the starting $HCOO^-$ concentration. This buffer is dominated by the HCOOH, since the $HCOO^-$ concentration is too low to repress the acid ionization successfully. Addition of 0.0005 mol of OH^- to one liter of this buffer changes the relative concentrations of the buffer components sufficiently to raise the pH to 2.6. Clearly, this is not a very effective buffer. If, however, both the HCOOH and $HCOO^-$ were equal at 0.005 M, addition of the 0.0005 mol of OH^- would change the pH from

3.77 to 3.85. The buffer with equal component concentrations is far more effective.

EXAMPLE 13–6

Consider the opposite case, that of a buffer solution made of a weak base and its conjugate acid. Calculate the pH of a solution that is 0.20 M in aqueous ammonia and 0.10 M in ammonium ion.

Solution We can go back to our primary acid-base relationship from the equilibrium expression:

$$NH_3 + H_2O \rightleftharpoons NH_4^+ + OH^-$$

$$K_b = \frac{[NH_4^+][OH^-]}{[NH_3]}$$

$$[OH^-] = K_b \frac{[NH_3]}{[NH_4^+]} = (1.8 \times 10^{-5})\left(\frac{0.20}{0.10}\right) = 3.6 \times 10^{-5} \text{ M}$$

$$[H_3O^+] = \frac{K_w}{[OH^-]} = \frac{1.0 \times 10^{-14}}{3.6 \times 10^{-5}} = 2.8 \times 10^{-10} \text{ M}$$

$$pH = -\log [H_3O^+] = -\log 2.8 \times 10^{-10} = 9.55$$

The same result within the limits of significant figures can be obtained directly, using the Henderson-Hasselbalch equation for a weak base:

$$pOH = pK_b + \log \frac{[BH^+]}{[B]}$$

$$pOH = -\log(1.8 \times 10^{-5}) + \log \frac{[0.10]}{[0.20]} = 4.74 + (-0.30) = 4.44$$

$$pH = 14 - pOH = 14 - 4.44 = 9.56$$

EXERCISE 13–6(A)

What is the pH of a solution containing 0.010 M methylamine (CH_3NH_2) and 0.030 M methylammonium ion ($CH_3NH_3^+$)?
Answer: 10.17.

EXERCISE 13–6(B)

Calculate the pH of an aqueous solution that is 0.025 M in nitrous acid (HNO_2) and 0.015 M in potassium nitrite (KNO_2). Determine the pH of an aqueous solution that is 0.020 M in ammonia (NH_3) and 0.009 M in ammonium nitrate (NH_4NO_3).
Answer: 3.13; 9.61.

pH ranges: blood, 7.3–7.5; saliva, 6.5–7.5.

Body fluids such as blood and saliva function properly only over a very narrow pH range. Nature achieves this range by the use of buffers. For example, blood has two buffering systems. One is a phosphate buffer containing the ions $H_2PO_4^-$ and HPO_4^{2-}. The dihydrogen phosphate ion ($H_2PO_4^-$) is a weak acid, ionizing in aqueous solution as follows:

$$H_2PO_4^- \text{ (aq)} + H_2O \text{ (liq)} \rightleftharpoons H_3O^+ \text{ (aq)} + HPO_4^{2-} \text{ (aq)}$$

$$K_a = \frac{[H_3O^+][HPO_4^{2-}]}{[H_2PO_4^-]}$$

The equilibrium constant for this reaction is $K_a(2)$ for H_3PO_4, which is known to be 6.2×10^{-8}. If the average pH for blood is assumed to be 7.40, then

$$[H_3O^+] = 10^{-7.40} = 4.0 \times 10^{-8} \text{ M}$$

Rearrangement of the equilibrium expression gives

$$\frac{[H_2PO_4^-]}{[HPO_4^{2-}]} = \frac{[H_3O^+]}{K_a} = \frac{4.0 \times 10^{-8}}{6.2 \times 10^{-8}} = 0.64$$

which is the ratio of $[H_2PO_4^-]$ to $[HPO_4^{2-}]$ in the blood.

However, $H_2PO_4^-/HPO_4^{2-}$ is not the principal buffer in the blood. The total concentration of these two ions in the blood is only about one-twelfth of the total concentration of the components of the H_2CO_3/HCO_3^- buffer. Thus most of the buffering in blood is performed by the species in the carbon dioxide/bicarbonate ion equilibrium:

$$CO_2 \text{ (g)} + H_2O \text{ (liq)} \rightleftharpoons H_2CO_3 \text{ (aq)}$$
$$H_2CO_3 \text{ (aq)} + H_2O \text{ (liq)} \rightleftharpoons H_3O^+ \text{ (aq)} + HCO_3^- \text{ (aq)}$$

K for this reaction is $K_a(1)$ for H_2CO_3, which is 4.3×10^{-7}. The ratio of $[H_2CO_3]$ to $[HCO_3^-]$ can be calculated in the same way as for the $H_2PO_4^-/HPO_4^{2-}$ buffer system:

$$\frac{[H_2CO_3]}{[HCO_3^-]} = \frac{[H_3O^+]}{K_a} = \frac{4.0 \times 10^{-8}}{4.3 \times 10^{-7}} = 0.093$$

Note that the low ratio of $[H_2CO_3]$ to $[HCO_3^-]$ indicates that the buffer would have limited effectiveness, since addition of acid would build up $[H_2CO_3]$ and destroy the balance. However, the body has a mechanism to combat the buildup of H_2CO_3; decomposition of H_2CO_3 to H_2O and CO_2 is catalyzed by the enzyme carbonic anhydrase.

$$H_2CO_3 \text{ (aq)} \rightleftharpoons H_2O \text{ (liq)} + CO_2 \text{ (g)}$$

The CO_2 is expired from the lungs. Failure to remove enough CO_2 due to respiratory disfunction results in a buildup of H_2CO_3 in the blood and a lowering of the pH. Thus, monitoring of blood pH can be used as a diagnostic device for unsatisfactory respiration in premature babies and other patients.

13.4 ACID-BASE TITRATIONS

The classic analytical technique for determining the concentration of an acid or a base in an aqueous solution is titration. **Acid-base titrations** are convenient procedures that can be carried out to a high degree of accuracy and precision. In a typical titration, carefully measured quantities of a solution of a strong base of known concentration, such as sodium hydroxide, are progressively added to a solution of a weak acid of unknown concentration, such as acetic acid. As the base is added, the hydronium ion concentration decreases because of the stoichiometric **neutralization** reaction with hydroxide ions, forming water. To maintain the equilibrium, more of the weak acid ionizes, releasing protons, which are promptly neutralized by the hydroxide ions being added. The process continues to completion, at which point an **indicator** signals that the correct stoichiometric amount of hydroxide ions has been added to completely neutralize the hydronium ions from the acid, mole-for-mole. The indicator can be a visual indicator (a compound that changes color at the endpoint) or a pH meter, which gives direct readings from an electrochemical cell that is sensitive to hydronium ion concentration. The known quantities in the titration are

(1) the volume V_u of the unknown solution and (2) the volume V_k and concentration c_k of the known solution. Since the number of moles of hydroxide ions added equals the number of moles of hydronium ions present, for the simple 1:1 stoichiometry of sodium hydroxide and a monoprotic acid, we can write the following:

$$\text{moles of known} = \text{moles of unknown}$$

$$V \times c = \text{mol} = \text{L} \times \text{mol/L}$$

$$V_k c_k = V_u c_u$$

$$c_u = \frac{V_k c_k}{V_u}$$

EXAMPLE 13–7

What is the concentration of an unknown acid if a 20.0-mL sample of it is neutralized with precisely 33.4 mL of 0.250 M base. Assume the stoichiometry is 1:1.

Solution

$$c_u = \frac{V_k c_k}{V_u} = \frac{(0.0334\ \text{L})(0.250\ \text{mol/L})}{0.0200\ \text{L}} = 0.418\ \text{mol/L} = 0.418\ \text{M}$$

EXERCISE 13–7

How many milliliters of a 0.153 M aqueous sodium hydroxide solution must be added to a 20.0-mL sample of a solution of hydrochloric acid whose pH is known to be 0.747?
Answer: 23.4 mL.

In order for the stoichiometric coefficients to apply, the acid-base reactions must proceed as far toward completion as possible.

1. When the titration reaction is between a strong acid and a strong base, the net ionic equation is

$$H_3O^+\ (aq) + OH^-\ (aq) \longrightarrow 2\ H_2O\ (liq)$$

with $K = 1/K_w = 1.0 \times 10^{14}$. This reaction clearly proceeds essentially to completion.

2. When the titration reaction is between a strong acid and a weak base, the net ionic equation is

$$H_3O^+\ (aq) + B^-\ (aq) \longrightarrow HB\ (aq) + H_2O\ (liq)$$

where $K = 1/K_a(HB)$. Since K_a for weak bases ranges approximately from 10^{-4} to 10^{-10}, this gives K for the reaction greater than 10^4. Again, this proceeds essentially to completion.

3. If the titration reaction is between a weak acid and a strong base, the net ionic reaction is

$$HA\ (aq) + OH^-\ (aq) \longrightarrow H_2O\ (liq) + A^-\ (aq)$$

with $K = K_a(HA)/K_w$. With K_a ranging from 10^{-4} to 10^{-10}, this again gives K for the reaction of 10^4 and greater, and the reaction can be assumed complete.

4. A special case arises when both the acid and the base in the reaction are weak:

$$HA\ (aq) + B^-\ (aq) \rightleftharpoons HB\ (aq) + A^-\ (aq)$$

FIGURE 13–1   Although many of the pieces of volumetric glassware shown in this figure are "calibrated," the accuracy and precision with which they can be used varies considerably from the beakers and the erlenmeyer flask (low) to the 250-mL volumetric flask (high), in the background, left-of-center. When filled to the mark on the neck, the volume of liquid contained is known with a high degree of precision; the 10-mL pipet on its side in the foreground, when properly filled to the mark, can deliver the calibrated volume of liquid with a high degree of precision.

with $K = K_a(HA)/K_a(HB)$. Depending on the relative values of the two K_as, the value of K can range from 10^6 to 10^{-6} for the weak acids in Table 12–4. Thus, a weak acid and a weak base do not guarantee us a complete reaction.

Titrating a Strong Acid with a Strong Base

In an acid-base titration, an **aliquot** of the unknown solution, which is a carefully measured volume, is placed in a flask (Figure 13–1). The solution of known concentration is added with the aid of a **buret**, a piece of glassware that allows one to dispense controlled volumes of liquid into the flask (Figure 13–2). The acid or base in the flask may be either strong or weak and, in fact, its identity may be unknown. Hence the acid or base in the buret will be strong to insure that one of the reactants is a strong acid or base. The pH of the solution in the flask is monitored and a plot of the data—a **titration curve** showing pH versus volume or molar equivalents of acid (or base) dispensed—is prepared.

To illustrate, examine the titration of a 25.0-mL aliquot of a 0.100 M solution of hydrochloric acid with a 0.100 M solution of aqueous sodium hydroxide. From the volume of the aliquot of acid and the known concentrations of the acid and the base, we know that when 25.0 mL of base have been added, the equivalence point has been reached (Figure 13–3).

1. Initially, before any base has been added, pH of the aliquot of acid is

$$[H_3O^+] = 0.100 \text{ M}$$

$$pH = -\log 0.100 = 1.00$$

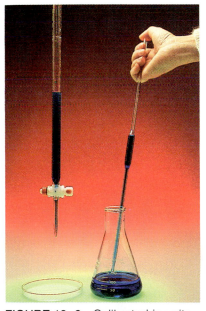

FIGURE 13–2   Calibrated in units of 0.1 mL in the range between 0 and 50 mL, the buret, shown on the left, can be used to deliver variable quantities of liquid with a high degree of accuracy and precision. The pipet, shown on the right, can be used to deliver an aliquot, or carefully measured fraction of the volume of the liquid in the erlenmeyer flask, in this case, 25.00 mL. Burets and pipets have been in use for over a century and continue to be important in the analysis of acids and bases. Modern versions are entirely automated. (See chapter-opening photograph and caption.)

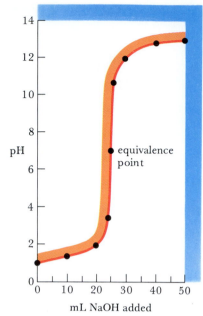

FIGURE 13–3 Titration curve for 25.0 mL of 0.100 HCl solution with 0.100 M NaOH solution. Note that the equivalence point is at pH 7.00, and that the "vertical" section of the curve, approaching and moving beyond the equivalence point, is long and steep. Titration curves for any other strong acid and strong base will look exactly the same as this one, as long as the concentrations of the acid and the base are the same, and both are "monoprotic."

Now, add 10.0 mL of the sodium hydroxide solution to the aliquot. The acid and base react:

$$H_3O^+ \text{ (aq)} + OH^- \text{ (aq)} \rightleftharpoons 2 \ H_2O \text{ (liq)}$$

The number of moles of H_3O^+ remaining is equal to (initial moles H_3O^+) − (moles OH^- dispensed):

$$(0.0250 \ L \times 0.100 \ mol/L) - (0.010 \ L \times 0.100 \ mol/L) = 0.00150 \ \text{mole}$$

The concentration of $[H_3O^+]$ can then be calculated, but be certain to take into account the dilution caused by the added volume. The total volume is now

$$25.0 \ mL \text{ initially} + 10.0 \ mL \text{ dispensed} = 35.0 \ mL$$

Therefore,

$$[H_3O^+] = \frac{0.00150 \ mol}{0.0350 \ L} = 0.0428 \ M$$

$$pH = -\log (0.0428) = 1.368$$

2. After 20.0 mL of base has been dispensed into the aliquot of hydrochloric acid,

$$[H_3O^+] = \frac{(0.0250 \ L)(0.100 \ mol/L) - (0.0200 \ L)(0.100 \ mol/L)}{(0.0250 + 0.0200) \ L}$$

$$= 0.011 \ M$$

$$pH = -\log (0.0111) = 1.95$$

We know the equivalence point for this titration occurs at 25.0 mL of dispensed base. To observe the behavior of the titration curve in this region, calculate pH values after 24.0, 25.0, and 26.0 mL of added base.

3. After 24.0 mL of base have been added:

$$[H_3O^+] = \frac{(0.0250 \ L)(0.100 \ mol/L) - (0.0240 \ L)(0.100 \ mol/L)}{(0.0250 + 0.0240) \ L}$$

$$= 0.00204 \ M$$

$$pH = 2.69$$

4. After 25.0 mL of base has been added to the original aliquot of hydrochloric acid, the acid has been neutralized and the pH is 7.000. Note that when no other source of H_3O^+ is present, the self-ionization of water determines the pH.

5. After 26.0 mL of added base has been dispensed, the net effect is the same as adding 1.0 mL of strong base to water. We can now calculate $[OH^-]$ in terms of mL of base dispensed beyond the equivalence point, and the total volume of the solution at that point.

$$[OH^-] = \frac{(0.0010 \ L)(0.100 \ mol/L)}{(0.0250 + 0.0260) \ L} = 0.0020 \ M$$

$$[H_3O^+] = \frac{K_w}{[OH^-]} = \frac{1.0 \times 10^{-14}}{0.0020 \ mol/L} = 5.0 \times 10^{-12} \ M$$

$$pH = -\log (5.0 \times 10^{-12}) = 11.30$$

Notice that our titration curve has jumped 8.6 pH units during the addition of only 2.0 mL of base, after moving only 1.69 pH units during the addition of the first 24 mL of base.

6. At 30 mL of base dispensed to the aliquot of hydrochloric acid:

$$[OH^-] = \frac{(0.0050\ \text{L})(0.100\ \text{mol/L})}{(0.0250 + 0.0300)\ \text{L}} = 0.00909\ \text{M}$$

$$[H_3O^+] = \frac{K_w}{[OH^-]} = \frac{1.0 \times 10^{-14}}{0.00909\ \text{mol/L}} = 1.1 \times 10^{-12}\ \text{M}$$

$$pH = -\log(1.1 \times 10^{-12}) = 11.96$$

7. At 40 mL of base dispensed:

$$[OH^-] = 0.0231\ \text{M}$$

$$[H_3O^+] = \frac{K_w}{[OH^-]} = \frac{1.0 \times 10^{-14}}{0.023\ \text{mol/L}} = 4.3 \times 10^{-13}\ \text{M}$$

$$pH = -\log(4.3 \times 10^{-13}) = 12.36$$

The titration curve is clearly leveling off again.

One of the interesting characteristics of the titration curve is that the equivalence point is the **inflection point**, the point where the curve changes direction. Titrating a strong base by a strong acid would give an equivalent titration curve, except that the pH would start at a high value and proceed to a low value (Figure 13–4).

Of course not every strong acid is monoprotic. Sulfuric acid is diprotic and gives one inflection point per proton when titrated with aqueous hydroxide ions. But because the successive K_a values for H_2SO_4 are of greatly different magnitude, the inflection points are well-separated along the titration curve (Figure 13–5).

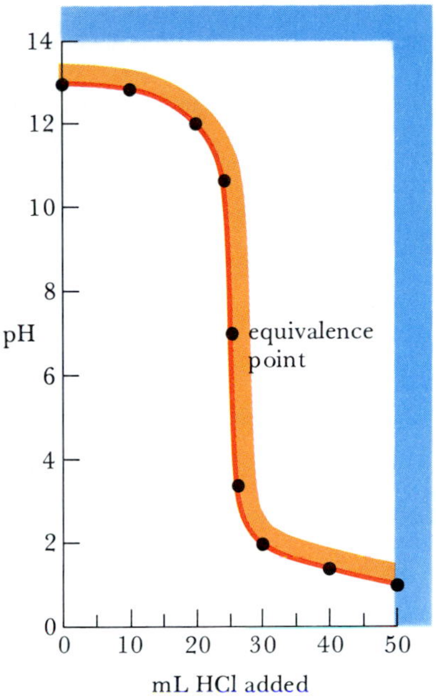

FIGURE 13–4 Titration curve for 25.0 mL of 0.100 M NaOH solution with 0.100 M HCl solution. Note that the curve is similar to that in Figure 13–3, only inverted with respect to pH, beginning at 13.0 and ending at 1.00.

Titrating a Weak Acid with a Strong Base

The general shape of the titration curve for a weak acid and a strong base (Figure 13–6) resembles that for a strong acid-base titration but with some very important differences.

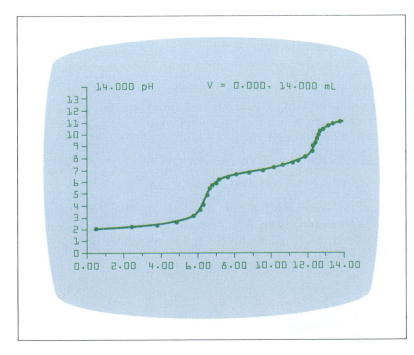

FIGURE 13–5 Titration curve for a weak diprotic acid, using sodium hydroxide and an automatic titrator, from data plotted in real time on a 9-inch CRT monitor. Note that there are two distinct inflection points (where the curve changes direction) at 6.25 mL and 12.5 mL, which mark the equivalence points for the titration of each proton.

FIGURE 13–6 Titration curves for three weak acids with different K_a values. Note the marked initial rise in pH for all three curves followed by the broad "buffer" range, and the steep rise near the equivalence point and beyond.

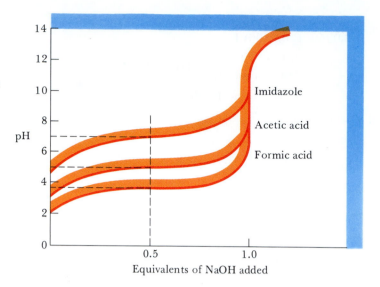

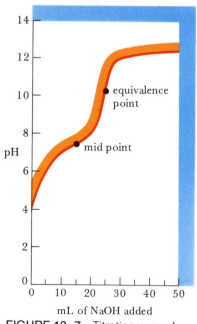

FIGURE 13–7 Titration curve for 25.0 mL of 0.100 M HOCl solution with 0.100 M NaOH solution. Note the initial rapid rise in pH and the broad buffer range that follows. Also note that the vertical section where the curve rises to the equivalence point and beyond, is much shorter (than for the strong acid in Figure 13–3) because of the fact that the buffer region for HOCl is closer to the endpoint region. The buffer solution is composed of varying concentrations of OCl⁻ and HOCl.

1. The weak acid curve has a shorter segment in the steep section around the equivalence point because the first part of the titration rises to a higher pH.

2. The equivalence point is at a higher pH for the weak acid because of hydrolysis of the conjugate base.

3. The first part of the weak acid curve rises relatively rapidly very early in the titration, and then flattens out as a buffer is formed from the weak acid and its salt; the strong acid curve (Figure 13–3) simply rises steadily from the start, as the acid is neutralized.

Figure 13–7 shows the titration curve for a 25.0-mL aliquot of a 0.100 M hypochlorous acid (HOCl) solution ($K_a = 3.7 \times 10^{-8}$) with 0.100 M NaOH solution. The equivalence point is reached on dispensing 25.0 mL of base. Note the differences between this curve and that for the titration of a strong acid with a strong base shown in Figure 13–3.

It is possible to calculate the K_a of a weak acid by knowing the pH at any point in the buffer region along the titration curve and the volume of base added at the equivalence point. Note the special case where pH = pK_a, the midpoint in the titration.

EXAMPLE 13–8

A sample of an unknown acid in an aqueous solution required 38.2 mL of 0.100 M NaOH solution to reach the equivalence point in a titration. After 14.3 mL of base had been added, the pH was 4.52. Find the K_a of the acid.

Solution   Begin with the general reaction of OH⁻ with a weak acid:

$$\text{HA (aq)} + \text{OH}^- \text{ (aq)} \rightleftharpoons \text{H}_2\text{O (liq)} + \text{A}^- \text{ (aq)}$$

Recognize that at the equivalence point, after adding 38.2 mL of base, all of the HA has been converted to A⁻. Therefore, when only 14.3 mL of base have been added, 14.3/38.2 of the HA has been converted to A⁻, and the fraction left as HA is

$$1 - \frac{14.3}{38.2} = \frac{38.2 - 14.3}{38.2}$$

The ratio of $\dfrac{[\text{A}^-]}{[\text{HA}]}$ will therefore be

$$\frac{14.3}{38.2 - 14.3} = \frac{14.3}{23.9}$$

Converting pH to $[H_3O^+]$, we find

$$[H_3O^+] = \text{antilog}(-pH) = \text{antilog}(-4.52) = 3.02 \times 10^{-5} \text{ M}$$

$$K_a = [H_3O^+]\frac{[A^-]}{[HA]}$$

and using the data from above

$$K_a = (3.02 \times 10^{-5})\frac{14.3}{23.9} = 1.8 \times 10^{-5}$$

This weak acid could well be acetic acid; note the K_a value in Table 12–4.

EXERCISE 13–8
The equivalence point for a solution of an unknown acid occurs after 46.7 mL of standard base has been added. Earlier in the titration, after 34.2 mL of base had been added, the pH was 4.64. What is K_a for the unknown acid?
Answer: 6.3×10^{-5}.

Acid-base titrations are commonly used to determine the nitrogen content of everything from the protein content of a grain to the polluting nitrogen oxides produced when coal is burned. The titration, originally developed by Kjeldahl, dates back to 1883. A sample is digested with sulfuric acid in the presence of a catalyst. Then an excess of strong base is added and the nitrogen in the sample is converted to NH_3, which is subsequently purified by distillation, then dissolved in water, and titrated with a strong acid (Figure 13–8).

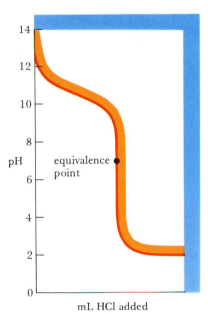

FIGURE 13–8 The Kjeldahl method for determining nitrogen content is ultimately based on titration of an aqueous ammonia solution with a strong acid. Hydrochloric acid is shown here; sulfuric or phosphoric acids are often used because they are less volatile and sometimes easier to handle on a routine analytical basis. Again, note the buffer range (pH 10–11).

13.5 INDICATORS

Often the purpose of a titration is simply to determine the concentration of an acid or a base in solution. In these cases it is not necessary to plot the entire titration curve; all that is needed is the volume of titrant (the standard base or acid) that has been added when the equivalence point is reached. Since the equivalence point is marked by a very rapid change of pH, a substance that would respond to a sharp rise in pH change by a change in color would provide a visual indicator. A number of dyes from plant sources have proved suitable for this purpose (Figure 13–9).

An acid-base indicator is itself a weak acid. The indicator acid and its conjugate base must have distinctly different colors. Furthermore, the colors must be intense enough that only a slight amount of the indicator needs to be added to the solution; thus, the presence of the indicator will not significantly affect the pH of the solution. The equilibrium expression for the indicator reaction can be represented as

$$\text{HIn (aq)} + H_2O \text{ (liq)} \rightleftharpoons H_3O^+ \text{ (aq)} + \text{In}^- \text{ (aq)}$$
acidic colorbasic color

and the hydronium ion concentration can be stated in the usual manner by rearranging the equation:

FIGURE 13–9 The juice of the red cabbage changes color with changing pH. At the left is pH = 1, and moving clockwise, 4, 7, 10, and 13. Red cabbage juice is a universal indicator of pH in aqueous solutions.

$$[H_3O^+] = K_a \frac{[HIn]}{[In^-]}$$

K_a is the ionization constant for the weakly acidic form of the indicator. For the special case where $[HIn] = [In^-]$, the hydronium ion concentration is equal to K_a for the indicator, and the color you see is intermediate between the colors of the acidic and basic forms. However, if the $[H_3O^+]$ is raised to $10K_a$, the ratio of $[HIn]$ to $[In^-]$ will be 10:1 and the color of acid HIn prevails. On the other hand, if $[H_3O^+]$ is dropped to $0.10K_a$, then $[In^-] = 10[HIn]$ and the color of the conjugate base (In^-) prevails. This means that the color change takes place over an indicator range of about 2 pH units. That is a factor of 100 in terms of $[H_3O^+]$, with the middle of the indicator range at pH = pK_a for the indicator acid, HIn.

Suppose we wanted to perform the titration in Figure 13–7. A change in color in the pH range from about 9 to 11 marks the equivalence point. Table 13–1 lists some common acid-base indicators and the pH ranges of their color changes. Thymolphthalein, with its color change between pH 9.3 and 11.0, would work well for this titration (Figure 13–10). However, you could not use any of the three indicators in Figure 13–11 since even phenolphthalein begins to change color with several percent of the titration still to be completed.

TABLE 13–1 Common Acid-Base Indicators and Their Effective pH Ranges

Indicator	Acidic Color	Basic Color	pH Range
thymol blue	red	yellow	1.2–2.8
methyl orange	orange	yellow	2.9–4.4
bromcresol green	yellow	blue	3.8–5.4
methyl red	red	yellow	4.4–6.0
bromthymol blue	yellow	blue	6.0–7.6
cresol red	yellow	red	7.2–8.8
phenolphthalein	colorless	red	8.3–10.0
thymolphthalein	colorless	blue	9.3–11.0
alizarin yellow	yellow	red	10.0–12.1

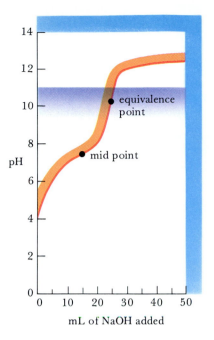

FIGURE 13–10 The indicator in this case is thymolphthalein which is colorless in acidic solutions and blue in alkaline solutions. The indicator's "range" is between 9.3 and 11.0, which places it squarely across the steepest part of the vertical segment of the titration curve and right on top of the equivalence point.

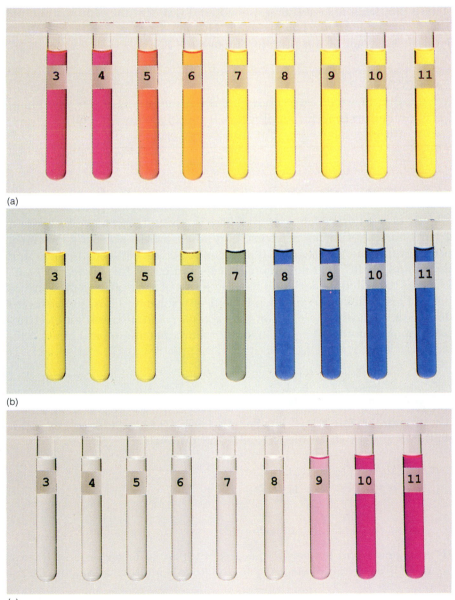

(a)

(b)

(c)

FIGURE 13–11 Three visual indicators: (a) methyl red changes from red (acidic form) to yellow (alkaline form) and its range is roughly 4.4–6.0; bromthymol blue is yellow at low pH and blue on the alkaline side of its range, which is roughly 6.0–7.6; phenolphthalein changes from colorless in acid solutions to pink, across its range of roughly 8.3–10.0.

EXAMPLE 13–9

What would be a good indicator for titrating a solution of NH_3 (aq) of unknown concentration using 0.100 M HCl (aq)?

Solution   To make that determination, note that at the equivalence point the solution would contain NH_4^+. We do not know the exact concentration of the NH_4^+, but if the original unknown solution were close to 0.1 M in NH_3, the concentration of NH_4^+ would be close to 0.050 M allowing for dilution. NH_4^+ would produce H_3O^+ according to

$$NH_4^+ \text{ (aq)} + H_2O \rightleftharpoons H_3O^+ \text{ (aq)} + NH_3 \text{ (aq)}$$

Letting x be the $[H_3O^+]$ formed in this reaction and assuming $x \ll 0.05$,

$$\frac{x^2}{0.050} = K_a = \frac{K_w}{K_b(NH_3)} = \frac{1.0 \times 10^{-14}}{1.8 \times 10^{-5}} = 5.6 \times 10^{-10}$$

$$x = [H_3O^+] = 5.3 \times 10^{-6}$$

$$pH = -\log 5.3 \times 10^{-6} = 5.28$$

This is the pH at the equivalence point, and it is near the center of the pH range for methyl red. This indicator changes from yellow to red at the equivalence point of this titration (Figure 13–12). ∎

EXERCISE 13–9

What indicator would you choose for the titration of a water solution of HCN (unknown concentration) with a 0.100 M solution of NaOH?
Answer: Alizarin yellow. ∎

FIGURE 13–12   The automatic titrator has done its work on half the analytical samples. Although no visual indicator is needed for these "potentiometric" titrations, methyl red was added to enhance the photograph. Which solutions are alkaline?

13.6   LEWIS ACIDS AND BASES

The Bronsted-Lowry definition of a base is a proton-acceptor. A proton is a bare nucleus with no electrons, and when a base accepts a proton it provides the pair of electrons that form the bond between the base and the proton. A different definition of acids and bases, formulated by G. N. Lewis in 1923, defines acids and bases in terms of the donation of electron-pairs. This definition is independent of any solvent. Thus, in the Lewis definition, an acid is an electron-pair acceptor and a base is an electron-pair donor. Consider the protonation of ammonia as a typical Lewis acid-base reaction:

$$\underset{\substack{\text{Lewis} \\ \text{acid}}}{H^+} + \underset{\substack{\text{Lewis} \\ \text{base}}}{NH_3} \longrightarrow \underset{\substack{\text{Lewis} \\ \text{adduct}}}{NH_4^+}$$

The ammonium ion, NH_4^+, is called the Lewis acid-base adduct. Let's draw the Lewis structures for the reactants and products:

$$\underset{\text{acid}}{H^+} + \underset{\text{base}}{H-\overset{\displaystyle H}{\underset{\displaystyle H}{N}}:} \longrightarrow \underset{\text{adduct}}{H-\overset{\displaystyle H}{\underset{\displaystyle H}{N^+}}-H}$$

H^+ is a Lewis acid because it can accept a pair of electrons and form a bond. NH_3 has a pair of electrons to donate to bond formation, and therefore it is a Lewis base. The halides of boron, BF_3, BCl_3, BBr_3, and BI_3, are Lewis acids. These compounds form Lewis acid-base adducts because each boron is surrounded by only six electrons and bonded to only three substituents. Thus, in a typical reaction, ammonia donates an electron pair to bond with boron trichloride:

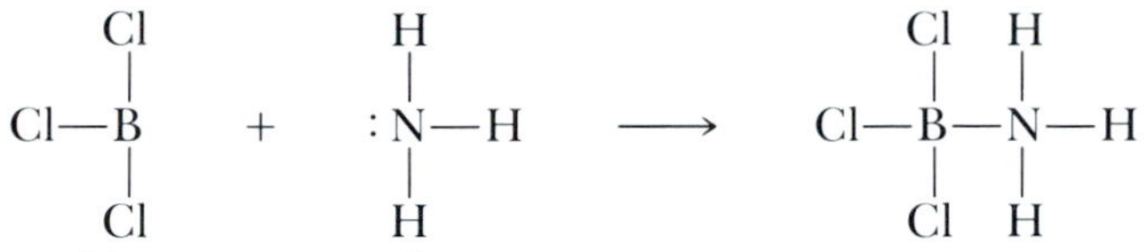

In order to comply with the octet rule, we can also write three equivalent resonance forms for the BCl_3 structure. Here is one:

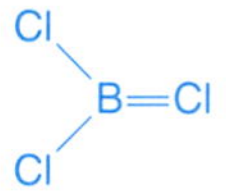

SUMMARY

A number of different definitions of acids and bases were discussed. The first really useful definition, and the one we used in the previous chapter, was the **Arrhenius** definition, that an acid is a compound that increases $[H_3O^+]$ in water solution. A base is a compound that increases $[OH^-]$ in water solution.

In the context of this chapter, it is important to remember the **pH scale**, introduced in Chapter 12. It is a way of expressing the acid concentration of a water solution as the negative logarithm of the hydronium ion concentration:

$$pH = -\log[H_3O^+]$$

A useful definition of acids and bases for considering a wide variety of acid-base reactions is the **Bronsted-Lowry** definition. An acid is a proton donor. A base is a proton acceptor. Since the proton has no independent existence in any chemically significant environment, Bronsted-Lowry acid-base reactions all involve a proton transfer in which an acid and a base react to form a new acid and a new base. The acid on one side of the reaction and the base on the other differ only in the presence of one proton and are called a **conjugate acid-base pair**.

The conjugate base of an uncharged weak acid is the anion of the acid. When added to water, such an anion increases the OH^- concentration by hydrolysis. The general reaction is

$$B^- \text{ (aq)} + H_2O \text{ (liq)} \rightleftharpoons HB \text{ (aq)} + OH^- \text{ (aq)}$$

The equilibrium constant for a **hydrolysis** reaction is a special case of K_b, the constant for the reaction of the conjugate base with the solvent; K_b is related to K_w and to K_a by

$$K_b = \frac{K_w}{K_a}$$

K_w is the ionization constant for water and K_a is the ionization constant for the weak acid.

Buffers are solutions containing a weak acid and its conjugate base. A buffer does not change its pH on dilution, and can absorb reasonable amounts of either acid or base with a minimal change of pH. The Henderson-Hasselbalch equation is very useful in dealing with buffers:

$$pH = pK_a + \log \frac{[A^-]}{[HA]}$$

where K_a is the ionization constant of the weak acid in the buffer. The concentrations of the acid and its conjugate base in a buffer should be within a factor of 10 of each other for effective buffering.

Titration techniques provide convenient means for determining the concentration of an unknown acid (or base) and its ionization constant if it is weak. Titration curves are obtained by graphing the pH of the solution versus the volume of added base (or acid); the shapes of these curves are characteristic and provide considerable information about the substances being analyzed. Indicators can be used to identify the endpoint of a titration. The ionization constant for a weak acid or base can be determined using buffer calculations.

Finally, the **Lewis** definition of an acid and a base was introduced. An acid is an electron-pair acceptor. A base is an electron-pair donor.

QUESTIONS AND PROBLEMS

QUESTIONS

1. Suggest a Bronsted-Lowry acid/base reaction in which each of the following would be an acid and another reaction in which each of the following would be a base. Give a reason for each choice.
(a) $H_2PO_4^-$
(b) HCO_3^-

2. Why is acid rain a less serious problem in lakes in which the water is exposed to limestone?

3. Why is blood pH a useful measure of effectiveness of respiration?

4. Explain how the same species (H_2O, for example) can be an acid in one Bronsted acid-base reaction and a base in another. Give some examples.

5. The following reaction can be described as a "half-reaction." Briefly explain.

$$HCN\ (aq) \rightleftharpoons H^+\ (aq) + CN^-\ (aq)$$

6. Why is the pH of a buffer unchanged if the buffer is diluted? Do you think that this will be true no matter how much water is added? Why or why not?

7. Why must the concentrations of the components of a buffer be large compared with the concentrations of strong acid or strong base added to the buffer if the pH is to remain reasonably constant?

8. Why are the concentrations of the components of a buffer normally within about a factor of ten of each other?

9. What kind of information can be obtained from an acid-base titration?

10. Why does one normally not use a weak acid or weak base to titrate an unknown solution in an acid-base titration?

11. What properties are necessary for a substance to be a satisfactory indicator for an acid-base titration?

12. What happens to the pH (rises, falls, stays the same) when you add each of the following to an aqueous solution of acetic acid?
(a) water
(b) sodium hydroxide
(c) HCl
(d) sodium acetate

13. Identify the Lewis acid and Lewis base in each of the following adduct-forming reactions:
(a) $H_2O + H^+ \rightleftharpoons H_3O^+$
(b) $H^+ + F^- \rightleftharpoons HF$
(c) $BF_3 + F^- \rightleftharpoons BF_4^-$
(d) $NH_3 + H^+ \rightleftharpoons NH_4^+$
(e) $BF_3 + NH_3 \rightleftharpoons F_3B{-}NH_3$

14. Name some species that are Lewis acids but are not Bronsted acids. Can you name any species that are Lewis bases without being Bronsted bases? Why or why not?

15. A first-year chemistry student was given 0.01 mol of a white crystalline, water-soluble, weakly acidic organic substance and told to determine the K_a of the acid. The student successfully solved the problem using nothing but water, a standardized sodium hydroxide solution, a few drops of phenolphthalein indicator solution, a few strips of indicator (pH) paper, and readily available laboratory glassware, all in about an hour. Briefly, but in a clear and orderly fashion, relate how the student most likely would have solved the problem, arriving at a reasonable approximation of K_a for the acid.

16. Briefly explain why you could (or could not) use any (or all) of the three indicators in Figure 13–11 to determine the endpoint in the titration of an aqueous ammonia solution with a "standardized" hydrochloric acid solution.

17. Do the same (as in question 16) for an aqueous acetic acid solution with a standardized sodium hydroxide solution.

PROBLEMS

Acid-Base Reactions [1–10]

1. Write balanced net ionic equations to show that each of the following is an acid, a base, or neither according to the Arrhenius definition:

(a) HCl
(b) NaCl
(c) NH_4Cl

2. Write balanced net ionic equations to show that each of the following is an acid, a base, or neither according to the Arrhenius definition:
(a) HCN
(b) NaCN
(c) NH_3

3. Write balanced net ionic equations to show that each of the following is an acid, a base, or neither according to the Bronsted-Lowry definition:
(a) $NaHSO_4$
(b) $Ca_3(PO_4)_2$
(c) HOCl

4. Write balanced net ionic equations to show that each of the following is an acid, a base, or neither according to the Bronsted-Lowry definition:
(a) NaOCl
(b) HF
(c) NaF

5. Give the conjugate base of each of the following acids:
(a) HCN
(b) HSO_4^-
(c) NH_4^+

6. Give the conjugate acid of each of the following bases:
(a) SO_3^{2-}
(b) F^-
(c) CH_3NH_2

7. In each of the following, predict which one of the pair of bases is the stronger:
(a) NO_2^- or NO_3^-
(b) HCO_3^- or CO_3^{2-}

8. In each of the following, predict which one of the pair of bases is the stronger:
(a) HSO_3^- or HSO_4^-
(b) F^- or Cl^-

9. Predict the direction favored by the equilibrium of each of the following reactions. Explain why.
(a) $NH_3 + H_3O^+ \rightleftharpoons NH_4^+ + H_2O$
(b) $CN^- + H_2O \rightleftharpoons HCN + OH^-$

10. For each of the following reactions, predict the direction favored by the equilibrium and offer a brief explanation.
(a) $H_2O + H_2O \rightleftharpoons H_3O^+ + OH^-$
(b) $HF + CN^- \rightleftharpoons HCN + F^-$

Bronsted-Lowry Acid-Base Reactions [11–18]

11. Calculate the equilibrium constants of the following reactions.
(a) $HCOOH (aq) + OH^- (aq) \rightleftharpoons$
$$H_2O + HCOO^- (aq)$$
(b) $CN^- (aq) + H_2O \rightleftharpoons HCN (aq) + OH^- (aq)$
(c) $F^- (aq) + H_2O \rightleftharpoons HF (aq) + OH^- (aq)$

12. For each of the following reactions, calculate the equilibrium constant.
(a) $NH_4^+ (aq) + OH^- (aq) \rightleftharpoons NH_3 (aq) + H_2O$
(b) $HCN (aq) + NO_2^- (aq) \rightleftharpoons$
$$HNO_2 (aq) + CN^- (aq)$$

13. The pH of a 1.0 M solution of aqueous acetic acid has a pH of 2.37. Calculate the hydronium ion concentration and the percent ionization of acetic acid at this concentration.

14. The pH of a 0.10 M solution of aqueous acetic acid has a pH of 2.87. Calculate the hydronium ion concentration and the percent ionization of acetic acid at this concentration.

15. Calculate $[OH^-]$, $[H_3O^+]$, and the pH when 1.000 g of CH_3NH_3Cl is added to sufficient water to produce 1.00 L of solution. CH_3NH_3Cl is composed of ions, $CH_3NH_3^+$ and Cl^-.

16. If the pH of an NH_4Cl solution is 5.30, what is the ammonium ion concentration? What is the pH of a 0.10 M NH_4Cl solution?

17. A sample of river water is contaminated with 0.137 g of NaF per liter. Calculate the pH of the water based on this information alone.

18. Calculate the pH of a water sample containing 6.32 g of formic acid per liter.

Buffered Solutions [19–26]

19. After diluting 25.0 mL of a 0.250 M aqueous hydrochloric acid solution to 33.3 mL, what is the pH of the resulting solution? What was the initial pH, before dilution?

20. After 25.0 mL of a 0.250 M solution of aqueous acetic acid is diluted to 33.3 mL, what is the pH of the resulting solution? What was the initial pH?

21. Calculate the pH of an aqueous solution that is 0.10 M in acetic acid and 0.10 M in sodium acetate.

22. Calculate the pH of an aqueous solution that is 0.10 M in acetic acid and 0.25 M in sodium acetate.

23. What is the ratio of concentrations in a CH_3COOH/CH_3COO^- buffer with a pH of 7.5? Is this an effective buffer? Why, or why not?

24. Find the ratio of concentrations for an NH_4^+/NH_3 buffer with a pH of 10.0. Comment on the effectiveness of this solution as a buffer in this range of pH.

25. A student has prepared 1.00 L of a buffer solution containing 0.10 M HCN and 0.12 M CN^-.
(a) What is the pH of this buffer?
(b) What is the pH if 0.01 mol of HCl is added to the buffer?
(c) What is the pH if 0.02 mol of NaOH is added to the original buffer?

26. Suppose you have 100 mL of a solution that is 0.10 M in HF and 0.10 M in NaF, to which your lab instructor now adds 10 mL of 0.10 M HCl. How does the pH change? Make a similar calculation for addition of 10 mL of 0.10 M NaOH instead of HCl. K_a for HF = 6.8×10^{-4}.

Acid-Base Titrations [27–36]

27. A 25.00-mL sample of an unknown base requires 18.34 mL of 0.100 M HCl to reach the equivalence point in a typical titration experiment. What is the concentration of the unknown base, assuming the acid-base stoichiometry is 1:1?

28. A 50.00-mL aliquot of an unknown acid is titrated with standard 0.250 M NaOH solution, and it is found

that 42.26 mL of the base is needed to reach the equivalence point. What is the concentration of the unknown acid, assuming it is monoprotic?

29. When 243.2 mg of an unknown solid, known to be a monoprotic acid, was dissolved in water and titrated with a 0.0760 M aqueous sodium hydroxide solution, 40.0 mL was required to neutralize the acid to a phenolphthalein endpoint. What is the formula weight of the dissolved acid?

30. A 310-mg sample of a weak, monoprotic acid was dissolved in sufficient water to prepare 100.00 mL of solution. It was found that 25.15 mL of a standard 0.10 M sodium hydroxide solution neutralized the acid to the equivalence point. What is the apparent molecular weight of the unknown acid?

31. An unknown weak acid is being titrated with 0.100 M NaOH. The equivalence point for a particular sample occurs when 28.2 mL of base has been added. After 19.2 mL of base had been added the pH was 7.76. What is the K_a of this acid?

32. A 25.00-mL sample of an unknown weak base is titrated with 0.100 M HCl. When 9.62 mL of HCl has been added, the pH is 9.37; the equivalence point is reached at 22.13 mL. What is the concentration of the weak base? What is the K_a of the *conjugate acid* of the weak base? Assume the stoichiometry is 1:1.

33. Plot the titration curve for 25.00 mL of an aqueous 0.10 M HCN solution being titrated with a 0.10 M solution of aqueous NaOH. Select a visual indicator to determine the equivalence point of this titration.

34. Plot the titration curve for a 0.10 M aqueous ammonia solution being titrated with 0.10 M aqueous HCl solution. Select a visual indicator for this titration. What is the color change observed through the equivalence point?

35. 2.00 g of a weakly acidic, monoprotic acid (molecular weight = 122 g/mol) was dissolved to form 100.0 mL of solution and titrated with NaOH solution. It was determined that the endpoint had been reached after addition of 41.00 mL of 0.200 M NaOH solution to 50.3 mL of the original unknown weak acid solution. The pH at that point was 8.76. with respect to the prepared solution of the unknown acid:

(a) What is the K_a for the acid?
(b) What is the percent dissociation of the acid before the first NaOH is added?
(c) What was the initial pH of the solution?
(d) What was the pH at the midpoint in the titration?
(e) Comment on the choice of thymolphthalein (pH range 9.3 to 10.5) versus methyl red (pH range 4.2 to 6.3) as a visual indicator for the titration.
(f) In what way can the general shape of the titration curve serve to confirm the initial assumption that the acid is monoprotic?

36. A solution of a certain weakly acidic substance was prepared by dissolving and diluting 2.344 g to a final volume of 100.00 mL. In a titration, 42.60 mL of 0.2500 M NaOH solution was required to reach a successful indicator endpoint. The shape of the titration curve was used as the basis for an assumption, namely that the acid question was monoprotic. The pH at the endpoint was 9.40.

(a) Calculate the apparent molecular weight of the unknown acidic substance.
(b) Calculate K_a for the acidic substance.
(c) Calculate the pH of the original 100.00 mL of solution (prior to titration).
(d) Calculate the pH at the midpoint of the titration (after addition of 21.30 mL of the sodium hydroxide solution).
(e) Using the calculated and given pH data, and your general understanding of acid-base titrations, carefully construct a graph of pH versus mL of base added for the titration.

Additional Problems [37–48]

37. Arrange the following bases in order from weakest to strongest in aqueous solution. Explain the reasons for the order you selected.

$$F^-, NH_3, NH_2^-, CN^-, H_2O, OCl^-, OH^-, NO_2^-$$

38. A 0.10 M solution of a weak acid HX is known to be 1.0% ionized. Calculate the hydronium ion concentration and the equilibrium constant K_a.

39. Compare the difference between the results in Problems 13 and 14 and offer an explanation of why there isn't a ten-fold difference in the H_3O ion concentrations, as you might expect on the basis of differences in concentration alone.

40. Suppose you need to prepare 1.00 L of a buffer solution with a pH of 5.00, which can absorb 0.05 mol of either strong acid or strong base with a minimal change of pH. What acid and salt would you select? What are the masses of your starting materials? What are the concentrations of the species in solution in your buffer?

41. The K_a for HF is 6.8×10^{-4}. Determine the pH of a solution made by mixing 0.125 mol of HF and 0.250 mol of NaF in enough water to make 200 mL of solution.

42. A 9.33-g sample of coal reacts with sulfuric acid in the presence of a mercury catalyst. The reaction mixture is then cooled and a strong base is added. The nitrogen in the coal is converted to ammonia in these reactions, and the ammonia is purified by distillation and dissolved in distilled water. This solution is titrated to a phenolphthalein endpoint with 58.2 mL of 0.100 M HCl. What is the percent of nitrogen in the coal sample?

43. You are given an acid-base indicator in which the acidic form is yellow and the basic form is purple. It appears to you that the solution of the indicator is completely yellow when the indicator ratio is 6:1, and completely purple when the ratio is 1:8. If K_a for the indicator is 5.0×10^{-6}, determine the pH range over which the color change appears to take place. State whether this indicator would provide an accurate endpoint for the following titrations (HOAC = acetic acid):

(a) HOAc/NaOH
(b) NH_3/HCl
(c) HCl/NaOH
(d) HOAc/NH_3

44. Calculate the pH of each of the following:
(a) pure water
(b) a 0.10 M sodium chloride (NaCl) solution
(c) a 0.10 M ammonia (NH_3) solution ($K_b = 1.8 \times 10^{-5}$)

(d) a 0.10 M acetic acid (HOAc) solution ($K_a = 1.8 \times 10^{-5}$)

(e) a 0.10 M ammonium chloride (NH_4Cl) solution

(f) a 0.10 M sodium acetate (NaOAc) solution

(g) a 0.10 M ammonium acetate (NH_4OAc) solution

(h) a 0.10 M ammonia (NH_3) solution that is also 0.10 M in NH_4Cl

45. Calculate $[OH^-]$, $[H_3O^+]$, and the pH when 0.050 mol of HCN is dissolved in sufficient water to produce 1.00 L of solution.

46. Calculate $[OH^-]$, $[H_3O^+]$, and the pH when 0.050 mol of HOCl is dissolved in sufficient water to produce 1.00 L of solution.

47. Hydrogen sulfide (H_2S) is a weak, diprotic acid in water. Write the equations for the two-stage equilibrium. Determine the pH and the sulfide ion concentration of a 0.10 M solution. Clearly indicate all simplifying assumptions made in your calculations. $K_a(1) = 9.1 \times 10^{-8}$; $K_a(2) = 1.2 \times 10^{-15}$.

48. Water that has been saturated with carbon dioxide gas can be considered 0.02 M in the weakly acidic species H_2CO_3, for which $K_a(1) = 4.3 \times 10^{-7}$ and $K_a(2) = 4.4 \times 10^{-11}$. Accordingly, by setting the hydrogen ion concentration, the carbonate ion concentration can be controlled.

(a) Determine the pH necessary to just prevent precipitation of $NiCO_3$ ($K_{sp} = 1.4 \times 10^{-7}$) during the precipitation of $FeCO_3$ ($K_{sp} = 2.1 \times 10^{-11}$) from a solution that was initially 0.10 M in each of the repective ions, namely Fe^{2+} and Ni^{2+}.

(b) Do you think that a 1:1 NaOAc/HOAc solution would provide effective buffering action in that range? Briefly explain your answer. (HOAc = acetic acid)

(c) How would you keep the H_2CO_3 concentration nearly constant at 0.02 M?

MULTIPLE PRINCIPLES   [49–54]

49. Consider the following acid-base reaction in aqueous solution:

$$HF\ (aq) + Cl^-\ (aq) \rightleftharpoons HCl\ (aq) + F^-\ (aq)$$

What can you say about K and $\Delta G°$ for this reaction? Is $\Delta H°$ or $\Delta S°$ the larger contributor to $\Delta G°$ at 25°C? What is the likely sign of $\Delta H°$?

50. At moderate gas pressures, the solubility of a gas in a liquid is proportional to the pressure of the gas in contact with the liquid (Henry's Law), which can be expressed in terms of the following equation:

$$P = K_{HL}X$$

where X is the mole fraction of the dissolved gas in the solution, K_{HL} is the Henry's Law constant for the gas and the liquid, and P is the equilibrium pressure of the gas. The value of K_{HL} for CO_2 (g) and H_2O (liq) is 1640 atm.

(a) Calculate the concentration (in mol/L) for CO_2 in water in equilibrium with a pressure of CO_2 (g) of 1.00 atm. For this calculation, assume that the concentration of water in the solution is 55.55 M and that the number of moles per liter of CO_2 in solution is negligible by comparison.

(b) Calculate the pH of the solution, assuming that CO_2 (aq) converts completely to H_2CO_3 (aq).

51. A saturated solution of CsCl in liquid ammonia contains 0.4 g CsCl per 100 g of ammonia. What is the solubility product of CsCl in liquid ammonia? The density of liquid ammonia is 0.817 g/cm^3.

52. Vinegar is a dilute solution of acetic acid and its pH varies between 2.4 and 3.4. What range of concentrations of acetic acid occur in vinegar?

53. The freezing point of 0.2000 molal aqueous benzoic acid solution is $-0.3785°C$. Calculate the percent ionization and the equilibrium constant K_a for benzoic acid. Assume that the solution has a density of 1.0 g/mL. Benzoic acid is monoprotic and has the formula C_6H_5COOH.

54. For each of the following, predict which one of the pair dissociates or dissolves in water, forming the stronger acid in aqueous solution:

(a) SO_2 or SO_3

(b) P_4O_6 or P_4O_{10}

(c) H_2O or H_2S

APPLIED PRINCIPLES   [55–57]

55. The water in a 10,000-gallon tank is found to have a pH of 2.9. At a cost of $1.60/kg for NaOH (s), what is the cost of raising the pH in the tank to 7.0?

56. The sulfur in coal can be converted to H_2S, SO_2, or SO_3, or to a mixture of these, depending on how the coal is processed and used. All of these products are acidic when dissolved in water. Calculate the pH values for the resulting solutions if the sulfur in 1.00 kg of coal containing 1.10% sulfur by mass is converted completely to each of the three possible products and then dissolved in 10.0 L of water.

57. The waste water from an industrial process has acetic acid as the principal possible contaminant. As a quick test of the acetic acid concentration, a sample of the water can be tested with a pH indicator. Which indicator would you choose to test whether the acetic acid concentration is about 0.01 M or lower? What color will this indicator be if the acetic acid concentration is too high?

PART TWO
Atomic & Molecular Structure, and Chemical Reactivity

The reverse side of the Nobel Prize medal awarded for
chemistry and physics.

Atomic Structure I:
Theories of the Atom

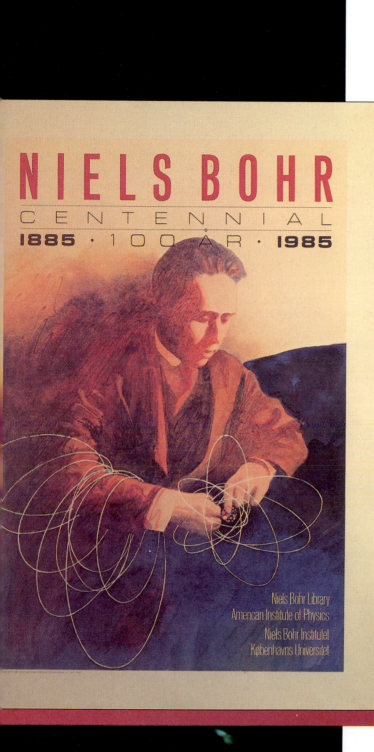

NIELS BOHR

CENTENNIAL

1885 · 100 ÅR · 1985

Niels Bohr Library
American Institute of Physics

Niels Bohr Institutet
Københavns Universitet

A "discontinuous" rainbow of colors is produced by a prism-like
device called a diffraction grating that disperses the 46 distinct
wavelengths put out by a helium-selenium metal vapor ion laser.
Included are visible blues, blue-green, greens, yellows, oranges and
reds, and many invisible infrared wavelengths. The importance of
quantum discontinuity to chemistry is nowhere more evident than in
the work of Niels Bohr, who was only 27 years old when he
published three monumental papers in 1913 on "The Constitution of
Atoms and Molecules." This original painting was created in 1985 in
celebration of the centennial of his birth.

14.1   THE MICROSCOPIC UNIVERSE AND THE DIVISIBLE ATOM

Few things illustrate the progress of scientific thought and practice in chemistry as well as the growth and development of our ideas about atoms and molecules. The concept of an atom as an indivisible unit, of which all matter is built, was considered philosophically sound for 2500 years. The idea of indivisible atoms gained scientific credibility with the formulation of the laws of definite and multiple proportions, and of Dalton's atomic theory. Of particular importance here are the parts of atomic theory that proposed the indivisibility and indestructibility of atoms. Both notions were useful in advancing scientific ideas in the 19th century; but by the 20th century they proved to be inadequate.

The first crack in the notion of the indivisible and indestructible atom came from the work of Michael Faraday on electrically conducting solutions. His electrochemical experiments suggested (1) that current is carried by positively and negatively charged ions; (2) that there is a particle of charge common to all atoms; and (3) that the flow of these carriers of charge is in direct proportion to the numbers of atoms and ions reacting. A subatomic idea had been quietly injected into 19th-century scientific thought when Faraday inferred that each atom is associated with a fixed quantity of electricity.

At the turn of the century, experiments with highly evacuated discharge tubes and radioactive materials led to the discovery that discrete particles are ejected from atoms; alpha rays (helium nuclei, He^{2+}), beta particles (electrons), and gamma rays (high energy x-rays). Ernest Rutherford's famous alpha particle scattering experiments provided compelling evidence for the existence of the nucleus, a positively charged center of mass within the atom, surrounded by negatively charged electrons in sufficient numbers to balance the nuclear charge. The idea of an indivisible atom was gone forever.

Our goal in this chapter is to unify a great many chemical phenomena in terms of a few simple principles. Not only do we have to come to terms with the nuclear atom, but with experiments conducted at the end of the 19th century that undermined the classical theories of energy and how it is transferred among atoms in material systems. In fact, our ideas about the very nature of energy, not just the transfer of energy, needed to be entirely redefined. The idea that energy was continuous, that a system could possess any amount of energy, had to be rejected. That became understandable through the work of a German physicist named Max Planck and a now famous hypothesis that bears his name: Planck's quantum hypothesis.

14.2   ELECTROMAGNETIC RADIATION AND THE QUANTUM HYPOTHESIS

At first glance, radio waves and light waves might appear to be entirely different phenomena. However, that is not the case. Both are manifestations of the same thing, namely electromagnetic waves that differ only in their wavelengths. Electromagnetic waves radiate at the speed c and have a wavelength λ. The speed c divided by the wavelength λ is the number of waves passing any point, per unit time. That is the frequency ν:

$$\nu = \frac{c}{\lambda}$$

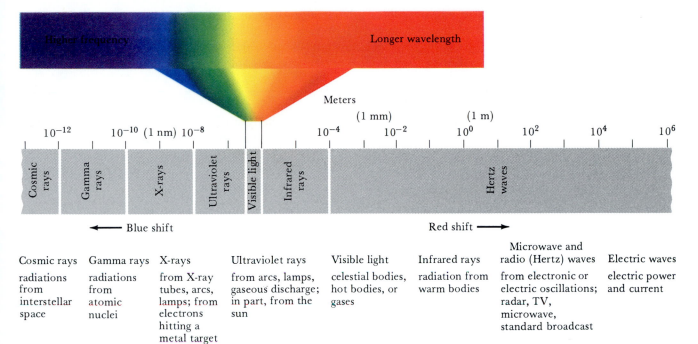

FIGURE 14–1   The range and regions of the electromagnetic spectrum, from the shorter wavelength, high energy X-rays and gamma radiations to the longer wavelength, lower energy infrared and microwave regions, and beyond. The "exploded" region shows the range of visible light; the blue end is higher frequency; the red end is longer wavelength; white light is the combination of all wavelengths.

TABLE 14–1   The Electromagnetic Spectrum

Wavelength*	Common Name
From beyond 10^6 to about 10^4 meters	long electric waves
10^4 to 10^{-2} meters	radio waves
10^0 to 10^{-3} meters	microwaves
10^{-4} meters to 8000 Å	heat waves/infrared radiation
8000 Å to 4000 Å	visible light
4000 Å to 10^{-8} meters	ultraviolet light
10^{-8} to 10^{-10} meters	x-rays
10^{-10} meters and beyond	gamma/cosmic rays

*As a point of reference, keep in mind that $1 \text{ m} = 10^9 \text{ nm} = 10^{10} \text{ Å}$ and that the distances between atoms in molecules are on the order of nm and Å.

Scientists have explored an enormous spectrum of electromagnetic radiations, covering the widest possible range of wavelengths; all move at the same speed of 3.00×10^{10} cm/s as they propagate through vacuum. This constant is known as the speed (or propagation velocity) of light. The units of frequency are reciprocal time:

$$\nu = \frac{c}{\lambda} = \frac{\text{cm/s}}{(\text{cm})} = \frac{1}{\text{s}} = \text{s}^{-1}$$

The frequency unit s^{-1} is called the hertz (Hz).

The light we are capable of seeing is a very small portion of the enormous range of the electromagnetic spectrum (Figure 14–1, Table 14–1). Without the aid of spectroscopic or photographic instruments, the human eye is incapable of detecting radiations that lie outside a narrow band. Our ability to detect red light fails for wavelengths longer than 8000 angstroms ($1\ \text{Å} = 10^{-10}$ m), and our ability to detect blue light fails for wavelengths shorter than 4000 Å.

EXAMPLE 14–1

The frequency of a certain wavelength of light is 6×10^{14} Hz. A change to 5×10^{14} Hz (a lower energy frequency) is referred to as a "red shift" (to a longer wavelength). A "blue shift" (to a shorter wavelength) corresponds to a change toward higher energy. Calculate the corresponding wavelengths λ in angstroms (Å) and nanometers (nm).

Solution   Using the fundamental relationship between ν and λ,

$$\lambda_1 = \frac{c}{\nu_1} = \frac{3.00 \times 10^{10}\ \text{cm/s}}{(6 \times 10^{14}\ \text{Hz})(s^{-1}/\text{Hz})}\ (10^8\ \text{Å/cm}) = 5000\ \text{Å}$$

$$5000\ \text{Å}\left(\frac{1\ \text{nm}}{10\ \text{Å}}\right) = 500\ \text{nm}$$

For the second case, the corresponding calculation yields

$$6000\ \text{Å} = 600\ \text{nm}$$

Both wavelengths lie roughly in the middle of the visible region (4000 to 8000 Å) of the electromagnetic spectrum. ∎

EXERCISE 14–1

Two wavelengths of electromagnetic radiation are respectively 1000 nm and 1000 Å. Which has the higher frequency? In what regions of the electromagnetic spectrum would radiation of these wavelengths be found? *Answer:* 1000 Å; 1000 nm in infrared, 1000 Å in ultraviolet. ∎

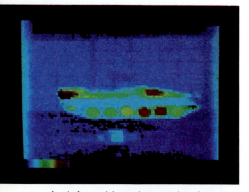

An infrared imaging seeker has produced this thermal image of a tank. Such "thermograms" are used in development of advanced imaging long range heat-seeking anti-tank missiles. Darker colors indicate which parts of the tank are hotter.

Newton correctly surmised that even though nothing more could be seen, the spectra produced by shining light through prisms and gratings hadn't just ended. The ultraviolet region lies before the violet (or blue) end of the visible spectrum, at shorter wavelengths, where radiations can easily be detected by their ability to expose photographic plates. At still shorter wavelengths lie the x-ray region, gamma radiations, and cosmic rays. The infrared region lies beyond the red region, at longer wavelengths, and its radiations can often be detected by their ability to warm thermometers. Infrared radiation is often referred to as heat radiation, since it is emitted by all warm bodies that are not hot enough to be luminous. At still longer wavelengths, we find the microwave and radiofrequency ranges of the electromagnetic spectrum.

In the first year of the 20th century, Planck introduced the then bizarre notion that there is a minimum value of the energy possessed by an oscillator, a vibrating particle, which is proportional to the frequency ν of oscillation:

$$E \propto \nu$$

The constant of proportionality h is known as **Planck's constant**. Thus,

$$E = h\nu$$

and the product $h\nu$ is a package of energy called a **quantum**. Planck further

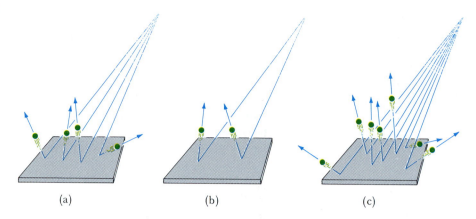

(a)   (b)   (c)

proposed that all transfers of energy between the particle and its surroundings must occur in integer multiples of this quantum; that is, the transfer of energy is *quantized*. This led eventually to the idea that energy itself is quantized, in packages called quanta. In the macroscopic world of everyday experience, we don't see these quanta because they are individually too small. We don't see discreteness in the flow of energy on the scale of cars and trucks colliding, or the transfer of momentum from a bat to a ball. The magnitude of Planck's constant is just too small:

$$h = 6.626 \times 10^{-34}\,\text{J·s}$$

But evidence for this quantization of energy is everywhere on the microscopic scale, when we consider the small energy changes that accompany collisions between atoms and subatomic particles or the excitation of a single atom on absorption of energy. We shall describe two important sets of results from experiments confirming this notion of quantization of energy at the microscopic level; data from photoelectric experiments and from the absorption and emission spectra of atoms.

14.3   THE PHOTOELECTRIC EFFECT

Planck's major pronouncement on quanta, published late in the year 1900, is considered by many scientists to be the opening statement of the modern era in the physical sciences. His equation was subjected to the most careful scrutiny by experimentalists of the day, including Planck himself. A persuasive and profoundly important application of Planck's quanta was introduced by Einstein a few years later as part of his explanation of the photoelectric effect—the ability of various materials to expel free electrons when irradiated by ultraviolet light.

As early as 1887, photoelectric experiments had been reported in the scientific literature in which ultraviolet light of certain wavelengths caused electrons to be expelled from the surface of some metals. Because the phenomenon involved both light and electricity, it was named the **photoelectric effect**. Many careful experiments (Figure 14–2) were performed using monochromatic light—light of a single wavelength—in which the number of expelled electrons and their kinetic energy were measured. The principal results are summarized in the following rules:

1. For incident light of a particular frequency, the number of photoelectrons produced is directly proportional to the intensity of the radiation, while their kinetic energy remains the same regardless of the intensity.

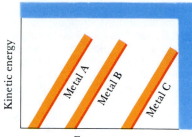

Kinetic energy

Metal A   Metal B   Metal C

Frequency

FIGURE 14–3 Plot of kinetic energy of ejected photoelectrons versus frequency of the illuminating radiation for three different metals. Note the threshold frequencies are all different, but that the slopes for each of the three lines are the same—all three metals give the same value, the Planck constant h.

Robert Millikan, the experimentalist who designed the ingenious oil-drop experiment for determining the charge on the electron, verified the Einstein photoelectric equation (in 1916) with his careful measurements of q_e. See Chapter 3.

2. Only certain frequencies of light falling on particular metals cause a photoelectric response, expelling electrons. For a given metal, only light of frequencies higher than some minimum or threshold value works. Below the threshold frequency, no response is observed no matter what the intensity of the radiation.

3. When radiation of increasing frequency is used, the energy of the expelled electrons increases as the frequency. A graph of kinetic energy (for the expelled electrons) versus frequency is linear. Plots of the kinetic energy versus frequency for different metals (Figure 14–3) give different straight lines, but they all display the same slope.

4. No atoms are torn off the metal surface by the incident light. But particles are being expelled, and they are identifiable as electrons.

5. For a photosensitive metal at any frequency beyond the threshold value, the response is instantaneous. The moment the light is turned on, charge begins to leak off the metal plate, no matter how low the intensity of the light. But when the light source is turned off, the photoelectric response abruptly stops.

Einstein's explanation begins with his now famous photoelectric equation, which encoded all the known experimental data:

$$\text{K.E.} = \tfrac{1}{2} mv^2 = h\nu - W$$

It says that the kinetic energy of an expelled electron is equal to Planck's energy quantum $h\nu$ for the incident light, less the energy W required to separate the electron from the metal surface. W is a characteristic constant for a given metal, sometimes referred to as the **work function**. To state it another way, if an electron is held to a metal surface with a certain energy W, then the quantum of light $h\nu$ required to expel it must have at least this much energy, defining the threshold condition as $h\nu_0 = W$ where ν_0 is the threshold frequency. If the incident light quanta carry more energy than this threshold energy, then the excess appears as the kinetic energy of the (expelled) photoelectrons. Work functions and threshold wavelengths needed to expel electrons from several metals are given in Table 14–2.

TABLE 14–2   Work Functions and Threshold Wavelengths for Several Metals*

Metal	W, eV	λ, nm
cesium (Cs)	1.94	639
rubidium (Rb)	2.13	582
potassium (K)	2.25	551
sodium (Na)	2.29	543
lithium (Li)	2.46	504
zinc (Zn)	4.01	310
copper (Cu)	4.48	277
silver (Ag)	4.52	274
gold (Au)	4.80	—
platinum (Pt)	5.38	231
palladium (Pd)	5.62	221

*It is worth noting the characteristically small work functions for the alkali metal elements in the IA family of the periodic table. We'll try to explain that in the next chapter in our discussion of electronic structure and periodic properties of the elements.

At the lower terminus of any of the lines in Figure 14–3, corresponding to zero kinetic energy, $h\nu$ is equal to the work function for that metal. Data plots of kinetic energy versus frequency that compare different photosensitive metals result in a family of parallel lines with slopes equal to Planck's constant. Because the use of the Einstein equation in this manner consistently reproduced Planck's constant to within a fraction of a percent of the predicted value, it was all but impossible to ignore Einstein's explanation and its essential implication—the quantized nature of light. In his classic paper on the subject, Einstein stated with great confidence that photoelectric behavior can easily be understood. One need only follow Planck's original proposal and assume that electromagnetic radiation propagating through space is composed of individual packets of energy called quanta, or photons. Every successful encounter of a photon with an electron on the metal surface that results in a photoelectron being expelled must have been accompanied by complete transfer of the corresponding energy of the photon to the electron (Figure 14–4).

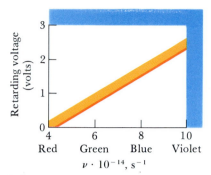

FIGURE 14–4 Graph adapted from Millikan's original paper showing the linear relationship between the opposing (or stopping) potential V_o needed to just prevent the emission of photoelectrons and the frequency of the illuminating light source: R.A. Millikan, *Physical Review*, Vol. 7, 355 (1916).

EXAMPLE 14–2

Given the threshold wavelength for potassium metal in Table 14–2, verify the corresponding value for the work function. Calculate the maximum kinetic energy for the photoelectrons expelled from potassium metal by light from a 500-nm source.

Solution   Using the Planck relationship and the proper conversion factors (in Table 14–3), we find the work function:

$$W = \frac{hc}{\lambda} = \frac{(6.626 \times 10^{-34} \text{ J·s})(3.00 \times 10^{10} \text{ cm/s})}{(551 \text{ nm})(10^{-7} \text{ cm/nm})(1.6 \times 10^{-19} \text{ J/eV})}$$

$$W = 2.25 \text{ eV}$$

To calculate the kinetic energy for the expelled electrons, use the Einstein equation:

$$\text{K.E.} = h\nu - W = \frac{hc}{\lambda} - W$$

$$= \frac{(6.626 \times 10^{-34} \text{ J·s})(3.00 \times 10^{10} \text{ cm/s})}{(500 \text{ nm})(10^{-7} \text{ cm/nm})(1.6 \times 10^{-19} \text{ J/eV})} - W$$

$$\text{K.E.} = 2.48 \text{ eV} - 2.25 \text{ eV} = 0.23 \text{ eV}$$

EXERCISE 14–2(A)

Light of a certain wavelength falls on a metal surface, causing electrons to be expelled.
(a) What happens as the intensity of the incident light increases?
(b) What happens as its wavelength increases?
(c) What happens as its frequency increases?
(d) Compare the photon energy of the incident light to the maximum kinetic energy of the expelled photoelectrons.

EXERCISE 14–2(B)

Determine the wavelength of light that is just capable of expelling electrons from a gold metal surface. In which region of the electromagnetic spectrum does it fall?
Answer: 259 nm; ultraviolet region.

EINSTEIN AND THE PHOTOELECTRIC EFFECT

Continuing Planck's radical departure from classical theory, Albert Einstein (1879–1955) reasoned that if the transfer of energy could be viewed as quantized, why not energy itself? Why not consider light itself to be atomic in character?

Einstein had already noted the experimental fact that some minimum frequency was necessary before one observed ionization of a gas, say in a discharge tube experiment. He also noted that light emitted by a fluorescent substance always turned out to be of a lower frequency than the incident light causing it to fluoresce, two results that (in retrospect) are clearly suggestive of quantization and discrete, rather than continuous, properties. But it was Einstein's explanation of the photoelectric effect that changed our way of thinking about the nature of light.

Einstein was recognized for this work by receipt of the Nobel Prize for 1921. His lifelong friend and colleague, Max Planck, had been so honored three years earlier. The general acceptance of these two theories, particularly that of Einstein, took the better part of two decades. This suggests how ingrained the classical ideas about the wave nature of light were in the scientific consciousness of the day. Albert Einstein was 26 years old at the time, a scientific unknown, working days at the Swiss patent office in Berne, supporting his wife and child, and doing theoretical physics (at home in the evenings) of such proportions that it literally shook the world. Besides the explanation of the photoelectric effect, he wrote critical papers on Brownian motion, the specific heats of solids, and the special theory of relativity, which made him famous for all time.

(Left) Einstein at age 26, newly married and newly employed by the Swiss patent office in Berne, Switzerland. This was about the time he was seriously considering what became known as the photoelectric effect; (right) Much later in life, and in a different part of the world, Einstein was photographed in his sailboat on Long Island Sound.

14.4 QUANTUM THEORY AND SPECTROSCOPY

Energy States and Spectral Lines

Because Einstein so successfully used Planck's hypothesis in his explanation of the photoelectric effect, scientists began to pay attention to the quantum concept in other areas involving optical phenomena. The product of this

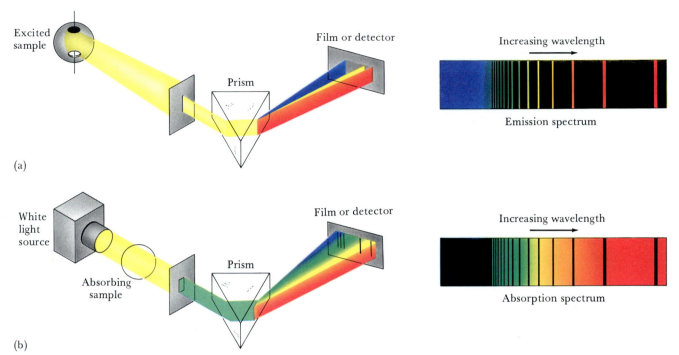

FIGURE 14–5 A hot solid emits all wavelengths in an overlapping, continuous spectrum that, depending on the temperature, may be visibly dull red to incandescent, white hot: (a) Discrete sets of lines referred to as "line spectra," or discontinuous spectra, are produced when a gas of excited atoms (of any element) emits light as the atoms fall to lower energy states. On passing through a prism, the emission lines are separated and the characteristic "fingerprint," an emission spectrum, is displayed. (b) When white light from an incandescent source is passed through a gas of unexcited (or ground state) atoms, certain characteristic wavelengths are absorbed. The transmitted light presents a rainbow spectrum with certain wavelengths missing, producing a second kind of "fingerprint," an absorption spectrum.

activity was a primary link between our understanding of the electronic structure and composition of atoms and the time-honored study of spectroscopy. That in turn led to Niels Bohr's explanation of atomic spectra and his quantized model of the atom.

Our most important sources of information at the atomic and molecular level have been optical spectra in the visible, infrared, ultraviolet, x-ray, microwave, and radiofrequency spectral regions. Optical spectra can be classified according to whether they are characteristically continuous, band, or line spectra. **Continuous spectra** are emitted by radiating solids and high-density gases. **Band spectra** are composed of large numbers of fine spectral lines that happen to be very close to each other and are generally associated with molecules. **Line spectra**, consisting mostly of single lines that can often be ordered into characteristic series, are typical of single atoms, usually in gases. Optical spectra can be observed by either absorption or emission (Figure 14–5). Table 14–3 lists some typical spectral quantities and conversion factors you are likely to encounter.

The existence of line spectra was explained by the radical notion that light is emitted or absorbed as discrete packets of radiant energy—as Planck's light quanta. When a quantum of radiant energy (a photon) is absorbed or emitted by an atom, a transition between two atomic states of different energies must take place. According to the general principles of energy conservation, the energy gained by the atom in a quantum jump to an

TABLE 14–3 **Quantities and Units for Spectroscopy**

Quantity	Unit/Conversion Factor
wavelength, λ	$1\ \text{Å} = 10^{-10}\ \text{m} = 10^{-8}\ \text{cm} = 10^{-1}\ \text{nm}$
	One wavelength standard is the yellow ^{86}Kr line, for which $\lambda = 605.78211\ \text{nm}$
wavenumber, $\bar{\nu}$	$\nu/c = 1/\lambda$; units of cm^{-1}
frequency, ν	The units are reciprocal time (s^{-1}) or Hertz (Hz, equivalent to cycles/s).
	Note that the product of frequency and wavelength is a universal constant, the propagation velocity of light: $\lambda\nu = c$.
energy, E	The electron-volt is often used as the energy unit: $1\ \text{eV} = 1.602 \times 10^{-19}\ \text{J} = 1.602 \times 10^{-12}\ \text{erg}$ $E = h\nu = hc/\lambda$
electron mass, m_e	$9.109 \times 10^{-28}\ \text{g} = 9.109 \times 10^{-31}\ \text{kg}$
electron charge, q_e	$1.602 \times 10^{-19}\ \text{C} = 4.80 \times 10^{-10}\ \text{esu}$
Planck constant, h	$6.626 \times 10^{-27}\ \text{erg·s} = 6.626 \times 10^{-34}\ \text{J·s}$

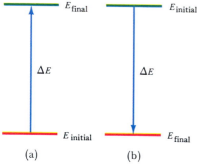

FIGURE 14–6 The change in energy levels for (a) radiation absorption and for (b) radiation emission.

excited state is equal to the energy of the absorbed photon. The reverse is also true, that the photon energy of emitted radiation equals the energy lost by the atom in the quantum relaxation to the lower energy state (Figure 14–6). In terms of Planck's photon hypothesis,

$$\Delta E = E_{\text{final}} - E_{\text{initial}} = h\nu$$

$$\nu = \frac{E_{\text{final}} - E_{\text{initial}}}{h} = \frac{\Delta E}{h}$$

The E terms are the initial and final energy states for the atom; h is Planck's constant; and ν is the photon frequency. When ΔE is negative, the frequency corresponds to emission; a positive value corresponds to absorption. In general, the energy is reported as an absolute value, and the sign is adjusted according to whether the process is emission or absorption. Finally, the equation gives ΔE for a single atom. To obtain energies on a per mole basis, multiply ΔE by the Avogadro number.

EXAMPLE 14–3

Calculate the energy associated with red light at 600 nm, in the middle of the visible spectrum.

Solution

On a mole basis,

$$\Delta E = N_A h\nu = \frac{N_A hc}{\lambda}$$

$$= \frac{(6.022 \times 10^{23}\ \text{mol}^{-1})(6.626 \times 10^{-34}\ \text{J·s})(3.00 \times 10^{10}\ \text{cm/s})}{(600\ \text{nm})(10^{-7}\ \text{cm/nm})}$$

$$\Delta E = 2.00 \times 10^5\ \text{J/mol}$$

EXERCISE 14–3

Calculate the energy in eV per atom and kJ/mol for the standard yellow line in the optical spectrum of ^{86}Kr. See Table 14–3 for the necessary wavelength for emission line of the krypton isotope.

Answer: $\Delta E = 2.05\ \text{eV}$; $\Delta E = 198\ \text{kJ/mol}$.

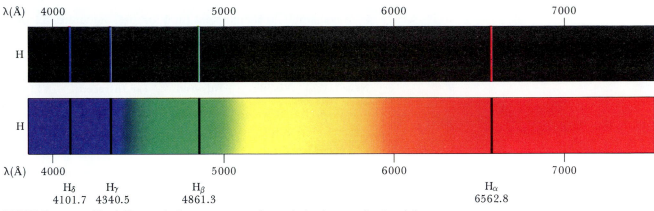

FIGURE 14–7 (Top) The emission spectrum of atomic hydrogen; (bottom) the absorption spectrum.

The Optical Spectrum of Atomic Hydrogen

Of all the optical spectra, it is the line spectrum of atomic hydrogen that has proved most important to our understanding of atomic theory. Robert Bunsen and Gustav Kirchhoff were first to discover that each element has its own characteristic spectrum. Hydrogen is the lightest element, and the simplest, consisting of a proton nucleus and an electron. The emission spectrum of atomic hydrogen shows three lines in the visible region at 656.3, 486.1, and 434.0 nm. They are referred to as the $H_{(\alpha,\beta,\gamma)}$ lines, with H_α (discovered by Anders Ångström in 1853) being the most intense. These three lines are followed by an entire series of lines in the near ultraviolet that fall closer to each other as they approach a short wavelength limit H_∞ (Figure 14–7). Johann Balmer (1855) found that the wavelengths of these lines could be accurately predicted by an empirical relationship. Johannes Rydberg (1899) found Balmer's relationship to be a special case of a more general equation that could be used to represent the entire spectrum for atomic hydrogen:

$$\bar{\nu} = \frac{\nu}{c} = \frac{1}{\lambda} = R_H \left[\frac{1}{n_1^2} - \frac{1}{n_2^2} \right]$$

The **Rydberg equation** states that the frequency of any line in a particular spectral series for atomic hydrogen can be represented as the difference between two terms. R_H is called the Rydberg constant and has the value 109,677.58 cm^{-1}; n_1 and n_2 are dimensionless integers (we now call them principal quantum numbers) where $n_1 < n_2$. For the Balmer lines, $n_1 = 2$; wavelengths for successive lines are found by setting $n_2 = 3, 4, 5$, and so on. Comparing calculated with observed values for the visible spectral lines (Table 14–4) shows that this empirical relationship is more than just good. It describes the visible series of hydrogen lines with great precision, and predicts the wavelengths in other spectral series for atomic hydrogen with equal success.

Bunsen and Kirchhoff, considered to be the founding fathers of spectroscopy, proved that the frequencies of spectral lines depend upon the composition of the material body from which they emanate.

EXAMPLE 14–4

Calculate λ for the signature lines for atomic hydrogen in the visible, or Balmer, series: $n_1 = 2$ and $n_2 = 3, 4$, and 5, respectively.

Solution Substituting the appropriate n values into the Rydberg equation gives the red line at $\lambda = 656.3$ nm:

TABLE 14–4   Wavelengths in Nanometers (nm) for the "Signature" Lines in the Spectral Series for Atomic Hydrogen

$n =$	1 Lyman Near UV	2 Balmer Visible	3 Paschen Infrared	4 Brackett Far Infrared	5 Pfund* VFIR
$n_2 = 2$	121.6	—	—	—	—
3	102.6	656.3	—	—	—
4	97.3	486.1	1875.1	—	—
5	95.0	434.0	1281.8	4050.0	—

*The wavelength for the first line in the series beginning with $n_1 = 5$ (discovered in 1924 by Pfund) is 7400 nm; VFIR = very far infrared.

$$\frac{1}{\lambda} = R_H \left[\frac{1}{n_1^2} - \frac{1}{n_2^2} \right] = 109{,}677.58 \text{ cm}^{-1} \left[\frac{1}{2^2} - \frac{1}{3^2} \right]$$

$$\lambda = 656.5 \text{ nm}$$

For the blue-green line at 486.1 nm, $n_1 = 2$ and $n_2 = 4$:

$$\frac{1}{\lambda} = 109{,}677.58 \text{ cm}^{-1} \left[\frac{1}{2^2} - \frac{1}{4^2} \right]$$

$$\lambda = 486.3 \text{ nm}$$

For the blue line at 434.0 nm, $n_1 = 2$ and $n_2 = 5$:

$$\frac{1}{\lambda} = 109{,}677.58 \text{ cm}^{-1} \left[\frac{1}{2^2} - \frac{1}{5^2} \right]$$

$$\lambda = 434.2 \text{ nm}$$

There are more than three lines in the Balmer series. These just happen to be the most pronounced, or signature lines. But they all fit the Rydberg equation with great precision. The spectral series in the lower energy infrared region, known as the Paschen series, is described by $n_1 = 3$ and $n_2 = 4, 5, 6$, and so on. The spectral series in the more energetic near-ultraviolet region is described by $n_1 = 1$ and $n_2 = 2, 3, 4, \ldots$. All told, there are five spectral series for hydrogen. All follow the Rydberg equation, from which we can conclude that the frequencies of the spectral lines can be represented as differences in two terms of the form R_H/n^2. These are the energy levels, or "allowed" states, for the electron in the hydrogen atom. The spectral lines representing the differences between allowed states have been carefully documented in the form of energy level diagrams.

EXERCISE 14–4(A)
Evaluate the energy associated with each of the signature lines for the near ultraviolet (Lyman) spectral series for atomic hydrogen. Record your answers alternately in ergs for the first, electron-volts for the second, kilojoules for the third, and finally cycles per second (Hz) for the fourth line.
Answer: 1.63×10^{-11} erg (first line); 12.1 eV (second line); 2.04×10^{-21} kJ (third line); 3.16×10^{15} Hz (fourth line).

ORIGINS OF MODERN SPECTROSCOPY

Robert Bunsen (German chemist, 1811–1899) and Gustav Kirchhoff (German physicist, 1824–1887) are considered to be the founding fathers of spectroscopy. They discovered that the frequencies of spectral lines depend upon the composition of the material body from which they emanate, to the point of being "fingerprints." For example, two distinct yellow lines at $\lambda = 5890$ nm and $\lambda = 5896$ nm known as the D lines, when observed in the emission spectrum of a "Bunsen" flame, definitively establish the presence of sodium atoms. Fifty years before, Johann von Fraunhofer (German physicist, 1787–1826) had observed these same two lines as dark lines, along with hundreds of others superimposed upon the continuous rainbow produced by a prism spectroscope as it dispersed light from the Sun. He lettered the most prominent lines, including the bright yellow D lines in the emission spectrum of sodium vapor. But what did it all mean? Fraunhofer had died before his fortieth birthday of tuberculosis, leaving the analysis to his famous successors, Kirchhoff and Bunsen. They reported detecting spectral lines characteristic of strontium and barium while observing a raging fire in Mannheim from their laboratory some distance away in Heidelberg. If one could analyze fires here on Earth this way, wouldn't it be possible to analyze the furnace of the Sun? If sunlight possessed the sodium D lines, then sunlight must have passed through sodium vapor on the way here, and the only reasonable place where that sodium vapor could exist was in the Sun's atmosphere. Armed with that hypothesis, they identified half a dozen elements in the Sun.

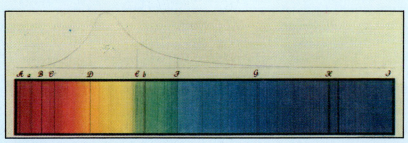

There are actually close to 600 distinct, dark lines in the visible region of Fraunhofer's original solar spectrum of the Sun. In particular, he identified the strongest lines in the red (A), orange (B and C), yellow (D), green (E), blue (F and G) and violet (H and I) regions.

EXERCISE 14–4(B)

Identify the n states for the first line in the very long wavelength Pfund series of spectral lines. Calculate the energy for the second line in the series in kJ/mol.

Answer: First line: $n_1 = 5$, $n_2 = 6$. Second line: $E = 25.7$ kJ/mol. ∎

14.5   THE BOHR HYDROGEN ATOM

Quantum Restrictions and the Rutherford Model

Niels Bohr was the first to apply the new quantum theory successfully to the problem of atomic structure and the frequencies of the spectral lines. While the Rydberg equation precisely predicted the emisson spectrum for

Niels Bohr (Danish, 1885–1962) became a Nobel Laureate (1922) for the quantum model of the atom he proposed as a student, working with Rutherford in Manchester. He was Rutherford's most famous student and one of the towering giants of 20th-century scientific thought.

atomic hydrogen, it offered little insight into the theoretical basis for the relationship. The Rydberg equation was an empirical relationship in need of explanation. The question of what structure to assign the nuclear atom was profoundly more difficult. Following Rutherford's lead, Bohr assumed that the electrons orbit about the nucleus as planets do the Sun. That is a plausible arrangement for a mechanical system involving large bodies in motion, held in place by a gravitational force, but it is hardly appropriate for an electrodynamical system of oppositely charged particles. Maxwell's electromagnetic theory of light, which had been thoroughly confirmed for macroscopic states, predicted that a charged planetary particle such as an orbiting electron would continuously accelerate toward the nucleus, radiating energy all the while, and eventually self-destruct as it crashed into the nucleus. Rutherford's model and Maxwell's theory predicted atomic collapse in an instant. Yet the facts are otherwise—the stability of the hydrogen atom is well known.

Bohr started with the fundamental idea that planetary electrons surround and move about an oppositely charged nucleus. In his system the coulombic force was just that required for circular motion of the electron as Rutherford had suggested. But now the difference! Taking into account the sequential regularity observed in the spectral emission lines for atomic hydrogen, Bohr imposed Planck's quantum restriction on the system. He proposed that only certain orbits are stable, or "allowed." Here are the requirements (or postulates) for the behavior of electrons in atoms according to the Bohr theory:

1. **Quantization of orbital angular momentum.** Only those orbits (or states) for planetary electrons are allowed for which the angular momentum is a whole number multiple n of $h/2\pi$.
2. **Radiationless motion for electrons.** As long as a planetary electron stays in its current orbit, the atom does not absorb or emit energy.
3. **Atomic spectra.** Emission of light occurs only during spontaneous (or stimulated) transitions between allowed orbits (or states).

Bohr was not discarding Newtonian mechanics. The classical equations of motion were entirely valid for electrons in atoms; but only certain orbits, and therefore only certain discrete energies, were to be allowed. These are the energy levels or energy states of the atom. Spectral lines result from transitions from one such allowed energy level to another, and the energies can be calculated according to the Planck equation:

$$\Delta E = E_{\text{final}} - E_{\text{initial}} = h\nu$$

Furthermore, it is interesting to note that Bohr made no statement or assumption about the processes that might be involved. He was concerned only with the energy states. Nothing is assumed about the electron as a function of time, only its initial and final states and the interpretation of the line spectra that herald transitions between states (Figure 14–8).

Quantitative Features of the Bohr Theory

Assume a hydrogen atom to be composed of a proton nucleus and a lone electron orbiting at a distance. The nucleus has a positive charge Ze^+. Around it an electron with negative charge e^-, mass m_e, and velocity v orbits at a radial distance r (Figure 14–9). For the electron's orbit about

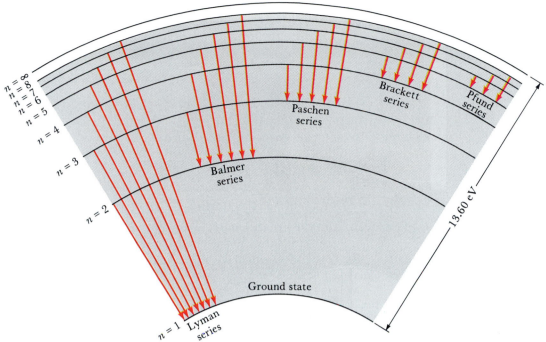

FIGURE 14–8   Energy level diagram for the hydrogen atom, with some of the predicted transitions between energy states. There are five principal sets of lines, beginning with the Lyman series in the ultraviolet and moving out to the Pfund series in the very far infrared.

the nucleus to be stable, the coulombic force (of attraction) between the nucleus (Ze^+) and electron (e^-) must be exactly the force required to cause circular motion of the orbiting electron:

coulomb force = force for circular motion

$$\frac{(Ze^+)(e^-)}{r^2} = \frac{Ze^2}{r^2} = \frac{m_e v^2}{r}$$

Solving for the radial distance r separating the electron charge and the nuclear charge:

$$r = \frac{Ze^2}{m_e v^2} \qquad (14\text{–}1)$$

Bohr quantized the angular momentum for the orbiting electron, dividing it into n integral packets of $h/2\pi$ units:

$$m_e v r = n\left(\frac{h}{2\pi}\right)$$

Solving for the velocity v of the orbiting electron:

$$v = \frac{nh}{2\pi m_e r}$$

Substituting the value for v into Eq. (14–1) for r:

$$r = \frac{Ze^2}{m_e(nh/2\pi m_e r)^2}$$

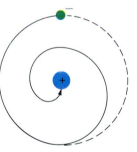

FIGURE 14–9   The hydrogen atom would presumably collapse if the circular motion of the electron about the oppositely charged nucleus was not balanced by the coulombic interaction between the two.

This simple equation is based upon Coulomb's Law, $F = q_1 q_2/r^2$, where F is in dynes or g·cm/s², r is in cm, and q is in electrostatic units (esu). Since an esu is very small, a much larger unit of charge known as the coulomb, equal to approximately 3×10^9 esu, is usually used.

$$r = \frac{n^2 h^2}{4\pi^2 m_e e^2 Z}$$

For the hydrogen atom, $Z = 1$ and the equation becomes

$$r = \frac{n^2 h^2}{4\pi^2 m_e e^2}$$

Finally, regrouping terms gives the result for hydrogen:

Keep in mind that an erg is an g·cm²/s², and therefore the units are consistent with the electrostatic unit (esu) of charge.

$$r = \left[\frac{h^2}{4\pi^2 m_e e^2}\right] n^2 = a_0 n^2 \qquad (14\text{–}2)$$

where a_0 equals the collected constants $[h^2/4\pi^2 m_e e^2]$, which is the radius of the hydrogen atom in the **ground state**, where $n = 1$.

To calculate a_0 in cm, substitute the appropriate values for Planck's constant h in erg·s, the electron mass m_e in g, and the charge e on the electron in electrostatic units (esu):

$$r = \frac{(6.626 \times 10^{-27} \text{ erg·s})^2}{4\pi^2 (9.109 \times 10^{-28} \text{ g})(4.803 \times 10^{-10} \text{ esu})^2}$$

$$r = (5.29 \times 10^{-9} \text{ cm})(10^8 \text{ Å/cm}) = 0.529 \text{ Å}$$

Expressing the electron charge in esu is consistent with energy units in ergs and results in units of length in cm. On the atomic scale, it has been traditional to convert units of length to angstroms (or nanometers).

EXAMPLE 14–5

Calculate the speed of an electron in the first Bohr orbit for atomic hydrogen.

Solution   According to the Bohr theory, $mvr = nh/2\pi$, or

$$v = \frac{nh}{2\pi mr} = \frac{(1)(6.626 \times 10^{-27} \text{ erg·s})}{2\pi(9.109 \times 10^{-28} \text{ g})(5.29 \times 10^{-9} \text{ cm})}$$

$$v = 2.19 \times 10^8 \text{ cm/s}$$

EXERCISE 14–5

By what factor is the speed of an electron in the second Bohr orbit different from that in the first?
Answer: 0.500.

Having defined the basic unit of length on the atomic scale as the radius for the ground state hydrogen atom, Bohr turned his attention to the energy states themselves. How did this model explain the empirical Rydberg equation and the discrete lines in the emission spectrum of atomic hydrogen? The electron in its orbit has a total energy equal to the sum of the kinetic energy and the potential energy:

$$E_{\text{total}} = \text{K.E.} + \text{P.E.}$$

But the kinetic energy is $\frac{1}{2} mv^2$. The potential energy, due to the electrostatic attraction between oppositely charged nuclei and electrons, is $(Ze^+)(e^-)/r$. Therefore,

$$E_{total} = \tfrac{1}{2} mv^2 + \frac{(Z)(+e)(-e)}{r} = \tfrac{1}{2} mv^2 - \frac{Ze^2}{r}$$

Since $mv^2 = Ze^2/r$, the equation for the total energy can be rewritten:

$$E_{total} = \tfrac{1}{2} \left[\frac{Ze^2}{r}\right] - \frac{Ze^2}{r} = -\tfrac{1}{2}\left[\frac{Ze^2}{r}\right] = -\frac{Ze^2}{2r}$$

Finally, substitute the value for the Bohr radius r into the equation for the total energy:

$$E_{total} = -\frac{2\pi^2 m_e e^4 Z^2}{n^2 h^2}$$

When $Z = 1$, the equation gives the total energy for the hydrogen atom. The **principal quantum number** n is an integer. For a given energy level

$$E_{total} \propto \frac{1}{n^2}$$

Therefore, the level for which $n = 1$ is four times lower in energy than the $n = 2$ state, nine times lower in energy than the $n = 3$ state, sixteen times lower than the $n = 4$ state, and so forth. The transition between states is discrete and abrupt.

The discrete nature of the emission lines in the optical spectrum for atomic hydrogen and the Rydberg relationship follow directly from the equation for E_{total}. Consider an electron in a higher energy n_2 state changing to a lower energy n_1 state:

$$E(n_2) = -\frac{2\pi^2 m_e e^4 Z^2}{n_2^2 h^2}$$

$$E(n_1) = -\frac{2\pi^2 m_e e^4 Z^2}{n_1^2 h^2}$$

The photon energy for the emission line is given by the Planck equation:

$$\Delta E = h\nu = E(n_1) - E(n_2)$$

$$\Delta E = -\frac{2\pi^2 m_e e^4 Z^2}{n_1^2 h^2} - \left[-\frac{2\pi^2 m_e e^4 Z^2}{n_2^2 h^2}\right]$$

$$\Delta E = \frac{2\pi^2 m_e e^4 Z^2}{h^2}\left[\frac{1}{n_2^2} - \frac{1}{n_1^2}\right]$$

Solving for the frequency term in the equation $\Delta E = h\nu$ and substituting for ΔE,

$$\nu = \frac{2\pi^2 m_e e^4 Z^2}{h^3}\left[\frac{1}{n_2^2} - \frac{1}{n_1^2}\right]$$

Since $\nu = c/\lambda$ and $Z = 1$ for H, we can rewrite the equation in its more familiar form:

$$\frac{1}{\lambda} = \frac{2\pi^2 m_e e^4}{h^3 c}\left[\frac{1}{n_2^2} - \frac{1}{n_1^2}\right] = \bar{\nu} = \frac{\nu}{c}$$

$$\bar{\nu} = R_H\left[\frac{1}{n_2^2} - \frac{1}{n_1^2}\right] \tag{14-3}$$

The Bohr theory allows us to derive the empirical Rydberg equation that predicted the emission lines for atomic hydrogen with such precision. Calculated values for the proportionality constant R_H from known values of m_e, e, h, and c come within 0.1% of the value obtained from experiment and are in good agreement with the value used today, namely 109,677.58 cm^{-1}:

$$R_H = \frac{2\pi^2 m_e e^4}{h^3 c}$$

$$R_H = \frac{(2)\pi^2(9.107 \times 10^{-28} \text{ g})(4.80 \times 10^{-10} \text{ esu})^4}{(6.626 \times 10^{-27} \text{ erg·s})^3(3.00 \times 10^{10} \text{ cm/s})}$$

$$= 1.09 \times 10^5 \text{ cm}^{-1}$$

Furthermore, the optical spectrum for atomic hydrogen was entirely consistent with Bohr's ideas about quantum restrictions, energy states, and the Rydberg equation.

EXAMPLE 14–6

For any particular atom, the ground state (or lowest-lying energy state) is assigned the principal quantum number $n = 1$. Emission lines in the Lyman series of hydrogen correspond to transitions between $n = 2$ and $n = 1$, between $n = 3$ and $n = 1$, between $n = 4$ and $n = 1$, and so forth. Calculate the energy for the limiting value.

Solution   The limiting value in the emission spectrum corresponds to a transition from $n = \infty$ to the ground state ($n = 1$). The calculated energy per atom is

$$\Delta E = h\nu = \frac{hc}{\lambda} = hc \left(\frac{1}{\lambda}\right)$$

$$= hcR_H \left[\frac{1}{1^2} - \frac{1}{\infty^2}\right] = hcR_H$$

$$= \frac{(6.626 \times 10^{-27} \text{ erg·s})(3.00 \times 10^{10} \text{ cm/s})(1.09 \times 10^5 \text{ cm}^{-1})}{1.602 \times 10^{-12} \text{ erg/eV}}$$

$$= 13.6 \text{ eV}$$

On a per mole basis, the energy is

$$\Delta E = (13.6 \text{ eV})(6.022 \times 10^{23} \text{ mol}^{-1})(1.602 \times 10^{-22} \text{ kJ/eV})$$

$$= 1.31 \times 10^3 \text{ kJ/mol}$$

For the reverse case (absorption rather than emission), the limiting value is the energy required to remove the electron from any influence due to the proton nucleus. This is the **ionization energy** (ΔE_{IE}) of the H atom:

$$H \longrightarrow H^+ + e^-$$

EXERCISE 14–6(A)

The ionization energy for the hydrogen atom corresponds to the Rydberg limit for the Lyman series of spectral lines:

$$(n_1 = 1) \longrightarrow (n_2 = \infty)$$

Calculate ΔE_{IE} in eV/atom and kJ/mol.
Answer: 13.6 eV/atom; 1.31×10^3 kJ/mol.

EXERCISE 14–6(B)
Beginning with the Rydberg equation, show that the ionization energy for the hydrogen atom is equal to $e^2/2a_0$.   ■

Experimental Confirmation of the Bohr Theory

The Bohr theory was an instant success because it was possible to derive the empirical Rydberg equation from theoretical principles, precisely reproducing the values for the emission lines in the spectrum of atomic hydrogen. But the hard question remained. Could one show from experiment that discrete quantum states exist in atoms? Could it be shown experimentally that there are gaps between energy levels within an atom?

In the **Franck-Hertz experiment**, heavy atoms were bombarded with electrons, and the energy lost by the electrons when they collided with these atoms was measured. A chamber filled with mercury vapor at low pressure served as a source of heavy atoms. The velocity of the electrons, and therefore the kinetic energy of the electrons, was measured as they left the source and immediately after passing through the gas of mercury atoms. Under these conditions, it was assumed that any lost energy would be due to collisions with mercury atoms (Figure 14–10a).

When the kinetic energy of the electrons emerging from the gun was small, little difference in energy could be detected after passing through the fog of mercury atoms. That is expected, since a mercury atom is about as likely to be moved by an electron striking it as a bowling ball by a pellet from a BB-gun. The electrons just bounce off after an elastic collision. However, when the kinetic energy of the electrons is raised beyond a certain threshold value, a sudden and quite dramatic change is observed. Below 3.4 eV nothing happened, but at 3.5 eV and beyond, the electrons slowed significantly, losing a major chunk of their kinetic energy; the K.E. of the

FIGURE 14–10   (a) The Franck-Hertz experiment established the existence of "gaps" between the energy states within atoms. (b) Data from a repeat of the original experiment, which showed that the electrons lost only energies amounting to multiples (because in this experiment, several consecutive collisions could occur) of 5 eV, indicating a discrete excitation energy of 5 eV in mercury.

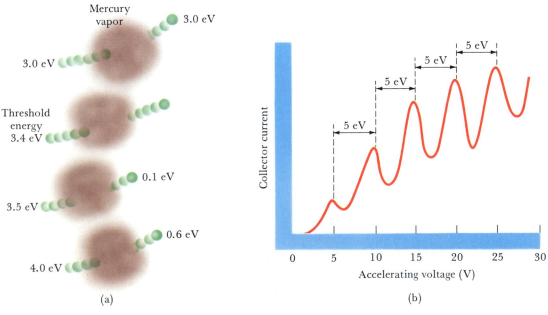

(a)

(b)

exiting electrons dropped to 0.1 eV. If 4.0-eV electrons were directed at the gas of mercury atoms, they still lost 3.4 eV and the beam of electrons exited with a kinetic energy of 0.6 eV. At 4.5 eV, the beam of electrons exited with 1.1 eV, 3.4 eV of their entering kinetic energy having been absorbed by the mercury atoms.

The clear implication from the experiment was that mercury atoms could not accept less or more than 3.4 eV. If more was available, the mercury atoms simply took the 3.4 eV, discarding the rest; when less was available, no energy was subtracted from the electron beam at all. Furthermore, it seemed reasonable to assume that the threshold 3.4 eV was being added to the internal energy of the atoms, since it was unlikely that a particle the size of an electron could alter the translational energy of giant-sized atoms of elements such as mercury. The conclusion was that mercury must have an energy state 3.4 eV above the ground state energy and that there is no stable state between them.

Finally, if the Bohr theory was to be believed, the extra energy absorbed by the mercury atoms must be emitted as photons whose frequency and wavelength correspond to an energy of 3.4 eV. In the emission spectrum of hot mercury vapor there is indeed a line at 3650 Å, or 3.4 eV. Franck and Hertz looked for that spectral line in their experiment and found it, concluding that excited mercury atoms emitted 3650 Å photons as they relaxed back to their ground state energy.

More experiments were subsequently carried out, and an abrupt absorption of kinetic energy was also observed at 4.86 eV. As expected, the corresponding emission line in the spectrum of hot mercury vapor was observed at 2560 Å. Energy gaps existed. That had been shown experimentally, just as Bohr had predicted (Figure 14–10b).

EXAMPLE 14–7

How much kinetic energy will a beam of electrons still possess after passing through a chamber of mercury vapor if they left the electron gun with 3.0 eV? If they left with 4.0 eV, by what factor did the velocity of the electrons change?

Solution   To begin with, 3.0 eV is less than the minimum value of 3.4 eV for mercury, so no energy can be lost. Electrons leaving still possess the 3.0 eV they entered with. However, at 4.0 eV, they lose all but

$$(4.0 - 3.4)\,\text{eV} = 0.6\,\text{eV}$$

Finally, we can calculate the factor by which the speed of the electrons changed:

$$\tfrac{1}{2}\,m_1 v_1^{\,2} = \tfrac{1}{2}\,m_2 v_2^{\,2}$$

$$\frac{m_f v_f^{\,2}}{m_i v_i^{\,2}} = \frac{0.6\,\text{eV}}{4.0\,\text{eV}}$$

$$\frac{v_f}{v_i} = 0.4$$

EXERCISE 14–7(A)

Using the Planck equation, show that a photon energy of 3.4 eV corresponds to a radiation wavelength of 3650 Å. In what region of the electromagnetic spectrum does this wavelength fall?
Answer: Ultraviolet.

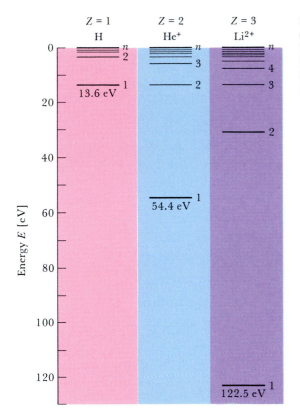

FIGURE 14–11 Selected energy states for H and for the hydrogen-like He$^+$ and Li^{2+} ions. Actual energies are negative; absolute energies are given here.

EXERCISE 14–7(B)

There is an emission line in the spectrum of hot mercury vapor at 1850 Å. What energy (in eV/atom) does that correspond to? In what region of the electromagnetic spectrum will you find this transition?
Answer: 6.72 eV; ultraviolet. ■

Hydrogen-Like Atoms

If the Bohr theory is correct, then the spectrum for any system composed of a nucleus and one electron should be the same as that of hydrogen, except for the Z^2 term and the (Rydberg) constant of proportionality. Shouldn't the logic and equations that predict the spectrum for H explain equally well the spectra for He$^+$, Li^{2+}, or any other nucleus with one electron? They should, and they do (Figure 14–11). For example, astronomers have been able to identify a **Pickering** series of spectral lines for He$^+$ in the solar spectrum. The Pickering lines can be obtained from the Rydberg equation in the form

$$\bar{\nu} = Z^2 R_H \left[\frac{1}{n_1^2} - \frac{1}{n_2^2} \right]$$

$$= 2^2 R_H \left[\frac{1}{4^2} - \frac{1}{n_2^2} \right]$$

$$= 4 R_H \left[\frac{1}{4^2} - \frac{1}{n_2^2} \right]$$

TABLE 14–5	**The Signature Lines in the Balmer Series for H and in the Pickering Series for He⁺, Measured in Å**						
H	6563 (H_α)	—	4861 (H_β)	—	4340 (H_γ)	—
He⁺	6560	5412	4859	4562	4339	4200

Data tabulated in Table 14–5 provide a comparison of the Balmer lines for H and the Pickering lines for He⁺. Note how every other Pickering line corresponds to a Balmer line, just as the Bohr theory equation predicts if we rewrite the series equation in the form

$$\bar{\nu} = R_H \left[\frac{1}{2^2} - \frac{1}{(n_2/2)^2} \right]$$

The Bohr Theory and Beyond

For all its success, the Bohr theory was not the last word. In describing the spectrum of even the simplest atom, the basic assumption of the Bohr theory that each spectral line is a single entity proved false. As closer examination of spectral lines became possible, it was observed that many of the boldest lines were in fact multiple lines close together. This **fine structure** among the spectral lines in atomic hydrogen turned out to be a common phenomenon. The Bohr theory successfully accounted for the lines in the hydrogen spectrum, but it could not account for their fine structure.

Others tried to bend, stretch, and patch the theory. Sommerfeld and Bohr jointly suggested that the existence of fine structure might mean that energy levels are composed of sublevels of slightly different energy. Therefore, any transition from an upper to a lower level would lead to a series of close lines representing the various sublevels the electron could fall into. Sommerfeld also introduced the idea that perhaps these discrete energy levels could be represented by eccentric ellipses as well as concentric circles. That is to say, an electron moving around a nucleus that does not happen to be at the atom's geometric center may trace an elliptical path. It all proved to be little more than a temporary fix, as matters continued to get ever more complex. There was the monumental problem of accounting for still further splitting of spectral lines into multiplets in atoms with two or more electrons, such as helium and lithium.

The Bohr theory turned out to be a curious hybrid of classical and quantum theory, and it has monumental significance as a bridge between the two. It took more than a decade to sort out the theory and to establish the basis for our contemporary quantum mechanical model of the structure of the atom. The legacy of the Bohr atom that remains in contemporary atomic theory is composed of four key points:

1. Stationary states. Bohr suggested that at any given time, electrons in atoms are in various states of motion, that each state of motion is distinct, and that each state can be identified by a definite energy. To put that another way, the energy in an atom is a quantized variable, limited to certain discrete values.

2. Quantum jumps. Bohr suggested that atoms are subject to sudden transitions, or quantum jumps, between stationary states (or energy levels). Here was the reduction to practice of Planck's and Einstein's ideas about the

Arnold Sommerfeld (German, 1868–1951) succeeded Boltzmann at Munich in 1906. He modified Bohr's theory (in 1916) to include elliptical orbits for electrons. In doing so, he applied relativity theory to electrons in atoms.

transfer of energy and the photon theory of light. Electrons were depicted as altering their states of motion all at once, emitting or absorbing packets of energy as they did so.

3. Conservation Principle. Bohr maintained that the classical principle of energy conservation must be rigorously observed when electrons leap between stationary states. That made possible the widespread use of energy level diagrams (which are commonly employed in explaining emission and absorption of radiation).

4. Correspondence Principle. The Bohr theory provided a link between the classical world of Newtonian mechanics and electromagnetic theory, and the quantum mechanical world. The link is called the Correspondence Principle. It states that quantum mechanics must observe a classical limit: wherever classical theory correctly accounts for the facts, quantum mechanics must agree with classical theory. Classical mechanics was not to be abandoned, and Newton had not been overthrown. The classical world of Newtonian mechanics had simply been reduced to special cases of a newer, more general, and more powerful theory.

Watching the Quantum ... JUMP!

Although the optical properties of matter have been increasingly well described by scientists for nearly a century, the discontinuity that exists between energy levels in atoms and the instantaneous switching between them has only recently been witnessed directly. Sightings of quantum jumps have been made by scientists at the University of Washington (Seattle) and at the National Bureau of Standards (Boulder) on single ions confined in an electromagnetic trap at very low temperatures and under very high vacuum. By cooling the ions to a few thousandths of a degree above absolute zero, their motion is damped almost to a stop; and under high vacuum, the ordinary activities of large numbers of particles—which have previously obscured quantum events amid average statistical behavior—are effectively removed.

The quantum jump itself is produced by energizing atoms with a unique photon frequency obtained from the light of a laser. Before the photon is absorbed, the atom is in one state; after the photon has been absorbed, the atom is in another of higher energy. But even when a single atom can be isolated and activated, detecting the quantum jump is still a difficult business. As we will see next chapter, one of the paradoxes that needs to be understood is the uncertainty principle, which warns us that it is impossible to measure a quantum state without disturbing the state we are trying to measure. As a result, there has always been the nagging question of whether anyone would ever be able to witness directly the event that has played such a critical role in modern theory. But the question has recently been answered. The quantum jump has been observed, visible to the eye with the aid of a simple microscope, as a single bright spot of light, blinking on and off.

The key to the experiment is the choice of Ba^{2+} or Hg^{2+} ions because they have two easily accessible excited states, one of which is fairly long-lived. Excited by the laser light, the atom jumps to one or the other of the two excited states, depending on the energy of the photons absorbed. A while later, it returns to the ground state. Keep in mind that the average time the atom waits before it returns to the ground state is significantly different for the two possible excited states. The wait is very short for one level; when the atom is irradiated, it jumps up and down hundreds of millions of time each second. Because a photon is given off with each jump down, the atom becomes fluorescent, visibly glowing; and because the jumps cannot be distinguished—they happen so rapidly—the glow appears to be steady.

The other excited state is many times more stable, waiting many seconds on average before emitting a photon as it returns to the ground state. However, this photon cannot be detected, having traded brightness for the much longer interval between jumps. Still, the observer can tell when the atom has made the second kind of jump because it is no longer available to make the first kind, which gives rise to a continuous glow; and so to the outside observer, the light switches off. It is this occasional switching off that is the evidence the eye-witness sees as confirmation of the quantum jump. As reported by the Universisty of Washington team:

> You have to hold yourself steady and look for many minutes at a time, and then you'll see it switch. You see the trapped ion blinking on and off, and each blink is a quantum jump. It's a striking illustration that things occur discontinuously in nature.

SUMMARY

The concept of definite atomic and molecular weights originated in the observation that the chemical elements combine in fixed proportions by weight, and as gases they combine in fixed proportions by volume. These 19th century ideas were based on an essential premise—indivisible atoms. But data and ideas began to accumulate that suggested otherwise; at first, electrochemical studies hinted at the existence of carriers of charge; later, discharge tube experiments led to the characterization and identification of these carriers of charge as electrons; then X-rays and radioactivity were discovered. A successful nuclear model for a "divisible" atom was proposed, based on the work of Rutherford and his students. And at about the same time, it began to appear to many scientists that light wasn't what they had thought, that certain experiments could be better explained if radiation was assumed to be a discontinuous, or particle, phenomenon, not the continuous wave-like phenomenon required by the prevailing theories of the 19th century.

A pivotal discovery was announced by Planck during Christmas week in the year 1900, opening the 20th century and marking the beginning of modern chemistry and physics. A correct formula, precisely describing the spectral distribution for radiations emitted from an incandescent source, could be obtained only by assuming that the transfer of energy from such a collection of hot, jittering atoms (material oscillators) had to be quantized; the transfer of energy took place in discrete events.

Einstein's explanation of the photoelectric effect powerfully confirmed Planck's work while extending it. Not just the transfer of energy among jittering atoms, but light energy as well, was quantized. Henceforth, light was to be understood as a storm of photons of discrete energy, frequency and wavelength:

$$\Delta E = h\nu = \frac{hc}{\lambda}$$

Planck's constant of proportionality $h = 6.626 \times 10^{-27}$ erg·s (with its unusual units of energy times time) assumed universal significance.

Rutherford's scattering experiments had indicated a dense, positively charged center of mass around which small, negatively charged electrons moved in circular orbits. But such a planetary atom would be unstable by any theory then known, radiating energy as the electron spiraled in toward the nucleus and finally collapsed, all in a brief instant. Bohr offered the first successful explanation for how this state of affairs could exist within the confines of classical electrodynamic theory.

Electrons could move only in certain allowed orbits. These circular pathways were to be governed by Newton's laws of motion and Coulomb's law, but with Planck's newly discovered quantum hypothesis as an added wrinkle. The angular momentum for these circular orbits was to be quantized:

$$mvr = n \left(\frac{h}{2\pi} \right)$$

That led to a determination of the Bohr radius and the speed of the electron in a Bohr orbit. Bohr orbits were to be radiationless pathways. As long as the electron maintained its prescribed state of motion, energy was neither gained nor lost.

Radiation of frequency ν was absorbed (or emitted) when electrons "jumped" up (or down) from some initial energy state or Bohr orbit:

$$\Delta E = E_{final} - E_{initial} = h\nu = \frac{hc}{\lambda}$$

The Franck-Hertz experiment confirmed the existence of discrete energy gaps within anions.

The Bohr theory was not the last word. It did successfully explain the emission spectrum for the hydrogen atom and hydrogen-like ions, and served as a theoretical bridge from the classical nuclear atom to the quantum mechanical models that prevail today.

QUESTIONS AND PROBLEMS

QUESTIONS

1. Why do atoms scatter streams of α-particles directed at them? When scattering is observed, the scattering angle is mostly small, but on occasion the angle is unexpectedly large. Briefly explain.

2. Light of a certain wavelength falls on a metal plate, causing a photoelectric response.
(a) What happens as the light intensity increases?
(b) What happens as the wavelength of the source increases?
(c) What happens as the frequency increases?

3. What major limitation did Einstein's explanation of the photoelectric effect remove from Planck's original suggestions about light quanta?

4. How do each of the following originate, and what are the characteristics of their appearance?

(a) continuous spectra
(b) band spectra
(c) line spectra
(d) visible spectrum

5. Light from an incandescent solid is allowed to pass through a haze of sodium vapor under greatly reduced pressure, and then examined spectroscopically. What does the spectrum look alike?

6. In a follow-up to the experiment described in Question 5, the incandescent light source is turned off, the sodium vapor is heated until it glows, and the spectrum is again observed. What does it look like?

7. A discharge tube filled with gaseous hydrogen emits a bright line spectrum in the visible region, while ionized hydrogen observed in the Sun's solar spectrum emits one continuous line. Why?

8. (a) What were the essential results of Rutherford's famous "scattering" experiments?

(b) What conclusions were drawn from these results?

(c) What was the reason for the failure of the Rutherford atom model?

(d) How did Bohr modify the Rutherford atom model to resolve the dilemma?

9. What kind of theoretical and experimental proof supported the Bohr theory?

10. In what ways did optical spectra at first give strong support to, and later help to undermine support for, the Bohr theory?

11. If the lone electron in the hydrogen atom were governed strictly by classical 19th-century electrodynamics (instead of quantum restrictions), would it emit a continuous or line spectrum? Why?

12. In comparing an electron moving in the first Bohr orbit to one moving in the fifth, which one:

(a) has the greater velocity?

(b) is in the orbit of smaller radius?

(c) radiates the greater energy?

(d) has the greater ionization energy?

(e) has the greater potential energy?

13. What was the Franck-Hertz experiment and what did it prove?

14. A stream of 10.2 eV electrons is effectively brought to rest as it passes through a gas of hydrogen atoms.

(a) Explain the result in terms of the Franck-Hertz experiment.

(b) What is the corresponding photon energy for the emission line that should be observed?

(c) You should expect a poorer fit between theory and practice for a gas of hydrogen atoms than for mercury vapor. Briefly explain why.

15. Where, according to the Bohr theory, is the orbiting electron in a ground state hydrogen atom after absorption of 13.6 eV? Briefly explain.

16. Can a ground state hydrogen atom absorb less than 13.6 eV? Briefly explain.

17. Why does each element have a characteristic emission spectrum that is as unique as a "fingerprint"?

18. Which is more energetic, blue light or red light? Which is of longer wavelength? Which is of higher frequency?

19. Suppose you lived in a universe of positively charged electrons and negatively charged nuclei, an "antiworld." Would Bohr's orbits be the same as in our world? Briefly explain.

PROBLEMS

Radiation and Planck's Constant   [1–8]

1. The limits of the visible region of the electromagnetic spectrum are roughly 4000 Å and 7500 Å.

(a) Identify the wavelength that corresponds to the blue end and calculate its frequency, wavenumber, and energy.

(b) Do the same for the wavelength limit on the red end.

2. The wavelength of X-rays from a highly evacuated dis-

charge tube having a tungsten anode is 0.21 Å. These X-rays are commonly used for medical purposes. What is their frequency?

3. Find the energy per photon for each of the following radiations:

(a) a microwave photon whose wavelength is 1.00 inch

(b) an infrared photon whose wavenumber is 1678 cm^{-1}

(c) a visible photon in the yellow region at 540 nm

(d) an ultraviolet photon whose frequency is 1.50×10^{15} Hz

(e) an x-ray photon at 1.00×10^{-10} m

4. Show that the emission lines for Hg at 253.5 and 184.4 nanometers correspond to photon energies of 4.9 and 6.7 eV, respectively.

5. Calculate the photon energies corresponding to each of the following:

(a) 0.5 Å x-rays

(b) 10^4 Å infrared radiation

(c) 10^{20} Hz gamma rays

(d) the first emission line in the fourth spectral series for atomic hydrogen

6. A solution of a certain biological compound absorbs light at both 3400 and 2000 angstroms. Determine the energy spacing between the levels and the wavelength of light corresponding to that energy transition.

7. Molecular oxygen in the upper atmosphere is split into a pair of atoms by radiation from the Sun. If the lowest frequency light that can initiate the process is 1.21×10^{15} Hz, what is the dissociation energy for molecular oxygen (in eV/molecule)?

8. The energy required to split chlorine molecules into atoms is 247 kJ/mol. What wavelength of electromagnetic radiation is needed to perform this reaction? In which part of the electromagnetic spectrum is this radiation found?

Photoelectric Effect   [9–18]

9. What is the threshold wavelength for a metal whose characteristic work function (W) is known to be 4.8 eV?

10. The minimum energy necessary to expel electrons from a certain metal surface was found to be 4.33 eV. Calculate the wavelength of light that will just initiate a photocurrent.

11. What is the minimum energy required to expel electrons from a silver foil? From a platinum surface?

12. Determine the maximum kinetic energy for expelled photoelectrons from a palladium surface illuminated by light from a 2000. Å source.

13. A gold foil is illuminated with a monochromatic UV source of wavelength 181 nm.

(a) What is the minimum energy required to expel electrons from gold?

(b) What is the energy of the incident photons?

(c) What are the maximum energy and the velocity of the photoelectrons?

14. A certain metal has a characteristic work function of 2 eV. When light of a certain wavelength falls on this metal, photoelectrons with kinetic energy of 2 eV are expelled. What is the wavelength of the irradiating photons?

15. The work function for a tungsten filament in a photoelectric experiment is 4.58 eV.
(a) Find the threshold frequency.
(b) Find the opposing (or stopping) potential that would just prevent a photoelectric response for incident light of 2000 Å wavelength.
(c) Find the kinetic energy of the photoelectrons ejected in part (b).

16. In a photoelectric experiment, incident light of 360 nm is allowed to fall on a potassium-coated cathode. Calculate:
(a) the opposing (or stopping) potential for the photoelectrons
(b) their maximum kinetic energy
(c) their maximum velocity

17. A photoelectric experiment is carried out by placing a metal plate near a monochromatic light source. The data shown by the accompanying graph were obtained. Using this graph, determine:
(a) the characteristic work function (W)
(b) Planck's constant
(c) the characteristic threshold frequency (ν) and the corresponding wavelength (λ)

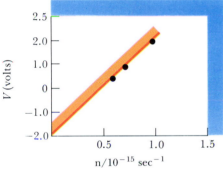

18. A metal surface is flooded with light of variable wavelength, the opposing (or stopping) potential measured, and the data tabulated as follows:

λ, nm	366	405	436	492	546	579
V, volts	1.48	1.15	0.93	0.62	0.36	0.24

Plot the opposing (or stopping) potential versus frequency. For the resulting curve, determine the threshold frequency, the work function, and the ratio h/q_e.

Bohr Theory   [19–26]

19. Calculate the radius for the first Bohr orbit for atomic hydrogen, and the radius for the next three orbits as multiples (or fractions) of the first.

20. Calculate the velocity of the electron in the first Bohr orbit for atomic hydrogen, and the velocities in the next three orbits as multiples (or fractions) of the first.

21. Compare the radius for the first Bohr orbit, a_0, of a ground state H atom to the radius of the fifth Bohr orbit. What is the relationship of one to the other? Derive a general expression for finding the Bohr radius of an orbit in terms of a_0.

22. Compare the velocity of an electron in the first Bohr orbit with that of one in the fifth Bohr orbit. What is the relationship of one to the other? Derive a general expression for the velocity of an electron in any Bohr orbit, in terms of its principal quantum number n.

23. The Rydberg relationship can be written as follows:

$$\bar{\nu} = \frac{2\pi^2 m e^4 Z^2}{h^3 c}\left[\frac{1}{n_1{}^2} - \frac{1}{n_2{}^2}\right] = RZ^2\left[\frac{1}{n_1{}^2} - \frac{1}{n_2{}^2}\right]$$

(a) Determine the shortest wavelength emitted by an electron falling from the $n = 4$ state for a hydrogen atom, in units of R.
(b) Determine the shift (red or blue) for the Balmer (visible) series spectral lines observed when comparing ground state He^+ ions to H atoms.
(c) If the wavelength of the lowest energy Balmer line for ground state H atoms is 6564 Å, what is the wavelength for the corresponding line for ground state He^+ ions?

24. Write the general form of the Rydberg equation that applies to the hydrogen-like Li^{2+} and Be^{3+} ions.

25. The singly charged helium ion (He^+) is isoelectronic with the hydrogen atom and should therefore be described by the Bohr theory. Calculate the wavelength for the first spectral line in the Lyman and Balmer series for such a species.

26. There is a spectral series for Li^{2+} that begins with $n_1 = 9$.
(a) Show that according to the Rydberg equation it would be expected that every third line corresponds to a line in the Lyman series for atomic hydrogen.
(b) Calculate the wavelengths for the first three lines and verify the expectation in part (a) by comparing the values with the data in Table 14–4.

Additional Problems   [27–37]

27. A certain wavelength of light in the visible region of the electromagnetic spectrum is known to be 654.3 nm.
(a) Express this value as frequency ν in Hz.
(b) Convert the wavelength to angstroms.
(c) Calculate the energy in kJ for Avogadro's number of photons of this wavelength.

28. Calculate the velocity of photoelectrons expelled from a sodium surface by incident light of 4000 Å.

29. Derive Bohr's equation for the radius of a ground state hydrogen atom, a system consisting one planetary electron moving about a proton nucleus, beginning with the basic notion that the Coulomb force of attraction must account for the electron's orbital motion.

30. Show that Planck's constant has units of angular momentum, and briefly explain why that is significant.

31. (a) What is the energy required to ionize a hydrogen atom in which the electron is in the fifth Bohr orbit?
(b) What is the energy released when the electron in the same hydrogen atom decays (or relaxes) to the fourth Bohr orbit?

32. If the value of the principal quantum number doubles, by what factor does the speed of an electron in a Bohr orbit change?

33. Ultraviolet lamps often make use of the 254 nm line in the emission spectrum of mercury. There is a green line in the visible spectrum of mercury at 544 Å. What is the energy spacing between these two lines in kJ/mol?

34. When a light source of wavelength λ_1 is used to illuminate a metal surface, the kinetic energy of the ejected photoelectrons is 1 eV. A second light source λ_2 has half the wavelength of λ_1, and when it illuminates the same metal surface, the kinetic energy of the ejected photoelectrons is 4 eV. What is the characteristic work function of the metal?

35. Show that the ionization energy for a hydrogen-like Bohr ion is given by the following equation:

$$\Delta E_{IE} = Z^2(\Delta E_{IE(H)})$$

36. Carry out the necessary algebraic manipulations to show that the following two forms of the Rydberg equation are equivalent.

$$\frac{1}{\lambda} = 4R_H \left[\frac{1}{4^2} - \frac{1}{n_2^2} \right]$$

$$\frac{1}{\lambda} = R_H \left[\frac{1}{2^2} - \frac{1}{(n_2/2)^2} \right]$$

37. Astronomers have observed the **Fowler series** for He$^+$ ions in the Sun's solar spectrum:

$$\frac{1}{\lambda} = 4R_H \left[\frac{1}{3^2} - \frac{1}{n_2^2} \right]$$

Calculate the wavelengths for the first three lines in angstroms, given that $R_H = 109,677.58$ cm^{-1}.

MULTIPLE PRINCIPLES [38–41]

38. Calculate the frequency and wavelength of a quantum of light having the same energy as the average translational energy of a gas molecule at room temperature. In what region of the electromagnetic spectrum will this wavelength be found? Based on this finding, would you expect most atoms to be in their ground state at room temperature?

39. Calculate the wavelength of the line in the spectral series for hydrogen in the visible region, corresponding to $n_1 = 2$ and $n_2 = 4$. Do the same for $n_2 = 8, 9, 10$, and so forth, until it is obvious that you are approaching a series limit. What is that limit? Compare the value to that obtained by considering $n_2 = \infty$. Briefly explain the significance of the result.

40. An electron in the $n = 2$ state in chromium relaxes

to the $n = 1$ state, but no energy is emitted. Instead, the excess energy in the atom is transferred to an outer electron in the $n = 4$ state, which is then ejected. Using the Bohr theory, calculate the kinetic energy of the ejected electron. *Note:* This is called an "Auger" process, and the ejected electron is called an Auger electron. This delayed photoemission is important in studying surface properties of materials.

41. Based on the Bohr model for a ground state hydrogen atom, how many times per second does the electron circumscribe the proton nucleus and how long does it take to complete each revolution?

APPLIED PRINCIPLES [42–46]

42. A biochemistry laboratory scientist is interested in finding the energy difference between two different states of reduced nicotinamide adenine dinucleotide (NADH), known to absorb light at 340.2 and 259.7 nm. Calculate the energy difference in kJ/mol.

43. Spectral lines corresponding to values of n as large as 50 have been observed in astronomical studies of hydrogen in our galaxy. What is the wavelength corresponding to the difference between the 49th and 50th n states for hydrogen? In what region of the electromagnetic spectrum does this energy transition lie? *Note:* This macroscopic wavelength lies in the domain of radio astronomy. Give the speed of the electron and the radius of its orbit.

44. In intergalactic space, what is the quantum state for an electron orbiting a proton if the radius is 1 meter?

45. (a) In radio astronomy, what is the energy equivalent (in kJ/mol) for the 21-cm radiation arising from the interaction of the electron and the proton in hydrogen atoms?

(b) A radio station announces that it is broadcasting at a frequency of 1250 Hz. Given that the "Hertz" (Hz) is cycles/s, or simply reciprocal seconds, what is the corresponding wavelength in nanometers?

46. Consider a macroscopic orbital system. In applying Bohr's quantization principles and theory to the Earth and its satellite, the Moon, use the following facts: (1) The radius of the Moon's orbit is approximately 385,000 km. (2) The time required to complete one orbit is about 30 days. (3) The mass of the Moon is close to 7.4×10^{22} kg. Estimate the following:
(a) the circumference of the Moon's orbit
(b) the average speed of the Moon in its orbit
(c) the Moon's angular momentum
(d) the n state of the Moon
(e) What does the magnitude of your answer in (d) suggest about applying Bohr theory to macroscopic orbital systems?

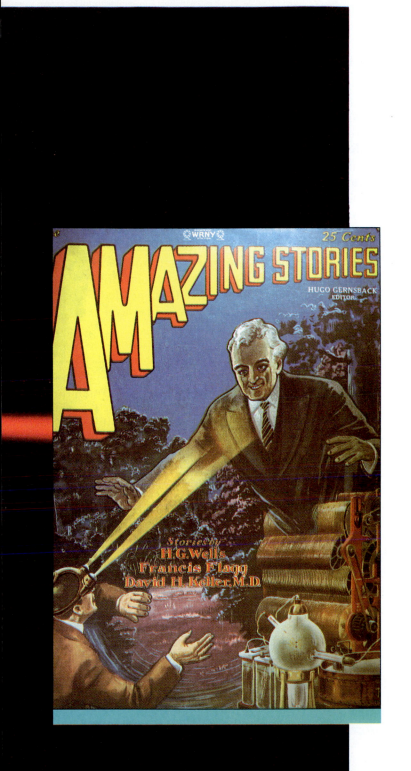

CHAPTER
15

Atomic Structure II: Atomic Structure and Quantum Theory

Barium sodium niobate crystal, like the one shown in the photograph, has the ability to convert infrared (invisible) laser light into visible light. This process for changing frequencies, known as harmonic generation, has already proved to be very important for the telecommunications industry, as well as in chemical research. The photograph was made possible with the aid of a card treated with an infrared-sensitive phosphor. Long before the operation of the first ruby laser in 1960, science fiction supplied us with imaginative anticipations of the practice of this science, as indicated by the cover of this 1928 edition of the science fiction magazine *Amazing Stories.* This issue contained the original publication of the H.G. Wells short story, "The Invisible Man."

15.1   ATOMIC STRUCTURE AND QUANTUM THEORY

The Bohr model for the hydrogen atom was responsible for the general acceptance of the quantum theory by a wide audience of scientists. Bohr's interpretation of the emission spectrum for the hydrogen atom was brilliantly contrived yet beautifully simple, and it faithfully reproduced every spectral line. But the Bohr theory was not useful for anything more complicated than a one-electron system. Furthermore, attempts at explaining chemical bonding on the basis of the Bohr theory were similarly frustrating. There was no evident way to calculate the binding energy for even the simplest neutral molecule, H_2, all but destroying the hope that Bohr's newly won understanding of atomic structure might provide insight into chemical problems.

In spite of its introduction of discrete energy state and quantum jumps, Bohr's atomic model remained a product of the physical theories of the 19th century. It was still a classical system. What was needed was a more general description of nature at the atomic and molecular level than classical mechanics and electrodynamics could provide.

The new mechanics of the 20th century became known as **quantum mechanics**, and our study of it begins with the resolution of a paradox, the so-called wave-particle duality. We'll see how scientists solved the problem of calculating quantum states without the cumbersome qualifications and conditions required by the Bohr theory, and how the new mechanics was put into general use. Quantum mechanics not only provides the means for quantitative determination of energy terms of quantum numbers, but allows us to paint qualitative pictures by means of reasonable approximations of the structure and composition of atoms and molecules. Finally, in the next chapter, we'll see how quantum mechanics was used by a new generation of young chemists to explore the nature of the chemical bond. In quantum mechanics, scientists found the theory they were looking for—one that gave results identical to those of the Bohr theory for atomic hydrogen while being generally applicable to atoms and molecules.

15.2   PARTICLES AND WAVES

Wave-Particle Duality

A young French scientist named Louis de Broglie (Figure 15–1) first stated that matter could manifest itself as either particles or waves, depending upon experimental circumstances. Einstein's brilliant interpretation of the photoelectric effect and the undeniable results of Millikan's photoelectric experiments validated the particle properties of radiation. Why not then propose wave properties for material particles such as electrons? If that proved reasonable, certain basic relationships should exist between energy, momentum, and wavelength. Assume as de Broglie did that the energy of the photon is proportional to its frequency while the energy of any material particle is proportional to its mass. For radiation particles, according to Planck and Einstein,

$$E = h\nu$$

For material particles, according to Einstein

$$E = mc^2$$

Setting the two expressions for the energy E equal to each other:

$$mc^2 = h\nu = \frac{hc}{\lambda}$$

$$mc = \frac{h}{\lambda}$$

Recognizing that the propagation velocity of light c is only a special velocity v gives the de Broglie relationship, which is a general expression for the wavelength of a material particle whose momentum is mv:

$$mv = \frac{h}{\lambda}$$

Wave behavior for walkers and runners, cars and trucks, jet aircraft and speeding bullets have all gone undetected because the corresponding wavelengths are too small to be measured by any known means. Consider the wavelength of a golf ball of mass 1.62 ounces, propelled at an average speed of 150 mi/hr:

$$\lambda = \frac{h}{mv} = \frac{6.626 \times 10^{-27} \text{ erg·s}}{(45.9 \text{ g})(6.71 \times 10^3 \text{ cm/s})} = 2.15 \times 10^{-32} \text{ cm}$$

The wavelength λ is an immeasurably short distance.

FIGURE 15–1 Emerging from a royal French family and a long line of scientists and engineers, 32-year old Louis Victor de Broglie presented a remarkable dissertation to the Faculté des Sciences in Paris in 1924.

EXAMPLE 15–1

According to de Broglie, the wavelength of the golf ball cannot be measured, being on the order of 10^{-32} cm. Can you measure the wavelength of an electron moving at a speed of 3.00×10^9 cm/s?

Solution   The mass of an electron is 9.11×10^{-28} g. Thus,

$$\lambda = \frac{h}{mv} = \frac{6.626 \times 10^{-27} \text{ erg·s}}{(9.11 \times 10^{-28} \text{ g})(3.00 \times 10^9 \text{ cm/s})} = 2.42 \times 10^{-9} \text{ cm}$$

The de Broglie wavelength for this particle is in the x-ray region of the electromagnetic spectrum and should therefore be measurable. ∎

EXERCISE 15–1

Compare the de Broglie wavelength of an official baseball (weighing 5.25 ounces) moving at a relatively ordinary speed of 95 mi/hr with that of an electron moving at 4.50×10^9 cm/s, ignoring any relativistic mass effects. *Answer:* The electron's wavelength is 1.54×10^{23} times that of the baseball. ∎

As radiation travels to us from the Sun and distant stars, we accept that the photons coming at us have characteristic wavelengths. Now we must also include within this framework the wave nature of material particles. But light waves and matter waves are different, even though we can calculate frequencies and wavelengths for both. Wavelengths for photons have dimensions within our experience. Although matter can be thought of as radiation, matter waves cannot move at the speed of light; although matter has mass at rest, light does not.

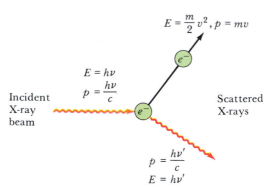

FIGURE 15–2   In explanation of the **Compton Effect**, consider the incident X-ray quantum with energy $E = h\nu$ and momentum $p = mv = h\nu/c$ as it collides with an electron. Energy and momentum are transferred to the electron. The net result is that the scattered X-ray photon has been reduced to $E = h\nu'/c$.

The Compton Effect

Having proposed that just as radiation can possess particle properties, electrons can display wave properties, it became a matter of considerable interest and concern to demonstrate just how true the de Broglie relationship $mv = h/\lambda$ really was. Convincing experiments were needed. One of the most important is known as the **Compton effect**, the name given to the scattering of x-radiations and the accompanying change in wavelength due to collisions with weakly bound electrons in atoms.

According to classical theory, when a beam of x-radiation strikes atoms in a thin metal foil, the radiation will scatter in all directions but without any alteration in wavelength. If radiation of a given wavelength happened to be absorbed, then of course it could emit radiation of different wavelengths. But if the radiation were simply deflected or scattered, there should be no change in wavelength. Compton's results were otherwise. He observed scattering accompanied by change in wavelength to lower energy. In explanation, Compton proposed a two-particle encounter between x-ray photons and target atoms. When a very small mass bumps into a very large stationary mass, there should be an elastic collision with negligible change in velocity, energy, or momentum. However, if the colliding mass is about the same size as the resting mass, quite a lot of energy transfer should be expected on collision.

If the mass of an x-ray photon is small compared to that of a metal atom, little change should be observed in the wavelength and momentum of the scattered x-rays. But if the photon struck an electron, which we know has a small mass, then an inelastic collision might be expected; and that should be accompanied by scattered x-rays at longer wavelengths (Figure 15–2). In Compton's experiment, a beam of x-rays is scattered as it passes through a metal foil. The radiation emerges along two paths. One beam is at the same wavelength as the incident beam. The other appears at slightly longer wavelengths. The x-rays scattered without change in wavelength are scattered by collisions with the much larger, stationary nuclei, while the scattering that is accompanied by a shift to longer wavelengths is the result of collisions with electrons. Compton's x-rays displayed the properties of particles.

The Davisson-Germer Experiment

If x-ray photons behave as particles about the size of electrons, might not one expect to observe the reverse phenomenon, electrons displaying wave properties? The first demonstration of the wave nature of electrons was made by Clinton Davisson and Lester Germer (Figure 15–3). On reflecting

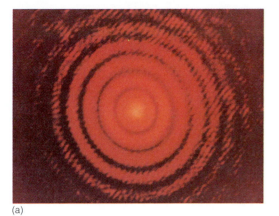

(a)

FIGURE 15–3 (a) A classic diffraction pattern formed by passing light from a helium-neon laser through a circular aperture. That light behaved this way was well known in the early 1920s; that particle beams such as electrons could behave this way was totally unexpected and provided revolutionary insight into the nature of matter. (b) In the Davisson-Germer experiment, electrons generated by a filament emerge from a "gun" that has accelerated them to some voltage, say 50 V, and directed them at a nickel crystal. The distribution-in-direction of the scattered electrons is determined with a movable collector.

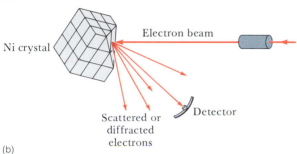

(b)

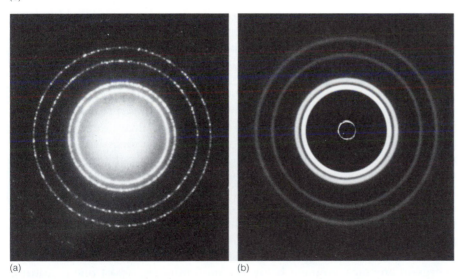

(a) (b)

FIGURE 15–4 (a) Beam of X-rays diffracted by a layer of aluminum filings ($\lambda = 0.71$Å); (b) beam of electrons diffracted by a thin foil of aluminum ($\lambda = 0.50$Å).

electrons from crystals, they observed typical interference (diffraction) phenomena, maxima and minima in the intensities of the electrons reflected from a nickel target. On analyzing these data, they were able to measure wavelengths for electrons, which compared favorably with predictions using the de Broglie relationship.

Since wavelengths for x-ray photons are on the order of 10^{-8} cm or about an angstrom, it would be useless to try to observe diffraction due to an electron beam by ordinary optical means, say with a diffraction grating. The best optical gratings have line spacings of several thousand angstroms. But nature lends a hand by providing suitable natural gratings in the form of crystals (Chapter 19), where the spacing is just right for diffracting x-rays or electrons, producing interference patterns. The result can be seen on comparison of the pattern created by an x-ray beam reflecting from aluminum crystals with the pattern of an electron beam reflected from an aluminum foil (Figure 15–4).

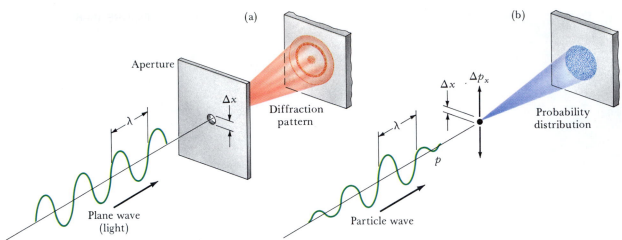

FIGURE 15–5 (a) Wave properties of a photon beam; (b) particle properties of an electron beam. Each can be determined in separate experiments, but neither can be adequately measured, simultaneously.

15.3 THE UNCERTAINTY PRINCIPLE

The Davisson-Germer experiment and the Compton effect served to demonstrate an important principle: The intrinsic nature of an electron or an x-ray photon does not change with the type of experiment performed, but what you are able to sense experimentally does. Electrons could behave like classical particles in certain experiments, while in others they behave nonclassically, like waves. X-rays can behave classically like electromagnetic radiations in certain experiments, while in others they behave nonclassically, like particles.

The **Heisenberg Uncertainty Principle** brings together two unconventional ideas: (1) the concept of a minimum amount of energy for quantum events, and (2) the existence of a fundamental uncertainty concerning the position and motion of electrons in atoms. Both ideas have an impact upon experiments at the quantum mechanical level. In the practical world of science and engineering, it was always assumed that one could correct for disturbances due to the measurement process by running a "blank" experiment, enabling us to "subtract out the background," in a manner of speaking. This is not the case when dealing with the very smallest systems, such as electrons and nuclei in atoms. According to Planck, energy has discrete structure at the atomic level, and we cannot reduce the amount of energy in any measurement below the minimum energy of one quantum, $h\nu$.

Consider the problem of trying to locate the position of a particle and measure its momentum simultaneously with absolute precision. Whenever the position is known precisely, the momentum is known less precisely, and vice versa. To locate the position of the particle, it must be considered as a point mass. To determine its momentum, its exact wavelength must be known, requiring a unique frequency operating through an infinite time span. Each quantity can be determined in a separate experiment, designed in the former case for the particle property and in the latter case for the wave property (Figure 15–5). However, determining both properties simultaneously with absolute certainty is not possible because the method of measurement influences the quantity measured.

To locate an electron somewhere in an atom, you might illuminate it with a radiation with a wavelength of the same order of magnitude as the size of the atom. But if you choose photons of such short wavelength, they

FIGURE 15–6 Werner Heisenberg, seated left, chatting with Niels Bohr at lunch during a 1930s conference at the Bohr Institute in Copenhagen. Not shown on the table is Carlsberg beer; the Carlsberg Foundation has supported much of the work of this institute to this day.

will have a correspondingly large momentum and consequently, a lot of energy:

momentum $\qquad mv = \dfrac{h}{\lambda}$

energy $\qquad mv^2 = \dfrac{hc}{\lambda}$

From the Compton effect we know that short wavelength x-ray photons give the electron quite a boot when they collide. Therefore, the best we can hope for is to determine where the electron was before the photon got to it by examining the scattered photons after the collision. All we can know is where the electron *was*, because in trying to locate the electron, our experiment altered its speed and direction. We would of course minimize the disturbance by using less energetic, longer wavelength photons, but that would produce greater uncertainty in the electron's position. Furthermore, as we are more successful at locating the electron, we can expect to be correspondingly less successful at determining its velocity. At the atomic level we have to live with the knowledge that uncertainty plays a fundamental role.

Heisenberg (Figure 15–6) was able to show that the uncertainty in the determination of the momentum p and the simultaneous uncertainty in determination of its location x is governed by the following equation:

$$\Delta p \Delta x \geq \dfrac{h}{4\pi}$$

Note that $p = mv$ and $\Delta p = m\Delta v$, so

$$[m\Delta v]\Delta x \geq \dfrac{h}{4\pi}$$

The product of the two uncertainties is close to or greater than $h/4\pi$. As one approaches the perfection of measurement, $h/4\pi$ is approached as a limit.

A second application of the uncertainty principle can be written for energy and time. Restating the equation,

$$\Delta E \Delta t \geq \frac{h}{4\pi}$$

As an electron "jumps" to a lower energy state, a quantum of energy is emitted; but the precise moment of the event cannot be known with certainty. Energy is associated with particle properties—mass and speed. Time is associated with wave properties—frequency. Trying to mark the exact moment of an event limits knowledge of the frequency of the wave. In this form, the uncertainty principle has proved useful in estimating the sharpness of spectral lines.

In a sense, Δv and Δx, or ΔE and Δt, represent the "smeared" nature of our knowledge. Yet the world around us seems crisp and free of such vagueness and limitations. In our everyday experience, the uncertainties are vastly overridden by general inaccuracies and other experimental difficulties long before the built-in limitations implied by the uncertainty principle become evident. In the world of subatomic particles that's not the case. Keep in mind that experiment cannot yield greater accuracy than the uncertainty principle allows.

The uncertainty in our knowledge of the coordinate x and the momentum p for a particle can be written as $x \pm \Delta x$ and $p \pm \Delta p$. That is to say, the particle is located between $x - \Delta x$ and $x + \Delta x$, and the momentum of the particle lies between $p - \Delta p$ and $p + \Delta p$. If we apply the uncertainty principle to a particle of classical mass, say about a milligram:

$$(\Delta p)(\Delta x) \geq \frac{h}{4\pi}$$

$$(m\Delta v)(\Delta x) = \frac{h}{4\pi}$$

$$\Delta v \Delta x = \frac{h}{4\pi m} \approx \frac{10^{-27}}{10^{-3}} \approx 10^{-24}$$

which says that when we know the position of the particle to within $\pm 10^{-12}$ cm, our ability to know the particle's velocity is limited to $\pm 10^{-12}$ cm/s. In dealing with milligram-sized particles (or larger), uncertainties of such small dimensions are of no real consequence.

EXAMPLE 15–2

Are the uncertainties in dealing with electron-sized particles of real consequence?

Solution   The mass of an electron is on the order of 10^{-27} g. Using the uncertainty principle,

$$(\Delta p)(\Delta x) \geq \frac{h}{4\pi}$$

$$(m\Delta v)(\Delta x) \geq \frac{h}{4\pi}$$

$$\Delta v \Delta x \geq \frac{h}{4\pi m} \approx \frac{10^{-27}}{10^{-27}} \approx 1$$

When trying to find the exact location of the electron, say to an uncertainty of only 10^{-6} cm, the fundamental requirements of quantum mechanics and the Heisenberg principle prevent us from knowing its velocity any better than to within $\pm 10^6$ cm/s, which is large enough to invalidate the classical picture of electrons riding about in Bohr's orbits. ■

EXERCISE 15–2(A)

Consider a 1-microgram mass traveling at a speed of 1 cm/sec. If the uncertainty in its velocity is 1 percent, what is the magnitude of the minimum uncertainty in its position?

Answer: 10^{-19} cm. ■

EXERCISE 15–2(B)

The energy of a certain nuclear state can be measured with an uncertainty of 1 eV. What is the magnitude of the minimum uncertainty in the lifetime of the state?

Answer: 10^{-15} s. ■

15.4 WAVE MECHANICS

Given everything said so far, we should be able to define quantum mechanics as "the theories and experiments that impose h upon us." Physicists call h Planck's quantum of action. With less mystery, chemists simply refer to it as Planck's constant, the proportionality constant in his famous equation, $E = h\nu$. The photoelectric effect first suggested a connection between the particle of light and its corresponding wave, a duality built into theory by the de Broglie relationship $mv = h/\lambda$. This hypothesis was established experimentally by the Compton effect and the Davisson-Germer experiment. Finally, the uncertainty principle reconciled this dualistic point of view, with far-reaching implications.

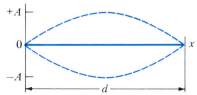

FIGURE 15–7 Imagine a guitar string between the two fixed points zero and x. The maximum displacement along the y-coordinate is the amplitude A.

Wave mechanics also reconciled particle and wave behavior, quantizing the wave and its corresponding energy. But it had an added benefit in that it could be applied in ways that made the fundamental nature of wave-particle duality much more accessible to a wider audience of scientists. Although highly mathematical in its pure form, wave mechanics has given us very useful qualitative portraits of electrons in atoms.

Picture a guitar string fixed between two endpoints in the x-direction (Figure 15–7). Assign a value d to that distance and an amplitude $\pm A$ to the maximum distance a point on the string may move in the y-direction. Now imagine this plucked guitar string, vibrating in a fundamental mode, moving up and down all along its length, except at the anchored endpoints. The maximum distance along the y-direction occurs at the center of the string for the case where the wavelength $\lambda = 2d$. Other modes of vibration, called **overtones**, are possible, the first occurring when $\lambda = d$ (Figure 15–8). In this mode there is a vibrationless point called a **node** at the center of the string, with two loops on either side. A second overtone arises when $\lambda = \frac{2}{3}d$; now there are two nodes and three loops. Higher overtones are constructed in the same way. Half-wavelengths correspond to integer values for n, where $n = 1, 2, 3, 4, \ldots$ for the fundamental vibration and the overtones, according to the general equation

(a) $\lambda = 2d$ or
$\quad\lambda/2 = d$

(b) $\lambda = d$ or
$\quad\lambda/2 = d/2$

(c) $\lambda = 2d/3$ or
$\quad\lambda/2 = d/3$

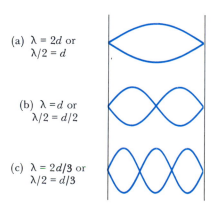

FIGURE 15–8 The fundamental mode of vibration, and the first and second overtones, are given at (a) $\lambda = 2d$, (b) $\lambda = d$, and (c) $\lambda = 2d/3$.

$$\lambda = \frac{2d}{n} \quad \text{or} \quad n\left(\frac{\lambda}{2}\right) = d$$

FIGURE 15–9 The mathematical relationships derived by Erwin Schrödinger, sometimes referred to as wave mechanics, sometimes as quantum mechanics, placed Planck's quantum theory (postulated a quarter century before) on a firm foundation.

An integral number of half-wavelengths is $n(\lambda/2)$. Note that because the string is anchored at the ends, only an integral number of half-wavelengths can fit along the length, d. Nothing between is allowed. The waves are quantized.

In order to apply this wave mechanical picture to atomic structure, imagine—as Schrödinger did (Figure 15–9)—bringing the endpoints of a distance $2d$ together into a circle to form an orbit (Figure 15–10). The standing waves exist as for the plucked guitar string, moving up and down, in place. Again, you may encounter the fundamental vibration, with no nodes, and overtones with one node, two nodes, three nodes, and so forth. You never encounter a non-integral number of wavelengths as you move through one complete turn around the circle ($n\lambda = 2d$). Why? Because that would give rise to overlapping waves, interference, and ultimately cancellation.

Now for the most important feature of our analogy. Imagine that this vibrating circle of guitar string represents the orbiting electron as it moves about the nucleus in an atom. If the orbit is far from the nucleus, the radius r of the circle is larger and there is a correspondingly longer path $2\pi r$ around the circle for any set of standing waves. Remember our quantum restriction, that an electron in its orbit about the nucleus must describe a path that is an integral number of complete wavelengths, not half-wavelengths; so $2\pi r = 2d$. Otherwise, interference and cancellation result. Substitutions lead directly to the equation Bohr could only assume fifteen years earlier. The de Broglie equation is

$$\lambda = \frac{h}{mv}$$

From wave mechanics we know that

$$\lambda = \frac{2d}{n}$$

Therefore we may write:

$$\frac{h}{mv} = \frac{2d}{n} = \frac{2\pi r}{n}$$

Finally, rearranging terms gives a familiar equation from Bohr theory:

$$mvr = n\left(\frac{h}{2\pi}\right)$$

Consider the wave patterns representing cases for $n = 1$, 2, and 6 in Figure 15–10. All three cases are consistent with the de Broglie relationship and the quantum restriction for a standing wave. Each depicts a radiationless electron wave, independent of time. Only when the electron jumps to a larger orbit is energy absorbed; only when it jumps to a smaller orbit is energy emitted. Finally, imagine the electron wave in three dimensions, not two, and imagine enough of these orbits to create the surface of a sphere of radius r, not just one lonely orbit about the nucleus.

The Bohr theory was good as far as it could go, and it was very important at the time. It pictured the electron as a small charged particle revolving about the nucleus in a quantized orbit. The wave mechanical view is much more powerful, and of a fundamental nature. It pictures the electron as an extended wave packet or matter wave.

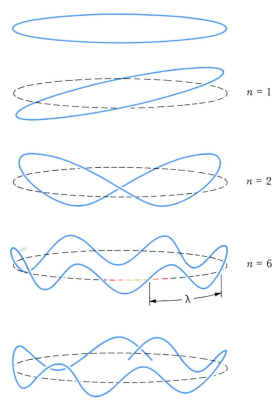

FIGURE 15–10 Imagine the fixed endpoints in Fig. 15–8 coming together in a circle—an orbit. Only the bottom illustration is out of phase on each circuit of the circle. The others, representing the cases for $n = 1$, $n = 2$, and $n = 6$, are stationary states with the circumference being divisible into an integral number of wavelengths—allowed states.

$n = 1$

$n = 2$

$n = 6$

λ

15.5 QUANTUM NUMBERS AND THE EXCLUSION PRINCIPLE

The hydrogen atom consists of only two particles, a proton and an electron. The force acting between the two particles is coulombic, and the potential energy term is given by the Bohr theory as $-e^2/r$, where r is the distance separating them. The position of the electron relative to the proton nucleus can be specified by a "wave function," traditionally given the Greek letter designation Ψ (psi), in the usual space coordinates (x, y, and z) in three dimensions. This mathematical function describing an electron (matter) wave can be written as a product of two parts. One part depends only on r and is referred to as the *radial* component; the other part depends on two *angular* components. Note the similarity to the Bohr theory for the hydrogen atom, which predicted the Bohr radius r and the quantization of angular momentum.

We are now in a position to consider qualitatively the wave functions and energy levels for hydrogen and hydrogen-like atoms, and for poly-electronic species. It is difficult enough to treat the simplest possible system, consisting of only a nucleus and one electron, let alone a nuclear atom with several electrons. Fortunately, for more complicated atoms we can, to a good approximation, neglect the interactions between the electrons and obtain results similar to those of hydrogen. These hydrogen-like wave functions—and the atomic electrons they represent—can be characterized by four quantum numbers: n, ℓ, m_ℓ, and m_s. The first three characterize the wave function Ψ; the orbital angular component is determined by ℓ and m_ℓ, while n is related to the radial component and represents the distance dependence. These three quantum numbers are referred to respectively as the **principal quantum number** (n), the **angular momentum quantum number** (ℓ), and the **magnetic quantum number** (m_ℓ). The fourth quantum number is the **spin quantum number** (m_s).

In the Bohr theory for the hydrogen atom, only one quantum number (n) was needed to characterize each stationary state.

The Four Quantum Numbers

Principal Quantum Number (n): As the name suggests, this quantum number is the most important of the four. It specifies the major energy level, or shell, to which an electron can be assigned. It can take on any positive integral value except zero:

$n = 1, 2, 3, 4, 5, \ldots$

Chemists also refer to the energy levels or shells in atoms by capital letters that correspond to the numerical values of n. You need to know both:

Letter	K	L	M	N	O	P
n Value	1	2	3	4	5	6

Quantum states of equal n for hydrogen atoms are degenerate or have the same energy. But equal n states are not degenerate for polyelectron atoms.

All the electrons in an atom possessing the same principal quantum number populate the same shell, or energy level. For hydrogen and hydrogen-like wave functions, the principal quantum number n determines the energy, which is quantized because the electron is confined in the radial direction; and it determines the radius of the shell. The quantum number n and the corresponding energy should be familiar to you as one of the surviving features of the Bohr theory:

$$E = -\left(\frac{1}{n^2}\right)\left(\frac{Z^2 e^2}{2a_0}\right)$$

Angular Momentum Quantum Number (ℓ): This quantum number determines the orbital angular momentum ℓ. It can take the following values for any given value of n:

$\ell = 0, 1, 2, 3, 4, 5, \ldots, (n - 1)$

There are n different states; for a given n value, there are a fixed number of ℓ states, with a maximum value of $n - 1$. So, for the ground state of the hydrogen atom, $n = 1$, and $\ell = 0$. Excited states of the hydrogen atom, which are characterized by $n > 1$, always exhibit an interesting multiplicity. For the energy state defined by $n = 2$ (the final energy for Balmer transitions) there are two distinct stationary states: one with $\ell = 0$, and therefore zero angular momentum, and the other with $\ell = 1$, and therefore one unit of angular momentum. These two states are said to be **degenerate states**, which is another way of saying they have equal energy. As a consequence, they do not appear as separate spectral lines in the emission spectrum for an excited hydrogen atom. For $n = 3$, there are three states of equal energy, corresponding to ℓ quantum numbers of 0, 1, and 2. Each state within a given energy level or shell is called a subshell.

Specifying n and ℓ for each electron in an atom specifies the **electronic configuration**. Chemists use a lowercase letter or number for the ℓ quantum number:

The corresponding lowercase designations are historical in their origins and stand for sharp, principal, diffuse, and fundamental.

Letter	s	p	d	f	g
Number Value	0	1	2	3	4

As the principal quantum number n determines the size of the shell, the angular momentum quantum number ℓ is related to the shape of the subshell and accounts for the existence of fine structure in the emission spectrum of atoms.

Magnetic Quantum Number (m_ℓ): The angular momentum of the electron is responsible for inducing a magnetic field. That should not be surprising if we assume the orbiting electron is equivalent to a circular electric current. From classical electrodynamics, circular electric currents generate magnetic fields. Orbiting electrons should behave in the same way. The effect is to produce further splitting of the emission lines in atomic spectra. It takes an external magnetic field to observe this induced magnetic field. Otherwise we see degeneracy—if the only difference between states is their orientation, they are of equal energy. In empty space, where all possible directions are equivalent, it should make no difference to the atom in what direction its angular momentum is pointing. The magnetic quantum number can take on any of the integral values from $-\ell$ to $+\ell$, including zero. So, for any ℓ value, there are $2\ell + 1$ values for m_ℓ, expressing the orientation of the subshells in space:

$$m_\ell = -\ell, (-\ell + 1), (-\ell + 2), \ldots, 0, \ldots, (\ell - 1), \ell$$

There are $2\ell + 1$ different values, including zero.

Spin Quantum Number (m_s): A fourth quantum number is needed to specify fully an electron's state. We can imagine that this number describes the direction in which an electron spins about its own axis. The spin quantum number can have one of two values: $m_s = +\frac{1}{2}$ or $-\frac{1}{2}$. The two different values ($+$ and $-$) represent the possible spin orientations:

$$m_s = -\frac{1}{2}, +\frac{1}{2}$$

The accompanying $+$ and $-$ signs are used to distinguish the possible spin orientations in much the same way as we speak of "clockwise" and "counterclockwise."

The state of the lone electron in a hydrogen atom is specified by its set of four quantum numbers. In the ground state, the electron has two distinct states that differ only in their spin orientation about a chosen axis. Both states have $n = 1$, $\ell = 0$, and $m_\ell = 0$; one state has $m_s = -\frac{1}{2}$, and the other has $m_s = +\frac{1}{2}$ (Figure 15–11).

The suggestion that the electron itself is capable of exhibiting magnetic behavior due to some intrinsic angular momentum, or spin, was established in a series of experiments that have come to be known collectively as the Stern-Gerlach experiment. An atom with an odd number of electrons ($Z = 47$, for Ag) would experience a deflecting force in a nonuniform magnetic field. As a consequence, a single beam of such atoms entering the field of the magnet emerges split in two. Lithium ($Z = 3$) and sodium ($Z = 11$) give similar results; so does a gas of hydrogen atoms ($Z = 1$) in a discharge tube. No such splitting occurs for atoms with even numbers of electrons.

The key feature of the experiment can best be understood in terms of a statistical distribution of spin orientation ($m_s = +\frac{1}{2}$ and $m_s = -\frac{1}{2}$) for the unpaired electron; in 50% of the atoms it has a "clockwise" orientation, and in the other 50% it has a "counterclockwise" orientation. In atoms with even numbers of electrons, all of the spins are paired, and the magnetic forces on them cancel out.

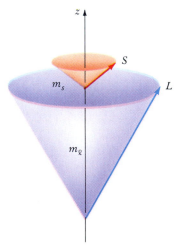

FIGURE 15–11 The orbital angular momentum L and spin S precess about the z-axis. The quantum numbers are n, ℓ, m_ℓ and m_s where $m_s = \pm 1/2$ and determines the component of the spin S along the chosen axis. The m_ℓ quantum number determines the component of orbital angular momentum along the chosen axis.

EXAMPLE 15–3

For the first excited state of the hydrogen atom, corresponding to transitions giving rise to the Balmer lines, how many different states are there?

Solution   There are eight states, which are listed in Table 15–1.   ■

TABLE 15–1 The Four Quantum Numbers

	n	ℓ	m_ℓ	m_s
ground state	1	0	0	$-\frac{1}{2}$
hydrogen	1	0	0	$+\frac{1}{2}$
first excited	2	0	0	$-\frac{1}{2}$
state for	2	0	0	$+\frac{1}{2}$
hydrogen	2	1	-1	$-\frac{1}{2}$
	2	1	-1	$+\frac{1}{2}$
	2	1	0	$-\frac{1}{2}$
	2	1	0	$+\frac{1}{2}$
	2	1	$+1$	$-\frac{1}{2}$
	2	1	$+1$	$+\frac{1}{2}$

EXERCISE 15–3

Using a format similar to that in Table 15–1, specify the states for a hydrogen atom in the second excited state, for which $n = 3$, corresponding to the Paschen series of lines in the emission spectrum. ◼

The Pauli Exclusion Principle

In 1925, Wolfgang Pauli formulated what is now called the **exclusion principle**. One statement of the exclusion principle is that no two electrons in the same atom can have identical states. Since the state of an atomic electron is specified by a set of quantum numbers, another way of wording the exclusion principle is that no two electrons can have identical sets of quantum numbers. Not only is there the coulombic repulsion between like-charged particles such as a pair of electrons or a pair of nuclei, but they avoid each other in a quantum mechanical sense. It is a fact of nature for which there is no counterpart in classical theory. Although the quantum mechanical basis for this principle is beyond the scope of our treatment of atomic structure, its impact could not be more important. The exclusion principle was originally put forward as a response to questions scientists were asking. For example: "Why is the population of the shells in atoms limited?" Or: "Why can some orbits accommodate more electrons than others?" The answers, in the form of this strange principle, are building blocks of the theoretical structure of the entire material world.

One of the consequences of the exclusion principle is that the number of electrons in a particular n shell is restricted to a maximum of $2n^2$ electrons, one in each of the possible states. That being the case, only two electrons are allowed for the $n = 1$ state, since

$$2n^2 = 2(1)^2 = 2$$

EXAMPLE 15–4

How many electrons are allowed in an atom for the n state corresponding to the L shell?

Solution Since $n = 2$, $2n^2 = 2(2)^2 = 8$. ◼

EXERCISE 15–4

Calculate the maximum number of electrons allowed in the n states corresponding to the M, N, and O shells.

Answer: 18, 32, and 50. ◼

TABLE 15–2 Composite Symbols Representing $n + \ell$ Quantum Numbers

Shell	n	ℓ	Symbol	Maximum Number of Pairs
1st shell	1	0	$1s$	1 pair
2nd shell	2	0	$2s$	1 pair
	2	1	$2p$	3 pairs
3rd shell	3	0	$3s$	1 pair
	3	1	$3p$	3 pairs
	3	2	$3d$	5 pairs
4th shell	4	0	$4s$	1 pair
	4	1	$4p$	3 pairs
	4	2	$4d$	5 pairs
	4	3	$4f$	7 pairs

15.6 ELECTRONIC STRUCTURES OF ATOMS

One of the great triumphs of wave mechanics is that calculations performed on hydrogen can be extended to all of the other atoms in the periodic table. In fact, we can "build" the other atoms by (1) adding the necessary electrons in their proper quantum states, according to the four quantum numbers, and (2) adding the appropriate numbers of neutrons and protons, up to the required mass and charge. It is, of course, a thought experiment, but with very interesting and useful results. For a ground state hydrogen atom, the single electron is in the lowest energy, or $n = 1$, state. Consequently:

$$\ell = 0 \text{ and } m_\ell = 0$$

The value of m_s can take on either of the two degenerate, or equal energy, values:

$$m_s = \pm\tfrac{1}{2}$$

A ground state helium atom has two electrons. According to the exclusion principle, both can assume the same values for n, ℓ, and m_ℓ as the one hydrogen electron, because they can have different m_s values: $+\frac{1}{2}$ for one and $-\frac{1}{2}$ for the other. That completes the first shell; by the $2n^2$ rule, for the $n = 1$ state there at most two ground state electrons. The next element is Li with three electrons, one of which must have $n = 2$ in the ground state atom. As shown in Table 15–1, eight electrons are possible for which $n = 2$; according to the $2n^2$ rule, $2(2)^2 = 8$. There can be up to eight electrons in the $n = 2$ state, plus the two in the $n = 1$ state, for a total of 10 electrons when we reach neon. Beyond neon, a third n state must be available for the buildup to continue.

Building up the elements in the periodic table in this way is known as the "Aufbau Prinzip," or **buildup principle**. It has given rise to a useful shorthand notation for the electron configurations of ground state atoms. Each electron in an atom has a composite symbol representing the n and ℓ quantum numbers. The symbol is the value of n, followed by the letter for the value of ℓ. Note that because of the spin orientations, two electrons can have the same symbol. Table 15–2 lists the composite $n + \ell$ symbols for the electrons from $n = 1$ through $n = 4$. Each set (or pair) of electrons having the same symbol represents what is called a set (or pair) of orbital electrons. The word *orbital* is not to be confused with *orbits*. An **orbital** has exact meaning in the mathematical sense: it is a one-electron wave function. But for our introduction to these ideas, we use the term "orbital" loosely and need only say that (1) the orbital electrons are distinguished by their m_ℓ values; (2) in the hydrogen atom, both orbital electrons have the same energy—that is, orbital electron pairs are degenerate; and (3) each orbital electron is assigned one of the two possible m_s values.

The **electron configuration** of an atom is a list of the orbitals that contain electrons and the number of electrons in each of those orbitals. The number of electrons in a particular orbital is specified by using superscript notation. Thus, the electron configuration for a ground state H atom is $1s^1$; for a ground state helium atom it is $1s^2$. According to the exclusion principle, the maximum number of electrons that can go into each set of orbitals depends on the possible values of m_ℓ and m_s. The number of orbital sets (or pairs) equals the number of m_ℓ states, which equals $2\ell + 1$. Since orbital electrons are paired according to their spin states, $m_s = +\frac{1}{2}$ or $-\frac{1}{2}$, the total number of electrons that can fit into a set of orbitals is $2(2\ell + 1) = 4\ell + 2$. Thus, we speak of a $1s$ orbital, which can hold

$$(4 \times 0) + 2 = 2 \text{ electrons}$$

We refer to three $2p$ orbitals, which can hold

$$(4 \times 1) + 2 = 6 \text{ electrons}$$

For the five $3d$ orbitals, there can be up to

$$(4 \times 2) + 2 = 10 \text{ electrons}$$

The seven $4f$ orbitals can have

$$(4 \times 3) + 2 = 14 \text{ electrons}$$

Beyond helium and the filled first shell, it is necessary to know the order in which the electrons fill the orbitals. The order of electron filling happens to follow diagonal lines in Figure 15–12, starting from the top of the diagram and moving downward. For example, the electron configuration for lithium, with three electrons, is $1s^2 2s^1$. The ordering goes according to the $\boldsymbol{n + \ell}$ **rule**, which states that the orbital with the lowest value of $n + \ell$ has the lowest energy. Where two orbital symbols give the same $n + \ell$ value, the lower n value takes precedence in determining the lowest energy state. For example, here are electron configurations for ground state atoms of five different elements, in proper order according to the $n + \ell$ rule. Note especially calcium and titanium:

O $1s^2 2s^2 2p^4$
Al $1s^2 2s^2 2p^6 3s^2 3p^1$
Cl $1s^2 2s^2 2p^6 3s^2 3p^5$
Ca $1s^2 2s^2 2p^6 3s^2 3p^6 4s^2$
Ti $1s^2 2s^2 2p^6 3s^2 3p^6 4s^2 3d^2$

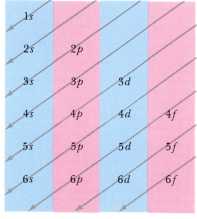

FIGURE 15–12   A memory device for the $n + \ell$ rule. Follow the arrows, starting at the top and proceeding down the page.

EXAMPLE 15–5

Using the $n + \ell$ rule, determine the ordering of the energy states for the orbital electron pairs in the ground state calcium atom.

Solution   The result is the same as that shown above.   ■

EXERCISE 15–5

Work out the $n + \ell$ rule values for the orbital pairs in the ground state titanium atom and confirm your answer by consulting Figure 15–12. ∎

We still have more to say later in this chapter about orbital energies and the order in which electrons are allowed to fill orbitals. The memory aids and rules presented to this point give correct electron configurations up to vanadium, and correct or nearly correct electron configurations for most of the elements beyond. Certain electron configurations are very stable, with very important chemical implications.

Shell Structure and Periodicity

Quantum mechanics provided a framework for understanding the ordering and the properties of the elements in the periodic table in terms of shell structures. If an electron shell is the set of all electrons in a polyelectronic atom having the same principal quantum number n, then helium and neon can be described as elements with closed (or filled) shells. That is a theoretical description. The corresponding experimental observation is that they are the first two members of a family of elements, known as the noble gases, all of which are chemically unreactive. Furthermore, it is interesting to note how many of the ions we commonly encounter in chemical bonding have the same electron shell configuration as the nearest noble gas atom in the periodic table. Figure 15–13 shows the shells for the ions of the first 60 elements in the periodic table. The number of electrons, plotted along the y-axis, is compared to the atomic number Z for the corresponding neutral atoms and their most common ions, plotted along the x-axis. Look at how the shells for the atoms fill or empty to the atomic number of a noble gas when ions form, as indicated by the arrows. With few exceptions (such as nickel and palladium), the preferred atomic numbers are those of the noble gases.

When cations form due to loss of electrons, the electrons most easily lost are those in excess of the atomic number of the nearest noble gas element. For anions resulting from gaining electrons, the electrons most easily acquired match the nearest noble gas element as well. Consider the elements reasonably close to the inert gas neon in the periodic table (for which $Z = 10$). The next element is sodium ($Z = 11$), which forms a unipositive cation, Na^+, with its neon-like electron shell structure. The neon atom and the sodium ion are said to be **isoelectronic**—having the same electron shell structures.

Magnesium ($Z = 12$), the next element in the row, forms a dipositive Mg^{2+} ion—also isoelectronic with neon. Looking on the lighter side of neon, fluorine ($Z = 9$) and oxygen ($Z = 8$) form F^- and O^{2-} on gaining one and two electrons, respectively—and both are isoelectronic with neon. Similar arguments can be made for the elements surrounding argon, krypton, and xenon.

You may also have noticed that Figure 15–13 suggests that stable electron shell structures are not limited to the noble gases. Palladium shares preferred status with members of the noble gas family of elements in the sense that nearby elements acquire isoelectronic shell structures when they become ions. Yet palladium is not a noble gas, nor is it chemically unreactive. What this means is that the simple view of closed shells and isoelectronic

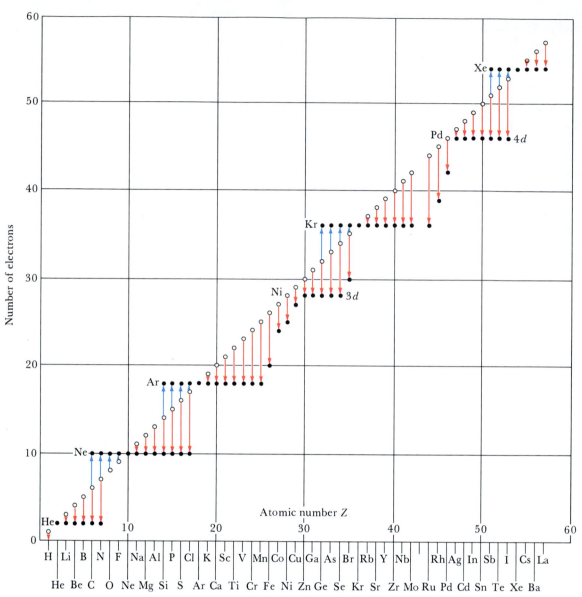

FIGURE 15–13   Electron shells for the ions of the atoms of the first sixty elements of the periodic table. (After Kossel, 1916.)

structures is incomplete for elements with higher Z values, and further explanation is needed. Note first that the electron configurations for helium, neon, and palladium are:

He $1s^2$

Ne $1s^2 2s^2 2p^6$

Pd $1s^2 2s^2 2p^6 3s^2 3p^6 3d^{10} 4s^2 4p^6 4d^{10}$

Since all three qualify as atoms with filled electron shells, what is it that is uniquely associated with especially stable electron configurations? It is the completion of a partial shell, upon occupation of all states with the same ℓ value for a given n value. In fact, this is the case for the noble gases beyond neon as well. Argon, for example, has this partially filled electron configuration:

Ar $1s^2 2s^2 2p^6 3s^2 3p^6$

TABLE 15–3 Electron Configurations for the Highest Occupied Shells or Partially Filled Shells* for the Noble Gases

State	Element (Z)	IE$_1$ eV
$1s^2$	He (2)	24.58
$2s^2 2p^6$	Ne (10)	21.56
$3s^2 3p^6$	Ar (18)	15.76
$4s^2 3d^{10} 4p^6$	Kr (36)	14.00
$5s^2 4d^{10} 5p^6$	Xe (54)	12.13
$6s^2 4f^{14} 5d^{10} 6p^6$	Rn (86)	10.75

*Beyond neon (and the second shell), it is the completed $s^2 p^6$ electron configuration that effectively closes the subshells associated with the characteristic chemical properties of the noble gas elements.

TABLE 15–4 Ionization Energies for the Transitions from Atoms to Ions, eV/atom

Atom	IE$_1$	IE$_2$	IE$_3$	IE$_4$
H	13.59			
He*	_24.5_	54.1		
Li	5.4	_75_	122	
Be	9.3	18.2	_154_	217
B	8.3	25.1	38	_259_
C	11.3	24.5	48	64.5
N	14.6	29.6	47	77.4
O	13.6	35.2	55	77.4
F	17.4	34.9	62.7	87.3
Ne*	_21.6_	41.0	63.9	96.4
Na	5.14	_47.3_	71.7	98.9
Mg	7.64	15.0	_80.2_	109.3
Al	5.97	18.8	28.5	_120_
Si	8.15	16.4	33.5	44.9
P	10.9	19.7	30.2	51.4
S	10.4	23.4	35.1	47.1
Cl	12.9	23.7	39.9	53.5
Ar*	_15.8_	27.5	40.7	61
K	4.3	_31.7_	45.5	60.6
Ca	6.1	11.9	_51_	67

*Note that the ionization energy is always particularly large for the noble gas electron configurations due to disruption of a closed shell. It is especially low for the element just beyond, with one electron more than the noble gas configuration.

The $3s$ and $3p$ orbitals are filled in the third shell, but there are no $3d$ electrons. The remaining noble gases, krypton, xenon, and radon, have similar electron configurations (Table 15–3). Thus, for elements that immediately follow the noble gases, removal of an electron to form the unipositive ($+1$) ion with all shells filled (closed) is particularly easy and requires very little energy.

In the previous chapter we mentioned the ionization energy, the energy needed to remove one electron from an atom or ion. This is called IE$_1$ for the first electron, IE$_2$ for the second, and so on. Comparing ionization energies between neighbors in the periodic table, the alkali metals have very low ionization energies, while the ionization energies for the noble gases are very high. For example, the value for helium is 24.46 eV, while neighboring lithium has an ionization energy of only 5.40 eV. The ionization energies of the heavier alkali metals are lower still. The energy required to remove the second electron from an alkali metal ion must be very large, because the electron configuration of the singly charged ion is the same as that of a chemically unreactive noble gas (Table 15–4). A table for electron configurations of all the elements is provided in Table 15–5.

Before we continue, let us plainly state that the shell structures we have been discussing, which are so important to our understanding of the periodic properties of the elements, account only for the assignment of "the last added electron." Beginning with hydrogen, as successive electrons

TABLE 15–5 Electron Configurations of the Elements

1	H	$1s^1$	37	Rb	$[Kr]\ 5s^1$	71	Lu	$[Xe]\ 6s^24f^{14}5d^1$
2	He	$1s^2$	38	Sr	$[Kr]\ 5s^2$	72	Hf	$[Xe]\ 6s^24f^{14}5d^2$
3	Li	$[He]\ 2s^1$	39	Y	$[Kr]\ 5s^24d^1$	73	Ta	$[Xe]\ 6s^24f^{14}5d^3$
4	Be	$[He]\ 2s^2$	40	Zr	$[Kr]\ 5s^24d^2$	74	W	$[Xe]\ 6s^24f^{14}5d^4$
5	B	$[He]\ 2s^22p^1$	41	Nb	$[Kr]\ 5s^14d^4$	75	Re	$[Xe]\ 6s^24f^{14}5d^5$
6	C	$[He]\ 2s^22p^2$	42	Mo	$[Kr]\ 5s^14d^5$	76	Os	$[Xe]\ 6s^24f^{14}5d^6$
7	N	$[He]\ 2s^22p^3$	43	Tc	$[Kr]\ 5s^24d^5$	77	Ir	$[Xe]\ 6s^24f^{14}5d^7$
8	O	$[He]\ 2s^22p^4$	44	Ru	$[Kr]\ 5s^14d^7$	78	Pt	$[Xe]\ 6s^14f^{14}5d^9$
9	F	$[He]\ 2s^22p^5$	45	Rh	$[Kr]\ 5s^14d^8$	79	Au	$[Xe]\ 6s^14f^{14}5d^{10}$
10	Ne	$[He]\ 2s^22p^6$	46	Pd	$[Kr]\ \ \ \ 4d^{10}$	80	Hg	$[Xe]\ 6s^24f^{14}5d^{10}$
11	Na	$[Ne]\ 3s^1$	47	Ag	$[Kr]\ 5s^14d^{10}$	81	Tl	$[Xe]\ 6s^24f^{14}5d^{10}6p^1$
12	Mg	$[Ne]\ 3s^2$	48	Cd	$[Kr]\ 5s^24d^{10}$	82	Pb	$[Xe]\ 6s^24f^{14}5d^{10}6p^2$
13	Al	$[Ne]\ 3s^23p^1$	49	In	$[Kr]\ 5s^24d^{10}5p^1$	83	Bi	$[Xe]\ 6s^24f^{14}5d^{10}6p^3$
14	Si	$[Ne]\ 3s^23p^2$	50	Sn	$[Kr]\ 5s^24d^{10}5p^2$	84	Po	$[Xe]\ 6s^24f^{14}5d^{10}6p^4$
15	P	$[Ne]\ 3s^23p^3$	51	Sb	$[Kr]\ 5s^24d^{10}5p^3$	85	At	$[Xe]\ 6s^24f^{14}5d^{10}6p^5$
16	S	$[Ne]\ 3s^23p^4$	52	Te	$[Kr]\ 5s^24d^{10}5p^4$	86	Rn	$[Xe]\ 6s^24f^{14}5d^{10}6p^6$
17	Cl	$[Ne]\ 3s^23p^5$	53	I	$[Kr]\ 5s^24d^{10}5p^5$	87	Fr	$[Rn]\ 7s^1$
18	Ar	$[Ne]\ 3s^23p^6$	54	Xe	$[Kr]\ 5s^24d^{10}5p^6$	88	Ra	$[Rn]\ 7s^2$
19	K	$[Ar]\ 4s^1$	55	Cs	$[Xe]\ 6s^1$	89	Ac	$[Rn]\ 7s^2\ \ \ \ 6d^1$
20	Ca	$[Ar]\ 4s^2$	56	Ba	$[Xe]\ 6s^2$	90	Th	$[Rn]\ 7s^2\ \ \ \ 6d^2$
21	Sc	$[Ar]\ 4s^23d^1$	57	La	$[Xe]\ 6s^2\ \ \ \ 5d^1$	91	Pa	$[Rn]\ 7s^25f^26d^1$
22	Ti	$[Ar]\ 4s^23d^2$	58	Ce	$[Xe]\ 6s^24f^15d^1$	92	U	$[Rn]\ 7s^25f^36d^1$
23	V	$[Ar]\ 4s^23d^3$	59	Pr	$[Xe]\ 6s^24f^3$	93	Np	$[Rn]\ 7s^25f^46d^1$
24	Cr	$[Ar]\ 4s^13d^5$	60	Nd	$[Xe]\ 6s^24f^4$	94	Pu	$[Rn]\ 7s^25f^6$
25	Mn	$[Ar]\ 4s^23d^5$	61	Pm	$[Xe]\ 6s^24f^5$	95	Am	$[Rn]\ 7s^25f^7$
26	Fe	$[Ar]\ 4s^23d^6$	62	Sm	$[Xe]\ 6s^24f^6$	96	Cm	$[Rn]\ 7s^25f^76d^1$
27	Co	$[Ar]\ 4s^23d^7$	63	Eu	$[Xe]\ 6s^24f^7$	97	Bk	$[Rn]\ 7s^25f^9$
28	Ni	$[Ar]\ 4s^23d^8$	64	Gd	$[Xe]\ 6s^24f^75d^1$	98	Cf	$[Rn]\ 7s^25f^{10}$
29	Cu	$[Ar]\ 4s^13d^{10}$	65	Tb	$[Xe]\ 6s^24f^9$	99	Es	$[Rn]\ 7s^25f^{11}$
30	Zn	$[Ar]\ 4s^23d^{10}$	66	Dy	$[Xe]\ 6s^24f^{10}$	100	Fm	$[Rn]\ 7s^25f^{12}$
31	Ga	$[Ar]\ 4s^23d^{10}4p^1$	67	Ho	$[Xe]\ 6s^24f^{11}$	101	Md	$[Rn]\ 7s^25f^{13}$
32	Ge	$[Ar]\ 4s^23d^{10}4p^2$	68	Er	$[Xe]\ 6s^24f^{12}$	102	No	$[Rn]\ 7s^25f^{14}$
33	As	$[Ar]\ 4s^23d^{10}4p^3$	69	Tm	$[Xe]\ 6s^24f^{13}$	103	Lr	$[Rn]\ 7s^25f^{14}6d^1$
34	Se	$[Ar]\ 4s^23d^{10}4p^4$	70	Yb	$[Xe]\ 6s^24f^{14}$	104		$[Rn]\ 7s^25f^{14}6d^2$
35	Br	$[Ar]\ 4s^23d^{10}4p^5$				105		$[Rn]\ 7s^25f^{14}6d^3$
36	Kr	$[Ar]\ 4s^23d^{10}4p^6$				106		$[Rn]\ 7s^25f^{14}6d^4$

are added, and accompanied by increasing nuclear charge, the relative energies of the inner shells and subshells change because each electron experiences the presence of the nucleus and the other electrons. Figure 15–14 provides a summary of the relative energies, which are reflected in both the shell and subshell quantum numbers. All of this either is known from or has been confirmed by x-ray spectroscopy, and more recently by photoelectron spectroscopy, a modern technique for determining binding energies for the inner electrons in atoms. Because each atom has a unique arrangement of energy levels for the quantum states, diagrams of these levels have proved to be useful descriptors.

EXAMPLE 15–6

Using the $n + \ell$ rule, pick the lower energy state in each pair: (a) $4s$ and $4p$; (b) $5s$ and $4p$.

Solution (a) Between $4s$ and $4p$, the rule selects $4s$ since

$$4s = 4 + 0 = 4 \quad \text{versus} \quad 4p = 4 + 1 = 5$$

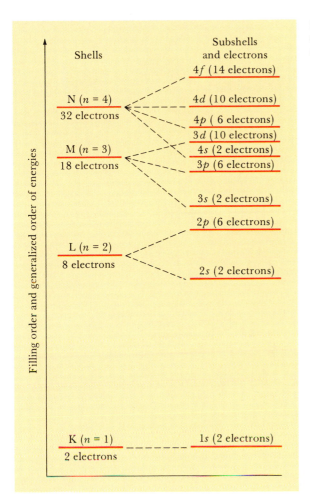

FIGURE 15–14 Shell structure and the atomic energy levels. The order is according to the last electron added and for the inner electrons.

(b) Between $5s$ and $4p$, the rule selects $4p$ since the $n + \ell$ values are the same in both cases; when that situation exists, the lower n value prevails:

$$5s = 5 + 0 = 5 \qquad \text{versus} \qquad 4p = 4 + 1 = 5$$

EXERCISE 15–6(A)

According to the $n + \ell$ rule, which subshell energy in each pair is lower?
(a) $5p$ or $4f$ (b) $5d$ or $4p$
Answer: $5p$; $4p$.

EXERCISE 15–6(B)

Why is the first ionization energy for helium so much greater than that for hydrogen? Why are both so much greater than the first ionization energy for lithium?

Hund's Rule

We have discussed the electronic configuration of ground state atoms with respect to the n and ℓ quantum numbers, and it remains to state the case for the m_ℓ and m_s numbers. Figure 15–15 shows the ground states and the electron configurations for atoms of the first eleven elements in the periodic table. The first excited states for beryllium and carbon are also included:

FIGURE 15–15 The manner in which electrons are accommodated in energy levels and sublevels for ground state atoms from H to Na. The two atoms noted with an asterisk are excited states.

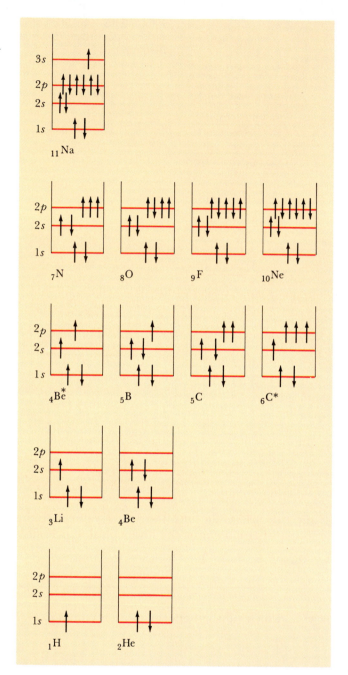

- Hydrogen and helium have $1s^1$ and $1s^2$ electrons, completing population of the K shell.
- Lithium contains the helium core plus an s electron in the second shell, and we write the electron configuration as $[He]2s^1$.
- The $2s$ subshell is closed with beryllium, which has the ground state electron configuration $[He]2s^2$. Because the $2p$ subshell lies very close (in energy) to the $2s$ state, beryllium can easily be excited to the lowest excited state, Be*, $[He]2s^12p^1$.
- The $2p$ subshell starts to fill, beginning with boron, $[He]2s^22p^1$.
- For carbon, $[He]2s^22p^2$, from spectroscopic data we know that the spins of the two $2p$ electrons are the same (parallel); that is, either both are $+\frac{1}{2}$ or both are $-\frac{1}{2}$.

- In nitrogen, the three $2p$ orbital electrons all have parallel spins.
- Spin pairing of the $2p$ electrons begins with oxygen, proceeds through fluorine, and ends with the closing of the subshell and the shell at neon, the noble gas: $[He]2s^2 2p^6$.
- Population of the M shell begins with metallic sodium and magnesium, first filling $3s$; then the $3p$ orbital electrons fill, until we again encounter the $s^2 p^6$ filled subshell (partially closed shell) arrangement, characteristic of all the remaining noble gas elements, at argon: $[Ne]2s^2 2p^6$.

As we discussed earlier, electrons having the same ℓ values are degenerate, or of equal energy. They are divided among the possible m_ℓ sublevels in ground state atoms in such a way that the maximum number of parallel spins result. That arrangement is called maximum multiplicity. The higher the multiplicity (the more parallel spins), the lower is the energy of a given state. Consequently, we must assign one electron to each m_ℓ state—spins parallel—before pairing of the spin states occurs. This is **Hund's Rule**: A second electron will not populate any m_ℓ state within a given subshell until each such state contains one electron; single electrons in their separate m_ℓ states display maximum multiplicity of spins. That means parallel spins occur whenever possible.

EXAMPLE 15–7

Consider how one might populate the three m_ℓ states in the L shell of nitrogen with three $2p$ electrons. The arrows in the following diagram indicate spin orientations, which have been placed parallel or antiparallel:

m_ℓ	-1	0	$+1$
case 1	$\uparrow$	$\uparrow$	$\uparrow$
case 2	$\uparrow$	$\downarrow$	$\uparrow$
case 3	$\uparrow\downarrow$	$\uparrow$	
case 4	$\uparrow\uparrow$	$\uparrow$	

According to Hund's rule, case 1 represents the ground state, or lowest energy arrangement; cases 2 and 3 are excited states; and case 4 represents an impossible state, with both electrons in a given orbital having parallel spins—that condition is forbidden. Here are the quantum numbers for all seven electrons in the ground state nitrogen atom:

	n	ℓ	m_ℓ	m_s	$n + \ell$
K shell	1	0	0	$+\frac{1}{2}$	1
	1	0	0	$-\frac{1}{2}$	1
L shell	2	0	0	$+\frac{1}{2}$	2
	2	0	0	$-\frac{1}{2}$	2
	2	1	-1	$+\frac{1}{2}$	3
	2	1	0	$+\frac{1}{2}$	3
	2	1	$+1$	$+\frac{1}{2}$	3

Note that the m_s values for the three $2p$ states could just as well have been designated $-\frac{1}{2}$.

EXERCISE 15–7(A)

Write the electronic configurations for the ground state aluminum and phosphorus atoms; the ground state chlorine and argon atoms; and the first excited states for magnesium and silicon.

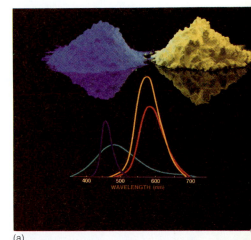
(a)

(b)

In a fluorescent lamp, a gas discharge produces ultraviolet light that is converted into visible light by a phosphor coating on the inside of the tube. (a) The two-component phosphor of praseodymium and yttrium fluorides, shown here, produces the visible soft violet-pink light flooding the face of the researcher in (b) from a process that begins with praseodymium ions absorbing ultraviolet radiation of 185 nm wavelength as they are excited into an upper energy state. After excitation, the praseodymium ions return to their normal state by releasing the absorbed energy, producing quanta that are characteristically in the visible region of the electromagnetic spectrum, 400–700 nm.

EXERCISE 15–7(B)

For each of the species in Exercise 15–7(A), use the pictorial scheme of parallel and antiparallel arrows to represent the electronic configurations. ∎

Ionization Energy and Shell Structure

The **periodic law** states that the properties of the elements recur periodically when the elements are arranged in order of increasing atomic number. A striking example can be found on examination of ionization energy, the energy required to remove an electron from the field of the nucleus and the other electrons of a gas-phase atom (Table 15–4). In the case of the hydrogen atom, 13.60 eV is required to produce the proton:

$H \longrightarrow H^+ + e^-$	13.60 eV	$1s^1$

The lithium atom is far less stable with respect to ionization, and the stability continues to fall off, going down through the list of alkali metal elements in Table 15–4:

$Li \longrightarrow Li^+ + e^-$	5.39 eV	$2s^1$
$Na \longrightarrow Na^+ + e^-$	5.14 eV	$3s^1$
$K \longrightarrow K^+ + e^-$	4.34 eV	$4s^1$
$Rb \longrightarrow Rb^+ + e^-$	4.18 eV	$5s^1$
$Cs \longrightarrow Cs^+ + e^-$	3.89 eV	$6s^1$

As we move down through a family of elements in the periodic table, the decreasing stability of the outermost electron is a consequence of the greater distance and shielding brought about by introduction of successive shells of electrons between the nucleus and the outer electron.

Going across the periodic table, the data aren't as simple to interpret, but the trend is clear:

Li	5.39 eV	$2s^1$
Be	9.32 eV	$2s^2$
B	8.30 eV	$2s^2 2p^1$
C	11.26 eV	$2s^2 2p^2$
N	14.54 eV	$2s^2 2p^3$
O	13.61 eV	$2s^2 2p^4$
F	17.42 eV	$2s^2 2p^5$
Ne	21.56 eV	$2s^2 2p^6$

Stability can be measured by the added energy required to remove an outer shell electron, which increases with the number of electrons populating the outer shell. But the manner of filling is interesting. There is an obvious increase on closing the $2s$ subshell at beryllium. Addition of one $2p$ electron creates the first of the three equivalent m_ℓ states, in boron; this results in a decrease, followed by increasing ionization energy as the second and third $2p$ electrons are added at carbon and nitrogen. There is a decrease at oxygen, followed by increasing ionization energy to the end of the period and the closing of the L shell.

The two irregularities that occur at boron and oxygen are caused by the requirement that the last electron be placed in a relatively high energy level for these two elements. In the case of boron, the last electron is placed in a $2p$ rather than a $2s$ orbital. In the case of oxygen, the last electron must spin-pair; it must share an orbital with another electron.

In any polyelectronic atom, the $2p$ sublevel is at a higher energy than the $2s$ sublevel, and spin-pairing always increases the energy of a system due to additional electron-electron repulsion. An electron that is in a higher energy level is easier to remove and has a lower ionization energy. It is useful to remember these irregularities by realizing that they occur immediately after closing the $2s$ subshell and immediately after half-filling the $2p$ subshell.

Down the periodic table or across it, ionization energies do not follow a simple empirical relationship, but they do help make a strong case for periodicity among chemical properties. Metallic character, for example, increases as one moves down through a family of elements and decreases as one moves from left to right across a period; in a very general way, it increases diagonally from the upper right-hand corner of the table to the lower left. Metallic character, as we shall verify later, is associated with the ability of an atom to give up an outer shell electron. The smaller the ionization potential, the more metallic the element. Sodium is more metallic than lithium. Lithium is more metallic than beryllium. More will be said about the relationships between structure and properties in Chapter 17.

15.7   QUANTUM MECHANICS AND THE HYDROGEN ATOM

In the beginning, there was the Bohr theory of the hydrogen atom, which was simple enough. The electron was a particle with a certain location and a definite velocity at a given instant. Unfortunately, it quickly proved to be far too simple a picture, being largely limited to one-electron systems such as the hydrogen atom or hydrogen-like ions. Quantum mechanics presents the electron as a pattern of unconfined standing waves, restricted by coulombic forces of attraction between the negative electron and the positive nucleus. Since coulombic forces become progressively weaker as the distance from the nucleus increases, the probability of finding the electron just anywhere in the universe is small. For the hydrogen atom in its lowest energy state, the probability is greatest at a distance $r = a_0 = 0.529$ Å, where the matter wave is at a maximum. We know a_0 from quantum mechanics; we also know a_0 from the classical Bohr theory. Both give the same result.

An enormously important part of the task ahead of us is to develop "pictures" that are consistent with the important properties of atoms and molecules. First, in the remaining pages of this chapter, we will look at how the distribution of the electrons in atoms can be understood in simple terms using quantum mechanics. Our goal here is to establish the means to understand the properties of molecules. In subsequent chapters, we will examine the chemical bonds that hold atoms in place in molecular arrangements, how the distribution of charge produces dipole moments and polar molecules, and what that means for intermolecular forces such as hydrogen bonding and hydration of ions and molecules in solution. We will also discover why some compounds are colored or why they are susceptible to chemical change under the influence of heat and light, and even look at the delicately turned biochemistry of the cell. These explanations must ultimately be quantum mechanical because they involve the distribution of the electrons. But if even the simplest molecules can't be understood exactly

It is always difficult to express abstract ideas. Even theorists and higher mathematicians try to use simple expressions of the physical reality their work represents. In a sense, we always try to "picture" ideas and connect them to things.

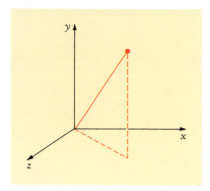

FIGURE 15–16 Spherical coordinate system for an atom, assuming the origin to be the nucleus.

(a)

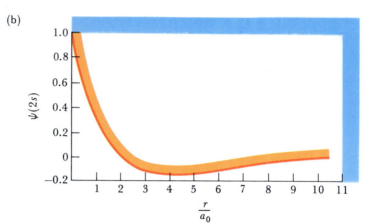

(b)

(c)

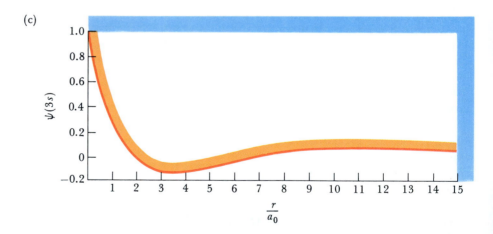

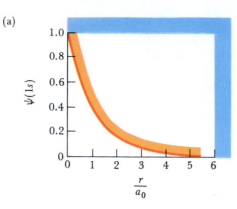

FIGURE 15–17 Plots of wavefunctions versus distance for spherically symmetric $\Psi 1s$, $\Psi 2s$ and $\Psi 3s$ states for hydrogen. The distance scale has units of $a_0 = 0.529$ Å.

using quantum mechanics, what hope do we have here of establishing some introductory level of understanding? The answer lies in good working models—pictures—based on approximations to an acceptable limit of accuracy.

The wave function Ψ is a particular mathematical solution of the equation describing a given state for an atomic electron. Although this function is abstract, the *square* of the wave function Ψ can be interpreted as the **probability density** for the electron at any point where the origin of the space coordinates is set at the nucleus (Figure 15–16). Mainly, we want to picture hydrogen wave functions and their squares because they will in turn be used in subsequent chapters to represent electron distributions. The graphs in Figure 15–17 are plots of Ψ for the hydrogen atom 1s, 2s,

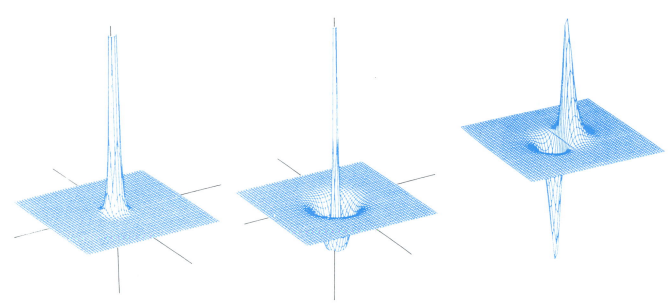

FIGURE 15–18 Reading left to right, computer plots of $\Psi 1s$, $\Psi 2s$ and $\Psi 2p_x$. (After Streitwieser and Owens.)

and $3s$ atomic orbitals. Note that each has a maximum at $r = 0$, the position of the nucleus, and each falls away as the distance from the nucleus increases. For the lowest energy, or ground state, for which $n = 1$, the electron is closest to the nucleus. The first excited state is $n = 2$, then $n = 3$ and so forth, with the electron being separated from the nucleus by more and more space. Computer graphics can be used to add some dimension to our picture of the hydrogen atom wave functions. The $\Psi 1s$ and $\Psi 2s$ pictures in Figure 15–18 were obtained by rotating the corresponding plots in Figure 15–17(a) and (b) about the origin. The $\Psi 2p_x$ picture represents a wave function that is not symmetrical about the origin.

For a given value of the distance r from the nucleus to the immediate vicinity of the electron, the chance of the electron being found is given by a **radial distribution function**, P_r. The radial distribution function is a measure of the probability of locating the electron on a thin, spherical shell of radius r surrounding the nucleus. The probability is proportional to the volume of the spherical shell. Since the volume of a spherical shell is directly proportional to r^2, the probability P_r of finding the electron some distance r from the nucleus is proportional to $r^2\Psi^2$. In Figure 15–19, this is plotted versus r/a_0, again for the $1s$, $2s$, and $3s$ states for the hydrogen atom. Note that the highest probability of finding the electron in the $1s$ ground state hydrogen atom ($n = 1$ and $\ell = 0$) is a maximum at $r = a_0$, corresponding to the first Bohr radius:

$$r = a_0 = 0.529 \text{ Å}$$

For the $2s$ orbital ($n = 2$ and $\ell = 0$), the first excited state for the hydrogen atom, the most probable value for r is $5.2a_0$; for the $3s$ orbital ($n = 3$ and $\ell = 0$), the most probable value is at a distance from the nucleus of about $13a_0$.

Generalizing, we can say that the radial probability density will be largest in the region where the Bohr radius is $r_n = n^2a_0/Z$. These calculations give an estimate of the extent of the electron clouds. As the principal quantum number n increases, the electron clouds expand, the average

FIGURE 15–19   Probability (which is proportional to $r^2\Psi^2$) of finding the electron in a hydrogen atom on the skin of a spherical shell of radius r from the origin, where the nucleus is located. Coordinates give $r^2\Psi^2$ versus distance in units of $a_0 = 0.529$ Å.

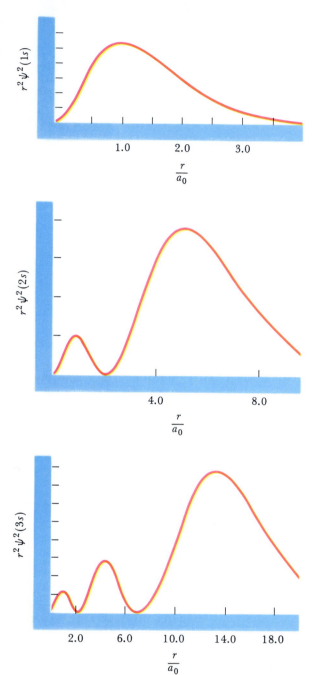

distance from the nucleus being proportional to n^2. It is also interesting to note that for atoms of elements beyond hydrogen (for which $Z = 1$), the electron clouds contract with increasing atomic number. In good agreement with the Bohr theory, quantum mechanics also predicts the proper relationship between atomic size and values of n and Z, except that the concept of atomic size is less precise for the quantum mechanical model because of the diffuseness of the electron cloud.

EXAMPLE 15–8

Confirm that the Bohr radius for the 2s state in an H atom is consistent with the graph in Figure 15–19.

Solution   We make the appropriate substitutions into the general equation for the Bohr radius and calculate the distance r:

$$r = (2)^2(0.529 \text{ Å}) = 2.12 \text{ Å}$$

The actual value (obtained from quantum mechanical theory) places the maximum close to 2.8 Å, as measured on the abscissa of the graph in Figure 15–19 as $r/a_0 = (2.77 \text{ Å}/0.529 \text{ Å}) = 5.24$. Recall that we can also calculate the energy corresponding to the $2s$ state for the H atom. For a Bohr atom:

$$E = -\frac{E_0}{n^2} = -\left(\frac{13.6}{n^2}\right)\text{eV} = -3.4 \text{ eV}$$

This represents the first excited state for hydrogen.   ■

EXERCISE 15–8

Find the precise values for r and E predicted by the Bohr theory for the $2s$ states in the hydrogen-like helium ion.
Answer: $r = 1.06 \text{ Å}$; $E = -13.6 \text{ eV}$.   ■

For the $3s$ orbital, we learn from the graph that the most likely location for the electron is 6 to 8 Å from the nucleus. The two smaller maxima, corresponding to about 0.5 and 1.5 Å, suggest the possibility that the electron can come fairly close to the nucleus. The first standing wave corresponds to the first Bohr orbit, at a radial distance of 0.529 Å. It should not escape your attention that there are three values for r where the electron has a zero probability, which means there is zero likelihood of finding the electron at the origin where $r = 0$, at $r = 0.9$ Å, and at $r = 3.8$ Å.

Because there is always a small but finite probability of finding the electron at large distances from the nucleus, we cannot draw a picture of a volume that "contains" the electron. However, it is a reasonable and convenient approximation to represent a boundary surface that contains 90% of the electron density. For all the s-orbital electrons, the electron distribution is uniform in all directions, and the boundary surfaces are simple spheres centered at the nucleus. For $1s$, $2s$, and $3s$ electrons, these representations are shown in Figure 15–20.

In addition to the radial distribution function, **angular distribution functions** for different quantum states yield three-dimensional representations for p, d, and f orbitals. The pictures are not spherical as is the case for s-orbital electrons. Instead, they are multi-lobed surfaces, with the lobes directed in space. The curves in Figure 15–21 are three-dimensional

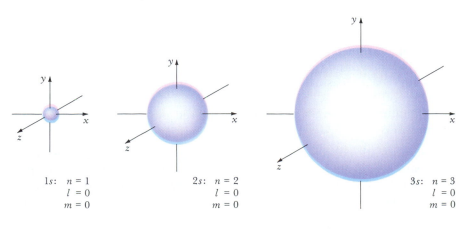

FIGURE 15–20   Spheres representing boundary surfaces for $1s$, $2s$, and $3s$ electrons.

$1s$: $n = 1$
    $l = 0$
    $m = 0$

$2s$: $n = 2$
    $l = 0$
    $m = 0$

$3s$: $n = 3$
    $l = 0$
    $m = 0$

FIGURE 15–21 Distribution functions for the electronic states of the H atom: s ($\ell = 0$), p ($\ell = 1$), and d ($\ell = 2$) electronic states. The f electronic states (for which $\ell = 3$) are shown on the next page.

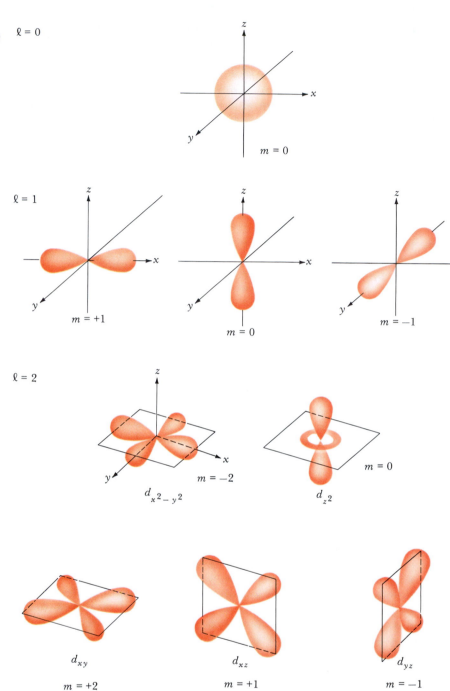

representations of the angular distribution function for s, p, d, and f electrons. The magnitude of a straight line from the origin to a point on a given curve is a measure of the probability of locating the electron in the direction of that line. .

Quantum mechanics has radically altered our view of reality in different ways:

1. Characteristic concepts of length and energy have been introduced that dominate atomic phenomena at the submicroscopic level. The electrostatic attraction between the electron and the proton nucleus of the hydrogen atom on the one hand, and the kinetic energy of such an electrostatically

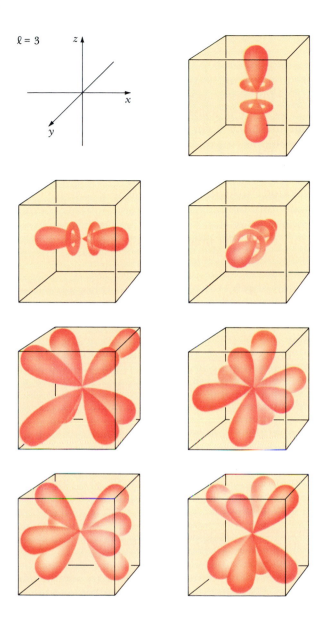

confined electron on the other hand, serve to define the Bohr radius as the unit of length and the Rydberg constant as the carrier of the unit of energy.

2. The use of quantum numbers for characterizing the states of electrons in atoms has reduced the qualities of atomic matter to quantities and magnitudes. As chemists, we attempt to explain all the properties of an atom in a given state in terms of the number of electrons and the quantum numbers of that state.

3. Fundamental shapes of atoms are directly determined from the forces binding the electrons. These shapes have only probabilistic meaning for any given atom. Nonetheless, understanding this intrinsic uncertainty has provided the most refined description of atomic reality to date, namely quantum mechanics (although we certainly have not heard the last word on this topic). Collectively, these wave functions and their orbital representations are important because they describe the structures of atoms, and ultimately, molecules.

15.8 LASER LIGHT

The Bohr atom model created all kinds of problems and challenges for chemistry, including such questions as "How does an electron in a given state know it should jump? How does an electron decide what state to choose? Under what circumstances does it absorb or emit energy?" Answers to those questions eventually took the form of probabilities that could be determined by quantum mechanical computations. But long before that was possible, Einstein opened the way. In a classic scientific paper published in 1917, he described three transition probabilities for an electron in a given energy state. The first kind Einstein referred to as spontaneous emission—the probability that an atom in an excited state spontaneously emits a photon and falls to a lower energy state. The second kind he referred to simply as absorption—the probability that an atom in any lower energy state absorbs a photon and jumps to a higher energy state when illuminated by light of the proper frequency.

Neither of these suggestions was surprising, even then. They accounted for the excitation of atoms, and for their emission and absorption spectra. It was the third probability that was surprising: If an atom in an *excited* state is illuminated with light of the proper frequency, causing it to emit an identical photon and fall to a lower energy state, the probability of such an event would be described as **stimulated emission**. The fall to a lower energy state, which would eventually have happened spontaneously, was triggered by a photon identical to the one the atom was about to emit, crashing into it.

You could of course argue that stimulated emission does little more than accelerate the process of spontaneous emission. Why then the special interest? Consider what happens when photons of proper frequency are directed at a substance containing atoms, each of which is in the same excited state. As atom after atom is stimulated to emit a photon, the number of photons begins to grow geometrically—one photon produces two; the two lead to four; then eight, sixteen, thirty-two, sixty-four, and so forth, until the process that began with a single photon has produced a flood of identical photons.

Now for the important difference! Stimulated photons are identical in every way to spontaneously emitted photons: they have the same energy and the same frequency. But whereas a million excited atoms spontaneously emitting photons would eject them in a million different directions, the same million excited atoms in a cascade of stimulated emissions produce an intense, directed beam of radiation—a million photons moving in parallel and, most importantly, almost perfectly in phase with each other. Such a beam is said to be **coherent**.

If the concept seems simple enough, reduction to practice proved to be quite another matter. It was nearly forty years (1955) before a chain reaction of stimulated emission of photons was achieved for microwave radiation, and a bit longer (1960) for visible light (Figure 15–22). The devices are known respectively as masers and lasers—acronyms for **m**icrowave (or **l**ight) **a**mplification by **s**timulated **e**mission of **r**adiation. The optical (or light) maser, or laser, has become the light source without which modern spectroscopy could not exist.

You may ask, "How can atoms be kept in the required excited state, ready to be stimulated into photon emission?" Not easily! Think of the problems working against this state of affairs. First, whenever an atom emits a stimulated photon, it falls to a lower excited state and becomes a potential absorber of exactly the photon emitted. That, of course, would be counterproductive. Secondly, in any normal state of thermal equilib-

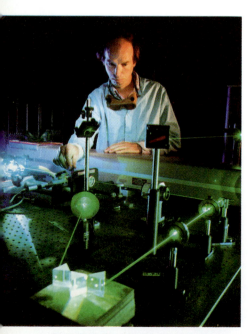

FIGURE 15–22 Using laser spectroscopy to measure the impurities in a GaAs semiconductor.

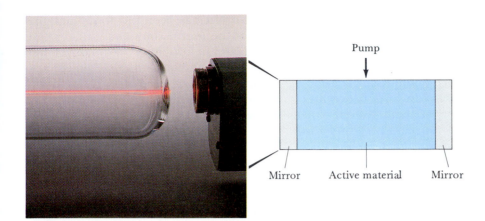

FIGURE 15–23 The helium-neon laser generates a coherent beam of light at approximately 6400 Å, in the orange-red region of the electromagnetic spectrum.

Pump

Mirror   Active material   Mirror

FIGURE 15–24 Using a special dye laser, short pulses of light—on the order of femtoseconds—can be generated and used to provide slow motion measurements of the twisting of chemical bonds.

rium, more atoms are photon absorbers, lying in the lower energy state, than higher energy photon emitters. What's required of the laser device, then, is the means (1) to create and maintain a large population of excited-state atoms (a situation called a **population inversion**), and (2) to pump the energy-deflated atoms back to the excited state by a process other than photon absorption, usually by electron bombardment in a discharge tube, or optical pumping, which is irradiation with light.

One of the earliest and still the most popular examples of a laser device is the helium-neon laser (Figure 15–23). Stimulated emission takes place among the neon atoms. The helium atoms are there to excite (or pump) the neon atoms back to their upper level so they may again emit photons. To begin, energetic electrons—produced with the aid of an external power supply—flow through the gas, exciting many helium atoms to a long-lived (metastable) state 20.61 eV above the ground state. Here, one of the helium atom's two electrons is in the $1s$ state; the other is in the higher energy $2s$ state. In this excited state the helium atom roams through the gas until it collides with a neon atom. By design, the helium excitation energy happens to match an excitation energy of neon ($2s \rightarrow 3s$). Energy is transferred, and the neon atom ends up 20.66 eV above its ground state. When more neon atoms populate the upper than the lower level, the laser is ready to work. By stimulated emission, the neon atom falls 1.96 eV to a state 18.70 eV above its ground state. This energy is dissipated by other means, usually collisions with the walls of the laser tube.

Mirrors at the ends of the laser tube cause most of the radiation to be reflected back and forth many times so that each photon stimulates the emission of many photons, all in the same direction. But 1 or 2% leaks through the mirror at one end, because it has been purposely designed not to be totally reflective, and that is the useful component of the laser beam. In the case of the helium-neon laser, a low-energy beam of red light escapes. It is monochromatic because the photons stimulate emission of more photons of the same frequency, and it is coherent because standing waves can form between two mirrors only if a multiple of $\lambda/2$ exactly fits between them. Note also that only the photons moving in the axial direction are reflected back and forth, while those moving in any other direction must quickly leave the laser.

In a few short years since the operation of the first ruby laser in 1960, it has become possible for engineers and scientists, in fields as diverse as ophthalmic surgery and automobile manufacturing, to perform an almost unlimited variety of new and unexpected functions made possible by these devices that generate or amplify coherent radiation frequencies in the infrared, visible, or ultraviolet (and most recently, X-ray) regions of the electromagnetic spectrum (Figure 15–24). In one recent study originally carried out on butterfly wings, the chemists reporting the work noted that instead of burning away living tissue (as most laser beams do), the ultraviolet beam of a particular type of laser caused tissue to be chemically decomposed. Incredibly clean cuts, only a few millionths of an inch deep, could be made in living tissue without charring or otherwise damaging underlying or adjacent tissue. That discovery has caught the attention of ophthalmologists seeking ways to reshape the cornea of the eye in order to correct human vision. In early experimental tests, scientists have found that laser cuts heal cleanly, without scarring. One possible application involves shaving off a layer of the cornea, flattening it to correct nearsightedness or elevating it to correct farsightedness. It may soon be possible to sit in front of a laser for a few brief moments, and never have to wear glasses again.

SUMMARY

Under certain circumstances, electrons exhibit classical properties of waves; and under certain circumstances, radiation can be shown to exhibit the classical properties of particles. It all depends on the choice of experiments. Both views are correct. At the quantum mechanical level the wave-particle duality must be accepted as a fundamental part of our understanding. It is succinctly stated in the de Broglie relationship, relating frequency and wavelength to momentum and energy:

$$\text{momentum} = \frac{h}{\lambda} = mv$$

The **Compton effect** showed the inertial properties of x-radiation, recoiling after a particle-like collision with atoms in a metal foil, while the **Davisson-Germer experiment** verified the wave nature of electrons by the diffraction pattern produced on passing through a crystal in which the spacing between particles was on the order of atomic dimensions.

A limiting factor operating at the quantum mechanical level is the fundamental impossibility of simultaneously measuring a particle's position and momentum with absolute accuracy. Known as the **Uncertainty (or Heisenberg) Principle,** the statement

$$(\Delta x)(m\Delta v) \geq \frac{h}{2\pi}$$

allows us to understand better the wave-particle duality. When an experiment has been designed to bring out the particle nature of matter, wave properties are less in evidence, and vice versa.

Although our introduction to quantum mechanics is not concerned with the **Schrödinger equation** in any great detail, it should be recognized that solutions to the Schrödinger wave equation for a nuclear atom lead to the three quantum numbers n, ℓ, and m_ℓ, and we are concerned with them in some detail. The ℓ and the m_ℓ quantum numbers are related to the orbital angular momentum. There are n values of ℓ, in the range $\ell = 0, 1, 2, \ldots, (n-1)$. There are $2\ell + 1$ values for m_ℓ from $-\ell$ to $+\ell$, in integer steps (including zero). The n quantum number, which specifies the radial component of the wave function, has to do with the size of the shell and the total energy. The ℓ quantum number has to do with the shape of the subshells and characterizes the orbital angular momentum; the m_ℓ quantum number describes the spatial orientation of the subshells.

In addition to its orbital angular momentum, an electron has an intrinsic spin angular momentum, $m_s = \pm\frac{1}{2}$. Consequently, for electrons in atoms, four quantum numbers describe the state of each electron, one for each of the three space coordinates, and one for the spin. The **Pauli Exclusion Principle** requires that no two electrons in any electronic system have the same set of four quantum numbers.

The **periodic table** of the elements can be viewed as built up, starting with hydrogen, by adding one electron to the preceding atom, increasing the charge on the nucleus by one at the same time, in the quantum state of lowest energy allowed by the exclusion principle and the $n + \ell$ rule. Lowest n state takes precedence in determining lowest energy when the $n + \ell$ rules does not distinguish between states.

Hund's rule states that for atoms with orbitals of equal energy—an orbital is a one-electron wave function—the order in which electrons appear for the ground state, or lowest energy condition, is such that the maximum number of electrons have unpaired spins with $m_s = \pm\frac{1}{2}$. Excited states are normally due to excitation of one of the outer-shell electrons into orbitals of higher energy, or into spin states other than those predicted by Hund's rule.

Ionization energies (or potentials) provide good supporting evidence for the relationship between chemical properties and electronic structure. That is especially in evidence across the second period elements, from metallic lithium to nonmetallic fluorine, and down through the members of the alkali metal family of elements from least metallic to most. Fluorine would be expected to be the best electron-acceptor or nonmetallic element; cesium, on the other hand, would be strongly electron-donating and metallic.

According to the quantum mechanical view, atoms do not have definite boundaries. The portrait we have is that of an electron cloud based on a probability distribution function Ψ^2. For the ground state H atom, the radial probability is a maximum at the first Bohr radius, $r = a_0$ (or 0.529 Å), leading to a picture of a spherically symmetric $1s$ orbital. Angular distribution functions lead to three-dimensional representations for p, d, and f orbitals, which are not spherically symmetric.

QUESTIONS AND PROBLEMS

QUESTIONS

1. Reconsider the photoelectric effect and the Franck-Hertz experiment in terms of a quantum mechanical universe. What was their collective significance?

2. Describe the Davisson-Germer experiment, the Compton effect, and the Stern-Gerlach experiment. What did each contribute to our understanding of the structure of the universe at the submicroscopic level?

3. The photoelectric effect and the Compton effect both have to do with the interaction of photons and electrons. How are the processes different?

4. What would be the implications of reducing Planck's constant to zero?

5. In our common experience we are unaware of the wave nature of particles. Why?

6. In our daily experience, we are equally unaware of the particle nature of electromagnetic radiation. Why?

7. Why is it not inconsistent to say that the wave nature of matter is irrelevant in the macroscopic world of daily experience, yet the entire structure of the macroscopic world depends on the wave nature of matter?

8. What's wrong with the statement that the de Broglie wavelength increases with the particle's speed?

9. Why is the Compton effect essentially unobservable for visible light?

10. Why do you think an electron microscope might be preferred to a light microscope for the study of matter at the atomic-molecular level?

11. What's wrong with the statement that the uncertainty principle limits the precision with which the position of a particle can be measured?

12. The speed of a photon (c, the speed of light) is known with great certainty. Does it follow therefore that nothing can be known about the photon's position?

13. Temperature measurements are used to record changes in heat content. Consider measuring temperature changes in terms of the uncertainty principle:

(a) What is the limiting uncertainty if one chooses to measure temperature and time simultaneously?

(b) Why would this be of no particular consequence in a macroscopic system—say, in experiments designed to measure the rate of cooling of the Sun?

(c) What are the implications for a microscopic system—say, the temperature change that occurs when one electron strikes a metal plate in a cathode ray tube?

14. Explain why an electron wave following a circular orbit about the nucleus of a stable atom must be divisible into an integral number of wavelengths within the circumference. Illustrate with a simple sketch.

15. What is your understanding of each of the following?
(a) the $n + \ell$ rule
(b) the exclusion principle
(c) Hund's rule
(d) the $2n^2$ rule
(e) the $2\ell + 1$ rule

16. Identify the four quantum numbers that characterize the state of an atomic electron. What are the permitted values for each?

17. What is the rationalization for the experimental fact that many apparently simple spectral lines are seen to be rather more complex when viewed under high resolution?

18. How can you explain the experimental fact that the lowest energy level for the hydrogen atom splits in two in the presence of an external magnetic field?

19. Use the $n + \ell$ rule to describe the electron configuration for the ground state of iron.

20. In the fourth row of the periodic table of the elements there is a block of ten transition metal elements stretching from scandium to zinc. Why are there ten?

21. The 14 lanthanide (or rare earth) elements (from $Z = 58$ to $Z = 71$) have been squeezed into one slot in the periodic table because their chemical and physical properties are all so similar. Offer an explanation based on their electronic structure.

22. What does it mean to say that according to the exclusion principle the total angular momentum for noble gas elements such as He, Ne, and Ar must be zero?

23. What do you understand by each of the following terms?
(a) ionization energy
(b) atomic number
(c) families of elements
(d) periods of elements

24. Consider the family of elements immediately preceding the noble gas family of elements, and the family immediately following. One easily develops a positive charge, the other a negative charge. Which is which, and why?

25. Offer an explanation for the relative ease of forming cations of
(a) sodium versus cesium
(b) sodium versus magnesium
(c) sodium versus calcium

26. The ionization energy for the ground state hydrogen atom is 13.61 eV.
(a) Is it possible for a ground state hydrogen atom to absorb less energy than that?
(b) Is it possible to absorb more? Explain.

27. The second ionization potential for potassium metal is much greater than the first, namely 31.7 eV versus 4.34 eV. Is that reasonable? Why?

28. Would you expect removal of an electron from a cation to be energetically less favorable than from an atom? Why?

29. A set of harmonic oscillators populate an excited state. Suddenly, one jumps down to a lower excited state, emitting a photon in the process. Is a flood of photons likely to be produced by stimulated emission as a result? Why or why not?

30. Why is there a difference in energy between the $3s$ and $3p$ states for sodium whereas in hydrogen these two states have essentially the same energy?

31. The following graph is a plot of the radial distribution function P_r for given values of r, the radial distance from the nucleus to the immediate vicinity of the electron, for the $n = 3$ and $\ell = 0$ energy state. Briefly explain the peaks and valleys in the curve.

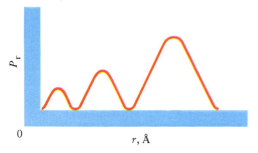

32. Indicate whether the following statements are true or false:
(a) Only particles of rest mass zero can move at the speed of light.
(b) The amplitude of a matter wave at some radial distance r from the nucleus is directly proportional to the probability of finding an electron at that radius.
(c) The first ionization potential for helium should be about the same as the second.
(d) F^- and Ne are isoelectronic.

33. Which of the following ground state atoms (or ions) exhibit spherical electron distribution. Briefly explain: Rb^+; C^+; C^-; Cr^{3+}; Mn^{2+}.

34. Briefly explain the "nonexistence" of $2d$ orbital electrons. Are $7f$ orbitals possible? Explain.

PROBLEMS

Wave-Particle Duality   [1–10]

1. Determine the de Broglie wavelength of an oxygen molecule moving at thermal speeds of 50,000 cm/s.

2. Find the de Broglie wavelength of a "speeding" 200-gram turtle moving across a road at a relatively slow speed, 4 mi/h.

3. Calculate the de Broglie wavelength for each of the following:
(a) a 22-caliber bullet, with a mass of 5.0 g, moving at a muzzle velocity of 25,000 cm/s.
(b) a thoroughbred racehorse, which weighs 600 pounds, moving with a speed of 40 mi/h.

4. Use the de Broglie relationship to calculate the wavelength associated with each of the following particles:
(a) a hydrogen molecule whose speed is 2×10^5 cm/s
(b) a 45-g golf ball moving at 3×10^3 cm/s
(c) a 150-pound person walking at a steady 3 mi/h
(d) a 5000-pound communications satellite moving at 5000 mi/h

5. Calculate the de Broglie wavelength for a very small, hardly measurable, but still clearly macroscopic particle of mass 9.1×10^{-9} g, moving at the rapid but nonrelativistic velocity of 3.0×10^5 cm/s. Comment on the quantum mechanical significance of the result.

6. Determine the wavelength associated with an electron moving at the velocity of light. Comment on the quantum mechanical significance of the result.

7. A neutron, an electron, and a photon all have the same wavelength, namely $\lambda = 10^{-10}$ m.
(a) Determine the speed (in cm/s) and the energy (in eV) for the neutron.
(b) Determine the energy of the electron (in eV).
(c) Determine the energy of the photon (in eV) and state where in the electromagnetic spectrum it lies.

8. Calculate the de Broglie wavelength for
(a) thermal neutrons with 0.02 eV kinetic energy
(b) 1 MeV protons
(c) 100 eV electrons

9. Thermal neutrons can be cooled by allowing them to pass through liquid helium at a temperature of 3 K. If they emerge with KE equal to $\frac{3}{2}kT$, what is their wavelength?

10. Ordinary thermal neutrons have kinetic energies two orders of magnitude greater than in Problem 9. What are their temperature and their wavelength?

Uncertainty Principle   [11–14]

11. What is the uncertainty in the location of a 100-g turtle crossing a road at a speed of 1 mm/s if the uncertainty in its speed is 10^{-4} mm/s? Is the uncertainty in the turtle's position negligible?

12. What is the minimum uncertainty in the position of a 10^{-6} g particle moving at a speed of 1 cm/s if the uncertainty in its speed is 0.01%?

13. What is the apparent wavelength of an electron moving at an approximate speed of 10^6 m/s? If the uncertainty in its speed is 3×10^5 m/s, what is the approximate uncertainty in its position?

14. The kinetic energy of an electron happens to be 25 eV and it has an uncertainty in its momentum of 10%. What is the minimum uncertainty in its position?

Quantum Numbers; Electron Configurations   [15–38]

15. List the values for all four quantum numbers for each of the electrons in a ground state beryllium atom.

16. List all four quantum numbers
(a) for a lone electron in a ground state hydrogen atom
(b) for a $3p$ electron

17. Provide the information requested for each of the following:
(a) all possible ℓ values for an $n = 3$ state
(b) all possible m_ℓ values for the $\ell = 3$ state
(c) all possible m_s values for the $m_\ell = 3$ state

18. How many electrons can be described by
(a) the quantum designation $3p$ in a polyelectron atom?
(b) the designation $5s$?
(c) the designation $4d$?
(d) the designation $6f$?

19. List the electron configurations of all the elements from hydrogen through scandium in their proper order according to the $n + \ell$ rule, lowest energy first. In what way is periodicity a consequence of such behavior?

20. List the values of all four quantum numbers for each of the 20 electrons in the ground state calcium atom, in proper order according to the $n + \ell$ rule. Into what quantum state would the twenty-first electron enter? What is that element, and why is its position in the periodic table significant (or unusual)?

21. Write the electronic configuration
(a) for the ground state arsenic atom ($Z = 33$),
(b) for the ground state titanium atom ($Z = 22$), and
(c) for the iodine atom ($Z = 53$).

22. Write the electron configuration for the ground state of each of the following atoms:
(a) Zn, Cd, and Hg
(b) Mg, Cu, and Ba
(c) Be, Ca, and Y

23. Give a complete quantum number description
(a) for the 18 electrons in the M shell;
(b) for the 10 electrons that can be described as $4d$.

24. List the 18 unique states possible for an H atom in which $n = 3$.

25. Name the element:
(a) $[Kr]5s^2 4d^7$
(b) $[He]2s^2 2p^1$

26. Name the element:
(a) $[Ne]3s^2 3p^3$
(b) $[Ar]4s^2 3d^{10} 4p^4$

27. Identify the element:

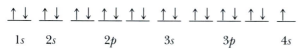

$1s$ $2s$ $2p$ $3s$ $3p$ $4s$

28. Identify the element:

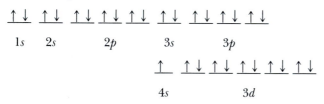

$1s$ $2s$ $2p$ $3s$ $3p$

$4s$ $3d$

29. Indicate the higher of the two energy states in each of the following pairs:
(a) $3d$ or $4s$
(b) $4s$ or $4p$
(c) $4p$ or $5s$

30. Determine which of the two energy levels in each of the following pairs has the lower energy:
(a) $5s$ or $4d$
(b) $4d$ or $4s$
(c) $6s$ or $6p$

31. List the proper ordering of the energy levels for the 47 electrons in a ground state silver atom.

32. Using arrows to represent the parallel and antiparallel spins of the electrons, order the energy levels for the electrons in a ground state copper atom.

33. Compare the electronic structures of sodium, potassium, silver, and gold. On that basis, explain why silver and gold are found in the free state in nature, but sodium and potassium never are.

34. List the occupied quantum states for the 36 electrons in atoms of the element krypton, and indicate why Kr atoms might be chemically inert.

35. Classify each of the following species according to charge (neutral, anion, or cation) and state (ground, excited, or impossible):
(a) $^1H(1s^1)$
(b) $^{16}O(1s^2 2s^2 2p^6)$
(c) $^{14}N(1s^2 2s^2 2p^3)$
(d) $^{19}F(1s^2 2s^2 2p^4 2d^1)$
(e) $^{12}C(1s^2 2s^2 2p^3)$

36. Classify each of the following atoms according to charge (anion, cation, neutral) and state (ground, excited, impossible):

Species	Designation	Charge	State
^{4}He	$1s^0$		
^{20}Ne	$1s^2 2s^2 2p^5 2d^1$		
^{32}S	$1s^2 2s^2 2p^6 3s^2 2p^6$		
^{34}K	$1s^2 2s^2 2p^6 3s^2 3p^6 3d^1$		

37. How many d electrons with parallel spins does a ground state iron atom possess?

38. How many electrons are paired in d-orbitals in of a ground state cobalt atom?

Ionization Energies   [39–42]

39. Calculate the energy required to remove both electrons from a ground state helium atom.

40. How much energy does it take to remove the electrons from a ground state Li^+ ion?

41. Calculate the energy required to form one mole of sodium ions from a gas of sodium atoms.

42. How much energy does it take to form one mole of Ne^+ (g)?

Wave Functions   [43–46]

43. Carefully draw boundary surfaces specifying approximately 90% probability for $1s$, $2s$, and $3s$ states for the hydrogen atom.

44. Carefully draw a boundary surface representing a "90% probability" picture for a $2p$ electronic state.

45. Referring to Figure 15–17(a), place a line representing a similar plot for He^+ on the graph. Would your line lie to the left of the line for H, to the right of it, or in approximately the same place? Briefly explain.

46. Referring to Figure 15–21, explain how representations of the $2p_y$ and $2p_z$ electronic states differ from the $2p_x$ state. Sketch all three together.

Additional Problems   [47–60]

47. Explain the general increase in ionization energies observed in traversing the second period of the periodic table from Li to Ne. Account for the alteration in ionization energies that is observed between Be and B; between N and O.

48. Write the electronic configuration for gold ($Z = 79$) and briefly explain why it might be chemically unreactive, with properties not unlike those of the noble gases. Why is it neither inert nor gas-like?

49. Given that the uncertainty in the position of an electron in an atom cannot be greater than the diameter of the atom:

(a) Determine the minimum uncertainty in momentum if the uncertainty in position is 10^{-8} cm.

(b) Find the kinetic energy for an electron if it has a momentum equal in magnitude to the uncertainty found in part (a).

50. (a) State the n and ℓ quantum numbers for the valence electron in a ground state lithium atom.

(b) What are the values for the n and ℓ quantum numbers after enough energy has been absorbed to excite lithium atoms to the first possible higher unoccupied state?

(c) Answer (a) and (b) for sodium.

51. Explain the existence of 10 transition metal elements (from yttrium at $Z = 39$ through cadmium at $Z = 48$) across the fifth period of the periodic table in terms of atomic structure and electronic configuration.

52. There are small dips in an otherwise increasing plot of first ionization energies, occurring after gallium ($Z = 31$) and indium ($Z = 49$). Write the complete ground state electronic configurations for gallium and indium and for the elements that follow each one. Draw a pictoral representation that shows the spins as parallel and anti-parallel arrows. Briefly explain the fall-off in ionization energy that is observed.

53. Using data in Table 15–4 for the first and second ionization energies, compare calcium and potassium and explain the differences based on their respective electronic configurations.

54. Identify the atoms among the elements in the first row of transition metal elements (from scandium to zinc) that have either the $3d^5$ or $3d^{10}$ electronic configuration for the ground state. Draw a pictoral representation, using arrows to represent the electrons and their spins.

55. The properties of iron and cobalt, elements adjacent to each other in the periodic table, are similar, whereas the properties of neon and sodium, also adjacent elements, are not. Based on their respective electronic configurations, briefly explain.

56. Into how many levels is a p state split in a magnetic field? Into how many levels is a d state split? Compare the spacing between p and d levels in a strong magnetic field.

57. (a) What is the degeneracy of s atomic orbitals? Of p orbitals? Of d orbitals?

(b) What are the ground state electron configurations for Se^{2-}, for Ag^+, and for Cs?

58. Write down the ground state electronic configuration for each of the following species:

(a) S, S^{6+}, and S^{2-}

(b) Cl, Cl^{5+}, and Cl^-

59. Sketch (from memory) a graphical representation for the radial distribution function for a ground state hydrogen atom in a $1s$, $2s$, and $3s$ state by plotting P_r versus r. Do the same for a plot of Ψ^2 versus r/a_0. Then draw the "fuzzy-cloud" representation for an s and a p orbital.

60. Consider the following hypothetical worlds and discuss each briefly:

(a) A world in which the exclusion principle operates, but electrons are spinless.

(b) A world in which the exclusion principle operates, and electron spins are $+\frac{3}{2}(h/2\pi)$ and $-\frac{3}{2}(h/2\pi)$.

(c) A world in which the exclusion principle does not operate.

MULTIPLE PRINCIPLES [61–62]

61. With the help of ionization energy data from Table 15–4, contrast the characteristic properties of the typically nonmetallic family of elements known as the halogens (such as F and Cl) and those of the typically metallic alkali metal family of elements (such as Li, Na, and K). In what way are ionization energies associated with nonmetallic and metallic character?

62. Estimate the energy in kJ/mol required to convert a mole of sodium at its melting point to a gas of positively charged Na^+ ions.

APPLIED PRINCIPLES [63–64]

63. Suppose that you measure the position of a red blood cell (as a smear on a slide) on the stage of an optical microscope. Assume that the mass of a red blood cell is about 10^{-10} g.

(a) If the position uncertainty is 100 nm, what is the limit of resolution for the optical microscope?

(b) Calculate the resulting uncertainty in the velocity of the blood cell.

(c) Explain the significance of the result of part (b).

64. If a "safe light" in a photographic darkroom turns out a steady supply of 660-nm visible photons at a rate of 5×10^{19} photons per second, what amount of power output is that equivalent to? Power is defined as energy delivered per unit time: 1 watt = 1 J/s. Compare that result to a water molecule falling over Niagara Falls, estimated to be 0.00015 eV; burning one carbon atom to carbon dioxide, estimated to be 4 eV; and fission of one uranium-235 nucleus, estimated to be 200,000,000 eV.

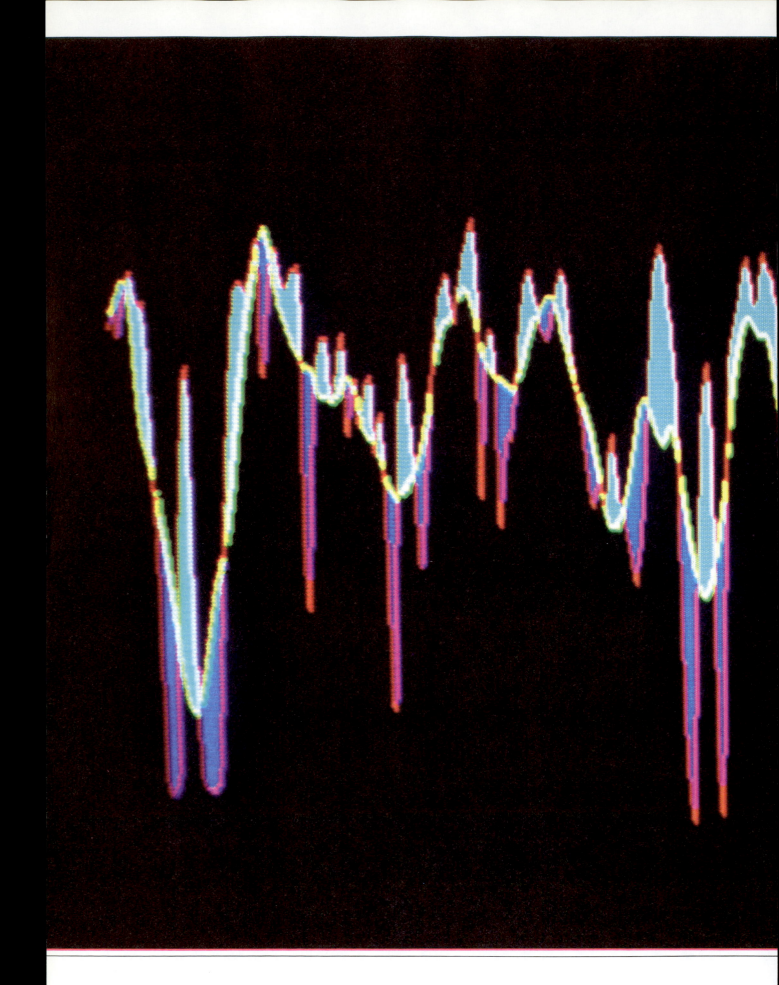

CHAPTER 16

Bonding I: The Properties of Bonds

16.1 THE NATURE OF THE CHEMICAL BOND

16.2 THE IONIC BOND AND ELECTROVALENCE

16.3 THE COVALENT BOND AND COVALENCE

16.4 THE PROPERTIES OF CHEMICAL BONDS

16.5 BREAKING CHEMICAL BONDS; BOND ENTHALPIES

16.6 POLAR MOLECULES, ELECTRONEGATIVITY, AND ELECTRON AFFINITIES

16.7 DOCUMENTING CHEMICAL BONDS BY INFRARED SPECTROSCOPY

Infrared (IR) spectroscopy is one of the most insightful tools for understanding the nature of the chemical bond and the structure of matter at the molecular level. Shown here are two infrared spectra, displayed in an overlay mode so an analytical chemist can qualitatively and quantitatively compare them. In a long professional career that first reached widespread prominence with the publication of his benchmark treatise on "The Nature of the Chemical Bond" in the 1930s, Linus Pauling stands today as he nears his 90th birthday as one of the great American chemists of the 20th century. Pauling, shown here with his wife, won the Nobel Prize in 1954.

Linus Pauling (American), Nobel Prize for Chemistry in 1954 for his research into the nature of the chemical bond and the structure of complex substances.

For our purposes, the word *valence* will refer to either the electrovalence or the covalence of an atom. The number of charges on an ion of an element is the *electrovalence* of the element. The number of covalent bonds that an atom of an element can form is the *covalence* of the element.

16.1 THE NATURE OF THE CHEMICAL BOND

It is not possible to understand fully the behavior of a molecule until its structure is known; that is, until we can measure its size, know its shape, and say something about the nature of the bonds that hold its atoms together. The explanation of how atoms combine to form molecules and why the resulting molecules have the unique shapes that identify them is one of the major triumphs of quantum mechanics. You might correctly guess from our earlier discussions that quantum mechanical descriptions are complicated; however, it is possible to extract the essential ideas in a qualitative way that can give us an appreciation of the factors responsible for molecular size and shape, and the nature of the chemical bond. In its simplest terms, what we are trying to do is to answer the question: "What are the properties of chemical bonds?"

The 20th century has witnessed superb experimental measurements that have provided an immense amount of information about chemical bonds. Bonding theories have established patterns out of which the modern principles of molecular architecture have emerged. As you will see, detailed mathematical analyses are not necessary in order to understand the arguments because they lend themselves quite nicely to visual representations.

To begin with, atoms are collections of positively charged nuclei and surrounding clouds of negatively charged electrons. Chemical bonding between atoms results from the mutual attractions and repulsions of these classical positive and negative charges. Coulomb's Law governs the balance of electrical forces; quantum mechanics governs the positions and motions of electrons. The main bonds of which we are about to speak are ionic bonds and covalent bonds. Whatever their source, these bonds are very strong, typically on the order of 400 kJ/mol. Understanding the nature of the main chemical bonds will in turn help us make sense of the van der Waals force and the weak interactions among molecules.

16.2 THE IONIC BOND AND ELECTROVALENCE

One or more electrons can be transferred from one neutral atom to another, producing electrically charged ions that are roughly spherical and that attract one another because of their opposite charges. As the force of attraction between the oppositely charged ions increases, their electron clouds penetrate each other until a balance of coulombic forces is struck between attractive forces between electrons and nuclei, and repulsive forces between electrons and between nuclei. The result of this process is an **ionic bond**. The number of charges on a stable ion is called the electrovalence—or simply the **valence**—of the atom. Crystals of common table salt, for example, are gigantic arrays (in three dimensions) of univalent Na^+ ions surrounded by six univalent Cl^- ions, and vice versa. Only at elevated temperatures in the vapor state can one actually observe the isolated properties of Na^+ and Cl^- ion pairs, held together by the electrostatic interaction characteristic of electrovalent bonds.

A sodium atom, with 10 of its 11 electrons in the filled K and L shells, and only the lone 11th electron in the outermost M shell, can achieve the noble gas configuration by giving up the M-shell electron, becoming a univalent sodium ion, Na^+:

$$Na\cdot \longrightarrow Na^+ + e^-$$

An atom of chlorine has seven valence (M-shell) electrons. It can reach the stable noble gas configuration by gaining one electron:

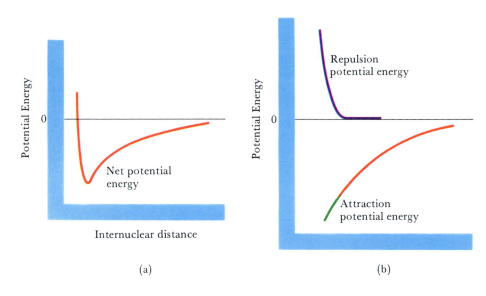

$$: \ddot{\underset{..}{Cl}} \cdot + e^- \longrightarrow : \ddot{\underset{..}{Cl}} :^-$$

Sodium and chlorine can react to form sodium chloride by transfer of one electron from one atom to the other. The energy required to remove an electron from a neutral atom is the **ionization energy**, and the amount of energy given off when an electron is added to a gaseous atom is known as the **electron affinity**:

Ionization: $Na \longrightarrow Na^+ + e^-$

Electron affinity: $: \ddot{\underset{..}{Cl}} \cdot + e^- \longrightarrow : \ddot{\underset{..}{Cl}} :^-$

In this case, the net process is the spontaneous formation of a sodium chloride ion-pair:

$$Na\ (g) + Cl\ (g) \longrightarrow Na^+\ (g) + Cl^-\ (g)$$

A complete transfer of the valence shell electrons from one atom to the other has taken place. The ions that are formed can be understood to be held together by ionic (or electrovalent) bonds, set up by the electrostatic forces that exist between the oppositely charged particles.

The potential energy associated with such bonds can be represented as the net potential energy due to the balance of two factors (Figure 16–1): (1) coulombic forces of attraction between two oppositely charged ions within some reasonable distance of each other, and (2) mutual repulsion between a pair of like-charged nuclei, and between a pair of like-charged electron clouds that are close enough to overlap each other:

$$\begin{matrix} \text{potential} \\ \text{energy} \end{matrix} = \begin{matrix} \text{electrostatic} \\ \text{attraction of} \\ \text{oppositely} \\ \text{charged ions} \end{matrix} + \begin{matrix} \text{electrostatic repulsion of} \\ \text{like-charged nuclei and} \\ \text{electron-cloud repulsion} \end{matrix}$$

Figure 16–1(a) is a plot of binding energy for sodium ions and chloride ions as a function of the distance between the nuclei. The generalized curve in Figure 16–1(b) shows the minimum energy, corresponding to the most stable internuclear distance. At very short distances, an extremely unstable state arises because the strong repulsive forces overcome the attraction. At large distances, the probability of bond formation approaches zero because of the inability of the ions to exert any influence on each other. They are

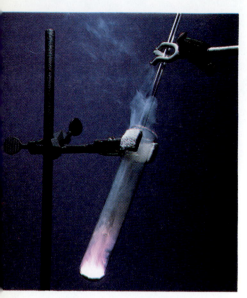

FIGURE 16–2 Burning potassium in a chlorine atmosphere, forming KCl. The chemistry of sodium is similar.

just too far apart for any attractive forces to be effective. The zero binding energy line on the graph corresponds to the completely separated ions.

Now let's take a closer look. For a sodium atom with a lone $3s$ electron beyond the neon core, the ionization energy is low: about 5.1 eV. Ionization forms a spherically symmetric sodium ion that is isoelectronic with a neon atom:

$$Na^+ = [He]2s^22p^6$$

The electron affinity for a chlorine atom, only one electron shy of the argon core, is 3.8 eV. In this case, ionization forms a spherically symmetric chloride ion:

$$Cl^- = [Ne]3s^23p^6$$

By convention, the electron affinity reaction is written:

$$: \overset{..}{\underset{..}{Cl}} \cdot \, (g) + e^- \longrightarrow : \overset{..}{\underset{..}{Cl}} :^- \, (g)$$

The ΔH of this process is the negative of the electron affinity, or -3.8 eV. Thus this process is exothermic and the net result—the energy for the overall electron-transfer process leading to formation of Na^+ and Cl^- ions—is

$$IE + EA = 5.1 \text{ eV} - 3.8 \text{ eV} = 1.3 \text{ eV}$$

Since the value is positive, the existence of the ions might seem to be a logical inconsistency. However, from classical physics, we can calculate the potential energy of the two ions a distance r apart:

$$PE = \frac{kZ_1Z_2e^2}{r}$$
$$= \frac{k(+1)(-1)e^2}{r}$$
$$= -\frac{ke^2}{r}$$

And when the separation of the ions is less than 11 Å, the negative PE of attraction is greater than the energy needed to create the ions from the atoms. Such reactions take place spontaneously (Figure 16–2).

16.3   THE COVALENT BOND AND COVALENCE

The Covalent Bond

What's needed for ionic bond formation is a pair of stable, oppositely charged ions. This requirement is achieved by dissimilar atoms from the extremes of the periodic table as they transfer electrons to and from each other, forming ions. But how shall we explain bonding between a pair of identical atoms such as hydrogen, and how shall we explain why atoms such as carbon and oxygen form stable molecules? Carbon and oxygen are near neighbors in the periodic table and would be expected to have similar properties. Similarly, there is no evidence for the existence of carbon ions or hydrogen ions in methane and, therefore, no obvious electrostatic interaction. There isn't any immediately obvious reason for two hydrogen atoms to form a hydrogen molecule, or any obvious explanation why four hydrogen atoms and one carbon atom come together to form a stable methane molecule.

The kind of bonding that does explain such molecules is called **covalent bonding**. Typically, when two atoms are bonded covalently, each one contributes an electron to a pair, which positions itself between the atoms in such a way as to increase the charge density in that space. When this happens, the like-charged nuclei are shielded from each other and are attracted to the oppositely charged electron clouds. Furthermore, the spins of the electrons are opposite, as would be required by the Pauli exclusion principle for bonds in stable molecules. Finally, the number of covalent bonds a single atom of an element can form is equal to the covalence of the element, and is a function of the electron configuration of the atom.

Consider two hydrogen atoms, each with its lone K-shell electron. If one atom shares its electron with another, each could be viewed as possessing both electrons and having acquired the stability associated with the two-electron, filled K-shell configuration of the helium atom:

$$H\cdot + H\cdot \longrightarrow H:H$$

With four L-shell bonding electrons on the carbon and four hydrogen atoms with their four separate electrons, the methane molecule can be viewed in the same way. The carbon atom assumes the eight-electron outer configuration and becomes neon-like. Each hydrogen atom becomes helium-like by sharing two electrons:

$$4H\cdot + \cdot\overset{\displaystyle\cdot}{\underset{\displaystyle\cdot}{C}}\cdot \longrightarrow \begin{array}{c} H \\ H:\overset{\displaystyle\cdot\cdot}{C}:H \\ H \end{array}$$

When carbon and oxygen combine to form carbon dioxide, our theory of the covalent bond must be expanded to include the concept of multiple bonds to accommodate all electrons:

$$\cdot\overset{\displaystyle\cdot}{\underset{\displaystyle\cdot}{C}}\cdot + 2\cdot\overset{\displaystyle\cdot\cdot}{\underset{\displaystyle\cdot\cdot}{O}}\cdot \longrightarrow :\overset{\displaystyle\cdot\cdot}{\underset{\displaystyle\cdot\cdot}{O}}::C::\overset{\displaystyle\cdot\cdot}{\underset{\displaystyle\cdot\cdot}{O}}:$$

The Coordinate-Covalent Bond

Gilbert Lewis suggested that covalent bonds could as well be formed by the sharing of an electron pair supplied by only one of the two atoms. Consider what takes place when a proton interacts with an ammonia molecule, forming an ammonium ion:

$$\begin{array}{c} H \\ H:\overset{\displaystyle\cdot\cdot}{N}: \\ H \end{array} + H^+ \longrightarrow \left[\begin{array}{c} H \\ H:\overset{\displaystyle\cdot\cdot}{N}:H \\ H \end{array}\right]^+$$

The nitrogen atom in the ammonia molecule possesses a nonbonding (free) electron pair, which it shares with the approaching proton. Once the bond is formed, the four H-to-N bonds—the three original ones and the new one—are indistinguishable, and the positive charge is assigned to the entire ion as a net charge. Covalent bond formation by electron-sharing, in which both electrons are supplied by one atom, results in a **coordinate covalent** (or donor-acceptor) bond.

EXAMPLE 16–1

Write Lewis structures for the formation of the coordinate covalent bond in the fluoborate ion on reaction of boron trifluoride with fluoride ion.

Solution   The fluoride ion supplies both electrons in the new bond, which is then identical to the existing bonds:

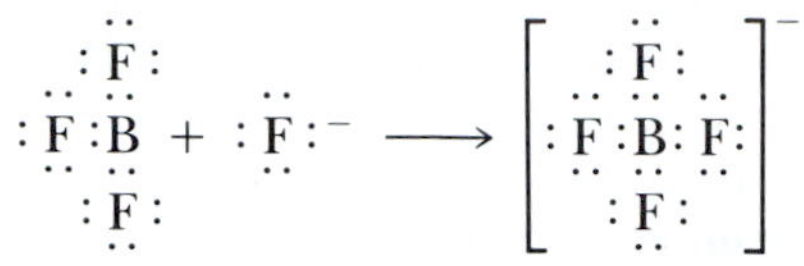

$$
\begin{array}{c}
:\ddot{F}: \\
:\ddot{F}:B \\
:\ddot{F}:
\end{array}
+ \;:\ddot{F}:^{-} \;\longrightarrow\;
\left[
\begin{array}{c}
:\ddot{F}: \\
:\ddot{F}:B:\ddot{F}: \\
:\ddot{F}:
\end{array}
\right]^{-}
$$

The fluoborate ion fully satisfies the stable neon configuration both for boron and for each fluorine. Compounds such as this make up a large and important class known as coordination compounds, about which we'll have much more to say in a later chapter.   ◼

EXERCISE 16–1

By drawing Lewis electron-dot structures, demonstrate that each of the following is consistent with the "octet" rule:

(a) H_2O	(d) Na_2SO_4	(g) S_8
(b) H_3O^+	(e) NO_3^-	(h) HCN
(c) H_2O_2	(f) O_2	(i) NH_4Cl

The Weakest Bonds

Why are real gases not ideal? How is it that even helium can be condensed? The answers must lie in the attractive interactions between neutral, covalent molecules. Otherwise, molecular substances would exist as gases at all temperatures. **van der Waals forces** were first introduced in our discussion of gas laws and solutions to account for deviations from ideal behavior. These forces can be viewed as due to the positively charged nucleus within a neutral molecule affecting the behavior of electrons lying beyond its normal radius of influence. Normally, interatomic distances are 1 to 2.5 Å, while the nearest intermolecular distances are 2.5 to 4.0 Å. The magnitude of the van der Waals forces is a tenth that of the usual covalent bond energies: 10 to 50 kJ/mol versus 100 to 500 kJ/mol. These weakest forces can be classified according to three principal types of interactions:

Dipole-Dipole Interactions: An electrostatic force exists between the oppositely charged ends of dipolar molecules, that is, molecules with nonuniform charge distributions (see Section 16.6).

Dipole-Induced Dipole Interactions: In a mixture of molecules—some polar, some nonpolar—the polar molecules can induce momentary dipoles in nearby nonpolar molecules. This gives rise to electrostatic attractions.

Dispersion Forces: These exist in all molecules, regardless of the presence or absence of dipoles. A specific polar configuration of an electron in one molecule can induce an instantaneous dipole in a second molecule. At some single instant in time, the electrons may lie within the internuclear distance, nearer one nucleus than the other. A temporary dipole results. This in turn induces a dipole in a neighbor, and so on. The mutual effects will tend to orient the molecular dipoles, leading to attraction. Dispersion forces, therefore, depend on the total number of electrons; as the size of the molecule increases, so does the size of the dispersion forces (Figure 16–3).

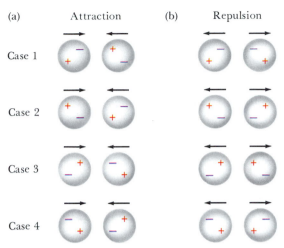

FIGURE 16–3   van der Waals attractions among molecules with zero dipole moments: (a) Possible orientations of instantaneous dipole moments at different times, leading to attraction; (b) possible orientations leading to repulsions. The electric field of the instantaneous dipole moment of one molecule tends to polarize the other. Thus, the orientations leading to attraction as in (a) are more likely than those leading to repulsion as in (b).

FIGURE 16–4   Cross section of electrically conducting copper wire embedded in insulating plastic materials. Cables constructed in this fashion are much lighter and more resistant to corrosion than lead cable, insulated with paper, that had been used traditionally. Copper is an especially good conducting material for electrical applications.

Metallic Bonds

A simple way of thinking about bonding in metals is to consider the crystal as a three-dimensional matrix of cations of the metallic element immersed in a sea of electrons. What holds the crystal together is the attraction between the positively charged ions of the metal and the negatively charged electrons. Because some of the electrons are free to move about the entire metal, the metal is a good conductor of electricity (Figure 16–4). We will discuss the metallic bond more thoroughly in Chapter 19.

Hydrogen Bonds

Although they were introduced earlier, and mentioned in several different contexts in our discussions of liquids and solutions, we briefly mention hydrogen bonds here in order to present a more complete discussion of chemical bonds. The hydrogen bond is the weak electrostatic interaction between a hydrogen atom in one covalently bonded H-X molecule, where X is O, N, or F, and the X atom (O, N, or F) in another H-X molecule. Such atoms are said to be electronegative. For example, in the network of bonds in water (Figure 16–5), it is always the internal H-to-O bond that is stronger; the external H-to-O bond—the intermolecular hydrogen bond—is weaker.

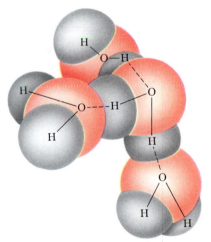

FIGURE 16–5   An association of water molecules brought about by the weaker, intermolecular (external) H-to-O bonds. Within the hydrogen bonded water molecules there are strong intramolecular (internal) H-to-O bonds.

16.4   THE PROPERTIES OF CHEMICAL BONDS

An essential feature of any chemical bond is the distance between the two nuclei, the **bond length**. If an atom is simultaneously involved in chemical bond formation with two (or more) other atoms, there must be a characteristic **bond angle** between the bonds at the central atom. The bond angle is also an essential structural feature of molecules. **Bond enthalpies** (or **energies**) are usually measured as the enthalpy change associated with the

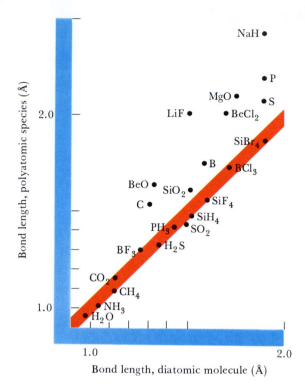

FIGURE 16–6 Comparison of bond lengths between atoms in simple diatomic molecules and in many-atom (polyatomic) systems. The 45° line represents a 1:1 ratio for bond lengths in aggregated species compared to bond lengths in single molecules and ion-pairs.

dissociation of any particular pair of connected atoms in a molecule or ion. Perhaps the most important reason why scientists are interested in bond lengths, bond angles, and bond enthalpies is that they determine the sizes and shapes of molecules, and many of their physical and chemical properties.

Bond Lengths

Some care needs to be taken when discussing the sizes and shapes of molelcules. First recognize that nuclei are always vibrating. Even at the absolute zero of temperature they are not free of what is known as their **zero point energy**. Thus, bond length is the average distance between nuclei that are in motion. Also, the electrons are best understood as an electron cloud or distribution of charge whose density varies from place to place. So the shape and direction of a bond have to do with the cloud of charge associated with the bond. Let's leave those ideas aside for a moment and examine how we measure and characterize bond lengths and directions, operationally.

Internuclear distances are determined experimentally by electron diffraction, x-ray diffraction, or spectroscopic techniques. These data are of two kinds:

- **interionic distances**, the distance between nuclei in ionic solids.
- **interatomic distances**, the distance between nuclei in molecules.

The latter are bond lengths. As we said, there can be no one permanent, fixed bond length or internuclear distance between atoms in molecules because the nuclei are constantly moving, and because the nature of measurement itself affects the measured distance. There is, however, an average or mean distance. Consider a sodium chloride ion pair at elevated tem-

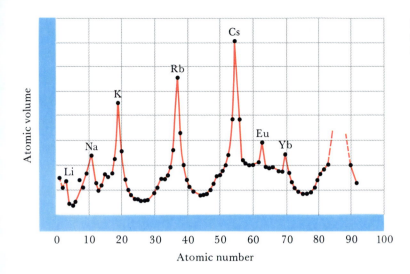

FIGURE 16-7 Atomic volume as a function of atomic number for the elements in their standard states.

peratures in the gas phase, for which the bond length is known to be about 2.4 Å—the bond length for a simple diatomic molecule. In crystalline sodium chloride, where each ion is surrounded by six others, the interionic distance is 2.8 Å. Clearly, the internuclear distance is a function of the immediate environment. Figure 16–6 is a plot of the internuclear distance for a number of gas-phase covalent molecules and ion-pairs versus the measured distances in the more complicated, aggregated state (as in a crystal lattice). Because the deviation from the 45° line is small for covalent species, we can conclude that there is little difference between isolated and aggregated covalent molecules; but great differences exist between isolated ion pairs and aggregate ionic crystals.

One of the most remarkable facts to emerge from the experimental determination of bond lengths is that in many different compounds the same type of bond always exhibits approximately the same average internuclear distance. In literally thousands of organic compounds, the C-to-C single bond is very close to 1.54 Å. In many compounds containing the O-to-H single bond, the internuclear distance is 0.96 Å. Many sulfur compounds show S-to-F bonds of 1.57 Å. However, multiple bonds are much shorter than single bonds between the same pair of atoms. For example, C-to-C double bond lengths (C=C) are close to 1.33 Å; C-to-C triple bond lengths (C≡C) commonly are close to 1.20 Å. Multiple bonds show up frequently between other atoms as well: C≡N, C=O, and S=O.

Atomic and Ionic Radii

A simple calculation of atomic volume can be made by dividing the atomic weight of an element by its density and the Avogadro number. When the result is plotted against the atomic number (Figure 16–7), a periodic relationship is evident. The size of an atom clearly increases from element to element within a group, as in:

Na < K < Rb < Cs

The valence electrons lie at greater distances from the nucleus, which means that the atom gets bigger as one moves down a group of elements. However, within a given period, atomic size shows a progressive decrease with increasing atomic number. Here, the principal quantum number stays the

TABLE 16–1   Covalent radii, Å

H 0.37						
Li 1.23	Be 0.89	B 0.80	C 0.77	N 0.74	O 0.73	F 0.72
		Al 1.25	Si 1.17	P 1.10	S 1.04	Cl 0.99
			Ge 1.22	As 1.21	Se 1.17	Br 1.14
			Sn 1.41	Sb 1.40	Te 1.37	I 1.33

TABLE 16–2   Interionic Distances Among the Alkali Metal Chlorides and Iodides, Å

cation	anion	Interionic Distances	
		by summing of the radii	by direct observation
Li^+	Cl^-	2.49	2.57
	I^-	2.87	3.02
Na^+	Cl^-	2.79	2.81
	I^-	3.17	3.23
K^+	Cl^-	3.14	3.14
	I^-	3.52	3.53
Rb^+	Cl^-	3.29	3.29
	I^-	3.67	3.66
Cs^+	Cl^-	3.48	3.47
	I^-	3.86	3.83

same, but the charge upon the nucleus increases, exerting a greater attractive force upon the electrons. Thus, the size of the atom shrinks.

A comparison of experimentally determined bond distance values and those calculated from the simple method of summing the atomic radii shows the latter to be consistently too high. The discrepancy arises because the sum of atomic radii does not account for distortion due to the unequal charge distribution between dissimilar atoms involved in the bond. Covalent atomic radii are generally obtained from homonuclear molecules such as H_2 or Li_2 (Table 16–1).

Ionic size is a consequence of the attractive force exerted on the outer electrons by the positive charge of the nucleus, which is shielded by the negative charge of the inner shells of electrons. When a neutral atom becomes a positive ion, the atomic size decreases because the shielding decreases and the *effective* nuclear charge increases. Similarly, a negative ion is always larger than its neutral atomic parent because there is a decrease in effective nuclear charge.

Measurable internuclear distances between a positive ion and a negative ion, often referred to as ionic bond lengths, are best called interionic

TABLE 16–3	**Ionic radii, Å**			
Li$^+$	Be^{2+}		O^{2-}	F$^-$
0.68	0.30		1.45	1.33
Na$^+$	Mg^{2+}	Al^{3+}	S^{2-}	Cl$^-$
0.98	0.65	0.45	1.90	1.81
K$^+$	Ca^{2+}	Ga^{3+}	Se^{2-}	Br$^-$
1.33	0.94	0.60	2.02	1.96
Rb$^+$	Sr^{2+}	In^{3+}	Te^{2-}	I$^-$
1.48	1.10	0.81	2.22	2.19
Cs$^+$	Ba^{2+}	Tl^{3+}		
1.67	1.29	0.91		

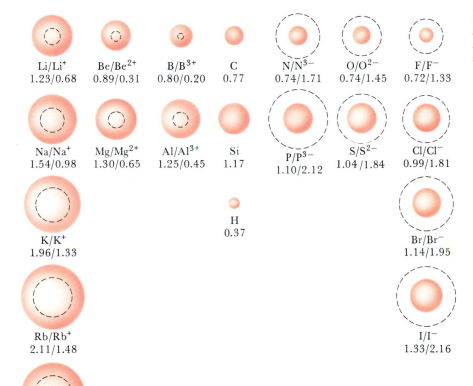

Li/Li$^+$ 1.23/0.68 Be/Be^{2+} 0.89/0.31 B/B^{3+} 0.80/0.20 C 0.77 N/N^{3-} 0.74/1.71 O/O^{2-} 0.74/1.45 F/F$^-$ 0.72/1.33

Na/Na$^+$ 1.54/0.98 Mg/Mg^{2+} 1.30/0.65 Al/Al^{3+} 1.25/0.45 Si 1.17 P/P^{3-} 1.10/2.12 S/S^{2-} 1.04/1.84 Cl/Cl$^-$ 0.99/1.81

K/K$^+$ 1.96/1.33 H 0.37 Br/Br$^-$ 1.14/1.95

Rb/Rb$^+$ 2.11/1.48 I/I$^-$ 1.33/2.16

Cs/Cs$^+$ 2.35/1.67

FIGURE 16–8 Comparing relative atomic radii (Å)—for a few metals and nonmetals—with the corresponding filled valence shell ions.

distances. Several values are listed in Table 16–2. Experimental values for internuclear distances between adjacent ions in a crystal are obtained from x-ray measurements. Ionic radii are then obtained by assigning a percentage contribution to the anion and to the cation (Table 16–3). Using ionic radii obtained this way, we can get useful approximations of interionic distances. The trends evident in Figure 16–8 are worth summarizing:

1. The ionic radius increases as you move down through a family:

Li$^+$ at 0.68 Å to Cs$^+$ at 1.67 Å

This is in keeping with the fact that electrons of successively higher n quantum levels lie at greater distances from the nucleus.

TABLE 16–4 Variations in measured bond angles, degrees

Hydride	$\angle$H—X—H	Substituted methanes	$\angle$H—C—H
CH_4	109.5	CH_3Cl	110.5
SiH_4	109.5	CH_3Br	111.2
GeH_4	109.5	CH_3I	111.4
SnH_4	109.5	CH_2Cl_2	112.0
NH_3	107.3	CH_3OH	107.9
PH_3	93.3		
H_2O	104.5		
H_2S	92.3		

2. A comparison of positive and negative ions containing equal numbers of electrons shows the negative ion to be the larger:

$$F^- \text{ at } 1.33 \text{ Å} > Na^+ \text{ at } 0.98 \text{ Å}$$

The negative ion has a smaller nuclear charge per electron, accounting for an expanded electron cloud.

3. Increasing nuclear charge within a period results in a decreasing atomic radius:

$$Na^+ \text{ at } 0.98 \text{ Å} > Mg^{2+} \text{ at } 0.65 \text{ Å} > Al^{3+} \text{ at } 0.45 \text{ Å}$$

The more positive ion has the larger nuclear charge per electron, accounting for a contracted electron cloud.

Once an atomic or ionic radius has been assigned, an atomic volume, consistent with a sphere of that radius, can be calculated. But keep in mind that the electron cloud has no exact physical boundary. The atomic radius corresponds to our view of the atomic orbital as represented by a contour map of about 90% probability that the electron will be found within.

EXAMPLE 16–2
Place the boron, beryllium, and lithium ions in proper order according to increasing size.

Solution B^{3+}, with three more nuclear charges than electrons, would be the smallest. Li^+ would be the largest. ■

EXERCISE 16–2
Order the following sets of ions according to increasing size.
(a) Br^-, F^-, Cl^-
(b) Cl^-, K^+, Br^- ■

Bond Angles

We measure values of bond lengths, not atomic radii.

When a given atom is bonded to two or more atoms in a molecule, the bonds are directed in space, creating characteristic bond angles. For example, the H—C—H bond angle is close to 110° in many different molecules; the H—N—H bond angle is often observed to be about 107°. The importance of this phenomenon in determining molecular structure should be self-evident. Like bond lengths, bond angles are average values rather than rigidly fixed (due to atomic vibrations), but the average values are well defined. Table 16–4 lists some typical examples.

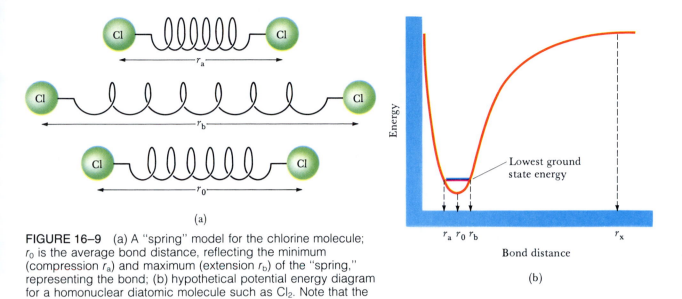

FIGURE 16–9 (a) A "spring" model for the chlorine molecule; r_0 is the average bond distance, reflecting the minimum (compression r_a) and maximum (extension r_b) of the "spring," representing the bond; (b) hypothetical potential energy diagram for a homonuclear diatomic molecule such as Cl_2. Note that the distance r_x corresponds to "free" atoms.

16.5 BREAKING CHEMICAL BONDS; BOND ENTHALPIES

If the two atoms in a diatomic molecule were pulled far enough apart, the bond between them would break. Imagine two chlorine atoms connected by a spring representing the chemical bond. As the two chlorine atoms are drawn apart and the spring is extended, a counter-force tries to pull them back to their original positions. Now let them spring back together; they not only return to their original position but they overshoot it, and only eventually vibrate back into the equilibrium position. That is just what a pair of chlorine atoms do in a chlorine molecule; they vibrate about an equilibrium position. Figure 16–9 is plot of bond distance versus the energy needed to displace the pair of chlorine atoms from the equilibrium position to that distance. With increasing separation, the curve flattens out until no attractive force remains and the bond is broken.

Because the spring is oscillating between two points, the lowest ground state energy in Figure 16–9 lies above the minimum, corresponding to the distance r_0 between the atoms. The **bond dissociation enthalpy** corresponds to the distance along the energy coordinate from the ground state level to the point where the two atoms have been displaced the distance r_x corresponding to complete nuclear separation. Using the compression and expansion of a spring as a model for internuclear motion, r_a and r_b are the classical turning points for the oscillating motion; but remember that on a molecular level these motions are quantized. Molecules have discrete vibrational states; as we shall see, that turns out to be very important in the spectroscopic examination of molecules.

In the case of the chlorine molecule, the bond dissociation enthalpy is about 243 kJ/mol, providing the Cl—Cl bond separates homolytically. **Homolytic bond dissociation** means that each atom is left with one of the two electrons that made up the bonding electron pair:

$$Cl:Cl \longrightarrow Cl\cdot + \cdot Cl$$

For **heterolytic bond dissociation**, in which one atom leaves with both bonding electrons, the enthalpy of dissociation is 1138 kJ/mole:

$$Cl:Cl \longrightarrow Cl:^- + Cl^+$$

Heterolytic dissociation is a special case. Generally, when we speak of bond dissociation the process is homolytic and accompanied by absorption of energy. The reverse process, bond formation, is energy-releasing.

The bond dissociation enthalpy is the minimum enthalpy needed to break the bond between a pair of atoms in a gaseous molecule, separating them completely. For the diatomic hydrogen molecule, we know from spectroscopic data that 431 kJ/mol is required:

$$H_2 \; (g) \longrightarrow 2 \; H \; (g) \qquad \Delta H = 431 \; kJ/mol$$

For anything more complex than homonuclear diatomic molecules, it is rather a more difficult task to obtain exact bond dissociation enthalpies. However, reasonable approximations have been obtained from thermodynamic data. For example, take the complete thermal dissociation of methane as calculated from Hess's Law:

$$CH_4 \; (g) \longrightarrow C \; (g) + 4 \; H \; (g) \qquad \Delta H = 1662 \; kJ/mol$$

According to the equation, four C-to-H bonds have been broken. The average dissociation enthalpy for the separate C-to-H bonds is

$$\Delta H = \frac{1662 \; kJ/mol}{4 \; C\text{—}H \; bonds} = (416 \; kJ/mol)/C\text{—}H \; bond$$

It is probably unreasonable to assume all four bonds break simultaneously. What is more likely is that there are intermediate stages in this process, and different bond dissociation enthalpies. Spectroscopic experiments yield the following data for methane:

$$CH_4 \; (g) \longrightarrow CH_3 \cdot \; (g) + \cdot H \; (g) \qquad \Delta H = 427 \; kJ/mol$$

For the O-to-H bond in the water molecule:

$$H_2O \; (g) \longrightarrow H \cdot \; (g) + \cdot OH \; (g) \qquad \Delta H = 502 \; kJ/mol$$
$$\underline{\cdot OH \; (g) \longrightarrow \cdot O \cdot \; (g) + \cdot H \; (g) \qquad \Delta H = 423 \; kJ/mol}$$
$$H_2O \; (g) \longrightarrow 2 \; H \cdot \; (g) + \cdot O \cdot \; (g) \qquad \Delta H = 925 \; kJ/mol$$

When the two O-to-H bonds in the water molecule thermally dissociate, 925 kJ/mol is given up, and the average O-to-H bond dissociation enthalpy is

$$925/2 = (463 \; kJ/mol)/O\text{—}H \; bond.$$

That represents the simultaneous breaking of the bonds. For the more likely successive dissociation of the O-to-H bonds, the enthalpies are 502 and 423 kJ/mol (Table 16–5).

EXAMPLE 16–3

Bond dissociation enthalpies are widely used to obtain fair approximations for heats of reaction. Calculate the bond dissociation energy for methyl alcohol vapor.

Solution Consider the heat of reaction associated with the formation of methyl alcohol vapor from a stoichiometric collection of carbon, hydrogen, and oxygen atoms (at 273 K), according to the equation

$$C \; (g) + 4 \; H \; (g) + O \; (g) \longrightarrow CH_3OH \; (g)$$

Based on Hess's Law, ΔH_{rxn} could be obtained from any set of thermo-

TABLE 16–5   Bond Enthalpies at 298 K*

Bond	kJ/mol	Bond	kJ/mol	Bond	kJ/mol
H—H	431	Sn—Sn	163	N—Cl	201
C—C	345	Sb—Sb	121	O—Cl	218
C=C	610	C—H	416	Si—Cl	381
C≡C	835	N—H	391	P—Cl	326
N—N	163	O—H	464	S—Cl	339
N≡N	945	F—H	565	As—Cl	293
O—O	146	Si—H	318	Se—Cl	243
O=O	494	P—H	322	Sn—Cl	318
F—F	155	S—H	347	Sb—Cl	310
Cl—Cl	242	Cl—H	431	C—N	305
Br—Br	193	As—H	247	C≡N	890
I—I	151	Se—H	276	C—O	358
Si—Si	222	Br—H	366	C=O	745
P—P	201	Te—H	239	C—S	272
S—S	226	I—H	299	C=S	536
Se—Se	209	C—Cl	339	S=O	498

*Adapted from T. L. Cottrell, *The Strengths of Chemical Bonds*, 2nd ed., Butterworths, London, 1956, pp. 270–289.

chemical equations that add up to the desired net equation. Note that the reaction is equivalent to the formation of three C-to-H bonds, one C-to-O bond, and one O-to-H bond. This is the reverse of the dissociation of the molecule:

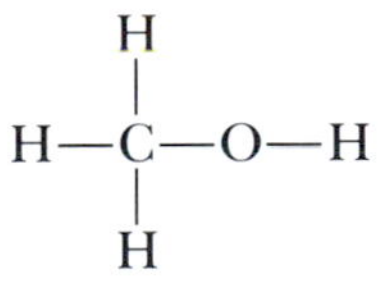

$$
\begin{array}{c}
\quad\;\; H \\
\quad\;\; | \\
H\!-\!C\!-\!O\!-\!H \\
\quad\;\; | \\
\quad\;\; H
\end{array}
$$

Breaking three C—H bonds	= 3(416) kJ/mol	= 1248 kJ/mol
Breaking one C—O bond	= 1(358) kJ/mol	= 358 kJ/mol
Breaking one O—H bond	= 1(464) kJ/mol	= 464 kJ/mol
Net result		= 2070 kJ/mol

Since the reaction describes bond formation, not bond dissociation, $\Delta H = -2070$ kJ/mol. Therefore, 2070 kJ is given off when one mole of methyl alcohol vapor forms from a gas of atoms.   ■

EXERCISE 16–3(A)

When CCl_4 is decomposed into its constituent atoms at 25°C, 1356 kJ/mol is absorbed. Calculate the average bond dissociation energy for the C—Cl bond.
Answer: 339 kJ/mol.   ■

EXERCISE 16–3(B)

Using the bond energy data in Table 16–5, calculate the heat required to decompose a mole of ammonia into a gas of atoms.
Answer: 1.17×10^3 kJ/mol.   ■

16.6 POLAR MOLECULES, ELECTRONEGATIVITY, AND ELECTRON AFFINITIES

Polarity and Dipole Moments

The characteristic properties of most chemical bonds lie somewhere between the ionic behavior of salts such as sodium chloride, which is essentially electrovalent, and the covalent character of petroleum hydrocarbons such as methane. For example, consider the hydrogen chloride molecule. There is little evidence of any chloride ion. HCl is a gas at room temperature (m.p. $-115°C$; b.p. $-85°C$); in the liquid state, it is a poor conductor of electricity. Both properties are distinctly covalent and characteristic of chemical bonds formed by atoms sharing electrons:

$$H\cdot + \cdot \overset{..}{\underset{..}{Cl}}: \longrightarrow H:\overset{..}{\underset{..}{Cl}}:$$

Still, hydrogen chloride is not entirely covalent. It dissolves in water to an almost unlimited degree (almost 800 g/L at 25°C), and dilute aqueous solutions of hydrogen chloride are good conductors of electricity. The implications are that in aqueous solution, the covalent bond between the hydrogen and the chlorine atom can break heterolytically, dispensing charge-carrying ions:

$$H\!-\!Cl \longrightarrow H^+ + Cl^-$$
$$H_2O + H^+ \longrightarrow H_3O^+$$

$$\overline{H\!-\!Cl + H_2O \longrightarrow H_3O^+ + Cl^-}$$

This activity, which is surprising for covalent bonds, can be written as a chemical equation:

$$HCl\ (g) + H_2O\ (liq) \longrightarrow HCl\ (aq) \longrightarrow H_3O^+\ (aq) + Cl^-\ (aq)$$

Continuing the development of this line of reasoning, consider the hydrogen molecule. It has a limited solubility in water, yet to the extent that it does dissolve, there is little evidence of heterolytic fission into hydronium (H_3O^+) ions and hydride (H^-) ions. Since the hydrogen molecule is symmetric, the shared electron pair in the covalent bond ought to be shared equally by both atoms. In hydrogen chloride, however, the two atoms aren't identical, and as a result the electron sharing isn't equal. The bonding electron pair in the hydrogen chloride molecule is more strongly attracted to the chlorine atom than to the hydrogen atom. Consequently, the hydrogen atom is assigned a partial positive charge and the chlorine atom is assigned a partial negative charge. A Greek letter delta, δ, is usually used to designate these partial charges on the atoms of the molecule:

$$H^{\delta +}\!-\!Cl^{\delta -}$$

Molecules such as these, in which the bonding electron pairs are unequally distributed, are **polar molecules**; the bonds themselves are **polar bonds**; and the atom that carries the partial negative charge is the more **electronegative** of the two. In general, when two opposite electric charges are separated at a distance, an electric **dipole** has been established (Figure 16–10).

The polarity of a bond is quantitatively measured by its **dipole moment**—the product of the magnitude of the separated charge and the distance between the centers of charge. Qualitatively, the polarity of an

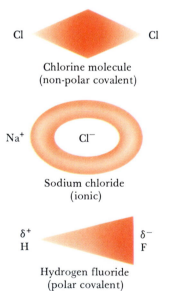

Cl ——— Cl

Chlorine molecule
(non-polar covalent)

Na⁺ Cl⁻

Sodium chloride
(ionic)

δ^+ δ^-
H F

Hydrogen fluoride
(polar covalent)

FIGURE 16–10 Ionic and covalent electron distributions. For polar molecules, the lowercase Greek letter delta (δ) denotes a partial charge. H—F (as H—Cl) is polar covalent, with the charge distribution shifted toward the electronegative halogen.

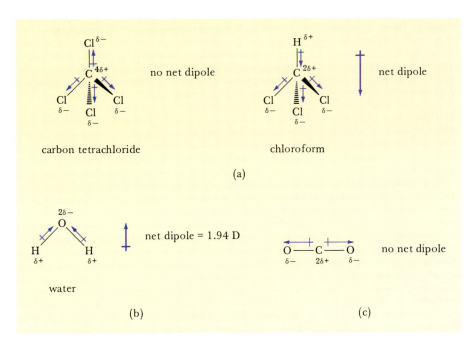

FIGURE 16–11   (a) The symmetric carbon tetrachloride molecule. Although the individual C—Cl bonds are polar, the resultant of the lines of force equals zero. The chloroform molecule has a net dipole; (b) the dipolar water molecule; (c) CO_2, being linear and symmetric, has no net dipole moment.

X—Y bond increases with increasing separation of X and Y in the periodic table of the elements. As one moves across the periodic table from the electropositive elements on the left-hand side to the electronegative elements on the right-hand side, the difference in electronegativity increases, along with the dipole moment and the polarity of the bond.

Dipole moments are actually vector quantities, directed from the more positive end of the bond toward the more negative end. Like forces, dipole moments in polyatomic molecules are added by vector addition. This means that the resultant (the sum of the dipole moments of all the bonds in the molecule) may point in a direction different from any of the bond directions. In fact, if the dipole moment vectors cancel each other, a molecule that contains polar bonds may have a total dipole moment of zero—that is, it can be nonpolar. For example, in CCl_4, the C—Cl bond is indeed polar because the chlorine atom attracts the bonding electron pair $(C^{\delta+}—Cl^{\delta-})$. However, the molecule is tetrahedral and symmetrical, so the vector addition gives a net zero dipole moment (Figure 16–11).

As a rule, unsymmetrical molecules are generally polar. The presence of one hydrogen atom changes the nonpolar carbon tetrachloride molecule into the highly polar chloroform ($CHCl_3$) molecule. Likewise, the presence of one chlorine atom changes the nonpolar methane (CH_4) molecule into the polar methyl chloride (CH_3Cl) molecule. The presence of strongly electron-withdrawing atoms of elements such as N, O, S, and the halogens (F, Cl, Br, and I) markedly enhances the effect by increasing the transfer of charge and the dipole moment.

Clearly, symmetry considerations impose restrictions on polarity. The converse is also true: Polarity in a molecule indicates asymmetry. Consider the carbon dioxide molecule. Is it linear or nonlinear? Because the molecule contains two highly electron-withdrawing oxygen atoms, one would expect two polar O-to-C bonds. Experimental data show that the molecule is nonpolar. Therefore, the vector addition of the two dipole moments must yield zero, which means that the molecule must be linear so that the vectors cancel. If the molecule were bent, the vector addition could not yield zero, and the carbon dioxide molecule would be polar. Furthermore, what if the

TABLE 16–6   Dipole Moment Data for Some Representative Gaseous Molecules, debyes

Diatomic molecules		Triatomic molecules		Polyatomic molecules	
HF	1.98 D	H_2O	1.86 D	NH_3	1.47 D
HCl	1.03 D	H_2S	1.11 D	PH_3	0.56 D
HBr	0.79 D	N_2O	0.15 D	AsH_3	0.21 D
HI	0.38 D	NO_2	0.33 D		
ClF	0.87 D	O_3	0.52 D	CH_3OH	1.70 D
NO	0.15 D				
CO	0.13 D				

D = Debye unit, the resultant electric moment $\times$ 10^{18} esu.

carbon atom were not the central atom—if it were C—O—O instead of O—C—O? Then the molecule would be polar no matter what the internal bond angle, because of the difference in dipole moments between the C—O and O—O bonds. Thus, dipole moment data give us information about polarity in a molecule, which in turn helps us understand the arrangement and geometry of the atoms in the molecule.

EXAMPLE 16–4

The water molecule has a large dipole moment, and the boron trifluoride molecule is nonpolar. What does that suggest about their respective geometries?

Solution   If the water molecule were linear, the dipole moments of the two O—H bonds would cancel; therefore, the water molecule must be bent. In fact, the internal H—O—H angle is 104.5° (Figure 16–11b). Boron trifluoride must be symmetrical and planar.

EXERCISE 16–4

The CO_2 molecule is symmetrical and linear; the SO_2 molecule is symmetrical and bent. What does this suggest about their respective dipole moments?

Answer: The CO_2 molecule has no observable dipole moment. The SO_2 molecule has a dipole moment.

Finally, dipole moment data can be used to determine the electron distribution and degree of ionic character of a bond. Examine the dipole moments for the hydrogen halides in Table 16–6. Note the decrease in percent ionic character for the H—X bond in proceeding down through the halogen family.

Electron Affinities

The ionization energy associated with removal of an electron from an atom M is of basic importance to any discussion of atomic structure and chemical bonding. Ionization energies are determined by recording the frequency at the threshold (onset) of absorption of energy corresponding to photoionization of a gas of atoms of the element being studied:

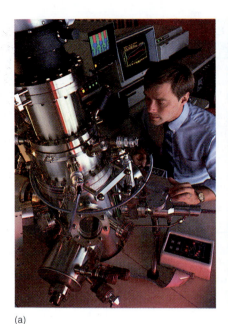

(a) (b)

A scanning Auger Microscope (a) produces secondary electron images (b) as a first step in guiding scientists to where to search for further analyses of the fractured metal surface of a failed part. The color images are computer-enhanced representations of the actual data obtained from the surface scans due to emission of a second electron after high energy radiation has ejected the first—Auger spectroscopy.

$$M + h\nu \longrightarrow M^+ + e^-$$

The energy of the photon corresponding to the threshold frequency is the ionization energy.

The energy released when an electron is captured by an atom M is also important and is called the **electron affinity**:

$$M\,(g) + e^- \longrightarrow M^-\,(g) \qquad \Delta H = -(\text{electron affinity})$$
$$Cl\,(g) + e^- \longrightarrow Cl^-\,(g) \qquad \Delta H = -361.5\ \text{kJ/mol}$$
$$EA = 361.5\ \text{kJ/mol}$$

Unfortunately, electron affinity data are more difficult to obtain, although the theory and practice are analogous to those described for ionization energies. A threshold frequency corresponding to a photodetachment reaction is the essential observation:

$$M^-\,(g) + h\nu \longrightarrow M\,(g) + e^-$$

In one method, an intense "tunable" laser light source is used; one searches for the wavelength corresponding to detachment. Photoelectron spectroscopy (to be described later) can also be used.

Electron affinities are smallest for elements having only a few outer shell electrons, or for filled shell elements such as neon; electron affinities are largest for elements with nearly filled shells such as fluorine. Taken together, electron affinities and ionization energies reflect the balance between nuclear charge and the shielding of the nucleus (Table 16–7).

The Electronegativity Scale

A useful definition of electronegativity is "the power of an atom in a molecule to attract electrons to itself." A very large number of chemical and physical properties have been explained, predicted, and correlated on the

TABLE 16–7	Experimental Values for Electron Affinities for the Atoms of Several Elements, kJ/mol				
H					
72.9					
Li	B	C	N	O	F
59.8	29	121	133	142	334
Na	Al	Si	P	S	Cl
52.7	50	134	72	197	360
K					Br
48.4					343
					I
					314

basis of electronegativity. Table 16–8 is one compilation of electronegativity values that have proved especially useful. Values have been compiled by Pauling.

The electronegativity scale is important in providing a qualitative statement of differences between the elements. For example, alkali metals, with their low ionization energies and small electron affinities, have low electronegativities. In fact, we more commonly refer to them as electropositive elements. Across any given period, the effective nuclear charge increases and the atomic volume decreases, both of which contribute to the tendency of an atom to attract electrons. Hence electronegativities increase. As we move down through a family of elements, the atomic volume increases as more shells are added. At the same time, there is a corresponding decrease in the atom's ability to attract the electrons in a bond, so electronegativities decrease with increasing atomic weight.

EXAMPLE 16–5
Using data taken from Table 16–8, predict whether CsCl and CH_4 would more likely be classified as ionic or covalent, assuming that a pair of combined elements with very different electronegativities would form a predominantly ionic combination.

TABLE 16–8	Pauling Electronegativity Scale for the Common Oxidation States of the Elements															
H																
2.1																
Li	Be											B	C	N	O	F
1.0	1.5											2.0	2.5	3.0	3.5	4.0
Na	Mg											Al	Si	P	S	Cl
0.97	1.2											1.5	1.8	2.1	2.5	3.0
K	Ca	Sc	Ti	V	Cr	Mn	Fe	Co	Ni	Cu	Zn	Ga	Ge	As	Se	Br
0.90	1.0	1.3	1.5	1.6	1.6	1.5	1.8	1.8	1.8	1.9	1.6	1.6	1.8	2.0	2.4	2.8
Rb	Sr	Y	Zr	Nb	Mo	Tc	Ru	Rh	Pd	Ag	Cd	In	Sn	Sb	Te	I
0.89	1.0	1.2	1.4	1.6	1.8	1.9	2.2	2.2	2.2	1.9	1.7	1.7	1.8	2.0	2.1	2.5
Cs	Ba	La	Hf	Ta	W	Re	Os	Ir	Pt	Au	Hg	Tl	Pb	Bi	Po	At
0.7	0.97	1.1	1.3	1.5	1.7	1.9	2.2	2.2	2.2	2.4	1.9	1.8	1.8	1.9	2.0	2.2
Fr	Ra	Ac	Th	Pa	U	Np										
0.7	0.9	1.1	1.3	1.5	1.7	1.3										

Solution   Cesium fluoride, according to the values in the table, would most likely be ionic in character, with a maximum difference of

$$3.0 - 0.7 = 2.3$$

Methane would be best described as covalent, based on the similar electronegativity values found in the table:

$$2.5 - 2.1 = 0.4$$ ■

EXERCISE 16–5
What are the relative differences between the electron affinities in each of the following pairs? Briefly explain the differences.
(a) argon and potassium   (c) copper and zinc
(b) argon and chlorine   (d) sodium and magnesium   ■

16.7   DOCUMENTING CHEMICAL BONDS BY INFRARED SPECTROSCOPY

Our understanding of molecular structure and bonding is largely based upon our knowledge of the interaction of matter with electromagnetic radiation. Instruments designed to measure these interactions are called spectrometers. For the study of individual atoms, **emission spectra** are extensively used. Of more general value with regard to molecules are **absorption spectra**, typically in the infrared region of the electromagnetic spectrum. The infrared region is particularly important for studying the structure of matter at the molecular level because the natural vibrational frequencies of the atoms in molecules fall in that region. Figure 16–12 shows the characteristics of the infrared (IR) spectra. Each of the successive minima in the absorption spectrum—we call them peaks, although they are clearly valleys—corresponds to a particular interaction between the molecules and the absorbed radiation; each represents a particular wavelength or photon of energy that is capable of exciting a response in the molecules. The molecular spectrum is highly characteristic and has often been referred to as a molecular "fingerprint."

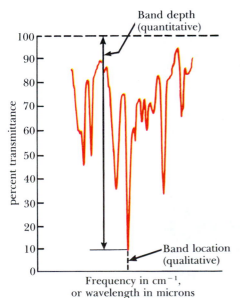

FIGURE 16–12   Significant characteristics of the infrared spectrum. Band depth can be used quantitatively; band location provides qualitative information.

When the frequency of the incident radiation corresponds to a natural frequency of the molecule, radiation is absorbed and the transmitted intensity is decreased, which is what the spectrograph detects and records. There are other motions as well, each characterized by its own frequency. In addition to stretching modes (frequencies), there are many bending modes and rotational modes. A compound exposed to radiation in the infrared region will absorb only those frequencies corresponding to such vibrations and rotations within the molecule. Taken together, they constitute a unique molecular code called the infrared spectrum that characterizes the molecule (Figure 16–13).

Fourier transform infrared spectroscopy is the latest addition to this family of techniques for recording rotational and vibrational spectra. It is fundamentally different and worth mentioning before leaving the subject because of its potential importance. In this technique, all the IR frequencies irradiate the sample simultaneously for an instant, and the data are stored in a computer. The absorption spectrum is then obtained by mathematically manipulating the stored data pattern. It is a very rapid technique that gives

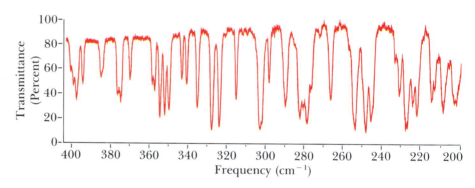

FIGURE 16–13 Infrared spectrum of water vapor in the region between 400 and 200 cm^{-1}.

FIGURE 16–14 Fourier-transform infrared (FTIR) spectroscopy.

high resolution and accurate wavelength assignments from very small samples; for all those reasons, it promises great benefits in basic research and process development for scientists and engineers (Figure 16–14).

SUMMARY

Stable anions and cations can be thought of as being formed by the transfer of electrons, giving rise to electrostatic forces called **ionic** chemical bonds that hold these particles together in the crystalline, solid state.

Among the more important features of the shared electron-pair or **covalent** bond is its multiplicity: there are single, double, and triple bonds. Homonuclear diatomic molecules such as H_2 and Cl_2 are typical examples of pure covalent single bonds. There are well-established values for carbon-to-carbon, carbon-to-hydrogen, and carbon-to-oxygen single and double bonds, and there are many molecules that contain carbon-to-carbon and carbon-to-nitrogen triple bonds. All represent distinct average distances between nuclei.

Ionic and covalent chemical bonds are documented by their characteristic bond lengths, bond angles, and bond enthalpies. **Bond lengths** are stated as **interionic distances** between nuclei in ionic solids, or **interatomic distances** between nuclei in molecules. **Covalent radii** can be calculated from atomic volumes and added together to give approximations of bond distances in molecules. The same can be done with **ionic radii** to obtain bond distances in salt-like substances. Both bond distances and bond angles are average values, reflecting the fact that a pair of connected atoms vibrate about an equilibrium position. Bond dissociation enthalpies are also tabulated as average values. Molecular vibrations exist at all temperatures down to the absolute zero of temperature.

Due to differences in the power of atoms to attract electrons, which we call their **electronegativities**, shared electron pairs are not generally shared equally. As a result, chemical bonds exhibit permanent dipoles and many molecules are polar; but the presence of polar bonds does not in itself guarantee **polarity** in a molecule. If the addition of the vector quantities representing the **dipole moments** yields zero, the molecule is nonpolar. That is the situation in molecules such as CH_4 and CCl_4, where symmetry exists. Using the **Pauling scale of electronegativities**, it is possible to determine the degree of ionic or covalent character of a bond.

The regularly repeating properties discussed in this chapter are sizes of atoms and ions, ionization energy and electron affinity, and electronegativity. The sizes of atoms increase proceeding down a group, and as a rule decrease proceeding across a period. The first ionization energy, the energy required to remove an electron from an atom, is generally greatest for small atoms. Thus, the first ionization energies can be expected to increase proceeding down a group and increase going across a period. Some irregularities in this order can be explained in terms of removing electrons from a newly populated orbital or from a half-filled shell. Second, third, and higher ionization energies result from removal of electrons from cations. These increase progressively for any element.

The electron affinity of an element is the energy given off when an electron is added to an atom to form a negative ion. These values are more irregular than ionization energies but tend to increase in the same general pattern. The electronegativity is a measure of the overall ability of an atom to draw electrons to itself. The electronegativity increases going upward and to the right in the periodic table. Fluorine is the most electronegative element.

Infrared spectroscopy is an analytical tool for studying molecular structure based on the absorption of energy at frequencies that perturb the rotational and vibrational motion of the atoms. The infrared spectrum is a molecular fingerprint.

QUESTIONS AND PROBLEMS

QUESTIONS

1. Distinguish between ionic, covalent, and coordinate bonds.

2. When speaking of HCl, it is reasonable to speak of "one molecule" of the compound. Why is that not the case for NaCl?

3. Based on the bonding theories developed in this chapter, explain each of the following:
(a) the electrovalence of potassium and chlorine in KCl
(b) the electrovalence of magnesium and oxygen in MgO

4. Briefly offer an explanation of each of the following:
(a) Neon and argon form no known compounds.
(b) Sodium chloride is invariably NaCl, while copper chloride can be CuCl or $CuCl_2$.

5. Write the electronic structure for BF_3. Would you predict that BF_3 should react with NH_3? If a compound does form, is the bond that is formed covalent? And if so, in what way does this bond differ from any other covalent bond?

6. Why should heterolytic cleavage of a homonuclear bond between two like atoms require so much more energy than homolytic cleavage?

7. Which member of each of the following pairs of bonds would you expect to exhibit the larger dipole moment?
(a) H—O and H—N
(b) H—F and H—Br
(c) C—O and C—S
(d) C—N and C—F

8. Briefly explain each of the following terms:
(a) bond length
(b) bond polarity
(c) bond enthalpy
(d) dipole moment
(e) electronegativity
(f) atomic radii

9. Distinguish between the terms ionization energy and electron affinity. What do we mean by electronegativity?

10. In what useful way(s) can you predict the type of bond—ionic or covalent—formed between two elements from their electronegativities?

11. Distinguish between metallic and nonmetallic elements in terms of electronegativities.

12. What is a dipole moment? What effect does the presence of a dipole moment have on the properties of chemical bonds and chemical compounds?

13. How would you expect the magnitude of the dipole moment for the X—Cl bond to change in moving from X = Al to X = S among the following chlorides: $AlCl_3$ to $SiCl_4$ to PCl_3 to SCl_2?

14. Explain why HCl has a dipole moment but H_2 and Cl_2 have none.

15. Briefly explain why the C—F bond is polar yet CF_4 has a zero dipole moment.

16. Which bond is most polar: CF, NF, or OF? Compare your answer with the observed dipole moments and briefly explain: F_2O $(0.21\ \mu)$; NF_3 $(0.21\ \mu)$; CF_4 $(0\ \mu)$.

17. Suggest a likely arrangement of the three atoms in the following molecules:
(a) The BeH_2 molecule is linear and nonpolar.
(b) The N_2O molecule is linear and polar.

18. The water molecule is polar. Is the molecule bent? Is the molecule symmetric?

19. The CO_2 molecule is nonpolar. Is the molecule bent? Is the molecule symmetric?

20. Predict the geometry of each of the following:
(a) $HgCl_2$ has a zero dipole moment.
(b) BF_3 has a zero dipole moment.
(c) NF_3 is a polar molecule.

PROBLEMS

Lewis Structures; Ionic and Covalent Bonds [1–10]

1. On the basis of the Lewis concept of an octet of electrons, show the outer electron configuration for each of the following substances:
(a) OH^-
(b) H_3O^+
(c) NH_4^+
(d) NH_2^-

2. On the basis of the Lewis concept of an octet of electrons, show the outer electron configuration for each of the following substances:

(a) P_4 (tetrahedron) and S_8 (ring)

(b) HCN

(c) NH_4Cl

(d) Na_2SO_4

3. Draw a Lewis electron-dot structure for each of the following ions, clearly indicating the formal charge on each atom and the net charge on the entire species:

(a) fluoborate ion (BF_4^-)

(b) hydrogen peroxide ion (HOO^-)

(c) ammonium hydrogen sulfate (NH_4^+, HSO_4^-)

4. Draw a probable Lewis structure for each of the following covalent molecules, showing all of the valence electrons required to comply with the octet rule. Write in the charges where necessary:

(a) N_2O_5

(b) NCl_3 and PCl_3

5. Write electronic structures for the following covalent molecules that are in keeping with the "octet" rule:

(a) H_2O

(b) H_2O_2

(c) CO_2

(d) H_2CO_3

(e) CH_4O (methyl alcohol)

(f) C_2H_6O (ethyl alcohol or dimethyl ether are possible)

6. Write electronic structures for the following covalent molecules that are in keeping with the "octet" rule:

(a) N_2

(b) NH_3

(c) N_2H_4 (hydrazine)

7. Write electronic structures for the following covalent molecules that are in keeping with the "octet" rule:

(a) HN_3 (hydrazoic acid)

(b) H_2N_2 (diimide)

(c) CH_3NH_2 (methylamine)

8. Draw the electronic structure for each of the following:

(a) CH_2 and CH_4

(b) O_2 and O_3

(c) N_2 and N_3^-

Would you expect CH_2, O_3, and N_3^- to be unusually reactive or very stable species?

9. The sulfate ion consists of a central sulfur atom with four equivalent oxygen atoms attached in a tetrahedral arrangement. Draw the electronic structure for the ion in keeping with the octet rule.

10. Draw electron-dot structures of the BF_4^- ion and the NH_4^+ ion.

Atomic and Ionic Radii   [11–14]

11. Place the following in proper order according to increasing size:

(a) Al^{3+}, Na^+, Mg^{2+}

(b) Br^-, F^-, Cl^-

(c) Cl^-, Na^+, F^-

12. Place the following in proper order according to decreasing size:

(a) K^+, Na^+, Li^+

(b) Ne, He, Ar

(c) F^-, Ne, Na^+

13. Order each of the following sets according to increasing atomic radius.

(a) Na, Li, Rb

(b) F, B, N

(c) P, Ga, Cl

14. Place each of the following sets of atoms in order of decreasing atomic radius.

(a) Ca, Mg, Rb

(b) S, O, Al

(c) Ar, Kr, Ne

Bond Enthalpies   [15–18]

15. Use the table of bond enthalpies to calculate the change in enthalpy that takes place when a mole of ethylene and a mole of hydrogen are allowed to react, forming a mole of ethane according to the reaction:

$$CH_2{=}CH_2 \text{ (g)} + H_2 \text{ (g)} \longrightarrow CH_3{-}CH_3 \text{ (g)}$$

16. Calculate the enthalpy change for the following reaction using the table of bond enthalpies:

$$N_2 \text{ (g)} + 3 H_2 \text{ (g)} \longrightarrow 2 NH_3 \text{ (g)}$$

17. The heat of formation of ethyl alcohol vapor has been found to be -56.3 kcal/mol for the equation as written:

$$2 C \text{ (s)} + \tfrac{1}{2}O_2 \text{ (g)} + 3 H_2 \text{ (g)} \longrightarrow C_2H_5OH \text{ (g)}$$

Using the data available in the table of bond enthalpies (Table 16–5), determine the heat for the reaction:

$$2 C \text{ (g)} + O \text{ (g)} + 6 H \text{ (g)} \longrightarrow C_2H_5OH \text{ (g)}$$

18. The bond dissociation energies for fluorine and chlorine are 37 and 58 kcal/mol, respectively. When chlorine monofluoride, Cl—F, is formed from its elements according to the following reaction, 26 kcal/mol of heat is given up:

$$F_2 + Cl_2 \longrightarrow 2 ClF$$

Calculate the dissociation energy of the Cl—F bond.

Polarity; Electronegativity; Electron Affinity   [19–26]

19. Predict whether the binary compound formed from each of the following pairs of elements is likely to be predominantly ionic or covalent:

(a) Cs and F

(b) O and S

(c) O and N

(d) H and Cl

20. Predict whether the binary compound formed from each of the following pairs of elements is likely to be predominantly ionic or covalent:

(a) Li and N

(b) Be and I

(c) Ca and S

(d) Ba and Br

21. Write the chemical formula for each of the binary compounds in Problem 19.

22. Give the chemical formula for the simplest binary compound that would be formed by each of the pairs of elements in Problem 20.

23. Which of the following has the largest dipole moment?

   HF, $SiCl_4$, H_2Se, $SbCl_5$

24. The following gaseous molecules have no dipole moment. What might their structures be?
(a) BCl_3
(b) SiF_4
(c) CO_2

25. Using the table of electronegativities, determine which of the atoms in each pair has a partial positive charge and which a partial negative charge.
(a) the O—F bond
(b) the O—N bond
(c) the O—S bond

   Using the data in the same table, identify three covalent bonds formed between dissimilar elements that are essentially nonpolar.

26. In the following pairs of molecules, which one would you anticipate to be the more polar? Explain your choices.
(a) HCl and HF
(b) PH_3 and NH_3
(c) H_2S and H_2O

Additional Problems   [27–31]

27. Write at least three (there are more) formulas for sulfuric acid that would be considered acceptable Lewis structures.

28. The diatomic nitrogen molecule has 10 valence shell electrons. In its ground state, the molecule has a zero magnetic moment, and all of the electrons are spin paired. The most probable electron-pair notation is the linear, triple-bond structure with a lone pair of each nitrogen, and the bond length is 1.1 Å:

   $: N \equiv N :$

Three other diatomic species—carbon monoxide, cyanide ion, and nitrosyl ion—all show no magnetic moment and are characterized by the following bond lengths:

CO	1.1 Å
CN^-	1.2 Å
NO^+	1.1 Å

Draw probable valence bond structures for these three species. Acetylene dianion ($C \equiv C^{2-}$) is similarly related. Draw its valence bond structure. Such species as these are said to be isoelectronic. Explain.

29. In the following reaction, the number of ClF bonds remains constant while one Cl—Cl bond is destroyed:

   Cl_2 (g) + ClF_3 (g) $\longrightarrow$ 3 ClF (g)

(a) Find ΔH_{rxn} from the standard bond dissociation enthalpy data in Table 16–5.
(b) Briefly explain why the answer to part (a) is likely to differ from the experimental value.

(c) Would the given equation serve as a suitable preparation for chlorine monofluoride? Briefly explain why or why not.
(d) Can you explain why there is a favorable entropy change for the reaction?

30. Based upon the following thermochemical data, show that ozone (O_3) is considerably more stable than a cyclic structure would suggest. The enthalpy for the O—O bond is approximately 35 kcal/mol.

$$\tfrac{3}{2} O_2 \longrightarrow O_3 \qquad \Delta H = +34.5 \text{ kcal}$$
$$O_2 \longrightarrow 2 O \qquad \Delta H = +119 \text{ kcal}$$

(b) Using the empirical data available in the table of bond enthalpies, what might you suspect to be a more reasonable structure?

31. Molecular chlorine exhibits continuous absorption of energy at wavelengths in the visible region of the electromagnetic spectrum above 4800 Å. At 4800 Å, chlorine gas is dissociated into a ground state atom and an atom in an excited state, some 10.5 kJ/mol higher. Determine the bond dissociation energy for chlorine.

MULTIPLE PRINCIPLES   [32–37]

32. Experimentally, electron-affinities can be obtained using intense laser light sources that have been "tuned" to the wavelength corresponding to detachment of the electron from the anion:

   M^- (g) + $h\nu$ $\longrightarrow$ M (g) + e^-

Using data taken from Table 16–7, calculate the photon wavelength corresponding to the electron affinity values for hydrogen and chlorine. Note in what region of the electromagnetic spectrum these energies fall.

33. Using bond enthalpy data available in Table 16–5, (a) calculate the enthalpy of reaction for the isomerization of ethyl alcohol (C_2H_6O) to dimethyl ether (C_2H_6O) at 298 K:

   CH_3CH_2OH (g) $\longrightarrow$ CH_3OCH_3 (g)
   ethyl alcohol   dimethyl ether

(b) With the help of Table 16–4, estimate the C—O—H and C—O—C bond angles in ethyl alcohol and dimethyl ether, respectively; (c) With the help of Table 16–6, suggest which molecule has the greatest dipole moment.

34. Consult Table 16–5 and estimate the heat of reaction for the trimerization of acetylene (C_2H_2) to benzene (C_6H_6) in the gas phase at 298 K:

3 H—C≡C—H (g) $\longrightarrow$

(g)

Compare your answer to that obtained from thermochemical data obtained from Table 7–4 and briefly suggest some possible basis for the differences observed.

35. Use bond enthalpies and an estimation of the entropy change to predict at what temperatures (high, low, all, none) the following reactions are likely to take place:

(a) NCl_3 (g) $+ 3 H_2$(g) $\longrightarrow NH_3$ (g) $+ 3$ HCl (g)

(b) CH_4 (g) $+ 4 Cl_2$ (g) $\longrightarrow CCl_4$ (g) $+ 4$ HCl (g)

(c) $HC{\equiv}CCl$ (g) $+ 2 Cl_2$ (g) $\longrightarrow HCl_2C{-}CCl_3$ (g)

36. For the following reaction at 298 K, $\Delta S°$ is 322 J/mol·K:

$$2 F_2 \text{ (g)} + 2 H_2O \text{ (g)} \longrightarrow O_2 \text{ (g)} + 4 \text{ HF (g)}$$

Use the table of bond enthalpies to calculate the enthalpy change of this reaction and then use this value to calculate K, the equilibrium constant at 298 K.

37. In Chapter 9 (on solutions), we learned that similar molecules tend to be soluble in each other. One way that molecules can either resemble or differ from each other is in the extent of their dipole moment. On this basis, predict the solubility of the following gases in water, and which one has a large dipole moment: O_2, N_2, HF, HCl, BF_3 (symmetrical and planar), and SO_2 (bent).

APPLIED PRINCIPLES   [38–40]

38. Referring back to Problem 34, if the $\Delta G_f°$ values for acetylene (g) and benzene (g) are respectively 209.2 kJ/mol and 124.5 kJ/mol, what is the equilibrium constant for the reaction? Make a recommendation as to the possibilities of using the process for the commercial production of benzene.

39. A copper smelter produces 25 tons of copper per hour, using chalcopyrite, $CuFeS_2$, as a starting material. The principal stack gas is sulfur dioxide (SO_2): (a) Write a Lewis structure for the polar sulfur dioxide molecule; (b) by chemical reaction(s), explain the origins of sulfuric acid rains due to sulfur dioxide in the atmosphere; (c) estimate the sulfur dioxide emissions from this particular smelter, assuming no emissions controls and that all the stack gas does indeed exit the stack; (d) will your estimate be too high or too low if the chalcopyrite is contaminated with iron pyrites, FeS_2?

40. Infrared radiation (heat) can be absorbed by molecules which have a dipole moment or can be bent into some shape which has a dipole moment. Such gases in the atmosphere will contribute to the "greenhouse" effect. Emissions of which of the following gases should be limited because of possible greenhouse effect contributions: O_2, N_2, CO, CO_2, NO_2.

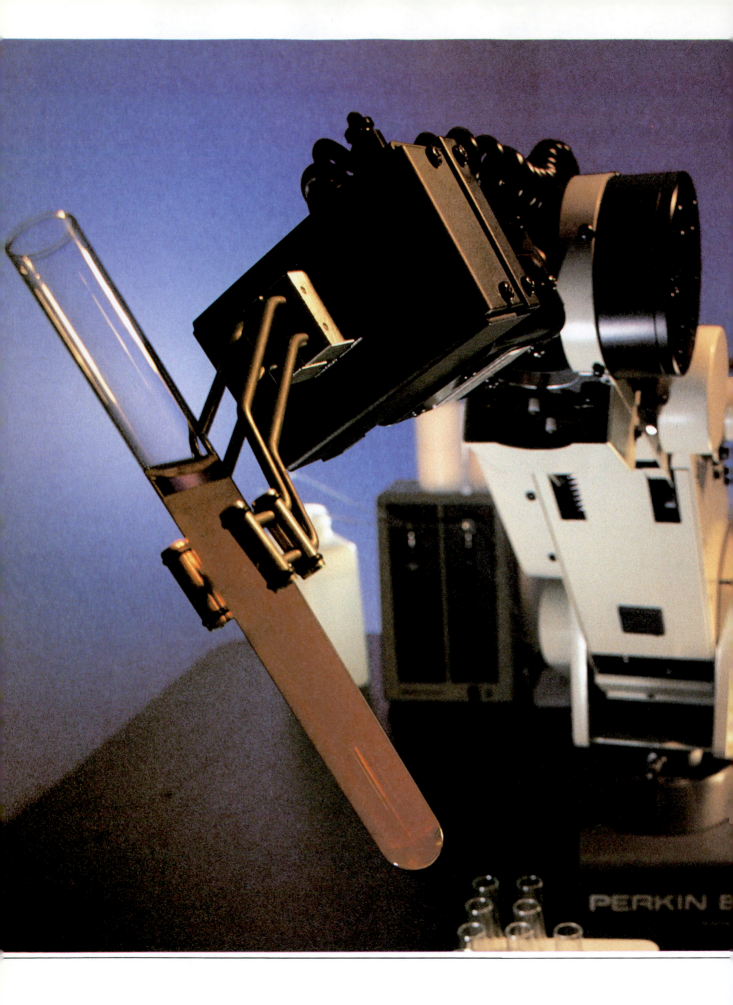

Main Group Elements; Periodic Properties

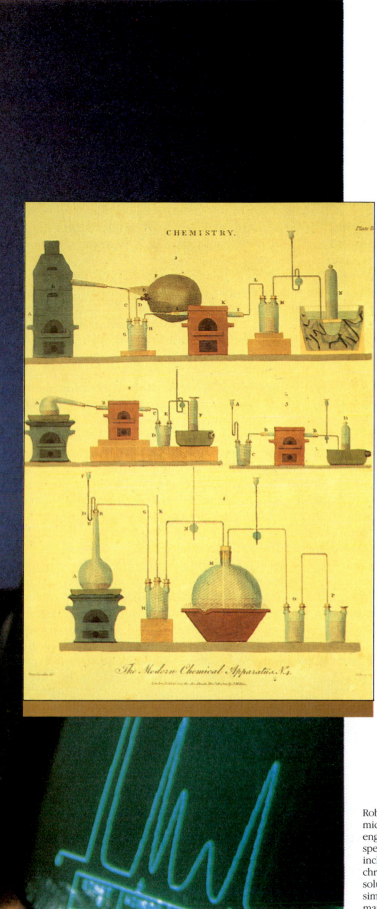

CHEMISTRY.

Plate II

The Modern Chemical Apparatus. N.4.

Robotics technology, aided by "smart" software and efficient microprocessors, has provided the laboratory scientist and the field engineer with improved methods of chemical analysis. Here, high-speed automated analysis of the progress of a chemical reaction includes sampling, centrifuging to remove the solid substrate, and chromatographic analysis (on the CRT screen) of the resulting solution. In earlier times, the progress of science was supported by simpler apparatus, as in this equipment from a London manufacturer's catalogue for the year 1800.

17.1   TRENDS IN GROUPS OF COMPOUNDS

In the last three chapters, we discussed the nature of atoms and chemical bonds. The theories of chemistry that were introduced allow us to understand some of the "Why?" questions of chemistry. Electrons in atoms, and the orbitals that they occupy, were described. Lattice energies were used to give a picture of the forces holding ionic compounds together, and the concepts of electronegativity and bond polarity were introduced to help describe covalent bonds. We are now in a position to consider some of the practices of chemistry, what chemical elements and compounds are like, and which reactions occur. Our discussion centers on the main group elements; the transition elements are considered in a later chapter. To begin with, let's look at a small group of compounds formed by one important element, chlorine.

Some elements such as chlorine, hydrogen, and oxygen form compounds with a very large number of other elements. Because of the great variation in properties of other elements, the compounds formed are very different. As an example, we can look at the group of compounds containing chlorine and one other element of the third period of the periodic table, from NaCl through Cl_2. Some of the physical properties of these compounds are shown in Table 17–1.

In these compounds, the number of chlorine atoms per other atom reaches a maximum of four at silicon and falls off at a rate of one chlorine per atom in either direction. Several trends are readily observed in the properties of these compounds. In particular, the melting and boiling points drop steadily through the group except for a slight irregularity at SCl_2. Thus, at room temperature NaCl and $MgCl_2$ are high melting solids, Al_2Cl_6 is a low melting solid, $SiCl_4$, PCl_3, and SCl_2 are liquids, and Cl_2 is a gas. Also, the heats of fusion and vaporization show general decreases. We are observing a progression from ionic to covalent compounds, from solids to gases, and from compounds that are melted and vaporized with difficulty to compounds that are melted and vaporized with ease.

When we look at the next series of chlorine compounds with elements of the fourth period the same trends repeat all over again. The repeating properties of elements and their compounds follow what is known as the periodic law; the periodic table, which we have used from time to time in our previous discussion, is the ultimate manifestation of the periodic law and an essential tool for our study of descriptive chemistry. Many periodic properties, such as sizes of atoms and ions, ionization energy, electron affinity, and electronegativity, have already been introduced in the three

TABLE 17–1   Some Physical Properties of the Chlorides of the Elements of the Third Period

Compound	Density (g/cm³)	Melting Point (°C)	Boiling Point (°C)	ΔH_{vap} (kJ/mol)	ΔH_{fus} (kJ/mol)
NaCl	2.165	801	1413	228	30.2
$MgCl_2$	2.316	708	1412		
Al_2Cl_6	2.44	sublimes, 178	120		
$SiCl_4$	1.48	−70	57.6	25.7	7.7
PCl_3	1.57	−91	75.5	29.5	
SCl_2	1.62	−78	59		
Cl_2	3.21	−103	−34.6		6.7

preceding chapters. In this chapter we will discuss a few more periodic properties and then proceed to a discussion of the chemistry of the main group elements.

We now understand the periodic law in terms of the electronic structure of the atom, which was covered in Chapters 14 and 15. However, the fact that the properties of elements repeat in a regular way was recognized before atomic structure was even vaguely comprehended.

17.2 BEGINNING OF THE PERIODIC TABLE

At the end of the 1700s and the beginning of the 1800s, as the existence of the chemical elements came to be recognized, it also became apparent that some elements have striking similarity to others. For example, sodium and potassium are both light, silvery conductors of electricity that react violently with water. Chlorine and bromine in the gaseous state are both colored, with pungent choking odors. They react rapidly with sodium and potassium to form stable, water-soluble salts. By 1830 a number of groups of closely related elements were recognized, including calcium, strontium, and barium, and sulfur, selenium, and tellurium.

No electronic theory of atoms would be available for many decades to explain why elements appeared to exist in families. However, the atomic weight was the one numerical quantity known to be attached to every element, and relationships between atomic weights and element properties were explored. Chemists did detect properties of elements that appeared to repeat in a regular fashion, but their theories were all tentative and they did not risk making any predictions. Finally, in 1869 Dmitrii Mendeleeff (in Russia) and Lothar Meyer (in Germany) hit on the nature and consequences of the regularly repeating properties of elements—the **periodic law**—which became apparent when elements were arranged in the order of the **periodic table**.

Mendeleeff's study of the relationships among the chemical elements was performed by preparing a card for each known element, which listed its properties. These he permuted in many ways. When he arranged them in order of increasing atomic weight he was able to construct the periodic table in a form only trivially different from the form we use today. Mendeleeff was definitely not afraid to go out on a limb with predictions based on his periodic table. He left spaces in his periodic table for elements that he anticipated would be discovered, and he predicted their properties. Where the properties of some elements did not fit in with the order of the atomic weights, he predicted that the atomic weights were incorrect and suggested the correct order. A surprising number of his predictions proved correct and a new level of chemical rationality emerged. Being able to place the large number of known elements into the structure provided by the periodic table greatly simplified our understanding and further study of chemistry.

In Mendeleeff's periodic table, the elements were arranged in the order of increasing atomic weight. Careful determination of atomic weights, however, showed that the order could not be strictly followed. Inversions were necessary at Ar and K, at Co and Ni, and Te and I. This problem was solved in 1913 by H.G.J. Moseley from x-ray data obtained by bombarding metals with electrons. His results showed that the wavelengths of the x-rays produced from specific target metals decreased in a regular way

Dmitrii Ivanovitch Mendeleeff (1834–1907) was denied entrance to the University of Moscow because he was a Siberian, but he finally entered a teacher's training college in St. Petersburg. His textbooks on chemistry became famous in their time.

FIGURE 17–1 The long form of the periodic table of the elements.

Group IA																	VIII
1 **H** 1.0079	IIA											IIIA	IVA	VA	VIA	VIIA	2 **He** 4.0026
3 **Li** 6.941	4 **Be** 9.0122											5 **B** 10.811	6 **C** 12.011	7 **N** 14.0067	8 **O** 15.9994	1 **H** 1.0079	10 **Ne** 20.1797
11 **Na** 22.9898	12 **Mg** 24.3050	IIIB	IVB	VB	VIB	VIIB	VIIIB			IB	IIB	13 **Al** 26.9815	14 **Si** 28.0855	15 **P** 30.9738	16 **S** 32.066	9 **F** 18.9984	18 **Ar** 39.948
19 **K** 39.0983	20 **Ca** 40.078	21 **Sc** 44.9559	22 **Ti** 47.88	23 **V** 50.9415	24 **Cr** 51.9961	25 **Mn** 54.9380	26 **Fe** 55.847	27 **Co** 58.9332	28 **Ni** 58.69	29 **Cu** 63.546	30 **Zn** 65.39	31 **Ga** 69.723	32 **Ge** 72.61	33 **As** 74.9216	34 **Se** 78.96	17 **Cl** 35.4527	36 **Kr** 83.80
37 **Rb** 85.4678	38 **Sr** 87.62	39 **Y** 88.9059	40 **Zr** 91.224	41 **Nb** 92.9064	42 **Mo** 95.94	43 **Tc** (98)	44 **Ru** 101.07	45 **Rh** 102.9055	46 **Pd** 106.42	47 **Ag** 107.8682	48 **Cd** 112.411	49 **In** 114.82	50 **Sn** 118.710	51 **Sb** 121.75	52 **Te** 127.60	35 **Br** 79.904	54 **Xe** 131.29
55 **Cs** 132.9054	56 **Ba** 137.327	57 *****La** 138.9055	72 **Hf** 178.49	73 **Ta** 180.9479	74 **W** 183.85	75 **Re** 186.207	76 **Os** 190.2	77 **Ir** 192.22	78 **Pt** 195.08	79 **Au** 196.9665	80 **Hg** 200.59	81 **Tl** 204.3833	82 **Pb** 207.2	83 **Bi** 208.9804	84 **Po** (209)	53 **I** 126.9045	86 **Rn** (222)
87 **Fr** (223)	88 **Ra** (226)	89 ****** Ac** (227)	104 **Unq** (261)	105 **Unp** (262)	106 **Unh** (263)	107 **Uns** (262)	108 **Uno** (265)	109 **Une** (266)								85 **At** (210)	

*Lanthanide Series

58 **Ce** 140.115	59 **Pr** 140.9076	60 **Nd** 144.24	61 **Pm** (145)	62 **Sm** 150.36	63 **Eu** 151.965	64 **Gd** 157.25	65 **Tb** 158.9253	66 **Dy** 162.50	67 **Ho** 164.9303	68 **Er** 167.26	69 **Tm** 168.9342	70 **Yb** 173.04	71 **Lu** 174.967

**Actinide Series

90 **Th** 232.0381	91 **Pa** 231.0359	92 **U** 238.0289	93 **Np** (237)	94 **Pu** (244)	95 **Am** (243)	96 **Cm** (247)	97 **Bk** (247)	98 **Cf** (251)	99 **Es** (252)	100 **Fm** (257)	101 **Md** (258)	102 **No** (259)	103 **Lr** (260)

Metals

Metalloids

Nonmetals

Note: Atomic masses are 1987 IUPAC values (up to four decimal places).

when moving from element to element along the rows of the periodic table. Rutherford recognized that Moseley's work identified the number of positive charges in the nucleus, and he determined that the elements in the periodic table must be arranged by *atomic number*, not *atomic weight*.

17.3 THE PERIODIC TABLE TODAY

A number of different forms of the periodic table have been prepared, but all give the same information. The most common form in use today is called the long form (Figure 17–1). Horizontal *rows* of the periodic table are called **periods**; vertical *columns* are called **groups** or **families**. In general, elements have properties most like those of other elements near them in the periodic table. Members of the same group have related properties, which change gradually upon moving up or down a column. The properties of the elements of the same period also change in a regular fashion, proceeding along a row. In addition, there is often a close relationship along the diagonals between an element and the element in the next greater column and row. Thus, Li has much in common with Mg, and B is similar to Si.

The nature of the periodic table is, of course, a result of the order in which the electrons fill atomic orbitals. Thus, the increasing lengths of the periods reflect the increasing number of orbitals available as the principal quantum number n increases. Only the $1s$ orbital is being filled in period 1 so there are only two elements, H and He. In period 2, both $2s$ and $2p$ are being filled, and eight elements are observed. He from period 1 is clearly similar to Ne in period 2; both are inert gases. However, H in period 1 is not like to any element in period 2. As the first element in the periodic table, hydrogen is chemistry's unique element, it is not really a member of any of the families of elements, although it is in some ways similar to Li, C, and F. It is customarily (though almost arbitrarily) placed in the same column with Li.

Each of the elements of period 2 is representative of a distinct family of elements. These eight families, labeled IA to VIIIA, make up what are called **main group** or representative elements. In Groups IA and IIA, s orbitals are being filled, so the block of the periodic table made up of these two groups is called the **s-block**. The orbitals being filled in Groups IIIA through VIIIA are p orbitals, so these groups form the **p-block** of the peroiodic table. Note that the space between Be and B exists only to allow room for the d-block elements that start in period 4. Some forms of the periodic table have been constructed to remove this artificial separation. Thus, Be and B are actually next-door neighbors and a diagonal relationship exists between Be and Al.

In period 4, the $3d$ orbitals start filling after Ca. The elements from Sc through Zn and the parallel elements in the later periods form the **d-block**. The elements of this block all have relatively similar properties and do not divide into families as clearly as do the main group elements. Group designations IIIB through IIB have been assigned as indicated in Figure 17–1. The elements of the d-block are also called **transition elements** or transition metals (since they are all metallic).

In period 6, after La, the $4f$ orbitals begin to fill, initiating the **f-block**, or lanthanide, elements. If those elements were placed in their rightful place in the periodic table, the d-block would be split between Groups IIIB and IVB and the table would become too elongated for practical printing.

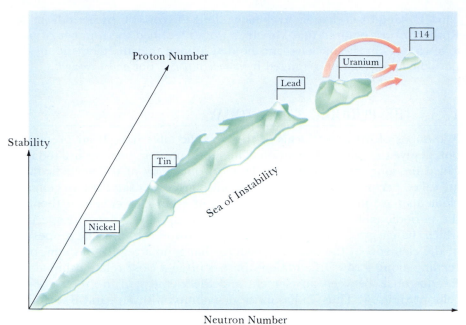

FIGURE 17–2   The nuclear landscape depicts islands of stability. Instead of being located by coordinates of lattitude and longitude, they are measured by binding energy per nucleon as a function of neutron number (*N*) and proton number (*Z*).

Therefore it is customary to split off the f-block elements into a separate part of the table. This means that the principal part of the table includes jumps from atomic numbers 57 to 72 and from 89 to 104.

The periodic table is still growing as new elements are synthesized, Currently, the d-block in period 7 is being filled. These elements, which are extremely unstable and produced in vanishingly small quantities, are the result of intense work in major national and international laboratories. Although elements $Z = 110$ to $Z = 118$ have not yet been reported, it is expected that elements around 114 may be relatively stable (Figure 17–2). The traditional rule that the discoverer of an element is allowed to name it has produced considerable controversy. As a means of eliminating battles on names of new elements, the International Union of Pure and Applied Chemistry has proposed that elements of atomic number 104 and greater be named by a new system, in which the name merely gives the atomic number. Thus, the element 104 is called unnilquadium (symbol Unq) where un = 1, nil = 0, and quad = 4. Element 105 is unnilpentium (Unp). These elements have been variously named in the East and West; kurchatovium and rutherfordium for 104 and nielsborium and hahnium for 105.

One of the premier applications of the periodic table has been in the search for new elements. In his periodic table, Mendeleeff noted the open spaces beneath aluminum and silicon (as well as in several other places). He was confident that elements would be found to fill these spaces, and called these then unknown elements eka-aluminum and eka-silicon. He predicted a number of the properties of these elements, thus simplifying the search for them. Both elements were discovered within twenty years of Mendeleeff's first publications of the periodic table. Eka-aluminum was called gallium and eka-silicon was called germanium. A comparison of some of the predicted and actual properties for germanium shows how close Mendeleeff's predictions really were.

	Eka-Silicon (Predicted)	Germanium (Measured)
atomic weight	72 g/mol	72.56 g/mol
density	5.5 g/cm³	5.35 g/cm³
maximum oxidation state (Sec. 17.5)	4	4
heat capacity	0.0073 cal/g	0.0074 cal/g
density of oxide	4.70 g/cm³	4.23 g/cm³
boiling point of chloride	less than 100°C	84°C

The predictive power of the periodic table has aided in finding a number of other elements. Until 1925 both of the positions under Mn were blank. The search for elements resembling manganese led to the eventual discovery of technetium ($Z = 43$) and rhenium ($Z = 75$).

Another blank in the periodic table existed at $Z = 72$. Searches for this element were complicated by a lack of knowledge of the total number of lanthanide elements. Thus, if element 72 was a fifteenth lanthanide, it would be expected to have properties similar to these elements and be present in their ores. Several research groups made this unfortunate assumption. Bohr, however, argued that quantum theory required that there be only fourteen lanthanides, and that element 72 be in the same group as Ti and Zr. This suggestion was followed, and the new element hafnium was subsequently found in zirconium ores. Discovery of the radioactive elements promethium, astatine and francium filled the remaining holes in the periodic table.

One whole family was omitted from Mendeleeff's periodic table. The noble gases were unknown until Ramsay and Rayleigh discovered argon in 1894. Helium was discovered a year later. The existence of two similar elements with no family suggested that a new group with more new members needed to be added to the periodic table. Krypton, neon, and xenon were discovered soon after as components of the air; and radon, the last member of the family, was discovered during the early explorations of radioactive materials.

Because the properties of elements in the first few periods change in a uniform fashion, an element may have properties similar to the averages of the properties of the elements on either side of it. That is particularly true for the main group elements among the first 36, and particularly for an element in the middle of a period. Thus, a compound containing equivalent proportions of B and N might be similar to C. In fact, the compound BN does form two structures, a diamond-like structure and a graphite-like structure—just like carbon. The total number of valence electrons is the same also, since boron has 3 and nitrogen has 5, averaging out to the 4 valence electrons of carbon.

Species with different atoms but the same number of electrons are said to be **isoelectronic**. Other examples of this isoelectronic principle are aluminum phosphide (AlP) as a substitute for silicon and gallium arsenide (GaAs) as a substitute for germanium. Both silicon and germanium are important semiconductor materials. Their isoelectronic substitutes are similar but with important subtle differences that lead to important applications in microelectronics.

The discovery of the predicted element 43 was erroneously reported in 1925. It was actually discovered in 1937 by Perrier and Segrè, in a sample of molybdenum that had been bombarded with deuterons by Lawrence. It was called technetium because it was the first artificially produced element.

17.4 STRUCTURE AND CONDUCTIVITY: METALS AND NONMETALS

Now let's look at some of the properties of the elements and the way these properties are reflected by their positions in the periodic table. First, the overwhelming majority of the elements are solids at room temperature. Only eleven are gases, and two are liquids. Two more elements are almost liquids—gallium and cesium have melting points a few degrees above room temperature, 29.8°C and 28.5°C respectively. The two elements that are liquids and the two that are nearly liquid are widely separated in the periodic table, and no useful generalization has emerged. Six are noble gases—helium through radon. These exist in the gas phase as independent single atoms. The other five elements that are gases under standard state conditions are diatomic molecules—H_2, N_2, O_2, and two members of Group VIIA, F_2 and Cl_2. The other common members of Group VIIA, Br_2 and I_2, also form diatomic molecules; however, their large electron clouds provide sufficient dispersion forces that at standard conditions Br_2 is a liquid and I_2 is a solid. Apart from these few elements, no other elements form stable diatomic molecules under standard conditions.

Although most of the elements are solids, the solid elements exhibit quite a large range in their other properties. For example, we can measure the ability of the solid to conduct electricity. Those that are good conductors are the **metals**; they start at the left side of the periodic table and include the d-block and the f-block, and extend into the p-block.

The conductivity of an element is usually a calculated quantity. The property that is actually measured is the resistance R (in ohms) of a sample of the element. This is related to the resistivity or specific resistance of the element, r (in ohm·cm), by

$$r = \frac{RA}{l}$$

where A is the uniform cross-sectional area of the sample (in cm^2) and l is its length (in cm). The conductivity or specific conductance of the element is defined as

$$K = \frac{1}{r} = \frac{l}{RA}$$

with units of $ohm^{-1}\ cm^{-1}$. The specific conductances of metals range from about $6 \times 10^3\ ohm^{-1}\ cm^{-1}$ for vanadium to $6.8 \times 10^5\ ohm^{-1}\ cm^{-1}$ for silver. An important feature of the conductivity of metals is that it decreases with increasing temperature.

A rough break in the conductivity of metals occurs in the p-block of the periodic table. Along the green zig-zag diagonal from boron to astatine (Figure 17–3), the electrical behavior changes sharply. On the blue side, Al, Ga, In, Sn, Sb, and Bi are all metallic conductors. The conductivity of Po is uncertain because of the element's low abundance. However, B, Si, Ge, As, and Te show very different behavior. These elements—the green area in Figure 17–3—are called **semiconductors** (or semimetals or **metalloids**). Their specific conductances are much lower, from about 10^{-5} to $10^1\ ohm^{-1}\ cm^{-1}$. Also, unlike the metals, their conductivities increase with increasing temperature. In recent years the use of semiconductors in electronic devices has become a major industry.

The remaining elements to the right of and above the metals and semiconductors are, for the most part, **insulators**—materials with no tendency to conduct electricity. These are the **nonmetallic elements**. However, classification of elements close to the semiconductor region of the periodic table is complicated in some cases by the existence of **allotropes**, different structures of the same element with quite different properties. Several illustrations of this phenomenon will be discussed later in this chapter.

Copper and magnesium (left) are metals; aluminum (top center) is considered to be a metal by some and a metalloid by others; silicon (bottom center) is a metalloid; bromine and carbon (right) are nonmetallic elements.

					He
B	C	N	O	F	Ne
Al	Si	P	S	Cl	Ar
Ga	Ge	As	Se	Br	Kr
In	Sn	Sb	Te	I	Xe
Tl	Pb	Bi	Po	At	Rn

FIGURE 17–3 The p-block elements of the periodic table, showing the zig-zag line that separates the metals from nonmetals.

(a)   (b)

Elements with metallic conductance also have a number of other properties in common. They are generally solids that are malleable (capable of being pounded into fine sheets), ductile (capable of being drawn into wires), lustrous, structurally strong, and attacked by strong acids. There are exceptions, but these properties of metals hold quite well. The crystal structures of metals will be discussed in Chapter 19. All have close packed structures for the atoms, with a large number of nearest neighbors. The number of nearest neighbors is called the **coordination number** and is most commonly eight or twelve.

The semiconductor elements have quite different structures. In general they have a covalent network structure held together by covalent bonds, and the bonded network extends over huge numbers of atoms. Their coordination numbers are much lower than for metals; four is a typical number.

The element carbon is not particularly easy to classify. It has two structures or allotropic modifications: diamond and graphite (Figure 17–4). The diamond structure is also adopted by the semiconductors Si and Ge. However, diamond is an insulator. The graphite structure is strongly anisotropic—different in different directions—being an insulator perpendicular to the carbon planes and a semiconductor in directions parallel to these planes. More will be said about these structures in the next chapter. The remainder of the nonmetals have structures composed of small molecules. Examples are P_4, S_8, and diatomic molecules of the gases (except for the noble gases, which are monatomic).

Examples of elemental allotropes that bridge the classifications between metals and nonmetals are black phosphorus and grey tin, which have covalent network structures and are semiconductors. The more common form of phosphorus is the white nonmetallic (P_4) form, whereas that of tin is a metallic structure.

Red and white phosphorus, two allotropic modifications of the element, differing only in structure. The white form is molecular (P_4).

17.5 OXIDATION STATES

An important periodic property is the **oxidation state**—a number which can be applied to any atom in an element or compound. The oxidation state is most easily defined in compounds made up of simple monatomic ions. In such cases the oxidation state of the element is the charge of the ion. For example, NaCl is composed of Na^+ and Cl^- ions. Therefore, in NaCl the oxidation state of Na is $+1$ and the oxidation state of Cl is -1.

Simple monatomic ions are significant in the chemistry of some elements, particularly those that are either strongly electropositive or electronegative. The elements in Group IA exist in compounds principally as 1^+

Crystalline tin.

ions and those of Group IIA as 2^+ ions. For Group VIIA, 1^- ions are very important, and the O^{2-} ion plays a big part in oxygen chemistry.

Even though most compounds do not consist simply of monatomic ions, it is still convenient to define oxidation states for the elements involved. But in order to do so, a number of rules are necessary:

The oxidation state of any atom in a pure element is zero.

The sum of the oxidation states of all atoms in a neutral compound is zero.

The sum of the oxidation states of all atoms in an ion is equal to the charge on the ion.

The oxidation state of all Group IA elements in compounds is $+1$, and that of the Group IIA elements in compounds is $+2$.

The oxidation state of H in compounds is $+1$ except in metal hydrides of very electropositive elements, where it is -1. Examples are NaH and CaH_2.

The oxidation state of F is always -1.

The oxidation state of O is always -2 except where two atoms are bonded together or O is bonded to F. Such compounds are quite rare. An example is hydrogen peroxide, H_2O_2.

EXAMPLE 17–1

Determine the oxidation states of the atoms in Na_3N.

Solution   Na in compounds is an oxidation state of $+1$. Let $x =$ oxidation state of N in Na_3N. Since the sum of the oxidation states for a neutral compound must be 0,

$$3(+1) + x = 0$$
$$x = -3 = \text{oxidation state of N}$$

EXERCISE 17–1

Determine the oxidation states of the atoms in MgB_2.
Answer: Mg, $+2$; B, -1.

EXAMPLE 17–2

Determine the oxidation states of the atoms in $Cr_2O_7^{2-}$.

Solution   Remember that the oxidation state of O is almost always -2. Let $y =$ oxidation state of Cr. Since the sum of the oxidation states of this ion will be equal to the ion charge of -2,

$$2y + 7(-2) = -2$$
$$y = +6 = \text{oxidation state of Cr}$$

EXERCISE 17–2

Determine the oxidation states of the atoms in Na_2SO_4.
Answer: Na, $+1$; S, $+6$; O, -2.

Because of the stability of the noble gas configuration, it should not be surprising that particularly stable oxidation states for the elements involve either removal of or addition of sufficient electrons to result in an s^2p^6 shell configuration. Furthermore, elements on the left side of the periodic table with relatively few valence electrons tend to lose them, to form positive ions with an s^2p^6 shell configuration. Elements on the right side of the periodic table with relatively many valence electrons tend to add electrons and form negative ions. Elements with approximately half-filled

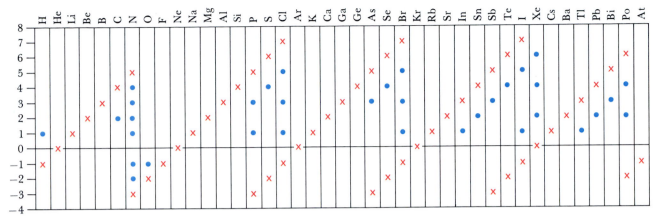

FIGURE 17–5 Oxidation states among the main group elements. Those equivalent to "closed shell" electron configurations are marked by x's.

valence shells find equal stability in either direction—gaining or losing electrons.

Figure 17–5 gives the observed oxidation states of the main group elements. Oxidation states equivalent to noble gas configurations are marked by x's. Note that these x's inscribe a series of parallel diagonal lines, one per period. The elements of the p-block have additional positive oxidation states corresponding to losing fewer than the maximum number of electrons. These oxidation states are indicated by dots in Figure 17–5. They too can be seen to form a pattern, generally being positive numbers that differ from the maximum oxidation state by 2, 4, or 6. Thus, the observed oxidation states for Sn are +4 (the maximum) and +2. For Cl they are +7, +5, +3, +1, and −1, where −1 represents the gain of an electron to fill the valence shell.

The oxidation states observed for the transition metals (d-block) and lanthanides (f-block) are rather more complex and are discussed in Chapter 22.

17.6   HYDROGEN

Hydrogen may well be the most important element in the periodic table for the following reasons. It forms compounds with almost every other element except the noble gases, and forms many compounds with some elements such as carbon, silicon, and boron. Its compounds span the range of chemical properties, including ionic salts such as sodium hydride (NaH), volatile covalent gases such as hydrogen sulfide (H_2S), very stable compounds such as water (H_2O), and compounds that burst into flame on being exposed to air, such as diborane (B_2H_6). Included are the strong acids—HCl, H_2SO_4, and HNO_3—as well as most of the strong bases like NaOH and $Ca(OH)_2$. The most important compounds in life and commerce are mostly hydrogen compounds.

The varied nature of the compounds and reactions of hydrogen can be explained in terms of the unique and versatile $1s^1$ electron configuration. Viewed as having a shell holding a single electron, hydrogen has a tendency to lose this electron to form the H^+ ion or proton. On the other hand, hydrogen is one electron short of a full shell, which it can complete to form the hydride (H^-) ion. Finally, because it has a half-filled valence shell, hydrogen can share electrons with other atoms to form covalent bonds. This is by far the most common circumstance for hydrogen and characterizes the bonding in most of its compounds.

TABLE 17–2	Energies of Bonds between Hydrogen and Some Nonmetals	
Element	**Bond Energy (kJ/mol)**	
B	290	
C	411	
N	391	
O	459	
S	364	
F	561	
Cl	428	
Br	362	
I	295	

Covalent compounds of hydrogen tend to have low melting and boiling points, indicative of the relatively weak intermolecular attractions holding them in the solid and liquid phases. However, NH_3, H_2O, and HF have abnormally high boiling points as a result of hydrogen bonding.

Average bond dissociation energies in some covalent hydrogen compounds are given in Table 17–2. Note that oxygen and fluorine form the strongest bonds with hydrogen, and such bonds tend to form, given the opportunity. Thus, CH_4 and NH_3 burn in air or oxygen,

$$CH_4 \text{ (g)} + 2\ O_2 \text{ (g)} \longrightarrow CO_2 \text{ (g)} + 2\ H_2O \text{ (g)}$$
$$4\ NH_3 \text{ (g)} + 3\ O_2 \text{ (g)} \longrightarrow 2\ N_2 \text{ (g)} + 6\ H_2O \text{ (g)}$$

to form water. The formation of water also occurs when hydrogen is used to reduce a metal oxide at elevated temperature:

$$CuO \text{ (s)} + H_2 \text{ (g)} \longrightarrow Cu \text{ (s)} + H_2O \text{ (g)}$$
$$Fe_2O_3 \text{ (s)} + 3\ H_2 \text{ (g)} \longrightarrow 2\ Fe \text{ (s)} + 3\ H_2O \text{ (g)}$$

Fluorine generally reacts with covalent compounds of hydrogen to form the very strongly bonded HF molecule,

$$CH_4 \text{ (g)} + 4\ F_2 \text{ (g)} \longrightarrow CF_4 \text{ (g)} + 4\ HF \text{ (g)}$$
$$2\ H_2O \text{ (liq)} + 2\ F_2 \text{ (g)} \longrightarrow O_2 \text{ (g)} + 4\ HF \text{ (g)}$$

Formation of H^+ is possible because hydrogen has a valence shell containing only one electron. Because it is an extremely small, positively charged species, the proton binds to any species possessing available electrons and exists independently only in a near vacuum. Therefore, reactions of protons actually involve the transfer of the proton from one base to another, as in

$$H_2SO_4 \text{ (aq)} + H_2O \text{ (liq)} \longrightarrow H_3O^+ \text{ (aq)} + HSO_4^- \text{ (aq)}$$

Some hydrogen-containing species are very strong donors of protons to the solvent water. Such compounds are strong acids in water:

sulfuric acid

$$H-\overset{..}{\underset{..}{O}}-\overset{\overset{\displaystyle :\overset{..}{O}:}{|}}{\underset{\underset{\displaystyle :\overset{..}{O}:}{|}}{S}}-\overset{..}{\underset{..}{O}}-H$$

nitric acid

$$H-\overset{..}{\underset{..}{O}}-\overset{\overset{\displaystyle}{\underset{\underset{\displaystyle :\overset{..}{O}:}{\|}}{N}}}{}-\overset{..}{\underset{..}{O}}$$

perchloric acid

$$H-\overset{..}{\underset{..}{O}}-\overset{\overset{\displaystyle :\overset{..}{O}:}{|}}{\underset{\underset{\displaystyle :\overset{..}{O}:}{|}}{Cl}}-\overset{..}{\underset{..}{O}}:$$

hydrochloric acid

$$H-\overset{..}{\underset{..}{Cl}}:$$

hydrobromic acid

$$H-\overset{..}{\underset{..}{Br}}:$$

In each case, hydrogen is bonded to oxygen or to a halogen. However, HF is not a strong acid in aqueous solution because of the very strong bond (Table 17–2) that must be broken to form the F^- and H^+ ions. Pure HCl,

HBr, and HI are all gases and nonconducting in the condensed state. In water they are all strong acids.

The other common strong acids all have hydrogen bonded to oxygen, and are called **oxy acids**. However, hydrogen bonded to oxygen is not in itself a sufficient condition to produce a strong acid. For example, consider these weakly acidic species:

water

$$\overset{\cdot\cdot\;\cdot\cdot}{O}$$
$$H \diagup \qquad \diagdown H$$

nitrous acid $H—\overset{\cdot\cdot}{\underset{\cdot\cdot}{O}}—N{=}\overset{\cdot\cdot}{O}{:}$

acetic acid CH_3COOH

In strongly acidic oxy acids, the O bearing an H is bonded to an electro-negative atom such as N, Cl, or S. This atom is itself bonded to at least two strong electronegative atoms, usually oxygen, and the oxygen atoms must not have their electron-withdrawing capabilities shared with another atom such as hydrogen. The result of such a structure is a concerted withdrawal of electrons away from the oxygen-bearing the hydrogen, facilitating the release of this hydrogen as the electron-free H^+.

In each of the three strongly acidic oxy acids, there are oxygen atoms on the central element that are not bonded to hydrogen, meeting the requirement for a strong acid in each case:

Sulfuric acid has two such oxygens.
Nitric acid has two such oxygens.
Perchloric acid actually carries three such oxygen atoms on the central element.

Thus, perchloric acid should be a stronger proton donor than either sulfuric or nitric acid, and this is borne out by experiments in suitable solvents (ones in which the leveling effect does not obscure the results).

Hydrogen gas is an important reagent for producing useful hydrogen compounds. Examples abound: ammonia (NH_3), methane (CH_4), and methanol (CH_3OH), to name a few. Ammonia is produced from hydrogen and nitrogen by the Haber process. Carbon monoxide is the carbon source for reactions leading to methanol and methane. These reactions have a good deal in common. Look at the methanol reaction, for example:

$$2\ H_2\ (g) + CO\ (g) \longrightarrow CH_3OH\ (liq)$$

The enthalpy change for this reaction is negative (as you can determine using the data in Table 7–4); the entropy change is also negative, since three moles of reactants lead to one mole of products. Le Chatelier's principle predicts that high pressure and relatively low temperatures are needed to optimize the yield of methanol at equilibrium. However, at low temperatures, the reaction is slow. In order to promote the reaction at a reasonable rate, a somewhat elevated temperature and a catalyst are needed. Typical conditions for this reaction are 100 atm and 250°C, with a metal catalyst such as cerium, chromium, cobalt, manganese, or molybdenum.

17.7   THE S-BLOCK ELEMENTS

The elements of the s-block all contain one or two electrons in the orbital of a new valence shell. Having a small number of rather loosely bound electrons determines a number of the important properties of these elements:

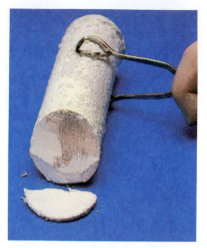

FIGURE 17–6 A piece of sodium showing a freshly cut surface. Note that oxide formation has obscured the metallic luster.

1. All are metals.
2. All are electrical conductors, and most are malleable and ductile.
3. The atoms are large because the valence electrons are effectively shielded by core electrons. Therefore, these metals tend to have low densities and are relatively soft and easily deformed.
4. The valence electrons are weakly held and easily removed. Note the low ionization energies listed in Table 15–4. Therefore, these elements tend to form cations easily in reactions with nonmetals, acids, and other oxidizing agents. The heavier members of each group are the most reactive and have an almost exclusively ionic chemistry.
5. Such covalent compounds as are formed by s-block elements occur with the earliest members, particularly lithium, beryllium, and magnesium.

Group IA: The Alkali Metals

The alkali metals are soft and silver-colored (except for cesium, which has a light, golden-yellow appearance). However, the oxide coating that forms as soon as a fresh surface on all these metals is exposed to air obscures the shiny surface (Figure 17–6). All are soft enough to be easily cut with a spatula or a pen knife, although lithium does offer some resistance. Their densities (Table 17–3) are low for metals. Because of their extreme reactivity, alkali metals are stored under mineral oil to prevent reaction with oxygen. Rubidium and cesium must be handled in an inert atmosphere at all times to prevent reaction with oxygen, nitrogen, or moisture.

None of these metallic elements occur free in nature because they are so reactive; instead, they are commonly found in the form of their extremely stable +1 ions. For example, sodium occurs widely as NaCl in salt deposits, underground brines, and seawater. The reduction of an alkali metal +1 ion to the neutral element cannot generally be accomplished by ordinary chemical means but can be readily performed by electrolysis of the neutral salt. For example,

$$2 \text{ NaCl (liq)} \longrightarrow 2 \text{ Na (liq)} + \text{Cl}_2 \text{ (g)}$$

The cell must be constructed in such a way that the Na and Cl_2 that are produced do not mix, or they will instantly react to re-form the NaCl. The melting point of the NaCl can be lowered somewhat by adding another salt such as CaCl_2 to form a **eutectic**.

A eutectic temperature is the lowest melting point of a solution of two or more substances. Eutectic mixtures behave as though they were pure substances with sharp melting points. More in our discussion of metal alloys of Chapter 20 (Materials Science).

Preparation of K, Rb, and Cs by the electrolysis of the molten salts is frustrated by the low melting and boiling points of the metals, which tend to vaporize away at the reaction temperature. However, sodium vapor can be used:

$$\text{KCl (liq)} + \text{Na (g)} \longrightarrow \text{NaCl (liq)} + \text{K (g)}$$

The unfavorable equilibrium constant of this reaction can be circumvented by using a counter-current fractionating tower which removes the desired product metal as it is formed.

The Group IA metals all have relatively low melting and boiling points (Table 17–3), which generally decrease as one proceeds down through the group. Because of its low melting point, liquid sodium is used as a coolant in nuclear reactors. Alloys of the metals show quite low melting points, and some are liquids at room temperature. A sodium-potassium alloy containing 22.8% Na is a eutectic that melts at −12.3°C and has been proposed as a substitute for pure sodium as a reactor cooling fluid.

TABLE 17–3	Melting and Boiling Points and Densities of the Group IA Metals		
Element	Melting Point (°C)	Boiling Point (°C)	Density (g/cm³)
Li	180.5	1326	0.53
Na	97.8	883	0.97
K	63.7	756	0.86
Rb	39.0	688	1.53
Cs	28.6	690	1.90

All of the Group IA metals dissolve in very pure ammonia to form blue solutions that conduct electricity. The most reasonable explanation for these observations is that the metal ionizes to form a cation and an electron. For example,

$$K \text{ (s)} \longrightarrow K^+ + e^-$$

The electron so produced becomes attached to NH_3 solvent molecules—that is, it becomes solvated just as a normal ion. This solvated electron produces the blue color of the solution and is principally responsible for the observed electrical conductivity.

The Group IA metals react with nonmetals to form ionic salts. Their reactions with hydrogen have been discussed in Chapter 6. Typical reactions are

$$2 \text{ K (s)} + Br_2 \text{ (liq)} \longrightarrow 2 \text{ KBr (s)}$$

$$2 \text{ Rb (s)} + I_2 \text{ (s)} \longrightarrow 2 \text{ RbI (s)}$$

$$2 \text{ Na (s)} + S \text{ (s)} \longrightarrow Na_2S \text{ (s)}$$

$$3 \text{ K (s)} + P \text{ (s)} \longrightarrow K_3P \text{ (s)}$$

$$2 \text{ Na (s)} + H_2 \text{ (g)} \longrightarrow 2 \text{ NaH (s)}$$

The Group IA metals react with oxygen to form normal oxides, M_2O, peroxides, M_2O_2, and superoxides, MO_2. Lithium forms principally the normal oxide. Sodium forms the oxide and the peroxide. Potassium, rubidium, and cesium form all three—oxide, peroxide, and superoxide.

In most of the reactions with nonmetals, the reactivity of the Group IA metals increases as one proceeds down the group to larger atoms and lower ionization energies. The exception is the reaction with the generally unreactive nitrogen—lithium is unusually and uniquely reactive:

$$6 \text{ Li (s)} + N_2 \text{ (g)} \longrightarrow 2 \text{ Li}_3N \text{ (s)}$$

The loss of electrons in a chemical reaction is called oxidation; the gain of electrons is called reduction. Because the electrons in these elements are so easily removed, the alkali metals are powerful agents for the reduction of other compounds. Reactions with acids, in which the hydrogen ions are reduced to hydrogen, proceed with explosive ferocity to give ionic salts; for example,

$$2 \text{ Na (s)} + 2 \text{ HCl (aq)} \longrightarrow 2 \text{ Na}^+ \text{ (aq)} + 2 \text{ Cl}^- \text{ (aq)} + H_2 \text{ (g)}$$

However, much weaker acids also react with Group IA metals. Water can be viewed as a weak acid that produces a very low concentration (1×10^{-7} M) of H^+ or H_3O^+ ions. These will be reduced to H_2 vigorously (Figure 17–7) leaving K^+ (aq) and OH^- (aq) in solution:

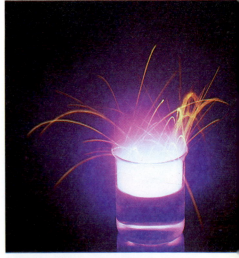

FIGURE 17–7 Potassium reacts vigorously with water. All the light necessary for this photograph was produced when a small piece of the metal was carefully added to the beaker of water.

$$2 \text{ K (s)} + 2 \text{ H}_2\text{O (liq)} \longrightarrow 2 \text{ K}^+ \text{ (aq)} + 2 \text{ OH}^- \text{ (aq)} + \text{H}_2 \text{ (g)}$$

The hydroxide compounds of the Group IA metals, particularly NaOH, are exceptionally important as strong bases. However, sodium hydroxide is not normally prepared by reaction of the metal with water, but rather by the less expensive electrolysis of NaCl solutions.

Similar reactions of alkali metals with compounds that can be viewed as very weak acids occur with NH_3 and with alcohols:

$$2 \text{ Na (s)} + 2 \text{ NH}_3 \text{ (g)} \longrightarrow 2 \text{ NaNH}_2 \text{ (s)} + \text{H}_2 \text{ (g)}$$
$$2 \text{ K (s)} + 2 \text{ CH}_3\text{OH (liq)} \longrightarrow 2 \text{ K}^+ + 2 \text{ CH}_3\text{O}^- + \text{H}_2 \text{ (g)}$$

Because their $+1$ ions form so easily, most of the chemistry of Group IA elements is ionic. However, covalent compounds do form, especially those of lithium. The greater tendency for lithium to form covalent bonds than the other Group IA metals is due to its higher ionization energy (which makes forming the cation more difficult) and its small size (which results in a stronger covalent bond). Lithium covalently bonded to carbon results from reactions such as

$$\text{CH}_3\text{CH}_2\text{Cl (liq)} + 2 \text{ Li (s)} \longrightarrow \text{CH}_3\text{CH}_2\text{Li (liq)} + \text{LiCl (s)}$$

The driving force for the reaction is principally the stability of the solid LiCl product.

Organic compounds of sodium and potassium can also be prepared. However, the compounds are extremely sensitive to air, moisture, or other reagents. They are not soluble to any great extent in organic solvents and are thought to be essentially ionic.

EXAMPLE 17–3

Complete and balance the following equations:
(a) K (s) + S (s) →
(b) Cs (s) + Cl_2 (g) →
(c) Li (s) + HCl (aq) →

Solution   To complete the equations, you need to know that:
(a) potassium forms a simple sulfide,

$$2 \text{ K (s)} + \text{S (s)} \longrightarrow \text{K}_2\text{S (s)}$$

(b) cesium burns in chlorine, forming a chloride,

$$2 \text{ Cs (s)} + \text{Cl}_2 \text{ (g)} \longrightarrow 2 \text{ CsCl (s)}$$

(c) lithium liberates hydrogen gas from the aqueous acid,

$$2 \text{ Li (s)} + 2 \text{ HCl (aq)} \longrightarrow 2 \text{ LiCl (aq)} + \text{H}_2 \text{ (g)}$$ ■

EXERCISE 17–3

Complete and balance the following:
(a) Na (s) + H_2O (liq) →
(b) K (s) + NH_3 (g) →
(c) CH_3Cl (liq) + Li (s) →   ■

Lithium is used in lightweight alloys with magnesium and aluminum. Because it reacts so readily with nitrogen and oxygen, lithium is used as a degasifier for producing high conductivity copper castings that must be exceptionally pure. Lithium salts—lithium carbonate, for example—are widely used for the treatment of depression, although the reason for their apparent effectiveness is unknown.

Sodium has been used for more than half a century to produce tetraethyl lead, a compound that raises the octane rating of gasoline, improving the combustion properties of the fuel. The sodium is melted with lead to produce an alloy that contains 10% Na by weight. This is formed into pellets, which are heated with ethyl chloride under pressure to favor this reaction:

$$4 \, NaPb \, (s) + 4 \, C_2H_5Cl \, (liq) \longrightarrow (C_2H_5)_4Pb \, (liq) + 4 \, NaCl \, (s) + 3 \, Pb \, (s)$$

Lead compounds in gasoline have been gradually phased out over the past decade because of environmental concerns about the volatile combustion products, and by 1990 they will no longer be commercially available in the United States.

The sodium $+1$ ion is an inexpensive, stable cation that forms a large number of water-soluble salts with useful anions. These salts are produced from NaCl, the standard sodium source. For example, NaOH is produced by the electrolysis of salt solutions. It is used for making soap and rayon, in the pulping of paper, and in the rubber, textile, and petroleum industries. Sodium carbonate (Na_2CO_3), which is used in manufacturing glass as well as in uses similar to those of NaOH, is prepared by the **Solvay process**. The first step in this process is a reaction that occurs at 15°C in a saturated aqueous solution of sodium chloride, ammonia, and carbon dioxide:

$$Na^+ \, (aq) + NH_3 \, (aq) + H_2O + CO_2 \, (aq) \longrightarrow$$
$$NaHCO_3 \, (s) + NH_4^+ \, (aq)$$

This reaction is promoted by the low solubility of sodium hydrogen carbonate under these conditions. Gentle heating of the solid product gives sodium carbonate:

$$2 \, NaHCO_3 \, (s) \longrightarrow Na_2CO_3 \, (s) + CO_2 \, (g) + H_2O \, (g)$$

The ammonia is regenerated for reuse through the reaction

$$2 \, NH_4Cl \, (s) + Ca(OH)_2 \, (s) \longrightarrow 2 \, NH_3 \, (g) + CaCl_2 \, (s) + 2 \, H_2O$$

Limestone ($CaCO_3$) serves as the source of the $Ca(OH)_2$.

The hydroxides of the Group IA elements are used for making soaps from animal or vegetable fats, for example,

$$\text{glyceryl tristearate} + 3 \, NaOH \longrightarrow 3 \text{ sodium stearate} + \text{glycerol}$$

Glyceryl tristearate is a typical animal fat, and sodium stearate is a soap. Soaps made from different alkali metals have different properties, particularly the temperatures at which they soften. Softening temperatures are highest for lithium. Because of their high softening points lithium soaps are often used as thickening agents for greases that must keep their viscosity at high temperatures. Sodium soaps have the most convenient softening points for general use and are the best known because sodium hydroxide is the least expensive alkali reagent. Potassium soaps are liquids at room temperature.

Group IIA: The Alkaline Earth Metals

The elements of Group IIA, generally called the alkaline earth metals (Table 17–4), are denser and have higher melting and boiling points than the alkali metals. They are also much harder than the alkali metals. Beryllium will scratch glass, and barium is slightly harder than lead.

TABLE 17–4   Melting and Boiling Points and Densities of the Group IIA Metals

Element	Melting Point (°C)	Boiling Point (°C)	Density (g/cm³)
Be	1278	2970	1.85
Mg	651	1107	1.74
Ca	850	1240	1.55
Sr	757	1150	2.6
Ba	850	1140	3.5
Ra	700	1140	5

These metals exhibit the same kinds of trends we saw in Group IA. Throughout most of the group, the chemistry is dominated by the tendency to lose the valence electrons in the outer s orbital to form +2 ions. As in Group IA, this tendency is weakest at the top of the group where the ionization energies are the highest. In this group the two lightest members, beryllium and magnesium, both tend to form covalent compounds. Magnesium is very similar to lithium in this respect, and a good example of the diagonal relationship in the periodic table. Beryllium has higher ionization energies than magnesium and, therefore, its compounds are predominately covalent. As an example, the melting point of beryllium chloride is only 440°C—very low for a metal halide—indicating a strong covalent character even with an element as electronegative as chlorine. The melting points of magnesium chloride and calcium chloride are 708°C and 772°C, respectively.

Beryllium is obtained principally from the mineral beryl, $Be_3Al_2(SiO_3)_6$, which when properly colored by impurities is known and valued as emerald (Figure 17–8). Covalent bonds between beryllium and carbon are readily formed. For example, organolithium compounds such as ethyl lithium react readily with beryllium chloride:

$$2\ CH_3CH_2Li\ (liq)\ +\ BeCl_2\ (s)\ \longrightarrow\ (CH_3CH_2)_2Be\ (liq)\ +\ 2\ LiCl\ (s)$$

This reaction is driven by the stronger tendency of beryllium than of lithium to be covalently bonded.

Magnesium occurs in important minerals such as dolomite, $CaCO_3 \cdot MgCO_3$. However, the most interesting source of magnesium is the almost limitless suppy of ocean water, where its ions occur at 1.3 grams per liter. Magnesium ions are commercially precipitated as the hydroxide,

FIGURE 17–8   (a) Beryl $[Be_3Al_2(SiO_3)_6]$ is the principal ore of the element beryllium. (b) This X-ray powder pattern for the mineral beryl displays the regularity of the arrangement of the atoms in the mineral's crystal structure. Note the flower-like symmetry. (c) An atom model of the characteristic crystal structure of the metal itself.

(a)

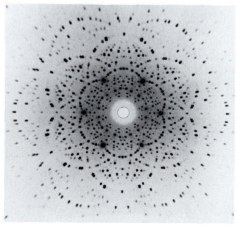

(b)

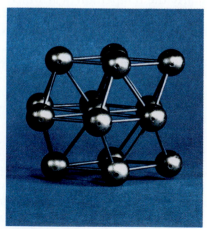

(c)

and then converted to the chloride with hydrochloric acid. The $MgCl_2$ is melted and then electrolyzed to produce the metal. Magnesium is a strong, low-density metal. It is used to produce important light-weight alloys with such metals as aluminum, zinc, and manganese, which are used in aircraft and a host of other applications. Of the covalent compounds of magnesium, the most important contain magnesium bonded to carbon, the so-called Grignard reagents.

The remaining elements in Group IIA can be produced by electrolysis of their molten salts or reaction of their chlorides with sodium, as for example:

$$CaCl_2 \text{ (s)} + 2 \text{ Na (liq)} \longrightarrow 2 \text{ NaCl (s)} + \text{Ca (liq)}$$

However, these elements are rarely produced in their metallic forms. Calcium is very abundant, occurring as $CaCO_3$ in seashells and in sedimentary rocks such as limestone and chalk (Figure 17–9). Metamorphic rocks such as marble are also largely calcium carbonate. Strontium and barium occur principally as their sulfates. Radium was first isolated (by Marie and Pierre Curie) from the uranium ore called pitchblende, in which it occurs at a concentration of 1 part in 3,000,000. All of the isotopes of radium are radioactive, the longest lived being ^{226}Ra with a half-life of about 1600 years. This isotope is a product of the decay series of ^{238}U.

In their elemental forms, calcium, strontium, barium, and radium are highly reactive. All three react with water,

$$Ba \text{ (s)} + 2 H_2O \text{ (liq)} \longrightarrow Ba^{2+} \text{ (aq)} + 2 \text{ OH}^- \text{ (aq)} + H_2 \text{ (g)}$$

with the Group VIIA elements,

$$Sr \text{ (s)} + Br_2 \text{ (liq)} \longrightarrow SrBr_2 \text{ (s)}$$

and with oxygen. Their compounds are generally salts. The four heavier members of Group IIA form salt-like hydrides,

$$Ba \text{ (s)} + H_2 \text{ (g)} \longrightarrow BaH_2 \text{ (s)}$$

The hydrides of Be and Mg are complex structures with covalent bonds.

The oxides of these metals are normally produced by heating the carbonates:

$$CaCO_3 \text{ (s)} \longrightarrow CaO \text{ (s)} + CO_2 \text{ (g)}$$

This reaction is endothermic but is favored by entropy because of the production of carbon dioxide. It is carried out on an industrial scale; the calcium oxide product is known as lime or quicklime. Quicklime reacts in a strongly exothermic manner with water to produce calcium hydroxide,

$$CaO \text{ (s)} + H_2O \text{ (liq)} \longrightarrow Ca(OH)_2 \text{ (s)}$$

The calcium hydroxide or slaked lime is reasonably soluble in water and is the cheapest source of hydroxide ions for preparing a strong base.

Examples of Grignard reagents include methylmagnesium bromide (CH_3MgBr) and phenylmagnesium bromide (C_6H_5MgBr).

FIGURE 17–9 Crystals of calcite, a form of calcium carbonate.

EXAMPLE 17–4

Complete and balance the following equations:
(a) $Ca \text{ (s)} + H_2 \text{ (g)} \rightarrow$
(b) $BaO \text{ (s)} + H_2O \text{ (liq)} \rightarrow$

Solution To complete the equations, you need to know that:
(a) calcium forms a hydride,

$$Ca \text{ (s)} + H_2 \text{ (g)} \longrightarrow CaH_2 \text{ (s)}$$

FIGURE 17–10 High voltage tube for the production of X-rays, and a close-up view of one of the beryllium windows.

(b) barium oxide is alkaline, forming the hydroxide in water,

$$BaO \text{ (s)} + H_2O \text{ (liq)} \longrightarrow Ba(OH)_2 \text{ (aq)}$$

EXERCISE 17–4

Complete and balance the following:

(a) CH_3Li (liq) + $BeCl_2$ (s) →

(b) Ca (s) + H_2O (liq) →

(c) Ba (s) + Cl_2 (g) →

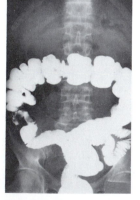

FIGURE 17–11 Gastro-intestinal (GI) series, using barium sulfate to provide an opaque background for the X-ray photograph.

Beryllium and barium are both used in x-ray technology, but for opposite reasons. The extent to which x-rays are absorbed by an atom is proportional to its number of electrons—the atomic number. Therefore, beryllium with only four electrons will not absorb x-rays very well. Beryllium is, in fact, the element of lowest atomic number out of which a physically strong and chemically stable wafer can be constructed. Such beryllium wafers are used as windows for vacuum tubes that produce x-rays, since the rays can pass through with minimum absorption (Figure 17–10). On the other hand, barium compounds absorb x-rays very well. Barium sulfate is used for x-ray diagnosis of the stomach and intestinal tract. The patient swallows a slurry of very sparingly soluble $BaSO_4$, and the x-ray picture shows a dark shadow where the rays are absorbed by barium. Although barium ions at any significant concentration would be extremely toxic, the very low solubility of barium sulfate prevents any problem (Figure 17–11).

Besides its use in alloys, magnesium is employed in the production of titanium metal,

$$TiCl_4 \text{ (liq)} + 2 \text{ Mg (s)} \longrightarrow Ti \text{ (s)} + 2 \text{ MgCl}_2 \text{ (s)}$$

Titanium ores are plentiful and the metal is strong, light, and corrosion-resistant. Unfortunately, the process is expensive because of the cost of magnesium, and full utilization of titanium awaits a less expensive alternative.

Calcium compounds have many uses. For example, slaked lime is mixed with sand and water to make plaster. Carbon dioxide in the air reacts with the calcium hydroxide to form hard, rigid, and insoluble calcium carbonate.

$$Ca(OH)_2 \ (s) + CO_2 \ (g) \longrightarrow CaCO_3 \ (s) + H_2O \ (liq)$$

The mineral gypsum is the dihydrate of calcium sulfate, $CaSO_4 \cdot 2H_2O$. When gypsum is warmed, the hemihydrate, $CaSO_4 \cdot \frac{1}{2}H_2O$, is formed as a powder. When water is added, heat is evolved and the dihydrate reforms. The hemihydrate is known as **plaster of paris**, and it is used for wallboard, orthopedic plaster casts, and statuary.

Relatively small ions with a +2 charge like Mg^{2+} and Ca^{2+} have a considerable electrostatic attraction for water. Thus, their salts are hygroscopic (they withdraw water from the air), and the chlorides of both of these metals dissolve into a pool of water when left out in the open on a relatively humid day. Calcium chloride and calcium sulfate are frequently used as desiccants to provide a dry atmosphere.

Strontium has few uses. When heated in an arc, it emits a bright red light, and its salts are used for highway lights and fireworks. The radioactive isotope ^{90}Sr is a product of nuclear fission weapons. Because calcium and strontium are chemically similar, ^{90}Sr falling on grazing land will be incorporated into milk along with calcium. In turn, the radioisotope is incorporated into the bones of the milk consumer, where it continues to emit biologically destructive (ionizing) beta particles.

The radioactivity of radium has been used to advantage in tumor therapy. Because tumor cells are reproducing rapidly, they are more susceptible to the destructive effects of radiation than are normal cells. Tiny vials of radium salts can be inserted in tumorous tissue to destroy the malignant cells. However, isotopes of other elements produced in nuclear reactors have replaced radium in most of these applications.

17.8 THE P-BLOCK ELEMENTS

In the p-block elements, s and p electrons are the valence shell electrons. The block divides diagonally into metals and nonmetals. Because ionization energies are higher here than in the s-block and because formation of a stable cation would involve the loss of at least three electrons, none of the p-block metals form compounds that are ionic to the extent seen in Groups IA and IIA. For the electronegative elements with 5, 6, and particularly 7 valence electrons, gaining electrons to form anions is an important part of their chemistry.

Group IIIA

This group includes the clearly nonmetallic element boron at the top of the column. It is hard, brittle, and nonconductive. Four metals follow—aluminum, gallium, indium, and thallium. Because of its excellent metallic properties of strength, ductility, and high conductivity, aluminum is in high demand for many purposes including foils, wire, nails, containers, and such high-performance applications as airplane parts and engine components (Figure 17–12). Gallium, indium, and thallium are for the most part curiosities, although gallium has been gaining applications in the microelectronics industry. Gallium has an enormous liquid range, from 30°C to 2250°C. Thallium is a particularly deadly toxic substance.

Among Group IIIA elements, the maximum oxidation state of +3 corresponds to loss of all three valence electrons. Compounds with this oxidation state range from the covalent BCl_3 to the largely ionic $TlCl_3$. The

FIGURE 17–12 Rolls of sheet aluminum, one of the many forms in which the metal is used.

FIGURE 17–13   (a) Jadeite, a rare stone that has been fashioned into fine jewelry and carved into sculptures for thousands of years, has recently been prepared synthetically by loading sodium and silicon oxides along with powdered alumina (Al_2O_3) into a small cylindrical furnace (b) and subjecting the mixture to extremely high temperatures and pressures. The technique is similar to that used to prepare diamonds and other gemstones synthetically.

more stable thallium chloride, however, is TlCl with Tl in the +1 oxidation state. Thus, the reaction

$$TlCl_3 \text{ (s)} \longrightarrow TlCl \text{ (s)} + Cl_2 \text{ (g)}$$

occurs at 40°C and above. There are other examples in which the heavier members of a group tend to adopt an oxidation state two units lower than the group maximum.

The principal ore of boron is borax, $Na_2B_4O_5(OH)_4 \cdot 8H_2O$, found naturally in arid regions such as the desert southwest in the United States, particularly Death Valley, California. The element can be prepared by the reaction of its oxide with magnesium

$$B_2O_3 \text{ (s)} + 3 \text{ Mg (s)} \longrightarrow 2 \text{ B (s)} + 3 \text{ MgO (s)}$$

Separation of the pure element from this mixture is difficult.

It has been found that filaments of elemental boron have exceptional tensile strengths, rivaling those of carbon filaments. Boron filaments can be prepared by passing an electric current through an extremely fine carbon filament core in a diborane (B_2H_6) atmosphere. The carbon heats up and B_2H_6 in contact with it decomposes to the elements, leaving the boron behind:

$$B_2H_6 \text{ (g)} \longrightarrow 2 \text{ B (s)} + 3 \text{ H}_2 \text{ (g)}$$

Aluminum is the most abundant metallic element in the Earth's crust, being an important component of the extremely common clays and feldspars. Economic separation of the metal from these sources has not yet been achieved, and bauxite, a form of Al_2O_3, is the only useful ore. Gallium and indium are found as trace impurities in aluminum ores. As with most of the active metals, these metals are prepared by electrolysis.

Aluminum is highly reactive and is rapidly dissolved by dilute solutions of acids such as hydrochloric. In fact, the reaction of aluminum with water is spontaneous at room temperature:

$$2 \text{ Al (s)} + 6 \text{ H}_2O \text{ (liq)} \longrightarrow 2 \text{ Al(OH)}_3 + 3 \text{ H}_2 \text{ (g)}$$

However, the instant formation of a very thin, tough, and adhering film of aluminum oxide "passivates" the surface, preventing reaction with water.

$$4 \text{ Al (s)} + 3 \text{ O}_2 \text{ (g)} \longrightarrow 2 \text{ Al}_2O_3 \text{ (s)}$$

The passive surface turns out to be a highly useful feature; aluminum containers can hold even the likes of concentrated nitric acid. Electrical treatment of the oxide suface makes it impassive in other ways, and it can be used to hold dyes. Such anodized aluminum surfaces are popular for decorative purposes.

Metallic aluminum is widely used in construction and fabrication of countless items. Aluminum would be useless in all of these applications were it not for the passive oxide surface. The oxide of aluminum shows up in a number of diverse applications (Figure 17–13). In a form known as corundum, it is highly refractory, chemically resistant, and second only to diamond in hardness. It is an excellent abrasive and is used to make containers and vessels for corrosive materials. It may be colored with trace amounts of transition metal impurities; it is known as sapphire if blue and ruby if red. Artificial gem-quality sapphires and rubies are prepared synthetically by melting corundum in the flame of an oxyhydrogen torch and adding trace amounts of coloring agents (Figure 17–14).

Except for gallium in the microelectronics industry, the other three metals in the group have found only limited applications. Gallium has also been used occasionally as thermometer fluid because of its extraordinarily large liquid range. Some indium alloys conduct heat very well and have found uses in bearings for jet engines.

Group IVA

Group IVA is headed by a clearly nonmetallic element, carbon. Chapter 25 discusses the wide range of strong covalent bonds that carbon forms with carbon, nitrogen, oxygen, sulfur, the halogens, and many other elements—the organic chemistry of carbon. The immense scope of the subject of the organic and inorganic chemistry of carbon is based on the ability of the carbon atom to form rings and long chains by bonding to other carbon atoms. This bonding phenomenon is called catenation, and although it can also be observed for silicon, it is not a dominant aspect of silicon chemistry. Silicon forms covalent bonds, but mainly to oxygen and halogens. Silicon and germanium are both very important semiconductor elements. Tin and lead are metals. Lead forms very stable compounds in the $+2$ oxidation state, two less than group maximum of $+4$.

Diamond is a transparent insulator, the hardest material known. Graphite is a soft, opaque conductor used for "pencil lead" and electrodes. (See Figure 17–4.) The structures and properties of the allotropes will be considered in some detail in Chapters 19 and 20. At standard temperature, graphite is the more stable form, and the reaction

$$C \text{ (diamond)} \longrightarrow C \text{ (graphite)}$$

has a standard free energy change of -2.89 kJ/mol. The change from diamond to graphite is extremely slow because of the large number of exceptionally strong bonds that must be broken. Thus, although a diamond is not absolutely forever, its change to pencil lead is not humanly perceptible. Synthesis of diamond from graphite has been accomplished at high temperature and pressure by using a suitable metal catalyst.

There are several important inorganic compounds of carbon. We have already discussed the carbonate ions in limestone. The reaction of calcium oxide with carbon at 3000°C in an electric furnace gives calcium carbide,

$$CaO \text{ (s)} + 3 \text{ C (s)} \longrightarrow CaC_2 \text{ (s)} + CO \text{ (g)}$$

The use of calcium carbide to prepare calcium cyanamide was discussed in Chapter 6. When treated with water, calcium carbide generates acetylene gas:

$$CaC_2 \text{ (s)} + 2 \text{ H}_2\text{O (liq)} \longrightarrow C_2H_2 \text{ (g)} + Ca(OH)_2 \text{ (s)}$$

Acetylene burns with a bright flame and was formerly used extensively for lighting. Acetylene generators, which dripped water onto calcium carbide, were used for home lighting, on automobiles and bicycles, and for miners' lanterns.

Silicon is surpassed only by oxygen in abundance in the Earth's crust. There is a large number of silicon minerals, the simplest having the formula SiO_2 as in quartz, the essential component of sand. The element is prepared by reaction with carbon in an electric furnace. For semiconductor use, extremely pure silicon is required. The crude silicon from the electric furnace is usually treated with chlorine to produce $SiCl_4$,

$$Si \text{ (s)} + 2 \text{ Cl}_2 \text{ (g)} \longrightarrow SiCl_4 \text{ (liq)}$$

The silicon tetrachloride can be purified by fractional distillation. It is then converted back to silicon by reaction with magnesium or zinc:

FIGURE 17–14 Artificial sapphires formed by melting aluminum oxide. Trace quantities of metal compounds are added to give the desired colors.

The garlic odor of crude acetylene from calcium carbide and water is due to the presence of phosphine (PH_3) derived from calcium phosphide (Ca_3P_2).

$$SiCl_4 \text{ (liq)} + 2 \text{ Mg (s)} \longrightarrow \text{Si (s)} + 2 \text{ MgCl}_2 \text{ (s)}$$

The final ultrapure silicon is prepared by **zone refining**. In this process a heat source is repeatedly moved slowly along a bar of the element. In this way a zone of the bar is melted, and this zone is repeatedly moved toward one end of the bar. This end eventually accumulates the remaining impurities in the silicon, since they are soluble in the melt and are dragged along with it. The impure end is discarded, leaving a bar of ultrapure silicon (Figure 17–15).

Germanium is found as a minor component of a number of silver, lead, tin, antimony, and zinc ores. It can be isolated as the oxide, which can then be reduced to the metal with carbon:

$$GeO_2 \text{ (s)} + \text{C (s)} \longrightarrow \text{Ge (s)} + CO_2 \text{ (g)}$$

Tin is found as the mineral cassiterite, SnO_2, which can be reduced to the metal in the same manner—by reduction with carbon. The principal lead ore is the simple sulfide, galena (PbS). This is roasted in the presence of air to give lead(II) oxide,

$$2 \text{ PbS (s)} + 3 O_2 \text{ (g)} \longrightarrow 2 \text{ PbO (s)} + 2 SO_2 \text{ (g)}$$

Note the oxidation state—again, two less than the maximum at the bottom of the group. PbO can be reduced with carbon to give the metal.

Silicon, germanium, tin, and lead are all capable of bonding directly to carbon. Important examples range from "silicone" polymers to the antiknock compounds in "leaded" gasolines. The tetrachlorides of these Group IVA elements all display covalent character and tetrahedral geometry about the central atom. Those of silicon, germanium, and tin can be prepared directly:

$$\text{Ge (s)} + Cl_2 \longrightarrow GeCl_4 \text{ (liq)}$$

On heating lead in a chlorine atmosphere, however, lead(II) chloride forms. The tetrachloride must be prepared from the oxide and acid:

$$PbO_2 \text{ (s)} + 4 \text{ HCl (aq)} \longrightarrow PbCl_4 \text{ (liq)} + 2 H_2O \text{ (liq)}$$

Carbon tetrachloride can be prepared from carbon disulfide and chlorine, or by the photochlorination of methane, as will be described in some detail in Chapter 25. All the tetrachlorides except carbon tetrachloride undergo hydrolysis in water:

$$GeCl_4 \text{ (liq)} + 2 H_2O \text{ (liq)} \longrightarrow GeO_2 \text{ (s)} + 4 \text{ HCl (g)}$$

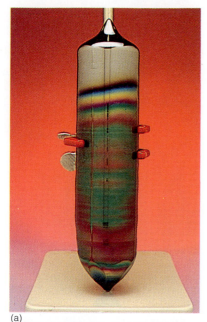

(a)

(b)

FIGURE 17–15 (a) A solid cylinder of ultrapure silicon, prepared for the semiconductor industry. (b) Using a special furnace, large, pure single crystals of many inorganic substances can be drawn directly from a melt of the substance. A mechanically rotated "seed" crystal is placed in the melt and then slowly withdrawn. Since the temperature falls rapidly above the crucible, the liquid "freezes" onto the seed as it is withdrawn.

EXAMPLE 17–5

(a) Like lead, germanium and tin also form dichlorides. All are characteristically ionic solids. Write the reaction for dissolving tin in acid, producing Sn(II) chloride.

(b) Lead(II) chloride is a sparingly soluble salt. Write the reaction for preparing it by precipitation from aqueous solutions of the ions.

Solution

(a) The balanced equation is

$$\text{Sn (s)} + 2 \text{ HCl (aq)} \longrightarrow SnCl_2 \text{ (aq)} + H_2 \text{ (g)}$$

(b) The balanced equation is

$$Pb^{2+} \text{ (aq)} + 2 \text{ Cl}^- \text{ (aq)} \longrightarrow PbCl_2 \text{ (s)}$$

EXERCISE 17–5
Complete and balance the following reactions:
(a) SnO_2 (s) + HCl (aq) →
(b) Ge (s) + HCl (aq) →
(c) Pb^{2+} (aq) + I^- (aq) →   ∎

The strength of the silicon-oxygen bond is used in synthetic polymers composed of chains of alternating silicon and oxygen atoms. To complete silicon's required valence of four, each is also bonded to two carbon atoms, usually as CH_3— groups. Here is what a small portion of a silicone polymer chain looks like:

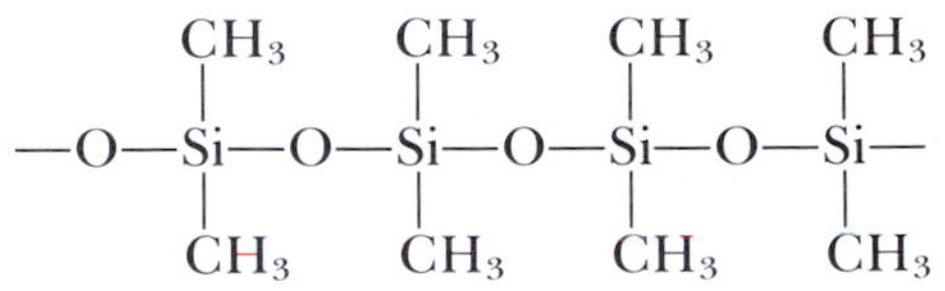

$$\begin{array}{ccccccc} & CH_3 & & CH_3 & & CH_3 & & CH_3 \\ & | & & | & & | & & | \\ -O- & Si & -O- & Si & -O- & Si & -O- & Si- \\ & | & & | & & | & & | \\ & CH_3 & & CH_3 & & CH_3 & & CH_3 \end{array}$$

The starting material for silicone polymers is silicon prepared from SiO_2. This reacts with methyl chloride, CH_3Cl, at 300°C over a copper catalyst to form dimethyldichlorosilane:

$$Si \text{ (s)} + 2\ CH_3Cl \text{ (g)} \longrightarrow (CH_3)_2SiCl_2 \text{ (g)}$$

Reaction of this product with water yields a silicone polymer,

$$n\ (CH_3)_2SiCl_2 \text{ (liq)} + n\ H_2O \text{ (liq)} \longrightarrow$$
$$-[(CH_3)_2SiO]_n- +\ 2n\ HCl \text{ (g)}$$

where n is a large number. These polymers can be oils or rubbery elastic materials, depending on chain length and crosslinking. More will be said about both those ideas in our later discussions of materials science (Chapter 20) and organic chemistry (Chapter 25). Silicone polymers show up in a variety of applications from seals to antifoaming agents in cooking oils and antacids.

Group VA

This group is headed up by the very important gaseous element nitrogen. As one of the $2p$ elements, nitrogen is capable of multiple bonding and forms the triply bonded N_2 molecule. Because of the extreme strength of this bond, the chemistry of nitrogen is dominated by a considerable reluctance to enter into chemical reactions. We have already discussed some of the important simple compounds of nitrogen, including ammonia and the oxides of nitrogen. Nitric acid is prepared by dissolving NO_2 in water:

$$3\ NO_2 \text{ (g)} + H_2O \text{ (liq)} \longrightarrow 2\ HNO_3 \text{ (aq)} + NO \text{ (g)}$$

Nitric acid is a strong acid. It is also an oxidizing acid, dissolving metals like copper and silver that do not dissolve in hydrochloric acid:

$$4\ HNO_3 \text{ (aq)} + 3\ Ag \text{ (s)} \longrightarrow 3\ AgNO_3 \text{ (aq)} + 2\ H_2O \text{ (liq)} + NO \text{ (g)}$$

Phosphorus is prepared by heating phosphate rock with sand (SiO_2) and coke (C) in an electric furnace.

$$Ca_3(PO_4)_2 \text{ (s)} + 5\ C \text{ (s)} + 3\ SiO_2 \text{ (s)} \longrightarrow$$
$$3\ CaSiO_3 \text{ (s)} + 5\ CO \text{ (g)} + 2\ P \text{ (g)}$$

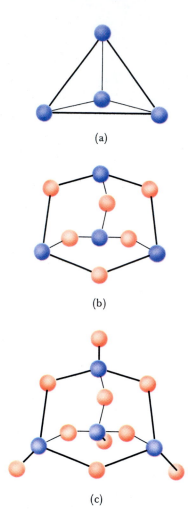

FIGURE 17–16 (a) The P_4 molecule typical of white phosphorus; (b) P_4O_6; (c) P_4O_{10}.

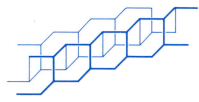

FIGURE 17–17 Black phosphorus is composed of very large interlocking layers. A segment cut from two layers is shown here.

Phosphorus does not form multiple bonds. Consequently, its elemental structures are entirely different from those of nitrogen. Furthermore, no one form is clearly more stable than any other, and, in fact, phosphorus forms three different allotropes: white, red, and black. The vaporized phosphorus from the preceding reaction condenses to give the white, waxy P_4 molecules of white (sometimes yellow) phosphorus (see p. 579). The structure of this form is shown in Figure 17–16. In this form, the individual molecules dissolve in solvents such as benzene and chloroform. The bond angles in the P_4 molecule are forced into a highly strained configuration with an internal bond angle of 60°, accounting in part for its exceptional reactivity. White phosphorus, for example, reacts rapidly with the oxygen of the air and is used in incendiary and tracer ammunition. Because of its extreme flammability, white phosphorus is stored under water. The oxidation products of phosphorus are P_4O_6 and P_4O_{10}. In P_4O_6 the addition of the oxygen atoms relieves the strain in the bond angles. This release of strain energy results in a strongly exothermic reaction. Addition of further oxygen to P_4O_6 results in addition of one oxygen atom to each phosphorus.

When white phosphorus is heated under very high pressure, the black allotrope forms. The structure is composed of infinite double layers, interconnected in a complex manner (Figure 17–17). The bond angle strain of white phosphorus does not exist in the black form, which is stable in air and ignites only with great difficulty. Because the black allotrope does not consist of discrete molecules, its solubility in solvents in solvents is extremely limited.

Red phosphorus forms on heating white phosphorus at 400°C for several hours. (See p. 579.) This allotrope of the element exists in several different modifications, each of which displays a characteristically complex and largely unknown structure. Red phosphorus is also stable in air and generally insoluble in solvents. The general properties of red phosphorus are consistent with a polymeric structure similar to that of black phosphorus.

Tetraphosphorus pentoxide (P_4O_{10}) has an extreme affinity for water. Accordingly, it is a very efficient drying agent, removing water from other compounds. For example, it will dehydrate H_2SO_4, leaving SO_3. Reaction of P_4O_{10} with sufficient water gives phosphoric acid:

$$P_4O_{10} \text{ (s)} + 6\ H_2O \text{ (liq)} \longrightarrow 4\ H_3PO_4 \text{ (aq)}$$

Arsenic, antimony, and bismuth are obtained by roasting the sulfide ores to give the oxides, As_4O_6, Sb_4O_6, and Bi_2O_3. These are then reduced to the element by treatment with carbon.

Applications of nitrogen and phosphorus as plant nutrients have been discussed previously. Arsenic is used in lead shot, where it serves to harden the alloy. Bismuth is used in some low-melting alloys like Wood's metal (50% Bi, 25% Pb, 12.5% Sn, and 12.5% Cd). Its melting point is a modest 65.5°C. Such alloys are used in sprinkler systems, fire alarms, and oil shutoff valves, where they melt and activate the system when the temperature rises to about that of a hot cup of coffee.

Group VIA

Included in this group are the nonmetals oxygen and sulfur, the semiconductors selenium and tellurium, and the metal polonium. Polonium occurs in uranium minerals and can be obtained by neutron bombardment of bismuth in a nuclear reactor:

$$^{209}\text{Bi} + n \longrightarrow {}^{210}\text{Bi} \longrightarrow {}^{210}\text{Po} + \beta$$

^{210}Po has a half-life of 138 days, but its study is hampered by the fact that the isotope is an intense alpha-emitter that endangers the experimenter and materials alike.

Because of its very high electronegativity, oxygen occurs in compounds in the -2 oxidation state. The only exceptions to this are compounds with O-to-O bonds such as hydrogen peroxide (H—O—O—H) and compounds with O-to-F bonds such as oxygen difluoride. Oxygen easily forms O^{2-} ions with the electropositive metals. Such compounds are high-melting ionic compounds. In contrast, oxygen forms covalent oxides with less electropositive elements. For example, CO and NO_2 are gases. Because of its high reactivity and great abundance, compounds of oxygen are everywhere, from the H_2O of the oceans to the $CaCO_3$ of limestone and the CO_2 present in the atmosphere.

Sulfur is also clearly nonmetallic. However, it commonly displays positive oxidation states, such as the $+6$ state in H_2SO_4. Other very important compounds of sulfur are sulfur trioxide, SO_3 (S in $+6$ oxidation state), sulfur dioxide, SO_2 (S in $+4$ oxidation state), and hydrogen sulfide, H_2S (S in -2 oxidation state). Selenium and tellurium have chemical properties similar to those of sulfur. For example, both form compounds analogous to SO_3, SO_2, and H_2S.

Elemental sulfur exists in a number of different forms. The common forms of solid sulfur contain eight-membered rings of sulfur atoms (Figure 17–18). This includes the most stable, orthorhombic form of sulfur, which exists in large yellow crystals in volcanic areas. Orthorhombic sulfur melts at about 113°C to give a mobile, transparent yellow liquid, that also contains S_8 rings. However, if heating is continued the liquid gradually turns brown and becomes much more viscous at about 160°C. The viscosity continues to increase until about 200°C. Within this temperature range, S_8 rings are breaking and linking to form long chains. The average chain length reaches a maximum at approximately 200°C. At higher temperatures the viscosity decreases and the color of the liquid becomes red as the average chain length decreases. If the viscous liquid is rapidly cooled in ice water, amorphous or plastic sulfur is obtained. This rubbery material contains the same long sulfur chains and can be stretched and molded (Figure 17–19).

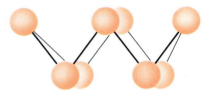

FIGURE 17–18   The crown-shaped S_8 ring of orthorhombic sulfur.

FIGURE 17–19   (a) Sulfur melts to form a mobile liquid which turns red. (b) On further heating, it becomes viscous and darkens. (c) From this viscous liquid, plastic sulfur forms on rapid cooling.

(a)

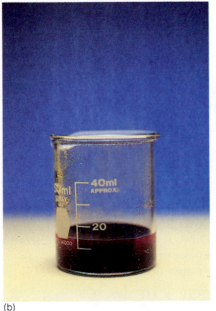

(b)

(c)

TABLE 17–5 Melting and Boiling Points of the Halogens

Element	Melting Point (°C)	Boiling Point (°C)
F_2	−233	−118
Cl_2	−103	−34.6
Br_2	−7.2	58.76
I_2	113.5	184.35

TABLE 17–6 Melting and Boiling Points of the Hydrogen Halides

Compound	Melting Point (°C)	Boiling Point (°C)
HF	−83.36	19.74
HCl	−114.25	−85.09
HBr	−86.92	−66.78
HI	−50.85	−35.41

Sulfur, selenium, and tellurium form the hydrogen compounds H_2S, H_2Se, and H_2Te. These are all gases with increasingly disagreeable odors. They are all weak acids when dissolved in water. Most metals react with S, Se, and Te to form metal salts.

Selenium exists in two allotropic forms, a nonconducting red form analogous to S_8 and a gray form that is a semiconductor. The red form is used for coloring glass. Because the electrical conductivity of the gray form is about 1000 times greater in light than in the dark, it has been used for light-sensing devices.

Group VIIA: The Halogens

The elements of Group VIIA, most commonly known by the name **halogens**, are one of two groups in the periodic table composed entirely of nonmetals. The other is the Group VIIIA elements—the noble gas elements—which are generally inert and have a very limited reaction chemistry. All of the halogens are clearly nonmetals, existing in the elemental form as diatomic molecules without conductivity or other metallic properties. Their melting points and boiling points (Table 17–5) become progressively lower as the size of their electron clouds and the attractions by dispersion forces increase. Thus, F_2 and Cl_2 are gases, Br_2 is a liquid, and I_2 is a solid at room temperature. However, even I_2 is volatile and is clearly visible as a violet vapor in equilibrium with the solid at room temperature; the vapor has a pungent odor (Figure 17–20).

All of the halogens have high electron affinities; therefore, they easily form salts in which the halide ion is in the −1 state. The halogens also form a wide range of typically covalent compounds. Because the valence shell requires only one electron to reach the noble gas configuration, this can be accomplished by forming one single bond to another element, which provides the needed electrons. Examples of covalent halogen compounds are:

FIGURE 17–20 The volatility of iodine is clearly demonstrated by gentle heating, causing the iridescent solid to sublime to a purple gas.

Among the most important covalent compounds of the halogens are the four hydrogen halides: HF, HCl, HBr, and HI. The physical properties of these four compounds are given in Table 17–6. All are gases at room temperature. In the liquid state, all undergo some autoionization, especially hydrogen fluoride:

$$HF + HF \rightleftharpoons H_2F^+ + F^- \qquad K = 10^{-10}$$

In water, all except HF ionize completely to form hydronium ion solutions. The acid strength in water increases going from HF to HCl. All are strong acids except HF, which has $K_a = 6.8 \times 10^{-4}$. Hydrogen fluoride has the unique ability to dissolve glass (largely SiO_2) by the reaction,

$$4\ HF\ (aq)\ +\ SiO_2\ (s)\ \longrightarrow\ 2\ H_2O\ (liq)\ +\ SiF_4\ (g)$$

There are a number of important compounds with Cl, Br, and I in positive oxidation states, particularly the $+1$, $+3$, $+5$, and $+7$ states. These are represented by the following four oxo acids of chlorine:

hypochlorous acid	HClO	H—O—Cl:
chlorous acid	HClO₂	H—O—Cl—O:
chloric acid	HClO₃	H—O—Cl—O: ¦O:
perchloric acid	HClO₄	:O: H—O—Cl—O: :O:

FIGURE 17–21 The bleaching power of chlorine gas has caused the coloring matter in the red rose to yellow dramatically.

Their acid strengths increase with the number of oxygen atoms in the formula. Chloric ($HClO_3$) and perchloric ($HClO_4$) acids are both strong acids in water.

Fluorine reacts vigorously with just about everything. For example, a stream of fluorine gas directed onto the surface of water will cause the water to "burn":

$$2\ H_2O\ (liq)\ +\ 2\ F_2\ (g)\ \longrightarrow\ 4\ HF\ (g)\ +\ O_2\ (g)$$

The great reactivity of fluorine makes its preparation very difficult. It was the last of the halogens to be prepared (not including astatine, the isotopes of which are known only as short-lived decay products of uranium and thorium). Fluorine was finally prepared by the electrolysis of a HF—KF mixture. Chlorine is prepared commercially in great quantities—millions of pounds per annum—by electrolysis of brine solutions. It is a powerful bleaching agent (Figure 17–21).

Each of the halogens can displace all of those below it in the periodic table from their chemical combinations. Thus bromine can be prepared by the reaction of chlorine with a bromide salt,

$$Cl_2\ (g)\ +\ 2\ Br^-\ (aq)\ \longrightarrow\ Br_2\ (liq)\ +\ 2\ Cl^-\ (aq)$$

Bromide salts occur in seawater (Figure 17–22) and in some salt brines, particularly near Saginaw, Michigan. That fact led to the creation of one of the major chemical companies, the Dow Chemical Company.

Some salts of the halide oxo acids are of great importance. Laundry bleach, for example, is a solution of sodium hypochlorite (NaClO), and solid calcium hypochlorite, $Ca(ClO)_2$, is added to swimming pools to kill bacteria and other microorganisms. Potassium chlorate is a powerful oxidizing agent (Figure 17–23).

The high reactivity of fluorine and the equally great stability of bonds to fluorine make for very stable fluorine compounds. The Freons are com-

FIGURE 17–22 Bromine can be obtained from seawater by first concentrating the solution by boiling and then bubbling in chlorine gas. The bromide ions are converted to bromine, which appears yellow in the saline solution.

FIGURE 17–23 Potassium chlorate reacts explosively with organic materials such as common table sugar.

pounds of carbon, chlorine, and fluorine—for example, dichlorodifluoromethane, CCl_2F_2. Stable and nontoxic, this and other Freons are useful as refrigerants and foaming agents. However, as we previously discussed, some freons are hazardous to the ozone layer of the atmosphere (Chapter 6). The very inert polymer Teflon contains long chains of carbon atoms bonded only to fluorine atoms (Chapter 25):

$$\begin{array}{ccccccc}
 & F & F & F & F & F & F & F \\
 & | & | & | & | & | & | & | \\
-C & -C- & C- & C- & C- & C- & C- & C- \\
 & | & | & | & | & | & | & | \\
 & F & F & F & F & F & F & F
\end{array}$$

Iodine is required in human nutrition to produce thyroxin, an iodine-containing growth-regulating hormone. This hormone is produced in the thyroid gland. An iodine insufficiency in the diet leads to enlargement of the thyroid in an attempt to gather iodide ions. This condition, called goiter, can be prevented by the addition of 0.01% NaI or KI to table salt.

17.9   THE OTHER ELEMENTS

The **transition elements** fall within the ten columns in the periodic table headed by scandium through zinc. These elements are all metals, and many—iron, copper, silver, gold, chromium, and mercury, for example—have tremendous commercial value. Characteristic properties of these metals include a large number of stable oxidation states, formation of many colored compounds, and formation of coordination complexes. The transition elements will be discussed in more detail as part of a general discussion of coordination chemistry in Chapter 22.

There are fourteen **lanthanide elements**, starting with cerium at $Z = 58$. Lanthanum itself is element number 57 and is often also considered a lanthanide. Known also as the **rare earths**, these elements have very similar chemical properties. Because the metals themselves are quite reactive, the lanthanides have classically been known through their compounds; the metals themselves, though, have become more readily available in recent years. The most important oxidation state of the lanthanides is the +3 state, and like the transition elements before them, they have a strong tendency to form complexes.

The **actinide elements** are the fourteen elements starting with thorium at $Z = 90$. Actinium is the preceding element, $Z = 89$. All the isotopes of these elements are radioactive. The first three actinides—thorium, protactinium, and uranium—occur naturally and have isotopes with long half-lives. The remaining actinides have been produced in nuclear reactors. In common with the transition metals, the first six actinides have a large number of oxidation states; but for the last six, the principal oxidation state is +3.

SUMMARY

Early versions of the periodic table of the elements listed them in order of increasing atomic weights, and a periodic repetition of properties was observed. Mendeleeff was confident enough in his ordering of the elements according to their periodic properties that he bypassed the few exceptions to the ordering by weight in favor of ordering by properties. Subsequently, it was discovered that indeed the fundamental order, fully in keeping with periodic properties, was according to atomic number.

The commonly used form of the periodic table is called the long form. Horizontal rows of the table are called periods; vertical columns are called groups, or families. The nature of the periodic table results from the order in which electrons are added to the atomic orbitals of atoms. The periods increase in length in the order 2, 8, 8, 18, 18, 32, and 32 as more atomic orbitals become available with the increase in the principal quantum number n. The periodic table can be divided into four blocks, s, p, d, and f, depending on the orbital being filled. The s- and p-blocks make up the representative or main group elements; the d-block spans the transition elements; and the f-block contains the lanthanides and actinides.

Except for mercury, which is a liquid at room temperature, all the elements on the left side of the periodic table are solids; most are good conductors of electricity. These elements are called metals and extend (from left to right) across the periodic table to a zigzag line through the p-block. Close to the zigzag line the elements are semiconductors; to the right of it they are nonmetals, which include several that are gases, one liquid, and several nonmetallic solids (at room temperature).

Because a few common elements have reliably fixed oxidation numbers in their chemical combinations, the oxidation state of an atom in a compound can be calculated; furthermore, the sum of the oxidation states in a molecule or ion is equal to its overall charge. The maximum oxidation state of an element is equal to its group number; the minimum is equal to the group number minus eight.

The chemical properties of the main groups were discussed. Groups with all metals—the alkali metals in Group IA and the alkaline earth metals in Group IIA—have quite similar properties. That is also true for groups with all nonmetals—the halogens in Group VIIA have quite similar properties. The other main groups (IIIA to VIA) have a wider range in properties as they proceed from nonmetals to metals down a group.

QUESTIONS AND PROBLEMS

QUESTIONS

1. How can you show that the periodic table is based on increasing atomic number rather than increasing atomic weight?

2. Mendeleeff was not the first to notice the periodic nature of the properties of elements. What use did Mendeleeff make of his periodic table that made his contribution so much more important than those of previous investigators?

3. Mercury and bromine are both liquids. How could you show that one is a metal and one is a nonmetal?

4. What is the general difference between the structures of metals and semiconductors?

5. Why is it impossible to assign some elements such as phosphorus and tin to one of the three groups of elements—metals, semiconductors, or nonmetals?

6. The atomic radius of Hf is actually less than that of Zr, even though they are in the same group and Hf is in a greater period. Briefly explain why.

7. Why do successive ionization energies increase for a given element?

8. Briefly describe and explain the observed trend in the boiling points of the Group VIIA halogen elements.

9. Give examples of two compounds in which H is not in the typical +1 oxidation state.

10. Write formulas for two compounds in which O is not in its commonly observed −2 oxidation state.

11. Chlorine need not always be in the common −1 oxidation state. Give examples of three types of compounds in which chlorine is in a positive oxidation state.

12. What are some of the advantages and disadvantages of using sodium versus sodium-potassium alloys for the coolants in nuclear reactors?

13. What kind of evidence can you give to show that beryllium forms largely covalent compounds?

14. Why is aluminum resistant to attack by water (and some strong acids) even though the reactions are thermodynamically favorable?

15. Which is the more stable form of carbon under ambient environmental conditions? Why isn't all carbon in this form?

16. List the names of nitrogen compounds with the nitrogen in as many different oxidations states as you can.

17. Describe the behavior of molten sulfur when it is heated, and offer a brief explanation.

PROBLEMS

Periodic Table   [1–18]

1. Without referring to a periodic table, use the atomic numbers of the following elements to identify the family to which each belongs:
(a) Ge, atomic number = 32
(b) Sr, atomic number = 38

2. Use the atomic numbers of the following elements to identify their family in the periodic table.
(a) Nb, atomic number = 41
(b) Tb, atomic number = 65

3. A metal wire 50. cm long and 0.10 mm² in cross-sectional area has a resistance of 0.217 ohm. What is the specific conductance of this metal?

4. What would be the resistance in ohms of a wire made of copper 100.0 cm long and 0.10 mm in diameter? What would be the diameter of an aluminum wire with the same resistance? Which wire would have greater mass? Use the following specific conductances:

$$K(Cu) = 65 \times 10^4 \ ohm^{-1}cm^{-1}$$

$$K(Al) = 40 \times 10^4 \ ohm^{-1}cm^{-1}$$

5. Arrange the following elements in increasing order of atomic radius, consulting only the periodic table: Al, Ca, Cs, Rb, Sr

6. Arrange the following elements in order of increasing atomic radius without consulting a table of atomic radii: As, Ba, In, Ra, Tl.

7. Arrange the following groups of elements in order of increasing first ionization energy, consulting only the periodic table.
(a) Be, K, Mg
(b) Na, Mg, Al

8. Use only a periodic table to arrange the following sets of elements in order of increasing first ionization energies.
(a) N, O, F
(b) Cs, Na, K

9. Arrange the following elements in order of increasing second ionization energy, consulting only the periodic table: Na, Mg, Al.

10. Use only a periodic table to arrange Ca, K, and Sr in increasing order of their second ionization potentials.

11. Arrange the following elements in order of increasing electron affinities: Cl, I, K, Li. Consult only the periodic table.

12. Using only a periodic table, arrange Al, Cl, S, Ga in order of increasing electron affinity.

13. Predict which of the following compounds will be ionic salts: BaO, CO, ClF, NO, KF, SrO, HI.

14. Predict which of the following binary compounds will be covalent: NaBr, CS_2, $FeCl_2$, PCl_3.

15. Determine the oxidation states of all elements in each of the following compounds and ions.
(a) NH_4^+
(b) $KMnO_4$
(c) BrO_3^-
(d) $CaCO_3$
(e) $NaClF_4$

16. Give the oxidation state of each element in the following compounds.
(a) CaH_2
(b) Na_2SO_3
(c) HNO_3
(d) H_2O_2
(e) IF_5

17. Arrange the following ions in order of increasing ionic radius: Al^{3+}, Ca^{2+}, Cl^-, F^-, Mg^{2+}, O^{2-}, S^{2-}. What were your reasons for selecting the order you did?

18. List the following ions in order of increasing ionic radius: Ca^{2+}, Cl^-, K^+, Se^{2-}.

Chemical Reactions   [19–24]

19. Write the balanced net ionic equations for the following reactions:
(a) potassium metal reacts with water
(b) cesium metal reacts with liquid bromine
(c) sodium metal reacts with hydrogen gas
(d) rubidium hydride reacts with water

20. Write a balanced net ionic equation for each of the following reactions:
(a) strontium metal reacts with water
(b) HF autoionizes
(c) HCl is added to water
(d) thallium(III) chloride is decomposed to thallium(I) chloride.

21. Write the balanced equation for each of the following compounds burning in oxygen gas.
(a) propane, C_3H_8 (g)
(b) hydrazine, N_2H_4 (g)

22. Write the balanced equation for the reaction that occurs when each of the following reacts with oxygen.
(a) diborane, B_2H_6 (g)
(b) silane, SiH_4 (g)

23. Predict whether the following acids are strong or weak.
(a) H_2SeO_4
(b) H_3PO_4
(c) H_3BO_3

24. Predict the relative acid strength of each of the following compounds.
(a) HNO_2
(b) $HClO_4$
(c) H_2CO_3

Additional Problems   [25–33]

25. Write a balanced equation for the reaction of each of the alkali metals with oxygen.

26. Write balanced equations for the preparation of magnesium metal from seawater.

27. Write balanced equations for the preparation of slaked lime from limestone.

28. Write equations for the preparation of the illumination gas known as acetylene, starting with only limestone, water, and coal (as carbon source).

29. Write equations for the preparation of ultrapure silicon, starting with sand.

30. Write the equations for the preparation of a silicone polymer.

31. Determine whether each of the following hydrides is a gas, liquid, or solid at room temperature and whether each is best described as ionic or covalent. Offer a few words of explanation in support of each answer.

(a) NaH
(b) HF
(c) HCl
(d) SiH_4
(e) SnH_4
(f) NH_3
(g) B_2H_6
(h) $NaBH_4$
(i) H_2S

32. Using balanced chemical equations, illustrate the difference between the chemistry of the ionic hydrides and that of the covalent hydrides with the reactions of two different examples of each with water.

33. Give balanced chemical equations that suggest a reasonable scheme for preparing $PbCl_4$ from the metal.

MULTIPLE PRINCIPLES   [34–39]

34. In the gas phase (as in water solution), HBr is a stronger acid than HCl. What does this say about the relative strength of the bond between hydrogen and chlorine compared to the bond between hydrogen and bromine?

35. $Ca(OH)_2$, prepared from $CaCO_3$ from limestone, chalk, or oyster shells, has a solubility of 1.60 g per 100 g of water at 25°C. What is the maximum pH that can be achieved using $Ca(OH)_2$ dissolved in water?

36. What is the change in pH if 10.0 g of $CaCl_2$ is added to 1.00 L of a saturated solution of $Ca(OH)_2$ in water at 25°C? Use the data in Problem 35.

37. ^{90}Sr is a radioactive isotope with a half-life of 25 years that emits beta particles. What isotope will be the product of this decay?

38. Predict as many as you can of the chemical and physical properties of the next element that will eventually be discovered in Group IIIA (that is, element 113). Briefly describe a few of the characteristic properties of the dichloride and tetrachloride of element 114.

39. Silicon is very high on the curve of nuclear binding energy. Calculate the mass defect and the binding energy per nucleon for Si and compare the values to those of oxygen and iron by consulting Figure 3–12. Assume the isotopes in question are ^{28}Si, ^{16}O, and ^{56}Fe.

APPLIED PRINCIPLES   [40–44]

40. You have a 100,000-L tank of water with a pH of 1.5. How many kg of $CaCO_3$ will be necessary to produce sufficient $Ca(OH)_2$ to neutralize this water to a pH of 7.0?

41. Hydrogen gas can be stored under pressure in a tank or as an easily decomposed compound such as CaH_2. The density of CaH_2 is 1.7 g/cm^3. At what pressure would H_2 at 25°C have to be stored in order to occupy the same volume as the same mass of hydrogen stored in the form of CaH_2?

42. How much Al_2O_3 would be necessary to prepare 10,000. m of an aluminum wire with a resistance of 0.0200 ohm? The specific conductance of aluminum is 3.54×10^5 ohm^{-1}cm^{-1} and its density is 2.702 g/cm^3.

43. What volume of SiF_4 gas at 1.00 atm and 298 K will be produced by the reaction of 4.00 L of 0.750 M HF with excess SiO_2 if the yield is 85.0%?

44. How much $Ca_3(PO_4)_2$ is required to produce 100.0 kg of phosphorus if the yield is 92.5% and the phosphate mineral is 78.1% calcium phosphate?

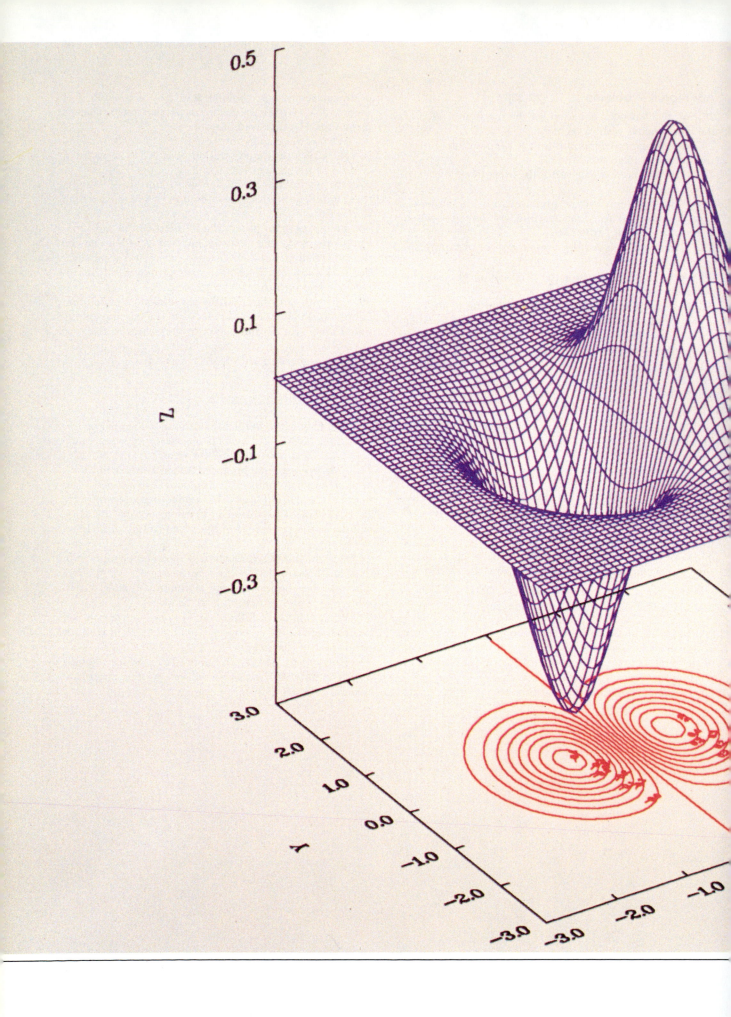

CHAPTER 18

Bonding II:
Molecular Structure

One of the most powerful theories in chemistry is the orbital theory, which views molecular structure *not* as atoms connected by bonds, but as electronic clouds interspersed with nuclear stars. Shown here is the σ^* molecular orbital for the hydrogen molecule, in the plane of the internuclear axis (in purple), with contour lines of constant electron density projected below (in red). More than a century ago, Friedrich Kekulé, then a young architecture student turned chemist, revolutionized chemical theory by popularizing the notion that the atoms in chemical formulas could be attached by little dashes. He suggested that carbon had four "tetrahedral" dashes and later, added the unique structure of benzene to our knowledge.

18.1   CHEMICAL BONDS AND BONDING THEORIES

A chemical bond exists between a pair of atoms when the forces between them are strong enough to be documented. We shall describe and measure several important kinds of bonds: ionic bonds, covalent bonds, and the weak bonds known as van der Waals forces, to name a few. The feature they share is their inherently electronic nature. As we have seen, the electrons surrounding nuclei determine the essential character of atoms, and that is equally true of the forces that chemically bond atoms together in molecules and ions. As a result, any description of the nature of the chemical bond must be based on a knowledge of charge density and electron distribution.

In this chapter, we wish to develop some basic ideas about chemical bonds using two different yet related theories—valence bond theory and molecular orbital theory. **Valence bond theory** is perhaps the more useful in beginning chemistry, having its origins in Lewis structures and simple orbital representations. **Molecular orbital theory**, or the quantum theory of the chemical bond, with its origins in quantum mechanics, is more difficult to apply although potentially more powerful. Through approximations, we can also make the molecular orbital theory useful at the introductory level. For all the accomplishments and successes of the valence bond and molecular orbital bonding theories, though, physics and chemistry are both still far away from anything like a complete solution to the fundamental problems in bonding. We shall address the reasons why that is the case, too.

We have already introduced the simplest form of valence bond theory in several places in the text, as early as Chapter 4. As atoms and their valence shell electrons come together, bonds form by pairing the electrons. This makes for an effective, yet simple model, one that works very well with little more than the chemical symbols for the bonded atoms, lines representing the bonding (or shared) electron pairs, and pairs of dots representing the nonbonding electron pairs. Based as it is on the electron pair, valence bond theory has come to be known as the "localized" orbital theory of bonding. Advanced valence bond concepts to be discussed in this chapter include the following:

1. Hybridized orbitals and **valence shell electron pair repulsion (VSEPR)** for predicting geometries of molecules, especially in methane (CH_4), ammonia (NH_4), water (H_2O), and hydrogen fluoride (HF), which are all hydrogen compounds of second period elements.
2. Resonance theory for describing bond lengths and bond energies in inorganic and organic carbon compounds such as sodium carbonate and benzene.

As successful as applications of valence bond theory can be, the theory paints an incomplete picture for some of the molecules it tries to describe. For example, predictions of magnetic properties and bond distances for the oxygen molecule do not agree with data from experiment. The simple Lewis structure by itself is unsatisfactory for these purposes. In some cases, the computations needed for successful applications of valence bond theory and its localized electron-pair bonds have been all but impossible until very recently. Furthermore, whereas valence bond theory is quite good at dealing with ground or lowest energy states, it has less often been the theory of choice in treating excited states.

Molecular orbital (MO) theory assumes that molecular orbitals arise from combinations of atomic orbitals from the bonded atoms. All the electrons of the molecule can be placed in molecular orbitals, starting with those of lowest energy and working up in energy. This procedure is much the same as the one we used when we added electrons to atomic orbitals of atoms to build up the periodic table. Molecular orbital theory has come to be known as the "delocalized" orbital theory of bonding because its fundamental assumption is that the pair of electrons involved in bonding lies in an orbital that is spread over more than one atomic nucleus. In many respects, molecular orbital theory has proved to be the more powerful theory, but at the expense of simplicity of application. As we shall see in the course of the chapter, molecular orbital theory is able to explain successfully many facts that valence bond theory could not.

It is also worthwhile remembering that measured properties of the molecules of which we speak are independent of what either theory (or any theory, for that matter) expects of their behavior. The molecules simply act according to their nature. Finally, the goal of this chapter is to describe the properties of molecules in the best way possible, regardless of what we call the theory or model we happen to be using.

18.2   SIMPLE MOLECULES OF THE FIRST, SECOND, AND THIRD PERIOD ELEMENTS

The Hydrogen Molecule and Beyond

The minimum requirement for the formation of a covalent bond is two electrons with opposite spins, and a stable "bonding orbital" in which to keep them. Generally, each bonding orbital is in an outer shell so that overlapping, or "interpenetration," of the two bonded atoms is possible. For hydrogen atoms in the first period, the only available bonding orbital is the $1s$ atomic orbital, so hydrogen atoms typically form only one covalent bond. Each hydrogen atom in an H_2 molecule has a $1s$ orbital and a lone electron. The two orbitals have overlapped, or interpenetrated, and the pair of electrons (with their spins paired) are equally shared by both atoms. This is the valence bond model for the **single bond**, so called because one pair of electrons and one pair of orbitals are used in bonding the two hydrogen atoms together (Figure 18–1).

Fluorine, oxygen, nitrogen, and carbon in the second period typically form up to four covalent bonds by making use of the single $2s$ orbital and the three $2p$ orbitals in the L shell. These orbitals are occupied by shared and unshared pairs of electrons in any combination totalling four. Thus, the carbon atom in methane forms four bonds, using all four orbitals on carbon; the nitrogen atom in ammonia has one of the four orbitals occupied by a nonbonding electron pair; the water molecule has two nonbonding electron pairs; and hydrogen fluoride has one shared electron pair and three nonbonding electron pairs.

Let's take a closer look at water and ammonia, polyatomic molecules that have been used traditionally to illustrate bonding to oxygen and nitrogen, two key second period elements. One of the simplest polyatomic molecules is the triatomic water molecule. To begin with, think of it as two hydrogen atoms and one oxygen atom, all separated beyond the point of influencing each other, and all in their lowest ground state energy levels. No bonds, yet! Both hydrogen atom orbitals are $1s^1$. The ground state electronic configuration for the oxygen atom is $1s^2 2s^2 2p^4$, but we need be concerned only with the four $2p$ electrons in the valence shell:

$$\underline{\uparrow\downarrow} \qquad \underline{\uparrow} \qquad \underline{\uparrow} \qquad \qquad 2p_{x,y,z}$$

This restriction is the basis for the Lewis octet rule, which states that second period elements are stabilized by achieving an eight-electron valence shell configuration.

Lewis structures for the second period elements, showing shared electron-pair bonds and nonbonding electron pairs:

$$
\begin{array}{c}
\text{H} \\
| \\
\text{H}-\textbf{C}-\text{H} \\
| \\
\text{H}
\end{array}
$$

$$
\begin{array}{c}
\text{H} \\
| \\
\text{H}-\textbf{N}: \\
| \\
\text{H}
\end{array}
$$

$$
\begin{array}{c}
\text{H}-\overset{\displaystyle ..}{\underset{\displaystyle |}{\textbf{O}}}: \\
\text{H}
\end{array}
$$

$$\text{H}-\overset{..}{\underset{..}{\textbf{F}}}:$$

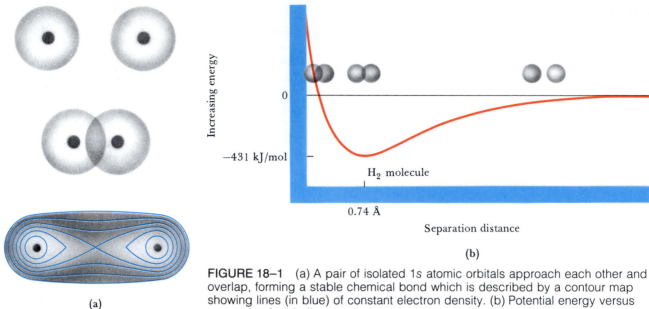

(a)

(b)

FIGURE 18–1 (a) A pair of isolated 1s atomic orbitals approach each other and overlap, forming a stable chemical bond which is described by a contour map showing lines (in blue) of constant electron density. (b) Potential energy versus separation for binding two hydrogen atoms together, based on a quantum mechanical model. The potential energy is at a minimum that corresponds to an average bond distance of 0.74 Å and an average bond enthalpy of 435 kJ/mol. There is no separation distance for H^+ and $H:^-$ for which the potential energy is less than zero, however, and so an ionic model such as we used in an earlier discussion in Chapter 16 to explain bonding in substances like NaCl cannot be used.

Note that combination along the *x* and *z* axes, or along the *y* and *z* axes, would give the same result.

The spin-paired electrons result in one atomic orbital having slightly higher energy than the other two that contain single electrons. These two lower-energy atomic orbitals overlap with the two hydrogen atomic orbitals to form what are known as molecular orbitals (Figure 18–2). As the atomic orbitals come together and overlap along the *x* and *y* axes, the symmetry of the dumbell-shaped 2*p* orbitals is lost as the charge density shifts toward the 1*s* orbitals of the hydrogen atoms. Finally, we are left with two single bonds. Why isn't the triatomic water molecule a simple linear arrangement of H—O—H? One suggestion is that the two molecular orbitals are initially perpendicular to each other, since the oxygen 2*p* atomic orbitals are at right angles to each other. But electrostatic repulsion between the clouds of charge that have been added by the interpenetrating hydrogen atom orbitals forces the molecular orbitals apart, from 90° to the observed bond angle of 104.5°.

The ammonia molecule has three hydrogen atoms bonded in a similar fashion to the central nitrogen atom. With three bonding 2*p* orbitals and electrons (one less than in oxygen), one nitrogen atom can bond with three hydrogen atoms:

 $2p_{x,y,z}$

Orbital overlap of the 1*s* and 2*p* atomic orbitals along the *x*, *y*, and *z* axes produces a molecule whose geometry is best described as a trigonal pyramid, but with a central bond angle expanded from the 90° of three mutually perpendicular 2*p* atomic orbitals to the experimentally observed value of 107° for the molecular orbital combination (Figure 18–3).

EXAMPLE 18–1

Briefly explain the driving force for opening the interior bond angle in the ammonia molecule from 90° to 107°.

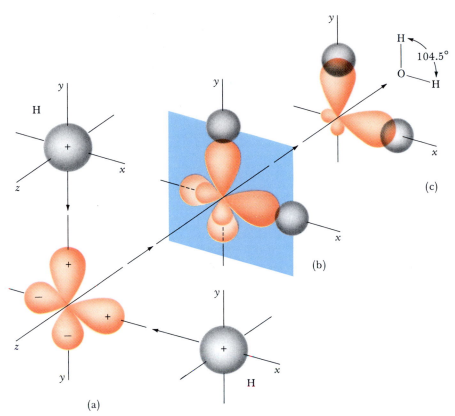

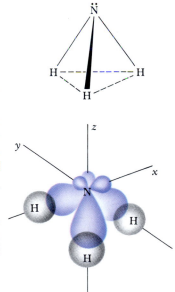

FIGURE 18–2 Formation of a water molecule by linear combination of atomic orbitals to give molecular orbitals, followed by opening of the internal H—O—H bond angle from 90° to the observed value of 104.5°. The s orbitals (a) approach along the x and y axes, overlap (b and c), and (c) the 90° formed by the overlapping orbitals opens up to the known bond angle in the water molecule. The molecular shape is bent.

Solution As with our picture of the water molecule, note the shift in the electron distribution from a symmetric structure as the bonding atomic orbitals become molecular orbitals. The charge density has moved in the direction of the $1s$ orbitals. As the $1s$ orbital penetrates the $2p$ orbital, the nearer lobe increases at the expense of the farther one. The repulsions between these electron clouds force the molecular orbitals apart. ■

EXERCISE 18–1(A)
Predict the likely bond angles for the other Group VA hydrides, PH_3, AsH_3, and SbH_3. Do the same for the Group VIA hydrides, H_2S and H_2Se. Check your predictions against the experimental values in Table 16–4. ■

EXERCISE 18–1(B)
Based on our discussion of H_2O and NH_3, is the hydride of the next element across the second period (carbon) predictably CH_4? Why does this suggest that our model is incomplete, or at least limited? Briefly explain.
Answer: The model predicts CH_2. Of course, methane is in fact CH_4. ■

Among the third period elements, covalent bonds and nonbonding electron pairs sometimes number five, six, or more. But the $3d$ orbitals that are needed to expand the valence shell beyond four orbitals are much less stable than the $3s$ and $3p$ orbitals and so even here, four bonds are most common. The notable exceptions are the transition elements—the big blocks

FIGURE 18–3 The formation of the ammonia molecule can be imagined in the same way as that of the water molecule (in Figure 18–2). Three s orbitals approach the nitrogen p orbitals, overlapping to form the molecular orbitals, after which the internal H—N—H bond angle opens to the measured value of about 107°. The molecular shape is pyramidal.

Compounds of third period elements using only 3s and 3p orbitals: HCl, H₂S, PH₃, and SiH₄. Compounds of third period elements using 3d orbitals as well: SF₄, SF₆, and PCl₅.

Double bonds in ethylene and formaldehyde:

$H_2C{=}CH_2$ and $H_2C{=}\ddot{O}:$

Triple bonds in acetylene and nitrogen:

$HC{\equiv}CH$ and $:N{\equiv}N:$

Forming the coordinate-covalent bond:

$:NH_3 + H^+ \longrightarrow NH_4^+$

of elements in the middle of the periodic table beginning in the fourth period with the $3d$ orbitals—where the use of the d orbitals in bond formation is the rule rather than the exception. In fact, an entire chapter on coordination chemistry (Chapter 22) is devoted to these compounds and their chemical bonds.

Double and triple bonds are also possible. Also, if one of the atoms contributes the pair of electrons to the bond while the other lends an empty orbital, the result is a coordinate-covalent bond.

18.3 HYBRID ATOMIC ORBITALS AND VSEPR THEORY

Consider the methane molecule (Figure 18–4). There are four C-to-H bonds, laid out in a perfectly symmetric tetrahedral arrangement, with an average bond angle of 109.5°. The ground state electronic structure for the carbon atom is $1s^2 2s^2 2p^2$, and we can represent the orbitals as follows:

$$\underset{1s}{\uparrow\downarrow} \quad \underset{2s}{\uparrow\downarrow} \quad \underset{}{\uparrow} \quad \underset{2p_{x,y,z}}{\uparrow} \quad \underline{}$$

On that basis, you might wrongly conclude that carbon is divalent. After all, we just got through establishing the structures for ammonia and water based on the number of unpaired electrons on nitrogen and oxygen. Nitrogen is trivalent and gives rise to NH_3; oxygen is divalent and gives rise to H_2O. Why not divalent carbon, combined with two hydrogen atoms, giving rise to the so-called methylene molecule, $:CH_2$? Nature tells us otherwise. Carbon is tetravalent and gives rise to CH_4, leaving us with the task of adjusting our theory to be consistent with the observed facts.

By uncoupling the paired $2s$ electrons and promoting one of them into a $2p$ orbital, we can establish the necessary number of unpaired valence shell electrons and orbitals—four in the L shell:

$$\underset{1s}{\uparrow\downarrow} \quad \underset{2s}{\uparrow} \quad \underset{}{\uparrow} \quad \underset{2p_{x,y,z}}{\uparrow} \quad \uparrow$$

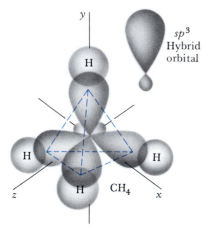

FIGURE 18–4 The methane molecule requires a different explanation than that used for water and ammonia (Figures 18–2 and 18–3). Since there are four equivalent C—H bonds, s atomic orbitals on hydrogen overlap with four hybrid sp³ atomic orbitals on carbon. Hybrid sp³ orbitals are depicted as vastly distorted p orbitals, with a very large lobe and a very small one.

But that alone leads to an incorrect prediction for the shape of the molecule. One of the C-to-H bonds (the one formed from the $2s$ orbital) would have a different energy than the other three. Therefore, the molecule would be unsymmetrical—but that is not the case. To get around that objection, we propose that the orbitals **hybridize**, or average the four states. With four equivalent, hybrid "sp^3" orbitals—so called because they are composed of three p orbitals and one s orbital—we can establish the correct geometry.

$$\underset{1s}{\uparrow\downarrow} \quad \underset{2sp^3}{\uparrow \quad \uparrow \quad \uparrow \quad \uparrow}$$

The four equivalent hybrid bonding orbitals project into space from the central carbon atom, toward the corners of a regular tetrahedron. When carbon bonds with hydrogen, there are four equivalent atomic orbitals on carbon to combine with the atomic orbitals of four hydrogen atoms, forming the molecular orbitals in methane. Note that the regular tetrahedron is the geometry that allows four equivalent hybrid orbitals and the electrons contained in them to separate from each other as much as possible.

Hybrid atomic orbitals have successfully accounted for the bonding in CH_4, but we must also examine the other second period hydrides. Let's consider the water molecule now in terms of hybridization. Looking at the

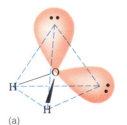

central oxygen atom in the water molecule before the hydrogen atoms are added, there are eight electrons, of which six are in the valence shell:

$$\underset{1s}{\underline{\uparrow\downarrow}} \qquad \underset{2s}{\underline{\uparrow\downarrow}} \qquad \underset{2p_{x,y,z}}{\underline{\uparrow\downarrow}\;\underline{\uparrow}\;\underline{\uparrow}}$$

Again we propose the formation of four sp^3 hybrid orbitals. Two contain paired electrons, and two contain lone electrons of parallel spin:

$$\underset{1s}{\underline{\uparrow\downarrow}} \qquad \underset{2sp^3}{\underline{\uparrow\downarrow}\;\underline{\uparrow\downarrow}\;\underline{\uparrow}\;\underline{\uparrow}}$$

One hydrogen atom can interact with each of the lone oxygen electrons, forming two covalent bonds. This leaves us with four pairs of electrons surrounding the central oxygen atom: two bonding pairs and two non-bonding pairs. All four pairs seek the maximum separation provided by a tetrahedral geometry (Figure 18–5). However, the geometry is not the regular tetrahedral arrangement with all possible internal bond angles equal to 109.5° because all four electron pairs are not equivalent. Good agreement between our structural model and data from experiment is obtained if we assume that bonding electron pairs are more tightly localized in the molecule than are the nonbonding pairs. Thus, the bonding pairs are squeezed in from the theoretical 109.5° internal H—O—H bond angle to the observed value of 104.5°.

This is the essence of **VSEPR** (Valence Shell Electron Pair Repulsion) theory. Applying VSEPR theory to ammonia gives similar results. The central nitrogen atom has five valence shell electrons in four hybrid orbitals, forming one lone pair and three bonding pairs when they bond with three hydrogen atoms (Figure 18–6):

$$\underset{1s}{\underline{\uparrow\downarrow}} \qquad \underset{2sp^3}{\underline{\uparrow\downarrow}\;\underline{\uparrow}\;\underline{\uparrow}\;\underline{\uparrow}}$$

The tetrahedral bond angle of 109.5° for the internal H—N—H bond is reduced, but not as much as in the water molecule because there is only one lone pair instead of two. Experimentally, the value is 107.3°.

Applying this discussion to the elements in the second period that lie to the left of carbon, we can rationalize the covalent bonding behavior in BF_3 and BeF_2. The covalent bonds between boron and fluorine in BF_3 (Figure 18–7) are formed by combination of three hybrid sp^2 orbitals on boron. Their preferred geometry is trigonal planar since this geometry places each orbital as far from the other two as possible. The internal bond angle is 120°.

$$\underset{1s}{\underline{\uparrow\downarrow}} \quad \underset{2s}{\underline{\uparrow\downarrow}} \quad \underset{2p_{x,y,z}}{\underline{\uparrow}\;\underline{\quad}\;\underline{\quad}} \qquad \text{before hybridization}$$

$$\underset{1s}{\underline{\uparrow\downarrow}} \quad \underset{2sp^2}{\underline{\uparrow}\;\underline{\uparrow}\;\underline{\uparrow}} \quad \underset{2p}{\underline{\quad}} \qquad \text{after hybridization}$$

BeF_2 is linear and covalent. The ground state electronic structure for beryllium atoms is $1s^2 2s^2$:

$$\underset{1s}{\underline{\uparrow\downarrow}} \qquad \underset{2s}{\underline{\uparrow\downarrow}} \quad \underset{2p_{x,y,z}}{\underline{\quad}\;\underline{\quad}\;\underline{\quad}}$$

There are no unpaired electrons initially, and presumably no opportunity

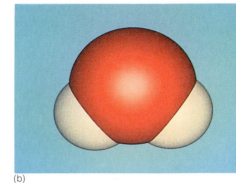

FIGURE 18–5 Structures for the water molecule: (a) the artist's sketch; (b) a computer-generated space-filling model for the water molecule, drawn to scale.

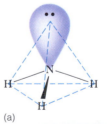

(a)

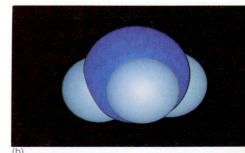

(b)

FIGURE 18–6 The ammonia molecule, based on the hybrid orbital model described in Figure 18–5: (a) artist's sketch and (b) computer model.

FIGURE 18–7 Overlap of 2p fluorine atomic orbitals on boron, forming the trigonal planar BF₃ molecule.

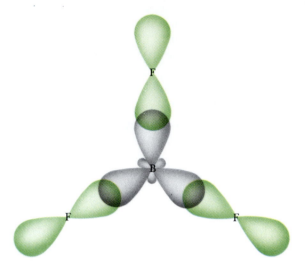

to form covalent bonds by overlapping of atomic orbitals. That is, there are no unpaired electrons until the pair of valence shell $2s$ electrons is uncoupled, promoted, and hybridized. The two hybrid orbitals assume a linear geometry, which separates the electrons contained in them as much as possible. These hybrid orbitals are referred to as sp orbitals, being composed of one s and one p orbital (Figure 18–8):

$$\underset{1s}{\underline{\uparrow\downarrow}} \qquad \underset{2sp}{\underline{\uparrow}\ \underline{\uparrow}} \qquad \underset{2p}{\underline{\quad}\ \underline{\quad}}$$

FIGURE 18–8 Overlap of 2p fluorine atomic orbitals on beryllium, forming the linear BeF₂ molecule.

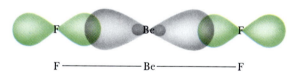

18.4   VALENCE SHELL EXPANSION

Sulfur forms three important fluorides, SF_2, SF_4, and SF_6, in which it appears that the central atom is respectively divalent, tetravalent, and hexavalent. The bonding in SF_2 (sulfur difluoride) can be understood in terms of the normal electronic state, for which we can write the orbitals as follows:

$$\underset{3s}{\underline{\uparrow\downarrow}} \qquad \underset{3p_{x,y,z}}{\underline{\uparrow\downarrow}\ \underline{\uparrow}\ \underline{\quad}\ \underline{\uparrow}} \qquad \underset{3d}{\underline{\quad}\ \underline{\quad}\ \underline{\quad}\ \underline{\quad}\ \underline{\quad}}$$

The hybridization on sulfur would be sp^3 (Figure 18–9), with two orbitals contributing the bonding electron pairs.

$$\underset{3sp^3}{\underline{\uparrow\downarrow}\qquad\underline{\uparrow\downarrow}\qquad\underline{\uparrow}\qquad\underline{\uparrow}}$$

Sulfur tetrafluoride is rather more difficult to treat. For one thing, it is not possible to write a Lewis structure for the molecule without expanding the valence shell beyond a simple octet. In order to arrive at the four atomic orbitals needed for bonding, plus one for the s lone pair, one of the $3d$

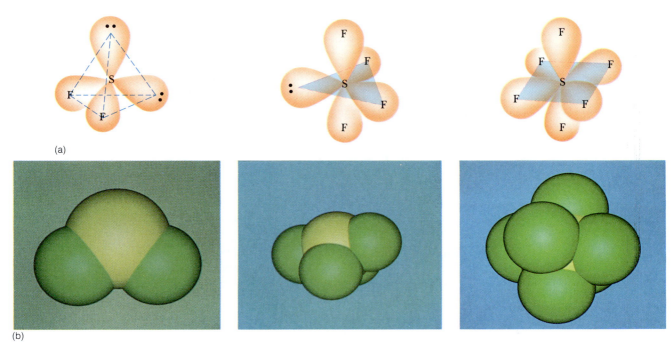

(a)

(b)

FIGURE 18–9 (a) Orbital overlap in SF_2 leads to a tetrahedral arrangement of the bonding and the nonbonding electron pairs in the valence shell and a molecule that is bent; the SF_4 structure has the bonding and nonbonding electron pairs displayed in a trigonal bipyramidal arrangement and the molecular structure looks something like a sawhorse; SF_6 is an octahedral molecule. (b) Computer-generated space-filling models of the three sulfur fluorides.

orbitals must be combined with the $3s$ and $3p$ orbitals, raising the hybridization on sulfur to sp^3d (Figure 18–9):

$$\underline{\uparrow\downarrow} \quad \underline{\uparrow} \quad \underline{\uparrow} \quad \underline{\uparrow} \quad \underline{\uparrow} \qquad \underline{\quad} \quad \underline{\quad} \quad \underline{\quad} \quad \underline{\quad}$$
$$3s \qquad\qquad 3sp^3d \qquad\qquad\qquad\qquad 3d$$

In sulfur hexafluoride, there are six equivalent hybrid sp^3d^2 hybrid orbitals (Figure 18–9):

$$\underline{\uparrow} \quad \underline{\uparrow} \quad \underline{\uparrow} \quad \underline{\uparrow} \quad \underline{\uparrow} \quad \underline{\uparrow} \qquad \underline{\quad} \quad \underline{\quad} \quad \underline{\quad}$$
$$3sp^3d^2 \qquad\qquad\qquad\qquad 3d$$

Expansion of the orbitals used to include d, as well as s and p orbitals, to form an atom with more than four bonds is referred to as **valence shell expansion**.

EXAMPLE 18–2(A)

Consider the bonding orbitals on the central phosphorus atom in phosphorus trichloride to be hybrid orbitals, and determine the geometry.

Solution   Including the nonbonding $3s$ orbital electron pair, there are four orbitals involved, and the hybridization on phosphorus is sp^3. Thus, the geometry is tetrahedral and the overall shape of the molecule is pyramidal, as in NH_3.

$$\underline{\uparrow\downarrow} \quad \underline{\uparrow} \quad \underline{\uparrow} \quad \underline{\uparrow}$$
$$3sp^3$$

In all these examples of hybridization, keep in mind that the energy requirement for promotion of an electron is more than compensated for by the extra stability of the bonding molecular orbitals that result.

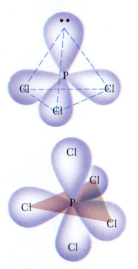

FIGURE 18–10 The orbitals used in constructing the PCl₃ structure are tetrahedrally arranged and the basic molecular design is pyramidal. PCl₅ has a trigonal bipyramidal structure.

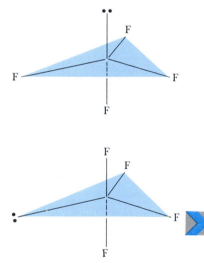

FIGURE 18–11 The SF₄ structure can be trigonal bipyramidal in two different ways, with the favored geometry placing the nonbonding electron pair in the central plane rather than above it.

EXAMPLE 18–2(B)
Determine the geometry about the central phosphorus atom in phosphorus pentachloride.

Solution   In order to provide five bonding orbital electrons (to share with five Cl atoms), the $3s$ orbital pair is uncoupled and one electron is promoted into an available $3d$ orbital, leading to five equivalent sp^3d orbitals (on hybridization of the energies). The overall shape is that of a trigonal bipyramid (Figure 18–10).

$$3sp^3d \qquad\qquad 3d$$

EXERCISE 18–2
Determine the general molecular shapes for arsenic(III) fluoride and arsenic(V) fluoride, using the hybridization principle and VSEPR theory. *Answer:* pyramidal and trigonal bipyramidal.

18.5   VSEPR THEORY AND STRUCTURE NUMBER

The molecular shapes that result from the overlapping of stable orbitals in the valence shells of atoms, hybridization of atomic orbitals, and VSEPR theory can be used collectively to establish a principle, namely that bonding and nonbonding electron pairs settle into whatever spatial arrangement provides minimum electron-pair repulsion. By counting the number of nonbonding electron pairs and the number of shared electron pairs used to form single bonds between atoms, we determine the number of physical objects surrounding the central atom in a polyatomic molecule, the so-called steric number or **structure number (SN)**. For sp, sp^2, and sp^3 hybrid orbitals, the SN values are 2, 3, and 4, respectively. Combination of one s, three p, and one d orbital leads to trigonal bipyramidal geometry for the hybrid orbitals, and SN = 5. For the octahedral geometry of SN = 6, the hybrid orbitals are sp^3d^2 (Table 18–1).

EXAMPLE 18–3
Predict the molecular shape of the SF₄ molecule by using the structure number (SN).

Solution   Counting the nonbonding electron pair, the hybridization is sp^3d, or trigonal bipyramidal, and SN = 5 (Figure 18–11). But SF₄ (in contrast to PCl₅) could be trigonal bipyramidal in two different ways. If the nonbonding electron pair were perpendicular to the plane of the triangle, it would have one 180° and three 90° repulsions from the F atoms. But should the nonbonding electron pair be one corner of the triangle, it would experience two 90° interactions and two 120° interactions. Recall that a nonbonding pair is larger (less localized) than a bonding pair, and that electrostatic interactions fall off abruptly with distance. Thus, it is fair to say the 90° interactions must be of higher energy and less stable. The electron pair will settle into a position at a corner of the triangle of the trigonal bipyramid, and the resulting molecular shape, the shape formed by the five nuclei, will look something like a seesaw.

TABLE 18–1   VSEPR Structures

Formula (Example)	Geometric Configuration	Formula (Example)	Geometric Configuration
AB_2 (CO_2) (BeF_2)	Linear	AB_4 (SF_4)	Seesaw
AB_3 (SO_3) (CO_3^{2-}) (BF_3)	Trigonal planar	AB_3 (ClF_3)	T-shaped
AB_2 (S_2Cl_2) (SO_2)	Angular	AB_2 (XeF_2)	Linear
AB_4 $Pb(C_2H_5)_4$ (CH_4) ($Ni(CO)_4$)	Tetrahedral	AB_6 $Mo(CO)_6$ $Fe(CN)_6^{3-}$ $Fe(CN)_6^{4+}$ $Co(NH_3)_6^{3+}$ (SF_6)	Octahedral
AB_3 (NH_3)	Pyramidal	AB_5 (IF_5)	Square based pyramid
AB_2 (H_2O)	Angular	AB_4 (XeF_4)	Square planar
AB_5 ($Fe(CO)_5$) (PCl_5)	Trigonal bipyramidal	AB_7 (IF_7)	Pentagonal bipyramidal

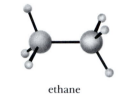

EXERCISE 18–3(A)

Determine the number of nonbonding electron pairs and the steric number for SF_6, IF_5, and XeF_4.

Answer: none/SN = 6; one/SN = 6; two/SN = 6 and the lone pairs are located at opposite ends. ■

EXERCISE 18–3(B)

What would the most likely geometric shape for each of the molecules in Exercise 18–3(A)?

Answer: octahedral; square pyramidal; square planar. Whether you say square planar or octahedral depends on whether you consider the nonbonding electron pairs in your determination of the molecular geometry. ■

ethane

propane

n-butane

iso-butane

Keep in mind that descriptions may vary depending on whether you are referring to the way electron pairs are arranged or the geometric shape of the molecule: XeF_4 has octahedral electron pairs but the geometric shape of the molecule is square planar.

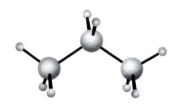

18.6 MULTIPLE BONDS IN CARBON COMPOUNDS

We explained the existence of four equivalent C-to-H bonds in methane (CH_4) on the premise that the valence shell electrons were hybridized. That gave us four equivalent bonding orbitals on carbon, which overlapped with bonding orbitals on hydrogen. Because the orbitals involved are directed along the internuclear axis, these particular single bonds are called by a special name, sigma (σ) bonds. If one H atom in CH_4 is replaced with a CH_3 group, the ethane molecule (C_2H_6) is obtained. It is useful to write the molecular formula in such a way as to indicate the C-to-C bond, which is a sigma bond, formed by overlap of two hybrid sp^3 orbitals along their common axis. The C-to-H bonds are the same sigma bonds as in methane, and the bond angle is very close to the expected tetrahedral angle of 109.5°. The bond angle is not exactly that, however, because the structure is not perfectly symmetrical as is methane:

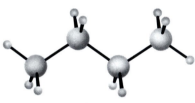

ethane $CH_3—CH_3$

A family of related hydrocarbons can be generated by systematically replacing H atoms with CH_3 groups. Propane, *n*-butane, and iso-butane can be constructed in this way, and they are all characterized by sigma bonds. They are said to be "saturated" hydrocarbons, since all four of the bonding orbitals on carbon are being used in bonds that are directed along the internuclear axes to carbon or hydrogen atoms:

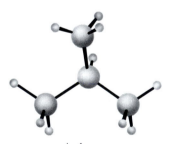

propane $CH_3—CH_2—CH_3$

n-butane $CH_3—CH_2—CH_2—CH_3$

iso-butane $CH_3—\underset{\displaystyle CH_3}{\underset{|}{CH}}—CH_3$

An expansion of this popular and successful model to include sp^2 and sp hybrid atomic orbitals can be used to explain the existence of "unsaturation," double and triple bonds, found among the compounds of carbon. Consider the case of ethylene (C_2H_4), the simplest possible hydrocarbon with a carbon-to-carbon double bond structure that satisfies the octet rule. Intead of promoting a $2s$ valence shell electron and then creating four

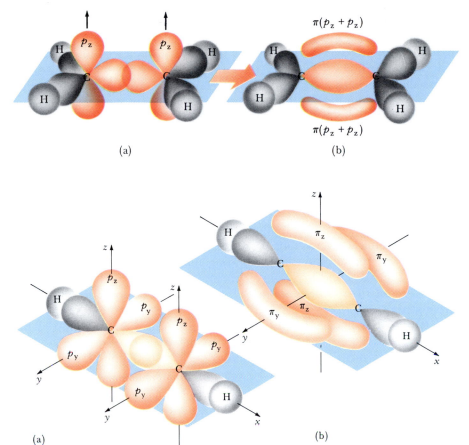

FIGURE 18–12 Ethylene has a "double bond" between the two carbon atoms: (a) a sigma bond forms as hybrid sp^2 and sp^2 orbitals overlap by combination along a common axis; (b) the pi bond forms as unhybridized p orbitals on the adjacent carbon atoms interact, leading to increased electron density above and below the internuclear plane.

FIGURE 18–13 Acetylene has a "triple" bond; (a) a sigma bond forms as hybrid sp and sp orbitals overlap by combination along a common axis; (b) two pi bonds form as unhybridized p orbitals on the adjacent carbon atoms interact, leading to increased electron density above and below the plane of the internuclear axis.

hybrid sp^3 atomic orbitals, create only three, and call them hybrid sp^2 atomic orbitals. That leaves one unhybridized $2p$ atomic orbital. Now, when the two carbon atoms come together with the four hydrogen atoms, the three equivalent sp^2 hybrid orbitals on each carbon atom overlap two $1s$ hydrogen orbitals and one sp^2 carbon orbital. The result is five sigma bonds. The unhybridized p orbitals of the carbon atoms form a pi (π) bond by concentrating the electron distribution of the orbital electrons above and below the plane of the internuclear axis. A requirement of sp^2 hybridization is a trigonal planar arrangement of the bonding orbitals on carbon. The carbon atom and its orbitals are no longer tetrahedral, and the resulting H—C—H bond angle is 120°. Figure 18–12 shows the three sigma bonds lying in a plane, with the π bond (due to the overlap of the unhybridized p atomic orbitals) placed above and below the plane of the internuclear axis. The sp^2-to-sp^2 sigma bond plus the p-to-p pi bond constitute the double bond between the carbon atoms.

In our model, triple bonds arise between adjacent carbon atoms by hybridization of only two of the four valence shell orbitals on each carbon atom. The result is a pair of sp hybrid atomic orbitals, leaving two unhybridized p atomic orbitals. The simplest example is acetylene (C_2H_2), in which there are two C-to-H sigma bonds. One of the three C-to-C bonds, lying at the core of the molecule and along the internuclear axis, is also a sigma bond, resulting from the overlap of sp orbitals directed along the internuclear axis. The p orbitals, lying perpendicular to the C-to-C sigma bond along the y and z axes, form pi bonds. They concentrate the electron density above, below, and on both sides, effectively surrounding the internuclear axis (Figure 18–13).

For the most part, multiple bonds are confined to the elements oxygen, nitrogen, and carbon. The differences between σ and π bonds depicted by these descriptions and drawings show up in the respective bond energies and the properties they impart to the entire molecule. The σ bond displays an essentially cylindrical shape, with its cloud of charge distributed symmetrically along and about the internuclear axis. If one were to attempt to rotate the two carbon nuclei in ethylene about the internuclear C-to-C σ bond, it would not change the overlapping electron density patterns of the sp^2-sp^2 molecular orbitals. That is not true for the π bond. In attempting to twist the internuclear axis, one would be forcing the p-p interaction into an increasingly unacceptable configuration. This imparts a considerable degree of rigidity to multiple bonds, strongly affecting chemical structure and reactivity. In addition, because the pi bonds lie external to the sigma bond, they are generally of higher energy and are more susceptible to a variety of chemical interactions.

However, the addition of pi-bonding makes double bonds more resistant to thermal bond breaking than single bonds, and triple bonds are more resistant than double bonds. Thus, the average bond enthalpies of carbon-carbon single, double, and triple bonds are 346, 610, and 834 kJ/mol, respectively. (See Table 16–5 for other comparisons.) The addition of pi-bonding also draws the atoms together, giving shorter bond lengths. The interatomic distances for carbon atoms bonded by single, double, and triple bonds are 1.54 Å, 1.34 Å, and 1.20 Å.

18.7   RESONANCE THEORY

Two thematic principles of valence bond theory have successfully expanded and strengthened the theory: hybridization of atomic orbitals and resonance. Our discussions of hybrid orbitals assumed a kind of "averaging" of atomic orbitals in order to account for the known tetrahedral geometry of methane despite an orbital electron configuration that would seem to favor the structure of $:CH_2$. Resonance theory deals with the tendency of the structure of a molecule to be a kind of a "sum" of the different distinct ways the orbitals can be arranged. It is one of the important ideas of modern bonding theory and deserves special attention.

In some cases, an adequate representation of an ion or molecule cannot be obtained from a single formula. The true electron distribution is better understood as a composite of two or more structures. For example, sodium carbonate exists as sodium ions and carbonate ions in the crystal lattice and in solution: Na_2CO_3 (s) $\longrightarrow$ 2 Na$^+$ (aq) + CO_3^{2-} (aq).

Although the sodium ion is reasonably represented as Na$^+$, the carbonate ion presents a more difficult problem for us:

The four valence bonds between the central carbon atom and the three oxygen atoms represent ordinary covalent bonds resulting from atomic orbital overlap, giving rise to molecular orbitals. We add the right number of dots about the oxygen atoms to represent the additional electrons required to satisfy the octet rule. Then we finish off the structure with the requisite formal charges. Here is the result: each of oxygen atoms 1 and 3 has six unshared electrons, and can be credited with a seventh from the pair shared with carbon. That gives each of those two oxygen atoms one electron in excess of its normal complement, resulting in a formal charge of -1. Oxygen atom 2 possesses four unshared electrons and is credited with two more from the pairs shared with carbon, giving it a formal and neutral complement of six.

The problem is that electrons are not just the stationary dots and lines we use to represent them. An improved picture would show the π-bond electron pair as **delocalized**, or spread across all three C-to-O bonds, giving each of them something between pure single-bond character and pure double-bond character. This picture of delocalized electrons can be satisfactorily represented by three equivalent forms called **resonance structures** or resonance hybrids:

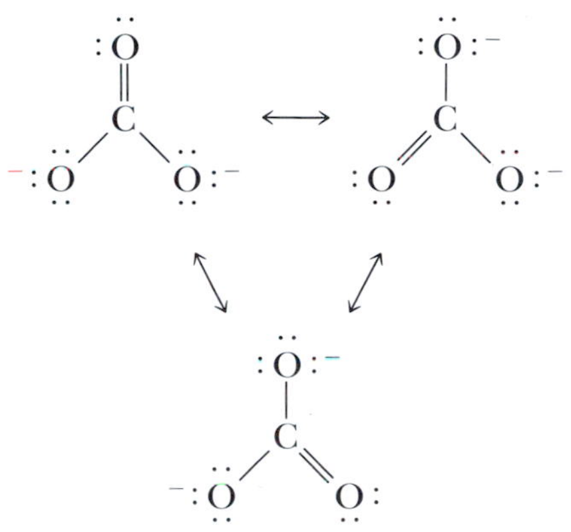

The carbonate ion is not an equilibrium mixture of the three forms, nor does any one of the three forms exist independently. All three of the C-to-O bonds are identical. If we compare the carbonate C-to-O bond distance with C-to-O double bond and single bond distances in other molecules, we find that it is shorter than a single bond and longer than a double bond. The double-headed arrow indicates that no one of these hypothetical **canonical structures** is truly representative, but that the real state of affairs is closer to some average of the canonical structures.

EXAMPLE 18–4

Briefly explain the following observations in terms of resonance theory. (a) If the carbonate ion actually contains one or more of the canonical structures, then it cannot be strictly trigonal planar (oxygen atoms at the corners of an equilateral triangle). (b) The resonance structure predicts that the carbonate ion is strictly trigonal planar. (c) What is the structure number?

Solution   (a) We know from spectroscopic evidence that single bonds are longer than double bonds. If the ion had one of the canonical structures (or an equilibrium mixture of them), then the oxygen atom that was doubly bonded to carbon would be closer to the center of the molecule than the other two oxygens. The molecular spectrum would indicate two types of bonds. In fact, though, it indicates only one type of bond.

(b) Resonance theory predicts that one electron pair (the pair shown in the canonical forms as the pi-bonding half of the double bond) is delocalized over the entire molecule. This electron pair contributes *partial* double-bonding character equally to all three C-to-O bonds. We predict that the three bonds have equal lengths, and that the molecule is strictly planar. This prediction is supported by experimental observations.

(c) The structure number is 3, because the delocalized electron pair is not counted. (It is not localized like the nonbonding pair in ammonia.) ◼

FIGURE 18–14 Benzene: (a) shows the orbitals giving rise to the sigma bonds, and the unhybridized p orbitals from which the pi bonds form; (b) the properties of the molecule suggest that the electron density associated with the pi bonds is uniformly distributed, or "delocalized," above and below all six carbon atoms in the hexagonal ring; (c) sigma and pi bonds, showing complete delocalization, along with the planar arrangement of the atoms in the molecule.

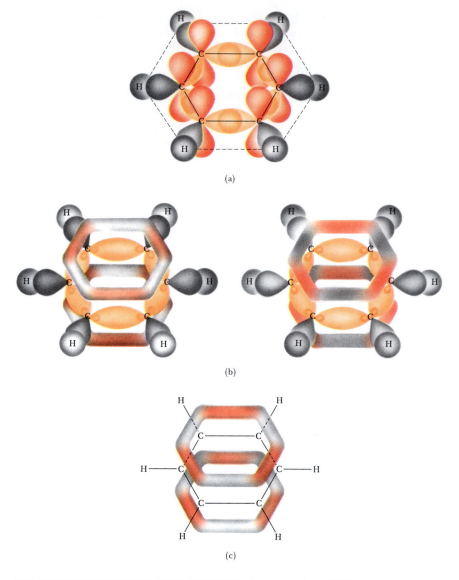

(a)

(b)

(c)

EXERCISE 18–4(A)
Write all of the hypothetical canonical structures for each of the following, using dots for lone electron pairs and lines for the bonding electron pairs:
(a) nitrate ion, NO_3^- (b) sulfur trioxide, SO_3 ■

EXERCISE 18–4(B)
Here are three canonical structures for carbon dioxide:

$$:\!\overset{..}{O}\!=\!C\!=\!\overset{..}{O}\!:\qquad\quad {}^+O\!\equiv\!C\!-\!\overset{..}{\underset{..}{O}}\!:^-\qquad\quad {}^-\!:\!\overset{..}{\underset{..}{O}}\!-\!C\!\equiv\!O^+$$

Which one best represents the correct structure of carbon dioxide? Briefly explain.
Answer: $:\!\overset{..}{\underset{..}{O}}\!=\!C\!=\!\overset{..}{\underset{..}{O}}\!:$ ■

The best-known example of resonance is the benzene molecule, C_6H_6. Picture the structure as a hexagonal ring of six carbon atoms, each bonded to two other carbon atoms and a hydrogen atom. That should suggest an overlapping arrangement of sp^2 hybridized orbitals, establishing the sigma bonds between the carbon atoms in the ring as well as the sigma bonds to the hydrogen atoms. The carbon atoms' hybridized p orbitals form pi bonds just as they would in isolated ethylene molecules. However, because of the alternating arrangement of the pi bonds, the electron clouds above and below the plane of the internuclear axes in the ring are not localized. Instead, they are delocalized, or smeared, around the entire ring above and below the carbon atoms (Figure 18–14).

An important aspect of resonance theory is that molecules that exhibit resonance have extra stability that they would not have if their structures were simple arrangements (in this case, alternating single and double bonds). The experimental evidence of this stability is that the bonds forming the carbon ring of benzene are more difficult to break than are single carbon-carbon single bonds. As a result, the chemistry of benzene and its relatives is primarily the substitution of other groups in place of the hydrogen atoms.

Benzene can be adequately represented by two canonical forms called Kekulé structures:

In a bold and imaginative moment, the German chemist Auguste Kekulé (1829–1896) pictured benzene's structure as an oscillating combination of two distinct hexagonal structures. The achievement was one of 19th-century chemistry's most cherished events.

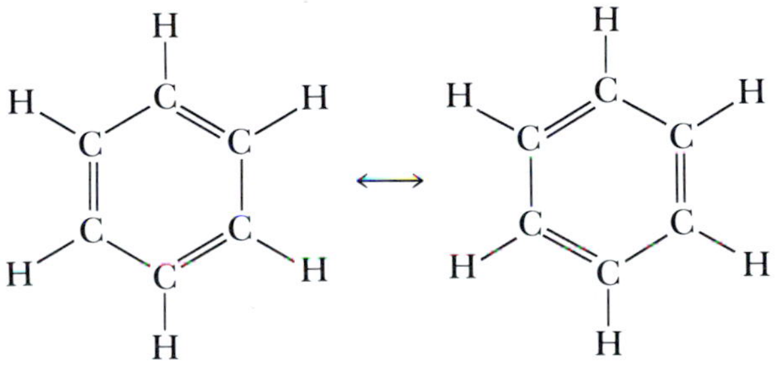

Remember, though, that we do not have three C-to-C single bonds and three C-to-C double bonds in the benzene ring, but rather six identical C-to-C bonds of intermediate character. There is only one bond length, a distance between the ordinary single and double carbon-to-carbon bond lengths. As we will show in later chapters on organic chemistry and biochemistry, this state of affairs has far-reaching consequences.

EXAMPLE 18–5

Toluene, $C_6H_5CH_3$, is derived from benzene by substituting a methyl (CH_3) group for one of the hydrogen atoms. Draw the canonical structures for toluene.

Solution   The bond between the ring carbon atom and the methyl group carbon atom does not take part in the resonance. The canonical structures of toluene are simply those of benzene, with a methyl group in place of a hydrogen atom:

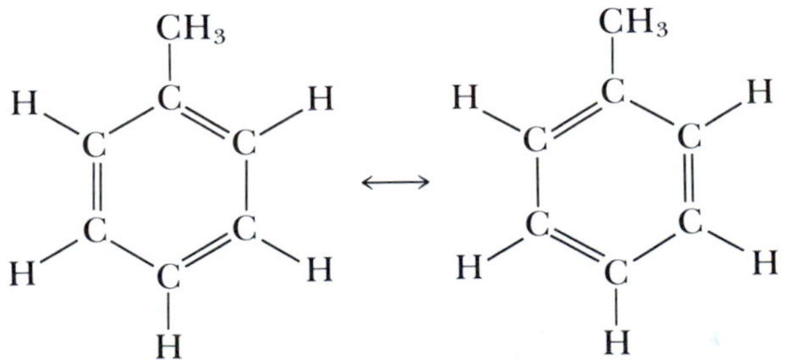

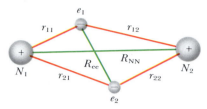

FIGURE 18–15 The hydrogen molecule: N_1 and N_2 are nuclei; e_1 and e_2 are electrons. Repulsive forces (by like-charged particles) are indicated by R; attractive forces (by opposite-charged particles) are indicated by r.

$\psi_H(1s) + \psi_H(1s)$

(a)

(b)

(c)

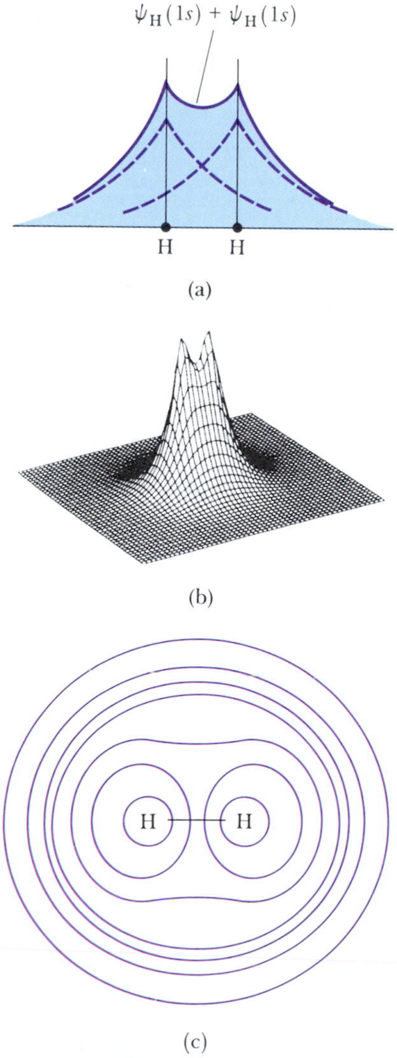

FIGURE 18–16 (a) Sum of two $1s$ atomic orbitals gives a sigma bonding molecular orbital for the hydrogen molecule; (b) computer plot maps the ground state in three dimensions; (c) contour map showing lines of constant electron density.

EXERCISE 18–5
One of the Kekulé structures for the naphthalene molecule ($C_{10}H_8$), a close relative of benzene, is shown. Draw another.

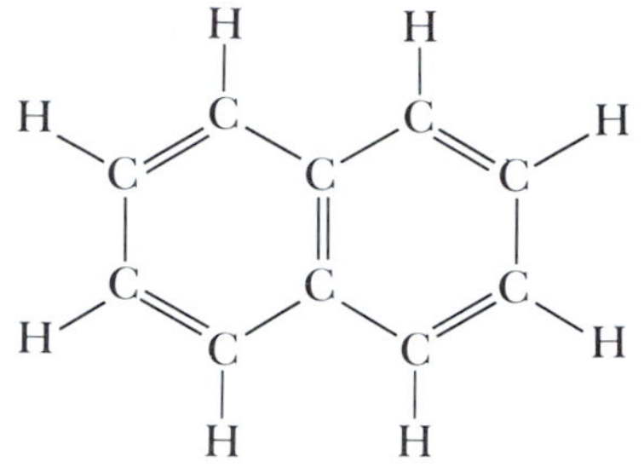

18.8   MOLECULAR ORBITAL THEORY OF THE CHEMICAL BOND

With the theories that we have examined, it is now possible to explain the forces holding together the simple H_2 molecule. Our model for the formation of ionic bonds in compounds such as sodium chloride makes good sense. The interaction between Na^+ and Cl^- involves a pair of closed-shell, spherically symmetric ions of dissimilar atoms with opposite charges. The purely coulombic interactions are easily understood from classical physics. However, a completely different mechanism must be responsible for the bonding in H_2. If we calculate the energy needed to form H^+ and H^- by transfer of an electron from one atom to another (as we did for sodium and chlorine atoms in Chapter 16), and then add this energy term to the electrostatic potential energy released on bringing the resulting ions together, here is what we find. There is no internuclear separation distance for which the total energy is negative (see Figure 18–1b). Either the bond can't form, which we know is not the case, or the bond isn't ionic, which must be the case. The attraction of two hydrogen atoms is a quantum mechanical effect.

When two H atoms come together, they form a H_2 molecule in which a pair of electrons move in the fixed field of a pair of protons. We can develop an approximation for such a four-particle system in terms of the individual one-electron orbitals. Each orbital in the molecular combination represents an electron caught in the field of a proton nucleus and an electron cloud of negative charge, the orbital of the other electron (Figure 18–15). The potential for this system is a composite of six terms, four due to opposite charge interactions between electrons and nuclei, and two due to like charge interactions between electrons or between nuclei:

$$\left[\left(-\frac{e^2}{r_{11}}\right) + \left(-\frac{e^2}{r_{22}}\right) + \left(-\frac{e^2}{r_{12}}\right) + \left(-\frac{e^2}{r_{21}}\right)\right] + \left[\left(+\frac{e^2}{R_{NN}}\right) + \left(+\frac{e^2}{R_{ee}}\right)\right]$$

$$\underbrace{\qquad\qquad\qquad\qquad\qquad}_{\substack{\text{attractive forces by} \\ \text{electrons for nuclei}}} \qquad \underbrace{\qquad\qquad\qquad}_{\substack{\text{replusive forces between} \\ \text{nuclei and between electrons}}}$$

The observed result is the net electrostatic interaction due to the time-averaged distribution of two electrons about the two nuclei.

Thus, we may realistically assume that electron e_1 in the hydrogen molecule is moving in a field not very different from that in a single hydrogen atom, and consequently the molecular orbital it occupies must resemble an atomic orbital of the first H atom in the vicinity of nucleus N_1. Similarly, in the vicinity of nucleus N_2, the occupied molecular orbital must resemble the atomic orbital of the second H atom. That line of reasoning suggests that the wave function representing the molecular orbital (MO) for the hydrogen molecule H_2 can be reasonably approximated as a linear combination of the two atomic orbitals (AO) for the separated H atoms in two different ways—the in-phase combination and the out-of-phase combination:

in-phase $\qquad \Psi_{MO}^{+} = \Psi_{AO_1} + \Psi_{AO_2} = \Psi_{A+A}$

out-of-phase $\qquad \Psi_{MO}^{-} = \Psi_{AO_1} - \Psi_{AO_2} = \Psi_{A-A}$

The molecular orbital resulting from the in-phase combination of atomic orbitals produces a pattern of high electron density between the two nuclei. The favorable case of the hydrogen molecule results; the Ψ_{MO}^{+} function has produced a charge cloud that effectively screens the two nuclei from each other. Bond formation takes place because that is a lower energy state than the energies for the separated atoms. This is referred to as the **bonding molecular orbital** (Figure 18–16).

If the two wave functions impinging upon each other happen to be out-of-phase

$$\Psi_{MO}^{-} = \Psi_{AO_1} - \Psi_{AO_2} = \Psi_{A-A}$$

then the linear combination of atomic orbitals leads to an **antibonding molecular orbital** (Figure 18–17). This is the unfavorable case for the hydrogen molecule. Midway between the nuclei, the wave function Ψ_{MO}^{-} has a value of zero, and the electron density is zero. It represents a higher state of energy, where the electrons in the region between the two nuclei have been displaced. In other words, when the orbitals combine we may get either constructive interference of the hydrogen atom orbital wave functions as they impinge on each other in-phase, or we may get destructive interference of the hydrogen atom orbital wave functions as the wave functions impinge on each other out-of-phase.

Since the electron density is given by the square of the wave function Ψ^2, we can write the first case as follows:

$(\Psi^{+})^2 = (\Psi_1 + \Psi_2)^2$
$(\Psi^{+})^2 = \Psi_1^2 + 2\Psi_1\Psi_2 + \Psi_2^2 \qquad$ constructive interference

Note that the combination of the atomic orbitals differs from the separated atomic orbitals by the term $2\Psi_1\Psi_2$, which represents the increased electron density in the region between the two nuclei due to the wave mechanical behavior of the electrons. The result is a larger electron-proton attraction, lowering the energy of the H_2 molecule with respect to a pair of isolated hydrogen atoms.

The second case leads to a reduction of the electron density in the critical region between the two nuclei in the hydrogen molecule due to destructive interference of electron waves, again by an amount equal to $2\Psi_1\Psi_2$:

$(\Psi^{-})^2 = \Psi_1^2 - 2\Psi_1\Psi_2 + \Psi_2^2 \qquad$ destructive interference

Consequently, the energy of the combined molecule is higher than that calculated for the isolated atoms. The Ψ^{+} molecular orbital corresponds to a situation in which combining electrons produces decreased energy and a favorable bonding effect between atoms. That's called the **bonding orbital**, or bonding MO. The Ψ^{-} molecular orbital, resulting in a higher energy and therefore less favorable combination of electrons, is called the **antibonding orbital** or antibonding MO. For the H_2 molecule, since the two electrons can both go into the bonding orbital, the system of two H

"Linear combination" is a mathematical term describing addition or subtraction of one quantity from another—in this case, atomic orbital wave functions.

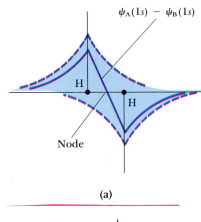

(a)

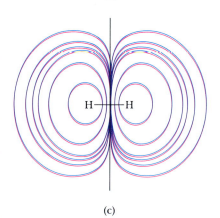

(b)

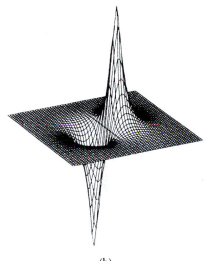

(c)

FIGURE 18–17 (a) The difference of two 1s atomic orbitals gives a sigma star antibonding molecular orbital for the hydrogen molecule; (b) computer plot maps the first excited state in three dimensions; (c) contour map shows lines of constant electron density.

FIGURE 18–18   An energy level diagram depicting formation of hydrogen molecular orbitals by combination of atomic orbitals.

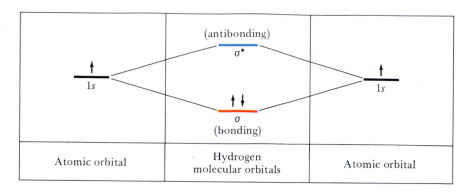

atoms is more stable when they are combined—close together—rather than when they are just a pair of isolated atoms.

These conditions can be illustrated with the aid of energy level diagrams. Figure 18–18 shows the separated hydrogen atoms in their free state, each with its lone electron in a $1s$ atom orbital, and the bonding and antibonding molecular orbitals. In-phase overlap of a pair of s atomic orbitals gives rise to a **sigma** (σ) or bonding molecular orbital. Out-of-phase overlap of a pair of s atomic orbitals gives rise to a **sigma star** (σ^*) or antibonding molecular orbital. Note that the energy level of the sigma molecular orbital is lower than that of the atomic orbitals, while the sigma star molecular orbital energy level is higher. Figure 18–18 shows the two electrons represented by arrows of like spin (direction) in the atomic orbitals, and by arrows of opposite spin (direction) in the molecular orbital of lowest energy.

The energy balance between bonding and antibonding molecular orbitals in any given bond can be expressed in terms of its **bond order**. This is defined as the *net* number of electron pairs in bonding orbitals. Each pair of electrons added to a bonding MO increases the bond order by 1; each pair added to an antibonding MO decreases the bond order by 1. A bond order of 1 is equivalent to a single bond; a bond order of 2, a double bond; and so forth. Fractional bond orders can occur in molecules or ions in which resonance structures occur; for example, one electron pair in the carbonate ion is delocalized over three C-to-O bonds, giving each bond a bond order of 4/3.

18.9   MOLECULAR ORBITAL THEORY IN MOLECULES OF FIRST AND SECOND PERIOD ELEMENTS

Because an s atomic orbital has a spherical shape, symmetrically distributed about the origin in three-dimensional space, there is no preferred orientation when two hydrogen atoms are brought together to form a hydrogen molecule. Each direction is the same as every other, and only a sigma bond can form. In our discussion of valence bonds, we described the bond in terms of overlapping or interpenetrating orbitals along the internuclear axis. The bonding sigma molecular orbital may be depicted as the symmetrical structure shown in Figure 18–19. Note the essential difference between MO theory and the valence bond view. Here the electron density is distributed over both nuclei, effectively surrounding them. The electron-pair bond is no longer localized as it was in our earlier model.

The three p atomic orbitals are depicted as two-dimensional shapes (lobes) projected along the x, y, and z axes; hence the p_x, p_y, and p_z atomic orbitals pictured in Figure 18–20. They depict a zero probability of finding

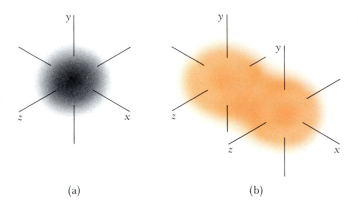

FIGURE 18–19 (a) A sigma-atomic orbital; (b) a sigma-bonding molecular orbital.

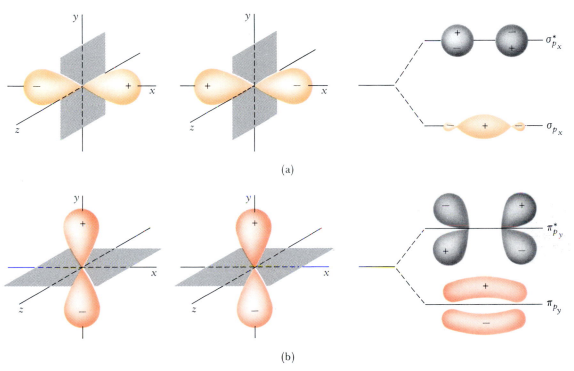

FIGURE 18–20 Combination of p atomic orbitals: (a) sigma (bonding) and sigma star (antibonding); (b) pi (bonding) and pi star (antibonding) molecular orbitals.

the electrons in a plane through the origin. In order to describe the molecular orbitals that arise from linear combinations of these atomic orbitals, we need to use π and π^* molecular orbitals, as well as σ and σ^* molecular orbitals. Remember from our earlier discussions of atomic orbitals and valence bonds that the three p orbitals are mutually perpendicular, with the origin at the nucleus. Therefore, we cannot combine more than one p orbital by overlapping or interpenetrating a p orbital from another atom along the internuclear axis. Instead, any further p orbital interaction leads to increased electron density above and below the internuclear axis.

Figure 18–21 shows two generalized energy level diagrams for the molecular orbitals resulting from combination of identical atomic orbitals in the homonuclear diatomic molecules having up to 20 electrons. That includes all ten elements across the first two periods of the periodic table: H_2, He_2, Li_2, Be_2, B_2, C_2, N_2, O_2, F_2, and Ne_2. Most, but not all, of these

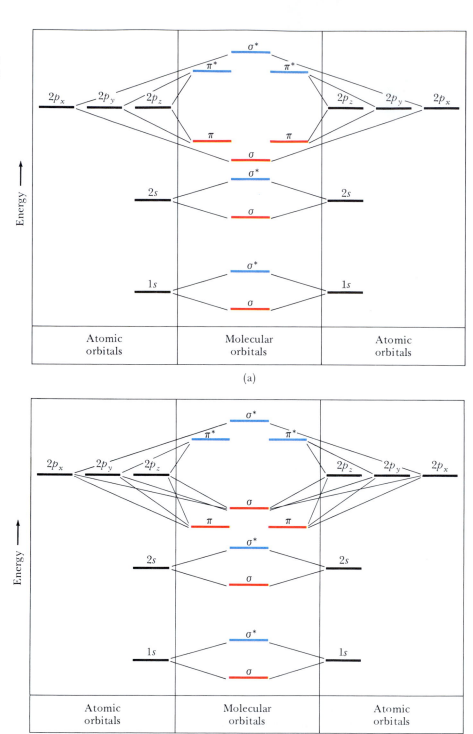

FIGURE 18–21 Generalized energy level diagram for hydrogen-like molecules: (a) large separation of s and p valence levels, with little or no mixing; (b) small separation of the s and p valence levels, with extensive mixing.

diatomic molecules are known. Figure 18–21a represents the order of energy levels one would expect when the separation between s and p valence levels is large. It constitutes the simplest case for such molecular orbital diagrams. The fluorine molecule is a likely example. If, however, the relative energies of the molecular orbitals happen to lie close to each other (as is often the case), mixing of the levels can take place, reversing the order. Just as the 4s atomic orbital had to be assigned a lower energy level

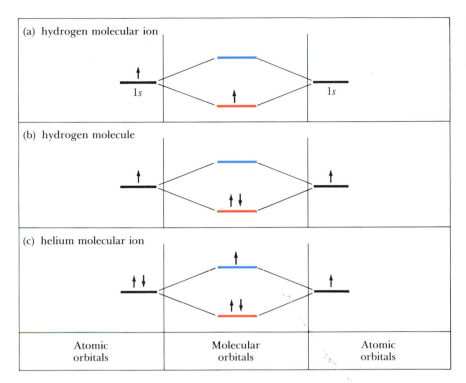

FIGURE 18–22 (a) The hydrogen molecular ion, H_2^+; (b) the hydrogen molecule, H_2; (c) the helium molecular ion, He_2^+.

(a) hydrogen molecular ion

$1s$ $1s$

(b) hydrogen molecule

(c) helium molecular ion

| Atomic orbitals | Molecular orbitals | Atomic orbitals |

than the $3d$ atomic orbital in ground state atoms, Figure 18–21b shows a reversal of the positions of the energy levels with respect to the σ_{2p} and π_{2p} levels. Studies of the spectroscopic behavior and magnetic properties of molecules give us the ultimate ordering of the energy levels, lowest to highest. Carbon has this reversed ordering of molecular orbitals; it is also probable for nitrogen and oxygen.

One of the triumphs of molecular orbital theory is its ability to predict the magnetic behavior of molecules. Most substances are not magnetic, and are not attracted by an external magnetic field. This behavior is called **diamagnetism**. Some substances, however, are attracted into a magnetic field. This behavior, called **paramagnetism**, is caused by the presence of one or more unpaired electron spins. The more unpaired electrons there are, the stronger the attraction is. Molecular orbital theory, as we shall see next, can predict the number of unpaired spins and thus the strength of the paramagnetic effect.

The Simplest Molecules: H_2^+, H_2, and He_2^+

Since molecular orbital theory assumes that the molecular orbitals are not much different from the atomic orbitals from which they are derived, we can begin to construct some simple molecules by placing electrons in orbitals, lowest energy first, according to the Pauli principle and Hund's rule. Using only K-shell orbitals, the ground state electronic structures for the hydrogen molecular ion (H_2^+), the hydrogen molecule (H_2), and the helium molecular ion (He_2^+) are obtained by using the bottom part of the extended energy level diagram in Figure 18–22. Having only one electron, H_2^+ is said to have only half a bond, or to be of bond order $\frac{1}{2}$; and as expected for a particle with an unpaired electron spin, the hydrogen molecular ion is paramagnetic. The experimental bond length is 1.06 Å and the bond

TABLE 18–2   Comparing Characteristics of MO Structures for First Period Elements

Molecular Species	Number of Bonding Electrons	Number of Antibonding Electrons	Bond Order	Bond Length (Å)	Bond Enthalpy (kJ/mol)
H_2^+	1	0	$\frac{1}{2}$	1.06	255
H_2	2	0	1	0.74	431
He_2^+	2	1	$\frac{1}{2}$	1.08	230
He_2	0	0	0	—	—

enthalpy is 255 kJ/mol. For H_2, which has a pair of electrons in the sigma bonding molecular orbital and no others, the bond order is one, and the molecule is diamagnetic because there are no unpaired spins. The experimental values for the bond length and bond energy are 0.74 Å and 431 kJ/mol, respectively. In some ways, He_2^+ is like H_2^+, and that's to be expected. The bond order is $\frac{1}{2}$, in this case because the antibonding orbital contains the third electron. Half the energy benefit from the bonding orbital electron pair is given back by populating the antibonding oribtal with one electron. The molecular ion is paramagnetic, having an unpaired electron spin. Finally, the theory predicts correctly that He_2 is unstable. Populating both the bonding and antibonding orbitals with pairs of electrons leaves a net zero bond order (Table 18–2). In fact, there has never been a verified report of the existence of He_2.

EXAMPLE 18–6

Compare the bond dissociation enthalpies for H_2^+ and H_2 and explain why the value for the later isn't simply twice the value for the former.

Solution   If the two electrons in the sigma bonding molecular orbital in H_2 were independent of each other, you might expect H_2 to have twice the bond enthalpy of H_2^+. However, the value for H_2 is not quite double that for H_2^+, 255 kJ versus 431 kJ, suggesting that the effects of electron-electron repulsion (Figure 18–15) destabilize the molecule to a small extent. That might not be an unreasonable explanation, since the electrons can be viewed as occupying the same region in space. ■

EXERCISE 18–6

Compare the bond dissociation energies for H_2^+ and He_2^+ and briefly explain why the values are similar, but not the same. *Hint:* Consider the difference between the nuclear charges. ■

Before moving on among the second period elements, let's pause to summarize the molecular orbital method we have employed in examining the homonuclear diatomic molecules of the first period elements:

1. Linear combination of the ground state atomic orbitals gives us the required molecular orbitals, which are always equal in number to the atomic orbitals. Thus, two atomic orbitals yield two molecular orbitals, a sigma MO and a sigma star MO.

2. Using the basic premise that bonding MOs are more stable than antibonding MOs, an MO diagram is created that orders the energies of the orbitals. That is simple enough for the first period elements, since we are dealing only with sigma and sigma star orbitals. Later, among the second period elements, we will have to consider the relative order of energies for both sigma and pi molecular orbitals.

3. Beginning with the lowest energy orbitals, the electrons are placed in the molecular orbitals in a way that is consistent with Hund's rule. The importance of that restriction becomes more evident among the second period elements. In the first period, it is necessary only to insure antiparallel spins for electrons sharing the same MO.

4. Bond distances and bond strengths are determined by analyzing the electron populations and the relative orbital energies. The model, to this point, is consistent with data on bond lengths and bond dissociation enthalpies from experiment. A net bond order of 1 is equivalent to a single bond in a Lewis structure, and to a localized shared electron pair bond in the valence bond model.

5. As we move into the second period and across the period from left to right, the effect of increasing nuclear charge must be considered.

6. Ground state electronic configurations for the molecules can be written in a short form, representing the energy level diagram:

H_2^+	$(1s\sigma)^1$
H_2	$(1s\sigma)^2$
He_2^+	$(1s\sigma)^2(1s\sigma^*)^1$
He_2	$(1s\sigma)^2(1s\sigma^*)^2$

Proceeding into the second period, we would expect Li_2 to be similar to H_2, but with some differences. The bond order is 1, as expected for a molecule created by combining lone s orbital electrons, but the bond length (2.67 Å) is much longer and the bond dissociation enthalpy is much less (110 kJ/mol) since longer bonds are generally weaker. The much larger effective radius of the $2s$ orbital versus the $1s$ orbital is the principal reason, but there is a secondary factor: the mutual repulsion of the $1s$ orbital electron pairs, which was not present in the H_2 molecule. The diatomic lithium molecule is diamagnetic, and the ground state electronic configuration is

Li_2	$(1s\sigma)^2(1s\sigma^*)^2(2s\sigma)^2$

Using the same line of reasoning we applied to He_2, we conclude that the bond order for Be_2 is zero and the species is unstable. Here is the ground state electronic configuration:

Be_2	$(1s\sigma)^2(1s\sigma^*)^2(2s\sigma)^2(2s\sigma^*)^2$

EXAMPLE 18–7

Construct a simple energy level diagram for Li_2^+, and analyze the structure according to bond order, bond length, and bond enthalpies, relative to the diatomic lithium molecule.

Solution   Remember that the structure should be analogous to that of the hydrogen molecular ion in the first period. Therefore the diagram should follow the electronic configuration for the ground state of the molecular ion,

Li_2^+	$(1s\sigma)^2(1s\sigma^*)^2(2s\sigma)^1$

On that basis, the bond order should be $\frac{1}{2}$, which means that the bond length should be longer and the bond dissociation enthalpy should be less than for Li_2. The molecular ion should be paramagnetic, since there is an unpaired electron in the half-filled orbital. ■

EXERCISE 18–7

Modify the energy level diagram in Example 18–7 for Be_2^+ and analyze the structure with respect to Be_2 and $He2^+$.

Answer: $(1s\sigma)^2(1s\sigma^*)^2(2s\sigma)^2(2s\sigma^*)^1$; bond order $\frac{1}{2}$; comparisons of bond length and bond dissociation energy cannot be made since Be_2 is unstable and not known to exist; compared to He_2^+, the bond length is longer and the bond dissociation energy is lower because of the larger size of $2s$ compared to the $1s$ orbitals. ■

Boron atoms have three valence shell electrons. For the diatomic B_2 molecule, there are six electrons to be placed in orbitals of lowest energy. If we assume there has been mixing of the sigma (p_x) and pi (p_y and p_z) orbitals, we get a better fit with experimental data, which shows some evidence of paramagnetic character. That would not be the case without mixing. The bond order is 1, counting $\frac{1}{2}$ for each $2p$ orbital electron; that is analogous to Li_2, but the bond distance is much shorter (1.59 Å), suggesting some double bond character. For example, one might expect a shorter bond distance if there is a greater electron density accumulated between two nuclei because of the added shielding of the nuclei that is provided. The bond dissociation energy is consistent with the shorter bond length (272 kJ/mol).

B_2	$(1s\sigma)^2(1s\sigma^*)^2(2s\sigma)^2(2s\sigma^*)^2(2p_y\pi)^1(2p_z\pi)^1$

One of the features of this analysis by molecular orbital theory, in contrast to Lewis structures and valence bond theory, is that there is no definitive way of determining the bond order. Writing the pi bond in ethylene as $CH_2=CH_2$, as valence bond theory suggests, has the stamp of authority: "There is the double bond." However, bond distances are a function of the electron density between nuclei and are never uniquely defined as chemical formulas often imply.

Following the same line of reasoning, our analysis of diatomic carbon suggests a bond order of 2, which is consistent with the shortened bond distance (1.24 Å) and higher bond dissociation enthalpy (602 kJ/mol). The energy level diagram implies diamagnetic behavior because of *s-p* mixing of the levels, and that is found experimentally. The ground state electronic configuration is

C_2	$(1s\sigma)^2(1s\sigma^*)^2(2s\sigma)^2(2s\sigma^*)^2(2p_y\pi)^2(2p_z\pi)^2$

For the familiar nitrogen molecule, a Lewis structure suggests a triple bond, or bond order of 3, giving a very short internuclear distance and a large bond dissociation energy. The MO diagram (Figure 18–23) is consistent with these data and is quite independent of whether we use *s-p* mixing. The picture isn't clear. We show mixing without being strongly committed to that position. As expected from this analysis, the molecule is diamagnetic:

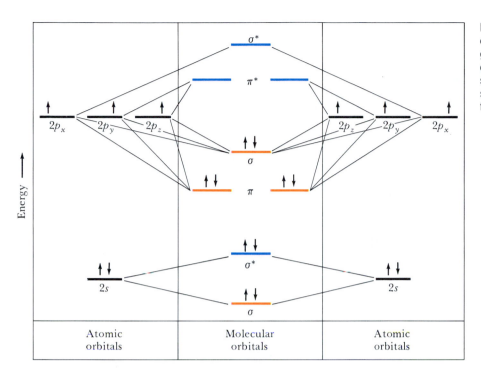

FIGURE 18–23 Molecular orbital energy representation for the nitrogen molecule. Note that all the electrons are paired, which is consistent with the experimentally observed diamagnetic behavior of the molecule.

| N_2 | $(1s\sigma)^2(1s\sigma^*)^2(2s\sigma)^2(2s\sigma^*)^2(2p_y\pi)^2(2p_z\pi)^2(2p_x\sigma)^2$ |

Writing a Lewis structure for the oxygen molecule suggests a double bond and paired electrons:

$$:\overset{..}{O}=\overset{..}{O}:$$

Its magnetic properties—the molecule *is* paramagnetic (Figure 18–24)— suggest otherwise; unpaired electrons must be present. However, the measured bond distance and bond dissociation energy are consistent with the presence of a double bond. The molecular orbital analysis (Figure 18–25) accurately explains both the magnetic properties and the measured bond distance. Note the two half-filled π^* molecular orbitals; the sum of the bonding and antibonding orbitals, five bonding and three antibonding, equals bond order 2. Again, the analysis is independent of whether there is s-p mixing. Based on the continued separation and increased stability as the shells fill, we show no mixing in the diagram for oxygen:

| O_2 | $(1s\sigma)^2(1s\sigma^*)^2(2s\sigma)^2(2s\sigma^*)^2(2p_x\sigma)^2(2p_y\pi)^2(2p_z\pi)^2(2p_y\pi^*)^1(2p_z\pi^*)^1$ |

The fluorine atom has seven valence shell electrons, and the diatomic molecule has fourteen. Analysis of the MO diagram and data from experiment suggest bond order of 1 and diamagnetic behavior. At this point, the L shell orbitals are nearly filled and the energy levels are clearly separated. There is no evidence of s-p mixing:

| F_2 | $(1s\sigma)^2(1s\sigma^*)^2(2s\sigma)^2(2s\sigma^*)^2(2p_x\sigma)^2(2p_y\pi)^2(2p_z\pi)^2(2p_y\pi^*)^2(2p_z\pi^*)^2$ |

Neon has the closed shell electronic configuration characteristic of all the noble gases. The diatomic Ne_2 molecule would be unstable if it existed at all. From the MO diagram, the bond order is zero:

Ne$_2$	$(1s\sigma)^2(1s\sigma^*)^2(2s\sigma)^2(2s\sigma^*)^2(2p_x\sigma)^2(2p_y\pi)^2(2p_z\pi)^2(2p_y\pi^*)^2(2p_z\pi^*)^2(2p_x\sigma^*)^2$

EXAMPLE 18–8

Summarize the relative bond order for the diatomic molecules of the second period from boron to fluorine.

FIGURE 18–24 Molecular O$_2$ is paramagnetic and so it is not surprising that it "sticks" to the poles of a powerful magnet. Liquid oxygen was poured between the poles where it got stuck; liquid nitrogen, on the other hand, would simply pour through the space between the poles.

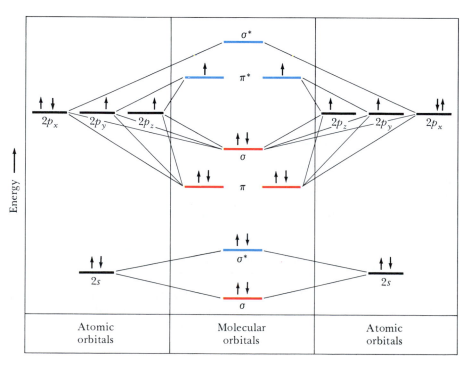

FIGURE 18–25 Molecular orbital energy representation for the oxygen molecule. Note that two of the electrons are unpaired, which is consistent with the experimentally observed paramagnetic behavior of the molecule.

Solution   Based on our discussions of MO theory, we would expect to find

$$B_2 < C_2 < N_2 > O_2 > F_2$$ ■

EXERCISE 18–8(A)
Write the relative ordering of the bond dissociation enthalpies for the diatomic molecules of the second period from boron to fluorine.
Answer: $B_2 < C_2 < N_2 > O_2 > F_2$.   ■

EXERCISE 18–8(B)
Write the relative ordering of the bond lengths for the diatomic molecules of the second period from boron to fluorine.
Answer: $B_2 > C_2 > N_2 < O_2 < F_2$.   ■

The next step would be to expand this discussion to some familiar heteronuclear diatomic molecules such as HF and CO. Water, ammonia, and methane can be treated similarly, as can the other molecules we analyzed using valence bonds earlier in the chapter—ethylene, acetylene, and benzene, for example. The strength of any of these models lies in the accuracy with which it predicts the actual properties of the molecules, and in the insight it gives us into the nature of the chemical bond. Although we have hardly scratched the surface, this is a good stopping point, and a good jumping off point into the last quarter of the book. Beginning with solid state and materials science, and ending with organic chemistry, coordination chemistry, and biochemistry, there is a strong dependence on these early ideas about bonding and bonding theories.

SUMMARY

Valence bond theory is particularly useful to chemists because it makes visualization easy. Structures can be drawn or modeled. Since 1930, we have been playing with these models, and newer versions of them, to help us see what we can't really see directly. The valence bond concept developed out of a quantum mechanical treatment of Lewis' shared electron pair bonds. As two atoms approach each other, chemical bonds form through interactions between atomic orbitals. The approaching atomic orbitals can have one electron each, or there may be two (spin-paired) electrons in one and none in the other. With overlap of the two atomic orbitals, the electron pairs become localized between the two nuclei, which can then move closer to each other due to electrostatic interactions with the electron clouds. As the nuclei approach each other to within some optimum distance, the electrons become indistinguishable and we say that a stable chemical bond exists.

The valence bond approach to chemical bonding depends upon choosing particular stable arrangements of electrons and nuclei. Molecular orbital theory is more general in that regard but does not supply the simple chemical formula that chemists are used to. In both theories, we speak of sigma and pi bonds, distinguished by whether the orbitals interpenetrate along the internuclear axis or lead to increased electron density above and below the internuclear axis. Resonance theory helped valence bond theory out of an apparent dilemma having to do with localization restrictions on electron pair bonds. For example, we are now better able to deal with structures such as benzene and the carbonate ion. But here, too, the simplicity of the formula was lost in a new requirement—hypothetical canonical, or resonance, structures representing the different distinct ways the orbitals can be arranged. Clearly, both valence bond and molecular orbital theories are important in that they can predict properties and behavior that are consistent with observation. As our assumptions about bonding become better founded, these theories converge.

One of the main elements of valence bond theory is hybridization of atomic orbitals, giving shape to otherwise spherical orbitals and thereby accounting for the known geometry of simple molecules. For the second row elements, a good fit between theory and practice was obtained for the simple hydrides—methane, ammonia, water, and hydrogen fluoride. The atomic orbitals on the central atom in each case were described as sp^3. For the left side elements, boron and beryllium, the hybridizations on the central atom were sp^2 and sp, respectively. The bonding in ethylene was explained on the basis of hybrid sp^2 atomic orbitals, and the bonding in acetylene was based on sp hybrid atomic orbitals on carbon.

The illustrations and examples of this chapter should help you understand why the H_2 molecule cannot be explained with an ionic model based on electron transfer and why He_2 does not exist. Simple energy level diagrams are useful in predicting bond orders, bond lengths, and bond energies for the homonuclear diatomic molecules of the first ten elements. Oxygen and nitrogen are especially interesting because of the first appearance of the double and triple bonds among important molecules. Keep in mind the mixing of s and p orbitals in the second period, and the π molecular orbitals that allow us to properly explain the magnetic properties of these diatomic molecules.

QUESTIONS AND PROBLEMS

QUESTIONS

1. Would you expect the H_2S molecule to be reasonably similar in shape to the H_2O molecule? Why, or why not?
2. How might you account for the fact that the H—S—H bond angle in H_2S is 93° while the H—O—H bond angle in H_2O is 104.5°?
3. Do you think it is a reasonable assumption that NH_3 is approximately tetrahedral? Explain.
4. Briefly explain why the CH_2F_2 molecule is not a regular tetrahedron (i.e., the internal bond angles between the central carbon and the other atoms aren't equal).
5. Identify the shapes that are generally associated with the following hybrid orbitals. Give an example of each. (a) sp; (b) sp^2; (c) sp^3; (d) sp^3d^2; (e) sp^3d.
6. Choose the strongest carbon-to-carbon bond among the following compounds and then explain the reasoning for that choice: ethane, ethylene (ethene), acetylene (ethyne).
7. Explain the meaning of resonance by describing differences between the chemical bonding in carbon dioxide and benzene.
8. What effect does the resonance phenomenon have on the "normal" carbon-to-carbon single bond length? The "normal" double bond length?
9. Two separate types of resonance structures can be written for the benzene molecule that properly represent the experimental facts. The first are the Kekulé formulas in (a): the second are the so-called Dewar formulas in (b). Explain why the Dewar formulas are less likely representations than the Kekulé formlas.

(a)

(b)

10. The noble gases and nonpolar molecules such as H_2, N_2, O_2, and F_2 can all be liquified. Account for the existence of forces of attraction that will allow these species to exist in the liquid state. Qualitatively compare these forces to hydrogen bonds and normal dipolar forces.

11. What is your understanding of each of the following terms? (a) atomic orbital; (b) molecular orbital; (c) hybrid orbital; (d) multiple bond; (e) resonance.

12. Distinguish between:
(a) a sigma bond and a pi bond
(b) a sigma bond and a sigma molecular orbital
(c) an sp^2 and an sp^3 hybrid atomic orbital

13. Why can we represent the four bonds on methane by one type of interaction—a bonding MO localized between a carbon atom and a hydrogen atom?

14. Upon what basic assumptions is the theory of formation of molecular orbitals by linear combination of atomic orbitals based?

15. Using atomic and molecular orbitals, describe the formation of the chemical bonds in H_2O and explain the observed 104.5° internal bond angle.

16. Draw a simple sketch of the molecular orbitals for ammonia and methane. Show how the linear combination of atomic orbitals leads to the molecular orbital picture.

17. Return to Chapter 2 and review the three fundamental laws of chemistry—conservation of matter, constant composition, and multiple proportions. Then, briefly discuss to what extent one may consider these laws valid in light of modern chemical bonding theory.

PROBLEMS

Valence Bond Structures of Simple Molecules   [1–4]

1. Which of the following molecules obey the octet rule?

NH_3, CH_4, PF_5, BF_3

2. The octet rule is not obeyed for which of the following compounds?

SF_4, $SbCl_5$, F_2O, CF_4, XeF_4

3. Write Lewis structures for formaldehyde (HCHO) and acetate ion (CH_3COO^-). Include resonance structures where appropriate.

4. Write Lewis structures for phosgene (Cl_2CO) and formate ion ($HCOO^-$).

Hybrid Orbitals and VSEPR Theory   [5–8]

5. Use the VSEPR theory to predict the geometries of the following species in the gas phase: (a) SF_2; (b) SF_4; (c) SF_6; (d) PF_3; (e) PF_5; (f) BrF_2^-.

6. Give the structure predicted by VSEPR for the following species in the gas phase:
(a) SeF_4; (b) $SbCl_3$; (c) $TeCl_2$; (d) SbF_5; (e) ICl_4^-.

7. Predict the geometries of the following compounds in the gas phase: (a) XeF_2; (b) XeF_4; (c) ClF_3; (d) IF_5; (e) $BeCl_2$.

8. What structures would the VSEPR theory predict for the following species in the gas phase? (a) MgH_2; (b) IF_4^-; (c) BrF_3; (d) TeF_4.

Multiple Bonds and Resonance   [9–12]

9. Determine the number of sigma and pi bonds in each of the following molecules: (a) H_2; (b) HCN; (c) CH_3CN.

10. Give the total number of sigma and pi bonds in each of the following molecules:
(a) $HC{\equiv}C{-}CH{=}CH_2$
(b) $CH_2{=}CH{-}COOH$
(c) $(CN)_2C{=}C(CN)_2$

11. Write resonance forms for the following: (a) SO_3; (b) SO_2; (c) O_3 and S_3.

12. Give reasonable resonance structures for the following: (a) PO_4^{3-}; (b) HPO_3^{2-}; (c) N_3^- and P_3^-.

Molecular Orbital Theory   [13–22]

13. Draw the energy level for the hypothetical Be_2 molecule and explain why it doesn't exist as a stable species.

14. Draw the hypothetical Ne_2 energy level diagram. Comment on the stability of the Ne_2 molecule.

15. Describe the formation of the nitrogen molecule from both the valence bond view and the molecular orbital view.

16. Use both the molecular orbital and valence bond theories to describe the formation of the fluorine molecule.

17. (a) Write the correct orbital configuration for the electrons in O_2^+.
(b) What is the bond order for the O_2^+ molecular ion?
(c) Is the O_2^+ molecular ion diamagnetic or paramagnetic?

18. Consider the three diatomic species, OF, OF^+, and OF^-. Provide a brief analysis of their respective bond orders, bond lengths, magnetic properties (diamagnetic or paramagnetic), and their relative bond dissociation energies.

19. Which of the following are paramagnetic species?

H_2, He_2^+, Li_2, Be_2, N_2, O_2

20. Arrange the following diatomic molecules and ions according to decreasing bond order.

B_2^+, C_2, F_2, H_2, He_2^+, N_2, O_2^-

21. Sketch a molecular orbital energy level diagram for a BeH_2 molecule. Show only the valence shell AO's and MO's.

22. Draw the molecular orbital diagram for the unstable species BH_3 showing only valence shell AO's and MO's.

Additional Problems   [23–29]

23. Draw an energy level diagram for a ground state Li_2 molecule showing all AO's and MO's involved, all appropriately labeled.
(a) Should the species exist?
(b) What is the apparent bond order?
(c) Is it diamagnetic or paramagnetic?
(d) Is the bond length longer, shorter, or the same as that in Li_2^+?
(e) Is the bond length longer, shorter, or the same as that in H_2?
(f) Is the energy released greater than, less than, or the same as that released during the formation of Na_2?

(g) Is the energy released greater than, less than, or the same as that released during the formation of B_2?

24. Treating HF as a simple heteronuclear diatomic molecule, draw and label an energy level diagram for the ground state, showing all AO's and MO's involved.

25. Carbon monoxide presents a somewhat more complicated case for a diatomic molecule than either Li_2 or HF (in the two preceding problems), but it can be treated similarly. Draw and label an energy level diagram for the ground state molecule.

26. Sulfur dioxide has 18 electrons involved in bonding orbitals. Write the electron dot structures, determine the structure number, and give an explanation for the probable structure.

27. Triodide ion forms when iodine is dissolved in a potassium iodide solution:

$$I^- \ (aq) \ + \ I_2 \ (s) \ \longrightarrow \ I_3^- \ (aq)$$

Viewing the structure as a central iodine atom bonded to two others, draw a satisfactory valence bond picture using electron dot structures. Give a molecular orbital explanation.

28. Use the VSEPR theory to predict the structures of the following: (a) ClF_2^-; (b) BF_4^-; (c) PCl_5; (d) PF_4^+.

29. Write two possible structures for P_4 in which all four phosphorus atoms are equivalent. Which is the more likely structure of the two?

MULTIPLE PRINCIPLES   [30–35]

30. Predict which of the following compounds would be most soluble in a polar solvent. Which would be least soluble? Give reasons for your answers.

NH_3, XeF_4, ClF_3, BCl_3, PCl_5

31. Which of the following compounds would you expect to be capable of acting as Lewis bases? For each likely Lewis base, give a possible reaction with a Lewis acid.

PCl_3, PCl_5, SF_2, SF_4, SF_6, BF_3

32. Which of the following compounds would you expect to be capable of acting as a Lewis acid? Give a possible reaction with a Lewis base for each likely Lewis acid.

PCl_5, BCl_3, SF_6

33. Sulfur tetrafluoride boils at $-40°C$ whereas sulfur hexafluoride vaporizes (actually it sublimes) at a lower temperature, $-63.8°C$. We normally expect larger molecules to have higher boiling points since induced dipole interactions will be larger. Explain the lower boiling point of the hexafluoride in this case.

34. Compare the bonding in diatomic phosphorus and acetylene dianion, P_2 and C_2^{2-}, specifying the number of sigma and pi bonds, nonbonding electron-pairs and possible hybridization. Acetylene dianion is very strongly basic. Suggest why.

35. NF_3 is known but NF_5 is not, while both PF_3 and PF_5 are known. Briefly explain.

APPLIED PRINCIPLES   [36–39]

36. Sulfur hexafluoride is a very stable gas that can be safely used as an insulator in high voltage equipment. Sulfur tetrafluoride is deadly to breathe since it reacts instantly with moisture to produce the toxic products, HF and SO_2. Explain the unusual stability of sulfur hexafluoride.

37. Two shock-sensitive compounds that are used as percussion caps in explosives and firearms are mercury fulminate $Hg(CNO)_2$ and lead azide $Pb(N_3)_2$. The fulminate and axide ions are isoelectronic and isostructural with each other. Write out possible contributional resonance forms for each.

(a) CNO^- fulminate ion

(b) N_3^- azide ion

38. The linear cyanate ion is obtained on mild oxidation of aqueous cyanide solutions:

$$PbO \ (s) \ + \ KCN \ (aq) \ \longrightarrow \ Pb \ (s) \ + \ KOCN \ (aq)$$

Write the Lewis structure for the cyanate ion, $[NCO]^-$, and suggest why the negative charge is most often formally assigned to oxygen.

39. Halogen cyanides are well known and fairly stable, but do tend to easily form cyclic structures called cyanuric halides:

$$CN^- \ (aq) \ + \ Cl_2 \ (aq) \ \longrightarrow \ ClCN \ (aq) \ + \ Cl^- \ (aq)$$

3 ClCN
cyanogen chloride $\longrightarrow$

cyanuric chloride

Cyanamide, an important agricultural chemical for its high nitrogen content, is prepared by direct reaction of cyanogen chloride (ClCN) with ammonia. By analogy to the chemistry above, write down the chemical reaction for the preparation of cyanamide and suggest a probable structure for the "cyclic" product, called Melamine, which is widely used in preparing home products such as Formica kitchen countertops and Melmac dishes and plates. Compare the Melamine and cyanuric chloride ring structures to the essential features of the benzene ring—resonance and geometry.

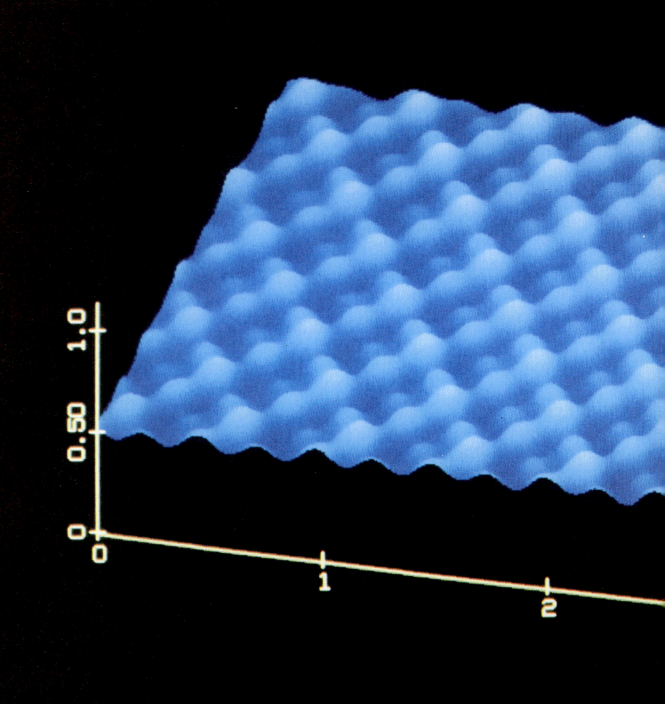

CHAPTER
19

The Solid State

OFFICIAL FIRST DAY COVER

25 YEARS OF TRANSISTORS

WALTER BRATTAIN, WILLIAM SHOCKLEY AND JOHN BARDEEN, AWARDED NOBEL PRIZE FOR TRANSISTOR INVENTION AT BELL LABORATORIES

Progress in Electronics

SERIES OF 1973

Discovered in 1981, scanning tunneling microscopy (STM) allows researchers to produce images of surfaces in which individual atoms are resolved. That capability—to "see" atoms, and features even smaller than atoms—has already impacted on research in fields as diverse as surface science, microelectronics, and biochemistry. Shown here is a three-dimensional surface plot of potassium intercalated graphite. In the potassium-graphite intercalation compound, potassium ions are arranged in ordered layers between graphite sheets. Another monumentally significant discovery of this century occurred more than 40 years ago: the transistor. Today, transistors are everywhere! John Bardeen, who shared the 1956 Nobel Prize for that discovery, was the first person to win a second Nobel Prize in physics, in 1972, for his work on superconductivity.

Ernst Ruska (German), Gerd Binnig (German), and Heinrich Rohrer (Swiss), Nobel Prize for Chemistry in 1986 for their design of the scanning tunneling microscope.

19.1 CRYSTALLINE SOLIDS AND THE AMORPHOUS STATE

Liquids and gases are fluid states. It is a matter of common experience that water takes the shape of its container, and we know that water vapor fills its space uniformly. This fluidity, or ability to flow, that characterizes the liquid and gaseous states suggests that the individual particles are not locked into definite places. Particles in the solid state, on the other hand, need no container to set or hold their shape.

Drawing the line between the solid state of a substance and its liquid state is not as straightforward as you might imagine. We have known for some 50 centuries that heating mixtures of soda, lime, and sand to the melting point produces a smooth "glassy" substance as rigid as a regular crystalline solid. But it was only when scientists began to understand the structure of the atom at the beginning of this century that a clearer understanding of the differences between glasses and crystals became possible. The glassy state is not as distinctly defined, and we know a lot less about it than we do about the crystal state. A **glass** is actually a liquid that has such a high viscosity that it will hold its shape for a very long time.

Window glass seems solid enough; yet there is evidence that the glass in very old windows is flowing. We can describe window glass as a very viscous liquid. In fact, it is so viscous at room temperature that we can think of it as a frozen liquid, rather than a solid.

A useful criterion for distinguishing between crystalline solids and glassy substances is that crystalline solids have definite melting points. Unless a pure compound decomposes on heating, it can be characterized by its melting point, which is uniquely and precisely defined. Glass, on the other hand, does not pass through a clearly defined transition between its solid and liquid states. Instead, its viscosity simply decreases with increasing temperature, until it visibly softens and finally flows into a puddle. Lower viscosity at higher temperatures is typical behavior for liquids; glass is really behaving as a liquid should across the entire temperature range.

If a solid melts precisely at a unique temperature, it freezes at the same temperature. When water freezes, the molecules rearrange themselves in order to maintain thermodynamic equilibrium, balancing enthalpy and entropy to a new point of lowest free energy. Glasses fail to reach the lowest possible free energy state. The glassy state occurs because the liquid state cools rapidly and becomes so viscous that the molecules do not have a chance to relax into the crystalline state. Molecules may have a tendency to join into an ordered, neighborly arrangement, but each has to wait for others to take their proper places, and many are impeded by the movements of still others as they move about. Finally, at some temperature below the temperature at which a glass—when cooled more slowly—would have settled into an ordered crystalline state, the glassy liquid has become so viscous that it is frozen solid. It is stuck in an unstable state. Glasses are also known as **amorphous** solids.

19.2 CRYSTALS

For most of the discussion in this chapter, we shall describe crystalline collections of atoms, molecules, and ions. The amorphous or glassy state—because of its greater complexity—is an advanced topic for us, but will receive some special attention in the next chapter on materials science. Furthermore, we shall limit our discussion of crystals to the simpler structures. Although there is no definitive way of classifying crystalline solids, it has been found convenient to discuss metallic, ionic, and covalent structures according to the following general features:

1. Crystals of the same substance, when prepared in the same way, have the same shape whether the crystals are large or small.

(a)

(b)

(c)

FIGURE 19–1 (a) The near-perfect cubic symmetry in the crystal of barium sodium niobate is clearly seen in this picture. This is the same crystal that produced the "harmonic generation" responsible for the optical effect shown in the chapter 15 opening photo; (b) a chip from a synthetically grown yttrium iron garnet. The faces, natural and unpolished, are defined by cleavage planes within the crystal; (c) synthetic quartz crystals.

2. Crystals are characterized by their faces and edges (Figure 19–1a). Each face normally has another face parallel to it; each edge normally has other edges parallel to it. The edges intersect at fixed angles.

3. Crystals have a high degree of symmetry.

4. When crystals shatter, they generally tend to fracture along planes parallel to the faces of the original crystal, producing small crystals shaped just like the larger crystals from which they broke away (Figure 19–1b). Quartz is an exception (Figure 19–1c). The planes along which the fracture occurs are called **cleavage planes**.

Since cleavage occurs along specific directions, the properties of a crystal are not the same in all directions; that is, crystals are **anisotropic**. On the other hand, amorphous materials are **isotropic**, which means they have the same properties in all directions. Thus, when glass is shattered, splintering occurs in all directions from the point of impact, producing sharp shards.

Because crystals have characteristic external shapes, the constituent particles that make up the crystal must be fixed in specific positions. Furthermore, since small crystals of a substance are the same shape as large crystals of the same substance, we can conclude that the basic arrangement within the crystal is repeated over and over in a regular way as the size of the crystal increases. Accordingly, a simple two-dimensional model of a crystal might look like Figure 19–2. The dotted lines suggest possible directions of cleavage in this hypothetical model.

Within the crystal, each atom or ion is uniformly surrounded by its neighbors. At the surface, however, the situation is different; surface atoms or ions are not regularly coordinated with their neighbors. Thus, for example, the particles of a gas can bond to the surfaces of metals, and in this way metal surfaces can provide temporary sites for atoms and molecules in the course of gas phase reactions. This provides a new pathway for the reaction, which often speeds the reaction greatly. Thus, the metal surface acts as a **catalyst**—a substance that alters the rate of a reaction without being consumed in the reaction. In this case, the metal is a **heterogeneous** catalyst, because the catalyst (solid) and reactants (gases) are in different phases. Precious metals such as palladium, platinum, and rhodium are often used for heterogeneous catalysis. An example is the automotive catalytic converter, which catalyzes the oxidation of the carbon monoxide to carbon dioxide:

$$2\ CO\ (g) + O_2\ (g) \longrightarrow 2\ CO_2\ (g)$$

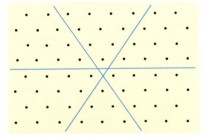

FIGURE 19–2 Model of a hypothetical two-dimensional crystal. Favored directions of cleavage are shown by lines.

Unfortunately, what catalysts are used and how they function are largely proprietary information. But metals such as Pt and Cr are likely candidates.

FIGURE 19–3   Two-dimensional crystal lattice. The circles represent identical atoms; lattice points are taken to be at the centers of each atom. The primitive cell takes the shape of a parallelogram; the nonprimitive cell has the additional unshared center lattice point.

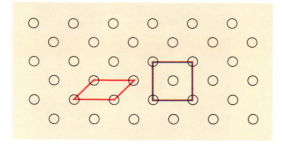

FIGURE 19–4   Movement and translation of the primitive cell (in Figure 19–3), along the directions indicated by the arrows, showing how the entire crystal lattice can be generated.

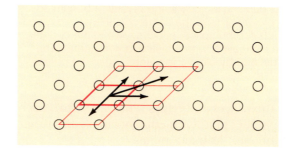

19.3   LATTICE STRUCTURE AND UNIT CELLS

To simplify our introduction to crystal structure and the solid state, imagine a two-dimensional crystal that might be analogous to a layer of atoms adsorbed onto a surface at high density. Consider the two-dimensional arrangement in Figure 19–3, in which the circles represent identical atoms. Any point in this structure can be defined as a **lattice point**, but for convenience we place the lattice point in the center of an atom. Now define the other lattice points in this structure as the collection of all points that have an environment indistinguishable from that of the first lattice point. Remembering that our picture is just a small part of a huge repeating structure, we can imagine a tiny being placed on any of the lattice points; the view in all directions would always be the same. Note that the first lattice point can be placed anywhere in the structure, and then the complete set of lattice points consists of all other points with identical environments. Generally, we choose the first lattice point to coincide with a real particle. The **crystal lattice** is made up of the complete set of lattice points.

The crystal lattice brings us to the concept of the unit cell. The unit cell in a structure has as many dimensions as the structure itself. In a three-dimensional structure the unit cell has three dimensions, whereas in a two-dimensional structure the unit cell has two dimensions. The unit cell is a very small part of the structure, formed by connecting lattice points in such a way that the entire structure can be generated from the unit cell simply by translation of the unit cell. **Translation** is a simple shift of position involving no other movements such as rotations. The unit cell of smallest possible volume for a given space lattice is the **primitive cell**. The primitive cell for a two-dimensional lattice is outlined in Figure 19–3 and takes the shape of a parallelogram. Movement of this cell in the directions shown in Figure 19–4 generates the entire structure. The geometry of the unit cell is specified by the lengths of the two intersecting sides and the angle between them.

The nonprimitive cell shown in Figure 19–3 has greater symmetry. Primitive and nonprimitive unit cells may be distinguished from each other by determining the total number of lattice points per unit cell. For the

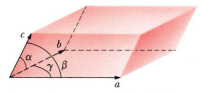

FIGURE 19–5   The dimensions of a three-dimensional unit cell.

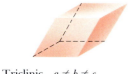

Triclinic $a \neq b \neq c$

$\alpha \neq \beta \neq \gamma \neq 90°$

Rhombohedral $a = b = c$

$\alpha = \beta = \gamma \neq 90°$

Monoclinic $a \neq b \neq c$

$\alpha \neq \gamma \neq 90°$ $\beta = 90°$

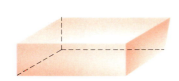

Orthorhombic $a \neq b \neq c$

$\alpha = \beta = \gamma = 90°$

Hexagonal $a = b \neq c$

$\alpha = \beta = 90°$ $\gamma = 120°$

Tetragonal $a \neq b = c$

$\alpha = \beta = \gamma = 90°$

Cubic $a = b = c$

$\alpha = \beta = \gamma = 90°$

FIGURE 19–6 Seven geometries for primitive cells are possible.

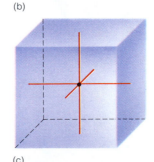

(a)

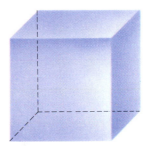

(b)

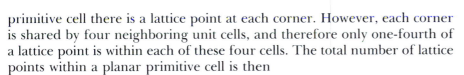

(c)

FIGURE 19–7 (a) The primitive, or simple cubic unit cell; (b) the face-centered cubic unit cell; (c) the body-centered cubic unit cell.

primitive cell there is a lattice point at each corner. However, each corner is shared by four neighboring unit cells, and therefore only one-fourth of a lattice point is within each of these four cells. The total number of lattice points within a planar primitive cell is then

$$4 \text{ corners} \times \frac{1}{4} \frac{\text{lattice point}}{\text{corner}} = 1 \text{ lattice point}$$

The nonprimitive or centered cell has the additional unshared lattice point at its center. Its total number of lattice points within the primitive cell is thus 2.

For three-dimensional lattices and unit cells, we need only extend these principles. The unit cell will have a lattice point at each corner. Three lengths and three angles are now necessary to specify the geometry of the unit cell. As shown in Figure 19–5, these include the lengths of the three intersecting edges a, b, and c and the three angles between these edges. The angles are represented by α, β, and γ. Angle α is the angle between edges b and c, β is between a and c, and γ is between a and b. The possible geometries for three-dimensional primitive cells are given in Figure 19–6.

19.4 CRYSTAL LATTICES BASED ON THE CUBE

Of the possible crystal lattices, three have a cubic unit cell (Figure 19–7). One is the simple or primitive cubic with lattice points only at the corners of the cell. There are two nonprimitive cubic lattices: face-centered cubic

(FCC) has extra lattice points in the centers of all faces, while body-centered cubic (BCC) has an extra lattice point in the center of the cell. These lattices have proved to be very important in understanding structures of metals and salts. Consider the number of lattice points in the unit cell for the three cubic lattices just as we did for the planar (two-dimensional) lattice. In the body-centered cubic lattice, the lattice point located within the cube is unshared. Lattice points located in faces, in the face-centered cubic lattice, are shared by two unit cells. Lattice points located at unit cell corners are shared by eight unit cells. Thus, the number of lattice points for each kind of unit cell are

- Primitive cubic:

$$8 \text{ corners} \times \frac{1}{8} \frac{\text{lattice point}}{\text{corner}} = 1 \text{ lattice point}$$

- Body-centered cubic:

$$\left[8 \text{ corners} \times \frac{1}{8} \frac{\text{lattice point}}{\text{corner}} \right] + 1 \text{ lattice point} = 2 \text{ lattice points}$$

- Face-centered cubic:

$$\left[8 \text{ corners} \times \frac{1}{8} \frac{\text{lattice point}}{\text{corner}} \right] +$$
$$\left[6 \text{ faces} \times \frac{1}{2} \frac{\text{lattice point}}{\text{face}} \right] = 4 \text{ lattice points}$$

19.5 METALLIC CRYSTALS

Closest-Packing of Spheres

In many ways the structure of a solid metallic element is the simplest of structures, since all the atoms are identical and regularly arranged. A regular arrangement of atoms suggests a fixed distance between adjacent atoms. Such a situation is most readily described in terms of the atoms being identical rigid spheres packed as close to each other as possible. A layer that forms naturally when hard spheres of equal diameter are placed close together on a planar surface, as shown in Figure 19–8, is called a **closest-packed** layer.

In a three-dimensional structure, closest-packed layers are stacked on top of each other. Just as the spheres within the layers are as close together as possible, the adjacent layers are also as close together as possible. This results in what is called a closest-packed crystal lattice. A second layer sits on top of the first so that spheres in the second layer are positioned directly

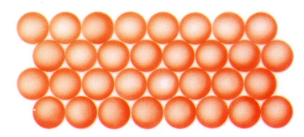

FIGURE 19–8 One layer of closest-packed spheres.

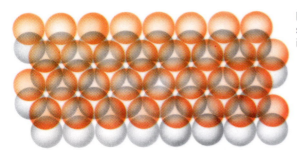

FIGURE 19–9 Two layers of closest-packed spheres, showing the second layer sitting over half the indentations between the spheres in the first layer.

FIGURE 19–10 Three layers of closest-packed spheres displayed (a) hexagonally in an ABA arrangement and (b) in an ABC cubic closest-packed arrangement.

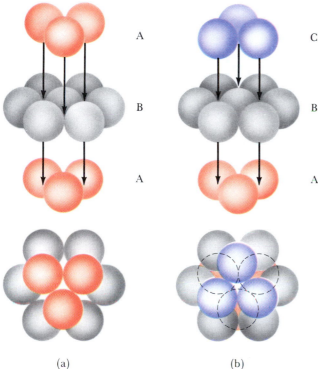

(a)   (b)

over indentations in the first layer, as shown in Figure 19–9. A third layer of spheres sits over indentations in the second layer.

There are, however, two possible arrangements here. The third layer can be positioned directly over the first layer, an *ababab.* . . arrangement, producing a hexagonal unit cell called **hexagonal closest-packing** (HCP) (Figure 19–10a). Alternatively, the third layer can align itself so that it is directly above neither layer one nor layer two, and the fourth layer is over layer one (Figure 19–10b). This is called an *abcabcabc.* . . arrangement, producing a cubic unit cell and **cubic closest-packing** (CCP). For example, copper has the CCP structure. Study of models reveals that the actual structure within this cubic unit cell is that of a face-centered cubic lattice with a metal atom centered at every lattice point. Thus the CCP structure is often called the face-centered cubic or FCC structure.

The FCC or CCP structure is adopted by a number of metals including silver, aluminum, calcium, lead, and copper. The other closest-packed arrangement, HCP, is observed for magnesium, osmium, ruthenium, and others. The third of the relatively simple metal structures, **body-centered cubic**, employs a body-centered lattice with an atom at every lattice point. The BCC structure is observed for iron, potassium, tungsten, and several other metals (Figure 19–11). Note that BCC is not closest-packed.

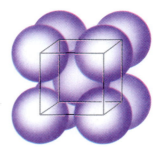

Body-centered

FIGURE 19–11 Packing of spheres in one of the three cubic systems: in this case, the body-centered arrangement of atoms (spheres) is presented.

Since the unit cell is simply the building block for the crystal, the unit cell must have the same density as the bulk sample. The density of a solid can normally be easily determined. Furthermore, the density of a unit cell can be calculated from the number of atoms it contains and its volume. Thus,

$$\text{density of unit cell} = d = \frac{m}{V}$$

where m is the total mass of the atoms in the unit cell and V is the volume of the unit cell. The mass of the atoms in the unit cell is

$$m = \frac{ZM}{N_A}$$

where Z is the number of atoms per unit cell, M is the atomic weight of the atom in g/mol, and N_A is Avogadro's number, 6.022×10^{23} atoms per mole. The volume of the cubic unit cell will be a^3, where a is the length of the edge of the unit cell, and the density can be calculated:

$$d \; \frac{\text{g}}{\text{cm}^3} = \frac{Z \text{ atoms} \times M \text{ g/mol}}{N_A \text{ atoms/mole} \times (a \text{ cm})^3} \qquad (19\text{–}1)$$

Now d is determined easily by experiment and a can be calculated from x-ray data.

EXAMPLE 19–1

X-ray diffraction experiments show that copper crystallizes in an FCC unit cell with a cell edge of 3.608 Å. In a separate experiment, copper is determined to have a density of 8.92 g/cm³. Calculate the atomic weight of copper.

Solution   Begin by rearranging Eq. (19–1) in order to solve for M. Since the atoms sit on lattice points, the number of atoms per unit cell, Z, is determined in exactly the same manner as the number of lattice points per unit cell was determined in Section 19.4. Thus

$$Z = \left[8 \text{ corners} \times \frac{1}{8} \frac{\text{atom}}{\text{corner}} \right] + \left[6 \text{ faces} \times \frac{1}{2} \frac{\text{atom}}{\text{face}} \right]$$
$$= 4 \text{ atoms}$$

To find the answer requires converting angstroms to centimeters to get density in g/cm³:

$$M = \frac{dN_A a^3}{Z}$$

$$= \frac{8.92 \text{ g/cm}^3 \times 6.022 \times 10^{23} \text{ atoms/mol} \times (3.608 \times 10^{-8} \text{ cm})^3}{4 \text{ atoms}}$$

$$= 63.1 \text{ g/mol} \qquad \blacksquare$$

EXERCISE 19–1

Tantalum metal crystallizes in a BCC arrangement with a unit cell edge of 3.281 Å. The density of tantalum has been measured to be 16.6 g/cm³. Calculate the atomic weight of tantalum as indicated by these data. *Answer:* 176 g/mol. ∎

Making use of the idea that the structure of a metal consists of hard spheres in direct contact, it is possible to calculate the relationship between the size of the spheres and the size of the unit cell. The radius of the sphere is the atomic radius r of the metal, and the size of the cubic unit cell is determined by the length of the cell edge a. Consider the FCC unit cell of a metal, using a diagram in which the atoms are shown at their full space-filling size (Figure 19–12). Corner atoms do not touch each other. However, atoms on the corners do touch atoms in the centers of faces, and the diagonal of a face (face diagonal, fd) is equal to four atomic radii. The Pythagorean theorem gives

$$fd = a\sqrt{2}$$

Thus, in a metal that crystallizes in an FCC lattice,

$$a\sqrt{2} = 4r$$

and solving for r gives

$$r = \frac{a\sqrt{2}}{4} \tag{19–2}$$

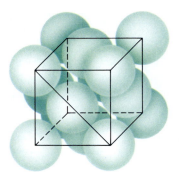

FIGURE 19–12 The face-centered cubic lattice of closest-packed spheres (CCP), showing the face diagonal to be four times the radius of the spheres.

In the BCC structure of a metal (Figure 19–13) the corner atoms again do not touch. However, the corner atoms do touch the atom in the body center of the unit cell, and in this case the body diagonal bd is equal to four atomic radii. But bd is the hypotenuse of a right triangle, the other two sides of which are a and fd. Thus, according to the Pythagorean theorem,

$$(bd)^2 = (fd)^2 + a^2$$

Substituting for fd,

$$(bd)^2 = (a\sqrt{2})^2 + a^2 = 2a^2 + a^2 = 3a^2$$
$$bd = a\sqrt{3}$$

Thus,

$$a\sqrt{3} = 4r$$

and solving for r gives

$$r = \frac{a\sqrt{3}}{4} \tag{19–3}$$

for a metal that crystallizes in a BCC crystal lattice.

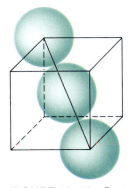

FIGURE 19–13 Part of the body-centered cubic lattice, showing the body diagonal to be four times the radius of the spheres.

EXAMPLE 19–2

Copper crystallizes in an FCC lattice with a unit cell edge of $a = 3.608$ Å. Calculate the atomic radius of copper.

Solution   Applying Eq. (19–2) and substituting for a,

$$r = \frac{(3.608 \text{ Å})\sqrt{2}}{4} = 1.276 \text{ Å}$$ ■

EXERCISE 19–2

Potassium crystallizes in a BCC structure. The unit cell edge length is 5.333 Å. Calculate the atomic radius of potassium.
Answer: 2.309 Å. ■

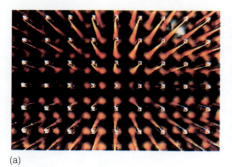

(a)

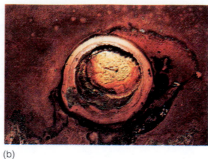

(b)

(c)

FIGURE 19–14   (a) Head-on view of some 10,000 pins on a connector backplane for complex electronic equipment. A computer-controlled system has aligned the pins with an accuracy of nine-thousandths of an inch so that the connector can readily be inserted in the receptacle on a circuit pack. (b) Electroplated gold on connector pins should be smooth, not rough and irregular, as in this test sample. (c) The "cauliflower-like" features are a result of current density variations during electroplating. This scanning electron micrograph is at about 18,000× magnification.

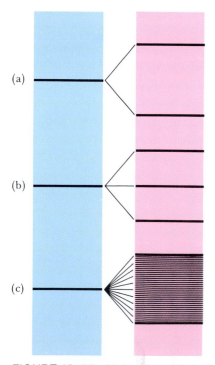

(a)

(b)

(c)

FIGURE 19–15   Molecular orbital diagram for the bonding of (a) two, (b) three, and (c) many sodium atoms.

Palladium and Gold Lattices

Both palladium and gold crystallize in the FCC arrangement. Because a face-centered cube represents the smallest repeating unit of the structure—the unit cell—only one value, the length along the edge of the cube, is needed to define the structure. This value is called the **lattice constant**. In general, lattice constants are on the order of 3 to 6 Å and both palladium and gold have similar constants. Thus, the number of atoms in a cube of equal length along an edge must be the same for both metals. However, the atomic mass of palladium is about 60% that of gold. This results in a large density difference between the two metals, which turns out to have very practical implications.

Palladium and gold have been widely used in the microelectronics industry for electroplating what are commonly known as "interconnects." All microelectronic devices contain interconnects that enable an electrical signal, and hence a message, to be transmitted from one component to another. While copper and silver are excellent interconnects because of their conductive characteristics, they corrode too easily. On the other hand, although gold is not as conductive as silver or copper, it is good enough, and extremely resistant to corrosion. However, since 1980, gold has cost an average of $500 per troy ounce. At a U.S. consumption level of 22.5 metric tons, or 724,000 troy ounces, that is some 375 million dollars worth of gold for microelectronic devices, some 60% of which was used in connector contacts (Figure 19–14). Another 25% was used in printed circuit wiring board fingers. At a recent price of $200 per troy ounce for palladium, the economic advantages of palladium over gold are obvious.

Bonding and Conduction in Metals.

Having discussed the general features of the structures of metals, we are now in a position to consider how the atoms in these structures are bonded together. In turn, this can provide some insight into the characteristic properties of metals such as conductivity of electricity and heat, malleability, and ductility, all of which depend on the presence of mobile electrons.

The metallic elements occupy the left side and lower portion of the periodic table of the elements. Compared to the nonmetallic elements on the right, their ionization energies and numbers of electrons in the valence shell are low. Low ionization energies indicate a weak hold on the valence shell electrons, as well as a low tendency to add electrons to their valence shells. As a result, not only do these atoms fail to form anions, but formation of shared electron-pair bonds—covalent bonds—would provide little stabilization. Therefore, it should not be surprising that species such as the diatomic Na_2 and Pb_2 molecules are not observed under any commonly experienced conditions; even under favorable conditions, these molecules are very unstable. The molecular orbital diagram for the ground state diatomic sodium molecule (Figure 19–15a) shows the two $3s$ orbital electrons in one molecular orbital. That should be favorable to bonding; but the bonding molecular orbital is little stabilized relative to the atomic orbitals, and the covalent bond is weak.

Metals, then, do not have a significant existence as diatomic molecules; instead, they exist principally as crystalline solids. In the solid state, each atom is surrounded by many near neighbors, usually eight or twelve. Metals have relatively small numbers of valence electrons—for example, sodium has one, and lead has four. No metal comes even close to having the eight or twelve valence electrons needed for ordinary covalent bonding to its nearest neighbors. Clearly, the available valence electrons have to be deployed in a delocalized way to hold the structure together. Thus, the electrons must be moving among the pairs of atoms to bond the structure together, in marked contrast to ordinary covalent bonding where the pairs of electrons are fixed in definite molecular orbitals.

By adding more atoms to the molecular orbital diagram of diatomic sodium, the situation that exists in the structure of the solid can be approached. Suppose we bond three atoms together as in Figure 19–15b. Three atomic orbitals form three molecular orbitals, and two of the valence electrons occupy the lowest molecular orbital while the third electron occupies the middle (nonbonding) molecular orbital. As the number of atoms increases, the number of molecular orbitals produced increases accordingly. For a very large number n of atoms, there are n molecular orbitals at very close intervals (Figure 19–15c). The valence electrons occupy the lower portion of this band of orbitals. Movement of electrons through the metal for conduction of electricity or heat is easy, since the electrons near the top of the band can jump to a slightly higher level that is empty and move about the structure. The highest occupied or half-occupied level for these mobile electrons is called the **Fermi level**.

The mobility of the electrons in metals such as aluminum, copper, gold, and lead helps to explain malleability and ductility. In each case, the metal maintains its strength and integrity even after severe deformation. Under such conditions, a nonmetal such as sulfur would simply crumble. However, the highly mobile electrons in a metal can readily form new bonds as the atoms are moved about and so the structure continues to be held firmly together.

The band structure of the molecular orbitals of a metal also accounts for the fact that the metals are opaque. Having a wide range of very closely spaced bands means that a wide range of electronic transitions is possible. Thus, light of most frequencies is absorbed, and little passes through.

19.6 SALTS, IONIC CRYSTALS, AND LATTICE ENERGY

The determination of the crystal structure of salts such as sodium chloride, revolutionized the study of solids. The structure of NaCl (Figure 19–16) consists of an array of positive (Na^+) and negative (Cl^-) ions. These exist in a lattice of unit cells. There is no evidence of pairing up of ions into NaCl molecules.

Solid salts consist of ionic crystals held together by very strong forces, as evidenced by their very high melting points (Table 19–1). The forces holding an ionic crystal together are simply coulombic forces of attractions between unlike charges. The force is described by Coulomb's law,

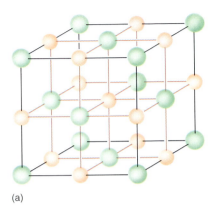
(a)

(b)

FIGURE 19–16 The NaCl structure. (a) In this model, green spheres represent Cl^- ions; yellow spheres represent Na^+ ions. (b) Single crystals, grown in the laboratory.

TABLE 19–1	Melting Points of Some Salts
Salt	**Melting Point (°C)**
AgCl	455
$CaCl_2$	772
LiF	870
NaCl	801
NaI	651
KCl	776

From first discovery by von Laue (1914) and the Braggs (father and son, 1915), to Watson, Wilkins, and Crick (DNA), in 1962, crystallography continues to be an essential tool for uncovering the secrets of structure at the atomic and molecular level.

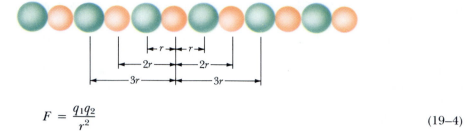

$$F = \frac{q_1 q_2}{r^2} \qquad (19\text{–}4)$$

where q_1 is the magnitude of one of the charges, q_2 the magnitude of the other, and r the distance between them. Since any ionic crystal structure contains like as well as unlike ions, a balance needs to be struck between attractive and repulsive forces. Stable crystals hold together because the overall attractive forces are greater than the overall repulsive forces at the interionic distance, r, characteristic of the crystal.

At this point we could examine the total forces holding a crystal together, but it will prove more useful to consider the energy that stabilizes the crystal structure. That is the case because we are not generally interested in disrupting the structure by applying a force to it, but rather by applying energy, for example in the form of heat. This energy, in the form of work required to move one charged particle in the electric field of another, is force multiplied by distance or

$$E = Fr$$

Substituting for F from Eq. (19–4) gives

$$E = \frac{q_1 q_2}{r} \qquad (19\text{–}5)$$

The potential energy for a whole crystal can be calculated by employing Eq. (19–5) for each pair and totalling all of the attractions and repulsions. This energy is called the **lattice energy** and is given by the symbol U. Attractions will make a negative contribution to U since q_1 and q_2 will have different signs, and repulsions will make positive contributions. Thus the crystal lattice is stable if there is a negative value for U.

Now consider the lattice energy for a hypothetical one-dimensional crystal of alternating positive and negative ions separated by a repeating distance r as shown in Figure 19–17. We can begin by calculating the attractions and repulsions for the positive ion at the center of the drawing. The net potential energy for this ion includes the following:

1. two attractions of $-|q_1 q_2|/r$
2. two repulsions of $+|q_1 q_2|/2r$
3. two more attractions of $-|q_1 q_2|/3r$

Carried out a few more steps

$$U = -\frac{2|q_1 q_2|}{r} + \frac{2|q_1 q_2|}{2r} - \frac{2|q_1 q_2|}{3r} + \frac{2|q_1 q_2|}{4r} - \frac{2|q_1 q_2|}{5r} \cdots$$

$$= -2\left(\frac{|q_1 q_2|}{r}\right)\left(1 - \frac{1}{2} + \frac{1}{3} - \frac{1}{4} + \frac{1}{5} \cdots\right)$$

The infinite sum within the parentheses converges to 0.693147 or ln 2:

$$U = -2(0.693147)\left(\frac{|q_1 q_2|}{r}\right) = -1.38629 \left(\frac{|q_1 q_2|}{r}\right) \qquad (19\text{–}6)$$

The number 1.38629 is called the **Madelung constant** M, for this particular linear lattice. Each type of lattice has its own specific value for the Madelung constant. Thus, for any crystal, the lattice energy can be calculated from the general equation

$$U = -M\left(\frac{|q_1 q_2|}{r}\right) \qquad (19\text{-}7)$$

Within a lattice type such as our linear lattice, the only variables are the charges on the ions q_1 and q_2 and the interionic distance, r. Consequently, the lattice energy becomes more negative and the lattice is more tightly held together if the ionic charges, q_1 and q_2, increase or the interionic distance, r, decreases.

EXAMPLE 19-3

For sodium chloride, the Madelung constant is 1.748 and the interionic distance is 2.81 Å. Calculate the value for the lattice energy (taken as the electrostatic potential energy of a pair of separated point charges).

Solution   We express the electron charge, e, in esu and the interionic distance in centimeters; this results in the energy having units of ergs per formula:

$$U = -M\frac{e^2}{r}$$

$$= -1.748\,\frac{(4.80 \times 10^{-10})^2}{(2.81 \times 10^{-8})} \text{ ergs/formula}$$

$$= -\frac{1.43 \times 10^{-11} \text{ ergs}}{\text{formula}} \times \frac{6.022 \times 10^{23}}{\text{mol}} \times \frac{1 \text{ kJ}}{10^{10} \text{ erg}}$$

$$= -861 \text{ kJ/mol} \qquad \blacksquare$$

EXERCISE 19-3

The Madelung constant for KCl is 1.763. Calculate the lattice energy for K^+ and Cl^- considering the ions to be point charges and the interionic distance to be 3.14 Å.
Answer: $U = -777$ kJ/mol. ■

Examples of Some Ionic Structures

The CsCl structure: CsCl has the simplest ionic crystal structure. The same structure is observed in a few other salts, such as CsBr, NH_4Cl, and TlBr. It is also found in a few of the so-called stoichiometric alloys that contain strict one-to-one molar ratios of two metals, such as CuZn, AuZn, and MgHg.

The CsCl structure is based on the primitive cubic lattice. This lattice has points only at the corners of a cubic unit cell, requiring an identical environment at each corner. For CsCl this can be accomplished by placing a Cl^- ion at each unit cell corner. Cubic symmetry is preserved by placing a Cs^+ ion at the center of the cube (Figure 19-18). The number of Cl^- ions per unit cell is

$$8 \text{ corners} \times \frac{1}{8}\frac{Cl^-}{\text{corner}} = 1 \ Cl^-$$

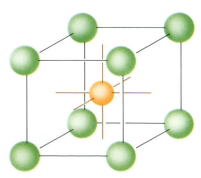

FIGURE 19-18 Body-centered cubic, one of three different space lattices in the cubic crystallographic system.

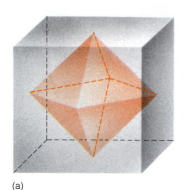

(a)

(b)

FIGURE 19–19 (a) Artist's sketch of the octahedral hole in the center of the face-centered (FCC) lattice of the cubic crystallographic system. (b) Crystals of alum, commonly a double salt of aluminum and potassium or ammonium sulfates, showing how the external form changes on passing from the cubic (left) to the octahedral (on the right). Notice that the progression consists of successively "shaving" the corners of the original cube.

The Cs^+ in the center of the unit cell is not shared with any other unit cell, so each unit cell contains one Cs^+ ion and one Cl^- ion. The ratio of the number of Cs^+ ions to Cl^- ions is therefore one-to-one throughout the crystal; the CsCl structure can accommodate only ions of equal charges, resulting in a $1:1$ ion ratio. In fact, the few salts that crystallize in the CsCl structure have cations with a charge of $+1$ and anions with a charge of -1.

If we extend the CsCl lattice to several adjacent unit cells, we find that the structure could be viewed just as well by unit cells with Cs^+ ions at the corners and Cl^- ions at the unit cell centers. One viewpoint is as good as the other, since the unit cell is devised merely to help describe the existing crystal structure. Using these two possible descriptions of the CsCl structure, we can determine the **coordination number** of each ion. The coordination number is the number of nearest neighbors that a particular type of ion has. Thus, if we draw a unit cell with the Cs^+ ion in the center, we can see that the Cs^+ ion has eight nearest neighbor Cl^- ions, giving the Cs^+ ion a coordination number of eight.

Similarly, if we draw the unit cell with the Cl^- atom in the center, we see that the Cl^- has eight nearest neighbor Cs^+ ions and thus also has a coordination number of eight.

The NaCl Structure: The NaCl, or rocksalt, structure is very common for $1:1$ salts. Among the many solids that adopt this structure are AgCl, KCl, KBr, LiCl, and NH_4I with $+1$ cations and -1 anions; BaO, BaS, and MgO with $+2$ cations and -2 anions; VN with a $+3$ cation and -3 anion; and ZrC with a cation and anion that are formally $+4$ and -4.

The NaCl structure (Figure 19–16) is based on an FCC lattice, but also uses spaces or holes in the FCC structure. These holes are situated in the center of the unit cell and halfway along each edge. All of these holes are equivalent in the overall structure, but we will consider the hole in the center of the FCC lattice first. As can be seen in Figure 19–16, an ion located in this hole has six nearest neighbors, and a coordination number of six. Furthermore, these six nearest neighbors can be thought of as placed in space at the vertices of a solid (Figure 19–19). The particular solid figure in this case is an octahedron, one of nature's regular solids. The octahedron has eight equal faces, each of which is an equilateral triangle, twelve equal edges, and six equivalent vertices. The hole thus described is called an **octahedral hole**. The Na^+ ions located halfway along the unit cell edges are also in octahedral holes, but it is necessary to consider neighboring unit cells in order to see the octahedra. In the NaCl structure, all of the octahedral holes in the lattice are occupied by Na^+ ions.

Just as the CsCl structure could be drawn having either Cs^+ or Cl^- ions at the unit cell corners, the NaCl lattice can be viewed in an alternative way—with Na^+ ions in an FCC lattice and Cl^- ions in the octahedral holes.

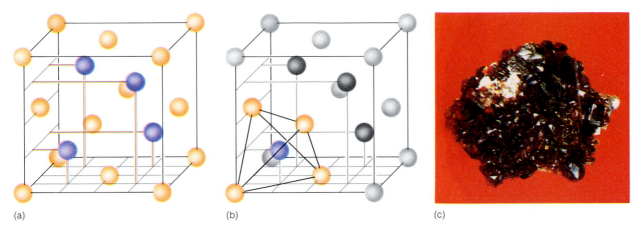

FIGURE 19–20 The zinc blende structure. (a) Blue spheres represent Zn^{2+} ions; orange spheres represent S^{2-} ions. (b) The tetrahedral hole from the front lower left, in (a). The Zn^{2+} ion is in a tetrahedron formed by four S^{2-} ions. (c) Zinc sulfide crystals in a chunk of the mineral.

The coordination number of the Cl^- must also be six. The ratio of Na^+ to Cl^- ions can be verified by considering the number of each in a unit cell. If we draw the cell with Cl^- ions at the lattice points, the number of Cl^- ions is

$$\left[8 \text{ corners} \times \frac{1}{8} \frac{Cl^- \text{ ion}}{\text{corner}} \right] + \left[6 \text{ faces} \times \frac{1}{2} \frac{Cl^- \text{ ion}}{\text{face}} \right] = 4 \ Cl^- \text{ per unit cell}$$

In order to count the Na^+ ions properly, we must consider that an ion on the unit cell edge is shared by four adjacent unit cells. Thus the number of Na^+ ions per unit cell is

$$\left[12 \text{ edges} \times \frac{1}{4} \frac{Na^+ \text{ ion}}{\text{edge}} \right] + \text{unshared ion at cell center} = 4 \ Na^+ \text{ per unit cell}$$

The cell thus contains ions in the ratio of 4:4 or to 1:1 for NaCl, indicating that the numbers of Na^+ ions and Cl^- ions in a sample of sodium chloride of any size are equal.

The Zinc Blende (ZnS) Structure: Zinc sulfide also crystallizes into an FCC lattice. In this case the S^{2-} ions are located at the lattice points of a FCC lattice. Each Zn^{2+} ion is located in a hole, but not an octahedral hole; instead, it is in a hole that has four nearest neighbors (Figure 19–20a). The four nearest neighbors can be considered to be at the vertices of another regular polyhedron, the tetrahedron (Figure 19–20b). There are eight **tetrahedral holes** in the FCC unit cell, and half of them are occupied by Zn^{2+} ions in the zinc blende structure.

The coordination number of each ion in the zinc blende structure is four. Each unit cell of the structure contains four S^{2-} ions a the FCC positions, just as the unit cell of the NaCl structure contains four Cl^- ions. The four Zn^{2+} ions are not shared with adjacent unit cells, so the ratio of Zn^{2+} and S^{2-} ions is 4:4 or 1:1. Other salts that adopt this structure are BeS, HgS, CuCl, CuBr, and AgI.

The Fluorite (CaF$_2$) Structure: In the fluorite structure (Figure 19–21) the unit cell can be drawn so that Ca^{2+} ions are situated in an FCC lattice with the F^- ions in all of the tetrahedral holes. This gives a ratio of Ca^{2+} to F^- ions of 4:8; which reduces to 1:2 or CaF$_2$. Other 1:2 salts that adopt this structure are BaF$_2$, SrCl$_2$, ZrO$_2$, and HfO$_2$. The coordination numbers for the ions in fluorite are eight for the Ca^{2+} and four for the F^-.

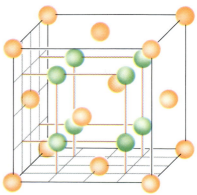

FIGURE 19–21 The fluorite, or calcium fluoride, structure. Green spheres represent F^- ions and orange spheres represent Ca^{2+} ions.

TABLE 19–2 Madelung Constant for 1:1 Salts

Structure	M	Coordination Numbers	
		Cation	Anion
CsCl	1.763	8	8
NaCl	1.746	6	6
ZnS (zinc blende)	1.638	4	4

The Madelung constant increases somewhat with increasing coordination number since the sum will include more shortest distance attractions.

Comparing Ionic Structures

The four structures we have described represent some simpler, relatively common ionic crystals. These structures are typically found among salts consisting of monatomic ions. Now we ask, "Why does a salt adopt a particular ionic structure?" To answer this question, we must consider four particularly important factors:

1. The stoichiometry of the salt; that is, the ratio of cations to anions.
2. The lattice energy.
3. The coordination number of each of the ions.
4. The ratio of the radii of the ions.

The stoichiometry of the salt determines what ionic structures are possible. Thus the NaCl structure can accommodate only a 1:1 salt. This requires that the magnitude of the charge be the same for both anion and cation. We have described three structures for 1:1 salts; it is important to point out that although these are all important structures, this is not a complete list.

Similarly, 1:2 salts or salts in which the cation has a charge with a magnitude twice that of the anion can adopt the fluorite structure. Again, this is only one of a number of possibilities. Salts with a 2:1 ratio or with an anion charge of twice the magnitude of the cation charge, such as Na_2O and Na_2S, can adopt an **antifluorite** structure (which is the same as the fluorite structure but with the positions of the anions and cations reversed).

The lattice energy can be calculated. Madelung constants for the structures of 1:1 salts we have described are listed in Table 19–2.

For a particular salt the values of q_1 and q_2 are fixed by the nature of the ions. Therefore, the most negative value of U will occur if r is as small as possible and the structure chosen has the greatest possible value of M resulting from the highest possible coordination number. This situation would represent the most stable possible crystal held together by the greatest possible energy. Note, however, that the differences among the Madelung constants for different structures are relatively small. In addition, minimizing r and maximizing M cannot be done independently, as we will next consider.

We can determine the value of r by considering the ions in an ionic crystal to be hard spheres of definite ionic radius, just as we considered the metal atoms in a closest-packed metal structure to be hard spheres of definite atomic radius. These models of metallic or ionic solids are very useful even though not perfectly accurate. Minimizing the value of r in an

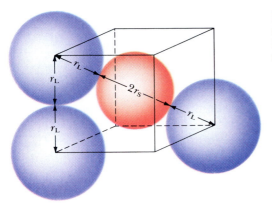

FIGURE 19–22 Part of the CsCl structure, showing the relationship between r_L and r_S at the minimum of r_S.

ionic solid depends on having a structure in which the spheres of the cation and adjacent anions can actually touch. But this places restrictions on the coordination number, the Madelung constant, and the crystal structure that is adopted.

Let us consider the CsCl structure and determine the smallest ion that can be placed in the hole in the center of the unit cell that will still be in contact with all of its eight nearest neighbors. Call the ionic radius of the small atom in the center of the unit cell r_S, and the radius of its eight larger nearest neighbors r_L. In order to make the holes as small as possible, draw the larger ions so they actually touch along the unit cell edges (Figure 19–22). Thus,

$$a = 2r_L$$

where a is the length of the edge of the unit cell. If the smaller ion just fits into the hole formed by its eight nearest neighbors, the body diagonal of the cubic unit cell will pass from the center of one of the large ions to its surface, through the smaller ion, and then from the surface of another of the larger ions to its center, so

$$bd = 2r_L + 2r_S$$

The length of the body diagonal is

$$bd = a\sqrt{3}$$

Therefore

$$a\sqrt{3} = 2r_L + 2r_S$$

Substitution of a from above gives

$$2r_L\sqrt{3} = 2r_L + 2r_S$$

Solving for r_S, we have

$$r_S = r_L\sqrt{3} - r_L = r_L(\sqrt{3} - 1) = 0.732r_L$$

Within the limitations of our ionic model of spheres of fixed radius, this calculation shows that if the radius of the smaller ion is less than 0.732 times the radius of the larger ion, then spheres of unlike charge do not touch whereas those of like charge do. This result is an unsatisfactory structural situation, and we can expect that a different structure will be observed. Similar calculations for the NaCl and zinc blende structures give the minimum r_S/r_L ratios:

TABLE 19–3	Ionic Radii Å for Some Common Monatomic Ions		
Cation	**Radius**	**Anion**	**Radius**
Li^+	0.90	F^-	1.19
Na^+	1.16	Cl^-	1.67
K^+	1.52	Br^-	1.82
Rb^+	1.66	I^-	2.06
Cs^+	1.88	O^{2-}	1.24
Mg^{2+}	0.86	S^{2-}	1.70
Ca^{2+}	1.14		
Sr^{2+}	1.32		
Ba^{2+}	1.49		
Zn^{2+}	0.74		

TABLE 19–4	Ionic Radius Ratios for Several 1:1 Salts		
Salt	r_S/r_L	**Predicted**	**Observed**
LiCl	0.54	NaCl	NaCl
LiI	0.44	NaCl	NaCl
NaCl	0.69	NaCl	NaCl
RbCl	0.99	CsCl	NaCl
CsCl	0.89	CsCl	CsCl
BaO	0.83	CsCl	NaCl
ZnS	0.44	NaCl	ZnS
CuCl	0.44	NaCl	ZnS
AgCl	0.68	NaCl	NaCl
AgI	0.55	NaCl	ZnS

Structure	Coordination number	Minimum r_S/r_L
CsCl	8	0.732
NaCl	6	0.414
zinc blende	4	0.225

The general result of these calculations is that the lower the value of r_S/r_L, the fewer the number of larger ions that can be coordinated about the smaller ion. Now, higher coordination numbers give higher Madelung constants and greater lattice energies if other factors remain equal. Therefore, it is reasonable to postulate that the structure will assume the highest coordination number possible consistent with the ratio of r_S/r_L. Table 19–3 gives some accepted ionic radii of monatomic ions. Table 19–4 lists several ratios of r_S/r_L and the predicted and observed structures for a few 1:1 salts.

Comparing predicted and observed structures clearly has its successes and its failures. The approximation of considering ions as spheres of fixed radii is an oversimplification. In fact, some books list ionic radii that vary with the coordination number. The CsCl structure, which does not have a closest packed lattice, appears rather seldom—and considerably less frequently than would be predicted by radius ratio calculations. The NaCl structure appears to be extremely stable, clearly dominating the structure of 1:1 salts, particularly those with +1 and −1 ions.

Covalent bonding is not favored by the six-coordinate NaCl structure. Such bonding becomes significant with ions of charges greater than +1 and −1. The zinc blende structure has a coordination number of four about each ion, and each ion finds its nearest neighbors in a tetrahedral arrangement. This is a common and stable arrangement that is favored when there is some measure of covalent bonding.

Calculating ionic radii from crystal structures follows principles similar to those for calculating atomic radii from crystal structures of metals. The basic assumptions are that the ions are spheres of fixed radius and that the ions are in contact with nearest neighbor ions of opposite charge, in order to maximize the lattice energy. However, the fact that two radii are involved in each structure prevents a unique solution. For example, in the CsCl structure we assume that unlike ions touch along the body diagonal of the cube. Thus,

$$bd = 2r_{Cl^-} + 2r_{Cs^+}$$

where r_{Cl^-} and r_{Cs^+} are the ionic radii of Cl^- and Cs^+. We also know that

$$bd = a\sqrt{3}$$

where a is the length of the unit cell edge. Thus

$$2r_{Cl^-} + 2r_{Cs^+} = a\sqrt{3}$$

and

$$r_{Cl^-} + r_{Cs^+} = \frac{a\sqrt{3}}{2}$$

If we know the value of a, we can calculate $r_{Cl^-} + r_{Cs^+}$ but we cannot calculate a unique value for either one. To compile tables of ionic radii, we require information in addition to unit cell lengths in order to fix the lengths of key ions. Only then can a set of ionic radii (consistent with each other) be constructed from crystal data.

EXAMPLE 19–4

CdO has the NaCl structure and a unit cell edge length of 4.689 Å. Calculate the ionic radius of Cd^{2+} using the crystal data and ionic radii given in Table 19–3.

Solution   Examination of the NaCl structure (Figure 19–16) shows that unlike ions touch along a unit cell edge, so that for CdO

$$2r_{O^{2-}} + 2r_{Cd^{2+}} = a = 4.689 \text{ Å}$$

and

$$r_{O^{2-}} + r_{Cd^{2+}} = \frac{4.689 \text{ Å}}{2} = 2.345 \text{ Å}$$

Table 19–2 gives $r_{O^{2-}}$ as 1.24 Å. Therefore,

$$r_{Cd^{2+}} = 2.345 \text{ Å} - 1.24 \text{ Å} = 1.10 \text{ Å}$$

EXERCISE 19–4

FeO has an NaCl structure and a unit cell edge length of 4.294 Å. Use the value of the ionic radius given of O^{2-} given in Table 19–3 to calculate the ionic radius of Fe^{2+}.
Answer: 0.91 Å. ■

19.7   COVALENT NETWORK CRYSTALS

Diamond and Graphite Structures

In the previous section we mentioned the suitability of the four-coordinate zinc blende structure for salts that have some covalent character. Diamond (Figure 19–23) is an example of a purely covalent crystal that adopts the same structure as zinc blende, except that every location of a Zn^{2+} or S^{2-} ion now contains a carbon atom. Each carbon atom is covalently bonded by a shared pair of electrons to four neighboring carbon atoms. This produces a network extended by a huge number of unit cells in each direction, forming a covalent network crystal.

The structure of diamond, which involves strong bonds between the carbon atoms and a rigidly linked three-dimensional network, makes it resistant to deformation and therefore extremely hard. Diamond is the hardest substance known, and it finds enormous industrial use for cutting

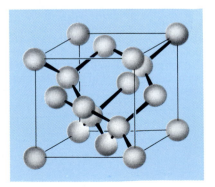

FIGURE 19–23   One unit cell of the diamond crystal lattice.

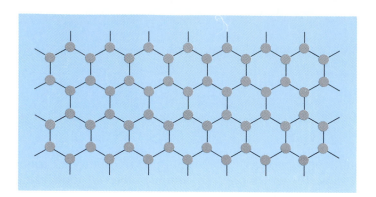

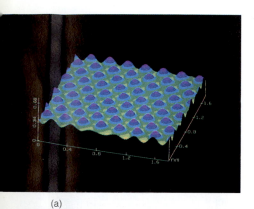

(a)

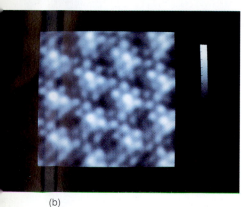

(b)

FIGURE 19–25   (a) Three-dimensional surface plot of graphite. Only alternate carbon atoms are observed due to the asymmetry in the electronic properties of these two structurally distinct sites.
(b) The hexagonal closest-packed sulfur plane in the layered material TaS_2. Bright clusters of sulfur atoms are observed in a regular pattern.

other materials. The diamond structure is also found in germanium, silicon, and the gray form of tin. However, because the bonds between the atoms in these elements are weaker than those between carbon atoms in diamond, they are all softer.

A number of other elements and compounds form covalent network crystals, including a variety of important minerals containing silicon. A few of these generally complex structures will be considered later.

Graphite, the other form of elemental carbon, displays a very different covalent network crystal with very different properties. This structure (Figure 19–24) is made up of flat layers of carbon atoms in a network of six-membered rings like chicken wire. The graphite crystal consists of these carbon layers stacked parallel to each other at a fixed interplanar distance of 3.35 Å. The bonding within the planes is very strong, whereas the bonding between planes is very weak. Thus, the planes slide easily over each other, giving graphite its slippery, greasy feel. This structure makes graphite useful as a lubricant and in pencils, where layers of graphite slide off the surface and adhere to the paper. Because of its very different structure in different directions, graphite is strongly anisotropic, being a conductor in the direction of the planes and an insulator perpendicular to them. Molybdenum sulfide also has a layered structure; it has been used as a lubricant additive in motor oil, especially because it is a solid and shows no appreciable change of viscosity with temperature. Scanning tunneling micrographs of graphite and tantalum disulfide, which also has a layered structure, are shown in Figure 19–25.

Diamond and Synthetic Diamonds

Applications of solid materials can be understood and extended through knowledge of their structures. For example, diamond is the supreme example of an abrasive, being harder than any other material. Natural and synthetic diamonds are widely used for grinding, cutting, and boring metals, rock, and other substances. Diamond is the most wear-resistant material for the tip of the stylus of a record player.

Natural diamond is the high-pressure allotrope of elemental carbon. Although it is the hardest substance known, it is also quite brittle and can be readily crushed into a powdery grit, in which form it is widely used in industry as an abrasive. Small single crystals are used for turning and boring tools in oil-well drilling and mining operations, dies for drawing wire, dressing tools for grinding wheels, and single point cutting tools. Diamonds of near flawless quality have been prized as gem stones for thousands of years.

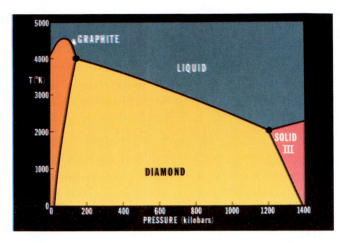

The phase diagram for carbon. Note that graphite, not diamond, is the standard state form of the element.

In 1955, scientists at General Electric prepared the first synthetic diamonds, completing a task that had stimulated scientific thought and engineering practice since the last years of the 18th century, when it first became known that diamonds were indeed elemental carbon (Figure 19–26). In the General Electric process, a charge of graphite and a catalyst metal are heated to melt temperatures at a pressure high enough for diamond to be stable. The graphite dissolves in the metal, and diamond is produced from it. Iron was the first molten metal used, but many are now known to serve as effective catalysts: Cr, Mn, Co, and Ni, for example. But this is equally an engineering problem, requiring a specially designed "belt" apparatus to produce the required temperature and pressure, in the range of 1400 to 2400°C and 55,000 to 130,000 atmospheres.

The General Electric scientists described those first small dark synthetic diamond crystals by well-known characteristics of natural diamond: (1) Readily scratched silicon carbide, which is, next to diamond, the hardest abrasive in common use. It scratches tungsten carbide, too. (2) Burned cleanly in pure oxygen without leaving a residue. (3) Had a density of 3.52 g/cm^3 at 25°C.

In 1970, General Electric announced the creation of carat-sized gemstone-quality diamonds. The process starts with tiny seed crystals—usually synthetic diamonds no bigger than the period at the end of this sentence. This seed crystal, along with a metal catalyst and a powder charge of synthetic diamond, is subjected to high pressures and temperatures in a special press. At the high temperatures of the molten metal, the diamond powder dissolves, but the end of the tube containing the seed crystal is kept cool enough so the seed does not dissolve. By carefully controlling the pressures and temperatures, the carbon atoms from the diamond powder can be made to migrate through the molten metal catalyst, finally redepositing themselves upon the seed crystal, and eventually building to a large diamond.

Scientists at Pennsylvania State University, in a 1986 report, described a way of coating objects with thin films of synthetic diamond, which promises to revolutionize major segments of electronics, optics, machine tools, chemical processing, and military technology. Diamond film is likely to become a commonplace material, and although such materials may have the appearance of a lacquer coating, they promise to radically improve the quality of the objects they coat.

Most recently (August, 1988), General Electric scientists have been able to photograph polycrystalline diamond balls grown at low pressures, at which it is the less stable form of carbon.

(a)

(b)

FIGURE 19–26 (a) This very large General Electric "belt" apparatus for making synthetic diamonds is designed to produce temperatures in the range of 2000 K and pressures on the order of 10^5 atmospheres. (b) A synthetic diamond, still in the jaws of a cell just removed from the apparatus.

The structure and hardness of diamond can be mimicked by a form of boron nitride (BN) that has the zinc blende structure. The bonds in this structure are principally covalent and about as strong as those in diamond. This gives a covalent network structure with hardness comparable to diamond.

Abrasives less expensive than diamond or boron nitride are desirable. One particular success has been silicon carbide, SiC, most commonly known by the trade name "Carborundum." Silicon carbide crystallizes in a number of modifications. Among these are zinc blende and several closely related structures. The bonds are somewhat weaker than those in diamond, so silicon carbide is not quite as hard. It is, however, inexpensively made from sand and coal.

19.8 NONCONDUCTORS AND SEMICONDUCTORS

A solid such as diamond, composed of nonmetallic carbon atoms, has properties very different from those of a metallic substance. For example, diamond is a very poor conductor of electricity and heat. It is not opaque. It cannot be seriously stressed and still maintain its basic structure.

In the diamond structure, each carbon atom has four hybrid sp^3 orbitals pointing toward the corners of a tetrahedron. There are also four valence electrons on each carbon atom with the electron configuration $2s^2 2p^2$. A large assemblage of n carbon atoms in a diamond structure has a total of $4n$ hybrid orbitals, of which $2n$ (or one-half) are bonding and $2n$ are antibonding. There are $4n$ valence electrons, and these just fit in the $2n$ bonding molecular orbitals, two to an orbital. Thus, the bonding orbitals are completely full and the antibonding orbitals are empty.

Because the ionization energy of carbon is high, bonding results in considerable stabilization and very strong bonds. These strong bonds in the rigid, three-dimensionally linked structure result in diamond's extreme hardness. It also results in a considerable energy difference, or band gap, between the two sets of orbitals (Figure 19–27a).

The valence electrons in the diamond structure are all held tightly in bonding orbitals with no opportunity for movement from one atom to another. In order for this substance to conduct electricity, an electron in the valence band (or bonding orbital band) would have to jump up to the conduction band (or antibonding orbital band) where it would be free to move. At higher temperatures the number of electrons excited to the point of making this jump increases. However, because of the large band gap in diamond, the temperature would have to be extremely high to excite a significant number of electrons, so diamond is effectively an **insulator**. If the band gap were narrower, conduction of electricity would be a more likely possibility at reasonable temperatures. This is observed with silicon ($3s^2 3p^2$) and germanium ($4s^2 4p^2$), for which the ionization energy is not so high. Substances such as these have the diamond structure but smaller band gaps (Figure 19–27b). Electrons can jump to the conduction band and move through the structure.

Once an electron has jumped, electrons are free to move in the valence band as well. The electron jumping out of the valence band leaves behind a hole—a missing electron location—and as an electron moves out of this hole, the hole appears to be moving. Thus, conduction is possible as soon as an electron leaves the valence band. At higher temperatures, more electrons make the jump; as a consequence, substances such as silicon and germanium are better conductors at higher temperatures. In contrast, met-

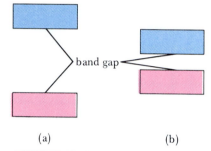

(a)   (b)

FIGURE 19–27 Valence and conduction bands for (a) an insulator and (b) a semiconductor.

als become poorer conductors at higher temperatures because the increased motion of the metal atoms inhibits the movement of the electrons. Pure substances that exhibit conductive behavior like that of silicon and germanium are called **intrinsic semiconductors**.

As a practical matter, the conductivities of pure silicon and germanium are too low at room temperature to be of much use. In order to increase the number of electrons (or the number of holes) that are free to move, the pure substances are carefully **doped** with impurities. Silicon can be doped with phosphorus atoms, for which the valence electron configuration is $3s^2 3p^3$. Each phosphorus atom occupies a location in the structure equivalent to a silicon atom. However, each phosphorus atom brings along an extra electron that occupies an energy level just below the conduction band. The electron easily jumps into the conduction band and moves about the crystal. Because the current is carried by the negatively charged electrons, phosphorus-doped silicon is called an *n*-type semiconductor.

On the other hand, when silicon is doped with aluminum, for which the valence electron configuration is $3s^2 3p^1$, the foreign atoms enter the silicon structure and there is now one electron less per aluminum atom. Holes have been introduced into the valence band. These holes are free to move in the direction opposite the direction of electron flow. Thus, the holes can be regarded as positively charged current carriers, and the result is a *p*-type semiconductor.

The transistor industry, which revolutionized modern electronics, depends on semiconductors of these types. The simplest application of such semiconductors is in current rectification, that is, allowing current flow in only one direction in a circuit. In Figure 19–28 an *n*-type and a *p*-type semiconductor are in contact. This produces a *p-n* transistor, each side of which can be connected to the terminal of a battery. If the negative terminal of the battery is connected to the *p*-type semiconductor and the positive terminal is connected to the *n*-type semiconductor, two things happen. Electrons from the battery fill all of the holes in the *p*-type semiconductor, and the donor electrons in the *n*-type semiconductor all drain away to the positive terminal of the battery. Thus, each side of the transistor becomes an insulator and current flow ceases. If the battery is connected in the opposite manner, however, electrons from the negative terminal of the battery can flow into the *n*-type semiconductor. The excess electrons on this side of the transistor can flow across the *p-n* junction into the holes in the *p*-type semiconductor. These can flow into the positive terminal of the battery, and a continuous current flows. If the rectifier is connected to an alternating current source, current flow in one direction will be stopped, giving a half-waved rectified (direct) current.

Another use of the *p-n* junction is in solar cells (Figure 19–29). A thin *n*-layer is placed on top of a *p*-layer by the controlled diffusion of phosphorus or another element that supplies electrons. Electrical contact strips are connected to each layer. Sunlight shining on the *n*-layer causes some electrons to jump to the conduction band, leaving holes in the valence band

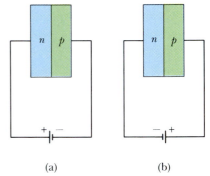

FIGURE 19–28 A rectifier constructed of *n*- and *p*-type semiconductors. In (a) the current is blocked, whereas in (b) the current can flow through the rectifier.

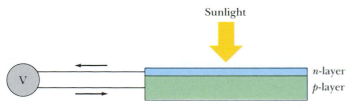

FIGURE 19–29 A solar cell employing a *p-n* junction. Sunlight striking the *n*-layer causes a flow of electrons as indicated through the voltmeter V.

FIGURE 19–30 The unusual character of these "chips" is the use of high-performance gallium arsenide integrated circuits. GaAs has the advantage over silicon in that it operates much faster, consumes less power, and can withstand radiation better.

in that layer. The electrons in the conduction band can flow off into the conductor, leaving the n-layer with a positive charge. This is neutralized by the holes in the n-layer flowing across the junction and through the p-layer, where they meet up with electrons coming through the outside circuit. The overall result is a flow of current through the circuit.

Very pure silicon is needed as a starting material for doping, since the number of conductor electrons (or holes) must be very carefully controlled. Methods for producing ultrapure silicon were discussed in Chapter 17. Purities up to 99.9999999%—the purest substances ever produced—have been realized. Single crystals of pure silicon weighing several hundred grams are pulled from the melted substance. Germanium with the diamond structure is also widely used in semiconductor devices. Substances with the same structure and number of valence electrons as silicon or germanium can be prepared from equal amounts of an element with one fewer valence electron and an element with one more valence electron. These compounds are known as the III-V compounds because they consist of an element from Group IIIA and an element from Group VA. Examples are aluminum phosphide and gallium arsenide.

Gallium arsenide semiconductors (Figure 19–30) have very fast responses and are revolutionizing the design of semiconductor devices. Integrated circuits based on gallium arsenide have already achieved operating speeds up to five times that of the fastest silicon chips. Solar cells based on the photovoltaic properties of alternating thin films of GaAs and GaAlAs have achieved a surprising milestone (in 1988) of 30% efficiency—at Sandia National Laboratory in New Mexico, scientists have demonstrated a solar cell that converts more than 30% of the light that strikes it into electricity.

19.9 MOLECULAR CRYSTALS

None of the structures that we have discussed so far involve compounds consisting of discrete molecules. Compounds that are made of molecules can also form crystals, although generally at much lower temperatures than

FIGURE 19–31 The regular, hexagonal arrangement of the water molecules in ice crystals. Molecules are held in place within the crystal structure mainly by hydrogen bonds and weaker dipole-dipole interactions and dispersion forces.

metals, salts, and covalent network structures. In ice, for example, water molecules are arranged in a regular repeating pattern (Figure 19–31). The forces between the molecules that hold them in the crystal structure include hydrogen bonds, other dipole-dipole interactions, and dispersion forces. In general, stronger intermolecular forces lead to higher melting points. However, these intermolecular forces are much weaker than the ionic attractions of salts, or covalent bonds in network structures. Therefore, molecular crystals tend to be broken up more easily by increasing temperature and have relatively low melting points. In addition, they are relatively soft and often do not have clearly defined cleavage planes.

19.10 DEFECTS IN CRYSTALS

The most noticeable aspect of crystals is the regularity and perfection in their design. But crystals do have their imperfections or defects. We must remember that the number of unit cells in a crystal is huge. Thus, if only a small percentage have defects there still will be a considerable number of defects in the crystal. For example, a cubic crystal of NaCl that is 0.1 mm on a side contains about 5×10^{18} unit cells. Even if there are defects in only 0.01% of these unit cells, there are still 5×10^{14} defects in the crystal.

There are several kinds of crystal defects. **Point defects** involve only one or a very few lattice sites. **Line defects** involve dislocation of rows of lattice sites. **Plane defects** deal with dislocations of entire planes of atoms or ions. We will consider only point defects.

The simplest point defect in a crystal is a vacancy where an atom or ion is missing. If the missing species is an ion, electrical neutrality and the proper stoichiometry can be maintained by also having a nearby ion or ions of opposite charge missing. Such paired defects are called **Schottky defects** (Figure 19–32a). For example, an NaCl crystal with a Schottky defect would be missing one Na^+ and one Cl^-, whereas a $CaCl_2$ crystal with such a defect would be missing one Ca^{2+} ion and two Cl^- ions.

If an atom or ion is out of its usual position, occupying a normally vacant hole, we have a **Frenkel defect** (Figure 19–32b). Frenkel defects are most common when the smaller ion, normally the cation, is much smaller than the anion. Thus, Ag^+ is a very small cation, and Frenkel defects in the silver halides AgCl, AgBr, and AgI are so common that Ag^+ ions can diffuse quite freely through the crystal. Both Frenkel and Schottky defects preserve the regular stoichiometry of the crystal compound.

Some point defects do not maintain the regular stoichiometry of the

Cholesteric liquid crystals, so named because their characteristic helical structure was first observed among esters of cholesterol, are useful for thermometer displays and thermal mapping. When the length of a single helical twist (in the phase) is comparable to the wavelengths in the visible light spectrum, iridescent colors are observed. Temperature differences cause the length to vary, changing the colors. The cholesteric mixture painted on the plate of the iron responded between 30°–35°C. Note the uneven heating indicated by the range of colors.

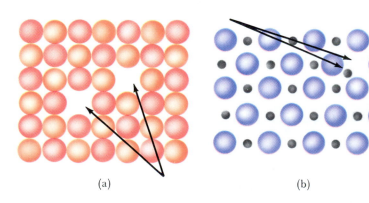

(a)   (b)

FIGURE 19–32 Point defects in crystals: (a) Schottke defect; (b) Frenkel defect.

compound. In these cases an impurity of some type either is substituted for a regular constituent atom or ion or occupies a vacant hole in the crystal lattice. Substitutions involving ions of the same charge and similar size are quite common. For example, Ba^{2+} and Sr^{2+} have similar ionic radii. Therefore, precipitation of $BaSO_4$ in the presence of Sr^{2+} ions will result in some Sr^{2+} ions occupying Ba^{2+} sites in the $BaSO_4$ lattice.

Sometimes an electron will be substituted for an anion in a crystal lattice. This situation can be caused by irradiating a salt such as NaCl with ultraviolet light. The energy of the ultraviolet light can remove an electron from a Cl^- ion, forming a Cl atom.

$$Cl^- \longrightarrow e^- + Cl$$

The uncharged (and much smaller) Cl atom is no longer held in the crystal structure, so it migrates to the surface and is lost. The electron with its negative charge is held in a vacant anion site by the neighboring positive ions. The salt is now nonstoichiometric, with more Na^+ ions than Cl^- ions. However, electrical neutrality is maintained (as it must be) because an electron replaces each Cl^- ion that was removed. The presence of the trapped electron causes a blue color in the NaCl crystal, and the defect is called an F-center (from *farbe*, the German word for color).

Knowledge of defects in metal structures allows us to add small amounts of impurities in order to modify properties. For example, metals are weakened by the presence of line defects in their crystal structures, which allow the structure to bend too easily or break. Metal impurities that have smaller atoms than the principal metal tend to occupy sites at the line defects and help hold the structure together. Thus 10% zinc increases the strength of copper by 30%.

19.11 NONSTOICHIOMETRIC COMPOUNDS

Some compounds commonly have nonstoichiometric ratios of their elements as a result of crystal defects. For example, the compound written as FeO actually varies in composition from $Fe_{0.95}O_{1.00}$ to $Fe_{1.00}O_{1.00}$. The lattice in this case is the NaCl structure, but some Fe^{2+} sites are normally vacant. In order to achieve the required electrical neutrality some Fe^{3+} ions are substituted for Fe^{2+} ions.

EXAMPLE 19-5

A sample is found to have the formula $Fe_{0.97}O_{1.00}$. What fraction of the metal ions are Fe^{3+}?

Solution   We can consider our sample to be $(Fe^{2+})_x(Fe^{3+})_yO$ and set up two simultaneous equations. According to the given formula

$$x + y = 0.97$$

Since the O ion has a charge of -2 and because the total charge on Fe must be $+2$,

$$2x + 3y = 2$$

Subtracting twice the first equation from the second yields

$$y = 0.06$$
$$x = 0.91$$

The formula of the compound is therefore $(Fe^{2+})_{0.91}(Fe^{3+})_{0.06}O$, and the fractions of Fe^{2+} and Fe^{3+} ions are

$$\frac{0.91}{0.97} = 0.94 \ Fe^{2+}$$

$$\frac{0.06}{0.97} = 0.06 \ Fe^{3+}$$

■

EXERCISE 19–5

Another "FeO" sample is found to have the actual formula $Fe_{0.95}O_{1.00}$. What fraction of the metal ions are Fe^{2+}?
Answer: 0.89. ■

NiO also has the NaCl structure, and its composition can vary from $Ni_{0.97}O_{1.0}$ to $Ni_{1.0}O_{1.0}$. The nonstoichiometric form with the low nickel content and some missing Ni^{2+} ions can be prepared in an excess of oxygen. In this case Ni^{3+} ions replace some Ni^{2+} ions to preserve the electrical neutrality. Transfer of an electron converts Ni^{3+} to Ni^{2+},

$$Ni^{3+} + e^- \longrightarrow Ni^{2+}$$

Thus in nonstoichiometric "NiO" electrons can jump from Ni^{2+} to Ni^{3+} through the structure, making it a modest conductor (actually a semiconductor), whereas the stoichiometric compound has no ready mechanism for electron transfer and is an insulator.

19.12 LIQUID CRYSTALS

Order of a crystalline nature is possible in liquids under certain circumstances. Such stituations are called **liquid crystals**, and they depend on dissolved molecules that have a rigid, rod-like structure At higher temperatures these rods will be oriented in a random fashion giving a typical disordered liquid (Figure 19–33a). However, as the solution is cooled, some ordering of the random structure is possible. In these ordered solutions

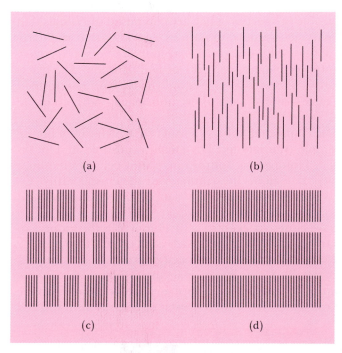

(a)   (b)   (c)   (d)

FIGURE 19–33 Four degrees of orientation of rod-shaped molecules. (a) Disordered, or only short-range order, of a solution; (b) nematic phase of a liquid crystal; (c) smectic phase of a liquid crystal; and (d) full order of a molecular crystal.

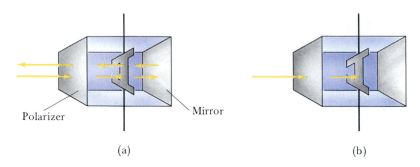

the molecules will show preferred orientations, which will predominate over other orientations. Thus changes are possible to several different ordered phases. The simplest of these is the **nematic** phase (Figure 19–33b). In this phase the molecules point in a preferred direction, but the centers of the molecules are distributed at random. **Smectic** phases, of which one is shown in Figure 19–33c, are more ordered than nematic ones because the molecules not only have the same preferred direction, but their centers are located on planes that run through the structure. Finally, the ordering of a molecular crystal is shown in Figure 19–33d. Other types of liquid crystals are also possible, including phases with planes of molecules oriented in different directions.

The orientation of the molecules in a liquid crystal is very sensitive to electric and magnetic fields. This is useful in liquid crystal devices, particularly liquid crystal displays. Figure 19–34 shows the schematic of a liquid crystal display as used in a digital watch or a calculator. A slight change in the electric field in a part of the display causes that part, such as the shape of a letter or a number, to be darkened.

SUMMARY

Crystalline solids are identified by their characteristic shapes, their symmetry, and their anisotropic properties. They are composed of crystals that consist of a lattice of particles—atoms, molecules, or ions—that regularly repeat themselves in a space-filling pattern. Crystals are very effectively studied by crystallographic techniques based on x-ray diffraction. The glassy state of a substance is best understood as a liquid that is frozen—trapped in an unstable energy state—and subject to extraordinarily slow, creeping change over time.

Lattice points can be connected to form a unit cell, which can generate the entire crystal by simple translations. The smallest possible unit cell is called the primitive cell. However, a nonprimitive unit cell is sometimes useful in order to incorporate all of the symmetry of the crystal lattice. Of particular interest to us are the cubic unit cells: simple, body-centered, and face-centered.

Metals normally have closest-packed structures in which the atoms can be approximated as hard spheres in contact with each other. Two regular closest-packed structures are possible, cubic closest packing (CCP) and hexagonal closest packing (HCP). Cubic closest packing is identical with the face-centered cubic lattice having an atom at every lattice point. Atomic radii of metals can be calculated from their crystal structures and unit cell dimensions.

Salts have structures in which the lattice sites are occupied by the positive and negative ions. Electrostatic interactions can be seen as the principal bonding forces, and lattice energies can be calculated. Various crystal structures are possible for salts having the same general formulas. The actual structure adopted depends on considerations of ion size and coordination number.

Crystals containing networks of atoms connected by covalent bonds are called covalent network crystals. Diamond and graphite, the two crystalline forms of carbon, exemplify two very different covalent network structures with very different properties.

Molecules also form crystals; these usually have relatively low melting points as a result of their being held together by relatively weak intermolecular interactions.

All crystals can be expected to have some defects. Defects in crystals account for the formation of nonstoichiometric compounds.

QUESTIONS AND PROBLEMS

QUESTIONS

1. How can you most easily distinguish between a glassy solid and a crystalline solid?

2. Can you determine the atomic weight of an unknown metal if you know its density, the dimensions of its unit cell, and the number of atoms in the unit cell? Why or why not?

3. The more negative the value of the lattice energy, the more stable the crystal. Arrange the following salts in order of stability, from least to most: NaCl, NaF, NaBr, NaI.

4. The stability of a crystal is reflected in the magnitude of its melting point. More stable crystals require more heat energy to break the ionic attractions holding the lattice together. Arrange the four salts in Question 3 in order of increasing melting point.

5. Look up the melting points of water, ethyl alcohol, diethyl ether, and methane. What can you say about the intermolecular forces between these molecules?

6. How does the bonding theory of metals explain why most metals can be bent without breaking?

7. What kind of semiconductor results if germanium is doped with gallium? If gallium arsenide is doped with phosphorus? If aluminum phosphide is synthesized from a slight excess of aluminum?

8. Explain why a semiconductor device consisting of a sandwich of two n-type semiconductors on either side of a p-type semiconductor would pass electric current in neither direction.

9. Briefly distinguish between the following pairs of terms:
(a) isotropic and anisotropic solids
(b) hexagonal close packing and cubic close packing of spheres
(c) crystal lattice and unit cell

10. Explain the differences between the properties of diamond and graphite on the basis of their fundamentally different space lattices.

11. Using the apparent properties of solid carbon dioxide and table salt, justify the contention that the former is a molecular crystal and the latter is an ionic crystal.

12. Explain why you would expect ionizing radiation to damage a covalent layer as a cell membrane but have no effect on a metallic film or layer.

13. Explain the basis for the similarities and differences between metallic and ionic crystals.

14. Comment on each of the following:
(a) A p-type semiconductor can be formed by substituting an aluminum atom for a silicon atom.
(b) An n-type semiconductor can be formed by substituting an arsenic atom for a silicon atom.

15. Gallium arsenide is referred to as a III-V semiconductor. Briefly explain.

16. Describe the effect of "doping" silicon crystals with impurities of phosphorus; with impurities of aluminum.

17. Offer a structural explanation for the differences between conductors, semiconductors, and insulators.

18. (a) Show how sodium chloride can be generated by filling octahedral holes in a face-centered cube. (An octahedron is a solid figure with eight equivalent faces).
(b) Explain why the unit cell is not simply a cube of four sodium ions and four chlorine ions in an alternating arrangement.

19. Why do computers "crash" if the temperature gets too high? To put it another way, why do semiconductors lose their effectiveness with increasing temperature?

PROBLEMS

Lattices   [1–2]

1. Determine the number of lattice points per unit cell in each of the following lattices.

(a) primitive tetragonal
(b) body-centered tetragonal
(c) end-centered orthorhombic

2. How many lattice points are there in one unit cell of each of the following lattices?
(a) face-centered orthorhombic
(b) primitive triclinic
(c) body-centered cubic

Crystal Density and Atomic Radii   [3–14]

3. Use the following data to calculate the value of the Avogadro number N_A. Tungsten crystallizes in a body-centered cubic structure with a lattice constant (unit cell edge) $a = 3.1583$ Å. The density of tungsten metal is 19.3 g/cm³ and its atomic weight is 183.85 g/mol.

4. Calculate Avogadro's number from the following data on the metal iridium. Iridium has a face-centered cubic structure with a unit cell edge of 3.823 Å. Its density is 22.42 g/cm³.

5. Silver crystallizes in an FCC lattice with lattice constant $a = 4.0776$ Å. The density is experimentally determined to be 10.5 g/cm³. Calculate the atomic weight of silver.

6. Vanadium has a body-centered cubic unit cell with a unit cell edge of 3.011 Å. Its density is 5.96 g/cm³. Calculate its atomic weight.

7. If copper crystallizes into a face-centered cube with a lattice constant of 3.61 Å, show that the calculated density is in agreement with a measured value of 8.92 g/cm³.

8. Chromium has a body-centered cubic arrangement for its unit cell with a length along an edge of 2.89 Å and a measured density of 7.0 g/cm³. What would be the calculated density if the unit cell were mistakenly assumed to be a face-centered cube?

9. Calcium has a face-centered cubic lattice with a lattice constant $a = 5.56$ Å. Determine the atomic radius of calcium.

10. Iron has a body-centered cubic lattice and a density of 7.86 g/cm³. Calculate the atomic radius of iron.

11. Copper has a face-centered cubic structure with a density of 8.92 g/cm³. Calculate its atomic radius.

12. Calculate the atomic radius of niobium from its atomic weight, its density of 8.55 g/cm³, and the fact that it crystallizes in a body-centered cubic structure.

13. Calculate the fraction of empty space in an FCC structure of a metal.

14. Calculate the fraction of free space in a body-centered cubic structure of a metal. Compare your results with those of Problem 13. Tell why your results seem reasonable or not.

Lattice Energy   [15–16]

15. The Madelung constant for CsCl is 1.763. Calculate the lattice energy for this salt, assuming the ions to be point charges separated by the interionic distance of 3.55 Å.

16. Calculate the lattice energy for CaF_2, for which the Madelung constant is 2.519 and the interionic distance is 2.36 Å. Assume the ions to be point charges.

Ionic Radii and Radius Ratios   [17–22]

17. Thallium(I) bromide has the CsCl structure with a lattice constant $a = 3.97$ Å. The ionic radius of Br^- is 1.82 Å. Calculate the ionic radius of Tl^+.

18. Iron(II) oxide has the NaCl structure with a unit cell length of 4.294 Å. The oxide ion has a radius of 1.24 Å. Calculate the ionic radius of Fe^{2+} in iron(II) oxide.

19. Calculate the minimum radius ratio of the smaller ion to the larger ion in the NaCl structure.

20. Calculate the minimum ratio of the radius of the smaller ion to the radius of the larger ion in the zinc blende (ZnS) structure.

21. The formula of sodium chloride is 58.443 g/mol. A sample of the pure salt has a density of 2.167 g/cm³. Using these data and Table 19–3, calculate Avogadro's number.

22. Cesium bromide has the CsCl structure. Use the ionic radii in Table 19–3 to calculate the closest distance between unlike ions and then use this to calculate the density of CsBr.

Nonstoichiometric Compounds   [23–24]

23. A sample was analyzed and found to have the formula $Ni_{0.98}O_{1.00}$. What fraction of the nickel ions are Ni^{2+} and what fraction are Ni^{3+}?

24. A sample of "FeO" contained 96.10% of the total Fe as Fe^{2+} and 3.90% of the total Fe as Fe^{3+}. Determine the actual formula of the sample.

Additional Problems   [25–31]

25. Silicon has the diamond structure. How many silicon atoms are there in one unit cell?

26. Tantalum metal has a body-centered cubic structure with a unit cell edge of 3.281 Å. Its atomic weight is 180.9. Calculate its density.

27. Crystalline lithium chloride is known to have the sodium chloride structure, a density of 2.07 g/cm³, and a Madelung constant of 1.75.
(a) Calculate the interionic distance in the crystal lattice.
(b) Estimate the lattice energy as the potential energy of attraction for the ions.
(c) Would you expect the measured lattice energy to be larger, smaller, or the same as the estimated value?

28. Magnesium oxide has the NaCl structure. Use the ionic radii in Table 19–3 to calculate the length of the edge of the unit cell of MgO.

29. Calculate the fraction of nickel ions that are Ni^{2+} in a sample of "nickel(II) oxide" having the formula $Ni_{0.97}O_{1.00}$.

30. Determine the average bond distance between K^+ and Cl^- ions in crystalline potassium chloride, assuming a density of 1.984 g/cm³ and the NaCl structure.

31. Given the interionic distance between Li^+ and Cl^- ions in LiCl to be 0.257 nm, determine the density of lithium chloride. Assume LiCl has the NaCl structure.

MULTIPLE PRINCIPLES   [32–35]

32. A compound of mercury and chlorine has a crystal structure with two formula weights per unit cell. The unit cell edges are 4.47, 4.47, and 10.89 Å and the unit cell angles are all 90°. The density of the substance is 7.15 g/cm³. What is the formula of the compound?

33. An unknown substance that is a gas at room temperature can be condensed to a solid at −80°C. X-ray analysis of its crystals can then be performed. As a solid, it is found to have a cubic unit cell, 5.15 Å on each side, containing four molecules. The density of the solid is 0.73 g/cm³. What is the density of the substance as a gas at 25°C and a pressure of 1.00 atm?

34. A compound of cobalt and sulfur has a cubic structure with four formula weights per unit cell. Its density is 4.269 g/cm³. What mass of sulfur would be required to produce a 1.50 kg of this compound?

35. The structure of graphite is described in Section 19.7. The parallel planes of carbon atoms are 3.35 Å apart. Each plane is composed of six-membered rings of carbon atoms connected together (Figure 19–24). The density of graphite is 2.25 g/cm³. Use these data to calculate the carbon-carbon distance within the carbon planes. What does this result tell you about the bonding within the carbon planes?

APPLIED PRINCIPLES [36–38]

36. Semiconducting, nonstoichiometric nickel(II) oxide can be prepared by reaction of the stoichiometric compound with oxygen gas. What volume of $O_2(g)$ at 25°C and 2.50 atm would be consumed in converting 1.00 kg of NiO to the appropriate amount of $Ni_{0.97}O_{1.00}$? What mass of the semiconductor would be produced?

37. Preparation of silicon for semiconductor uses requires the purification of the silicon up to 99.9999999% pure. At this level of purity, approximately how many silicon atoms per gram have been replaced by some other element?

38. What is the relative cost advantage to plate a surface to a given thickness with palladium rather than gold? Use the prices noted in the discussion and the density values at 25°C: Au = 19.3 g/cm³; Pd = 12.0 g/cm³.

CHAPTER
20

Materials Science

PLASTICS MAKE A MAGIC LENS

Designed as a living laboratory, the 3000-square-foot concept house known as "Living Environments" in Pittsfield, Massachusetts, is an imaginative introduction to home design, manufacturing and construction, based on high-performance materials and processes for the 1990s. Fifty years ago, innovative use of heat-conducting plastic materials solved the World War II aviator's problem of fogged goggle lenses and precipitated the rear window defogger found in most automobiles on the road today.

"We have labeled civilizations by the main materials which they have used: The Stone Age, the Bronze Age and the Iron Age. . . . a civilization is both developed and limited by the materials at its disposal. Today, Man lives on the boundary between the Iron Age and a New Materials Age."— Sir George Paget Thomson, Nobel Laureate in Physics (1937)

20.1 INTRODUCTION TO MATERIALS SCIENCE

The development and selection of appropriate materials for enormously different applications is one of the triumphs of modern science and engineering technology, and it continues to provide major challenges. Discoveries and inventions include tonnage steel, magnesium, aluminum, and their alloys; synthetic textile fibers such as nylon and polyester; technology for manufacturing paper and such plastic materials as cellophane, Styrofoam, rubber O-rings and RTV bathroom sealants and caulks; pressure-sensitive tape and those ubiquitous yellow stick-on notes; transistors and microprocessor chips, and biological and electrochemical membrane materials. In the automobile, everything from the tires and bumpers to components under the hood and inside the passenger compartment is a testimony to materials science (Figure 20–1). These are towering achievements in this century!

As a discipline, materials science brings together knowledge gained from physics, from a variety of areas of theoretical and practical chemistry, and from important areas of engineering practice including electrical, thermal, mechanical, chemical, and fabrication and processing technology. Thus, materials scientists need to be knowledgeable in many related fields; the combinations have produced interesting and often astonishing results. This is the age of stress and strain. We operate more often than not near the limits of high and low temperature, of elasticity and rigidity, where we need to be concerned with flow, fatigue, and failure. When accidents happen, were they caused by materials failure? Materials science is a fascinating world. We begin discussion of this subject by surveying the range of useful materials.

20.2 TYPES OF MATERIALS

Planet Earth has been blessed with a bounty of raw materials. For most of our history, these materials have been used as they were found: stone implements, wood and forest products, a rare mineral or two, air and water. Civilization's advance often coincided with the introduction of new uses of materials and the discovery of new materials. The winning, refining, alloying, and fabricating of metals is most notable. In particular, the met-

FIGURE 20–1 Automobiles are designed with a range of plastic body parts—bumpers, wheels and guards, quarter panels, hood and roof, and tail assemblies. The BMW Z-1 is an all-plastic automobile.

allurgy of iron, copper, tin, and zinc and the discovery of the first alloys—bronze and brass—allowed the production of the first "advanced" weapons and other material objects. However, it is only within the last century and a half that people began to synthesize, formulate, and fabricate "secondary materials," based on Earth's primary resources, in ways we recognize today as modern.

Although there seem to be a nearly infinite number of useful materials for engineering purposes, they can be grouped into five major categories that include both natural and synthetic materials:

1. Metals and alloys
2. Ceramics and glasses
3. Polymers and plastic materials
4. Composite materials
5. Semiconductor materials

Metals and **alloys** (Figure 20–2) find their uses in the elemental form, that is, as structures composed of atoms, not molecules. Of all metals, the one that comes to mind first is structural steel, an alloy containing iron as its principal component. It can readily be fabricated into useful shapes and has outstanding qualities of strength combined with flexibility. It can be formed into girders strong enough to support modern skyscrapers, but is capable of deforming when subjected to sudden forces. Thus, in an earthquake the building can give a little, instead of crumbling as it would if it were constructed of a brittle material that was unable to bend. Steel shares the qualities generally found in other metals. It is lustrous, a good conductor of heat and electricity, and (unfortunately) subject to corrosion by acids and other oxidizing compounds that can convert its atoms to ions and diminish the favorable structural properties of the material.

An outstanding characteristic of most metals is their ability to form alloys, or solid solutions, with each other. Alloys may actually include small amounts of intermetallic compounds and compounds of the metals with nonmetals such as carbon. Since the properties of an alloy depend on its composition, and since their compositions can be varied almost infinitely, alloys with a huge range of properties can be prepared. In many cases the alloys have properties superior to those of the pure metal; iron has few uses, yet iron alloys are used more than any other.

Ceramics (Figure 20–3) are composed of compounds that exist in their structures as molecules or ions. The compounds may be metallic or non-metallic, and very often contain oxygen. Simple examples of ceramics are materials fabricated from silica (SiO_2, silicon dioxide), magnesia (MgO, magnesium oxide), and alumina (Al_2O_3, aluminum oxide). Since these are already oxygen compounds, they are not susceptible to reaction with oxygen or with compounds such as the oxidizing acids, so they are highly resistant to corrosion. Many ceramics have high melting points and can be fabricated into materials that can perform at high temperatures—furnaces, for example. Such materials are called **refractories**.

The range of compounds possible for ceramics is broad. In general, ceramics contain a metal or a metalloid such as silicon, along with one of the elements C, N, O, P, or S. Examples of nonoxide ceramics are silicon nitride (Si_3N_4), boron carbide (B_4C), and tungsten carbide (WC). In addition, there are many ceramics composed of various mixtures of compounds such as those formed from natural clays. Thus, the number of different ceramics is comparable to the number of alloys: large, indeed.

The principal difference between ceramics and **glasses** is that ceramics have a regular crystalline structure that repeats in a very regular way,

FIGURE 20–2 Molten steel pouring from a furnace. Recall from Chapter 14 that the temperature range of the white-hot liquid material can be determined directly by the color.

Intermetallic compounds can be loosely defined as metal systems (prepared with gallium and other metals) having semiconducting properties.

FIGURE 20–3 Samples of a high-temperature, high-strength silicon carbide (SiC) "ceramic."

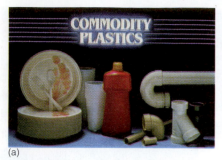
(a)

(b)

(c)

FIGURE 20–4 (a) Lower-cost "commodity" plastics are generally used for applications where the material will not have to endure stressful environments; (b) colorful plastic pellets contain the desired pigments (coloring agents) and other additives that enhance their properties; (c) typical "engineering plastics" are polymeric materials that outperform traditional materials, especially metal, wood, and glass.

whereas glasses are amorphous. Silica is the basic ingredient for most glasses; a glass can be formed by simply melting quartz and then cooling it sufficiently rapidly that it does not have time to crystallize. Such glass is chemically resistant and generally useful at elevated temperatures because of its high melting range. However, it is expensive to fabricate because of the high temperatures needed to produce it, and it is difficult to shape. Adding the basic oxides Na_2O and CaO lowers the melting point and gives the familiar soda-lime-silicate glasses used in typical window panes and bottles. Other oxides can be added to impart special properties. Borosilicate glasses contain B_2O_3, which improves the thermal shock resistance of the glass; these products carry famous trade names like Pyrex. Lead oxide makes the glass softer and heavier and increases the index of refraction, giving the glass more sparkle. Such glass is used for fine decanters and tableware.

Important new materials called **glass ceramics** start with a glass, which is then crystallized in a network of microscopic crystals. The starting glass in these cases often contains lithium and aluminum oxides along with silica. The object is shaped in the glassy state; then, a heat treatment converts the glassy material into a form with very fine-grained crystals and no porosity. The resulting products have exceptional mechanical strength and heat resistance and are excellent for cooking ware.

In marked contrast to metals and metal alloys, the main shortcoming of ceramics and glasses is their brittleness, which limits their use in applications, such as engines, where their excellent heat resistance would be extremely useful. Efforts are being made to find ceramics with properties that will permit such uses in spite of their inherent brittleness. They cannot be easily modified to bend instead of break, in response to load and impact stress. However, **structural ceramics**, compositions that include silicon nitride, are being tested for high temperature automobile engines. As we discussed in Chapter 11, the higher the temperature at which an engine can operate, the more efficient it can be. Structural ceramics open a way to very efficient engines by allowing construction of engines that can operate at high temperatures for protracted periods of time. For the kitchen and the bathroom, unusually break-resistant tableware and tile, dishes that won't shatter when dropped on a ceramic floor, and ceramic floor tiles that won't shatter when struck by a falling object have been introduced by DuPont and Dow Corning.

The principal structural elements of **polymers** and **plastic materials** (Figure 20–4) are long chain molecules. In polyethylene, the simplest polymeric structure, the monomer unit from which the polymer is made is ethylene;

$$CH_2{=}CH_2 \qquad -(CH_2-CH_2)_n-$$

ethylene   polyethylene

(monomer)   (polymer)

and *n*, the number of repeating ethylene monomer units in the polymer, can easily range up into the thousands. Thus, the molecular weights of these giant polymer molecules are in the tens of thousands (and higher). For metals and ceramics, the nature of their crystal structures is responsible for characteristic properties. In polymers, the long chain molecules are the primary features.

The chemistry of polymers and their preparation from monomer units will be discussed in Chapter 25 (Organic Chemistry). Here, we consider only the general properties of polymers and plastic materials. The starting monomeric materials from which polymers are prepared are generally relatively inexpensive coal and petroleum products, and the means of production in large quantities are well developed and economically favorable processes. Thus, the cost of plastics is low compared with that of alternate materials. Because polymers are mostly composed of light atoms such as hydrogen, carbon, oxygen, and nitrogen, the materials themselves tend to be of low density.

Plastic materials can generally be molded and extruded, and compared to ceramics and glasses they have considerable strength. These useful properties make them low-cost alternatives for metal and glass. A plastic container has many of the favorable properties of glass, but it will not break when dropped, and its shipping weight is much less. A plastic frame for a backpack has all the strength of its metal counterpart, yet is much lighter in weight. The shortcomings of plastics are generally poor resistance to heat and their reactivity to a variety of chemicals, including common solvents such as carbon tetrachloride, toluene, and gasoline.

Combination of materials yields **composites** and offers an attractive way to enhance the useful properties of each one. "Fiberglass" is probably the best known man-made composite. In this material, a polymer is reinforced by impregnating it with fine glass fibers. The flexibility and lightness of the polymer is enhanced by the high strength of the thin glass fibers, producing a structural material used in boat hulls, auto bodies, fuel storage tanks, and many other applications. Carbon fibers have great strength and are used to reinforce polymer composites. Wood is a naturally occurring composite that is reinforced by fibers. Its unusually wide range of uses and applications results directly from its lightness, flexibility, and overall strength. Multi-layered materials are made of successive sheets of plastic films; each contributes special barrier properties against harmful substances such as moisture and oxygen, or thermal stability. The layers are glued together into composites that are revolutionizing the packaging industry. The modern toothpaste tube or potato chip bag, for example, consists of anywhere from five to nine layers or more.

Semiconductor materials make up a category of materials in their own right. In our discussion of the periodic table in Chapter 17, we mentioned that electrical conductivity is a periodic property and that the change between metals and nonmetals (that is, conductors and nonconductors) occurs along a zigzag line that splits the p-block of the table. The elements along the divide are intermediate in conductivity between conductors and insulators and are called metalloids, semimetals, or semiconductors. The conducting mechanism for semiconductors is different than for metals, as we discussed in Chapter 19. At higher temperatures metals become poorer conductors, whereas semiconductors have higher conductivities. The usefulness of semiconductors in microelectronic circuits has created a revolution that has changed our entire way of life.

As we noted in Chapter 19, the major semiconductor elements are silicon (Figure 20–5) and germanium in Group IVA. Each uses its four

Although Babe Ruth might have insisted on 52 ounces of Northern ash for his Louisville Slugger, a graphite-filament polyester composite can simulate many of the attributes of the real thing. It sounds like wood, at the "crack" of the bat. But, unlike wood, it won't splinter.

FIGURE 20–5 In a vacuum chamber, fine, intricate patterns are etched into silicon wafers with an ion discharge. The violet glow is emitted by the ion plasma. The plasma etching process makes possible the small geometries needed for very large scale integration in silicon chips.

valence electrons to form a diamond-like structure. In this structure each atom is surrounded by four other atoms in a regular tetrahedron. By introducing small amounts of elements having three or five valence electrons, holes are produced in the valence band or electrons in the conduction band. This makes it possible for current to flow through the material.

The diamond-like structure of germanium and silicon can be duplicated by combining elements having an average of four valence electrons per atom—for example, gallium with three valence electrons and arsenic with five. Such compounds are often referred to as III-V compounds. The resulting gallium arsenide (GaAs) is isoelectronic and isostructural with germanium; as we mentioned in Chapter 19, it is used in very high speed microelectronic devices and photovoltaic cells. Aluminum phosphide (AlP) is another III-V compound with semiconductor applications. There are II-VI semiconductor compounds, including CdS and ZnO. Finally, silicon carbide (SiC)—a IV-IV compound—exists in a number of structural modifications, some of which have semiconductor applications.

20.3 SELECTING MATERIALS

What is the basis for selecting a material from one of these five categories? Bear in mind that cost is always a factor, often the deciding factor when other properties are similar. In some cases, low-cost materials are absolutely necessary because a highly competitive, established market exists and a somewhat better or new product at a higher price may not be given a chance to establish itself. Substitutes for wood in construction or window glass in automotive applications must be competitively priced. On the other hand, applications requiring extremely high performance standards, such as space vehicles, race car engines, or electronics, can sometimes tolerate costly materials.

Within all of the categories, except perhaps semiconductors, there is a considerable range of cost. Winning metals from their ores is fairly expensive in all cases. Metals range from moderate in cost, for example iron and aluminum, to very expensive, such as tungsten, iridium, and platinum. Ceramics range from low-cost fired clays used in flower pots and dishes to high-cost silicon nitride and related specialty ceramic materials suitable for fabricating blades for gas turbines and jet engines. Polymers are generally less expensive materials, ranging from phenolics used in Formica for kitchen countertops at fifty cents a pound to high-temperature polybenzimidazoles that sell currently at thirty dollars a pound.

Anticipating 21st-century materials of construction and modular designs, the exteriors and interiors of this house have been largely built of engineering plastics—materials that perform beyond the capabilities of traditional materials of construction and protection: metal, wood, and glass.

The blades of a gas turbine or jet engine elongate due to creep and may fail by striking the engine casing. Accordingly, the original choice of material is critical as is regular monitoring of performance characteristics.

Semiconductor materials do not usually compete with the other categories, since their principal applications are the highly specialized needs of microelectronics circuits.

EXAMPLE 20–1

Select the category of materials that you think would be best for the construction of a propeller blade for a small boat to be used in salt water. Useful properties are the following: (1) Sufficient strength so the load from the boat's engine can be transferred to the water. (2) Flexibility, so the propeller does not shatter if it strikes an object in the water. (3) Resistance to corrosion by salt water. (4) Resistance to wear due to rotating through the water at high speeds for long periods.

Solution   The brittleness of ceramics rules them out. Either a metal or a composite appears to be a reasonable choice. Metals have the advantage of greater strength and resistance to wear. However, metals are subject to corrosion by the combination of salt water and dissolved oxygen. Lower cost and corrosion resistance might favor the choice of a composite material in this application.

EXERCISE 20–1

Consider the following applications. For each, list the desired characteristics and select a possible material from the five major categories. In some cases, there may be more than one choice. When multiple possibilities exist, give the advantages and disadvantages of each choice.
(a) The hull of a canoe.
(b) A tank for a gas under pressure.
(c) The heating elements of an electric stove.
(d) A pipe to carry molten iron.

20.4   BONDING IN MATERIALS

The properties of materials are a direct consequence of the chemical bonds they contain—ionic bonds, metallic bonds, covalent bonds, and van der Waals bonds. It is useful to summarize our earlier discussions of bonding (in Chapters 16, 17, 18, and 19) in the context of this chapter.

The **ionic bond** is a direct consequence of the attraction between oppositely charged particles—the ions that make up the structure. It is a nondirectional bond. Generally, it is satisfactory to assume that these ions are hard spheres and that the spheres representing ions of opposite charge are in contact. This maximizes the energy of the structure and establishes specific distances between the ions. Lattice energies for crystalline substances such as sodium chloride are on the order of 10^3 kJ/mol. When compared with examples of other types of bonds, these turn out to be among the very strongest. The ionic bond is the principal bond in ceramic materials.

Metallic bonds are also nondirectional. The structures of metals are characterized by the high numbers of nearest neighbors (high coordination numbers). Metal atoms are held together by their valence electrons, which are highly delocalized (in bands) composed of large numbers of overlapping bonding orbitals. This makes for characteristically strong, malleable, ductile, and conducting materials.

Covalent bonds, contrary to the previous two types, are strongly directional. They involve the sharing of pairs of electrons between atoms,

TABLE 20–1	Melting Points of Some Substances Related to Type of Bonding	
Substance	**Bonding**	**Melting Point (°C)**
NaCl	ionic	800
CaO	ionic	2580
Fe	metallic	1535
W	metallic	3370
C (diamond)	covalent	~3500
B	covalent	2300
Br_2	covalent and van der Waals	−7.2
NH_3	covalent and van der Waals	−77.7
$—(CH_2—CH_2)_n—$	covalent and van der Waals	~120 (softens)

with very specific bond angles between the other atoms to which an atom is bonded. The strengths of covalent bonds can be measured as the average bond dissociation enthalpy (Table 16–5). On the whole, covalent bond energies are reasonably strong, and in many cases are comparable to ionic lattice energies on a per mole basis. Thus, we should not be surprised to find many materials held together by covalent bonds to be strong and stable. Yet such covalent materials are uncommon, with the exception of diamond and diamond-like structures (which are indeed very hard, stable materials). The problem is the inclusion of some weaker van der Waals bonds, which affects even the strongest, most stable covalent materials—plastics.

van der Waals bonds are the weakest of the bonds. They do not involve transfer of electrons between atoms or into bonding orbitals. Instead, they result from attractions between permanent and induced dipole moments. They are not directional to the same extent as covalent bonds, although they are not purely undirectional either. The hydrogen bond is the best known and generally the strongest of the bonds of this group. This is the result of the relatively powerful attraction between dipoles containing hydrogen and one of the very electronegative elements nitrogen, oxygen, or fluorine.

A measure of the strength of the bonds that hold a substance together is the melting point. The higher the melting point, the more thermal energy must be put in to break the bonds, and the stronger the bonds must be (Figure 20–6). If a material has more than one kind of bond, the weakest break first, and this will be noted by melting or softening of the material. Table 20–1 gives the types of bonding present in some substances and their melting points.

EXAMPLE 20–2

Discuss the relative melting points of polyethylene (a polymer), lithium chloride, potassium chloride, chlorine, and ethylene.

Solution Polymer molecules have a backbone of covalent bonds. However, one molecule is held to another by weaker van der Waals bonds. Thus, polymers' resistance to heat is relatively low. Lithium chloride and potassium chloride would be expected to display the high melting points typical of ionic solids. As a substance composed of small molecules, chlorine is a greenish-yellow gas at room temperature. Ethylene is a colorless gas at room temperature, but polyethylene—a major material for transparent food wrapping—is a relatively low-melting polymer; note what happens if a bread wrapper accidentally falls against a hot oven surface. ■

Estimate the melting (or softening) points of the following materials. Then check in a chemistry handbook to see how close you came.
(a) Si   (b) KBr   (c) polyvinyl chloride   (d) Cr   (e) BCl_3 ■

20.5 PERFORMANCE CHARACTERISTICS OF MATERIALS

Stress and Strain

Materials science has achieved such a high degree of sophistication that it is no longer possible to provide a summary of materials properties in a short space. Still, it is important to introduce the basis for the notion of **performance characteristics** of materials and their measurement. Because stress/strain measurements are the most commonly performed of the many standard tests and measurements, it makes good sense for us to discuss this particular performance characteristic here.

The words stress and strain have particular meaning. **Stress** is the load (force) imposed on a test specimen divided by the original cross-sectional area of the specimen. Thus, if you exert a towing force on a nylon line with a cross-sectional area of $0.50 \ in^2$ while hauling in a 50-lb anchor, the stress is

$$\frac{50 \ lb}{0.5 \ in^2} = 100 \ lb/in^2$$

Strain is the increase in length of the specimen divided by the original length. If the anchor line was 12 feet long, and under the stress of 100 lb/in^2 it stretched 0.10 inch, the strain is

$$\frac{0.10 \ in}{(12 \ ft)(12 \ in/ft)} = 0.00069 \ in/in$$

Strain is sometimes reported as a dimensionless number, but it is best to keep the units here to remind us that we are dealing with a relative change in length.

There are three classic types of stresses—tension, compression, and shear. A shear stress is a twisting or tearing stress, as in cutting a metal sheet or twisting a bolt at high torque. Compression is squeezing and tension is pulling. The strength of a material is lower in shear stess than in tension; for most materials, especially brittle materials, the strength under tension is less than under compression.

On removing the load (or stress), materials exhibit two kinds of behavior.

(a)   (b)

FIGURE 20–6 Thermal stability is a function of many factors, not the least of which is "bonding." Here, a molded plastic part (a) is subjected to elevated temperatures; (b) although the test temperature was not high enough to cause charring or discoloration, the basic design structure has collapsed.

FIGURE 20–7 Bumper deformation and recovery testing. When the stressing force equivalent to a 5 mph collision is removed from the plastic bumper, elastic recovery will be 100%; a steel bumper part would be permanently deformed.

Since 1984, most cars in the American market have had plastic composite bumpers that can withstand collisions of up to 5 mph. The force of such an impact is within the elastic limit of the material. Contrast that to the old, much heavier, and always dented chrome-plated steel bumpers.

1. The material may be permanently deformed. Such strain behavior is termed **plasticity** or **plastic deformation**. Wet sand and freshly poured concrete, which retain an imprint, are good examples. Stretching the plastic top of a soft drink six-pack is plastic deformation—the material is permanently deformed.

2. The material may recover its original size and shape. When the deformation disappears as the stress is removed—as in stretching a rubber band—the strain behavior is termed **elasticity** or **elastic deformation**. Such materials are said to have a "memory."

A wide variety of materials display elastic deformation when subjected to limited stress, but only a few are perfectly elastic up to the stress at which they fail completely. Glass is one such material. It will recover its original shape completely if you bend it without breaking it. Brittle materials are, in general, elastic since they are incapable of accepting any significant degree of strain. Ductile and malleable materials are capable of a high degree of plastic deformation. Copper and aluminum, for example, can be drawn into wires and hammered and rolled into sheets.

Creep is long-term plastic deformation. It is a familiar effect. Metals, woods, paper products, and plastics such as polyethylene sag when stressed under relatively light loads for long periods of time. Although few metals creep at room temperature, creep is a cause of failure of metal parts under severe or prolonged conditions of operation.

Both kinds of strain—elastic and plastic—are useful performance characteristics. For example, since dimensional stability is required of manufactured products, they must be designed in materials that can sustain the expected stresses in any anticipated use. That is to say, under "use" conditions, the stress-strain characteristics must fall within the **elastic limit** of the material. However, in order to mold or form a material into an object of desired size and shape, it must be plastic under fabricating conditions. The traditional fender material for automobiles has been sheet steel, and fender parts for the Pontiac Fiero are being made from flat plastic composite sheets. Both the steel sheets and the plastic sheets must be plastically deformed. However, once the fender is installed on the frame of the car, only elastic deformations are acceptable. Of course, that has rarely been achieved with steel fenders. Even a small stone, spun off the road by another car, is enough to leave a permanent dent. But external plastic parts, having a much greater elastic range, can withstand even minor accidents without being permanently deformed (Figure 20–7).

Modulus of Elasticity

Many metals, rubber elastic materials, and any number of other compositions of matter display elastic behavior at load (stress) levels below the elastic limit. In this range the strain ϵ is (ideally) always proportional to the stress τ and is independent of the rate of deformation. The constant of proportionality is known as Young's modulus Y, and the general relationship is known as Hooke's law (if we are referring to a tensile stretch or deformation). In general, we speak of the **modulus of elasticity**, or simply the modulus E:

$$\text{Modulus of elasticity:} \qquad E = \frac{\text{stress}}{\text{strain}} = \frac{\tau}{\epsilon}$$

High E values mean rigid materials, since the numerator (load or stress) will be very large with respect to the denominator (deformation or strain). A rubber band may have a modulus of elasticity in the range of only a few

TABLE 20–2 Modulus of Elasticity Values of Some Typical Materials

Material	E (psi)*	Material	E (psi)*
carbon steel	29×10^6	nickel	30×10^6
glass	10×10^6	iron	30×10^6
concrete	3×10^6	copper	17×10^6
wood	2×10^6	gold	12×10^6
nylon-66 fiber	1×10^5	silver	11×10^6
polyethylene	4×10^4	aluminum	9×10^6

*1 psi = 6.89×10^4 dyne/cm^2 = 6.89×10^3 N/m^2 (Pa).

hundred pounds per square inch (psi); metals have moduli on the order of 10^6 psi (Table 20–2).

Actual measurements are made on a standardized test piece mounted in a specially designed hydraulic system capable of loading materials in tension or compression. In a tension (tensile) test of a polymeric material, a parallel-sided strip is held by two clamps that are separated by the hydraulic system at constant speed, and the force (stress) needed to carry this out is recorded as a function of clamp separation. The test strips are usually dogbone-shaped to promote deformation between the clamps and deter flow in the clamped section; the two-inch center of the test strip, called the gauge length, is the active material region in the test (Figure 20–8).

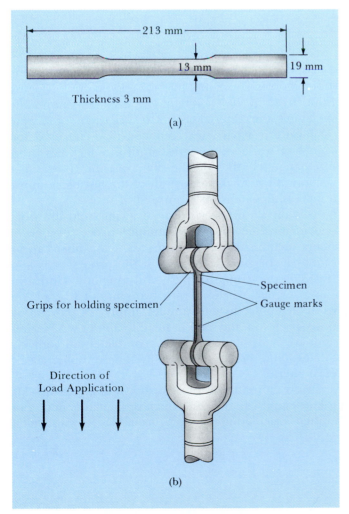

213 mm

13 mm   19 mm

Thickness 3 mm

(a)

Grips for holding specimen — Specimen

Gauge marks

Direction of Load Application

(b)

FIGURE 20–8 Tensile properties are the most important single indication of strength in a material. The force necessary to pull the specimen apart is determined, along with how much the material stretches before breaking. Both ends of the standard specimen (a) are clamped in the jaws (b) of a testing machine. The jaws then move apart at controlled rates, pulling the sample from both ends and the recorded stress (load) is plotted against the strain (deformation).

FIGURE 20–9 Tensile history of a test specimen which has deformed by "cold-drawing." Along the segment *oabc*, Hooke's Law holds: stress is proportional to strain. This is the elastic (linear) region of the curve and the slope (*bx/ax*) is the modulus (of elasticity). Point *c* is the yield point, the highest stress the material can withstand and still "recover." The segment *cde*, beyond the yield point, is the region of plastic (permanent) deformation; and finally, the break point is observed at *e*. See Figure 20–10.

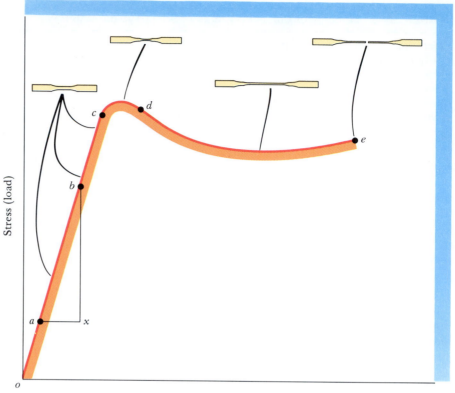

Stress (load)

Strain (deformation)

The modulus of elasticity is computed from data in the ideal region where the stress-to-strain ratio is constant—point *b* on the curve (Figure 20–9). At the top of the curve, in the vicinity of the stress maximum and beyond the proportional limit, the material begins to **neck down**, or narrow. The necked region stabilizes at a particular reduced diameter as deformation continues at more or less constant stress until the neck has propagated across the entire gauge length or to the point of rupture—point *e*. Note how the cross section of the necking portion of the test piece in the figure decreases with increasing extension (Figure 20–10).

The necking process is also known as yielding, drawing, cold-drawing, or cold-flow. It is an essential element in the orientation processes used to stengthen synthetic fibers such as nylon. An undrawn nylon fiber can be elongated to several times its original length. If in the process it is also twisted or warped, it takes on the strength typical of fishing line. Undrawn nylon can be pulled apart or necked down with two pairs of pliers. You can perform a simple test on a polyethylene sample: simply grasp two close points in a soft drink six-pack top and pull it steadily apart. It stretches some 80%. Rubber bands are made of materials that completely and quickly recover from stresses that extend them dimensionally. These rubber elastic materials are called **elastomers**.

A plot of stress versus strain for a steel piece is shown in Figure 20–11. Note that the strain is linear up to 42,000 psi, which is the elastic limit and the end of the elastic range. Calculating the modulus of elasticity at 30,000 psi,

$$E = \frac{30,000 \text{ lb/in}^2}{0.001 \text{ in/in}} = 30 \times 10^6 \text{ psi}$$

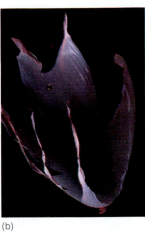
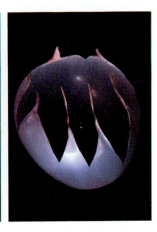

(a)

(b)

FIGURE 20–10 (a) Tensile test bars, displaying the kind of behavior described in the graph in Figure 20–9. The percent elongation (deformation) is about 25% in this case. (b) At the extreme end of the load-deformation curve for a rubber elastic material, brittle failure is often observed. For example, if you stick a pin into a distended balloon, the failure is dramatic. At the moment of impact, the balloon explodes and "shatters" into several long finger-like projections, before beginning to relax back into rubber elastic failure.

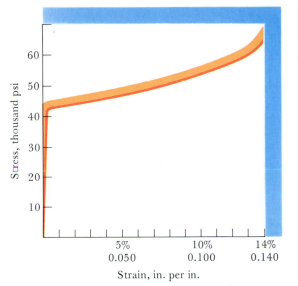

FIGURE 20–11 Stress/strain plot for a steel specimen. While not brittle, this piece is not especially ductile.

The piece ruptured at 63,000 psi at an elongation of 14%, ending the test. While not brittle, this steel piece is not especially ductile. A more ductile metal might stretch as much as 50%.

EXAMPLE 20–3

A mass of 100.00 kg is suspended by a wire that has a cross-section with a radius of 0.500 mm. Before the mass was attached, the length of the wire was 100.00 cm. With the mass in place the length of the wire was 100.62 cm. When the mass was removed, the wire returned to its original length. Calculate the stress and strain of the wire, and its modulus of elasticity.

Solution The force or load on the wire is

$$f = ma = (100.00 \text{ kg})(1000 \text{ g/kg})(980.7 \text{ cm/sec}^2)$$
$$= 9.807 \times 10^7 \text{ g·cm/sec}^2$$
$$= 9.807 \times 10^7 \text{ dynes}$$

The stress τ is this force divided by the wire's cross-sectional area,

$$\tau = \frac{9.807 \times 10^7 \text{ dynes}}{\pi (0.500 \text{ mm/10 mm/cm})^2} = 1.25 \times 10^{10} \text{ dynes/cm}^2$$

FIGURE 20–12 One of the earliest direct observations of the "links" in plastic crystals is seen in this photomicrograph by scientists at the Bell Telephone Laboratories, published in 1966. Note the fine fibrous links bridging the gaps between radial arms of crystalline regions in a polyethylene sample.

The strain ϵ is

$$\epsilon = \frac{100.62 \text{ cm} - 100.00 \text{ cm}}{100.00 \text{ cm}} = 0.0062 \text{ cm/cm}$$

The modulus of elasticity E is thus

$$E = \frac{\tau}{\epsilon} = \frac{1.25 \times 10^{10} \text{ dynes/cm}^2}{0.0062 \text{ cm/cm}} = 2.02 \times 10^{12} \text{ dynes/cm}^2$$

EXERCISE 20–3

A wire with a cross-sectional radius of 1.00 mm has an elastic modulus of 1.50×10^{12} dynes/cm². How much will the length of a 5.00-m length of this wire increase if a mass of 125.00 kg is suspended from it?
Answer: 1.3 cm.

Strain Mechanisms and Relief of Strain

In elastic deformation, a reasonable strain mechanism is the stretching and relaxing of the interatomic spacing. The greater elasticity or memory of many rubbery, elastomeric materials is due to the effects of chemical cross-linking, in which a few weak bonds connect the major chains. When the stressing force is removed, these crosslinks "remind" the molecules where they belong in the structure (Figure 20–12). However, if the degree of cross-linking is extensive, a three-dimensional matrix is established that forms the material into a rigid rather than elastic material.

Plastic deformation in metals and alloys is governed by a more com-

FIGURE 20–13 (a) Schematic diagram of a single crystal metal sample oriented for plastic deformation with the least applied stress. (b) Slip planes in the FCC, BCC, and HCP metals. These planes are those of greatest atomic density.

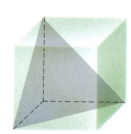

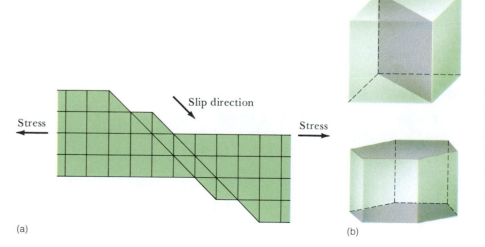

(a)

(b)

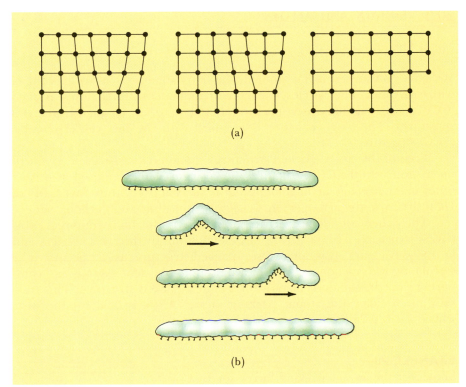

(a)

(b)

FIGURE 20–14 (a) Plastic permanent deformation of a crystal caused by sliding one plane of atoms past an adjacent plane. In contrast to a perfect crystal (where high stress is required to "slip" the planes), the presence of a dislocation lowers the required slip stress. (b) In the simplest sense, the dislocation "slips" according to the caterpillar principle.

plicated mechanism than simple extension of the interatomic spacing (Figure 20–13a). We say **slippage** has occurred when atoms move a unit (interatomic) distance along the slip plane. When the stress is removed, there is no driving force to return atoms that lie below the slip plane to their original position. Slippage occurs along planes of greatest atomic or molecular density. They are shown for typical FCC, BCC, and HCP structures in metals in Figure 20–13b. Since the FCC lattice has the greatest number of slip planes—eight in all—such metals are the most ductile. Copper and aluminum are examples.

In principle, if metals crystallized from the melt "perfectly," without **dislocations**, all the bonds between the atoms that lie across the slip plane would have to break simultaneously for slippage to occur. Because of that, metals would be perhaps hundreds of times stronger than any ever tested. In practice, the strengths of metals are limited by irregularities, or dislocations. Slippage occurs as the dislocations move bond-by-bond along the slip plane. Eventually, the dislocation is consumed and the crystal becomes more perfect (with fewer dislocations) and of greater strength. It becomes **work-hardened** or strain-hardened. At the end of this sequence, when no more dislocations are available to slip in response to a stress—because they have all run off the edge of the map or run into each other in a tangle of dislocations—the stress required to slip any further equals the stress required to rupture the structure, and the structure fails (Figure 20–14).

To understand the brittle character of ceramic materials, consider the typical magnesium oxide (magnesia) structure and the idea of slippage along a slip plane. Slippage cannot occur (Figure 20–15) because shifting one interatomic unit along a plane would force like ions to align with each other, destroying the ionic arrangement. Imposing a large enough stress to force such a distortion results in shattering the structure. That would not be the case in a metal, where all the atoms are the same.

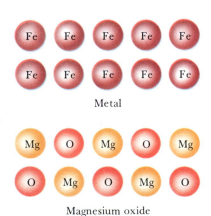

Metal

Magnesium oxide

FIGURE 20–15 Composition of the cubic crystal structures of a metal (Fe) and a metal oxide (MgO).

20.6 METALS AND ALLOYS

Metal Structures

We have considered metal atoms to be hard spheres that are in contact with each other. The two closest-packed arrangements, cubic closest-packed (CCP) and hexagonal closest-packed (HCP) both occur frequently in metal structures. The CCP structure is identical with the face-centered cubic (FCC) structure.

Each of the closest-packed structures contains closest-packed layers of metal atoms. In the CCP structure, the layers are superimposed over each other in an *abcabcabc* arrangement that repeats at every third layer. In the HCP structure the arrangement is *ababab*, repeating every other layer.

The other common structure of metals is the body-centered cubic (BCC) structure. This structure has metal atoms at the cube corners and in the body center. It is not a closest-packed structure, as can be shown by calculating the fraction of the unit cell occupied by actual atoms and comparing with the same calculation for a closest-packed structure. This fraction, often called the atomic packing factor (APF), is 0.68 for the BCC whereas it is 0.74 for the closest-packed structures.

EXAMPLE 20–4

Show that the atomic packing factor for BCC is 0.68.

Solution From Chapter 19, recall Eq. (19–3):

$$r = \frac{a\sqrt{3}}{4}$$

Also recall that the BCC unit cell contains a total of two atoms. Thus,

$$\text{volume of atoms} = 2 \times \left(\frac{4}{3}\pi r^3\right)$$

$$= 2 \times \left(\frac{4}{3}\pi \times \frac{3\sqrt{3}}{64}a^3\right)$$

$$= 0.68\,a^3$$

$$\text{volume of the cell} = a^3$$

$$\text{APF} = 0.68$$

EXERCISE 20–4

Calculate the APF for the simple cubic metal structure. Why do you think that this structure is not observed in actual metals?
Answer: APF = 0.52. Too little of the space in the unit cell is utilized.

The properties of a metal or alloy depend on heat treatment and subtle differences in composition, including very low levels of impurities. Ductility depends on the slippage of closest-packed or high density planes across each other. If one builds models of the CCP and HCP structures, it is possible to determine that there are twelve different directions in which high density planes can slip against each other in the CCP structure, compared with three for the HCP structure. A good comparison between these

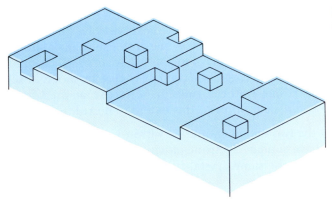

FIGURE 20–16 The Hirth-Pound model of the surface of a crystal.

two structures can be obtained examining aluminum (CCP) and titanium (HCP). In agreement with their structures, aluminum-based alloys tend to be ductile and titanium-based alloys tend to be brittle.

Crystals and Grains

So far, we have been considering structure in the interior of the crystal, taking the position that the crystal contains countless millions of atoms that seem to extend almost infinitely from any atom under consideration. However, we also know that crystals have finite sizes and they are, in fact, usually quite small. Except in the semiconductor industry, which depends on single crystals, crystalline materials are composed of many small crystals held together in a polycrystalline structure.

Before we can discuss polycrystalline materials, we need to answer the question: "What occurs at the boundary between one crystal and another?" There is a defect where the regular arrangement of one crystal ends and a new regular arrangement, oriented in a different direction, begins. Ideally, the surface of a crystal would be simply a close-packed layer of atoms. Note that atoms on the surface do not have exactly the same properties as atoms within the bulk of the structure, since their coordination numbers are different. That is, surface atoms have empty coordination sites available. Furthermore, even though the face of a crystal may appear extremely flat, we do not really expect to find an absolutely flat surface of a perfect plane of atoms. Rather, the surface has irregularities in the form of steps or terraces as layers of atoms start or stop. An attractive visualization of this is the Hirth-Pound model of a crystal surface shown in Figure 20–16.

Polycrystalline materials are considered to be composed of grains. Each grain is a single crystal with a regular arrangement of atoms within it. However, as the grains are packed tightly together, these crystals are not able to form the characteristic flat faces of crystals that are formed with space to grow. Thus, the shapes of grains are irregular, much more irregular than the Hirth-Pound model. Grain boundaries exist between the grains as shown in the schematic diagram in Figure 20–17.

Solid Solutions

Many metals dissolve in each other when they are melted, resulting in a **solid solution** when the melt cools. In some cases atoms of one metal simply

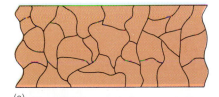

(a)

(b)

FIGURE 20–17 (a) Model of grains in a metal. (b) Photomicrograph of aluminum crystals, with reflected polarized light. The sample was a piece of annealed aluminum wire that was ground, polished, and etched to reveal its crystal structure. Each colored region is a separate crystal. The variety of color results from the interaction of the polarized light with the metal surface.

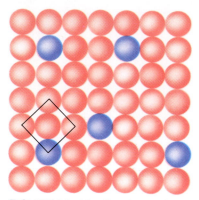

FIGURE 20–18 One layer of a substitutional solid solution in the FCC (CCP) metal structure. The edges of one face of the FCC cube are indicated.

fit into sites that would otherwise be occupied by the other. This is called a **substitutional solid solution**. As an example, nickel dissolves in copper in this manner, and one plane of the CCP structure of this solution is shown in Figure 20–18. Only some combinations of metals will form solid solutions in which the mole fraction of the lesser component is 10% or more. A set of three rules called the Hume-Rothery rules are effective at predicting which metals will dissolve in each other to this extent. The rules are:

1. The atomic radii of the two metals should be within 15% of each other.
2. The metals should have similar electronegativities.
3. The metals should have the same crystal structures in order to form a complete range of solid solutions.

The second rule can be checked in a table of electronegativities or simply by choosing metals close to each other on the periodic table. The crystal structures and unit cell lengths of some important metals are given in Table 20–3. The relationship between unit cell length and atomic radius was discussed in Chapter 19. Note that for any two metals with the same crystal structure, the ratio of their atomic radii and unit cell lengths will be identical.

EXAMPLE 20–5

Which metals from Table 20–3 would you expect to form substitutional solid solution with iron to the extent of 10% or more?

Solution   Table 20–3 shows that Fe has a BCC structure in common with Cr, K, Na, and W. As we showed in Chapter 19, the atomic radius in the BCC case is

$$r = \frac{a\sqrt{3}}{4}$$

and thus the atomic radii of these metals are all in the same proportion to the unit cell length a. The percentage differences between the unit cell of Fe and the BCC metals are:

$$\text{Cr:} \quad 100\% \times \frac{|2.861 - 2.878|}{2.861} = 0.6\%$$

$$\text{K:} \quad 100\% \times \frac{|2.861 - 5.333|}{2.861} = 86\%$$

$$\text{Na:} \quad 100\% \times \frac{|2.861 - 4.24|}{2.861} = 48\%$$

$$\text{W:} \quad 100\% \times \frac{|2.861 - 3.158|}{2.861} = 10\%$$

W and especially Cr fall within the atomic radius requirement. The Pauling electronegativities of Fe, Cr, and W are 1.8, 1.6, and 1.7 respectively, all of which are relatively close. Therefore we would expect both Cr and W to be able to form solid solutions with iron quite freely.   ■

EXERCISE 20–5

Which metals of Table 20–3 would you expect to be capable of forming substitutional solid solutions with Cu to the extent of 10% or more?
Answer: Ag, Al, Au, and Ni all satisfy the atomic radius requirement. Of these, all but (probably) Al satisfy the electronegativity requirement.   ■

TABLE 20–3	Crystal Structures and Lattice Parameters of the Stable Forms of Some Important Metals	
Metal	**Structure**	**Unit Cell Length(s) (Å)**
Ag	CCP	4.078
Al	CCP	4.041
Au	CCP	4.070
Be	HCP	2.283, 3.607*
Cd	HCP	2.973, 5.606
Co	HCP	2.514, 4.105
Cr	BCC	2.878
Cu	CCP	3.608
Fe	BCC	2.861
K	BCC	5.333
Mg	HCP	3.203, 5.196
Na	BCC	4.24
Ni	CCP	3.514
Pb	CCP	4.941
W	BCC	3.158
Zn	HCP	2.658, 4.934

*The two parameters given for hexagonal closest-packed structures are the values for a and c (see Figure 19–6).

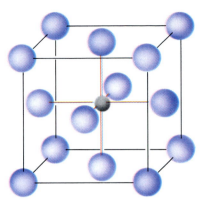

FIGURE 20–19 An interstitial solution of carbon, the small solid sphere, in the CCP form of iron, the larger spheres. Equivalent locations for the carbon atoms are at the centers of the unit cell edges. The carbon atoms are too large to fit into interstitial sites without causing some strain in the structure and the carbon content is limited to about 2%. The carbon atoms are distributed at random.

Another kind of solid solution involves the insertion of atoms into holes or interstices of the metal structure. Such interstitial solid solutions are possible only with small atoms such as hydrogen, boron, carbon, and nitrogen. An example of a solid solution composed of carbon inserted into tetrahedral holes of the CCP form of iron is shown in Figure 20–19.

Properties of Simple Metal Structures

So far we have considered pure metals and solid solutions. The solid solution is composed of crystals of the same kind and often is called a **single phase alloy**. Recall that a phase is a homogeneous and physically distinct region of matter. In this section we will consider a few properties of pure metals and single phase alloys, which we will group as **simple metals**.

Substitution of atoms of a pure metal for those of another metal normally increases the strength and hardness of a metal while decreasing its ductility. There are only a few exceptions to this. The general reason for these effects is that when atoms in a structure have been replaced by atoms of a somewhat different size, the planes of atoms in the structure are no longer so flat and cannot slide across each other so easily. Interstitial atoms tend to have the same kind of effect, locking the atomic network in a more rigid structure.

When stress is applied to a metal it deforms. The elastic modulus for a single metal crystal depends on the direction in which the force is applied. This is a consequence of the fact that most crystals are **anisotropic**, having different properties in different directions. For example, a single crystal of iron has an elastic modulus of 2.8×10^{12} dynes/cm^2 along the body diagonal direction and only 1.2×10^{12} dynes/cm^2 along the direction parallel to the unit cell edge. In Figure 20–20 we can see that in the BCC structure of iron the density of atoms is greatest in the body diagonal

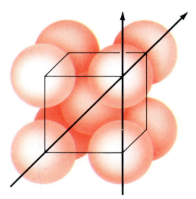

FIGURE 20–20 The body-centered cubic structure, showing the effect of strain along the edge and the diagonal of the cell.

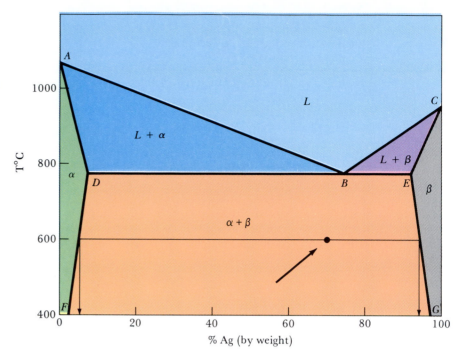

direction, accounting for the higher elastic modulus. However, in an ordinary polycrystalline sample of iron, the elastic modulus is very close to 2.2×10^{12} dynes/cm². This results from the grain structure with the grains all oriented in random directions, so that an intermediate elastic modulus is observed.

Polyphase Structures and Phase Diagrams

In contrast to the subject of the previous discussion (in which all of the grains were of the same phase), many important alloys contain grains of different phases. The existence of these different crystals in the material has enormous effects on the properties of an alloy. For example, an alloy of copper and tin can contain two different kinds of crystals. The principal crystals, composed of a copper-rich solid solution of Cu and Sn, are rounded. However, needle-shaped tin-rich crystals are also present. These long crystals act like reinforcing rods in concrete and give the alloy increased strength.

Carbon is involved in the metallurgy of iron and steel. The number of phases in the cast irons and carbon steels is large, including the CCP and BCC forms of the pure metal, the compound Fe_3C, graphite, and a variety of solid solutions of these components. The hard and brittle material called cast iron has a high carbon content and includes large, coarse crystals of Fe_3C that confer the material's properties. Low carbon steels contain two phases, BCC iron and a solid solution of BCC iron and Fe_3C.

A phase diagram for copper and silver is given in Figure 20–21. In this figure, L is a liquid phase, α is a copper-rich solid solution and β is a silver-rich solution. Above the line ABC, the sample exists entirely as a liquid. The line ABC represents the melting point as a function of composition. The melting point of pure copper occurs at point A and the melting point of pure silver occurs at point B. Note that the melting point of each of the pure metals is lowered by the addition of the dissolved solute.

This agrees with our earlier discussions of colligative properties and the lowering of melting points (in Chapter 9). Point *B* is the **eutectic** point, corresponding to the composition of minimum melting point, which occurs at about 75% silver.

Below the line *ABC*, except at the three points *A*, *B*, and *C*, a mixture of two phases exists. Each region contains a liquid solution and a solid solution in a heterogeneous mixture. Thus, if a liquid with any composition but pure copper, pure silver, or the eutectic mixture is cooled, it will start to solidify when it crosses the *ABC* line but will not be fully solid until the *ADBEC* line is crossed. Pure solid solutions exist at high copper or high silver compositions. These are denoted by the α and β regions. For most compositions, however, the solid consists of a two-phase mixture of α and β. This is the region below the *FDEG* line. Within this region, the alloy contains two kinds of grains, which strongly affect the properties of the alloy.

Phase diagrams provide useful information about what phases are present for any composition and temperature. From Figure 20–21 we can tell that at 40% Ag and 800°C two phases are present, liquid and copper-rich solid solution α. Also, phase diagrams allow us to find the compositions of phases present in two-phase regions. To do that, locate the point on the diagram corresponding to the given temperature and composition. If this point is in a two-phase region, draw horizontal lines in either direction until the boundary with a one-phase region is reached. Then drop vertical lines directly to the percent composition line and read off the compositions of the two phases.

EXAMPLE 20–6

Using Figure 20–21, determine the composition of the phases present for a sample at 600°C that is 70% Ag and 30% Cu.

Solution   First locate the point on the phase diagram corresponding to the given conditions. This point is marked by a solid circle on the diagram and pointed out by an arrow. The point is clearly in the solid two-phase region containing the solid solutions α and β. A horizontal line is drawn to the boundaries with the pure α phase and the pure β phase. Dropping verticals to the % Ag line tells us that the composition of the α phase is about 5% Ag and 95% Cu and the composition of the β phase is about 95% Ag and 5% Cu.   ■

EXERCISE 20–6

Determine the compositions of the phases present for a composition of silver and copper at 800°C that is 40% Ag and 60% Cu. Use Figure 20–21. *Answer:* A copper-rich solid solution containing about 8% Ag and a liquid solution containing about 71% Ag.   ■

Rapid cooling of certain compositions of alloys can result in nonequilibrium conditions that can persist indefinitely. On cooling a silver-copper alloy with 5% Ag from 800°C, it should convert from a single-phave solid solution (α) to a two-phase mixture of the solid solutions α and β. However, rearrangement of atoms in a solid is slow, especially at lower temperatures. Thus, if the alloy is cooled fairly rapidly, it is possible to have pure solid solution α at room temperature containing 5% Ag, and it can persist for a long time. Obtaining nonequilibrium alloys is an important aspect of metallurgy.

20.7 CERAMIC MATERIALS

If it is inorganic and nonmetallic, and fashioned into a material with high performance properties through high temperature processing, it is safe to say: "It is a ceramic!" Modern ceramic materials are indispensible to an industrialized society, especially the refractory materials that tolerate the high processing temperatures required of the basic manufacturing industries.

Beyond that, ceramics encompass the range of materials from the single ruby crystal in a laser to dense polycrystalline materials, pigments in paints, and porcelain enamel finishes. As a class of materials, compared to the metals, ceramics are thermal insulators and electrical insulators, and are chemically more stable. Furthermore, ceramic materials are appreciably more stress-resistent in compression and display greater rigidity and thermal stability than polymeric materials. Glasses are ceramics or ceramic components that are amorphous. Concretes and cements are ceramics, as are bricks and tiles.

Ceramics are among our oldest materials, and they are also among the most durable we know. Archeologists have found pottery specimens that are thousands of years old and still in excellent condition. Glassmaking probably dates back to 2500 B.C., and the use of crystalline ceramics for glazing is older still, with commercial pottery dating back to about 4000 B.C. Some simple earthenware vessels date from 15,000 B.C.

Amorphous Ceramics—Glasses

The most important aspect of glasses is that, contrary to metals and crystalline ceramics, their structures are amorphous. Long-range order is absent. This difference is clearly evident in the behavior of glasses on heating. Crystalline substances have a definite melting point. Below the melting point they are solids with regular crystalline structures; above the melting point they are liquids, displaying the typical random or disordered structure of the liquid state. The behavior of glasses upon heating or cooling is more complex. There is no melting point. However, as the temperature of a specific glass is raised, it is possible to identify a number of points for which a significant change in the properties can be detected.

For a glass at a low temperature, say ordinary soda-lime-silicate window glass, it is generally true that no permanent or plastic deformation is possible without fracture. That is, bending of the glass without breakage is limited to elastic strain from which it can spring back to its original shape. This behavior is observed at temperatures up to the **strain point**. Above the strain point small amounts of permanent deformation are possible. When the temperature is increased further, the **anneal point** is reached. At this temperature any stresses that exist in the glass are readily removed by movement of the atoms in the structure to new locations. Newly fabricated glass objects are normally held at the anneal point long enough to remove all stresses in the glass. Otherwise the glass object would be excessively susceptible to mechanical or thermal shock.

If the glass is heated still higher, the **softening point** is reached. At this temperature the glass will deform under its own weight. Finally, if the glass is heated still further, the **working point** is reached. At this temperature the glass can be readily drawn, blown, or pressed into the desired shape. The fact that glass can be heated to the point of easy working makes

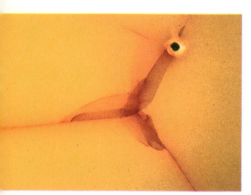

A thin ceramic plate of the type used in integrated circuits has been deliberately cracked in a study of its mechanical properties. An injection of red dye and illumination from below highlight the hairline crack.

There is of course slow cold flow in glasses, and one does not have to wait a century or two in order to observe it. If you support a 4-ft glass rod or tube at the ends and hang a weight on the center, after two or three years there is measurable evidence of permanent deformation.

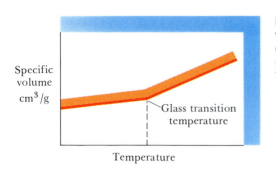

FIGURE 20–22 A plot of specific volume versus temperature for a glass. The glass transition temperature is noted as the point where the slope of the curve changes.

it especially valuable for fabricating items of all conceivable shapes. Approximate values of the four temperatures for soda-lime-silicate glass are 550°C, 590°C, 750°C and 1050°C.

The **glass transition temperature** (Figure 20–22) is the temperature at which the behavior of the glass changes from being more like a solid to being more like a liquid. This temperature can be determined by plotting the specific volume (1/density) of the glass as a function of temperature. As the temperature increases the specific volume increases. However, the specific volume of a solid-like material does not increase as rapidly as the specific volume of a liquid-like material, and a change in the slope of the curve is noted at the glass transition temperature. This occurs at about 530°C for soda-lime-silicate glass.

Silica (SiO_2) can be thought of as the starting material for most glasses. It is readily available in the crystalline form as quartz and silica sand. In these structures, each silicon atom is surrounded by four oxygen atoms in a regular tetrahedral arrangement. Each oxygen atom bridges two silicon atoms, giving a regular three-dimensional structure held together by covalent bonds. These can be fused by heating to around 1700°C. In the molten state, the atoms are freed from their regular positions and movement is possible. On rapid cooling, the atoms do not have time to find their regular positions and an amorphous, glassy material is formed. Referred to as **fused silica**, this material has exceptional resistance to heat and thermal shock but requires extremely high temperatures for working. Consequently, only small objects can be fabricated from it and they are very expensive. Adding impurities lowers the softening point of the glass. For example, sodium carbonate (soda ash) can be added. At the high temperatures of the glass manufacturing process, carbon dioxide is evolved, giving a glass with the components Na_2O and SiO_2:

$$n\ Na_2CO_3 + m\ SiO_2 \longrightarrow n\ Na_2O \cdot m\ SiO_2 + n\ CO_2$$

The sodium ions in the glass are called **modifiers**, and they break up the regular crystal structure. Figure 20–23 shows schematic two-dimensional representations of crystalline silica and a noncrystalline form containing modifier ions. Calcium oxide is also added to glass to form the common soda-lime-silicate glass. Other metal oxides such as B_2O_3 and PbO confer special properties.

Fiber optics is an important modern application made possible in part by glass technology. A beam of light is transmitted along a thin glass fiber with very little loss through the fiber walls (Figure 20–24). Special glass is used for the fiber, and the walls of the fiber are coated with another glass with a lower index of refraction, which keeps reflecting the light back into the fiber instead of allowing it to escape. Bunches of these fine fibers are

FIGURE 20–23   Two-dimensional representations of the silica structure. In (a), the structure is crystalline, and in (b), the structure is amorphous, containing a modifying ion such as Na^+. Silicon atoms are shown in red, oxygen atoms in grey, and sodium ions in yellow.

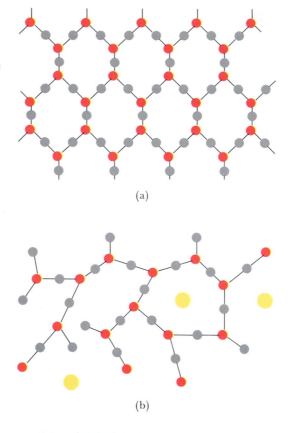

(a)

(b)

FIGURE 20–24   (a) Hair-thin optical fibers of ultrapure glass such as this spool now transmit voice, data, and video communications in the form of digital signals emitted by semiconductor lasers the size of a grain of salt. In 1988, the first fiberoptic transatlantic cable, capable of transmitting more than 24,000 calls over a pair of fibers, went into operation. (b) The photograph shows optical fibers for the telecommunications industry, protected by loose cable buffer tubes of plastic resins, which provide chemical resistance, dimensional stability and "color" codability.

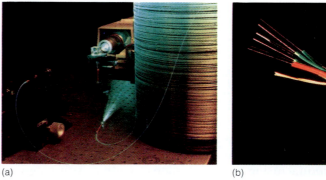

(a)

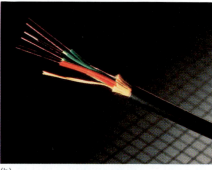

(b)

now being used by the telephone companies to transmit signals and messages instead of using electrical wires. Fiber optics has an advantage in the number of messages that can be carried simultaneously by one carrier.

Another exciting area of current research and development is the formation of **glassy metals** by extremely rapid cooling of molten metals. The absence of grain boundaries gives them very interesting mechanical and electrical properties, and resistance to corrosion.

Crystalline Ceramics

Ceramics are held together by ionic and covalent bonds and the weaker van der Waals forces. When a ceramic is fractured, it can be expected that the weakest bonds break first. For example, clay has layers that are held together by ionic and covalent bonds, but the layers are held to each other

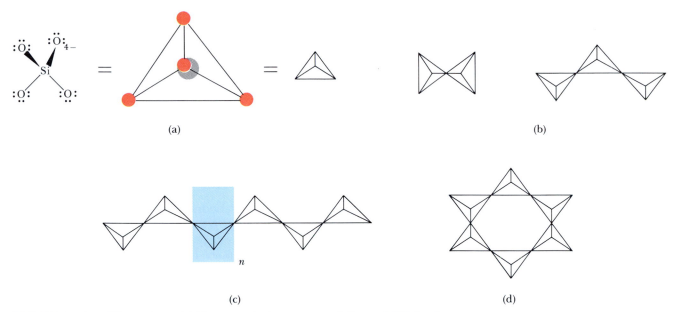

(a)

(b)

(c)

(d)

FIGURE 20–25 Silicate ions. (a) Three equivalent representations of SiO_4^{4-}, the orthosilicate ion; (b) disilicate, $Si_2O_7^{6-}$, and trisilicate, $Si_3O_{10}^{8-}$; (c) an "infinite" single chain silicate, $(SiO_3^{2-})_n$; (d) a ring silicate $Si_6O_{18}^{12-}$.

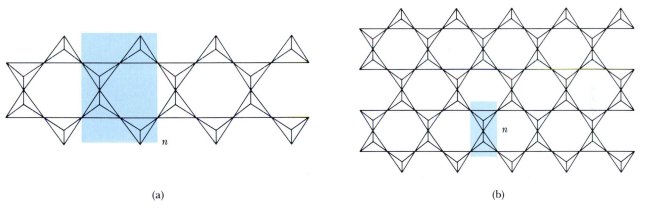

(a)

(b)

FIGURE 20–26 Complex silicate ions. (a) An "infinite" double chain silicate, $(Si_4O_{11}^{6-})_n$; (b) an "infinite" sheet silicate ion, $(Si_2O_5^{2-})_n$.

by van der Waals bonds. Therefore, clays cleave in a direction parallel to these layers. Examples of ceramics with ionic bonding are metal oxides such as Al_2O_3 and MgO. Magnesium oxide has the sodium chloride structure (which we discussed in Chapter 19). The bonding in diamond is purely covalent. Quartz, a crystalline form of silica (SiO_2), tungsten carbide (WC), and many other ceramics have bonding that is intermediate between purely ionic and purely covalent. Ceramics from fired clays contain ionic and covalent bonds and van der Waals forces.

The most common ceramics are the silicates, which include fired clays, and Portland cement. A tetrahedron of four oxygen atoms surrounding a central silicon atom makes up the basic building block of the structure. One such unit has the formula SiO_4^{2-} and is called orthosilicate. Its structure and those of several other silicate ions are given in Figures 20–25 and 20–26. The Lewis structure of SiO_4^{4-} and two other representations of this ion are given in Figure 20–25a. Note that the Lewis structure requires 32 electrons, whereas only 28 electrons are available in the valence orbitals

The term "asbestos" is generic, referring to a group of impure magnesium silicate minerals that occur in fibrous form. Typical colors are white, gray, green, and brown. It is now well known that asbestos dust particles are highly toxic and carcinogenic by inhalation and as a result, substitutes are being sought; one promising material is glass fiber made from slate and limestone.

of the elements. Hence, the ion must have a charge of -4. In general, for each terminal oxygen atom—that is, an oxygen atom not shared by another silicon—a charge of -1 is necessary. The reason for this is that the oxygen needs to be surrounded by eight electrons for a stable configuration; six of these electrons come from the oxygen, one comes from the silicon it is bonded to, and one is extra and accounts for the -1 charge.

For the other silicate ions in Figures 20–25 and 20–26, the SiO_4 unit is simply shown as a tetrahedron. It is assumed an oxygen atom sits at each of the four corners of the tetrahedron with a silicon atom at the center. As we go through these figures, count up the numbers of silicon atoms, oxygen atoms, and negative charges. The disilicate ion has two silicon atoms, one in the center of each tetrahedron. There are six terminal oxygen atoms and one shared between the two tetrahedra, for a total of seven. The six terminal oxygen atoms impart a charge of -6. The disilicate ion's formula is thus $Si_2O_7^{6-}$. Draw the Lewis structure of this ion, showing that a charge of -6 is correct. Similar reasoning shows that the formula for trisilicate is $Si_3O_{10}^{8-}$.

Longer and longer single chain silicate ions can be formed. To determine the formula of a polymeric silicate, it is necessary to locate the building block unit that, when repeated over and over, generates the whole structure. In Figure 20–25c, a building block has been marked off; n is a large number that gives the total length of the ion. Within the building block there is a single silicon atom, plus two terminal oxygen atoms and two oxygen atoms shared with neighboring building blocks. Any atom shared between two neighboring blocks can be counted only as one-half in each block. Thus, there are three oxygen atoms in each block and the formula is $(SiO_3^{2-})_n$.

Figure 20–25d shows a ring silicate ion with the formula $Si_6O_{18}^{12-}$. Note that the proportional numbers of atoms and charges are the same as for the infinite chain silicate in Figure 20–25c. A double chain silicate is shown in Figure 20–26a. Counting the atoms in the building block that has been marked gives the formula $(Si_4O_{11}^{6-})_n$. This structure occurs in some important minerals called amphiboles. In these minerals, the very long ions are separated by cations such as K^+ and some water molecules. The bonding that holds parallel silicate ions to each other is much weaker than the bonds within the ion itself. Thus, the mineral cleaves between the ions, giving mineral fibers such as those in asbestos.

An infinite sheet silicate ion with the formula $(Si_2O_5^{2-})_n$ is shown in Figure 20–26b. Such ions occur in micas and talcs. In these minerals the sheets are separated by cations and water molecules, and cleavage takes place between the sheets, yielding flat platelets. When micas are subjected to rapid heating, the water between the sheets is vaporized quickly, puffing it up and giving the familiar packing material called "vermiculite."

EXAMPLE 20–7
Give the formula for tetrasilicate, a single chain containing four SiO_4 groups.
Solution Four SiO_4 groups contain four silicon atoms and 16 oxygen atoms, with a total charge of -16. Three oxygen atoms and six negative charges are removed to link the structure together, so the formula is $Si_4O_{13}^{10-}$.

EXERCISE 20–7
Give the formulas for the following silicate ions.
(a) A ring silicate containing eight tetrahedra.
(b) An infinite double chain silicate based on rings of four tetrahedra each.
Answer: $Si_8O_{24}^{16-}$, $(Si_2O_5^{2-})_n$.

So far, our silicate ions have included isolated ions and polymers that extend in one and two dimensions. Is it possible to form a silicate ion that extends in three dimensions? Various ways of attaching the SiO_4 tetrahedra to form a three-dimensional network can be imagined. However, extending the network in three dimensions requires that all four of the oxygen atoms of each tetrahedron be shared with another tetrahedron. (Recall that to form a sheet extended in two dimensions, three oxygen atoms per tetrahedron must be shared; see Figure 20–26b.) If all four oxygen atoms are shared, there is no excess charge; the silicate is no longer an ion but rather a neutral species. Furthermore, since each silicon atom will be surrounded by four oxygen atoms, each of which is shared, the formula must be SiO_2, the same as silica in quartz or sand.

However, it is possible to construct an ionic, three-dimensional silicate structure by replacing some of the silicon atoms by aluminum atoms. Since aluminum is similar to silicon in size and electronegativity, this is a reasonable substitution. However, the aluminum atom has only three valence electrons compared to four for silicon. Therefore, for every silicon atom that is replaced by an aluminum atom, an extra electron and a charge of -1 is also gained. The general formula of these aluminum-substituted silicas or **aluminosilicates** is $Si_mAl_nO_{2(m+n)}{}^{n-}$. Examples of aluminosilicates with structures that extend in all three directions are feldspars and zeolites.

Aluminosilicates also form layered structures. A structure can be formed of two sheets (as shown in Figure 20–26b) that share the formerly unshared oxygen atom at the apex of each tetrahedron. Without replacing silicon atoms by aluminum, this would be a neutral species with the formula SiO_2. Replacing some silicon atoms by aluminum atoms leads to an ionic layer structure. Such layer structures appear in micas and talcs, and in the very important precursors of ceramics, the clays. Because of their charged layer structures, clays can absorb large amounts of water. This and the sheetlike cleavage are what give clays their greasy feel. The wet mass is easily shaped to the desired configuration and then will dry to a rigid mass. The product at this point, called "greenware," is weak and brittle and has no practical uses. Slow heating to a high temperature or "firing" is necessary to produce the desired ceramic properties.

Mica is the name given to several silicate minerals of varying chemical composition but similar physical properties and crystalline structure. All micas characteristically cleave into thin sheets which are flexible and elastic. Uses range from electrical equipment, vacuum tubes and incandescent lamps, to lubricants, cosmetics, and dusting powders.

During the firing process the ceramic shrinks and is strengthened as a number of chemical changes take place:

1. Water is driven off. This includes water that was merely adsorbed as well as water that is the product of chemical reactions.
2. Calcium and magnesium carbonates decompose:

$$CaCO_3\ (s) \longrightarrow CaO\ (s)\ +\ CO_2\ (g)$$

3. The metal oxides produced react with silica to produce silicates:

$$m\ CaO\ +\ n\ SiO_2 \longrightarrow Ca_mSi_nO_{m+2n}$$

4. Organic matter in the clay is decomposed and oxidized.
5. Quartz is transformed to a different structure called crystobolite.
6. A material called mullite ($3Al_2O_3 \cdot 2SiO_2$) is formed with long needle-like crystals that reinforce the structure.

Portland cement is a ceramic of exceptional importance and utility. It reacts with water to harden slowly, forming a hard durable solid. The fact that it hardens even when submerged in water makes it useful for bridge abutments and other applications where such conditions are encountered. Properly reinforced, Portland cement can be used as a substitute for steel beams and even for the hulls of boats.

Portland cement is manufactured by firing a mixture of clay and ground limestone ($CaCO_3$) in a cement kiln. The limestone is converted to calcium oxide, and then a number of reactions finally lead mainly to $2CaO \cdot SiO_2$ (dicalcium silicate) and $3CaO \cdot SiO_2$ (tricalcium silicate). This mixture is then cooled and mixed with gypsum ($CaSO_4 \cdot 2H_2O$), which regulates the hardening rate, and the mixture is ground to a fine powder. A number of slow and complex reactions take place when the powder is mixed with water. These continue as the cement slowly hardens for more than a year.

Other important ceramic materials include alumina, Al_2O_3, and magnesia, MgO. Alumina melts at over 2000°C, so it is an important refractory. In practice, grains of alumina are pressed together (sintered) at high temperatures with a small amount of flux. The flux melts and also lowers the melting point of the alumina at the surface of the grains so that they stick together. Alumina tubes are used to contain the sodium in sodium vapor lamps because the material is resistant to sodium, and it is translucent when formed with a microstructure that is almost pore-free. Magnesia has the properties of being an electrical insulator and yet a good conductor of heat. It is used to insulate the heating elements of electric stoves.

Still other ceramics are the carbon compounds of metals, including tungsten carbide (WC), silicon carbide (SiC, known by the trademark "Carborundum"), and boron carbide (B_4C). All are very hard with high melting points.

20.8 POLYMERS

As has so often been the case, craft tradition and technology led the way in advance of polymer science by a century. Anecdotal history has it that Columbus found the natives of the West Indies playing games with rubber balls, and rubber artifacts have been found among the Mayan ruins in Mexico's Yucatan Peninsula. Natural rubber from South American and Southeast Asian plantations was mechanically worked (masticated) to make it suitable for dissolving or coating cloth at the beginning of the 19th century. The first patents were issued to Charles Goodyear for the curing or vulcanization of rubber in the 1840s. That discovery eliminated the stickiness of the natural product and led to commercialization.

Natural fibers of plant and animal origin such as cotton and wool have, of course, been used for thousands of years. Cellulose nitrate, a modified plant product, was discovered in the mid-19th century and used by John

The name "rubber" has been attributed to Priestley, who first discovered the ability of the material to "rub out" pencil marks.

Leo Baekelund

BAKELITE

Leo Baekelund (1863–1944), a young Dutch emigré chemistry professor, inventor, and entrepreneur, was in search of a synthetic substitute for shellac. While working in his Bronx, New York, garage laboratory, he came upon a phenol-formaldehyde formulation that, once formed, was very hard and intractable, could be cut and machined, was an excellent electrical insulator, and at the same time, water- and organic solvent-resistant. He recognized potential applications for such a material, refined his product and process to maximize its unique properties, and began to market his invention in 1909 under the TM "Bakelite," after himself. It was not the first commercial plastic material—celluloid retains credit for that—but it was of major industrial importance, and nearly a century later, is still manufactured and used at a level of multi-millions of pounds per year, worldwide. It was Bakelite that kicked off the development of modern plastics.

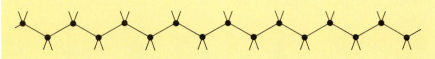

Hyatt as early as 1869. Hyatt was seeking a substitute for ivory and found a product that was truly revolutionary. This Celluloid, as it came to be known, showed up in everything from men's shirt collars to the first motion picture films. But Leo Baekelund's discovery of the phenolic plastic resin known as Bakelite probably was the event that pushed us into the Age of Plastics. It was the first major commercial polymeric material. Nylon, dis- covered some twenty years later, was the second.

Finally, in the 1930s, the science caught up with the technology as it became generally accepted that natural rubber and other long-chained collections of carbon atoms we now call polymers are composed of covalent bonds of the usual kind, and that they differ from ordinary molecules only in size. We now recognize that "the size" makes the difference. The waxy petroleum products used in making candles are composed of chains of carbon atoms as well, but perhaps only 30 to 40 atoms in length, giving molecular weights of 300 to 400. That is not what we mean by "giant" molecules of long chain length. When the length of the chain results in molecular weights on the order of 30,000 to 40,000—and beyond—then we begin to see the materials properties that make collections of these molecules suitable for applications as widely varied as Saran Wrap, phon- ograph records, and pipe for plumbing.

The typical backbone of a carbon-based polymer has two atoms or groups of atoms attached to each carbon atom (Figure 20–27). Poly- ethylene, $H—(C_2H_4)_n—H$, is the simplest polymer. The all-carbon chain is relatively flexible. Polymers in which some of the backbone atoms are oxygen, nitrogen, or silicon, or rings of carbon atoms, have stiffer backbone chains. Because of the simplicity of the polyethylene structure, it is possible to grow this material with a high degree of crystallinity, provided the back- bone chain is not branched to any significant degree. However, this is unusual for polymers, most of which are noncrystalline, amorphous struc- tures. Polyethylene can be prepared both ways. The amorphous form is transparent; the crystalline form is translucent because of the grain boun- daries. Polyesters such as Mylar, Lucite, and Plexiglas are crystalline poly- mers; polycarbonates such as Lexan are amorphous polymers.

There are two important groups of plastic polymers—thermoplastics (TP) and thermosets (TS):

- A **thermoplastic polymer** softens and can be made to flow when heated. On cooling, it hardens and retains the shape established by molding or extrusion at the higher temperature, and the heating and cooling cycle can be repeated. Polyethylene, polystyrene, and polyvinyl chloride are typical examples.

- A **thermoset polymer** hardens (or cures), becoming fixed in a given shape, by a chemical reaction that produces a three-dimensional network. Heat and pressure are generally used to kick off the reaction. Once formed, the cured polymer is generally insoluble and cannot be heated to deform it without causing chemical decomposition. Familiar examples are the Bake- lite and Formica materials, the epoxy resins used to glue things perma- nently, and the vulcanized rubber so widely used in tire materials.

Construction of Formica, a lami- nate, with an overlay, a decorative sheet, and a barrier sheet, which are dip-coated with melamine- phenol/formaldehyde resin and pressed together at high tempera- ture, producing a thermoset com- posite that is widely used for deco- rative applications as surfacing (kitchen counter and table tops) and as adhesives for bonding lam- inated plastics to other surfaces.

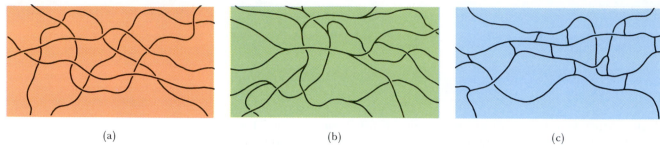

(a)   (b)   (c)

FIGURE 20–28 Three types of polymers: (a) chains arranged in a linear (though spaghetti-like) collection; (b) with branches along the main chains; (c) with cross-links between the main chains.

The different properties of these materials can be understood in terms of the model polymer structures shown in Figure 20–28. In Figure 20–28a the individual molecules are linear and in Figure 20–28b they are branched. van der Waals bonds hold one molecule to another. In either of these cases, heating the polymer provides enough energy to break the van der Waals bonds, allowing the molecules to begin to slide and separate from each other. This thermoplastic behavior increases with the temperature, increasing the plasticity of the material. In the normal case of an amorphous polymer, no definite melting point is observed, but there is an approximate softening point like that of a glass.

Our model structure for a thermosetting resin is shown in Figure 20–28c. Here, the long-chain polymer molecules are held together by crosslinking the chains. These crosslinks give the material additional strength and prevent softening and melting. Normally the crosslinks are formed when the original material is heated.

The actual chemical structures of important polymers and their preparation will be discussed in some detail in Chapter 25. However, we can make a few observations about the effects of their chemical makeup on their important properties. Polyethylene has a relatively low tensile strength. This can be improved by replacing some of the hydrogen atoms on the carbon backbone with other atoms or groups of atoms. For example, if one hydrogen on every other carbon is replaced by a chlorine atom, the resulting polymer is polyvinyl chloride. This polymer has a tensile strength two to three times that of polyethylene. Imagine polyethylene under stress and picture the long molecules being pulled past each other. The larger chlorine atoms tend to catch on each other as the molecules pull past, thus increasing the tensile strength. Furthermore, the chlorine atom is at the negative end of the carbon-chlorine dipole whereas hydrogen is at the positive end of the carbon-hydrogen bond dipole. These opposite poles will attract each other to form stronger van der Waals bonds, thus also increasing the strength of the material.

In the remarkable fluorocarbon polymer known as Teflon, all of the hydrogen atoms in polyethylene have been replaced by fluorine atoms. In this polymer, each fluorine atom is at the strongly negative end of the carbon-to-carbon fluorine dipole. These negative poles repel each other; as a result, the material has little strength and creeps readily at room temperature. But this opaque, soft, and waxy white crystalline polymer has marvelous electrical resistance, is virtually immune to chemical attack, and has the lowest (unlubricated) coefficient of friction of any known material. Teflon has become famous as a coating for kitchen utensils. Although it is

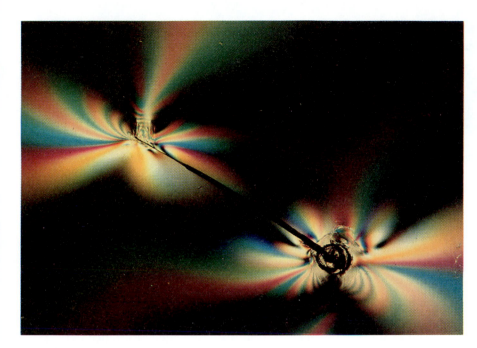

Polarized light has been used to reveal the stresses (lying perpendicular) along a crack in a sheet of plastic material. Two smaller cracks have propagated from the lower end of the large one, creating additional stress patterns. Smaller circular flaws surround the upper end of the large crack.

defined as a thermoplastic because it is not crosslinked, its use characteristics and processing properties make it virtually a thermoset. It can be extruded or injection-molded only by applying the "sintering" techniques used to fabricate powders. But once fabricated, it has service temperatures across a huge range from near 0 K to several hundred Kelvins (Figure 20–29).

Silicones are ceramic-organic polymers, sharing some of the characteristics of both. The polymer is not carbon-based; it consists of a backbone of alternating silicon-to-oxygen bonds instead. If the chain lengths are relatively short, the products are silicone fluids that are widely used as high temperature oils and greases, and as dielectrics or fluid insulators. They are serviceable to several hundred Kelvins, which is well beyond the useful range of petroleum-based oils and greases; also, the fluids do not congeal at low temperatures as petroleum lubricants do. Longer chain lengths give solid silicones that have excellent insulating properties, make water-repellent coatings for fabrics and finishes, and have good elastic properties. The well-known caulking materials such as the RTV polymers—cross-linked silicone elastomers—can be used in showers, bathtubs, and sinks because of their water-repellent character.

Polymers are generally susceptible to chemical and environmental damage. Nonpolar solvents such as carbon tetrachloride and benzene can penetrate between the polymer chains, swelling the polymer and even dissolving it. In fact, polyethylene can be recrystallized from benzene-like solvents. As you might expect, thermosetting resins are far more resistant to solvent action because of their crosslinks. These problems limited the use of plastic materials in automobiles until recently because of contact with oils, greases, and gasolines.

The molecules in polymers are generally susceptible to heat and degradation by oxygen of the air. Applying excessive heat results in breaking covalent bonds, and oxidative degradation occurs at elevated temperatures, resulting in shortening of the chain lengths and eventual loss of characteristic polymeric properties. Because fluorine does not react with oxygen

FIGURE 20–29 Plasma techniques can be used to coat with many modern materials—Teflon is one. The coating material, in powdered form, is blown into the spray nozzle at elevated temperatures and is instantly carried to the target—in this case, a metal bucket—by a jet of hot gases. The spray is passed back and forth across the surface several times to achieve the desired coating buildup.

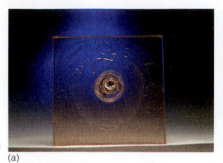

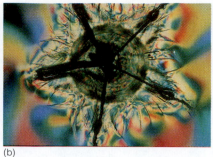

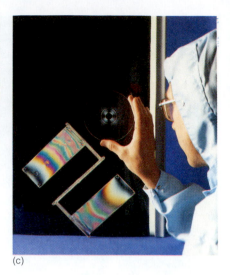

FIGURE 20–30 (a) LEXAN™ polycarbonate, in the form of LEXGARD™. The left photo shows a 1-inch-thick composite piece that has successfully stopped a .357 magnum round—a "Dirty Harry" round; (b) a close-up view under polarized light shows stresses along the fracture lines. (c) By contrast, the material-of-construction of the compact disk is all but taken for granted, yet the requirements the polycarbonate polymer must meet for this application are stringent.

FIGURE 20–31 Barrier bottles and jars are in the process of revolutionizing the packaging of foods and the supermarket shelf, once almost entirely the domain of metal, paper, and glass packages and containers. One of the first was the 5-ply ketchup bottle composed of three different polymers, polycarbonate to provide strength during hot-filling and breakage-protection, a moisture barrier of polypropylene and an oxygen barrier of polyethylene/vinyl alcohol copolymer, all to extend shelf life indefinitely, and two adhesive "tie" layers to bond the three materials together. The bottle is co-blow-molded, with all five layers coming together as the bottle is formed.

to form many stable molecules, Teflon polymers have good oxygen resistance. Inclusion of already oxidized groups such as the Si—O—Si—O—Si—O backbone of silicone polymers also improves oxygen resistance.

Some polymers, especially thermoplastics, are depolymerized by ultraviolet radiation, so they weather poorly in out-of-doors applications. That shows up as a yellowing or general discoloring, which can usually be overcome by adding UV-absorbers or by surface-coating the material.

In an era that is notable for its innovative use of materials, it is interesting to note some of the special features of engineering plastics such as polycarbonate polymer products. LEXGARD™ provides unique protection of the most demanding kind, typically against handguns (Figure 20–30); and although not nearly as dramatic a success, the multilayered ketchup bottle draws strength and character from polycarbonates, while other layers deflect oxygen in the air from entering and prevent moisture from leaving (Figure 20–31).

SUMMARY

Materials for engineering purposes can be divided into five important categories: metals and alloys, ceramics and glasses, polymers, composites, and semiconductors. Each of these categories has a distinct range of structures that result in particular properties. Knowledge of these properties is essential in selecting materials for specific uses. The bonds in materials include all of those we have previously encountered—ionic, metallic, covalent, and van der Waals. The strength of a material and its performance characteristics depend on the bonding within and on the surface properties.

In a general way, materials made of metal, wood, glass, and plastics have particular responses to mechanical, thermal, environmental, and electrical demands, which in turn dictate the choice of material for a particular application. Stress, strain, and modulus of elasticity are the most widely measured performance characteristics. Comparisons among metals and plastics are particularly important because, in many ways, they resemble each other. It is therefore not surprising that plastics have successfully replaced metals in applications where they are superior.

Metals are normally made up of crystals or grains. The grains are frequently solid solutions of metals. The Hume-Rothery rules allow us to predict which metals will dissolve significantly in each other. Alloy composition and crystal defects are important in determining the strength of a metal. Many important alloys have polyphase structures with different kinds of crystal grains present. These affect the properties of the metal significantly. Phase diagrams are used to help understand polyphase systems.

Glasses have amorphous structures and are most commonly based on silica. Silicate ions are also very important in crystalline ceramics. Structures for a large number of silicate ions can be drawn. Substitution of aluminum for silicon atoms leads to aluminosilicate ions. Such ions in sheet-like forms are the basis of clays. Many important ceramics result from clays, which are shaped into the desired configuration and then fired.

Polymers contain very long molecules, most usually consisting of long chains of carbon atoms. These molecules may be linear or branched. If the molecules are not crosslinked, the polymer is thermoplastic and will soften when heated. If the molecules are crosslinked, the polymer is a thermosetting resin that does not soften. A number of important aspects of polymer behavior can be related to chemical structure.

QUESTIONS AND PROBLEMS

QUESTIONS

1. Assign each of the following materials to one of the five major materials categories.
(a) "Plexiglas" or "Lucite"
(b) asphalt used for roads
(c) paper
(d) rope
(e) blackboard chalk
2. Why do pure metals find limited use, although applications for alloys abound?
3. Why are metals corroded by oxidizing agents such as acids and oxygen, whereas ceramics are resistant?
4. What properties allow polymers to replace metals for some applications? In what general aspects are polymers inferior to metals? In what respects are they superior?
5. What does it mean to say that a bond is directional? Which kinds of bonds are directional and which are not?
6. How can you show whether a structure is closest-packed?
7. What is the difference between a crystal and a grain? Why do grains grow in the shapes they do?
8. The Hume-Rothery rules say that in order to form a solid solution, metals should have about the same electronegativity. What is the reason for this? If, say, potassium and tin were compatible in the other respects, what effect would the difference in their electronegativities have?
9. Explain the difference between a substitutional and an interstitial solid solution.
10. What is the definition of a simple metal? A glass? A ceramic? A plastic?
11. Briefly explain the effect on the properties of a metal of substituting some different atoms into the structure.
12. Differentiate between elastic and plastic strain. Explain this in terms of macroscopic properties as well as in atomic terms.

13. Explain why tensile strengths for metals are only about one-tenth of the theoretical values.
14. Why does work hardening increase the tensile strength of a metal? What other properties are also affected by cold working?
15. Explain under what conditions a polyphase structure can increase the strength of a metal.
16. Describe the behavior of a glass on heating. How does this differ from the behavior of a metal or a crystalline ceramic?
17. Why is fused silica in very limited use as a glass? How is it modified to make it more usable?
18. Give examples in which the structures of silicates and aluminosilicates affect the properties of the material in which they are contained.
19. Why is clay able to hold so much water?
20. How could you distinguish between crystalline and amorphous polyethylene?
21. Describe the differences between thermoplastics and thermosets in terms of their molecular structures. List as many differences in properties as you can that distinguish between these two groups of polymers.
22. Why is it so difficult for synthetic polymers to crystallize?
23. In what way is crystallinity associated with greater density in a thermoplastic?
24. Summarize the performance characteristics of Teflon. What's good? What's not so good?
25. What distinguishes silicone polymers from the usual polymeric materials?
26. Offer a brief explanation of why window glass is transparent to visible light.
27. What is a eutectic?

28. In what ways have the elastic and plastic performance characteristics of steel established its universal value?

29. Imagine a material that is elastic in the solid phase under all conditions. How might you fabricate it into a useful product? Suggest more than one way.

30. Distinguish between asphalts and cements. What is the principal difference?

31. From your experience, name several materials—metals, plastics, and wood products—that creep at room temperature over a long time.

PROBLEMS

Types of Materials [1–2]

1. For each of the following applications, list the properties required and suggest which category or categories of materials would be applicable. Do not hesitate to suggest possibilities that are not currently in use.
(a) paper clip
(b) phonograph record
(c) rain coat
(d) chair
(e) insulator for carrying a high voltage conductor

2. List the properties required and suggest which categories of materials would be applicable for each of the following applications.
(a) stock for a rifle
(b) fender of a car
(c) bumper of a car
(d) driveshaft of a car
(e) cover of a golf ball

Bonding in Materials [3–4]

3. Predict approximate melting points for the following materials:
(a) SiC
(b) MgO
(c) Cu

4. Give your best estimate of the melting point of each of the following substances.
(a) I_2
(b) polypropylene
(c) LiF

Stress and Strain [5–6]

5. Determine the strain (measured as in/in) in a magnesium bar stressed in tension to 10,000 psi if the modulus of elasticity is 6.5×10^6 psi.

6. A mass of 50.00 kg is found to increased the length of a 1000.00 mm wire by 0.423 mm. The cross-sectional area

of the wire was 0.10 mm². What is the modulus of elasticity of the wire?

Solid Solutions [7–8]

7. What metals would you expect to form substitutional solid solutions with beryllium?

8. What metals would you expect to form solid solutions with nickel?

Phase Diagrams [9–10]

9. Using Figure 20–21, tell what phases would be present in the copper-silver system and what the composition of each phase would be for each of the following sets of conditions.
(a) 70% Ag; 900°C
(b) 90% Ag; 800°C

10. Use the copper-silver phase diagram to tell what phases would be present and what would be the composition of each phase for each of the following temperatures and weight percentages.
(a) 98% Cu; 800°C
(b) 10% Cu; 500°C

Ceramic Materials [11–14]

11. Draw silicate ions that are consistent with each of the following formulas.
(a) $(Si_4O_{11}{}^{6-})_n$. Try to come up with a structure different than the one we have mentioned.
(b) $(Si_2O_5{}^{2-})_n$. Again, try to come up with a new structure.

12. Draw silicate ions that would fit the following formulas.
(a) $Si_5O_{16}{}^{12-}$
(b) $Si_4O_{12}{}^{4-}$

13. Calculate the correct value for x and y in each of the following aluminosilicate ions.
(a) $Si_3AlO_x{}^{y-}$
(b) $Si_2Al_xO_8{}^{y-}$

14. Determine x and y for the following ions.
(a) $Si_xAl_4O_{12}{}^{y-}$
(b) $Si_4Al_xO_{12}{}^{y-}$

Additional Problems [15–18]

15. Calculate the atomic packing factor for the face-centered cubic structure.

16. Which metals do you think would be able to form substitutional solid solutions with chromium?

17. Write the balanced equation for the complete reaction of polyethylene with air.

18. Determine the 0.2 percent yield stress and the modulus of elasticity for the two materials in the accompanying figures:

(For Problem 18)

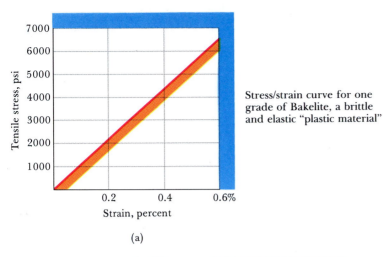

Stress/strain curve for one grade of Bakelite, a brittle and elastic "plastic material"

(a)

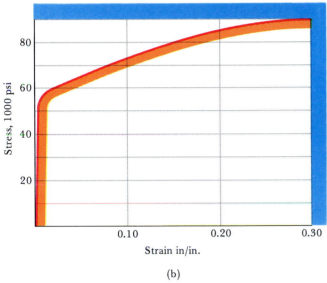

Stress/strain curve for a commercially pure titanium sheet, aircraft quality

(b)

MULTIPLE PRINCIPLES [19]

19. A sample that contains 55.0% by weight of Ag_2O and 45.0% by weight of CuO reacts with hydrogen to produce the metals, which are then heated to form a liquid-liquid solution and cooled. At what temperature will the solution start to solidify? How many phases will be present and what will be their compositions when the temperature has dropped to 700°C?

APPLIED PRINCIPLES [20–25]

20. Calculate the stress on a wire with a cross-sectional radius of 0.10 mm if a mass of 10.0 kg is supported by it.
21. Calculate the strain for the wire in Problem 20 if it increases elastically in length by 4.35% when stressed by the mass. What is the elastic modulus of the wire?
22. A 3.500-m length of a wire that has an elastic modulus of 1.2×10^{12} dynes/cm² increases elastically to a length of 3.582 m when a load of 50.00 kg is suspended by it. What is the diameter of the wire?
23. Copper is malleable and can be "hammered" flat at room temperature. What is the effect on the number of defects and dislocations in the crystal structure?
24. One of the hot new semiconductor materials is gallium arsenide (GaAs).
(a) What does it mean to say that it is "isoelectric" with Group IV elements such as silicon and germanium?
(b) If a small number of germanium atoms replace some gallium atoms, what type of semiconductor material results?
(c) If a small number of arsenic atoms are replaced by germanium, what type results?
25. For the purpose of suspending a heavy load from a cable, does it make a difference whether you use cable of pure iron, or one containing a small percentage of carbon? Why? If cables of equal diameter are used, which will support the greater load to failure?

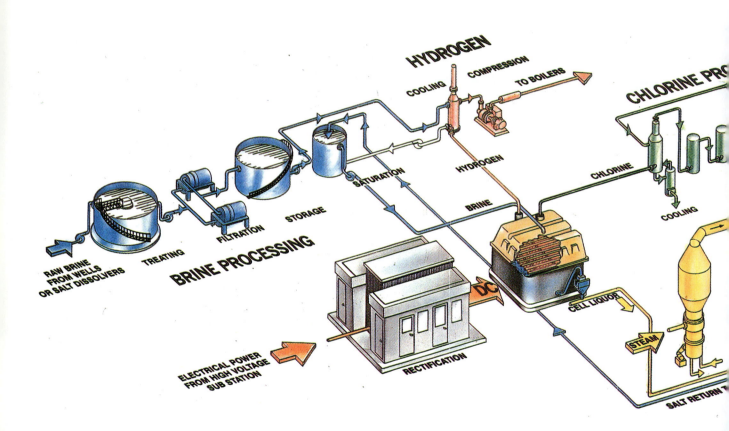

HYDROGEN

COOLING   COMPRESSION   TO BOILERS

CHLORINE PRO

SATURATION   HYDROGEN   CHLORINE

STORAGE   BRINE   COOLING

FILTRATION

RAW BRINE
FROM WELLS   TREATING   BRINE PROCESSING
OR SALT DISSOLVERS

CELL LIQUOR

ELECTRICAL POWER   DC
FROM HIGH VOLTAGE   STEAM
SUB STATION   RECTIFICATION

SALT RETURN

C H A P T E R
21

Electrochemistry

Two primary products of the chloralkali industries are produced by the electrolysis of saturated salt solutions—chlorine at one electrode, sodium hydroxide (caustic) at the other. The electrolytic chloralkali cell in the center of the flow chart contains an asbestos diaphragm that separates the oxidation (anode) process, producing chlorine, from the reduction (cathode) process, producing caustic. Following the blue, green, and yellow color codes of the schematic diagram takes you through the process. Today, newer cells fitted with porous, chemically active polymer membranes in place of asbestos are used—but the process is much as in this early chloralkali photo, dating back to the turn of the century.

21.1 INTRODUCTION TO ELECTROCHEMISTRY

Electricity plays a very important role in chemistry because electrons, the common carriers of electric current, are also the source of chemical reactivity in atoms and molecules. Bonds break and bonds form as electrons are shared and transferred. Reactions that involve the flow of electrons through a conductor are called **electrochemical reactions**, and their study constitutes **electrochemistry**. This chapter deals with these ideas and more. In earlier discussions, we introduced the free energy ΔG and described its role in predicting the spontaneous direction of chemical processes. The simplest yet most precise method for obtaining ΔG values is based on the battery—an electrochemical cell—and the value of its cell voltage. In turn, that is very useful for everything from the simple determination of pH to understanding the biological processes involved in respiration and to solving the complex problems of corrosion and rusting.

Applied electrochemistry has given us portable sources of electric current in a huge range of "dry cells" and storage batteries. Electrochemical reactions are also useful because of the products they can produce. Aluminum, sodium, and chlorine—all articles of commerce—are all prepared by electrochemical processes. Electrochemical methods are used to purify other metals of industrial importance such as copper and silver. Nylon is produced on an industrial scale by an electrochemical process. Many materials are metal-coated by electrodeposition for uses ranging from car bumpers and hub caps to compact disks. Fuel cells are a relatively new application of electrochemistry that may well store our energy in the 21st century.

21.2 OXIDATION AND REDUCTION REACTIONS

Electrochemical reactions involve the flow of electrons and ions through a conducting chemical system. In order for this to happen, electrons must be transferred from one chemical species to another. Such reactions involving electron transfer are called **oxidation-reduction reactions**. (The abbreviated term **redox reactions** is often used.) In order for a redox reaction to be an electrochemical reaction, it must take place under quite specific circumstances. For the moment we will simply concern ourselves with redox reactions; we will study these reactions under electrochemical conditions later.

The concept of oxidation states is particularly useful in analyzing redox reactions. For monatomic ions, the ion charge is the oxidation state. Neutral atoms of an element are said to be in the zero oxidation state. Thus, for a simple reaction of atoms to produce monatomic ions such as Na^+ and Cl^-,

$$2 \text{ Na (s)} + Cl_2 \text{ (g)} \longrightarrow 2 \text{ Na}^+ + 2 \text{ Cl}^-$$

the transfer of electrons can be traced by following the change in oxidation states. In this example, Na (oxidation state = 0) is converted to Na^+ (oxidation state = +1), signifying a loss of one electron. Similarly, Cl_2 (oxidation state = 0) is transformed to 2 Cl^- ions (oxidation state of each ion = −1), showing that formation of each Cl^- ion involves the gain of one electron. At this point we can introduce a few definitions:

Oxidation involves a loss of electrons.

Oxidation corresponds to an increase in oxidation state.

Reduction involves a gain of electrons.

Reduction corresponds to a decrease in oxidation state.

It is also customary to label the reactants in an oxidation-reduction reaction. In every such reaction, one reactant is oxidized and another is reduced. The reactant that is oxidized is called the **reducing agent** or **reductant**, since it causes the other reactant to be reduced. The reactant that is reduced is called the **oxidizing agent** or **oxidant**. Thus, in the reaction of Na with Cl_2, Na is oxidized and therefore is the reducing agent; and Cl_2 is reduced and is the oxidizing agent.

The use of oxidation states to analyze redox reactions goes far beyond simple reactions of elements to form monatomic ions. We have previously (Sec. 17.5) discussed calculating the oxidation states of the elements in a wide variety of ionic and covalent compounds as well as ions themselves. In the following aqueous redox reaction, the oxidation state of each element is shown beneath the equation.

$$H_2O + SO_3^{2-} + 2\ Fe^{3+} \longrightarrow SO_4^{2-} + 2\ Fe^{2+} + 2H^+$$
$$+1-2 \qquad +4-2 \qquad +3 \qquad\qquad +6-2 \qquad +2 \qquad +1$$

Notice that two elements have changed their oxidation states. S has been oxidized from $+4$ in SO_3^{2-} to $+6$ in SO_4^{2-}. Fe has been reduced from $+3$ in the Fe^{3+} ion to $+2$ in the Fe^{2+} ion. In such cases, we say that SO_3^{2-} is oxidized to SO_4^{2-}, and thus SO_3^{2-} is the reducing agent or reductant. Similarly, Fe^{3+} is said to be the oxidizing agent or oxidant.

EXAMPLE 21–1

Identify the oxidizing and reducing agents in the following balanced reaction in aqueous acidic solution:

$$ClO^- (aq) + 2\ H^+ (aq) + Cu (s) \longrightarrow$$
$$Cl^- (aq) + H_2O (liq) + Cu^{2+} (aq)$$

Solution Begin with the oxidation state of each atom,

$$ClO^- (aq) + 2\ H^+ (aq) + Cu (s) \longrightarrow$$
$$+1-2 \qquad\qquad +1 \qquad\qquad 0$$
$$Cl^- (aq) + H_2O (liq) + Cu^{2+} (aq)$$
$$-1 \qquad\quad +1-2 \qquad\quad +2$$

The oxidation state for Cu is observed to increase from 0 to $+2$. Therefore, Cu is oxidized and acts as the reducing agent, reducing the Cl in ClO^- from $+1$ to -1. Thus, ClO^- is reduced and is the oxidizing agent. ∎

EXERCISE 21–1

Identify the oxidizing and reducing agents in the following reaction:

$$7\ H_2O + 2\ Cr^{3+} + XeF_6 \longrightarrow Cr_2O_7^{2-} + 14\ H^+ + Xe + 6\ F^-$$

Answer: Oxidizing agent, XeF_6; reducing agent, Cr^{3+}. ∎

21.3   HALF-REACTIONS IN REDOX PROCESSES

It is often useful in studying chemical processes to break the process into **half-reactions**. We have already done this in the Bronsted definition of

The pyrotechnic display was produced by introducing finely divided iron filings into the Bunsen flame. Oxygen is the oxidizing agent and the "redox" reaction is straightforward:

$$4\ Fe\ (s) + 3\ O_2\ (g) \longrightarrow$$
$$2\ Fe_2O_3\ (s)$$

Though no less dramatic in quality, the chemistry accompanying the oxidation of finely divided zinc filings by ammonium nitrate is not direct, nor immediately obvious:

$$Zn\ (s) + NH_4NO_3\ (s) \longrightarrow$$
$$N_2\ (g) + ZnO\ (s) + 2\ H_2O\ (g)$$

Visually, and otherwise, the combustion of aluminum foil in contact with liquid bromine is an oxidation, though oxygen itself is clearly absent:

$$2 \text{ Al (s)} + 3 \text{ Br}_2 \text{ (liq)} \longrightarrow \text{Al}_2\text{Br}_6 \text{ (s)}$$

acid-base reactions, in which a typical half-reaction is

$$\text{NH}_4^+ \text{ (aq)} \longrightarrow \text{NH}_3 \text{ (aq)} + \text{H}^+$$

This is a half-reaction because of the proton (H^+), which is a highly reactive species. We could calculate the energy of the reaction if the state of the proton were defined. Two half-reactions can be combined into a complete reaction, provided that the reactive species produced in one half-reaction is used up in the other.

Half-reactions are particularly useful in the study of redox reactions and electrochemistry. The reactive species in this case is the electron (e^-):

$$2 \text{ Cl}^- \text{ (aq)} \longrightarrow \text{Cl}_2 \text{ (g)} + 2 \ e^-$$

To produce a reaction it is necessary to combine this half-reaction with one that uses up electrons. For example:

$$\text{Au}^{3+} \text{ (aq)} + 3 \ e^- \longrightarrow \text{Au (s)}$$

In order that the number of electrons produced equal the number of electrons consumed, multiply the first half-reaction by 3 and the second by 2 before combining. The complete reaction is thus

$$3[2 \text{ Cl}^- \text{ (aq)} \qquad \longrightarrow \text{Cl}_2 \text{ (g)} + 2 \ e^-]$$
$$2[\text{Au}^{3+} \text{ (aq)} + 3 \ e^- \longrightarrow \text{Au (s)}]$$
$$\overline{6 \text{ Cl}^- \text{ (aq)} + 2 \text{ Au}^{3+} \text{ (aq)} \longrightarrow 3 \text{ Cl}_2 \text{ (g)} + 2 \text{ Au (s)}}$$

and the number of electrons transferred in the reaction (as written) is six.

A half-reaction that is written so as to produce electrons is called an **oxidation half-reaction** since the chemical species involved is oxidized. If the direction of the half-reaction is reversed; it becomes a **reduction half-reaction**.

21.4   BALANCING OXIDATION-REDUCTION EQUATIONS

Some redox reactions can be balanced easily by inspection—by trial-and-error. However, many redox reactions are complex and difficult to balance. The situation is further complicated by the fact that the balancing of many redox equations in aqueous solution involves incorporation of H_2O, H^+, or OH^- from the water solvent into the reaction. Generally, in oxidation-reduction reactions, H^+ (aq) is written instead of H_3O^+ (aq). This simplifies the balancing process a little, but it should always be remembered that H^+ (aq) and H_3O^+ (aq) are equivalent. As an example, consider the following redox reaction:

$$\text{I}_2 \text{ (s)} + \text{PbO}_2 \text{ (s)} \longrightarrow \text{IO}_3^- \text{ (aq)} + \text{Pb}^{2+} \text{ (aq)}$$

Although it is a typical redox reaction, simple inspection probably will not easily lead to a balanced equation. Bear in mind that the charges as well as the elements must balance.

The key to success in balancing redox equations is to be sure that the electrons are balanced as well as the elements and charges. That is, the number of electrons produced in the oxidation half of the reaction must be the same as the number of electrons used up in the reduction half. There are several different methods for balancing redox equations, and we will look at two of these. One of these methods uses oxidation states and the other uses half-reactions.

In either of these methods, the first step is to write out as much of the reaction as is known. Then the oxidation state of every element on each side of the reaction is indicated. Use the rules for determining oxidation states listed in Section 17.5. In this way it is possible to identify what is oxidized and what is reduced. For example, in the preceding reaction, the oxidation states are

$$\underset{0}{I_2\,(s)} + \underset{+4-2}{PbO_2\,(s)} \longrightarrow \underset{+5-2}{IO_3^-\,(aq)} + \underset{+2}{Pb^{2+}\,(aq)}$$

We can see that I_2 is oxidized (I goes from 0 to $+5$) and PbO_2 is reduced (Pb goes from $+4$ to $+2$).

At this point the two methods take different routes to getting the equations balanced and, in particular, ensuring that the electrons are evenly exchanged. One is the **oxidation state method** and the other is the **half-reaction** or **ion-electron method**.

Oxidation State Method: In the oxidation state method we keep in mind that a change in oxidation state is a result of the gain or loss of the same number of electrons. Thus, if the oxidation state goes up by three, three electrons have been lost. If the oxidation state decreases by two, two electrons have been gained. Therefore, making sure that the number of electrons lost in the oxidation equals the number gained in the reduction can be accomplished by keeping the total increase in oxidation state of one reactant equal to the total decrease in oxidation state of another. In the example above, iodine has an oxidation state increase of 5, whereas lead has an oxidation state decrease of 2. We can equalize the gain and loss if we allow five Pb atoms to be reduced for every two iodine atoms that are oxidized. Adding the appropriate coefficients to the equation gives us

$$I_2\,(s) + 5\,PbO_2\,(s) \longrightarrow 2\,IO_3^-\,(aq) + 5\,Pb^{2+}\,(aq)$$

Note that we already had two iodine atoms at the left, but the coefficient of 2 had to be added at the right. After the oxidized and reduced atoms have been balanced, it is necessary to balance the other atoms—in this case, oxygen and hydrogen. In order to do this, we can add certain common species to the equation. It may be necessary to add H_2O (the usual solvent) and H^+ if the reaction is taking place in an acidic solution, or to add H_2O and OH^- if the reaction is taking place in a basic solution. The rules for these additions are as follows:

In Acidic Solution:

1. Balance O by adding the appropriate number of H_2O molecules to one side of the equation.
2. Balance H by adding the appropriate number of H^+ ions to one side of the equation.

In Basic Solution:

1. Balance O by adding OH^- and H_2O to the equation. For each O needed, add two OH^- ions to the deficient side of the equation and one H_2O to the other side.
2. Balance H by adding OH^- and H_2O to the equation. For each H needed, add one H_2O to the deficient side of the equation and one OH^- to the other side.

EXAMPLE 21–2(A)
Balance the preceding reaction, assuming that it occurs in acidic solution.

Solution Note there are four too few O's on the right. Step 1 gives us

$$I_2 \text{ (s)} + 5 \text{ PbO}_2 \text{ (s)} \longrightarrow 2 \text{ IO}_3^- \text{ (aq)} + 5 \text{ Pb}^{2+} \text{ (aq)} + 4 \text{ H}_2\text{O (liq)}$$

Now there are eight H's too few on the left; Step 2 gives us

$$I_2 \text{ (s)} + 5 \text{ PbO}_2 \text{ (s)} + 8 \text{ H}^+ \text{ (aq)} \longrightarrow$$
$$2 \text{ IO}_3^- \text{ (aq)} + 5 \text{ Pb}^{2+} \text{ (aq)} + 4 \text{ H}_2\text{O (liq)}$$

This is now the balanced equation in acidic solution. If you are balancing this for an assignment or a test, you should check that all atoms and charges actually do balance. ■

EXAMPLE 21–2(B)
Balance this reaction in basic solution.

Solution Again there are four too few O's on the right; Step 1 gives us

$$I_2 \text{ (s)} + 5 \text{ PbO}_2 \text{ (s)} + 4 \text{ H}_2\text{O (liq)} \longrightarrow$$
$$2 \text{ IO}_3^- \text{ (aq)} + 5 \text{ Pb}^{2+} \text{ (aq)} + \mathbf{8} \text{ OH}^- \text{ (aq)}$$

The H's already balance and it is unnecessary to use Step 2. We now have the balanced equation in basic solution. ■

EXERCISE 21–2
Complete and balance the following reaction, which takes place in acid solution:

$$\text{Cl}^- \text{ (aq)} + \text{MnO}_4^- \text{ (aq)} \longrightarrow \text{Cl}_2 \text{ (g)} + \text{Mn}^{2+} \text{ (aq)}$$

Answer: $10 \text{ Cl}^- \text{ (aq)} + 2 \text{ MnO}_4^- \text{ (aq)} + 16 \text{ H}^+ \text{ (aq)} \longrightarrow$
$$5 \text{ Cl}_2 \text{ (g)} + 2 \text{ Mn}^{2+} \text{ (aq)} + 8 \text{ H}_2\text{O (liq)}$$ ■

Half-Reaction Method: In the half-reaction method the first thing to do is to separate the original reaction into those species involved in the oxidation and those species involved in the reduction. For the reaction of Example 21–2 those are

oxidation: $I_2 \text{ (s)} \longrightarrow \text{IO}_3^- \text{ (aq)}$
reduction: $\text{PbO}_2 \text{ (s)} \longrightarrow \text{Pb}^{2+} \text{ (aq)}$

Next balance all elements except O and H by adding suitable coefficients. This gives us

$$I_2 \text{ (s)} \longrightarrow 2 \text{ IO}_3^- \text{ (aq)}$$
$$\text{PbO}_2 \text{ (s)} \longrightarrow \text{Pb}^{2+} \text{ (aq)}$$

Now balance O and H in each half-reaction for an acidic or basic solution, using Steps 1 and 2 above. For an acidic solution this would give

oxidation: $6 \text{ H}_2\text{O (liq)} + I_2 \text{ (s)} \longrightarrow 2 \text{ IO}_3^- \text{ (aq)} + 12 \text{ H}^+ \text{ (aq)}$
reduction: $4 \text{ H}^+ \text{ (aq)} + \text{PbO}_2 \text{ (s)} \longrightarrow \text{Pb}^{2+} \text{ (aq)} + 2 \text{ H}_2\text{O (liq)}$

Now add electrons to the half-reactions to balance the charges. In the oxidation, 10 negative charges are needed on the right side.

$$6 \text{ H}_2\text{O (liq)} + I_2 \text{ (s)} \longrightarrow 2 \text{ IO}_3^- \text{ (aq)} + 12 \text{ H}^+ \text{ (aq)} + 10 \, e^-$$

In the reduction, two negative charges are needed on the left.

$$4 \text{ H}^+ \text{ (aq)} + \text{PbO}_2 \text{ (s)} + 2 \, e^- \longrightarrow \text{Pb}^{2+} \text{ (aq)} + 2 \text{ H}_2\text{O (liq)}$$

These half-reactions can be added together if the number of electrons produced in the first half-reaction equals the number consumed in the second. This is accomplished if the second half-reaction is multiplied by 5:

$$[6\ H_2O\ (liq)\ +\ I_2\ (s)\ \longrightarrow\ 2\ IO_3^-\ (aq)\ +\ 12\ H^+\ (aq)\ +\ 10\ e^-]$$
$$5[4\ H^+\ (aq)\ +\ PbO_2\ (s)\ +\ 2\ e^-\ \longrightarrow\ Pb^{2+}\ (aq)\ +\ 2\ H_2O\ (liq)]$$

$$6\ H_2O\ (liq)\ +\ I_2\ (s)\ +\ 20\ H^+\ (aq)\ +\ 5\ PbO_2\ (s)\ \longrightarrow$$
$$2\ IO_3^-\ (aq)\ +\ 12\ H^+\ (aq)\ +\ 5\ Pb^{2+}\ (aq)\ +\ 10\ H_2O\ (liq)$$

Note the presence of H_2O and H^+ on both sides of the equation. We cancel the excess of each:

$$I_2\ (s)\ +\ 8\ H^+\ (aq)\ +\ 5\ PbO_2\ (s)\ \longrightarrow$$
$$IO_3^-\ (aq)\ +\ 5\ Pb^{2+}\ (aq)\ +\ 4\ H_2O\ (liq)$$

This result is identical to the result obtained by the oxidation state method.

EXAMPLE 21–3

Use the half-reaction method to balance the following reaction, which occurs in basic solution:

$$ClO_3^-\ (aq)\ +\ Zn\ (s)\ \longrightarrow\ ZnO_2^{2-}\ (aq)\ +\ ClO_2^-\ (aq)$$

Solution Assignment of oxidation states gives us the start of two half-reactions,

$$\text{reduction:}\quad \underset{+5-2}{ClO_3^-}\ (aq)\ \longrightarrow\ \underset{+3-2}{ClO_2^-}\ (aq)$$

$$\text{oxidation:}\quad \underset{0}{Zn}\ (s)\ \longrightarrow\ \underset{+2-2}{ZnO_2^{2-}}\ (aq)$$

Application of Steps 1 and 2 to balance O and H in basic solutions gives

$$ClO_3^-\ (aq)\ +\ H_2O\ (liq)\ \longrightarrow\ ClO_2^-\ (aq)\ +\ 2\ OH^-\ (aq)$$
$$Zn\ (s)\ +\ 4\ OH^-\ (aq)\ \longrightarrow\ ZnO_2^{2-}\ (aq)\ +\ 2\ H_2O\ (liq)$$

Addition of electrons yields

$$ClO_3^-\ (aq)\ +\ H_2O\ (liq)\ +\ 2\ e^-\ \longrightarrow\ ClO_2^-\ (aq)\ +\ 2\ OH^-\ (aq)$$
$$Zn\ (s)\ +\ 4\ OH^-\ (aq)\ \longrightarrow\ ZnO_2^{2-}\ (aq)\ +\ 2\ H_2O\ (liq)\ +\ 2\ e^-$$

These half-reactions can be added directly because each involves two electrons. Addition and cancellation of redundant H_2O and OH^- gives

$$ClO_3^-\ (aq)\ +\ Zn\ (s)\ +\ 2\ OH^-\ (aq)\ \longrightarrow$$
$$ClO_2^-\ (aq)\ +\ ZnO_2^{2-}\ (aq)\ +\ H_2O\ (liq)\quad ■$$

EXERCISE 21–3

Complete and balance the following redox reaction, which takes place in basic solution:

$$Al\ (s)\ +\ IO_3^-\ (aq)\ \longrightarrow\ H_2AlO_3^-\ (aq)\ +\ I^-\ (aq)$$

Answer: $2\ Al\ (s)\ +\ 2\ OH^-\ (aq)\ +\ IO_3^-\ (aq)\ +\ H_2O\ \longrightarrow$
$$2\ H_2AlO_3^-\ (aq)\ +\ I^-\ (aq)\quad ■$$

There is one variety of redox reaction in which an ion or molecule will oxidize an identical ion or molecule, and itself be reduced. This is called a **disproportionation**.

EXAMPLE 21–4

Complete and balance the following redox equation for the reaction of $HS_2O_3^{2-}$ ion in acid solution:

$$HS_2O_3^- \text{ (aq)} \longrightarrow S \text{ (s)} + HSO_4^- \text{ (aq)}$$

Solution   The oxidation states of sulfur can be seen to change from $+2$ in $HS_2O_3^-$ to 0 in S, and to $+6$ in HSO_4^-. Therefore, $HS_2O_3^-$ is both oxidized and reduced and is the reactant in each half-reaction:

$$HS_2O_3^- \text{ (aq)} \longrightarrow S \text{ (s)}$$
$$HS_2O_3^- \text{ (aq)} \longrightarrow HSO_4^- \text{ (aq)}$$

Completing and balancing each half-reaction gives

$$4\ e^- + 5\ H^+ \text{ (aq)} + HS_2O_3^- \text{ (aq)} \longrightarrow 2\ S \text{ (s)} + 3\ H_2O \text{ (liq)}$$
$$5\ H_2O \text{ (liq)} + HS_2O_3^- \text{ (aq)} \longrightarrow 2\ HSO_4^- \text{ (aq)} + 9\ H^+ \text{ (aq)} + 8\ e^-$$

We multiply the reduction half-reaction by 2, add the two half-reactions, and then remove excess H_2O and H^+. This gives

$$3\ HS_2O_3^- \text{ (aq)} + H^+ \text{ (aq)} \longrightarrow 4\ S \text{ (s)} + H_2O \text{ (liq)} + 2\ HSO_4^- \text{ (aq)}$$

The products of this reaction are the result of two $HS_2O_3^-$ ions oxidizing a third.   ■

EXERCISE 21–4
Complete and balance the equation for the disproportionation reaction of phosphorus in basic solution:

$$P_4 \text{ (s)} \longrightarrow PH_3 \text{ (g)} + H_2PO_2^- \text{ (aq)}$$

Answer: $3\ OH^- \text{ (aq)} + P_4 \text{ (s)} + 3\ H_2O \longrightarrow 3\ H_2PO_2^- \text{ (aq)} + PH_3 \text{ (g)}$ ■

21.5   ELECTROCHEMICAL CELLS

In the previous sections we have discussed oxidation-reduction reactions, reactions that involve the transfer of electrons. Such reactions can occur under conditions such that the electrons are transferred directly from the reductant to the oxidant, just as protons are transferred directly in an acid-base reaction. However, because electrons are the carriers of electric current and can be conducted from place to place, it is possible to separate the oxidation and reduction halves of the reaction, with conduction of electrons (through a wire) between the halves. When this is done our oxidation-reduction reaction becomes an electrochemical reaction, and the apparatus is called an **electrochemical cell**.

Electrochemical cells can be used in two ways. If the reaction is spontaneous, the flow of electrons in the conductor can be used as a source of energy. This kind of cell is called a **galvanic cell**, and familiar examples are cells of the lead-acid battery in an automobile, the common dry cell for flashlights, and numerous other types of cells used for cameras, calculators, and hundreds of other applications. If the reaction is not spontaneous, an outside source of electrical energy can be placed in the cell circuit to drive the reaction to the desired products. Such a cell is called an **electrolytic cell**, and the production of many industrial materials such as aluminum and chlorine depend on its use.

A few electrical terms should be reviewed before we discuss actual cells. The SI derived unit of charge is the coulomb (C). The faraday ($\mathscr{F}$) is the total charge of one mole of electrons or 96,485 C. The flow of current is measured in amperes (A), where $1\ A = 1\ C/s$. The energy per coulomb of charge is called the electrical potential or just the potential. The only recognized unit of electrical potential is the volt (V), where 1 V = J/C.

21.6 GALVANIC CELLS

If a bar of Zn is placed in a solution containing Cu^{2+} ions, a redox reaction takes place (Figure 21–1):

$$Zn \ (s) + Cu^{2+} \ (aq) \longrightarrow Zn^{2+} \ (aq) + Cu \ (s)$$

This reaction can be separated into two half-reactions:

$$Zn \ (s) \longrightarrow Zn^{2+} \ (aq) + 2 \ e^-$$
$$2 \ e^- + Cu^{2+} \ (aq) \longrightarrow Cu \ (s)$$

There is a transfer of electrons, but under the described conditions, the transfer occurs directly from Zn to Cu^{2+} right at the surface of the Zn bar. Thus, there is clearly no flow of electrons along a conductor and this is a redox reaction, but not an electrochemical reaction. It can become an electrochemical reaction if the oxidation and reduction halves of the reaction are separated so that the transferred electrons flow from one half to the other by conduction along a wire. Such an arrangement is shown in Figure 21–2 and is called a galvanic cell.

Let's examine the various parts of this cell. It is composed of two compartments, one of which contains a bar of Zn metal dipping into a solution containing 1 M Zn^{2+} ions (from $ZnCl_2$, for example). The other compartment contains a bar of Cu metal dipping into a solution containing 1 M Cu^{2+} ions (from a dissolved salt such as $CuSO_4$). The bars dipping

FIGURE 21–1 Zinc displaces copper ions in aqueous solution in a spontaneous oxidation-reduction (or "redox") reaction. A strip of zinc is placed in the blue solution of copper(II) sulfate on the left. As the reaction proceeds, copper "plates out," depleting the supply of Cu^{2+}, and the original blue color, which was due to the presence of hydrated copper ions, fades away (on the right).

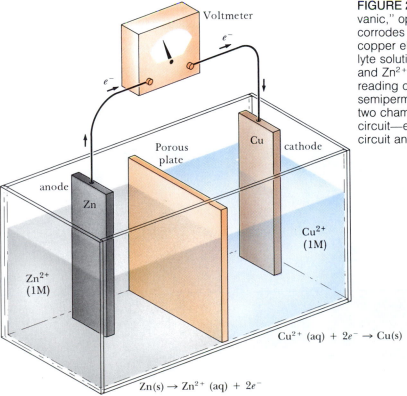

FIGURE 21–2 This electrochemical cell is "galvanic," operating spontaneously as the zinc electrode corrodes and copper plates out of solution onto the copper electrode. If the concentrations of the electrolyte solutions in each compartment are 1 M in Cu^{2+} and Zn^{2+}, respectively, the cell should produce a reading on the voltmeter of 1.10 V. A clay plate or semipermeable barrier must be used to separate the two chambers, or half-cells, in order to complete the circuit—electrons flow along the wire in the external circuit and ions move across the barrier internally.

into the compartments are called **electrodes**. The solutions in the compartments are **electrolyte solutions**.

In the Zn compartment, the half-reaction

$$Zn \text{ (s)} \longrightarrow Zn^{2+} \text{ (aq)} + 2 \, e^-$$

occurs. The electrons produced by this half-reaction make this electrode negative. In the other compartment, the half-reaction

$$Cu^{2+} \text{ (aq)} + 2 \, e^- \longrightarrow Cu \text{ (s)}$$

consumes electrons, making that electrode positive. Electrons from the Zn electrode flow along the wire to the Cu electrode to equalize the charges, and the difference in the electrical potential between the two electrodes registers on the voltmeter in the circuit.

The compartments are separated by a porous plate. Without contact between the electrolyte solutions through the porous plate, there would be no current flow in the circuit. The problem is that the half-reaction in the Zn compartment would immediately cause the buildup of an excess of positively charged ions, and that in the Cu compartment would create a shortage of positively charged ions. The porous plate permits migration of positive ions from the Zn compartment to the Cu compartment (and negative ions in the opposite direction), which balances the charges and allows current to flow through the wire. A **salt bridge** (Figure 21–3), a tube containing a solution of a salt such as ammonium nitrate or potassium chloride, can be used in place of the porous plate. Glass wool or cotton prevents the salt solution filling the bridge from running freely into the electrolyte compartments.

The terms **anode** and **cathode** are used to name the two electrodes, and the following definitions are correct under all circumstances:

> **The anode is the electrode at which oxidation takes place.**
> **The cathode is the electrode at which reduction takes place.**

Electrochemical Cell Notation

Before continuing, it is necessary to introduce our notational scheme for describing electrochemical cells. It is not universal, but it is widely used. To describe a half-cell reactaion, we write:

FIGURE 21–3 In this version of the galvanic cell described in Figure 21–2, the two half-reactions have been established in physically separated beakers. A "salt bridge" was carefully filled with an ammonium nitrate solution, then plugged at the ends with tightly packed cotton swabs, and placed across the two beakers, allowing ions to pass freely from one "half-cell" to the other, completing the electrical circuit, allowing "current" to flow. The reading on the voltmeter is 1.10 V.

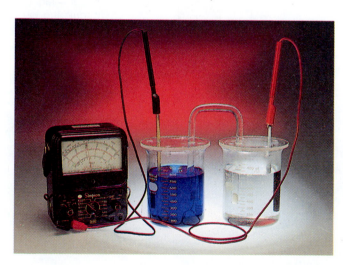

half-cell reactants | half-cell products

The vertical line indicates a phase boundary. Thus, when we place a copper electrode in a $CuSO_4$ solution, we would write the reduction half-cell reaction taking place in the cathode compartment as

Cathode half-cell reaction: Cu^{2+} (aq) $+ 2\,e^- \longrightarrow Cu$ (s)

Cathode half-cell notation: Cu^{2+} (aq) $|$ Cu (s)

But to describe a galvanic cell completely, a pair of half-reactions needs to be combined, a reduction half-cell as well as an oxidation half-cell. Thus, if the reaction taking place in the anode compartment involved zinc metal and a $ZnSO_4$ solution, we write:

Anode half-cell reaction: Zn (s) $\longrightarrow Zn^{2+}$ (aq) $+ 2\,e^-$

Anode half-cell notation: Zn (s) $| Zn^{2+}$ (aq)

The complete notational scheme for the cell would be written (by convention) with the anode on the left, and a double vertical line denoting the salt bridge. A single vertical line denotes a boundary between phases, for example, solid | aqueous solution. Finally, the concentrations of ions in solution or pressures of gases are given. Thus:

Cell notation: Zn (s) $| Zn^{2+}$ (1 M) $\| Cu^{2+}$ (1 M) $|$ Cu (s)
                  anode   cathode
          (oxidation half-cell)   (reduction half-cell)

In contrast to situations such as this—where electrodes such as zinc and copper are chemically involved in the cell processes—we commonly encounter situations where an inert electrode material is used that serves only as a surface across which electrons are transferred. Our notational scheme would be adjusted as follows:

inert material | half-cell reactants | half-cell products

A typical example (which will be described in detail shortly) is the hydrogen gas/hydrogen ion electrode:

Half-cell notation: Pt (s) $| H_2$ (1 atm) $| H^+$ (1 M)

Note that we have included the pressure of the gaseous hydrogen and the concentration of the hydronium ions. This example has been (arbitrarily) written as an oxidation half-cell. As a reduction, the same half-cell reaction would be written as

H^+ (1 M) $| H_2$ (1 atm) $|$ Pt (s)

Cell Potentials

Galvanic cells run spontaneously, and the potential measured by the voltmeter indicates the extent of this spontaneity. The symbol for the electrical potential is $\mathscr{E}$, and its units are volts. If, as in our example, the concentrations of all ions involved in the reaction are 1 M and the pressures of gases are 1 atm, the potential observed is called the **standard potential**, which has the symbol $\mathscr{E}°$. The value of $\mathscr{E}°$ for the Cu-Zn cell we have described is 1.100 V.

The potential of a cell is an intrinsic property that does not depend on the size of the cell or the total number of electrons transferred. Thus, if we were to write

$$2 \text{ Zn (s)} + 2 \text{ Cu}^{2+} \text{ (aq)} \longrightarrow 2 \text{ Cu (s)} + 2 \text{ Zn}^{2+} \text{ (aq)}$$

the standard cell potential would still be 1.100 V.

The reaction of Zn (s) with Cu^{2+} (aq) at standard conditions is spontaneous, with $\mathscr{E}° = 1.100$ V. If we were to write the reaction in the opposite direction,

$$\text{Cu (s)} + \text{Zn}^{2+} \text{ (aq)} \longrightarrow \text{Zn (s)} + \text{Cu}^{2+} \text{ (aq)}$$

we would have a process that is not spontaneous. In fact, the degree of its nonspontaneity depends directly on the degree of spontaneity of the reverse reaction. Therefore, the value of $\mathscr{E}°$ for a reaction in one direction is simply the negative of its value in the opposite direction, and the $\mathscr{E}°$ for this nonspontaneous reaction is -1.100 V.

In our electrochemical cell the processes in the two half-cells occur essentially independently. Each half-cell has its own degree of spontaneity and its own $\mathscr{E}°$. The $\mathscr{E}°$ for the overall cell process is simply the sum of the values of $\mathscr{E}°$ of the half-cells, one oxidation and one reduction. Thus,

$$\mathscr{E}° = 1.100 \text{ V} = \mathscr{E}°_{\text{Zn/Zn}^{2+}} + \mathscr{E}°_{\text{Cu}^{2+}/\text{Cu}}$$

However, although it is easy to determine the sum of the $\mathscr{E}°$ values of two half-reactions, one cannot determine $\mathscr{E}°$ for a single half-reaction independently. The problem is that in a half-reaction such as

$$\text{Zn (s)} \longrightarrow \text{Zn}^{2+} \text{ (aq)} + 2 \, e^-$$

the state of the two electrons is not specified. For the overall reaction this is not a problem, since the two electrons generated in this reaction are simply picked up by the other half-reaction

$$2 \, e^- + \text{Cu}^{2+} \text{ (aq)} \longrightarrow \text{Cu (s)}$$

and the states of all reactants and products are known.

In order to determine useful values of $\mathscr{E}°$ for half-reactions, it is necessary to set one half-reaction as an arbitrary standard. Then we can determine the values of $\mathscr{E}°$ for all other half-reactions relative to this standard. The universally agreed upon standard half-reaction is the hydrogen half-cell:

$$\text{H}_2 \text{ (1 atm)} \longrightarrow 2 \text{ H}^+ \text{ (1 } M\text{)} + 2 \, e^-$$

or

$$2 \text{ H}^+ \text{ (1 } M\text{)} + 2 \, e^- \longrightarrow \text{H}_2 \text{ (1 atm)}$$

and its $\mathscr{E}°$ is assigned to be exactly 0 V for either direction. The electrode most often used for this half-reaction is called the hydrogen electrode, and it is shown as the left-hand electrode in Figure 21–4. In the hydrogen electrode a piece of platinum acts as the conductor and the surface for the electron transfer. However, platinum is inert under these circumstances because the H_2 (g) is oxidized much more easily. The overall reaction for the cell in Figure 21–4 is

$$\text{H}_2 \text{ (g)} + \text{Cu}^{2+} \text{ (aq)} \longrightarrow 2 \text{ H}^+ \text{ (aq)} + \text{Cu (s)}$$

and the value measured for $\mathscr{E}°$ is 0.337 V. Since $\mathscr{E}° \, (\text{H}_2/\text{H}^+)$ is 0 V and

$$\mathscr{E}° = \mathscr{E}°_{\text{H}_2/\text{H}^+} + \mathscr{E}°_{\text{Cu}^{2+}/\text{Cu}}$$

then

$$\mathscr{E}°_{\text{Cu}^{2+}/\text{Cu}} = 0.337 \text{ V}$$

For the cell in Figure 21–4, the complete notational scheme is

$$\text{Pt (s)} \mid \text{H}_2 \text{ (1 atm)} \mid \text{H}^+ \text{ (1 } M\text{)} \parallel \text{Cu}^{2+} \text{ (1 } M\text{)} \mid \text{Cu (s)}$$

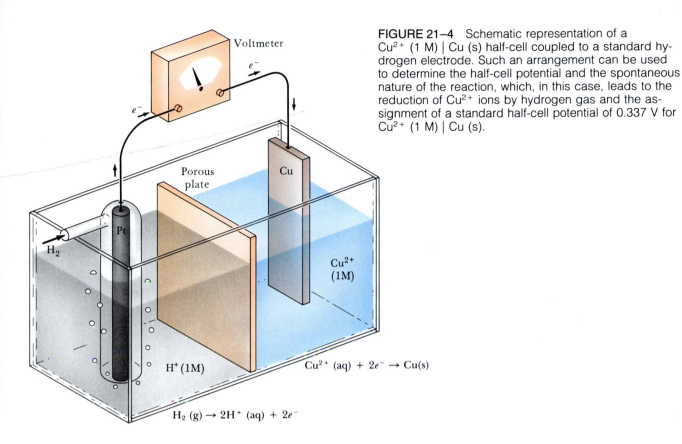

FIGURE 21–4 Schematic representation of a Cu^{2+} (1 M) | Cu (s) half-cell coupled to a standard hydrogen electrode. Such an arrangement can be used to determine the half-cell potential and the spontaneous nature of the reaction, which, in this case, leads to the reduction of Cu^{2+} ions by hydrogen gas and the assignment of a standard half-cell potential of 0.337 V for Cu^{2+} (1 M) | Cu (s).

We can substitute $\mathscr{E}°_{Cu^{2+}/Cu}$ into the sum of $\mathscr{E}°$ values for the Cu-Zn cell, for which the overall $\mathscr{E}°$ is 1.100 V. We solve for $\mathscr{E}°_{Zn/Zn^{2+}}$ and obtain

$$\mathscr{E}°_{Zn/Zn^{2+}} = \mathscr{E}° - \mathscr{E}°_{Cu^{2+}/Cu}$$

$$\mathscr{E}°_{Zn/Zn^{2+}} = 1.100 \text{ V} - 0.337 \text{ V} = 0.763 \text{ V}$$

We now have $\mathscr{E}°$ values for three half-reactions:

	$\mathscr{E}°$
$Zn \text{ (s)} \longrightarrow Zn^{2+} \text{ (aq)} + 2\,e^-$	0.763 V
$Cu^{2+} \text{ (aq)} + 2\,e^- \longrightarrow Cu \text{ (s)}$	0.337 V
$H_2 \text{ (g)} \longrightarrow 2\,H^+ \text{ (aq)} + 2\,e^-$	0.000 V

Two are oxidations, and one is a reduction. For purposes of comparison it is useful to have all reactions written in the same direction, and the present practice is to tabulate half-reactions as reduction potentials. This is simply done. Since the $\mathscr{E}°$ for a half-reaction represents its degree of spontaneity, reversing the direction of the half-reaction merely requires changing the sign of its $\mathscr{E}°$. Thus:

	$\mathscr{E}°$
$Cu^{2+} \text{ (aq)} + 2\,e^- \longrightarrow Cu \text{ (s)}$	0.337 V
$2\,H^+ \text{ (aq)} + 2\,e^- \longrightarrow H_2 \text{ (g)}$	0.000 V
$Zn^{2+} \text{ (aq)} + 2\,e^- \longrightarrow Zn \text{ (s)}$	−0.763 V

and $\mathscr{E}°$ in each of these cases is called the half-reaction standard reduction potential. These reductions are arranged in decreasing order of spontaneity. Note, however, that the negative value of the $\mathscr{E}°$ for the reduction

FIGURE 21–5 In this modification of the basic galvanic cell design in Figure 21–2, an inert electrode— platinum, in this case, or carbon could be used—provides a surface across which electrons are transferred during the Fe^{3+}/Fe^{2+} half-cell reaction taking place in the right-hand chamber. The electrode is chemically inert, in contrast to the zinc electrode in the other chamber, which corrodes as the cell reaction proceeds.

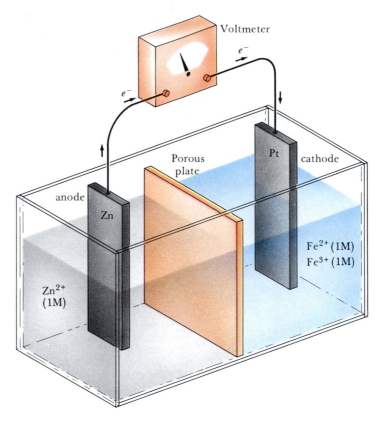

of Zn^{2+} only means that the reduction of Zn^{2+} is 0.763 V less spontaneous than the reduction of H^+.

EXAMPLE 21–5

Figure 21–5 shows a cell in which one of the compartments contains a platinum electrode dipping into a solution containing 1 M Fe^{3+} and 1 M Fe^{2+}. As in the H_2 gas electrode, the platinum serves only as an inert surface for the electron transfer, which in this case is between Fe^{2+} and Fe^{3+}. The $\mathscr{E}°$ for this cell is measured to be 1.533 V and the Zn electrode is negative. Determine the half-reactions taking place in each compartment.

Solution   First identify the anode and the cathode, and determine the reduction potential for $\mathscr{E}°_{Fe^{3+}/Fe^{2+}}$. Since the Zn electrode is negative and the cell is galvanic, the half-reaction taking place in that compartment is

$$Zn \ (s) \longrightarrow Zn^{2+} \ (aq) + 2 \ e^-$$

A reduction is taking place in the other compartment, which will be

$$Fe^{3+} \ (aq) + e^- \longrightarrow Fe^{2+} \ (aq)$$

So the Pt electrode is the cathode. The overall cell reaction therefore is

$$Zn \ (s) + 2 \ Fe^{3+} \ (aq) \longrightarrow Zn^{2+} \ (aq) + 2 \ Fe^{2+} \ (aq)$$

Adding up the standard potentials for the half-reactions—one written for oxidation, the other for reduction—gives the measured standard potential for the cell:

$$\mathscr{E}°_{cell} = \mathscr{E}°_{Zn/Zn^{2+}} + \mathscr{E}°_{Fe^{3+}/Fe^{2+}}$$

We mentioned before that $\mathscr{E}°$ is an intrinsic property, not dependent upon the amount of matter undergoing reaction. Therefore, $\mathscr{E}^\circ_{Fe^{3+}/Fe^{2+}}$ is not multiplied by a coefficient even though two moles of Fe^{3+} must be reduced for every one mole of Zn oxidized:

$$\mathscr{E}^\circ_{Zn/Zn^{2+}} = -\mathscr{E}^\circ_{Zn^{2+}/Zn} = -(-0.763 \text{ V}) = +0.763 \text{ V}$$

Now we can solve for $\mathscr{E}^\circ_{Fe^{3+}/Fe^{2+}}$:

$$\mathscr{E}^\circ_{Fe^{3+}/Fe^{2+}} = \mathscr{E}^\circ_{cell} - 0.763 \text{ V} = 1.533 \text{ V} - (+0.763 \text{ V})$$

$$\mathscr{E}^\circ_{Fe^{3+}/Fe^{2+}} = +0.770 \text{ V} \quad ■$$

EXERCISE 21–5

A cell contains one compartment with a Pt electrode dipping into a solution containing Hg^{2+} and Hg_2^{2+}, each at 1 M concentration. The other compartment contains a Cu electrode in a $1 \text{ M } Cu^{2+}$ solution. The spontaneous cell reaction gives an $\mathscr{E}^\circ$ of 0.583 V and the Pt electrode is positive. State the half-reactions taking place at anode and cathode, the overall cell reaction, and the standard reduction potential, $\mathscr{E}^\circ_{Hg^{2+}/Hg_2^{2+}}$:

Answer:

Anode (oxidation): $Cu \text{ (s)} \longrightarrow Cu^{2+} \text{ (aq)} + 2 \, e^-$

Cathode (reduction): $2 \, Hg^{2+} \text{ (aq)} + 2 \, e^- \longrightarrow Hg_2^{2+} \text{ (aq)}$

Overall: $Cu \text{ (s)} + 2 \, Hg^{2+} \text{ (aq)} \longrightarrow Cu^{2+} \text{ (aq)} + Hg_2^{2+} \text{ (aq)},$

$$\mathscr{E}^\circ_{Hg^{2+}/Hg_2^{2+}} = 0.920 \text{ V} \quad ■$$

The half-reaction standard reduction potentials that we have gathered thus far are:

	$\mathscr{E}^\circ$
$2 \, Hg^{2+} \text{ (aq)} + 2 \, e^- \longrightarrow Hg_2^{2+} \text{ (aq)}$	0.920 V
$Fe^{3+} \text{ (aq)} + e^- \longrightarrow Fe^{2+} \text{ (aq)}$	0.770 V
$Cu^{2+} \text{ (aq)} + 2 \, e^- \longrightarrow Cu \text{ (s)}$	0.337 V
$2 \, H^+ \text{ (aq)} + 2 \, e^- \longrightarrow H_2 \text{ (g)}$	0.000 V
$Zn^{2+} \text{ (aq)} + 2 \, e^- \longrightarrow Zn \text{ (s)}$	-0.763 V

Many more have been tabulated. Lists of important half-reactions in acidic and basic aqueous solutions are given in Table 21–1. With these half-reactions you can construct a large number of oxidation-reduction reactions for which the standard cell potentials are known.

EXAMPLE 21–6

Consider a cell consisting of two compartments. One has a Pb electrode dipping into a 1 M solution of Pb^{2+}. The other has a Pt electrode dipping into a solution that is 1 M in both Cr^{3+} and Cr^{2+}. The two compartments are joined by a salt bridge and the electrodes are connected through a voltmeter. (a) Write the balanced equation for the spontaneous cell reaction and the half-reactions as they take place in each cell. (b) Tell which electrode is the anode and which the cathode. (c) What are the respective polarities (charges) of the electrodes? (d) In which direction are electrons flowing? (e) Determine $\mathscr{E}^\circ$ for the cell. The required half-reactions—written as reductions—are to be found in Table 21–1:

$$Cr^{3+} \text{ (aq)} + e^- \longrightarrow Cr^{2+} \text{ (aq)} \qquad \mathscr{E}^\circ = -0.41 \text{ V}$$

$$Pb^{2+} \text{ (aq)} + 2 \, e^- \longrightarrow Pb \text{ (s)} \qquad \mathscr{E}^\circ = -0.126 \text{ V}$$

TABLE 21–1 Standard Reduction Potentials

Acid Solutions

Half-Reaction	$\mathscr{E}°$ (V)	Half-Reaction	$\mathscr{E}°$ (V)
$Li^+ + e^- \rightleftharpoons Li$	-3.045	$I_3^- + 2\,e^- \rightleftharpoons 3\,I^-$	0.536
$Ca^{2+} + 2\,e^- \rightleftharpoons Ca$	-2.866	$MnO_4^- + e^- \rightleftharpoons MnO_4^{2-}$	0.558
$Na^+ + e^- \rightleftharpoons Na$	-2.714	$H_3AsO_4 + 2\,H^+ + 2\,e^- \rightleftharpoons$	0.560
$La^{3+} + 3\,e^- \rightleftharpoons La$	-2.52	$\qquad HAsO_2 + 2\,H_2O$	
$Mg^{2+} + 2\,e^- \rightleftharpoons Mg$	-2.36	$O_2\,(g) + 2\,H^+ + 2\,e^- \rightleftharpoons H_2O_2$	0.695
$AlF_6^{3-} + 3\,e^- \rightleftharpoons Al + 6\,F^-$	-2.07	$PtCl_4^{2-} + 2\,e^- \rightleftharpoons Pt + 4\,Cl^-$	0.73
$Al^{3+} + 3\,e^- \rightleftharpoons Al$	-1.66	$Fe^{3+} + e^- \rightleftharpoons Fe^{2+}$	0.77
$SiF_6^{2-} + 4\,e^- \rightleftharpoons Si + 6\,F^-$	-1.24	$Hg_2^{2+} + 2\,e^- \rightleftharpoons 2\,Hg$	0.788
$V^{2+} + 2\,e^- \rightleftharpoons V$	-1.19	$Ag^+ + e^- \rightleftharpoons Ag$	0.799
$Mn^{2+} + 2\,e^- \rightleftharpoons Mn$	-1.18	$2\,Hg^{2+} + 2\,e^- \rightleftharpoons Hg_2^{2+}$	0.920
$Zn^{2+} + 2\,e^- \rightleftharpoons Zn$	-0.763	$Br_2 + 2\,e^- \rightleftharpoons 2\,Br^-$	1.087
$Cr^{3+} + 3\,e^- \rightleftharpoons Cr$	-0.744	$2\,IO_3^- + 12\,H^+ + 10\,e^- \rightleftharpoons$	1.19
$Fe^{2+} + 2\,e^- \rightleftharpoons Fe$	-0.44	$\qquad I_2 + 6\,H_2O$	
$Cr^{3+} + e^- \rightleftharpoons Cr^{2+}$	-0.41	$O_2 + 4\,H^+ + 4\,e^- \rightleftharpoons 2H_2O$	1.23
$PbSO_4 + 2\,e^- \rightleftharpoons Pb + SO_4^{2-}$	-0.359	$Cr_2O_7^{2-} + 14\,H^+ + 6\,e^- \rightleftharpoons$	1.33
$Co^{2+} + 2\,e^- \rightleftharpoons Co$	-0.277	$\qquad 2\,Cr^{3+} + 7\,H_2O$	
$Ni^{2+} + 2\,e^- \rightleftharpoons Ni$	-0.250	$Cl_2 + 2\,e^- \rightleftharpoons 2\,Cl^-$	1.36
$Sn^{2+} + 2\,e^- \rightleftharpoons Sn$	-0.138	$PbO_2 + 4\,H^+ + 2\,e^- \rightleftharpoons$	1.45
$Pb^{2+} + 2\,e^- \rightleftharpoons Pb$	-0.126	$\qquad Pb^{2+} + 2\,H_2O$	
$Fe^{3+} + 3\,e^- \rightleftharpoons Fe$	-0.037	$Au^{3+} + 3\,e^- \rightleftharpoons Au$	1.50
$2\,D^+ + 2\,e^- \rightleftharpoons D_2$	-0.0034	$MnO_4^- + 8\,H^+ + 5\,e^- \rightleftharpoons$	1.51
$2\,H^+ + 2\,e^- \rightleftharpoons H_2$	0 (definition)	$\qquad Mn^{2+} + 4\,H_2O$	
$Cu^{2+} + e^- \rightleftharpoons Cu^+$	0.153	$Ce^{4+} + e^- \rightleftharpoons Ce^{3+}$	1.61
$AgCl + e^- \rightleftharpoons Ag + Cl^-$	0.222	$MnO_4^- + 4\,H^+ + 3\,e^- \rightleftharpoons$	1.679
$Hg_2Cl_2 + 2\,e^- \rightleftharpoons 2\,Hg + 2\,Cl^-$	0.2676	$\qquad MnO_2 + 2\,H_2O$	
$Cu^{2+} + 2\,e^- \rightleftharpoons Cu$	0.337	$H_2O_2 + 2\,H^+ + 2\,e^- \rightleftharpoons 2\,H_2O$	1.776
$Fe(CN)_6^{3-} + e^- \rightleftharpoons Fe(CN)_6^{4-}$	0.36	$O_3 + 2\,H^+ + 2\,e^- \rightleftharpoons O_2 + H_2O$	2.07
$Cu^+ + e^- \rightleftharpoons Cu$	0.521	$F_2 + 2\,e^- \rightleftharpoons 2\,F^-$	2.87
$I_2 + 2\,e^- \rightleftharpoons 2\,I^-$	0.535	$H_4XeO_6 + 2\,H^+ + 2\,e^- \rightleftharpoons$	3.0
		$\qquad XeO_3 + 3\,H_2O$	

Basic Solutions

Half-Reaction	$\mathscr{E}°$ (V)	Half-Reaction	$\mathscr{E}°$ (V)
$H_2AlO_3^- + 2\,H_2O + 3\,e^- \rightleftharpoons$	-2.33	$IO_3^- + 3\,H_2O + 6\,e^- \rightleftharpoons$	0.26
$\qquad Al + 4\,OH^-$		$\qquad I^- + 6\,OH^-$	
$CrO_2^- + 2\,H_2O + 3\,e^- \rightleftharpoons$	-1.27	$ClO_3^- + H_2O + 2\,e^- \rightleftharpoons$	0.33
$\qquad Cr + 4\,OH^-$		$\qquad ClO_2^- + 2\,OH^-$	
$ZnO_2^{2-} + 2\,H_2O + 2\,e^- \rightleftharpoons$	-1.21	$ClO_4^- + H_2O + 2\,e^- \rightleftharpoons$	0.36
$\qquad Zn + 4\,OH^-$		$\qquad ClO_3^- + 2\,OH^-$	
$Sn(OH)_6^{2-} + 2\,e^- \rightleftharpoons$	-0.93	$O_2 + H_2O + 4\,e^- \rightleftharpoons 4\,OH^-$	0.40
$\qquad HSnO_2^- + 2\,H_2O + 3\,OH^-$			
$HSnO_2^- + H_2O + 2\,e^- \rightleftharpoons$	-0.91	$HO_2^- + H_2O + 2\,e^- \rightleftharpoons 3\,OH^-$	0.88
$\qquad Sn + 3\,OH^-$			
$2\,H_2O + 2\,e^- \rightleftharpoons H_2 + 2\,OH^-$	-0.8277	$ClO^- + H_2O + 2\,e^- \rightleftharpoons$	0.89
$HPbO_2^- + H_2O + 2\,e^- \rightleftharpoons$	-0.54	$\qquad Cl^- + 2\,OH^-$	
$\qquad Pb + 3\,OH^-$		$HXeO_4^- + 3\,H_2O + 6\,e^- \rightleftharpoons$	0.9
$Co(OH)_3 + e^- \rightleftharpoons Co(OH)_2 + OH^-$	0.17	$\qquad Xe + 7\,OH^-$	

Solution   In order to produce the complete cell reaction, one of these half-reactions will have to be reversed, and the resulting overall $\mathscr{E}°$ must be positive for the reaction to be spontaneous. Reversal of the Pb^{2+}/Pb half-reaction would give $\mathscr{E}° = -0.41 + 0.126 = -0.28$ V, which is not the potential for a spontaneous reaction. However, reversal of the Cr^{3+}-Cr^{2+} half-reaction gives $\mathscr{E}° = 0.41 - 0.126 = +0.28$ V. Thus, the overall spontaneous reaction results from reversing the Cr^{3+}/Cr^{2+} half-reaction to make it an oxidation and multiplying it by 2 to balance the electrons:

$$Pb^{2+} (aq) + 2\ Cr^{2+} (aq) \longrightarrow Pb (s) + 2\ Cr^{3+} (aq) \qquad \mathscr{E}° = 0.28\ V$$

The oxidation is taking place at the Pt electrode, making it the anode. The half-reaction taking place at the anode is

$$Cr^{2+} (aq) \longrightarrow Cr^{3+} (aq) + e^-$$

and the electrons produced there cause it to be negatively charged. Similarly, the Pb electrode is the cathode with the half-reaction

$$Pb^{2+} (aq) + 2\ e^- \longrightarrow Pb (s)$$

and a positive charge. The electrons in the external circuit flow from the Pt electrode through the voltmeter to the Pb electrode. ■

EXERCISE 21–6

A galvanic cell has two compartments connected by a salt bridge and an external circuit through a voltmeter. One compartment contains a piece of Mn dipping into a $1\ M$ solution of Mn^{2+}, and the other contains a Pt electrode in a solution that is $1\ M$ in both Cu^{2+} and Cu^+. Give the spontaneous cell reaction, its $\mathscr{E}°$, and the half-reaction in each compartment. Tell which electrode is the anode, which the cathode, what are the electric charges on the electrodes, and what is the direction of electron flow in the external circuit.
Answer: $Mn (s) + 2\ Cu^{2+} (aq) \rightarrow Mn^{2+} (aq) + 2\ Cu^+ (aq)$, $\mathscr{E}° = 1.33$ V; $Mn (s) \rightarrow Mn^{2+} (aq) + 2\ e^-$; $Cu^{2+} (aq) + e^- \rightarrow Cu^+ (aq)$; Mn electrode, anode, negative; Pt electrode, cathode, positive; electron flow in external circuit from Mn to Pt. ■

The Electrochemical Series

The reduction half-reactions in Table 21–1 are part of a still longer list called the **electrochemical series**, in Tables 21–2, 21–3, and 21–4. This series can be used in at least three different ways to make predictions about redox reactions in aqueous solutions, as shown in the next three examples.

EXAMPLE 21–7   Displacements of One Metal by Another

Will aluminum displace copper from aqueous solutions of Cu^{2+} ions?

Solution   From Table 21–1 we see that aluminum atoms lose electrons more easily than do copper atoms:

$$2\ [Al (s) \longrightarrow Al^{3+} (aq) + 3\ e^-] \qquad \mathscr{E}° = +1.66\ V$$
$$3\ [Cu (s) \longrightarrow Cu^{2+} (aq) + 2\ e^-] \qquad \mathscr{E}° = -0.34\ V$$

$$2\ Al (s) + 3\ Cu^{2+} (aq) \longrightarrow 2\ Al^{3+} (aq) + 3\ Cu (s)$$
$$\mathscr{E}° = 1.66 + 0.34 = 2.00\ V$$

Thus, aluminum metal will displace Cu^{2+} ions from solution. ■

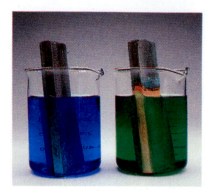

When a nickel bar is placed in a solution of copper(II) ions (left), copper deposits on the bar. As the copper ions are depleted and replaced by nickel(II) ions, the solution turns from characteristic blue to green (right). That is the "spontaneous" direction, never the other way. If "wired up," as in Figure 21–3, the cell could produce 0.59 V.

TABLE 21–2 Some of the Important Elements and the Electrochemical Series		
lithium		
potassium		
calcium		
sodium		
magnesium	**METALS**	Increasing tendency
aluminum		to act as reducing agents
zinc		and give up electrons,
iron		forming cations
nickel		
lead		
hydrogen		
copper		
oxygen		
iodine		
mercury		Increasing tendency
silver		to act as oxidizing agents
bromine		and gain electrons,
chlorine	**NONMETALS**	forming anions
fluorine		

EXERCISE 21–7 Will lead atoms displace zinc ions from solution?
Answer: No. The reverse process is spontaneous, zinc atoms displace lead ions.   ∎

EXAMPLE 21–8 Reactivity of Metals with Hydrogen Acids
Will aluminum metal generate hydrogen gas from aqueous solutions of hydrogen acids?

Solution Again, from Table 20–1 we see that aluminum atoms lose electrons more easily than do hydrogen atoms:

$$2 \,[Al \,(s) \longrightarrow Al^{3+} \,(aq) + 3 \,e^-] \qquad \mathscr{E}° = +1.66 \text{ V}$$
$$3 \,[H_2 \,(g) \longrightarrow 2 \,H^+ \,(aq) + 2 \,e^-] \qquad \mathscr{E}° = 0$$

Thus, aluminum atoms will displace hydrogen ions, generating hydrogen from aqueous acids such as hydrochloric and sulfuric acids:

$$2 \,Al \,(s) + 6 \,H^+ \,(aq) \longrightarrow 2 \,Al^{3+} \,(aq) + 3 \,H_2 \,(g)$$ ■

EXERCISE 21–8
Will copper atoms displace hydrogen from aqueous solutions of HCl?
Answer: No. The spontaneous process is the reduction of Cu^{2+} ions by hydrogen.   ∎

EXAMPLE 21–9 Displacement of One Halogen by Another
Will chlorine molecules displace bromide ions in aqueous solutions?

Solution Again, from Table 21–1 we see that electrons are most easily lost by bromide ions:

$$2 \,Br^- \,(aq) \longrightarrow Br_2 \,(aq) + 2 \,e^- \qquad \mathscr{E}° = -1.09 \text{ V}$$
$$2 \,Cl^- \,(aq) \longrightarrow Cl_2 \,(aq) + 2 \,e^- \qquad \mathscr{E}° = -1.36 \text{ V}$$

TABLE 21–3   Activity of Some Metals Toward Water, and the Electrochemical Series

Relative Order of the Metal in the Series	Activity with Water, and the Rate of H_2 (g) Displacement
Li	Generates hydrogen and forms alkaline solution. $2\ Li\ (s)\ +\ 2\ H_2O\ (liq)\ \longrightarrow$ $2\ Li^+\ (aq)\ +\ 2\ OH^-\ (aq)\ +\ H_2\ (g)$
K	Rapid generation of hydrogen. Ignites the gas, which burns with characteristic purple-red flame. Alkaline solution forms. $2\ K\ (s)\ +\ 2\ H_2O\ (liq)\ \longrightarrow$ $2\ K^+\ (aq)\ +\ 2\ OH^-\ (aq)\ +\ H_2\ (g)$
Ca	Moderate rate of hydrogen gas generation. Alkaline solution forms. $Ca\ (s)\ +\ 2\ H_2O\ (liq)\ \longrightarrow$ $Ca^{2+}\ (aq)\ +\ 2\ OH^-\ (aq)\ +\ H_2\ (g)$
Na	Rapid generation of hydrogen. Occasionally ignites the gas, which burns with characteristic yellow flame. Alkaline solution forms. $2\ Na\ (s)\ +\ 2\ H_2O\ (liq)\ \longrightarrow$ $2\ Na^+\ (aq)\ +\ 2\ OH^-\ (aq)\ +\ H_2\ (g)$
Mg	Slow generation of hydrogen. Alkaline solution forms. $Mg\ (s)\ +\ 2\ H_2O\ (liq)\ \longrightarrow$ $Mg^{2+}\ (aq)\ +\ 2\ OH^-\ (aq)\ +\ H_2\ (g)$
Al and all the metals below in the series	No reaction

TABLE 21–4   Activity of the Halogens Toward Water, and the Electrochemical Series

Relative Order of the Halogens in the Series	Activity with Water, and Rate of Oxygen Displacement
I_2	Slightly soluble in water. No reaction.
Br_2	Soluble in water. Very slow reaction. $Br_2\ (aq)\ +\ H_2O\ (liq)\ \longrightarrow\ HOBr\ (aq)\ +\ H^+\ (aq)\ +\ Br^-\ (aq)$
Cl_2	Soluble in water. Slow reaction. $Cl_2\ (aq)\ +\ H_2O\ (liq)\ \longrightarrow\ HOCl\ (aq)\ +\ H^+\ (aq)\ +\ Cl^-\ (aq)$
F_2	Soluble in water. Rapid reaction. $2\ F_2\ (aq)\ +\ 2\ H_2O\ (liq)\ \longrightarrow\ 4\ HF\ (aq)\ +\ O_2\ (g)$

Thus, molecular chlorine will displace bromide ions in aqueous solution:

$$Cl_2\ (aq)\ +\ 2\ Br^-\ (aq)\ \longrightarrow\ Br_2\ (aq)\ +\ 2\ Cl^-\ (aq)$$
$$\mathscr{E}°\ =\ +0.27\ V\ ■$$

EXERCISE 21–9
What will happen if aqueous iodine is added to an aqueous solution of bromide ions?
Answer: Nothing. The electrochemical series tells us that electrons are more readily lost by iodide ions, so the spontaneous process is bromine displacing iodide ions. ∎

21.7 THERMODYNAMICS OF ELECTROCHEMICAL CELLS

The thermodynamics that we discussed in Chapters 7 and 11 are very useful in further discussion of electrochemical cells. In particular, we can use thermodynamics to learn about the relationship between the standard potential of a cell and the equilibrium constant of the reaction involved. All the cells we have discussed so far have been at standard conditions. The concentrations have been 1 M and the gas pressures (such as in the hydrogen electrode) have been 1 atm. The temperatures have all been 298 K. It is perfectly possible and very useful to construct cells that operate under other conditions. We can use the thermodynamics we studied in Chapter 11 to give us the understanding we need for cells at nonstandard conditions.

The standard potential for an electrochemical reaction is clearly related to its spontaneity. That is, a positive value of the standard potential means that the process is spontaneous under standard conditions. The standard free energy, $\Delta G°$, is also related to the spontaneity. Therefore, $\mathscr{E}°$ and $\Delta G°$ must be related to each other. By looking at the thermodynamics of an electrochemical reaction we can see the relationship between the two. Electrochemical reactions are particularly susceptible to thermodynamic analysis because, by placing a large resistance in the circuit, the process can be slowed to near reversible conditions. Also, an electrochemical process is capable of doing work. This work can be of the ordinary pressure-volume type where volume changes are involved. But, more important in these cases, the potential generated in an electrochemical process can be used for electrical work like running a motor. Thus, for an electrochemical process,

$$w = w_{\text{elec}} + w_{\text{PV}} \tag{21–1}$$

where w is the total work for the process. We have considered pressure-volume work in Chapter 7. How do we calculate electrical work?

A galvanic cell that is proceeding spontaneously has a positive potential. Such a cell can do work on the surroundings, giving a negative w_{elec}. Electrical work equals the negative of charge times potential, or

$$w_{\text{elec}} = -\mathscr{E}Q$$

If $\mathscr{E}$ is in volts (J/C) and Q is in coulombs, w_{elec} is calculated in joules. Under standard conditions $\mathscr{E}$ becomes $\mathscr{E}°$. For an electrochemical reaction, the charge is equal to the number of moles of electrons transferred per mole of reaction, n, times the Faraday constant $\mathscr{F}$, 96,485 C/mol:

$$Q = n\mathscr{F}$$

Therefore,

$$w_{\text{elec}} = -n\mathscr{F}\mathscr{E}° \tag{21–2}$$

Typically, 96,500 is used in problem-solving.

Now we can examine electrical work in terms of the laws of thermodynamics.

For an electrochemical cell, the First Law of Thermodynamics is

$$\Delta E = q + w = q + w_{PV} + w_{elec}$$

At constant pressure,

$$\Delta H = \Delta E + P\Delta V = \Delta E - w_{PV}$$

Substituting for ΔE gives

$$\Delta H = q + w_{elec}$$

The free energy at constant temperature is

$$\Delta G = \Delta H - T\Delta S$$

Substituting for ΔH gives

$$\Delta G = q + w_{elec} - T\Delta S$$

$T\Delta S$ equals q_{rev}, and if the process is run reversibly

$$\Delta G = +w_{elec}$$

Combining this result with Eq. (21–2),

$$\Delta G = -n\mathscr{F}\mathscr{E}$$

which becomes

$$\Delta G° = -n\mathscr{F}\mathscr{E}° \tag{21–3}$$

under standard conditions.

> Note that an equation we used in Chapter 7, namely $\Delta H = q_p$, is valid only when the only work done is PV work. This equation does not hold for electrochemical cells.

EXAMPLE 21–10

Calculate $\Delta G°$ for the reaction

$$Cu\ (s) + 2\ Ag^+\ (aq) \longrightarrow Cu^{2+}\ (aq) + 2\ Ag\ (s)$$

Solution Oxidation of one mole of copper in this reaction requires two moles of electrons. The standard cell potential for this process is 0.462 V. Use of Eq. (21–3) gives

$$\Delta G° = -(2\ mol\ e^-/mol\ Cu)(96{,}500\ C/mol\ e^-)(0.462\ J/C)(1\ kJ/1000\ J)$$
$$= -89.2\ kJ/mol\ Cu$$
$$\Delta G° = -89.2\ kJ\ \text{for one mole of the reaction as written} \qquad\blacksquare$$

EXERCISE 21–10

Calculate $\Delta G°$ for the reaction

$$2\ Cr\ (s) + 3\ PbSO_4\ (s) \longrightarrow 2\ Cr^{3+}\ (aq) + 3\ Pb\ (s) + 3\ SO_4^{2-}\ (aq)$$

Answer: -223 kJ. $\blacksquare$

Since $\Delta G°$ is related to the equilibrium constant K by Eq. (11–20),

$$\Delta G° = -RT \ln K \tag{11–20}$$

the link between the standard potential and the equilibrium constant can be made by combining Eq. (21–3) with Eq. (11–20) to give

$$n\mathscr{F}\mathscr{E}° = RT \ln K \qquad (21\text{--}4)$$

$$\mathscr{E}° = \frac{RT \ln K}{n\mathscr{F}}; \ln K = \frac{n\mathscr{F}\mathscr{E}°}{RT}$$

$$K = e^{n\mathscr{F}\mathscr{E}°/RT}$$

Thus, a spontaneous process under standard conditions has $\mathscr{E}° > 0$ and $K > 1$. A nonspontaneous process has $\mathscr{E}° < 0$ and $K < 1$; a reaction for which $\mathscr{E}° = 0$ would be a situation where equilibrium prevails at standard conditions.

EXAMPLE 21–11

Determine the equilibrium constant at 298 K for the reaction

$$V^{2+} (aq) + 2 Cr^{2+} (aq) \longrightarrow V (s) + 2 Cr^{3+} (aq)$$

Solution   The standard potential for this reaction is -0.78 V; this is not a spontaneous reaction. Rearrangement of Eq. (21–4) gives

$$K = e^{n\mathscr{F}\mathscr{E}°/RT}$$

$$= \exp\left[\frac{(2 \text{ mol } e^-/\text{mol V})(96{,}500 \text{ C/mol } e^-)(-0.78 \text{ J/C})}{(8.314 \text{ J/mol·K})(298 \text{ K})}\right]$$

$$= 4.1 \times 10^{-27}$$

Note that the modestly negative potential of -0.78 V leads to the very small K of 4.1×10^{-27}. ■

EXERCISE 21–11

Calculate the equilibrium constant for the following reaction taking place at 298 K:

$$2 Fe^{3+} (aq) + 3 I^- (aq) \longrightarrow 2 Fe^{2+} (aq) + I_3^- (aq)$$

Answer: $K = 8.25 \times 10^7$. ■

If we enter the values of the constants and a temperature of 298 K into Eq. (21–4) and convert from natural to common logarithms (multiplying by 2.303), we get

$$\mathscr{E}° = \frac{0.0591}{n} \log K$$

$$K = 10^{n\mathscr{E}°/0.0591}$$

EXAMPLE 21–12

Determine the equilibrium constant K for this reaction at 298 K:

$$2 Cr^{3+} (aq) + Fe (s) \longrightarrow 2 Cr^{2+} (aq) + Fe^{2+} (aq)$$

Solution $\mathscr{E}°$ is calculated from Table 21–1 to be 0.03 V, and 2 mol of electrons are transferred per mole of reaction. Thus,

$$K = 10^{2 \times 0.03/0.059} = 1 \times 10^1$$ ■

EXERCISE 21–12

Determine K for this reaction at 298 K:

$$Hg_2Cl_2 \text{ (s)} + 2 \text{ Cu}^+ \text{ (aq)} \longrightarrow$$
$$2 \text{ Hg (liq)} + 2 \text{ Cl}^- \text{ (aq)} + 2 \text{ Cu}^{2+} \text{ (aq)}$$

Answer: 7.79×10^3. ∎

In Section 11.9 we discussed the value of ΔG at nonstandard conditions for the generalized reaction,

$$a\text{A} + b\text{B} \longrightarrow c\text{C} + d\text{D}$$

For the case in which all reactants and products are gases, Eq. (11–19) gave us

$$\Delta G = \Delta G^\circ + RT \ln \left[\frac{(P_C)^c(P_D)^d}{(P_A)^a(P_B)^b} \right]$$

For dissolved species, the partial pressures, P, would be replaced by concentrations. We can recognize the quantity within the brackets, $(P_C)^c(P_D)^d/(P_A)^a(P_B)^b$, as the reaction quotient, Q. Thus,

$$\Delta G = \Delta G^\circ + RT \ln Q$$

For nonstandard partial pressures or concentrations, Eq. (21–3) will be

$$\Delta G = -n\mathscr{F}\mathscr{E}$$

Combining the preceding equations gives the expression for calculating potentials of cells and half-cells under nonstandard conditions,

$$n\mathscr{F}\mathscr{E} = -n\mathscr{F}\mathscr{E}^\circ + RT \ln Q$$

$$\mathscr{E} = \mathscr{E}^\circ - \frac{RT}{n\mathscr{F}} \ln Q \qquad (21\text{–}5)$$

Entering the values of the constants and a temperature of 298 K and converting to common logarithms gives

$$\mathscr{E} = \mathscr{E}^\circ - \frac{0.0591}{n} \log Q$$

This is called the **Nernst equation**, and it was determined by experiment before the thermodynamic reason for it was understood.

We can apply the Nernst equation to the reaction

$$\text{Pb}^{2+} \text{ (aq)} + 2 \text{ Cr}^{2+} \text{ (aq)} \longrightarrow \text{Pb (s)} + 2 \text{ Cr}^{3+} \text{ (aq)} \qquad \mathscr{E}^\circ = 0.28 \text{ V}$$

In this reaction, two moles of electrons are transferred per mole of reaction, so $n = 2$. The Nernst equation for this reaction is

$$\mathscr{E} = \mathscr{E}^\circ - \frac{0.059}{2} \log \frac{[\text{Cr}^{3+}]^2}{[\text{Pb}^{2+}][\text{Cr}^{2+}]^2}$$

Note that Pb (s) does not appear in the expression for Q. This is the case for all pure solids and pure liquids, which do not appear in expressions for equilibrium constants. If we do have standard conditions with all concentrations equal to 1 M, then Q will equal 1, log Q will equal 0, and $\mathscr{E}$ will simply equal $\mathscr{E}^\circ$ as expected. If we increase the concentrations of the reactants, Q will be less than 1, log Q will be negative, and $\mathscr{E}$ will be greater than $\mathscr{E}^\circ$. Thus the change in $\mathscr{E}$ will reflect an increased spontaneity of the process, in agreement with Le Chatelier's principle. If instead the concentration of the product is increased, Q will be greater than 1, log Q will be positive, and $\mathscr{E}$ will be less than $\mathscr{E}°$.

EXAMPLE 21–13

Calculate $\mathscr{E}$ for the above reaction at 298 K if $[Cr^{3+}] = 1.25 \times 10^{-3}$, $[Pb^{2+}] = 1.00 \times 10^{-1}$, and $[Cr^{2+}] = 9.00 \times 10^{-1}$.

Solution

$$\mathscr{E} = 0.28 \text{ V} - \frac{0.0591}{2} \log \frac{(1.25 \times 10^{-3})^2}{(1.00 \times 10^{-1})(9.00 \times 10^{-1})^2}$$

$$= 0.28 \text{ V} - \frac{0.0591}{2} \log 1.93 \times 10^{-5}$$

$$= 0.42 \text{ V}$$

EXERCISE 21–13

One of the two compartments of a galvanic cell contains a strip of Co dipping into a 0.00235 M solution of Co^{2+}. The other compartment has a bar of Ag dipping into a 0.61 M solution of Ag^+. What is the value of $\mathscr{E}$ for this cell at 298 K?

Answer: 1.141 V.

Concentration Cells

Since the potential for a cell reaction is simply the sum of the potentials of its half-reactions, the Nernst equation applies to half-cell reactions as well as to complete cell reactions. Thus, two half-cells having the same components but different ion concentrations will have different half-cell potentials. At standard conditions the reduction potential for the half-reaction

$$Ag^+ \text{ (aq)} + e^- \longrightarrow Ag \text{ (s)}$$

is $\mathscr{E}°$ or 0.799 V. However, if the Ag^+ concentration is reduced, say to 1.00×10^{-3} M, the half-cell potential is

$$\mathscr{E} = \mathscr{E}° - \frac{0.0591}{n} \log \frac{1}{[Ag^+]}$$

$$= 0.799 \text{ V} - \frac{0.0591}{1} \log \frac{1}{1.00 \times 10^{-3}}$$

$$= 0.622 \text{ V}$$

Furthermore, because these two half-cells have different potentials, they could be connected together to produce a cell with a net positive potential—a spontaneous process. Such a cell is called a **concentration cell**, and it is shown in Figure 21–6. The two half-reactions involved are

$$Ag^+ \text{ (1 }M\text{)} + e^- \longrightarrow Ag \text{ (s)} \qquad \mathscr{E}° = 0.799 \text{ V}$$
$$Ag^+ \text{ (1.00} \times 10^{-3} M\text{)} + e^- \longrightarrow Ag \text{ (s)} \qquad \mathscr{E}° = 0.622 \text{ V}$$

A complete and spontaneous reaction will result if the second half-reaction is reversed and the two are added together:

$$Ag^+ \text{ (1 }M\text{)} \longrightarrow Ag^+ \text{ (1.00} \times 10^{-3} M\text{)}$$
$$\mathscr{E} = [0.799 - 0.622] \text{ V} = 0.177 \text{ V}$$

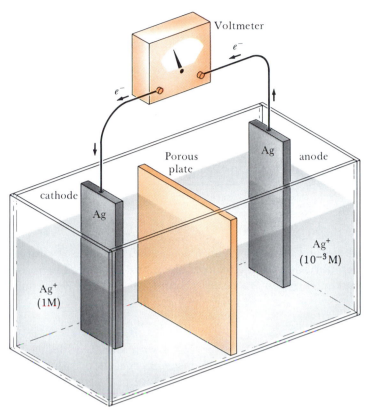

Voltmeter

Porous plate

Ag   anode

cathode

Ag

Ag^+ $(10^{-3}M)$

Ag^+ $(1M)$

FIGURE 21–6   As the name implies, a "concentration" cell is capable of producing a cell voltage based only on a difference in concentration of the ions in two otherwise identical half-cells.

An alternative approach to calculating the potential for this cell is to use the Nernst equation for the complete reaction. In this case, $\mathscr{E}°$ will be zero since the reaction products and reactants are the same, and

$$\mathscr{E} = 0 - \frac{0.0591}{1} \log \frac{1.00 \times 10^{-3}}{1} = 0.177 \text{ V}$$

As the reaction taking place in the concentration cell proceeds, the more dilute solution increases in concentration while the more concentrated decreases. $\mathscr{E}$ approaches zero as Q moves closer and closer to unity. Finally, when the concentrations are equal, the cell process stops. The overall spontaneous process that occurs in the concentration cell is the same as would result from mixing the two solutions direcly and obtaining a solution of intermediate concentration. Furthermore, it is interesting to note the thermodynamic consistency of the result. Predictions based on an understanding of entropy and the Second Law are in full agreement since one would expect the spontaneous process to be dilution of the more concentrated solution, which is the equivalent of a tendency toward randomization and disorder.

EXAMPLE 21–14

Determine the potential for a concentration cell containing Co electrodes dipping into solutions of $0.500 \ M \ Co^{2+}$ and $1.25 \times 10^{-5} \ M \ Co^{2+}$.

Solution

$$\mathscr{E} = 0 - \frac{0.0591}{2} \log \frac{0.0000125}{0.500} = 0.136 \text{ V}$$

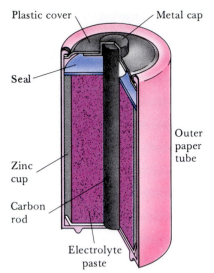

FIGURE 21–7 The standard flashlight-type of dry cell (or Le Clanché cell) consists of a graphite rod in a moist $MnO_2/ZnCl_2/NH_4Cl$ paste, in a zinc wrapper. The electrodes are the zinc outer wrapper (for the oxidation half-reaction) and the graphite center rod (for the reduction half-reaction).

EXERCISE 21–14

What is the potential of a concentration cell containing Cr^{3+} at 0.100 M and 2.00×10^{-3} M and Cr electrodes? Give the equation for the overall process.

Answer: 0.0335 V.

$$Cr^{3+} \ (0.100 \ M) \longrightarrow Cr^{3+} \ (0.00200 \ M)$$ ∎

21.8 APPLICATIONS OF GALVANIC CELLS

Batteries

Galvanic cells, either used singly or coupled together in series to form batteries, are very important portable sources of electric current. The most familiar of these is the common "dry cell" (Figure 21–7). The Zn casing serves as the anode:

$$Zn \ (s) \longrightarrow Zn^{2+} \ (aq) \ + \ 2 \ e^-$$

The electrolyte is a paste of MnO_2 and NH_4Cl with starch added as a thickener. A graphite rod serves as an inexpensive inert cathode, for which we may write the following half-reaction at its surface:

$$2 \ NH_4^+ \ (aq) \ + \ 2 \ MnO_2 \ (s) \ + \ 2 \ e^- \longrightarrow$$
$$Mn_2O_3 \ (s) \ + \ 2 \ NH_3 \ (aq) \ + \ H_2O \ (liq)$$

This cell produces about 1.5 V.

A modern version of this classic cell, marketed as the "alkaline battery," has a paste of NaOH or KOH for the electrolyte rather than NH_4Cl. Here are the half-reactions:

$$Zn \ (s) \ + \ 2 \ OH^- \ (aq) \longrightarrow Zn(OH)_2 \ (s) \ + \ 2 \ e^-$$
$$2 \ MnO_2 \ (s) \ + \ H_2O \ + \ 2 \ e^- \longrightarrow Mn_2O_3 \ (s) \ + \ 2 \ OH^- \ (aq)$$

Most batteries generate electricity by a chemical reaction in which a metal such as lead reacts with an electrolyte like sulfuric acid, or by the alkaline chemistry of the flashlight battery. In one new and novel approach, the Duracell zinc-air battery (a) generates electricity as air reacts with the zinc anode and a highly conductive KOH electrolyte. The surrounding air is the source of the needed oxygen. With about the same capacity as a typical "C" size alkaline battery, the new Duracell zinc-air battery is less than one-third the size (b). Note the air intake holes in the outer surface.

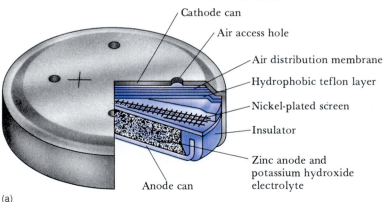

(a)

(b)

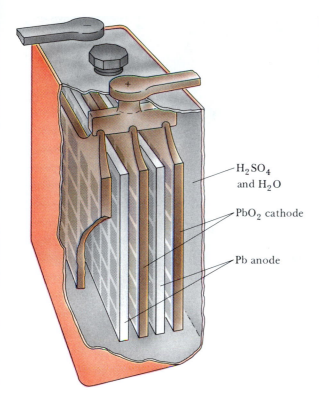

FIGURE 21–8 The essentials of the lead storage battery are alternating plates of lead and lead dioxide, immersed in a sulfuric acid solution. The positive plates consist of a lead grid filled with lead dioxide; the alternating negative plates are lead grids filled with "spongy" lead.

H_2SO_4 and H_2O

PbO_2 cathode

Pb anode

This alkaline version lasts longer than the NH_4Cl or acid version because the Zn electrode corrodes more slowly under the alkaline conditions.

The galvanic cells described so far cannot be recharged for further use once their components are used up. Such cells are called primary cells. Rechargeable or secondary cells are very useful since the same cell can be used over and over, after recharging. The best known of the rechargeable cells is the familiar **lead-acid storage battery**, which has been known since 1859. Since about 1915 when self-starters began to appear in automobiles, the lead-acid storage battery has played a major role in transportation. It is still in use, powering literally millions of internal combustion engines in automobiles. Each cell produces a potential of about 2 V, at the electrolyte concentration employed. Six cells are connected in series (anode to cathode) to produce a 12-volt battery. One battery cell is shown in Figure 21–8. The anode consists of Pb in a porous form so as to maximize its surface area. The cathode is composed of PbO_2 (s) and the electrolyte is a solution of sulfuric acid. Here are the electrode reactions:

Anode: $Pb \text{ (s)} + SO_4^{2-} \text{ (aq)} \longrightarrow PbSO_4 \text{ (s)} + 2\ e^-$

Cathode: $PbO_2 \text{ (s)} + 4\ H^+ \text{ (aq)} + SO_4^{2-} \text{ (aq)} + 2\ e^- \longrightarrow$
$$PbSO_4 \text{ (s)} + 2\ H_2O$$

Overall: $Pb \text{ (s)} + PbO_2 \text{ (s)} + 4\ H^+ \text{ (aq)} + 2\ SO_4^{2-} \text{ (aq)} \longrightarrow$
$$2\ PbSO_4 \text{ (s)} + 2\ H_2O$$

Current is drawn from the cell as the spontaneous reaction proceeds. Note that this uses up H_2SO_4, lowering its concentration, the cell potential, and the density of the electrolyte. Measuring the electrolyte density is a common way of detecting a run-down battery. The $PbSO_4$ (s) produced in each half-reaction precipitates and deposits onto the electrodes. This allows the cell to be operated as an electrolytic cell in the recharging process, where each of the half-reactions is made to run in the opposite direction.

(a)

(b)

This one-of-a-kind "hybrid" automobile (a) was built for the U.S. Department of Energy by research scientists and automotive engineers. It features an electric motor and a gasoline engine and offers the fuel savings of an electric, but without the major drawback: limited range. Ten 12-volt lead-acid batteries provide the energy pack (b) for the electric motor. The idea is to run around town on batteries and recharge at night, but to use the gasoline engine for longer trips, without the inconvenience of having to stop for recharging.

Lead-acid storage batteries can undergo many cycles of charging and discharging before flaking off of the $PbSO_4$ (s) or internal short circuits cause irreversible failure.

Recently, there has been a great deal of interest in electric vehicles with batteries that can be recharged during periods of nonuse. Because of its relatively low cost and high dependability, the lead-acid storage battery is a candidate for such applications. However, it is very heavy for the amount of energy it can generate. The ratio of energy to mass for batteries is generally called specific energy, given units of watt-hours per kilogram, W·h/kg.

We can calculate the specific energy for the lead-acid storage battery, considering only the reactants in the overall chemical process. The total mass of one mole of Pb (s), one mole of PbO_2 (s), and two moles of H_2SO_4 is 642 g. This mass of starting materials will produce 2 mol of electrons or 2 mol × 96,500 C/mol = 193,000 C. If this charge is allowed to flow over 1 h (or 3600 s), the current will be 193,000 C/3600 s = 53.6 A.

The average power in watts is current times average potential. Because the potential of the cell steadily decreases as sulfuric acid is used, the average potential is 1 V. Thus, the average power is 53.6 A × 1.0 V = 53.6 W over the 1 hour, in this case. The specific energy is thus 53.6 W·h/0.642 kg = 83.5 W·h/kg. This number, however, considers only the actual reacting material. Inclusion of the casing, electrolyte, and supporting material in the electrodes increases the mass enough to lower the specific energy to 35 W·h/kg. In general, 60 W·h/kg is considered a satisfactory specific energy for batteries for general use.

Higher specific energies can be achieved by using lighter electrode materials. The nickel-cadmium battery has a higher specific energy and produces 1.3 V per cell. Its electrode reactions are

$$Cd\ (s)\ +\ 2\ OH^-\ (aq)\ \longrightarrow\ Cd(OH)_2\ (s)\ +\ 2\ e^-$$
$$NiO_2\ (s)\ +\ 2\ H_2O\ +\ 2\ e^-\ \longrightarrow\ Ni(OH)_2\ (s)\ +\ 2\ OH^-\ (aq)$$

Note that OH^- is not used up as the cell is discharged. Therefore, there is no voltage drop through most of the useful life of the battery as it discharges, in marked contrast to the lead storage battery, which uses up sulfuric acid.

Very high specific energies require very light electrode materials. The sodium-sulfur cell delivers about 2.0 V and achieves a value near 200 W·h/kg. The electrode reactions are

$$\text{Na (liq)} \longrightarrow \text{Na}^+ + e^-$$
$$\text{S (liq)} + 2\,e^- \longrightarrow \text{S}^{2-}$$

However, an operating temperature above 300°C is required (to obtain liquid sodium and sulfur), limiting the usefulness of the cell.

Fuel Cells

A fuel cell is a galvanic cell to which the chemical reactants are constantly added as the electrochemical reaction proceeds. The species that is oxidized is a substance commonly recognized as a fuel; methane (CH_4), the principal component of natural gas, and other hydrocarbons such as propane (C_3H_8) and butane (C_4H_{10}) as well as carbon monoxide, methanol (CH_3OH), ammonia, and hydrazine ($H_2N{-}NH_2$) can be used. However, most work has been performed using hydrogen as the fuel. The species that is reduced is normally oxygen. Most fuel cells produce low voltages, commonly less than one volt, so a number of them are connected together in series in "fuel batteries." The attractive aspects of fuel cells are their ability to convert fuel to electric energy without moving parts and their theoretical high efficiencies. Conversion of fuel to mechanical work by means of a heat engine is subject to efficiency limits. Conversion of this mechanical work to electrical energy using a generator involves further energy losses. On the other hand, if its electrodes were perfect and its electrolyte had a negligible resistance, a fuel cell would operate at an efficiency of nearly 100%.

The fuel cell also has been proposed as a part of a system for storing electrical energy. Hydrogen and oxygen gases produced by the electrolysis of water could be stored in high pressure gas cylinders, and then react in a fuel cell when the energy was needed. Calculations have been presented which show that this method of energy storage is far more compact than the use of rechargeable batteries.

A hydrogen-oxygen fuel cell is shown in Figure 21–9. The electrolyte is concentrated KOH. The anode is porous nickel, and hydrogen is forced through it, yielding the half-reaction

$$\text{H}_2\text{ (g)} + 2\,\text{OH}^-\text{ (aq)} \longrightarrow 2\,\text{H}_2\text{O} + 2\,e^-$$

The cathode is also porous nickel but coated with a NiO catalyst. Oxygen is forced through this electrode, and the half-reaction here is

$$\text{O}_2\text{ (g)} + 2\,\text{H}_2\text{O} + 4\,e^- \longrightarrow 4\,\text{OH}^-\text{ (aq)}$$

The overall reaction

$$2\,\text{H}_2\text{ (g)} + \text{O}_2\text{ (g)} \longrightarrow 2\,\text{H}_2\text{O}$$

is equivalent to the simple combustion of hydrogen. Although the hydrogen-oygen fuel cell is relatively expensive, it has already found important uses in small-scale applications such as spacecraft.

Fuel cells using hydrazine have been found to function rapidly and efficiently and have been given trial uses in propelling small vehicles. Such a fuel cell can be constructed to use a 3% solution of hydrazine in 25% KOH solution as the fuel. Porous nickel sheets can be used as the electrodes

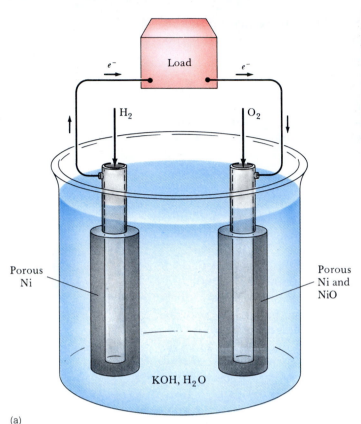

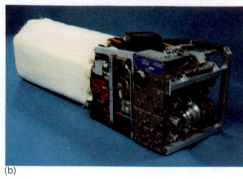

(b)

FIGURE 21–9 (a) Schematic drawing depicting the design of a typical hydrogen/oxygen fuel cell; (b) an actual fuel cell used in the U.S. Manned Space Program.

and catalyst. The disadvantages of this system are the high cost of hydrazine and the fact that the pure substance is capable of detonation.

Biochemical fuel cells have reactions at one or both electrodes catalyzed by enzymes, the highly efficient catalysts of biochemical processes. Such fuel cells could utilize the vast quantities of plant material available as fuel. They could also serve to use up organic waste from environments such as spacecraft. Problems that are encountered in this work are the susceptibility of enzymes to heat and to "poisoning" by even traces of substances containing elements such as arsenic, mercury, and silver.

Advances in fuel cells are particularly aimed at improving the materials used in the electrodes that catalyze the processes. Increasing the efficiency in the direction of the high theoretical limit is extremely important. The fuels vary considerably in cost, and it is desirable that a fuel cell not require expensive, highly purified fuels. At the moment, cells with relatively inexpensive fuels require expensive catalysts, whereas those with relatively inexpensive catalysts require expensive fuels.

EXAMPLE 21–15

What standard cell potential can be expected from a fuel cell that consumes hydrogen and oxygen gases and produces water vapor?

Solution For the overall reaction,

$$H_2 (g) + \frac{1}{2} O_2 (g) \longrightarrow H_2O (g)$$

the value of $\Delta G°$ from Table 11–2 is -228.6 kJ/mol.

$$\Delta G^\circ = -n\mathcal{F}\mathcal{E}^\circ$$

$$\mathcal{E}^\circ = \frac{-\Delta G^\circ}{n\mathcal{F}} = \frac{(228.6 \text{ kJ/mol})(1000 \text{ J/kJ})}{2(96,500 \text{ C/mol})} = 1.18 \text{ V}$$

EXERCISE 21–15

What would be the cell potential for a fuel cell that converted methane and oxygen to carbon dioxide and water vapor, with the partial pressures of all gases at 1 atm? (*Hint:* First write half-reactions to show that $n = 8$ for one mole of methane.)
Answer: 0.97 V.

pH Determination

A galvanic cell in which differences in concentration produce a potential is the basis for some interesting and especially important applications involving the precise determination of cell concentrations. For example, consider two cells, one of which contains an electrolyte of known concentration while in the other—a hydrogen gas electrode—the electrolyte is a solution of unknown hydrogen ion concentration. The pH of this solution can be determined.

EXAMPLE 21–16

In this example, a galvanic cell is constructed of a Cu electrode dipping into a 1 M solution of Cu^{2+} and a hydrogen gas electrode [$p(H_2) = 1$ atm] in a solution of unknown pH. The Cu electrode is positive and the potential measured is 0.573 V. Calculate the pH in the solution in the hydrogen electrode compartment.

Solution Since the Cu electrode is positive and the cell is galvanic, its half-reaction must be using up electrons. The two electrode reactions are therefore

$$Cu^{2+} \text{ (aq)} + 2\,e^- \longrightarrow Cu \text{ (s)} \qquad \mathcal{E}^\circ = 0.337 \text{ V}$$
$$H_2 \text{ (g)} \longrightarrow 2\,H^+ \text{ (aq)} + 2\,e^- \qquad \mathcal{E}^\circ = 0.000 \text{ V}$$

The overall reaction is

$$Cu^{2+} \text{ (aq)} + H_2 \text{ (g)} \longrightarrow Cu \text{ (s)} + 2\,H^+ \text{ (aq)} \qquad \mathcal{E}^\circ = 0.337 \text{ V}$$

$[H^+]$ can be calculated from the Nernst equation,

$$\mathcal{E}_{cell} = 0.573 \text{ V} = \mathcal{E}^\circ_{cell} - \frac{0.0591}{n} \log \frac{[H^+]^2}{[Cu^{2+}]p(H_2)}$$

$$0.573 \text{ V} = 0.337 \text{ V} - \frac{0.0591}{2} \log \frac{[H^+]^2}{(1)(1)}$$

Rearranging terms and solving for $-\log[H^+]$ gives

$$-\log[H^+]^2 = -2 \log[H^+] = \frac{2(0.573 - 0.337)}{0.0591} = 7.99$$

$$-\log[H^+] = 4.00$$
$$pH = 4.00$$

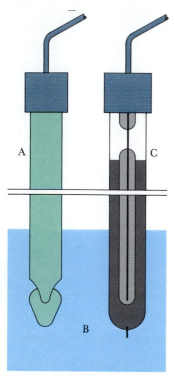

A. Glass electrode
B. Electrolyte solution
C. Calomel reference electrode
(a)

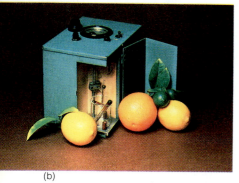

(b)

FIGURE 21–10   (a) The "glass electrode," on the left, consists of a specially constructed thin-walled glass bulb, containing an 0.10 M HCl solution about a silver wire/silver chloride electrode; the "calomel electrode," pictured in the simplified schematic diagram on the right, consists of a Pt wire that bridges the surrounding electrolyte solution and a pool of mercury that is layered with a paste of Hg_2Cl_2 (calomel) and a saturated KCl solution. (b) An original Beckman pH meter as it appeared in a recent advertising piece.

EXERCISE 21–16
A galvanic cell with a measured potential of 0.98 V has one compartment consisting of a positive Pt electrode dipping into a solution that is 0.100 M in Fe^{3+} and 0.0100 M in Fe^{2+}. The other compartment contains a hydrogen gas electrode [$p(H_2)$ = 1 atm] in a solution of unknown pH. Determine this unknown pH.
Answer: pH = 2.6.   ∎

Actual determination of pH with the half-cells described above would be very cumbersome. In the early 1930s Arnold Beckman invented a pH meter that was convenient and easily portable. His invention enjoyed immense success, and a company to produce these meters was formed. The schematic of the pH meter is shown in Figure 21–10. Each of the half cells is contained in an electrode that dips into the solution of unknown pH. The glass electrode is sensitive to the pH of the solution into which it is dipping. The calomel electrode is simply a standard electrode that provides a reference half-cell potential (at a constant temperature).

Determination of Solubility Products

The very sensitive response of cell potentials to changes in concentrations makes possible the determination of equilibrium constants that would be very difficult to achieve in other ways. The determination of solubility products of sparingly soluble salts, where the concentrations of dissolved ions are low, is an important example.

EXAMPLE 21–17
Determine the solubility product of AgCl. Make use of data from the following experiment. A galvanic cell has one compartment with a Zn electrode that dips into a solution of 1.00 M Zn^{2+}. The other compartment contains AgCl (s) at the bottom of the vessel in equilibrium with a solution of 1.00 M Cl^-. The electrode in this compartment is Ag. The Zn electrode is negative, and the measured potential is 1.00 V.

Solution   Since the Zn electrode is negative, the two half-reactions are:

$$Zn\ (s) \longrightarrow Zn^{2+}\ (aq) + 2\ e^- \qquad \mathscr{E}° = 0.763\ V$$
$$Ag^+\ (aq) + e^- \longrightarrow Ag\ (s) \qquad \mathscr{E}° = 0.799\ V$$

The overall reaction is

$$Zn\ (s) + 2\ Ag^+\ (aq) \longrightarrow Zn^{2+}\ (aq) + 2\ Ag\ (s) \qquad \mathscr{E} = 1.562\ V$$

Substitution into the Nernst equation gives

$$1.00\ V = 1.562\ V - \frac{0.0591}{2} \log \frac{1}{[Ag^+]^2}$$

Rearrangement to solve for [Ag^+] gives

$$\log[Ag^+]^2 = \frac{2(1.00\ V - 1.562\ V)}{0.0591}$$

$$\log[Ag^+] = \frac{(1.00\ V - 1.562\ V)}{0.0591} = -9.51$$

$[Ag^+] = 3.09 \times 10^{-10}$

Since $[Cl^-]$ is $1.00\ M$,

$$K_{sp} = [Ag^+][Cl^-] = 3.09 \times 10^{-10}$$

EXERCISE 21–17

A cell has one compartment with a Cu electrode in a $1.00\ M$ solution of Cu^{2+}. The other compartment contains AgBr (s) in equilibrium with $0.500\ M$ Br^- and a Ag electrode. The Cu anode is negative and the potential measured for the cell is 0.244 V. What is the solubility product of AgBr as determined from this experiment?

Answer: 5.6×10^{-13}. ■

Corrosion of Metals

In the corrosion of a metal (Figure 21–11), the metal is oxidized

$$M\ (s) \longrightarrow M^{n+}\ (aq) + ne^-$$

and is thus slowly eaten away. The electrons produced by this half-reaction have to be taken up by a reduction reaction that occurs at the same time. In some cases this can take place right at the point where the corrosion is occurring, as in the reaction with an acid:

$$2\ H^+\ (aq) + 2\ e^- \longrightarrow H_2\ (g)$$

This represents the simple case in which an acid dissolves a metal such as iron or aluminum. In other cases of corrosion an electrochemical cell can be set up; the corrosion can occur through a process that involves an oxidation half-cell where the metal is oxidized and a reduction half-cell where some chemical species is being reduced. The species that is reduced can be the proton of an acid. It can be dissolved oxygen,

$$O_2\ (aq) + 2\ H_2O\ (liq) + 4\ e^- \longrightarrow 4\ OH^-\ (aq)$$

Metal ions present in solution can be reduced,

$$M^{n+}\ (aq) + ne^- \longrightarrow M\ (s)$$

These can be ions of the metal being corroded or the ions of a different metal. Thus corrosion can take place in the presence of water that contains dissolved oxygen or dissolved electrolytes such as acids and the cations of salts. In all cases of corrosion of metals there is an oxidation reaction, with the metal being oxidized and some species being reduced.

The first step in avoiding corrosion is in choosing the metal to be used. The table of standard reduction potentials (Table 21–1) is the reasonable starting point. Metals that have large negative reduction potentials such as calcium, vanadium, and zinc have high oxidation potentials and are easily oxidized. As a first approximation we expect such metals to corrode easily. On the other hand, metals with positive reduction potentials such as copper, silver, platinum, and gold have low oxidation potentials and can be expected to be resistant to corrosion.

Some metals such as aluminum and chromium, which have positive electrode potentials for corrosion, are protected by a "passivated" surface. Each of these metals is immediately oxidized upon exposure to the air. The thin oxide coating is tough and adheres very tightly, preventing additional

FIGURE 21–11 Rust and corrosion are electrochemical effects that cost Americans upwards of $10 billion per year for maintenance, especially painting, protecting exposed surfaces, and massive replacement of structural components. This photograph was taken near the Brooklyn Battery Tunnel in New York; it could have been taken just about anywhere where moisture, oxygen, and iron come together.

corrosion from taking place. Thus chromium can be used as a protective coating on steel, and aluminum has myriad uses, including containers for concentrated nitric acid! For other metals, most notably iron, the oxide coating or rust flakes off and gives no protection to the metal.

Some metals such as lead have reduction potentials that indicate that they should dissolve in acids but are, in fact, quite resistant. For example, lead does not react appreciably even with moderately strong sulfuric acid. Such resistance is attributed to "overvoltage," a term that describes a situation in which the thermodynamics predicts a spontaneous process but the reaction is so slow that it essentially does not take place.

Metals with reduction potentials more positive than that of H^+ are not corroded by pure nonoxidizing acids of 1 M concentration at 298 K. Such metals as copper, silver, gold, and platinum require stronger oxidizing conditions for corrosion. Copper will dissolve in hydrochloric acid if oxygen of the air is present,

$$2 \; Cu \; (s) + 4 \; H^+ \; (aq) + O_2 \; (aq) \longrightarrow 2 \; Cu^{2+} \; (aq) + 2 \; H_2O \; (liq)$$

Copper and silver dissolve in nitric acid,

$$3 \; Ag \; (s) + NO_3^- \; (aq) + 4 \; H^+ \; (aq) \longrightarrow$$
$$3 \; Ag^+ \; (aq) + NO \; (g) + 2 \; H_2O \; (liq)$$

Gold and platinum can be dissolved in mixed HCl and HNO_3, also called aqua regia:

$$Au \; (s) + 5 \; H^+ \; (aq) + 4 \; Cl^- \; (aq) + NO_3^- \; (aq) \longrightarrow$$
$$HAuCl_4 \; (aq) + NO \; (g) + 2 \; H_2O \; (liq)$$

Corrosion of a metal can occur upon exposure to the ions of a less reactive metal. Thus Cu^{2+} ions will corrode Fe,

$$Fe \; (s) + Cu^{2+} \; (aq) \longrightarrow Fe^{2+} \; (aq) + Cu \; (s)$$

You can easily calculate that at standard conditions the cell potential for this reaction is 0.78 V.

The junction of two dissimilar metals provides the makings of a galvanic cell if an electrolyte is present. An example of this is a steel bolt in a copper plate with exposure to sea water. The steel is the anode with the half-reaction

$$Fe \; (s) \longrightarrow Fe^{2+} \; (aq) + 2 \; e^-$$

and the copper is the cathode,

$$Cu^{2+} \; (aq) + 2 \; e^- \longrightarrow Cu \; (s)$$

We would expect that this second half-reaction would be severely limited by the low concentration of Cu^{2+} that would be available. However, if the area of the steel anode is very small, as we would expect for a bolt, and the copper plate cathode has a relatively large area, a very small deposit per unit area of copper on the cathode can result in considerable corrosion per unit area of the anode. In general, corrosion will be most serious when the anode has a small area and the cathode has a large area.

Such two metal corrosion can also occur in polyphase alloys. For example, brass (an alloy of zinc and copper) can have zinc-rich and copper-rich phases. These can be distributed in such a way on the surface that they set up minute galvanic cells that will result in corrosion.

An example of corrosion resulting from dissolved oxygen gas is shown in Figure 21–12. A water drop on the surface of a piece of iron serves as

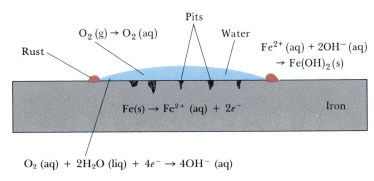

Rust

$O_2 (g) \rightarrow O_2 (aq)$

Pits

Water

$Fe^{2+} (aq) + 2OH^- (aq)$
$\rightarrow Fe(OH)_2 (s)$

$Fe(s) \rightarrow Fe^{2+} (aq) + 2e^-$

Iron

$O_2 (aq) + 2H_2O (liq) + 4e^- \rightarrow 4OH^- (aq)$

FIGURE 21–12 Corrosion of iron at a water drop begins as dissolved oxygen comes in close contact with metal, usually in pits and cavities in the surface, particularly along the edge of the water drop where oxygen concentration is greatest. Electrons are carried to the site, ions diffuse to the edge of the drop where they come in contact with hydroxide ions and there combine to form Fe^{2+}, and subsequently, Fe^{3+}, hydroxides. Rust is the name given hydrated Fe(III) oxides and hydroxides.

the electrolyte. The anode of the cell consists of pits in the iron surface where iron is being oxidized. The cell cathode is at the edge of the water droplet where dissolved oxygen is being reduced. The electrons flow through the iron itself to complete the circuit. Corrosion will take place particularly at pits that are close to the edge of the water drop, which minimize the distance of electron conduction and thus minimize the resistance. A line of rust can be seen at the water level of a steel tank partially filled with water. Pitting of the steel in the tank will be most severe close to the rust line.

Pits in a metal surface are particularly susceptible to continued corrosion and eventual failure. The pit can start as an inclusion of an impurity or a grinding mark. As the pit grows, positively charged metal ions accumulate in it. These attract chloride ions into the pit if they are available, forming a metal chloride. A number of metal chlorides such as iron(II) chloride can react with water to form hydrochloric acid,

$$FeCl_2 (aq) + 2 H_2O (liq) \longrightarrow Fe(OH)_2 (s) + 2 HCl (aq)$$

The hydrochloric acid can attack the iron, resulting in further corrosion.

Grain boundaries in metals are particularly susceptible to corrosion because the surface metal at the grain boundary is more reactive than the bulk metal. Thus grain boundaries can be observed by etching with acid. In regions of high stress in a metal the grain boundaries are separated to some extent, exposing more of the reactive surface. This can act as the anode in electrochemical corrosion.

There are a number of ways of preventing or limiting corrosion. In Figure 21–13(a), a steel sample has been coated with the more active metal zinc. The zinc is slowly corroded, making any exposed steel surface cathodic and preventing it from becoming the anode. In this case the large area of the zinc anode compared with the small area of the exposed steel means that the corrosion per unit area of the zinc surface will be very small.

Some metal surfaces such as those of buried pipes, bridge abutments, piers, and the hulls of ships are continually exposed to electrolyte solutions. In order to protect against their constant susceptibility to corrosion, such structures are attached by a conductor to a **sacrificial anode**. This anode is a metal more easily oxidized than the metal to be protected (Figure 21–13b). This forms a complete cell, with the electrolyte environment acting as the salt bridge, and with the connection between the metals completing the circuit. If magnesium is used to protect iron, the half-reactions of interest are

$$Mg^{2+} (aq) + 2 e^- \longrightarrow Mg (s) \qquad \mathscr{E}° = -2.36 \text{ V}$$
$$Fe^{2+} (aq) + 2 e^- \longrightarrow Fe (s) \qquad \mathscr{E}° = -0.44 \text{ V}$$

FIGURE 21–13 Preventing the corrosion of metals: (a) Here, a steel surface has been coated with Zn, a more active metal, which is preferentially corroded, making any exposed steel "cathodic" instead of anodic, and therefore not subject to oxidation and corrosion; (b) an iron pipe is shown in this schematic diagram, protected by a "sacrificial" magnesium anode, a metal more easily oxidized than iron.

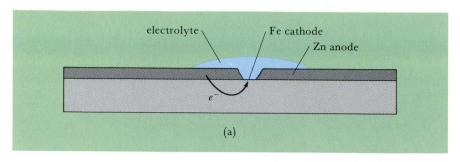

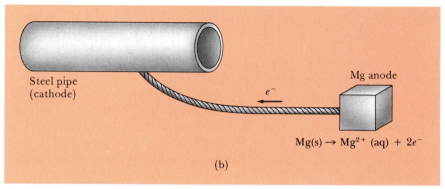

The spontaneous cell reaction under any reasonable range of concentrations is

$$Mg \text{ (s)} + Fe^{2+} \text{ (aq)} \longrightarrow Mg^{2+} \text{ (aq)} + Fe \text{ (s)}$$

Thus the Mg anode will be used up slowly and the Fe will not corrode. Periodic replacement of the sacrificial anode is far less expensive than replacement of the pipe, hull, or structure.

21.9 ELECTROLYSIS AND ELECTROLYTIC PROCESSES

All of the electrochemical reactions we have discussed so far take place spontaneously, in a galvanic cell. However, because such reactions produce a flow of electrons in the conductor connecting the electrodes, it is possible to oppose this flow by applying an external voltage source so as to stop the reaction or force it in the opposite direction. The voltage source could be another electrochemical cell, a battery of such cells, or a direct current power supply. Thus, at standard conditions, the reaction

$$2 \text{ Ag}^+ \text{ (aq)} + Cu \text{ (s)} \longrightarrow 2 \text{ Ag (s)} + Cu^{2+} \text{ (aq)}$$

is spontaneous with $\mathscr{E}° = 0.462$ V. If this process is carried out in a galvanic cell, the Cu electrode is negatively charged and is the anode. The Ag electrode is positively charged and is the cathode. The electron flow in the external circuit is from Cu to Ag. If we insert a voltage source into the external circuit so as to oppose the spontaneous electron flow (Figure 21–14), the reaction will stop if the imposed potential is 0.462 V and will reverse if the imposed potential is greater than 0.462 V. A cell in which an external voltage source forces a reaction in the nonspontaneous direction is called an **electrolytic cell**, and the process taking place is called **electrolysis**. In this case the electrolysis reaction is

$$2 \text{ Ag (s)} + Cu^{2+} \text{ (aq)} \longrightarrow \text{Ag}^+ \text{ (aq)} + Cu \text{ (s)}$$

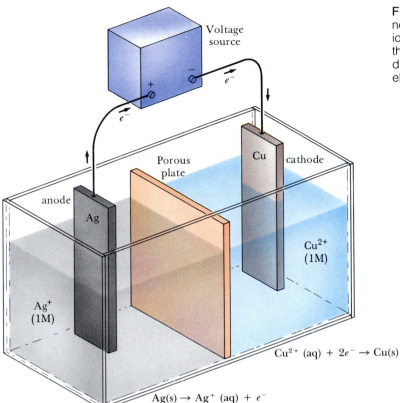

Voltage source

Porous plate

Cu cathode

anode

Ag

Cu^{2+} (1M)

Ag^+ (1M)

Cu^{2+} (aq) $+ 2e^- \rightarrow Cu(s)$

$Ag(s) \rightarrow Ag^+$ (aq) $+ e^-$

FIGURE 21–14 As a galvanic cell, the spontaneous process would be the reduction of silver ions by copper. An opposing voltage source in the external circuit can reverse the process, driving it in the other direction, making the cell electrolytic. The process is called *electrolysis*.

Each half-reaction has been reversed. Now the Cu electrode is the cathode because reduction is taking place at its surface. It is still negatively charged. The Ag electrode is positively charged, and it is the anode.

The potentials needed to perform an electrolysis at a reasonable rate are normally somewhat higher than would be calculated by the standard potentials and sometimes are a volt or more higher. Three factors contribute to the higher potential needed:

1. The internal resistance of the cell, R_{int}, leads to a potential drop of $\mathcal{E}_{int} = iR_{int}$ where i is the current in amperes.

2. Lower concentrations of reactants near the electrodes result from the cell reaction. This is called concentration polarization and can be minimized by vigorous stirring.

3. The remainder of the extra potential needed is called *overvoltage* and is mainly the result of factors that slow the process until the potential is sufficiently high. Overvoltages tend to be particularly high when gases such as hydrogen and oxygen are evolved. The overvoltages in these situations are found to be strongly dependent on the nature of the electrode. Thus it takes an extra volt to evolve hydrogen from water using a mercury electrode compared to using a platinum electrode with a large surface area. The choice of electrode materials in electrolysis reactions is usually critical although not necessarily thoroughly understood.

One important use of electrolysis is electrodeposition of metals, or **electroplating**. For example, silver plating of tableware is shown in Figure 21–15. The surface of each electrode can be considered to be Ag (s) and the electrolyte solution contains Ag^+ (aq). The overall reaction

$$Ag\ (s) + Ag^+\ (aq) \longrightarrow Ag^+\ (aq) + Ag\ (s)$$

FIGURE 21–15 Schematic representation of the silver-plating of tableware. The bar of silver is consumed and the utensil is plated.

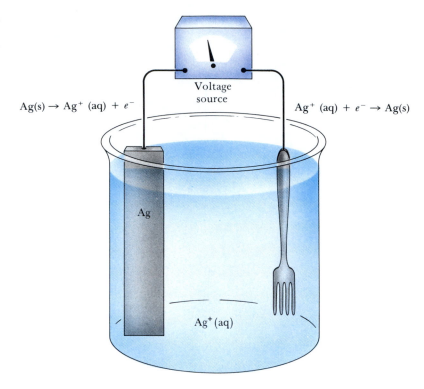

$$Ag(s) \rightarrow Ag^+ (aq) + e^-$$

Voltage source

$$Ag^+ (aq) + e^- \rightarrow Ag(s)$$

Ag

$$Ag^+ (aq)$$

In early 1989, a process was described for plating metal on plastic, an application with important ramifications for the microelectronics industries. Shown here is the deposition of copper electrical conductors on circuit boards molded in a polyetherimide polymer resin. The chemical bond provided by the metalization process results in a "peel" strength nearly double that of the traditional technique in which copper is mechanically attached, by deposition in microscopic cracks created in the plastic by etching. The plating bath is a copper sulfate solution.

has no driving force because the concentration of Ag^+ (aq) is the same for both reactions. As external voltage is applied to give the electrode at which plating is desired a negative charge. The half-reaction at this electrode (the fork in this case) is

$$Ag^+ \text{ (aq)} + e^- \longrightarrow Ag \text{ (s)}$$

The other electrode, a bar of silver, is used up:

$$Ag \text{ (s)} \longrightarrow Ag^+ \text{ (aq)} + e^-$$

The total charge on one mole of electrons is 96,500 coulombs, or a faraday of charge. This number is used for calculating the total electron charge needed for electrolysis. Charge passed per unit time is the current, usually given in amperes:

$$I \text{ amperes} = \frac{Q \text{ coulombs}}{t \text{ seconds}}$$

EXAMPLE 21–18

Calculate the total electron charge needed to plate 10.0 g of Cu metal onto a metal electrode from a solution of $CuSO_4$.

Solution   First state the half-reaction involved in this process:

$$Cu^{2+} \text{ (aq)} + 2 e^- \longrightarrow Cu \text{ (s)}$$

Note that two moles of electrons are required to reduce one mole of Cu^{2+}, so the required charge is

$$(10.0 \text{ g Cu}) \left(\frac{1 \text{ mol Cu}}{63.5 \text{ g Cu}} \right) \left(\frac{2 \text{ mol } e^-}{1 \text{ mol Cu}} \right) \left(\frac{96,500 \text{ C}}{\text{mol } e^-} \right) = 30,400 \text{ C}$$

To put it another way, the passage of 30,400 C of charge deposits 10.0 g of Cu metal. Since a coulomb is an ampere-second, that could be arranged by allowing a 10-A current to flow for 3040 s, or any other combination of amperes and seconds whose product is 30,400. ◼

EXERCISE 21–18

Calculate the mass of Ni (s) that can be plated out of a solution of $NiSO_4$ by 100,000 C of charge.

Answer: 30.4 g. ◼

EXAMPLE 21–19

How long would it take to deposit 1.00 g of Zn from a solution of $ZnCl_2$ by applying a current of 100 amps?

Solution   The half-reaction is

$$Zn^{2+} (aq) + 2\,e^- \longrightarrow Zn\,(s)$$

and

$$(1.00 \text{ g Zn}) \left(\frac{1 \text{ mol Zn}}{65.4 \text{ g Zn}} \right) \left(\frac{2 \text{ mol } e^-}{1 \text{ mol Zn}} \right) \left(\frac{96,500 \text{ C}}{1 \text{ mol } e^-} \right) \left(\frac{1 \text{ s}}{100 \text{ C}} \right) = 29.5 \text{ s}$$

◼

EXERCISE 21–19

What current is needed to deposit 2.00 g of Cr (s) from a solution of $Cr(NO_3)_3$ in 1.00 minute?

Answer: 186 A. ◼

Electrolysis reactions provide an important tool for obtaining elements and compounds that would be very difficult to obtain in other ways. For example, the reaction

$$2\,NaCl \longrightarrow 2\,Na + Cl_2$$

would convert a very common chemical into two very useful products. However, NaCl is a very stable compound and there is no reasonable chemical process for performing this decomposition. Besides, it is necessary to produce the products, Na and Cl_2, separate from each other since they would instantly react to reform NaCl. If we consider this reaction electrochemically, the two half-reactions are

$$Na^+ + e^- \longrightarrow Na \qquad \mathscr{E}° = -2.714 \text{ V}$$
$$2\,Cl^- \longrightarrow Cl_2 + 2\,e^- \qquad \mathscr{E}° = -1.36 \text{ V}$$

Thus for the overall process

$$2\,Na^+ + 2\,Cl^- \longrightarrow 2\,Na + Cl_2 \qquad \mathscr{E}° = -4.07 \text{ V}$$

Clearly this is not spontaneous. However, application of 5 V or so would force the reaction to proceed as written, and 5 V is not a high electrical potential. Thus, the general preparation of reactive metals such as Na, K, Mg, Ca, Al, and others is by electrolysis, usually of their molten salts. Note that the standard potentials in aqueous solution given here for NaCl would not be correct under molten conditions, but they do give an indication of the potentials involved.

The production of aluminum is almost entirely by electrolysis procedures. Alumina, Al_2O_3, is the pure derivative of bauxite, an important

ELECTROLYTIC ALUMINUM—THE HALL-HEROULT PROCESS

Aluminum is the major non-ferrous metal. It is produced today by a modification of the original process developed simultaneously and independently by Charles Martin Hall (American) and Paul Heroult (French), based on the electrolysis of bauxite (Al_2O_3) in a bath of molten cryolite (Na_3AlF_6). Although the modern cells are huge, boxy containers, careful formulation and control are critical. The melting point of pure cryolite is about 1010°C. Because the electrolyte solution contains CaF_2 (fluorspar) and AlF_3, and dissolved aluminum, the electrolytic cells can operate at somewhat lower temperatures of about 950°C. But that is considerably lower than the normal 2030°C melting point of bauxite. Here are the principle electrochemical reactions:

cathode: $AlF_6^{3-} + 3\,e^- \longrightarrow Al\,(liq) + 6\,F^-$

anode: $2\,Al_2O_3\,(liq) + 24\,F^- + 3\,C\,(s) \rightleftharpoons$
$$4\,AlF_6^{3-}\,(liq) + 3\,CO_2\,(g) + 12\,e^-$$

net: $2\,Al_2O_3\,(liq) + 3\,C\,(s) \rightleftharpoons 4\,Al\,(liq) + 3\,CO_2\,(g)$

It is interesting to note that the carbon reduction suggested by the net reaction is not possible without the molten cryolite because Al_4C_3 is formed and a back reaction between aluminum and carbon dioxide quickly leads to re-formation of the original oxide:

$2\,Al\,(liq) + CO_2\,(g) \rightleftharpoons Al_2O_3\,(liq)$

Aluminum produced by the Hall-Heroult process is 99.9% pure at best. This may seem very high, but for some space-age and modern electronics applications, metal of much higher purity is needed. Fractional crystallization (zone refining) is capable of producing aluminum of more than 99.9995% purity. Trace quantities of by-product hydrogen fluoride (HF) gas are produced by reaction of water with AlF_3 present in the electrolyte solution:

$2\,AlF_3\,(liq) + 3\,H_2O\,(g) \rightleftharpoons Al_2O_3\,(liq) + 6\,HF\,(g)$

If not adequately scrubbed, HF vapors exiting the manufacturing facility in the off-gases would severely damage local vegetation downwind.

Because of the huge power demands for the Hall-Heroult process, less energy-intensive processes (see Table) for the preparation of aluminum have been considered and are still being sought, but none is currently in use to any significant degree. Here is one such process that has been tried commercially; it is an interesting application of LeChatelier's principle:

$AlCl_3\,(g) + 2\,Al\,(s) \rightleftharpoons 3\,AlCl\,(g)$

At 700–800°C, the volatilized AlCl is transferred out of the active reaction chamber, causing the reaction to shift to the right, allowing more AlCl to form. Meanwhile, the AlCl that has been removed from the original equilibrium frees aluminum metal as the reaction now reverses itself in re-establishing the equilibrium:

$3\,AlCl\,(g) \rightleftharpoons 2\,Al + 3\,AlCl_3\,(g)$

A second process that is in use on a modest commercial scale is the Alcoa chlorine process. The second step is electrolytic:

aluminum ore. Unfortunately, the melting point of Al_2O_3 is over 2000°C, making electrolysis of the molten substance impractical. Until a way around this problem was discovered, aluminum was a very expensive metal used only in the rarest circumstances. This was all changed in 1886 when Charles

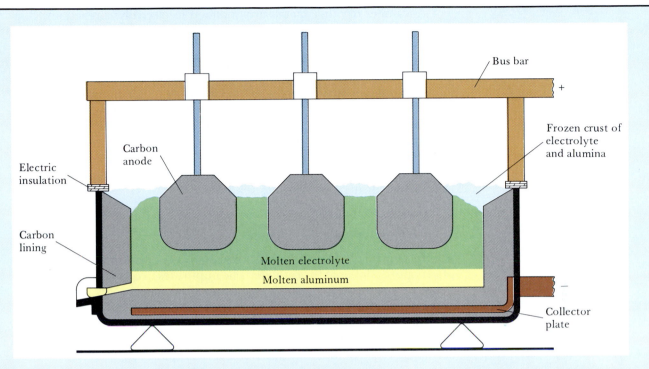

$$2\ Al_2O_3\ (s)\ +\ 3\ C\ (coke)\ +\ 6\ Cl_2\ \longrightarrow\ AlCl_3\ (g)\ +\ 3\ CO_2\ (g)$$

$$AlCl_3\ (liq)\ \longrightarrow\ Al\ (liq)\ +\ Cl_2\ (g)$$

Write chemical equations for the following aluminum production processes, beginning with bauxite. The name should suggest the process chemistry:

(1) Carbothermic reduction.
(2) Metallothermic reduction:
    a) Sodiothermic reduction:
    b) Manganothermic reduction:

If a typical Hall-Heroult cell were operating at 88–95% efficiency, what is the theoretical maximum ratio of Al_2O_3 to Al?

Using the data in the Table, determine how many tons of coal are required to produce a ton of aluminum.

Approximate Energy Requirements for Aluminum Produced by the Hall-Heroult Process, MJ/metric ton* of Al			
Operation	**Thermal Energy**	**Electrical Energy**	**Total**
Mining/refining	30,000	5,000	35,000
Recovering	30,000	157,000	187,000
Processing	30,000	18,000	48,000
Total	90,000	180,000	270,000

*A metric ton is approximately 2200 pounds.
1 MJ ≅ 12,000 BTU ≅ 1 lb coal

Martin Hall and Paul Heroult independently discovered that a mixture of alumina and cryolite (Na_3AlF_6) could be melted and electrolyzed at 800° to 1000°C. As a result of this discovery, the price of aluminum plummeted and this strong, light, durable, and attractive metal has found virtually endless uses.

A number of other metals are obtained in electrolysis reactions. Copper, cadmium, gallium, and some other metals can be electrolyzed from aqueous solutions of their salts. More active metals such as magnesium and lithium must be electrolyzed from their fused salts as sodium and aluminum are.

Electrolysis reactions are also important in purifying metals. For example, copper for power transmission must be extremely pure in order to approach its maximum conductivity. In the electrorefining of copper, an impure bar of copper is the anode and the cathode is started with a thin sheet of pure copper. Both electrodes dip into the same Cu^{2+} and so the cell potential is zero. A potential is applied so that the bar of impure copper is positive, causing the following reactions:

Impure Cu (anode): $\quad Cu\ (s) \longrightarrow Cu^{2+}\ (aq) + 2\ e^-$

Pure Cu (cathode): $\quad Cu^{2+}\ (aq) + 2\ e^- \longrightarrow Cu\ (s)$

Careful control of the potential imposed on this cell results in the deposition of very pure copper at the cathode. The impurities in copper include metals that are more difficult to oxidize than Cu, such as silver, gold, and platinum, and some that are easier to oxidize such as iron, zinc, lead, and nickel.

Keeping the imposed potential sufficiently low will ensure that the Ag, Au, and Pt will not be oxidized since their electrode potentials for oxidation are more negative than that for Cu. For example,

$$Cu\ (s) \longrightarrow Cu^{2+}\ (aq) + 2\ e^- \qquad \mathscr{E}° = -0.337\ V$$
$$Ag\ (s) \longrightarrow Ag^+\ (aq) + e^- \qquad \mathscr{E}° = -0.799\ V$$

Thus, as the anode disintegrates these valuable metals simply fall to the bottom of the cell for later reclaiming. The other impurities, Fe, Zn, Pb, and Ni, are oxidized along with the Cu since they are more easily oxidized. However, a sufficiently low imposed potential will insure that they are not reduced back to the metal at the cathodes, since their electrode potentials for reduction are lower than copper. For example,

$$Zn^{2+}\ (aq) + 2\ e^- \longrightarrow Zn\ (s) \qquad \mathscr{E}° = -0.763\ V$$
$$Ni^{2+}\ (aq) + 2\ e^- \longrightarrow Ni\ (s) \qquad \mathscr{E}° = -0.250\ V$$

These metals remain in the electrolyte solution, which therefore has to be changed from time to time. Proper control of this cell can result in 99.9% pure copper at the cathode. Other metals that can be refined by electrolysis include nickel, cobalt, lead, and plutonium.

Electrolysis of water solutions is of considerable importance. In these cases it is necessary to consult the table of electrode potentials to see what will be oxidized and what will be reduced. For example, suppose we pass a direct current through a water solution of Na_2SO_4 (Figure 21–16). The electrode reactions are

anode: $\quad 2\ H_2O\ (liq) \longrightarrow O_2\ (g) + 4\ H^+\ (aq) + 4\ e^-$

cathode: $\quad 2\ H^+\ (aq) + 2\ e^- \longrightarrow H_2\ (g)$

The H^+ (aq) comes from the self-ionization of water,

$$H_2O\ (liq) \longrightarrow H^+\ (aq) + OH^-\ (aq)$$

and the overall cell reaction is

$$2\ H_2O\ (liq) \longrightarrow 2\ H_2\ (g) + O_2\ (g)$$

The Na^+ and SO_4^{2-} ions of the dissolved Na_2SO_4 increase the conductivity of the water, aiding the electrolysis. Neither of these ions is affected by the

FIGURE 21–16 The electrolysis of water is carried out here using two inert electrodes and sodium sulfate as an electrolyte. Water is oxidized at the anode, producing oxygen and hydronium ions; water is reduced at the cathode, producing hydrogen and hydroxide ions. Note that just as the balanced chemical equation would predict, twice as much hydrogen is produced in the left tube as oxygen in the right.

electrolysis since Na^+ is more difficult to reduce than H^+ (see Table 21–1) and SO_4^{2-} is more difficult to oxidize than H_2O.

When a reasonably concentrated solution of NaCl is electrolyzed, the situation becomes more complex. The possible half reactions are:

anode:
$$2\ H_2O\ (liq) \longrightarrow O_2\ (g) + 4\ H^+\ (aq) + 4\ e^-$$
$$\mathscr{E}° = -1.23\ V$$
$$2\ Cl^-\ (aq) \longrightarrow Cl_2\ (g) + 2\ e^- \qquad \mathscr{E}° = -1.36\ V$$

cathode:
$$2\ H^+\ (aq) + 2\ e^- \longrightarrow H_2\ (g) \qquad \mathscr{E}° = 0.000\ V$$
$$Na^+\ (aq) + e^- \longrightarrow Na\ (s) \qquad \mathscr{E}° = -2.714\ V$$

If overvoltage were not an issue, this cell would be driven by the lowest necessary potential of 1.23 V to give hydrogen and oxygen gases. However, the choice of the electrode materials controls the substances that are actually produced in this electrolysis. Chlorine is produced exclusive of any oxygen in the commercially important Hooker chloralkali process (Figure 21–17, Chapter opening display).

In the chloralkali industry, an anode with a high overvoltage for the production of oxygen is chosen so that chlorine is produced instead. Graphite is the common anode material, and its high overvoltage for oxygen is postulated to be the result of the formation of transient carbon oxides on the graphite surface. Furthermore, the cathode is normally a flowing mercury surface. The overvoltage for hydrogen production on mercury is quite high, and sodium ions are actually preferentially reduced at this electrode. Reduction of the Na^+ ions does not lead to the Na (s) indicated above, but rather to a solution of the sodium in the mercury called an **amalgam**. Production of the amalgam of sodium does not have as low a reduction potential as production of the Na (s), and this further favors the Na^+ reduction. The sodium amalgam is treated with water outside the cell to produce sodium hydroxide and by-product hydrogen, and to regenerate the mercury for further use.

$$2\ Na{\cdot}Hg + 2\ H_2O\ (liq) \longrightarrow$$
$$2\ Na^+\ (aq) + 2\ OH^-\ (aq) + H_2\ (g) + 2\ Hg\ (liq)$$

Electrolysis of water provides a preparation of hydrogen for fuel from a very inexpensive source. However, the electrical energy involved makes this preparation costly; unless the electric cost is relatively low, chemical means of hydrogen preparation will be preferred.

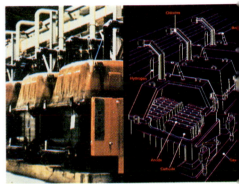

FIGURE 21–17 The photograph (left) shows one of two hundred chloralkali cells, banded together in an on-line commercial scale chloralkali synthesis of chlorine and caustic from the electrolysis of saturated salt solutions; the schematic (right) is a view inside the cell. See also the chapter-opening display.

SUMMARY

Electrochemical reactions involve the flow of electrons and ions through a conducting chemical system. This requires an oxidation-reduction or redox reaction in which electrons are transferred from one species to another. An oxidation-reduction reaction requires a change of oxidation states. A substance is oxidized if it loses electrons and its oxidation state increases. It is reduced if it gains electrons and its oxidation state decreases. In a redox reaction, the species that is reduced is called the oxidizing agent or oxidant, and the species that is oxidized is called the reducing agent or reductant.

Oxidation-reduction reactions are often difficult to balance, and several balancing methods are used. In all methods of balancing these equations, an accounting has to be made of the electrons transferred. The chemical equation depends on whether it is taking place in acidic or basic solution.

The half-reactions in an oxidation-reduction reaction can be separated in an electrochemical cell so that the electrons are transferred between the two halves by an electrical conductor (e.g., a metal wire). A cell in which the overall reaction proceeds spontaneously can produce a useful electric current. It is called a **galvanic cell**. A cell upon which a current is imposed so as to produce a reaction that would otherwise not be spontaneous is called an **electrolytic cell**.

In any cell the anode is the electrode at which oxidation takes place and the cathode is the electrode at which reduction takes place. The measured potential $\mathscr{E}$, of a galvanic cell is the sum of the electrode potentials of the two half-cells. These are called standard potentials, $\mathscr{E}°$, if concentrations of dissolved species are 1 M and pressures of gases are 1 atm. The sign convention for cell potentials is such that a spontaneous process has a positive potential. Half-cell potentials are defined so that the hydrogen electrode at standard conditions has an electrode potential, $\mathscr{E}°$, of zero. Half-cells are normally tabulated as reductions. Changing the direction of a reaction or half-reaction simply changes the sign of its potential. Multiplying a reaction (or a half-reaction) by a constant has no effect on its potential.

The position of a half-cell reaction in the **electrochemical series** suggests the tendency to give up or take on electrons. In turn, the electrochemical series can be used to make predictions about electron-transfer reactions in aqueous solutions.

Potentials for cells (or half-cells) that are not operating under standard conditions (of pressure or concentration) can be determined from the standard potential by means of the **Nernst equation**:

$$\mathscr{E} = \mathscr{E}° - \frac{0.0591}{n} \log Q \qquad \text{at 298 K}$$

It is an expression of the thermodynamic relationships that exist between cell potential, electrical work, and the change in free energy for a chemical reaction. In this form of the equation, n is the number of electrons transferred in the equation and Q is the reaction quotient. And one should not forget the thermodynamic foundations of the Nernst equation, namely that

$$-w_{\text{elec}} = \Delta G° = -n\mathscr{F}\mathscr{E}°$$

under reversible and standard conditions. The Nernst equation can be used to calculate potentials for concentration cells in which the half-cells differ only in the concentrations of reactants. This equation also leads to the relationship between $\mathscr{E}°$ and the equilibrium constant, K:

$$\mathscr{E}° = \frac{0.0591}{n} \log K$$

Electrolysis cells can be used to carry out a number of useful reactions. The quantitative relationships in electrolysis depend on the number of moles of charge that pass through the cell during the process. The total charge passed in an electrochemical reaction can be determined by knowing that one mole of electrons carries a total charge of 96,485 coulombs. This information is stored in the Faraday constant, $\mathscr{F} = 96,485$ C/mol.

QUESTIONS AND PROBLEMS

QUESTIONS

1. Describe the distinction between an oxidation-reduction (redox) reaction and an electrochemical reaction.
2. What is the difference between an electrolytic cell and a galvanic cell? Give an example of each.
3. Define the terms

(a) oxidation
(b) reduction
(c) electrode
(d) anode

(e) cathode
(f) electrolyte
(g) oxidizing agent
(h) oxidant

$$\overset{+2\ -4\ -2}{Cl} \overset{-2}{O_2}$$

4. Explain the function of the salt bridge in a galvanic cell.

5. (a) What is meant by the term **standard electrode potential**?
(b) How can you distinguish between oxidation and reduction potentials for the same electrode?
(c) What is a reference electrode?

6. Describe the essential differences between passage of electricity through aqueous sodium chloride solutions and through molten sodium chloride. Why are they different?

7. Explain how standard electrode potentials are used for measuring each of the following:
(a) standard free energy change ($\Delta G°$) for a chemical process
(b) equilibrium constant for a chemical reaction
(c) direction of spontaneous change

8. Distinguish between each of the following pairs of terms:
(a) oxidation and reduction
(b) anode and cathode
(c) electrolytic and galvanic cells
(d) faradays and coulombs
(e) $\mathscr{E}° > 0$ and $\mathscr{E}° < 0$

9. Under what conditions does the equation $\Delta G° = -n\mathscr{F}\mathscr{E}°$ hold true?

10. In what way(s) can the Nernst equation be viewed as an example of Le Chatelier's principle?

11. Why would the sodium-sulfur cell be unsatisfactory for powering the starter of the family automobile? What uses can you imagine for this storage cell (which has a very high specific energy)?

12. What are the advantages of using a fuel cell to produce electric energy compared with burning the fuel in a conventional power plant?

13. (a) Explain why copper piping might be preferable to iron piping for standard plumbing applications.
(b) Explain the basis for the plumbing problems that arise in copper piping when well water containing significant $[Fe^{3+}]$ is used.

14. Certain types of brass are rapidly corroded by sea water as zinc dissolves from the alloy, leaving behind a spongy mass of very nearly pure copper. Why is zinc more easily attacked than copper? Why does the corrosion process proceed rapidly once started?

15. Do you think anything of a chemical nature will occur to a pure gold coin dropped into a $1\ M$ ferrous ammonium sulfate [$FeSO_4 \cdot (NH_4)_2SO_4$] solution? Why or why not? What about a copper wire placed in a $1\ M$ silver nitrate solution? Write electrochemical equations for both cases.

PROBLEMS

Oxidation-Reduction Reactions   [1–10]

1. For each of the following, give the apparent oxidation state.
(a) the chlorine atom in Cl_2O, ClO_2, and Cl_2O_7
(b) the sulfur atom in HS^-, HSO_3^-, $HS_2O_8^-$, $HS_4O_6^-$, and HSO_4^-
(c) the phosphorus atom in P_4, PH_3, P_2H_4, H_3PO_2, H_3PO_3, H_3PO_4, and $(NH_4)_2P_2O_7$

2. Give the oxidation state for each of the following:

(a) the uranium atom in UF_6, UO_2Cl_2, and UO_3
(b) the manganese atom in MnO, MnO_2, Mn_2O_3, Mn_3O_4, MnO_4^{2-}, MnO_4^- and Mn_2O_7
(c) the chromium atom in $CrCl_2$, CrF_3, CrO_2, CrO_3, CrO_3Cl^-, CrO_4^{2-}, and $Cr_2O_7^{2-}$

3. For each of the following processes, identify the oxidizing agent and the reducing agent:
(a) $2\ Na\ (s) + Cl_2\ (g) \longrightarrow 2\ NaCl\ (s)$
(b) $Zn\ (s) + H_2SO_4\ (aq) \longrightarrow ZnSO_4\ (aq) + H_2\ (g)$

4. Identify the oxidizing agent and the reducing agent in each of the following processes:
(a) $Br_2\ (aq) + 2\ I^-\ (aq) \longrightarrow I_2\ (aq) + 2\ Br^-\ (aq)$
(b) $2\ KClO_3\ (s) \longrightarrow KClO_2\ (s) + KClO_4\ (s)$

5. Which of the following correspond to oxidations? Which are reductions?
(a) a ferrous salt is converted to a ferric salt: Fe^{2+}/Fe^{3+}
(b) a sulfide is converted to sulfur: S^{2-}/S
(c) a sulfite is converted to a sulfate: SO_3^{2-}/SO_4^{2-}
(d) a persulfate is converted to a sulfite: $S_2O_8^{2-}/SO_3^{2-}$
(e) a hydrogen sulfate becomes a sulfate: HSO_4^-/SO_4^{2-}

6. Tell what is oxidized and what is reduced in the following reactions.
(a) $Hg_2^{2+}\ (aq) + Pt\ (s) + 4\ Cl^-\ (aq) \longrightarrow$
$$2\ Hg\ (liq) + PtCl_4^{2-}\ (aq)$$
(b) $5\ MnO_2\ (s) + 4\ H^+\ (aq) \longrightarrow$
$$2\ MnO_4^-\ (aq) + 3\ Mn^{2+}\ (aq) + 2\ H_2O$$

7. Complete and balance the following oxidation-reduction reactions, which take place in acid solution.
(a) $Cr_2O_7^{2-}\ (aq) + Cl^-\ (aq) \longrightarrow Cr^{3+}\ (aq) + Cl_2\ (g)$
(b) $S_2O_3^{2-}\ (aq) + I_2\ (aq) \longrightarrow I^-\ (aq) + S_4O_6^{2-}\ (aq)$
(c) $MnO_2\ (s) + Hg\ (liq) + Cl^-\ (aq) \longrightarrow$
$$Mn^{2+}\ (aq) + Hg_2Cl_2\ (s)$$
(d) $Ag\ (s) + NO_3^-\ (aq) \longrightarrow Ag^+\ (aq) + NO\ (g)$
(e) $H_3AsO_4\ (aq) + Zn\ (s) \longrightarrow AsH_3\ (g) + Zn^{2+}\ (aq)$
(f) $Au^{3+}\ (aq) + I_2\ (aq) \longrightarrow Au\ (s) + IO_3^-\ (aq)$
(g) $IO_3^-\ (aq) + I^-\ (aq) \longrightarrow I_3^-\ (aq)$
(h) $HS_2O_3^-\ (aq) \longrightarrow S\ (s) + HSO_4^-\ (aq)$
(i) $MnO_4^{2-}\ (aq) \longrightarrow MnO_2\ (s) + MnO_4^-\ (aq)$
(j) $O_2^{2-}\ (g) \longrightarrow O_2\ (g) + H_2O\ (liq)$

8. Balance the following equations for the reactions in aqueous acid solution:
(a) $Cr_2O_7^{2-}\ (aq) + I_2\ (aq) \longrightarrow Cr^{3+}\ (aq) + IO_3^-\ (aq)$
(b) $S_2O_3^{2-} + I_2\ (aq) \longrightarrow S_4O_6^{2-}\ (aq) + I^-\ (aq)$
(c) $MnO_4^-\ (aq) + H_2O_2\ (aq) \longrightarrow$
$$Mn^{2+} + H^+\ (aq) + O_2\ (g)$$
(d) $Hg_2Cl_2\ (s) + NO_2^-\ (aq) \longrightarrow Hg^{2+}\ (aq) + NO\ (g)$
(e) $MnO_4^{2-}\ (aq) \longrightarrow MnO_2\ (s) + MnO_4^-\ (aq)$
(f) $Pb\ (s) + PbO_2\ (s) + SO_4^{2-}\ (aq) \longrightarrow PbSO_4\ (s)$

9. Complete and balance the following equations for reactions that take place in basic solution:
(a) $Co(OH)_3\ (s) + Sn\ (s) \longrightarrow$
$$Co(OH)_2\ (s) + HSnO_2^-\ (aq)$$
(b) $ClO_4^-\ (aq) + I^-\ (aq) \longrightarrow ClO_3^-\ (aq) + IO_3^-\ (aq)$
(c) $PbO_2\ (s) + Cl^-\ (aq) \longrightarrow ClO^-\ (aq) + Pb(OH)_3^-\ (aq)$
(d) $NO_2^-\ (aq) + Al\ (s) \longrightarrow NH_3\ (g) + AlO_2^-\ (aq)$
(e) $ClO^-\ (aq) \longrightarrow Cl^-\ (aq) + O_2\ (g)$
(f) $HXeO_4^-\ (aq) + Pb\ (s) \longrightarrow Xe\ (g) + HPbO_2^-\ (aq)$
(g) $Ag_2S\ (s) + CN^-\ (aq) + O_2\ (g) \longrightarrow$
$$S\ (s) + Ag(CN)_2^-\ (aq)$$
(h) $MnO_4^-\ (aq) + S^{2-}\ (aq) \longrightarrow MnS\ (s) + S\ (s)$
(i) $Cl_2\ (g) \longrightarrow ClO^-\ (aq) + Cl^-\ (aq)$

10. Balance the following equations for the reactions in aqueous basic solution:

(a) $MnO_4^- (aq) + H_2O_2 (aq) \longrightarrow MnO_2 (s) + O_2 (g)$
(b) $ClO_2 (aq) \longrightarrow ClO_2^- (aq) + ClO_3^- (aq)$
(c) $CrO_4^{2-} (aq) + N_2H_4 (aq) \longrightarrow Cr^{3+} (aq) + N_2 (g)$
(d) $Ag (s) + CN^- (aq) + O_2 (g) \longrightarrow$
$$Ag(CN)_2^- (aq) + OH^- (aq)$$
(e) $Co (s) + ClO^- (aq) \longrightarrow Co^{2+} (aq) + Cl^- (aq)$
(f) $Cd (s) + H_2O (liq) + Ni_2O_3 (s) \longrightarrow$
$$Cd(OH)_2 (s) + NiO (s)$$

Standard Potentials; Electrochemical Series [11–16]

11. Using the data in Table 21–1, arrange the following substances in their proper order as reducing agents: Al, Co, Ni, Ag, H_2, Na.

12. List the following in order as oxidants: Fe^{3+}, F_2, Pb^{2+}, I_2, Sn^{4+}, O_2.

13. Using the standard reduction potentials listed in Table 21–1:

(a) Pick an oxidizing agent that could cause the following to happen spontaneously:

$$Fe (s) \longrightarrow Fe^{3+} (aq)$$
$$Fe (s) \longrightarrow Fe^{2+} (aq)$$
$$Fe^{2+} (aq) \longrightarrow Fe^{3+} (aq)$$
$$2 F^- (aq) \longrightarrow F_2 (g)$$

(b) Pick an appropriate agent (or agents) for the following:

$$Ni^{2+} (aq) \longrightarrow Ni (s)$$
$$Cl_2 (aq) \longrightarrow 2 Cl^- (aq)$$

14. Making use of the table of Table 21–1, select a reagent to perform the indicated task:

(a) $Zn (s) \longrightarrow Zn^{2+} (aq)$
(b) $F_2 (g) \longrightarrow 2 F^- (aq)$
(c) $Mn^{2+} (aq) \longrightarrow MnO_4^- (aq)$
(d) $H_2O_2 (liq) \longrightarrow H_2O (liq)$
(e) $H_2O_2 (liq) \longrightarrow O_2 (g)$
(f) $2 I^- (aq) \longrightarrow I_2 (s)$
(g) $Cr^{3+} (aq) \longrightarrow Cr_2O_7^{2-} (aq)$

15. For each of the following unbalanced redox reactions, calculate the standard cell potential and determine whether the reaction is spontaneous for 1 M solutions of ionic reactants, gaseous reactants at 1 atm partial pressures, and solids in their standard states, all at 298 K:

(a) $Fe^{2+} (aq) + MnO_4^- (aq) + H^+ (aq) \longrightarrow$
$$Fe^{3+} (aq) + Mn^{2+} (aq) + H_2O (liq)$$
(b) $Ag (s) + H^+ (aq) + Cl^- (aq) + O_2 (g) \longrightarrow$
$$AgCl (s) + H_2O (liq)$$
(c) $Sn (s) + Hg_2Cl_2 (s) \longrightarrow$
$$Sn^{2+} (aq) + Hg (liq) + Cl^- (aq)$$
(d) $H_2O_2(aq) + IO_{3-} (aq) + H_+ (aq) \longrightarrow$
$$O_2 (g) + I_2 (g) + H_2O (liq)$$
(e) $H_3AsO_2 (aq) + 2MnO_4^- (aq) \longrightarrow$
$$H_3AsO_4 (aq) + 2MnO_4^{2-} (aq) + 2H^+ (aq)$$

16. Determine which of the following reactions would take place spontaneously as written under standard state conditions:

(a) $Hg_2^{2+} (aq) \longrightarrow Hg (liq) + Hg^{2+} (aq)$
(b) $Sn (s) + Fe^{2+} (aq) \longrightarrow Sn^{2+} (aq) + Fe (s)$
(c) $Sn^{2+} (aq) + 2 Fe^{3+} (aq) \longrightarrow Sn^{4+} (aq) + 2 Fe^{2+} (aq)$
(d) $2 Ce^{4+} (aq) + 2 Cl^- (aq) \longrightarrow 2 Ce^{3+} (aq) + Cl_2 (g)$
(e) $2 Fe(CN)_6^{4-} (aq) + 2 H^+ (aq) \longrightarrow$
$$2 Fe(CN)_6^{3-} (aq) + H_2 (g)$$

Galvanic Cells [17–24]

17. Draw a complete galvanic cell in which the electrodes are a Ni bar in a 1.00 M Ni^{2+} solution and a Mn bar in a 1.00 M Mn^{2+} solution. The Mn electrode is negative. Label all parts of the cell including the anode and the cathode. Give the half-reaction taking place at each electrode.

18. A galvanic cell consists of a Co electrode in a 1.00 M Co^{2+} solution and a Pt electrode in a solution containing Fe^{3+} and Fe^{2+}, each at 1.00 M. The two compartments are separated by a salt bridge. Draw the cell and label all components. Indicate the anode and the cathode, the positive and negative electrodes, and the direction of electron flow. Give the half-reaction taking place at each electrode and the overall spontaneous cell reaction. Calculate the potential of the cell.

19. Devise a cell with all concentrations at 1.00 M in which one electrode is a Cu bar in a Cu^{2+} solution. Choose the other half-cell so that the standard cell potential is about 0.4 V. Label the cell completely.

20. Draw a cell containing a Cu^{2+}/Cu^+ half-cell that has a standard cell potential of close to 1.20 V. Is the Cu^{2+}/Cu^+ electrode negative? Describe fully.

21. Calculate the potential of a galvanic cell containing a Pt electrode dipping into a solution 0.00235 M in both Hg^{2+} and Hg_2^{2+} and a Pb electrode dipping into a 0.0936 M solution of Pb^{2+}. Which is the positive electrode?

22. A galvanic cell with a measured potential of 0.11 V contains a Pt electrode in a solution 0.0135 M in Cr^{2+} and 0.000216 M in Cr^{3+}. In the second compartment a Ni electrode dips into a solution of Ni^{2+} of unknown concentration. The Ni electrode is positive. Determine $[Ni^{2+}]$.

23. A miniaturized battery designed for use in hearing aids is composed of Zn and a paste of KOH, water, mercury(II) oxide and mercury. Write the separate half-reactions and the overall cell process (as chemical equations) if zinc and potassium hydroxide are consumed, mercury is deposited, and potassium zincate (K_2ZnO_2) is formed.

24. Devise a cell for the essential reaction in the corrosion of steel, a process that may be viewed as the conversion of Fe to Fe_2O_3 in an aqueous electrolyte environment. Write a complete equation for the process and draw a diagram for the cell.

Electrical Work; Free Energy; Equilibrium Constants [25–36]

25. For the following reaction at 298 K, determine the equilibrium constant and the standard free energy change $\Delta G°$:

$$Sn (s) + Pb^{2+} (aq) \rightleftharpoons Sn^{2+} (aq) + Pb (s)$$

26. Use the table of standard reduction potentials to determine the equilibrium constant of the reaction

$$2 Hg^{2+} (aq) + 2 Br^- (aq) \rightleftharpoons Hg_2^{2+} (aq) + Br_2 (liq)$$

27. If 1.00 mol of silver is oxidized to Ag^+ by a stoichiometric quantity of Br_2 (liq), what is the maximum electrical work that can be done by this process?

28. How much electrical work can be performed if 10.0 g of zinc metal is oxidized to Zn^{2+} by a sufficient amount of Ni^{2+}?

29. A galvanic cell is constructed of two half-cells connected by a salt bridge. The first consists of an inert Pt electrode dipping into a solution that is 1 M in Fe^{2+} and 1 M in Fe^{3+}. In the other half-cell, a Zn electrode dips into a 1 M Zn^{2+} ion solution.
(a) Write the cell reaction for the spontaneous process.
(b) Which is the negative electrode? The positive electrode?
(c) Determine the cell voltage and the equilibrium constant.
(d) How will the cell voltage change if $[Fe^{3+}]$ increases?

30. A galvanic cell consists of a platinum wire dipping into a solution that is 1.0 M in Ce^{3+} and Ce^{4+}. The other electrode is silver, dipping into 1.0 M silver nitrate. Making use of Table 21–1, determine each of the following:
(a) the cell polarity
(b) the anode and the cathode
(c) the standard cell potential
(d) the equilibrium constant
(e) the effect on the cell voltage of increasing the silver ion concentration

31. A solution in a beaker contains Zn^{2+} and Cr^{3+} in contact with Zn and Cr metals. The concentration of Zn^{2+} is 0.132 M at equilibrium. What is the equilibrium concentration of Cr^{3+}?

32. The chlorate ion, ClO_3^-, can disproportionate in basic solution according to the reaction

$$2\ ClO_3^-\ (aq) \rightleftharpoons ClO_2^-\ (aq) + ClO_4^-\ (aq)$$

What are the equilibrium concentrations of the ions resulting from a solution initially 0.100 M in ClO_3^-?

33. Use the following data to calculate the solubility product of PbF_2. A galvanic cell consists of a hydrogen gas electrode ($p(H_2) = 1$ atm, $[H^+] = 1\ M$) and a Pb electrode in a 1 M F^- solution in equilibrium with PbF_2 (s). The cell potential is 0.562 V and the hydrogen gas electrode is positive.

34. A galvanic cell contains one Pb electrode in a 1 M Pb^{2+} solution and the other Pb electrode in a saturated solution of $PbSO_4$. The cell potential is 0.235 V and the electrode in the saturated $PbSO_4$ is negative. What is the solubility product of $PbSO_4$ calculated from these data?

35. The solubility product for $PbSO_4$ can be calculated using the half-reactions

$$PbSO_4\ (s) + 2\ e^- \longrightarrow Pb\ (s) + SO_4^{2-}\ (aq)$$
$$\mathscr{E}° = -0.359\ V$$
$$Pb^{2+}\ (aq) + 2\ e^- \longrightarrow Pb\ (s) \qquad \mathscr{E}° = -0.126\ V$$

Calculate K_{sp} for $PbSO_4$ (s) from these data.

36. From data in Table 21–1, calculate the equilibrium constant for the following reaction:

$$2\ Ag^+\ (aq) + 2\ Hg\ (liq) \rightleftharpoons Hg_2^{2+}\ (aq) + 2\ Ag\ (s)$$

With all substances in their standard states, determine the direction in which the cell would operate spontaneously.

Concentration Cells   [37–40]

37. A concentration cell contains Pb electrodes in equal volume solutions of Pb^{2+} of 0.432 M and 0.000149 M concentration. Draw the cell and label all components. Write the spontaneous half-reaction at each electrode and the overall cell process. What will be the cell potential when the cell is first connected? What will happen to the cell potential as current is allowed to flow? What will be the final concentration of Pb^{2+} in each compartment?

38. The following half-cells are coupled together to form a concentration cell:

$$H_2\ (1\ atm) \longrightarrow 2\ H^+\ (0.10\ M) + 2\ e^-$$
$$H_2\ (1\ atm) \longrightarrow 2\ H^+\ (1.0\ M) + 2\ e^-$$

(a) Sketch the cell diagram, showing anode, cathode, flow of electrons, movement of ions, and the net cell reaction for the spontaneous process.
(b) Calculate the maximum cell voltage and comment on the source of the cell's driving force.
(c) Determine the change in the free energy and the electrical work the cell is capable of performing.

39. A galvanic cell contains two hydrogen gas electrodes. One is in a 1.00 M H^+ solution and is positive. The other is in a solution of unknown pH. The cell potential is 0.0251 V. What is the unknown pH?

40. A galvanic cell has a Cu electrode in a 1.00 M Cu^{2+} solution and a hydrogen electrode in a solution of unknown pH. The cell potential is 0.7205 V. Why do these data allow calculation of two possible pH's for the unknown solution? What are the two possible solutions? What are the two possible pH's? Which can be ruled out as unreasonable?

Electrolytic Cells; Faraday's Laws   [41–56]

41. How many grams of Cr (s) would be deposited in an electrolytic cell from Cr^{3+} (aq) by 150,000 coulombs of charge?

42. What mass of Pb would be deposited by electrolysis of Pb^{2+} using a current of 100. A for 8.0 hours?

43. What current would be necessary to plate 10.0 g of Co (s) from Co^{2+} (aq) in 24.0 hours?

44. How long would it take to plate a 0.100 mm coating of Cr (s) on an automobile bumper having a total surface area of 1.00 m^2 from a solution of Cr^{3+} (aq) employing a current of 100. A? The density of Cr (s) is 7.20 g/cm^3.

45. Calculate the total energy in joules needed to produce 1.00 kg of Cl_2 by electrolysis from a 1.00 M solution of NaCl.

46. Calculate the maximum number of grams of cadmium deposited at the cathode when a $CdCl_2$ solution is electrolyzed in a cell using inert electrodes, by passage of 0.450 faraday of charge.

47. In each of the following cases, calculate the maximum weight in grams of vanadium that can be deposited on the cathode when a solution of $VO(NO_3)_2$ is electrolyzed with:
(a) 1.00×10^4 C of electricity.
(b) a 1.00-A current flowing for 60.0 minutes.
(c) 1.00 faraday of electricity.

48. A 0.750-A current passes for 1.00 h through a cell that contains an aqueous sodium sulfate solution and nickel electrodes. At the cathode, water decomposes to hydrogen gas and hydroxide ions. At the anode, hydronium ions

and nickel oxide are produced. Calculate each of the following, after writing out the electrode reactions:
(a) the number of moles of H_3O^+ and OH^- produced
(b) the number of moles of NiO deposited
(c) the number of moles of Ni dissolved
49. (a) An aqueous silver nitrate solution is electrolyzed using a 2.50-A current for a period of 1.00 h. Calculate the volume of oxygen gas collected at STP. The anode reaction is as follows:

$$2\ H_2O\ (liq) \longrightarrow O_2\ (g) + 4\ H^+\ (aq) + 4\ e^-$$

(b) In a second experiment, the cathode reaction resulted in an increase in weight of the silver electrode of 0.523 g. How long had current flowed through the cell?
50. A liter of a 1.0 *M* aqueous permanganate ion (MnO_4^-) solution is reduced at the cathode of an electrolytic cell. Determine how many faradays of electricity would be required to bring about the formation of each of the following:
(a) a 0.010 *M* solution of manganate ion (MnO_4^{2-})
(b) 1.00 g of MnO_2 (s)
(c) 1.00 g of Mn
If the current in the cell was 10.0 A, how much time would be required to bring about each of these transformations?
51. The coulometer, invented by Michael Faraday, is a device for measuring the total charge passing through a circuit. This is done simply by placing an electrolytic cell in series with the circuit and measuring the chemical result of the electrolysis. Faraday's coulometer measured the amount of H_2 (g) produced from the electrolysis of water. How many coulombs of electric charge would have passed through a circuit producing 3.62 L of H_2 (g) at 1.00 atm pressure at a temperature of 35°C?
52. A battery was used to supply a constant current of what was believed to be exactly 0.450 A as read on a meter in the external circuit. The cell was based on the electrolysis of a copper sulfate solution. During the 30.0 minutes that current was allowed to flow, a total of 0.3000 g of copper metal was deposited at the cathode. Determine the extent to which the meter was inaccurate.
53. Four electrolytic cells are connected in series. In the first, silver ions are reduced to metallic silver; in the second, copper is oxidized to copper(II) ions. In the third, a chromium(III) nitrate solution deposits chromium. In the fourth, water is electrolyzed. If 1.000 g of silver was deposited in the first cell, what are the weights of copper dissolved and chromium deposited? Determine the volumes of hydrogen and oxygen gas evolved (at STP).
54. Six different cells are connected in series so that they can be electrolyzed between inert electrodes, producing pure metal at each of the respective cathodes. In the first of the six cells, a 2.15 mL pool of mercury metal (d = 13.59 g/mL) was obtained from $Hg_2(NO_3)_2$. Calculate how many grams of Fe, Au, Co, or Cr could be deposited from solutions of the following:
(a) $K_4Fe(CN)_6$
(b) $K_3Fe(CN)_6$
(c) $Au(NO_3)_3$
(d) $Co(NO_3)_3$
(e) $Cr_2(SO_4)_3$

55. A lead storage battery is allowed to discharge until 23.92 g of PbO_2 have been reduced at the cathode. Determine each of the following:
(a) the number of grams of $PbSO_4$ formed
(b) the time required to recharge the battery to its original state at a current of 3.0 A
56. Two different types of batteries discharge according to the following reactions:

$$Fe\ (s) + Ni_2O_3\ (s) + 3\ H_2O\ (liq) \longrightarrow$$
$$Fe(OH)_2\ (s) + 2\ Ni(OH)_2\ (s)$$

$$Cd\ (s) + Ni_2O_3\ (s) + 3\ H_2O\ (liq) \longrightarrow$$
$$Cd(OH)_2\ (s) + 2\ Ni(OH)_2\ (s)$$

Beginning with exactly 100.0 g of pure Fe and Cd, determine each of the following:
(a) Which battery will have undergone the greater chemical change per unit time?
(b) Assuming both batteries generate 1.40 V, which converted the greater quantity of chemical energy to electrical energy?
(c) If a 10.-A recharging current is used, which is restored to its original condition faster?

Additional Problems   [57–70]

57. For each of the following half-reactions, determine the number of electrons transferred and the direction of the transfer. Complete the balancing of the equation for stoichiometry and charge.
(a) VO^{2+} (aq) $\longrightarrow VO_3^-$ (aq) (in basic solution)
(b) Cr^{3+} (aq) $\longrightarrow Cr_2O_7^{2-}$ (aq) (in acid solution)
(c) Mn^{2+} (aq) $\longrightarrow MnO_2$ (s) (in acid solution)
(d) NO (g) $\longrightarrow NO_3^-$ (aq) (in basic solution)
(e) Fe^{3+} (aq) $\longrightarrow Fe^{2+}$ (aq) (in acid solution)
58. From among the following reactions, and with the aid of the reduction potentials listed in Table 21–1, determine the strongest oxidant and the strongest reductant:
(a) Na^+ (aq) $+ e^- \longrightarrow Na$ (s)
(b) Br_2 (liq) $+ 2\ e^- \longrightarrow 2\ Br^-$ (aq)
(c) O_2 (g) $+ 2\ H^+$ (aq) $+ 2\ e^- \longrightarrow H_2O_2$ (liq)
(d) Ce^{4+} (aq) $+ e^- \longrightarrow Ce^{3+}$ (aq)
(e) Sn^{2+} (aq) $+ 2\ e^- \longrightarrow Sn$ (s)
59. Devise a cell with all standard concentrations and gas pressures (1 *M* or 1 atm) that has a standard cell potential as close as possible to 1.40 V. Describe the cell fully.
60. For a galvanic cell composed of a silver electrode immersed in a 1.0 *M* silver nitrate solution and an aluminum electrode in a 1.0 *M* aluminum nitrate solution, calculate the standard cell potential and the value for $\Delta G°$.
61. Using standard reduction potentials from Table 21–1, determine the electrode potential for each of the following half-reactions at the concentrations given:
(a) Fe^{2+} (aq) $\longrightarrow Fe^{3+}$ (aq) $+ e^-$

$$[Fe^{2+}] = 0.10\ M,\ [Fe^{3+}] = 0.20\ M$$

(b) Ni (s) $\longrightarrow Ni^{2+}$ (aq) $+ 2\ e^-$

$$[Ni^{2+}] = 0.001\ M$$

(c) MnO_4^- (aq) $+ 8\ H^+$ (aq) $+ 5\ e^- \longrightarrow$
$$Mn^{2+}\ (aq) + 4\ H_2O\ (liq)$$

$$[Mn^{2+}] = 0.10\ M,\ [MnO_4^-] = 1.00\ M,\ pH = 1.00$$

62. A galvanic cell composed of a Zn electrode in a 1.00 *M* zinc sulfate solution coupled to a hydrogen electrode [H_2(1 atm)/H^+(1 *M*)] is connected in series to a second cell composed of a silver electrode dipping into a 1.00 *M* silver nitrate solution coupled to a hydrogen gas electrode [H_2(1 atm)/H^+(1 *M*)].
(a) Write the half-cell reaction occurring at each electrode.
(b) Write the equation for the net reaction and calculate the standard cell potential.
(c) Sketch the cell, being sure to include the anodes, cathodes, electron flow, migration of ions, and all other significant features.
(d) If the two cells in series are now operated as an electrolytic cell rather than a galvanic cell by placing an opposing voltage of 1.50 V against the cell voltage, what will happen? If the opposing cell voltage is increased to 1.60 V, what happens?

63. Calculate the approximate specific energies of the nickel-cadmium and sodium-sulfur cells, considering only the reacting species in the overall reactions.

64. The lead-acid storage battery is composed of galvanic cells that do not contain salt bridges. Why is this possible in this case? Use the table of standard reduction potentials to devise two other galvanic cells that do not require salt bridges. Give the half-reaction taking place at each electrode and calculate the standard cell potential.

65. (a) For the cell process described by the following equation, find the standard cell potential, the cell potential, and the concentration ratio at which the potential generated by the cell is zero:

$$Fe \text{ (s)} + Co^{2+} \text{ (0.5 } M) \longrightarrow Fe^{2+} \text{ (1.0 } M) + Co \text{ (s)}$$

(b) Using the Nernst equation and the data in Table 21–1 (giving standard electrode potentials), calculate the cell potential for the following reaction:

$$Fe \text{ (s)} + MnO_4^- \text{ (0.01 } M) + 4 \text{ } H^+ \text{ (0.10 } M) \longrightarrow$$
$$MnO_2 \text{ (s)} + Fe^{3+} \text{ (0.01 } M) + 2 \text{ } H_2O \text{ (liq)}$$

66. Two electrodes are both constructed of an inert platinum gauze supporting a paste of MnO_2, and they both operate according to the following half-cell reaction, with $[MnO_4^-] = 0.01$ *M*:

$$MnO_4^- \text{ (aq)} + 4 \text{ } H^+ \text{ (aq)} + 3 \text{ } e^- \text{ (aq)} \longrightarrow$$
$$MnO_2 \text{ (s)} + 2 \text{ } H_2O \text{ (liq)}$$

The difference between the two half-cells is that the pH in one is 1.0 while the pH in the other is 2.0.
(a) Write the net equation for the spontaneous process.
(b) What is the cell potential?
(c) What is the equilibrium constant?

67. Derive the equation for the value of $\mathscr{E}$ for the reaction in the lead-acid storage battery as a function of pH if all other concentrations are held at 1 *M*.

68. Aluminum metal can be won from alumina (Al_2O_3) in an electrolytic process. What is the maximum number of grams of pure metal that can be obtained from a ton of crude ore containing 1.5% alumina?

69. Draw a simple schematic diagram of a cell for the electrolytic decompositon of water. What gases will be found and in what relative proportions? At which electrode will each collect? Write the half-cell reaction for each electrode process and note the direction in which ions and electrons are flowing.

70. Exactly 1000. mL of a 0.100 *M* NaCl solution was electrolyzed for 100. minutes with a 1.0-A current. Calculate the pH of the solution in the cell due to the hydroxide ion concentration that accumulates there.

MULTIPLE PRINCIPLES   [71–74]

71. When XeO_3 is reduced, the product is Xe gas. A 5.00-g sample containing Na_2SO_3 is treated with XeO_3. Under these conditions the Na_2SO_3 is oxidized to Na_2SO_4. The Xe gas produced is collected and measured, and is found to occupy 175 mL at 35.0°C and 0.950 atm. What is the percentage of Na_2SO_3 in the sample?

72. A concentration cell has Zn electrodes. The electrolyte in each of the half-cells is a solution of $ZnCl_2$ dissolved in water. One has a freezing point of -2.00°C. The other has a freezing point of -0.90°C. What is the potential of the cell at 25°C?

73. A balloon is being filled with hydrogen produced by the electrolysis of an aqueous solution of an acid. How long will it take to generate enough hydrogen to lift 1.50 kg by using a current of 8.5 A?

74. A galvanic cell consists of one half-cell with a Zn electrode in a 0.100 *M* Zn^{2+} solution and a second half-cell with a Pt electrode in a 0.100 *M* solution of HIO_3. Write the equation for the spontaneous cell reaction and calculate the cell potential. The dissociation constant (K_a) for HIO_3 is 1.9×10^{-1}.

APPLIED PRINCIPLES   [75–79]

75. How much energy is required to electroplate 1.00 g of silver metal onto a metal surface from a 1.00 *M* solution of $AgNO_3$? How does the amount of energy required change if the $AgNO_3$ solution is 2.00 *M*?

76. A galvanic cell consists of a Cu electrode in 0.500 L of 1.00 *M* $CuSO_4$ and a Zn electrode in 0.500 L of 1.00 *M* $ZnCl_2$. A steady current of 1.00 A is drawn from the cell. Plot the cell potential as a function of time at 1-hour intervals from 0 to 10 hours. Use this graph to calculate the approximate power output during this time.

77. Chlorine is one of the products of the electrolysis of molten sodium chloride. Use the van der Waals equation to calculate the pressure at 35°C that would result if the chlorine produced with a yield of 96.0% from 10.00 kg of sodium chloride were compressed into a tank with a volume of 75.0 L.

78. Calculate the standard cell potential for a fuel cell in which ammonia and oxygen gas react to produce nitrogen gas and water vapor.

79. If the fuel cell in Problem 78 operates at an efficiency of 40.0%, how many coulombs of charge will be produced if 100 g of ammonia are consumed?

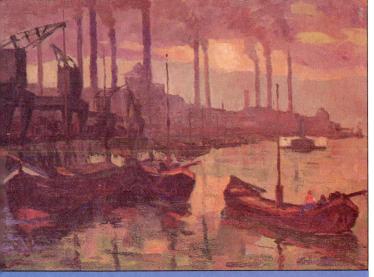

CHAPTER 22

Transition Metals and Coordination Chemistry

Using air oxidation to remove excess carbon and other impurities, "open hearth" steel has been the modern backbone of steel manufacturing technologies. Newer methods (shown here) inject pure oxygen directly into the molten iron for that purpose. In the 19th century, "smokestack" industries such as steel manufacturing, became the benchmarks of industrialization. Fortunately, the smokestacks have largely given away to cleaner, cheaper and more efficient process technologies, leaving us with this artist's view of what was one of the great "smokestack" steel manufacturing industrial centers in the world—"Ludwigshafen-am-Rhein, 1910."

22.1 THE TRANSITION METALS

The metals of Groups IA and IIA, the alkali and alkaline earth metals, may be typical metals in their properties, but they would not be the first metals that would come to most people's minds. When asked to name a few important metals, you would probably name iron, copper, gold, and silver. Other metals that would come to mind quickly are nickel, chromium, and platinum, all transition metals. The transition metals and their alloys such as steel, alnico, carbaloy, monel, brass, bronze, nichrome, German silver, and many others are of enormous importance in practical and decorative applications.

We classify the transition metals as including the d-block elements, the lanthanides or rare earths, and the actinides. The Group IIB elements zinc, cadmium, and mercury are sometimes not included in the transition metals, but are treated here more conveniently than elsewhere. The number of elements in this classification is 58 (plus the elements numbered 104 and above)—the majority of the periodic table. All are metals and generally exhibit typical metallic properties including luster, conductivity, opacity, malleability, and ductility. In addition, these elements have some distinguishing characteristics of their own. Most of them take on different oxidation states in the compounds they form; a large fraction of these compounds are colored, and many of them have the property of paramagnetism—they are attracted to a magnetic field. In addition, they form coordination complexes. (Figure 22–1).

As we shall discuss later, transition metals tend to form coordination complexes because their ions have relatively small sizes and high charges. Their other distinguishing properties, multiple oxidation states, colored compounds and ions, and paramagnetism, are the result of partially filled d shells in the elements themselves or some of their oxidation states. Elemental iron with an electron configuration of $[Ar]4s^2 3d^6$ clearly has a partially filled d shell, since the shell is filled when it holds ten electrons. Elemental copper has an electron configuration of $[Ar]4s^1 3d^{10}$. This does not include a partially filled d shell. However, the Cu^{2+} ion has the electron configuration $[Ar]3d^9$. This does have the partially filled d shell, and all

FIGURE 22–1 (a) Aqueous solutions of transition metal ions display a rich assortment of colors across the entire visible range of the electromagnetic spectrum, except for Zn^{2+}. From left to right: Cu^{2+} (blue); Fe^{3+} (yellow); Zn^{2+} (colorless); Cr^{3+} (purple); Co^{2+} (pink); Ni^{2+} (green); Mn^{2+} (pale pink). (b) The two flasks on the left are both +3 states of chromium, violet for the chloride (Cl^-) and green for the nitrate (NO_3^-). Changing the counter ion generally alters the color of the solution. The two flasks on the right are +6 states of chromium, the yellow chromate (CrO_4^{2-}) and the orange dichromate ($Cr_2O_7^{2-}$).

(a)

(b)

The display of transition metal sulfates is characteristically highly colored and typically "glassy." From left to right: $CuSO_4 \cdot 5H_2O$ (blue); $FeSO_4 \cdot 7H_2O$ (green); $CoSO_4 \cdot 7H_2O$ (red); $NiSO_4 \cdot 6H_2O$ (dark green).

the distinguishing properties of a transition metal. Zinc, one of the three metals of Group IIB, has the electron configuration $[Ar]4s^2 3d^{10}$, and its only significant ion, Zn^{2+}, has the configuration $[Ar]3d^{10}$. Zinc and the other metals of Group IIB do not have atoms or ions with partially filled d shells. Consequently their compounds are not paramagnetic, and multiple oxidation states and colored compounds are rare. That is why these three metals are often not considered along with the transition metals. We shall consider them briefly in a separate section in this chapter.

So the electrons in d orbitals play a key role in transition metal chemistry. As we discussed in Chapter 15, the $3d$ orbitals are very well shielded by inner orbitals, so their energy does not drop significantly as the nuclear charge is increased while the $1s$, $2s$, $2p$, $3s$, and $3p$ orbitals are being filled. When the filling of the $3p$ orbitals is completed, the next orbital to be filled is the $4s$ in potassium and calcium. The $4s$ orbital is sufficiently far from the nucleus that it does not completely shield the $3d$ orbitals, and these finally can drop in energy as the nuclear charge increases. Thus, after two electrons have been placed in the $4s$ orbital, the next ten electrons fill the five $3d$ orbitals, giving us the first transition series from scandium to zinc. After that the $4p$ orbitals fill for the elements gallium to krypton. In a similar way $4d$ fills after $5s$, and $5d$ starts to fill after $6s$.

The f orbitals are shielded even more effectively than d orbitals, and only when the $6s$ orbital is being filled do the $4f$ orbitals begin to show a significant drop in energy. At this point the $5d$ and $4f$ orbitals drop to competitive values. One electron goes into a $5d$ orbital to give lanthanum and then the seven $4f$ orbitals begin to fill, giving rise to the next fourteen elements. Because electrons in f orbitals do not contribute significantly to chemical properties, the fourteen elements are all similar to lanthanum and are called lanthanides (or rare earths). In the next period, a similar situation occurs to produce the actinides. Here the competition between $6d$ and $5f$ orbitals is even keener and the filling is more irregular.

22.2 THE FIRST TRANSITION SERIES

The elements Sc to Zn, in which the $3d$ shell is filling, make up the first transition series. Within this series, the properties change in a relatively smooth manner with increasing atomic number. Some properties of the first transition series are shown in Table 22–1. Note the steady increase in density from scandium through copper. This can be directly correlated to the decreasing size of the metal atoms as more and more electrons are

TABLE 22–1   Properties of the First Transition Series

Metal	m.p. (°C)	b.p. (°C)	d (g/cm³)	I_1 (eV)	I_2 (eV)	I_3 (eV)
Sc	1200	2400	2.5	6.54	12.80	24.45
Ti	1677	3277	4.51	6.83	13.57	27.47
V	1919	3400	6.1	6.74	14.65	29.31
Cr	1903	2642	7.14	6.76	16.49	30.95
Mn	1244	2095	7.44	7.43	15.64	33.69
Fe	1535	3000	7.87	7.90	16.18	30.64
Co	1493	3100	8.90	7.86	17.05	33.49
Ni	1453	2732	8.91	7.63	18.15	35.16
Cu	1083	2595	8.95	7.72	20.29	36.83
Zn	420	908	7.14	9.39	17.96	39.7

TABLE 22–2   Electron Configurations of the First Transition Series

Orbital	Number of Electrons									
	Sc	Ti	V	Cr	Mn	Fe	Co	Ni	Cu	Zn
$4s$	2	2	2	1	2	2	2	2	1	2
$3d$	1	2	3	5	5	6	7	8	10	10

added to the d shell. As we have discussed before, this effect is seen as any shell is filled. It is a result of incomplete shielding of the nucleus by electrons in the same shell. Thus the effective nuclear charge felt by these electrons increases, drawing them in closer to the nucleus. The increasing values of the first ionization energy also reflect the increased attraction of the electrons for the nucleus as the nuclear charge to which they are subjected increases. Zinc has a particularly high first ionization energy because both its $3d$ and $4s$ shells are complete.

Look at the electron configurations of the neutral metals of the first transition series (Table 22–2). Notice that the filling of the orbitals is irregular at Cr and Cu. In each case an electron has been removed from the $4s$ orbital, giving $3d$ configurations of 5 and 10 electrons, respectively. These half-filled and filled d shells are more stable arrangements than the $4s^2$ configurations, which have less favorable spin pairing energy. The electron configurations for the two series below the first transition series are more irregular.

Removal of electrons from these metals produces cations. The smaller number of electrons in cations increases the nuclear charge felt by the outer electrons and reduces shielding effects. Therefore, the energy of the $3d$ orbitals drops below that of the $4s$ orbitals, and **in all first transition series cations the $4s$ orbital is empty.** That is, the s electrons are lost first in forming cations. In addition, one or more d electrons may also be missing. However, higher ion charges are successively less stable, and therefore the total number of missing $4s$ and $3d$ valence electrons is generally limited. For most of the metals, several different ions are known.

Starting with the first of the elements, we note that Sc loses all three valence electrons to form the Sc^{3+} ion. Titanium has the possibility of losing all four $4s$ and $3d$ electrons to form the Ti^{4+} ion. However, such a high charge cannot be stabilized in a pure ionic compound, and Ti^{4+} compounds such as $TiCl_4$ (melting point = −30°C) are largely covalent. More ionic compounds occur for Ti^{3+} and Ti^{2+}. Proceeding along the first transition

TABLE 22–3   Common Oxidation States of the First Transition Series

	+1	+2	+3	+4	+5	+6	+7
Sc			x				
Ti		x	x	x			
V		x	x	x	x		
Cr		x	x	x	x	x	
Mn		x	x	x	x	x	x
Fe		x	x	x	x	x	
Co		x	x				
Ni		x					
Cu	x	x					
Zn		x					

series through Fe, successively higher oxidation states become possible although actual cations are generally limited to +2 and +3 ions. Cobalt forms stable compounds only as Co^{2+} and Co^{3+}, Ni as Ni^{2+}, and Cu as Cu^+ and Cu^{2+}. For Zn, formation of an ion greater than +2 would involve breaking into the very stable filled d shell, so Zn^{2+} is the only Zn ion. The oxidation states commonly observed for the first transition series are indicated in Table 22–3. For transition metals with multiple oxidation states, those that have high states tend to be good oxidizing agents while those that have low states tend to be good reducing agents.

EXAMPLE 22–1

Determine the valence electron configuration of the Fe^{3+} ion.

Solution   Note first that the valence electron configuration of Fe is $4s^2 3d^6$. Removal of the two s electrons gives the Fe^{2+} ion. It is necessary to remove one d electron to produce the Fe^{3+} ion, giving a valence electron configuration of $3d^5$. ■

EXERCISE 22–1

Determine the electron configurations of the Cu^{2+}, Cr^{3+}, Mn^{2+}, and V^{4+} ions. ■

Metal Compounds of the First Transition Series

The compounds formed by these metals reflect the oxidation states that the metals are capable of attaining. Strongly oxidizing elements like oxygen and fluorine tend to form stable compounds with the metals in the higher oxidation states. On the other hand, the lower oxidation states of the metals tend to be the most stable in the presence of a strongly reducing ion like I^- or S^{2-}. For example, copper forms Cu(I) and Cu(II) compounds. CuF_2 is the only copper fluoride. Both chlorides and bromides exist, but CuI is the only copper iodide. Both oxides and sulfides exist. However, Cu_2S (which melts at 1100°C) is much more stable than CuS (which decomposes at 220°C).

For each metal the compounds at the higher oxidation states are more covalent. That is, at the higher oxidation states the ions are not stable as independent entities as they are when the oxidation states are lower. Mn^{2+}

FIGURE 22–2 Chromium forms a green oxide, with the metal in the +3 oxidation state, and a red oxide where the metal is in the +6 state.

is a stable entity, and the oxide MnO has the NaCl structure of a typical ionic compound. However, Mn^{7+} cannot exist in the same way. The charge of +7 is much too high for an independent ion; such a species, if formed, would immediately seize electrons from any available source. The oxidation state of +7 can be achieved only in a covalent compound, and Mn_2O_7 is an explosive green *liquid* with the covalent structure

$$\begin{array}{ccc}
O & O & O \\
\backslash & | & | \\
O-Mn & Mn-O \\
/ & & \backslash \\
O & & O
\end{array}$$

Another way of looking at this is to consider the electronegativity of a metal as a function of its oxidation state. The higher its oxidation state, the greater its demand to draw electrons toward itself. This means that the electronegativity increases with oxidation state, and metals in higher oxidation states behave more like nonmetals. In the preceding example, MnO has the ionic structure of a metal oxide whereas Mn_2O_7 has the covalent structure of a nonmetal oxide.

Another example of the increase in nonmetal character as the oxidation state of a metal increases is the reaction of the metal oxide with water. As we have mentioned before, metal oxides are basic when dissolved in water and nonmetal oxides are acidic under the same circumstances. The oxides of chromium are a good example (Figure 22–2). Chromium(II) oxide is basic and will dissolve in water when acid is present,

$$CrO \text{ (s)} + 2\ H_3O^+ \text{ (aq)} + 3\ H_2O \text{ (liq)} \longrightarrow [Cr(H_2O)_6]^{2+} \text{ (aq)}$$

Chromium(III) oxide is **amphoteric**; that is, it will dissolve in water solutions of either acid or base:

$$Cr_2O_3 \text{ (s)} + 6\ H_3O^+ \text{ (aq)} + 3\ H_2O \text{ (liq)} \longrightarrow 2[Cr(H_2O)_6]^{3+} \text{ (aq)}$$

$$Cr_2O_3 \text{ (s)} + 2\ OH^- \text{ (aq)} + 7\ H_2O \text{ (liq)} \longrightarrow 2[Cr(OH)_4(H_2O)_2]^- \text{ (aq)}$$

Chromium(VI) oxide is highly acidic. It will dissolve in neutral water to form chromic acid, H_2CrO_4, which dissociates to form protons (hydronium ions) and the chromate ion:

$$CrO_3 \text{ (s)} + 3\ H_2O \text{ (aq)} \longrightarrow 2\ H_3O^+ \text{ (aq)} + CrO_4^{2-} \text{ (aq)}$$

22.3 THE SECOND AND THIRD TRANSITION SERIES

The metals of the second and third transition series are distinctly different from those of the first. Those of the second and third transition series tend to be hard, high-melting, and unreactive. They form compounds in a wide range of oxidation states and tend to higher oxidation states than the first transition series metals. The compounds they form have a high degree of covalent character with few simple ionic compounds. Typical compounds are WCl_5, which has a melting point of 244°C, and WCl_6, which melts at 280°C.

The similarities between the pairs of equivalent elements in these two series are very strong. For example, molybdenum is very similar to tungsten, the element just below it, and these have little similarity to chromium, the element above them. Chromium forms very stable salts and complexes in the +2 and +3 oxidation states, whereas molybdenum and tungsten do not. Molybdenum and tungsten form a large series of complex oxygen-containing anions that chromium does not form. These anions consist of octahedra of MoO_6 and WO_6 fused into structures with formulas like $[Mo_7O_{24}]^{6-}$ and $[W_{12}O_{42}]^{12-}$.

The differences between the behaviors of the first and second transition series can be attributed mainly to the large differences in atomic and ionic radii between equivalent members of the two series. This is the result of the electrons being in orbitals with a higher value of n for the second series. However, the radii for the second and third series are very close for equivalent elements, and the properties are very similar. This rather surprising fact is the result of the filling of the $4f$ orbitals of the lanthanides prior to filling of the $5d$ orbitals of the third transition series. The incomplete shielding by the $4f$ electrons results in a decrease in the atomic and ionic radii of subsequent elements. This decrease almost exactly balances the increase in radii that would be expected because the quantum number n increases on going from the second to the third transition series. This effect is called the "lanthanide contraction." Typical results are the ionic radii of V^{3+}, Nb^{3+}, and Ta^{3+}, which are 0.78, 0.86, and 0.86 Å, respectively. With equal charges and ionic radii, we can expect Nb^{3+} and Ta^{3+} to behave very similarly.

22.4 REFINING PROCESSES

A number of chemical principles appear in the production and refining of the transition metals from their ores. We will consider four important and typical examples: iron, copper, titanium, and tungsten.

Refining of Iron

Iron is a soft metal that, in its pure state, has almost no uses. Its alloys, however—particularly steels—are among the mainstays of industrial civilization. The principal ores of iron are:

- **hematite** (Fe_2O_3) and hydrated Fe^{3+} oxides such as $2Fe_2O_3 \cdot 3H_2O$, which are red.

- **magnetite** (Fe_3O_4), which is brown to black and strongly magnetic. The apparent fractional oxidation state of Fe is the result of a crystal structure that contains both Fe^{2+} and Fe^{3+} ions.

(a)

(b)

(a) Among the principal iron ores are hematite, basically a red to red-brown Fe_2O_3 oxide, and magnetite, a darker Fe_3O_4 oxide. The nearly black ore sample shown here is mostly magnetite. (b) The beguiling "fool's gold" is an iron sulfide, FeS_2.

FIGURE 22–3 Reduction of iron ore in a "blast" furnace is largely a carbon monoxide-based reduction process, with the molten iron being drawn off at the bottom after separation of the unwanted calcium silicate "slag."

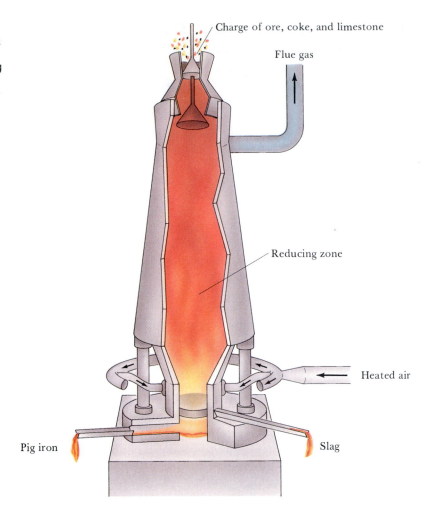

- **siderite** (FeCO$_3$), which is gray to black.
- **taconite**, which contains all of the other iron ores. It is commonly green, strongly magnetic, and relatively low grade because of the presence of significant quantities of silica, which must be removed at the mining site.

Iron ore, normally containing considerable SiO$_2$ and silicate impurities, is charged into a blast furnace (Figure 22–3) along with limestone (CaCO$_3$) and coke. Coke is produced from heating soft coal in the absence of air and is composed of carbon and the fused coal ash. A blast of hot air at the bottom of the furnace results in temperatures up to 1300°C as the coke burns:

$$2 \text{ C (s)} + \text{O}_2 \text{ (g)} \longrightarrow 2 \text{ CO (g)}$$

The carbon monoxide progressively reduces the iron oxides to lower oxides and finally to the metal,

$$3 \text{ Fe}_2\text{O}_3 \text{ (s)} + \text{CO (g)} \longrightarrow 2 \text{ Fe}_3\text{O}_4 \text{ (s)} + \text{CO}_2 \text{ (g)}$$

$$\text{Fe}_3\text{O}_4 \text{ (s)} + \text{CO (g)} \longrightarrow 3 \text{ FeO (s)} + \text{CO}_2 \text{ (g)}$$

$$\text{FeO (s)} + \text{CO (g)} \longrightarrow \text{Fe (liq)} + \text{CO}_2 \text{ (g)}$$

The liquid iron is drawn off at the bottom of the furnace.

Hot carbon in the coke can react with the carbon dioxide evolved from these reactions to regenerate carbon monoxide,

The open hearth process is used to produce a steel with a lower carbon content and very little other nonmetallic impurities, which makes it more flexible, stronger, harder, and malleable. Open hearth furnaces are usually lined with magnesium oxide or mixed magnesium and calcium oxides.

$$CO_2 \text{ (g)} + C \text{ (s)} \longrightarrow 2\ CO \text{ (g)}$$

The heat of the furnace decomposes the limestone,

$$CaCO_3 \text{ (s)} \longrightarrow CaO \text{ (s)} + CO_2 \text{ (g)}$$

The calcium oxide helps to remove the silicon-containing impurities by forming calcium silicate slag, which is a liquid at the temperatures involved:

$$CaO \text{ (s)} + SiO_2 \text{ (s)} \longrightarrow CaSiO_3 \text{ (liq)}$$

This slag is less dense than the molten iron and floats on top of it. From time to time it is drained off. Some of it is blown with air to make a fluffy, nonflammable insulation known as rock wool.

The iron produced by the blast furnace is called *cast iron* or *pig iron*. It contains considerable impurities:

C, 2.0–4.5%;   Si, 0.7–3.0%;   S, 0.1–0.3%;   P, 0–3.0%;
Mn, 0.2–1.0%

As a result of these impurities, pig iron is brittle and suitable only for producing castings that will not be subjected to shock. Reduction of the carbon content to 0.05 to 2.0% and removal of almost all of the other nonmetallic impurities leads to steel—alloys with more desirable qualities of flexibility, hardness, strength, and malleability. This is normally achieved in an open hearth furnace, consisting of a shallow pool of molten iron heated by gas flames over the surface. The furnace is lined with magnesium oxide or mixed magnesium and calcium oxides. Sufficient iron oxides are added to oxidize the sulfur, phosphorus, and most of the carbon. The acidic oxides produced react with the basic oxides of the furnace lining.

EXERCISE 22–2

Write balanced equations for the processes going on in the conversion of pig iron to steel in the open hearth furnace.   ∎

FIGURE 22–4 To free, or "win" copper, sulfide ores are first enriched in the "flotation" process by removing lighter metal sulfide particles, caught up in soapy bubbles floating on water, while heavier "gangue" settles to the bottom.

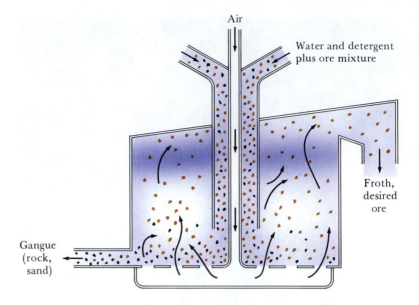

Refining of Copper

Copper is an attractive, reddish, durable metal. It is second only to silver in conductivity of heat and electricity. Bronze, brass, and others of its alloys have long been among the most useful materials. Copper occurs in the two important classes of ores.

1. Sulfide ores such as $CuFeS_2$, Cu_3FeS_3, and Cu_2S.
2. Oxide ores such as CuO, $Cu_2(OH)_2CO_3$, and $Cu_3(OH)_2(CO_3)_2$.

The copper ores now available are mainly low grade, containing a large proportion of sand and rock or "gangue." The ore is separated from the gangue by flotation. First the impure ore is ground with oil and water. The ore particles are wetted by the oil, while the gangue is wetted by the water. This mass is then added to a larger amount of water containing a foaming agent, and the mixture is blown with air and beaten until a froth forms and rises to the top (Figure 22–4). The ore particles stick to the surface of the bubbles of the froth and are carried off at the surface. The gangue particles, weighed down with the water, settle to the bottom. This method will remove 95% of the pure copper ore from an impure ore that is 98% waste.

The purified ore is then roasted in the presence of air. This performs a number of useful functions, including oxidizing arsenic and antimony impurities to volatile oxides that distill off, and oxidizing iron in sulfide ores to FeO:

$$2\ CuFeS_2\ (s)\ +\ 3\ O_2\ (g)\ \longrightarrow\ 2\ CuS\ (s)\ +\ 2\ FeO\ (s)\ +\ 2\ SO_2\ (g)$$

Sand (SiO_2) and limestone ($CaCO_3$) are then added to the mixture, which is heated in a reverberatory furnace (Figure 22–5). The limestone and sand form calcium silicate, which is liquid at the furnace temperature and acts as a flux for removal of the FeO,

$$FeO\ (s)\ +\ SiO_2\ (s)\ \xrightarrow{\ CaSiO_3\ (liq)\ }\ FeSiO_3\ (liq)$$

The iron silicate produced in this reaction is a form of slag. Sulfur still present in the mixture reduces the CuS to Cu_2S, which is a liquid at this

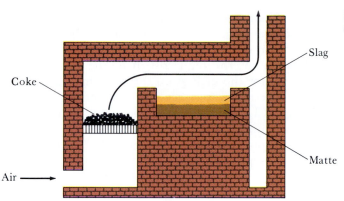

temperature and is called copper matte. The slag is less dense than the matte and floats on top of it.

The matte is drawn off and run into a copper converter. There, air is blown through the molten Cu_2S and the free metal is produced by the reaction

$$Cu_2S \text{ (liq)} + O_2 \text{ (g)} \longrightarrow 2 \text{ Cu (liq)} + SO_2 \text{ (g)}$$

Most of the copper produced is used for electrical transmission. The high electrical conductivity of copper is not realized unless it is very pure. Electrolytic purification of the copper obtained from the copper converter was discussed in Chapter 21.

Refining of Titanium

Titanium is abundant, strong, light, and corrosion-resistant. It can maintain its strength at high temperatures. Unfortunately, it is expensive because of its method of preparation, and the metal finds use only in specialized aircraft engine and airframe applications where relatively high cost can be tolerated. Ilmenite, $FeTiO_3$, is the chief ore of titanium. The major impurity is silica, SiO_2, from which the ilmenite can be separated magnetically. The ore is heated with coke in the presence of chlorine,

$$FeTiO_3 \text{ (s)} + 3 Cl_2 \text{ (g)} + 3 C \text{ (s)} \longrightarrow$$
$$3 CO \text{ (g)} + FeCl_2 \text{ (s)} + TiCl_4 \text{ (g)}$$

The $TiCl_4$ is a volatile compound and can be purified by fractional distillation. The pure $TiCl_4$ is then reduced by reaction with an active metal. Magnesium is most commonly used for this. The $TiCl_4$ vapor is passed over molten magnesium and the metallic titanium forms as a spongy mass,

$$TiCl_4 \text{ (g)} + 2 \text{ Mg (liq)} \longrightarrow Ti \text{ (s)} + 2 \text{ MgCl}_2 \text{ (liq)}$$

Alternatively, molten sodium can be used as a reducing agent,

$$TiCl_4 \text{ (g)} + 4 \text{ Na (liq)} \longrightarrow Ti \text{ (s)} + 4 \text{ NaCl (s)}$$

In either case the solid titanium is washed with water to remove the chloride salt. The metal can then be pressed into the shape of an electrode and melted and cooled in this shape under high vacuum. This prepares it for electrolytic purification.

The high cost of titanium results from the need to use magnesium or sodium in the preparation. This can be avoided by producing TiI_4 instead of $TiCl_4$ in the first step by reaction with I_2 instead of Cl_2. The TiI_4 can be decomposed directly to titanium, making the use of magnesium or sodium unnecessary. In fact, very pure titanium crystals can be produced by decomposition of TiI_4 on an electric filament by what is known as the van Arkel process. Unfortunately, the high cost of iodine compared to chlorine more than makes up for the savings on the reducing agent.

The electrolyte in the electrolytic purification of titanium is molten sodium chloride containing $TiCl_2$. The solution is a liquid at the operating temperature of 850°C. The impure titanium is oxidized to Ti^{2+} at the anode and then redeposited as purified metal at the cathode. Cell potentials of 0.3 to 2.5 V are applied. The purity of the metal depends on a number of factors including applied potential, cell design, rate of deposition (current), and electrolyte composition.

Refining of Tungsten

In a study of materials properties, a sample of heat-treated steel—shown here in cross section—was immersed in a liquid in which high-speed vapor bubbles were generated, simulating cavitation and other mechanical stresses. The force of the small bubbles impacting on the steel at high speed created the "pit" visible as a black area extending from the bottom into the center of the photograph, causing the failure of the part. On adding small amounts of tungsten to the steel—a few tenths of a percent—such mechanical effects are greatly minimized.

Tungsten is a very dense metal (19.3 g/cm^3) with extremely high melting and boiling points (3370°C and 5900°C). Its very low vapor pressure at high temperatures makes it useful in lamp filaments, x-ray tube targets, electrical contacts, and furnaces. Its main use, however, is in making steel alloys in which small amounts of tungsten result in enormous increases in hardness and strength. Tungsten and chromium are used to make steels for high speed cutting tools that remain hard at red heat.

The main ores of tungsten are tungstates, $FeWO_4$, $MnWO_4$, $CaWO_4$, and $PbWO_4$. After mechanical and magnetic separation, the ore reacts with aqueous NaOH to produce Na_2WO_4,

$$CaWO_4 \text{ (s)} + 2 \text{ NaOH (aq)} \longrightarrow Na_2WO_4 \text{ (aq)} + Ca(OH)_2 \text{ (s)}$$

The aqueous tungstate ion then reacts with acid to produce WO_3:

$$WO_4{}^{2-} \text{ (aq)} + 2 \text{ } H_3O^+ \text{ (aq)} \longrightarrow WO_3 \text{ (s)} + 3 \text{ } H_2O \text{ (liq)}$$

The WO_3 is dried and then reduced to the metal with hydrogen gas:

$$WO_3 \text{ (s)} + 3 \text{ } H_2 \text{ (g)} \longrightarrow W \text{ (s)} + 3 \text{ } H_2O \text{ (g)}$$

Hydrogen is used for the reduction in order to obtain a high purity product. In particular, the use of carbon as the reducing agent woud lead to the formation of tungsten carbide, WC. The temperature is held at 800°C during the reduction. The hydrogen is used as a flowing gas so that the water that is formed will be swept away without reacting with the metal.

Before it is exposed to air, the metal is cooled to room temperature under an atmosphere of hydrogen to prevent oxidation. The metal is a powder at this point, and it is compacted into porous bars at room temperature and very high pressure. It is then slowly heated to 1300°C under a hydrogen atmosphere. In this process, called sintering, the powder grains are stuck together by movements of the atoms at their surfaces. An electric current is then passed through the bar to heat it to 3000°C. The sintering continues and spaces between the original powder grains decrease until the full density of the metal is approached. The metal can then be worked mechanically at a temperature of about 900°C into the desired shape.

22.5 COORDINATION COMPLEXES

A **coordination complex** is a compound in which one or more neutral molecules are bonded to a metal atom or ion. Coordination complexes of

FIGURE 22–6 On adding aqueous ammonia to a solution containing "hydrated" Ni^{2+} ions in solution $[Ni(H_2O)_6^{2+}]$, a greenish precipitate of the hydroxide $[Ni(OH)_2]$ forms, followed by dissolution of the precipitate as excess ammonia is added and the water-soluble, dark blue ammonia complex forms $[Ni(NH_3)_6^{2+}]$.

the transition metals can be prepared from their salts. As an example, chlorine will react with nickel metal to produce the yellow salt, $NiCl_2$.

$$Ni\ (s)\ +\ Cl_2\ (g)\ \longrightarrow\ NiCl_2\ (s)$$

If this salt is added to water, it will evolve considerable heat and form a green solution. Evaporation of the water from this solution gives a green solid, $NiCl_2 \cdot 6H_2O$ (s). If ammonia is added to the green solution, it turns deep blue (Figure 22–6). If this solution is evaporated to dryness, the compound that is isolated has the formula $NiCl_2 \cdot 6NH_3$. This chemistry is much more complicated than that of main group metals like sodium or calcium. The difference is that nickel, in common with other transition metal ions, forms coordination complexes.

Formation of coordination complexes by transition metal ions has been known for a long time. Pioneering chemical studies were carried out by Alfred Werner, who prepared ammonia complexes named ammines a century ago. Ammines of Ni^{2+}, Cu^{2+}, Co^{3+}, Pt^{2+}, and Pt^{4+} were well known (if poorly understood) at that time. Werner made careful and complete studies of families of complexes. For example, when water solutions of $PtCl_4$ reacted with different amounts of ammonia, five different ammines formed, with formulas ranging from $PtCl_4 \cdot 2NH_3$ to $PtCl_4 \cdot 6NH_3$. Werner prepared pure solutions of all five complexes and performed two key experiments on each:

- First, by treating each solution with a silver nitrate solution and measuring the mass of silver chloride produced, he was able to calculate the number of "free" chloride ions in each complex.

- Second, by measuring the conductance of the ammine solutions, he was able to determine the total number of ions in each complex.

His results were:

Complex	Formula	Free Cl$^-$ Ions	Total Ions
1	$PtCl_4 \cdot 6NH_3$	4	5
2	$PtCl_4 \cdot 5NH_3$	3	4
3	$PtCl_4 \cdot 4NH_3$	2	3
4	$PtCl_4 \cdot 3NH_3$	1	2
5	$PtCl_4 \cdot 2NH_3$	0	0 (nonionic)

Here is how Werner interpreted these results. In complex 1, four of the five ions were the free chloride ions, and the remaining ion was a complex ion consisting of all the other atoms. In each successive complex, one more Cl atom is incorporated into the complex ion and becomes unavailable for reaction. He recognized the actual formulations as follows:

FIGURE 22–7 (a) If the coordination about the central metal ion took on a planar, hexagonal arrangement, three isomers would be possible. (b) Octahedral symmetry with respect to the central metal ion leads to only two possible isomers. (c) The two isomers of the octahedral $[Co(NH_3)_4Cl_2]^+$ ion.

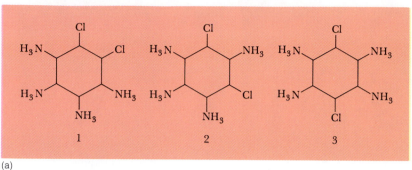

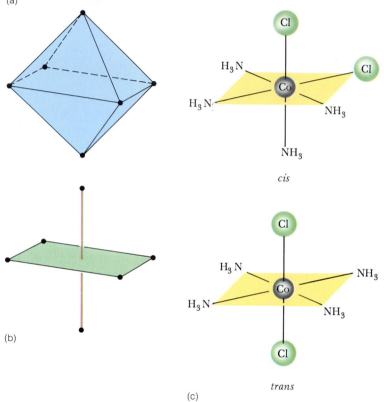

Complex	Formulation	
1	$Pt(NH_3)_6^{4+}$	$4\ Cl^-$
2	$PtCl(NH_3)_5^{3+}$	$3\ Cl^-$
3	$PtCl_2(NH_3)_4^{2+}$	$2\ Cl^-$
4	$PtCl_3(NH_3)_3^+$	Cl^-
5	$PtCl_4(NH_3)_2$	

Two more members of the series, $PtCl_5(NH_3)^-$ and $PtCl_6^{2-}$, were discovered later. Werner established that in each of these Pt^{4+} complexes, there were six species bonded to the central ion. These species are called **ligands**, and the number of ligands about an atom or ion is called its **coordination number**. Platinum has a coordination number of six in these examples.

On the basis of the results of similar sets of experiments, Werner was able to show that Co^{3+} and Cr^{3+} formed complexes with coordination number six, whereas Pt^{2+} and Pd^{2+} formed complexes with coordination number four. Some transition metal ions such as Ni^{2+} and Cu^{2+} formed some complexes with coordination number four and some with coordination number six.

Werner was able to infer the geometry of the four- and six-coordinate complexes by counting the number of **isomers**—compounds of identical composition but with different arrangements of atoms, resulting in different properties. For example, the compound of Co^{3+} with a coordination number of six and the formula $[Co(NH_3)_4Cl_2]Cl$ has just two isomers. What does this tell us about the coordination geometry? At this point it is useful to get out a scrap of paper and try to draw a few structures of possible coordination geometries. For example, if the coordination about the central metal ion were in the form of a planar hexagon, there would be three isomers (Figure 22–7a). Since the planar geometry is not satisfactory, it will be necessary to use a solid or three-dimensional figure for the coordination model. If we are to end up with as few as two isomers, we must have a very regular or symmetric solid. The octahedron is such a solid (Figure 22–7b). It has six equivalent corners or vertices, eight equilateral triangles for the faces, and twelve edges, all of the same length. Because this geometry occurs fairly regularly in chemical structures, a skeleton notation is used that shows four vertices in a square plus one vertex above and one below. We will use this convenient notation, although it creates the illusion that two of the vertices are distinct from the other four when, in fact, all six are equivalent. Figure 22–7c shows the two isomers of the $[Co(NH_3)_4Cl_2]^+$ ion. The isomer with the two chlorines adjacent (or 90° apart) is referred to as the *cis* form; the one with the two chlorines opposite (or 180° apart) is the *trans* isomer.

Similar reasoning was used to deduce the coordination geometry about Pt^{2+} in $Pt(NH_3)_2Cl_2$. Two isomers exist, leading to a prediction of the square planar geometry as shown in Figure 22–8a. But not all four-coordinate complexes are square planar. Some, particularly those of Co^{2+} and Zn^{2+}, exhibit four-coordination with tetrahedral geometry (Figure 22–8b). Note that if $Pt(NH_3)_2Cl_2$ had been tetrahedral instead of square planar, there would be only one structure for the complex rather than the two that were found.

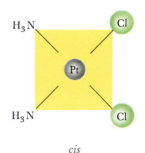

cis

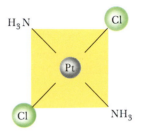

trans

(a)

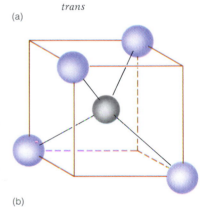

(b)

FIGURE 22–8 (a) The two isomers of the square planar $Pt(NH_3)_2Cl_2$ coordination complex. (b) The geometry of a regular tetrahedron.

22.6 GEOMETRIES OF COORDINATION COMPLEXES

Although the octahedral geometry in coordination complexes occurs extremely frequently for the more common metals and has been the most thoroughly studied, many other geometries are important. Drawings of significant coordination geometries are shown in Figure 22–9. These are:

a. Linear, 2-coordination; observed for Ag(I) and Au(I) complexes.
   Examples, $[Ag(NH_3)_2]^+$, $[Au(CN)_2]^-$
b. Trigonal, 3-coordination; rare, observed for Ag(I) and Hg(II).
   Example, $[HgI_3]^-$
c. Tetrahedral, 4-coordination; common, observed for Cu(I), Co(−I,0,I,II), Ni(0,II), Zn(II), Cd(II), etc.
   Examples, $[Zn(CN)_4]^{2-}$, $[CoCl_4]^{2-}$
d. Square planar, 4-coordination; common, observed for Ni(II), Pd(II), Pt(II) (d^8 electron configurations).
   Examples, $[Ni(CN)_4]^{2-}$, $[PtCl_4]^{2-}$
e. Trigonal bipyramidal, 5-coordination; rare, observed for a few cases such as Cu(II), Fe(0), Co(I).
   Examples, $Fe(CO)_5$, $[CuCl_5]^{3-}$
f. Square pyramidal, 5-coordination; very rare, observed in Ni(II), Fe(II).
   Example, $[Ni(CN)_5]^{3-}$

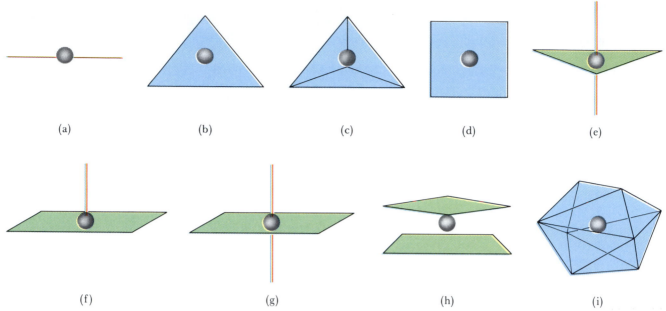

(a) (b) (c) (d) (e)

(f) (g) (h) (i)

FIGURE 22–9 Coordination geometries: (a) linear; (b) trigonal; (c) tetrahedral; (d) square planar; (e) trigonal bipyramidal; (f) square pyramidal; (g) octahedral; (h) square antiprismatic; (i) dodecahedral.

g. Octahedral, 6-coordination; very common, observed for most transition metals in most oxidation states.
Examples, $[PdCl_6]^{2-}$, $[Fe(H_2O)_6]^{2+}$, $W(CO)_6$

h. Square antiprismatic, 8-coordination; higher oxidation states, third transition series.
Examples, $[ReF_8]^{2-}$, $[TaF_8]^{3-}$

i. Dodecahedron, 8-coordination; more common 8-coordination geometry.
Example, $[Mo(CN)_8]^{4-}$

22.7 NAMING COORDINATION COMPLEXES

Coordination complexes are named as a central cation, most often of a transition metal atom, surrounded by two or more ligands. The ligands are either anions or neutral molecules that can donate electrons; the highly unusual cases of cationic ligands will not be considered here.

In naming coordination complexes, the anionic ligands are identified by names ending in "o." The important examples are:

We first noted the effect of the ligand on the color of transition metal ions in Figure 22–1(b). Here, five different ligands are associated with Co^{3+}.

F^-	fluoro	NO_2^-	nitro
Cl^-	chloro	$C_2O_4^{2-}$	oxalato
Br^-	bromo	SCN^-	thiocyanato
I^-	iodo	OH^-	hydroxo
CN^-	cyano		

$[Co(NH_3)_6]^{3+}$ $[Co(NCS)_6]^{3-}$ $[Co(H_2O)_6]^{3+}$ $[Co(Cl)_6]^{3-}$ $[Co(NH_3)_4Cl_2]^+$

The names of neutral compounds acting as ligands are not systematic, and the following four important examples should be learned:

H_2O	aqua
CO	carbonyl
NH_3	ammine
NO	nitroso

Some coordination complexes are neutral molecules, but many are either anions or cations; the latter are called **complex ions**. The formula of the complex molecule or ion is enclosed in [square] brackets. If a salt contains a complex ion, the usual convention is to name the cation before the anion. In naming a complex ion (or molecule), the ligands are named in alphabetical order before the central atom. The prefixes di-, tri-, tetra-, penta-, hexa-, and so on are used to indicate the number of each ligand present if that number is more than one. If the ligand itself contains such a prefix (for example, ethylene *di*amine), a different set of prefixes (bis, tris, tetrakis, pentakis, etc.) are used. The metal is then named, with the suffix "-ate" added to the metal name if the metal is part of a complex anion. Finally, the oxidation state of the metal is indicated by a Roman numeral enclosed in parentheses. As we have mentioned before, the oxidation states in transition metals tend to be variable and must be calculated for each case by considering the charges of the ligands and other ions. The correct oxidation state of the metal is the one that is consistent with an overall charge of zero for a salt or neutral complex and gives the correct charge to a complex ion. As in simple salts (for example, $FeCl_2$, iron(II) chloride), it is not necessary to specify how many ions of each type are present, as long as the oxidation state of the metal is specified.

To name the neutral coordination compound $[Co(NH_3)_3Cl_3]$, note that the cobalt ion in this complex is surrounded by six ligands, including three neutral compounds (NH_3) and three -1 ions (Cl^-). The cobalt ion must be in the oxidation state $+3$ to balance the charge. Alphabetical order puts ammine before chloro, and the name is triamminetrichlorocobalt(III).

To name the salt $Na_4[Fe(CN)_6]$, note that the Fe must be in the oxidation state $+2$ to balance the Na^+ and CN^- ions. The suffix "-ate" must be added to iron to show that the iron is part of the anion. The name is sodium hexacyanoferrate(II).

EXAMPLE 22–2

Name the following:
(a) $K_3[Al(OH)_6]$
(b) $[Cr(NH_3)_4(H_2O)Cl]Cl_2$

Solution
(a) The three K^+ ions indicate that the complex ion has a charge of -3. Each of the six OH— groups supplies -1 charge, so the Al atom has a $+3$ oxidation state. The name is potassium hexahydroxoaluminate(III).
(b) The two chlorine ions outside the brackets indicate that the complex ion has a charge of $+2$. The chromium atom is surrounded by six ligands, five of them neutral species. The sixth ligand is a chloride ion, so the chromium atom must have an oxidation state of $+3$. The name is tetraammineaquachlorochromium(III) chloride. ■

EXERCISE 22–3

Give the formula for hexamminechromium(III) hexachlorocobaltate(III).
Answer: $[Cr(NH_3)_6][CoCl_6]$. ∎

22.8 BONDING IN COORDINATION COMPLEXES

A coordination complex consists of a central ion surrounded by ligands. The ligands may be negatively charged ions such as Cl^-, F^-, or CN^-, or they may be neutral but polar molecules such as NH_3, H_2O, or CO. In complexes with ligands that are polar molecules, the ligands are oriented so that the negative end of the dipole points toward the metal cation. For the moment we shall view the tendency of a cation to form complexes as being based principally on electrostatic attraction, that is, having a small ionic radius and a fairly high charge (at least +2 electron units). As an example of a metal ion that does not form complexes, the large size and small charge of Na^+ would result in only a weak electrostatic attraction that would be easily broken by molecular collisions.

In Chapter 16 we discussed the effect of position in the periodic table on atomic and ionic radius. The transition metals have relatively small atoms and ions as a result of incomplete shielding by electrons in the valence shell. As discussed above, the charges of +2 are the result of losing the valence s electrons. Ions with higher charges result from loss of d electrons as well. We shall look at a purely electrostatic model of the bonding in transition metal complexes, called the **crystal field theory**.

The Crystal Field Theory

Crystal field theory assumes a purely electrostatic model of bonding in coordination complexes. Furthermore, in order to produce easily interpretable results, the negative charge of each ligand is assumed to be concentrated at a point. The arrangement of these negative point charges about the central positively charged metal ion produces an electric field, which is called the **crystal field**. We will examine closely the six-coordinate octahedral geometry, the most commonly encountered geometry, particularly in the first transition series. The octahedral crystal field is shown in Figure 22–10. All transition metal ions have lost their s valence electrons. Therefore the only valence electrons and orbitals that we need to consider in this model are in the d shell. The principal question in this theory is, "What effect does the presence of the crystal field have on the relative energy levels of the d orbitals?"

The shapes of the five d orbitals are shown in Figure 15–21. In the absence of any disturbing influences (such as electric charges) all five are said to be **degenerate**, having the same energy. However, the presence of the crystal field breaks the degeneracy. Consider the octahedral crystal field superimposed on the d orbital diagrams. Two of the orbitals, $d_{x^2-y^2}$ and d_{z^2}, point directly at the negative charges. Repulsion between the point charges and electrons in these two orbitals will destabilize the orbitals somewhat and raise the orbital energy. The other three orbitals, d_{xy}, d_{xz}, and d_{yz}, are directed between the point charges. Their energies would not be raised so much by the presence of the negative charges. The degeneracy of the

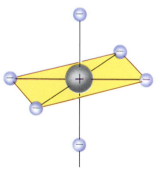

FIGURE 22–10 The octahedral crystal field. Six negative point charges are arranged symmetrically about a positively charged central ion.

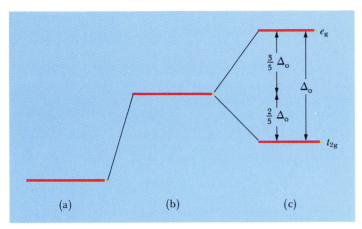

FIGURE 22–11 An energy level diagram showing relative energies of d orbitals for a transition metal ion: (a) in the absence of ligand charges; (b) with ligand charges spread equally about a sphere that is concentric about the metal ion; and (c) in an octahedral crystal field.

five orbitals has therefore been broken into two groups, a group of two at higher energy and a group of three at lower energy.

In order to construct an energy level diagram we must imagine three situations. (1) We start with an isolated metal ion in which all five d orbitals are degenerate. Then we will move in the total charge of the six ligands. (2) To begin with, the total charge of the six points in the octahedral crystal field will be spread evenly over a sphere that is concentric with the positively charged metal ion. This will raise the energy of all five orbitals because the repulsion of the negative charge will cause each orbital to be a less stable location for an electron. However, it will affect all five d orbitals equally and they will still all be degenerate. (3) Then we will move this same total charge so that it is concentrated at the six octahedral crystal field points, at the same distance from the metal ion. The total energy will remain the same since the total charge and distance from the metal ion are unchanged.

Now split the orbitals into two sets. The result is shown in Figure 22–11. The $d_{x^2-y^2}$ and d_{z^2} orbitals have been increased in energy. They now form a set called e_g, where e is a symbol used to designate two orbitals of the same energy. (The "g" refers to the symmetry of the orbitals.) The d_{xy}, d_{xz}, and d_{yz} orbitals have been decreased in energy and now form a set called t_{2g}, where the symbol t is used for three orbitals of the same energy. The splitting between the two sets is called Δ_o, where the "o" stands for octahedral. Since the total energy is unchanged before and after splitting the orbitals, we can calculate the increase in energy of the e_g orbitals, Δ_e, and the decrease in energy of the t_{2g} orbitals, Δ_t, from the following two simultaneous equations:

$$\Delta_e + \Delta_t = \Delta_o$$
$$2\Delta_e - 3\Delta_t = 0$$

Solution of these equations gives

$$\Delta_e = +\tfrac{3}{5}\Delta_o$$
$$\Delta_t = +\tfrac{2}{5}\Delta_o$$

as shown in Figure 22–11.

Next it is necessary to populate the orbitals—which have been split in an octahedral field—with electrons. Consider this on a one-by-one basis (Figure 22–12). First, three electrons enter the lower energy t_{2g} orbitals. Each occupies a different orbital according to Hund's rule, and the spins are not paired. However, the fourth electron has two alternatives:

FIGURE 22–12   Possible electron configurations for one to ten d electrons in an octahedral crystal field.

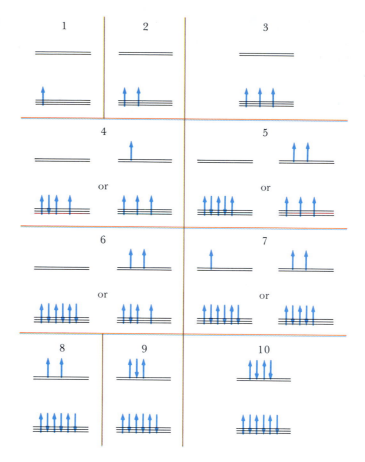

It can enter a t_{2g} orbital that already contains another electron, pairing its spin with the second electron.

Or it can enter an empty e_g orbital.

Each alternative requires more energy than was expended for any one of the first three electrons.

If it enters a t_{2g} orbital, it must overcome the electrostatic repulsion of the other electron, in addition to the spin pairing energy.

If it enters an e_g orbital, there is the added cost of Δ_o—it must overcome the octahedral splitting energy—but along with that cost comes the credit of "exchange" energy if its spin is parallel to the first three.

The net result depends on which energy is greater. The spin pairing energy depends principally on the metal ion. The octahedral splitting energy depends on both the metal ion and the ligand and will be discussed later. Two alternatives occur for 4, 5, 6, or 7 electrons. For 1, 2, 3, 8, 9, or 10 electrons, only one possibility exists.

Important properties, particularly regarding magnetism, depend on the number of unpaired electrons in a coordination complex. From Figure 22–12 it is clear that for 1, 2, or 3 electrons, all will be unpaired. For 4 electrons there will be 2 unpaired electrons if the spin pairing energy is less than the octahedral splitting energy, and 4 unpaired electrons in the opposite situation. The configuration with fewer unpaired electrons is called the **low spin** configuration and that with more unpaired electrons is called the **high spin** configuration. Thus, in an octahedral crystal field, low and high spin configurations exist for 4, 5, 6, and 7 d electrons. Only one possible ground state configuration exists for ions with 8, 9, or 10 d electrons.

EXAMPLE 22–3

Determine the possible numbers of unpaired electrons for the Fe^{3+} ion in an octahedral crystal field.

Solution Fe^{3+} has a d^5 electron configuration. According to Figure 22–12, this can result in either a low spin configuration with one unpaired electron or a high spin configuration with five unpaired electrons. ■

EXERCISE 22–4

Determine the possible numbers of unpaired electrons for each of the following ions in an octahedral crystal field: Cr^{2+}, Co^{2+}, Cu^+, Fe^{2+}, Zn^{2+}, and Sc^{3+}.

Answer: 4 or 2; 3 or 1; 0; 4 or 0; 0; 0. ■

Magnetic Properties of Complexes

Magnetism is a phenomenon common to all kinds of matter. There are several different kinds of magnetism. The most familiar is the extremely strong form known as **ferromagnetism**, which is possessed by horseshoe magnets, the magnets in electric motors, and so forth. This is a complicated form of magnetism involving "domains" of magnetic ions all oriented in the same direction. Another form of magnetism, called **diamagnetism**, is possessed by all matter. Diamagnetism is a result of the small degree of magnetism that is induced whenever matter is placed near either pole of a magnet. Since the polarity induced is the same as the pole of the magnet that induced it, this results in a very weak repulsion of the matter from the magnetic field. Diamagnetism is an extremely weak effect, about 10^{-12} as strong as ferromagnetism. Intermediate in effect between ferromagnetism and diamagnetism is **paramagnetism**, which is due to the presence of unpaired electrons in matter with random orientations of spin. Unlike substances that have only diamagnetic behavior, paramagnetic substances are pulled into a magnetic field.

Classically, we can view the electron as a tiny charged mass spinning on an axis and moving in a closed path around the nucleus. Each movement of a charged object results in a magnetic moment, and the total magnetic moment of the electron is the sum of a spin moment and an orbital moment. However, the effects of the orbital moment tend to be "quenched" as a result of interactions with neighboring atoms or ions, and the magnetic moment of an electron can usually be considered from the "spin only" viewpoint.

Remember from our earlier discussions on the structure of the atom that two electrons can be accommodated in a single orbital so long as they have different values for the spin quantum number, $m_s = \pm\frac{1}{2}$. That is, one electron will have m_s equal to $+\frac{1}{2}$ and the other will have m_s equal to $-\frac{1}{2}$. These two values of m_s are equivalent to electron spins in opposite directions, and we commonly say that such electrons have their spins paired. Electrons are shown as arrows in energy diagrams, and the two values of m_s are represented by the directions in which the arrows point. Such spin-

FIGURE 22–13 A Gouy balance for measuring magnetic susceptibilities of a paramagnetic substance. The mass of the sample is first determined with the power in the electromagnet turned off. Then, the power is turned on and the change in the mass (determined by the balance) is a measure of the extent to which the sample is drawn into the magnetic field.

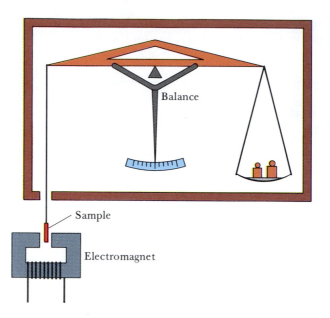

Balance

Sample

Electromagnet

paired electrons are like bar magnets oriented in opposite directions, each canceling the effect of the other. Thus paramagnetism in a substance depends on having one or more orbitals containing an unpaired electron. Such situations are particularly common in transition metal chemistry, where there are partially filled d orbitals.

Paramagnetism can be measured by determining the force by which a sample is drawn into a magnetic field. The more unpaired electrons in a given mass of sample, the stronger this effect will be. An instrument used for this purpose is the Gouy balance (Figure 22–13). The difference in the measured mass of the sample with the electromagnet off and on is determined, from which one can calculate the number of unpaired electrons for each ion.

EXERCISE 22–5

Which numbers of d electrons will always result in diamagnetic complexes in an octahedral crystal field? Which numbers will always have paramagnetic complexes in an octahedral crystal field? Which can be either diamagnetic or or paramagnetic in this field. ∎

Spectral Properties of Complexes

One of the characteristic properties of transition metals is the prevalence of colored ions (Figure 22–14). Also, addition of reagents like ammonia to solutions of transition metals can cause color changes. We now need to talk about the source of these colors.

The Ti^{3+} ion has one d electron. If Ti^{3+} is placed in water, it will form the $Ti(H_2O)_6^{3+}$ complex ion with an octahedral ligand environment. As shown in Figure 22–12, this one electron will reside in a t_{2g} orbital in the most stable or ground state. However, absorption of an amount of energy equal exactly to Δ_o would result in the d electron jumping up to the e_g orbital, producing an excited state.

The exact amount of energy for excitation of the $Ti(H_2O)_6^{3+}$ complex is most conveniently delivered with electromagnetic radiation of the proper frequency according to

$$E = h\nu$$

$Cu(H_2O)_4^{2+}$ $Cu(NH_3)_4^{2+}$

FIGURE 22–14 In contrast to the sky blue color of aqueous solutions of $Cu(H_2O)_4^{2+}$ ions, ammoniacal solutions are an intense, almost midnight blue, due to the presence of $Cu(NH_3)_4^{2+}$ ions.

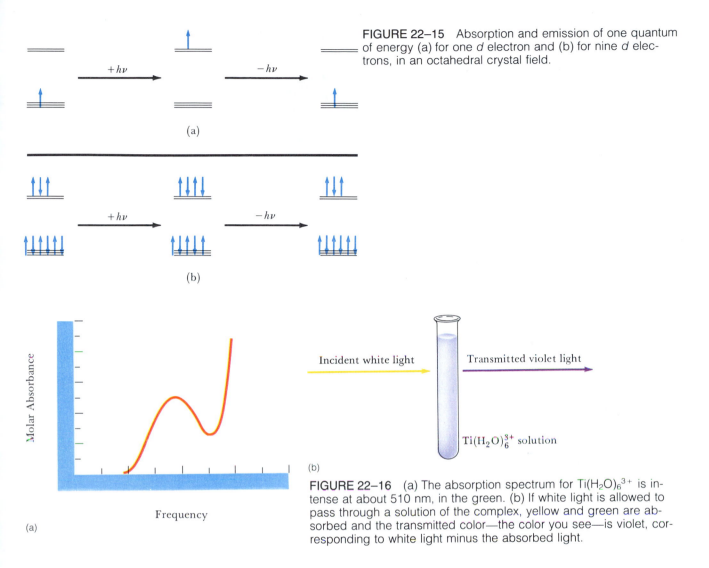

FIGURE 22–15 Absorption and emission of one quantum of energy (a) for one d electron and (b) for nine d electrons, in an octahedral crystal field.

(a)

(b)

Incident white light

Transmitted violet light

$Ti(H_2O)_6^{3+}$ solution

(b)

Molar Absorbance

Frequency

(a)

FIGURE 22–16 (a) The absorption spectrum for $Ti(H_2O)_6^{3+}$ is intense at about 510 nm, in the green. (b) If white light is allowed to pass through a solution of the complex, yellow and green are absorbed and the transmitted color—the color you see—is violet, corresponding to white light minus the absorbed light.

Since the energy desired is Δ_o, the frequency is related to the octahedral crystal field splitting by

$$\nu = \frac{\Delta_o}{h}$$

The excited $Ti(H_2O)_6^{3+}$ will emit radiation of the same frequency when it returns to the ground state. These processes for a d^1 system are shown in Figure 22–15a. The crystal field splitting for $Ti(H_2O)_6^{3+}$ is such that the radiation absorbed and emitted is in the green and yellow region of the visible spectrum. The absorption spectrum for this complex is shown in Figure 22–16a. If white light, say from the sun, passes through a solution of this complex (Figure 22–16b), the yellow and green will be absorbed; the transmitted light is violet, which is the color you see. The absorbed light will be re-emitted equally in all directions but will go unnoticed by the eye.

An equally simple situation occurs with d^9 configurations such as Cu^{2+} (Figure 22–15b). Here the ground state has one space or "hole" in the e_g orbitals, and the excited state has a hole in the t_{2g} orbitals. It is useful to view the hole as jumping to the lower energy orbitals when radiation is absorbed. Spectra for the configurations d^2 to d^8 are more complex but follow the same general ideas.

FIGURE 22–17 Remembering that what you see is white light minus the absorbed light, suggest likely colors for the absorbed light for these two classic coordination complexes: (a) Prussian blue, $Fe_4[Fe(CN)_6]_3$; (b) iron(III) thiocyanate, $Fe(SCN)_6$.

(a)

(b)

Early in our discussion of complexes we mentioned that addition of simple reagents can cause drastic changes in their colors. Ni^{2+} ions, for example, dissolve in water to form a green solution. Addition of aqueous ammonia to this solution turns the color to purple. Clearly, the frequency of the light being absorbed is changing, and since

$$\nu = \frac{\Delta_o}{h}$$

Δ_o must be changing with the ligand exchange that is going on. Thus, the value of Δ_o depends on the identity of the ligands (Figure 22–17).

The Spectrochemical Series

The list of ligands in order of increasing Δ_o values for a given metal ion is called the **spectrochemical series**. The order of the more common ligands in the series is

$$I^- < Br^- < Cl^- < F^- < OH^- < C_2O_4^{2-} < H_2O < \text{pyridine}$$
$$\approx NH_3 < NO_2^- < CN^-$$

When the ligands on a transition metal complex are exchanged, the change in Δ_o will result in a change in color. For example, adding NH_3 to a water solution of Ni^{2+} ions exchanges NH_3 ligands for H_2O ligands. According to the spectrochemical series, this increases the value of Δ_o. This is observed as a change in color (or the transmitted light) from green to purple. To fully understand the reasons for the order of the spectrochemical series, one has to go beyond the crystal field theory.

Effects of the Metal on the Value of Δ

In addition to the effect of the ligand on the value of Δ, giving the spectrochemical series, the value of Δ for a particular complex also depends on the identity of the metal ion and its charge. Two important observations are as follows:

- Higher oxidation states result in higher values of Δ. The higher the charge on the ion, the stronger the attraction between the metal and the ligands. This will pull the ligands in more closely so that the perturbation of the metal orbitals and the splitting of the energy levels by the ligands will be greater.

- The value of Δ increases going from the first to the second to the third

TABLE 22–4 **Enthalpies of Hydration of Some of the Gaseous Ions of the Metals from Potassium to Zinc in kJ/mol.**

K^+	-322
Ca^{2+}	-1577
Cr^{2+}	-1904
Mn^{2+}	-1841
Fe^{2+}	-1946
Co^{2+}	-1996
Ni^{2+}	-2105
Cu^{2+}	-2100
Zn^{2+}	-2046

transition series. The observed result of this effect is that practically all complexes of second and third transition series metals are low spin.

Geometries other than octahedral also have different values of Δ. For example, for tetrahedral coordination the value of Δ is about half that for octahedral coordination.

22.9 REACTIONS OF SOME COORDINATION COMPLEXES

Of the many reactions of coordination complexes, let's look at the formation of complexes from uncomplexed metal ions, exchange of one ligand for another in a complex, and oxidation-reduction reactions.

Formation of Aqua Complexes

A transition metal ion that is dissolved in water can already be considered an aqua complex, where water molecules are the ligands. A strong tendency toward formation of the aqua complex can be seen in the enthalpies of hydration for gaseous metal ions plunged into water. Table 22–4 lists enthalpies of hydration for some of the elements of the first transition series as well as potassium and calcium, illustrating the ease with which aqua complexes are formed by transition metal ions.

Aqua complexes of metals are all acidic to some extent because of the following sort of ionization:

$$[Cr(H_2O)_6]^{3+} + H_2O \longrightarrow [Cr(H_2O)_5(OH)]^{2+} + H_3O^+$$

The driving force for this reaction is the stronger coordination of the metal cation to the negative OH^- ligand compared to the neutral H_2O ligand. The acidity of these ions is comparable to that of other weak acids; in this case, $K = 1.26 \times 10^{-4}$.

Ligand Exchange Reactions

As a result of the very high enthalpies of hydration of transition metal ions, any water solution of a transition metal ion can be considered to be an aqua complex. For all $+2$ and $+3$ ions of the first transition series, the aqua ions are octahedral $M(H_2O)_6^{2+}$ or $M(H_2O)_6^{3+}$ species, although there are distortions from perfect octahedral geometry in some cases. When in water solution, the complexed water molecules are usually not identified explicitly and the complexed ion is normally given simply as M^{n+} (aq). The water molecules can be replaced in a stepwise manner with other ligands, and an equilibrium constant expression can be written for each step. For example,

$$M^{n+} \text{ (aq) } + NH_3 \text{ (aq) } \longrightarrow M(NH_3)^{n+} \text{ (aq) } + H_2O$$

$$K_1 = \frac{[M(NH_3)^{n+}]}{[M^{n+}][NH_3]}$$

$$M(NH_3)^{n+} \text{ (aq) } + NH_3 \text{ (aq) } \longrightarrow [M(NH_3)_2{}^{n+}] \text{ (aq) } + H_2O$$

$$K_2 = \frac{[M(NH_3)_2{}^{n+}]}{[M(NH_3)^{n+}][NH_3]}$$

$$M(NH_3)_2{}^{n+} \text{ (aq) } + NH_3 \text{ (aq) } \longrightarrow [M(NH_3)_3{}^{n+}] \text{ (aq) } + H_2O$$

$$K_3 = \frac{[M(NH_3)_3{}^{n+}]}{[M(NH_3)_2{}^{n+}][NH_3]}$$

$$M(NH_3)_3{}^{n+} \text{ (aq) } + NH_3 \text{ (aq) } \longrightarrow [M(NH_3)_4{}^{n+}] \text{ (aq) } + H_2O$$

$$K_4 = \frac{[M(NH_3)_4{}^{n+}]}{[M(NH_3)_3{}^{n+}][NH_3]}$$

$$M(NH_3)_4{}^{n+} \text{ (aq) } + NH_3 \text{ (aq) } \longrightarrow [M(NH_3)_5{}^{n+}] \text{ (aq) } + H_2O$$

$$K_5 = \frac{[M(NH_3)_5{}^{n+}]}{[M(NH_3)_4{}^{n+}][NH_3]}$$

$$M(NH_3)_5{}^{n+} \text{ (aq) } + NH_3 \text{ (aq) } \longrightarrow [M(NH_3)_6{}^{n+}] \text{ (aq) } + H_2O$$

$$K_6 = \frac{[M(NH_3)_6{}^{n+}]}{[M(NH_3)_5{}^{n+}][NH_3]}$$

The values of the stepwise formation constants decrease somewhat as the substitution progresses. This is the result of the decreasing availability of sites for substitution—a statistical factor—and ligand crowding if the new ligand is more bulky than the old.

Also due to the buildup of charge, for charged ligands.

The stepwise reactions can be summed to give an overall reaction,

$$M^{n+} \text{ (aq) } + 6\ NH_3 \text{ (aq) } \longrightarrow [M(NH_3)_6{}^{n+}] \text{ (aq) } + 6\ H_2O$$

The overall equilibrium constant for the formation for the complex is

$$K_{\text{overall}} = \frac{[M(NH_3)_6{}^{n+}]}{[M^{n+}][NH_3]^6} = K_1 \cdot K_2 \cdot K_3 \cdot K_4 \cdot K_5 \cdot K_6$$

EXAMPLE 22–4

The overall formation constant for $Ni(NH_3)_6{}^{2+}$ is 4.07×10^8. Calculate the concentration of Ni^{2+} (aq) when $[NH_3] = 0.120\ M$ and $[Ni(NH_3)_6{}^{2+}] = 0.501\ M$.

Solution   Substitution into the overall formation constant expression gives

$$[Ni^{2+}] = \frac{[Ni(NH_3)_6{}^{2+}]}{K_{\text{overall}} \times [NH_3]^6}$$

$$= \frac{0.501}{(4.07 \times 10^8)(0.120)^6} = 4.12 \times 10^{-4}\ M$$

EXERCISE 22–6

Cu^{2+} forms the ammine complex, $Cu(NH_3)_4{}^{2+}$. The overall formation constant for this complex is 6.8×10^{12}. Calculate the concentration of

Cu^{2+} (aq) in equilibrium with NH_3 (aq) at 0.331 M and $Cu(NH_3)_4^{2+}$ at 0.413 M.

Answer: $[Cu^{2+}] = 5.06 \times 10^{-12} M$. ∎

Complexes with Chelating Ligands

Ammonia is a **monodentate** ("one-toothed") ligand, having one coordinating (nitrogen) site. Ethylenediamine is similar to NH_3 but has two coordinating (nitrogen) sites, tied together in one molecule. It is a **bidentate** ligand. It is often represented by the symbol **en** and has the formula:

$$H_2N-CH_2CH_2-NH_2$$

ethylenediamine (en)

Because of the flexibility of the molecule, the lone pairs of electrons on both nitrogens are available for complexing to the same metal ion, forming five-membered rings. The complex of ethylenediamine (analogous to $Ni(NH_3)_6^{2+}$) is shown in Figure 22–18. Such ligands as ethylenediamine are called **chelating** ligands, from the Greek word for "claw." The magnitude of the overall formation constant increases surprisingly from the nonchelating, monodentate NH_3 ligand to the chelating, bidentate $H_2NCH_2CH_2NH_2$ ligand. For Ni^{2+}, the increase is from 4.07×10^8 to 1.91×10^{18}. The tendency expressed in these results is known as the **chelate effect**.

Several factors contribute to the chelate effect. We shall consider entropy. The formation of $Ni(en)_3^{2+}$ from the aquated ion is

$$Ni(H_2O)_6^{2+} \text{ (aq)} + 3 \text{ en (aq)} \longrightarrow Ni(en)_3^{2+} + 6 H_2O$$

The reaction is favored by entropy because four reactant particles change seven product particles. The equivalent reaction of NH_3 has no entropy advantage, since seven reactant molecules form seven product molecules. Only three ethylenediamine molecules need to find each metal ion in order to form the complex, compared with six NH_3 molecules. Once one end of the ethylenediamine molecule has attached to the metal ion, the other end is close by and it is a relatively convenient and favorable matter to complete the five-membered ring.

Another colorful and especially interesting example of a six-coordinate bidentate ligand is the $Co(en)_2Cl_2^+$ complex ion, which exists as a pair of *cis/trans*-isomers (Figure 22–19).

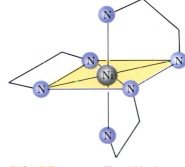

FIGURE 22–18   The $Ni(en)_3$ complex. Hydrogen atoms have been omitted (for simplicity in illustration), and the carbon atoms are located at the bends in the lines linking the nitrogen atoms.

Ethylenediamine = en
en = $H_2NCH_2CH_2NH_2$

Optical isomerism will be discussed along with organic and biomolecules in Chapter 26.

FIGURE 22–19   On the left is the *cis*-isomer, the purple cation, *cis*-$Co(en)_2Cl_2^+$; on the right is the *trans*-isomer, the green cation, with the same chemical composition but different spatial orientation of the chlorine atoms about the central metal ion, *trans*-$Co(en)_2Cl_2^+$.

EXAMPLE 22–5

Chelating ligands with more than two complexing sites show even greater chelate effects. Ethylenediaminetetraacetate ion, widely known as EDTA, is a notable example:

$$O=\overset{\overset{\displaystyle O^-}{|}}{C}-CH_2 \qquad\qquad CH_2-\overset{\overset{\displaystyle O^-}{|}}{C}=O$$

$$:N-CH_2-CH_2-N:$$

$$O=\underset{\underset{\displaystyle O^-}{|}}{C}-CH_2 \qquad\qquad CH_2-\underset{\underset{\displaystyle O^-}{|}}{C}=O$$

Both nitrogen atoms and the single-bonded oxygen anions are capable of coordinating with a metal ion. Such a ligand would be hexadentate (six-toothed), and this is a potent chelating ligand. In the case of Fe^{3+}, an octahedral complex is formed in which these electron-pairs occupy the six bonding sites on the metal ion. Adding the sodium salt of EDTA to solutions containing Fe^{3+} ions results in the complete complexation of the Fe^{3+} ions. Once complexed, or sequestered, such ions can be maintained in the presence of other ions in solution that would normally have caused their precipitation, or the sequestered ions can be removed from solution. Hydroxide ions, for example, normally precipitate Fe^{3+} ions as the insoluble hydrated iron oxides, but the EDTA complex is soluble in the presence of hydroxide.

Use of EDTA as a sequestering agent in the treatment of heavy metal poisoning has proved to be one of the most important practical applications of the coordination chemistry of the transition metal ions. After EDTA is administered as an antidote in lead and mercury poisoning, the sequestered ions are excreted by the body. That is essential since even small quantities of lead, from peeling paint or improperly glazed pottery, for example, can affect certain enzyme systems involved in the manufacture of hemoglobin for red blood cells. Minor symptoms such as headache and irritability may be followed by gastrointestinal symptoms—loss of appetite, constipation, nausea, vomiting, and abdominal pain. The central nervous system may be involved. However, the treatment is not without its side effects, since EDTA removes calcium ions from body tissue, along with several essential trace metal ions (which must be replenished when EDTA is used therapeutically).

EXERCISE 22–7

(a) Pt^{2+} ions form a four-coordinate, square planar complex. Draw a sketch of its ammonia and ethylenediamine complexes.

(b) Oxalic acid (HOOC—COOH) is a diprotic acid that forms a divalent anion capable of acting as a bidentate chelating agent. Draw a sketch of the octahedral complex it forms with Cr^{3+} ions.

(c) Sketch the octahedral complex formed between EDTA and Fe^{3+} ions.

Oxidation-Reduction Reactions

Formation of a coordination complex has a significant effect on the oxidation-reduction potential of a metal ion. For example, the relative stabilities of Co^{2+} and Co^{3+} complexes are strongly ligand-dependent. Oxidation of the aqua complex of Co^{2+} to the Co^{3+} complex is very unfavorable, and the reverse reaction is spontaneous:

$$[Co(H_2O)_6^{3+}] + e^- \longrightarrow [Co(H_2O)_6^{2+}] \qquad \mathscr{E}° = 1.84 \text{ V}$$

However, in the presence of NH_3 ligands the stability of the Co^{3+} ions is greatly increased:

$$[Co(NH_3)_6^{3+}] + e^- \longrightarrow [Co(NH_3)_6^{2+}] \qquad \mathscr{E}° = 0.1 \text{ V}$$

The oxidation-reduction chemistry of iron also depends on the ligand. For example, Fe^{3+} is more stable in the presence of cyanide ion:

$$[Fe(H_2O)_6^{3+}] + e^- \longrightarrow [Fe(H_2O)_6^{2+}] \qquad \mathscr{E}° = 0.77 \text{ V}$$
$$[Fe(CN)_6^{3-}] + e^- \longrightarrow [Fe(CN)_6^{4-}] \qquad \mathscr{E}° = 0.36 \text{ V}$$

The oxidation-reduction chemistry of copper depends strongly on the ligands present. Note the standard reduction potentials in pure water and in the presence of cyanide ion:

$$Cu^{2+} (aq) + e^- \longrightarrow Cu^+ (aq) \qquad \mathscr{E}° = 0.153 \text{ V}$$
$$Cu^{2+} (aq) + 2 CN^- + e^- \longrightarrow Cu(CN)_2^- (aq) \qquad \mathscr{E}° = 1.103 \text{ V}$$

The disproportionation of Cu^+ is also interesting:

$$Cu^+ (aq) + e^- \longrightarrow Cu (s) \qquad \mathscr{E}° = 0.52 \text{ V}$$
$$Cu^{2+} (aq) + e^- \longrightarrow Cu^+ (aq) \qquad \mathscr{E}° = 0.153 \text{ V}$$

Reversing the second half-reaction and adding gives

$$2 Cu^+ (aq) \longrightarrow Cu (s) + Cu^{2+} (aq) \qquad \mathscr{E}° = 0.37 \text{ V}$$

The standard reduction potential corresponds to $K = 2 \times 10^6$, implying that this disproportionation of Cu^+ to Cu metal and Cu^{2+} is spontaneous. In fact, the only Cu^+ compounds that can exist in the presence of water are highly insoluble species such as CuCN and CuCl, or the cyanide complex $Cu(CN)_2^-$ mentioned earlier.

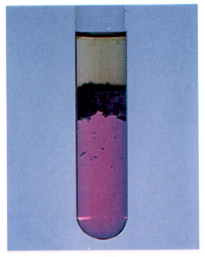

Aqueous cobalt(II) solutions $[Co(H_2O)_6^{2+}]$ are pink, and can be seen in the lower portion of the test tube; on careful addition of aqueous ammonia, a blue precipitate of the hydroxide $[Co(OH)_2]$ forms (middle), which dissolves easily in excess ammonia as a pale yellowish-orange $[Co(NH_3)_6]^{2+}$ solution (top). The thin red layer, just at the surface where the solution is in contact with oxygen in the air, is due to:
$[Co(NH_3)_6]^{2+} \rightarrow [Co(NH_3)_6]^{3+} + e^-$

22.10 ZINC, CADMIUM, AND MERCURY

Zinc, cadmium, and mercury have filled d and s subshells. The electron configuration of zinc is $[Ar]3d^{10}4s^2$; that of cadmium is $[Kr]4d^{10}5s^2$; and that of mercury is $[Xe]4f^{14}5d^{10}6s^2$. In forming ions, the two outer s-electrons are lost, giving the M^{2+} ions. These ions have none of the properties of transition metals that depend on a partially filled d shell. The only other significant ion formed by these metals is the Hg_2^{2+} ion. Known as the mercurous ion, it contains Hg in the $+1$ oxidation state. In this ion, two mercury atoms are bonded together at a distance ranging from about 2.49 to 2.69 Å. Zinc and cadmium have quite similar chemistry; both are electropositive metals that are easily oxidized. Both dissolve easily in nonoxidizing acids such as HCl. Mercury, however, is quite different and is not very reactive. Mercury is inert to nonoxidizing acids.

The $+2$ ions of all three metals form coordination complexes. The most important coordination number is four, with a tetrahedral arrangement of the ligands. A typical example is $[Zn(CN)_4]^{2-}$. Much higher formation constants are observed for Hg^{2+} than for the other two ions. This can be partly ascribed to the relatively small size of the Hg^{2+} ion resulting from the lanthanide contraction. Thus the ionic radius of the Cd^{2+} ion is 0.92 Å and that of Hg^{2+} is only 0.01 Å greater.

(a)   (b)

FIGURE 22–20   (a) Cinnabar (vermilion) is natural mercury(II) sulfide. It is a red to scarlet volcanic rock, found widely in Southern California and Nevada, and is the only important mercury ore. (b) Droplets of mercury can be obtained directly by simply heating the ore in air: $HgS (s) + O_2 (g) \rightarrow Hg (liq) + SO_2 (g)$

Important ores of zinc are $ZnCO_3$ and ZnS. The carbonate can be treated with sulfuric acid to form the sulfate, and the zinc metal is then produced by electrolysis. This is a particularly good route for electroplating zinc onto steel, a major use of zinc. The ZnS can be roasted in the presence of oxygen to form the oxide, which is then reduced with carbon,

$$2 \ ZnS \ (s) + 3 \ O_2 \ (g) \longrightarrow 2 \ ZnO \ (s) + 2 \ SO_2 \ (g)$$

$$ZnO \ (s) + C \ (s) \longrightarrow Zn \ (liq) + CO \ (g)$$

Coating of iron for the prevention of corrosion (as we discussed in Chapter 20) is the major use of zinc. It can be applied by electrolysis, or the object can be dipped in or sprayed with molten zinc. Other uses include dry cells, where Zn forms the case and anode, and preparation of alloys, particularly brass (which contains copper and zinc). Zinc oxide is used as a white pigment in paints and in the vulcanizing of rubber. $ZnCrO_4$ is the active ingredient in many anti-rust paints. ZnS fluoresces when struck by x-rays or an electron beam, and is used for television screens and oscilloscopes. Zinc is also required in nutrition for manufacturing the hormone insulin.

Cadmium occurs in zinc ores, where it substitutes for about 0.4% of the zinc atoms. It is recovered by distilling out the relatively volatile salt $CdCl_2$. Cadmium has limited uses, the principal ones being electroplating on steel for corrosion resistance and making the yellow paint pigment CdS. Mercury occurs naturally as cinnabar, HgS, and as HgO. Either one heated in air gives the metal:

$$HgS \ (s) + O_2 \ (g) \longrightarrow Hg \ (liq) + SO_2 \ (g)$$

$$2 \ HgO \ (s) \longrightarrow 2 \ Hg \ (liq) + O_2 \ (g)$$

Mercury is the one metal that is a liquid at room temperature (Figure 22–20b). Because it is an inert, liquid conductor of electricity, it has some important applications. In particular, it is used in switches and as the flowing electrode in electrolysis cells, the most important being the chlor-alkali cell for the electrolysis of brine. Excited mercury atoms emit intensely in the green and ultraviolet regions of the electromagnetic spectrum. Therefore mercury is used in mercury vapor lamps on streets and highways and in fluorescent tubes. In the fluorescent tubes, the ultraviolet radiation that is emitted by mercury causes the fluorescent coating on the tube to give off visible light.

Mercury and many of its compounds are extremely toxic. Because it has an appreciable vapor pressure even at room temperature, mercury can be accidentally inhaled; it can also be absorbed through the skin.

22.11   THE LANTHANIDE AND ACTINIDE ELEMENTS

The fourteen elements following lanthanum resemble it strongly and are therefore called lanthanides. An alternative and still widely used name for these elements is the "rare earth elements." Because of their similarity to lanthanum, it is convenient to discuss lanthanum along with these elements. In fact, scandium, the element above lanthanum, also has very similar chemical behavior.

The electron configuration of lanthanum is $[Xe]5d^1 6s^2$. The first three ionization energies for lanthanum are reasonably low, leading to easy formation of the La^{3+} ion with the $[Xe]$ electron configuration. At this point in the filling of the atomic orbitals, the energy of the $4f$ orbitals has dropped to the point that these can start to fill. The fourteen lanthanide elements occur as the $4f$ shell is being filled. The principal and most stable ion of all of the lanthanides is M^{3+}. For some of these elements the $+2$ or $+4$ ions will form, particularly if they lead to the half-filled shell, $4f^7$, or the completely filled shell, $4f^{14}$. Examples of the relatively stable $+2$ ion are Eu, $[Xe]4f^7 6s^2$, which forms Eu^{2+}, $[Xe]4f^7$; and Yb, $[Xe]4f^{14}6s^2$, which forms Yb^{2+}, $[Xe]4f^{14}$. Likewise, Ce, $[Xe]4f^1 5d^1 6s^2$, forms Ce^{4+}, $[Xe]$; and Tb, $[Xe]4f^9 6s^2$, forms Tb^{4+}, $[Xe]4f^7$. Because the $+3$ ions are the most stable, the $+2$ ions are reducing agents and the $+4$ ions are oxidizing agents. As with lanthanum, the lanthanides form ions easily, and their chemistry is almost entirely ionic.

These so-called rare earth elements are not particularly rare; thulium is the least common of the group, and it is more common than arsenic, cadmium, mercury, and selenium. The most important lanthanum and lanthanide mineral is monazite sand, which contains MPO_4 in which M represents uranium, thorium, lanthanum, and the lanthanides as $+3$ ions. Europium, which tends to form the $+2$ ion (see above), is poorly represented in monazite sands and is often found in Group IIA minerals.

Because of the similarity of their chemistry, lanthanum and the lanthanides are difficult to separate. The principal difference in their $+3$ ions is their ionic radii. These decrease steadily from 1.061 Å for La^{3+} to 0.848 Å for Lu^{3+}. Thus the ability to bind to anions will increase from La^{3+} to Lu^{3+}. The ions are separated on an **ionic exchange resin**. This is a polymer with negatively charged sites. The mixed $+3$ ions are complexed to a polydentate ligand and poured into the top of a column (Figure 22–21) containing the resin. Some of the ions will detach from the ligand and bind to the anionic sites on the resin. The Lu^{3+} will be most strongly bound to the ligand and therefore will progress through the column the most quickly; it will leave (be eluted) first. The other ions will follow and the La^{3+} will be eluted last. This **ion exchange chromatography** is used on a commercial scale.

The lanthanides have a few important uses, especially as phosphors for lamps and TV screens. An alloy of cerium with iron gives off sparks when struck and is used as the "flint" in cigarette lighters.

The actinide elements are the fourteen elements following actinium, which they resemble to some extent. All of the isotopes of actinium and the actinides are radioactive. The electron configurations of the actinides

FIGURE 22–21 Ion exchange chromatography: (a) The column is packed with an ion exchange resin; (b) schematic representation of an anionic exchange resin.

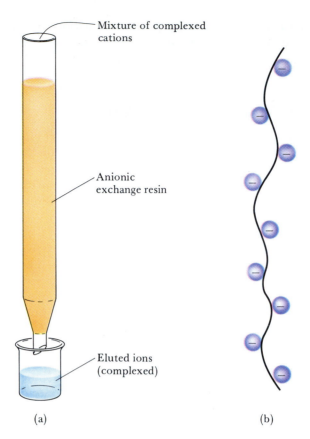

Mixture of complexed cations

Anionic exchange resin

Eluted ions (complexed)

(a)

(b)

are somewhat irregular. Actinium is [Rn]$6d^17s^2$, analogous to lanthanum. However, thorium is [Rn]$6d^27s^2$, giving the impression that the $6d$ orbitals are going to fill in preference to the $5f$ and that a series parallel to the lanthanides is not going to occur. Then protactinium has the configuration [Rn]$5f^26d^17s^2$ or [Rn]$5f^16d^27s^2$, showing the beginning of filling of the $5f$ shell. The remaining actinides do have $5f$ electrons, although the extent of occupation of the $6d$ shell is irregular.

The metals are all electropositive, and the +3 ion is the most common for the elements after plutonium. Thorium is the most plentiful of the actinides since its natural isotope, ^{232}Th, has a half-life of 1.4×10^{11} years. The others of these elements that occur naturally, actinium, protactinium, and uranium, all have isotopes with sufficiently long half-lives to last since the elements were formed. Thorium(IV) oxide, ThO_2, or thoria, is used in mantles for gasoline lanterns and for refractory crucibles and insulators. Its use in mantles stems from its ability to catalyze the combustion reaction at a high temperature and the brilliant white light it emits when heated to a high temperature. The uses of uranium and plutonium in nuclear fission were discussed in Chapter 3.

SUMMARY

The transition elements include the d-block elements, the lanthanides, and the actinides. They are all metals and characteristically exhibit variable oxidation states and colored compounds, many of which are paramagnetic. Much of their chemistry involves coordination complexes and electrons in d orbitals. The electron configurations of the elements show a preference for half-filled and filled d shells.

Many of the transition metals are of great commercial importance, as are the processes whereby they are won from natural resources.

Coordination complexes are composed of a central metal ion and co-ordinating ligands. Of the several different coordination numbers, four and six are very common.

Crystal field theory assumes a model in which the positively charged metal ion is surrounded by ligands represented by point charges. These charges split the otherwise equal-energy d orbitals. For octahedral coordination, the d orbitals are split into two groups, and magnetic and spectral properties can be explained accordingly. The list of ligands in order of degree of splitting is called the spectrochemical series.

Chemical equilibria in solution are among the most important reactions involving exchange of ligands. Particularly stable complexes are formed with chelating ligands, which have more than one site that can complex a single metal ion. The oxidation-reduction chemistry of transition metal ions depends strongly on the nature of the complexing ligands.

QUESTIONS AND PROBLEMS

QUESTIONS

1. Explain why chromium and copper have electron configurations with only one s electron in the valence shell whereas their neighboring elements have two.
2. What establishes the maximum oxidation state for a transition metal?
3. Why is limestone included in the blast furnace charge for the smelting of iron ore?
4. What is the function of the open hearth furnace in the production of useful iron alloys?
5. Write the reactions for producing copper from a copper ore.
6. Write the equations for preparing tungsten from $PbWO_4$ ore.
7. Why does the fact that $Pt(NH_3)_2Cl_2$ has two isomers eliminate the possibility that this particular complex might have the geometry of a tetrahedron?
8. Octahedrally coordinated Cr^{3+} is particularly stable. In terms of the crystal field theory, why is this?
9. Why do all substances exhibit diamagnetism?
10. Why do water solutions of many transition metal ions change color when ammonia is added?
11. Why is the crystal field theory uninformative regarding the actual colors of complexes?
12. Are water solutions of transition metal ions basic or acidic? Explain.
13. Explain why a chelating ligand forms complexes so much more readily than the comparable nonchelating ligand.
14. What ligand would you want to have present when oxidizing Co^{2+} to Co^{3+}? Why?
15. EDTA can be successfully used to titrate Cu^{2+} ions to a sharp endpoint. Briefly explain why ammonia cannot be used.
16. Suppose a transition metal gives colored complexes

with all ligands in the spectrochemical series. How will the colors of the complexes tend to change proceeding from one end of the spectrochemical series to the other?

PROBLEMS

Electron Configurations   [1–2]

1. Give the electronic configurations of Ni^{2+}, Cu^+, Zn^{2+}, and Ti^{2+}.
2. Write the electronic configuration of each of the following ions: Cr^{3+}, Au^{3+}, Co^{3+}, and V^{3+}.

Preparation of Metals   [3–4]

3. Copper ores now considered workable are of very low purity. What is the maximum amount of pure copper metal that can be extracted from an ore that is 2.0% $CuFeS_2$ and 98.0% waste if the flotation process is able to separate out 95.0% of the pure ore?
4. Suppose the current price of magnesium metal is $3.00 per pound. Using the current method for the production of titanium, what is the minimum price that can be charged for a pound of titanium?

Coordination Complexes: Structures, Names   [5–14]

5. $PdCl_2$ will react with two, three, and four molecules of NH_3 to give three different complexes. Propose the formula of each of these comounds, the number of "free" chloride ions each contains, and the total number of ions in each formulation.
6. The Cr^{3+} ion has a coordination number of 6. Give the formulas for all possible combinations that would result from adding NH_3 (aq) to an aqueous solution of $CrCl_3$.

For each formula, predict the number of "free" Cl^- ions and the total number of ions per formula unit.

7. How many isomers would you expect for the complex $[Ni(NH_3)_3Cl_3]^-$ if the coordination geometry is octahedral? Draw the isomers.

8. Draw the isomers that would be observed for $[Co(NH_3)_4Cl_2]Cl$ if the coordination had the geometry of a triangular prism. Do the same for a pentagonal pyramid. How many isomers would there be in each case?

9. Name the following compounds:
(a) $K_2[Ni(CN)_4]$
(b) $Na_3[Cr(C_2O_4)_3]$
(c) $[Pt(NH_3)_5Cl]Cl_3$
(d) $[Fe(NH_3)_4(NO_2)_2]_2SO_4$
(e) $[Co(H_2O)_6]I_3$
(f) $[Ru(NH_3)_5Cl]Br_2$
(g) $K_2[CoCl_4]$

10. Give the formulas of the following compounds:
(a) pentammineiodochromium(III) bromide
(b) sodium tetrachloroferrate(III)
(c) potassium tetracyanonickelate(II)
(d) pentaquachlorochromium(III) chloride
(e) sodium tetrahydroxozincate(II)
(f) hexaquachromium(III) sulfate
(g) trichlorotrithiocyanatochromium(III)

11. Give both the coordination number and the oxidation number for the metal atom in each of the following complexes:
(a) $CuCl_4^{2-}$
(b) $Ag(CN)_2^-$
(c) $Zn(NH_3)_4^{2+}$

12. Give the coordination number and oxidation state of the metal in each of the following complexes.
(a) $Fe(CO)_5$
(b) $Cr(H_2O)_6^{3+}$
(c) $Fe(CN)_6^{3-}$

13. For each of the following geometric forms, state the number of faces, the number of corners, and coordination number of a central atom in such an arrangement:
(a) a tetrahedron
(b) a trigonal bipyramid

14. Give the number of faces, edges, and corners for each of the following geometric solids.
(a) an octahedron
(b) a trigonal prism
(c) a pentagonal pyramid

Coordination Complexes: Bonding, Magnetic Properties, and Spectra   [15–22]

15. Determine the possible numbers of unpaired electrons in complexes of the following ions in an octahedral crystal field: Mn^{2+}, Co^{3+}, and Ni^{2+}.

16. Give the possible numbers of unpaired electrons in octahedral complexes of Cu^{2+}, Zn^{2+}, and Fe^{3+}.

17. If an octahedral Fe^{2+} complex is paramagnetic, what do you know about its electron configuration?

18. Give the configuration of the $3d$ electrons in an octahedral complex of Co^{2+} that has one unpaired electron.

19. Calculate the value of Δ_o for a d^9 octahedral complex with a strong visible absorption centered at 600 nm. What color would you expect a solution of this complex to be?

20. What is the value of Δ_o for an octahedral d^1 complex with an absorption centered at 550 nm? Predict the color of the solution.

21. Which of the following species when aquated would you expect to exhibit paramagnetic behavior?
(a) Cr^{3+}
(b) Ni^{2+}
(c) Co^{2+}
(d) Fe^{2+}
(e) Fe^{3+}
(f) Cu^{2+}

22. Which of the following ions would you expect to contain unpaired d electrons?
(a) $Cr(NH_3)_6^{3+}$
(b) $Fe(CN)_6^{3-}$
(c) $Fe(CN)_6^{4-}$
(d) $Cu(NH_3)_4^{2+}$

Reactions of Coordination Complexes   [23–26]

23. Calculate the pH of the resulting solution if 10.0 g of $CrCl_3$ is dissolved in sufficient water to produce 1.00 L of solution.

24. The first acid ionization constant of $Al(H_2O)_6^{3+}$ is 1.07×10^{-5}. Calculate the pH of a solution initially $0.100\ M$ in $Al(H_2O)_6^{3+}$.

25. The overall formation constant for $[Co(NH_3)_6]^{3+}$ is 5.0×10^{31}. What will be the concentration of Co^{3+} (aq) at equilibrium if the concentrations of $[Co(NH_3)_6]^{3+}$ (aq) and NH_3 (aq) are respectively 0.100 and $0.120\ M$?

26. The overall formation constant for $[Cu(NH_3)_4]^{2+}$ is 6.8×10^{12}. What concentration of NH_3 (aq) would be present at equilibrium with $[Cu^{2+}]$ at $0.001\ M$ and $[Cu(NH_3)_4]^{2+}$ at $0.633\ M$?

Additional Problems   [27–33]

27. Give the name of each of the following compounds and give the oxidation state of every metal in each compound.
(a) $[Cu(H_2O)_6]Cl_2$
(b) $Ni(NH_3)_4Cl_2$
(c) $Na_2[PtCl_4]$
(d) $K_4Fe(CN)_6$

28. Draw all possible isomers for the complex ion $[Cr(H_2O)_3(NH_3)_2Cl]^+$, assuming octahedral coordination. (Remember how symmetrical the octahedron is; there are probably fewer isomers than you will first think.)

29. The valence bond interpretation of covalency, coordination number, and coordination complexes is based upon involvement of available bonding orbitals on the metal atom in formation of coordinate-covalent bonds.

Size and charge are also determining factors. With that in mind, complete the following table.

Coordination Number	Hybrid σ Orbitals	Geometric Configurations	Examples
2	sp	linear	—
3	sp^2	—	BF_3, NO_3^-
4	—	tetrahedral	$Ni(CO)_4$, MnO_4^-
4	sp^2d	—	$Ni(CN)_4^{2-}$, $Pt(NH_3)_4^{2+}$
5	sp^3d	—	$CuCl_5^{2-}$, $Fe(CO)_5$
6	sp^3d^2	—	—
—	sp^3d^4	dodecahedral	$Mo(CN)_8^{4-}$

30. The values of the first three stepwise formation constants for $[Cu(NH_3)_4]^{2+}$ are 1.7×10^4, 3.2×10^3, and 8.3×10^2. The overall formation constant is 6.8×10^{12}. What is the value of K_4?

31. Silver forms the sparingly soluble salt AgCl ($K_{sp} = 1.6 \times 10^{-10}$) and the soluble complex $[Ag(NH_3)_2]^+$ ($K_{formation} = 1.5 \times 10^7$). Calculate the solubility of AgCl in $1.00\ M$ NH_3.

32. A tetrahedral crystal field splits the d orbitals into two groups, a lower energy group of two and a higher energy group of three. Draw a diagram and determine which two d orbitals will end up in the lower energy group.

33. Ions with seven d electrons such as Co^{2+} tend to form some tetrahedral complexes. How many unpaired electrons will such complexes have?

MULTIPLE PRINCIPLES   [34–37]

34. Suggest an oxidizing agent that will convert Fe^{2+} to Fe^{3+} in water solution but not if excess CN^- is present. (Assume the concentrations of the oxidizing agent and the iron ions are all $1.00\ M$).

35. Magnetite, Fe_3O_4 has a density of $5.18\ g/cm^3$ and a cubic unit cell with a cell edge of 8.37 Å. How many formula weights of Fe_3O_4 are there per unit cell? How many Fe^{2+} ions and how many Fe^{3+} ions are there per unit cell?

36. Determine the equilibrium constant for the reaction

$$[Ni(NH_3)_6]^{2+}\ (aq)\ +\ 6\ H_3O^+\ (aq) \longrightarrow Ni(H_2O)_6^{2+}\ (aq)\ +\ 6\ NH_4^+\ (aq)$$

37. Determine $\Delta G°$ for the reaction

$$[Cr(H_2O)_6]^{3+}\ +\ CN^- \longrightarrow [Cr(H_2O)_5(OH)]^{3+}\ +\ HCN$$

APPLIED PRINCIPLES   [38–39]

38. A charge of 3000 kg of pig iron is placed in an open hearth furnace to remove some of the impurities. The pig iron contains 0.22% sulfur (by mass), which we want to remove completely, and 3.83% carbon, which we want to cut down to 0.75% for the production of steel. The furnace is lined with magnesium oxide. What mass of magnesium oxide will be used up in removing the impurities?

39. Plating parts of automobile trim with chromium serves to protect as well as add attractiveness. Suppose you are electroplating a bumper with a total area of $1.50\ m^2$ in a bath containing Cr^{3+} in acid solution. You want a coating exactly 0.12 mm thick. Exactly how would you arrange the conditions so that this would happen?

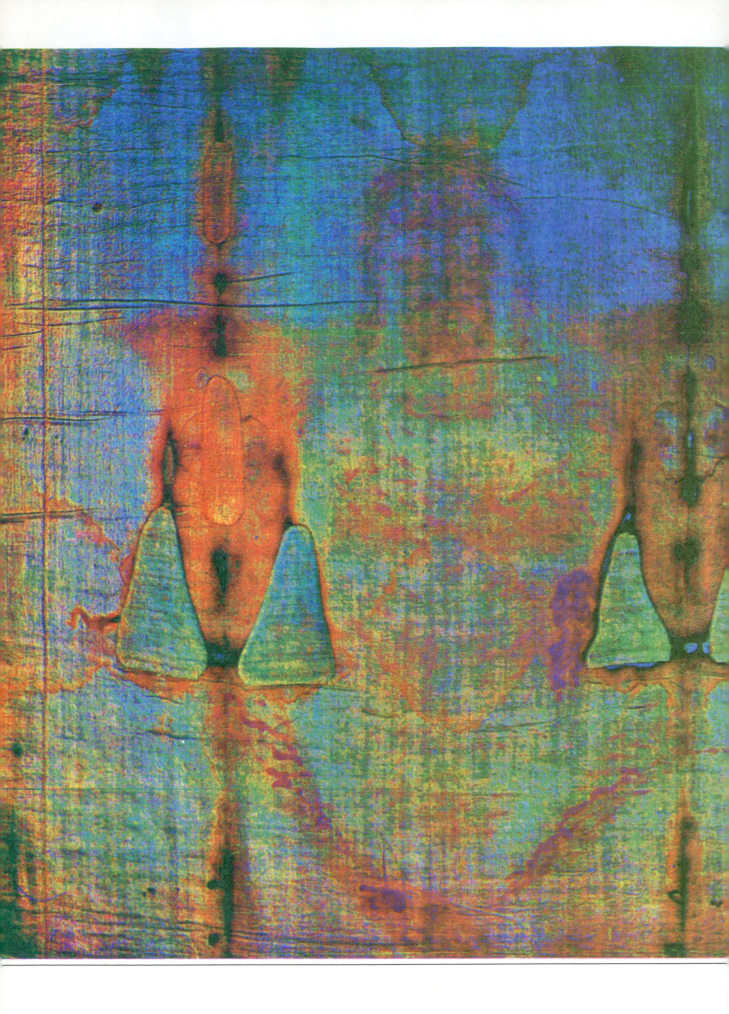

Kinetics I

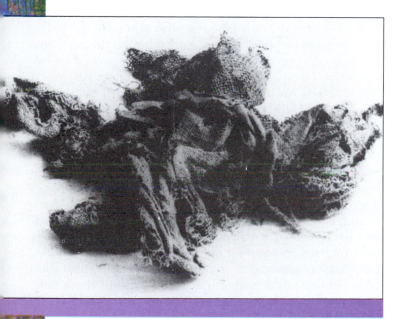

Based on carbon-14 studies, independent test results by three research groups have recently dated the cloth known as the "Shroud of Turin" between 1260 and 1390 A.D. For many, these data laid to rest the mystery of the shroud, with its faint image (here color enhanced and remarkably well defined by computer graphics analysis) of the front and back of a hollow-eyed, bearded man, and with what appear to be scourge marks and blood stains corresponding to New Testament accounts of the crucified Jesus. Father of the carbon-14 method of dating carbonaceous materials, J. Willard Libby examined the linen wrappings from the Dead Sea Scroll containing the Book of Isaiah in 1947, shortly after its discovery. He found the sack cloth wrappings to be 1917 ± 200 years old.

23.1   REACTION DYNAMICS

Thermodynamics is concerned with initial and final states at equilibrium and with the direction of spontaneous change for chemical reactions and other processes. However, it is not concerned with the time required for a system to pass from one equilibrium state to another, nor with the pathway the chemical reaction or process must follow. **Kinetics**, on the other hand, is concerned with the rate at which chemical reactions happen and, accordingly, with the details of the chemistry of the process, all of which thermodynamics can conveniently ignore. Thermodynamics tells you the extent of the reaction. Kinetics tells you the rate of the reaction and, by so doing, tells you useful things about how reactions happen.

The details of a chemical reaction are of great interest for very practical reasons. Imagine a reaction that we want to happen, and that in fact is known to be thermodynamically favorable. Such a reaction may not be very useful if it is too slow or too fast. If the reaction takes place too slowly, it may well prove to be uneconomic. As a manufacturer you won't profit, or as a researcher your time will be poorly spent. And although the very slow conversion of vegetation into petroleum and coal has indeed proved useful, the study of reactions that take place over geologic time frames isn't often practical. Very fast reactions can be of limited value because they often prove difficult to control and study. But many chemical reaction are fast indeed—and important, too. Most elementary steps involving enzymes and neurons are very rapid, and there are modern techniques available for their study, ranging from resonance phenomena to "stopped flow."

Understanding kinetic factors can lead to control of a chemical reaction by minimizing competing reactions and formation of undesirable secondary products, and by accelerating slow reactions or moderating the fast ones. Corrosion is an example of a slow process that can be made still slower if we first understand the details of the process. That's obviously desirable. Engine knock in the internal combustion engine is the result of very fast processes that can be largely eliminated in the modern high compression engine. That's desirable, too. By learning a great deal about the kinetic details, we know how to design the explosion chamber of the automobile and the combustion characteristics of the fuel.

In the chapters on thermodynamics and equilibrium, we explored what nature demands of chemistry. In this chapter on chemical kinetics, we'll see the extent to which we have been able to exercise a measure of control over nature, and how we learn the details of chemistry in progress.

Under normal conditions, precipitation of a sparingly soluble salt is governed by purely thermodynamic considerations. Consider the precipitate that forms when a small amount of sulfide ion (S^{2-}) is added to an equimolar aqueous solution of Zn^{2+} and Fe^{2+} ions:

$$Zn^{2+} + S^{2-} \longrightarrow ZnS\ (s) \qquad K_{eq} = 1/K_{sp} = 2.2 \times 10^{23}$$

$$Fe^{2+} + S^{2-} \longrightarrow FeS\ (s) \qquad K_{eq} = 1/K_{sp} = 2.7 \times 10^{18}$$

It is the larger equilibrium constant for the less soluble ZnS that governs this competition between Fe^{2+} and Zn^{2+} for the S^{2-} ions, and the reaction that occurs *first* as the S^{2-} ion is added is selective precipitation of zinc sulfide. The reaction is said to be thermodynamically controlled, and as a practical matter affords us an efficient way of separating the cations by selective precipitation as the sparingly soluble sulfides.

Now consider two gas phase reactions that appear to be similar but with puzzling differences:

$$2 \text{ NO (g)} + \text{O}_2 \text{ (g)} \longrightarrow 2 \text{ NO}_2 \text{ (g)}$$
$$2 \text{ CO (g)} + \text{O}_2 \text{ (g)} \longrightarrow 2 \text{ CO}_2 \text{ (g)}$$

Although the equilibrium constants for both reactions are large numbers and both reactions are spontaneous, the atmospheric oxidation of NO is very fast while the atmospheric oxidation of CO is very slow. Different molecular pathways are available for CO and NO oxidations, and we say the reactions are kinetically controlled. There are clear practical implications here, too. Note the sickly yellowish cast to the air during temperature inversions in urban environments due to the rapid conversion of colorless NO to the obnoxious, brownish fumes of NO_2. Note too the headaches and nausea experienced in urban traffic jams due to steadily increasing concentrations of long-lived CO molecules.

An elegant example of thermodynamic versus kinetic control is the reaction of ethyl alcohol with acid. At 140°C in dilute sulfuric acid solution, the principal reaction is dehydration of two moles of alcohol, yielding a mole of diethyl ether and a mole of water; at 180°C in concentrated sulfuric acid, the principal reaction is dehydration of one mole of ethyl alcohol, yielding one mole of ethylene and one mole of water.

At 180°C in concentrated H_2SO_4:

$$\underset{\text{ethyl alcohol}}{\text{C}_2\text{H}_5\text{OH}} \longrightarrow \underset{\text{ethylene}}{\text{C}_2\text{H}_4} + \underset{\text{water}}{\text{H}_2\text{O}}$$

At 140°C in dilute H_2SO_4:

$$2 \text{ C}_2\text{H}_5\text{OH} \longrightarrow \underset{\text{diethyl ether}}{\text{C}_2\text{H}_5\text{—O—C}_2\text{H}_5} + \text{H}_2\text{O}$$

Success in the competition between the two reactions is not due to differences in equilibrium constants but rather because one of the reactions is favored by the particular conditions. This is kinetic control. We see the products of the faster reaction.

23.2 FACTORS INFLUENCING REACTION RATES

There are a number of controllable features of chemical reactions that can be used to influence the course of events in our favor. As we have just learned, temperature and concentration are two such factors. Depending on our particular interest, by altering the reaction temperature and the sulfuric acid concentration we can favor the preparation of ethylene or diethyl ether. Both happen to be important articles of commerce; 33 billion pounds of ethylene were manufactured in 1985, though by the more economical catalytic process of cracking petroleum.

Let us now consider the reaction of mossy zinc with hydrochloric acid. This reaction is genuinely simple to follow since one of the products is a gas:

$$\text{Zn (s)} + 2 \text{ H}^+ \text{ (aq)} \longrightarrow \text{Zn}^{2+} \text{ (aq)} + \text{H}_2 \text{ (g)}$$

We begin our reaction with the addition of the metal to 1 M aqueous hydrochloric acid at 20°C. By observing the volume of gas produced, as measured by the piston (shown in Figure 23–1) rising from the syringe, we can track the course of the reaction. Since it is a rapid reaction, observations have to be made at relatively short intervals of no more than ten seconds each. Data are recorded, tabulated, organized conveniently, and finally plotted. During the first minute, hydrogen is produced at a steady

FIGURE 23–1 A simple method for following the course of a chemical reaction when one product happens to be a gas. In this case, the tube contains "mossy" zinc and hydrochloric acid:

$$\text{Zn (s)} + 2 \text{ HCl (aq)} \longrightarrow$$
$$\text{ZnCl}_2 \text{ (aq)} + \text{H}_2 \text{ (g)}.$$

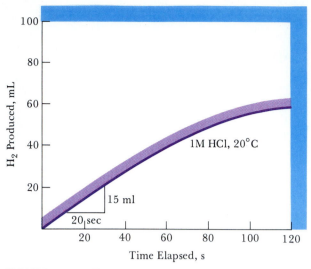

FIGURE 23–2 The slope of the line represents the time rate of change in concentration. Here, the acid is 1 M and the evolution of milliliters of hydrogen gas per unit time is measured.

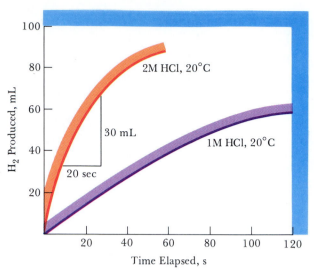

FIGURE 23–3 Data plot demonstrating concentration effects at 1 M and 2 M HCl, at constant temperature.

rate of 15 mL/20 s, after which the rate noticeably decreases, eventually falling to zero with depletion of the supply of zinc (Figure 23–2).

Concentration

Repeating the experiment without altering any essential feature except concentration gives the results shown in Figure 23–3. When 2 M HCl is used, hydrogen is again produced smoothly, but at a more rapid rate during the early part of the reaction: about 30 mL/20 s versus the 15 mL/20 s produced at the lower concentration (Figure 23–2). The effect is easily explored by adding different concentrations of acid to zinc metal in a beaker.

Dilute mineral acids such as hydrochloric and sulfuric acids react slowly (left) with zinc; more concentrated solutions (right) react rapidly; in both cases, bubbles of hydrogen gas are generated, and in addition, the appearance of steam (on the right) suggests the added vigor of that reaction.

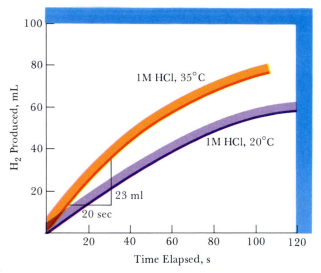

FIGURE 23–4 Data plot demonstrating temperature effects at 20°C and 35°C for 1 M HCl, at constant temperature.

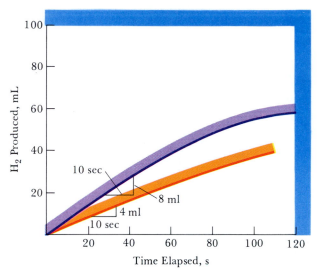

FIGURE 23–5 Data plot demonstrating effects of increased surface area, at fixed concentration and temperature.

Temperature

Repeating the initial experiment a third time, without altering any essential feature except temperature, gives the data shown in Figure 23–4. This time the temperature of the solution is raised from 20°C to 35°C. Again, hydrogen is smoothly produced during the early part of the reaction; as the figure shows, the rate at the higher temperature has increased from 15 mL/20 s to about 23 mL/20 s. This reaction clearly speeds up at the elevated temperature. However, it can be dangerous to make wide-ranging generalizations here; there are well-known reactions that are insensitive to temperature changes.

Surface Area

Again we can repeat the initial experiment with one essential difference. Instead of using several small pieces of mossy zinc, adding up to our 1-gram sample, we've chosen a single piece of zinc of nearly identical weight. The effect is to reduce the surface area (with respect to the original zinc samples). Hydrogen is once again produced smoothly during the course of the reaction, though the rate is considerably diminished (Figure 23–5). On the other hand, if the metal had been finely powdered, the reaction might have been too vigorous to carry out safely. Thus the reaction rate increases as the surface area of the metal is increased. This is generally observed in reactions of solids with liquids or gases (heterogeneous reactions) and the rate is defined in terms of moles of reactant or product per unit area of interface.

Catalytic Agents

The rates of many reactions are altered by the presence of catalysts—substances that participate in the events that bring about chemical change without undergoing permanent change themselves. Whether they serve to

(a)

(b)

Potassium permanganate oxidizes organic reactants such as glycerin. (a) Using the oxidizing agent in the usual granular form produces reaction products slowly (if at all) at room temperature; (b) however, if the potassium permanganate is finely powdered, increasing the surface area in contact with the glycerin, the rate of reaction under the same room-temperature conditions, is dramatically increased.

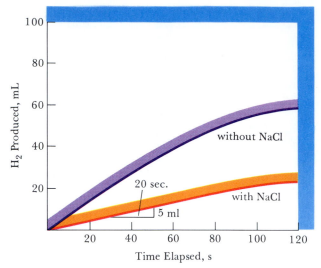

FIGURE 23–6 Data plot demonstrating the inhibiting effect of added salt, in this case NaCl.

inhibit or accelerate reaction rates, they are considered catalysts only if this happens without their being a part of the stoichiometry of the overall reaction. In the case of the Zn/HCl reaction, the presence of dissolved salts such as NaCl inhibits the rate of the reaction. You can see that by going back to our original experiment once more. Comparing it to one in which the hydrochloric acid has been saturated with common salt demonstrates the catalytic effect, reducing the rate to 5 mL/20 s (Figure 23–6).

Sodium chloride inhibition requires that the inhibiting ions be present in high concentration in order to have a significant effect. However, many catalysts, inhibitors and accelerators alike, produce considerable change in reaction rates when present in only very small or even trace quantities, which makes them particularly important. The subject is of such importance as to warrant our full attention later in this chapter.

23.3 REACTION RATES AND RATE LAWS

Variation of Rate with Time

There is a great deal of experimental evidence demonstrating that the greater the amount of reactant(s) present, the greater the extent of reaction per unit time, and consequently, the greater the rate of reaction. Remembering the law of mass action and earlier discussions of equilibrium, that should not be a surprising qualitative statement. It is a much more difficult task, however, to arrive at a quantitative statement for the rate of reaction, since the rate is different every time you look. The concentration at first changes rapidly, then slows with time, and eventually approaches some limiting value that corresponds to completion of the reaction or reaching equilibrium. And as the concentration becomes less and less, so does the slope of the line, the rate of change in the concentration with respect to time.

If a change in concentration Δc occurs in a small time interval Δt, then the rate of change in concentration during that time interval is the ratio $\Delta c/\Delta t$. When the time interval Δt is made sufficiently small, $\Delta c/\Delta t$ can be

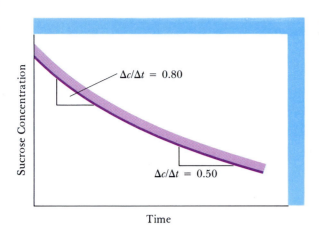

FIGURE 23–7 Time rate of change of sucrose concentration. The rate of change, $\Delta c/\Delta t$, is greater earlier in the reaction.

$\Delta c/\Delta t = 0.80$

$\Delta c/\Delta t = 0.50$

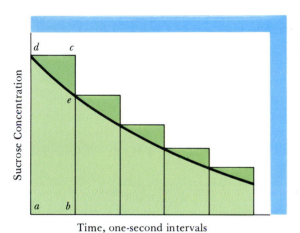

FIGURE 23–8 Changing sucrose concentration, measured over five one-second intervals. The error is the sum of the colored triangles.

regarded as the change in concentration per unit time at that particular instant and at that concentration. As an example, consider the hydrolysis of sucrose:

$$\text{sucrose} + \text{water} \longrightarrow \text{glucose} + \text{fructose}$$

The rate of the reaction is the change in sucrose concentration (mol/L) per unit time:

$$\text{rate} = \frac{\text{sucrose reacted}}{\text{observation time}} = \frac{c_2 - c_1}{t_2 - t_1} = \frac{\Delta c}{\Delta t}$$

The experimental data described in Figure 23–7 indicate that as sucrose hydrolysis proceeds, it proceeds more slowly. The rate of reaction is not constant. Consequently, the average rate does not accurately represent the rate of change at a single instant. Strictly speaking, Δt should be infinitesimally small if we want to express the instantaneous rate of change of concentration. We must keep in mind that the plot of c versus t is not linear and that $\Delta c/\Delta t$ changes with time.

To illustrate the difference between average and instantaneous values, we can construct a set of graphs using rectangular partitioning. This method approximates the area under the curve by filling the space in question with narrow vertical rectangles. In the time interval ab from 0 to 1 second (Figure 23–8), the amount of sucrose that reacts can be approximated by the area $abcd$. But the rate does not fall suddenly at $t = 1$ second from c to e. Instead,

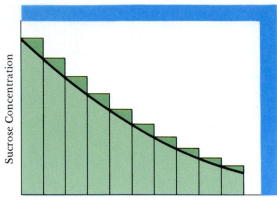

FIGURE 23–9   Changing sucrose concentration, measured over ten half-second intervals. Individually and collectively, the error—the sum of the colored triangles—is less than in Figure 23–8.

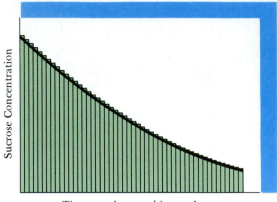

FIGURE 23–10   Changing sucrose concentration, measured over twice as many still smaller tenth-second intervals. The clear suggestion is that as the measured interval approaches the "instantaneous" change of concentration with time, the error in the measurement approaches zero.

the rate changes smoothly along the curve *de*, and the error that has been introduced corresponds to the shaded area *dce*. The error results from the fact that the indicated calculations have not taken into consideration the moment-to-moment change of concentration with time, the instantaneous change of concentration with time.

Now shorten the intervals from five 1-second intervals to ten half-second intervals and note how the error shrinks (Figure 23–9). Finally, as the time interval becomes smaller still (Figure 23–10), the summation procedure becomes more and more accurate, approaching the true value as the interval becomes infinitesimally small. At that point we can say the sucrose hydrolyzed during the time interval from zero to five seconds is the summation of all the Δc terms for an infinite number of time intervals Δt. Hence, the rate of the reaction is

$$\text{rate} = -\frac{\Delta c}{\Delta t} \text{ (very small intervals of } t\text{)}$$

with the negative sign indicating disappearance of a reactant, sucrose in this case. If instead we had chosen to examine the increasing concentration of either fructose or glucose—the formation of product rather than disappearance of reactant—the expression for the slope would be written with the sign changed (Figure 23–11):

$$\text{rate} = +\frac{\Delta c}{\Delta t}$$

Please note two important features: (1) by convention, the rate is always positive, even when describing the decrease in concentration of a reactant; (2) the particular choice of which substance to follow is based exclusively on which is most simply and accurately measured.

Rate Laws and Rate Constants

By measuring the concentration of a given component as a function of time, we get some of the information needed to determine the rate of a chemical reaction. But you'll also need to know how this change is altered

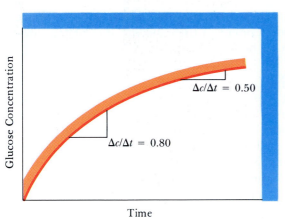

FIGURE 23–11 The rate of sucrose decomposition can be studied in terms of product formation, in this case, by observing the increasing concentration of glucose (or fructose).

by changing the concentrations of the other substances in the balanced chemical equation. There are other chemical factors that do not appear in the balanced chemical equation but that turn out to be part of the reaction mechanism, such as pH; they might also affect the rate, and they need to be included in this discussion.

We first need to say something about balanced chemical equations. More often than not, they involve multi-step pathways. Consider the following reaction. Under no circumstances can it occur in a single step:

$$6 \text{ Fe}^{2+} \text{ (aq)} + \text{Cr}_2\text{O}_7^{2-} \text{ (aq)} + 14 \text{ H}^+ \text{ (aq)} \longrightarrow$$
$$6 \text{ Fe}^{3+} \text{ (aq)} + 2 \text{ Cr}^{3+} \text{ (aq)} + 7 \text{ H}_2\text{O (liq)}$$

It is inconceivable to imagine 21 particles colliding at the same instant. The reaction must take place by a series of steps, of which the stoichiometry is only an expression of the net result.

This concept of simultaneous collisions of reacting particles is the essence of molecularity and the elementary process. There is no known elementary process that is greater than termolecular.

When the chemical equation really is a single-step reaction in which reactants go directly to products, we give it a special name, an **elementary reaction**. A classic example is the proton transfer reaction between hydronium and hydroxide ions:

$$\text{H}_3\text{O}^+ \text{ (aq)} + \text{OH}^- \text{ (aq)} \longrightarrow 2 \text{ H}_2\text{O (liq)}$$

On the other hand, apparent simplicity is no guarantee of an elementary process. The formation of water from its elements seems simple enough:

$$2 \text{ H}_2 \text{ (g)} + \text{O}_2 \text{ (g)} \longrightarrow 2 \text{ H}_2\text{O (liq)}$$

The equation suggests that water forms on collision of two hydrogen molecules and one oxygen molecule. Detailed kinetics studies have demonstrated otherwise.

To establish a general statement for the rate of a chemical reaction, let's consider the concentration of the product P that forms when A, B, and C react at a specific temperature:

$$\text{A} + \text{B} + \text{C} \longrightarrow \text{P}$$

We can write the rate law, an equation relating the reaction rate and the concentrations of reactants A, B, and C:

$$\text{rate} = \frac{\Delta[\text{P}]}{\Delta t} = k[\text{A}]^a[\text{B}]^b[\text{C}]^c$$

where the exponents a, b, and c are yet to be determined.

Suppose reactant C is excluded because we have found out (experimentally) that the rate is independent of the concentration of reactant C. Then C can be excluded from the expression, which reduces to

$$\text{rate} = \frac{\Delta[P]}{\Delta t} = k[A]^a[B]^b$$

The proportionality constant k in this generalized rate law expression is always a positive number, and it is called the **specific reaction rate constant**. It is specific to this reaction at this temperature. We could of course write a similar rate law for one of the two reactants, but that would require us to include a minus sign since the concentration of reactants would be decreasing and the rate must be a positive number by convention:

$$\text{rate} = -\frac{\Delta[A]}{\Delta t} = -\frac{\Delta[B]}{\Delta t} = k[A]^a[B]^b$$

Rate Laws and the Order of a Reaction

The **overall order** of the reaction described by a particular rate law is defined as the sum of the exponents. This would be $a + b$ for our generalized rate example. Consider the decomposition of N_2O_5

$$N_2O_5\ (g) \longrightarrow N_2O_4\ (g) + \tfrac{1}{2} O_2\ (g)$$

for which the rate law has been experimentally determined to be

$$\text{rate} = -\frac{\Delta[N_2O_5]}{\Delta t} = k[N_2O_5]$$

The rate is directly proportional to the concentration of N_2O_5. The exponent total is 1 and so the reaction is first order.

For HI decomposition, the reaction is said to be second order:

reaction: $\qquad 2\ HI\ (g) \longrightarrow H_2\ (g) + I_2\ (g)$

rate law: $\qquad \text{rate} = -\dfrac{\Delta[HI]}{\Delta t} = k[HI]^2$

The rate is directly proportional to the square of the concentration of HI. The exponent total is 2, and so the reaction is second order. Note that for this reaction the concentrations of the reactants and the products change at different rates, and this must be reflected in the rate laws as well. Every time a mole of HI disappears, half a mole of H_2 and half a mole of I_2 appear. In other words, the rate of change in the concentration of HI must be twice the rate of change of either H_2 or I_2. We can express this as follows:

$$\text{rate} = -\frac{\Delta[HI]}{2\Delta t} = +\frac{\Delta[H_2]}{\Delta t} = +\frac{\Delta[I_2]}{\Delta t}$$

For this reaction in particular we end up with

$$-\frac{\Delta[HI]}{\Delta t} = 2\,\frac{\Delta[H_2]}{\Delta t} = k[HI]^2$$

which leads to

$$\frac{\Delta[H_2]}{\Delta t} = \frac{k}{2}[HI]^2 = k'[HI]^2 \qquad \text{where } k' = \frac{k}{2}$$

In a general way, then, the rate of reaction and the specific reaction rate constant both depend on how you define them. It will facilitate matters in more complicated cases than this to note the relationship between rates of change and stoichiometric coefficients. Consider a reaction such as

$$3\ A\ +\ 2\ B\ \longrightarrow\ C\ +\ 4\ D$$

$$-\frac{\Delta[A]}{3\Delta t}\ =\ -\frac{\Delta[B]}{2\Delta t}\ =\ +\frac{\Delta[C]}{\Delta t}\ =\ +\frac{\Delta[D]}{4\Delta t}$$

For the general reaction

$$a\mathrm{A}\ +\ b\mathrm{B}\ \longrightarrow\ c\mathrm{C}\ +\ d\mathrm{D}$$

$$\mathrm{rate}\ =\ -\left(\frac{1}{a}\right)\frac{\Delta[A]}{\Delta t}\ =\ -\left(\frac{1}{b}\right)\frac{\Delta[B]}{\Delta t}\ =\ \left(\frac{1}{c}\right)\frac{\Delta[C]}{\Delta t}\ =\ \left(\frac{1}{d}\right)\frac{\Delta[D]}{\Delta t}$$

The rate of reaction is the rate of change of concentration divided by the appropriate stoichiometric coefficient for the species in question. Suppose we have a general reaction that obeys the following stoichiometric relationship:

$$A\ +\ 2\ B\ \longrightarrow\ 3\ C$$

If experimentally it was found that the reaction is directly proportional to the concentration of A, then we could write the following relationships:

$$-\frac{\Delta[A]}{\Delta t}\ =\ k[A]$$

$$-\frac{\Delta[B]}{2\Delta t}\ =\ k[A]$$

$$+\frac{\Delta[C]}{3\Delta t}\ =\ k[A]$$

or restating these equations,

$$-\frac{\Delta[A]}{\Delta t}\ =\ k[A]$$

$$-\frac{\Delta[B]}{\Delta t}\ =\ 2\ k[A]\ =\ k'[A]$$

$$+\frac{\Delta[C]}{\Delta t}\ =\ 3\ k[A]\ =\ k''[A]$$

where $k'\ =\ 2\ k$, $k''\ =\ 3\ k$, and $k'\ =\ \frac{2}{3}\ k''$.

For reactions in which more than one kind of chemical entity is found to the affect the rate, it is useful to speak of the order for each particular species, as well as the overall order. The following is known about the oxidation of nitrogen dioxide with ozone:

reaction: $\mathrm{NO_2\ (g)\ +\ O_3\ (g)\ \longrightarrow\ NO_3\ (g)\ +\ O_2\ (g)}$

rate law: $\mathrm{rate}\ =\ \dfrac{\Delta[O_2]}{\Delta t}\ =\ k[\mathrm{NO_2}][\mathrm{O_3}]$

The reaction is first order in NO_2, first order in O_3, and second order overall.

Nitrogen oxide (NO) is oxidized to nitrogen dioxide (NO_2) in a reaction with O_2:

reaction: $\mathrm{2\ NO\ +\ O_2\ \longrightarrow\ 2\ NO_2}$

rate law: $\mathrm{rate}\ =\ -\dfrac{\Delta[O_2]}{\Delta t}\ =\ k[\mathrm{NO}]^2[\mathrm{O_2}]$

The reaction is second order in NO, first order in O_2, and third order overall. As we'll have occasion to see later in the chapter, half-integer exponents can show up in rate law expressions and there are important examples of zeroth order reactions, too. And keep in mind that although the overall order is the same as the stoichiometry in all the examples so far, that is not generally true—rate laws are determined only by experiment

EXAMPLE 23–1

Consider the dichromate oxidation of iron(II). The reaction obeys a third order rate law, first order in each reacting species. Write the exponential rate law.

Solution

$$\text{rate} = k[Cr_2O_7^{2-}][Fe^{2+}][H^+]$$

■

EXERCISE 23–1(A)

For a reaction that obeys a stoichiometric equation of the type

$$2\,A + B \longrightarrow 3\,C$$

relate the rate of formation of component C to the rate of the disappearance of components A and B respectively.

Answer: Rate $= -\dfrac{\Delta[A]}{2\Delta t} = -\dfrac{\Delta[B]}{\Delta t} = \dfrac{\Delta[C]}{3\Delta t}.$ ■

EXERCISE 23–1(B)

Even reactions that display unusual stoichiometry are governed by these same rules. Determine the relationships that exist between the different ways of expressing the rate of reaction for the following chemical equation:

$$30\,CH_3OH + B_{10}H_{14} \longrightarrow 10\,B(OCH_3)_3 + 22\,H_2$$

Answer: For example,

$$\text{rate} = -\frac{\Delta[B_{10}H_{14}]}{\Delta t} = -\frac{\Delta[CH_3OH]}{30\,\Delta t} = \frac{\Delta[B(OCH_3)_3]}{10\,\Delta t} = \frac{\Delta[H_2]}{22\,\Delta t}.$$ ■

Reversibility, Molecularity, and the Order of a Reaction

When we speak of the order of a reaction, we are providing a statement about the experimental reaction rate. By contrast, molecularity is a theoretical statement about the reaction's mechanism. When we speak of **molecularity**, we mean the number of molecules that come together in an elementary process. Each step in a multi-step process has its own molecularity.

The exponents in the rate law are experimentally arrived at by testing the effect of concentration on rate. If the balanced equation happens to represent a direct reaction that occurs in one step (as written), we call it an elementary process, and these same exponents turn up as coefficients in the rate law. If a reaction is reversible, the rates forward and reverse must be taken into account. Remember, as reactant disappears it is being replaced as the product of the reverse reaction. For example, consider (Figure 23–12) dissociation of the diatomic HI molecule:

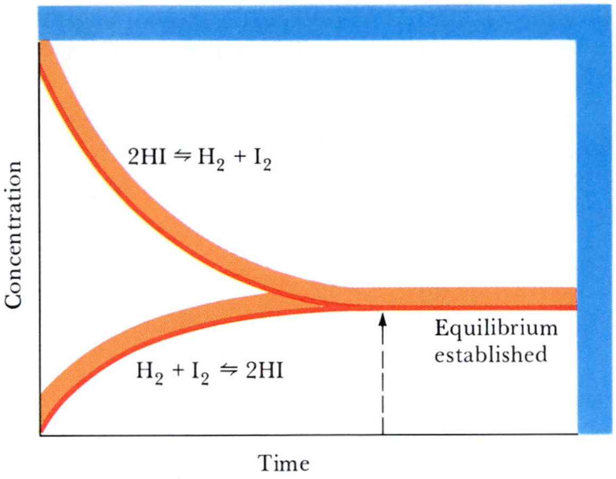

FIGURE 23–12 Graphical representation of the approach to equilibrium for the H_2 (g) + I_2 (g) $\rightleftharpoons$ 2 HI (g) reaction. The equilibrium can be approached from either direction.

$$2 \text{ HI (g)} \underset{k_{-1}}{\overset{k_1}{\rightleftharpoons}} \text{H}_2 \text{ (g)} + \text{I}_2 \text{ (g)}$$

The rate law is given by

$$\frac{\Delta[\text{H}_2]}{\Delta t} = \frac{\Delta[\text{I}_2]}{\Delta t} = k_1[\text{HI}]^2 - k_{-1}[\text{H}_2][\text{I}_2]$$

The forward (k_1) and reverse (k_{-1}) steps are elementary processes. Each involves a two-particle collision (HI + HI or H_2 + I_2); therefore, both are bimolecular reactions.

EXAMPLE 23–2

Consider the kinetic analysis of the oxidation of nitrogen oxide and determine the experimental rate law:

$$2 \text{ NO (g)} + \text{O}_2 \text{ (g)} \longrightarrow 2 \text{ NO}_2 \text{ (g)}$$

A termolecular collision takes place, leading directly to reaction products.

Solution   Remember that a termolecular reaction is always third order, but a third order reaction isn't necessarily termolecular:

$$2 \text{ NO (g)} + \text{O}_2 \text{ (g)} \longrightarrow 2 \text{ NO}_2 \text{ (g)}$$

The experimental third order rate law is

$$\text{rate} = k[\text{NO}]^2[\text{O}_2] \qquad \blacksquare$$

EXERCISE 23–2

The following reactions are elementary. Give the order and molecularity:
(a) 2 NO (g) + Cl_2 (g) $\longrightarrow$ 2 NOCl (g)

(b) 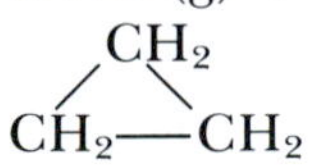 $\underset{\text{CH}_2-\text{CH}_2}{\overset{\text{CH}_2}{\diagup \diagdown}}$ (g) $\longrightarrow$ $CH_3CH{=}CH_2$ (g)

(c) CH_3I (g) + HI (g) $\longrightarrow$ CH_4 (g) + I_2 (g)
(d) The following reaction is second order. What can you state about the molecularity?

$$2 \text{ NO} + \text{Br}_2 \longrightarrow 2 \text{ NOBr} \qquad \blacksquare$$

In summary, we can say that:

1. Molecularity defines the rate law and the mechanism for an elementary reaction, giving the corresponding order directly: biomolecular is second order, and termolecular is third order. True unimolecular reactions involve only one particle, for example radioactive decay, and are rare. They are first order.

2. The converse of these statements is not true. Many kinetically second order reactions do not follow the simple bimolecular mechanism of an elementary process. Thus, we cannot use the form of the rate law to predict the mechanism of a reaction; however, knowing the mechanism does allow one to deduce the rate law.

3. We cannot determine the order by inspecting the balanced chemical equation for the reaction; order must be determined by experiment.

The Rate-Determining Step

On mixing an aqueous solution of hydrogen peroxide with iodide ion, the peroxide solution decomposes with evolution of oxygen. The stoichiometric equation represents the overall reaction:

$$2\ H_2O_2\ (aq) \longrightarrow 2\ H_2O\ (liq)\ +\ O_2\ (g)$$

But the reaction takes place in two steps, a slow step followed by a fast reaction:

slow first step: $\qquad H_2O_2\ +\ I^-\ \xrightarrow{k_1}\ H_2O\ +\ OI^-$

fast second step: $\qquad OI^-\ +\ H_2O_2\ \xrightarrow{k_2}\ H_2O\ +\ O_2\ +\ I^-$

The stoichiometry is satisfied because the sum of these two consecutive, essentially irreversible and elementary reactions is the balanced equation. As you might expect, if one step in a multi-step reaction is much slower than all the others, it is the **rate-determining step** in product formation. Thus the rate of oxygen production is determined by the rate of the slow first step.

To complete the kinetic analysis, the rate-determining step is bimolecular because two molecules collide. Only reactants appear in the rate law because each step is irreversible, so we do not need to consider either reverse reaction and we can write the rate law from that first, slow, rate-limiting step:

$$\text{rate}\ =\ -\frac{\Delta[H_2O_2]}{\Delta t}\ =\ k_1[H_2O_2][I^-]$$

The reaction is second order overall, first order in hydrogen peroxide and first order in iodide ion. If we had chosen to do so, we could have expressed the rate of reaction equally well in terms of the formation of product oxygen:

$$\frac{\Delta[O_2]}{\Delta t}\ =\ -\frac{\Delta[H_2O_2]}{2\Delta t}\ =\ \tfrac{1}{2}k_1[H_2O_2][I^-]$$

In many multi-step reactions, however, no one step is much slower than any other; and so we are left with the rate-determining step as the elementary reaction that happens to dominate the overall kinetics. Being "slowest" does not make it rate-determining.

The Order and the Rate Constant for a Chemical Reaction

The rate laws themselves offer suggestions on how to find the order and the rate constant for a chemical reaction. For a general reaction in which

$$aA + bB \longrightarrow products$$

to determine the order with respect to A, make up a series of solutions in which the concentration of A is varied while maintaining the concentration of B at some fixed value. Then measure the initial rate of reaction for each solution by finding the change in concentration of any one of the reactants occurring in the first, smallest time interval after mixing reactants together.

Since our small series of experiments contains only one variable, the initial concentration of A, the computed initial rates should be directly proportional to $[A]^m$. If the rate of reaction doubles on doubling the concentration of A, then the rate depends on the first power of the concentration of A; m equals 1; and we say the reaction is first order in A. Should the reaction rate increase by a factor of four on doubling the concentration of A, then m must equal two, and the reaction is said to be second order in A.

Once the order has been found with respect to A, the procedure is repeated, fixing the concentration of A while the concentration of B is varied. That allows us to deduce the order of the reaction with respect to B.

Finally, knowing the order with respect to each reagent, we can (1) calculate the overall order of the reaction as the sum of the exponents and (2) evaluate the specific reaction rate constant (k) as the ratio of the measured rate of reaction and the product of the concentrations of the reactants A and B, raised to the appropriate powers:

$$\text{rate} = -\frac{\Delta[A]}{\Delta t} = k[A]^m[B]^n$$

$$k = \frac{\text{rate}}{[A]^m[B]^n}$$

Remember that m and n equal the coefficients a and b in the overall chemical equation only if it happens to be an elementary reaction.

EXAMPLE 23–3

The data below are for the process A + B reacting directly to products. Based on the data provided, determine the overall order, the general rate law, and the specific reaction rate constant:

$$A + B \longrightarrow products$$

Experiment	#1	#2	#3	#4	#5
[A], M	0.01	0.01	0.01	0.02	0.03
[B], M	0.01	0.02	0.03	0.01	0.01
rate, M/min	0.05	0.10	0.15	0.20	0.45

Solution First compare experiments #1, #2, and #3. Since the concentration of A doesn't change, the observed two-fold and three-fold increase in the rate must be due to the two-fold and three-fold increase in the concentration of B. We describe the reaction as first order in component B.

Next we compare experiments #1, #4, and #5. Since the concentration of B doesn't change, the observed four-fold and nine-fold increase in the reaction rate must be due to the corresponding two-fold and three-fold increases in the concentration of A, leading to the conclusion that the rate of the reaction is directly proportional to the square of the concentration of A.

The overall reaction would be described as third order and written as a general rate law:

$$\text{rate} = k[A]^2[B]$$

For the specific reaction rate constant, any of the five experiments can provide needed data. From experiment #1,

$$k = \frac{\text{rate}}{[A]^2[B]} = \frac{5.0 \times 10^{-2} \text{ M/min}}{[0.01 \text{ mol/L}]^2[0.01 \text{ mol/L}]}$$
$$k = 5.0 \times 10^4 \text{ M}^{-2} \text{ min}^{-1}$$

EXERCISE 23–3(A)

Evaluate the overall order of the reaction from the following data:

$$A + B + C \longrightarrow \text{products}$$

Experiment	#1	#2	#3	#4
[A], M	1.0	2.0	1.0	1.0
[B], M	1.0	1.0	2.0	1.0
[C], M	1.0	1.0	1.0	2.0
rate, M/min	1.0	2.0	1.0	1.0

EXERCISE 23–3(B)

Again, consider the reaction of A, B, and C to give products, only this time it is found that the rate doubles when either the concentration of A or B doubles, but the rate is halved when the concentration of C doubles. Write a rate law that is consistent with these data. ∎

EXERCISE 23–3(C)

Show the relationships between the reaction rate constant and the rates of the reaction as expressed in terms of disappearance of A and of B, and the appearance of C, respectively, for a reaction

$$3A + B \longrightarrow 2C$$

known to be first order in A and first order overall. ∎

23.4 EXPERIMENTAL RATE LAWS

Virtually all the interesting information about reaction rates can be obtained by using rate laws. For the most part, this method depends on being able to derive simple algebraic expressions for the time dependence of concentration from the rate laws. In the next sections, we'll try to show you what that means for several kinds of chemical reactions.

The theory on which the method is based derives from the fact that the instantaneous reaction rate is proportional to a power of the concentrations. That can be restated as

$$\text{rate} = k[A]^n$$

Taking the logarithm of both sides of the equation gives

$$\log \text{rate} = \log k + n \log[A]$$

Plotting log rate versus log[A] gives a straight line for a reaction of simple order. From the point where the line intercepts the y-axis, we obtain the value for the log of the specific reaction rate constant k. The slope of the line gives the value of n, the order of the reaction. Several examples and exercises follow for kinetically first order and simple second order reactants.

Rate Law for a Reaction of the First Order

For a first order reaction, the rate depends on the first power of the concentration of only one reactant. The decomposition of dinitrogen pentoxide is a classic example of a first order reaction. It has been studied extensively in the gas phase and in solution. To describe the change that takes place, we can write the following stoichiometric equation:

$$2 \, N_2O_5 \, (g) \longrightarrow 4 \, NO_2 \, (g) \, + \, O_2 \, (g)$$

A great deal of experimental evidence has shown the rate to be accurately described by the first order rate law expression

$$-\frac{\Delta[N_2O_5]}{2\Delta t} = k[N_2O_5]$$

For the general case

$$A \longrightarrow \text{products}$$

the rate law for decomposition of A is

$$\text{rate} = -\frac{\Delta[A]}{\Delta t} = k[A]$$

k is the first order rate constant, and it can be shown that

$$\ln\left(\frac{[A]_0}{[A]}\right) = kt$$

where $[A]_0$ is the initial concentration at time $t = 0$, and $[A]$ is the concentration some time t later. The expression represents a linear function, indicating the constant nature of the fractional decrease of $[A]$ with time. A plot of $\ln([A]_0/[A])$ versus t gives a straight line with slope equal to k for any first order reaction. In other words, adherence to a general equation of this form is an essential criterion for a first order reaction.

The plots in the following figures are typical of kinetics data for a first order decomposition. Figure 23–13a shows concentration decreasing rapidly with time from some initial concentration $[A]_0$ but with the rate slowing all the while, and finally approaching zero or some equilibrium value $[A]_\infty$. A plot of $\ln[A]$ or $\log[A]$ versus time should be a straight line with a slope of $-k$ or $-k/2.3$, as is shown in Figure 23–13b. The specific reaction rate constant can be determined from Figure 23–13b by finding the slope of the line.

> There are some benefits in remembering the equation with the minus sign, $\ln([A]/[A]_0) = -kt$, because it shows that the logarithm of the concentration of A decreases with time, and if $t = 0$, the intercept is $\ln[A]_0$.

The Half-life for a First Order Reaction

The half-life $(t_{1/2})$ for a chemical reaction is the time required for half the reactant initially present to decompose. Start with the general rate law for a first order reaction,

$$\ln\left(\frac{[A]_0}{[A]}\right) = kt$$

FIGURE 23–13 (a) Plot of concentration versus time for a typical chemical reaction; (b) for a first order reaction, a plot of ln [A] versus time is a straight line whose slope is $-k$.

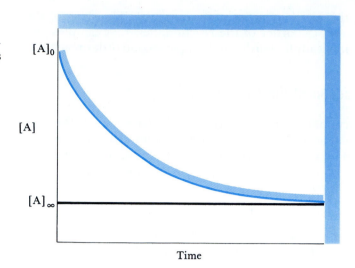

But we are interested in the special case where [A] is 0.50 at time $t_{1/2}$, which means that the rate law reduces to a simple statement:

$$\ln \frac{1}{0.5} = kt_{1/2} = \ln 2$$

$$t_{1/2} = \frac{0.693}{k}$$

To help illustrate the concept of half-life, consider the fate of a million molecules undergoing a first order decomposition, with $t_{1/2}$ equal to 1 minute (Figure 23–14a). During the first minute, 500,000 molecules disappear along the route to products, leaving an equal number of molecules that have not reacted. During the next one-minute interval, half the remaining molecules react, leaving 250,000 that have not reacted. After three minutes, or three half-life intervals, 125,000 remain. Then 62,500; 31,250; and so forth.

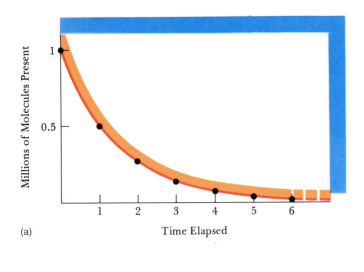

(a)

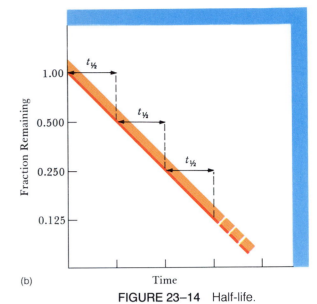

(b)

FIGURE 23–14 Half-life.
(a) During each interval, half the remaining molecules decay, decompose, or otherwise react.
(b) Here, three successive half-lives are marked off.

The concentration of reactant A remaining at any time t (not just integer multiples of the half-life) can be found by rearranging and exponentiating the general first order rate law to get

$$[A] = [A]_0 e^{-kt}$$

That is, **first order decay is an exponential decay process**.

Note that in our equation, there is no concentration term, simply:

$$t_{1/2} = \frac{0.693}{k}$$

The half-life is independent of concentration for a kinetically first order process and can be quickly determined from any pair of concentrations $[A]_1$ and $[A]_{1/2}$. Because concentration independence of the half-life does not occur for any other order reaction, it is the unique signature of a first order reaction. Graphically, this can be represented in an interesting fashion by modifying the preceding graph, turning it into a semilogarithmic plot (Figure 23–14b). As you can see, the half-life can be determined for any pair of concentrations that differ by a factor of two. Once the half-life is known (by whatever means), the reaction rate constant can be simply computed from $t_{1/2} = 0.693/k$.

EXAMPLE 23–4

Sulfuryl chloride spontaneously decomposes at 600 K according to the equation

$$SO_2Cl_2 \text{ (g)} \longrightarrow SO_2 \text{ (g)} + Cl_2 \text{ (g)}$$

The first order rate constant is 1.3×10^{-3} min^{-1}. After heating a certain sample for 30 minutes at 600 K, what fraction of the original sulfuryl chloride sample remains? How long will it take for 50% of the original sample to decompose to products under these same conditions?

Solution

$$\ln\left(\frac{[A]_0}{[A]}\right) = kt = (1.3 \times 10^{-3} \text{ min}^{-1})(30 \text{ min}) = 3.9 \times 10^{-2}$$

$$\frac{[A]_0}{[A]} = 1.04$$

$$\frac{[A]}{[A]_0} = \frac{1}{1.04} = 0.96$$

The half-life is

$$t_{1/2} = \frac{0.693}{k} = \frac{0.693}{1.3 \times 10^{-3} \text{ min}^{-1}} = 533 \text{ min}$$ ■

EXERCISE 23–4

If $t_{1/2}$ for a first order reaction is 600 seconds, *approximately* how long will it take for 99% of the reactant molecules to decompose to products?   ■

Rate Law for a Reaction of the Second Order

For a second order reaction, the rate of reaction must be affected by the concentrations of two species. Both may be the same, as in decomposition and dimerization, or they may be different as is the case for hydrolysis.

For the case in which both molecules are the same:

$$2 \text{ A} \longrightarrow \text{products}$$

the rate law is

$$\text{rate} = -\frac{\Delta[A]}{\Delta t} = k[A]^2$$

For a starting concentration $[A]_0$ and a concentration of $[A]$ at time t, it can be proved that

$$\frac{1}{[A]} - \frac{1}{[A]_0} = kt \qquad \text{or} \qquad \frac{[A]_0 - [A]}{[A]_0 [A]} = kt$$

To illustrate the application of this rate law, we shall consider two reactions, one in solution and the other in the gas phase. The first is the conversion of the inorganic salt ammonium cyanate (NH_4OCN) into the organic compound urea (NH_2CONH_2). Discovered in 1828 by Friedrich Wöhler, the proof that this reaction occurs marked the birth of modern organic and biochemistry (Chapters 25 and 26).

$$NH_4OCN \text{ (aq)} \longrightarrow NH_2CONH_2 \text{ (aq)}$$

The data in Table 23–1 have been reported by Barrow for a solution initially containing 22.9 g/L of NH_4OCN.

To confirm that the reaction is kinetically second order, either (1) plot $1/[A]$ versus t and evaluate the linearity of the best line drawn through the points, or (2) algebraically evaluate

$$\left(\frac{[A]_0 - [A]}{[A]_0[A]}\right)\left(\frac{1}{t}\right) = k$$

for each value of t and check the constancy of k for these solutions. To begin with,

$$[A]_0 = (22.9 \text{ g/L})\left(\frac{1 \text{ mol}}{60.0 \text{ g}}\right) = 0.381 \text{ mol/L} = 0.381 \, M$$

and since the number of grams of NH_2CONH_2 formed equals the number

TABLE 23–1 Conversion of Ammonium Cyanate to Urea in Aqueous Solution

time (min)	0	20	50	65	150
urea formed (g)	—	7.0	12.1	13.8	17.7
ammonium cyanate concentration	0.381	0.264	0.180	0.151	0.086
calculated rate constant, k	—	0.058	0.059	0.062	0.060

$$k = \frac{1}{t} \cdot \frac{[A]_0 - [A]}{[A]_0[A]}$$

of grams of NH_4OCN that reacted, we can list the NH_4OCN concentrations at the several times noted in Table 23–1, and calculate the k values accordingly. The calculated k values fall into a near-constant range with an average value of

$$k = 0.060 \ M^{-1} \ min^{-1}$$

and the general rate law is

$$rate = k[NH_4OCN]^2$$

which says that the rate of reaction is second order in ammonium cyanate.

Consider the decomposition of acetaldehyde, which takes place at 791 K:

$$CH_3CHO \ (g) \longrightarrow CH_4 \ (g) + CO \ (g)$$

Time, s	0	42	105	242	480	840
Pressure, torr	363	397	437	497	557	607

The second order rate constant can be calculated from these data:

1. The initial pressure (at the time $t = 0$) represents the concentration of acetaldehyde at the beginning of the experiment, before decomposition commences.

2. We can use the increasing pressure that we observe as a measure of the amount of acetaldehyde decomposed. This is the case because a mole of CH_4 and a mole of CO form for every mole of CH_3CHO decomposed. Thus the increase in total pressure is equal to the decrease in the partial pressure of acetaldehyde.

3. Let x be the measure of the acetaldehyde decomposed.

	CH_3CHO	$\longrightarrow$	CH_4	+	CO
Initial	p_0		0		0
Final	$p_0 - x$		x		x

$$p_{tot} = (p_0 - x) + x + x = p_0 + x$$
$$x = p_{tot} - p_0 = 397 - 363 = 34 \ torr \ at \ t = 42 \ s$$

4. The existing acetaldehyde partial pressure is

$$p = p_0 - x = 363 - 34 = 329 \ torr \ at \ t = 42 \ s$$

5. Now the rate constant from the expression for a second order reaction can be calculated in terms of partial pressure for a gas-phase reaction:

- At $t = 42$ s,

$$\frac{1}{p} - \frac{1}{p_0} = kt$$

$$\frac{1}{329} - \frac{1}{363} = k(42)$$

$$k = 6.8 \times 10^{-6} \text{ torr}^{-1} \text{ s}^{-1}$$

- At $t = 105$ s,

$$\frac{1}{289} - \frac{1}{363} = k(105)$$

$$k = 6.7 \times 10^{-6} \text{ torr}^{-1} \text{ s}^{-1}$$

- At $t = 242$ s, $k = 6.7 \times 10^{-6} \text{ torr}^{-1} \text{ s}^{-1}$
- At $t = 480$ s, $k = 6.6 \times 10^{-6} \text{ torr}^{-1} \text{ s}^{-1}$
- At $t = 840$ s, $k = 6.7 \times 10^{-6} \text{ torr}^{-1} \text{ s}^{-1}$

These data fit the empirical formula for a second order reaction as demonstrated by the constancy of k. That kind of information could not have been extracted from the simple equation representing the stoichiometry.

The half-life for a second order reaction is concentration-dependent. That is in marked contrast to reactions that follow first order chemical kinetics. For the reaction

$$2\,A \longrightarrow \text{products}$$

the rate law is

$$\text{rate} = -\frac{\Delta[A]}{\Delta t} = k[A]^2$$

for which we can write

$$kt = \frac{1}{[A]} - \frac{1}{[A]_0}$$

$$kt_{1/2} = \frac{2}{[A]_0} - \frac{1}{[A]_0}$$

and the half-life is

$$t_{1/2} = \frac{1}{k[A]_0}$$

EXAMPLE 23–5

Consider the dimerization or coupling of butadiene (C_4H_6):

$$2\,C_4H_6\,(g) \longrightarrow C_8H_{12}\,(g)$$

Butadiene is introduced into an empty flask at 600 K at some time $t = 0$. Dimerization begins, and we follow the progress of the coupling reaction by observing the pressure, which decreases as each pair of molecules of butadiene monomer becomes one molecule of dimer. Show that the data in Table 23–2 and Figure 23–15 support the rate law

TABLE 23–2   Gas Phase Dimerization of Butadiene

Time (s)	p_{tot} (torr)	$p_{C_4H_6} = p_0 - 2x$* (torr)	k (torr^{-1} sec^{-1})
0	632.0	—	—
367	606.6	581.2	3.77×10^{-7}
731	584.2	536.4	3.86×10^{-7}
1038	567.3	502.6	3.92×10^{-7}
1751	535.4	438.8	3.98×10^{-7}
2550	509.3	386.6	3.93×10^{-7}

*Can you explain why $p(C_4H_6) = p_0 - 2x$?

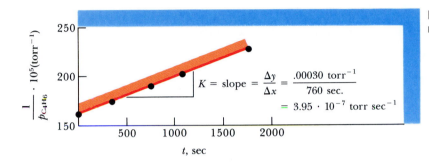

$$K = \text{slope} = \frac{\Delta y}{\Delta x} = \frac{.00030 \ \text{torr}^{-1}}{760 \ \text{sec.}}$$
$$= 3.95 \cdot 10^{-7} \ \text{torr sec}^{-1}$$

FIGURE 23–15   Second order plot for the dimerization of butadiene.

$$-\frac{\Delta[C_4H_6]}{2\Delta t} = k[C_4H_6]^2$$

Solution   Let $x = pC_8H_{12}$

	2 C$_4$H$_6$	$\longrightarrow$	C$_8$H$_{12}$
Initial	p_0		0
Final	$p_0 - 2x$		x

$$p_{tot} = (p_0 - 2x) + x = p_0 - x$$
$$x = p_0 - p_{tot}$$
$$p_{C_4H_6} = p_0 - 2x = p_0 - 2(p_0 - p_{tot}) = 2p_{tot} - p_0$$

At $t = 367$ s, the second order rate law predicts

$$\frac{1}{p} - \frac{1}{p_0} = kt$$

$$\frac{1}{581.2} - \frac{1}{632.0} = k(367)$$
$$k = 3.78 \times 10^{-7} \ \text{torr}^{-1} \ \text{sec}^{-1}$$

Similar computations for the other times yield the k values in Table 23–2. The constancy of the k values confirms that the reaction is second order. ▪

EXERCISE 23–5

Dinitrogen oxide (N$_2$O) was thermally decomposed at 650°C. With the initial N$_2$O pressure recorded at 300 torr, it was found that 50% reacted in 250 seconds. When the initial N_2O pressure was 350 torr, it only took 214 seconds for 50% to react.

(a) Is the reaction first or second order? Explain.

(b) With the initial N_2O pressure set at 1 atm, how much time would be required to achieve the same 50% decomposition?

Answer: (a) Second order because the half-life depends on the concentration or pressure. (b) 99 s.   ■

A Zero Order Reaction

If the rate of reaction is independent of the concentration of any component, the reaction is described as zero order. Consider the reaction

$$A \longrightarrow products$$

If the reaction is zero order we may write the rate law as

$$-\frac{\Delta[A]}{\Delta t} = k$$

This can be shown to lead to the relationship

$$[A]_0 - [A] = kt$$

The reaction will be zero order if a plot of $[A]_0 - [A]$ versus t yields a straight line whose slope is k, directed toward the origin. This is typical of many important industrial processes involving heterogeneous gas phase catalysis at surfaces. For example, the platinum metal-catalyzed addition of hydrogen to the double bond in ethylene (catalytic hydrogenation) is zero order.

The Quantitative Aspects of Radioactivity

Almost immediately after Becquerel's discovery of natural radioactivity, Rutherford observed that decay rates (measured by emission of α- and β-particles) were not constant. Rates decreased exponentially with time, and therefore must describe statistical processes.

Radioactive decay is a purely random happening. There is a statistical probability that a given atom will decay within a given time, without regard to origins or present environment of the nuclide in question. The decay rate for a particular atomic nucleus is solely a decreasing function of the number of nuclei present. If twice as many are present, there will be twice the number of disintegrations per unit time.

It is useful to review a few of the ideas presented in this chapter as they pertain to rates of decay of radioactive nuclei. The equation for radioactive decay is

$$\text{rate} = \frac{\Delta N}{\Delta t} = -\lambda N$$

where N is the number of radioactive atoms present at time t and λ is the disintegration (or decay) constant for the radioisotope—the specific reaction rate constant. Thus,

$$N = N_0 e^{-\lambda t}$$

where N_0 is the number of atoms present at time zero.

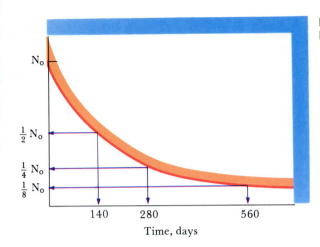

Note that the number of nuclei that have disappeared at time t follows an exponential decay plot. If we plot the number of radioactive nuclides at given times on linear graph paper, a classic exponential curve is produced (Figure 23–16). Similarly, a plot of $\ln(N_0/N)$ versus t should give a straight line, with slope equal to λ.

The half-life can be calculated as we did for a first order reaction:

$$\frac{N}{N_0} = e^{-\lambda t} \text{ (by simple rearrangement)}$$

$$\ln\left(\frac{N}{N_0}\right) = -\lambda t \text{ (taking the natural log)}$$

$$\ln\left(\frac{0.5\, N_0}{N_0}\right) = -\lambda t_{1/2} \text{ (for the special case where half the atoms present have decayed)}$$

$$\ln 2 = \lambda t_{1/2} \text{ (clearing terms)}$$

$$\frac{0.693}{\lambda} = t_{1/2}$$

Note as well that the number of atoms left after time $t_{1/2}$ has elapsed is half the original number present; after $2t_{1/2}$ has elapsed, half of half or a quarter of the original number are present. It should be apparent that the fraction remaining after n half-lives have elapsed is $(1/2)^n$.

EXAMPLE 23–6

A small particle of cotton cloth taken from an old painting had a decay rate that was 92% that of the newly spun cloth. The radionuclide responsible for the activity is ^{14}C, for which $t_{1/2}$ is 5720 yr. What is the approximate age of the painting?

Solution

$$\lambda = \frac{0.693}{t_{1/2}} = \frac{0.693}{5720 \text{ yr}} = 0.000121 \text{ yr}^{-1}$$

Then substituting back into the general relationship,

$$\ln\left(\frac{N}{N_0}\right) = -\lambda t$$

$$t = \left(\frac{-1}{\lambda}\right) \ln \left(\frac{N}{N_0}\right) = \left(\frac{-1}{0.000121}\right) \ln (0.92) = 6.9 \times 10^2 \text{ yr}$$

The age of the painting is approximately 690 years. ■

EXERCISE 23–6

After 16,000 years, what percentage of the ^{14}C radionuclide will remain? *Answer:* 14%. ■

Geochronology

With the discovery of radioactivity, it became possible in principle to determine the age of rock formations in the Earth's crust. In practice, modern isotopic geochronometric methods using potassium, rubidium, and uranium nuclides do provide reliable results, accurate to within a few percent, permitting reconstruction of the geologic history of the Earth's crust in many parts of the world. Carbon isotopes have been successfully used to establish the chronology of life on Earth.

In the simplest scenario, a mineral sample containing radioactive (R) parent atoms, but as yet none of the stable (S) daughter atoms, crystallizes from a silicate melt during formation of an igneous (volcanic) rock. The chronometer starts its count as R atoms steadily disintegrate into S atoms. Time passes, and the larger the interval between mineral formation and the moment of measurement, the larger the ratio of S to R atoms. Beginning with our earlier relationship for nuclear disintegrations, an age relationship formula can be derived:

$$\begin{aligned}
t &= \left(-\frac{1}{\lambda}\right) \ln \left(\frac{N}{N_0}\right) \\
&= \left(\frac{1}{\lambda}\right) \ln \left(\frac{N_0}{N}\right) \\
&= \left(\frac{1}{\lambda}\right) \ln \left[\frac{R + S}{R}\right] \\
&= \left(\frac{1}{\lambda}\right) \ln \left[1 + \frac{S}{R}\right]
\end{aligned}$$

where t is the age of the Earth, provided that the R (parent) and S (daughter) atoms are quantitatively separate at the moment of mineral formation, and that the mineral has remained a closed system through its history.

There are three very important chronometers—parent/daughter pairs—that are used in determining isotopic ages by this general approach. As one can see from the comparative data (Table 23–3), results are in good agreement among the different analyses.

The widespread occurrence of potassium minerals makes the K/Ar pair (potentially) one of the best chronometers. But ^{40}K decays by two pathways, beta-decay to ^{40}Ca and K-capture to ^{40}Ar, and that complicates the calculations, especially since ^{40}Ca is so widespread. The radioactive transformation of ^{87}Rb to ^{87}Sr by beta-decay, on the other hand, is straightforward and equally reliable.

$$^{87}Rb \longrightarrow {}^{87}Sr + \beta^- + \nu$$

Naturally occurring rubidium contains 27.85% ^{87}Rb, which has a half-life

TABLE 23–3 **Isotopic Ages of Unaltered Rocks from the Bearfoot Mountains of Montana**

Primary Geochronometer	Age (10^9 yr)
K/Ar	2.50
Rb/Sr	2.74
$^{238}U/^{206}Pb$	2.60
$^{235}U/^{207}Pb$	2.64

TABLE 23–4 **Half-Lives of Parent Nuclides Used for Geologic Dating**

R/S	$t_{1/2}$ for R	Suitable Minerals and Rocks
$^{87}Rb/^{87}Sr$	5.2×10^{10}	muscovite, biotite, K-feldspar, lepidolite, glauconite
$^{40}K/^{40}Ar$	1.3×10^9	muscovite, biotite, glauconite
$^{238}U/^{206}Pb$	4.51×10^9 ⎱	uraninite, monazite, zircon,
$^{235}U/^{207}Pb$	7.13×10^8 ⎰	black shale

of 5.2×10^{10} years. This is on the order of the age of the Earth. Consider a feldspar formation at the moment it solidified in the Earth's crust, when N_0 ^{87}Rb atoms are present. The chronometer begins to wind down to the present as beta-particles are emitted. The following example develops its chronometric equation.

EXAMPLE 23–7

Write a general equation that gives the geologic age of a rock based on the $^{87}Rb/^{87}Sr$ ratio.

Solution   The number N_0 of ^{87}Rb nuclides originally present must equal the sum of the ^{87}Rb and ^{87}Sr nuclides present in the sample today; N is the existing number of ^{87}Rb nuclides:

$N_0 = (^{87}Rb + {}^{87}Sr)$ present today

$N = {}^{87}Rb$ present today

Using our generalized equation for radioactive decay,

$$\ln\left(\frac{N_0}{N}\right) = \lambda t$$

$$\ln\left[\frac{{}^{87}Rb + {}^{87}Sr}{{}^{87}Rb}\right] = \lambda t$$

Since we know $t_{1/2}$ from experiment (Table 23–4), we can find the decay constant:

$$\lambda = \frac{0.693}{t_{1/2}} = \frac{0.693}{5.2 \times 10^{10} \text{ yr}} = 1.3 \times 10^{-11} \text{ yr}^{-1}$$

Finally, solving for t gives the age of the sample and (presumably) geologic time:

$$t = \left(\frac{1}{\lambda}\right) \ln \left[1 + \frac{^{87}Sr}{^{87}Rb}\right] = (7.5 \times 10^{10} \text{ yr}) \ln \left[1 + \frac{^{87}Sr}{^{87}Rb}\right]$$

EXERCISE 23–7

Even though formation of ^{40}Ca from ^{40}K cannot be used to date minerals, certain rocks are known to trap ^{40}Ar, and so reliable geochronometric measurements can be based on the ^{40}K-to-^{40}Ar ratio. Ignoring the fraction of ^{40}K that decays into ^{40}Ca, develop a generalized expression for this clock and find, for the samples in Table 23–3
(a) the $(^{40}Ar + {}^{40}K)/^{40}K$ ratio
(b) the $^{40}Ar/^{40}K$ ratio
(c) the isotopic % ^{40}Ar in the sample
Answer: (a) 3.8; (b) 2.8; (c) 10.1%.

The geochronometric method based on uranium is the oldest and most elegant, but because it depends on three radioactive isotopes (^{238}U, ^{235}U, ^{232}Th) it is also more complex. Suffice it to say the method consistently produces data establishing the age of the Earth as about 4.5×10^9 years. It is interesting to note as well that stone meteorites and moon rocks give identical ages since solidification.

EXAMPLE 23–8

A certain uranium core was found to have 0.865 g of ^{206}Pb per gram of ^{238}U. What is the minimum age of the deposit, and what assumptions must be made to give this estimate?

Solution   Since the atomic ratio is 1:1, half of the original ^{238}U must have decayed, and so the ore deposit must have been undisturbed for $t_{1/2}$ for ^{238}U, or 4.5×10^9 years. We assume:
(a) No ^{206}Pb was deposited in the beginning.
(b) None of the radioactive daughters of the several decay processes, particularly gaseous radon, leached out or were otherwise lost from the rock (and thus lost to later accounting).
(c) The half-lives of the daughter products are short compared to the half-life of the parent.

EXERCISE 23–8

The weight ratio of uranium-235 to lead-207 in a certain mineral deposit was found to be 1:1. Would you expect the mole ratio of ^{235}U to ^{207}Pb to be greater than, equal to, or less than 1:1?

Cosmic Rays and Radiocarbon Dating

Most of the ^3H in the atmosphere comes from nuclear weapons tests in the 1950s and 1960s.

The primary long-lived radioactive nuclides that have survived since nucleogenesis, along with their short-lived descendents, are not the only radioactivities in nature. High energy cosmic rays produce violent nuclear reactions near the top of the atmosphere. Neutrons, protons, alpha-particles, and many other fragments form. Among the fragments found in significant yields (by radiochemical methods) are ^3H and ^{14}C. Careful investigations indicate there is a small but unchanging concentration of tritium in surface water and rain. With a half-life of only 12.3 years, tritium has at least one interesting use as a chronometer—in wine-dating (Figure

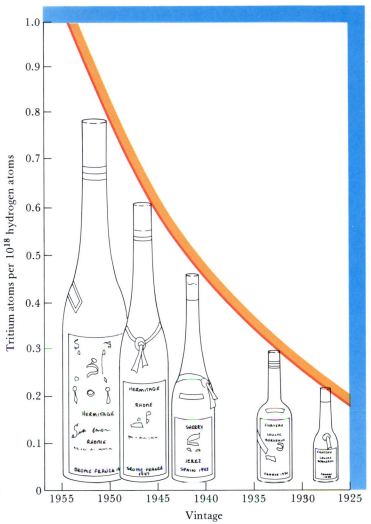

FIGURE 23–17 Age and tritium activity in various wines. Because of the relatively small half-life of tritium, 12.3 years, it is not possible to use tritium activity in dating very old wines.

TABLE 23–5	Carbons in Exchangeable Reservoirs
Source	**Exchangeable Carbon**
atmospheric CO_2	1.6%
dissolved HCO_3^- in oceans	88.1%
living organisms	0.8%
dissolved organic matter	6.8%
humus	2.7%

23–17). Is it really a Chateau Lafite Rothschild 1886? If there is any measurable concentration of tritium it is more likely a much more recent vintage. On a more serious note, one useful area of research concerns whether underground water supplies originate from recent rainfall or from underground reservoirs. You can check the tritium content to find out.

Most of the neutrons produced by cosmic rays and slowed to thermal energies produce ^{14}C, a beta-emitter with a half-life of 5720 years. The production rate, averaged over the entire atmosphere, is $1.97 \text{ s}^{-1} \text{ cm}^{-2}$ on the Earth's surface. Because of its long lifetime, the radioactive ^{14}C isotope thoroughly mixes with all the carbon in exchangeable reservoirs, equilibrating at sea level in a period estimated to be 500 years (Table 23–5).

Since the total carbon content of these reservoirs is known to be 7.88 g/cm^2 over the Earth's surface, the specific activity of ^{14}C is 1.97/7.88 = 0.25 disintegrations per second per gram, or about 15 disintegrations per minute per gram. All the carbon in the world's life cycle is thus kept uniformly radioactive. Recognition of this fact led to the invention of the very important ^{14}C dating method for determining the ages of carbon-containing specimens. The technique has provided a reliable chronometric record of human culture.

The decay rate per gram of carbon in living organisms can be confirmed independently since the ^{14}C/^{12}C ratio is known to be 1.3×10^{-12}. Begin with one gram of ^{12}C atoms:

$$(6.022 \times 10^{23} \text{ nuclei/mol}) \times \left(\frac{1 \text{ mol}}{12.0 \text{ g } ^{12}\text{C}} \right)$$

$$= 5.02 \times 10^{22} \ ^{12}\text{C nuclei/g}$$

The number of ^{14}C nuclei is 1.3×10^{-12} times that number, for which the decay rate can be calculated (remembering that the decay rate is related to the number N of radioactive nuclei) as follows:

$$\text{rate} = -\frac{\Delta N}{\Delta t} = \lambda N = \left(\frac{0.693}{t_{1/2}} \right) N$$

$$= \left[\frac{0.693}{5720 \text{ yr}} \right] (1.3 \times 10^{-12})(5.02 \times 10^{22}/\text{g})(1 \text{ yr}/5.26 \times 10^5 \text{ min})$$

$$= 15 \text{ decays/min/g}$$

which is the decay rate in living organisms. When the organism dies, carbon intake stops, equilibrium is no longer maintained, and the decay rate decreases. Measurement of the decay rate per gram of carbon per minute allows one to calculate the time of death, and therefore the isotopic age.

EXAMPLE 23–9

Consider a bone known to contain 200 grams of carbon, for which the decay rate is found to be 400 disintegrations/minute. How old is the bone?

Solution   If the bone was still part of a living system, the decay rate should be

$$(15 \text{ decays/g/min})(200 \text{ g}) = 3000 \text{ decays/min}$$

In other words, the decay rate has decreased by a factor of 400/3000. From an earlier discussion in this chapter, we know that after n half-lives the decay rate decreases by $(\frac{1}{2})^n$. Therefore,

$$(\tfrac{1}{2})^n = \frac{400}{3000}$$

$$2^n = \frac{3000}{400} = 7.5$$

$$n \ln 2 = \ln 7.5$$

$$n = \left(\frac{\ln 7.5}{\ln 2} \right) = \frac{2.01}{0.693} = 2.91$$

Finally, the age of the bone is given by

$$t = n t_{1/2} = (2.91)(5720 \text{ yr}) = 1.66 \times 10^4 \text{ yr}$$

The problem of dating the bone could also have been handled in a direct fashion using the basic relationship

$$\ln\left(\frac{N_0}{N}\right) = kt$$

where

$$(N_0/N) = 3000/400$$

since the decay rate is proportional to the number of nuclei. Then

$$\ln (3000/400) = [0.693/(5720 \text{ yr})]t$$
$$t = 1.66 \times 10^4 \text{ yr} \quad \blacksquare$$

EXERCISE 23–9(A)

A bone found in an archeological "dig" is believed to be 12,000 years old. What is the approximate decay rate for the ^{14}C radionuclide (if that is true)?

Answer: Approximately 3 to 4 decays/min/g. $\quad \blacksquare$

EXERCISE 23–9(B)

A wood sample contains 10 g of carbon and exhibits a decay rate of 10/min. How old is the sample?

Answer: 2.2×10^4 years. $\quad \blacksquare$

FIGURE 23–18 Based on carbon-12/carbon-13 isotope ratios, anthropologists were able to resolve a puzzle about this 11th-century South African villager.

Anthropologists have successfully used radioisotope content to date and study fossils. The skeleton of the 11th-century inhabitant of a South African village (Figure 23–18), for example, posed a significant problem. Physically, the man's skeleton was different from those of the other villagers, suggesting he was not a native of the area. However, when the skeleton was analyzed for isotopes, its carbon-12/carbon-13 ratio was found to be not unlike that of other skeletons recovered from the same village. Since different kinds of plants incorporate different proportions of isotopes into the food produced from them, this similarity of isotope content indicates that these individuals all ate the same foods. The anthropologists concluded that the man in question probably migrated to the village from a distant region and then spent most of his life after that in the village.

SUMMARY

The rate (or velocity) of a chemical reaction is measured as the change in the concentration (Δc) over a time interval (Δt), or $\Delta c/\Delta t$. For the general chemical reaction in which A reacts with B, yielding C and D as products,

$$a\text{A} + b\text{B} \longrightarrow c\text{C} + d\text{D}$$
$$\text{rate} = -\left(\frac{1}{a}\right)\frac{\Delta[\text{A}]}{\Delta t} = -\left(\frac{1}{b}\right)\frac{\Delta[\text{B}]}{\Delta t} = +\left(\frac{1}{c}\right)\frac{\Delta[\text{C}]}{\Delta t} = +\left(\frac{1}{d}\right)\frac{\Delta[\text{D}]}{\Delta t}$$

Rate laws, rate expressions, or rate equations—whatever you choose to call them—take the form of a product of concentrations raised to a power

$$-\frac{\Delta[\text{B}]}{\Delta t} = k[\text{A}]^m[\text{B}]^n$$

where the exponents m and n may or may not be equal to the stoichiometric coefficients a and b. The order of the reaction is said to be m in component A and n in component B, and the overall order is described as the sum of the exponents, $m + n$. If m is 1, then the rate of the reaction must change directly as the concentration of A; if m is 2, then the rate of the reaction must change directly as the square of the concentration of A; and so forth. First and second order reactions are commonplace; but third order reactions are uncommon. Although the exponents in the rate law are usually positive integers, negative values and even fractions occasionally show up.

Whereas the order is an empirical construct, molecularity is a theoretical concept. Unimolecular, bimolecular, and termolecular describe elementary processes, direct collisions resulting in reactions, in which one, two, or three particles are involved in a single kinetic event. Remember that a termolecular elementary reaction is always third order but a third order reaction isn't necessarily termolecular. Corresponding statements can be made for the terms "bimolecular" and "unimolecular."

The proportionality constant k in the rate law expression for a given reaction is characteristic, and the simplified equations we routinely use are in the integrated form:

$$\ln \left(\frac{[A]_0}{[A]} \right) = kt \qquad \text{(first order)}$$

$$\left(\frac{1}{[A]} \right) - \left(\frac{1}{[A]_0} \right) = kt \qquad \text{(simple second order)}$$

where $[A]$ and $[A]_0$ can be concentrations or pressures.

By fitting analytical data obtained in studies of concentration versus time (at a given temperature) to equations of these kind, we obtain the means to form theories on how the reactions occurred. The half-life ($t_{1/2}$) is the time required for half the reacting material present to react, and from the preceding equations, useful half-life equations can be obtained:

$$t_{1/2} = 0.693/k \qquad \text{(first order)}$$

$$t_{1/2} = \frac{1}{k[A]_0} \qquad \text{(second order)}$$

The concentration dependency of the reaction rate, along with its temperature dependence, the presence of a catalyst, and other factors, helps paint a picture of the simple elementary steps that occur as the initial reactants move toward final products. The portrait of theory that results is called a mechanism. There is usually one among the complex collection of processes that make up the proposed mechanism that is the rate-controlling, slowest step.

Radioactive nuclides decay by alpha-emission (^{4}He nuclei), beta-emission (nuclear electrons), gamma-emission (high energy photons), positron emission, and electron capture. Quantitatively, the number of nuclei spontaneously decaying and the rate of radioactive decay follow exponential laws:

$$\text{rate} = \lambda N = \lambda N_0 e^{-\lambda t}$$

where λ is the characteristic decay constant, N is the number of nuclei remaining at time t, and N_0 is the original number of nuclei (at time zero). A special case of this equation gives the half-life, the time required for half the nuclei originally present to decay:

$$t_{1/2} = \frac{0.693}{\lambda}$$

QUESTIONS AND PROBLEMS

QUESTIONS

1. What do we mean by the terms kinetic control and thermodynamic control, and why is it essential for chemists and engineers to understand the difference?

2. Explain why a heap of flour burns slowly, but the same quantity suspended in the air as dust can explode violently.

3. Which of the following reactions are not (likely to be) elementary processes?

(a) $A + B \longrightarrow AB$
(b) $2 A_2 + 2 B \longrightarrow 2 A_2 B$
(c) $A_2 + B + 2 C \longrightarrow AC + ABC$
(d) $A + B + C + D \longrightarrow ABCD$

4. What is the molecularity of the following elementary processes?

(a) $A_2 + B_2 \longrightarrow 2 AB$
(b) $A_2 + B \longrightarrow A + AB$
(c) $2 A + B_2 \longrightarrow A_2B_2$
(d) $2 A \longrightarrow A_2$

5. What graphical methods could you use to test whether a reaction in which A goes to products is first or second order?

6. If a reaction follows second order kinetics, is it necessarily second order? Explain.

7. Why are a "termomolecular" elementary process and processes of higher molecularities considered to be extraordinary?

8. Explain why the following reaction does not proceed by an apparent fifth order rate law:

$$4 HCl + O_2 \longrightarrow 2 H_2O + 2 Cl_2$$

9. Explain the differences between the stoichiometry of the balanced chemical equation and the concentration dependencies expressed in the corresponding rate law:

$$HCrO_4^- + 3 Fe^{2+} + 7 H^+ \longrightarrow$$
$$Cr^{3+} + 3 Fe^{3+} + 4 H_2O$$
$$rate = k[HCrO_4^-][Fe^{2+}][H^+]$$

10. Why is it inaccurate to speak about the lifetime of a sample of a certain radioisotope? Or of a single radioactive nucleus?

11. How much of a certain radioisotope will remain unchanged after a time equal to $4t_{1/2}$ has passed?

12. After many half-lives have passed, only two atoms of a certain radioisotope remain. What happens after one more half-life elapses?

13. Why does it make little sense to try to date the following by ^{14}C methods: coal deposits; Texas crude oil; the fossilized remains of dinosaurs?

14. What problems do you envision in trying to carbon-date a sample of recent vintage, say 1983 Bordeaux wine?

PROBLEMS

Rate Expression   [1–4]

1. For the following reactions, write the rate law in terms of the appearance of product, assuming that no intermediates are present.

(a) $A + 2 B \longrightarrow AB_2$
(b) $A_2 + 2 B \longrightarrow A_2B_2$

2. For the following reactions, write the rate law in terms of the appearance of product, assuming that no intermediates are present.

(a) $X + Y + Z \longrightarrow XYZ$
(b) $X_2 + Y_2 \longrightarrow 2 XY$

3. For each reaction, determine the relationship between the rate of reaction stated in terms of the reactant(s) and the rate stated in terms of the product(s)—that is,

$$\frac{\Delta[\text{reactant}]}{\Delta t} \quad \text{and} \quad \frac{\Delta[\text{product}]}{\Delta t}$$

(a) $A_2 + B_2 \longrightarrow 2 AB$
(b) $A_2 + B_2 \longrightarrow A_2B_2$

4. For each reaction, determine the relationship between the rate of reaction stated in terms of the reactant(s) and the rate stated in terms of the product(s)—that is,

$$\frac{\Delta[\text{reactant}]}{\Delta t} \quad \text{and} \quad \frac{\Delta[\text{product}]}{\Delta t}$$

(a) $X + Y_2 \longrightarrow XY + Y$
(b) $X_2 + 2 Y \longrightarrow X_2Y_2$

Reaction Orders and Rate Constants   [5–12]

5. A known termolecular reaction is the combination of nitric oxide and chlorine:

$$2 NO (g) + Cl_2 (g) \longrightarrow 2 NOCl (g)$$

(a) What is the order of the reaction?
(b) Express the rate of formation of NOCl in terms of the rate of disappearance of Cl_2.

6. Consider the following reaction, which takes place under mildly acidic conditions:

$$4 H_3O^+ (aq) + MnO_4^- (aq) + 2 Sb(OH)_3 (aq) \longrightarrow$$
$$Mn^{3+} (aq) + 2 H_3SbO_4 (aq) + 6 H_2O (liq)$$

It can be shown experimentally that the rate of permanganate disappearance doubles when either the permanganate concentration or the hydronium ion concentration doubles, but halves when the $Sb(OH)_3$ concentration is doubled. Write the experimental rate law for the reaction.

7. Consider the following chemical reaction and the corresponding kinetic data showing the initial reaction rate as a function of the initial concentrations of the reactants:

$$H_3AsO_4 (aq) + 2 H_3O^+ (aq) + 3 I^- (aq) \longrightarrow$$
$$HAsO_2 (aq) + I_3^- (aq) + 4 H_2O (liq)$$

Initial Rate (moles/liter·sec)	Molar Concentration		
	$[H_3AsO_2]$	$[H_3O^+]$	$[I^-]$
3.70×10^5	0.001	0.010	0.10
7.40×10^5	0.001	0.010	0.20
7.40×10^5	0.002	0.010	0.10
3.70×10^5	0.002	0.005	0.20

Using the data, establish the correct experimental rate law.

8. (a) Write the rate expression for the following reaction:

$$A + 3 B \longrightarrow 2 C$$

(b) For the reaction

$$2 A + 3 B \longrightarrow 2 C$$

determine the rate law from the following data.

Experiment	[A]	[B]	Rate (M/min)
I	1.0	1.0	0.20
II	1.0	2.0	0.40
III	2.0	1.0	0.80

9. Based on the accompanying experimental data:

Experiment	#1	#2	#3
[A]	0.500	0.500	1.000
[B]	0.050	0.100	0.050
observed rate (M/min)	5×10^{-4}	2.0×10^{-3}	1.0×10^{-3}

Determine:

(a) the overall order of the reaction

(b) the order with respect to each component

(c) the specific reaction rate constant

10. The following data show the effect of concentration on reaction rate for a given reaction:

[A]	[B]	[C]	Rate (M/s)
0.01	0.20	0.10	2.8
0.01	0.40	0.10	5.6
0.01	0.80	0.05	5.6
0.02	0.20	0.10	2.8

(a) Write the indicated rate law equation.

(b) Determine the specific reaction rate constant.

11. An investigation of the thermal decomposition of arsine (AsH_3) on a hot glass surface at 623 K gave the following results:

Time (hr)	0	3.77	16.0	25.5	37.7	44.8
P_{total} (torr)	392	403	436	454	480	488

Determine:

(a) the overall order of the reaction

(b) the specific reaction rate constant

(c) the half-life for the reaction under these circumstances

12. At 800 K, dimethyl ether decomposes smoothly to methane, carbon monoxide, and hydrogen, according to the equation

$$CH_3OCH_3 \text{ (g)} \longrightarrow CH_4 \text{ (g)} + CO \text{ (g)} + H_2 \text{ (g)}$$

The following data were obtained:

Time (min)	0	390	780	1200	3160
Pressure (torr)	310	410	490	560	780

What are the order of the reaction, the reaction rate constant, and the half-life?

First Order Rate Laws   [13–20]

13. If $t_{1/2}$ for a first order reaction is 10 minutes, approximately how long will it take for the reaction to be 99% complete?

14. The half-life for a first order reaction is known to be 600 seconds. What is the specific reaction rate constant, and what percentage of the reacting material remains after one hour of reaction time?

15. A particular reacting substance is shown to be disappearing at a rate of 1% per minute at a given temperature. If the reaction is known to be first order, what are the specific reaction rate constant and the half-life for the reactant (at the given temperature)?

16. Consider the first order decomposition of benzoyl peroxide in ether. The reaction is 60% complete in 220 seconds at 60°C. Calculate the specific reaction rate constant for the reaction, the half-life for the reaction, and the time required for the reaction to proceed 75% to completion.

17. Consider a hypothetical reaction in which A is observed to decompose to products according to the following equation:

$$A \longrightarrow B + C$$

(a) Write the rate law for such a first order reaction in terms of the disappearance of A.

(b) Develop a general expression for the time required for 90% of the original quantity of A present to be used up.

18. A certain organic peroxide thermally decomposes in solution via first order kinetics. The reaction was 72.5% complete after 600 seconds had elapsed.

(a) Calculate the specific reaction rate constant.

(b) Determine $t_{1/2}$ for the reaction.

19. The vapor phase decomposition of ethylene oxide according to the following equation has been studied at 400°C.

$$C_2H_4O \text{ (g)} \longrightarrow CH_4 \text{ (g)} + CO \text{ (g)}$$

The following data were recorded:

Time (min)	0	7	12	18
P_{total} (torr)	119	130.7	138.2	146.4

(a) Calculate the rate constant at 400°C and show that this is a reaction of the first order.

(b) Calculate the half-life of ethylene oxide at 400°C.

(c) Calculate the time required to reach a total pressure of 200 torr.

20. A group of investigators studied the thermal decomposition of phosphine (PH_3) to elemental phosphorus and hydrogen as the reaction takes place on a ceramic surface at 600 K according to the following reaction:

$$4 PH_3 \text{ (g)} \longrightarrow P_4 \text{ (g)} + 6 H_2 \text{ (g)}$$

time (min)	0	240	1000	1500	2700
P_{total} (torr)	393	404	436	454	495

Based on these results, the investigators proposed a first order mechanism for the phosphine decomposition.
(a) Verify that the reaction is first order with respect to phosphine.
(b) Calculate the total pressure when 50% of the original sample has decomposed.

Radioactivity [21–28]

21. A sample of radioactive silver foil was placed adjacent to a Geiger counter and an initial decay rate of 1000 counts per second was observed. The half-life for the silver isotope is known to be 2.4 minutes.
(a) What is the decay rate 4.8 minutes later?
(b) When will the decay rate be 30 counts per second?
22. The initial decay rate for a certain radioisotope is 8000 counts per second. After ten minutes have elapsed, the decay rate is observed to have fallen to 1000 counts per second.
(a) What is the half-life of the radioisotope?
(b) What is the decay constant for the isotope?
23. Calculate the number of disintegrations per second for a gram of ^{226}Ra, and calculate the decay rate. The half-life for the radioisotope is 1620 years.
24. One radioisotope in the uranium series has a decay rate such that after 53.6 minutes only 25% of the original sample was still present.
(a) Calculate the decay constant (λ).
(b) Calculate the half-life of the radioisotope.
25. A medically important isotope of iodine, ^{131}I, has a radioactive half-life of 193.2 hours, decaying to a stable xenon isotope.
(a) Write the nuclear reaction of the indicated decay process.
(b) If the iodine isotope has an initial decay rate of 1500 counts per minute, what would the count or rate be after 24.00 hours have elapsed?
26. The nuclide ^{40}K is known to decay to ^{40}Ca by beta-decay (89%) and to ^{40}Ar by electron-capture (11%).
(a) Write nuclear equations for the two processes.
(b) A particular mineral known to trap ^{40}Ar contains 10.1% of the nuclide. What is the age of the sample?
27. A sample taken from a wooden artifact was found to contain 10 g of carbon. The decay rate was found to be 100 counts per minute. How old was the specimen?
28. An animal bone, supposedly more than 10,000 years old, contained 15 g of carbon. Calculate the maximum value for the decay rate of the radiocarbon in this old bone.

Additional Problems [29–37]

29. If 1.0 g of ^{238}U gives off 1.3×10^4 alpha-particles per second, what is the half-life of the isotope?
30. The reaction

$$2\ Ti^{3+} \longrightarrow Ti^{2+} + Ti^{4+}$$

proceeds in a single elementary step in aqueous solution. At 25°C, k is 1×10^2 when expressed in the usual units. A solution is initially 0.015 M in Ti^{3+}.
(a) What are the "usual units" for k?
(b) What is the initial rate of the reaction in the solution?

(c) What is the initial rate of change of the concentration of Ti^{3+} in the solution?
(d) How long is required for the concentration of Ti^{3+} to drop to 10% of its initial value?
31. The disintegration rate for a certain radioactive source was measured every minute, with the following results: 1000, 820, 673, 552, 453, 305, and 250 counts per second:
(a) Plot decay rate versus time.
(b) Use your graph to estimate the half-life for the reactions.
(c) Again using your graph, determine the approximate initial rate of decay.
32. Alcohol was dehydrated to a mixture of diethyl ether and water at 140°C in a dilute sulfuric acid experiment. One particular set of experiments produced the following data:

Time (min)	Fraction of Alcohol Reacted
60	0.197
93	0.290
143	0.409
295	0.672
590	0.889

Determine the probable experimental order of the reaction based on these data.
33. For a second order reaction in which

$$2\ A \longrightarrow products$$

clearly show that $t_{1/2} = 1/k[A]_0$.
34. Consider the kinetic experiment described by the following graph of log concentration versus time for N$_2$O$_5$ decomposition.

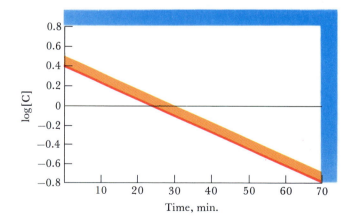

(a) Determine the order of the reaction.
(b) Evaluate the specific reaction rate constant.
(c) Determine the initial concentration of the reacting material.
35. At 415°C, ethylene oxide undergoes a first order gas phase decomposition, producing methane and carbon monoxide as the only products. The specific reaction rate constant was found to be 0.0123 min^{-1}.
(a) After 60 min at 415°C what percent of any given ethylene oxide sample will have decomposed to products?

(b) How much time will be required for three quarters of the starting material to react?

36. A compound named dimethyldiimide ($CH_3N{=}NCH_3$) is known to decompose according to the following reaction in the gas phase:

$$CH_3N{=}NCH_3 \text{ (g)} \longrightarrow CH_3CH_3 \text{ (g)} + N_2 \text{ (g)}$$

After 20 min, the total pressure in the reaction flask is 50% greater than the initial pressure. Assuming that nothing but $CH_3N{=}NCH_3$ molecules was initially present, and that the decomposition is first order, what is the specific reaction rate constant?

37. (a) Briefly explain why the following reaction is not likely to be an "elementary" reaction:

$$A + B + C + D \longrightarrow X$$

(b) If the reaction were elementary, state the order with respect to A, the overall order of the reaction, and the rate law.

MULTIPLE PRINCIPLES   [38–42]

38. If a certain uranium core contains 0.865 g of ^{206}Pb per gram of ^{238}U present, what is the minimum age of the sample?

39. The decay rate for a particular process is R_0 at $t = 0$, but falls to R at $t = 1$. If the decay constant is λ, show that

$$\lambda = \left(\frac{1}{t}\right) \ln \left(\frac{R_0}{R}\right)$$

$$t_{1/2} = \frac{0.693t}{\ln (R_0/R)}$$

40. Calculate the rate at which a gram of carbon in a living organism is decaying, given the ratio

$$\frac{^{14}C}{^{12}C} = 1.3 \times 10^{-12}$$

and a half-life for the radioactive parent isotope of 5720 years.

41. Given a third order reaction in which

$$3 \text{ A} \longrightarrow \text{products}$$

it can be shown that the integrated form of the rate law assumes the form

$$\frac{1}{[A]^2} = \frac{1}{[A]_0^2} + 2\,kt$$

State what you would do to obtain a graphical solution for the specific reaction rate constant, the initial concentration of A, and a test of the overall order of the reaction.

42. Data describing the first order decomposition of N_2O_5 in CCl_4 solution at 45°C are given in Table 23–6.

$$2 \text{ } N_2O_5 \text{ (g)} \longrightarrow 4 \text{ } NO_2 \text{ (g)} + O_2 \text{ (g)}$$

The analytical procedure depends on the fact that N_2O_5 and NO_2 are soluble in carbon tetrachloride (CCl_4) while O_2 is not. These data have been used to obtain the plots shown in the following figures.

TABLE 23–6	Decomposition of N_2O_5 in CCl_4 at 45°C*
Time Elapsed (s)	**N_2O_5 Concentration (mol/L)**
184	2.08
319	1.91
526	1.67
867	1.36
1198	1.11
1877	0.72
2315	0.55
3144	0.34

*Recorded in a classic kinetics paper by Eyring and Daniels in 1933.

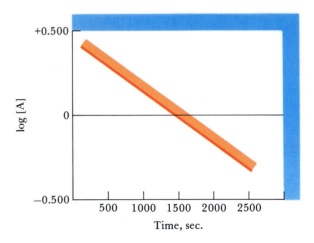

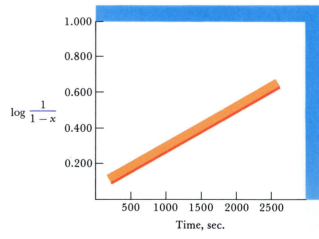

(a) How might you "follow" the progress of the reaction if you were conducting the experiment?

(b) Explain how these data points confirm the statement that the reaction is first order.

(c) Using both data plots, calculate the specific reaction rate constant and the initial N_2O_5 concentration.

(d) Determine the N_2O_5 concentration at the point where the reaction has diminished to half the initial rate.

(e) What is the half-life of N_2O_5?

(f) The time on the lab clock reads 3:00 PM and it is Friday afternoon. When should you return to shut down the reaction if it is desired to allow 87.5% of the original N_2O_5 sample to decompose to products?

(g) Take any two consecutive sets of numbers from the data table and determine the rate constant algebraically (in contrast to the graphical determination employed earlier in this problem).

APPLIED PRINCIPLES [43–45]

43. (a) If the Shroud of Turin is the sacred relic many believe, what would the approximate ^{14}C decay rate be today in a sample (assuming it is a cotton cloth)?

(b) Evidence suggests the relic is only about 800 years old. What is the decay rate?

44. (a) Calculate the decay constant for ^{226}Ra if the half-life of the isotope is 1620 years.

(b) If the unit of radiation (known as the curie) is defined as the number of disintegrations per second per gram of ^{226}Ra, what is its value?

45. The growth, multiplication, and expansion of colonies (cultures) of bacteria are classic kinetic processes. When a growing bacterium reaches a critical size, it divides in two; then into four, eight, sixteen, and the multiplication process continues. In one growth experiment, the following data were obtained:

Time (min)	—	60.0	120	180	240	300
Concentration [(bacteria/L) $\times 10^{-3}$]	3.50	7.00	14.0	28.0	56.0	112

(a) How many minutes must pass before the population of the colony doubles?

(b) What is the bacteria concentration at 6 hours?

(c) When the bacteria concentration has reached $22.4 \times 10^4/L$, how much time has elapsed?

(d) Calculate the growth constant for this multiplication process.

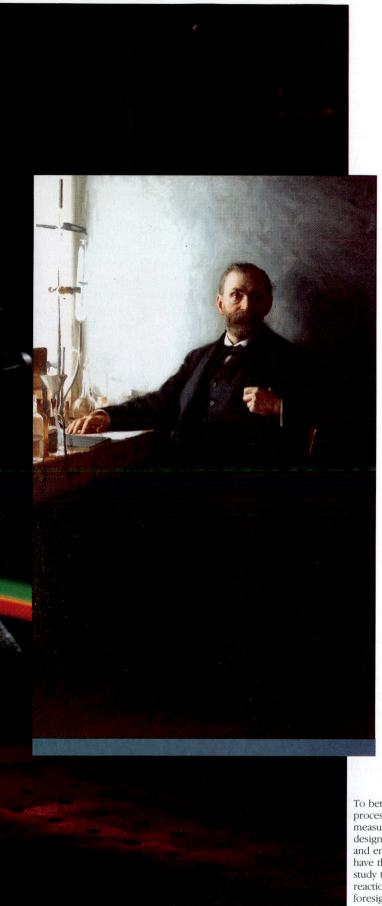

CHAPTER
24

Kinetics II

24.1 KINETICS OF MORE COMPLEX REACTIONS

24.2 TEMPERATURE DEPENDENCE OF THE RATE

24.3 KINETIC THEORIES

24.4 REACTION MECHANISMS

24.5 CATALYSTS

To better understand the reaction mechanisms for combustion processes, light-scattering experiments are being used here to measure temperature in an engine cylinder. The experiments are designed to shed light on the causes of autoignition in motor fuels and engine knock in gasoline-powered engines. Alfred Nobel did not have the scientific means at his disposal in the late 19th century to study the mechanisms of the complex combustion and explosion reactions of his marvelous invention, dynamite. He did have the foresight to motivate and provide for fundamental research by future generations of scientists.

Paul Sabatier and Victor Grignard (French), Nobel Prize for Chemistry in 1912 for their methods of synthesis, especially the technique of catalytic hydrogenation and the "Grignard" reaction.

24.1 KINETICS OF MORE COMPLEX REACTIONS

Having discussed the factors influencing rates of reactions and the characteristics of the rate laws for first order, simple second order, and zero order reactions, we can now go on to some advanced topics in reaction dynamics, especially multi-step reactions and the effect of temperature on reaction rates. Theories of reaction rates are important because they provide us with bridges to the microscopic scale of dynamic activity. Finally, we will discuss the special role of catalysis in chemical reactions, how catalysts alter the pathways by which reactions occur, and how that affects the rates of the reactions.

Up to this point, our discussion of chemical kinetics has dealt with the one-step elementary reactions, and with sequences of irreversible steps among which one can be found that is slow and therefore rate-determining. Although reactions that follow a simple kinetic course make good illustrations of the basic principles and methods, it is obviously important to pay some attention to harsher realities, if only qualitatively. What follows are a few examples of complex kinetics situations that are commonly encountered, along with some comments on their causes.

Competing Reactions and Reaction Pathways

Chemical reactions can occur simultaneously, in competition with each other, making experimental measurements of individual reaction rates more difficult. One possibility for such parallel reactions is that two or more processes take place at the same time. Consider the situation where A can react with B or C:

$$A + B \longrightarrow products$$
$$A + C \longrightarrow products$$

Another involves a single species decomposing by alternative competing pathways, yielding the same or different products.

$$A \longrightarrow B$$
$$A \longrightarrow C$$

As a rule, when parallel reactions compete, the major (and sometimes the only) product observed is that of the fastest reaction. That's important because it has a direct bearing on formation of unwanted side-products. Classic examples are the syntheses of Agent Orange and PCB's, where undesired side-products have produced disastrous biological effects. Often, the success of commercial processes depends on catalysts that can exert kinetic control, speeding up the rate of the desirable route.

Consecutive Reactions

The graph in Figure 24–1 shows concentration changing with time for consecutive first order reactions, where the product of one reaction becomes the reactant in the next:

$$A \longrightarrow B \longrightarrow C$$

The concentration of A steadily decreases with time, irrespective of anything to do with B or C; the concentration of C increases steadily, approaching a limiting concentration that is the initial concentration of A.

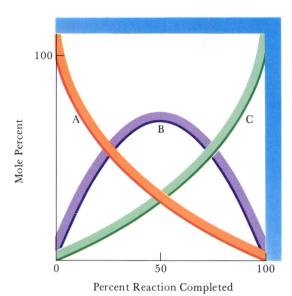

FIGURE 24-1 Typical concentration versus time plot for consecutive reactions where *B*, the product of the first reaction, is the reactant in a second reaction.

The concentration of B increases initially, and as it does, so does its rate of decomposition; [B] passes through a maximum before finally decreasing toward zero. Nuclear chemistry supplies us with numerous examples of consecutive first order reactions that carry a parent nuclide through several generations to a stable daughter nuclide. For example, we find the following sequences in the (artificially produced) radioactive decay series that begins with ^{137}Np:

$$^{225}\text{Ra} \xrightarrow[\beta]{14.8\ \text{d}} {}^{225}\text{Ac} \xrightarrow[\alpha]{10.0\ \text{d}} {}^{221}\text{Fr}$$

$$^{213}\text{Bi} \xrightarrow[\alpha]{47\ \text{m}} {}^{209}\text{Ti} \xrightarrow[\beta]{2.2\ \text{m}} {}^{209}\text{Pb}$$

The times above the arrows are half-lives, and the decay processes are indicated below the arrows.

The chemistry of blood coagulation provides an interesting illustration of complex reaction kinetics. The major sequence of events appears to be the conversion of an enzyme called prothrombin into another called thrombin, followed by the subsequent action of thrombin upon fibrinogen to form fibrin in the blood. The fibrin separates into long fibers in a three-dimensional mesh that retains the formed components of the blood. Fibrin threads are extremely adhesive, sticking to each other and to blood cells, tissue, and any other substance close by, causing the blood to clot. The clotted blood in turn holds firmly to injured tissue in a wound, stopping the flow of blood (Figure 24–2).

The prothrombin-thrombin reaction follows the kinetics of a system of successive first-order processes of the type A → B → C. Figure 24–1 can be used to show the thrombin concentration, observed by measuring the ability of the test solution to clot blood, plotted against time. Thrombin corresponds to B, the middle reactant in our scheme. Its concentration rises rapidly as A (prothrombin) decomposes. [B] reaches a maximum, and then decreases toward zero as it in turn decomposes to C (by reaction with fibrinogen) to form fibrin.

More often than not, intermediates such as B tend to be reactive. We find that their concentrations never rise to any significant level because

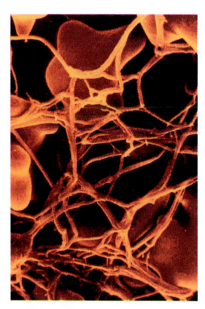

FIGURE 24-2 This scanning electron micrograph shows part of a blood clot. The red blood cells are enmeshed in a network of fibrin.

FIGURE 24–3 From the information presented by the graph, give a qualitative description of the events that might be taking place. Compare the processes in this graphical representation to those in Figure 24–1.

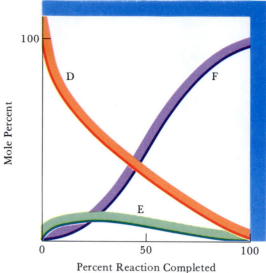

they disappear as fast as they form; and we can say that the intermediate has established a steady state concentration. Consider the example illustrated in Figure 24–3.

EXAMPLE 24–1

Using composition versus time plots, qualitatively describe the chemistry of the photochlorination of methylene chloride, producing chloroform and carbon tetrachloride, both of which are stable products:

$$CH_2Cl_2 \longrightarrow CHCl_3 \longrightarrow CCl_4$$
methylene chloride   chloroform   carbon tetrachloride

Solution   Figure 24–3 is similar to Figure 24–1. A considerable concentration of $CHCl_3$ builds up during the process and is then converted to CCl_4.   ■

EXERCISE 24–1

Qualitatively describe the progress of the hydrolysis of *tert*-butyl chloride, producing *tert*-butyl alcohol (a stable product) by means of an unstable cationic intermediate:

$$(CH_3)_3CCl \longrightarrow [(CH_3)_3C^+] \longrightarrow (CH_3)_3COH$$
tert-butyl chloride   *tert*-butyl alcohol

Answer: Similar to Figure 24–3.   ■

Equilibrium Reactions

What do we really mean when we say that a reaction goes to completion? After sufficient time has passed, the equilibrium will lie far to the right and the concentration of at least one reactant will have fallen almost to zero. However, there are many reactions in which significant concentrations of reactants remain even after a long time because a chemical equilibrium has been established between reactants and products that does not lie far to the right.

The equilibrium constant for an elementary reaction is the ratio of the rate constants for the forward and reverse directions. This follows from our earlier discussion of the equilibrium condition for a reversible chemical reaction, where the rates of the forward and reverse reactions are equal. Consider the dinitrogen tetroxide/nitrogen dioxide equilibrium:

$$N_2O_4 \text{ (g)} \underset{k_{-1}}{\overset{k_1}{\rightleftharpoons}} 2 \text{ NO}_2 \text{ (g)}$$

The forward reaction (k_1) is a first order decomposition for which the following rate law can be written:

$$\text{rate} = k_1[N_2O_4]$$

For the reverse reaction, a bimolecular recombination takes place according to the rate law

$$\text{rate} = k_{-1}[NO_2]^2$$

If we state the overall reaction rate in terms of disappearance of N_2O_4,

$$-\frac{\Delta[N_2O_4]}{\Delta t} = k_1[N_2O_4] - k_{-1}[NO_2]^2$$

net rate of N_2O_4 N_2O_4 N_2O_4 formed from

disappearance   disappeared NO_2 disappearance

Imposing the condition that the overall rate at equilibrium must be zero:

$$-\frac{\Delta[N_2O_4]}{\Delta t} = 0$$

it must be true that

$$k_1[N_2O_4] = k_{-1}[NO_2]^2$$

which gives us the value of the equilibrium constant as the ratio of the respective rate constants:

$$\frac{k_1}{k_{-1}} = \frac{[NO_2]^2}{[N_2O_4]} = K$$

24.2 TEMPERATURE DEPENDENCE OF THE RATE

The general rate law (equation) and the rate constant for a specific chemical reaction can be deduced from experiment by measurements of the rate of reaction at a fixed temperature. What happens when a set of kinetics experiments for a specific reaction is carried out at several different temperatures? The original concentration dependencies remain unchanged, but the specific reaction rate constant turns out to be very much a temperature-dependent characteristic. An increase in temperature usually results in an increase in the reaction rate. Rarely do reaction rates decrease with increasing temperature. A useful generalization among chemists and photographers is: in the vicinity of room temperature, a 10°C rise in temperature doubles the rate of reaction.

The first quantitative formulation of the sensitivity of reaction rates to temperature changes was made a century ago by Svante Arrhenius, the Swedish Nobel Laureate (Figure 24–4; see also Chapter 12). He observed that a plot of ln k versus $1/T$ gives a straight line with a negative slope if T is the absolute temperature. This led him to the still widely used form of the **Arrhenius equation,**

FIGURE 24–4 Although best known (to first-year chemistry students) for his electrolytic dissociation theory (Chapter 12), do not underestimate the theoretical and practical importance of Arrhenius's studies in the field of reaction dynamics. The portrait of the great man, in mid-career, is interesting in part for the suggestion of the laboratory equipment he used.

$$k = Ae^{-E_a/RT}$$

where A is the pre-exponential (or frequency) factor, E_a the activation energy, and R the universal gas constant.

Equally useful is the logarithmic form of the Arrhenius equation:

$$\ln k = -\frac{E_a}{RT} + \ln A$$

An **Arrhenius plot** of $\ln k$ versus $1/T$ gives values for the two reaction parameters E_a and A, obtainable from the slope and the intercept:

$$\text{slope} = -\frac{E_a}{R}$$

$$\text{intercept} = \ln A$$

Both terms are important, but the interpretation of E_a has physical significance, as we'll explain in the next section.

The activation energy E_a can also be determined algebraically from values for the rate constants k_1 and k_2 at temperatures T_1 and T_2. In the first experiment, at T_1,

$$\ln k_1 = \ln A - \frac{E_a}{RT_1}$$

In the second experiment, at T_2,

$$\ln k_2 = \ln A - \frac{E_a}{RT_2}$$

Subtracting the expression for the first experiment from that for the second:

$$\ln k_2 - \ln k_1 = \left(\ln A - \frac{E_a}{RT_2} \right) - \left(\ln A - \frac{E_a}{RT_1} \right)$$

Finally, rearranging and combining terms:

$$\ln \frac{k_2}{k_1} = -\frac{E_a}{R} \left[\frac{1}{T_2} - \frac{1}{T_1} \right]$$

$$= -\frac{E_a}{R} \left[\frac{T_1 - T_2}{T_1 T_2} \right]$$

$$\log \frac{k_2}{k_1} = -\frac{E_a}{2.303R} \left[\frac{T_1 - T_2}{T_1 T_2} \right]$$

EXAMPLE 24–2

The rate of decomposition of N_2O_5 was measured at several different temperatures. The data are tabulated in Table 24–1. Making use of an Arrhenius plot of $\log k$ versus $1/T$ for these data (Figure 24–5), evaluate the slope and obtain E_a.

Solution

$$\text{slope} = -\frac{E_a}{2.303R} = -5.3 \times 10^3 \text{ K}$$

$$E_a = 1.0 \times 10^5 \text{ J/mol}$$

TABLE 24–1	$2 N_2O_5 (g) \rightarrow 4 NO_2 (g) + O_2 (g)$
Temperature (K)	$k \times 10^5 \ (\text{min}^{-1})$
273	0.0787
298	3.46
318	49.8
338	487

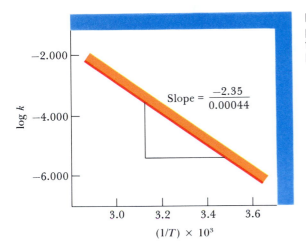

FIGURE 24–5 Arrhenius plot of log k versus $1/T$ data for the decomposition of N_2O_5 in the vapor state.

To see how the Arrhenius equation is treated algebraically, consider a second reaction in which reaction rate constants were determined at different temperatures:

$$k_1 = 9.67 \times 10^{-6} \quad \text{at } T_1 = 288 \text{ K}$$
$$k_2 = 6.54 \times 10^{-4} \quad \text{at } T_2 = 333 \text{ K}$$

Using the suitable form of the Arrhenius equation, we can calculate the activation energy E_a:

$$\log \frac{k_2}{k_1} = -\frac{E_a}{2.303R} \left[\frac{1}{T_2} - \frac{1}{T_1} \right]$$

$$\log \frac{6.54 \times 10^{-4}}{9.67 \times 10^{-6}} = -\frac{E_a}{2.303(8.314 \text{ J/mol·K})} \left[\frac{1}{333 \text{ K}} - \frac{1}{288 \text{ K}} \right]$$

$$E_a = 7.5 \times 10^4 \text{ J/mol}$$

Having obtained the value for E_a, the rate constant at 100°C (or any other reasonable temperature) can be calculated. For example, using data for k_1 and the value for E_a gives

$$\log \frac{k_3}{k_1} = -\frac{E_a}{2.303R} \left[\frac{1}{T_3} - \frac{1}{T_1} \right]$$

$$\log \frac{k_3}{9.67 \times 10^{-6}} = -\frac{7.5 \times 10^4}{(2.303)(8.314)} \left[\frac{1}{373} - \frac{1}{288} \right]$$

$$k_3 = 1.2 \times 10^{-2} \qquad \blacksquare$$

EXERCISE 24–2
(a) Use the data plot in Figure 24–6 to determine E_a by a graphical method.
(b) Then, taking the following rate constants from the same data plot,

FIGURE 24–6   Arrhenius plot of ln *k* versus 1/*T*.

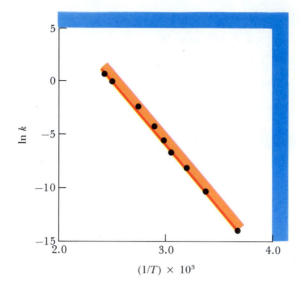

algebraically determine the value of the rate constant at 318 K:

$$k_1 = 4.5 \times 10^{-5} \quad \text{at 298 K}$$
$$k_2 = 1.0 \times 10^{-3} \quad \text{at 328 K}$$

Answer: (a) $E_a = 8.3 \times 10^4$ J; (b) $k = 4.1 \times 10^{-4}$.   ∎

24.3   KINETIC THEORIES

The primary goal of reaction kinetics is to find a theory that fits the facts, a mechanism that is consistent with reaction rate data. So far, we have been largely concerned with the operational elements of the science: measuring the rate of a reaction, establishing the experimental rate law, and finding the order. Essentially, the approach is phenomenological. But eventually we need to come to grips with hard questions about the underlying reality. What kind of molecular interpretation will support these facts? Can we suggest some theory of molecular interactions that is consistent with the dynamics we observe in the laboratory? In this section and the next (on reaction mechanisms) we'll try to introduce you to some conventional wisdom, but with the disclaimer that it may not be wisdom at all. Reaction kinetics is much more subject to interpretation than most other topics discussed in this first text on chemistry.

Collision Theory and Activation Energies

Let's consider the oxidation of CO and NO:

$$2 \text{ CO (g)} + \text{O}_2 \text{ (g)} \longrightarrow 2 \text{ CO}_2 \text{ (g)}$$
$$2 \text{ NO (g)} + \text{O}_2 \text{ (g)} \longrightarrow 2 \text{ NO}_2 \text{ (g)}$$

As a starting point, assume that molecules must collide in order to react. We could reason that simple collisions are controlling the rate of reaction. Then we could argue that the two processes should follow along comparable pathways from reactants to products. But as we noted in the introduction to the last chapter, the facts are contradictory. At ordinary temperatures, the oxidation of CO proceeds at a negligible rate while NO is subjected to rapid oxidation. Yet, the equilibrium constants for both reactions are so large that both oxidations should be essentially complete at equilibrium.

One can reasonably assume that both reactions require molecular collisions. It is also reasonable to assume that under the same conditions of temperature and concentration, comparable numbers of collisions occur. But even the faster NO oxidation is really a very slow reaction compared to the rate for a hypothetical model reaction in which we assume that each collision results in reaction. Consequently, the number of **effective collisions**—those leading to reaction—must be very small compared to the total number of collisions that occur. Any attempt at a general understanding of reaction rates and processes must be able to explain why so few collisions are effective, and why the fraction of collisions that are effective is greater for some reactions than for others.

The Arrhenius theory of reaction rates begins with the assumption that when two molecules react in a gas phase process such as the oxidation of NO, only exceptional collisions between the molecules, oxygen and nitric oxide, cause reaction. The molecules that reacted were "activated," and we say that these successful collisions passed through an energy-rich **activated complex**. According to the Arrhenius point of view, the fraction of collisions producing reaction is the ratio of the effective collisions to total collisions:

$$\frac{\text{effective collisions}}{\text{total collisions}} = e^{-E_a/RT}$$
$$N = N_0 e^{-E_a/RT}$$

where

N = the number of effective collisions per liter per second

N_0 = the total number of collisions per liter per second

$e^{-E_a/RT}$ = the fraction of molecules with kinetic energy E_a or greater

We may therefore say that the effective reaction rate is given by

$$\text{rate} \propto [N_0 \text{ collisions/L·s}]e^{-E_a/RT}$$

The rate constant is thus

$$k = Ae^{-E_a/RT}$$

When molecules collide:

- If their kinetic energy is less than E_a, then nothing happens—there is no reaction.
- If their kinetic energy is equal to or greater than E_a, then a reaction takes place.
- The larger the value of E_a, the smaller the value for N—note the negative exponent—and the slower the reaction at a given temperature. As T increases, more collisions are effective—as T increases, $e^{-E_a/RT}$ increases, too.

If E_a is high, only the occasional highest-energy collisions will successfully lead to reaction. On the other hand, an increase in temperature leads to larger numbers of molecules moving at greater speeds, increasing the rate of effective collisions and the rate of reaction.

The reaction rate theory sits squarely on this concept of the energy of activation (E_a), the energy barrier between reactants and products. In a

FIGURE 24–7 Relationships between the activation energies E_a for the forward reaction and the reverse reaction and the net change in internal energy, ΔE.

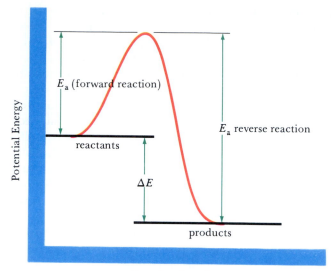

Reaction Coordinate

E_a/R can be thought of as the temperature at which the reaction occurs at the equivalent of $1/e$, or 37%, of its maximum rate. The e is 2.718, the base of natural logarithms.

general way the transition from reactants to products can be illustrated by following the energy of the reacting species along a **reaction coordinate**—a measure of the completeness of the process by which a set of reactant molecules is transformed into a set of product molecules.

Figure 24–7 shows the relation between the amounts of energy required for the forward and reverse reactions and the energy change for the overall process. Note that there is one value of the constant E_a for the forward reaction, and a different value for the reverse direction. In this case, reactants are shown with a higher energy than the products. That situation should be energetically favorable for reaction. However, thermodynamics considerations alone are insufficient. The reaction must move along the reaction coordinate, following the only "allowable" pathway, and passing through some activated state before finally decomposing to products. In order for that to happen, the additional energy E_a must be acquired.

We can express the probability that a given molecule will possess this added energy at a particular temperature by the exponential term $e^{-E_a/RT}$ in the Arrhenius equation. If the activation energy is roughly 80,000 J/mol, the value for $e^{-E_a/RT}$ at two temperatures, say 300 K and 400 K degrees, will be the following:

At 300 K:

$$e^{-E_a/RT} = e^{-[80,000/(8.314 \times 300)]} = 1.18 \times 10^{-14}$$

At 400 K:

$$e^{-E_a/RT} = e^{-[80,000/(8.314 \times 400)]} = 3.57 \times 10^{-11}$$

An increase in temperature of 100 degrees increases the value of the exponential by a factor of more than 10^3. The physical significance is this: As the temperature is raised, the proportion of molecules in the activated state has increased (Figure 24–8). And since reactants must move along the reaction coordinate through the activated complex, that must also favor product formation. In general, a bimolecular process that has an E_a of about 80,000 J/mol will proceed at a moderate rate at ordinary temperatures. Above 170,000 J/mol, elevated temperatures are required to produce a sufficient concentration of activated complexes to sustain the reaction.

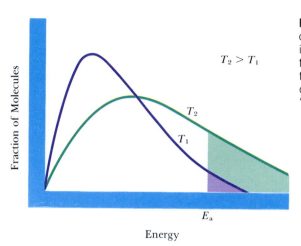

FIGURE 24–8 Distribution of energies for a set of reacting molecules at two different temperatures. At the higher temperature, a larger fraction of the molecules present are "activated."

The Theory of the Transition State

In spite of the plausibility of the Arrhenius collision theory and its general utility, there are many reactions for which agreement between theory and observation is less than satisfactory. For example, not all collisions between molecules with the necessary energy of activation result in product formation. Therefore our reaction rate theory must be viewed as being incomplete. The molecules are energized but still fail to get to the activated state. Apparently there must be more to chemical reaction than activation energy. The way in which molecules react must also have some bearing on whether anything happens.

Consider a collision between two molecules of A. Energy is transferred from one to the other, and we are left after the event with a molecule of below average energy and an another of above average energy. If enough energy is transferred, the molecule may be considered to be energized (which we designate with an asterisk as A*). Energized molecules can do one of two things. They can collide with inactive molecules in what we would regard as fruitless collisions because they accomplish little more than redistribution of their collective energies—a pair of inactive molecules result. Or one energized molecule can collide with another, in which case one of them will acquire the extra boost necessary to get beyond energized, to the activated state, and then on to reaction products:

Fruitless collisions

Molecular collisions, resulting in energized, but not **activated**, molecules— no products form:

$$A + A \longrightarrow A + A*$$

Molecular collisions, resulting in redistribution of energies—no products form:

$$A* + A \longrightarrow A + A*$$
$$A* + A \longrightarrow A + A$$

☐ Fruitless collisions
☐ Fruitful collisions

Fruitful collisions

Molecular collisions between energized molecules, leading to activated states— products form:

$$A^* + A^* \longrightarrow \text{products}$$

In another scenario, let's suppose some particular molecular arrangement or configuration is required before reaction AB + C can occur, producing BC. A particular collision with the right energy but the wrong configuration would be fruitless.

Fruitful reactions

$$A{-}B + C \longrightarrow [A{-}B{-}C^*]^{\neq} \longrightarrow A + B{-}C$$

Fruitless reactions

$$C + A{-}B \longrightarrow [C{-}A{-}B] \longrightarrow C + A{-}B$$

The symbol $\neq$ marks the **activated complex**, a mid-reaction geometry that is required in order to go on to the products. An ABC configuration can produce the desired BC product, but CAB cannot, leading to different products or no products at all.

In still another scene, imagine we have two reacting molecules that require a substrate to serve as a molecular organizer or template, bringing them into proper spatial orientation or alignment. Here again we have introduced a required element of reaction beyond simply energizing molecules. To attain the activated state, the reacting atoms and bonds need to be somehow properly directed. One possibility might begin with an equilibrium establishing the activated complex in a three-body event, followed by collapse of the complex to products.

$$M + A + B \longrightarrow [\text{complex}] \longrightarrow \text{products} + M$$

A more likely route is bimolecular rate-controlling formation of a complex, followed by a fast, second event leading to products:

$$M + A \longrightarrow [\text{complex}]$$
$$[\text{complex}] + B \longrightarrow \text{products} + M$$

Transition state theory is a statistical, more modern approach to reaction kinetics. In many situations the fit with data from experiment is better than for the older collision theory. Instead of bothering with details of events involving colliding molecules, transition state theory follows the reaction's potential energy profile. Reactant molecules approach each other from a great distance, pass through some activated state, and finally move apart as molecular products. The kinetic event leading to reaction can be characterized in terms of the activated complex, with the rate being a function of the number of complexes that succeed in advancing to the top and over to the other side of the potential energy curve per unit time.

24.4 REACTION MECHANISMS

"The primary goal of reaction kinetics is to find a theory that fits the facts, a mechanism that is consistent with reaction rate data." With those words we began the preceding section on kinetic theories. It is now left for us to deal with the mechanisms themselves, and with the thorny problem of using facts from experiment to deduce a molecular model for the probable pathway from reactants to products. Throughout this discussion, it is essential to keep in mind that as complex as the overall reaction pathway may seem to be, nature operates more modestly, in simple elementary processes. The task at hand is to uncover these mostly unimolecular and bimolecular one-step conversions and to patch them together so that we gain useful information about how chemical reactions happen.

EXAMPLE 24–3(A)   Decomposition of Acetaldehyde

If the thermal decomposition of acetaldehyde to methane and carbon monoxide in the gas phase is known from experiment to follow a second order rate law, what reasonable mechanism would you propose?

Solution

$$\text{rate} = -\frac{\Delta[CH_3CHO]}{\Delta t} = k_{exp}[CH_3CHO]^2$$

Because the acetaldehyde concentration appears to the second power in the rate law, one possible mechanism includes a rate-controlling bimolecular event in which there is some close approach, physical contact, or direct collision:

$$CH_3CHO \text{ (g)} + CH_3CHO \text{ (g)} \xrightarrow[\text{slow}]{k} 2\, CH_4 \text{ (g)} + 2\, CO \text{ (g)}$$

EXAMPLE 24–3(B)   Acetaldehyde and Iodine Vapor

When acetaldehyde is decomposed in the presence of iodine vapor,

$$CH_3CHO \text{ (g)} \longrightarrow CH_4 \text{ (g)} + CO \text{ (g)}$$

the kinetic description is also second order, but the experimental rate law is different:

$$\text{rate} = -\frac{\Delta[CH_3CHO]}{\Delta t} = k_{exp}[CH_3CHO][I_2]$$

What reasonable mechanism explains this observation?

Solution   There is no evidence of any iodine-containing products, suggesting the presence of an intermediate that enters into the reaction. Since I_2 appears in the experimental rate law, it must be a reactant in the slow step:

(1) $$CH_3CHO + I_2 \xrightarrow[\text{slow}]{k_1} CH_3I + HI + CO$$

(2) $$CH_3I + HI \xrightarrow[\text{fast}]{k_2} CH_4 + I_2$$

Adding the two elementary steps depicted in equations 1 and 2 cancels all the intermediate iodine-containing species, and regenerates the original iodine. The overall reaction (3) is simple decomposition, where step (1) is the slow, rate-controlling step:

(3) $$CH_3CHO \longrightarrow CH_4 + CO$$

If we could go into the laboratory in a separate experiment and show that the reaction of hydrogen iodide and methyl iodide is a fast reaction leading to methane and iodine as step 2 suggests, we would have a consistent mechanism with independent chemical support for our proposal. The mechanism is not proved, but it has been established beyond a reasonable doubt, which is all we can hope for.

EXAMPLE 24–3(C)   Persulfate Reduction by Iodide Ion

Consider the reduction of potassium persulfate by iodide ion in solution, a classic undergraduate kinetics experiment:

$$S_2O_8{}^{2-} \text{ (aq)} + 2 \text{ I}^- \text{ (aq)} \longrightarrow 2 \text{ SO}_4{}^{2-} \text{ (aq)} + I_2 \text{ (aq)}$$

The kinetic data support a second order rate law:

$$\text{rate} = k_{exp}[S_2O_8{}^2][I^-]$$

What mechanism would you suggest?

Solution   A possible mechanism might begin with the slow formation of an intermediate complex, followed by a rapid second step that instantaneously transforms it to products. In effect, that says there is never any measurable concentration of the intermediate. The sum of the two-step mechanism gives the stoichiometric equation:

$$S_2O_8{}^{2-} + I^- \underset{k_{-1}}{\overset{k_1}{\rightleftharpoons}} [\text{complex}]$$

$$[\text{complex}] + I^- \overset{k_2}{\longrightarrow} 2 \text{ SO}_4{}^{2-} + I_2$$

Beyond simple consistency with stoichiometry and rate law, little else is provided here. One would hope to be able to find experimental evidence concerning the nature of the intermediate complex. It is important to remember that kinetics data and the rate laws do not prove mechanisms, but are necessary conditions. Mechanisms are always subject to changes in our understanding of them. ■

EXAMPLE 24–3(D)   Peroxide Oxidation of Iodide Ion

When hydrogen peroxide is used to bring about the oxidation of iodide ion to iodine, the reaction is best carried out in acid solution:

$$H_2O_2 \text{ (aq)} + 2 \text{ H}^+ \text{ (aq)} + 2 \text{ I}^- \text{ (aq)} \longrightarrow I_2 \text{ (aq)} + 2 \text{ H}_2O \text{ (liq)}$$

A second order rate law is observed with concentration dependencies that are first order in peroxide and first order in iodide:

$$\text{rate} = k_{exp}[H_2O_2][I^-]$$

Propose a reasonable mechanism.

Solution   Our proposed mechanism begins with a rate-controlling reaction between peroxide and iodide, followed by a pair of fast reactions:

$$(1) \qquad H_2O_2 + I^- \underset{\text{slow}}{\overset{k_1}{\longrightarrow}} OH^- + HOI$$

$$(2) \qquad OH^- + H^+ \underset{\text{fast}}{\overset{k_2}{\longrightarrow}} H_2O$$

$$(3) \qquad H^+ + I^- + HOI \underset{\text{fast}}{\overset{k_3}{\longrightarrow}} I_2 + H_2O$$

Again, chemical information from experiment is required to confirm (or at least increase confidence in) the proposed steps. ■

EXAMPLE 24–3(E)   Dinitrogen Pentoxide Decomposition

The stoichiometric equation can be written as follows:

$$N_2O_5 \text{ (g)} \longrightarrow 2 \text{ NO}_2 \text{ (g)} + \tfrac{1}{2} O_2 \text{ (g)}$$

Kinetics data suggest the following experimental rate law:

$$\text{rate} = -\Delta[N_2O_5]/\Delta t = k_{\text{exp}}[N_2O_5]$$

and the following mechanistic scheme has been proposed:

(a) Establishment of a rapid equilibrium between N_2O_5, NO_2, and NO_3:

$$N_2O_5 \underset{k_{-1}}{\overset{k_1}{\rightleftharpoons}} NO_2 + NO_3$$

(b) That is followed by a rate-controlling slow step:

$$NO_2 + NO_3 \xrightarrow[\text{slow}]{k_2} NO_2 + O_2 + NO$$

(c) Then, two fast reactions:

$$NO + NO_3 \xrightarrow[\text{fast}]{k_3} 2\ NO_2$$

$$NO_2 + \tfrac{1}{2}\,O_2 \xrightarrow[\text{fast}]{k_4} NO_3$$

Note that the four elementary reactions (not including k_{-1}) add up to the stoichiometric equation. Show that the rate of oxygen formation, which is one of three experimental handles we might choose to use in studying the reaction—disappearance of N_2O_5, or formation of either nitrogen dioxide or oxygen—is directly proportional to $[N_2O_5]$.

Solution   Begin with the rate law, written in terms of oxygen formation, as established by the slow step:

$$\Delta[O_2]/\Delta t = k_2[NO_2][NO_3]$$

From the equilibrium we can write the following:

$$\text{rate forward} = k_1[N_2O_5]$$
$$\text{rate reverse} = k_{-1}[NO_2][NO_3]$$

At equilibrium, rate forward = rate reverse,

$$k_1[N_2O_5] = k_{-1}[NO_2][NO_3]$$

The equilibrium constant is the ratio of the rate constants for the forward and reverse reactions:

$$k_1/k_{-1} = [NO_2][NO_3]/[N_2O_5] = K_{\text{equil.}}$$

Now solve this expression for the $[NO_2]$ and $[NO_3]$ terms

$$[NO_2][NO_3] = K_{\text{equil.}}[N_2O_5]$$

and substitute back into the original rate law

$$\Delta[O_2]/\Delta t = k_2[NO_2][NO_3]$$
$$\Delta[O_2]/\Delta t = k_2 K_{\text{equil.}}[N_2O_5]$$
$$= k[N_2O_5], \text{ where } k = k_2 K_{\text{equil.}}$$

Finally, it should be a straightforward matter to equate k with k_{exp} by going back to the original equation and rate law:

$$-\Delta[N_2O_5]/\Delta t = k_{\text{exp}}[N_2O_5] = 2\,\Delta[O_2]/\Delta t = 2\,k[N_2O_5]$$

Therefore,

$$k_{\text{exp}} = 2\,k$$

EXERCISE 24–3(A)

Beginning with the statement of the equilibrium below, write an equation for the net rate of disappearance of N_2O_4 in terms of N_2O_4 and NO_2, and develop an expression for the equilibrium constant in terms of the rate constants for the elementary processes for the dissociation.

$$N_2O_4 \underset{k_{-1}}{\overset{k_1}{\rightleftharpoons}} 2\,NO_2 \qquad \textit{Answer:} \text{ See p. 835.} \qquad ■$$

EXERCISE 24–3(B)

There are two forms of molecular hydrogen, parahydrogen and ortho-hydrogen, and conversion takes place homogeneously at a convenient rate at 1000 K. The order of the reaction is $\frac{3}{2}$ in parahydrogen, and the following mechanism has been suggested:

$$p\text{-}H_2 \underset{k_{-1}}{\overset{k_1}{\rightleftharpoons}} 2\,H \qquad \text{fast, equilibrium}$$

$$H + p\text{-}H_2 \overset{k_2}{\longrightarrow} o\text{-}H_2 + H \qquad \text{slow}$$

Derive a rate law for the formation of $o\text{-}H_2$ that is consistent with the mechanism proposed above and with the experimentally observed order of the reaction. ■

These molecules are isomers with different nuclear spins, aligned in the case of para-H_2. Below 20 K, the ratio of ortho to para is about 1:100. In the vicinity of room temperature, it is about 1:4.

24.5 CATALYSTS

Early in the 19th century, Jöns Jakob Berzelius (Figure 24–9) described a number of laboratory situations in which chemical reactions were subject to the effects of certain substances that themselves remained unchanged. He described these reactions as **catalyzed**. In the modern world, industrial manufacture of just about everything from fuels to drugs is pushed along by catalyzing agents called accelerators. Others, called inhibitors, are added to slow the spoilage of foods and reduce the velocities of a host of important reactions to acceptable, manageable levels. Recall our introduction to kinetics at the beginning of Chapter 23. Simply stated, a **catalyst** is a substance that accelerates or inhibits—changes—the rate of a chemical reaction without being consumed in the process. If a substance is to be truly a catalyst, it should show up at the end of the reaction in the same form and quantity as at the beginning.

Some fifty years after Berzelius, Wilhelm Ostwald was able to demonstrate conclusively for the first time that catalysts do not increase the yields in chemical reactions because they possess some hidden, magical qualities. He recognized that a given chemical reaction must be subject to the usual equilibrium conditions and constraints. No catalyst could operate beyond the theoretical limits set by the second law of thermodynamics. However, the approach to the state of equilibrium might be so slow as to make realization of the theoretical yield of reaction product(s) all but unattainable under normal circumstances. That is the basis for our modern understanding of catalytic activity—speeding the approach to the equilibrium state (if the substance is an accelerator). Catalysts open alternate pathways with different activation energies. Catalysis offers hope for many thermodynamically feasible but kinetically unfavorable reactions.

Catalyzed reactions fall into two major categories, depending on whether they involve homogeneous processes occurring in solution, or heterogeneous processes taking place at the interface between two distinct phases. The mechanisms involved in catalyzed reactions are often poorly defined

FIGURE 24–9 Berzelius (Swedish) was perhaps the world's greatest authority on chemistry in the early decades of the 19th century. His contributions to analysis and nomenclature include confirming Dalton's work, establishing the symbols for the elements, and chemical formulas; he made major contributions to the new electrical science that was just emerging and opened the field of catalysis.

TABLE 24–2 Activation Energies for Some Catalyzed Reactions

Reaction	Catalyst	E_a, kJ/mol
Homogeneous		
decomposition of H_2O_2	uncatalyzed	75
	iodide ion	58
	Fe^{3+}	42
	catalase (enzyme)	4
hydrolysis of grape sugar	H_3O^+	109
	sucrase (enzyme)	46
hydrolysis of urea	H_3O^+	105
	urease (enzyme)	54
Heterogeneous		
hydrogenation of ethylene	uncatalyzed	188
	Pt (on carbon)	146
	Ni (finely divided)	84

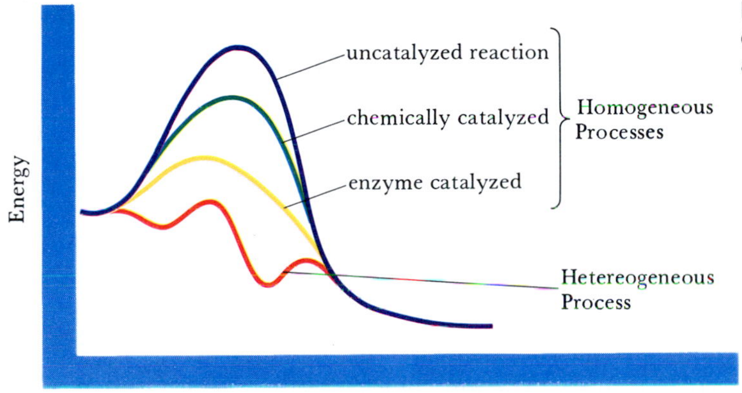

FIGURE 24–10   A family of generalized potential energy diagrams for a variety of catalyzed chemical reactions.

at best. In practice, a catalyst will be found to exhibit comparable effects on both the forward and reverse reactions; in principle, a catalyst must have identical acceleration or retarding effects in both directions. In many situations, the activation energy for a catalyzed reaction is indeed found to be different than that for the same process in the absence of the catalyst material, as you can see from the data in Table 24–2. The presence of the catalyst material has served to shift the mechanistic pathway to one having a lower activation energy, facilitating conversion of reactants to products (Figure 24–10).

Catalyzed and Uncatalyzed Reactions

In addition to modifying the rate of a given chemical reaction, catalysts are often used in situations where several reactions are possible; the catalyst favors one, effectively excluding the others. For example, depending on the nature of the catalyst, entirely different products can be obtained from the reaction of carbon monoxide and hydrogen (Table 24–3). Such selectivity obviously has great economic as well as scientific significance.

The essential feature of the catalyzed reaction is the entry of the catalyst into the process, followed by its regeneration at the end of the reaction cycle. Consequently, each catalyst unit can cause many reactant molecules to be converted to products. Consider a slow reaction in which reactant molecules (A) are converted into product molecules (C). On addition of

Another example of a homogeneous process, which is currently receiving a great deal of attention because of its environmental significance, is the decomposition of ozone. In the gas phase, it is known to be catalyzed by the presence of oxides of nitrogen (which have been widely introduced into the atmosphere as automobile emissions).

TABLE 24–3   The CO/H₂ Reaction

Conditions	Reaction Products
Ni catalyst, 100°C, 1 atm	$CH_4 + H_2O$
$Zn(CrO_2)_2$, 400°C, 500 atm	$CH_3OH + H_2O$
ThO_2, 400°C, 200 atm	branched hydrocarbons suitable for use in unleaded gasolines

catalyst molecules (X), an alternate, faster reaction pathway is available:

without the catalyst:

$$A \longrightarrow C$$

with the catalyst:

$$A + X \longrightarrow B$$
$$B \longrightarrow X + C$$

The entry of X into the catalyzed reaction scheme causes large numbers of reactant A molecules to be converted to product C molecules. The net result is still the same as the uncatalyzed process. What makes the overall process catalytic is the continuous regeneration of the X molecules.

Homogeneous Catalysis

When the catalyst is in the same phase (most often aqueous solution) as the reacting species, the process is **homogeneous catalysis**. A classic example is the electron-transfer reaction between Ce^{4+} and Tl^+ ions:

$$2\ Ce^{4+}\ (aq) + Tl^+\ (aq) \longrightarrow 2\ Ce^{3+}\ (aq) + Tl^{3+}\ (aq)$$

It is generally believed that the reaction is slow because Tl^+ ions lose two electrons in going to product ions while each Ce^{4+} ion undergoes a one-electron transfer:

$$Tl^+ \longrightarrow Tl^{3+} + 2\ e^-$$
$$2\ Ce^{4+} + 2\ e^- \longrightarrow 2\ Ce^{3+}$$
$$\overline{2\ Ce^{4+} + Tl^+ \longrightarrow 2\ Ce^{3+} + Tl^{3+}}$$

Mn^{2+} successfully catalyzes the reaction, perhaps because it separates the one-electron and two-electron transfers from each other:

$$Ce^{4+} + Mn^{2+} \longrightarrow Ce^{3+} + Mn^{3+}$$
$$Ce^{4+} + Mn^{3+} \longrightarrow Ce^{3+} + Mn^{4+}$$
$$Tl^+ + Mn^{4+} \longrightarrow Tl^{3+} + Mn^{2+}$$
$$\overline{2\ Ce^{4+} + Tl^+ \longrightarrow 2\ Ce^{3+} + Tl^{3+}}$$

The net result is the same, catalyzed or uncatalyzed. Mn^{2+} ions were consumed and then regenerated.

To find out the whole story of "drugstore" peroxide, you have to read the fine print on the label. Generally an inhibitor has been added. Organic molecules called quinones are commonly used for this purpose.

The fact that you can purchase an aqueous solution of hydrogen peroxide of various concentrations with a substantial shelf life is a reflection of just how slow the decomposition is at 25°C:

$$2\ H_2O_2\ (liq) \longrightarrow 2\ H_2O\ (liq) + O_2\ (g)$$

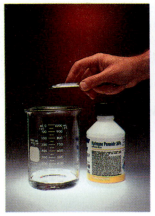
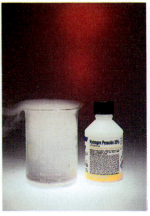

FIGURE 24-11 The catalytic decomposition of 30% hydrogen peroxide solutions by "trace" quantities of transition metal oxides is a rapid, exothermic reaction, heating the solution quickly to the boiling point of water, generating copious quantities of steam, and further accelerating the decomposition by the temperature increase.

But just a drop or two of a dilute solution of Fe^{3+} ions (Table 24–2) is sufficient to initiate the catalyzed peroxide decomposition reaction (Figure 24–11). Enzymes are very large, high molecular weight protein molecules generally containing a few active sites that serve as seeds for catalytic activity. These giant organic molecules are among the most effective and uniquely specialized catalyst materials we know. For example, one molecule of the Fe^{3+}-containing enzyme catalase brings about the decomposition of 5 million molecules of H_2O_2 per minute at 273 K. Since hydrogen peroxide (even in dilute solutions) is an acutely toxic substance, and since it is produced as a byproduct of a number of other enzyme reactions, catalase plays the role of protector, destroying any hydrogen peroxide (Figure 24–12).

Hydroformylation: The Oxo Process

This process uses homogeneous catalysis to convert the double bonds in olefins (R—CH=CH₂, where R represents the rest of the hydrocarbon) to aldehydes (R—CH₂CH₂CHO) by insertion of carbon monoxide (CO) in the presence of cobalt hydridotetracarbonyl (HCo[CO]₄), a transition metal complex. Propylene is primarily used as the olefin; *n*-butyraldehyde is the principal product, which is finally converted into detergents, lubricating oils, plasticizers, and solvents in Western Europe, Japan, the United States, and the USSR.

$$CH_3—CH=CH_2 \xrightarrow[HCo[CO]_4]{CO/H_2} CH_3—CH_2—CH_2—CHO$$
propylene $\qquad$ *n*-butyraldehyde

FIGURE 24-12 The bombardier beetle uses the enzyme-catalyzed decomposition of hydrogen peroxide as a defense mechanism. Oxygen gas, formed in the reaction, is used to expel water and other chemicals with explosive force, and since peroxide decomposition is very much an exothermic reaction, the water is ejected as steam.

Heterogeneous Catalysis

When the catalyst in a chemical reaction is (or is part of) a second phase, the process is described as **heterogeneous catalysis**. Many millions of tons of fertilizer ammonia are produced each year by the heterogeneous Haber process:

$$N_2 \text{ (g)} + 3 H_2 \text{ (g)} \longrightarrow 2 NH_3 \text{ (g)}$$

The process uses a catalyst consisting of an iron powder doped with a few

FIGURE 24–13 These finely divided catalyst materials of palladium on charcoal, supported on a ceramic material, are used for converting methane and steam to carbon monoxide and hydrogen—synthesis gas—which is a major industrial feed for commodity chemicals such as methanol (Chapter 6).

Development of multi-layered packaging materials have eliminated much of the need for inhibitors in food such as potato chips. See the discussion in Chapter 20.

percent K_2O and Al_2O_3 that act as promoters, improving the overall performance of the metal catalyst. Another important example of heterogeneous catalysis is the hydrogenation of ethylene on the surface of a finely divided nickel or palladium catalyst (Figure 24–13):

$$CH_2{=}CH_2 \text{ (g)} + H_2 \text{ (g)} \xrightarrow{\text{Ni}} CH_3CH_3 \text{ (g)}$$

The steps involved in a heterogeneous catalysis process can be generalized as follows:

1. Reactants diffuse to the catalyst surface.
2. Reactants adsorb onto the catalyst surface.
3. Adsorbed reactants are converted to adsorbed products.
4. Adsorbed products diffuse away.

Free Radical Processes

A **free radical** is a chemical species characterized by an unpaired electron, often present as a result of homolytic cleavage of a bond to yield two neutral species. The presence of the unpaired electron makes most radicals very reactive. The chemistry of free radicals is dominated by **chain reactions** in which a single radical molecule is consumed and regenerated many times.

There are many important chain reactions among industrial, commercial, and biological processes. Some are desirable, and others are not. Among the desirable ones are the large-scale syntheses of plastic polymers. Undesirable free radical chain reactions include the pinging or knock in internal combustion engines and the spoilage of foods; certain aspects of the aging process in humans and animals are also believed to be promoted by free radicals.

As we shall see, undesired chain reactions can often be minimized by introducing substances that interrupt the chain. Very small amounts of lead and manganese compounds in motor fuels prevent engine knock; trace quantities of antioxidants prevent spoilage of foods caused by air oxidation. In some investigations, vitamin E (alpha-tocopherol) appears to interfere with free radical oxidation reactions associated with aging processes.

Additives in Foods: A wide range of chemicals are added to many foods to retard the rate of oxidative breakdown of fats and oils. By inhibiting free radical chain reactions, they delay the onset and limit the degree of rancidity and the accompanying objectionable tastes and odors. In addition, they minimize oxidative destruction of certain vitamins and essential fatty acids. Their action is due to the ability to react with oxygen free radicals more readily than can the oxidizable substances in foods. For example, potato chips containing an antioxidant will remain fresh at 65°F for up to a month. Without the antioxidant, rancidity becomes evident within a few days at room temperature.

Knock Inhibitors in Motor Fuels: When uncontrolled burning occurs in a spark-ignition internal combustion engine, there is a sudden increase in pressure; this is accompanied by sound waves that produce a pinging or knocking. Over thousands of miles of driving, this phenomenon can damage the engine. It is due to free radical reactions that take place during the heating and compression of the fuel-air mixture prior to ignition. Certain gasoline-soluble organometallic compounds, such as tetraethyllead, inhibit the reactions that cause knocking.

However, the widespread use of tetraethyllead in motor fuels for more than half a century caused unacceptable levels of lead pollution in the environment. Ecological concern in the United States and elsewhere resulted in legislation phasing out leaded motor fuels, and gasolines are almost lead-free today. Chemical research has shown that the same knock inhibition can be accomplished by altering the structure and composition of the fuel itself. The cost of this "reforming" process is much greater than that of the lead antiknock additive, but the environmental benefits are greater still.

Gas Phase Chain Reactions: A gaseous mixture of Cl_2 and H_2 molecules will not form HCl at any appreciable rate, even though the thermodynamics are highly favorable. However, if we introduce only a few atoms of uncombined chlorine, the result is a large-scale conversion that proceeds at a dangerously explosive rate if proper care isn't taken:

$$H_2 \text{ (g)} + Cl_2 \text{ (g)} \longrightarrow 2 \text{ HCl (g)}$$

To explain these results, we propose a sequence of reactions known as a chain reaction:

1. To begin the reaction, we need to generate a few chlorine atoms. To do this, we expose molecular chlorine to light of a certain wavelength; sunlight will do. The resulting homolytic cleavage of the Cl—Cl bond produces a pair of neutral atoms:

$$Cl_2 \text{ (g)} + h\nu \longrightarrow 2 \text{ Cl} \cdot \text{ (g)}$$

The dot represents the free (unpaired) electron, and the species is a free radical. This reaction is called the **initiation step** of the chain reaction.
2. The Cl· free radicals created in the initiation step are highly reactive, attacking the bonds of H_2 molecules. This reaction creates an HCl molecule and an H· radical, which in turn can attack a Cl_2 molecule:

$$Cl \cdot + H_2 \longrightarrow HCl + H \cdot$$
$$H \cdot + Cl_2 \longrightarrow HCl + Cl \cdot$$

The radicals resulting from these steps can go on to attack a second molecule, and a third, and so on. These are referred to as **propagation reactions**. It would not be unusual for a light-induced chain reaction initiated by the pair of radicals from a single Cl_2 molecule to produce a million HCl molecules.
3. The chain reaction does not go on forever, or even to the point at which one of the reactants is used up. At some point, by chance, the chain carriers—the Cl· and H· free radicals—enter into **termination reactions**. There are three possibilities, and in each of them the free radicals required to continue the chain reactions are consumed without replacement:

Recombination of chlorine atoms:

$$Cl \cdot + Cl \cdot \longrightarrow Cl_2$$

Recombination of hydrogen atoms:

$$H \cdot + H \cdot \longrightarrow H_2$$

Combination of hydrogen and chlorine atoms:

$$H \cdot + Cl \cdot \longrightarrow HCl$$

In addition, other compounds that are present in the reaction mixture—

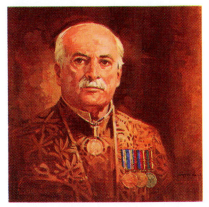

To better understand how nickel forms a volatile compound with carbon monoxide, Paul Sabatier chose to study the reaction of nickel with ethylene and discovered ethane among the product gases. Apparently, a cheap metal, nickel, had "catalyzed" the addition of hydrogen to the double bond, opening the way to large industrial-scale catalytic processes. An immediate result was the formation of edible fats such as margarine and shortenings from otherwise inedible oils such as cottonseed, on a world scale and with great economy.

such as the antioxidants discussed earlier—can react with the chain carriers to produce much less reactive species, thus disrupting the chain.

Among the interesting experimental features of chain reactions are auto-inhibition and fractional orders. Free radical-initiated chain reactions are often subject to **auto-inhibition** by the product of the reaction (as the product concentration increases), because the reverse reaction's rate increases:

$$H\cdot + HCl \longrightarrow H_2 + Cl\cdot$$

That presents us with a faithful test for the presence of a free radical chain process. Increase the concentration of the product and see what happens to the rate of the reaction. If the reaction is free radical in nature, it should be inhibited.

The complicated empirical rate law (below) with its **fractional exponents** does more than suggest hidden complexity for the apparently simple gas phase reaction between hydrogen and bromine:

$$H_2 \text{ (g)} + Br_2 \text{ (g)} \longrightarrow 2 \text{ HBr (g)}$$

It should remove the last shreds of doubt that the form of the rate law can be determined only by experiment. Clearly, it could not have been predicted from the stoichiometric equation for the overall reaction.

$$\frac{\Delta[HBr]}{\Delta t} = \frac{k'[H_2][Br_2]^{1/2}}{1 + k''([HBr]/[Br_2])}$$

One of the essential themes of chemical kinetics is to offer reasonable explanations for experimental rate laws. Obviously this reaction is particularly difficult, but it has been studied extensively and a great deal is known about it. For the time being, let's limit ourselves to the proposed steps leading to the overall reaction. They provide a review of the typical chemical events found in free radical gas phase processes.

EXAMPLE 24–4

Summarize the steps involved in a typical chain reaction by writing out the initiation, propagation, and termination steps for the HBr synthesis. Since it is known to be a free radical-initiated process, include an inhibition step, too.

Solution

Initiation:	$Br_2 \text{ (g)} + h\nu \longrightarrow 2 \text{ Br}\cdot \text{ (g)}$
Propagation:	$Br\cdot \text{ (g)} + H_2 \text{ (g)} \longrightarrow HBr \text{ (g)} + H\cdot \text{ (g)}$
	$H\cdot \text{ (g)} + Br_2 \text{ (g)} \longrightarrow HBr \text{ (g)} + Br\cdot \text{ (g)}$
Termination:	$H\cdot \text{ (g)} + Br\cdot \text{ (g)} \longrightarrow HBr \text{ (g)}$
	$H\cdot \text{ (g)} + H\cdot \text{ (g)} \longrightarrow H_2 \text{ (g)}$
	$Br\cdot \text{ (g)} + Br\cdot \text{ (g)} \longrightarrow Br_2 \text{ (g)}$
Inhibition:	$H\cdot \text{ (g)} + HBr \text{ (g)} \longrightarrow H_2 \text{ (g)} + Br\cdot \text{ (g)}$
	$Br\cdot \text{ (g)} + HBr \text{ (g)} \longrightarrow Br_2 \text{ (g)} + H\cdot \text{ (g)}$

SUMMARY

In this chapter, we considered more complex situations in our coverage of chemical kinetics, the study of the rates of chemical reactions. We began by looking at chemical processes where there are competing reactions or reactions that take place in consecutive steps. In the cases where a reaction is at equilibrium, we showed that the equilibrium constant, K, is the ratio of the rate constants for the forward and reverse processes.

As part of our mechanistic picture, we often include concepts such as activation energies and energy barriers, collision models, and transition states, in order to tune our kinetic theories to experimental rate data. The activation energy can be obtained by studying rates of reactions at different temperatures. The relationship between $\ln k$ and $1/T$ is linear, so the analysis can be algebraic or graphical. We assume activation energies to be constant over reasonable temperature ranges, so we can write

$$\ln k = \ln A - \frac{E_a}{RT}$$

$$\ln \frac{k_2}{k_1} = -\frac{E_a}{R}\left[\frac{1}{T_2} - \frac{1}{T_1}\right] = -\frac{E_a}{R}\left[\frac{T_1 - T_2}{T_1 T_2}\right]$$

Catalysts have gained considerable attention industrially and biologically because of the controls they provide for chemical reactions. Catalysts sometimes allow us to select the products formed in a reaction from among several different pathways, and often they give us the power to control the rate. Heterogeneous reactions depend on adsorption phenomena taking place on the catalyst surface or at the surface of the reacting material. Free radical processes are usually fast chain reactions.

Finally, and perhaps most importantly, in no other place is the experimental nature of the natural sciences more evident than in studies of kinetics and mechanisms of reactions.

QUESTIONS AND PROBLEMS

QUESTIONS

1. What do we mean by the term "activation energy"?

2. State two reasons why reaction rates increase with temperature.

3. What's wrong with the following logic? Reaction 1 is exothermic and reaction 2 is endothermic. Therefore reaction 1 has the smaller activation energy and is the faster of the two reactions.

4. What do you understand by the term "mechanism of a reaction"?

5. What is a zero order reaction? Explain why many heterogeneous catalytic processes are zero order.

6. Of what benefit (if any) is a catalyst in a reaction for which ΔG is known to be greater than zero?

7. To ignite sugar in air requires relatively high temperatures. Yet sugar burns easily and efficiently during cell metabolism in the human body at a modest 37°C. Explain.

8. Give a simple mechanistic accounting for the following facts: The intensity of the deep purple color of an acidic potassium permanganate solution can be used as an indicator of the extent to which oxidation has taken place. Often there is a considerable time lapse before any visual evidence of reaction is noted. But once begun, the process proceeds vigorously. On the other hand, if a small quantity of the essentially colorless manganous salt $MnCl_2$ is dissolved in the aqueous acidic medium, the process is im-

mediately vigorous, with no latent period.

9. State the essential difference between the following:

(a) order and molecularity

(b) an elementary process and a stoichiometric reaction

(c) the half-life and the time required for the reaction to go halfway to completion

(d) the rate constant and the rate law

(e) a transition state and a reaction intermediate

(f) a reaction that is first order in A and one that is second order in A

10. Indicate whether the following statements are best described as true (T) or false (F) and add explanatory comments where appropriate:

(a) The order of a chemical reaction is determined by inspection of the stoichiometry of the process.

(b) A bimolecular elementary process follows a second order rate law.

(c) A first order (overall) reaction is described by a concentration function that is exponential with respect to time.

(d) The rate of a reaction is proportional to the accompanying change in free energy.

(e) The rate constant for a unimolecular reaction has units of concentration divided by time.

(f) The slow step among the two or more that constitute a reaction mechanism is considered to be rate determining.

(g)   The Arrhenius Theory suggests that the rate of reaction is a function of the number of collisions between reacting molecules.

(h)   Catalysts are substances that can improve the yield of an otherwise poor reaction by favorably shifting the equilibrium without being altered in the overall process.

(i)   A given reaction mechanism is a true statement of the pathway from reactants to products.

(j)   Isolation of an activated complex is necessary if one is ultimately to be able to verify proposed mechanisms involving such species.

(k)   The sum of the exponents for the concentration terms in the experimental rate law gives the overall order of the reaction.

(l)   For a reaction occurring in solution that has no activation energy or special orientation requirements, the molecules should react as soon as they become neighbors and the rate of reaction should be controlled by the rate at which reacting molecules diffuse together.

(m) Diffusion-controlled reactions are very rapid and consequently it is difficult to measure their rates.

(n)   The value of the activation energy for an endothermic reaction is ΔE.

11.   For each of the following, answer the question from the selections of possible answers:

(a)   Consider the following graph and determine whether it is essentially (correct, incorrect), as drawn for a reaction in which the energy of the products is less than that of the reactants.

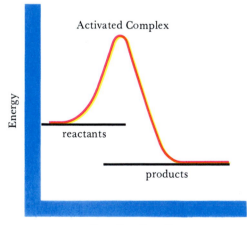

Reaction Coordinate

(b)   An increase in temperature will usually lead to a (greater, smaller, unchanged) rate of reaction.

(c)   At a given temperature, an increase in E_a will usually lead to a (greater, smaller, unchanged) rate of reaction.

(d)   An inhibitor or negative catalyst for a given chemical reaction can provide a reaction pathway leading to a (more, same, less) favorable equilibrium state.

(e)   The half-life of a chemical reaction is the time required for the reaction to proceed halfway to completion. (true, false)

(f)   For a reaction with activation energy E_a at temperature T, the fraction of the total number of molecules re-

acting to produce products is given by $e^{-E_a/RT}$ (true, false)

(g)   The heterogeneous gas phase hydrogenation of ethylene over a finely divided platinum catalyst would likely be described overall as (zero, first, second, indeterminate) order.

PROBLEMS

Activation Energy   [1–6]

1.   Here is a set of experimental rate constants for a particular first order decomposition:

temperature (°C)	20	40
$k \times 10^5$/sec	47.5	576

Determine the activation energy for the reaction in this temperature range and calculate the half-life for the reaction at 30°C.

2.   Here are some kinetic data reported for a particular proton-transfer reaction involving the ethoxide ion:

temperature (K)	273	283	293	303
$k(M^{-1}s^{-1})$	35	150	500	1800

(a)   Determine the order of the reaction from the units of k.

(b)   Using a graphical method, determine the activation energy, E_a.

(c)   Find the specific reaction rate constant at 300 K.

3.   The elementary reaction:

$$2 \text{ I} \longrightarrow \text{I}_2$$

has a rate constant of 4.0×10^9 L $M^{-1}s^{-1}$ at 25°C.

(a)   In a solution that is 10^{-4} M in I, what is the initial rate of the reaction?

(b)   What is the half-life of I?

(c)   At 35°C, the reaction rate constant is 6.1×10^9 $M^{-1}s^{-1}$. What is the activation energy of the reaction?

4.   For a first order reaction, the following experimental data were recorded:

temperature (K)	273	293	313
$k \times 10^5$ (min^{-1})	2.45	45.0	575

Calculate the activation energy and the rate constant at 303 K.

5.   The data in Figure 24–5 give the temperature dependence of the rate constant for the reaction

$$2 \text{ N}_2\text{O}_5 \longrightarrow 4 \text{ NO}_2 + \text{O}_2$$

By graphical analysis of the data:

(a)   Calculate the activation energy.

(b)   Determine the pre-exponential factor.

(c)   Determine the fraction of activated complexes that might be considered "activated enough" at 328 K, according to the Arrhenius Theory.

6.   A cyclic trimer of acetaldehyde, known as paraldehyde, decomposes into the monomer according to the following reaction:

$$\text{C}_6\text{H}_{12}\text{O}_3 \longrightarrow 3 \text{ C}_2\text{H}_4\text{O}$$

or, in a simplified, general form:

1 paraldehyde $\longrightarrow$ 3 acetaldehyde

The first order rate constant at a certain temperature has been determined as 0.00105 s^{-1}.
(a) Calculate the total pressure exactly 1000 seconds after a sample of paraldehyde at an initial pressure of 100 torr was introduced into an enclosed space.
(b) State your understanding of the half-life for a chemical reaction, and calculate $t_{1/2}$ for paraldehyde decomposition at 273°C.
(c) If the rate of the reaction were independent of concentration, what would the order of reaction be?
(d) Calculate the fraction of collisions that are effective in producing paraldehyde decomposition, and comment briefly on the significance of the magnitude of your numerical answer. E_a for the reaction is known to be $+44,975$ cal at 273°C.

Mechanisms   [7–16]

7. Consider a bimolecular reaction involving A and B molecules reversibly forming an intermediate complex C, which then rapidly breaks down to form product P.
(a) Write equations for the elementary processes that might form the mechanisms of the reaction.
(b) Graphically illustrate how the concentrations of A, B, C, and P might vary as functions of time, assuming the initial concentrations of A and B to be equimolar.
(c) From the information at hand, write the implied experimental rate law.

8. Consider the reaction

$$NO (g) + NO (g) + H_2 (g) \longrightarrow 2 \, NOH (g)$$

A mechanism has been proposed in which two bimolecular processes take place. Demonstrate that such a bimolecular mechanism would be entirely consistent with the experimental third order rate law.

9. Consider the following reaction and data:

$$2 \, H_2 (g) + 2 \, NO (g) \longrightarrow 2 \, H_2O (g) + N_2 (g)$$

P_{NO} (atm)	P_{H_2} (atm)	Initial Rate (atm/min)
0.006	0.001	0.025
0.006	0.002	0.050
0.006	0.003	0.075
0.001	0.009	0.0063
0.002	0.009	0.025
0.003	0.009	0.056

(a) Being sure to indicate your reasoning clearly, establish the correct experimental rate law.
(b) If chemical evidence indicates the fleeting presence of N_2O as a transitory intermediate, write a likely mechanism for the reaction that is consistent with the overall stoichiometry and kinetic information.

10. The following data were collected for the decomposition of hydrogen peroxide into water and molecular oxygen in the presence of iodide ion:

Time (min)	Fraction of Peroxide Decomposed	Iodide Ion Concentration
0	0	0.02 M
5	0.130	0.02 M
15	0.339	0.02 M
25	0.497	0.02 M
45	0.712	0.02 M
65	0.835	0.02 M

The rate of the reaction is proportional to the hydrogen peroxide concentration and it is also proportional to the iodide ion concentration.
(a) Write the experimental rate law.
(b) Evaluate the observed (experimental) rate constant, determine the specific reaction rate constant, and show that the data fit your rate law.
(c) Calculate the half-life for the reaction.
(d) Suggest a plausible mechanism that is consistent with all the information provided.

11. It has been observed that NO_2 can be oxidized by molecular fluorine according to the following reaction:

$$2 \, NO_2 (g) + F_2 (g) \longrightarrow 2 \, NO_2F (g)$$

The experimental data fit a second order rate law:

$$\text{rate} = k[NO_2][F_2]$$

Propose a reasonable mechanism indicating the rate-controlling step, based on this information and the possible presence of atomic fluorine during the reaction. Can you suggest any obvious complicating factors?

12. Show that the apparent third order dependence of the following stoichiometric reaction can be successfully explained by a rapid equilibrium, followed by a rate-controlling second order reaction:

$$2 \, NO (g) + F_2 (g) \longrightarrow 2 \, NOF (g)$$

Write your rate law in terms of the rate of NOF formation. It is known that NOF_2 is a chemical intermediate in the process.

13. For the photochemically initiated free radical chlorination of hydrogen in the gas phase, producing hydrogen chloride, write out the key steps in the mechanism and show how you might arrive at the indicated $\frac{3}{2}$ order rate law:

$$H_2 (g) + Cl_2 (g) \longrightarrow 2 \, HCl (g)$$

$$\frac{1}{2} \frac{\Delta[HCl]}{\Delta t} = k[H_2][Cl_2]^{1/2}$$

14. For the decomposition of ozone (O_3) into oxygen (O_2), according to

$$2 \, O_3 (g) \longrightarrow 3 \, O_2 (g)$$

the following rate law is believed to be an accurate representation:

$$-\frac{\Delta[O_3]}{\Delta t} = k \frac{[O_3]^2}{[O_2]}$$

Write a reasonable mechanistic scheme to explain this rate

law. (*Hint:* It is reasonable to expect oxygen atoms to be formed initially.)

15. Explain why the following statements or illustrations are best described as true (T) or false (F):

(a) The following reaction is known to be bimolecular elementary process:

$$CH_3CHO \ (g) \ + \ I_2 \ (g) \longrightarrow$$
$$CH_3I \ (g) \ + \ HI \ (g) \ + \ CO \ (g)$$

Therefore, rate $= k[CH_3CHO][I_2]$.

(b) The following sequence illustrates a catalyzed chemical reaction:

(1) $CH_3CHO + I_2 \longrightarrow CH_3I + HI + CO$ (slow)

(2) $CH_3I + HI \longrightarrow CH_4 + I_2$ (fast)

(overall) $CH_3CHO \longrightarrow CH_4 + CO$

(c) A compound C decomposes to products by a first order reaction:

$$\text{rate} = \frac{\Delta[C]}{\Delta t} = k[C]$$

16. Answer each of the following in as precise and clear a manner as you can. Give a short answer only:

(a) For a gas phase reaction in which

$$A \longrightarrow 3 \ B$$

what is the relationship between the pressure change and the existing concentration of A?

(b) Indicate the experimental order for each component and the overall order of the reaction from the following collected rate data:

$$A + B + C \longrightarrow \text{products}$$

Experiment	[A]	[B]	[C]	Rate
1	1	1	1	0.01
2	1	1	2	0.01
3	1	2	2	0.02
4	2	2	2	0.01

(c) For the following reaction, the kinetics can be followed equally well by studying the rate of N_2O_5 disappearance or the rate of O_2 appearance. Write an equation expressing the relationship between the two rates:

$$N_2O_5 \ (g) \longrightarrow N_2O_4 \ (g) \ + \ \tfrac{1}{2} O_2 \ (g)$$

(d) Given the following mechanistic proposal, indicate the stoichiometric reaction:

$OCl^- + H_2O \longrightarrow HOCl + OH^-$
                                    (rapid equilibrium)

$HOCl + I^- \longrightarrow HOI + Cl^-$ (slow)

$HOI + OH^- \longrightarrow H_2O + OI^-$ (fast)

Additional Problems   [17–23]

17. Calculate the temperature at which half the molecules present in gas having an activation energy of 1500 cal/mol would be activated.

18. For an *endothermic* reaction, state the minimum value that the activation energy can have. For an *exothermic* reaction, state the minimum value that the activation energy can have.

19. According to the Arrhenius Theory, at what temperature would you expect to find 10^{14} molecules in a mole of molecules to be activated in a first order gas phase reaction for which the activation energy is known to be 84 kJ/mol?

20. Draw and label a diagram, including ΔE, E_a, and a proposed transition state for any of the propagation steps leading to the principal product, the free radical chlorination of methane to methyl chloride:

$$2 \ CH_4 \ (g) \ + \ Cl_2 \ (g) \longrightarrow 2 \ CH_3Cl \ (g) \ + \ H_2 \ (g)$$

21. For a reaction for which

$$-\frac{\Delta[A]}{\Delta t} = -\frac{\Delta[B]}{\Delta t} = k$$

what is the overall order of the reaction?

22. For the oxidation of V^{3+} by Fe^{3+}, the following mechanistic pathway has been suggested:

$$Fe^{3+} + Mn^{2+} \longrightarrow Fe^{2+} + Mn^{3+}$$
$$Fe^{3+} + Mn^{3+} \longrightarrow Fe^{2+} + Mn^{4+}$$
$$V^{3+} + Mn^{4+} \longrightarrow V^{5+} + Mn^{2+}$$

Which element is the catalyst species? What would be the general rate law?

23. For the following reaction:

$$H \ (g) \ + \ Br_2 \ (g) \longrightarrow HBr \ (g) \ + \ Br \ (g)$$

draw a graph of energy versus distance along the reaction coordinate for the exothermic process and label the diagram completely, including:

(a) the activated complex
(b) the transition state
(c) the activation energy
(d) the change in internal energy
(e) the reaction pathway in the presence of a catalyst

How would the diagram be different if the process were endothermic?

MULTIPLE PRINCIPLES   [24–25]

24. For the imaginary reaction below, the rate constant at 298 K for the forward reaction k_f is 5.03×10^{-2} $\text{sec}^{-1} \cdot M^{-1}$, and the rate constant at 298 K for the reverse reaction k_r is 7.35×10^{-5} sec^{-1}. Calculate $\Delta G°$ of the reaction at 298 K.

$$2 \ A \ (g) \rightleftharpoons B \ (g)$$

25. Rate constants for the following hypothetical reaction at 298 K are taken to be $k_f = 2.44 \times 10^{-7}$ $\text{sec}^{-1} \cdot M^{-1}$ and $k_r = 3.19 \times 10^{-1}$ $\text{sec}^{-1} \cdot M^{-1}$. Calculate the equilibrium partial pressures of all species at equilibrium starting with $p_A = 0.10$ atm and $p_B = 0.15$ atm.

$$A \ (g) \ + \ B \ (g) \rightleftharpoons C \ (g) \ + \ D \ (g)$$

APPLIED PRINCIPLES   [26–28]

26. For the decomposition of nitramide according to the following reaction:

$$NH_2NO_2 \longrightarrow N_2O + H_2O$$

the experimental rate law is

$$\frac{\Delta[N_2O]}{\Delta t} = \frac{k[NH_2NO_2]}{[H^+]}$$

Show that the following is a reasonable mechanism, and evaluate the experimental rate constant in terms of the respective rate constants for the elementary molecular events indicated:

(1) $NH_2NO_2 \longrightarrow [NHNO_2^-] + H^+$
                                (rapid equilibrium)
(2) $[NHNO_2^-] \longrightarrow N_2O + OH^-$ (slow)
(3) $H^+ + OH^- \longrightarrow H_2O$ (rapid)

27. Consider the set of reactions

$$A + B \underset{k_{-1}}{\overset{k_1}{\rightleftharpoons}} C + D$$

$$C + E \overset{k_2}{\longrightarrow} F$$

proceeding in a situation where all concentrations are equal. What relationships between the magnitudes of the rate constants for the elementary processes will lead to the following rate laws?

(a) $\dfrac{\Delta[F]}{\Delta t} = k\dfrac{[A][B][E]}{[D]}$

(b) $\dfrac{\Delta[F]}{\Delta t} = k'[A][B]$

Show that this mehanism is consistent with the rate laws and express the experimental rate constant in terms of the rate constants for the elementary processes.

28. It is known that iodide ion reacts with hypochlorite ion in alkaline solution to produce hypoiodite and chloride according to the reaction

$$I^- (aq) + OCl^- (aq) \longrightarrow OI^- (aq) + Cl^- (aq)$$

The experimental rate law that was obtained is

$$rate = k\frac{[I^-][OCl^-]}{[OH^-]}$$

The following mechanism has been proposed:
(1) Rapid hydrolysis of the hypochlorite ion in a reaction that proceeds to equilibrium immediately.
(2) A slow, rate-controlling step involving iodide ion and a product of the prior equilibrium reaction.
(3) Iodite formation in a rapid reaction.

Show that the reaction mechanism is consistent with the stoichiometric reaction, and that the experimental rate law is consistent with the rate law written for the bimolecular, rate-controlling, slow step in the mechanism.

Theory and Practice of Organic Chemistry

HERE'S HOW   By Gus Edson

SILK MADE FROM CASTOR-OIL !

By Combining an Anti-Freeze Solution and Castor Oil, W. H. Carothers and J. W. Hill Produced a Fabric-Like Silk—Its Scientific Name Is Exademe-thylenedicarboxylic Acid.

CASTOR OIL

WANT A BUFFALO?

Uncle Sam Gives Away Almost One Hundred Buffaloes Every Year—There Is No Charge But the Animals Must Not Be Slaughtered.

STILL ME A GALLON. HERE'S THE FAMILY JEWELS!

GASOLINE IS TAXED 12 CENTS A GALLON IN PERU

Waxes are of considerable importance in the manufacture of candles, lipsticks and crayons—for pharmaceuticals and cosmetic preparations. Classified as lipids, waxes are thermoplastic materials of relatively low molecular weight. Nylon, the first synthetic textile fiber, was discovered by two Du Pont chemists, Wallace Carothers and Julian Hill. One of the first public disclosures of this profoundly important chemical event appeared in a 1931 newspaper cartoon (before any scientific publication). The author's caption—silk made from castor oil—didn't get it quite right, nor is it likely that any of his readers could have imagined what was afoot in modern organic chemistry.

Robert B. Woodward (American), Nobel Prize for Chemistry in 1965, for his contributions to synthetic organic chemistry, especially the synthesis of complex natural products.

25.1 BEGINNING ORGANIC CHEMISTRY

Organic chemistry is one of the most vigorous and exciting subjects in all of natural science and one of our most remarkable achievements. In this chapter, we shall try and explain just what organic chemistry is, what organic chemists do, what questions they ask of nature, and what kinds of answers they seek. It is worth exploring how organic chemistry differs from the other strains—inorganic and physical chemistry, biochemistry, polymer chemistry, and theoretical chemistry. Furthermore, we will explore the role played by engineering in organic chemistry, for here the social and economic implications of the work of the chemist are especially significant. In short, we need to seek answers to the question: "Why is there a separate branch of chemistry devoted exclusively to the 'organic' compounds of carbon?"

Until the early 19th century, chemists commonly distinguished between animal (or organic) matter and mineral (or inorganic) matter. **Organic substances** were those produced by living organisms—plant or animal—or obtainable from living organisms by chemical reactions. **Inorganic substances**, on the other hand, occurred in minerals and could be isolated from them by physical and chemical means. Rock salt and marble were classically understood to be inorganic compounds (as they are today); vinegar, grape sugar, and alcohol were known as organic chemicals then, as now.

It was not thought possible in those days to prepare organic compounds from inorganic substances, leading to the widely held view that there was a **vital force**, unique to living matter. Scientists believed organic compounds could not be prepared in the laboratory from *lifeless* inorganic materials. However, in 1828 Friedrich Wöhler laid that erroneous notion to rest when he obtained urea from ammonium sulfate and potassium cyanate, and succeeded in showing that his product was in all ways the same as that isolated from animal urine. The chemistry here is the formation of ammonium cyanate, followed by its thermal rearrangement to urea:

$$(NH_4)_2SO_4 \text{ (aq)} + 2 \text{ KOCN (aq)} \longrightarrow 2 \text{ NH}_4\text{OCN (aq)} + K_2SO_4 \text{ (aq)}$$

$$\underset{\text{ammonium cyanate}}{NH_4\text{OCN (aq)}} \longrightarrow \underset{\text{urea}}{NH_2\text{CONH}_2 \text{ (aq)}}$$

The year 1828 is now generally accepted as the starting point for modern organic chemistry. Although the distinction between organic and inorganic no longer retains its original significance—and in fact has become still more vague in recent years—chemistry (and chemists) continue to distinguish compounds of organic carbon from all others.

Organic compounds of carbon that are both interesting and useful to humans have been obtained from nature since time immemorial. Today, our main resources of organic starting materials are still (1) coal and petroleum, the fossilized remains of once living matter (Figure 25–1); (2) agricultural products such as potatoes and cereal crops (as sources of starches), cane and beet sugars, corn, rice and wheat, vegetable fats, and related tree and plant materials; and (3) animal fats and proteins, as well as traditional by-products such as milk and hides.

From these natural resources, organic chemists extract or construct thousands of products—from pharmaceuticals to plastics—either by direct isolation techniques or by synthetic methods. Historically, these resources were used directly. The properties of fermented fruit juices and grains have been used for social and medicinal purposes for as long as we have records. In the Old Testament we find evidence that people in biblical times knew that acetic acid was present in sour wine and recognized the action

(a)

(b)

FIGURE 25–1 (a) Pumping crude oil, an organic starting material, from the Texas Panhandle; (b) a platform rig in the Gulf of Mexico pumping crude oil.

of vinegar on chalk. The Egyptians wrapped mummified bodies in cloth dyed with indigo (Figure 25–2) and alizarin. Long before Socrates and Shakespeare lived, the effects of deadly nightshade (atropine) were well known. To the Indians of the American Southwest, the mind-altering constituents of the peyote cactus (mescaline) were part of their ceremony and ritual, and the effect of marijuana was found without any need to know about the chemistry of cannabis (tetrahydrocannabinol).

Positive identification and isolation of organic chemicals in pure form are clearly more difficult tasks for which techniques needed to be developed before purposeful synthesis became possible. Definitive preparations of alcohol and acetic acid were first reported only in the last decades of the 18th century. At the conclusion of the 19th century, we understood a great deal about many of the most complex natural products such as carbohydrates, fats, and proteins. By that time, local anesthetics such as aspirin and novocaine had been synthesized—not extracted from plants—along with colorful and important dyes, the first synthetic plastics, and chemotherapeutic agents. Still, for all our successes, it is still pretty hard to beat Mother Nature for the economical production of many of our chemical needs.

Fortunately, the immensity of organic chemistry has been simplified considerably by classification of organic molecules according to the recurring presence of small assemblies of atoms called **functional groups** that consistently behave in the same way. For example, hydroxide (OH^-) ions attack many organic halides in a similar fashion, displacing the halide ion and resulting in the synthesis of an alcohol (an organic compound with an —OH group attached to a carbon atom). Once the displacement reaction has taken place, the hydroxyl (OH) group behaves in essentially the same way whether it is part of the smallest, simplest alcohol, with only one carbon atom—wood alcohol—or part of a much larger alcohol such as lauryl alcohol with twelve carbon atoms. From a handful of functional groups and a few reaction mechanisms, we can learn a lot about the general chemistry of tens of thousands of organic compounds.

It is not sufficient, however, to know that the reaction of a particular compound depends on the number and kind of functional groups present. Organic chemists also need to know where the functional groups are in a molecule, and even the particular spatial relationships that exist among single atoms. Contributing to the richness and complexity we see in the world of carbon compounds is the existence of **isomers**—different compounds of the same atoms, having different spatial arrangements, resulting in different characteristic properties. Nature takes unusual advantage of **stereochemistry**, which exploits differences among isomers, in many of the molecules important to all living things.

Organic chemistry is everywhere—in the materials of construction of the automobile and the fuels that drive it, in the clothes we wear and the food we eat, in the drugs that heal animal diseases and control pests, in the drugs of abuse, and in the processes by which these compounds and composites are made.

atropine: gastrointestinal relaxant
mescaline: hallucinogen
tetrahydrocannabinol: mild hallucinogen

FIGURE 25–2 Indigo, a deep blue water-insoluble substance, may well be the oldest known dye. Egyptian mummy cloths known to be more than 4000 years old were dyed with it. It reached Europe as an article of commerce after the 12th century and was selling for the equivalent of $3 per pound in 1890. After that, the synthetic product appeared on the market. The photo is that of an original sample of the first synthetic indigo, from Baeyer's laboratory in Germany, 1890.

25.2 CARBON CHEMISTRY IS SPECIAL

Silicon and Carbon

The covalent chemistry of organic carbon shows every sign of being unique. A casual glance at the periodic table might suggest that silicon closely resembles carbon. Both elements commonly display four valence shell elec-

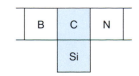

trons in four sp^3 hybrid atomic orbitals; both elements bond to oxygen and to hydrogen, to themselves, and to each other. But carbon does all those things with much greater versatility. That's because the C-to-C bond is much stronger than the Si-to-Si bond, while the C-to-O bond is much weaker than the Si-to-O bond. Consequently, whereas carbon atoms tend to link themselves together in chains and rings through C-to-C bonds, silicon atoms tend to link themselves together through Si-to-O bonds:

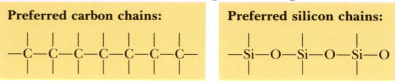

In nature, we see the preferred bonding of carbon in coal and petroleum, and in graphite and diamond. We have learned to translate nature's practice into synthetic analogs, such as the polymers of plastics—materials such as the ubiquitous polyethylene grip for a six-pack and the clear kitchen wrapping material made of polyvinyl chloride.

Si-to-O chains, on the other hand, make up the major part of the rocks, soils, and sands in the Earth's crust. We have also learned to translate nature's silicon chemistry into synthetic polymers called silicones (Chapter 20). These have proved to be important compounds because they impart water-repellent and heat-resistant properties, and they can be fabricated into very useful forms as fibers, films, and membranes.

The C-to-H bond is stronger than the Si-to-H bond. As a result, SiH_4 (silane) burns rapidly in oxygen, forming more stable Si-to-O bonds. But the C-to-H bonds in CH_4 (methane), being much more stable to begin with, are less susceptible to similar reactions with oxygen. For example, whereas methane is subject to oxidation only at elevated temperatures:

$$CH_4 \text{ (g)} + 2\,O_2 \text{ (g)} \longrightarrow CO_2 \text{ (g)} + 2\,H_2O \text{ (g)}$$

silane oxidizes rapidly at room temperature:

$$SiH_4 \text{ (g)} + 2\,O_2 \text{ (g)} \longrightarrow SiO_2 \text{ (s)} + 2\,H_2O \text{ (liq)}$$

So although the periodic table suggests that silicon should bear a close family resemblance to carbon because both are Group IVA members—four valence electrons in four sp^3 hybrid bonding atomic orbitals—it isn't that way at all. The essential difference can be understood in terms of silicon's extra inner shell of electrons. The shell effectively shields any one silicon atom from penetrating another's electron-cloud and forming Si-to-Si bonds. In effect, a repulsion that does not exist in carbon atom chemistry exists between silicon atoms because these extra electrons effectively minimize the otherwise strong influence of the positive nucleus. Hence, whereas C-to-C bonding is favorable, Si-to-Si bonding is much less so.

Boron and Carbon

The simplest compound of B and H is the boron hydride known as diborane (B_2H_6). It is analogous to ethane (C_2H_6) but with a total of 12 rather than 14 valence shell electrons. There are three electrons on each of two boron atoms and one electron on each of six hydrogen atoms.

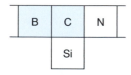

If boron were to form chain structures as carbon does, an ethane-like structure might be expected for diborane. However, for such a structure B_2H_6 would require seven bonds, and therefore 14 electrons. Since there are only 12 to go around, diborane is called an "electron-deficient" molecule; its structure is quite different from that of ethane.

Nitrogen and Carbon

If boron is different because it has too few electrons, in the same qualitative sense nitrogen can be seen as different because it has too many electrons. The result is excessive electrostatic forces of repulsion that lead to instability and unusual reactivity. For example, whenever chains of nitrogen atoms begin to form, highly reactive (and in many cases explosive, shock-sensitive, and otherwise unstable) molecules result. Hydrazine, for example, is a rocket propellent; hydrazoic acid is both heat- and shock-sensitive; triazine and tetrazine require even more delicate handling because of their explosive nature.

hydrazine	$H_2N—NH_2$
hydrazoic acid	HN_3
triazine	$H_2N—NH—NH_2$
tetrazine	$H_2N—N_2—NH_2$

On the other hand, there is an obvious stability associated with molecular nitrogen (N_2), a situation without counterpart in carbon chemistry. Interestingly enough, carbon binds readily to nitrogen and very stable C-to-N bonds can be found in natural and synthetic arrangements. The implication is that greater stability has somehow been achieved by nitrogen atoms paired together as nitrogen molecules—which helps to explain why 80% of the atmosphere near the Earth's surface is N_2—or paired up with carbon atoms in a number of important functional groups such as amines, amides, amino acids, and peptides.

25.3 ALKANES

Hydrocarbons Large and Small

The simplest organic compounds are those formed by combinations of carbon atoms with hydrogen atoms and nothing else. Such compounds are called **hydrocarbons** (Table 25–1). On the basis of their structures, hydrocarbons can be divided into two principal categories, aliphatic hydrocarbons and aromatic hydrocarbons. **Aliphatic hydrocarbons** can be subclassified as **alkanes**, **alkenes**, and **alkynes**, and their **alicyclic** (ring-containing) analogs. **Aromatic hydrocarbons** are characterized by the presence of the benzene ring with its alternating single and double bonds; they will be discussed in a later section.

Alkanes are hydrocarbons containing only sigma (single) bonds. Linear chain alkanes fit the general formula C_nH_{2n+2}, and cycloalkanes fit the general formula C_nH_{2n}. Because early chemists noted that alkanes (unlike alkenes and alkynes) fail to react with hydrogen, alkanes are sometimes called **saturated** hydrocarbons; they are also known as "paraffin" hydrocarbons.

TABLE 25–1 Basic Classification of Hydrocarbons

Hydrocarbons	
Aliphatic Hydrocarbons	*Aromatic Hydrocarbons*
alkanes $-CH_2-CH_2-$ alkenes $-CH=CH-$ alkynes $-C\equiv C-$ alicyclics 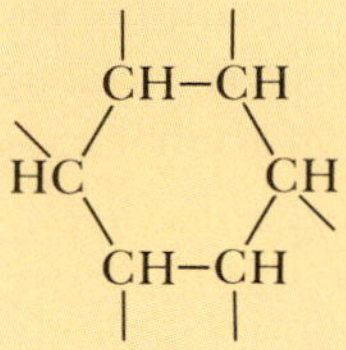 ![alicyclic ring structure]	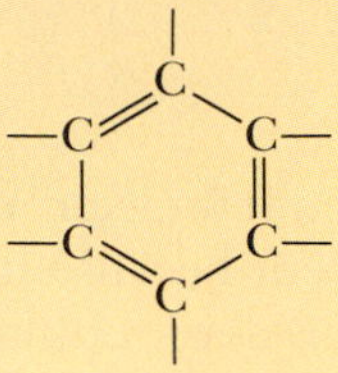 ![aromatic benzene ring structure]

TABLE 25–2 Normal Alkanes

Common Name	Condensed Formula	Structural Representations	m.p. (°C)	b.p. (°C)	Petroleum Fraction
Gases at room temperature					
methane	CH_4	CH_4	−183	−162	natural gas
ethane	C_2H_6	CH_3CH_3	−172	−89	
propane	C_3H_8	$CH_3CH_2CH_3$	−187	−42	
n-butane	C_4H_{10}	$CH_3(CH)_2CH_3$	−135	0	
Liquids at room temperature					
n-pentane	C_5H_{12}	$CH_3(CH_2)_3CH_3$	−130	36	petroleum ethers
n-hexane	C_6H_{14}	$CH_3(CH_2)_4CH_3$	−94	69	
n-heptane	C_7H_{16}	$CH_3(CH_2)_5CH_3$	−91	98	
n-octane	C_8H_{18}	$CH_3(CH_2)_6CH_3$	−54	126	
n-nonane	C_9H_{20}	$CH_3(CH_2)_7CH_3$	−30	151	
n-decane	$C_{10}H_{22}$	$CH_3(CH_2)_8CH_3$	−26	174	
Solid at room temperature					
n-octadecane	$C_{18}H_{38}$	$CH_3(CH_2)_{16}H_3$	28	308	lube oils

EXAMPLE 25–1

Write the formulas for the alkanes containing six, twelve, and fifteen carbon atoms.

Solution Using the general formula C_nH_{2n+2} gives C_6H_{14}, $C_{12}H_{26}$, and $C_{15}H_{32}$. ◼

EXERCISE 25–1

Write the formulas for cycloalkanes containing five, six, and eight carbon atoms.
Answer: C_5H_{10}, C_6H_{12}, C_8H_{16}. ◼

The first alkane (Table 25–2) is **methane** (CH_4), the ignitable gas known as fire damp that coal miners the world over fear. **Ethane** (C_2H_6) is next;

 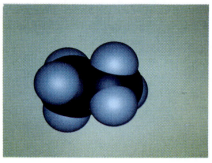 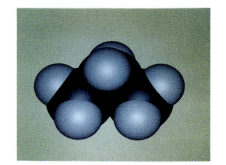

Computer-generated space-filling models of the methane molecule (CH_4), the ethane molecule (CH_3—CH_3), and the propane molecule (CH_3—CH_2—CH_3).

then **propane** (C_3H_8) and **butane** (C_4H_{10}). For butane and larger alkanes, there are families of isomers, each member of which has the same composition but different geometric arrangements and therefore different properties. All of the butanes are gases at room temperature. **Pentane** (C_5H_{12}), the next member of the alkane family, is the simplest liquid hydrocarbon. As the carbon chain lengthens and the molecular weight increases, melting and boiling points increase, resulting in progressively more viscous liquids; then oils and greases; and finally the familiar solid hydrocarbons called waxes, at about 25 carbons and beyond.

Methane and ethane together make up about 95% of what we call natural gas. Propane is bottled and sold for a variety of heating needs ranging from campfires and home heating to industrial use. Butane is the gas used in cigarette lighters (Figure 25–3). The liquid hydrocarbons from C_5 (pentanes) to C_8 (octanes), having a boiling range from 0°C to about 200°C, are called gasoline range hydrocarbons. Other petroleum distillation fractions are the kerosenes, turbine fuels, and jet fuels (boiling range 175°C to 275°C); gas oil, fuel oil, and diesel oil (boiling range 250°C to 400°C); and the lubricating oils, greases, and paraffins or waxes (Figure 25–4).

The *n* prefixed to most of the alkanes found in Table 25–2 is an abbreviation for "normal":

n-butane = normal-butane

n-pentane = normal-pentane

Normal means that the carbon atoms are all linked in a head-to-tail chain arrangement, with no branches or side-groups.

Branching first becomes possible when there are four carbons in the chain, giving rise to the very important phenomenon of **isomerism**. Two different C_4H_{10} structures—two isomeric butanes—are possible, and although they have the same chemical composition, their physical and chemical properties are different (Table 25–3). The branched isomer is called isobutane:

$$CH_3—CH_2—CH_2—CH_3 \qquad CH_3—\overset{\overset{\displaystyle CH_3}{|}}{C}H—CH_3$$

n-butane   isobutane

There are three structural isomers of the C_5H_{12} hydrocarbon, a normal (straight-chained) pentane and two branched pentanes, isopentane and neopentane:

FIGURE 25–3 "Flick-your-Bic." As soon as the pressure is released, the liquid butane (CH_3—CH_2—CH_2—CH_3), vaporizes and is ignited simultaneously as the flint wheel is struck.

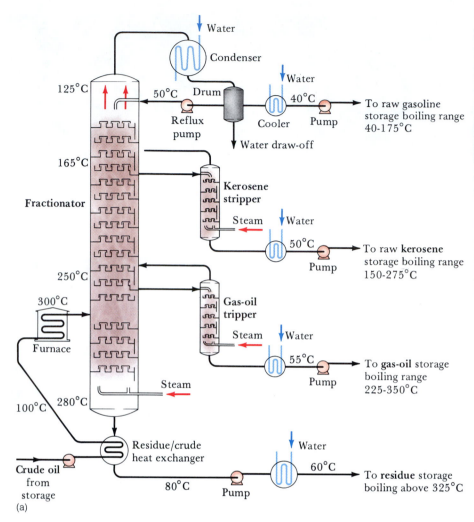

FIGURE 25–4 (a) An engineering diagram for the fractional distillation of crude oil. (b) Fractional distillation on an industrial scale.

TABLE 25–3 Characteristic Melting and Boiling Points of the Isomeric Butanes and Pentanes

	Butanes		Pentanes		
	n-	*iso-*	*n-*	*iso-*	*neo-*
m.p. (°C)	−138	−145	−130	−160	−20
b.p. (°C)	0	−10	36	28	10

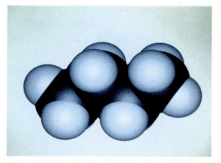

Computer-generated space-filling models of the *n*-butane molecule (CH₃—CH₂—CH₂—CH₃); and the *iso*-butane molecule (CH₃—CH—CH₃).
                                              |
                                              CH₃

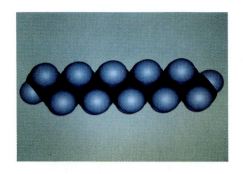 

Computer-generated space-filling models of the *n*-decane molecule (CH_3—$(CH_2)_8$—CH_3) and the *n*-eicosane molecule (CH_3—$(CH_2)_{18}$—CH_3).

TABLE 25–4   Cycloalkanes: Saturated Ring Hydrocarbons

Compound	Structural Formula	Skeleton Structures
Cyclopropane	CH_2 / CH_2–CH_2	△
Cyclobutane	H_2C——CH_2 / H_2C——CH_2	▢
Cyclopentane	CH_2 / H_2C CH_2 / H_2C——CH_2	⬠
		Boat form
Cyclohexane	CH_2 / H_2C CH_2 / H_2C CH_2 / CH_2	Chair form

n-pentane   isopentane   neopentane

Branching has a considerable effect upon physical properties.

There are five isomeric hexanes. The number of possible isomers increases exponentially with increasing number of carbon atoms. There are nine isomeric heptanes (C_7H_{14}); at ten carbons in the chain ($C_{10}H_{22}$), a total of 75 structural isomers are possible; at 20 carbons ($C_{20}H_{42}$), there are 366,319 possible arrangements of the atoms. By the time we get to C_{30}, computer calculations tell us that more than 4 billion isomers are possible, and just for the fun of it, at C_{40} there ought to be 62,491,178,805,831 possible isomers.

As the name suggests, cycloalkanes (Table 25–4) are hydrocarbons in which the carbon chains have been formed into rings. The fact that carbon

TABLE 25–5	Common Radicals Used in Systematic Naming of Hydrocarbons		
Parent		**Radical**	
CH_4	methane	$CH_3—$	methyl
CH_3CH_3	ethane	$CH_3CH_2—$	ethyl
$CH_3CH_2CH_3$	propane	$CH_3CH_2CH_2—$	propyl
		$CH_3\overset{\mid}{C}HCH_3$	isopropyl
C_6H_{12}	cyclohexane	$C_6H_{11}—$	cyclohexyl
C_6H_6	benzene	$C_6H_5—$	phenyl

normally has a tetrahedral bond angle of 109.5° suggests at least one basic difference between alkanes and cycloalkanes—ring strain. Cyclopropane is the smallest cycloalkane ring structure, and the internal C-to-C bond angle is 60°. The molecule exists but is highly strained. Cyclobutane at 90° is also highly strained relative to the normal alkanes. The five- and six-membered rings are almost free of strain. Cyclohexane has a "puckered" ring that is capable of existing in two "conformations," one boat-like and the other chair-like. The chair is the preferred geometry. Note that if the ring were not puckered, there would be strain because the internal bond angle would be 120° rather than the 109.5° tetrahedral bond angle. Cyclohexane is puckered because all the ring bonds are sigma bonds, with characteristic sp^3 hybridization. Compare that to benzene, with sp^2 hybridization, leading to a trigonal planar geometry and the hexagonal arrangement (Chapter 18).

Naming Organic Molecules

A common or **trivial** system for naming hydrocarbons has grown up with organic chemistry, beginning with methane, ethane, and propane. There are two butanes, three pentanes, five hexanes, nine heptanes, and eighteen octanes. You can see what's happening! Naming is going to get quite complicated, and rapidly out-of-hand if we rely on a trivial system of nomenclature. To help clear things up, a **systematic** nomenclature has been established, based on a main chain and on the location of side chains. This system is approved by the International Union of Pure and Applied Chemistry (IUPAC). In its simplest form, we find the longest straight chain of carbon atoms that can be followed, nonstop, through the entire molecule. The root of the compound's systematic name is the name of this "parent" chain. Beginning at the end carbon atom closest to where branching begins, number this main chain of carbon atoms. The longest chain in *n*-butane is four carbons. But in isobutane the longest chain is three, making it a propane derivative. Note that the branch, or side chain, is a methane molecule that has lost a hydrogen, and with it the -*ane* in its name, now replaced by -*yl*; a methane molecule (CH_4) becomes a methyl group ($CH_3—$) (Table 25–5). This methyl group is attached to the second carbon in the chain. Thus, 2-methylpropane is the systematic name for isobutane. Isooctane is the trivial name for the compound on which the gasoline industry's octane standard is based; this compound is systematically named 2,2,4-trimethylpentane. Note that when two alternative numbering schemes would provide different sets of numerical prefixes, we choose the one giving the smallest numbers. Thus, we number isooctane from the end that gives 2,2,4- rather than 2,4,4-:

$$CH_3\text{—}\underset{\underset{1}{}}{CH}\text{—}\underset{\underset{3}{}}{CH_3}$$

$$\underset{5}{CH_3}\text{—}\underset{4}{CH}\text{—}\underset{3}{CH_2}\text{—}\underset{2}{C}\text{—}\underset{1}{CH_3}$$

isobutane   (trivial)   isooctane   (trivial)
2-methylpropane (systematic)   2,2,4-trimethylpentane (systematic)

The large straight-chained hydrocarbon molecules such as *n*-triacontane ($C_{30}H_{62}$) and hexacontane ($C_{60}H_{122}$) are nothing more than polymethylenes, nature's $(CH_2)_x$ hydrocarbon polymers. The synthesis of polyethylene can be carried out so as to produce an almost linear product, tens of thousands of methylene (—CH_2—) units long. Polyethylene can also be prepared with a branched structure. Branching alters the properties of the large, polymeric hydrocarbons, as it does the smaller alkanes. Linear polyethylene is rigid and tough because the long strands can lie very close to each other in a regular, almost crystalline pattern. Branching results in a less rigid, less dense, less crystalline structure because the molecules can no longer arrange themselves in as orderly a fashion (Figure 25–5).

Whether we're talking about hydrocarbon chains of candle wax length—C_{25} to C_{30}—or polymer chains on the order of $C_{25,000}$, consider what we

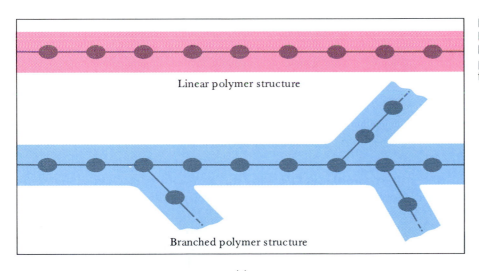

Linear polymer structure

Branched polymer structure

(a)

FIGURE 25–5 (a) Linear and branched polyethylene structures have characteristically different properties, (b) such as density and toughness.

Low density, flexible, branched

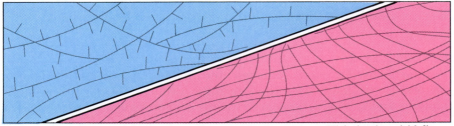

High density, rigid, linear

(b)

FIGURE 25–6 A linear hydro-carbon structure, showing (a) space-filling representation along a segment of the chain, and (b) the structural formula, with the main chain depicted in the plane of the page, the dashed lines extending below the plane of the page and the bold lines extending above the plane of the page. Note the "straight chain" hydrocarbons are really zig-zagged straight chains.

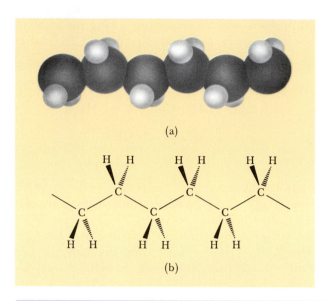

(a)

(b)

TABLE 25–6 The Five Isomeric Hexanes		
Structural Formula	**Name**	**Boiling Point (°C)**
$CH_3CH_2CH_2CH_2CH_2CH_3$	*n*-hexane	69
CH_3CH_2—CH—CH_2CH_3 $\quad\quad\quad$ | $\quad\quad\quad CH_3$	3-methylpentane	63
$CH_3CH_2CH_2$—CH—CH_3 $\quad\quad\quad\quad$ | $\quad\quad\quad\quad CH_3$	2-methylpentane	60
CH_3—CH—CH—CH_3 $\quad\quad$ |$\quad$ | $\quad\quad CH_3\ CH_3$	2,3-dimethylbutane	58
$\quad\quad\quad CH_3$ $\quad\quad\quad$ | CH_3CH_2—C—CH_3 $\quad\quad\quad$ | $\quad\quad\quad CH_3$	2,2-dimethylbutane (*neo*-hexane)	50

mean by **linear** and **branched**. First, because of the tetrahedral geometry about carbon, linear chains of C atoms aren't truly straight (180° linear) chains. The approximately 109° bond angle between carbon atoms results in a zigzag arrangement (Figure 25–6). Nevertheless, "linear" still provides a satisfactory general description for the overall interlocking arrangement of many connected carbon atoms. "Branched," on the other hand, refers to carbon atoms that stick out from the main chain. Branching generally produces impediments to molecular associations, disrupting the already weak van der Waals forces. That results in lower boiling points for the liquid hydrocarbons, and pliability in the polymeric hydrocarbons.

EXAMPLE 25–2

Beyond the five-carbon alkane compounds, it is necessary to use a systematic method of naming. Sketch the structural formulas for the five isomeric hexanes and give their systematic names in approximate order of their boiling points.

Solution The formulas are shown in Table 25–6. ■

TABLE 25–7 The Isomeric Heptanes

Structural Formula	Name	Boiling Point (°C)
$CH_3CH_2CH_2CH_2CH_2CH_2CH_3$	*n*-hexane	93.6
$CH_3CH_2—CH—CH_2CH_3$ $\quad\quad\quad\mid$ $\quad\quad CH_2CH_3$	3-ethylpentane (triethylmethane)	93.5
$CH_3CH_2CH_2—CH—CH_2CH_3$ $\quad\quad\quad\quad\mid$ $\quad\quad\quad CH_3$	3-methylhexane	92
$CH_3CH_2CH_2CH_2—CH—CH_3$ $\quad\quad\quad\quad\quad\mid$ $\quad\quad\quad\quad CH_3$	2-methylhexane (*iso*-heptane)	90
$CH_3CH_2—CH—CH—CH_3$ $\quad\quad\quad\mid\quad\mid$ $\quad\quad CH_3\ CH_3$	2,3-dimethylpentane	89.8
$\quad\quad\quad CH_3$ $\quad\quad\quad\mid$ $CH_3CH_2—C—CH_2CH_3$ $\quad\quad\quad\mid$ $\quad\quad\quad CH_3$	3,3-dimethylpentane (*neo*-hexane)	86
$\quad\quad\quad CH_3$ $\quad\quad\quad\mid$ $CH_3CH—C—CH_3$ $\quad\mid\quad\mid$ $\ CH_3\ CH_3$	2,2,3-trimethylbutane (triptane)	80.9
$CH_3—CH—CH_2—CH—CH_3$ $\quad\quad\mid\quad\quad\quad\mid$ $\quad\ CH_3\quad\quad CH_3$	2,4-dimethylpentane	80.5
$\quad\quad\quad\quad CH_3$ $\quad\quad\quad\quad\mid$ $CH_3CH_2CH_2—C—CH_3$ $\quad\quad\quad\quad\mid$ $\quad\quad\quad\quad CH_3$	2,2-dimethylpentane	79.2

FIGURE 25–7 Petroleum is generally a viscous brownish-black mixture of hydrocarbons.

EXERCISE 25–2

Draw the structural formulas for the nine isomeric heptanes, give their systematic names, and order them approximately according to their boiling points.
Answer: See Table 25–7. ∎

25.4 NATURAL GAS AND PETROLEUM RESOURCES

Natural gas and petroleum are the chief sources of hydrocarbons and will continue to be so well into the 21st century. **Natural gas** is mostly methane, and it is mostly used as a fuel, although lately it has become increasingly interesting to scientists and engineers as a resource for the synthesis of large numbers of organic compounds. **Petroleum** is a brownish-black, generally viscous mixture of organic compounds (Figure 25–7), chiefly hydrocarbons—aliphatic, alicyclic, and aromatic—in varying proportions, along with a few percent sulfur, nitrogen, and oxygen. It has been estimated that the petroleum fraction boiling roughly between 0°C and 200°C, called the gasoline range, contains more than 500 identifiable compounds, and it is impossible to separate them completely by fractional distillation into distinct chemical species.

Engine knock was discussed in Chapter 24. It is the audible ping caused by premature ignition of the fuel in an internal combustion engine. Knock-

The availability and price of energy and materials have always determined the technological base and therefore the expansion and development of industry. Nowhere is that more evident than in organic chemistry.

In the United States, the most important (1987) feedstocks available to the chemical industry are light naphthas, the gasoline-range hydrocarbons:

Hydrocarbon gases (LPG/LNG)	10%
Light naphthas	50%
Kerosenes/diesel fuel	25%
Heavy end oils, bitumens	15%

ing varies with the fuel and is a function of the hydrocarbon structures present. As a general rule, the more side-chains or branches on the parent hydrocarbon, the less pinging. To measure the knocking properties of gasolines, an **octane rating** system was established some fifty years ago. It is based on a zero rating for *n*-heptane, which has the greatest tendency to cause knocking, and a 100 rating for 2,2,4-trimethylpentane (or iso-octane), which was the fuel for the Spitfire aircraft that won the Battle of Britain in World War II. At the time, there was no better motor fuel for a high-compression internal combustion engine. By blending these two hydrocarbons, mixtures can be made that match the knocking characteristics of ordinary gasolines. For example, a fuel with a typical 87-octane rating has knocking equivalent to a mixture of 87% isooctane and 13% *n*-heptane.

For many years, the octane rating of gasolines was improved by addition of tetraethyllead, $Pb(C_2H_5)_4$, in a mixture with ethylene dibromide (which served to scavenge the lead by-products, mostly lead oxides and metallic lead, and prevent scarring of the cylinder walls in the engine). As you might expect, it is a severe environmental contaminant, especially so in densely populated, urban areas. It has been completely phased out (1988) of the current motor fuel mix.

The first distillation products from petroleum, in the boiling range from 0°C to 200°C, are referred to as "straight run" gasoline. Large amounts of gasoline range hydrocarbons are prepared by **cracking** higher boiling, longer-chain fractions down to size. At first, cracking of hydrocarbons was strictly thermal. But as new cracker reactors were built, thermal cracking was replaced by **catalytic cracking** in which petroleum vapors are passed over heated beds of aluminum silicate (Figure 25–8). Not only are the yields of gasoline increased, but the octane rating is also improved because the process tends to produce branches along the main hydrocarbon chain.

The chemistry of the alkanes is limited but very important. Combustion and halogenation are the most familiar processes. The complete combustion of any hydrocarbon eventually produces carbon dioxide and water. Chlorination and bromination (replacement of hydrogen atoms by chlorine or bromine) are the most common industrial halogenation processes.

FIGURE 25–8 Petroleum refineries are capital-intensive process facilities, designed to operate continuously. They are generally "sited" near sources of refinery feedstocks, near ocean tanker terminals of pipeline depots.

25.5 CHLORINATED HYDROCARBONS AND GEOMETRIC ISOMERISM

The Chlorination of Methane

Under appropriate conditions, a molecule of methane can react with a molecule of chlorine, yielding methyl chloride (CH_3Cl) and hydrogen chloride:

$$CH_4 \text{ (g)} + Cl_2 \text{ (g)} \longrightarrow CH_3Cl \text{ (g)} + HCl \text{ (g)}$$

The reaction is known to follow a free radical chain reaction mechanism (see Chapter 24). In the first step, thermal energy in excess of 242 kJ/mol is needed to break the Cl_2 bond homolytically:

initiation step $\qquad Cl\!-\!Cl \longrightarrow 2\ Cl\cdot$

Once chlorine free radicals are present, a number of very rapid propagation steps push the reactants along the way to products by first abstracting hydrogen atoms from methane, producing HCl and $CH_3\cdot$

ROBERT WILHELM BUNSEN

Bunsen's ingenious physical insights that led to the founding of the modern study of spectroscopy were mentioned in Chapter 14. It is his great contributions to organic chemistry and his famous burner that we wish to note here. After completing his graduate studies at Göttingen, Bunsen began studies on the organic compounds of arsenic, nearly lost an eye in a laboratory explosion, and collapsed twice in the laboratory, probably because of inhalation and slow absorption of the metal leading to arsenic poisoning. Safety in the laboratory was not easily or often practiced and certainly not well understood in those early days. His student Frankland (British, 1825–1899), was inspired to continue this work, which in turn provided the foundations needed by Ehrlich in the first decade of the 20th century to produce Salvarsan-606 (arsphenamine), an organoarsenic compound that proved effective against syphilis and opened the field of chemotherapy.

The Bunsen burner, first used in 1855, may be the one thing remembered by more people than any other feature of their introduction to science. It is a vertical tube perforated at the bottom, allowing air to be drawn in by the flowing gas used as the combustion fuel. The "carburated" fuel-air mixture burns steadily, providing uniform heat, little light, not much smoke, and hardly any flickering. It revolutionized laboratory work in chemistry. Although it was not the first of its kind—Faraday had designed something similar in 1830—it was Bunsen, one of the great early teachers of chemistry, who popularized it and justly gets proper credit for the "Bunsen burner."

Bunsen, in a portrait commissioned about 1860. He was at the height of his career and professor of physical chemistry at Heidelberg.

The typical laboratory version of the Bunsen burner in use today, burning a carbureted air-fuel mixture that is typically blue.

free radicals, which in turn are capable of producing $CHCl_3$ and $Cl\cdot$ free radicals.

Unfortunately, atomic chlorine is so reactive that all types of C-to-H bonds are attacked with equal vigor, and the replacement of one hydrogen in methane does not appreciably affect the replacement of a second. What that means for the thermal or photoinduced chlorination of methane is formation of a mix of four products, all of which are commercially important, and which must be separated from each other by distillation:

Methyl chloride (chloromethane)

$$CH_4 \text{ (g)} + Cl_2 \text{ (g)} \longrightarrow CH_3Cl \text{ (g)} + HCl \text{ (g)}$$

Methylene chloride (dichloromethane)

$$CH_3Cl \text{ (g)} + Cl_2 \text{ (g)} \longrightarrow CH_2Cl_2 \text{ (liq)} + HCl \text{ (g)}$$

Chloroform (trichloromethane)

$$CH_2Cl_2 \text{ (liq)} + Cl_2 \text{ (g)} \longrightarrow CHCl_3 \text{ (liq)} + HCl \text{ (g)}$$

Carbon tetrachloride (tetrachloromethane)

$$CHCl_3 \text{ (liq)} + Cl_2 \text{ (g)} \longrightarrow CCl_4 \text{ (liq)} + HCl \text{ (g)}$$

Industrially, the reaction is carried out by passing the mixture of methane and chlorine through hot tubes or special photochemical reactors fitted with mercury lamps to provide the necessary light quanta.

EXAMPLE 25–3

How would you go about maximizing the yield of methyl chloride, the monochlorination product in the photochlorination of methane, and minimizing the di-, tri-, and tetrachloromethanes in the industrial synthesis?

Solution The answer lies in having a very large excess of methane so that chlorine molecules almost invariably encounter CH_4 and not CH_3Cl molecules.

EXERCISE 25–3

Would the photobromination of methane require light of longer or shorter wavelength to initiate the reaction? Your answer should be based on the bond dissociation energy for Br_2. Briefly explain. ■

More than 10 million tons of chlorine were produced in the United States in 1986, three fourths of which was used in the synthesis of organic compounds. Methyl chloride is a gas at room temperature (b.p. −24°C), mostly used for the synthesis of silicone polymers and plastic materials such as butyl rubber and to a small extent as a pesticide. For many years it was used in the synthesis of tetraethyllead. It has also been used in the manufacture of methyl cellulose for the food industry. Methylene chloride (b.p. 40°C), the least toxic of the chlorinated methanes, is used extensively as a reaction medium and solvent (especially in the synthesis of triacetate films and fibers) and as a paint thinner and varnish remover. Chloroform (b.p. 61°C) is the principal intermediate in the synthesis of dichlorodifluoromethane and other "Freon" aerosol propellants (see Chapter 6). Carbon tetrachloride (b.p. 77°C) is the most important chlorinated methane. It is converted to tetrachloroethylene, the most widely used solvent in the dry cleaning industry, and into Freon aerosols. In common with some other chlorinated hydrocarbons, carbon tetrachloride is quite toxic, causing liver damage. Some contemporary environmental problems are believed to have arisen from the widespread use of Freons (or chlorofluorocarbons, "CFCs").

Structural Isomers and Chlorinated Alkanes

We encountered structural isomerism in our earlier discussion of coordination complexes (Chapter 22). For example, because the geometry around the central platinum atom in the anti-cancer drug cisplatin, cis-Pt(NH$_3$)$_2$Cl$_2$, is square planar, the NH$_3$ and Cl ligands can exist in two distinct arrangements (Figure 25–9). Both isomers are known, but only the cis isomer is biologically active. In contrast, dichloromethane cannot exist as $cis/trans$ isomers because the geometry about the central carbon atom is tetrahedral rather than square planar (Figure 25–10).

Only one monochlorination product is possible for ethane. However, if you remove an H atom from ethyl chloride and add another Cl atom in its place, two structural isomers are possible. Both chlorine atoms may be attached to the same carbon atom, or one chlorine may be attached to each carbon. The characteristic properties of the two isomers, such as melting and boiling points, are clearly different:

	1,2-dichloroethane	1,1-dichloroethane
m.p. (°C)	−35	−96
b.p. (°C)	84	57

Two structural isomers are possible for trichloroethane:

(1,1,1-trichloride)   (1,1,2-trichloride)

With four chlorine atoms in the molecule, two isomeric structures again are possible:

(1,1,2,2-tetrachloride)   (1,1,1,2-tetrachloride)

With five and six chlorine atoms, again only one bonding arrangement is possible in each case:

(pentachloride)   (hexachloride)

Finally, summing up the results for ethane, nine distinctly different chloroethanes are possible.

25.6 ALKENES, ADDITION POLYMERIZATION, AND THE CHEMISTRY OF RUBBER

Linear hydrocarbons containing one carbon-to-carbon double bond are known by the general name **alkenes** or **olefins**, and conform to the formula

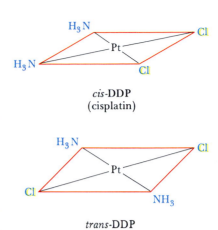

cis-DDP
(cisplatin)

trans-DDP

FIGURE 25–9 The square planar geometries of *cis*- and *trans*-Pt(NH$_3$)$_2$Cl$_2$, diaminodichloroplatinum, DDP. This *cis* isomer is the anticancer drug, cisplatin.

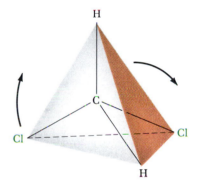

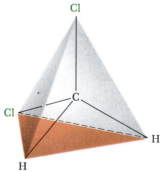

FIGURE 25–10 The tetrahedral geometry about the central carbon atom in dichloromethane accounts for the *nonexistence* of structural isomers.

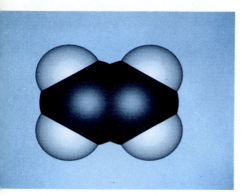

Computer-generated space-filling model of ethylene ($CH_2 = CH_2$).

C_nH_{2n}. Here too (unfortunately) you'll have to learn many common names, and (fortunately) a systematic method for naming. The first three members of the homologous series of alkenes are ethylene, propylene, and 1-butene. Ethylene and propylene are common names; the systematic names are ethene and propene.

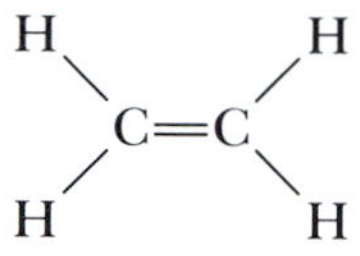

ethylene
(ethene)

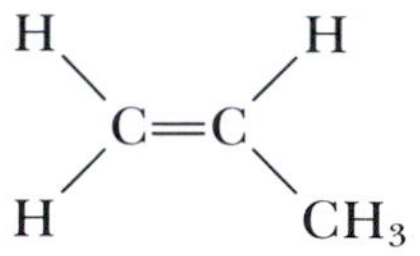

propylene
(propene)

Beginning with the butenes, alkenes are named systematically, in order to locate the double bonds unambiguously. Systematic nomenclature for the alkenes begins with the parent hydrocarbon chain of longest continuous length that contains the double bond. Drop the -*ane* ending of the corresponding alk*ane* name and replace it with the -*ene* ending for alk*ene*. Assign the numbering so that carbon atom at which the double bond begins has the lowest possible number:

$$CH_3 \underset{4}{-} CH_2 \underset{3}{-} CH \underset{2}{=} CH_2 \underset{1} \qquad CH_3 \underset{1}{-} CH \underset{2}{=} CH \underset{3}{-} CH_3 \underset{4}$$

1-butene   2-butene

Because double bonds impose restrictions on rotation (Chapter 18), two isomeric 2-butenes are possible; indeed, both isomers are well known:

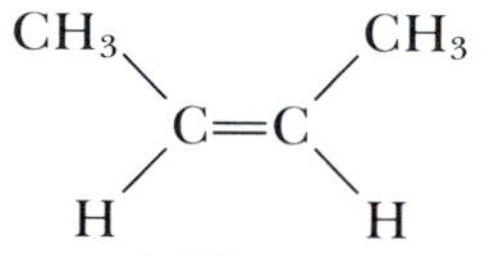

cis-2-butene 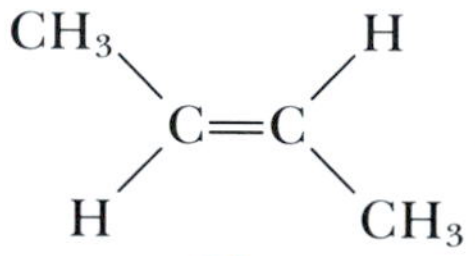   *trans*-2-butene

In order to distinguish between the geometric isomers of 2-butenes, we name then in a special way. Consider the methyl substituents on the two carbon atoms of the centrally located double bond. When the hydrogen atoms (or the methyl groups) are on the same side of an imaginary plane containing the double bond, we refer to the geometry as **cis**-. We refer to the geometry as **trans**- when they lie on opposite sides of the plane. Thus, we have *trans*-2-butene and *cis*-2-butene. There is yet another butene isomer, in which both methyl groups are attached to the carbon atom at one end of the double bond. It is named isobutene or 2-methylpropene. Note the difference in properties between ethylene, propylene, and the four possible butene isomers (Table 25–8).

Alkenes can be synthesized on an industrial scale in several ways:

TABLE 25–8	Melting and Boiling Points of the C_2–C_4 Alkenes	
	m.p. (°C)	b.p. (°C)
ethylene (ethene)	− 169	− 102
propylene (propene)	− 185	− 48
1-butene	− 195	− 6
trans-2-butene	− 106	1
cis-2-butene	− 139	4
iso-butene	− 141	− 7

1. By thermal cracking and dehydrogenation of the higher alkanes found in petroleum distillates. Ethylene and propylene are the major products. The synthesis of ethylene and propylene from wet natural gas, a mixture of ethane and propane from Texas, is typical:

At 850°C, in the presence of steam, on passing quickly through catalyst beds packed into 1-inch pipes in towers 100 feet high and 16 feet in diameter:

$$CH_3CH_3 \text{ (g)} \xrightarrow{\Delta} CH_2{=}CH_2 \text{ (g)} + H_2 \text{ (g) (dehydrogenation)}$$
$$CH_3CH_3 \text{ (g)} \xrightarrow{\Delta} C \text{ (s)} + CH_4 \text{ (g)} + H_2 \text{ (g) (cracking)}$$

At 750°C, under similar reactor conditions:

$$CH_3CH_2CH_3 \text{ (g)} \xrightarrow{\Delta} CH_3CH{=}CH_2 \text{ (g)} + H_2 \text{ (g) (dehydrogenation)}$$
$$CH_3CH_2CH_3 \text{ (g)} \xrightarrow{\Delta} CH_2{=}CH_2 \text{ (g)} + CH_4 \text{ (g)}$$
$$\text{(cracking and dehydrogenation)}$$

The delta over the arrow is the standard symbol for heat.

2. By elimination of HX (hydrogen and halogen atoms, typically HCl and HBr) from adjacent carbon atoms of halogenated alkanes, using an alcohol solution of KOH:

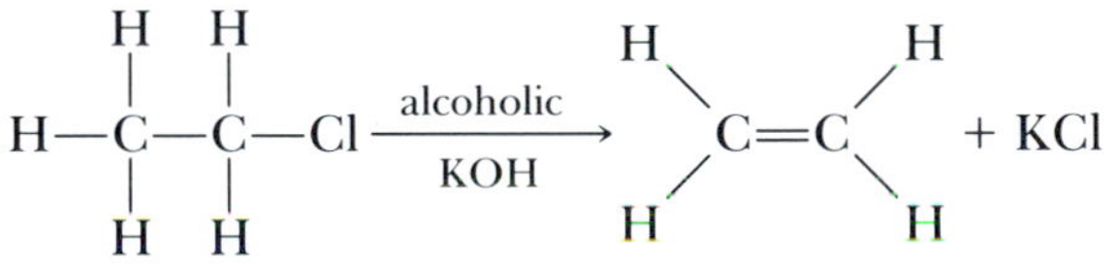

3. The double bond can also be introduced between adjacent carbon atoms by dehydration of alcohols, using strongly dehydrating agents such as sulfuric acid:

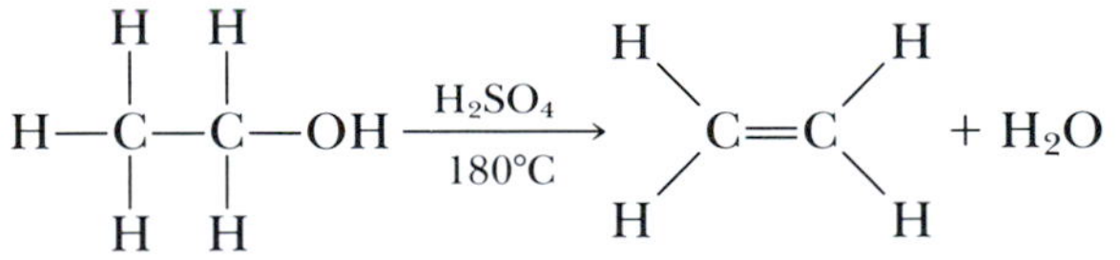

EXAMPLE 25–4

Synthesize propylene from propane by two different sets of chemical reactions.

Solution

1. Photochlorination of propane, followed by dehydrohalogenation:

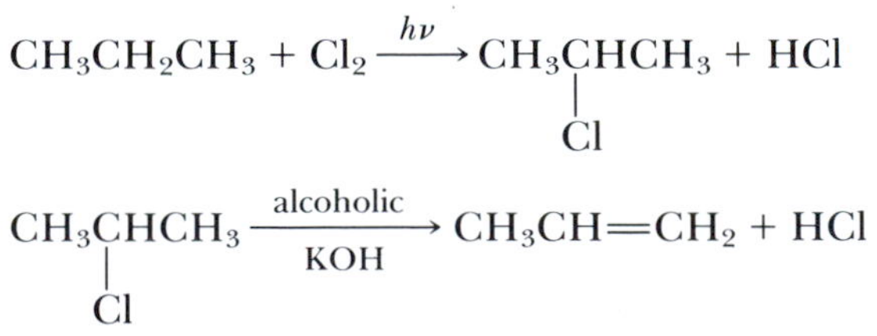

$$CH_3CH_2CH_3 + Cl_2 \xrightarrow{h\nu} \underset{\underset{Cl}{|}}{CH_3CHCH_3} + HCl$$

$$\underset{\underset{Cl}{|}}{CH_3CHCH_3} \xrightarrow[KOH]{\text{alcoholic}} CH_3CH{=}CH_2 + HCl$$

2. Dehydrogenation of propane:

$$CH_3CH_2CH_3 \xrightarrow[750°C]{\Delta,\ \text{catalyst}} CH_3CH{=}CH_2 + H_2$$

EXERCISE 25–4
Write equations for the synthesis of isobutene from 2-methylpropane.
Answer: Photochlorination of 2-methylpropane, followed by dehydrohalogenation. ∎

Alkenes are characterized by the ease with which reagents react with their double bonds. The net effect is addition to the double bond at the expense of the higher-energy π bond, leaving the σ bond intact. For example, a simple test for distinguishing between an alkane and an alkene is the fading of the characteristic red-brown color of bromine (dissolved in carbon tetrachloride), caused by addition of the bromine across the double bond:

$$\text{C=C} + \text{Br—Br} \xrightarrow{\text{CCl}_4} -\overset{\overset{\displaystyle Br}{|}}{\text{C}}-\underset{\underset{\displaystyle Br}{|}}{\text{C}}-$$

red-brown color   colorless

For example,

$$\text{CH}_3\text{CH}_2\text{CH}_2\text{CH=CH}_2 + \text{Br}_2 \xrightarrow{\text{CCl}_4} \text{CH}_3\text{CH}_2\text{CH}_2\overset{\overset{\displaystyle Br}{|}}{\underset{\underset{\displaystyle Br}{|}}{\text{CH}}}\text{CH}_2$$
1-pentene

$$\text{CH}_3\text{CH}_2\text{CH}_2\text{CH}_2\text{CH}_3 + \text{Br}_2 \xrightarrow{\text{CCl}_4} \text{no reaction}$$
pentane

Hydrogen, under high pressure and in the presence of catalysts such as nickel, palladium, or platinum, adds across double bonds:

$$\text{C=C} + \text{H}_2 \xrightarrow{\text{catalyst}} -\overset{|}{\underset{\underset{\displaystyle H}{|}}{\text{C}}}-\overset{|}{\underset{\underset{\displaystyle H}{|}}{\text{C}}}-$$

$$\text{CH}_3\text{CH}_2\text{CH}_2\text{CH=CH}_2 \text{ (g)} + \text{H}_2 \text{ (g)} \longrightarrow \text{CH}_3\text{CH}_2\text{CH}_2\text{CH}_2\text{CH}_3 \text{ (g)}$$

In fact, that is one of the best ways of synthesizing alkanes, providing the corresponding alkenes are available. It is because of this reaction that alkenes are known as **unsaturated** hydrocarbons, in contrast to saturated alkanes.

The hydrogenation reaction in turn can be used to measure the relative stabilities of alkenes by comparing respective ΔH values. Let's look at the heats of hydrogenation for *cis-* and *trans*-2-butene. There is a difference of 4 kJ/mol, even though both isomers yield *n*-butane, the same product, on adding one mole of hydrogen. Therefore, we can conclude that the *trans* isomer ($\Delta H_{\text{hyd}(trans)} = -115$ kJ/mol) must be stabilized, relative to the *cis* isomer ($\Delta H_{\text{hyd}(cis)} = -119$ kJ/mol) to the extent of the difference, or 4 kJ/mol.

Hydrogen chloride and hydrogen bromide can be added directly to the double bond, yielding the corresponding alkyl halides:

$$\text{C=C} + \text{HX} \longrightarrow -\overset{\overset{\displaystyle H}{|}}{\text{C}}-\underset{\underset{\displaystyle X}{|}}{\text{C}}-$$

X = Cl, Br, I, etc.

However, with the exception of purely symmetrical alkenes such as ethylene and 2-butene, a special problem arises. Which way do the elements of the hydrogen halides add? For example, H—Cl can add to propylene in two different ways:

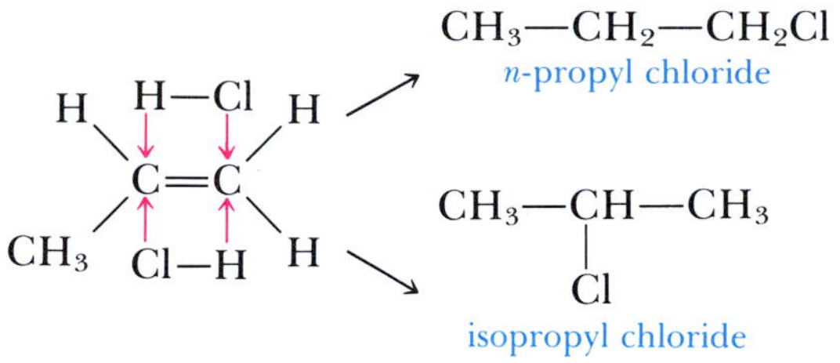

$$CH_3—CH_2—CH_2Cl$$
n-propyl chloride

$$CH_3—CH—CH_3$$
$$|$$
$$Cl$$
isopropyl chloride

The experimental facts strongly suggest a preference for one way over the other:

- Isopropyl chloride (2-chloropropane) is the dominant product on adding HCl to propylene.

- Tertiary butyl chloride is the only product that forms when HCl is added to isobutylene:

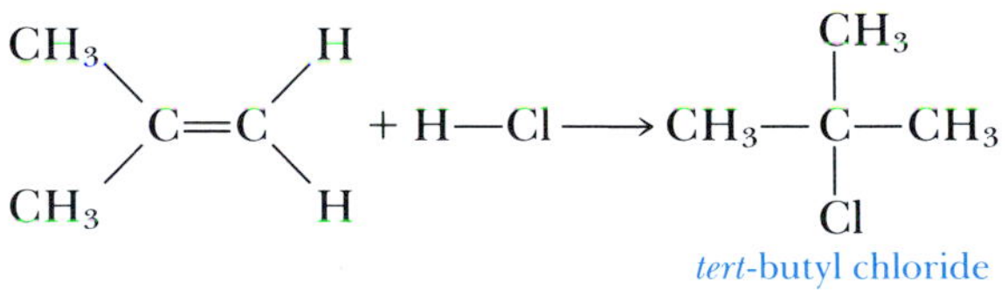

tert-butyl chloride

An empirical rule was established by a Russian chemist, Vladimir Markovnikov, after examining many such additions to alkenes:

> **The reaction of a polar molecule such as HCl with a carbon-carbon double bond results in the addition of the positive end of the polar molecule to the end of the double bond that has fewer alkyl substituents.**

Loosely stated, the H atom always goes to the carbon atom of the double bond that has the most H atoms already. [The reason behind the rule is that alkyl substituents are relatively electronegative and draw electrons away from the carbon atom to which they are attached. As a result, that carbon is more positive than the other carbon in the double bond. Electrostatic attractions thus orient the polar molecule so that Markovnikov's rule is obeyed.]

Another example of **Markovnikov addition** is the addition of water (HOH) to double bonds. Hydration of alkenes yields products called alcohols, in which an alkyl group now has an alcoholic OH group (instead of a halogen) attached to it:

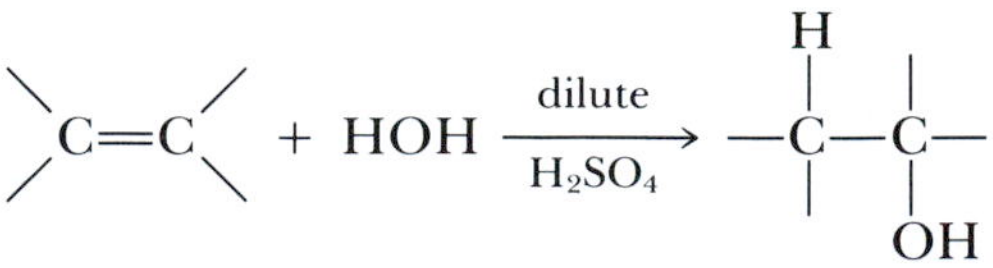

In the presence of peroxides, the mechanism changes (to free radical) and the addition is "anti-Markovnikov."

$$CH_3CH{=}CH_2 + HBr + peroxides \longrightarrow$$
$$CH_3CH_2CH_2Br$$

The reaction is catalyzed by acids, particularly sulfuric acid. Ethylene yields ethyl alcohol, and 2-butene gives rise to a molecule called secondary butyl alcohol (or 2-hydroxybutane). The direction of addition is not important in either case because of the symmetry of the starting material. But with propylene, Markovnikov's rule needs to be consulted in order to predict the product. It could be either *n*-propyl alcohol or isopropyl alcohol:

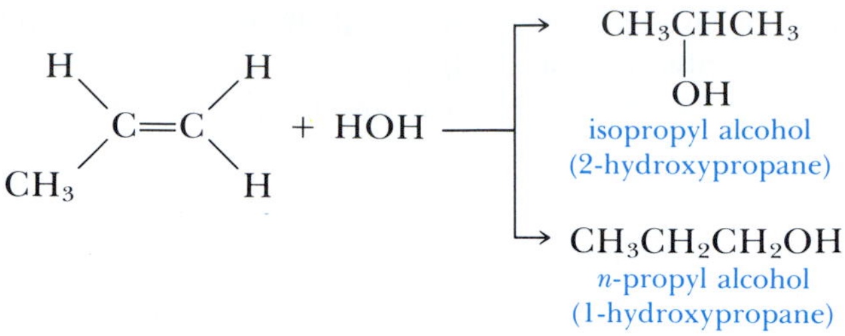

$$
\begin{matrix}
H & & H \\
& C=C & \\
CH_3 & & H
\end{matrix}
+ HOH \longrightarrow
\begin{cases}
\text{CH}_3\text{CHCH}_3 \\
\qquad | \\
\qquad \text{OH} \\
\text{isopropyl alcohol} \\
\text{(2-hydroxypropane)} \\[1em]
\text{CH}_3\text{CH}_2\text{CH}_2\text{OH} \\
\text{n-propyl alcohol} \\
\text{(1-hydroxypropane)}
\end{cases}
$$

The dominant product is isopropyl alcohol, just as Markovnikov would have predicted.

EXAMPLE 25–5

In light of Markovnikov's rule, predict the principal products of the reactions of HBr and H_2O with 1-butene and 3-hexene.

Solution   Remember, the H atom goes to the negative center, which generally turns out to be the carbon atom of the double bond having the most H atoms on it to begin with. Thus, for 1-butene,

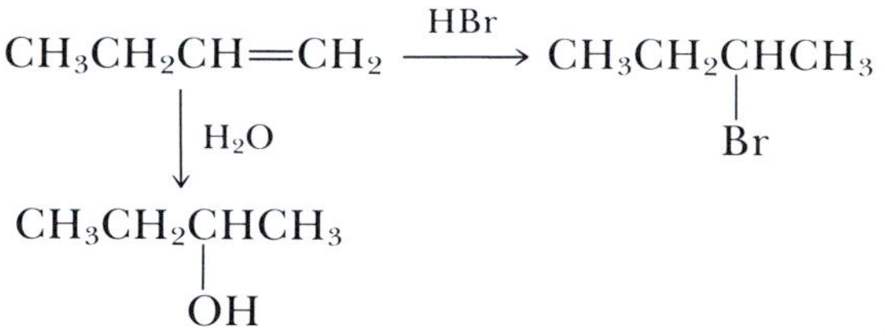

$$
\text{CH}_3\text{CH}_2\text{CH}=\text{CH}_2 \xrightarrow{\text{HBr}} \text{CH}_3\text{CH}_2\underset{\underset{\text{Br}}{|}}{\text{C}}\text{HCH}_3
$$

$$
\Big\downarrow \text{H}_2\text{O}
$$

$$
\text{CH}_3\text{CH}_2\underset{\underset{\text{OH}}{|}}{\text{C}}\text{HCH}_3
$$

For 3-hexene the same product is obtained regardless of the mode of addition because of the symmetry of the 3-hexene molecule:

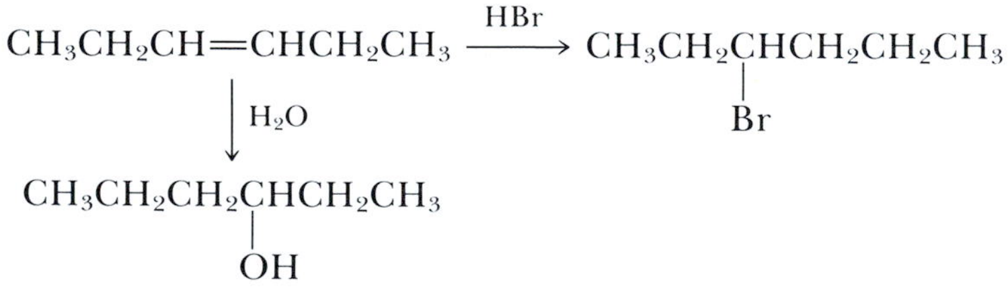

$$
\text{CH}_3\text{CH}_2\text{CH}=\text{CHCH}_2\text{CH}_3 \xrightarrow{\text{HBr}} \text{CH}_3\text{CH}_2\underset{\underset{\text{Br}}{|}}{\text{C}}\text{HCH}_2\text{CH}_2\text{CH}_3
$$

$$
\Big\downarrow \text{H}_2\text{O}
$$

$$
\text{CH}_3\text{CH}_2\text{CH}_2\underset{\underset{\text{OH}}{|}}{\text{C}}\text{HCH}_2\text{CH}_3
$$

EXERCISE 25–5

Name each of the structures in Example 25–5 unambiguously as a hydroxy- or bromoalkane.
Answer: 2-hydroxybutane, 2-bromobutane, 3-hydroxyhexane (or 3-hexanol), 3-bromohexane. ∎

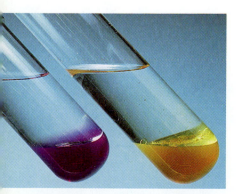

When aqueous $KMnO_4$ is shaken with hexane (left), it retains its purple color. The purple color is removed by reaction with the double bond of 1-hexene (right); and brown MnO_2 is formed.

Alkenes are much more susceptible to the action of oxidizing agents than are alkanes. Under mild conditions, alkenes turn the purple color of dilute aqueous permanganate solutions a murky brown due to formation of an insoluble precipitate of manganese dioxide. Alkanes won't do that (or much of anything else!) with permanganate solutions except under prolonged and more stringent conditions:

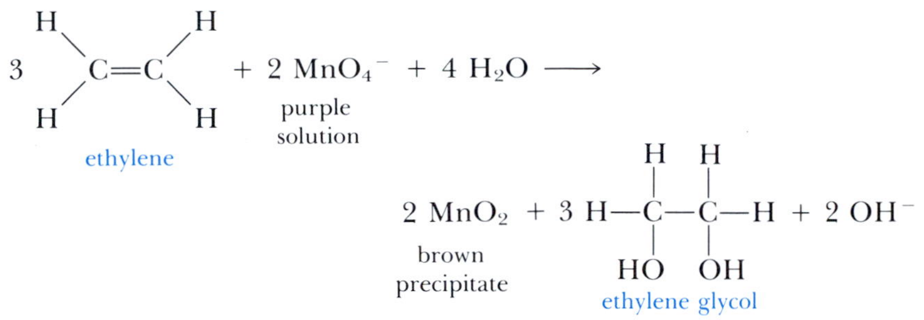

$$3 \quad \underset{H}{\overset{H}{>}} C = C \underset{H}{\overset{H}{<}} + 2 \, MnO_4^- + 4 \, H_2O \longrightarrow$$

ethylene
purple
solution

$$2 \, MnO_2 + 3 \, H-\underset{\underset{HO}{|}}{\overset{\overset{H}{|}}{C}}-\underset{\underset{OH}{|}}{\overset{\overset{H}{|}}{C}}-H + 2 \, OH^-$$

brown
precipitate

ethylene glycol

That suggests an interesting **permanganate test** for the presence of double bonds that can be used to distinguish between alkanes and alkenes.

One of the more important reactions alkenes undergo is **addition polymerization**, the combination of many small molecules called monomers into much larger molecules called polymers. When a monomer such as ethylene is heated in the presence of oxygen under pressure, a high molecular weight, alkane-like polymer called polyethylene is obtained. The process by which the monomer units are "snapped" together is an addition reaction in which the π bond electron pair is redistributd to the two carbon atoms. That alters the hybridization about carbon from sp^2 to sp^3, allowing formation of two new σ bonds between adjacent molecules. In that way, the monomer becomes a dimer, then a trimer, a tetramer, and eventually, a polymer:

$$n \, CH_2 = CH_2 \xrightarrow{\text{catalyst}} +CH_2 - CH_2 +_n$$

Vinyl chloride ($CH_2 = CHCl$) differs from ethylene ($CH_2 = CH_2$) by a single atom. The polymer produced is polyvinylchloride (PVC), which has uses as different as phonograph records, hubcaps, car bumpers, table cloths, curtains, and raincoats:

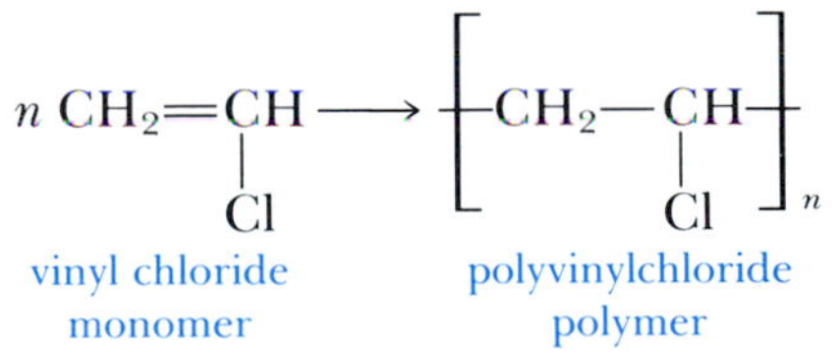

$$n \, CH_2 = \underset{\underset{Cl}{|}}{CH} \longrightarrow \left[CH_2 - \underset{\underset{Cl}{|}}{CH} \right]_n$$

vinyl chloride
monomer

polyvinylchloride
polymer

Saran Wrap and Glad Bags are typical examples of packaging products made from a similar polymer, polyvinylidene chloride. The monomer is vinylidene chloride ($CH_2 = CCl_2$):

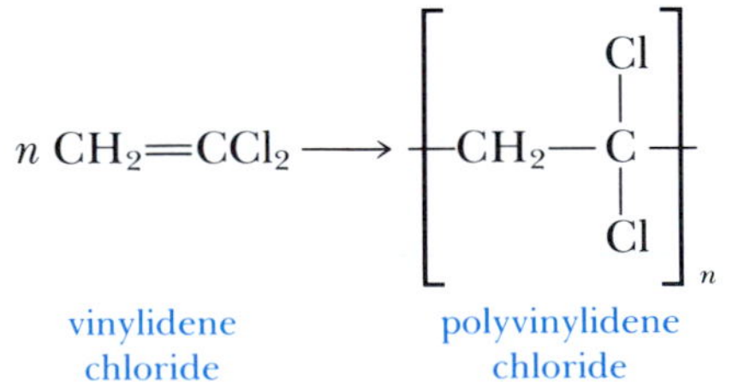

$$n \, CH_2 = CCl_2 \longrightarrow \left[CH_2 - \underset{\underset{Cl}{|}}{\overset{\overset{Cl}{|}}{C}} \right]_n$$

vinylidene
chloride

polyvinylidene
chloride

Orlon is used to make fibers, fabrics and the like. The monomer is called acrylonitrile ($CH_2 = CHCN$):

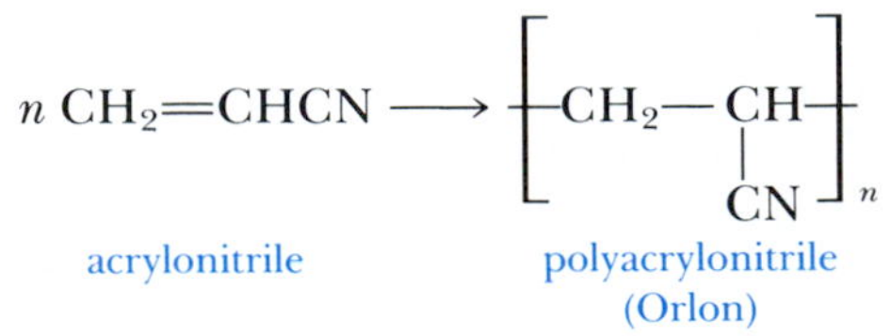

$$n \, CH_2 = CHCN \longrightarrow \left[CH_2 - \underset{\underset{CN}{|}}{CH} \right]_n$$

acrylonitrile

polyacrylonitrile
(Orlon)

Lucite and Plexiglas are acrylate polymers made from methyl methacrylate:

$$n \; CH_2{=}\underset{\underset{COOCH_3}{|}}{\overset{\overset{CH_3}{|}}{C}} \longrightarrow \left[CH_2{-}\underset{\underset{COOCH_3}{|}}{\overset{\overset{CH_3}{|}}{C}}{-}CH_2{-}\underset{\underset{COOCH_3}{|}}{\overset{\overset{CH_3}{|}}{C}} \right]_n$$

methyl methacrylate polymethylmethacrylate (Lucite/Plexiglas)

Natural rubber is composed of many hundreds of isoprene (2-methyl-1,3-butadiene) monomer units, added to each other, end-to-end, forming a high molecular weight polymeric structure with a *cis* geometry about the double bond. This is the soft, elastic material isolated from the sticky sap of the rubber tree:

$$n \; CH_2{=}C\overset{CH_3}{\underset{CH{=}CH_2}{}} \longrightarrow \left[CH_2{-}\overset{CH_3}{\underset{\parallel}{C}}\; CH{-}CH_2 \right]_n$$

the basic isoprene unit (2-methyl-1,3-butadiene)

natural rubber
(note the *cis* geometry about the double bond)

If the orientation about the double bond is trans, there is quite a change in the properties of the polymeric product. Gutta-percha is an all-*trans*-polyisoprene, and it has none of the elastic properties of natural rubber, the *cis* structure:

gutta-percha
(all *trans* structure)

One of the problems with natural rubber is that you can't use it directly for manufacturing tires or any other heat-sensitive products. On long trips or on warm days, natural rubber softens and eventually melts. Furthermore, if natural rubber is simply left standing in air for a period of time, it becomes hard and brittle with little elasticity. But that problem can be alleviated through a process called **vulcanization**, which introduces intramolecular polysulfide rings and intermolecular polysulfide bridges, connecting the long strands of polyisoprene units at points adjacent to the double bonds. There is no evidence that any new C-to-C bonds form. Although it would be nice to discuss the details of the reaction mechanism, after more than a century we still don't have all the answers. For example, the optimum vulcanization of rubber requires addition of four ingredients: sulfur, an accelerator, zinc oxide, and a zinc soap. In experiments involving all four ingredients in the absence of rubber, no less than 14 primary products and an unnumbered host of secondary products were formed. When rubber is present, only the desired product is found.

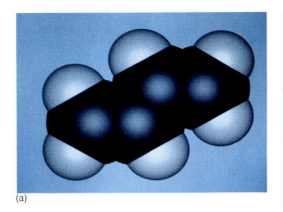
(a)

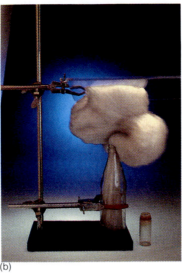
(b)

(c)

FIGURE 25–11 (a) Computer-generated space-filling model of the butadiene molecule used to produce a synthetic rubber. (b) In a cork-sealed soda bottle demonstration of the polymerization of butadiene, the elastomer product (left), swollen by the solvent, has "jumped" out. (c) After several hours, the solvent has evaporated, leaving (right) a "crepe-like" rubber, similar to that used on the soles of shoes.

In our automobile economy, rubber and the vulcanization process take on huge significance. Yet for all the interest, vulcanization technology is little changed in the 150 years since Charles Goodyear's pioneering experiments. In 1938, in the shadow of World War II and the isolation of the rubber plantations of Southeast Asia, the first commercially successful synthetic rubber was introduced. It made use of chlorobutadiene as the repeating monomer unit in the polymer. Although it is somewhat inferior to natural rubber as far as elasticity is concerned, the chloroprene that is formed is much less sensitive to the swelling and degrading effects of oils, greases, and organic solvents than is natural rubber. Figure 25–11 shows the preparation of a synthetic rubber made from butadiene.

25.7   ALKYNES

The distinguishing features of the alkynes are *sp*-hybridization and the resulting carbon-to-carbon triple bonds. The compounds fit the general formula C_nH_{2n-2}, and the simplest and best-known member of the family is acetylene, $H-C\equiv C-H$. Acetylene melts at $-82°C$ and boils at $-75°C$. Until recently, all acetylene was prepared by the reaction of calcium oxide and coke at elevated temperatures, followed by reaction of the calcium carbide product with water:

$$\text{C (coke)} + \text{CaO (lime)} \xrightarrow{2000°C} \underset{\text{calcium carbide}}{\text{CaC}_2} \xrightarrow{\text{H}_2\text{O}} \underset{\text{acetylene}}{\text{H}-\text{C}\equiv\text{C}-\text{H}}$$

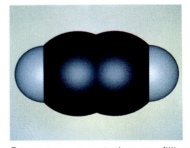
Computer-generated space-filling model of acetylene ($CH\equiv CH$).

Newer methods of preparing acetylene involve the cracking of natural gas or liquid hydrocarbon feedstocks:

$$2\ \text{CH}_4\ (g) \longrightarrow \text{C}_2\text{H}_2\ (g) + 3\ \text{H}_2\ (g)$$

Reaction conditions are critical. At 1600 K and beyond, C_2H_2 is more stable than other hydrocarbons that can form.

A general method for the laboratory preparation of alkynes is based on the same kind of elimination reaction used to prepare alkenes—dehydrohalogenation. But in order to generate the second π bond, two molecules of HX have to be removed:

$$-\overset{\overset{\displaystyle H}{|}}{\underset{\underset{\displaystyle X}{|}}{C}}-\overset{\overset{\displaystyle H}{|}}{\underset{\underset{\displaystyle X}{|}}{C}}- \xrightarrow[\text{KOH}]{\text{alcohol}} \left[-\overset{\overset{\displaystyle H}{|}}{C}=\overset{\underset{\underset{\displaystyle X}{|}}{}}{C}- \right] \longrightarrow -C\equiv C-$$

Just as alkene chemistry revolves around addition to the double bond, **alkyne chemistry** is largely the chemistry of additions to the triple bond. Although the chemistries are similar in many ways, a major distinguishing feature is the acidity of acetylenic hydrogen atoms. For example, acetylene will react with active metals, producing salts called acetylides:

$$2\ H-C\equiv C-H\ +\ 2\ Na\ \longrightarrow\ 2\ H-C\equiv \overset{\ominus}{C}:\overset{\oplus}{Na}\ +\ H_2$$

The acidic nature of the hydrogens can be demonstrated by reaction with $NaNH_2$, displacing NH_3 from solutions of sodamide in ether:

$$H-C\equiv C-H\ +\ NaNH_2 \Longrightarrow NH_3\ +\ H-C\equiv \overset{\ominus}{C}:\overset{\oplus}{Na}$$

Addition to the triple bond in alkynes is possible, but the intermediate double-bonded compound is hard to isolate because the double bond is actually more reactive than the triple bond.

$$HC\equiv CH \xrightarrow{\text{HI}} H_2C=CHI \xrightarrow{\text{HI}} H_3C-CHI_2$$

$$HC\equiv CH \xrightarrow{Cl_2} ClHC=CHCl \xrightarrow{Cl_2} Cl_2HC-CHCl_2$$

25.8   AROMATIC HYDROCARBONS

Resonance Stabilization in Benzene

The cyclic hydrocarbon known as benzene (C_6H_6) is one of the most important compounds in all of organic chemistry. The characteristic that set it apart from alkanes and alkenes is called its **aromatic character**. The principal feature of the aromaticity of benzene and benzene-like hydrocarbons has to do with the system of alternating single and double bonds. This is known as **conjugation**, and it allows delocalization of the π electrons throughout the system of alternating bonds. The delocalization of the electrons is continuous because of the ring structure, making for an extraordinarily symmetric molecule.

Benzene is flat and hexagonal, with all six carbon atoms lying in the same plane. The six carbon-to-carbon bonds are all 1.39 Å long, considerably shorter than the C-to-C sigma bond in ethane (1.54 Å) but longer than the C-to-C pi bond distance in ethylene (1.33 Å). All of the carbon-to-carbon bond angles are 120°. Recall our earlier discussion of resonance in Chapter 18: either we draw two equivalent canonical forms—Kekulé structures—or we simply show the π electrons delocalized over the entire six-membered ring by a circle within the hexagon. The π bonding electrons are not the property of any one carbon atom, but are delocalized about the entire ring.

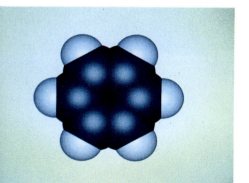

Computer-generated space-filling model of benzene (C_6H_6).

Aromatic Chemistry

To illustrate how aromaticity alters chemical behavior, let's compare the reactivity of the bonds in benzene to that of the double bond in cyclohexene and that of the single bonds in cyclohexane. Because of its alkane-like

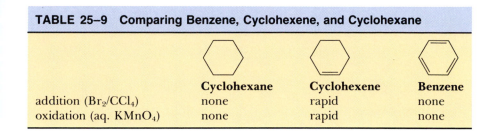

TABLE 25–9   Comparing Benzene, Cyclohexene, and Cyclohexane

	Cyclohexane	Cyclohexene	Benzene
addition (Br$_2$/CCl$_4$)	none	rapid	none
oxidation (aq. KMnO$_4$)	none	rapid	none

character, cyclohexane is notably unreactive. Cyclohexene, on the other hand, is alkene-like and typically undergoes addition across the double bond. The bonds in benzene, however, are not susceptible to the same addition reactions (Table 25–9). Resonance imparts a pronounced stability to benzene with respect to hydrogenation and oxidation. This added stability can be measured quantitatively in terms of a lower heat of hydrogenation (chemical addition) and heat of combustion (chemical oxidation).

While we are looking at these structures, note the convention for writing structural formulas of ring compounds. Each corner in the skeleton represents a carbon atom. Hydrogen atoms are omitted; it is understood that there are enough hydrogens attached to each carbon to bring the number of bonds on the carbon to four. Thus, the formula shown for cyclohexane represents six carbon atoms in a ring, each bonded to two hydrogen atoms.

Cyclohexane is considered to be an alicyclic hydrocarbon.

Substitution Reactions

The reactivity that benzene and benzene-like molecules exhibit can be understood in terms of preserving the integrity of the aromatic ring and aromatic character. Thus, substitution reactions, not addition reactions, are typical. For example, benzene reacts by undergoing ring-substitution rather than double bond addition in its behavior toward chlorine and bromine, reagents that we know add to the usual kinds of isolated double bonds.

Benzene also undergoes a number of other important ring substitution reactions, leading to many useful derivatives and intermediates (Table 25–10). Whole industries—especially drug and dye manufacturing—have been built upon these reactions, which were largely developed in Germany, Great Britain, and Switzerland in the late 19th century:

Nitration, with nitric acid:

$$\underset{HNO_3}{\overset{H_2SO_4}{\longrightarrow}} \quad \text{—NO}_2 \quad + \ H_2O$$

To help understand how the chemical reaction results in placing a —NO$_2$ group on the benzene ring, think of HNO$_3$ as HO—NO$_2$. Recall the Lewis structure.

Sulfonation, with sulfuric acid:

$$\underset{H_2SO_4}{\overset{SO_3}{\longrightarrow}} \quad \text{—SO}_3\text{H} \quad + \ H_2O$$

Think of H$_2$SO$_4$ as HO—SO$_3$H. Recall the Lewis structure.

Alkylation, with alkyl halides such as methyl chloride:

$$\underset{CH_3Cl}{\overset{AlCl_3}{\longrightarrow}} \quad \text{—CH}_3 \quad + \ HCl$$

Halogenation, with bromine or chlorine:

TABLE 25–10   Benzene Derivatives

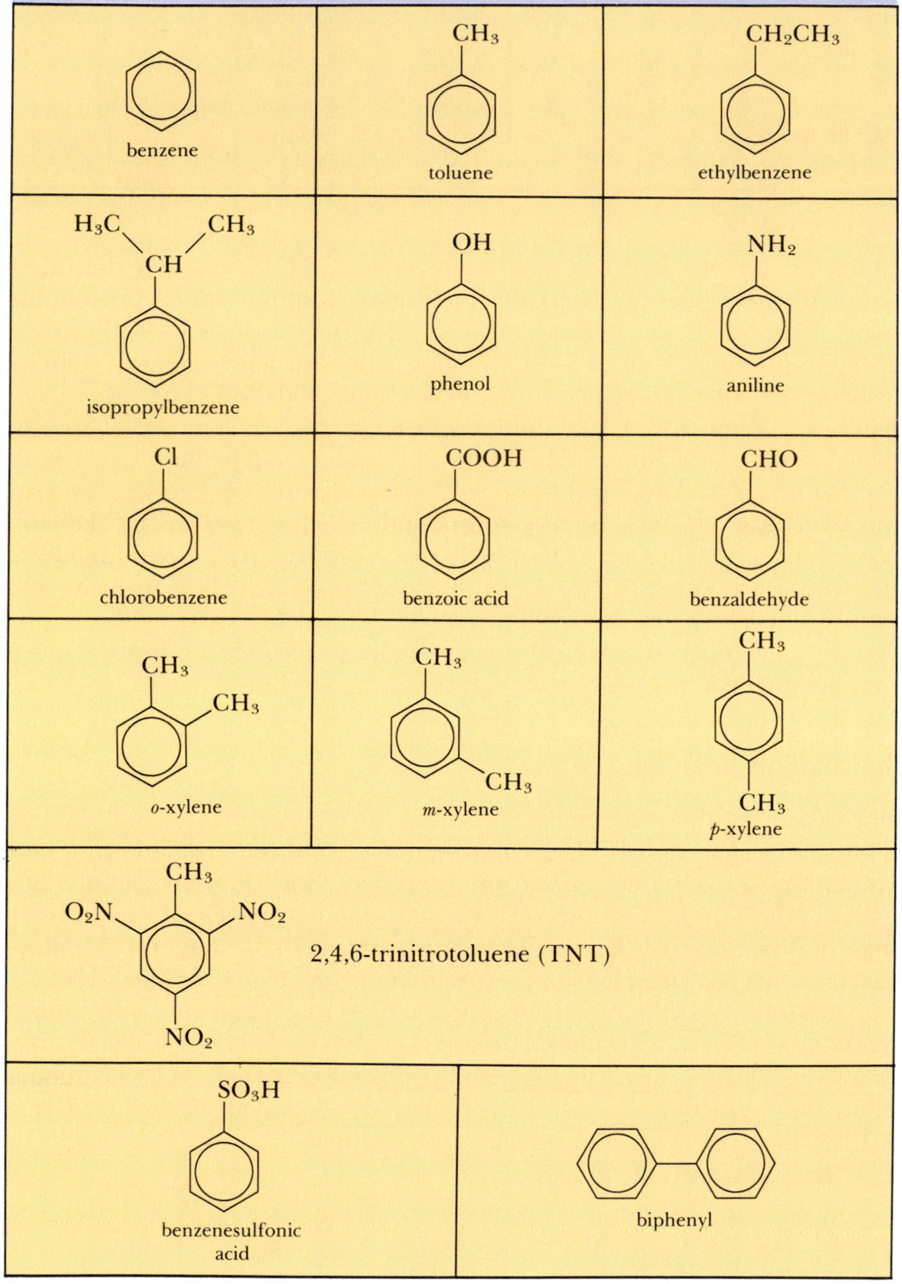

benzene	CH₃ toluene	CH₂CH₃ ethylbenzene
H₃C CH₃ CH isopropylbenzene	OH phenol	NH₂ aniline
Cl chlorobenzene	COOH benzoic acid	CHO benzaldehyde
CH₃ CH₃ o-xylene	CH₃ CH₃ m-xylene	CH₃ CH₃ p-xylene

CH₃
O₂N   NO₂

2,4,6-trinitrotoluene (TNT)

NO₂

SO₃H benzenesulfonic acid	biphenyl

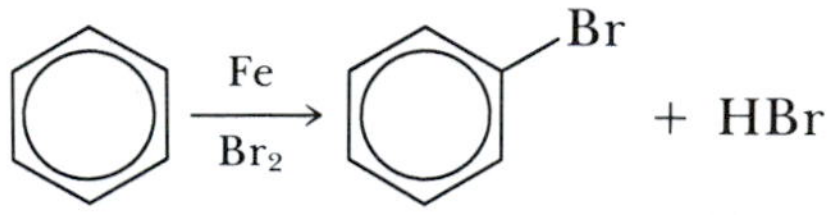

$$\text{benzene} \xrightarrow[\text{Br}_2]{\text{Fe}} \text{bromobenzene (Br)} + \text{HBr}$$

25.9   SOURCES OF AROMATIC COMPOUNDS

Coal, Coke, and Coal Tar

Coal is a complex mass of organic compounds derived from plants that have partially decayed (over millenia) due to the action of heat and pressure. The substances called peat, lignite, bituminous (or soft) coal, and anthracite

TABLE 25–11 Principal Components of Coal Tar	
Compound	**Percent**
benzene	0.1
toluene	0.2
mixed xylenes	1.0
naphthalene	10.9
α- and β-methylnaphthalenes	2.5
dimethylnaphthalenes	3.4
acenaphthalene	1.4
fluorene	1.6
phenanthrene	4.0
anthracene	1.1
carbazole	1.1
coal tar bases/pyridine (0.1)	2.0
coal tar acids/phenol (0.7)	2.5

(a)

(b)

FIGURE 25–12 Coal mining: (a) deep-mining operations in Pennsylvania; (b) strip-mining in Indiana.

(or hard) coal are successively more advanced stages in a metamorphosis leading to an ever higher ratio of carbon to all other elements present (Figure 25–12). For example, when bituminous coal is heated to temperatures near 1000°C in the absence of air (a process called **destructive distillation**), volatile products are given off and a residue of impure carbon, called **coke**, remains. On cooling, a fraction of the volatile products condenses to a black, viscous liquid called **coal tar**; the noncondensable gases are known as **coal gas**.

- Coal gas is a 1:1 mixture of hydrogen and methane. After it is properly treated to remove noxious trace compounds such as CO, H_2S, HCN, oxides of nitrogen, ammonia, and water vapor, it can be run directly into "mains" for use as illuminating gas, in domestic heating, and as a source of industrial energy.

- Coal tar varies in composition according to the carbonization process used to create it in the first place. Table 25–11 lists the chief products and those of commercial importance. The percentages are approximate; although the numbers may seem small, consider the huge quantities of coal that are processed annually, leading to nearly a billion gallons of coal tar in the United States in 1985.

- Coke is used for reduction of ores in blast furnaces and as a smokeless industrial fuel. As recently as the 1940s, it was a major fuel for home heating, but it has now been largely replaced by oil and natural gas.

Coal is consumed in appreciable amounts, for manufacturing benzene, naphthalene, anthracene, and (to a lesser extent) acetylene and carbon monoxide. However, all that still corresponds to less than 5% of organic chemicals produced.

Production of Benzene and Its Principal Derivatives

In 1986, in excess of 800 million gallons of benzene was produced in the United States, more than 85% of which came from petroleum, not coal. Of the 500 million gallons of toluene produced, more than 95% came from petroleum, as did almost all of the 400 million gallons of xylenes. Most of the 3 billion pounds of ethylbenzene (for use in the synthesis of styrene) was synthesized from benzene, along with 4 billion pounds of phenol. Phenol's principal use is in the synthesis of many of the monomers used in modern synthetic polymers.

Because of the demand for benzene, processes for its production from toluene have been developed. As you might expect from the following reaction, the process is called **hydrodemethylation**. It takes place on the

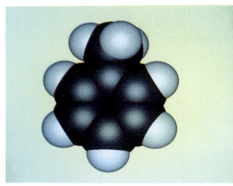
Computer-generated space-filling model of toluene ($C_6H_5CH_3$).

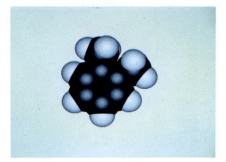 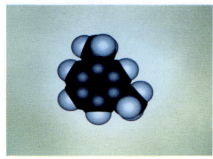 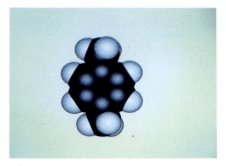

Computer-generated space-filling models of ortho-xylene, meta-xylene, and para-xylene: $[C_6H_4(CH_3)_2]$.

Hydrogen is obtained from fossil fuels, and especially from water, to fill all its process functions in industrial organic synthesis. Synthesis gas is the largest source:

$$CH_4 \text{ (g)} + H_2O \text{ (steam)} \longrightarrow CO \text{ (g)} + 3 H_2 \text{ (g)}$$

Du Pont is a major supplier of polyester fiber under the trade name Dacron; Du Pont's biaxially oriented film is called Mylar. Hoechst-Celanese is a major supplier of related polyester resins for manufacturing clear plastic soda bottles.

surface of a mixed Cr_2O_3—Al_2O_3—NaOH catalyst at 600°C and 800 psi:

$$C_6H_5CH_3 \text{ (g)} + H_2 \text{ (g)} \longrightarrow C_6H_6 \text{ (g)} + CH_4 \text{ (g)}$$

Because the benzene ring has six equivalent hydrogen atoms positioned around the ring, **positional isomers** arise when two substituents are added. The dimethylbenzenes, or xylenes (as they are commonly called), are important industrial solvents and intermediates. They can be conveniently distinguished by numbering the positions around the ring: thus, we would have 1,2-, 1,3-, and 1,4-dimethylbenzene. However, the trivial names *ortho-* for 1,2-disubstituted benzenes, *meta-* for 1,3-disubstituted benzenes, and *para-* for 1,4-disubstituted benzenes are commonly used.

ortho-xylene   *meta*-xylene   *para*-xylene

A typical example is *para*-dichlorobenzene, which is the substance in moth balls. The oxidation product of *para*-xylene, called terephthalic acid, is made by the hundreds of millions of pounds as an intermediate in the manufacture of plastics for monofilament fibers and blow-molded bottles:

1,4-dimethyl-
benzene
(*para*-xylene)
→ [O] oxidation →
terephthalic
acid
→ CH_3OH esterification →
dimethyl
terephthalate
→ fibers
→ films and sheets
→ molded products

Polycyclic Aromatic Hydrocarbons and Carcinogenesis

Considerable research is currently being devoted to the properties of condensed or polycyclic aromatic hydrocarbons. They are among a small but growing number of compounds known to induce cancer in humans. It was known in the 18th century that soot was a factor leading to a high incidence

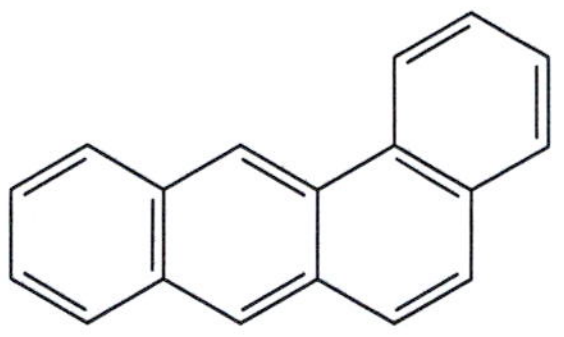

1,2-benzanthrene

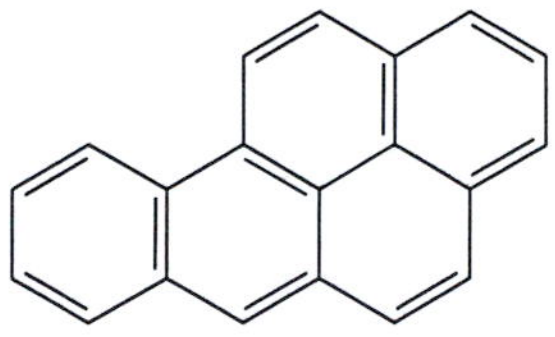

3,4-benzpyrene

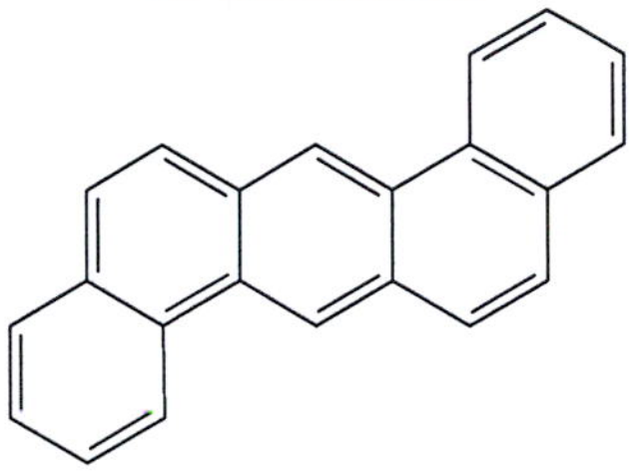

1,2,5,6-dibenzanthrene

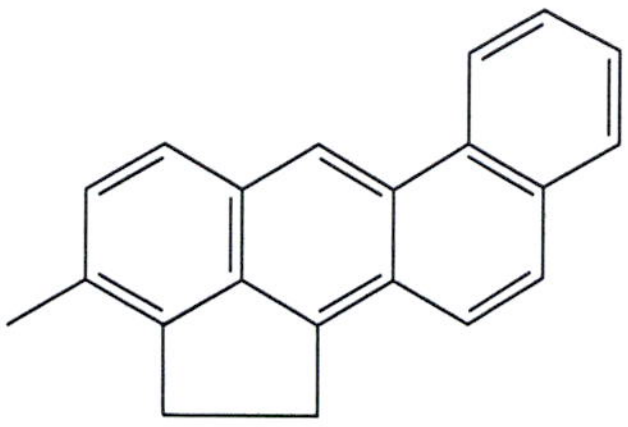

methylcholanthrene

of testicular cancer among chimney sweeps. Furthermore, workers in the coal tar industry frequently developed skin cancer. Recent studies conducted with laboratory animals under carefully controlled conditions have shown that cancers can be caused by a polycyclic aromatic compound, 3,4-benzpyrene, that is isolated from coal tar, tars from cigarette smoke, and the soots found in heavily polluted urban environments. These observations have led to the synthesis and study of a large number of related compounds, which have also turned out to be potent carcinogens.

25.10 ALCOHOLS AND ETHERS

The prospect of studying the chemical and physical properties of the 300,000 or so isomers of eicosane, $C_{20}H_{42}$, is beyond comprehension. Most hydrocarbons and their derivatives, however—eicosane and its isomers included—follow a general pattern of physical and chemical behavior. Thus, one need only study the group characteristics to know a fair amount about any one member of the group. When other atoms or groups of atoms (such as Cl, OH, or NH_2) are introduced, the resulting compound tends to exhibit patterns of reactivity that are characteristic of the **hetero** atom (or group). Since these groups react as functional units, they have come to be called **functional groups**. To know the chemistry of the variety of hydrocarbon derivatives containing a particular function group, one need only generalize from the study of the chemistry of that functional group in a few cases. A number of those most frequently encountered are listed in generalized molecular form, and with specific examples:

$CH_3—\overset{..}{\underset{..}{O}}H$
methyl alcohol
(alcohol)

$CH_3—\overset{..}{\underset{..}{O}}—CH_3$
dimethyl ether
(ether)

$CH_3\overset{\overset{..}{\overset{\displaystyle :O}{\|}}}{C}—H$
acetaldehyde
(aldehyde)

$CH_3\overset{\overset{\displaystyle :O}{\|}}{C}—CH_3$
dimethylketone
(ketone)

$CH_3\overset{\overset{\displaystyle :O}{\|}}{C}—\overset{..}{O}H$
acetic acid
(carboxylic acid)

$CH_3\overset{\overset{\displaystyle :O}{\|}}{C}—\overset{..}{O}CH_2CH_3$
ethyl acetate
(ester)

$CH_3\overset{\overset{\displaystyle :O}{\|}}{C}—\overset{..}{N}H_2$
acetamide
(amide)

$CH_3—\overset{..}{N}H_2$
methylamine
(amine)

Alcohols

Alcohols are compounds that have the general formula R—OH, where R is any alkyl or substituted alkyl group, and OH is the **hydroxyl** functional group. The properties of the alcohols depend on how the hydrocarbon chain is arranged. They are classified as primary (1°), secondary (2°), or tertiary (3°) alcohols depending on the number of carbon atoms bonded directly to the carbon atom to which the OH group is attached. A number of different systems of nomenclature are used for alcohols, but for our purposes the simplest is to name the fragment of the parent hydrocarbon (to which the OH has been added), and to add the word *alcohol*. Ethyl alcohol is a primary alcohol; isopropyl alcohol is a secondary alcohol; and tertiary butyl alcohol is a tertiary alcohol:

$$CH_3-CH_2-\ddot{O}H \qquad CH_3-\underset{\underset{\displaystyle :\ddot{O}H}{|}}{CH}-CH_3 \qquad CH_3-\underset{\underset{\displaystyle :\ddot{O}H}{|}}{\overset{\overset{\displaystyle CH_3}{|}}{C}}-CH_3$$

ethyl alcohol   isopropyl alcohol   *tert*-butyl alcohol
1°   2°   3°

The physical properties of the lower alcohols differ markedly from those of the corresponding hydrocarbons. They may be looked upon as derivatives of water. Because of the presence of the polar OH group, there is considerable hydrogen bonding. However, as the carbon chain becomes longer, the OH group becomes less and less of a factor on the composite molecular properties, and the character of the alcohol becomes increasingly hydrocarbon-like rather then water-like. The hydroxyl hydrogen is not readily ionizable—alcohols are not good proton donors—and as a result alcohols are only very weakly acidic.

One general method for preparing alcohols is the sulfuric acid-catalyzed hydration of alkenes. As we have already noted, addition takes place according to Markovnikov's rule. Hence the secondary alcohol is the principal product of propylene hydration:

$$CH_3CH{=}CH_2 + H_2O \xrightarrow{H_2SO_4} CH_3\underset{\underset{\displaystyle OH}{|}}{CH}CH_3$$

Hydrolysis of an alkyl halide with aqueous potassium hydroxide will also result in alcohol formation.

Some Specific Alcohols

Ethyl alcohol (grain alcohol or ethanol) can be produced industrially by the acid-catalyzed hydration of ethylene, although the fermentation of sugar residues (molasses) with yeast is still widely used:

$$\text{starch} \xrightarrow[\text{catalyst}]{\text{enzymes}} \underset{\text{(glucose)}}{C_6H_{12}O_6} \xrightarrow[\text{catalyst}]{\text{enzyme}} 2\,CH_3CH_2OH + 2\,CO_2$$

The carbon dioxide byproduct is important in baking and in the natural carbonation of beers and sparkling wines. Beer contains about 4% alcohol, wines near 12%, and gins, brandies, and whiskies about 40 to 50%. Ethyl alcohol is the only alcohol that can be consumed by the human body with a modicum of safety (though it, too, should be considered a toxic substance). Fortunately, in the case of ethyl alcohol, the human body has the capacity to metabolize the substance, although at a limited rate. It is a depressant to the central nervous system (contrary to popular belief).

Ethyl alcohol is an important industrial solvent for a variety of industries ranging from paints, inks, and dyes to perfumes, cosmetics, and pharmaceuticals. It is an important intermediate in the manufacture of many other chemicals. It can also be used as a fuel for internal combustion engines either neat, or as a gasoline extender.

Methyl alcohol (wood alcohol or methanol) is prepared by reduction of carbon monoxide with hydrogen in the presence of a mixed metal oxide catalyst:

$$CO \text{ (g)} + 2 H_2 \text{ (g)} \longrightarrow CH_3OH \text{ (g)}$$

The common name wood alcohol (Chapter 6) derives from an earlier method of preparation, the destructive distillation of wood. At about 250°C, in the absence of air, wood decomposes to charcoal and a volatile fraction that is 3% methyl alcohol. Widely used as a paint solvent, it is a particularly insidious posion, producing first blindness and eventually death upon prolonged exposure to the liquid or vapor. There has been considerable interest in methyl alcohol in recent years as an extender for gasolines, and even as a motor fuel by itself.

Methanol remains one of the world's most important industrial synthetic raw materials and process feedstocks—more than 20 billion pounds were manufactured in 1987. A typical catalyst system is modified $ZnO—Cr_2O_3$ or $CuO—ZnO$.

EXAMPLE 25–6

Until about 1965, methyl alcohol was produced on an industrial scale almost exclusively by the reaction of synthesis gas over a 3:1 mixture of ZnO and Cr_2O_3 at 350°C and 30.39 MPa, under which conditions the conversion is 19.5%:

$$CO \text{ (g)} + 2 H_2 \text{ (g)} \rightleftharpoons CH_3OH \text{ (g)} \qquad \Delta H = -94.5 \text{ kJ/mol}$$

(a) If the pressure drops to 10.13 MPa, the conversion drops to 3.5%. Briefly explain.
(b) Will the conversion be favored (or inhibited) by a temperature increase? Briefly explain.
(c) By this time, you should be comfortable with pressure units of pascals or atmospheres. To what pressure in atmospheres does 30.39 MPa correspond?

Solution (a) High pressure favors the products and low pressure favors the reactants, according to LeChatelier's principle.
(b) The exothermic reaction is inhibited by high temperatures, again according to LeChatelier.
(c) $\dfrac{30.39 \text{ MPa} \times 10^6 \text{ Pa/MPa}}{101{,}300 \text{ Pa/atm}} = 300 \text{ atm}$

EXERCISE 25–6(A)

In the process for the synthesis of methyl alcohol described in Example 25–6, the unconverted synthesis gas is recycled to the catalyst bed along with fresh reacting materials. Why can't the unreacted gaseous mixture be recycled indefinitely? *Hint:* From an earlier chapter, what competing reactions might be going on? ■

TABLE 25–12 Phenols and Phenolic Compounds

EXERCISE 25–6(B)

The industrial synthesis of ethyl alcohol, carried out in the gas phase over phosphoric acid on Celite, a form of clay, takes place at 300°C and 6.8 MPa and gives a 5% conversion. If the heat of reaction is -44.1 kJ/mol, how would you go about varying T and P in order to increase the conversion? ∎

Aromatic alcohols can be described in the same general way, except for the special case in which the hydroxyl group is substituted directly on one of the ring carbon atoms. In this situation, the delocalization of the electrons in the ring exerts a marked effect upon the OH group. Such aromatic alcohols are called **phenols** (Table 25–12).

The availability of phenol from coal is limited by the demand for other coal tar products and coal gas, and the demand has long since outstripped availability. Phenol may well be the highest volume organic chemical manufactured worldwide. More than 4 billion pounds were manufactured in 1986. Fortunately, phenol is easily (and therefore economically) available by means of some chemical technology left over from WW II—the British Petroleum/Hercules cumene process. Cumene happens to be a 92-octane aviation fuel, and in 1940 the British had a process for manufacturing it that gave their fighter aircraft a considerable advantage over their competition in the Battle of Britain. When the war ended, they had an enormous cumene capacity and nothing to do with it. It was quickly found that cumene could be air-oxidized to cumene hydroperoxide (CHP), an unstable intermediate that easily decomposes in the presence of trace quantities of acid to phenol and acetone:

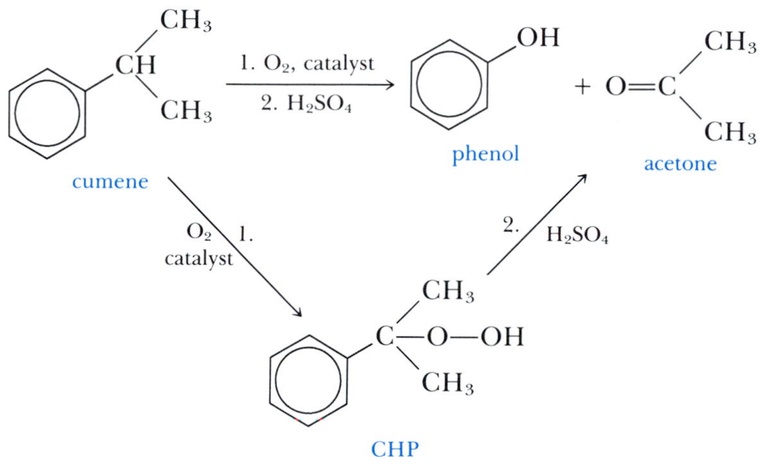

Phenols, as we have mentioned, are important intermediates in the synthesis of resins for use in the manufacture of plastics. They show up in the supermarket in alkylated form as antioxidants (in foods) and as surface-active agents (in detergent compositions). Chlorinated phenols have been used as antiseptics, plant growth regulators, and timber preservatives. A significant fraction of the worldwide synthesis of nylon comes from compounds obtained from phenol. Aspirin and photographic chemicals depend on phenol, too (Table 25–13).

Ethers

Compounds of the general form R—O—R′ are referred to as ethers. Again, R stands for an alkyl or an aryl group:

dimethyl ether	R = CH_3	CH_3—O—CH_3
diethyl ether	R = C_2H_5	C_2H_5—O—C_2H_5
methylethyl ether	R = CH_3, R′ = C_2H_5	CH_3—O—C_2H_5
anisole	R = C_6H_5, R′ = CH_3	C_6H_5—O—CH_3

Low molecular weight ethers tend to be volatile liquids, prepared by the loss of a molecule of water from two molecules of alcohol. This dehydration happens during distillation in the presence of sulfuric acid, under carefully controlled conditions:

$$2\ R{-}OH \xrightarrow{H_2SO_4} R{-}O{-}R + H_2O$$

The temperature is critical. For example, the preparation of diethyl ether takes place smoothly at 140°C, but at 180°C dehydration of the alcohol to ethylene is the preferred reaction:

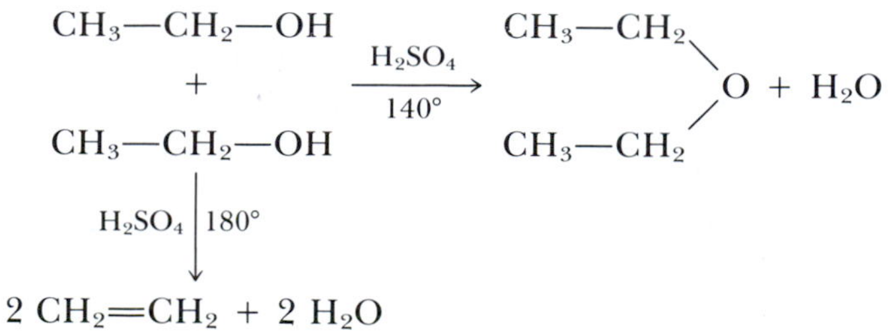

TABLE 25–13   Industrial Uses of Phenolic Compounds

Nylon from phenol via caprolactam

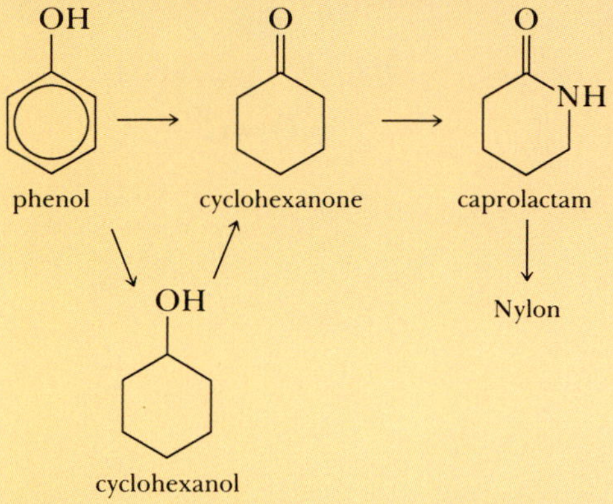

phenol → cyclohexanone → caprolactam → Nylon

cyclohexanol

Aspirin from salicylic acid, a phenol derivative

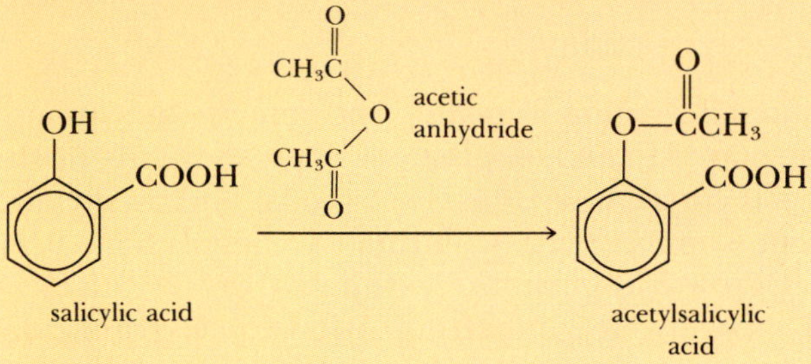

salicylic acid + acetic anhydride → acetylsalicylic acid

Hydroquinone as a photographic chemical

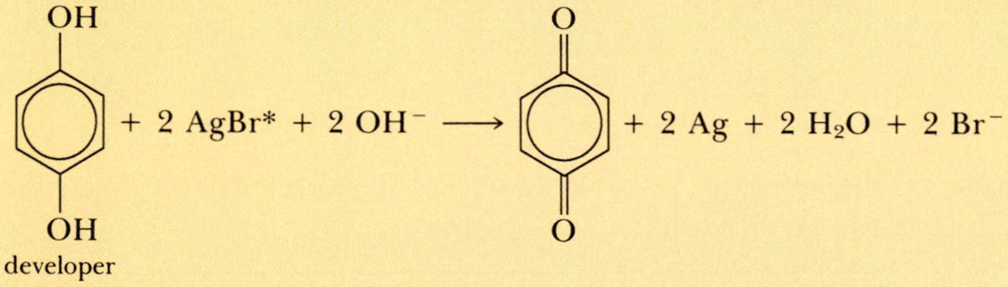

$$\text{developer} + 2\,AgBr^* + 2\,OH^- \longrightarrow \text{quinone} + 2\,Ag + 2\,H_2O + 2\,Br^-$$

developer

---

*Light-activated silver bromide crystals in the AgBr emulsion of the photographic film. The developer is oxidized to a quinone and the silver is reduced to the metal, which precipitates on the emulsion. The unactivated AgBr is removed with a "fixer" solution, often aqueous thiosulfate ($S_2O_3{}^{2-}$), leaving the precipitated silver as the black part of the "black-and-white" negative.

Diethyl ether is the best-known member of the family, because of its early use as an anesthetic and for its wide laboratory use as a solvent. Diethyl ether, or simply ethyl ether, is a notorious fire hazard because of its very high vapor pressure (b.p. 35°C), vapor density, and flammability. Having a density greater than that of air, the ether fumes will roll along bench tops and floors and be ignited by burners, hot surfaces, and electrical equipment long distances from the source. An equal (or perhaps greater) hazard is the tendency of ether to form a peroxide. These peroxygen compounds are often explosively unstable. Cumene hydroperoxide, the intermediate in the phenol/acetone synthesis, is just such compound.

TABLE 25–14   Aldehydes and Ketones

Characteristic Structure of the Carbonyl Group $\diagdown C {=} \ddot{O}:$

Aldehydic Carbonyl Group $\underset{H}{\overset{R}{\diagdown}} C{=}\ddot{O}:$

| $\underset{H}{\overset{H}{\diagdown}} C{=}\ddot{O}:$ formaldehyde | $\underset{H}{\overset{CH_3}{\diagdown}} C{=}\ddot{O}:$ acetaldehyde | benzaldehyde |

Ketonic Carbonyl Group $\underset{R'}{\overset{R}{\diagdown}} C{=}\ddot{O}:$

| $\underset{CH_3}{\overset{CH_3}{\diagdown}} C{=}\ddot{O}:$ acetone | acetophenone | benzophenone |

25.11   ALDEHYDES AND KETONES

The intermediate products in the controlled oxidation of primary alcohols are aldehydes. When secondary alcohols are oxidized, ketones result. The general formulas of aldehydes and ketones are shown in Table 25–14.

Aldehyde and ketone chemistry is predominantly **carbonyl group** chemistry. The properties of these two classes of compounds are similar, but differences do arise because of the hydrogen atom that is present in the former and replaced by an alkyl group in the latter. The aldehydic hydrogen atom leads to a marked susceptibility toward further oxidation (to carboxylic acids).

Formaldehyde, the simplest aldehyde and the only one with two hydrogen atoms on the carbonyl group, is prepared by the controlled air oxidation of methyl alcohol over a mixed oxide catalyst at about 600 K:

$$2\ CH_3OH\ (g)\ +\ O_2\ (g) \longrightarrow 2\ HCHO\ (g)\ +\ H_2O\ (g)$$
$$\Delta H\ =\ -154\ kJ/mol$$

At room temperature, formaldehyde is a gas with a particularly irritating and offensive odor. Because of its high solubility in water, it is often handled as a 40% aqueous solution (known as formalin). It is widely used in the synthesis of low-cost, durable plastic materials, the best known of which is Formica. It is also used to make a polyether polymer called Delrin that has largely replaced a number of metals in light engineering applications, including outdoor in-the-ground plumbing. In its clinical applications, formaldehyde is a general antiseptic and a widely used preservative agent, and the major active ingredient in most embalming fluids (because it stiffens protein).

Most of the other lower aldehydes and ketones are liquids with more pleasant odors. Acetaldehyde can be synthesized by the hydration of acetylene in the presence of sulfuric acid and mercuric sulfate, or by the controlled oxidation of ethyl alcohol:

$$HC \equiv CH \text{ (g)} + H_2O \text{ (liq)} \xrightarrow{\text{HgSO}_4} CH_3CHO \text{ (liq)}$$

Potassium dichromate can also be used for the alcohol oxidation. The trick here is to have the more volatile aldehyde distill as it forms so as to not be subjected to further oxidation to acetic acid.

One of the characteristic reactions of aldehydes is their oxidation by a silver-ammonia complex (known as Tollens's reagent), which is in turn reduced, depositing a telltale silver mirror on the walls of the reaction container:

$$RCHO \text{ (aq)} + 2 Ag(NH_3)_2{}^+ \text{ (aq)} + 2 H_2O \text{ (liq)} \longrightarrow$$

$$RCOO^- \text{ (aq)} + 2 Ag \text{ (s)} + 4 NH_4{}^+ \text{ (aq)} + OH^- \text{(aq)}$$

Ketones can be prepared from alcohols. Acetone, for example, can be prepared by oxidation of isopropyl alcohol,

$$\begin{matrix} CH_3 \\ \\ CH_3 \end{matrix}\!\!\!\!\diagdown\!\!\!\!\diagup CHOH \xrightarrow{\text{oxidation}} \begin{matrix} CH_3 \\ \\ CH_3 \end{matrix}\!\!\!\!\diagdown\!\!\!\!\diagup C{=}O$$

However, note that the commercial synthesis of acetone, at a rate of 2.5 billion pounds per year, is as a co-product (with phenol) in the cumene process described earlier. It is an industrial solvent and common laboratory reagent. Its most famous commercial use is in the synthesis of methyl methacrylate, the monomer used in the preparation of Lucite (by Du Pont) and Plexiglas (by Rohm & Haas) in the United States, and Perspex (by ICI) in the United Kingdom. All are acrylate polymers, characterized by unusual structural and optical properties.

In the human body, acetone is a by-product of certain metabolic pathways, showing up in the blood in very small concentrations. Diabetics produce excessive amounts that can be detected in the urine, and in severe cases, the characteristic odor of acetone is obvious on the patient's breath.

Both aldehydes and ketones are reduced to the corresponding alcohol or all the way back to the hydrocarbon itself, depending on reagents and conditions:

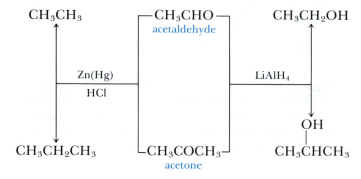

Starch and Cellulose

The formula for the sugar called glucose is $C_6H_{12}O_6$:

$$
\begin{array}{c}
\text{CHO} \\
| \\
\text{H—C—OH} \\
| \\
\text{OH—C—H} \\
| \\
\text{H—C—OH} \\
| \\
\text{H—C—OH} \\
| \\
\text{CH}_2\text{OH}
\end{array}
$$

It can form condensation polymers called **starches** (several hundred units long) and **celluloses** (several thousand units long) by elimination of water. Collectively, sugars, starches, and cellulose polymers are called **carbohydrates** because the hydrogen and oxygen atoms are in the same 2:1 ratio as in water. Under suitable conditions, starch and cellulose can be hydrolyzed back to simple glucose sugars once again:

$$(C_6H_{10}O_5)_n + H_2O \xrightarrow{H^+} n\ C_6H_{12}O_6$$

$$\text{starches} \quad + \quad \text{water} \quad \longrightarrow \quad \underset{\text{glucose sugars}}{\text{simple}}$$

25.12 CARBOXYLIC ACIDS, ESTERS, AND AMINES

The class of acidic organic substances characterized by the carboxyl functional group (—COOH) is referred to as **carboxylic acids**. They are weak as acids go, being only slightly dissociated in aqueous solution. For example, the K_a for acetic acid is about 10^{-5}. The acidity of carboxylic acids lies between that of phenol and inorganic acids such as HF and H_3PO_4. Formic acid (HCOOH) and acetic acid (CH_3COOH), the lower members of the family of organic acids, are soluble in water. But as we saw with the alcohols, extension of the carbon chain results in a diminishing contribution by the functional group and the solubility (in water) begins to drop off rapidly.

The low molecular weight carboxylic acids are liquids with sharp, unpleasant odors (Table 25–15):

- acetic acid—vinegar
- butyric acid—rancid butter
- caprylic and caproic acids—essence of goat (on a humid day)

Beyond 10 carbons in the chain, the carboxylic acids are all waxy solids with low vapor pressures. One of the most fascinating uses to which nature puts the carboxylic acids, especially formic acid, is as the active component in the chemical defense mechanisms of a variety of insects such as ants, cockroaches, beetles, and bees. Fatty acids, soaps, and detergents will be discussed briefly in Chapter 26.

Carboxylic acids have classically been prepared by oxidation. They are themselves the stable oxidation products of alcohols and hydrocarbons:

$$\text{hydrocarbon} \longrightarrow \text{alcohol} \longrightarrow \text{aldehyde} \longrightarrow \text{acid}$$

With two important exceptions, carboxylic acids are commercially synthesized by the oxidation of intermediates such as aldehydes (illustrated for acetic acid) and alkyl halides (illustrated by the formation of benzoic acid):

TABLE 25–15   Carboxylic Acids

HCOOH	formic acid
CH_3COOH	acetic (ethanoic) acid
$CH_3CH_2CH_2COOH$	butyric (butanoic) acid
$CH_3(CH_2)_4COOH$	caproic (hexanoic) acid
$CH_3(CH_2)_6COOH$	caprylic (octanoic) acid

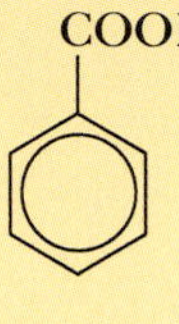

benzoic acid

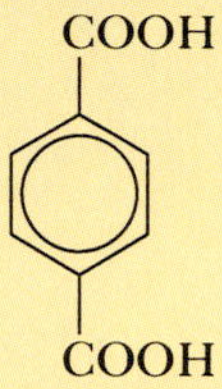

terephthalic acid

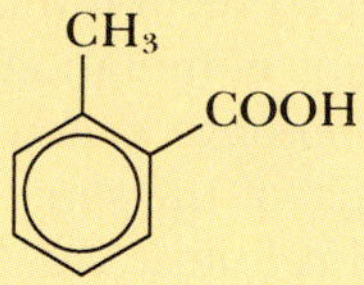

o-toluic acid

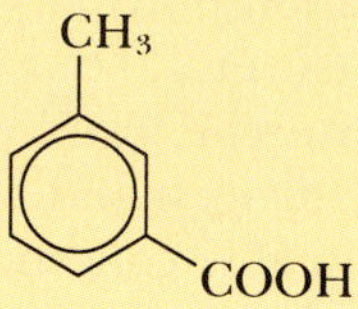

m-toluic acid

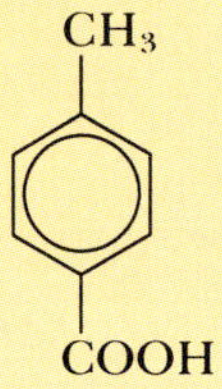

p-toluic acid

$$CH_3CHO \xrightarrow[Mn^{2+}]{O_2} CH_3COOH$$
acetic acid

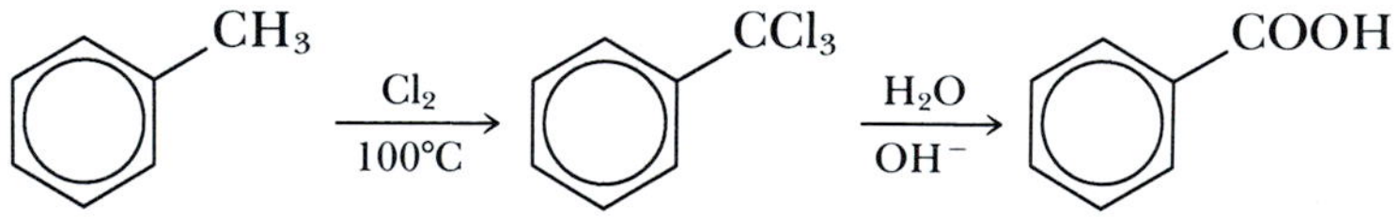

Formic acid is prepared by reaction of carbon monoxide with sodium hydroxide at elevated temperatures and pressures:

$$CO\ (g)\ +\ NaOH\ (aq) \longrightarrow HCOONa\ (aq) \xrightarrow{H^+} HCOOH\ (aq)$$

As is the case for both methyl alcohol and formaldehyde as the first members in a family, formic acid differs from all of the higher carboxylic acids. The most important reactions of carboxylic acids involve their ready conversion into one of a number of other functional group derivatives (Table 25–16).

One can generalize the structures in Table 25–16 in terms of the **acyl** group that is common to all:

$$R—C\!\!=\!\!O\ (acyl\ group)$$

The fourth bond to the carbonyl carbon is to an electronegative atom (rather than to C or H). The acid halides are used primarily as precursors to amides and esters. Acid chlorides are prepared by substituting Cl for the OH in the carboxyl group. $SOCl_2$ (thionyl chloride) and PCl_5 are the reagents commonly employed:

TABLE 25–16 Carboxylic Acid Derivatives

	General Formula	Specific Example	Parent Acid
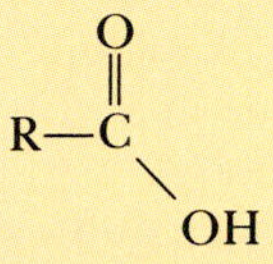	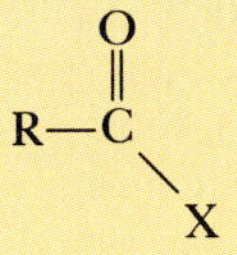 acid halide	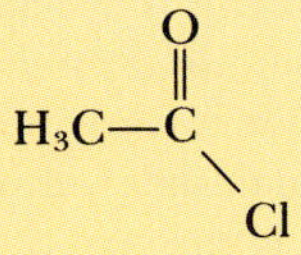 acetyl chloride	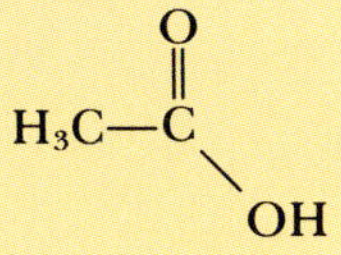 acetic acid
	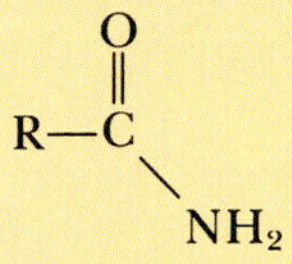 acid amide	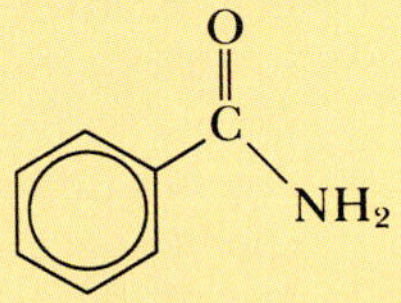 benzoic acid amide	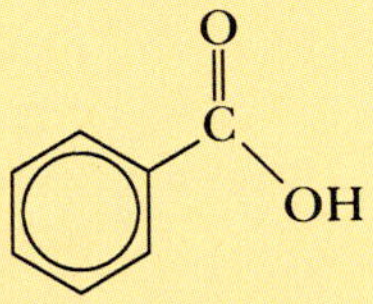 benzoic acid
	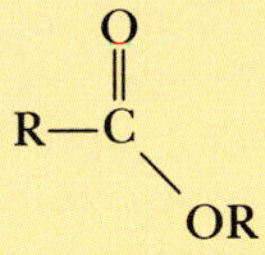 acid ester	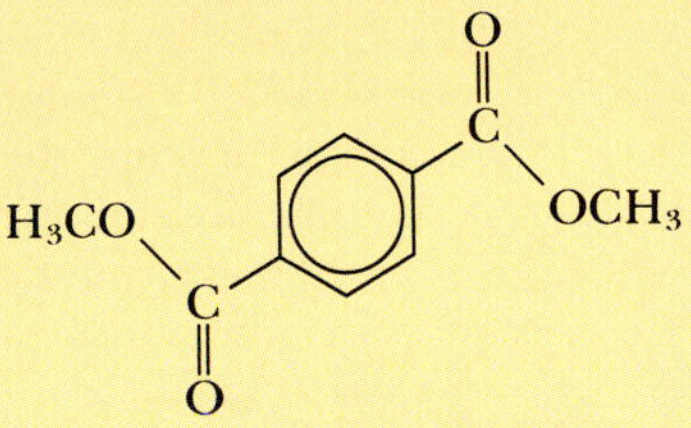 dimethyl terephthalate	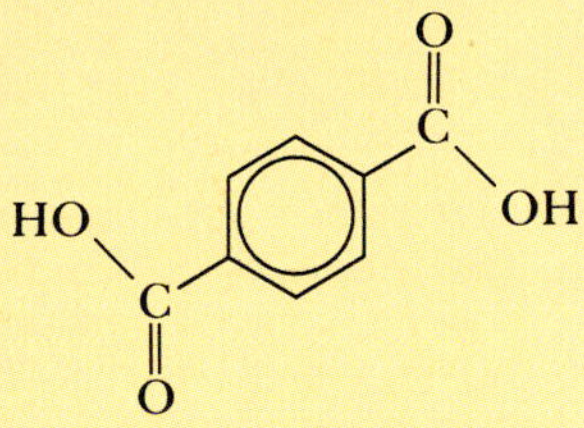 terephthalic acid

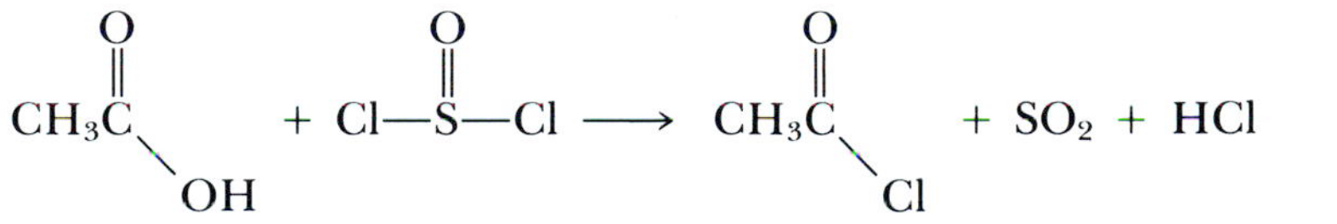

acetic acid acetyl chloride

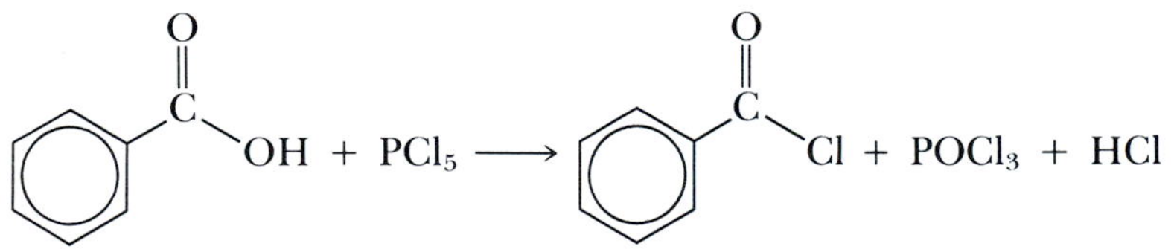

benzoic acid benzoyl chloride

Esters

Esters can be prepared by reaction of an acid chloride with an appropriate alcohol:

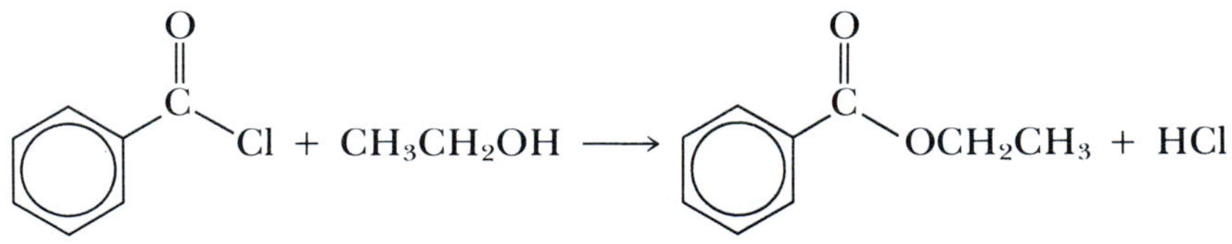

Alternatively, the ester can be prepared by direct reaction of alcohol and acid, but the equilibrium is not always as favorable for preparing esters as in the acid chloride reaction. The reaction is catalyzed by acids.

THE KRAFT PROCESS FOR PAPER AND THE
FIRST COMMERCIAL PLASTICS

Wood is a major material of construction and protection, and an important resource for the production of paper and paper products. The main components of wood are cellulose and lignins. There are two major cellulosic materials, an insoluble alpha-cellulose and an alkali-soluble hemicellulose. Lignins are the compounds that endow wood with its unique structural properties, but they must be separated from the cellulosic materials in manufacturing paper. Because lignins are highly complex and largely insoluble polymers, they must be chemically degraded to the point of solubility, but without degrading the cellulose fibers. That is usually done in a digester (after chipping logs into wood chips) with a caustic solution of chemical agents called "cooking liquor," which is mostly Na_2S, $NaHS$, and $NaOH$. The cooking converts the chips into a pulp and degrades the lignins, which are then separated. After washing and neutralizing—a critical step, since residual acid affects the lifetime of the paper product—the pulp is bleached in order to enhance its brightness. Before recycling the spent cooking liquors, they are neutralized, concentrated, and treated to remove the tall oils and rosin acids, which have commercial value.

Treatment of cellulose from wood pulp with concentrated solutions of nitric and sulfuric acids produces a variety of nitrated cellulose products, used as lacquers for coatings and finishes, for printing inks, in bookbinding applications, and as rocket propellants and explosives—just touch a flame to cellulose nitrate flash paper or flash powder and note the instantaneous "flash." In 1869, John Wesley Hyatt (American, 1837–1920) discovered that mixtures of cellulose nitrate and camphor could be molded and hardened. Under the trademark "Celluloid," this first commercial plastic material was fabricated into everything from billiard balls to shirt collars. However, its extreme flammability made it a hazardous material. For ex-

(a) A piece of cellulose nitrate "flash" paper is consumed almost instantly on ignition. (b) A line of cellulose nitrate "flash powder" is consumed in a flash.

(a)

(b)

$$CH_3\overset{\overset{O}{\|}}{C}\diagdown_{OH} + CH_3CH_3OH \xrightleftharpoons{H^+} CH_3\overset{\overset{O}{\|}}{C}\diagdown_{OCH_2CH_3} + H_2O$$

ethyl acetate

Many esters are pleasant-smelling substances used as perfumes and flavoring agents. Among the more important applications of compounds characterized by the presence of the ester group are industrial solvents and the universal analgesic known as aspirin, the sodium salt of acetylsalicylic acid (sodium acetylsalicylate).

ample, silk-like fabrics woven from these cellulosic fibers became known as "mother-in-law's silk." Early motion picture film used celluloid as a base material, and many cinema epics were totally destroyed by fire in studio warehouses.

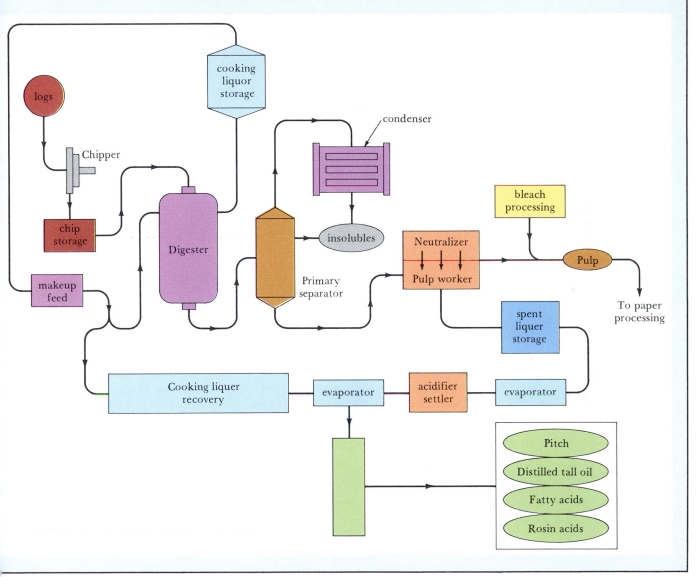

sodium acetylsalicylate
(aspirin)

Amides form when ammonia or an amine (see below) is caused to react with an acid chloride:

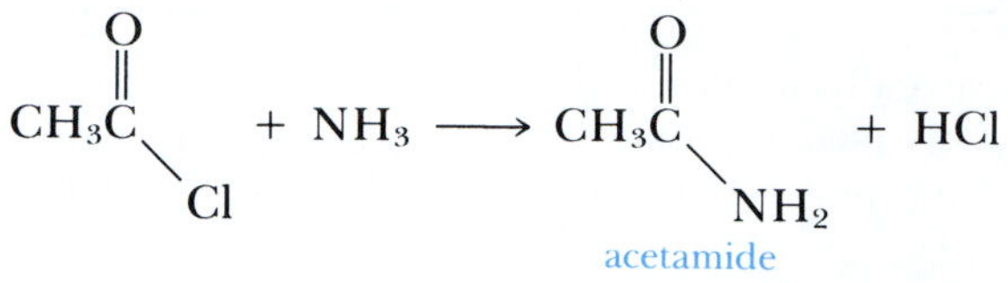

$$CH_3\overset{\overset{O}{\|}}{C}\diagdown Cl \;+\; NH_3 \longrightarrow CH_3\overset{\overset{O}{\|}}{C}\diagdown NH_2 \;+\; HCl$$

acetamide

They can also be prepared by making the ammonium salt of the carboxylic acid, followed by heating to drive off a molecule of water.

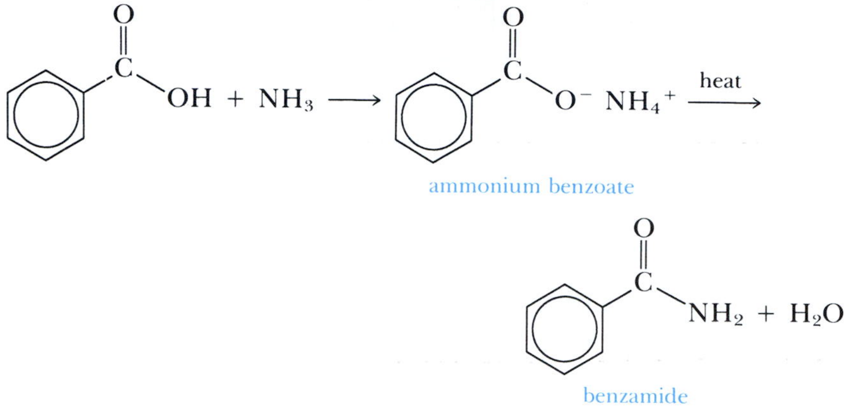

ammonium benzoate

benzamide

Amines

It is useful to think of amines as being derived from ammonia by replacing one of the amine N—H bonds by an N—C bond. As was true for alcohols, amines may be classified as primary (1°), secondary (2°), and tertiary (3°), according to the number of carbon atoms attached to nitrogen. The amines are weak bases with strengths generally on the order of that of NH_3, for which $K_b = 1.8 \times 10^{-5}$. They form amine and ammonium salts on addition of acids, including carboxylic acids.

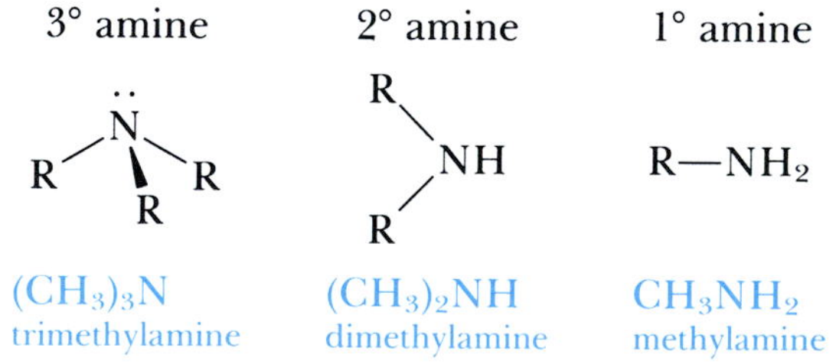

3° amine   2° amine   1° amine

$(CH_3)_3N$ $(CH_3)_2NH$ CH_3NH_2

trimethylamine   dimethylamine   methylamine

A general method of preparation for amines involves the action of an alkyl halide on ammonia or an amine:

$$NH_3 + CH_3Cl \longrightarrow CH_3NH_2 + HCl$$

1° amine

$$CH_3NH_2 + CH_3CH_2Cl \longrightarrow CH_3\overset{\overset{CH_2CH_3}{|}}{-}NH \;+\; HCl$$

2° amine

$$CH_3\overset{\overset{CH_2CH_3}{|}}{-}NH \;+\; \overset{\overset{CH_3CHCH_3}{|}}{Cl} \longrightarrow CH_3\overset{\overset{CH_2CH_3}{|}}{-}N-CH(CH_3)_2 \;+\; HCl$$

3° amine

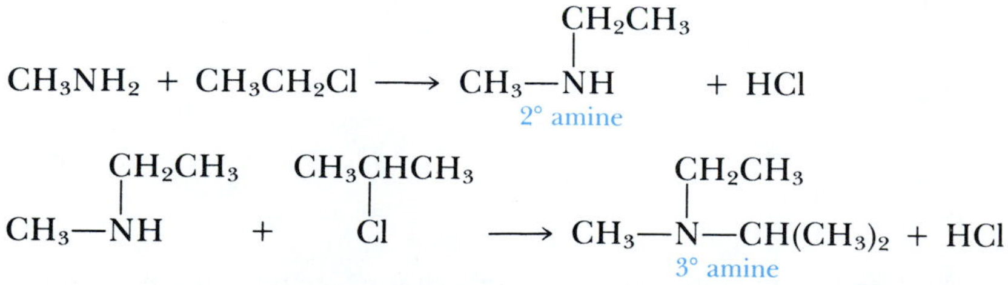

Lower amines are ammonia-like and water-soluble. They have disagreeable fishy odors. Putrescine is a four-carbon difunctional amine,

$H_2NCH_2CH_2CH_2CH_2NH_2$, and cadaverine has the same difunctional amine structure, but with one additional carbon atom in the chain; both are found in rotting flesh. The six-carbon diamine called hexamethylenediamine is used on a large scale for the synthesis of nylon.

EXAMPLE 25–7

Why do you think the direct reaction of ammonia or amines with alkyl halides might be of limited preparative value?

Solution   The reaction is not specific to a single reaction product because primary, secondary, and tertiary products can all form.   ■

EXERCISE 25–7

How would you go about minimizing the amounts of secondary and tertiary amine products? How would you produce largely the tertiary amine product?   ■

25.13   AMIDE AND ESTER CONDENSATION POLYMERS

If one of the hydrogen atoms attached to carbon (in the methyl group) of an acetic acid molecule is replaced by an amine functional group, the new compound is α-aminoacetic acid, a difunctional molecule better known as glycine:

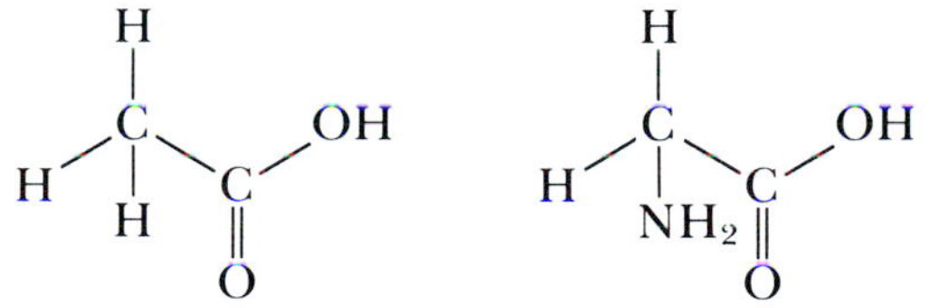

acetic acid   glycine (α-aminoacetic acid)

The designation α- refers to the location of substituents on the carbon atom adjacent to the carboxylic acid functional group. As we will see next chapter, α-amino acids are a very important class of organic compounds. They have the ability to polymerize by condensation reactions, linking consecutive units through **amide** (peptide) bonds (—NHCO—) into biologically important molecules called peptides (at the lower end of the molecular weight scale) and proteins (on the higher end):

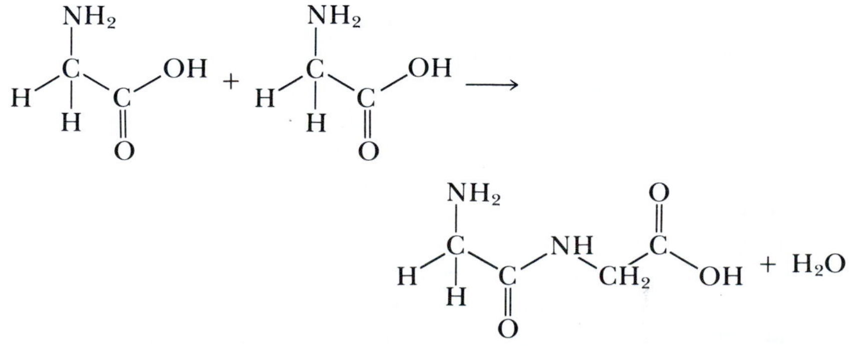

glycylglycine (dipeptide)

On reacting further by elimination of another water molecule between a third glycine monomer (or another α-amino acid) and the dimer from the first reaction, a trimer, then a tetramer, a pentamer, and eventually a

One way of distinguishing peptides and proteins—a polypeptide is an α-amino acid polymer; a polypeptide formed of many different amino acids is a protein.

(a)

(b)

FIGURE 25–13   (a) Blow-molded polyester bottles made from the injection-molded "preform" in the center (see Chapter 20); (b) polyester paint brush bristles with "exploded" monofilament tips.

polymeric peptide or protein will form. Hydrolysis of naturally occurring proteins leads to the formation of α-amino acids, about 20 of which turn out to be uniquely important in nature. Amino acids other than glycine have a second substituent (in addition to an amino group) attached to the α-carbon (see Chapter 26).

Note that in contrast to polyethylene and polymethylmethacrylate—which are addition polymers—proteins are **condensation polymers**. A molecule of water has been "condensed" or removed from between the functional (amino and carboxyl) groups. For obvious reasons, they are sometimes referred to as elimination polymers.

Dacron and Mylar

Dacron fibers and Mylar film are made of the same condensation polymer, poly(ethylene terephthalate). This polymer could in principle be produced directly by condensation of terephthalic acid and ethylene glycol. However, in practice, it is more conveniently prepared from dimethyl terephthalate and ethylene glycol by transesterification:

$$CH_3O-\overset{\overset{\displaystyle O}{\|}}{C}-\underset{\text{dimethyl terephthalate}}{\bigcirc}-\overset{\overset{\displaystyle O}{\|}}{C}-OCH_3 + n\ \underset{\text{ethylene glycol}}{HOCHCHOH} \longrightarrow$$

$$\left[\overset{\overset{\displaystyle O}{\|}}{C}-\bigcirc-\overset{\overset{\displaystyle O}{\|}}{C}-OCH_2CH_2O\right]_n + CH_3OH$$

poly(ethylene terephthalate)
Dacron or Mylar

These compounds are properly described as polyesters. Poly(ethylene terephthalate) and its near-twin, poly(butylene terephthalate), have become the materials of choice for the manufacture of plastic soda bottles (Figure 25–13).

Macromolecules and Chemical Bonds

In the 1920s, two chemists—one in Europe, the other in America—brought to light one of the most significant advances in our understanding of chemical bonding. In its own way, their work was as significant as Dalton's introduction of the atomic theory and Planck's quantum of action. Working in Zurich and Freiburg, Hermann Staudinger championed the idea that molecules composed of normal covalent bonds could indeed be grown to high molecular weights. At the time, it was thought that compounds such as peptides, the largest molecules known at the time, could not have molecular weights beyond a limit of about 5000.

Staudinger showed not only that was it possible to produce molecular weights well beyond that arbitrary limit, but that their physical properties were often a function of molecular weight. Furthermore, these polymeric or **macromolecular** structures could be fashioned into materials that in many cases had marvelous engineering properties, such that they could often be used to replace metal, wood, and glass, our traditional materials of construction and protection. The polymers of plastics were no longer to be understood as collections of smaller units (the aggregation theory). Rather, the regularly repeating monomer units were best understood as

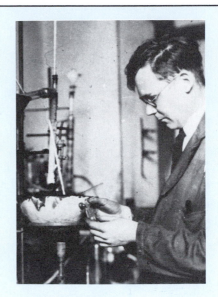

chemically bonded into chains by normal covalent bonds, the same kind that held hydrogen, nitrogen, oxygen, and carbon atoms together in simple molecules. These were molecules with molecular weights commonly in excess of 10,000, discrete molecules with chain lengths on the order of 10,000 Å rather than 10 Å. These were true macromolecules, or polymers. Staudinger literally forced the scientific community to change its view of what was meant by the word molecule.

Commercial synthesis of **nylon-6,6** (Figure 25–14) begins with elimination of water molecules between two synthetic monomer molecules, hexamethylenediamine and adipic acid. Both raw materials are obtained from cheap sources such as oat hulls, corncobs, and petroleum chemicals. Blending the difunctional amine with the difunctional acid produces first a nylon salt that, when heated to about 250°C, yields a condensed amide molecule containing all the original atoms of the two monomers, less one molecule of water. These condensed amide molecules themselves condense; about 45 to 50 units link head-to-tail, releasing water molecules at each condensation site and forming a polymer strand of about 10,000 molecular weight—a single, giant macromolecule.

FIGURE 25–14 Nylon polymer is removed as it is formed (a) at the interface between the two immiscible solutions of the monomers—hence the name "interfacial" polymerization; (b) by capping the jar and shaking it vigorously, crude nylon polymer is produced in bulk.

(a)   (b)

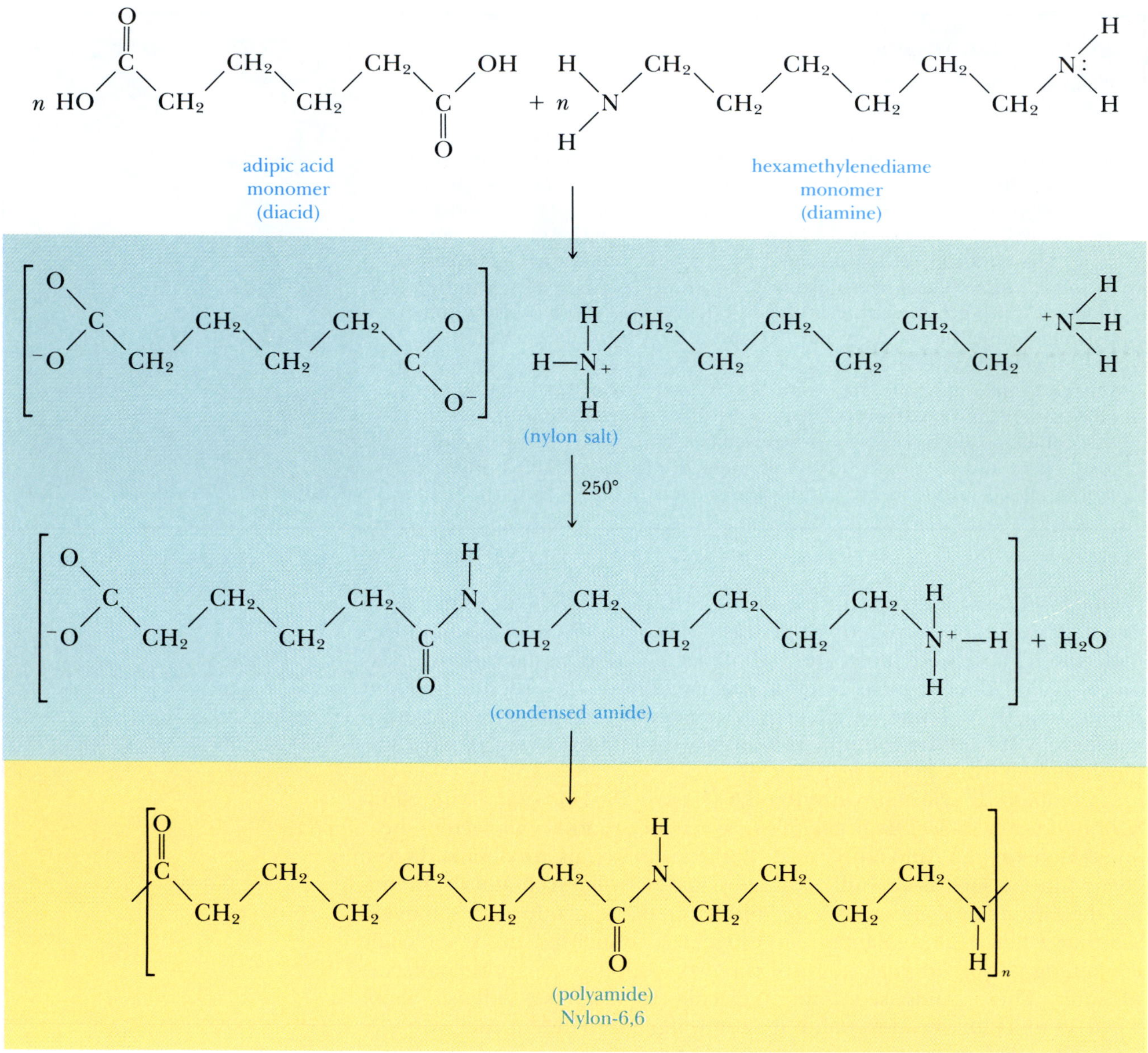

n HO — adipic acid monomer (diacid) $+ n$ hexamethylenediame monomer (diamine)

(nylon salt)

250°

(condensed amide) $+ H_2O$

(polyamide)
Nylon-6,6

One of the best-known polyamide fibers is Kevlar, the Du Pont trade-name for its aromatic polyamide. Because of its extremely high tensile strength, greater resistance (than steel) to elongation and high-energy absorption, it is an excellent material for belting radial tires (for which it was originally developed), in applications requiring high heat deflection, and in bulletproof vests (Figure 25–15).

SUMMARY

The chemistry of carbon is very different from that of its nearest neighbors in the periodic table, especially so because of its tendency to form stable covalent bonds with itself and with hydrogen, oxygen (and sulfur), nitrogen (and phosphorus), and the halogens. From a few natural sources such as petroleum and coal, and plant and animal matter, literally millions of well-

(a)

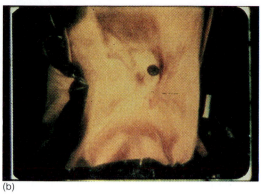

(b)

FIGURE 25–15 The welder (a) is wearing Kevlar gloves, which deflect the heat; (b) a woven vest of Kevlar fiber has deflected the bullet.

defined organic compounds have been isolated or synthesized, then characterized and identified, beginning with Wöhler's work on the conversion of ammonium cyanate into urea more than 150 years ago.

Aliphatic hydrocarbons can be saturated or unsaturated, straight-chained or cyclic. The straight-chain **alkanes**, or saturated hydrocarbons, having no multiple C-to-C bonds, are characteristically unreactive except to halogenation and air oxidation. They fall into a series with the general formula C_nH_{2n+2}, where n is the number of C atoms in the molecule. The hydrocarbon chains found in coal and petroleum range from C_1 to C_3 in the components of natural gases, C_5 to C_8 in the gasoline range, through the high teens in kerosenes and motor oils, and finally to lube oils and candle waxes in the 20s. Branching alters physical properties; thus, isobutane has a lower boiling point than n-butane. The IUPAC system of nomenclature is based on the number location of the alkyl branch along the main chain of the carbon skeleton; hence isobutane would be named 2-methylpropane. Synthetic procedures for preparing hydrocarbons are limited because of the inherent lack of reactivity of any necessary reagents and because hydrocarbons are readily available from natural sources.

Alkenes are unsaturated hydrocarbons. Sometimes referred to as olefins, their enhanced reactivity (over the alkanes) is due to the presence of the double bond. Bromine and hydrogen bromide typically add to the double bonds in ethylene and propylene. In the case of HBr and $CH_3CH=CH_2$, addition is according to **Markovnikov's rule**, leading to 2-bromopropane rather than 1-bromopropane.

An especially important reaction of the delocalized pi electrons in olefins is the self-addition or **polymerization** reaction, leading to high molecular weight, polymeric products such as polyethylene, polypropylene, and polymethylmethacrylate. Typically, **condensation polymers** form by elimination of small molecules such as water between monomer units. For example, peptide polymers such as proteins, nylons, and carbohydrate polymers are formed this way.

Alkynes contain triple bonds; the best-known example is acetylene, the first member of the family.

When **functional groups** are present in the molecule, replacing hydrogen atoms along the carbon chains in the backbone or in the branches, the number of possible isomeric forms increases dramatically, along with general reactivity. For example, there are two isomeric C_4H_9 hydrocarbons, but there are four isomeric C_4H_9OH alcohols. From alcohols, it is possible to synthesize alkenes or ethers (depending on the reaction conditions) by acid-catalyzed dehydration. Among the most widely encountered functional groups are the alcoholic and phenolic —OH, aldehydes (RCHO) and ketones (RCOR), acids (RCOOH) and esters (RCOOR'), ethers (ROR'), amines (RNH_2), and amides (RCONHR'). Industrial and biological applications of these compounds abound.

QUESTIONS AND PROBLEMS

QUESTIONS

1. What is your understanding of each of the following?
(a) C_nH_{2n+2}
(b) C_nH_{2n}
(c) coal gas and coal tar
(d) cracking and reforming
(e) alkanes, alkenes, and alkynes
(f) alicyclic and aromatic hydrocarbons
(g) isomeric pentanes
(h) *cis* and *trans* isomers
(i) octane rating
2. Distinguish between each of the following:
(a) resonance and conjugation
(b) an alkane and an alkyl group
(c) primary, secondary, and tertiary carbon atoms
(d) addition and substitution reactions
(e) addition and condensation reactions
(f) proteins and peptides
(g) starch and cellulose
3. Which of the following sets of structures represent identical compounds?

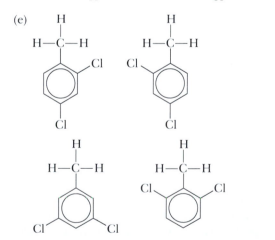

4. What simple chemical tests can be used to distinguish between *n*-hexane and 1-hexene?
5. How many different brominated benzenes are possible?
6. Briefly explain why the direct chlorination of ethane in sunlight may not be a useful way of preparing 1,1,1-trichloroethane, an important industrial solvent for paints and finishes.
7. What is Markovnikov addition, and why is it synthetically important?
8. How do aldehydes differ from ketones? How are amides different from amines?

9. Distinguish between each of the following functional groups:
(a) an alcoholic OH and a phenolic OH
(b) an alcoholic OH and an acidic OH
(c) an alcohol and an ether
(d) an acid and an ester
10. Write the formula for each of the following compounds:
(a) hydrazine
(b) isobutane
(c) acetylene
(d) isopropyl alcohol
(e) acetic acid
(f) formaldehyde
(g) acetone
(h) phenol
(i) chloroform
(j) glycine
11. What is the molecular formula for each of the following?
(a) methane
(b) ethylene
(c) trichloromethane
(d) ethyl alcohol
(e) urea and ammonium cyanate
(f) toluene
(g) glycerin
(h) the Plexiglas monomer
(i) the natural rubber monomer
(j) sodium stearate
12. An organic compound reacts readily with bromine to form the compound $CH_3CHBrCH_2Br$.
(a) What was the original compound?
(b) How would the original compound react with an alkaline permanganate solution?
13. An organic compound reacts readily with hydrogen bromide to form the compound $CH_3CHBrCH_3$.
(a) What was the original compound?
(b) What product would be formed when the original compound reacted with water in the presence of acid?
14. By what general process—oxidation, reduction, cracking, dehydration, halogenation, etc.—would you accomplish each of the following?
(a) isopropyl alcohol to acetone
(b) 2-butene to 2-bromobutane
(c) ethyl alcohol to ethylene
(d) kerosenes to gasolines
(e) coke to carbon monoxide
(f) coal to coke
(g) methane to chloroform
(h) glycine to glycylglycine
(i) cumene to cumene hydroperoxide
(j) starch to glucose

PROBLEMS

Structural Formulas   [1–4]

1. Draw structural formulas for the isomeric hexanes

(C_6H_{14}), and when you are finished, name the compounds by the IUPAC system.
2. Draw structural formulas for the isomeric pentenes (C_5H_{10}), and when you are finished, name the compounds by the IUPAC system.
3. There are nine possible isomeric heptanes.
(a) Write the structural formula for each.
(b) Name each by its systematic name.
(c) Which might be commonly called isoheptane?
(d) Which might be commonly called neoheptane?
(e) Are names like those used in (c) and (d) unambiguous?
4. There are 17 possible isomeric hexenes.
(a) Write the structural formula for each.
(b) Which is 3-hexene?
(c) Explain why there are two isomeric 3-hexenes.

Reactions and Preparation of Alkanes, Alkenes, and Alkynes   [5–8]

5. Using whatever inorganic reagents you might need, write a sequence of chemical reactions that would produce at least a modest overall yield of:
(a) isopropyl alcohol from propane
(b) propane from isopropyl alcohol
(c) n-butane from ethane
(d) propane from methane
6. Write chemical reactions, using necessary inorganic reagents, to produce the following transformations:
(a) acetylene from ethylene
(b) poly(vinyl chloride) from acetylene
(c) phenol from cumene
7. Write Lewis structures for acetylene, the acetylide anion, and the acetylene dianion. Write chemical reactions for the synthesis of the ions. Starting with methane, write chemical reactions that would produce a species that is isoelectronic with the dianion.
8. Explain why the carbon-to-carbon bond length changes from 1.54 Å in ethane to 1.33 Å in ethylene and 1.20 Å in acetylene.

Aromatic Derivatives   [9–10]

9. Write structures for all the possible bromobenzenes.
10. For the various substituted chlorobenzenes, fill in the chart.

Substitution	Number of Possible Isomers	Examples
mono-		
di-		
tri-		
tetra-		
penta-		
hexa-		

Alcohols and Ethers   [11–12]

11. Place the following compounds in order of their increasing solubility in water:

CH$_2$CH$_2$OH
CH$_3$(CH$_2$)$_5$OH
HO(CH$_2$)$_6$OH

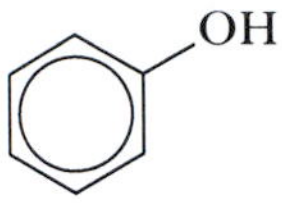

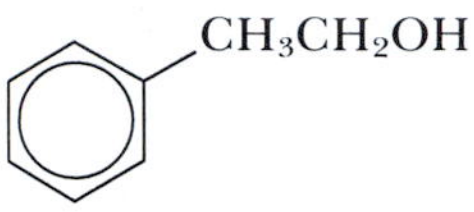

[benzene ring with OH]

[benzene ring with CH$_3$CH$_2$OH]

12. Consider the following data:

	Molecular Weight	b.p. (°C)
CH$_3$CH$_2$OH	46	78.5
CH$_3$CH$_2$SH	62	37.0

Explain why ethanethiol (CH$_3$CH$_2$SH) boils at an appreciably lower temperature than ethanol (CH$_3$CH$_2$OH) although its molecular weight is nearly 50% greater.

Aldehydes, Ketones, Carboxyl Acids, Esters, and Amines   [13–20]

13. There are a number of isomeric monobromobutanes.
(a) Draw structural formulas for all that are possible.
(b) Indicate by equations which one(s) will form primary alcohols on hydrolysis with aqueous alkali. Which one(s) will form secondary alcohols? Which one(s) will form tertiary alcohols?
(c) Indicate which of these alcohols:
  (i) can be oxidized to carboxylic acids
  (ii) can be oxidized to an aldehyde
  (iii) can be oxidized to a ketone
  (iv) cannot be further oxidized without disrupting the carbon skeleton
(d) Draw structural formulas for the alkenes that might form on dehydrohalogenation of the bromobutanes with alcoholic potassium hydroxide.
14. What are the basic structural units of aldehydic and ketonic molecules? How are they both structurally related to carboxylic acids? Explain why aldehydes are susceptible to oxidation but ketones are not.
15. The pK_a values for acetic acid and α-bromoacetic acid are, respectively, 4.7 and 2.5.
(a) Write out the structural formulas for both acids.
(b) What is the pH of a 0.10 M solution of each?
(c) How might you account for the considerable acidity of brominated acetic acid compared to acetic acid itself?
16. Consider a solution that is 0.10 M in acetic acid and 0.10 M in sodium acetate.
(a) Show by equations the effect of adding acid and base to the solutions.
(b) In what pH range might a solution of α-bromoacetic acid and its sodium salt be an effective buffer?
17. Complete the following:

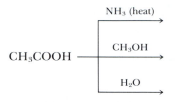

18. Write the products of the following reactions.

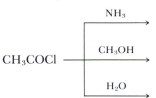

19. Write the structural formulas for all the possible isomeric butylamines.
20. Write the formulas of all the amines having the formula C$_3$H$_9$N. Label each amine as primary, secondary, or tertiary.

Organic Reactions   [21–24]

21. Using the organic material indicated and whatever inorganic reagents you require, show how the following conversions can be accomplished:
(a) CH$_3$CH$_2$CH$_2$OH $\longrightarrow$ CH$_3$CHCH$_3$
                                                              |
                                                             OH
(b) CH$_3$CHCH$_3$ $\longrightarrow$ CH$_3$CH$_2$CH$_2$Br
         |
         Cl
(c) CH$_3$CHCH$_3$ $\longrightarrow$ CH$_3$CHCH$_3$
         | |
        OH   NH$_2$
22. Perform the following transformations, using whatever inorganic substances are necessary, and write all reactions.
(a) CH$_3$CH$_2$OH $\longrightarrow$ CH$_3$COCH$_2$CH$_3$ (with O above the C)
(b) CH$_3$CH$_2$OH $\longrightarrow$ CH$_3$COOH
23. Consider each of the following and determine the lettered compounds in question. Write the equations taking place.
(a) A contains only C, H, and O, has a molecular weight of 30, and reduces Tollens's reagent.
(b) B has the molecular formula C$_3$H$_8$O. An alkaline permanganate solution oxidizes it to a new compound C, whose molecular formula is C$_3$H$_6$O. C is not further oxidized readily.
24. Determine the formulas of the compounds identified by letters. Write the reactions that take place.
(a) D is an aromatic compound that can be chlorinated at elevated temperatures, yielding E in a process in which a mole of HCl is produced for every mole of chlorine absorbed. D has a molecular formula of C$_7$H$_8$, and E is C$_7$H$_5$Cl$_3$.
(b) F, whose molecular formula is C$_2$H$_6$O, is treated with aqueous sulfuric acid to yield a new compound G, whose formula is C$_4$H$_{10}$O. If the reaction were carried out under somewhat more stringent conditions, H with the formula C$_2$H$_4$ is formed instead of G.

Additional Problems [25–30]

25. Draw structural formulas for each of the following:
(a) the nitrogen analogs of ethane and ethylene
(b) the silicon analog of methane
(c) the boron analogs of methane and ethane
26. Consider the following thermochemical data:

$$cis\text{-2-butene} \xrightarrow{H_2,\ catalyst} n\text{-butane}$$
$$\Delta H = -120 \text{ kJ/mol}$$

$$trans\text{-2-butene} \xrightarrow{H_2,\ catalyst} n\text{-butane}$$
$$\Delta H = -115 \text{ kJ/mol}$$

(a) Write the structural formula for both the *cis*- and *trans*-2-butenes.
(b) By examination of the structural models, which would you guess to be more stable? Why?
(c) Is your answer to (b) borne out by the thermochemical data above? Explain.
(d) When 1-butene is reduced to *n*-butane, ΔH is -127 kJ/mol. Compare the relative stabilities of the three isomeric butanes.
27. Inasmuch as both propylene and toluene can be characterized as having a methyl group adjacent to a C-to-C double bond, how do you account for the obvious differences in reactivity with respect to bromination and oxidation? Where reactions occur, write the equations.
28. Explain why the hydrolysis of an ester to yield alcohol and acid is best carried out in basic rather than acidic solution. Use equations to illustrate your answer.
29. List boiling point data (in °C) for the following saturated hydrocarbons from the appropriate table in the chapter: ethane, propane, hexane, octane, and nonane. The cracking of a saturated hydrocarbon that has a boiling point of $+174$°C gives, among the products, compounds with boiling points of -250°C, -162°C, and $+98$°C. None of these compounds is capable of being polymerized, and they have no effect on bromine water.
(a) Suggest what the original saturated hydrocarbon might have been.
(b) Suggest what the three products might have been,
being sure to show how you arrived at your conclusions.
(c) Some saturated hydrocarbons are gases, some are liquids, and some are solids at 25°C. Which of the liquids has the lowest mass for one mole of molecules?
30. By proper use of boiling point elevation data, the molecular weight of benzoic acid can be determined. When measurements are made in acetone solution, the molecular weight is close to 122; in carbon tetrachloride, it is nearly 244. How do you rationalize the data?

MULTIPLE PRINCIPLES [31–33]

31. A sample contains a mixture of propane and propylene (propene). A 10.0-g sample of this mixture is found to react exactly with 24.7 g of bromine. What is the percentage of propane in the mixture?
32. You are running a reaction of ammonia with methyl chloride, CH_3Cl, to produce a mixture of methyl amine, dimethyl amine and trimethyl amine. Suppose you know the mass of the ammonia and the mass of the methyl chloride that react and the total mass of the mixture of amines produced. Write three equations in three unknowns that could be used to determine the amount of each of the three amines in the mixture. It is not necessary to solve the equations.
33. 0.100 L of carbon monoxide at 4.50 atm and 25°C is reacted with excess sodium hydroxide to produce sodium formate, which is acidified to produce formic acid. The formic acid produced is titrated with 0.100 M HCl and it is found that the inflection point occurs at 123 mL. What was the percent yield of the formic acid synthesis?

APPLIED PRINCIPLES [34–35]

34. What volume of water vapor is produced for every kg of nylon-6,6 at the reaction temperature of 250°C?
35. How many liters of methanol, or CH_3OH, are produced as a biproduct for every 100. kg of poly(ethylene terephthalate), also known as Dacron or Mylar? The density of methanol at 25°C is 0.79 g/cm³.

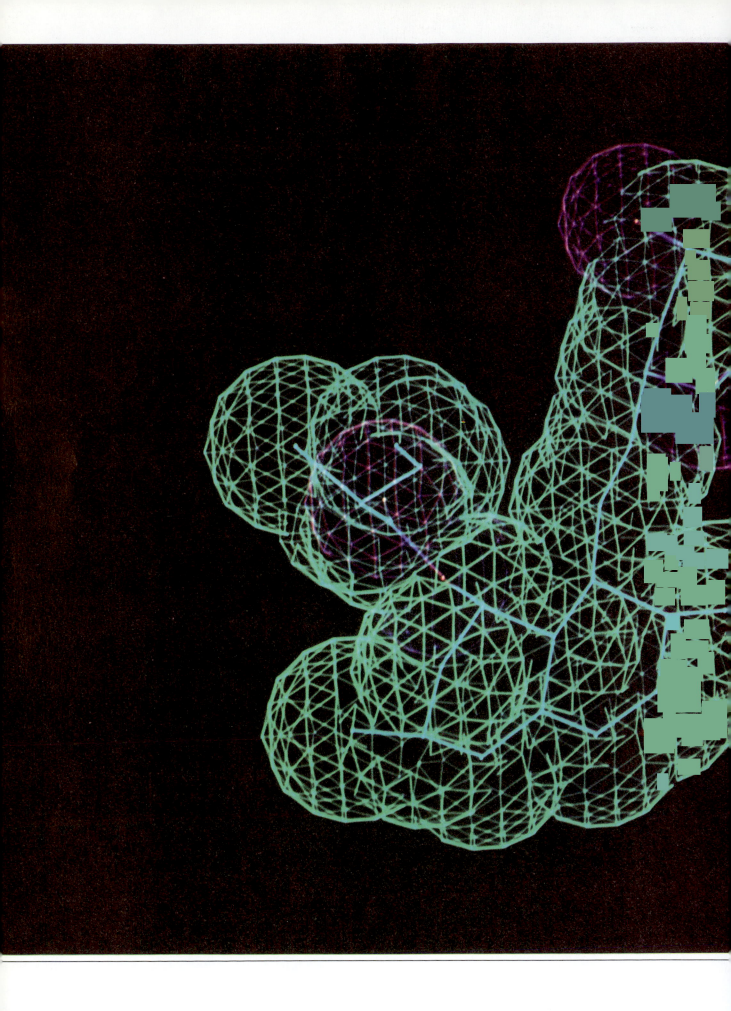

Biochemistry

The elevation of biochemistry to the level of public awareness is nowhere better demonstrated than in the products of pharmaceutical research and biomedical engineering—the drug industry. Shown here is a computer-generated model of the molecule that is the biologically active agent in Mevocor, a drug used to lower cholesterol levels. Nobel laureate Charles Pedersen's successes, over a long career, were not the result of a group of scientists working under a director. They were his alone. And those researches have had a major impact on organic and biochemistry. Shown here is the diploma recognizing his Nobel Prize-winning work on cage-like structures called "crown ethers."

Biochemistry has taken on a life of its own and is often found in academic departments as a separate discipline beyond biology and chemistry, with its own special techniques and methods.

Until about 20 years ago, the chemistry of the very complex biochemicals such as proteins and nucleic acids was thought to be impossible to duplicate in the laboratory. Now, however, such compounds have been prepared in the laboratory and there is reason to believe that it will be possible to synthesize living organisms from nonliving materials.

26.1 INTRODUCTION: UNIQUE ASPECTS OF BIOCHEMISTRY AND BIOCHEMICALS

Biochemistry is the study of the molecular basis of life, and it requires an understanding of both biology and chemistry. It is difficult to point to a specific event that signals the beginning of modern biochemistry. Many of the threads in the web were evident as early as 1835, when Berzelius included biological reactions in his list of catalyzed chemical changes. The biochemical process that attracted the most interest at that time was the fermentation that produced alcohols. Theodor Schwann recognized that alcoholic fermentation was a biological process when he described yeast as "a plant capable of converting sugars and starches into ethyl alcohol and carbon dioxide."

Louis Pasteur carried out important studies on fermentation and the role of aerobic and anaerobic microorganisms. Finally in 1897, two years after Pasteur's death, Edouard Büchner was able to show that the fermenting agents were indeed contained within living yeast cells, but that they could be extracted into a functional form, capable of fermenting sugar to alcohol in the absence of intact yeast cells. Büchner called these agents "ferments." Today we call them enzymes. Perhaps it is this work on enzyme-catalyzed reactions that opened the field of biochemistry.

How does the chemistry that takes place in living organisms differ from the chemistry that takes place in the organic chemistry laboratory? First, the conditions of biochemical reactions are extremely restrictive and not at all what an organic chemist would choose to stimulate reactions to a reasonable rate. Biochemistry commonly takes place in dilute aqueous solutions, at pH near 7, at temperatures very close to 37°C, and at a pressure of about 1 atm. All the usual tools and techniques for speeding the conversion of reactants to desired products—exotic solvents, concentrated solutions, and extremes of pH, temperature, and pressure—are gone. Yet biochemical reactions proceed with striking selectivity and at incredible speeds. For example, the reactions involved in transmitting a nerve impulse have been estimated to take place in a few microseconds. That is much faster than the typical reactions we are accustomed to running in the laboratory. Second, biochemicals are often very large and extremely complex molecules, and only recently have we been able to determine their structures.

There are about a million different kinds of living organisms, and the diversity among them is huge. As you might expect, the differences between organisms are based on the differences in their chemistry. However, because many of the same biochemical molecules are found in many very different organisms, surprising simplifications in our understanding are possible. For example, the same 20 amino acids are observed in all plant and animal proteins. The genetic code in DNA is universal in living things, and many biochemical reactions are seen repeatedly.

26.2 STEREOCHEMISTRY

Optical Isomerism and Chiral Molecules

Take two pairs of "Polaroid" sunglasses and set them together, one in front of the other. As you observe a light source through one lens of each pair, the maximum intensity occurs when they are in parallel. But if you rotate one lens so that it is at a right angle (perpendicular) to the other, the transmitted light diminishes to zero. You can use this simple **polarimeter**—

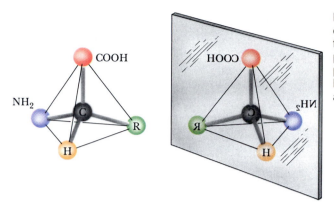

FIGURE 26–1 Nonsuperimposable mirror image isomers, created when four different atoms or groups of atoms are attached to the central *asymmetric* carbon center, are nonsuperimposable in the same sense as the left and right of a pair of gloves or shoes. If the four species about carbon are H, NH_2, CH_3, and COOH, the molecule is either an essential amino acid, or its mirror image isomer.

the fixed lens is called the *polarizer,* the rotating lens, the *analyzer*—to observe an interesting and unique phenomenon with far-reaching consequences for biochemistry. Put two heaping tablespoons of sucrose (common table sugar) into a small clear glass of water and place the glass between the two parallel lenses (polaroids)—between the polarizer and the analyzer—and rotate the one lens. Note that the orientation of the lenses for maximum and minimum transmission of light for the sucrose solution is different than when no sucrose solution was used. That is because sucrose is an **optically active** compound. It affects the rotation of polarized light.

To explain this observation, reconsider the tetrahedral geometry that is characteristic of sp^3 carbon centers in organic molecules. The bonds are directed along the axes of a regular tetrahedron in three-dimensional space. **Optical isomers** and optical activity typically arise when four different atoms or groups of atoms are attached to the central *asymmetric carbon* (Figure 26–1). The essential physical property that distinguishes optical isomers is the direction in which each rotates a special kind of light, **plane polarized light**—light whose radiations vibrate only in certain planes. If one isomer turns the plane of polarization in a clockwise (+) manner in the face of the beam, it is described as **dextrorotatory**; then the other should alter the passage of light through a solution of the sample in the polarimeter in a counterclockwise (−) manner, and that is described as **levorotatory**. In a casual way, we describe such molecules as "left-handed" (levorotatory) or "right-handed" (dextrorotatory).

This property of left- or right-handedness is called **chirality**. A molecule is said to be chiral (or to have a chiral center) if it cannot be superimposed upon its *mirror image* isomer. The pair—the isomer and its nonsuperimposable mirror image—are called **enantiomers**. For example, if the four different groups about the asymmetric carbon center in Figure 26–1 are —H, —NH_2, —CH_3 and —COOH, the molecule is an α-amino acid or its mirror image isomer. The enantiomers—the two nonsuperimposable mirror image amino acid isomers—carry distinguishing name tags, the origins of which we shall explain a page further on: one is the L-amino acid and the other, the D-amino acid. The far-reaching consequences we spoke of at the outset of this section are immediately evident when we examine nature's proteins because only L-amino acids are found to be biologically important.

EXAMPLE 26–1

Which of the following objects are chiral?

(a) a baseball bat

(b) a tennis racket

(c) a golf club

(d) a three-legged stool

(e) a coiled spring

(f) either die in a pair of dice

Solution   Each of the bat, the tennis racket, and the stool has symmetry about an axis through its center, so they are all superimposable on their mirror images and are not chiral. The golf club, the coiled spring, and the die are asymmetric and are not superimposable on their mirror images, so they are chiral.   ■

EXERCISE 26–1

Which of the following exist as enantiomers? For those that do, identify the chiral center.

(a) $CH_3\underset{\underset{\displaystyle OH}{|}}{C}HCOOH$

(b) $CH_3\underset{\underset{\displaystyle Cl}{|}}{C}HCH_2CH_3$

(c) $CH_2\diagdown \diagup CH_2 \! \rangle CHCOOH$

(d) $CH_3\underset{\underset{\displaystyle Cl}{|}}{\overset{\overset{\displaystyle CH_3}{|}}{C}}{-}COOH$

(e) $CH_3CH_2\underset{\underset{\displaystyle CH_3}{|}}{C}HCH_2CH_2CH_3$

(f) $CHBrClF$

Answer: (a), (b), (e), and (f) are chiral. The chiral center is the carbon atom bonded to four different groups.   ■

Tartaric acid ($C_4H_6O_6$) is probably the most famous organic compound for which stereochemical properties have been studied. It is present in grape juice as the potassium hydrogen salt. Pasteur originally discovered that there are two different tartaric hydrogen salts, one optically active (from wine) and the other optically inactive (by synthetic methods). As the sugar in the juice of the grape fermented, the increasing concentration of alcohol decreased the solubility of the tartrate salt, causing it to crystallize from solution. Acidification of the salt produced the free acid, which turned out to be dextrorotatory (or right-rotating). Using a simple hand-held magnifying glass, Pasteur noted that the tartrate crystals were all geometrically the same, suggesting a relationship between crystal structure and the rotation of plane polarized light by molecules in solution.

When Pasteur examined an optically inactive form of tartaric acid, there was clear evidence of a different geometry. As he picked through the pile of inactive crystals, he found not only the form he had noticed in the optically active tartrate, but a second kind. Pasteur painstakingly collected two piles of crystals, a left-handed pile and a right-handed pile (Figure 26–2). When he examined their respective solutions, he discovered that one rotated plane polarized light to the right (dextrorotatory) just as the optically active form isolated from wine sludge, while the other rotated light to the left (levorotatory). When equal masses of the two were combined, the resulting solution was optically inactive—the right-rotating compound cancelled the effect of the left-rotating one. Optically inactive tartaric acid had been **resolved**, by separating the crystals, into two optically active components: the dextro (D-tartaric acid) and the levo (L-tartaric acid). Pasteur had isolated two optical isomers; both were $C_4H_6O_6$, and they differed in just one property, the way in which they rotated a plane of polarized light (Figure 26–3).

Tartaric acid isolated from the lees of wine—the sludge remaining at the bottom of the barrel—is always right-handed, but the synthetic salt that Pasteur examined was a half-and-half mixture of left-handed and right-

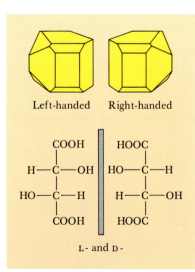

FIGURE 26–2   Tartaric acid enantiomorphs: (a) crystals; (b) structures, illustrating the L- and D-forms.

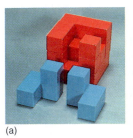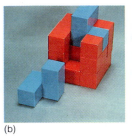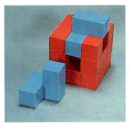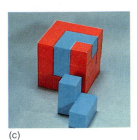

(a)   (b)   (c)

FIGURE 26–3 This puzzle is an illustration of what we mean by "handedness." (a) A piece and its mirror image are placed before the block. (b) A few moves will quickly convince you that (c) only one piece fits. So it is with chiral molecules. As Pasteur observed, having an inherent asymmetry endows such molecules with uniquely important spatial properties.

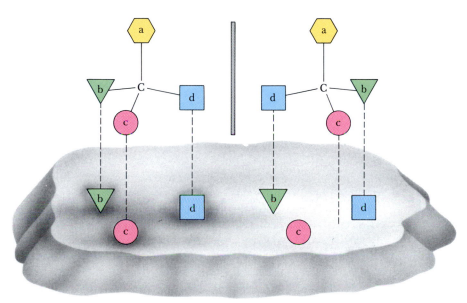

FIGURE 26–4 Protein specificity, depicted as a template. The protein substrate is keyed to a single reactant molecule having a particular stereochemical arrangement. Only one isomeric form (enantiomorph) is capable of reaction, because only it is capable of properly "landing" on the reaction sites.

handed. In looking for a nonmechanical way of separating the two, Pasteur discovered that a penicillin mold selectively destroyed the right-handed tartaric acid (D-tartaric acid) but would not touch the left-handed tartaric acid. Pasteur had discovered a sublime subtlety in nature: the properties of biomolecules are "keyed" to their chirality (Figure 26–4).

26.3 CLASSES OF BIOMOLECULES

Biomolecules can be divided into several broad classes according to their structures, functions, and properties. These classes include proteins, nucleic acids, carbohydrates, lipids, coenzymes, and nucleoside triphosphates.

Amino Acids and Proteins

Proteins make up the major structural components of animals: cartilage, skin, hair, feathers, horns, hoofs, muscles, and tendons. In addition, enzymes, the catalysts of bioreactions, are all proteins. Most hormones, antibodies for combating infectious agents, and major blood constituents are also proteins.

The word protein was first used by Berzelius; it comes from the Greek meaning "of the first order," suggesting the level of importance he perceived.

Proteins are macromolecules composed principally of long chains of amino acids connected together. An **amino acid** is any organic compound containing the amine (NH_2) and carboxylic acid (COOH) functional groups. However, of the limitless number of such compounds that can be formed, only twenty show up in all of the proteins in the plant and animal kingdoms. The names, abbreviations, and structures of these twenty amino acids are shown in Table 26–1.

The location of the amino group along the carbon chain in amino acids is specified by the Greek letters α, β, and so forth. Amino acids bearing the —NH_2 group on the carbon atom directly adjacent to the carboxyl group are α-amino acids, and all of the amino acids in Table 26–1 are in this category. Furthermore, in all of these amino acids the α-amino groups are primary (—NH_2) with the exception of proline, which is known as the α-imino acid. The general formulation for the amino acids in Table 26–1 is:

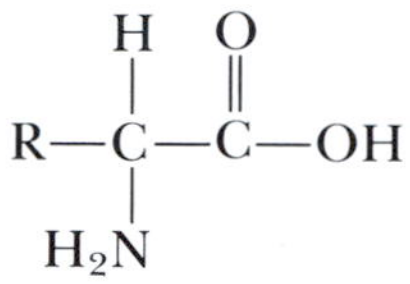

$$R-\underset{\underset{\displaystyle H_2N}{|}}{\overset{\overset{\displaystyle H}{|}}{C}}-\overset{\overset{\displaystyle O}{\|}}{C}-OH$$

where R can be hydrogen or a larger organic group. The R groups in these amino acids fall into several categories. In the first seven in the list, R is a nonpolar organic group (either aliphatic or aromatic). The next six have R groups that are polar, containing one of the electronegative elements S, O, or N. Asparagine and glutamine have the amide functionality ($CONH_2$) in their R groups. They are capable of forming hydrogen bonds with each other or with the polar R groups. Aspartic acid and glutamic acid have the carboxylic acid (—COOH) in their R groups and are capable of forming anions. The last three amino acids on the list have R groups containing basic nitrogen atoms that can be protonated to form cations. Finally, all of the amino acids in Table 26–1 (except glycine) have an optically active α-carbon and are therefore capable of exhibiting optical activity. Almost all known amino acids in plants and animals are L-amino acids. That is, they have a conformation analogous to L-glyceraldehyde:

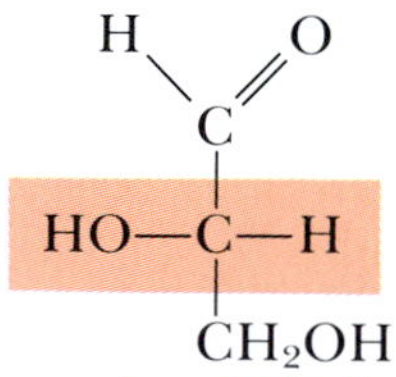

The left-handed form of the monosodium salt of glutamic acid (sold commercially as Accent) is a flavoring stimulant. The taste buds on the human tongue are insensitive to the right-handed isomer.

An amino acid, phenylalanine, is harmless in its left-handed form but produces symptoms of insanity in the right-handed configuration.

L-glyceraldehyde 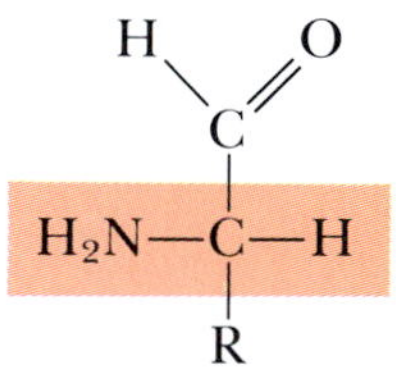   L-amino acid 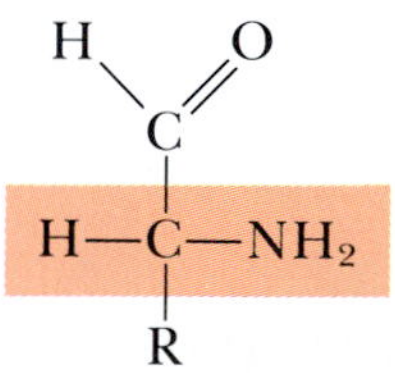   D-amino acid

Note that the letters D and L refer only to the absolute configuration of the molecule. Some L-amino acids rotate the plane of polarized light to the left (levorotatory) and some rotate it to the right (dextrorotatory).

The linkage between the amino acids in a protein chain involves the elimination of one molecule of water to form an amide:

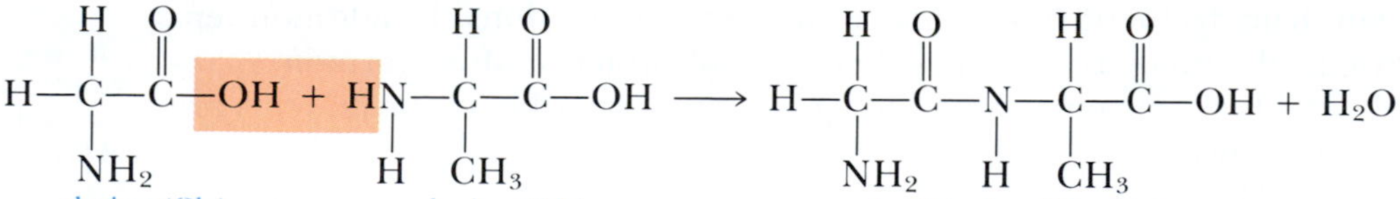

glycine (Gly)   alanine (Ala)   glycylalanine (Gly-Ala)

TABLE 26–1 Amino Acids: Name, Abbreviation, and Structure

Glycine (Gly)

$$\underset{\underset{NH_2}{|}}{H-\overset{\overset{H}{|}}{C}-}\overset{\overset{O}{\|}}{C}-OH$$

Alanine (Ala)

$$H_3C-\overset{\overset{H}{|}}{\underset{\underset{NH_2}{|}}{C}}-\overset{\overset{O}{\|}}{C}-OH$$

Valine (Val)

$$H_3C-CH-\overset{\overset{H}{|}}{\underset{\underset{NH_2}{}}{C}}-\overset{\overset{O}{\|}}{C}-OH$$
(CH₃ on the CH)

Leucine (Leu)

$$CH_3CH-CH_2-\overset{\overset{H}{|}}{\underset{\underset{NH_2}{}}{C}}-\overset{\overset{O}{\|}}{C}-OH$$
(CH₃ on the first CH)

Isoleucine (Ileu)

$$CH_3CH_2-CH-\overset{\overset{H}{|}}{\underset{\underset{NH_2}{}}{C}}-\overset{\overset{O}{\|}}{C}-OH$$
(CH₂ on the CH)

Phenylalanine (Phe)

$$C_6H_5-CH_2-\overset{\overset{H}{|}}{\underset{\underset{NH_2}{|}}{C}}-\overset{\overset{O}{\|}}{C}-OH$$

Proline (Pro)

Ring: H_2C, $N-H$, $CH-C(=O)-OH$, H_2C-CH_2

Serine (Ser)

$$HO-CH_2-\overset{\overset{H}{|}}{\underset{\underset{NH_2}{|}}{C}}-\overset{\overset{O}{\|}}{C}-OH$$

Threonine (Thr)

$$H_3C-\overset{\overset{HO}{|}}{\underset{\underset{H}{|}}{C}}-\overset{\overset{H}{|}}{\underset{\underset{NH_2}{|}}{C}}-\overset{\overset{O}{\|}}{C}-OH$$

Cysteine (Cys)

$$HS-CH_2-\overset{\overset{H}{|}}{\underset{\underset{NH_2}{|}}{C}}-\overset{\overset{O}{\|}}{C}-OH$$

Methionine (Met)

$$H_3C-S-CH_2-CH_2-\overset{\overset{H}{|}}{\underset{\underset{NH_2}{|}}{C}}-\overset{\overset{O}{\|}}{C}-OH$$

Tryptophan (Trp)

Indole ring (CH, HC, C, HC, C, CH, CH, NH) $-CH_2-\overset{\overset{H}{|}}{\underset{\underset{NH_2}{}}{C}}-\overset{\overset{O}{\|}}{C}-OH$

Tyrosine (Tyr)

$$HOC_6H_4-CH_2-\overset{\overset{H}{|}}{\underset{\underset{NH_2}{|}}{C}}-\overset{\overset{O}{\|}}{C}-OH$$

Asparagine (Asn)

$$H_2NCCH_2\overset{\overset{H}{|}}{\underset{\underset{NH_2}{|}}{C}}CO_2H$$
(O double bond on first C)

Glutamine (Gln)

$$H_2NCCH_2CH_2\overset{\overset{H}{|}}{\underset{\underset{NH_2}{|}}{C}}CO_2H$$
(O double bond on first C)

Aspartic acid (Asp)

$$HO-\overset{\overset{O}{\|}}{C}-CH_2-\overset{\overset{H}{|}}{\underset{\underset{NH_2}{|}}{C}}-\overset{\overset{O}{\|}}{C}-OH$$

Glutamic acid (Glu)

$$HO-\overset{\overset{O}{\|}}{C}-CH_2-CH_2-\overset{\overset{H}{|}}{\underset{\underset{NH_2}{|}}{C}}-\overset{\overset{O}{\|}}{C}-OH$$

Lysine (Lys)

$$H_2N-CH_2-CH_2-CH_2-CH_2-\overset{\overset{H}{|}}{\underset{\underset{NH_2}{|}}{C}}-\overset{\overset{O}{\|}}{C}-OH$$

Arginine (Arg)

$$H_2N-\overset{\overset{}{\underset{\underset{NH}{\|}}{C}}}-NH-CH_2-CH_2-CH_2-\overset{\overset{H}{|}}{\underset{\underset{NH_2}{|}}{C}}-\overset{\overset{O}{\|}}{C}-OH$$

Histidine (His)

Imidazole ring (HC, CH, N, NH, CH) $-CH_2-\overset{\overset{H}{|}}{\underset{\underset{NH_2}{|}}{C}}-\overset{\overset{O}{\|}}{C}-OH$

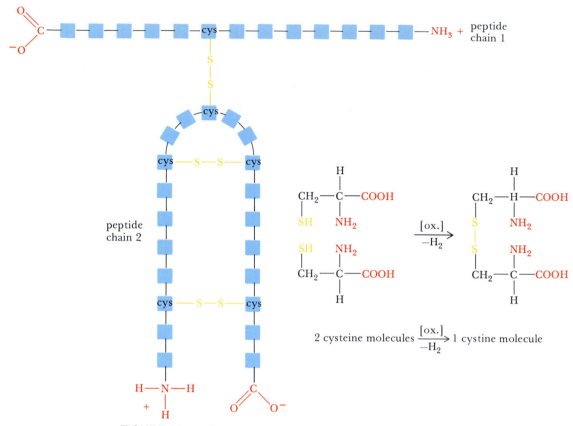

FIGURE 26–5 A pair of peptides, linked through an intermolecular disulfide bond. Note the two intramolecular disulfide bonds within one of the two peptides and the example of an oxidative coupling reaction producing an intermolecular disulfide bond.

The amide linkage between two amino acids is called the **peptide bond**, and the product of the combination of two or more amino acids is called a **peptide**. Glycylalanine is a **dipeptide**, one with three amino acids is a **tripeptide**, and so forth. Peptides with larger numbers of amino acids, from about 5 to 35, are called **oligopeptides**. Longer structures are called **polypeptides** or **proteins**. Peptides are named by the amino acid sequence starting at the end with the free amino group. The ending of each amino acid is changed to -yl, except for the amino acid with the free carboxylic acid group. Note the naming of glycylalanine, where "alanine" is unchanged.

EXAMPLE 26–2

Give the names for the tripeptides that can be constructed from one unit each of alanine, valine, and leucine.

Solution The possible combinations are alanylvalylleucine, valylalanylleucine, alanylleucylvaline, leucylalanylvaline, valylleucylalanine, and leucylvalylalanine.

EXERCISE 26–2

Give the names of all possible tetrapeptides that contain one unit each of serine, threonine, methionine, and tryosine.
Answer: The list will include 24 different tetrapeptides.

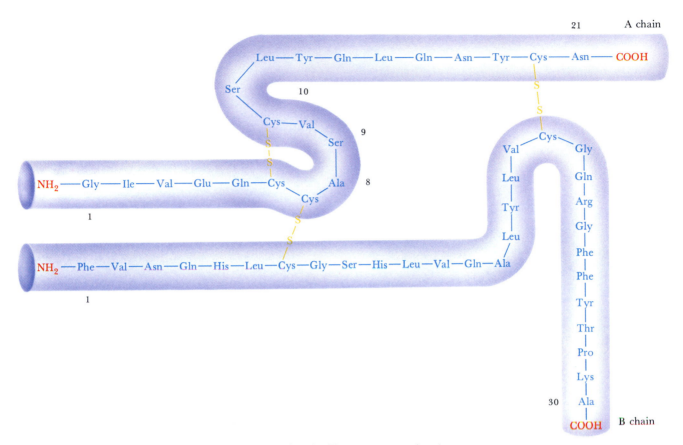

FIGURE 26–6 Primary structure of the insulin molecule. The sequence of amino acids is that found in bovine insulin.

Primary Protein Structure: All of the proteins that occur in nature are made up of one or more long peptide chains of α-amino acids. Connections *between* chains are made through bridges built by —S—S— linkages known as cystine, or "disulfide," bridges. The schematic representation in Figure 26–5 shows two peptide chains so joined; one of them is also internally joined to itself through two similar linkages. Note the terminal amine and carboxyl functions.

The sequence of amino acids in a protein is called its **primary structure**. With considerable skill and a great deal of patient effort, chemists have determined primary structures for a number of important proteins. The primary structure of insulin, a protein that is essential for the metabolic utilization of glucose, was proved by Sanger in 1950. Insulin has a basic unit with a molecular weight of 5700. Small and simple by comparison to most proteins, this fundamental unit consists of two peptide chains joined by two disulfide bridges. The smaller A-chain is composed of 21 amino acids and one internal disulfide linkage; the longer B-chain is made up of 30 amino acids (Figure 26–6). Use of abbreviations such as Val for valine and Cys for cysteine is typical. Ribonuclease, whose structure was worked out by Stein and Moore in 1960, is a single polypeptide chain composed of 124 amino acid residues and four disulfide bridges (Figure 26–7).

Secondary Protein Structure: Cysteine is capable of cross-linking different parts of the chain with disulfide bonds, forming cystine. But the long chain of protein does not just flop into a random configuration like a piece of string. The **secondary structure** of a protein describes the general shape

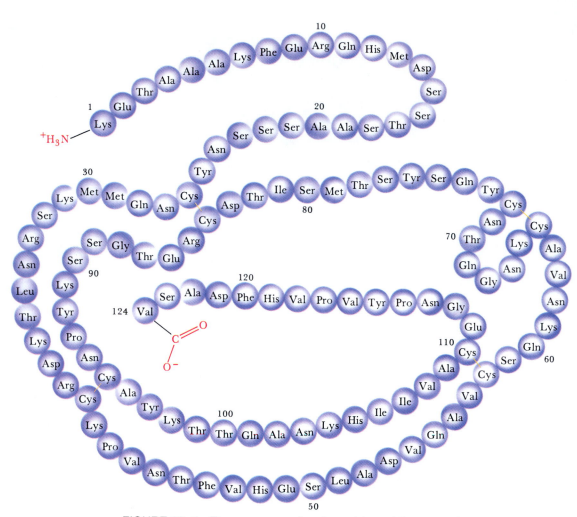

FIGURE 26–7 The sequence of amino acids and the secondary structure in bovine ribonuclease.

that the chain assumes. A spiral coil or helix is a common secondary protein structure. In the same sense as the turn of a screw or a spiral staircase, the coil can be either right- or left-handed, turning one way or the other. However, the L-configuration of naturally occurring α-amino acids forces the helix into the right-handed (α-helical) structure. Pauling and Corey first proposed a single helical structure in 1951 to explain the inherently stable spatial geometry observed in many proteins (Figure 26–8).

The nature of the α-helix is determined in large part by hydrogen bonding between $\diagdown$C=Ö: groups and nearby $\diagdown$N—H groups within the same protein chain. Fibrous proteins in structures such as hair, skin, and feathers have the α-helix structure.

Pauling and Corey also proposed an additional periodic protein structure, referred to as the β-pleated sheet. It is completely different from the α-helix. Whereas the α-helix is a twisted screw-like rod, the β-pleated sheet is just that, a sheet in which the peptide chains are fully extended. For example, silk fibers and muscles consist almost entirely of anti-parallel stacks of β-sheets, stabilized by hydrogen bonds between —NH groups and oxygen atoms in *adjacent* chains (Figure 26–9).

They named it β because it was the second structure they discovered that year, the first being the α-helix.

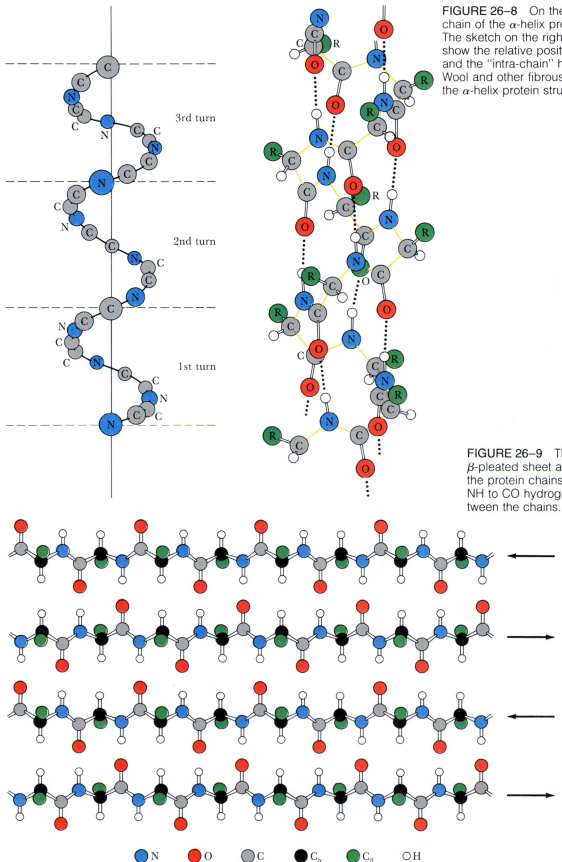

FIGURE 26–8   On the left is the main chain of the α-helix protein structure. The sketch on the right attempts to show the relative positions of the atoms and the "intra-chain" hydrogen bonds. Wool and other fibrous proteins display the α-helix protein structure.

3rd turn

2nd turn

1st turn

FIGURE 26–9   The anti-parallel β-pleated sheet arrangement of the protein chains in silk. Note the NH to CO hydrogen bonds between the chains.

● N   ● O   ● C   ● C_α ● C_β ○ H

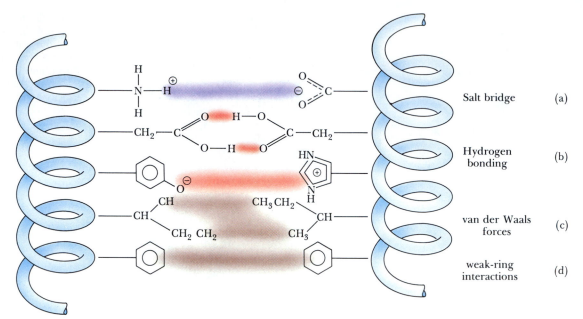

FIGURE 26–10 Types of bonding interactions that typically sustain and support the three-dimensional structures of proteins.

Salt bridge (a)

Hydrogen bonding (b)

van der Waals forces (c)

weak-ring interactions (d)

Tertiary Protein Structure: The actual three-dimensional structure of the protein is called the **tertiary structure**. Groups along the chain making up the backbone of the protein tend to interact with other groups in specific ways. Thus polar groups can link through dipole-dipole attractions and hydrogen bonds (Figure 26–10). The transfer of protons between acidic carboxyl groups and basic amine groups leads to ionic attractions. Finally, van der Waals attractions occur between nonpolar R groups. The role of the water that surrounds the protein in its environment is especially important to the protein's configuration. The polar groups tend to be turned to the outside of the tertiary structure, whereas the nonpolar groups are turned inward. The tertiary structure of a protein is best determined by single crystal x-ray diffraction analysis.

EXAMPLE 26–3

In the tertiary structure of a protein, state which amino acid R groups would tend to attract each of the following amino acids in the protein chain: valine, serine, aspartic acid, and lysine. What kinds of interactions would exist in each case?

Solution Beginning with valine: The R group is a nonpolar hydrocarbon. It would be expected to form van der Waals attractions with the other amino acids with nonpolar R groups: glycine, alanine, leucine, isoleucine, phenylalanine, and proline, as well as another valine. These groups would tend to be turned toward the interior of the structure.

With serine: The R group is neutral but polar, containing an OH group. It will form hydrogen bonds with threonine, tyrosine, another serine, aspargine, and glutamine, and dipole-dipole interactions with cysteine, methionine, and tryptophan. It will also react with the water in the environment.

With aspartic acid: The carboxylic acid COOH included in the R group can protonate the N atom in the amine NH_2 group of lysine, arginine, or histidine. For example,

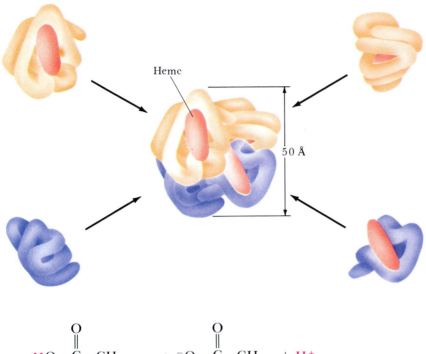

FIGURE 26–11 The hemoglobin protein consists of four uniquely folded polypeptide chains, shown here as "skeletons" so as not to obscure the intricate folding pattern by the amino acid R groups. Two are designated as α-chains and two, as β-chains. The heme groups are non-protein structures (attached to the polypeptide chains) containing iron atoms; their function is to bind oxygen atoms transported to them by hemoglobin.

$$HO-\overset{\overset{\displaystyle O}{\|}}{C}-CH_2- \longrightarrow {}^-O-\overset{\overset{\displaystyle O}{\|}}{C}-CH_2- + H^+$$

R group in
aspartic acid

$$H_2N-CH_2-CH_2-CH_2-CH_2- \quad + H^+ \longrightarrow$$

R group in
lysine

$$H_3\overset{+}{N}-CH_2-CH_2-CH_2-CH_2-$$

Anions and cations created in this way will form strong ionic bonds.

With lysine: This amino acid will form ionic attractions to aspartic acid and glutamic acid by the formation of ions. All of the R groups tending to form ions will also interact with water. ■

EXERCISE 26–3
State which amino acids tend to associate with the R groups of each of the following amino acids: isoleucine, cysteine, threonine, glutamic acid, and arginine. Also state the types of linkages that form. ■

Quaternary Protein Structure: Size and complexity in some proteins result in further differentiation due to **quaternary structure**. Such proteins have two or more polypeptide chains as **subunits** that fit together in a very specific fashion—the quaternary structure—to form the complete protein. The interactions holding the subunits are the same as in the tertiary structure, and are not regular covalent bonds. As an example, the protein myoglobin that is used to store oxygen in muscle cells has no quaternary structure, that is, no subunits. However, hemoglobin, an important protein of the blood, has four subunits, each of which is very similar to a single myoglobin molecule. The unique way that these four subgroups fit together constitutes the quaternary structure of hemoglobin (Figure 26–11).

Typical of nature's proteins is the subtle complexity of the four complementary peptide chains constructed from a unique arrangement of 574 amino acids. Change just one, switching valine for glutamic acid in the sixth position from the end of one chain, and you have the variant hemo-

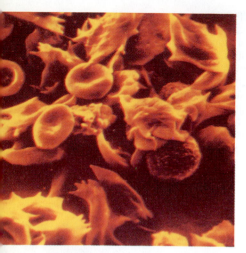

FIGURE 26–12 Normal red blood cells (doughnut-shaped) are seen here dispersed among abnormally shaped "sickle" cells. The sickle cells tend to get blocked up and do not flow freely through the blood stream, causing the painful symptoms of this genetic disease. (Magnification approximately 4000X)

globin responsible for the condition known as sickle cell anemia (Figure 26–12). A quarter of a million black Americans carry the gene for this chronic and eventually life-threatening disorder. Should the switch introduce lysine for glutamic acid, a much milder hemolytic anemia results.

Enzymes

Enzymes are the highly efficient and highly selective catalysts of bioreactions. All are proteins. Most act within the cell that produces them and are called **intracellular** enzymes. A few, such as the salivary and gastric enymes, act outside cells and are called **extracellular** enzymes. Enzymes act upon a **substrate**, converting it to one or more other compounds. Often an enzyme is named for the substrate that it acts upon and the reaction catalyzed. For example, alcohol dehydrogenases are oxidation enzymes that catalyze the dehydrogenations of alcohols into aldehydes. As mentioned in the previous section, bioreactions proceed rapidly under the very mild conditions of life. Enzymes catalyze these reactions with amazing specificity. For example, H_3O^+ ions are modestly effective catalysts for hydrolyzing sugars such as sucrose, maltose, and lactose into their smaller component sugars. On the other hand, the enzyme **sucrase** will effectively and selectively catalyze the hydrolysis of sucrose into glucose and fructose. Sucrase will not affect other sugars at all.

Some enzymes are simple proteins containing only amino acids. Others form generally noncovalent complexes with non-peptides. In such enzymes, the polypeptide part is called the **apoenzyme**. The non-peptide part is called the **coenzyme**, and it can often be dissociated from the rest of the enzyme. Without the coenzyme, the apoenzyme is inactive. Coenzymes or "enzyme helpers" are extremely important and will be discussed in more detail subsequently.

Nucleic Acids: Information Storage and Information Expression

Nucleic acids are macromolecules containing the information that determines the growth and development of all living things. They are of two types, deoxyribonucleic acid or DNA, and ribonucleic acid or RNA. The backbone of the macromolecular chain in each of the nucleic acids consists of alternating phosphoric acid and sugar molecules condensed together with the elimination of water—a phosphate ester linkage. In DNA the sugar is deoxyribose, and in RNA the sugar is ribose:

deoxyribose
DNA sugar

ribose
RNA sugar

Attached to each of the sugar residues is one of five organic bases. The **purine** bases are adenine and guanine, and the **pyrimidine** bases are cytosine, thymine, and uracil. Uracil is found only in RNA, and thymine is found only in DNA; adenine, guanine, and cytosine are common to both DNA and RNA.

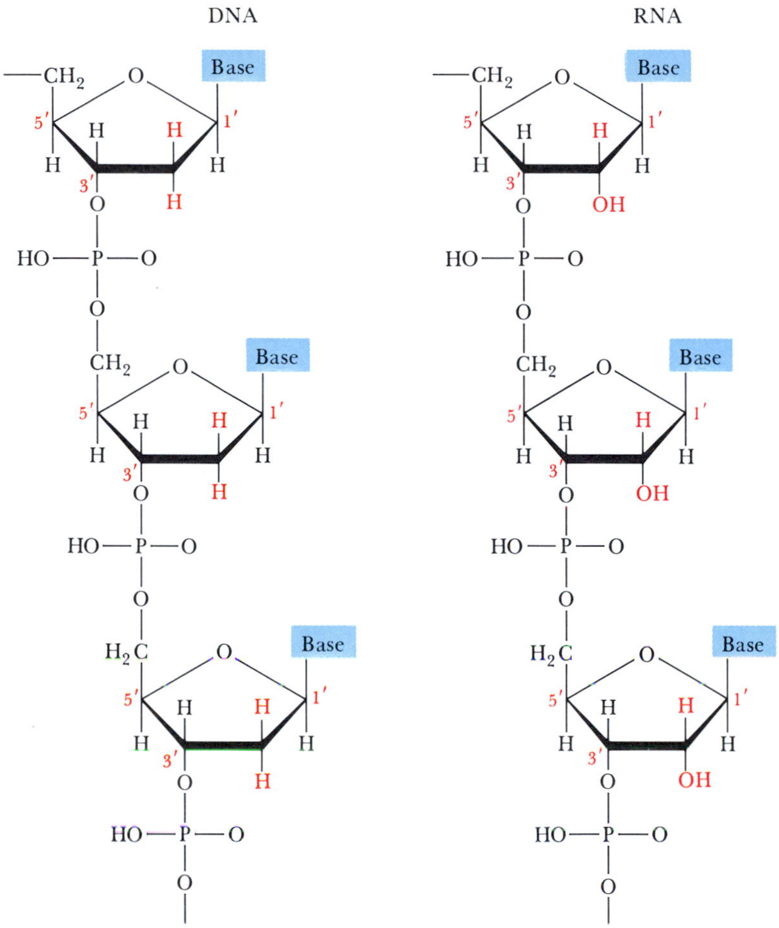
DNA
RNA
Base
CH_2
O
H
OH
HO—P—O
H_2C
5′
3′
1′

FIGURE 26–13 On the left, the typical deoxyribose sugar phosphate arrangement found in DNA; on the right, the ribose sugar phosphate structure of RNA. Only the positions of the bases are shown here.

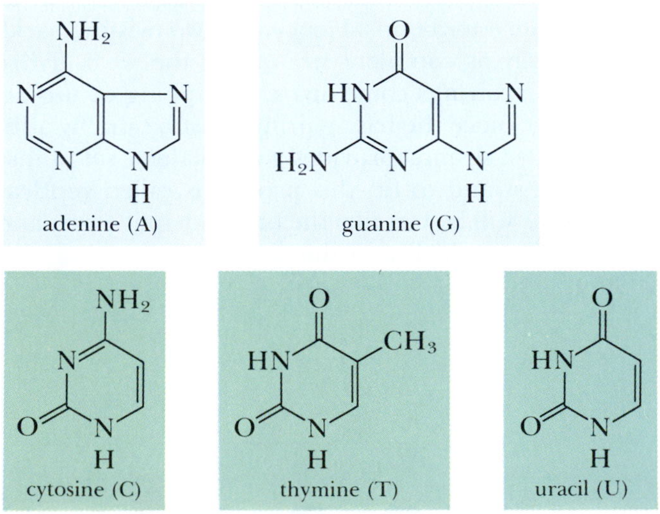
adenine (A)
guanine (G)
cytosine (C)
thymine (T)
uracil (U)

□ = purines
■ = pyrimidines

The structure of part of a DNA strand is shown in Figure 26–13. Each subunit (consisting of a phosphate group, a sugar residue, and a base) is called a **nucleotide**.

DNA varies greatly from species to species, and judgments on the closeness of the relationships of species can be made by comparing the similarities in their DNA. The determination of the double-stranded structure of DNA was the result of important discoveries made by several

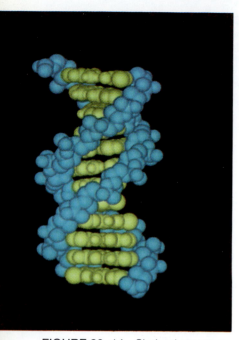

FIGURE 26–14 Skeletal structure of the DNA molecule, along the α-helix. The structure repeats at intervals of 34Å, corresponding to ten residues. Twelve residues are shown here.

investigators. Around 1950, it became known that the molar amount of adenine (A) is always equal to the molar amount of thymine (T) and that the amount of cytosine (C) is always equal to the amount of guanine (G). Thus A = T and C = G. X-ray diffraction data clearly showed that the DNA molecule is formed in a helix. The distance along the axis of the helix is 34 Å per complete turn and includes 10 subunits (Figure 26–14). This allows 3.4 Å per subunit, which is consistent with the size of a nucleotide.

Using analyses of the bases and the x-ray results of Maurice Wilkins and Rosalind Franklin, James Watson and Francis Crick proposed the double helix model of DNA. In their model, the two helical DNA strands lay parallel to each other and were held together by hydrogen bonds between specific base pairs. That is, adenine was always bonded to thymine, and cytosine was always bonded to guanine, which meant that the two strands had to have a definite relationship to each other.

The cytosine/guanine coupling, with three hydrogen bonds (Figure 26–15), is somewhat stronger than the thymine/adenine coupling with only two. These couplings are possible only in a double helix of unique dimensions with the base pairs directed inward; cytosine on one strand is opposite and bonded to guanine on the second strand, while thymine and adenine are bonded in corresponding positions. Why this is so can be clarified by looking at Figure 26–16. If adenine were paired with either adenine or guanine instead of thymine, the distance between the two backbone strands would have to be lengthened to accommodate the larger units. Using the same argument, base-pairing cytosine with cytosine or thymine would require a shortening of the distance between backbone units. But since we know that there is no variation in the length of the rungs (the base pairings), then each pair must be a purine/pyrimidine base pair—purine/purine pairs would be too long and pyrimidine/pyrimidine pairs would be too short. The hydrogen bonds fit perfectly only for the purine/pyrimidine base pairs (Figure 26–17).

In the complete DNA molecule, each unit is bonded to the one above through covalent phosphate ester bonds. Hence, the two twisting backbone helices tend to divide easily as complete strands, as the weak hydrogen bonds connecting the base pairings come apart. Unzipping or unstacking takes place (Figure 26–18). Since the base pairing must be strictly adhered to, each of the two unzipped strands provides a template or set of instructions for making another strand to fit; this process is called **replication**. The new strand that forms will be just like the one that split away, and the new DNA molecule is an exact replica of the old one.

FIGURE 26–15 The proper base pairings are adenine-thymine (left); and guanine-cytosine (right).

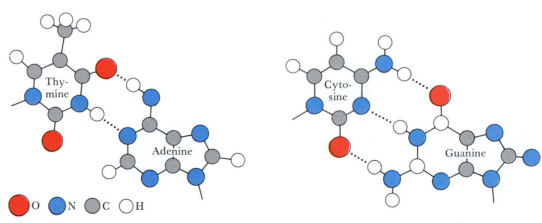

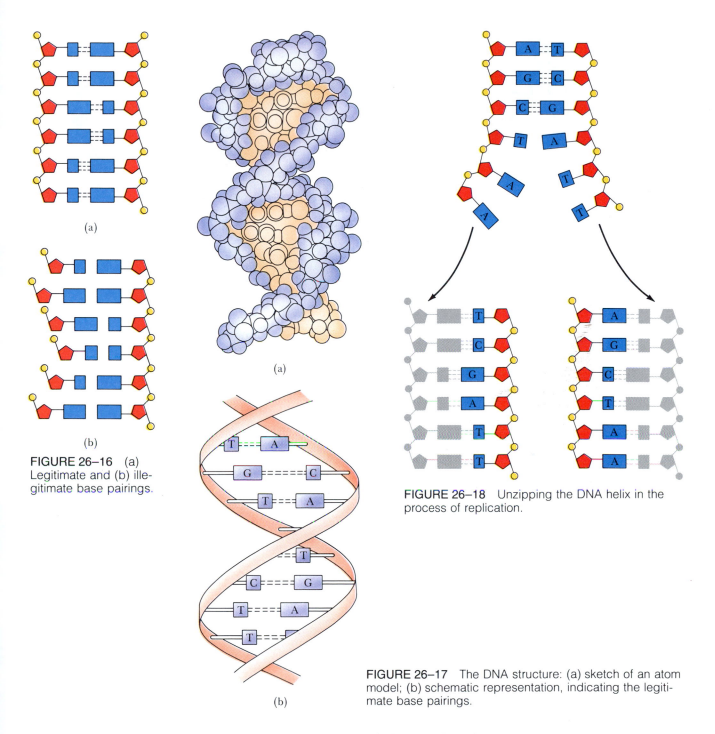

FIGURE 26–16 (a) Legitimate and (b) illegitimate base pairings.

FIGURE 26–18 Unzipping the DNA helix in the process of replication.

FIGURE 26–17 The DNA structure: (a) sketch of an atom model; (b) schematic representation, indicating the legitimate base pairings.

Protein synthesis doesn't occur where DNA is found. DNA is found in the cell nucleus while amino acids are snapped together into peptides and proteins in the cytoplasmic matter surrounding the nucleus. Therefore, to suggest that DNA is the template for protein synthesis is inaccurate. Instead, it is the single-stranded RNA molecule that is most closely associated with direct protein synthesis.

RNA strands are very similar to DNA strands. The principal difference is that one of the four nucleic acid residues has been switched—uracil for thymine. But even that isn't a big difference since both are pyrimidine bases, and uracil is capable of hydrogen bonding with adenine in exactly

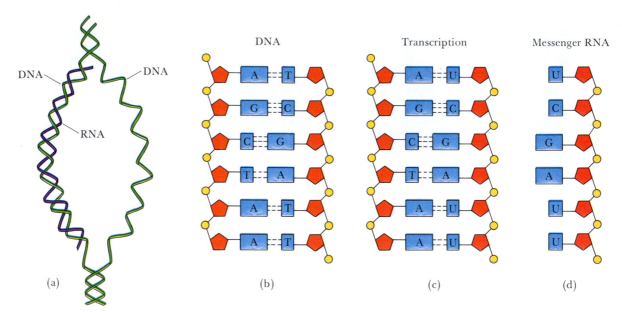

FIGURE 26-19 (a) Transcription, showing complementary strands. (b)-(d) show sequencing the DNA chain, the transcribed message (c) and the messenger RNA strand (d).

the same way thymine does. In fact, RNA is so closely related to DNA that a strand of one can coil around and pair with the other. Or, to put it another way, the two strands are complementary. Since the sequence of the base pairs is the coded message, a DNA message can be copied onto an RNA strand, after which the RNA strand is unzipped from the DNA strand, leaves the nucleus, and proceeds to make up the amino acid sequences for peptide and protein synthesis (Figure 26-19).

A typical DNA base sequence might be A G C T A A C A C A T T. Remembering that uracil replaces thymine (so the new base pair will be adenine/uracil), that transcribes into the following RNA complement (Figure 26-20 and Table 26-2):

U C G	A U U	G U G	U A A
serine	isoleucine	valine	*stop*

For each amino acid unit, the coding is a three-letter "word" consisting of three successive bases on the RNA strand. This group of bases is called a **codon**. There are 64 possible codons that can be used in coding 20 amino acids. Hence, some of the codons listed in Table 26-2 are unimportant, and some amino acids are represented by more than one codon. The valine code is either GUA or GUG, while glutamic acid is either GAA or GAG. If we replace the middle base (adenine for uracil) in both cases, a seemingly unimportant event when you consider the overall dimensions and complexity of amino acids, the molecular disease sickle cell anemia is produced.

Several types of RNA are involved in the chain of events leading to protein synthesis. The RNA that copies the code on DNA and carries it out of the nucleus is **messenger RNA**. In the cytoplasm, **transfer RNA** molecules select the correct protein monomer, bring it to the site of protein synthesis in the cell, then read the coded message on the messenger RNA, and finally place the amino acid in its place on the growing polymer chain. If the codon calls for the amino acid tryptophan, messenger RNA will

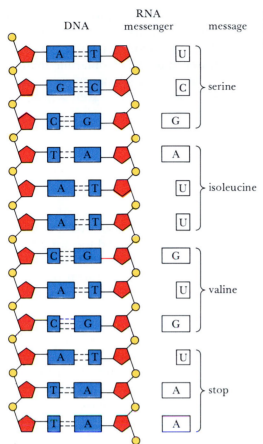

DNA

RNA messenger

message

serine

isoleucine

valine

stop

FIGURE 26–20 Codons identifying amino acids in a short segment along a peptide chain.

TABLE 26–2   Three-Letter Words Used in Coding Amino Acids

Amino Acid	Codon Assignments				
Alanine	GCC	GCU	GCA	GCG		
Arginine	CGC	CGU	CGA	CGG	AGA	AGG
Asparagine	AAU	AAC				
Aspartic acid	GAU	GAC				
Cysteine	UGU	UGC				
Glutamic acid	GAA	GAG				
Glutamine	CAA	CAG				
Glycine	GGU	GGC	GGA	GGG		
Histidine	CAU	CAC				
Isoleucine	AUU	AUC	AUA			
Leucine	UUA	UUG	CUU	CUC	CUA	CUG
Lysine	AAA	AAG				
Methionine	AUG					
Phenylalanine	UUU	UUC				
Proline	CCC	CCU	CCA	CCG		
Serine	UCU	UCC	UCA	UCG	AGU	AGC
Threonine	ACU	ACC	ACA	ACG		
Tryptophan	UGG					
Tyrosine	UAU	UAC				
Valine	GUU	GUC	GUA	GUG		
Terminate (stop)	UAA	UAG	UGA			

contain the three-letter word UGG. That means the transfer RNA must have the complementary word ACC. Protein synthesis continues in this systematic, programmed fashion until the *stop* codon is reached. At that point, the completed polypeptide is released into the cytoplasmic matter. It has been estimated that messenger and transfer RNA can snap together amino acids at the rate of 30 to 40 units per second, with rarely a mistake. That's quite an impressive feat of chemical synthesis!

Carbohydrates: Intermediate-Term Energy Storage and Structural Support

Carbohydrates are the principal constituents of plants and are crucial components of animal food chains. Plants manufacture carbohydrates from carbon dioxide and water by photosynthesis, using sunlight as an energy source. Cellulose is the main structural material of plants, and starch and sugars serve plants for energy storage. All three of these forms of carbohydrates provide food for animals. Glycogen is a carbohydrate that animals store in their cells as an energy reserve.

Some carbohydrates are long-chain macromolecules and some are much smaller units. The smallest units are called **monosaccharides**. They cannot be broken up by hydrolysis into smaller carbohydrate units. Carbohydrates with two or three monosaccharide units are called di- and trisaccharides. Carbohydrates with many monosaccharide units are called **polysaccharides**. Cellulose and starch are the most familiar polysaccharides.

Mono-, di-, and trisaccharides are known as sugars. All tend to have a sweet taste, although this varies considerably from sugar to sugar. The monosaccharide most common in nature is glucose:

$$
\begin{array}{c}
O \diagdown \quad \diagup H \\
C \\
| \\
H - C - OH \\
| \\
HO - C - H \\
| \\
H - C - OH \\
| \\
H - C - OH \\
| \\
CH_2OH
\end{array}
$$

This structure is typical of monosaccharides. All of these compounds are either aldehydes or ketones with a large number of OH groups. Glucose is an aldehyde and therefore would be expected to be easily oxidized. The OH groups confer a high solubility in water. The center four carbons in the chain are all chiral, so glucose is an optically active compound. The naturally occurring form is dextrorotatory (D-glucose), and is often called dextrose.

The glucose molecule is capable of reacting internally, producing either of two cyclic six-membered **pyranose rings**. These are the most stable structures under most conditions, although the open chain form must also be considered to account for all of the observed chemistry. The rings are designated β-D-glucopyranose (or simply β-D-glucose) and α-D-glucopyranose (or just α-D-glucose):

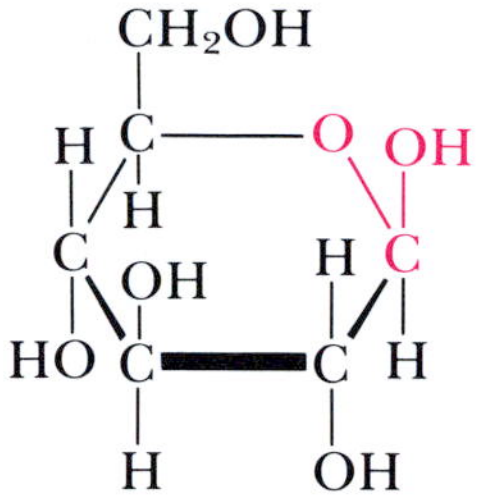

β-D-glucopyranose 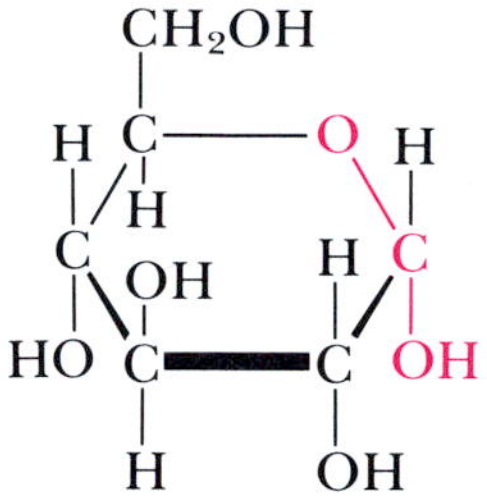 α-D-glucopyranose

"Sugar chemists" refer to these planar hexagonal representations as *Haworth projections*. The ring oxygen is placed in the upper right vertex and the OH at the "anomeric" carbon is up in the β-anomer and down in the α-anomer.

A second common monosaccharide is fructose, a sugar found in fruits. In its noncyclic form, this sugar contains the ketone functional group:

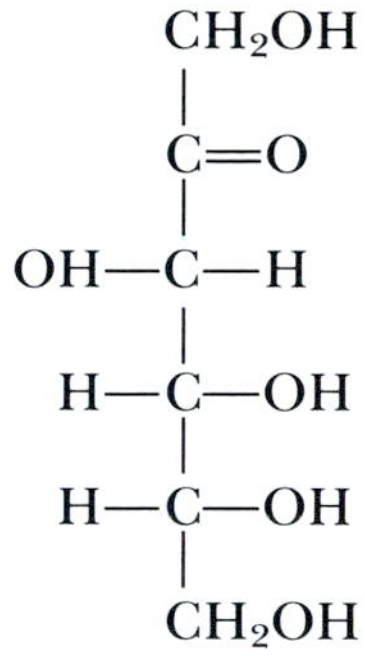

The cyclic forms are **furanose rings**:

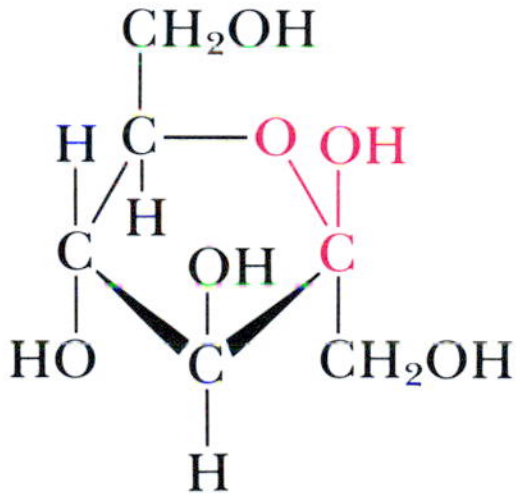

β-D-fructofuranose   α-D-fructofuranose
(β-D-fructose)   (α-D-fructose)

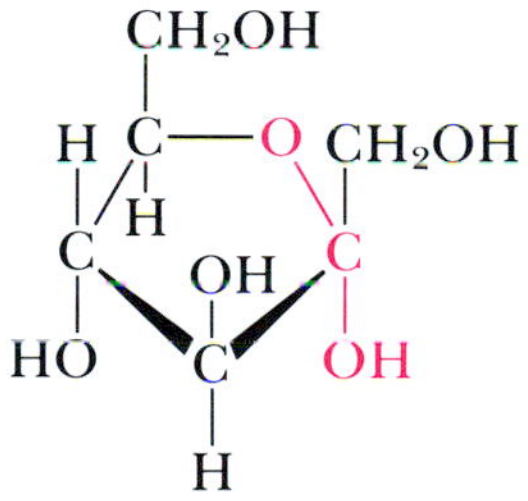

The condensation of a glucose molecule and a fructose molecule with the loss of water yields the disaccharide called sucrose (Figure 26–21).

Glucose units are the building blocks of the polysaccharide **starch**. Starch is found in most plants, particularly in the seeds. It can be broken down by the plant to its glucose units for reserve food during times when photosynthesis is insufficient for the plant's needs. The principal constit-

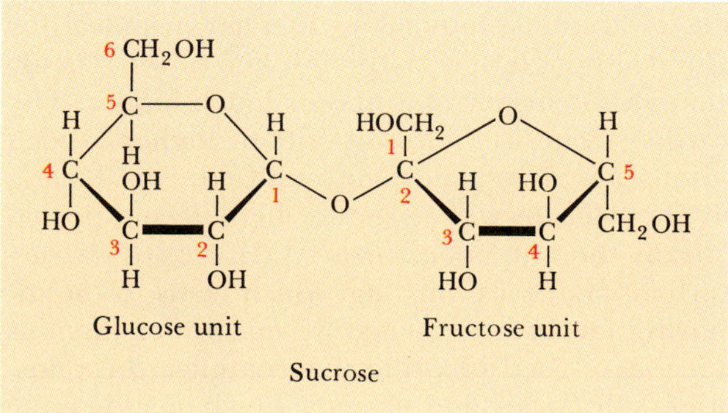

FIGURE 26–21 The sucrose molecule: In cane sugar and in beet sugar, sucrose consists of a glucose sugar joined to a fructose sugar at C-2.

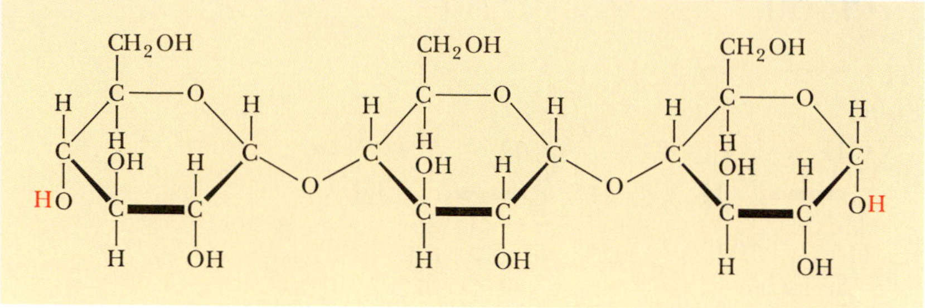

FIGURE 26–22 Starch is a mixture of two polysaccharides, one of which is amylose, shown here. The individual sugars are α-D-glucose and typically number 200–300 in the chain.

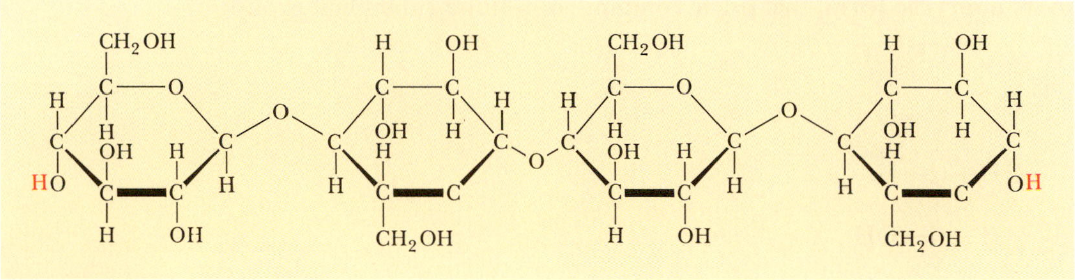

FIGURE 26–23
The small part of the cellulosic β-D-glucose chain.

Iodine serves as a test for starch. When the iodine is droppered on a potato, a deep blue complex is formed.

uent of starch, called **amylose**, is a straight chain of α-D-glucose units (Figure 26–22). However, the starch structure also contains a considerable amount of branching. Starch is an important food for humans and many animals. It is broken down into glucose units by a series of enzymes known as amylases.

Cellulose is also a glucose polymer. However, it is composed almost exclusively of straight chain β-D-glucose units (Figure 26–23). Cellulose is the main structural component of all plants. Many nonedible plant products such as paper and cotton are mainly cellulose. The enzymes necessary for breaking the cellulose linkages are not present in humans or carnivorous animals. Animals that can digest cellulose range from cows and sheep to termites. The stomachs of these animals contain microorganisms that have the necessary enzymes.

Lipids: Long-Term Energy Storage and Cell Membranes

The term **lipid**, unfortunately, is not clearly defined. Originally, it was used to describe natural products that were soluble in ether and insoluble in water. A biochemist might typically include in this category fatty acids and substances that yield fatty acids, as well as numerous other fat-soluble compounds. Fats are the lipids we are most familiar with. They are ideal for long-term energy storage, and although they are not broken down as easily as carbohydrates, their energy content per unit mass is much higher. The lower energy content of carbohydrates is the result of their higher oxygen content—that is, carbohydrates are already partially oxidized. Fats, having a much lower oxygen content to begin with, produce higher energies on oxidation. But that also means the water-solubilizing OH groups characteristically displayed by carbohydrates are missing, which leads to the interesting problem of removing fats and oils when we encounter them as stains. Other lipids include waxes used in protective coatings, the phospholipids found in membrane substances, and steroids, which include some regulatory hormones.

Fats and **oils** are esters of glycerol occurring in plants and animals:

$$
\underset{\text{glycerol}}{\begin{array}{c} CH_2OH \\ | \\ HO-C-H \\ | \\ CH_2OH \end{array}}
\qquad
\underset{\text{glyceryl tristearate}}{\begin{array}{c}
\qquad\qquad O \\
\qquad\qquad \| \\
CH_2O-C(CH_2)_{16}CH_3 \\
CH_3(CH_2)_{16}\overset{O}{\overset{\|}{C}}-O-C-H \quad O \\
| \qquad\quad \| \\
CH_2O-C(CH_2)_{16}CH_3
\end{array}}
$$

Long-chain acids of lipid esters are called **fatty acids**. Animal fats tend to be saturated; fatty acids of plant oils tend to be unsaturated. A fat or oil having only a single kind of fatty acid is called a simple **triglyceride**; one having two or three fatty acids is referred to as a mixed triglyceride. On reacting a fat or oil with aqueous sodium or potassium hydroxide solutions, a process known as **saponification**, glycerol is recovered along with the sodium or potassium **soaps**, the salts of the fatty acids. For example, if the triglyceride is glyceryl tristearate and the aqueous alkali is sodium hydroxide, the soap is sodium stearate:

$$
\underset{\text{glyceryl tristearate}}{\begin{array}{c}
\qquad\qquad\qquad O \\
\qquad\qquad\qquad \| \\
\qquad H_2C-O-C-(CH_2)_{16}CH_3 \\
CH_3(CH_2)_{16}-\overset{O}{\overset{\|}{C}}-O-C-H \\
\qquad H_2C-O-C-(CH_2)_{16}CH_3 \\
\qquad\qquad\qquad \| \\
\qquad\qquad\qquad O
\end{array}}
\quad + \ 3\ NaOH\ \longrightarrow
$$

$$
3\ CH_3(CH_2)_{16}\overset{O}{\overset{\|}{C}}-O^-Na^+ \ + \
\begin{array}{c} H_2C-OH \\ | \\ HO-C-H \\ | \\ H_2C-OH \end{array}
$$

Sodium and potassium salts of fatty acids in the C-12 to C-20 range are the most important, commercially. Stearic acid (in our example above) has 18 carbon atoms, lauric acid has 12, and palmitic acid has 16.

The familiar and important properties of a soap are due to a long-chain hydrocarbon tail of **hydrophobic** CH_2's on the one end, which is nonpolar, and a **hydrophilic** region on the other, due to a polar carboxylic acid anion. Butter stains, oils and greases, though not normally water-soluble, are easily removed from an article of clothing by an agitated soap solution. The hydrophobic ends of soap molecules dissolve into the butter fat, oil, or grease while the hydrophilic end extends into the wash water. The net effect is that the stain is dispersed and washed away.

You have probably noticed that soap solutions are cloudy, due to **micelle** formation. In a micelle, large numbers of soap molecules assemble with their hydrophobic tails directed toward the center and their hydrophilic ends directed outward. The resulting micelle structures are large enough to disperse light, giving soap solutions a generally cloudy (turbid) appearance (Figure 26–24).

Sodium stearate soaps are precipitated in "hard" water by the presence of calcium and magnesium ions, forming an undesirable "scum" of calcium or magnesium stearate that is responsible for the familiar ring around the collar (or the bathtub). The soap scum problem has now largely been overcome by the introduction of **detergents**, sulfuric acid analogs of the carboxylic acid soaps. But it has not been an easy conquest. For example, the first of these synthetic surface-active agents, or **surfactants**, were branched hydrocarbons, and although they formed more soluble salts (leaving no

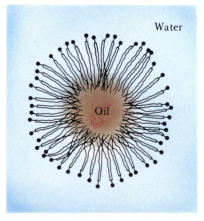

FIGURE 26–24 Soap molecules, arranged in a micelle, with their hydrophilic ends directed toward the surrounding water (shown as the smaller circles) and their hydrophobic ends directed toward an oil droplet.

residues), they were not easily biodegradable, persisting in the environment. Straight-chain hydrocarbon tails, it seemed, were digestible by bacteria in sewage plants but branched hydrocarbon tails were not. In our zeal to remove the bathtub ring, we had filled our rivers, lakes and streams with effluents that caused suds and foam. Straight-chain sulfuric acid soaps have helped mitigate and partially resolve that problem:

$$CH_3(CH_2)_{10}CH_2\!-\!O\!-\!\overset{\overset{\textstyle O}{\|}}{\underset{\underset{\textstyle O}{\|}}{S}}\!-\!O^-\ Na^+$$

sodium dodecanyl sulfate

But even detergents are not without problems, since they tend to lather and foam, causing esthetic as well as mechanical difficulties in their use and disposal.

Waxes are also esters of fatty acids, but the alcohols involved have long chains and generally have only one OH group unlike glycerol with three. An example is myricyl palmitate, a component of beeswax, which is an ester of myricyl alcohol ($C_{30}H_{61}OH$) and palmitic acid ($CH_3(CH_2)_{14}COOH$).

EXAMPLE 26–4

Write the chemical formulas for the fatty acid "soaps" obtained from saponification of the following triglyceride:

$$
\begin{aligned}
&\overset{\overset{\textstyle O}{\|}}{CH_2OC}(CH_2)_{14}CH_3\\
&\underset{|}{}\\
&CH_2O\overset{}{C}(CH_2)_7CH\!=\!CH(CH_2)_7CH_3\\
&\underset{|}{\underset{\textstyle O}{\|}}\\
&CH_2O\overset{}{C}(CH_2)_{16}CH_3\\
&\underset{\textstyle O}{\|}
\end{aligned}
$$

Solution   This is a mixed triglyceride. Therefore the fatty acid sodium salts, or soaps, from top to bottom are palmitic (hexadecanoic), $CH_3(CH_2)_{14}COO^-Na^+$; unsaturated oleic, $CH_3(CH_2)_7CH\!=\!CH(CH_2)_7COO^-Na^+$; and stearic (octadeconoic), $CH_3(CH_2)_{16}COO^-\ Na^+$.

EXERCISE 26–4

Write the structure for the mixed triglyceride of linoleic acid, $CH_3(CH_2)_4CH\!=\!CHCH_2CH\!=\!CH(CH_2)_7COOH$, and any other two fatty acids. ■

Phospholipids are similar in structure to fats and oils except that one of the OH groups in the glycerol has been substituted with phosphoric acid rather than a fatty acid. The basic structure of a phospholipid is

$$
\begin{array}{c}
\overset{\textstyle O}{\|}\\
\overset{\textstyle O}{\|}CH_2\!-\!O\!-\!C\!-\!R'\\
R''\!-\!C\!-\!O\!-\!\underset{|}{C}\!-\!H\overset{\textstyle O}{\|}\\
CH_2\!-\!O\!-\!\underset{\underset{\textstyle O^-}{|}}{P}\!-\!OX
\end{array}
$$

R′ and R″ represent the fatty acid chains, and X can be any of a large number of groups from H to some that are quite complex. The phosphate end of a phospholipid is hydrophilic and the fatty acid end is hydrophobic. Phospholipids are important in cell membranes, where the hydrophobic end points toward the center of the membrane and the hydrophilic end points toward the water medium on either side of the membrane (Figure 26–25).

Steroids have no structural relationship to the lipids mentioned previously. The steroid structure is based on the condensed four-ring structure shown in Figure 26–26. Important steroids are cholesterol, cortisone and the sex hormones testosterone, progesterone, estradiol, and estrone. These compounds play an enormous role in proper body function and development, and the applications of natural and synthetic steroids range from birth control to body building.

Coenzymes and Nucleoside Triphosphates: Enzyme Helpers and Energy Storage

In our discussion of proteins, we mentioned that enzymes are proteins consisting mainly of peptide chains. However, chains formed only of the twenty amino acids do not have nearly enough diverse functional groups to catalyze all bioreactions. Therefore, it is often necessary for an enzyme to contain a nonpeptide part at the site of catalysis. This is called a **coenzyme**.

The study of coenzymes was initiated as a result of the discoveries of the **vitamins**. Vitamins are organic compounds that are needed for the growth and survival of an organism but that the organism cannot synthesize for itself. Thus, vitamins must be included in the diet. Many vitamins are slightly modified by the organism to form a coenzyme. As an example, vitamin B_6 (pyridoxine) is the precursor of a coenzyme, pyridoxal phosphate.

There have been a number of recent episodes involving athletes who have abused the use of steroids to enhance their performance in body building and other sports.

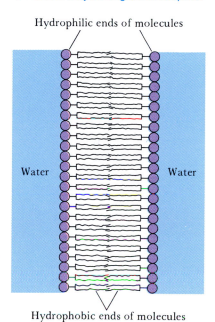

FIGURE 26–25 Simplified picture of a cell membrane.

pyridoxine

pyridoxal phosphate

FIGURE 26–26 (a) Four condensed carbon rings provide the fundamental steroid structure; (b) Cholesterol crystals viewed through a microscope using polarized light.

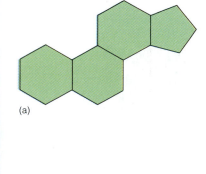
(a)

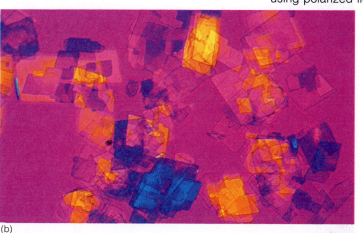
(b)

An important function of three of the coenzymes is to participate in oxidation-reduction processes. These coenzymes, nicotinamide adenine dinucleotide (NAD^+), nicotinamide adenine dinucleotide phosphate ($NADP^+$), and flavin adenine dinucleotide (FAD), undergo reversible gain and loss of electrons symbolized by the following reactions:

$$NAD^+ + H^+ + 2\,e^- \longrightarrow NADH$$
$$NADP^+ + H^+ + 2\,e^- \longrightarrow NADPH$$
$$FAD + 2\,H^+ + 2\,e^- \longrightarrow FADH_2$$

The form of the coenzyme bearing one or two hydrogens is called the **reduced coenzyme**. These coenzymes and their reduced forms serve as agents for the transfer of electrons (or hydrogen atoms) in biochemical reactions.

Nucleoside triphosphates such as adenosine triphosphate (ATP) have structures related to those of many of the coenzymes.

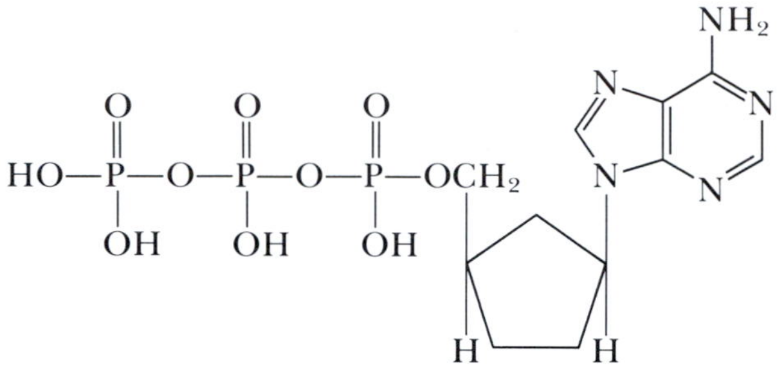

The phosphorus-oxygen bonds in ATP are often referred to as "high energy" or "energy rich" bonds. Hydrolysis of these bonds will produce two acids—phosphoric acid, and a phosphoric acid substituted with adenosine—and evolve much energy. Removal of phosphate groups from ATP yields adenosine diphosphate (ADP) first, and then adenosine monophosphate (AMP):

Ad = adenosine

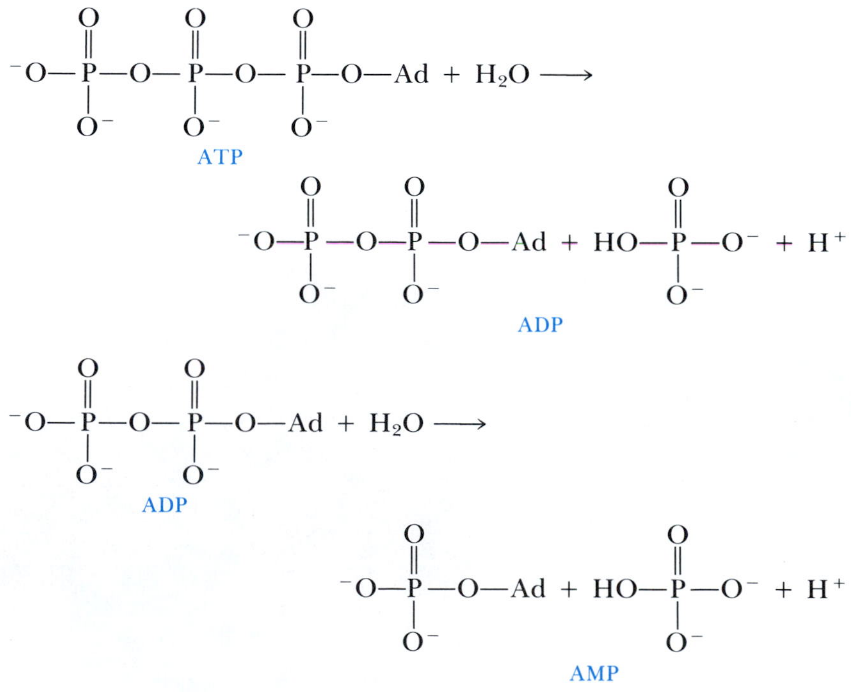

Considerable energy must be supplied to form these phosphate bonds in the first place, and considerable energy is given off when they are broken. The high-energy phosphate bonds in ATP and ADP represent the energy bank for bioreactions. Energy released from the hydrolysis process is used to drive other processes.

Of course, the storage of energy in the form of high-energy phosphate bonds means that exchange of phosphate groups assumes an important role in biochemical reactions. How does an ADP molecule get converted to an ATP? In some cases the phosphate is transferred to the ADP by a compound with an even higher-energy phosphate bond than ATP. Thus the overall reaction will be energy releasing and favorable. However, the explanation that seems to be most generally applicable for the formation of ATP involves existence of a proton gradient—a difference in proton concentrations—across a membrane in many biochemical reactions. The natural tendency for such an electrochemical potential (or gradient) to be equalized provides energy that can be harnessed for the production of ATP. The actual details of this mechanism are quite complex.

26.4 COMMERCIAL USES OF ENZYMES

Among the important applications of biochemistry are the commercial uses of enzymes to produce products and the use of recombinant DNA technology to cause alterations in organisms. In both cases, the life forms involved are most commonly microorganisms such as bacteria.

Enzymes are complex proteins that efficiently catalyze biochemical reactions. They have the additional advantage over other catalysts of a high degree of specificity. Therefore, it should not be at all surprising that their usefulness in chemical reactions has long been recognized:

- Since before the beginning of recorded history, enzymes are known to have been successfully used for production of wine, beer, and a variety of cheeses and related dairy products.

- Yeast contains the enzyme necessary for production of alcohol from sugars.

- Cheese is made by using the enzyme rennin, which is contained in rennet, a substance found in the stomach lining of calves. Yogurt, sour cream, and cultured buttermilk are all produced using enzymes.

- Penicillin is produced from a bread mold (Figure 26–27). It is claimed that the best mold for this process is an x-ray mutated version of a mold found originally on the surface of a cantaloupe in a Peoria, Illinois supermarket in 1942.

- Other substances produced by enzymes are citric acid, amino acids, and other organic compounds, including monosaccharides from the hydrolysis of starch or cellulose.

Processes in which the enzyme is mixed directly into the reaction are inherently expensive, since all or most of the enzyme will be lost. Recent efforts with immobilized enzymes offer a promising way around this problem. If the enzyme is attached to an inert support, it can catalyze a reaction without being mixed in with the reactants and products. Furthermore, no separation of the enzyme from the desired products is necessary. In effect, this is an example of heterogeneous catalysis in which the immobilized enzyme surface is filling the same role as the platinum surface in a catalytic converter. Enzymes have been immobilized on many materials, including synthetic macromolecules such as polystyrene and nylon, resins, cellulose,

FIGURE 26–27 Perhaps the most famous and widely used biochemical agent in modern times (other than aspirin) is penicillin. Responding to the pressures of World War II, the active agent found first in mold cultures discovered by Sir Alexander Fleming, was successfully scaled up into commercial production almost overnight.

FIGURE 26–28 Space-filling model of an alpha-cyclodextrin, with the six glucose monomers arranged in a doughnut-shaped ring. It is the hydrophobic nature of the internal cavity that makes the molecule into a molecular "guesthouse." Researchers have been especially interested in cyclodextrins for their control of biologically active substances and their ability to "mimic" enzyme-catalyzed reactions.

glass beads, stainless steel, and porous ceramics made of such metal oxides as ZrO_2, TiO_2, Al_2O_3, and NiO. Immobilized enzymes are being tested and used for the manufacture of foods and chemicals. In addition, they have promising uses in clinical analysis; an enzyme that reacts specifically with a certain compound can be immobilized and then used to test for that compound in blood, urine, or other fluids. Examples of this are tests for glucose and other sugars, and for cholesterol.

The conformation of an enzyme (or any protein) depends on the aqueous environment in the immediate vicinity of the protein. If the conditions are not optimal, the interactions that hold the tertiary structure together will be disrupted. However, if the enzyme is placed into a non-aqueous environment such as a liquid hydrocarbon, a very thin layer of water will remain attached to the enzyme and it will retain its conformation and its activity. A solvent such as ethanol is not satisfactory because it will dissolve away the water layer and denature (inactivate) the enzyme. In nonaqueous organic solvents, enzymes are finding uses as catalysts in situations previously unexplored. Furthermore, under these conditions enzymes tend to retain their activity but not their specificity. Thus the range of reactions that they will catalyze is enlarged.

Cyclodextrins are produced by a highly selective enzymatic synthesis. They consist of six, seven, or eight glucose monomers arranged in doughnut-shaped rings which are denoted alpha, beta or gamma cyclodextrin, respectively. The specific coupling of the glucose monomers gives the cyclodextrins a rigid, conical molecular structure with a hollow interior of specific volume (Figure 26–28). It is this internal cavity that is particularly interesting to researchers. Because of the hydrophobic nature of the internal cavity, the cyclodextrins can complex and contain a variety of "guest" molecules. The guests must, of course, satisfy the size criterion of fitting at least partially into the cyclodextrin's internal cavity, resulting in an "inclusion" complex. This unique molecular encapsulation leads to specific modifications of the properties of the guest, effecting control of biologically active substances and the means to "mimic" the behavior of nature's all-important enzyme-catalyzed reactions. These natural biopolymers offer a new dimension in basic and applied chemistry.

26.5 RECOMBINANT DNA TECHNOLOGY

Most of the genetic information of a cell or an organism is coded on the DNA in the cell nucleus. The molecules are massive; the DNA for the relatively simple bacterium *Escherichia coli* has a molecular weight of about 2.5×10^9, and a single molecule is 1.3 mm long! Actually changing the DNA in an organism to alter its specific properties is at once challenging and incredibly exciting. The great strides that have been made in understanding the structures and functions of the nucleic acids have made this possible. These changes are not the same as those produced in an organism by irradiating its cells or treating them with a chemical reagent. In these cases the changes will be random, and essentially all will produce organisms with inferior properties that do not survive. In recombinant DNA technology, the specific part of the DNA molecule of interest is identified and altered in the attempt to produce the desired result.

Modification of DNA was made possible by the discovery of two key families of enzymes, the **restriction endonucleases** (which will cleave the DNA at specific sites) and the **ligases** (which will rejoin the fragments). The restriction endonucleases are highly specific enzymes that can recog-

DEGRADATION OF PCBs

Polychlorinated biphenyls (PCBs) were widely used dielectric fluids in transformers. However, more than 25 years ago they were shown to be toxic environmental pollutants, especially to aquatic ecosystems, and generally slow to biodegrade. Their manufacture was discontinued in the United States in 1976, but they continue to be an environmental problem. The name PCBs is applied loosely to a family of aromatic compounds containing two connected benzene rings, with two or more chlorine atoms in place of hydrogen atoms on the rings:

biphenyl   polychlorinated biphenyl (PCB)

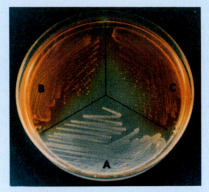

FIGURE 26–29 Mutants of a PCB-degrading bacterial strain accumulate a yellow intermediate of "biphenyl" metabolism in the areas marked B and C in the Petri dish—A is a control.

Research chemists have now isolated the genes encoding PCB bacterial degradation through the use of recombinant DNA technology. As we write, sequence analysis and biochemical studies are in progress to characterize the constituents in the multienzyme biodegradation pathway. A still more important outgrowth of this research is the design of recombinant strains with superior degradative activity (Figure 26–29).

---

nize sequences of four or more base pairs and cleave both strands of the DNA molecule near the recognized point. Simple organisms such as bacteria, yeasts, molds, and algae are the most easily modified by recombinant DNA technology. The DNA must first be withdrawn from the nucleus of the cell. The DNA molecule can then be cleaved at very specific points in the chain by the restriction endonucleases. The fragments produced by the cleavage can be separated by a technique called **electrophoresis**. In this technique, the fragments in an aqueous medium are subjected to an electric field. This field will draw charged fragments to the electrodes, and the rate of transport will depend on the fragment size. The desired fragments are carried by plasmid vectors, small circular pieces of DNA. Hybrid DNA molecules can then be produced by recombining the DNA fragments with a ligase.

Once recombined, the DNA can be reinserted into the cell of an organism. Here the ordinary division of the cell will cause the altered DNA molecule to be replicated many times. As we mentioned earlier, the information stored in the DNA molecule of an organism contains the information necessary for forming all of the proteins of the organism in the exactly correct sequences. The portion of the DNA molecule with the information for one specific protein is called a **gene**. The basic challenge of recombinant DNA technology is locating the desired gene and removing it or replacing it with a gene that will produce a protein modified in the desired way.

Recombinant DNA inserted into the nuclei of the cells of microorganisms and (recently) higher plants has resulted in many realized and potential products. Human insulin and hepatitis B vaccine produced by modified microorganisms are now on the market. Drugs produced by these methods for treating cancers and heart disease are now in the testing stages. Particularly promising for heart disease therapy are enzymes that dissolve blood clots.

Progress in modifying agricultural crops has been particularly impressive. Plants have been produced that have shown resistance to pesticides, drought, salt, and freezing. If plants that previously were unable to fix nitrogen are modified to be able to do so, the savings in artificial plant nutrients will be enormous. Current goals also include modifying plants so that their fruit or seeds, the commercially valuable parts, have improved nutritional or processing properties. Recent advances in determining the base sequence in DNA have led to new means of criminological identification. It is now possible to compare the DNA sequences in two tissue samples with essentially complete confidence that no two samples except from identical twins would be the same. Improvements in life brought about by recombinant DNA technology are sure to touch all of us.

SUMMARY

Biochemistry is distinguished from other forms of chemistry by (1) the very restricted conditions of reactions, (2) their unusual efficiency and specificity, and (3) the generally large size and great complexity of biomolecules. Stereochemistry, **optical isomerism** and activity, and **chiral behavior**—the property of "handedness"—also play essential roles. Based on the unique geometry of tetrahedral carbon, when four different substituent atoms or groups of atoms are appended, **asymmetry** is introduced and two-mirror image, isomeric forms can exist. In many instances, only one proves to be biologically important: for example, the L-amino acids in proteins.

Proteins are long-chained macromolecules that have been snapped together from a pool of twenty amino acid building blocks known as the **essential amino acids**. The **amide** linkage —NHCO— is the unique protein-connector, the sequence of amino acids along the chain is known as the **primary protein structure**, and the general shape the chain assumes—a spiral, coiled **helix**, for example, or a β-pleated sheet—is referred to as **protein secondary structure**. Other interactions and geometric considerations form the basis for **tertiary** and **quaternary protein structure**.

The **nucleic acids** include DNA and RNA. These compounds carry genetic information and vary greatly among plant and animal species. The building units of nucleic acids are called **nucleotides**, each consisting of the residues of a sugar, a phosphate group, and an organic base. In **DNA**, the molecules are in the form of a double helix with a specific base pairing pattern between the two strands of the helix. Protein synthesis depends on a unique mechanism for **transcription** and **replication** which produces the specific base-pairing.

Carbohydrates include monosaccharide sugars, disaccharide and trisaccharide sugars, still larger oligomeric sugars constructed from several monosaccharide units, and monosaccharide polymers. They are principal constituents of plants and are crucial components in the human food chain. **Lipids** include a number of compounds that are soluble in organic solvents: particularly, esters of glycerol, a trihydric alcohol, and fats and oils. As a class of compound, they are ideally suited for long-term energy storage, in contrast to carbohydrates, which are easily broken down and provide more readily accessible energy reserves through metabolism. **Soaps** are salts of long-chained carboxylic acids obtained from lipid esters; **detergents** are sulfuric acid analogs; and both categories serve as cleansing or surface-active agents—**surfactants**.

Coenzymes are helpers in enzyme-catalyzed reactions. Both reduced coenzymes and nucleoside triphosphates provide ways in which energy can be stored and rapidly recovered on demand.

QUESTIONS AND PROBLEMS

QUESTIONS

1. What aspects of biochemistry are very different from chemistry as practiced in the laboratory?

2. Alanine has the structure $CH_3\overset{\overset{\displaystyle NH_2}{|}}{C}HCOOH$. Identify the chiral center in this molecule and sketch its optical isomers.

3. Why are protein helices always observed to coil in a right-handed sense (that is, they are all α-helices)?

4. What are the alkaline hydrolysis products of a typical dipeptide? What are the acid hydrolysis products?

5. What are the principal differences between enzymes and the catalysts normally used in laboratory and industry?

6. Explain what is meant by the primary, secondary, tertiary, and quaternary structures of proteins.

7. The base pairings in DNA are always purine/pyrimidine, never purine/purine or pyrimidine/pyrimidine. Briefly explain.

8. Why do soaps not form clear solutions in water? Distinguish between a soap and a detergent. Why is "branching" such an important structural feature of surface active agents?

9. Explain how some coenzymes can be used for storing energy.

10. Why does hydrolysis of two but not three of the phosphate groups from ATP result in the evolution of large amounts of energy?

11. What is the advantage of immobilized enzymes in comparison to conventional use of enzymes in production of food and chemicals?

12. Why will an enzyme be denatured in a solvent like enthanol or methanol but not in a solvent like hexane (C_6H_{14})?

13. What were the discoveries that led to the development of recombinant DNA technology?

PROBLEMS

Stereochemistry   [1–4]

1. Draw three optically active compounds, each having a carbon atom located at a chiral center.

2. Draw three different chiral compounds, each having the formula $C_5H_{10}BrCl$.

3. When lactic acid ($CH_3\overset{\overset{\displaystyle OH}{|}}{C}HCOOH$) isolated from muscle is mixed with an equal quantity isolated from yeast, the resulting solution is optically inactive, yet each is optically active by itself. Sketch structures to explain this behavior.

4. Sketch structures to help you explain the optical activity observed in glyceraldehyde, $CH_2\overset{\overset{\displaystyle OH}{|}}{C}H\overset{\overset{\displaystyle OH}{|}}{C}HCHO$.

Peptides   [5–8]

5. How many tetrapeptides can be constructed from alanine, serine, proline, and leucine? You are not required to use all four in any one compound.

6. How many tetrapeptides can be constructed if you are restricted to the twenty amino acids in Table 26–1?

7. Draw the structure of a β-amino acid that is optically active. Draw one that is optically inactive.

8. Draw an α-amino acid having the formula $C_5H_{11}NO_2$. Draw one having the formula $C_9H_{11}NO_2$.

Carbohydrates   [9–14]

9. Consider the straight chain structure of D-glucose. Draw the structure of L-glucose, the mirror image of D-glucose.

10. Other sugars can be drawn with the same chain as glucose. For example, D-mannose is

Note that D-mannose is not related to D-glucose by a reflection. Therefore, mannose is a different sugar with different physical properties. Draw the structure of L-mannose. Draw the structures of the other sugars and their optical isomers that can be constructed on the glucose framework.

11. Draw the pyranose ring structure for α-D-glucose.

12. Draw the furanose ring structure for α-D-fructose.

13. A sugar like glucose, with an aldehyde functional group, can be oxidized to a carboxylic acid. Draw the product of such an oxidation of glucose. Glucose is called a reducing sugar because it will reduce an oxidizing agent. Write the balanced equation for the oxidation of glucose by Cu^{2+} in aqueous solution, where Cu^{2+} is reduced to Cu (s).

14. The aldehyde group in glucose can be reduced to an alcohol group. Draw the resulting compound.

Lipids   [15–18]

15. Write the equation for the formation of sodium palmitate, a soap.

16. A certain wax can be hydrolyzed to ethyl alcohol and stearic acid. Draw the structure of the wax.

17. A particular soap precipitates calcium myristate in hard water. Draw the structures of the salt and the free acid.

18. A triglyceride produces a trihydric alcohol—3 OH groups—and an unsaturated fatty acid on hydrolysis. Write a possible equation for the reaction.

MULTIPLE PRINCIPLES   [19–23]

19. What is the maximum number of gallons of fuel grade (100%) ethanol that could be made from the fermentation of 1.00 ton (2000 lb) of wheat, assuming that the grain is 80.0% starch? The density of ethanol is 0.789 g/cm^3.

20. Each turn of the DNA helix occupies a distance of 34 Å along the axis, and there are 10 subunits per turn. What would be the total length of the helix along this axis for a DNA molecule with a molecular weight of 10,000,000?

21. Consider the values of K_a for typical carboxylic acids such as acetic acid, and the values of K_b for typical organic bases such as methylamine. Predict the equilibrium form of an amino acid such as alanine in water solution. Write the reaction that would occur if an acid is added to alanine in aqueous solution. Write the reaction that would occur if a base is added to alanine in aqueous solution.

22. A cord of wood has a mass of about 1800 kg. If the wood is considered to be purely carbohydrate, what volume of oxygen gas at 30°C and 750 torr is liberated in forming a cord of wood?

23. The Calorie of nutrition is equal to 1000 ordinary calories. Calculate the energy in nutritional Calories that can be obtained by the complete metabolism of 1.00 g of glucose.

APPLIED PRINCIPLES [24–25]

24. The fermentation of grain is an important source of alcohol for industrial processes. The starch in the grain is converted to maltose, a sugar, by the enzymes α-amylase and amyloglucosidase. The maltose is then fermented with yeast to form ethyl alcohol and CO_2. Current technology obtains 2.6 gallons of 99.5% (by volume) ethyl alcohol from 1 bushel (56 lb) of grain. The average starch content of wheat is 55% by mass. Calculate the overall percent yield for the process.

25. How many kilograms of by-product glycerol would be yielded by the production of 2000 kg of soap from fats, in which the average composition of the fatty acids is $C_{17.3}H_{34.3}O_2$?

Appendices

APPENDIX A

Answers to Odd-Numbered Problems

CHAPTER 1

1.1. 1.31×10^{-1} nm, 1.31×10^{-10} m

1.3. 5.08×10^{-5} m

1.5. 1.10×10^{27} electrons

1.7. 7.9×10^2 g at 20°C

1.9. (a) 56.3 g (water + flask)
(b) 51.0 g (alcohol + flask)
(c) 371 g (mercury + flask)

1.11. 19 g/mL

1.13. 8.4×10^{25} atoms

1.15. (a) 28.02 μ (b) 32.00 μ (c) 38.00 μ

1.17. (a) 18.02 μ (b) 34.02 μ
(c) 70.06 μ, FW(SF_4) = 108.06 μ, FW(SF_6) = 146.1 μ

1.19. CF_2Cl_2

1.21. 228.3 μ

1.23. (a) 30 kcal   (b) 6.4 kcal   (c) 990 cal
(d) 13 kcal

1.25. $T_f = 313$ K

1.27. $T_f = 23$°C

1.29. (a) 161 km/hr   (b) 91.4 m   (c) 481 yd
(d) 10,000,000 m   (e) 2.99×10^{10} cm/s

1.31. (a) 100 m is longer than 100 yd.
(b) It will take less time for the same swimmer to swim 400.0 yd.

1.33. 5.87×10^{12} mi

1.35. \$1.13/kg

1.37. (a) 2.0500×10^3 (b) 2.050×10^{-5}
(c) 1.0×10^0 (d) 7.27×10^{-1}
(e) 6.35×10^1

1.39. 0.000000000300   625   0.0625
10,000   0.0001

1.41. (a) four   (b) seven   (c) three

1.43. (a) 0.10 mm   (b) 11.19% H, 88.809% O

1.45. 6×10^{-4}

1.47. 19.3 g/mL

1.49. (a) 165.8 μ (b) 1.625×10^3 kg/m³
(c) 70 lb

1.51. 1.3×10^{-10} m, 8.1×10^{-14} mi, $V = 9.2 \times 10^{-24}$ cm³

1.53. $T_f = 12$°C

1.55. \$15.53

1.57. 4.99 m

1.59. 14.6 flips

1.61. 54.9%

1.63. (a) 321 g NaCl/kg H_2O (b) 0.289 kg/L

CHAPTER 2

2.1. 3.6×10^{21} ergs

2.3. 3 hours

2.5. ≈2:1 ratio, which is a ratio of small whole numbers

2.7. (a) 35.5   (b) 23.0   (c) 16.0   (d) 12.0
(e) 32.0   (f) 32

2.9. 71.0 g/mol

2.11. 207.2 μ

2.13. 55.85 μ

2.15. (a) ≈190 g/mol   (b) 207.210 g/mol
(c) $PbCl_2$

2.17. Pb_2O_3

2.19. 14.01 g/mol

2.21. 238 g/mol, XF_6, XO_2

2.23. (a) one U atom   (b) six O atoms
(c) two N atoms   (d) 362 g/mol

2.25. (a) (1 g KOH)(1 mol KOH/56.1 g)
          = 1.78×10^{-2} mol KOH
(b) 1.743×10^{-1} kg K_2SO_4
(c) 1×10^3 mg $KClO_3$

2.27. 2.02×10^{21} atoms

2.29. $r = 4.68$ cm

2.31. 35.17, 35.17

2.33. (a) 1.00 mol compound
(b) 1.00 mol Na, 2.00 mol H, 1.00 mol P, 4.00 mol O
(c) 23.0 g Na, 2.02 g H, 31.0 g P, 64.0 g O
(d) 6.02×10^{23} atoms Na

2.35. 6.02×10^{24} molecules

2.37. 17.4 g/mol

2.39. (a) $2Al + 3Cl_2 \longrightarrow Al_2Cl_6$
(b) $N_2 + 3H_2 \longrightarrow 2NH_3$
(c) $C_3H_8 + 5O_2 \longrightarrow 3CO_2 + 4H_2O$
(d) $Fe_2O_3 + 3CO \longrightarrow 2Fe + 3CO_2$

(e) $Mg_3N_2 + 6H_2O \longrightarrow 2NH_3 + 3Mg(OH)_2$

(f) $Ca_3(PO_4)_2 + 4H_2SO_4 \longrightarrow$
$$Ca(H_2PO_4)_2 + 2Ca(HSO_4)_2$$

(g) $3K_2CO_3 + Al_2Cl_6 \longrightarrow Al_2(CO_3)_3 + 6KCl$

(h) $8KClO_3 + C_{12}H_{22}O_{11} \longrightarrow$
$$8KCl + 12CO_2 + 11H_2O$$

(i) $KOH + H_3PO_4 \longrightarrow KH_2PO_4 + H_2O$

2.41. (a) 28.98% Li   (b) 8.136% C   (c) 50.48% O
(d) 36.08% H_2O (e) 81.63% SO_3

2.43. C_3O_2

2.45. $CaCO_3$

2.47. (a) CrO_3 (b) Cr_2O_3 (c) CrO

2.49. (a) $FeC_{10}H_{10}$ (b) $FeC_{10}H_{10}$

2.51. C_3H_8

2.53. CH_2O

2.55. (a) $4NH_3 + 5O_2 \longrightarrow 4NO + 6H_2O$
$2NO + O_2 \longrightarrow 2NO_2$
$3NO_2 + H_2O \longrightarrow 2HNO_3 + NO$
$HNO_3 + NH_3 \longrightarrow NH_4NO_3$
(b) Two moles N   (c) 425.5 g NH_3

2.57. (a) $2C_8H_{18} + 25O_2 \longrightarrow 16CO_2 + 18H_2O$
(b) 3.08 g CO_2
(c) 115 g C_8H_{18}, 28 g CO_2, and 13 g H_2O

2.59. (a) 225.7 g C required
(b) 300.6 g O_2 required

2.61. LiOH is more effective on a weight basis than NaOH and Li_2O is more effective on a weight basis than LiOH.

2.63. (a) 0.75 g of metal   (b) 5.75 g MX_2

2.65. 24.0 g MnO_2 required

2.67. (a) 1 mol HCl can produce 0.5 mol H_2 and 0.5 mol Cl_2.
1 mol H_2O can produce 1 mol H_2 and 0.5 mol O_2.
1 mol NH_3 can produce 1.5 mol H_2 and 0.5 mol N_2.
(b) 2 g H_2 4 g H_2 6 g H_2
(c) 4.9 mol H_2 (9.9 g) and 0.28 mol HCl (10.2 g)
4.4 mol H_2 (8.9 g) and 0.62 mol H_2O (11.2 g)
3.9 mol H_2 (7.9 g) and 0.72 mol NH_3 (12.3 g)

2.69. (a) $3LiAlH_4 + 4BF_3 \longrightarrow 3LiF + 3AlF_3 + 2B_2H_6$
(b) BF_3 is the limiting reagent, 45.9 g B_2H_6

2.71. (a) 20 moles HCl   (b) 5 moles Cl_2
(c) 5 moles Cl_2

2.73. 57.45% Ag in the alloy

2.75. 50.0% K_2SO_4 and 50.0% Na_2SO_4

2.77. mixture of 0.43 g NaCl and 0.57 g KCl

2.79. 1.204×10^{24} molecules

2.81. 470. g SO_2

2.83. (a) $(2x + y)$ C $+ y$ $CO_2 + x$ $O_2 \longrightarrow$
$$(2x + 2y)$$ CO
(b) H_2O is the limiting reactant
(c) 64.5%   (d) 2.16 g H_2

2.85. 69 g HCl are produced.

2.87. 3.0 mol N_2, 5.0 mol H_2.

2.89. $Ni_{0.871}O_{1.00}$.

2.91. empirical formula is HSO_4
molecular formula is $(HSO_4)_2$ or $H_2S_2O_8$

2.93. 1.1×10^4 lb CaO
1.5×10^3 lb $Mg(OH)_2$

2.95. 7×10^{-6} mol benzoic acid

CHAPTER 3

3.1. 58 N

3.3. (a) 2.2×10^{-12} cm $= r_0$ (b) 1.68

3.5. 1.88×10^3 MeV

3.7. 2.3 MeV
1.2 MeV/nucleon
92.2 MeV
7.68 MeV/nucleon
492.4 MeV
8.793 MeV/nucleon

3.9. 0.1325 amu
7.714 MeV/nucleon

3.11. 17.5 MeV of energy are released

3.13. 5 protons, 5 electrons, and 5 neutrons

3.15. (a) ^{16}O: ^{17}O is an isotope. ^{15}N is an isotone.
^{208}Pb: ^{207}Pb is an isotope. ^{207}Tl is an isotone.
^{120}Sn: ^{118}Sn is an isotope. ^{121}Sb is an isotone.
(b) ^{14}N: ^{15}N is an isotope. ^{14}C is an isobar.
^{63}Cu: ^{65}Cu is an isotope. ^{63}Co is an isobar.
^{238}U: ^{235}U is an isotope. ^{238}Np is an isobar.

3.17. $^{22}_{11}Na \longrightarrow {}^{0}_{1}\beta + {}^{22}_{10}Ne$ or $^{22}_{11}Na + {}_{-1}^{0}\beta \longrightarrow {}^{22}_{10}Ne$
$^{24}_{11}Na \longrightarrow {}_{-1}^{0}\beta + {}^{24}_{12}Mg$

3.19. $^{226}_{88}Ra \longrightarrow {}^{4}_{2}He + {}^{222}_{86}Rn$

3.21. $^{14}_{6}C \longrightarrow {}_{-1}^{0}\beta + {}^{14}_{7}N$

3.23. (a) $^{10}_{5}B + {}^{4}_{2}He \longrightarrow {}^{13}_{6}C + {}^{1}_{1}H$
(b) $^{9}_{4}Be + {}^{1}_{1}H \longrightarrow {}^{8}_{4}Be + {}^{2}_{1}H$ or $^{9}_{4}Be + {}^{1}_{1}H \longrightarrow$
$${}^{8}_{4}Be + {}^{2}_{1}D$$
(c) $^{27}_{13}Al + {}^{1}_{0}n \longrightarrow \gamma + {}^{28}_{13}Al$

3.25. $^{235}_{92}U \longrightarrow {}^{90}_{38}Sr + {}^{143}_{54}Xe + 2\ {}^{1}_{0}n$.
$^{235}_{92}U + {}^{1}_{0}n \longrightarrow {}^{90}_{38}Sr + {}^{143}_{54}Xe + 3\ {}^{1}_{0}n$.

3.27. $^{14}_{7}N + {}^{4}_{2}He \longrightarrow {}^{1}_{1}p + {}^{17}_{8}O$

3.29. $^{231}_{90}Th \longrightarrow {}^{231}_{91}Pa \longrightarrow {}^{227}_{89}Ac \longrightarrow {}^{227}_{90}Th \longrightarrow$
$${}^{223}_{88}Ra \longrightarrow {}^{219}_{86}Rn \longrightarrow {}^{215}_{84}Po \longrightarrow {}^{211}_{82}Pb \longrightarrow$$
$${}^{211}_{83}Bi \longrightarrow {}^{211}_{84}Po \longrightarrow {}^{207}_{82}Pb$$

3.31. 1.6×10^4 to 2.4×10^4

3.33. $^{232}_{90}Th$ $^{233}_{90}Th$ $^{234}_{90}Th$
$^{233}_{91}Pa$ $^{234}_{91}Pa$
$^{232}_{92}U$ $^{233}_{92}U$ $^{234}_{92}U$ $^{235}_{92}U$ $^{236}_{92}U$ $^{237}_{92}U$
$^{237}_{93}Np$

3.35. 3×10^{-13} cm 4×10^{14} g/cm^3
6×10^{-13} cm 4×10^{14} g/cm^3

3.37. mass ^{120}Sn $= 119.90\ \mu$

3.39. Energy is released in this case (approximately 0.6 MeV/nucleon).

3.41. (a) $^{90}_{39}Y \longrightarrow {}_{-1}^{0}\beta + {}^{90}_{40}Zr$
$^{66}_{29}Cu \longrightarrow {}_{-1}^{0}\beta + {}^{66}_{30}Zn$
(b) $^{236}_{94}Pu \longrightarrow {}^{4}_{2}He + {}^{232}_{92}U$
$^{226}_{91}Pa \longrightarrow {}^{4}_{2}He + {}^{222}_{89}Ac$

3.43. $^{231}_{90}Th \longrightarrow {}^{231}_{91}Pa \longrightarrow {}^{227}_{89}Ac \longrightarrow {}^{223}_{87}Fr \longrightarrow$
$${}^{219}_{85}At \longrightarrow {}^{215}_{83}Bi \longrightarrow {}^{215}_{84}Po \longrightarrow {}^{211}_{82}Pb$$

3.45. (a) $20.18 \ \mu$

(b) $0.73\% \ ^{235}U$ and $99.27\% \ ^{238}U$.

3.47. $^{24}_{12}Mg + ^4_2He \longrightarrow ^{28}_{14}Si$

$^{28}_{14}Si + ^1_0n \longrightarrow ^{22}_{11}Na + ^7_3Li$

3.49. $^{238}_{92}U + ^1_0n \longrightarrow ^{239}_{92}U \longrightarrow ^{239}_{94}Pu + 2 \ ^0_{-1}\beta$

3.51. $2.5 \ kg \ ^{235}U$ $997.5 \ kg \ ^{238}U$

CHAPTER 4

4.1. (a) Ga: ·Te: Ba: ·Bi: Rb·

(b) Be: Mg: ·Si: :Kr:

4.3. Sr^{2+}, At^-, Fr^+, Se^{2-}, Sn^{4+} or Sn^{2+} or Sn^{4-}, N^{3-}, Ne^0, H^+ or H^-

4.5. (a) calcium bromide, $CaBr_2$

(b) potassium selenide, K_2Se

(c) aluminum chloride, $AlCl_3$

(d) sodium phosphide, Na_3P

4.7. (a) strontium oxide   (b) aluminum fluoride

(c) beryllium carbide   (d) rubidium hydroxide

(e) lithium nitride   (f) sodium hydride

(g) tin(II) chloride   (h) thallium(I) iodide

(i) potassium hydrogen selenide

4.9. (a) hydrogen fluoride, hydrofluoric acid

(b) hydrogen bromide, hydrobromic acid

(c) hydrogen sulfide, hydrosulfuric acid

(d) hydrogen selenide, hydroselenic acid

4.11. (a) KI   (b) $Ca(OH)_2$ (c) SrH_2 (d) $BiCl_3$

(e) BaHS

4.13. (a) $2Na + Cl_2 \longrightarrow 2NaCl$

(b) $2Ba + O_2 \longrightarrow 2BaO$

(c) $2Al + 3Br_2 \longrightarrow 2AlBr_3$

(d) $H_2 + Cl_2 \longrightarrow 2HCl$

4.15. (a) $2Al + 6HCl \longrightarrow 2AlCl_3 + 3H_2$

(b) $Mg + 2HBr \longrightarrow MgBr_2 + H_2$

(c) $2Li + 2HI \longrightarrow 2LiI + H_2$

4.17. (a) $H_2SO_4 + MgS \longrightarrow H_2S + MgSO_4$

(b) $H_2SO_4 + NaF \longrightarrow HF + NaHSO_4$

(c) $H_2SO_4 + NaBr \longrightarrow HBr + NaHSO_4$

4.19. (a) $KOH + HCl \longrightarrow KCl + H_2O$

(b) $Ca(OH)_2 + 2HF \longrightarrow CaF_2 + 2H_2O$

4.21. (a) H:F: (b) H:S:H   (c) :Cl:F:

(d) :F:N:F: (e) :C:::O:
      :F:

4.23. (a) arsenic trichloride

(b) iodine heptafluoride

(c) tetraphosphorus hexoxide

(d) dinitrogen tetroxide

(e) ammonia

(f) phosphine

(g) tetranitrogen tetrasulfide

4.25. (a) CS_2 (b) PCl_5 (c) Cl_2O_7

(d) CH_4 (e) Te_2O_{10} (f) SiH_4

4.27. (a) $Cl_2 + 3F_2 \longrightarrow 2ClF_3$

(b) $P_4 + 3O_2 \longrightarrow P_4O_6$

(c) $S + 3F_2 \longrightarrow SF_6$

4.29. (a) :O:Cl:O: (b) H:N:H
          :O: H

(c) C (with O atoms)   (d) :O:P:O:

(e) :O:S:O:

4.31. (a) sodium chlorate   (b) ammonium chloride

(c) sulfurous acid   (d) lead(II) nitrate

(e) ammonium nitrate   (f) nitric acid

4.33. (a) $Al(NO_3)_3$ (b) HNO_2 (c) NH_4ClO_3

(d) $Mg(NO_3)_2$ (e) $HClO_4$

4.35. (a) $Ca(OH)_2 + H_2CO_3 \longrightarrow CaCO_3 + 2H_2O$

(b) $2NH_4OH + H_2SO_4 \longrightarrow (NH_4)_2SO_4 + 2H_2O$

(c) $NaOH + HClO_2 \longrightarrow NaClO_2 + H_2O$

(d) $3KOH + H_3PO_4 \longrightarrow K_3PO_4 + 3H_2O$

4.37. (a) $2HClO_2 + 2Li \longrightarrow 2LiClO_2 + H_2$

(b) $H_2SO_4 + Sn \longrightarrow SnSO_4 + H_2$

(c) $2HNO_3 + Ca \longrightarrow Ba(NO_3)_2 + H_2$

4.39. (a) $S + O_2 + H_2O \longrightarrow H_2SO_3$

(b) $2Na + H_2 + 2C + 3O_2 \longrightarrow 2NaHCO_3$

(c) $N_2 + 4H_2 + Cl_2 \longrightarrow 2NH_4Cl$

4.41. $4.60 \ g \ H_2S$ required to produce CaS

$9.20 \ g \ H_2S$ required to produce $Ca(HS)_2$

4.43. 300 kg of sulfur are required

4.45. $6.32 \times 10^5 \ g \ H_3PO_4$

4.47. Every 100. g 40% oleum can produce 109 g H_2SO_4.

4.49. potassium selenide; magnesium iodide; aluminum bromide; tin(II) oxide; bismuth(III) chloride; nickel(III) bromide; copper(I) chloride; sodium hydride

4.51. (a) $6.79 \ g \ Na_2CO_3$

(b) $2.23 \ g \ Na_2CO_3$

(c) $108 \ g \ Na_2S_2O_3$

(d) $18.5 \ g \ Na_2S_2O_3$ produced, $3.9 \ g \ Na_2S$ remain, $5.86 \ g \ Na_2CO_3$ remain

(e) $K_2CO_3 + 2K_2S + 4SO_2 \longrightarrow 3K_2S_2O_3 + CO_2$

4.53. 220 tons Cl_2 produced, 6.4 tons H_2 produced

4.55. 52.2%

4.57. (a) 0.623 g HCl required

(b) 0.714 g HF required

(c) 0.680 g HBr required

(d) 0.870 g HI required

4.59. $0.0841 \ g \ O_2$, $0.126 \ g \ O_2$, $0.168 \ g \ O_2$

4.61. $102–108 \ kg \ H_2SO_4$

CHAPTER 5

5.1. $2.48 \ g/cm^3$

5.3. 715.3 torr

5.5. $357°C = 675°F$

5.7. 1.2 atm, 20 kPa

5.9. 6.33 g O_2

5.11. 439 K

5.13. 1. 10 atm   2. 1013 kPa   3. 1 m^3
   4. 25 atm

5.15. 3×10^{13} molecules

5.17. 212 L CO_2

5.19. 5.02 g/L

5.21. 131 g/mol

5.23. CH_2Cl, $C_2H_4Cl_2$.

5.25. 2140 K

5.27. $P(N_2) = 603$ torr $P(Ar) = 7.21$ torr

5.29. 29 g/mol

5.31. (a) 0.21 atm   (b) 0.79 atm   (c) 0.79 atm

5.33. $CH_4 = 8.51\%$ $C_2H_6 = 4.0\%$ $CO_2 = 87.5\%$

5.35. C_3H_8

5.37. 5.53 atm, 2.81 atm

5.39. 2.944 Å

5.41. $M = 44.02$ g/mol

5.43. $M(CO) = 27.986$ g/mol
   $AW(C) = 11.987$ g/mol

5.45. 3.72 kJ

5.47. 1.37×10^3 m/s

5.49. (a) same average kinetic energy
   (b) There must be twice as many B molecules as A molecules.
   (c) $v_A/v_B = 0.9995$

5.51. 1.41

5.53. 5.68 cm^3/s

5.55. compression ratio = 7
   volumetric efficiency = 0.54

5.57. $P^0 = P(17000 \text{ m} + H)/(17000 \text{ m} - H)$

5.59. 5.7 cubic feet

5.61. 1.4 cm

5.63. 4.89 atm, 4.02 atm

5.65. 4.

5.67. (a) 97 g/mol   (b) 4.0 g/L

5.69. (a) four times   (b) the same
   (c) one fourth, 0.25   (d) the same rate

5.71. $-254.4°C$

5.73. 1. 89.6 L   2. 471°C   3. 3.76 L
   4. 73.7 kPa

5.75. 0.261 L helium

5.77. $M_Y = 28.1$ g/mol, $M_X = 43.8$ g/mol

5.79. (a) 1.70 torr   (b) 1.586

5.81. 1.18

CHAPTER 6

6.1. 26.3 MeV, 2.530×10^9 kJ

6.3. (1) $^{28}_{14}Si \longrightarrow {}^4_2He + {}^{24}_{12}Mg$
   (2) $^{28}_{14}Si + {}^4_2He + {}^{24}_{12}Mg \longrightarrow {}^{56}_{28}Ni$
   (3) $2\,^{28}_{14}Si \longrightarrow {}^{56}_{28}Ni$

6.5. $NO + O_3 \longrightarrow NO_2 + O_2$
   $NO_2 + O \longrightarrow NO + O_2$, etc.

6.7. 0.0249 g Fe

6.9. 7360 L

6.11. 93 kg Chilean Saltpeter

6.13. 3.34 L

6.15. 400. kg Mn

6.17. (a) $Zn(s) + 2H^+(aq) \longrightarrow Zn^{2+}(aq) + H_2(g)$ or
   $3Fe(s) + 4H_2O(g) \longrightarrow Fe_3O_4(s) + 4H_2(g)$
   (b) $CH_4(g) + H_2O(g) \longrightarrow CO(g) + 3H_2(g)$
   (c) $N_2(g) + O_2(g) \longrightarrow 2NO(g)$
   (d) $CaCO_3(s) + 2HCl(aq) \longrightarrow$
   $CaCl_2(aq) + H_2O(l) + CO_2(g)$
   (e) $S(s) + O_2(g) \longrightarrow SO_2(g)$ followed by
   $2SO_2(g) + O_2(g) \longrightarrow 2SO_3(l)$

6.19. (a) $Zn(s) + 2H^+(aq) \longrightarrow Zn^{2+}(aq) + H_2(g)$ or
   $3Fe(s) + 4H_2O(g) \longrightarrow Fe_3O_4(s) + 4H_2(g)$
   (b) Synthesis gas: $CH_4(g) + H_2O(g) \longrightarrow$
   $CO(g) + 3H_2(g)$
   (c) $N_2(g) + O_2(g) \longrightarrow 2NO(g)$
   (d) $CaCO_3(s) + 2HCl(aq) \longrightarrow$
   $CaCl_2(aq) + H_2O(l) + CO_2(g)$
   (e) $S(s) + O_2(g) \longrightarrow SO_2(g)$ followed by
   $2SO_2(g) + O_2(g) \longrightarrow 2SO_3(g)$

6.21. (a) Haber: $N_2(g) + 3H_2(g) \longrightarrow 2NH_3(g)$
   Ostwald: $12NH_3(g) + 21O_2(g) + 4H_2O(l) \longrightarrow$
   $8HNO_3(l) + 4NO(g) + 14H_2O(g)$
   (b) $2NaN_3(s) \longrightarrow 2Na(s) + 3N_2(g)$
   (c) $CaC_2(s) + N_2(g) \longrightarrow CaCN_2(s) + C(s)$
   (d) $2NaCN(s) + H_2SO_4(l) \longrightarrow$
   $Na_2SO_4(s) + 2HCN(g)$

6.23. $2CO(g) + O_2(g) \longrightarrow 2CO_2(g)$
   $2NO_x(g) \longrightarrow N_2(g) + xO_2(g)$
   $S \text{ (in gasoline)} + O_2(g) \longrightarrow SO_2(g)$
   $2SO_2(g) + O_2(g) \longrightarrow SO_3(l)$

6.25. $N_2(g) + O_2(g) \longrightarrow 2NO(g)$
   $N_2(g) + 2O_2(g) \longrightarrow 2NO_2(g)$
   $N_2(g) + 2O_2(g) \longrightarrow N_2O_4(g)$

6.27. $6Li + N_2 \longrightarrow 2Li_3N$
   $2Li_3N + 3H_2O \longrightarrow 3Li_2O + 2NH_3$
   $3Mg + N_2 \longrightarrow Mg_3N_2$
   $Mg_3N_2 + 3H_2O \longrightarrow 3MgO + 2NH_3$

6.29. (a) $4FeS(s) + 7O_2(g) \longrightarrow 2Fe_2O_3(s) + 4SO_2(g)$
   (b) $2ZnS(s) + 3O_2(g) \longrightarrow 2ZnO(s) + 2SO_2(g)$
   (c) $NaNO_3(s) + H_2SO_4(l) + \text{heat} \longrightarrow$
   $HNO_3(l) + NaHSO_4(s)$

6.31. (a) $NaCN(s) + HCl(aq) \longrightarrow NaCl(aq) + HCN(g)$
   (b) $NH_3(aq) + HCl(aq) \longrightarrow NH_4Cl(aq)$
   (c) $CaCO_3(s) + 2HCl(aq) \longrightarrow$
   $CaCl_2(aq) + H_2O(l) + CO_2(g)$
   (d) $Zn(s) + 2HCl(aq) \longrightarrow ZnCl_2(aq) + H_2(g)$
   (e) $Na_2SO_3(s) + 2HCl(aq) \longrightarrow$
   $2NaCl(aq) + H_2O(l) + SO_2(g)$

6.33. $^{12}_6C + {}^1_1H \longrightarrow {}^{13}_7N + \gamma$
   $^{13}_7N \longrightarrow {}^{13}_6C + {}^0_1\beta + \gamma$
   $^{13}_6C + {}^1_1H \longrightarrow {}^{14}_7N + \gamma$
   $^{14}_7N + {}^1_1H \longrightarrow {}^{15}_8O + \gamma$
   $^{15}_8O \longrightarrow {}^{15}_7N + {}^0_1\beta + \gamma$
   $^{15}_7N + {}^1_1H \longrightarrow {}^{12}_6C + {}^4_2He$

6.35. (a) $SO_2(g) + MgO(s) \longrightarrow MgSO_3(s)$
   (b) $CaCO_3(s) + 2H^+(aq) \longrightarrow$
   $Ca^{2+}(aq) + H_2O(l) + CO_2(g)$

6.37. $NO_2 \longrightarrow NO + O$; $O + O_2 \longrightarrow O_3$.

6.39. (a)

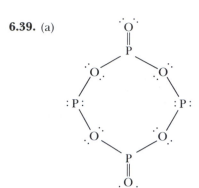

(b)

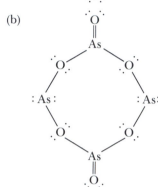

(c)

(d)

6.41. Element 118 should be the next noble gas.
6.43. 1.53 atm
6.45. 1.76 kg Li, 9 L
6.47. 2,500 L Cl_2, 15,000 L HCl, and 2,500 L N_2
6.49. $^{56}_{26}Fe \longrightarrow \, ^{46}_{22}Ti + 2\, ^{4}_{2}He + 2\, ^{1}_{0}n$
$^{56}_{26}Fe \longrightarrow \, ^{52}_{28}Ni + \, ^{4}_{2}He + 4\, ^{0}_{-1}\beta$
6.51. 0.3 g to 3.5 g CaO
6.53. 1×10^7 m^3 air required

CHAPTER 7

7.1. (a) K = °C + 273.15
(b) °C = (5/9)°F − 160/9
(c) °R = (9/5)K
7.3. −297.346°F = −182.97°C = 162.4°R = 90.18 K
32.0°F = 0.0°C = 491.7°R = 273.2 K
212.0°F = 100.0°C = 671.7°R = 373.2 K
832.28°F = 444.60°C = 1292.0°R = 717.75 K
1761.44°F = 960.80°C = 2221.1°R = 1233.95 K
1945.40°F = 1063.00°C = 2405.1°R = 1336.15 K
−109.3°F = −78.5°C = 350.4°R = 194.6 K
−37.97°F = −38.87°C = 421.7°R = 234.28 K
32.018°F = 0.0100°C = 491.7°R = 273.16 K

7.5. 77.08 kJ, work done = 8.59 kJ, ΔE = 68.49 kJ
7.7. $q = -650$ J
7.9. (a) 530°R = 70°F = 21°C = 294 K
(b) $\Delta E = 0$ (c) $w = 65$ cal
7.11. −3.1 kJ
7.13. (a) 68.26 kJ/mol   (b) 1.000 atm
7.15. (a) $\Delta E = 0, \Delta H = 0, q = 900$ J, $w = -900$ J
(b) $-w = q = \Delta E = \Delta H = 0$
7.17. (a) $\Delta E = 200.$ J   (b) $w = 0$ (c) $T_f = 33$°C
7.19. 12.8 J
7.21. (a) $C_6H_5CO_2H(s) + (7.5)O_2(g) \longrightarrow$
$7CO_2(g) + 3H_2O(l)$
(b) $w = 0, q = -32.27$ kJ, $\Delta E = -32.27$ kJ
(c) $\Delta H = -3228$ kJ/mol
7.23. −44.3 kJ/mol
7.25. (a) −283 kJ/mol   (b) −286 kJ/mol
(c) 44 kJ/mol   (d) $\Delta H_{rxn} = -41$ kJ/mol
7.27. 7640 J
7.29. −235 kJ
7.31. −984 kJ
7.33. 3528 kJ of heat are released
7.35. −298 kJ/mol
7.37. 75.32 kJ/mol K, 4.180 J/g·K
7.39. monatomic
7.41. (a) $\Delta H_f°(PbCl_2(s)) = -359$ kJ/mol
(b) $\Delta H_f°(C_{10}H_8(s)) = 256$ kJ/mol
7.43. $\Delta H = -175$ kJ
7.45. 10.8 kJ/°C, $q_v = -13.1$ kJ/g
7.47. (a) −500 J
(b) new pressure = 1 atm
(c) $R = 12.47$ J/mol·K, $C_p = 20.78$ J/mol·K,
$C_p/C_v = 1.667$

7.49. 12.3 J/mol·K (neon), 20.8 J/mol·K (hydrogen
chloride), 30.9 J/mol·K (ozone)
7.51. (a) $\Delta H° = -618$ kJ/mol
(b) 560 g C_2H_2
(c) $\Delta H° = -1198$ kJ/mol
7.53. 0.786 kg LPG
7.55. 11.00 atm., 360. kJ of heat is given off
7.57. −136.8 kJ/mol, −137 kJ/mol, $\Delta H° = -2097$
Btu/lb
7.59. $\Delta H_f°(WC(s)) = -34.5$ kJ/mol
7.61. 42.0 kJ/g
7.63. 1.9×10^{10} Btu, 1.9×10^{-5} quad
7.65. (a) −2204 kJ/mol
(b) $\Delta H_c° = -2220$ kJ/ft^3
(c) 40 ft^3
7.67. $\Delta T = 1 \times 10^{-2}$ K

CHAPTER 8

8.1. 77°C
8.3. 100°C
8.5. 97°C and 91°C
8.7. 23.8 torr
8.9. 3.65 g
8.11. $T_b = 353$ K = 80°C
8.13. (a) 335 kJ   (b) 2259 kJ
8.15 (a) 0°C   (b) 0°C

8.17. (a) 1 atm

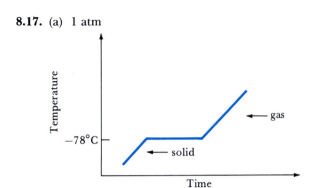

(b) 5.2 atm

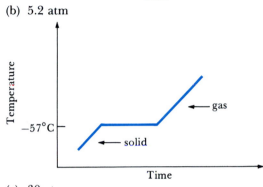

(c) 30 atm

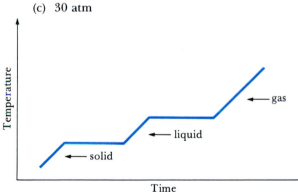

8.19. (a) Increasing the pressure will increase the melting point.
   (b) Decreasing the pressure will cause the liquid to boil.
   Decreasing the temperature at constant pressure will cause the liquid to freeze.
   (c) melting point: 345 K, boiling point: 390 K, vapor pressure: 0.25 atm at 300 K

8.21. (a) 95 atm   (b) will sink
   (c) 8°C and 25 atm   (d) not

8.23. (a) The three triple points occur at points B, C, and F.
   (b) The dotted lines represent extension of the "normal" phase diagram to the point where they would meet if the monoclinic phase of sulfur did not exist.
   (c) rhombic

8.25. 964 g ice and 36 g liquid water

8.27. It will take 115 kJ, 4260 kJ of heat must be removed

8.29. 14%.

8.31. 20.6%

8.33. 3.25%

8.35.

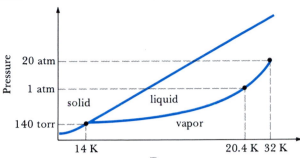

(a) At 25 K and 760. torr, hydrogen is a vapor.
(b) At 25 K and 100. torr, hydrogen is a vapor.
(c) At 75 K and 30. atm, hydrogen is a gas (since it is beyond the critical point, it is no longer considered a vapor).

8.37. (1) 1.36×10^6 J   (2) 5.46×10^5 J
   (3) 2.1×10^6 J   (4) 1.45×10^7 J
   total heat $= 1.85 \times 10^7$ J.
   $q_1 = 1.51 \times 10^6$ J
   $q_2 = 7.30 \times 10^5$ J
   $q_3 = 2.73 \times 10^6$ J
   $q_4 = 1.80 \times 10^7$ J
   total $= 2.30 \times 10^7$ J.
   This heat will only vaporize 8.56×10^3 g of gold.

8.39. 188 g/mol

8.41. 9.35×10^4 L of air

8.43. (a) The slope of the solid-liquid equilibrium line is very slightly positive. This means that the solid phase is somewhat more dense than the liquid phase, since an increase in pressure will favor formation of the solid phase. Increasing the pressure will increase the melting point a small amount, but the effect will be barely noticeable since the equilibrium line does not slant very much.
   (b) At 1.5 atm and 350. K, the substance is a liquid. If the pressure is increased at constant temperature (a vertical line on the graph), the liquid will eventually freeze.
   (c) At 0.5 atm and 350. K, the substance is a vapor. If the temperature is lowered at constant pressure (a horizontal line on the graph), the vapor will eventually condense to the solid phase.

CHAPTER 9

9.1. χ(ethanol) = 0.232, χ(water) = 0.595, χ(ethylene glycol) = 0.172

9.3. (a) 2.5 m NaOH   (b) 0.10 m CH_3OH
   (c) 0.10 m naphthalene

9.5. (a) 0.0574   (b) 0.780 m naphthalene
   (c) 94.3 torr

9.7. 26.4°C

9.9. −2.4°C

9.11. 100 g/mol

9.13. $-33.5°C$

9.15. (a) 4.20 torr   (b) 697 torr
  (c) $T_b = 102.6°C$ (d) $T_f = -9.30°C$

9.17. (a) $k_f = 40.5°C/m$ (b) 499 g/mol

9.19. (a) 174 g/mol
  (b) dissociation or aggregation, solute is reasonably volatile

9.21. $\pi = 2.71$ atm

9.23. (a) 2.0×10^4 g/mol
  (b) $\Delta T_b = 2.6 \times 10^{-4}°C$, $\Delta T_f = 9.3 \times 10^{-4}°C$
  (c) osmotic pressure is the most accurate technique

9.25. 3.5

9.27. $P_A = 16$ torr, $P_B = 16$ torr, 50:50 mole ratio of A:B

9.29. (a) χ(ethyl alcohol) = 0.51, χ(methyl alcohol) = 0.49
  (b) χ(ethyl alcohol) = 0.34 and χ(methyl alcohol) = 0.66

9.31. $P_B° = 0.70$ atm, $P_A° = 1.90$ atm

9.33. χ(acetic acid) = 0.0292, χ(H_2O) = 0.972, 1.66 m

9.35. χ(hydrogen) = 1.46×10^{-5}, $K = 6.83 \times 10^4$ atm

9.37. 23.7 torr

9.39. 180 g/mol

9.41. $P(CCl_4) = 87.4$ torr, $P(C_2H_4Cl_2) = 96.7$ torr, $P_{tot} = 184.1$ torr

9.43. $P°_{benz} = 400$ torr, $P°_{tol} = 200$ torr

9.45. Acetone-chloroform interactions are favored over acetone-acetone or chloroform-chloroform interactions.

9.47. Molality = 16 m ethanol, χ(ethanol) = 0.22, χ(H_2O) = 0.78
  Molarity of ethanol = 8.5 M, $T_f = -30.°C$

9.49. 0.080

9.51. (a) $T_f = -3.72°C$, van't Hoff's $i = 2$
  (b) $T_f = -2.79°C$

9.53. 140 g/mol

9.55. 500 m H_2SO_4

9.57. 2.54 times as cost-effective

9.59. 0.41 L

9.61. greater than 12.7 atm, $T_f = -0.97°C$

CHAPTER 10

10.1. (a) right   (b) right

10.3. (a) left   (b) left

10.5. (a) $K = [Zn^{2+}]P(H_2)/[H_3O^+]^2$
  (b) $K = [F^-]^2[Ca^{2+}]$
  (c) $K = 1/[Cl^-]^2[Cu^{2+}]$
  (d) $K = [V^{2+}]^3/[Cr^{3+}]^2$

10.7. (a) left   (b) right   (c) left
  (d) right   (e) right   (f) unchanged

10.9. (a) $K_2 = (K_1)^{1/2}$ (b) $K_2 = (K_1)^{-1}$

10.11. 0.111

10.13. $K_c = 0.078$

10.15. not at equilibrium and will proceed spontaneously to the left

10.17. 1.9, to the right

10.19. $P(PCl_3) = P(Cl_2) = 0.012428$ atm, $P(PCl_5) = 7.2 \times 10^{-5}$ atm

10.21. $P(SO_3) = 0.019$ atm, $P(O_2) = 0.091$ atm and $P(SO_2) = 0.081$ atm

10.23. $P(CO_2) = P(H_2) = 0.133$ atm, and $P(CO) = P(H_2O) = 0.067$ atm

10.25. (a) $K = P(H_2)P(CO_2)/P(H_2O)P(CO)$
  (b) 0.24 moles each of CO_2 and H_2, 0.76 moles each of CO and H_2O.
  (c) The reaction will favor CO formation.

10.27. (a) $K_c = 9.4 \times 10^{-3}$, $K_p = 2.6$, $K_c' = 110$, $K_p' = 0.38$
  (b) left   (c) left   (d) unchanged

10.29. (a) $K_p = P(UF_6)/P(F_2)$
  (b) $K_p = P(BrCl)^2/P(Cl_2)$
  (c) $K_p = P(SiF_4)$

10.31. $P(CO_2) = 5.38 \times 10^{-3}$ atm, $P(NH_3) = 4.87 \times 10^{-2}$ atm

10.33. not at equilibrium

10.35. $P(NO_2) = 0.0794$ atm, $P(N_2O_4) = 0.0556$ atm

10.37. $6.3 \times 10^{-8}\%$

10.39. $P(O_3) = 1.55 \times 10^{-6}$ atm

10.41. $P(NO_2) = 0.154$ atm, $P(N_2O_4) = 0.208$ atm

10.43. right

10.45. $K_p = 67$, $C_p/C_v = 1.62$

10.47. 0.52

10.49. (a) $K_p = P(K)/P(Na)$

CHAPTER 11

11.1. one-step: -67.6 J, two-step: -74.3 J, three-step: -76.8 J

11.3. 11 kJ

11.5. $\Delta S = 6.37$ J/mol·K, $\Delta S_{univ} = 0$

11.7. -5.06 J/K

11.9. $\Delta S = 2.36$ J/mol·K, $\Delta S_{univ} = 2.36$ J/mol·K

11.11. 0.851 J/K

11.13. (a) 72 J/K   (b) 36 J/K

11.15. $\Delta S = -0.502$ J/K, $\Delta S_{surr} = 0.502$ J/K

11.17. $\Delta S_{univ} = 0$, $\Delta S = 61.2$ J/K

11.19. 340 K

11.21. -12 J/K

11.23. 6.68×10^{-22} J/K

11.25. (a) $\Delta S° = -208.2$ J/mol·K, $\Delta G° = -1208$ kJ/mol, $\Delta H° = -1270$ kJ/mol
  (b) $\Delta S° = -269.6$ J/mol·K, $\Delta G° = -23.5$ kJ/mol, $\Delta H° = -103.8$ kJ/mol
  (c) $\Delta S° = 92.7$ J/mol·K, $\Delta G° = 403.9$ kJ/mol, $\Delta H° = 431.5$ kJ/mol
  (d) $\Delta S° = -161.5$ J/mol·K, $\Delta G° = -115.4$ kJ/mol, $\Delta H° = -163.5$ kJ/mol
  (e) $\Delta S° = 42.4$ J/mol·K, $\Delta G° = 28.5$ kJ/mol, $\Delta H° = 41.1$ kJ/mol
  (f) $\Delta S° = -184.3$ J/mol·K, $\Delta G° = -90.8$ kJ/mol, $\Delta H° = -145.6$ kJ/mol

11.27. (a) -1208 kJ/mol   (b) -23.5 kJ/mol
(c) 403.9 kJ/mol   (d) -115.4 kJ/mol
(e) 28.5 kJ/mol   (f) -90.8 kJ/mol

11.29. (a) -216 kJ/mol   (b) 14.3 kJ/mol
(c) -86.3 kJ/mol

11.31. (a) 7.29×10^{37} (b) 3.11×10^{-3}
(c) 1.34×10^{15}

11.33. $\Delta G°(683) = -264$ kJ/mol, $K = 1.55 \times 10^{20}$

11.35. (a) $\Delta H° = -1532$ kJ/mol, $\Delta E° = -1520$ kJ/mol
(b) -1525 kJ/mol

11.37. -42.1 kJ/mol

11.39. (a) 969 K   (b) 75 K

11.41. 81.6 kJ/mol

11.43. high temperatures, low pressures
low temperatures, not be affected much by
pressure variations

11.45. high pressure and low temperature

11.47. 33.0 kJ/mol

11.49. -610 J

11.51. $\Delta H = 44$ kJ, $\Delta S = 120$ J/mol·K, $\Delta G = 0$, $w = -3.1$ kJ, $\Delta E = 41$ kJ

11.53. $dS_{univ} < 0$

11.55. 34.6%

11.57. 325 J

11.59. $K = 1.3$, $P(O_2) = 0.577$ atm

11.61. (a) $K_p = P(SO_3)/P(SO_2)P(O_2)^{1/2}$
(b) $\Delta H° = -95.3$ kJ/mol, $\Delta S° = -90.5$ J/mol·K

(c)

$T(K)$	$1/T(K^{-1})$	$\ln K$
800	1.25×10^{-3}	3.444
850	1.18×10^{-3}	2.625
900	1.11×10^{-3}	1.879
950	1.05×10^{-3}	1.176
1000	1.00×10^{-3}	0.615
1100	9.09×10^{-3}	-0.465

11.63. 300°C: 66.0%, 400°C: 50.7%, 500°C: 39%,
600°C: 31%, 700°C: 25%, 800°C: 20.%

11.65. 1.15×10^4 kJ

CHAPTER 12

12.1. solubility $= 9.2 \times 10^{-9}$ moles/L

12.3. 3.9×10^{-7}

12.5. $K_{sp}(MnCO_3) = 3.75 \times 10^5$

12.7. solubility $= 1.44 \times 10^{-22}$ mol/L $[Ni^{2+}] = 1.44 \times 10^{-22}$ M; $[S^{2-}] = 0.010$ M

12.9. $[CO_3^{2-}] = 4.0 \times 10^{-4}$ M, $[Pb^{2+}] = 3.8 \times 10^{-9}$ M

12.11. 0.98

12.13. (a) $Mg^{2+}(aq) + 2\ OH^-(aq) \longrightarrow Mg(OH)_2(s)$
(b) $3\ Ca^{2+}(aq) + 2\ PO_4^{3-}(aq) \longrightarrow Ca_3(PO_4)_2(s)$

12.15. 0.257 M

12.17. 42 mL concentrated HCl and 958 mL water

12.19. (a) 17.9 M H_2SO_4 (b) 5.59 mL

12.21. (a) 1.89 (b) 1.636 (c) 7.10

12.23. (a) $[H_3O^+] = 0.06$ M, $[OH^-] = 2 \times 10^{-13}$ M
(b) $[H_3O^+] = 2 \times 10^{-7}$ M, $[OH^-] = 5 \times 10^{-8}$ M
(c) $[H_3O^+] = 4 \times 10^{-14}$ M, $[OH^-] = 0.25$ M

12.25. (a) $[H_3O^+] = 6.1 \times 10^{-5}$ M, $[OH^-] = 1.6 \times 10^{-10}$ M, pH $= 4.21$
(b) $[H_3O^+] = 6.7 \times 10^{-3}$ M, $[OH^-] = 1.5 \times 10^{-12}$ M, pH $= 2.17$

12.27. (a) $[OH^-] = 0.15$ M, $[H_3O^+] = 6.7 \times 10^{-14}$ M, pH $= 13.17$
(b) $[OH^-] = 8.1 \times 10^{-3}$ M, $[H_3O^+] = 1.2 \times 10^{-12}$ M, pH $= 11.92$

12.29. $[H_3O^+] = 6.5 \times 10^{-9}$ M, $[OH^-] = 1.5 \times 10^{-6}$ M

12.31. 4.2%

12.33. $[H_3O^+]$ decreased by 8.45×10^{-6} M

12.35. $[H_2SeO_3] = 0.46$ M, $[HSeO_3^-] = 0.037$ M, $[SeO_3^{2-}] = 5 \times 10^{-8}$ M, $[H_3O^+] = 0.037$ M

12.37. 0.096 mol/L, 5.6 g/L

12.39. $V = 302$ L

12.41. (a) solubility $= 0.457$ g/L, 2.07×10^{-3} M
(b) Adding lanthanum nitrate would decrease the solubility.

12.43. (a) solubility $= 7.3 \times 10^{-4}$ g/L
(b) solubility $= 0.019$ g/L
(c) $K_{sp} = 4.4 \times 10^{-13}$ (d) 18.3 g

12.45. (a) 7.00 (b) 7.00 (c) 11.11

12.47. 0.63

12.49. At 25°C: 0.687 g/L, at 60°C: 0.482 g/L, at 95°C: 0.362 g/L

12.51. (a) -0.372°C   (b) -0.201°C
(c) -0.400°C

12.53. 7.48

12.55. -251 kJ/mol, -2293 kJ/mol, -2163 kJ/mol

12.57. 2.6×10^{-7} M

CHAPTER 13

13.1. (a) Arrhenius acid   (b) neither
(c) Arrhenius acid

13.3. (a) Bronsted-Lowry acid
(b) Bronsted-Lowry base
(c) Bronsted-Lowry acid

13.5. (a) CN^- (b) SO_4^{2-} (c) NH_3

13.7. (a) NO_2^- is a stronger base.
(b) CO_3^{2-} is a stronger base.

13.9. (a) forward   (b) reverse

13.11. (a) 1.8×10^{10} (b) 1.4×10^{-5}
(c) 1.5×10^{-11}

13.13. 4.3×10^{-3} M, 0.42%

13.15. $[H_3O^+] = 5.8 \times 10^{-7}$ M, $[OH^-] = 1.7 \times 10^{-8}$ M, pH $= 6.24$

13.17. pH $= 7.29$

13.19. pH $= 0.726$, pH $= 0.602$

13.21. pH $= 4.74$

13.23. $[HOAc]/[OAc^-] = 0.002$

13.25. (a) pH $= 9.22$ (b) pH $= 9.14$
(c) pH $= 9.4$

13.27. 7.32×10^{-2} M

13.29. 80.0 g/mol

13.31. 3.7×10^{-8}

13.33.

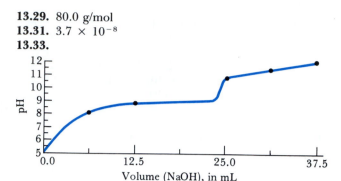

13.35. (a) 2.7×10^{-5} (b) 1.3%
(c) pH = 2.68   (d) pH = 4.57
(e) Thymolphthalein or Phenolphthalein
(f) There will be a sharp increase in slope at the equivalence point.

13.37. $H_2O < F^- < NO_2^- < OCl^- < CN^- < NH_3 < OH^- < NH_2^-$

13.39. The pH of 0.10M acetic acid is lower than you might predict.

13.41. pH = 3.47

13.43. (a) It would not provide an accurate endpoint.
(b) it would provide
(c) it would not provide
(d) it would not provide

13.45. $[H_3O^+] = 6.0 \times 10^{-6}$ M, $[OH^-] = 1.7 \times 10^{-9}$ M, pH = 5.22

13.47. pH = 4.02, 1.2×10^{-15} M

13.49. K will be a very small number, $\Delta G°$ will be large and positive.
$\Delta H°$ will probably be the larger contributor, $\Delta H°$ must be positive

13.51. 4×10^{-4}

13.53. 5.5%, $K_a = 6 \times 10^{-5}$

13.55. $3.20

13.57. Bromcresol green: yellow
Methyl orange: orange

CHAPTER 14

14.1. (a) 7.5×10^{14} s^{-1}, 5.0×10^{-12} ergs
(b) 4.0×10^{14} s^{-1}, 2.6×10^{-12} ergs

14.3. (a) 7.82×10^{-17} ergs   (b) 3.34×10^{-13} ergs
(c) 3.7×10^{-12} ergs   (d) 9.94×10^{-12} ergs
(e) 1.99×10^{-8} ergs

14.5. (a) 4.0×10^{-8} ergs   (b) 2.0×10^{-12} ergs
(c) 6.6×10^{-7} ergs   (d) 4.91×10^{-13} ergs

14.7. 5.00 eV/molecule

14.9. threshold wavelength = 260 nm

14.11. silver: 7.24×10^{-12} ergs, platinum: 8.62×10^{-12} ergs

14.13. (a) 7.69×10^{-12} ergs   (b) 1.10×10^{-11} ergs

14.15. (a) $\nu = 1.1 \times 10^{-15}$ s^{-1} (b) $V = 1.6$ volts
(c) 2.6×10^{-12} ergs

14.17. (a) $\omega = 2.0$ eV.   (b) 6.6×10^{-27} erg·s
(c) 4.9×10^{14} s^{-1}, 610 nm

14.19. $r_1 = 0.53$ Å $\int a_0$, $r_2 = 4a_0$, $r_3 = 9a_0$, $r_4 = 16a_0$

14.21. first Bohr orbit: a_0, fifth Bohr orbit: $25a_0$, any Bohr orbit, $r = n^2 a_0$.

14.23. (a) 16/15 R   (b) blue shift   (c) 1641 Å

14.27. (a) 4.58×10^{14} Hz   (b) 6543 Å
(c) 183 kJ/mol.

14.29. $r = a_0$.

14.31. (a) 0.543 eV   (b) 0.305 eV

14.33. $\Delta E = 1730$ kJ/mol

14.35. $\Delta E_{IE} = hcR_H Z^2 = (\Delta E_{IE}(H))Z^2$.

14.37. $1/\lambda_1$: 4689 Å, $1/\lambda_2$: 3205 Å, $1/\lambda_3$: 2735 Å

14.39. $n_1 = 2$ and $n_2 = 4$: 4863 Å,
$n_1 = 2$ and $n_2 = 8$: 3890 Å
$n_1 = 2$ and $n_2 = 9$: 3836 Å,
$n_1 = 2$ and $n_2 = 10$: 3799 Å
$n_1 = 2$ and $n_2 = 11$: 3772 Å,
$n_1 = 2$ and $n_2 = 12$: 3751 Å
$n_1 = 2$ and $n_2 = 13$: 3735 Å,
$n_1 = 2$ and $n_2 = 14$: 3723 Å
$n_1 = 2$ and $n_2 = 15$: 3713 Å,
$n_1 = 2$ and $n_2 = 16$: 3705 Å

14.41. 1.52×10^{-16} s, 6.56×10^{15} times.

14.43. 5.53 mm, radiowave, 1320 Å, $v = 4.37 \times 10^6$ cm/s

14.45. (a) 5.7×10^{-4} kJ/mol   (b) 2.40×10^{14} nm

CHAPTER 15

15.1. 0.25 Å

15.3. (a) 5.3×10^{-24} Å   (b) 1.4×10^{-27} Å

15.5. 2.4×10^{-24} cm

15.7. (a) $v = 4.0 \times 10^5$ cm/s, 0.08 eV
(b) 2×10^2 eV
(c) 1×10^4 eV, X-ray region

15.9. 1.45×10^{-7} cm

15.11. 5×10^{-24} mm

15.13. $\lambda = 7.3 \times 10^{-8}$ cm, $\Delta x \geq 1.8 \times 10^{-8}$ cm

15.15. $n = 1$, $\ell = 0$, $m_\ell = 0$, $m_s = +\frac{1}{2}$
$n = 1$, $\ell = 0$, $m_\ell = 0$, $m_s = -\frac{1}{2}$
$n = 2$, $\ell = 0$, $m_\ell = 0$, $m_s = +\frac{1}{2}$
$n = 2$, $\ell = 0$, $m_\ell = 0$, $m_s = -\frac{1}{2}$

15.17. (a) $\ell = 0, 1, 2$ (b) $-3, -2, -1, 0, 1, 2, 3$
(c) $+\frac{1}{2}, -\frac{1}{2}$

15.19. H, He: $(n + \ell) = 1$
Li, Be: $(n + \ell) = 2$
B, C, N, O, F, Ne: $(n + \ell) = 3$, $n = 2$
Na, Mg: $(n + \ell) = 3$, $n = 3$
Al, Si, P, S, Cl, Ar: $(n + \ell) = 4$, $n = 3$
K, Ca: $(n + \ell) = 4$, $n = 4$
Sc: $(n + \ell) = 5$

15.21. (a) As: $[Ar]4s^2 3d^{10} 4p^3$
(b) Ti: $[Ar]4s^2 3d^2$
(c) I: $[Kr]5s^2 4d^{10} 5p^5$

15.23. (a) two $3s$ electrons ($m_\ell = 0$, $m_s = \pm\frac{1}{2}$), six $3p$ electrons
($m_\ell = -1, 0, 1$ and $m_s = \pm\frac{1}{2}$) and ten $3d$ electrons
($m_\ell = -2, -1, 0, 1, 2$ and $m_s = \pm\frac{1}{2}$)

(b) The ten $4d$ electrons ($n = 4$, $\ell = 2$) have
$m_\ell = -2, -1, 0, 1, 2$ and $m_s = \pm\frac{1}{2}$.

15.25. (a) rhodium (Rh)   (b) boron (B)

15.27. potassium (K)

15.29. (a) $3d$ higher than $4s$ (b) $4p$ higher than $4s$
(c) $5s$ higher than $4p$

15.31. $1s^2 < 2s^2 < 2p^6 < 3s^2 < 3p^6 < 4s^2 < 3d^{10} < 4p^6$
$< 5s^1 < 4d^{10}$

15.33. Na: [Ne]$3s^1$, K: [Ar]$4s^1$, Ag: [Kr]$5s^14d^{10}$, Au:
[Xe]$6s^15d^{10}$

15.35. (a) neutral, ground state
(b) anion, ground state
(c) neutral, ground state
(d) impossible (no such thing as a $2d$ orbital)
(e) anion, ground state

15.37. five

15.39. 78.6 eV

15.41. 496 kJ

15.43.

     1s   2s   3s

15.45. left

15.47. There is a general increase in ionization energy,
the exceptions are B and O.

15.49. (a) 5×10^{-20} g·cm/s   (b) 0.96 eV

15.51. There are five possible values of m_ℓ and two
possible values of m_s, for a total of ten electrons;
therefore, ten transition metals occur in this
period.

15.53. It is more difficult to remove the first electron
from Ca than from K. It is more difficult to
remove the second electron from K than from
Ca.

15.55. Iron and cobalt are both transition metals; they
are both elements with partially filled $3d$ orbitals.
Ne has a closed-shell, noble gas configuration
whereas Na has a partially filled $3s$ orbital.

15.57. (a) In the absence of an external magnetic field,
s-orbitals are non-degenerate (or "one-fold
degenerate"), p-orbitals are three-fold
degenerate, and d-orbitals are five-fold
degenerate.
(b) Se^{2-}: [Kr], Ag$^+$: [Kr]$4d^{10}$, Cs: [Xe]$6s^1$

15.59.

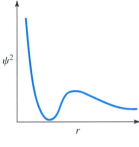 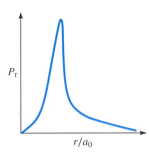

 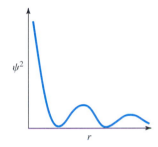 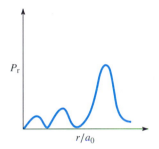

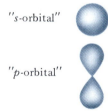

"s-orbital"

"p-orbital"

15.61. The metals (Li, Na, K) have much lower first
ionization energies than the nonmetals (F, Cl).

15.63. (a) 200 nm   (b) 5×10^{-13}
(c) A typical red blood cell will move many
orders of magnitude faster than this
calculated Δv. The red blood cell is still well
within the "macroscopic" range.

CHAPTER 16

16.1. (a) $\overset{\ominus}{\underset{..}{\overset{..}{:}}}\text{O} : \text{H}$ (b) $\text{H} : \overset{..}{\underset{..}{\text{O}}} \overset{\oplus}{:} \text{H}$ (c) $\text{H} : \overset{\overset{\text{H}}{\overset{\oplus}{|}}}{\underset{\underset{\text{H}}{|}}{\text{N}}} : \text{H}$

(d) $\text{H} : \overset{..}{\underset{..}{\text{N}}} \overset{\ominus}{:} \text{H}$

APPENDIX A   ANSWERS TO ODD-NUMBERED PROBLEMS

16.3. (a) :F:B:F: with F above and F below (boron center), each F with lone pairs; ⊖ charge   (b) H:O:O:⊖

(c) [H:N:H with H above and below, N⊕]   [H:O:S:O:H with O above (2+) and O below, ⊖ charges]

16.5. (a) H:O:H   (b) H:O:O:H

(c) :O::C::O: (d) C center with :O: above, O and O below bonded to H and H

(e) H:C:O:H with H above and below (C center)

(f) H:C:C:O:H with H's around   or   H:C:O:C:H with H's around

16.7. (a) H:N::N::N: (b) H:N::N:H

(c) H:C:N:H with H above and below

16.9. :O:S:O: with O above (2+) and O below, ⊖ charges (two ⊖ on side oxygens, one ⊖ below)

16.11. (a) $Al^{3+} < Mg^{2+} < Na^+$
(b) $F^- < Cl^- < Br^-$ (c) $Na^+ < F^- < Cl^-$

16.13. (a) $Li < Na < Rb$ (b) $F < N < B$
(c) $Cl < P < Ga$

16.15. -134 kJ/mol

16.17. $\Delta H = -766$ kcal/mol

16.19. (a) ionic   (b) covalent
(c) covalent   (d) covalent

16.21. (a) CsF (b) SO_3 (c) N_2O_5 (d) HCl

16.23. HF

16.25. (a) O is positive; F is negative
(b) N is positive; O is negative
(c) S is positive; O is negative
P—H, C—S, and N—Cl

16.27. H:O:S:O:H with :O: above and :O: below   H:O:S:O:H with O above and O below   H:O:S:O:H with O above and O below

16.29. (a) 242 kJ/mol
(b) Part (a) assumes that all Cl—F bonds are identical even if they are in different molecules.
(c) a suitable preparation for ClF (g) under conditions of high temperature and low pressure
(d) There is an entropy increase.

16.31. 239 kJ/mol

16.33. (a) 36 kJ/mol   (b) 105°
(c) Ethyl alcohol

16.35. (a) low temperatures   (b) low temperatures
(c) low temperatures

16.37. We would expect polar molecules to be more soluble in water.

16.39. (a) S center with :O. and .O: below

(b) $2\ SO_2 + O_2 \longrightarrow 2\ SO_3$
$SO_3 + H_2O \longrightarrow H_2SO_4$
(c) 50 tons SO_2/hour   (d) too low

CHAPTER 17

17.1. (a) Group IVA   (b) Group IIA

17.3. 2.3×10^5 ohm^{-1}cm^{-1}

17.5. $Al < Ca < Sr < Rb < Cs$

17.7. (a) $K < Mg < Be$ (b) $Na < Al < Mg$

17.9. $Mg < Al < Na$ for I_2

17.11. $Li < K < I < Cl$

17.13. BaO, KF and SrO

17.15. (a) H is $+1$, and N is -3.
(b) K is $+1$, O is -2, and Mn is $+7$.
(c) O is -2, and Br is $+5$.
(d) Ca is $+2$, O is -2, and C is $+4$.
(e) Na is $+1$, F is -1, and Cl is $+3$.

17.17. $Al^{3+} < Mg^{2+} < Ca^{2+} < F^- < O^{2-} < Cl^- < S^{2-}$.

17.19. (a) $K + H_2O \longrightarrow KOH + \frac{1}{2} H_2$
(b) $Cs + \frac{1}{2} Br_2 \longrightarrow CsBr$
(c) $Na + \frac{1}{2} H_2 \longrightarrow NaH$
(d) $RbH + H_2O \longrightarrow RbOH + H_2$

17.21. (a) $C_3H_8 + 5O_2 \longrightarrow 3CO_2 + 4H_2O$
(b) $N_2H_4 + O_2 \longrightarrow N_2 + 2H_2O$

17.23. (a) reasonably strong ($K_{a1} = 1.2 \times 10^{-2}$)
(b) weak ($K_{a1} = 7.5 \times 10^{-3}$)
(c) weak ($K_{a1} = 7.3 \times 10^{-10}$)

17.25. $4Li + O_2 \longrightarrow 2Li_2O$
$2Na + O_2 \longrightarrow Na_2O_2$
$K + O_2 \longrightarrow KO_2$
$Rb + O_2 \longrightarrow RbO_2$
$Cs + O_2 \longrightarrow CsO_2$

17.27. $CaCO_3 \longrightarrow CaO + CO_2$
$CaO + H_2O \longrightarrow Ca(OH)_2$

17.29. $SiO_2 + C \longrightarrow Si$ (crude) $+ CO_2$
Si (crude) $+ 2Cl_2 \longrightarrow SiCl_4$ (crude)
Distillation: $SiCl_4$ (crude) $\longrightarrow SiCl_4$ (pure)
$SiCl_4$ (pure) $+ 2Mg \longrightarrow Si$ (pure) $+ 2MgCl_2$
Zone refining: Si (pure) $\longrightarrow Si$ (ultrapure)

17.31. (a) ionic, solid   (b) covalent, gas
(c) covalent, gas   (d) covalent, liquid
(e) covalent, liquid   (f) covalent, gas
(g) covalent, liquid   (h) ionic, solid
(i) covalent, gas

17.33. $Pb + O_2$ (excess) $\longrightarrow PbO_2$
$PbO_2 + 4\ HCl \longrightarrow PbCl_4 + 2\ H_2O$

17.35. 13.635

17.37. $^{90}_{38}Sr \longrightarrow \; ^{0}_{-1}\beta + \; ^{90}_{39}Y$

17.39. $\Delta m = -0.253931$ amu, binding energy $= 8.448$ MeV/nucleon

$\Delta m = -0.5159$ amu, binding energy $= 7.3$ MeV/nucleon

binding energy per nucleon $= 8.2$ MeV/nucleon

17.41. 980 atm

17.43. 15.6 L SiF_4

CHAPTER 18

18.1. NH_3 and CH_4 obey the octet rule

CH_4: only C obeys the octet rule

PF_5: only F obeys the octet rule

BF_3: only F obeys the octet rule

18.3. H:C:H   H:C:C:O: ⟷ H:C:C::O:

18.5. (a) angular   (b) see-saw shaped
(c) octahedral   (d) trigonal pyramidal
(e) trigonal bipyramidal   (f) linear

18.7. (a) linear   (b) square planar
(c) bent-T shaped   (d) square pyramidal
(e) linear

18.9. (a) H_2 contains 1 σ-bond and no π-bonds.
(b) HCN contains 2 σ-bonds and 2 π-bonds.
(c) CH_3CN contains 5 σ-bonds and 2 π-bonds.

18.11. (a)

(b)

(c)

18.13. The bond order is zero so Be_2 does not exist

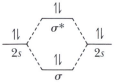

18.15. The valence bond view: A total of six electrons are shared. A triple bond results.

The molecular orbital view: There are eight bonding electrons and two antibonding electrons for a net bond order of 3. A triple bond results.

18.17. (a) O_2^+ has eleven valence electrons
$KK(\sigma_{2s})^2(\sigma^*_{2s})^2(\pi_{2p})^4(\sigma_{2p})^2(\pi^*_{2s})^1$
(b) 2.5   (c) paramagnetic

18.19. He_2^+ and O_2

18.21. BeH_2 has four valence electrons.

18.23. (a) The species does exist.
(b) The apparent bond order is 1.
(c) It is diamagnetic.
(d) The bond length is shorter than in Li_2^+.
(e) The bond length is longer than in H_2.
(f) The energy released is greater than in Na_2 formation.
(g) The energy released is less than in B_2 formation.

18.25. CO has ten valence electrons.

18.27. I_3^- has two single I—I bonds and one double bond. The structure number is five, and the molecule is linear. The central iodine is dsp^3 hybridized. The molecular geometry is linear.

18.29. It could have a cyclic structure with two double bonds, but a single-bonded tetrahedral structure is more likely.

18.31. PCl_3, SF_2, and SF_4

18.33. SF_4 has a permanent dipole moment whereas SF_6 does not. So SF_4 boils at a higher temperature than that at which SF_6 sublimes.

18.35. NF_3 and PF_3 have sp^3 hybridization and do not require the involvement of d-orbitals.

18.37. Both ions have sixteen valence electrons.

18.39. Cyanogen chloride can react with ammonia as follows.

$ClCN + NH_3 \longrightarrow NH_2CN + HCl$

The resulting cyanamide can form a cyclic product as follows.

$3\ NH_2CN \longrightarrow [(NH_2)—C=N]_3$

In the cyanuric chloride, the ring is planar (hexagonal), and it should exhibit resonance. In the Melamine, the ring twists into three dimensions, and there should be no resonance.

CHAPTER 19

19.1. (a) 1 lattice point   (b) 2 lattice points
(c) 2 lattice points

19.3. 6.08×10^{23}

19.5. 107 g/mol

19.7. 8.97 g/cm³
19.9. 1.96 Å
19.11. 1.28 Å
19.13. 25.6%.
19.15. −165 kcal/mol
19.17. 1.62 Å
19.19. 0.414.
19.21. $N_0 \approx 5.95 \times 10^{23}$
19.23. 96% of the nickel ions are Ni^{2+}, and 4% are Ni^{3+}
19.25. eight silicon atoms
19.27. (a) 2.59 Å   (b) −224 kcal/mol
    (c) The measured value will be smaller in absolute value than this estimated number.
19.29. 94% of the nickel ions are Ni^{2+}, and 6% are Ni^{3+}
19.31. 2.07 g/cm³
19.33. 0.613 g/L
19.35. 1.43 Å
19.37. 2.144×10^{13} atoms

CHAPTER 20

20.1. (a) A simple metal or a polymer
    (b) A composite with minimum tendency to creep
    (c) A polymer or composite material
    (d) A variety of materials (metals, polymers, composites)
    (e) A structural ceramic, a heat-resistant polymer, or a composite
20.3 (a) SiC forms a covalent lattice similar to that of diamond, so SiC should melt at a temperature somewhat less than 3500°C. Its actual melting point is around 2700°C.
    (b) MgO should have an ionic structure similar to CaO, so MgO should melt at a somewhat higher temperature due to the smaller size of the Mg^{+2} ion as compared with the Ca^{+2} ion. Its actual melting point is 2800°C.
    (c) Cu is held together by metallic bonding and is close to Fe on the periodic table, so Cu should melt at a temperature somewhat close to that of Fe. The actual melting point of copper is 1083°C.
20.5. 0.0015 in/in
20.7. cobalt
20.9. (a) liquid solution
    (b) heterogeneous mixture of solid solution and liquid solution

20.11. (a) The infinite double chain structure can be drawn as follows:

$(Si_4O_{11}^{6-})_n$

An alternative structure could be:

$(Si_4O_{11}^{6-})_n$

(b) The infinite sheet structure can be drawn as follows:

$(Si_2O_5^{2-})_n$

An alternative structure could be:

$(Si_2O_5^{2-})_n$

20.13. (a) $Si_3AlO_8^-$ (b) $Si_2Al_2O_8^{2-}$
20.15. 0.74
20.17. $-(CH_2{-}CH_2)_{\overline{n}} + 6n\ O_2 \longrightarrow$
$$2n\ CO_2 + 2n\ H_2O$$

20.19. 850°C; a silver-rich solid solution (containing roughly 93% Ag, 7% Cu), and a copper-rich solid solution (containing roughly 7% Ag, 93% Cu).

20.21. $\epsilon = 0.0435$, $E = 1.0 \times 10^7$ psi

20.23. (a) The number of flaws will decrease.
  (b) The number of flaws will decrease.
  (c) The number of flaws will decrease.

20.25. The carbon-substituted cable will support the greater load.

CHAPTER 21

21.1. (a) $+1$ in Cl_2O, $+4$ in ClO_2, and $+7$ in Cl_2O_7.
  (b) -2 in HS^-, $+4$ in HSO_3^-, $+3$ in $HS_2O_8^-$, $+2.5$ in $HS_4O_6^-$ and $+6$ in HSO_4^-
  (c) 0 in P_4, -3 in PH_3, -2 in P_2H_4, $+1$ in H_3PO_2, $+3$ in H_3PO_3, $+5$ in H_3PO_4 and $+6$ in $(NH_4)_2P_2O_7$.

21.3. (a) Na is the reducing agent. Cl_2 is the oxidizing agent.
  (b) Zn is the reducing agent. H_2SO_4 is the oxidizing agent.

21.5. (a) oxidized   (b) oxidized   (c) oxidized
  (d) reduced   (e) no change

21.7. (a) $14H^+ + Cr_2O_7^{2-} + 6Cl^- \longrightarrow$
$$2Cr^{3+} + 3Cl_2 + 7H_2O$$
  (b) $2S_2O_3^{2-} + I_2 \longrightarrow 2I^- + S_4O_6^{2-}$
  (c) $4H^+ + MnO_2 + 2Hg + 2Cl^- \longrightarrow$
$$Mn^{2+} + Hg_2Cl_2 + 2H_2O$$
  (d) $4H^+ + 3Ag + NO_3^- \longrightarrow$
$$3Ag^+ + NO + 2H_2O$$
  (e) $8H^+ + H_3AsO_4 + 4Zn \longrightarrow$
$$AsH_3 + 4Zn^{2+} + 4H_2O$$
  (f) $18H_2O + 10Au^{3+} + 3I_2 \longrightarrow$
$$10Au + 6IO_3^- + 36H^+$$
  (g) $6H^+ + IO_3^- + 8I^- \longrightarrow 3I_3^- + 3H_2O$
  (h) $H^+ + 3HS_2O_3^- \longrightarrow 4S + 2HSO_4^- + H_2O$
  (i) $4H^+ + 3MnO_4^{2-} \longrightarrow$
$$MnO_2 + 2MnO_4^- + 2H_2O$$
  (j) $4H^+ + 2O_2^{2-} \longrightarrow O_2 + 2H_2O$

21.9. (a) $OH^- + 2Co(OH)_3 + Sn \longrightarrow$
$$2Co(OH)_2 + HSnO_2^- + H_2O$$
  (b) $3ClO_4^- + I^- \longrightarrow 3ClO_3^- + IO_3^-$
  (c) $OH^- + H_2O + PbO_2 + Cl^- \longrightarrow$
$$ClO^- + Pb(OH)_3^-$$
  (d) $OH^- + H_2O + NO_2^- + 2Al \longrightarrow$
$$NH_3 + 2AlO_2^-$$
  (e) $2ClO^- \longrightarrow 2Cl^- + O_2$
  (f) $2OH^- + HXeO_4^- + 3Pb \longrightarrow$
$$Xe + 3HPbO_2^-$$
  (g) $2H_2O + 2Ag_2S + 8CN^- + O_2 \longrightarrow$
$$2S + 4Ag(CN)_2^- + 4OH^-$$
  (h) $8H_2O + 2MnO_4^- + 7S^{2-} \longrightarrow$
$$2MnS + 5S + 16OH^-$$
  (i) $2OH^- + Cl_2 \longrightarrow ClO^- + Cl^- + H_2O$

21.11. $Na > Al > Co > Ni > H_2 > Ag$

21.13. (a) I_2 could oxidize $Fe \rightarrow Fe^{3+}$; I_2 could oxidize $Fe \rightarrow Fe^{2+}$; Br_2 could oxidize $Fe^{2+} \rightarrow Fe^{3+}$; H_4XeO_6 could oxidize $2F^- \rightarrow F_2$.
  (b) Fe will reduce $Ni^{2+} \rightarrow Ni$; Br^- will reduce $Cl_2 \rightarrow 2Cl^-$.
  Note: There are many correct answers to this question. Each answer given is just one possibility.

21.15. (a) spontaneous   (b) spontaneous
  (c) spontaneous   (d) spontaneous
  (e) spontaneous

21.17. $Mn + Ni^{2+} \longrightarrow Mn^{2+} + Ni$, $E° = 0.93$ V

21.19. $Cu + PtCl_4^{2-} \longrightarrow Cu^{2+} + Pt + 4Cl^-$, $E° = 0.393$ V

21.21. 0.999 V, Pt electrode is positive.

21.23. Anode: $Zn + 2KOH + 2OH^- \longrightarrow$
$$K_2ZnO_2 + 2e^- + 2H_2O$$
Cathode: $H_2O + 2e^- + HgO \longrightarrow Hg + 2OH^-$
Overall: $Zn + 2KOH + HgO \longrightarrow$
$$K_2ZnO_2 + Hg + H_2O$$

21.25. $\Delta G° = -2.32$ kJ/mol, $K = 2.55$

21.27. 27.8 kJ

21.29. (a) $2Fe^{3+} + Zn \longrightarrow 2Fe^{2+} + Zn^{2+}$
  (b) Pt electrode is the cathode, and the Zn electrode is the anode.
  (c) $E° = 1.53$ V, $K = 10^{52}$
  (d) If $[Fe^{3+}]$ increases, Q decreases and E becomes more positive.

21.31. 5.19×10^{-3} M

21.33. 4.1×10^{-8}

21.35. 1.3×10^{-8}

21.37. 0.102 V, cell potential will drop, 0.216 M

21.39. 0.85

21.41. 26.9 g chromium

21.43. 0.380 amps

21.45. 3700 kJ

21.47. (a) 1.32 g V   (b) 0.475 g V   (c) 12.7 g V

21.49. (a) 0.522 L O_2 (b) 3.1 minutes

21.51. $27,600$ coul

21.53. 0.2945 g Cu, 0.1606 g Cr, 0.104 L H_2, 0.0519 L O_2

21.55. (a) 60.65 g $PbSO_4$ (b) 1.8 hours

21.57. (a) $4OH^- + VO^{2+} \longrightarrow VO_3 + 2H_2O + e^-$
  (b) $7H_2O + 2Cr^{3+} \longrightarrow$
$$Cr_2O_7^{2-} + 14H^+ + 6e^-$$
  (c) $2H_2O + Mn^{2+} \longrightarrow MnO_2 + 4H^+ + 2e^-$
  (d) $4OH^- + NO \longrightarrow NO_3^- + 2H_2O + e^-$

21.59. $Mn + 2AgCl \longrightarrow Mn^{2+} + 2Ag + 2Cl^-$, $E° = 1.402$ V, Electrons will flow from the magnesium to the silver.

21.61. (a) -0.79 V   (b) 0.338 V   (c) 1.43 V

21.63. Ni-Cd cell: 291 W·hr/kg, Na-S cell: 687 W·hr/kg.

21.65. (a) $E° = 0.16$ V, $E = 0.15$ V, $[Fe^{2+}]/[Co^{2+}] = 2.6 \times 10^5$
  (b) $E° = 1.716$ V, $E = 1.637$ V

21.67. $E = E° - 0.118$ pH

21.69. Cathode: $4H_2O + 4e^- \longrightarrow 2H_2 + 4OH^-$,
$E° = -0.828$ V
Anode: $2H_2O \longrightarrow 4e^- + 4H^+ + O_2$,
$E° = -1.23$ V

21.71. 49.6%

21.73. 15 days

21.75. 390 J

21.77. 21.7 atm

21.79. 6.80×10^5 coul

CHAPTER 22

22.1 Ni^{2+} is [Ar] $3d^8$ $\quad$ Cu^+ is [Ar] $3d^{10}$
Zn^{2+} is [Ar] $3d^{10}$ $\quad$ Ti^{2+} is [Ar] $3d^2$

22.3. 6600 g Cu

22.5. $Pd(NH_3)_2Cl_2$ has zero free chloride ions and zero total ions
$[Pd(NH_3)_3Cl]Cl$ has one free chloride ion and two total ions
$[Pd(NH_3)_4]Cl_2$ has two free chloride ions and three total ions

22.7. Two isomers are possible.

22.9. (a) potassium tetracyanonickelate(II)
(b) sodium trioxalatochromate(III)
(c) pentaamminechloroplatinum(IV) chloride
(d) tetraamminedinitroiron(III) sulfate
(e) hexaaquacobalt(III) iodide
(f) pentaammine chloro ruthenium(III) bromide
(g) potassium tetrachlorocobaltate(II)

22.11. (a) coordination number = 4, oxidation number = +2
(b) coordination number = 2, oxidation number = +1
(c) coordination number = 4, oxidation number = +2

22.13. (a) 4 faces, 4 corners, and coordination number of 4
(b) 6 faces, 5 corners, and coordination number of 5

22.15. Mn^{2+}: five or one, Co^{3+}: zero or four, Ni^{2+}: two

22.17. Fe^{2+} has six d-electrons; in an octahedral crystal field. The complex is high-spin (low value of Δ_0).

22.19. 3.3×10^{-12} ergs, blue

22.21. (a) paramagnetic $\quad$ (b) paramagnetic
(c) paramagnetic $\quad$ (d) paramagnetic
(e) paramagnetic $\quad$ (f) paramagnetic

22.23. 2.559

22.25. 6.7×10^{-28} M

22.27. (a) hexaaquacopper(II) chloride
(b) tetraamminedichloronickel(II)
(c) sodium tetrachloroplatinate(II)
(d) potassium hexacyanoferrate(II)

22.29.

Coordin. Number	Orbital	Geometry	Examples
2	sp	linear	$Ag(NH_3)_2{}^+$, BeH_2
3	sp^2	trigonal	BF_3, NO_3^-
4	sp^3	tetrahedral	$Ni(CO)_4$, MnO_4^-
4	dsp^2	square planar	$Ni(CN)_4^{2-}$, $Pt(NH_3)_4^{2+}$
5	dsp^3	trigonal bipyramidal	$CuCl_5^{2-}$, $Fe(CO)_5$
6	d^2sp^3	octahedral	$Fe(CN)_6^{4-}$, $Cr(NH_3)_6^{3+}$
8	d^4sp^3	dodecahedral	$Mo(CN)_8^{4-}$

22.31. 6.7 g/L

22.33. three

22.35. 8 formula weights, sixteen Fe^{3+} and eight Fe^{2+} per unit cell

22.37. $K = 1.8 \times 10^5$, $\Delta G° = -30$ kJ/mol

22.39. Electrolyze it with 10.0 amp current in 1.00M Cr^{3+} solution for 8.4 days.

CHAPTER 23

23.1. (a) rate $= \Delta[AB_2]/\Delta t$ $\quad$ (b) rate $= \Delta[A_2B_2]/\Delta t$

23.3. (a) $-\Delta[A_2]/\Delta t = -\Delta[B_2]/\Delta t = \frac{1}{2}\Delta[AB]/\Delta t$
(b) $-\Delta[A_2]/\Delta t = -\Delta[B_2]/\Delta t = \Delta[A_2B_2]/\Delta t$

23.5. (a) third order
(b) $\frac{1}{2}\Delta[NOCl]/\Delta t = -\Delta[Cl_2]/\Delta t$

23.7. $(3.7 \times 10^{13}$ L^3/mol^3·sec$)[H_3AsO_2][H_3O^+]^2[I^-]$

23.9. (a) third order overall
(b) first order in [A] and second order in [B]
(c) 0.4 M^{-2} min^{-1}

23.11. (a) first order in arsine
(b) 0.0154 hr^{-1} $\quad$ (c) 45.0 hr

23.13. 66.4 min

23.15. $k = 0.010$ min^{-1}; $t_{1/2} = 69.3$ min

23.17. (a) $-d[A]/dt = k[A]$ $\quad$ (b) $2.303/k$

23.19. (a) ≈ 0.0147 min^{-1}
(b) 47.1 min
(c) 77.6 min

23.21. (a) Decay rate = 250 counts per second
(b) twelve minutes

23.23. 3.61×10^{10} counts/sec

23.25. (a) $^{131}I \longrightarrow {}^{131}Xe + b^-$ $\quad$ (b) 1380 counts/min

23.27. 3400 yrs

23.29. 4.2×10^9 years

23.31. (a)

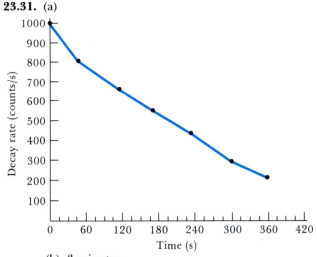

(b) 3 minutes.

(c) 1000 counts per second

23.33. $t_{1/2} = 1/k[A]_0$

23.35. (a) 52.2%   (b) 113 min

23.37. (a) Elementary processes are not likely to be greater than termolecular.
It is unlikely to have more than three particles collide simultaneously.

(b) first order in [A] and fourth order overall, $\Delta[X]/\Delta t = k[A][B][C][D]$.

23.39. $t_{1/2} = (0.693/t) \ln(R_0/R)$

23.41. $k = m/2$, $[A_0] = (1/b)^{1/2}$

23.43. (a) 12 decays/min/g   (b) 14 decays/min/g

23.45. (a) 60.0 minutes

(b) 2.24×10^5 bacteria/L, 2.28×10^5 bacteria/L

(c) six hours, 5.98 hours

(d) 0.0116 min^{-1}

CHAPTER 24

24.1. 95 kJ/mol, 403 sec

24.3. (a) 4.0×10^1 M/sec   (b) 2.5×10^{-6} sec

(c) 32.2 kJ/mol

24.5. (a) 25 kcal/mol   (b) 1×10^{13}

(c) 2.8×10^{-17}

24.7. (a) $A + B \longleftrightarrow C$ $C \longrightarrow P$

(b)

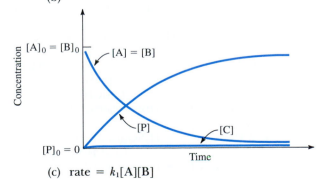

(c) rate = $k_1[A][B]$

24.9. (a) $d[N_2]/dt = k[NO]^2[H_2]$

(b) $2NO + H_2 \longrightarrow N_2O + H_2O$, slow
$N_2O + H_2 \longrightarrow N_2 + H_2O$, fast

24.11. $NO_2 + F_2 \longrightarrow NO_2F + F$ slow
$F + NO_2 \longrightarrow NO_2F$ fast
rate = $k_1[NO_2][F_2]$
Since NO_2 and F are both radicals, each species may recombine with itself rather than combining with the other.
$F + F \longrightarrow F_2$; $NO_2 + NO_2 \longrightarrow N_2O_4$

24.13. $Cl_2 \longleftrightarrow 2 Cl$ (rapid equilibrium)
$Cl + H_2 \longrightarrow HCl + H$ (slow step)
$d[HCl]/dt = k_2[H_2][Cl]$
We can calculate [Cl] from the initial equilibrium step:
$[Cl]^2/[Cl_2] = K_{eq} = k_1/k_{-1}$. $[Cl] = (k_1[Cl_2]/k_{-1})^{1/2}$
$d[HCl]/dt = k_2[H_2](k_1[Cl_2]/k_{-1})^{1/2}$
$\frac{1}{2} d[HCl]/dt = \frac{1}{2} k_2(k_1/k_{-1})^{1/2} [H_2][Cl_2]^{1/2}$
$\frac{1}{2} d[HCl]/dt = k[H_2][Cl_2]^{1/2}$, where $k = \frac{1}{2} k_2(k_1/k_{-1})^{1/2}$

24.15. (a) True   (b) True   (c) False

24.17. 1088 K

24.19. 447 K

24.21. zero order reaction

24.23.

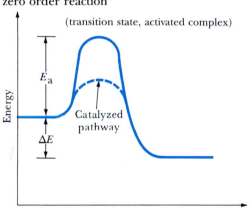

If the process were endothermic, the energy of the products would be higher than the energy of the reactants:

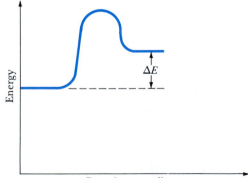

24.25. $P_A = 0.15$ atm, $P_B = 0.10$ atm, $P_C = P_D = 1.07 \times 10^{-4}$ atm

24.27. (a) rate = $\Delta[F]/\Delta t = k[A][B][E]/[D]$

(b) rate = $\Delta[F]/\Delta t = k'[A][B]$

CHAPTER 25

25.1 There are five isomeric hexanes. Their IUPAC names are *n*-hexane, 2-methyl pentane, 3-methyl pentane, 2,2-dimethyl butane and 2,3-dimethyl butane.

25.3. (a) There are nine isomeric heptanes.
(b) Their IUPAC names are *n*-heptane; 2-methylhexane; 3-methylhexane; 2,2-dimethylpentane; 2,3-dimethylpentane; 2,4-dimethylpentane; 3,3-dimethylpentane; 3-ethylpentane; and 2,2,3-trimethylbutane.
(c) It might commonly be called isoheptane.
(d) It might commonly be called neoheptane.
(e) The names used in (c) and (d) are emphatically *not* unambiguous (pardon the double negative).

25.5 (a) $CH_3CH_2CH_3$ + heat (750°C) $\longrightarrow$
$$CH_2=CHCH_3$$
$CH_2=CHCH_3$ + H_2O (in dilute acid) $\longrightarrow$
$$CH_3-CH(OH)-CH_3$$
(b) $CH_3-CH(OH)-CH_3$
     + H_2SO_4 (180°C) $\longrightarrow$ $CH_3CH=CH_2$
     + H_2O
$CH_3CH=CH_2$ + H_2 (and catalyst) $\longrightarrow$
$$CH_3CH_2CH_3$$
(c) CH_3CH_3 + Cl_2 + light $\longrightarrow$
$$CH_3CH_2Cl + HCl$$
$2CH_3CH_2Cl$ + $2Na$ $\longrightarrow$
$$CH_3CH_2CH_2CH_3 + 2NaCl$$
(d) CH_4 + Cl_2 + light $\longrightarrow$ CH_3Cl + HCl
$2CH_3Cl$ + $2Na$ $\longrightarrow$ CH_3CH_3 + $2NaCl$
CH_3CH_3 + Cl_2 + light $\longrightarrow$
$$CH_3CH_2Cl + HCl$$
CH_3CH_2Cl + CH_3Cl + 2 Na $\longrightarrow$
$$CH_3CH_2CH_3 + CH_3CH_3 + CH_3(CH_2)_2CH_3$$

25.7. C_2H_2 + $NaNH_2$ $\longrightarrow$ HC_2^- + Na^+ + NH_3
C_2H_2 + $2NaNH_2$ $\longrightarrow$ C_2^{-2} + $2Na^+$ + $2NH_3$
CH_4 + Cl_2 + light $\longrightarrow$ CH_3Cl + HCl
$2CH_3Cl$ + $2Na$ $\longrightarrow$ CH_3CH_3 + $2NaCl$
CH_3CH_3 + heat (850°C) $\longrightarrow$ $CH_2=CH_2$ + H_2
$CH_2=CH_2$ + Cl_2 $\longrightarrow$ CH_2ClCH_2Cl
CH_2ClCH_2Cl + alcoholic KOH $\longrightarrow$
$$C_2H_2 + 2HCl$$
C_2H_2 + $2NaNH_2$ $\longrightarrow$ C_2^{-2} + $2Na^+$ + $2NH_3$

25.9. There are twelve possible brominated benzenes: bromobenzene, ortho-dibromobenzene, meta-dibromobenzene, para-dibromobenzene, 1,2,3-tribromobenzene, 1,2,4-tribromobenzene, 1,3,5-tribromobenzene, 1,2,3,4-tetrabromobenzene, 1,2,4,5-tetrabromobenzene, 1,2,3,5-tetrabromobenzene, pentabromobenzene, and hexabromobenzene.

25.11. $C_6H_5(CH_2)_2OH$ < $CH_3(CH_2)_5OH$ < $HO(CH_2)_6OH$ < C_6H_5OH < CH_3CH_2OH

25.13. (a) There are four possible monobromobutanes. Their IUPAC names are 1-bromo butane, 2-bromo butane, 1-bromo-2-methyl propane and 2-bromo-2-methyl propane.

(b) $CH_3(CH_2)_3Br$ + OH^- $\longrightarrow$
$$CH_3(CH_2)_3OH + Br^-$$
$(CH_3)_2CHCH_2Br$ + OH^- $\longrightarrow$
$$(CH_3)_2CHCH_2OH + Br^-$$
$CH_3CH_2-CH(Br)-CH_3$ + OH^- $\longrightarrow$
$$CH_3CH_2-CH(OH)-CH_3 + Br^-$$
$(CH_3)_3CBr$ + OH^- $\longrightarrow$ $(CH_3)_3COH$ + Br^-

(c) Primary alcohols can be easily oxidized to carboxylic acids.
Sometimes secondary alcohols can also be oxidized to carboxylic acids.
Primary alcohols can be easily oxidized to aldehydes.
Secondary alcohols can be easily oxidized to ketones.
Tertiary alcohols cannot be further oxidized without disrupting the carbon skeleton.

(d) $CH_3(CH_2)_3Br$ + alcoholic KOH $\longrightarrow$
$$CH_3CH_2CH=CH_2$$
$CH_3CH_2-CH(Br)-CH_3$ + alc. KOH $\longrightarrow$
$$CH_3CH=CHCH_3 + CH_3CH_2CH=CH_2$$
$(CH_3)_2CHCH_2Br$ + alc. KOH $\longrightarrow$
$$(CH_3)_2C=CH_2$$
$(CH_3)_3C-Br$ + alc. KOH $\longrightarrow$
$$(CH_3)_2C=CH_2$$

25.15. (a)

$$\underset{\text{acetic acid}}{H_3C-\overset{\overset{O}{\|}}{C}-O-H} \qquad \underset{\underset{Br}{|}}{H-\overset{\overset{H}{|}}{\underset{}{C}}-\overset{\overset{O}{\|}}{C}-O-H}$$

α-bromoacetic acid

(b) Acetic acid: pH = 2.85, α-bromoacetic acid: pH = 1.78
(c) The electronegative bromine atom in α-bromoacetic acid stabilizes the α-bromoacetate anion that forms as protons dissociate. In general, electronegative substituents tend to increase the acidity of carboxylic acids. This is called the inductive effect.

25.17. CH_3CO_2H + NH_3 + heat $\longrightarrow$
$$CH_3CONH_2 + H_2O$$
CH_3CO_2H + CH_3OH $\longrightarrow$ $CH_3CO_2CH_3$ + H_2O
CH_3CO_2H + H_2O $\longrightarrow$ N.R.

25.19. There are four isomeric butylamines. Their IUPAC names are 1-amino butane, 2-amino butane, 1-amino-2-methyl propane and 2-amino-2-methyl propane.

25.21. (a) $CH_3CH_2CH_2OH$ + H_2SO_4(catalyst) $\longrightarrow$
$$CH_3CH=CH_2 + H_2O$$
$CH_3CH=CH_2$ + H_2O $\longrightarrow$ $(CH_3)_2CHOH$
(b) $(CH_3)_2CHCl$ + alcoholic KOH(catalyst) $\longrightarrow$
$$CH_3CH=CH_2 + HCl$$
$CH_3CH=CH_2$ + HBr + peroxides $\longrightarrow$
$$(CH_3)_2CHBr$$
(c) $(CH_3)_2CHOH$ + H_2SO_4(catalyst) $\longrightarrow$
$$CH_3CH=CH_2 + H_2O$$
$CH_3CH=CH_2$ + HCl $\longrightarrow$ $(CH_3)_2CHCl$
$(CH_3)_2CHCl$ + NH_3 $\longrightarrow$
$$(CH_3)_2CH-NH_2 + HCl$$

25.23. (a) Aldehydes reduce Tollen's reagent so "A" must be formaldehyde.

$$H_2CO(l) + 2Ag(NH_3)_2{}^+(aq) + 2H_2O \longrightarrow$$
$$HCO_2{}^-(aq) + Ag(s) + 4NH_4{}^+(aq)$$
$$+ OH^-(aq)$$

(b) "B" and "C" must be *n*-propanol and *n*-propanal, respectively.

$$5CH_3(CH_2)_2OH(l) + 2MnO_4{}^-(aq) \longrightarrow$$
$$5CH_3CH_2CHO(l) + 2Mn^{+2}(aq) + 6OH^-(aq)$$
$$+ 2H_2O$$

25.25. (a) $H_2N—NH_2$ and $HN=NH$
(b) SiH_4 (c) BH_3 and B_2H_6

25.27. Halogenation and oxidation of propylene result in addition across the double bond.

$$CH_3CH=CH_2 + Br_2 \longrightarrow CH_3CH(Br)—CH_2Br$$
$$CH_3CH=CH_2 + KMnO_4 \longrightarrow$$
$$MnO_2 + CH_3CH(OH)—CH_2OH$$

Halogenation and oxidation of toluene do *not* involve chemistry of the double bond. Instead, aromatic chemistry is predominantly substitution chemistry or the chemistry of appended functional groups.

$$C_6H_5CH_3 + Br_2 + Fe \longrightarrow \text{brominated toluenes}$$
(e.g., $C_6H_4BrCH_3$)
$$C_6H_5CH_3 + KMnO_4 \longrightarrow C_6H_5CO_2H$$

25.29. (a) decane, $CH_3(CH_2)_8CH_3$
(b) The three products could be H_2, CH_4, and $CH_3(CH_2)_5CH_3$, based on boiling point data. None of the products could be alkenes or else they would be capable of polymerization as well as reacting with bromine water.
(c) *n*-pentane, $CH_3(CH_2)_3CH_3$ is the liquid hydrocarbon with the lowest molar mass (MW = 72 g/mol).

25.31. 35.0%
25.33. 66.8%
25.35. 42 L

CHAPTER 26

26.1. Here are three examples of optically active compounds:

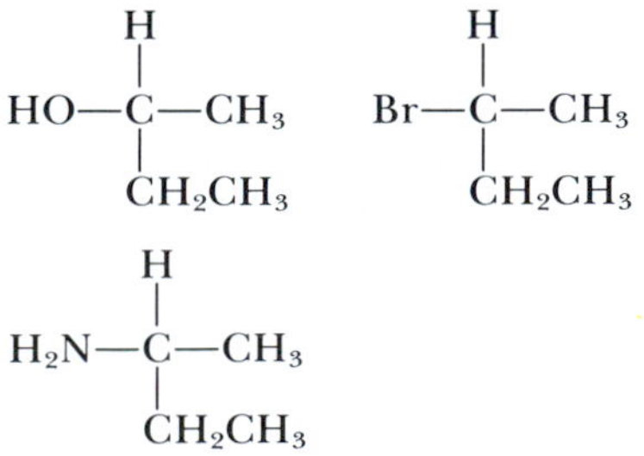

26.3. The two enantiomers of lactic acid are shown below.

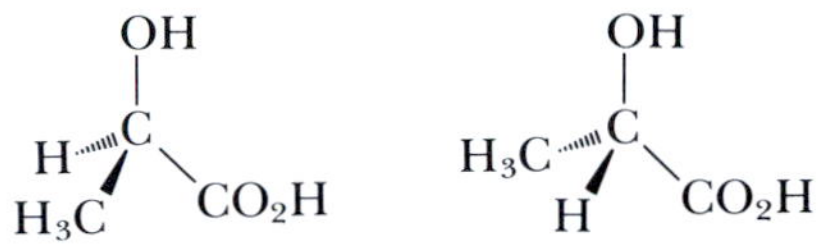

26.5. 256 possibilities
26.7.

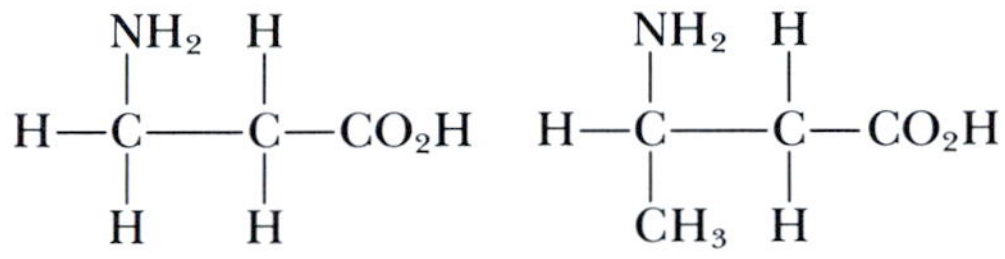

optically inactive   optically active

26.9.

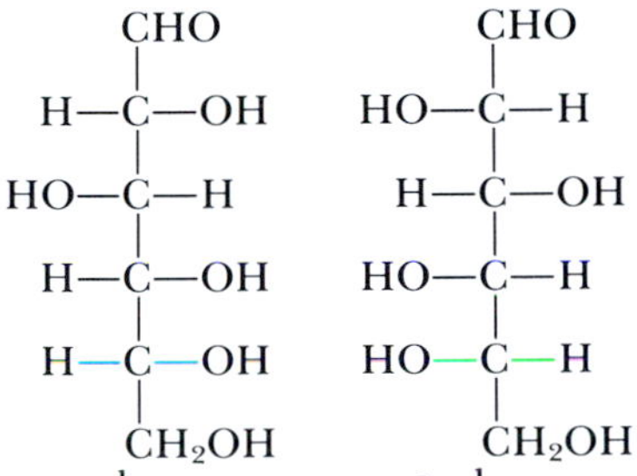

D-glucose   L-glucose

26.11.

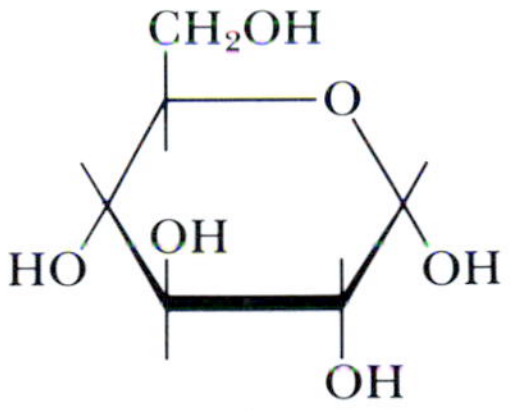

α-D-glucose

26.13. $C_6H_{12}O_6 + Cu^{+2} + H_2O \longrightarrow C_6H_{12}O_7 + 2H^+$
26.15. $CH_3(CH_2)_{14}COOH + NaOH \longrightarrow$
$$H_2O + CH_3(CH_2)_{14}COO^- Na^+$$
26.17. $2 CH_3(CH_2)_{12}COO^- Na^+ + Ca^{2+} \longrightarrow$
$$Ca[CH_3(CH_2)_{12}COO]_2 + 2Na^+$$
26.19. 123 gallons of ethanol
26.21. Alanine has a basic amino group and an acidic carboxylic acid group. The molecule as a whole will be neutral.

If acid is added to alanine in aqueous solution, the alanine will acquire a positive charge due to the ammonium cation.

If base is added to alanine in aqueous solution, the alanine will acquire a negative charge due to the carboxylate anion.

26.23. 3.72 Cal
26.25. Two hundred kilograms of glycerol

Appendix B Nobel Prize Winners

Year	Chemistry	Physics	Medicine/Physiology
1901	JACOBUS HENRICUS van't HOFF (*Dutch*)—Laws of chemical dynamics and osmotic pressure	WILHELM K. RÖNTGEN (*German*)—Discovery of X-rays	EMIL von BEHRING (*German*)—Diphtheria antitoxin
1902	EMIL FISCHER (*German*)—Studies on sugars, purine derivatives, and peptides	HENDRIK ANTOON LORENZ and PIETER ZEEMAN (*Dutch*)—Effect of magnetism on radiation—the Zeeman effect	Sir RONALD ROSS (*British*)—Malaria, the malarial parasite, and the mosquito
1903	SVANTE A. ARRHENIUS (*Swedish*)—Electrolytic dissociation theory	ANTOINE HENRI BECQUEREL and PIERRE and MARIE CURIE (*French*)—Natural (spontaneous) radioactivity	NIELS R. FINSEN (*Danish*)—The light treatment of disease, especially *Lupus vulgaris*
1904	Sir WILLIAM RAMSAY (*British*)—Discovery of inert gaseous elements in the atmosphere	LORD RAYLEIGH, JOHN WILLIAM STRUTT (*British*)—Discovery of argon and studies of gaseous densities	IVAN P. PAVLOV (*Russian*)—The physiology of digestion
1905	ADOLPH von BAEYER (*German*)—Organic dyes, aromatic compounds, and the synthesis of indigo	PHILIPP LENARD (*German*)—The properties of cathode rays	ROBERT KOCH (*German*)—Tuberculosis, discovery of tubercule bacillus and tuberculin
1906	HENRI MOISSAN (*French*)—Fluorine; the development of the electric furnace	JOSEPH JOHN THOMSON (*British*)—Electrical discharge through gases	SANTIAGO RAMON y CAJAL (*Spanish*) and CAMILLIO GOLGI (*Italian*)—Studies of the nervous system and the structure of nerve tissue
1907	EDUARD BUCHNER (*German*)—Cell-free fermentation	ALBERT A. MICHELSON (*American*)—Spectroscopic studies, optical instrumentation, and the speed of light	CHARLES L. A. LAVERAN (*French*)—Studies on protozoa in the generation of disease
1908	ERNEST RUTHERFORD (*British*)—Behavior of alpha rays; physics and chemistry of radioactive substances	GABRIEL LIPPMANN (*French*)—Color photography; the phenomenon of interference	PAUL EHRLICH (*German*) and ELIE METCHNIKOFF (*French*)—Immunity
1909	WILHELM OSTWALD (*German*)—Catalysis, chemical equilibrium, and rates of chemical reactions	GUGLIELMO MARCONI (*Italian*) and KARL FERDINAND BRAUN (*German*)—Wireless telegraphy	THEODOR KOCHER (*Swiss*)—The thyroid gland; physiology, pathology, and surgery
1910	OTTO WALLACH (*German*)—Alicyclic substances	JOHANNES D. van der WAALS (*Dutch*)—Studies of relationships between gases and liquids	ALBRECHT KOSSEL (*German*)—Cell chemistry, especially proteins and nucleic substances
1911	MARIE CURIE (*French*)—Radium and polonium and the compounds of radium	WILHELM WIEN (*German*)—Heat radiation by black bodies	ALLVAR GULLSTRAND (*Swedish*)—Studies in the refraction of light through the eye (the dioptrics of the eye)
1912	VICTOR GRIGNARD and PAUL SABATIER (*French*)—The Grignard reaction for synthesizing organic compounds	NILS G. DALÉN (*Swedish*)—Automatic gas regulators for coastal lighting	ALEXIS CARREL (*French*)—Studies on vascular seams and the grafting of organs and blood vessels
1913	ALFRED WERNER (*Swiss*)—Coordination theory for arrangements of atoms in molecules	HEIKE KAMERLINGH-ONNES (*Dutch*)—Properties of matter at low temperatures; liquefaction of helium	CHARLES RICHET (*French*)—Studies on allergies caused by foreign substances; anaphylactic test
1914	THEODORE W. RICHARDS (*American*)—Atomic weight determination of the elements	MAX T.F. von LAUE (*German*)—X-ray diffraction in crystals	ROBERT BARANY (*Austrian*)—Studies on the physiology and pathology of the human vestibular system
1915	RICHARD WILLSTATTER (*German*)—The coloring in plants, especially chlorophyll	Sir WILLIAM H. BRAGG and Sir WILLIAM L. BRAGG (*British*)—Crystal structure study by X-ray methods	*No award*
1916	*No award*	*No award*	*No award*
1917	*No award*	CHARLES G. BARKLA (*British*)—Studies on the diffusion of light and X-radiations from elements	*No award*

Year	Chemistry	Physics	Medicine/Physiology
1918	FRITZ HABER (*German*)—The Haber process for synthesizing ammonia	MAX PLANCK (*German*)—The quantum theory of light (the element of action)	*No award*
1919	*No award*	JOHANNES STARK (*German*)—The Stark effect of spectral lines in electric fields; the Doppler effect	JULES BORDET (*Belgian*)—Discoveries on immunity
1920	WALTHER NERNST (*German*)—Studies on heat changes in chemical reactions	CHARLES-EDOUARD GUILLAUME (*Swiss*)—The special properties of nickel alloys and their importance in precision physics	AUGUST KROGH (*Danish*)—Studies on the regulating action and behavior of the blood capillaries
1921	FREDERICK SODDY (*British*)—Radioactive substances and the origin and nature of isotopes	ALBERT EINSTEIN (*German*)—Contributions to mathematical physics and the photoelectric effect	*No award*
1922	FRANCIS W. ASTON (*British*)—The behavior of isotope mixtures, whole number rule on atomic weights, and the mass spectrograph	NEILS BOHR (*Danish*)—Atomic structure and atomic radiations	ARCHIBALD V. HILL (*British*) and OTTO MEYERHOF (*German*)—Discoveries on heat production in muscles and lactic acid production in muscles
1923	FRITZ PREGL (*Austrian*)—Methods of microanalysis for organic substances	ROBERT A. MILLIKAN (*American*)—The elementary electric charge and photoelectric effect	Sir FREDERICK G. BANTING (*Canadian*) and JOHN J. R. MacLEOD (*Scottish*)—Discovery of insulin
1924	*No award*	KARL M. G. SIEGBAHN (*Swedish*)—X-ray spectra	WILLEM EINTHOVEN (*Dutch*)—Discovery of the electrocardiogram
1925	RICHARD ZSIGMONDY (*German*)—Studies on the nature of colloids	JAMES FRANCK and GUSTAV HERTZ (*German*)—Laws governing the collision of an electron and an atom	*No award*
1926	THEODOR SVEDBERG (*Swedish*)—Dispersions and the chemistry of colloids	JEAN B. PERRIN (*French*)—The discontinuous structure of matter; measurements on sizes of atoms	JOHANNES FIBIGER (*Danish*)—Discovery of Spiroptera carcinoma, a cancer-producing parasite
1927	HEINRICH WIELAND (*German*)—Bile acids and related substances	ARTHUR H. COMPTON (*American*) and CHARLES T.R. WILSON (*British*)—Discovery of dispersion of X-rays reflected from atoms (the Compton effect)	JULIUS WAGNER-JAUREGG (*Austrian*)—Fever treatment (malaria vaccination) in treating paralysis
1928	ADOLPH WINDAUS (*German*)—Studies on sterols and their connection with vitamins	OWEN W. RICHARDSON (*British*)—Studies on the thermionic effect and electrons emitted by hot metals	CHARLES NICOLLE (*French*)—Typhus exanthematicus
1929	Sir ARTHUR HARDEN (*British*) and HANS von EULER-CHELPIN (*German*)—Fermentation and enzyme action	Prince LOUIS-VICTOR de BROGLIE (*French*)—The wave nature of the electron	CHRISTIAAN EIJKMAN (*Dutch*) and Sir FREDERICK G. HOPKINS (*British*)—Growth-promoting and antineuritic vitamins
1930	HANS FISCHER (*German*)—Chemistry of pyrrole and the synthesis of hemin	Sir CHANDRASEKHARA V. RAMAN (*Indian*)—The diffusion of light and the Raman effect	KARL LANDSTEINER (*American*)—The four main human blood types
1931	CARL BOSCH and FRIEDRICH BERGIUS (*German*)—High-pressure methods for chemical manufacture	*No award*	OTTO H. WARBURG (*German*)—Enzyme role in tissue respiration
1932	IRVING LANGMUIR (*American*)—Surface chemistry	WERNER K. HEISENBERG (*German*)—Quantum mechanics	EDGAR D. ADRIAN and Sir CHARLES S. SHERRINGTON (*British*)—Function of neurons
1933	*No award*	PAUL A. M. DIRAC (*British*) and ERWIN SCHRODINGER (*Austrian*)—New forms of atomic theory	THOMAS H. MORGAN (*American*)—The hereditary function of the chromosomes

APPENDIX B NOBEL PRIZE WINNERS IN CHEMISTRY, PHYSICS, AND MEDICINE/PHYSIOLOGY

Year	Chemistry	Physics	Medicine/Physiology
1934	HAROLD C. UREY (*American*)—Discovery of deuterium	*No award*	GEORGE R. MINOT, WILLIAM P. MURPHY, and GEORGE H. WHIPPLE (*American*)—Liver treatment for anemia
1935	FREDERIC and IRENE JOLIOT-CURIE (*French*)—Synthesis of new radioactive elements	JAMES CHADWICK (*British*)—Discovery of the neutron	HANS SPEMANN (*German*)—Studies in embryonic growth and development
1936	PETER J. W. DEBYE (*Dutch*)—Studies on dipole moments; diffraction of electrons in X-rays and gases	CARL D. ANDERSON (*American*) and VICTOR F. HESS (*Austrian*)—The positron and cosmic rays	Sir HENRY H. DALE (*British*) and OTTO LOWEI (*Austrian*)—Chemical transmission of nerve impluses
1937	Sir WALTER N. HAWORTH (*British*) and PAUL KARRER (*Swiss*)—Research on carbohydrates, vitamins C, A, and B₂, carotinoids, and flavins	CLINTON J. DAVISSON (*American*) and GEORGE P. THOMSON (*British*)—Diffraction of electrons by crystals	ALBERT SVENT-GYÖRGYI (*Hungarian*)—Research on biological oxidation, especially vitamin C and fumaric acid
1938	RICHARD KUHN (*German*)—Carotinoids and vitamins	ENRICO FERMI (*Italian*)—New elements beyond uranium; nuclear reactions by slow electrons	CORNEILLE HEYMANS (*Belgian*)—Regulation of respiration
1939	ADOLPH BUTENANDT (*German*) and LEOPALD RUZICKA (*Swiss*)—Sex hormones and polymethylenes	ENREST O. LAWRENCE (*American*)—The cyclotron and artificially radioactive elements	GERHARD DOMAGK (*German*)—Prontosil, the first sulfa drug
1940	*No award*	*No award*	*No award*
1941	*No award*	*No award*	*No award*
1942	*No award*	*No award*	*No award*
1943	GEORG VON HEVESY (*Hungarian*)—Isotopes as tracers in chemical studies	OTTO STERN (*American*)—Molecular beam method for the study of the atom; the magnetic moment of the proton	HENRIK DAM (*Danish*) and EDWARD DOISY (*American*)—Discovery and synthesis of vitamin K
1944	OTTO HAHN (*German*)—Atomic fission of heavy nuclei	ISADOR I. RABI (*American*)—Magnetic properties of atomic nuclei	JOSEPH ERLANGER and HERBERT S. GASSER (*American*)—Studies on single nerve fibers
1945	ARTTURI I. VIRTANEN (*Finnish*)—Discovery of new methods for agricultural biochemistry	WOLFGANG PAULI (*American*)—The exclusion principle (Pauli principle) of electrons	ERNST B. CHAIN (*German*), Sir ALEXANDER FLEMING, and Sir HOWARD W. FLOREY (*British*)—Penicillin
1946	JOHN H. NORTHROP, WENDELL M. STANLEY, and JAMES B. SUMNER (*American*)—Crystallizing enzymes; preparation of pure enzymes and virus proteins	PERCY W. BRIDGMAN (*American*)—High-pressure apparatus; studies at very high pressures	HERMANN J. MULLER (*American*)—X-ray–induced mutations
1947	Sir ROBERT ROBINSON (*British*)—Alkaloids and other plant substances	Sir EDWARD V. APPLETON (*British*)—Physical properties of the ionosphere	CARL F. and GERTY T. CORI (*American*) and BERNARDO A. HOUSSAY (*Argentinian*)—Animal metabolism and the study of the pituitary gland and pancreas
1948	ARNE TISELIUS (*Swedish*)—Nature of the serum proteins	PATRICK M. S. BLACKETT (*British*)—Cosmic radiation and nuclear physics	PAUL MULLER (*Swiss*)—Insect-killing properties of DDT
1949	WILLIAM F. GIAUQUE (*American*)—Chemical thermodynamics; effects due to extreme cold	HIDEKI YUKAWA (*Japanese*)—The meson	WALTER R. HESS (*Swiss*) and ANTONIO E. MONIZ (*Portuguese*)—Studies in brain control and brain surgery techniques
1950	OTTO DIELS and KURT ALDER (*German*)—Diene synthesis	CECIL F. POWELL (*British*)—Photographic techniques for atomic nuclei; discoveries concerning mesons	PHILIP S. HENCH, EDWARD C. KENDALL (*American*), and TADEUS REICHSTEIN (*Swiss*)—Cortisone and ACTH
1951	EDWIN M. McMILLAN and GLENN T. SEABORG (*American*)—Plutonium and other transuranium elements	Sir JOHN D. COCKCROFT (*British*) and ERNEST WALTON (*Irish*)—Transmutation of atomic nuclei through artificially accelerated atomic particles	MAX THEILER (*American*)—Discovery of yellow fever vaccine

Year	Chemistry	Physics	Medicine/Physiology
1952	ARCHER J. P. MARTIN and RICHARD L. M. SYNGE (*British*)—Partition chromatography	FELIX BLOCH and EDWARD M. PURCELL (*American*)—Magnetic moment method for atomic nuclei measurements	SELMAN A. WAKSMAN (*American*)—Streptomycin
1953	HERMANN STAUDINGER (*German*)—Studies in the synthesis of giant molecules	FRITZ ZERNIKE (*Dutch*)—Phase contrast microscope for cancer research	HANS A. KREBS (*British*) and FRITZ A. LIPMANN (*American*)—Metabolic studies and biosynthesis
1954	LINUS PAULING (*American*)—Chemical bonds in protein molecules and forces in matter	MAX BORN and WALTHER BOTHE (*German*)—Quantum mechanics and cosmic radiation	JOHN F. ENDERS, FREDERICK C. ROBBINS and THOMAS H. WELLER (*American*)—Method for growing polio viruses in test tubes
1955	VINCENT du VIGNEAUD (*American*)—Synthetic hormones	POLYKARP KUSCH and WILLIS E. LAMB (*American*)—The structure of the hydrogen spectrum; the magnetic moment of the electron	HUGO THEORELL (*Swedish*)—The nature of oxidation enzymes
1956	Sir CYRIL N. HINSHELWOOD (*British*) and NIKOLAJ N. SEMENOV (*Russian*)—Reaction kinetics and chemical reaction mechanics	JOHN BARDEEN, WALTER H. BRATTAIN, and WILLIAM SHOCKLEY (*American*)—The transistor	ANDRE F. COURNAND, DICKINSON W. RICHARDS, Jr., (*American*), and WERNER FORSSMANN (*German*)—The use of the catheter in heart research
1957	ALEXANDER R., LORD TODD (*British*)—Work on nucleotides and nucleotide coenzymes	TSUNG DAO LEE and CHEN NING YANG (*American*)—Disproving the laws of conservation of parity	DANIEL BOVET (*Italian*)—Antihistamines
1958	FREDERICK SANGER (*British*)—Structure of proteins, especially the insulin molecule	PAVEL A. CHERENKOV, ILYA M. FRANK, and IGOR TAMM (*Russian*)—Study of high-energy particles and the Cherenkov effect	GEORGE W. BEADLE, JOSHUA LEDERBERG, and EDWARD L. TATUM (*American*)—Genetic mechanisms and heredity
1959	JAROSLAV HEYROVSKY (*Czechoslavakian*)—Polarographic analysis and techniques	OWEN CHAMBERLAIN and EMILIO SEGRÈ (*American*)—The antiproton	ARTHUR KORNBERG and SEVERO OCHOA (*American*)—Artificial nucleic acids
1960	WILLARD F. LIBBY (*American*)—Radiocarbon dating	DONALD A. GLASER (*American*)—The bubble chamber for subatomic particles	Sir FRANK MACFARLANE BURNET (*Australian*) and PETER B. MEDAWAR (*British*)—Acquired immunological tolerance
1961	MELVIN CALVIN (*American*)—Photosynthesis	ROBERT HOFSTADTER and RUDOLPH MÖSSBAUER (*American*)—The nucleons and gamma ray research; the Mossbauer effect	GEORGE von BÉKÉSY (*American*)—Physical mechanisms of stimulation in the cochlea
1962	JOHN C. KENDREW and MAX F. PERUTZ (*British*)—Structure of complex globular proteins	LEV D. LANDAU (*Russian*)—Theories for condensed matter, especially liquid helium	FRANCIS H. C. CRICK, MAURICE H. F. WILKINS (*British*), and JAMES D. WATSON (*American*)—Study of structure of deoxyribonucleic acid (DNA)
1963	GIULIO NATTA (*Italian*) and KARL ZIEGLER (*German*)—Polymers of simple hydrocarbons; improved plastics	J. HANS JENSEN (*German*) and MARIA GOEPPERT-MAYER, and EUGENE P. WIGNER (*American*)—Structure of atomic nuclei and elementary particles	Sir JOHN C. ECCLES (*Australian*), ALAN L. HODGKIN, and ANDREW F. HUXLEY (*British*)—Ionic mechanisms and behavior of nerve impulses
1964	DOROTHY CROWFOOT HODGKIN (*British*)—Structures of vitamin B_{12} and penicillin by X-ray methods	NIKOLAY BASOV, ALEXANDER M. PROKHOROV (*Russian*), and CHARLES H. TOWNES (*American*)—Masers and lasers	KONRAD BLOCH (*American*) and FEODOR LYNEN (*German*)—Cholesterol and fatty acid metabolism
1965	ROBERT B. WOODWARD (*American*)—Contributions to synthetic organic chemistry	RICHARD P. FEYNMAN, JULIAN S. SCHWINGER (*American*), and SCHINICHIRO TOMONAGA (*Japanese*)—Basic studies in quantum electrodynamics	FRANCOIS JACOB, ANDRE LWOFF, and JACQUES MONOD (*French*)—Genetic control in the synthesis of enzymes and viruses
1966	ROBERT S. MULLIKEN (*American*)—Molecular orbital theory for chemical structure	ALFRED KASTLER (*French*)—Studies on the energy levels of atoms	CHARLES B. HUGGINS and FRANCIS P. ROUS (*American*)—Use of hormones in treating cancer and discovery of a cancer-producing virus

APPENDIX B NOBEL PRIZE WINNERS IN CHEMISTRY, PHYSICS, AND MEDICINE/PHYSIOLOGY

Year	Chemistry	Physics	Medicine/Physiology
1967	MANFRED EIGEN (*German*), RONALD G. W. NORRISH, and GEORGE PORTER (*British*)—Very fast chemical reactions	HANS ALBRECHT BETHE (*American*)—Theory of nuclear reactions, especially on the source of energy in stars	RAGNER GRANIT (*Swedish*), H. KEFFER HARTLINE, and GEORGE WALD (*American*)—Chemical and physiological visual processes in the eye
1968	LARS ONSAGER (*American*)—Various types of relationships for thermodynamic activity	LUIS W. ALVAREZ (*American*)—Studies on subatomic particles and techniques for detecting them	ROBERT W. HOLLY, H. GOBIND KHORANA, and MARSHALL W. NIRENBERG (*American*)—Role of the genes in cell function
1969	DEREK H. BARTON (*British*) and ODD HASSEL (*Norwegian*)—Conformations and the relationship between chemical shape and chemical reactivity	MURRAY GELL-MANN (*American*)—Contributions toward the understanding and classification of elementary particles	MAX DELBRÜCK, ALFRED D. HERSHEY, and SALVADOR E. LURIA (*American*)—Viruses, viral diseases, and the foundations of molecular biology
1970	LUIS F. LELOIR (*Argentinian*)—Sugar nucleotides	HANNES ALFVEN (*Swedish*) and LOUIS NEEL (*French*)—Theoretical basis for magnetohydrodynamics; antiferromagnetic materials and behavior	JULIUS AXELROD (*American*), ULF S. von EULER (*Swedish*), and BERNARD KATZ (*British*)—Discoveries related to the search for remedies for nervous and mental disturbances
1971	GERHARD HERZBERG (*Canadian*)—Contributions to electronic structure and geometry of molecules	DENNIS GABOR (*British*)—Invention and development of the holographic method of three-dimensional imagery	EARL W. SUTHERLAND, Jr. (*American*)—Discoveries concerning the mechanisms of the actions of hormones
1972	CHRISTIAN B. ANFINSEN, STANFORD MOORE, and WILLIAM H. STEIN (*American*)—Work on ribonuclease	JOHN BARDEEN, LEON N. COOPER, and JOHN R. SCHRIEFFER (*American*)—Jointly developed the theory of superconductivity	GERALD M. EDELMAN (*American*) and RODNEY R. PORTER (*British*)—Research on the chemical structure and nature of antibodies
1973	ERNST OTTO FISCHER (*West German*) and GEOFFREY WILKINSON (*British*)—Work on the chemistry of the organometallic, so-called sandwich compounds	IVAR GIAEVER (*American*), LEO ESAKI (*Japanese*), and BRIAN D. JOSEPHSON (*British*)—Experimental discoveries regarding tunneling phenomena in semiconductors and superconducters	KARL von FRISCH, KONRAD LORENZ (*Austrian*), and NIKOLAAS TINBERGEN (*Dutch*)—Studies of individual and social behavior patterns
1974	PAUL J. FLORY (*American*)—Fundamental achievements in the physical chemistry of the macromolecules	MARTIN RYLE and ANTHONY HEWISH (*British*)—Radio astrophysics and the discovery of pulsars	GEORGE E. PALADE, CHRISTIAN de DUVE (*American*), and ALBERT CLAUDE (*Belgian*)—Contributions to understanding the inner workings of living cells
1975	JOHN CORNFORTH (*Australian-British*) and VLADIMIR PRELOG (*Yugoslavian-Swiss*)—Stereochemistry of enzyme-catalyzed reactions	JAMES RAINWATER (*American*), BEN MOTTELSON (*American-Danish*), and AAGE BOHR (*Danish*)—Discovery of the connection between motion in atomic nuclei and the development of the theory of the structure of the atomic nucleus	DAVID BALTIMORE, HOWARD M. TEMIN, and RENATO DULBECCO (*American*)—Work on the interaction between tumor viruses and the cell's genetic material
1976	WILLIAM N. LIPSCOMB (*American*)—Studies on the structure of boranes illuminating problems of chemical bonding	BURTON RICHTER and SAMUEL C. C. TING (*American*)—Discovery of a heavy elementary particle of a new kind	BARUCH S. BLUMBERG and D. CARLETON GAJDUSEK (*American*)—Discoveries concerning new mechanisms for the origin and dissemination of infectious diseases
1977	ILYA PRIGOGINE (*Belgian*)—Contributions to nonequilibrium thermodynamics	JOHN H. van VLECK, PHILIP W. ANDERSON (*American*), and NEVILL F. MOTT (*British*)—Fundamental theoretical investigations of the electronic structure of magnetic and disordered systems	ROSALYN S. YALOW, ROGER C. L. GUILLEMIN, and ANDREW V. SCHALLY (*American*)—Research in the role of hormones in the chemistry of the body
1978	PETER MITCHELL (*British*)—Contribution to the understanding of biological energy transfer through formulation of the chemiosmotic theory	PYOTR KAPITSA (*Russian*), ARNO PENZIAS, and ROBERT WILSON (*American*)—Basic inventions and discoveries in the area of low-temperature physics	DANIEL NATHANS, HAMILTON SMITH (*American*), and WERNER ARBER (*Swiss*)—Discovery of restriction enzymes and their application to problems of molecular genetics

Year	Chemistry	Physics	Medicine/Physiology
1979	HERBERT C. BROWN (*American*) and GEORGE WITTIG (*German*)—Development of boron- and phosphorus-containing compounds	STEVEN WEINBERG, SHELDON L. GLASHOW (*American*), and ABDUS SALAM (*Pakistani*)—Contributions to the theory of the unified weak and electromagnetic interaction between elementary particles	ALLAN McLEOD CORMACK (*American*) and GODFREY NEWBOLD HOUNSFIELD (*British*)—Development computed axial tomography (CAT scan) X-ray technique
1980	PAUL BERG, WALTER GILBER (*American*), and FREDERICK SANGER (*British*)—Fundamental studies of the biochemistry of nucleic acids, with particular regard to recombinant DNA	JAMES W. CRONIN and VAL L. FITCH (*American*)—Discovery of violations of fundamental symmetry principles in the decay of neutral K-mesons	BARUJ BENECERRAF, GEORGE D. SNELL (*American*), and JEAN DAUSSET (*French*)—Discoveries that explain how the structure of cells relates to organ transplants and diseases
1981	KENICHI FUKUI (*Japanese*) and ROALD HOFFMANN (*American*)—Theories concerning the course of chemical reactions	NICOLASS BOEMBERGEN, ARTHUR SCHLAWLOW (*American*) and KAI M. SIEGBAHN (*Swedish*)—Contribution to the development of laser spectroscopy	ROGER W. SPERRY, DAVID H. HUBEL (*American*), and TORSTEN N. WIESEL (*Swedish*)—Studies vital to understanding the organization and functioning of the brain
1982	AARON KLUG (*South African*)—Development of crystallographic electron microscopy and structural elucidation of biologically important nucleic acid-protein complexes	KENNETH G. WILSON (*American*)—Theory for critical phenomena in connection with phase transitions	SUNE BERGSTROMM, BENGT SAMUELSSON (*Swedish*), and JOHN R. VANE (*British*)—Research in prostaglandins, a hormone-like substance involved in a wide range of illnesses
1983	HENRY TAUBE (*Canadian*)—Studies of the mechanisms of electron transfer reactions, particularly of metal complexes	SUBRAHMANYAN CHANDRASEKHAR and WILLIAM FOWLER (*American*)—Theoretical studies of the physical processes of importance to the structure and evolution of stars	BARBARA McCLINTOCK (*American*)—Discovery of mobile genes in the chromosomes of a plant that change the future generations of plants they produce
1984	BRUCE MERRIFIELD (*American*)—Methodology for chemical synthesis on a solid matrix	SIMON van der MEER (*Dutch*) and CARLO RUBBIA (*Italian*)—Discovery of field particles of W and Z, communicators of the weak interaction	CESAR MILSTEIN (*British-Argentinian*), NIELS K. JERNE (*British-Danish*), and GEORGES J. F. KOHLER (*West German*)—Work in immunology
1985	HERBERT A. HAUPTMANN and JEROME KARLE (*American*)—Development of direct methods for the determination of crystal structures	KLAUS von KLITZING (*West German*)—Discovery of the quantum Hall effect	MICHAEL S. BROWN and JOSEPH L. GOLDSTEIN (*American*)—Contributions to our understanding of cholesterol metabolism, prevention and treatment of atherosclerosis and heart attacks
1986	DUDLEY HERSCHBACH, YUAN T. LEE (*American*), and JOHN C. POLANYI (*Canadian*)—Fundamental contributions to the development of the dynamics of chemical reactions	ERNEST RUSKA (*German*), GERD BINNIG (*West German*), and HEINRICH ROHRER (*Swiss*)—Design of the scanning tunneling microscope	RITA LEVI-MOTALCINI (*American-Italian*) and STANLEY COHEN (*American*)—Contributions to understanding the substance that influences cell growth
1987	DONALD J. CRAM, CHARLES J. PEDERSON (*American*), and JEAN-MARIE LEHN (*French*)—Discovery of how to make relatively simple molecules that mimic the functions of much more complex molecules produced by living cells	K. ALEX MUELLER (*Swiss*) and J. GEORGE BEDNORZ (*West German*)—Discovery of superconductivity in a new class of ceramics at temperatures higher than had previously been thought possible	SUSUMU TONEGAWA (*Japanese*)—Discoveries of how the body can marshal its immunological defenses against millions of new and different disease agents
1988	HARTMUT MICHEL, ROBERT HUBER, and JOHANN DEISENHOFER (*West German*)—Mapping the structure of protein molecules essential in photosynthesis	LEON LEDERMAN, MELVIN SCHWARTZ, and JACK STEINBERGER (*American*)—Experiments with the subatomic neutrino particle	GEORGE H. HITCHINGS, GERTRUDE B. ELION (*American*), and Sir JAMES W. BLACK (*British*)—Work that had led to the introduction of drugs widely used to treat heart disease, ulcers, and leukemia
1989	SIDNEY ALTMAN and THOMAS CECH (*American*)—Discovered the active role played by RNA in catalyzing chemical reactions in cells	NORMAN RAMSAY (*American*)—Techniques for studying the structure of atoms, the hydrogen atom and hydrogen maser, and development of the cesium atomic clock; HANS DEHMELT (*American*) and WOLFGANG PAUL (*German*)—The ion-trap method for separating charged particles, especially the electron and ions	MICHAEL BISHOP and HAROLD VARMUS (*American*)—Studies leading to understanding of how normal genes that control cell growth can cause cancer

APPENDIX
C

Scientific Notation

Scientific notation of numbers is often also referred to as **exponential notation** since numbers are expressed in terms of exponents or powers of ten. This notation serves well in science since it is convenient for expressing the very large and very small numbers often encountered. In addition it gives a way of stating precision or number of significant figures without any ambiguities.

To begin with it is worthwhile to review some even powers of ten.

10000000000.	10^{10}
100000.	10^5
100.	10^2
10.	10^1
1.	10^0
0.01	10^{-2}
0.0000001	10^{-7}
0.000000000001	10^{-12}

For the numbers above, the exponent on the ten equals the number of places that the decimal point must be moved to reach the number 1. Movement of the decimal point to the left gives a positive exponent, whereas movement to the right gives a negative exponent.

In order to express numbers other than even powers of ten it is necessary to multiply ten raised to some power by a coefficient. This coefficient is normally chosen to be between one and ten. Thus,

$$269000 = 2.69 \times 100000 = 2.69 \times 10^5$$
$$0.000000318 = 3.18 \times 0.0000001 = 3.18 \times 10^{-7}$$

The convenience of scientific notation for very large and very small numbers is clear. Avogadro's number, the number of atoms or molecules in a mole of a substance, is always given in scientific notation, 6.022×10^{23}, to avoid having to write 602200000000000000000000. The mass of an electron in kilograms is always given as 9.1094×10^{-31} so that we don't have to write 0.00000000000000000000000000000091094.

Significant figures are very easily expressed in scientific notation. The rule is that all figures given in the coefficient are significant. As examples,

$3. \times 10^5$	1 significant figure
3.00×10^5	3 significant figures
3.00000×10^5	6 significant figures

When the number is expressed as 300000, there is no standard way of stating the number of significant figures it contains.

Addition or subtraction of numbers in scientific notation requires that all contain the same power of ten. This can involve moving of decimal points to obtain the same powers. Then the coefficients can be added or subtracted. For example,

$$
\begin{aligned}
4.23 \times 10^4 &= 0.423 \times 10^5 \\
+3.26 \times 10^5 &= \underline{+3.26 \times 10^5} \\
&3.68 \times 10^5
\end{aligned}
$$

In multiplication, the coefficients are multiplied in the ordinary way and then the exponents are added together.

$$(4.0 \times 10^2) \times (5.0 \times 10^4) = 20. \times 10^6 = 2.0 \times 10^7$$

For division the coefficients are divided and then one exponent is subtracted from the other.

$$(3.0 \times 10^3)/(4.0 \times 10^4) = 0.75 \times 10^{-1} = 7.5 \times 10^{-2}$$

Raising numbers in scientific notation to powers is accomplished by raising the coefficient to the desired power and multiplying that power by the exponent of the ten.

$$(2.0 \times 10^4)^3 = 8.0 \times 10^{12}$$

Taking roots of numbers in scientific notation follows the same rules. Remember that a root is the reciprocal of a power. Thus

$$\sqrt{a} = (a)^{1/2}$$

Often the decimal point in the coefficient has to be moved so that the power of ten in the answer is an integer.

$$(5.0 \times 10^{-3})^{1/2} = (50 \times 10^{-4})^{1/2} = 7.1 \times 10^{-2}$$

APPENDIX

D

Logarithms

The logarithm of a number is the power to which some base number must be raised to give the original number. The most familiar base is 10, giving common logarithms which have the symbol **log** or **log₁₀**. The other common base is **e** (2.71828), giving natural logarithms with the symbol **ln** or **logₑ**. Common logarithms have traditionally been the more usually tabulated. However, natural logarithms show up in scientific derivations and are now easily available on electronic calculators.

As two examples of common logarithms,

$\log 1000 = 3$, since $10^3 = 1000$

$\log 513 = 2.71$, since $10^{2.71} = 513$

An example of a natural logarithm is

$\ln 10 = 2.303$, since $e^{2.303} = 10$

As shown in these examples, the common or natural logarithm of a number greater than one has a positive value. The common or natural logarithm of exactly one is zero, since the zero power of any base equals one.

$\log 1 = 0$, since $10^0 = 1$

$\ln 1 = 0$, since $e^0 = 1$

Numbers less than one have negative common or natural logarithms.

$\log 0.1 = -1$, since $10^{-1} = 1$

$\log 0.0209 = -1.68$, since $10^{-1.68} = 0.0209$

$\ln 0.0030 = -5.81$, since $e^{-5.81} = 0.0030$

Conversion between common and natural logarithms uses the equation

$\ln a = 2.303 \log a$

Formerly, logarithms were determined from extensive tables, but either common or natural logarithms are now directly obtained using an electronic calculator.

The antilogarithm is the inverse of a logarithm. Thus

$\log n = x$

antilog $x = n$

$10^x = n$

Logarithms of numbers are convenient for multiplication or division of numbers and determining powers and roots of numbers. Either common or natural logarithms can be used for these purposes. Because logarithms are exponents, multiplication of two numbers can be accomplished by adding their logarithms and then taking the antilogarithm of the sum.

$m \cdot n$ = antilog (log m + log n)

m/n = antilog (log m − log n)

m^y = antilog ($y \cdot$ log m)

Similarly, taking the logarithm of each side of these three equations,

log ($m \cdot n$) = log m + log n

log (m/n) = log m − log n

log (m^y) = $y \cdot$ log m

In the logarithm of a number, the digits to the left of the decimal point merely determine the place of the decimal point in the original number. Therefore, the number of significant figures in the original number should equal the number of digits to the *right* of the decimal point in its logarithm.

APPENDIX

E

Quadratic Equations

A quadratic equation is one in which the unknown value to be solved for appears as the second power or square. These equations are easy to solve in cases such as

$$x^2 = 9.86$$
$$x = \sqrt{9.86}$$

In obtaining the solution to this equation it is only necessary to recall that the square root can be either a positive or a negative number. Thus there are two solutions to this equation,

$$x = +3.14$$
$$x = -3.14$$

If the equation is simply a pure, mathematical expression with no physical significance, it is impossible to make a choice between the two solutions. Either solves the equation. However, in actual problems it is often possible to discard one solution. For example, one might be solving for a pressure, a volume, or an absolute temperature for which a negative value is impossible.

Several methods of solution are available if the unknown appears in an equation to both the first and second power. The most dependable method uses the quadratic formula,

$$x = \frac{-b \pm \sqrt{b^2 - 4ac}}{2a}$$

To use this formula it is necessary to arrange the quadratic equation to be solved in the form

$$ax^2 + bx + c = 0$$

The values of a, b, and c can then be substituted into the quadratic formula. Use of the formula will give two solutions for x, one of which may possibly be discarded because of physical restrictions imposed by the nature of the problem. Examples of the use of this formula appear in Chapter 10.

Graphing experimental data is a useful technique for determining the relationships between measured variables and the values of parameters. The most useful form in which to graph an equation is that which produces a linear result, a straight line. The general form of a straight line is

$$y = mx + b$$

The variables are x and y with y plotted on the vertical axis, or ordinate, and x plotted on the horizontal axis, or abscissa. The slope of the line, $\Delta y/\Delta x$, is m and the intercept on the y-axis is b.

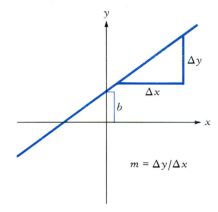

As an example, consider graphing the following equation as a straight line.

$$\ln K = -\frac{\Delta H^0}{RT} + \frac{\Delta S^0}{R}$$

The variables in this equation are K and T. The values of R, ΔH^0, and ΔS^0 are considered to be constants under the conditions involved. This equation will give a linear plot if

$$y = \ln K$$

$$x = \frac{1}{T}$$

$$m = -\frac{\Delta H^0}{R}$$

$$b = \frac{\Delta S^0}{R}$$

Thus experimental values for K and T can be plotted in a graph of $\ln K$ as a function of $1/T$. R is a standard constant which can be looked up. The values of ΔH^0 and ΔS^0 can be determined from the measured values of the slope and the intercept.

Derivations Using Calculus

A number of equations derived or given without proof in the chapters on thermodynamics and kinetics can be arrived at easily using fundamental calculus.

Thermodynamics

If a gas is contained in a container of variable volume and its volume is changed by the infinitesimal increment dV against an external pressure P_{ex}, the infinitesimal increment of work done is

$$dw = -P_{ex}dV \tag{1}$$

If the volume is changed from V_1 to V_2 against a constant external pressure, the work can be determined by integrating both sides of Eq. (1), giving

$$\int dw = w = -P_{ex}\int_{V_1}^{V_2} dV = -P_{ex}\Delta V \quad \text{(for constant } P_{ex})$$

This is the same result given in the text.

For the reversible change of volume of an ideal gas the external pressure will be changed by infinitesimal increments and nRT/V can be substituted for P_{ex} in Eq. (1). In this case integration between V_1 and V_2 gives

$$\int dw = w = -\int nRTdV/V = nRT\int_{V_1}^{V_2} dV/V$$
$$= -nRT(\ln V_2 - \ln V_1) = -nRT \ln (V_2/V_1)$$

This equation was given in the text without proof.

The first law of thermodynamics is

$$E = q + w$$

Differentiation of this gives

$$dE = dq + dw = dq - P_{ex}dV \tag{2}$$

If V is constant, then $P_{ex}dV = 0$ and

$$dE = dq_v$$

Integration of both sides of this equation gives

$$\Delta E = q_v$$

Enthalpy was defined as

$$H = E + PV$$

Differentiation of this expression gives

$$dH = dE + PdV + VdP \qquad (3)$$

Substitution of Eq. (2) into Eq. (3) gives

$$dH = dq - P_{ex}dV + PdV + VdP$$

If $P = P_{ex}$ and is constant, then

$$dH = dq_p$$

Integration of this expression gives

$$\Delta H = q_p$$

For the infinitesimal change of temperature of a substance, dT, the heat transferred is

$$dq = CdT \qquad (4)$$

where C is the heat capacity of the substance. Integration of the equation between T_1 and T_2 gives

$$\int dq = q = C \int_{T_1}^{T_2} dT = C\Delta T$$

For the infinitesimal transfer of heat at temperature T the entropy change is given as

$$dS = dq_{rev}/T \qquad (5)$$

If the temperature is held constant for a finite heat transfer

$$\int dS = \Delta S = \int dq_{rev}/T = q_{rev}/T$$

For the reversible change of temperature of a substance, substitution of Eq. (4) into Eq. (5) gives

$$dS = CdT/T$$

Integration of this between T_1 and T_2 gives

$$\int dS = \Delta S = \int_{T_1}^{T_2} CdT/T = C \int_{T_1}^{T_2} dT/T = C \ln (T_2/T_1)$$

The Gibbs free energy is defined by the equation

$$G = H - TS = E + PV - TS$$

Differentiation of this equation gives

$$dG = dE + PdV + VdP - TdS - SdT \qquad (6)$$

If all of the work is pressure-volume work,

$$dE = dq - P_{ex}dV$$

If the process is reversible,

$$dS = dq_{rev}/T \text{ and } dE = TdS - PdV$$

Therefore, under these conditions

$$dG = VdP - SdT$$

If T is constant, $SdT = 0$ and

$$dG = VdP$$

Substitution of $(RT)/P$ for one mole of an ideal gas gives

$$dG = (RT)dP/P$$

Integration of this expression between the standard pressure, $P^0 = 1$ atm, and P gives

$$\int_{G^0}^{G} dG = (RT) \int_{P^0}^{P} dP/P = G - G_0 = RT \ln P/P^0 = RT \ln P$$

For electrochemical cells

$$w = w_{pv} + w_{elec} \text{ and } dw = dw_{pv} + dw_{elec}$$

Substitution of $dE = dq + dw$ into Eq. (6) gives

$$dG = dq + dw + PdV + VdP - TdS - SdT$$

For a reversible process at constant pressure and temperature, VdP and SdT are zero and $TdS = q$. Thus,

$$dG = -dw + PdV$$

and for an electrochemical cell

$$dG = dw_{pv} + dw_{elec} + PdV$$
$$dG = dw_{elec}$$

Kinetics

Since the rate of a reaction is normally constantly changing, it is more accurately expressed in differential notation with infinitesimal increments of time and concentration rather than finite increments as expressed by the use of Δ's. Thus the rate for the disappearance of reactant A is given by

$$\text{rate} = -d[A]/dt$$

For a first-order reaction this gives a rate law of

$$-d[A]/dt = k[A] \text{ or } -d[A]/[A] = kdt$$

Integration of this expression between the initial time and the initial concentration, $[A_0]$, and the final concentration, $[A]$, gives

$$-\int_{[A_0]}^{[A]} (d[A]/[A]) = k \int_{0}^{t} dt = -\ln [A] + \ln [A_0] = kt$$

And

$$-\ln ([A]/[A_0]) = kt \quad \text{or} \quad \ln ([A_0]/[A]) = kt$$

For a second-order reaction the rate law is

$$-d[A]/dt = k[A]^2 \quad \text{or} \quad -d[A]/[A]^2 = kdt$$

A similar integration of this gives

$$-\int_{[A_0]}^{[A]} (d[A]/[A]^2) = k \int_{0}^{t} dt = (1/[A]) - (1/[A_0]) = kt$$

For a zero-order reaction the rate law is

$$-d[A]/dt = k \quad \text{or} \quad -d[A] = kdt$$

Integration of this expression gives

$$-\int_{[A_0]}^{[A]} d[A] = k \int_{0}^{t} dt = [A_0] - [A] = kt$$

APPENDIX
H

Physical Constants

Acceleration Due to Gravity	980.7 cm/sec^2
Atomic Mass Unit (μ or amu)	1.66×10^{-24} g
Avogadro's Number, N_A	6.022×10^{23} particles/mol
Boltzmann's Constant, k	1.38×10^{-16} erg/K; 1.38×10^{-23} J/K
Electron Charge	4.8033×10^{-10} esu
Electron Mass	9.1094×10^{-28} g
Faraday's Constant	9.6485×10^4 J/V
Gas Constant, R	0.0821 L·atm/mol·K 1.987 cal/mol·K 8.31451×10^7 erg/mol·K 8.31451 J/mol·K
Pi (π)	3.1416
Planck's Constant, h	6.626×10^{-27} erg·sec; 6.626×10^{-34} J·sec
Speed of Light	2.9979×10^{10} cm/s

APPENDIX

I

Conversion Factors

Density

$1 \text{ g/cm}^3 = 10^3 \text{ kg/m}^3$

$1 \text{ lb/ft}^3 = 16.0185 \text{ kg/m}^3$

Electrical Charge

$1 \text{ esu} = 3.33560 \times 10^{-10} \text{ C (coulombs)}$

$1 \text{ electron} = 4.8033 \times 10^{-10} \text{ esu}$

$1 \text{ Faraday} = 96500 \text{ C} = 6.022 \times 10^{23} \text{ electrons}$

Electrical Potential

$1 \text{ V} = 1 \text{ J/C}$

Energy

$1 \text{ J} = 1 \text{ N·m (newton·meter)} = 10^7 \text{ erg}$

$1 \text{ erg} = 1 \text{ dyne·cm} = 6.2415 \times 10^{11} \text{ eV}$

$1 \text{ cal} = 4.184 \text{ J}$

$1 \text{ L·atm} = 24.2 \text{ cal}$

$1 \text{ kWh (kilowatt hour)} = 3.6 \times 10^6 \text{ J}$

$1 \text{ Btu} = 1.055 \times 10^3 \text{ J}$

Force

$1 \text{ newton} = 1 \times 10^5 \text{ dynes} = 1 \text{ kg·m/sec}^2$

$1 \text{ dyne} = 1 \text{ g·cm/sec}^2$

Frequency

$1 \text{ Hz} = 1 \text{ cycle/s}$

Length

$1 \text{ mile} = 1609 \text{ m}$

$1 \text{ in} = 2.54 \text{ cm}$

$1 \text{ Å} = 10^{-8} \text{ cm}$

$1 \text{ } \mu \text{ (micron)} = 10^{-6} \text{ m}$

Mass

$1 \text{ metric ton} = 1000 \text{ kg}$

$1 \text{ ton (short ton)} = 2000 \text{ lb} = 907.2 \text{ kg}$

$1 \text{ amu} = 1 \text{ } \mu = 1.66 \times 10^{-24} \text{ g}$

$1 \text{ lb} = 454 \text{ g}$

Power

　　1 W (watt) = 1 J/s

Pressure

　　1 atm = 101,325 Pa (pascal)

　　1 atm = 760 torr = 760 mm of Hg

　　1 lb/in^2 = 6894 Pa

　　1 bar = 10^5 Pa

Volume

　　1 m^3 = 1000 L

　　1 cm^3 = 1 ml

　　1 ft^3 = 28.317 L

　　1 gal (US) = 3.785 L

APPENDIX

J

Vapor Pressure of Water as a Function of Temperature

T, °C	P_{vap}, torr	T, °C	P_{vap}, torr
0	4.58	35	42.2
5	6.54	40	55.3
10	9.21	45	71.9
12	10.52	50	92.5
14	11.99	55	118.0
16	13.63	60	149.4
17	14.53	65	187.5
18	15.48	70	233.7
19	16.48	80	355.1
20	17.54	90	525.8
21	18.65	92	567.0
22	19.83	94	610.9
23	21.07	96	657.6
24	22.38	98	707.3
25	23.76	100	760.0
26	25.21	102	815.9
27	26.74	104	875.1
28	28.34	106	937.9
29	30.04	108	1004.4
30	31.81	110	1073.6

Glossary

absolute temperature A temperature scale that has its zero value at absolute zero.

acceleration The rate of change of velocity per unit time.

accuracy The degree to which measured and true values correspond; how correct a measurement is. See **precision**.

acid rain Precipitation that has been made more than naturally acidic as a result of sulfur and nitrogen oxides from polluted air.

acids Substances noted for their sour taste and ability to dissolve some metals and metal oxides, change the colors of vegetable dyes, and neutralize bases (see **bases**). Acids are defined chemically as being proton donors or electron acceptors.

actinides The 14 elements, thorium through lawrencium, that follow actinium in the periodic table.

activation energy The energy barrier to reaction; the excess over the ground state level that has to be overcome for a chemical reaction to take place.

adhesion Molecular forces of attraction between unlike particles acting to hold them together.

adsorption Adhesion at the surface of a solid; "sticking" of a few monolayers of particles on the solid surface.

air pollution Fouling of the atmospheric environment by the introduction of natural and artificial contaminants.

alcohol A class of organic molecules generally consisting of a hydrocarbon portion and a hydroxy (OH) group; for example, ethyl alcohol (ethanol), CH_3CH_2—OH.

aldehyde A class of organic compounds resulting from oxidation of alcohols and having the characteristic —CHO group; for example, formaldehyde (H—CHO); acetaldehyde (CH_3—CHO).

alkali metals The IA family of metallic elements, from lithium through cesium (and francium); readily give up an electron, forming stable cations; are strongly reducing.

alkaline earth metals The group IIA elements.

allotropes Element substances that occur in more than one crystalline form: diamond and graphite, for example.

alloy A solid solution of two or more metals.

alpha particle (α), alpha ray Positively charged particle emitted by certain radioactive substances and present in certain emanations and in cathode ray tubes; composed of two neutrons and two protons, a doubly charged helium cation. An alpha ray is a stream of alpha particles.

amides A general class of organic nitrogen molecules characterized by the functional group —CONH— or peptide linkage.

amino acids Organic compounds containing amine and a carboxylic acid group. Specifically the 20 α-amino acids that are the constituents of proteins.

amines Organic ammonia derivatives in which successive hydrogen atoms have been replaced by hydrocarbon groups.

amorphous Noncrystalline; having neither definite shape nor structure.

ampere An electric current flow of one **coulomb** per second.

anion See **ion**.

anode The electrode where **oxidation** (removal of electrons) takes place in an electrochemical cell.

antibiotics Chemical substances that inhibit the action of bacteria. Most antibiotics are produced from microorganisms.

antibonding molecular orbital A molecular orbital higher in energy than the two atomic orbitals that were combined to form it. Electrons in an antibonding molecular orbital are not generally between the atoms involved.

antiparticles Pairs of particles that can annihilate each other, such as electrons and positrons.

aromaticity The general chemical properties and behavior characteristic of benzene and organic molecules containing the benzene ring or related benzene ring systems.

Arrhenius equation A relationship between the specific reaction rate constant and the activation energy for a given reaction: $k = Ae^{-Ea/RT}$.

atom Smallest particle of an element substance that still retains the characteristic properties of the element.

atomic mass number (A) An integer which gives the total number of nucleons (protons and neutrons) within the nucleus of an atom.

atomic number (Z) The number of protons in the atomic nucleus (unit positive charges on the nucleus); equal to the number of electrons in a neutral atom.

atomic pile An old-style nuclear reactor built by piling up bricks of graphite and natural uranium. Provides a self-sustaining fission process that releases a great deal of energy.

atomic weight, atomic mass The mass of an atom of an element relative to the mass of the ^{12}C isotope of carbon taken as 12.

autoionization, autoprotolysis An acid-base interaction between solvent and solute molecules involving proton transfer. In the case of water the resulting ions are hydronium and hydroxide.

atmospheric pressure See **pressure**.

Avogadro's hypothesis (law) Under the same conditions of temperature and pressure equal volumes of gases contain identical numbers of particles.

Avogadro's number (N) The number of carbon atoms in exactly 12 grams of the ^{12}C isotope of carbon, 6.022×10^{23}. The number of molecules in a mole of molecules; the ratio of the gas constant R and the Boltzmann constant k.

barbiturates A general class of organic nitrogen compounds related to barbituric acid; they act as sedatives, depressing the central nervous system.

bases Classically, ionic substances that dissociate into hydroxide ions in aqueous solutions; they are defined chemically as proton acceptors or electron donors. See also **acids**.

beta particle (β) Elementary particle emitted from a radioactively decaying nucleus, identical to an electron; also generated in certain vacuum tube discharges.

binary Containing atoms of only two elements. Said of salts, acids, and covalent compounds.

black body radiation Spectral distribution of radiant energies described by Planck's modified version of the distribution law for a perfect black body; a perfect radiator, for which the distribution of energies is strictly a function of the temperature of the radiation.

Bohr theory An atomic structure theory based on the existence of discrete electron orbits (energy levels) and emission and absorption of electromagnetic energy when electrons move between orbits or levels according to the Planck relationship $\Delta E = h\nu = E_2 - E_1$.

boiling point The temperature at which the vapor pressure of a liquid equals the pressure of the atmosphere. It is called the **normal boiling point** at a pressure of one standard atmosphere.

bonding molecular orbital For the hydrogen molecule, when two atomic orbitals are mixed in-phase, the electron density is concentrated between the nuclei and the energy of nuclei and electrons is lowered. In a manner of speaking, increased electron density is the nuclear "adhesive" holding things together. See **orbital**.

Boyle's law The relationship between pressure and volume for ideal gases; at constant temperature the product of pressure times volume is a constant.

buffers Solutions prepared in such a way that they are protected from drastic changes in acidity or basicity. A solution of a weak acid and its salt is the most common form of buffer.

calorie A common unit of heat, classically defined as the quantity of heat needed to raise the temperature of one gram of water by one degree Celsius or Kelvin. The calorie can be expected to be gradually replaced by the SI unit, the joule (1 cal = 4.184 joules).

carbohydrates Sugars and the more complex polysaccharides: cellulose, starch, glycogen.

catalyst A substance that changes the rate of a chemical reaction without being permanently consumed in the process; catalysts which slow a reaction are commonly called **inhibitors**.

catenation The formation of chains or rings of an element by means of covalent bonds between atoms. Carbon has the highest tendency to catenation.

cathode The electrode where **reduction** (loss of electrons) takes place in an electrochemical cell.

cation See **ion**.

ceramic A material composed of crystalline substances. Ceramics are characteristically brittle but capable of resisting high temperatures.

chain reaction A self-perpetuating series of reactions involving an initiation step, a series of chain-propagation reactions, and one or more chain-termination processes; for example, the photochemical chlorination of methane. Also, nuclear chain reactions involving neutron absorption and production.

Charles's law The relationship between volume and temperature of an ideal gas at constant pressure; the volume occupied by an ideal gas is proportional to the absolute temperature.

chemistry The science of substances, dealing with the investigation of all composition of matter and interaction between matter and energy in the universe.

chemical properties Properties of a substance which result in its being changed into a different substance.

chirality The characteristic quality of an asymmetric molecule that it cannot be superimposed on its reflection in a mirror.

chlorofluorocarbons Compounds containing chlorine, fluorine, and carbon, which are used as propellants, refrigerants, and foaming agents. They are generally stable and nontoxic but are a serious threat to the ozone layer. Also called CFC's and "Freons."

chromatography A technique for physically separating and purifying substances by selective adsorption, usually on a solid support, as components pass over.

clathrates Included or caged compounds, trapped within a cavity formed by a crystal lattice or present in a large molecule. Generally the properties of the clathrate are those of the enclosing species.

closed system A system limited to exchange of energy with its surroundings; no matter can be transferred.

cloud chamber A device for detecting nuclear particles based on the fact that their passage through a saturated water vapor atmosphere leaves a trail.

cohesion Intermolecular forces between particles of a substance that act to bind them together.

colligative properties Properties of a solution that depend upon the nature of the solvent and the number of solute particles present without concern for the nature of the solute particles themselves.

colloid A chemical system in which one phase is composed of particles whose dimensions range from 1 to 1000 nanometers.

combining capacity The relative amounts of different elements that will combine with one common element such as hydrogen or oxygen.

combining volumes, law of Gay-Lussac For reactions involving ideal gases as reactants and/or products, the ratios of the respective gas volumes can always be expressed as ratios of small whole numbers.

complementarity Where two different and conflicting descriptions of nature can both be made to partially fit the facts or data; for example, the classical and quantum mechanical descriptions of the atom.

Compton effect The scattering observed on collision of photons and electrons. Because collisions are elastic, the electron gains energy and recoils, while the photon loses energy, resulting in a change in wavelength that depends on the scattering angle.

compound Homogeneous chemical combination of two or more different chemical elements in definite proportions.

condensation Transformation from vapor (gaseous) to liquid state; the reverse of vaporization. Sometimes also used for the reverse of **sublimation**.

conductivity A measure of the effectiveness of a substance at transferring heat (**thermal conductivity**) or electricity (**electrical conductivity**).

constant composition, law of See **definite proportions, law of**.

conversion factor The ratio of one set of units to another, which can be used for converting between the two sets of units. The conversion factor has a numerical value of one and so multiplication by it does not change the numerical value of another quantity.

coordination complex (compound) A compound or an ion consisting of a central atom or ion surrounded by **ligands**.

coulomb A unit of electrical charge; a steady flow of one coulomb per second is called one **ampere**.

covalent Bonded by shared pairs of electrons in **covalent bonds**. Contrasted with **ionic**.

critical point The temperature and pressure on the phase diagram of a substance which represents the highest temperature at which the liquid and gaseous states can be distinguished. Above this temperature the substance is known as a **supercritical fluid**.

crystal, crystalline matter Matter consisting of atoms in systematically repeating, three-dimensional units periodically extended through space.

crystal field theory A bonding theory for **coordination complexes** that assumes bonding by purely electrostatic forces in which the **ligands** can be considered to be point charges.

crystal structure The arrangement of atoms in a **crystal**.

cyclotron An accelerator consisting of two **dees** for charged particles which take a circular path.

dee See **cyclotron**.

degenerate Having the same energy.

defect An irregularity in a crystal structure. Can lead to a **nonstoichiometric** compound.

definite proportions, law of All pure chemical compounds have a fixed, definite percent-mass composition, no matter the source of the substance. Also called law of constant composition.

density Ratio of mass to volume for a homogeneous substance.

diffusion The mixing process that takes place among particles in the fluid state because of their random thermal motions.

discreteness The concept that matter and energy consist of definite units (atoms, molecules, photons).

dispersion (of light) Separation of a light wave into its spectral components.

dissociation A usually reversible decomposition (often by heat or solvent interaction) resulting in separation of a given chemical composition into atoms, ions, or radicals.

distillation A method of separating components of a mixture by taking advantage of relative differences in volatility among the various components.

double bond Four pairs of shared electrons between two atoms. One of the two bonds is a **sigma bond** and the other is a **pi bond**.

ductility The ability to be drawn into a wire. A characteristic of metals.

effusion The passage of a gas through a small opening.

elasticity The ability of a material to be deformed and then return to its original dimensions.

electrochemical series A series of elements in order of their ability to displace each other in chemical reactions under **standard conditions**.

electrochemistry The study of reactions that involve flow of electrons along a conductor. The reactions studied in electrochemistry are oxidation-reduction reactions.

electrode The conducting material that draws electrons into and out of an electrolyte solution, a molten medium, a gas, or through a vacuum. Electrodes can be reactive (participate in the electrochemical process) or inert.

electrolysis Decomposition process brought about by passing an electric current through a compound substance in solution or in the molten state.

electrolyte Substance which dissolves in a solvent to produce a solution that conducts electricity, usually by movement of ions; acids, bases, and soluble salts are electrolyte substances in aqueous media.

electron An elementary, subatomic particle with unit negative electrical charge (4.8×10^{-10} esu) and mass $1/1837$ of a proton (9.1×10^{-28} g); electrons constitute the nonnuclear particles within the atom and determine chemical properties.

electron affinity The energy when an electron is added to an atom or an anion.

electronegativity The relative ability of each atom in a molecule to attract bonding electrons to itself.

electron shells The spheres or orbits, located at definite distances from the nucleus, in which electrons are said to be located in fixed numbers; the principal energy levels in the electronic structure model for the atom.

electron spin The intrinsic angular momentum or spin-orientation of an electron; the spin-quantization number, s, is $\pm 1/2$.

electrovalence The valence or combining capacity of an atom involved in ionic bonding; a theory of chemical combination based on electron-transfer.

element One of the 106 presently known, pure chemical substances that cannot be decomposed (divided) into any other pure substance by ordinary chemical processes. A substance which has atoms of all the same atomic number.

elementary particles The growing list of particles of matter and radiation, often referred to as "fundamental" particles, including electrons, protons, neutrons, and neutrinos and particles that do not exist independently under normal conditions, such as mesons, muons, baryons and antiparticles.

empirical formula A symbolic statement of the relative ratios of the elements in a molecular arrangement of atoms expressed as integers; the simplest possible molecular formula in keeping with the percent-by-weight relationships among the combined elements.

emulsion A more or less uniform suspension (dispersion) of two immiscible liquids.

endothermic Absorbing heat from the surroundings. Said of a system, reaction, or process.

energy The ability to do work.

energy conservation The principle that energy is conserved in any chemical process or that the total amount of energy in the universe is constant. See **mass conservation**.

enthalpy (H) A thermodynamic state function of a system ($H = E + PV$) that can be used to measure processes taking place at constant pressure; the change in enthalpy of a system, ΔH, is the heat of a process at constant pressure.

entropy (S) A measure of the disorder of a system; the change in entropy of a system is ΔS.

enzymes Biochemical catalysts; generally complex protein structures, operating in living systems at the cellular level.

equilibrium A system after it has reached a time-independent state with no changes in the composition of the system occurring; characterized by the fact that no spontaneous reaction is taking place.

equilibrium constant The constant value for an expression involving the reactants and products of a chemical process in terms of their concentrations or partial pressures. This expression will have a constant value for a particular chemical process at a particular constant temperature. The equilibrium constant has the symbol K. Symbols of equilibrium constants for specific processes are: K_w, ion product of water; K_a, acid dissociation constant; K_b, base dissociation constant; K_{sp}, solubility product constant; K_f, formation constant.

error The difference between the actual and measured value of a quantity. Usually of two types: systematic and random.

ester The class of organic compounds formed from organic acids and alcohols by elimination of a molecule of water; they are often characteristically pleasant, sweet-smelling substances.

ether Organic molecules in which an oxygen atom is bound to two carbon atoms as in diethylether (CH_3CH_2—O—CH_2CH_3), the classic anesthetic.

eutrophication Decay of organic matter that has accumulated in a body of water due to the presence of excessive amounts of dissolved nutrients, resulting in oxygen-depletion of the water.

evaporation The liquid-to-vapor transformation at temperatures below the boiling point of the liquid.

exclusion principle, Pauli exclusion principle The rule that no two electrons in a given atom can have the same set of quantum numbers.

exothermic Evolving heat to the surroundings. Said of a system, reaction, or process.

extraction A physical method of separation and purification that involves dissolving a substance in a solvent. Liquid-liquid extraction takes advantages of differences in solubilities of one substance between a pair of immiscible solvents.

faraday (F) The electric charge on a mole of electrons, or the charge associated with one gram-equivalent of electrochemical reaction; 96,500 coulombs (amp·sec).

fatty acid An organic acid with a long carbon chain; components of fats.

fermentation The enzyme-catalyzed reactions of certain organic substrates; for example, the hydrolysis of starches and the fermentation of the resulting sugars forming alcohol; in a sense, the biological equivalent of oxidation taking place in the absence of air.

filtration A physical method for separating solids from liquids or gases by allowing the liquid or gas to pass through a semipermeable membrane (for example, filter paper), leaving the solid phase behind.

fission Splitting of a heavy nucleus into nuclei of lighter elements, releasing considerable quantities of energy and usually emitting one or more neutrons; can occur spontaneously, but usually does not.

fluid Aggregated state of matter in which particles (usually molecules) flow past each other as in gases and liquids.

foam A colloidal system in which a gas phase is uniformly dispersed through a liquid phase.

formal charge The charge on an atom in a compound calculated on the assumption that the electrons in each bond are shared equally between the two atoms involved.

formula weight A relative weight calculated from the numbers of atoms given in a formula which is not necessarily the **molecular formula**.

freezing point The temperature at which the liquid and solid forms of a substance are at equilibrium. At a pressure of one atmosphere this is called the **normal freezing point**.

"freons" See **chlorofluorocarbons**.

frequency The number of cycles or repetitions of a periodic process that are completed per unit of time; usually repetitions per second, when dealing with electromagnetic radiations.

friction The resistance to the flow or motion of one object or substance over or through another.

fusion Synthesis of heavier nuclei by fusing lighter nuclei together; process occurs with release of energy.

galvanic cell An electrochemical cell that makes use of a spontaneous chemical reaction to generate a flow of electricity as in a battery.

Gay-Lussac, law of See **combining volumes**.

gamma rays (γ) High-energy and short-wavelength electromagnetic radiation; frequently accompanies alpha- and beta-emission; always accompanies fission.

gas A state of matter with particles moving independently at distances sufficiently great that interactions between them are minimal. Gases are characterized by great compressibility, high rates of diffusion, and the ability to fill their container completely.

Geiger counter A counter for high-energy particles which depends on the ability of these particles to ionize the gas in the counter.

gel An essentially solid or semisolid two-phase colloidal system composed of the solid phase and a liquid phase; the liquid component is generally the principal component of the system.

gene (DNA) The hereditary unit of the chromosome that controls the development of inherited traits and characteristics.

geochemistry The study of the chemistry of the earth and the adjacent portion of the atmosphere.

glass A liquid with viscosity so high that it appears not to flow.

greenhouse effect An atmospheric effect caused by infrared absorbers such as CO_2. The projected result is global warming.

ground state The state of an atom, atomic nucleus, molecule, or ion associated with its lowest energy, that is, with the electrons occupying their lowest energy levels.

half-life Time required for half of a substance to decompose or disintegrate. Measured half-lives of radioactive elements vary from billionths of a second to billions of years.

halogens The VIIA family of nonmetallic elements, including fluorine, chlorine, bromine, and iodine (and astatine); atoms easily gain one electron, forming halide anions; elements are strongly oxidizing.

harmonic motion Periodic motion that can be described as a sinusoidal function of time; a plucked guitar string; a swinging pendulum; the regular motion described by a sine curve.

heat Energy that can be transferred between a system and its surroundings because of an existing temperature differential between the two.

heat capacity The amount of heat needed to raise the temperature of a system one degree, usually measured either at constant volume or constant temperature.

hemoglobin The iron-protein complex found in red blood cells that is responsible for oxygen transport.

heterogeneous Not homogeneous; having a composition which is not constant over the entire sample.

homogeneous Having continuous properties; having the same composition throughout the entire sample.

humidity Moisture content of the air expressed as mass/volume is the **absolute humidity**. The ratio of the partial pressure of water in air to the vapor pressure of water at the given temperature is the **relative humidity**.

hybridization The equalization of energies to explain equivalency of chemical bonds formed by combination of orbitals of apparently different energies; for example, hybridization of s- and p-type orbitals on carbon in methane.

hydrate A compound containing water of hydration. See **hydration**.

hydration The binding of water molecules to a substance usually through hydrogen bonding and in crystalline hydrates, at definite lattice sites. Binding or association of water molecules with ions in solution.

hydrocarbons A general class of organic compounds consisting of only atoms of carbon and hydrogen. Subclasses are alkanes, alkenes, alkynes, aromatics, and alicyclics.

hydrogen bond A weak chemical bond formed by dipole-dipole attraction between a hydrogen atom in one molecule with an electronegative atom in the same or another molecule; the strongest hydrogen bonds form with F, O, Cl, and N atoms.

hydrolysis Decomposition or other chemical transformations brought about by the action of water; for example, hydrolysis of certain salts to form acidic and basic substances.

hydronium ions A hydrated proton of general formula H_3O^+ found in pure water and in all aqueous solutions.

hydrophilic Water-loving; attracted to water.

hydrophobic Water-hating; repelled by water; said of compounds or parts of compounds.

hypothesis A generalization made from observations but based on as yet incomplete evidence; not as certain as a theory.

ideal gas One which behaves according to the ideal gas law, $PV = nRT$; the particles in an ideal gas have no volume and neither attract nor repel each other.

immiscible Liquids not capable of forming a homogeneous solution; insoluble.

inertia The property of all matter in motion or at rest representing the resistance to any alteration in the existing state of affairs.

inhibitor See **catalyst**.

interference phenomena The result of the combination of two or more waves. If the waves are exactly in phase, the interference is said to be **constructive** and the amplitude of the resulting wave will be larger than that of the component waves. If the waves are out of phase, the interference is said to be **destructive** and the amplitude of the resulting wave will be smaller than that of the component waves.

ion An atom or molecule that has lost or gained one or more electrons, becoming electrically charged in the process: the loss of electrons produces a positive ion, or **cation**; the gain of electrons produces a negative ion, or **anion**.

ionic Consisting of ions. Contrasted to **covalent**.

ionization The process of removing or adding electrons from atoms and molecules, creating electrically charged particles called ions in the process. High temperatures, electrical discharges, or nuclear radiation can cause ionization. So can dissolving in water.

ionization energy, ionization potential The energy per unit charge required to remove an electron an infinite distance from an atom or molecule.

isolated system (thermodynamic) A system that cannot exchange matter or energy with its surroundings.

isomers One of two or more substances whose chemical compositions are identical but whose structures, and therefore properties, differ. See also **chirality**.

isomorphism Different compositions with the same crystalline form.

isotherm A curve representing some function at a constant temperature, for example, the relationship between pressure and volume at a constant temperature.

isotope One of two or more forms of an element with the same atomic number (Z) but different mass number (A); $^{12}_{6}C$, $^{13}_{6}C$ and $^{14}_{6}C$ are all isotopes of carbon and have very nearly identical chemical properties.

isotropic Having identical properties in all directions; no distinguishable direction through the material.

ketones Organic molecules characterized by the presence of the carbonyl group (—CO—): acetone, for example

$$\overset{\|}{\underset{O}{}}$$

(CH_3COCH_3).

kinetic energy The energy a body possesses as a result of its being in motion; a particle of mass m moves with speed s and has a kinetic energy equal to $1/2\ ms^2$.

kinetic (-molecular) theory A statistical explanation of the behavior of gases. Assumption: They are composed of very large collections of particles (atoms or molecules) in a state of random, ceaseless motion, involving collisions in which energy and momentum are conserved.

lanthanides A group of 14 metallic elements, cerium through lutitium, that follow lanthanum in the periodic table. Their properties are very similar to those of lanthanum.

lattice A system of identical points called **lattice points** located in a crystal structure. See **unit cell**.

Le Chatelier's principle Changes in a system at equilibrium will result in internal compensation (changes in the position of the equilibrium) in order to partially restore the system to its original stable (equilibrium) state.

ligand An atom, molecule, ion, or other complex group of atoms chemically bound to a central atom, often a metal atom; groups coordinated to central atoms.

linear accelerator An accelerator for charged particles in which the particle path is a straight line.

lipids Biomolecules that are soluble in organic solvents. The most familiar lipids are fats, triglyceride esters of glycerol, and fatty acids.

liquid A condensed form of matter without the regular long-range structure of a crystal. Liquids flow but have low compressibility and do not fill their containers completely.

lone pair A pair of valence electrons on an atom in a molecule not participating in bonding and confined to a single nonbonding orbital.

macroscopic Directly observable properties of a system.

main groups The eight groups of the periodic table in the **s-** and **p-blocks**.

malleability The ability to be pounded into a thin sheet; a characteristic property of metals.

mass The quantity of matter constituting any substantive body; the physical measure of the inertial property of a substance.

mass conservation The principle that the mass of a system is constant in the universe through any chemical transformation and that the total mass of the universe is constant. Mass-energy conservation broadens the interpretation of this law to include energy conservation as well.

mass defect The difference between the mass of a nucleus and the sum of the masses of its component nucleons. This difference is the mass equivalent of the **nuclear binding energy**.

material A combination of **matter**; called a **raw material** if used as a starting point for the production of substances or other materials.

matter Any collection of particles having mass and occupying space.

metalloid An element with properties intermediate between the metals and nonmetals. Also called **semimetal** and **semiconductor**.

metal One of the majority of elements on all but the upper right side of the periodic table, characterized in general by tendency to form cations, conductivity of electricity and heat, luster, ductility, malleability, and solubility in mineral acids.

melting point The temperature at which crystalline solid and liquid phases for pure substances, eutectics, and alloys are in thermodynamic equilibrium.

meniscus The curvature of a liquid surface in a container of narrow diameter due to surface tension; concave if the liquid wets the walls of the container; convex otherwise.

metathesis A substitution or decomposition reaction that does not involve the transfer of electrons with accompanying change in oxidation states.

micelle A highly charged, hydrated colloidal aggregate in which a polar group is oriented toward the aqueous medium and a nonpolar group toward the nonaqueous medium in solutions of electrolytes such as soaps, or in detergent systems.

mixture An aggregate collection of different substances and materials, separable by simple mechanical or physical means.

mole (mol) Avogadro's number of atoms or molecules of a substance. An amount of a substance equal to its atomic or **molecular weight**.

molecular weight The sum of the relative or atomic weights of the component atoms in a molecule; can be calculated directly from the **molecular formula**.

molecule The smallest particle of any chemical combination of atoms that still retains all the properties of the original collections of particles.

momentum For any moving body, the product of its mass and velocity.

monatomic Said of ions consisting of a single atom such as Na^+ or S^{2-}.

multiple proportions, law of The small whole-number relationship that exists between two elements when combined in different ways (i.e., H_2O and H_2O_2).

neutralization Classically, the reaction of equivalent amounts of an acid with a base, producing salt and water. Bronsted definition: an acid-base reaction forming a conjugate acid-base pair.

neutrino Particle of zero charge and near-zero mass.

neutron Neutral elementary particle of mass slightly greater than the proton; found in the nucleus of all atoms heavier than hydrogen. A highly penetrating, interactive form of matter.

nitrogen fixation The formation of nitrogen compounds from free, molecular nitrogen.

nonmetals The elements at the extreme right of the periodic table with properties contrasting to those of the **metals**.

nonstoichiometric Not having atoms in a strict ratio of small whole numbers. Normally encountered in **defect** structures.

nuclear binding energy See **mass defect**.

nuclear magnetic resonance spectroscopy (nmr) A spectroscopic technique based on a property of many nuclei that, when placed in a magnetic field, absorb characteristic energies from a radio frequency field superimposed upon them.

nucleic acids Polymeric strands composed of nucleotide units in which phosphoric acid is combined with a sugar (ribose, for example) and a nitrogen base (purine or pyrimidine derivatives); prominent cellular components, functioning in information storage and transfer. Ribonucleic acid (RNA) and deoxyribonucleic acid (DNA).

nucleon One of the components of the nucleus, particularly **protons** and **neutrons**.

nuclide A specific atomic nucleus.

octet rule The classical notion (from the Lewis-Kossel theory) that atoms tend toward stable bonds by electron-sharing or electron-transfer until the eight-electron, noble gas configuration is achieved.

open system (thermodynamic) A system that can freely transfer matter and energy with its surroundings.

orbital As close to an electronic orbit as you can come in quantum mechanics; a one-electron wave function expressing the probability of finding the electron in a given region in space.

osmosis Dilution of a solute by passage of solvent molecules through a semipermeable membrane due to osmotic pressure; a colligative property.

oxidation Chemical reactions involving removal of electrons, resulting in a higher **oxidation state**; **reduction**, or gain in electrons by another species must accompany the oxidation process.

oxidation state The charge on an atom in a molecule calculated on the assumption that bonding electrons between atoms are assigned to the more electronegative atom. Calculation is normally by a set of rules which assumes certain oxidation states for certain atoms.

oxo anions Anions containing one or more oxygen atoms around a central atom, for example, SO_4^{2-} and ClO^-.

oxyacids The acids of **oxoanions**, for example, H_2SO_4 and $HClO$.

ozone layer Layer in upper atmosphere containing a high ozone concentration which serves as a filter for ultraviolet radiation. Ozone is currently threatened by atmospheric pollutants, particularly chlorofluorocarbons.

Pauli exclusion principle See **exclusion principle**.

partial pressure The partial pressure p_a of component a in a mixture of gaseous components (real or ideal) can be stated as the product of the total pressure (P_T) and the mole fraction of $a(X_a)$: $p_a = X_a P_T$.

path function A quantity such as work or heat for which the value depends on the path taken between the initial and final states. See **state function**.

p-block The six columns of the periodic table farthest to the right; characterized by filling of p orbitals.

peptization Transformation of a colloidal gel into a sol.

percent yield The percentage of the stoichiometrically possible yield in a reaction that is actually obtained. % yield = 100%·actual yield/theoretical yield.

periodic table A systematic arrangement for the elements according to increasing atomic number and periodic recurrence of chemical properties that suggests family classification of the elements in vertical columns.

permeability The ability to allow passage of a substance through a membrane under given conditions.

pH The negative logarithm (base 10) of the hydrogen ion concentration; a measure of relative acidity in aqueous solutions: above 7, basic; below 7, acidic.

phase Visibly distinct portion of a heterogeneous system; one that is separated by a clearly defined surface.

photoelectric effect Liberation of electrons by a substance (usually a metal) on irradiation by an incident beam of electromagnetic radiation, usually light of short wavelength (UV, for example); the speed of ejected electrons is proportional to the frequency of radiation. Einstein photoelectric law: $E = hv - w$.

photon The carrier of the quantum of electromagnetic energy; photons have momentum but not mass or electrical properties.

photosynthesis The process whereby green plants effect conversion of carbon dioxide and water to carbohydrates and oxygen in the presence of sunlight.

physical methods Methods of separating, purifying, and otherwise processing a substance which do not change the identity of the substance.

physical properties Properties of a substance that are exhibited without any change in the identity of the substance, for example, melting point or viscosity.

pi-bond A bond formed by lateral (rather than linear) combination of p or d atomic orbitals; the second bond in a double bond, for example, in ethylene; the second and third bonds in a triple bond, for example, in acetylene and nitrogen.

Planck's constant The proportionality constant h in the Planck relationship between energy and the frequency (or wavelength) of radiation. See also **quantum hypothesis**.

plasticity The ability of a material to be deformed permanently without returning to its original dimensions.

polar molecule A molecule that has a permanent electrical dipole moment due to unequal distribution of electric charge between bonded atoms within the molecule.

polyatomic Having more than one atom present; said of ions.

polymer A compound formed of giant molecules generally constructed from one or two monomer units that have been chemically combined in polymerization processes.

polymorphism Same composition exhibiting two or more different crystalline forms.

positron The antiparticle of the electron having the same mass as the electron but a positive charge.

precipitate A solid that separates from a solution because the existing properties at that moment make it insoluble; often due to temperature change, loss of solvent through evaporation, or addition of a reactive reagent.

precision The degree to which measured observation can be judged reliable; the reproducibility of results.

pressure Uniform stress (in all directions) produced by a gas, measured as a force per unit area; the pressure exerted by a gas in a container results from collisions of the gas molecules with the container walls; **atmospheric pressure** is due to the weight of the atmosphere on the surface of the Earth.

proteins A biologically important class of complex, long-chained organic nitrogen compounds of high molecular weight made up of many alpha-amino acids linked through peptide (—CONH—) bonds.

proton Positively charged elementary particle in the nucleus of all atoms; the nucleus of the ^1H atom; carries a charge equal to the charge on the electron but of opposite sign. Atomic number (Z) of an atom equals the number of protons in its nucleus.

quantum hypothesis The principle that emission or absorption of radiant energy in atomic events can assume only discrete values; the units of energy of quanta are equal to the product of the frequency of the radiation and h, the Planck constant (6.626×10^{-27} erg·sec).

radical A classical name for a group of atoms (charged or uncharged) that enter into chemical combination as a unit: NH_4^+; CH_3; SO_4^{2-} (ammonium, methyl, and sulfate, respectively).

radioactive decay (disintegration) Spontaneous transformation of one nuclide into another or into a different energy state of the same nuclide by emission of alpha- or beta-particles, or gamma rays. The decay process follows a first-order rate law.

radioactivity Spontaneous disintegration or decay of an unstable atomic nucleus, usually accompanied by ionizing radiation.

rate law A mathematical expression of the rate of a reaction (shown as the rate of change of the concentration of a reactant or product) as a function of the concentration of one or more reactants and products.

recrystallization A physical method for separating and purifying substances which involves dissolving the substance and then allowing it to form crystals again slowly.

reduction Chemical reaction accompanied a gain in electrons by one or more participating reagents. See also **oxidation**.

resonance A stabilizing quality of certain molecules that can be represented by considering the electron distribution in an ion or molecule as a composite of two or more forms, in those cases where a single form is an inadequate representation; for example, benzene and the carbonate ion.

reversible Said of a process for which the direction can be changed by an infinitesimal change of an outside variable. In general, equilibrium constitutes a reversible situation.

Rydberg equation The empirical spectroscopist's equation which gave the frequencies for the lines in the spectrum for hydrogen and from which Bohr was able to derive considerable strength and support for his theory.

salt The ionic product of the reaction of an acid-base neutralization reaction, forming a salt and water; also produced by the displacement of acidic hydrogen atoms by metallic ions.

saponification A reaction that yields a soap. The reaction of an ester with a strong base in water yielding an alcohol and the salt of an organic acid.

saturation (1) The dissolution of the maximum quantity of solute in a particular solvent at a given temperature; results in a saturated solution under equilibrium conditions. (2) Having the full number of hydogen atoms possible; said of hydrocarbons and hydrocarbon radicals. Having no carbon-carbon double bonds.

s-block The first two groups of the periodic table, IA and IIA.

semiconductor Certain crystalline substances which have electrical conductivities between metals (conductors) and insulators (nonconductors). See **metalloid**.

semimetal See **metalloid**.

sigma bond Chemical bond formation involving linear combination of s- and p-type atomic orbitals between adjacent atoms. Such bonds have their bonding electron density concentrated directly between the bonding atoms and have molecular orbitals that do not change sign when crossing the internuclear axis.

significant figures The digits in an accurate measurement, typically including the first doubtful digit as a reasonable estimate.

single bond A single pair of shared electrons forming a covalent bond between two atoms. This is normally a **sigma bond**.

soap The salt of a fatty acid. It has **hydrophilic** and **hydrophobic** ends and forms **micelles** in water.

sol A colloidal solution of a gas, liquid, or solid in a suitable dispersion medium; usually applies to liquid systems; if the continuous phase is water, the term *hydrosol* is used.

solid An aggregated state of matter in which the substance exhibits definite volume and shape and is resistant to distorting forces that might alter its volume and shape. Most solids have regular crystalline structures.

solubility The property of substances that allows them to form homogeneous mixtures called **solutions**.

solubility product An equilibrium constant which is the product of the ion concentration in a saturated solution; it defines the degree of solubility of a given substance in a specific solvent at a specific temperature.

solution Mixture of substances forming a homogeneous phase.

specific gravity The ratio of the **density** of a substance or material to that of some standard substance. Water is the common standard for liquids and solids.

specific heat The quantity of heat required per unit mass per degree to change the temperature of a substance without causing a phase change; heat capacity per gram.

spectral lines Single wavelengths of electromagnetic radiations arising from electronic transitions (emissions and absorptions) in atoms.

spectrochemical series A series of **ligands** which occur in **coordination complexes** in order of the extent to which they split the energy of otherwise **degenerate** valence orbitals on the central atom.

spontaneous Occurring in a nonreversible manner. See **reversible**.

standard conditions A set of prescribed conditions for chemical processes, particularly all concentrations of reactants and products at 1 M and all partial pressures of reactants and products at 1 atm.

standard potential The potential in volts for a cell or a half cell under **standard conditions**. **Standard half cell potentials** are based on the assignment of 0.0000 V for the hydrogen half cell.

state function A quantity such as entropy or enthalpy for which the value depends only on the initial and final states of the system and is independent of the path taken. See **path function**.

steroid A lipid with a characteristic four-ring structure. Includes important hormones such as progesterone and estradiol.

stoichiometry That aspect of chemical science dealing with proportions of reactants and products involved in chemical transformations.

STP "Standard" temperature and pressure, defined as 273.15 K and 1 atm.

strain The change in a dimension divided by the original dimension for a sample put under **stress**.

stress The load put on a sample per unit cross section of the sample.

sublimation Direct transformation from solid to vapor without passing through the liquid state.

substance A homogeneous form of matter of definite composition; a pure chemical compound or element.

surface tension A tension at the surface of a liquid resulting from forces of attraction acting normally to the surface on molecules close by; this causes the liquid surface to behave as a tense membrane.

supercritical fluid See **critical point**.

system (thermodynamic) That sample or limited segment of the universe under investigation.

temperature A property of systems that establishes thermodynamic equilibrium: Two systems are in equilibrium with each other when they are at the same temperature. Temperature is proportional to the average kinetic energy for an ideal gas.

temperature inversion An atmospheric condition characterized by increasing temperature with altitude; little interchange and mixing of materials results in buildup of atmospheric pollutants.

theoretical yield The highest yield allowed by the stoichiometry of a reaction.

theory An underlying principle governing natural phenomena; experimentally defined doctrine that has been established by confirmation of predictions based on the theory.

thermochemistry The study of the heat exchanges that accompany chemical reactions and phase transformations.

thermodynamics The branch of physical science dealing with the transfer of heat and energy interconversions. The laws of thermodynamics are based solely on experience.

titration A technique for analysis of the composition of a solution by careful addition of a reagent solution of accurately known concentration (standard solution) until the indicator present confirms the end-point.

transition elements (transition metals) The block of metallic elements in the middle of the periodic table characterized by the progressive buildup of d electrons.

triple bond A covalent bond between two atoms consisting of three pairs of shared electrons. Of the three bonds, one is a **sigma bond** and two are **pi bonds**.

triple point The temperature and pressure at which all three phases of a pure substance can exist at equilibrium.

uncertainty principle The principle that the position and the velocity of a particle can be identified but not simultaneously. The same problem arises in the simultaneous determination of energy and time.

unit cell Basic repeating volume unit in a sample of crystalline matter which can be used to generate the entire structure by translation along the coordinate axes. The unit cell can be constructed by connecting specific **lattice points**.

units The actual quantity being measured, as kilograms when determining mass.

universal gas constant The proportionality constant, R, in the **ideal gas law**, $PV = nRT$. The value of R depends on its units.

valence A number representing the general combining capacity of an element in its chemical combination or the proportions in which atoms combine relative to hydrogen taken as 1.

valence electrons The electrons in an atom that are involved in chemical reactions. These are normally those outside the next lowest noble gas configuration.

valence shell electron pair repulsion (VSEPR) theory A system for predicting the geometries of covalent compounds based on the assumption that the principal determinent of structure is repulsions between bonded and nonbonded pairs of valence electrons.

van der Waals forces Attractive forces between atoms or among molecules which arise because of the presence of a dipole moment; in nonpolar molecules, weak intermolecular forces other than the usual valence forces.

vapor pressure The equilibrium pressure due to the vapor of a solid or liquid substance at a given temperature.

viscosity The internal resistance of fluids to flow.

volume The space occupied by a body of matter at any given temperature and pressure.

volt An electrical potential of one joule per coulomb.

wavelength The distance between two points that are said to be exactly in phase in consecutive cycles of a periodic wave, along the direction of propagation of the wave; usually measured in nanometers or angstroms when dealing with electromagnetic radiations in chemistry.

wax An ester of a fatty acid and an alcohol.

weight The force exerted on an object as a result of its mass and the prevailing acceleration of gravity. If two objects are subject to the same gravitational acceleration, their weights are proportional to their masses. Therefore, weight is commonly, if loosely, used as a synonym for **mass**.

work The classical definition is the product of a force and the distance through which the force operates.

X-rays Penetrating form of electromagnetic radiation with high energy and short wavelength emitted either when inner orbital electrons of an excited atom fall back to their normal state or when a metal target is bombarded with high-speed electrons. X-rays are nonnuclear in origin.

yield See **percent yield** and **theoretical yield**.

PHOTO CREDITS

CHAPTER 1

Chapter opening photo p. 2: Perkin-Elmer Corporation; Chapter opening photo p. 3, Fig. 1–5: Columbia University; Figs. 1–1a, 1–6, 1–7, 1–10a,b: General Electric Company; Figs. 1–1b, 1–8a,b, 1–9a,b: AT & T Bell Laboratories; Fig. 1–2a,b: Du Pont Company; Fig. 1–3, unnum. fig. p. 7: Burndy Library; Fig. 1–4: Deutcher Museum; Fig. 1–11: Adapted from an original drawing by Waisum Mok; Fig. 1–12: American Institute of Physics, Niels Bohr Library.

CHAPTER 2

Chapter opening photo p. 34, Fig. 2–4: General Electric Company; Chapter opening photo p. 35: AT & T Bell Laboratories; Figs. 2–1, 2–3a,b,c, 2–6a, 2–7c: Charles Steele; Fig. 2–2a: Royal Institution, London; Fig. 2–2b: Burndy Library; Fig. 2–6b: Richard Roese; Fig. 2–7a: Hal Levin; Figs. 2–7b, 2–8: Charles Winters.

CHAPTER 3

Chapter opening photo p. 68: University of Rochester, Laboratory for Laser Energetics; Chapter opening photo p. 69: Chicago Historical Society; Fig. 3–1: C. Winters; Fig. 3–2a: Science Museum (London); Fig. 3–3: Burndy Library; Fig. 3–5: Columbia University; Fig. 3–6: LWF Photos; Fig. 3–10: Cheodoke-McMaster Hospitals, Hamilton, Ontario; Fig. 3–13: Fermi Lab; Fig. 3–14b: Argonne National Laboratory; unnum. fig. p. 92: U.S. Army, White Sands Missile Range; unnum. fig. p. 94: Argonne National Laboratories; unnum. fig. p. 97: U.S. Council for Energy Awareness; Fig. 3–21b: Princeton Plasma Physics Laboratory, DOE; Fig. 3–22c: Lawrence Livermore National Laboratories.

CHAPTER 4

Chapter opening photo p. 108, unnum. figs. pp. 116, 117 (top), 119, Fig. 4–7a,b: AT & T Bell Laboratories; Chapter opening photo p. 109, unnum. figs. pp. 139, 144: LWF Photos; Fig. 4–1: NASA; Fig. 4–4: Burndy Library; unnum. fig. p. 117 (bottom): General Electric Company; Fig. 4–5a, unnum. figs. p. 130: C. Steele; Fig. 4–5b, unnum. figs. pp. 124, 128, Fig. 4–8, unnum. fig. p. 142a,b,c: C. Winters; Fig. 4–5c: Comstock; Fig. 4–6: Marna G. Clarke.

CHAPTER 5

Chapter opening photo p. 150: Dan Francis/Mardan Photography; Chapter opening photo p. 151: International Museum of Photography at George Eastman House; Fig. 5–1: Perkin-Elmer Corporation; Fig. 5–12: Douglas Faulkner/Photoresearchers; Fig. 5–15: Denver Bryan/Comstock.

CHAPTER 6

Chapter opening photo p. 194: NASA; Chapter opening photo p. 195: AT & T Bell Laboratories; Figs. 6–1a, 6–10, unnum. fig. p. 221: Columbia University; Fig. 6–3, unnum. fig. p. 217: NASA; Figs. 6–4, 6–6, 6–9, 6–13, unnum. fig. p. 216 (bottom): C. Winters; Fig. 6–5: H.C. Metcalfe, J.E. Williams, J.F. Castka; Fig. 6–7: Union Carbide; Fig. 6–8: Metcalfe, Williams, Castka & Walter O. Scott; unnum. fig. p. 211: C. Steele; Fig. 6–11: Royal Institution, London; Fig. 6–12: Marna G. Clarke; Fig. 6–14: U.S. Geological Survey; unnum. fig. p. 223: Argonne National Laboratories.

CHAPTER 7

Chapter opening photo p. 228, Fig. 7–1d: General Electric Company; Chapter opening photo p. 229: Columbia University, Chandler Collection; Figs. 7–6d, 7–14: LWF Photos; Fig. 7–7b: Andrew Davidhazy; Fig. 7–13: C. Steele.

CHAPTER 8

Chapter opening photo p. 266, Fig. 8–11b: Peter Aprahamian/Science Photo Library/Photoresearchers; Chapter opening photo p. 267: Akzochemie; unnum. fig. p. 271, Fig. 8–20: Marna G. Clarke; Figs. 8–6, 8–16a,b, unnum. figs. pp. 274, 287: C. Winters; Fig. 8–11a: C. Steele; Fig. 8–11c: Andrew Davidhazy.

CHAPTER 9

Chapter opening photo p. 296, Figs. 9–21b, 9–22: Dresser Industries; Chapter opening photo p. 297, unnum. fig. p. 304: Columbia University, Chandler Collection; Figs. 9–2, 9–5: C. Steele; Figs. 9–3, 9–6c, unnum. figs. pp. 303, 314 (bottom), Figs. 9–26, 9–28: C. Winters; Fig. 9–4, 9–7: Marna G. Clarke; unnum. fig. p. 311: Union Carbide Corporation; Fig. 9–15: Dan McCoy, Black Star.

CHAPTER 10

Chapter opening photo p. 336: Dresser Industries/Kellogg Company; Chapter opening photo p. 337: BASF; Figs. 10–1, 10–4: Marna G. Clarke; Fig. 10–3: C. Steele; Figs. 10–2, 10–10b: C. Winters.

CHAPTER 11

Chapter opening photo p. 370: Andrew Davidhazy; Chapter opening photo p. 371: General Electric Company; Fig. 11–1, unnum. fig. p. 374: C. Winters; Fig. 11–2: *New Yorker Magazine*; unnum. fig. p. 386: Gallerie der Stadt Stuttgart; unnum. fig. p. 398: Dresser Industries/Kellogg Company.

CHAPTER 12

Chapter opening photo p. 406: Brinkmann Instruments; Chapter opening photo p. 407: National Gallery, Berlin; Figs. 12–3, 12–4, unnum. fig. p. 418, Fig. 12–7: C. Winters; unnum. fig. p. 415: P.P. Berlow, D.J. Burton, J.I. Routh; unnum. fig. p. 420: Grant Heilman; unnum. fig. p. 421: C. Steele; Fig. 12–8: U.S. Environmental Protection Agency; Fig. 12–9: American Association for Advancement of Science.

CHAPTER 13

Chapter opening photo p. 440, Fig. 13–12: Brinkmann Instruments; Chapter opening photo p. 441: National Technical Laboratories; Figs. 13–1, 13–2: C. Winters; Fig. 13–9: C. Steele.

CHAPTER 14

Chapter opening photo p. 472: AT & T Bell Laboratories; Chapter opening photo p. 473: Liam Roberts; unnum. fig. p. 476: Perkin-Elmer Corporation; unnum. fig. p. 480: Neils Bohr Library; unnum. fig. p. 485: Deutsches Museum.

CHAPTER 15

Chapter opening photo p. 502, p. 507, Figs. 15–22, 15–23, 15–24: AT & T Bell Laboratories; Chapter opening photo p. 503: Columbia University, Chandler Collection; Figs. 15–1, 15–6, 15–9: American Institute of Physics, Neils Bohr Library; Fig. 15–4a,b: Burndy Library; unnum. fig. p. 525: General Electric Company.

CHAPTER 16

Chapter opening photo p. 542, unnum. fig. p. 561a,b: Perkin-Elmer Corporation; Chapter opening photo p. 543: Burndy Library; Fig. 16–2: C. Steele; Fig. 16–4: Ciba-Geigy.

CHAPTER 17

Chapter opening photo p. 570: Perkin-Elmer Corporation; Chapter opening photo p. 571: Columbia University, Chandler Collection; unnum. fig. p. 578, Fig. 17–10: C. Steele; Figs. 17–4a,b, 17–8a,b,c, 17–9, 17–14, 17–15b: AT & T Bell Laboratories; unnum. figs. pp. 579 (middle), 585, Figs. 17–15a, 17–19a,b,c, 17–20, 17–21, 17–22, 17–23: C. Winters; unnum.

fig. p. 579 (bottom): Kurt Nassau/AT & T Bell Laboratories; Fig. 17–6: Marna G. Clarke; Fig. 17–12: Aluminum Association of America; Fig. 17–13a,b: General Electric Company.

CHAPTER 18

Chapter opening photo p. 604: Systat, Inc.; Chapter opening photo p. 605: Columbia University, Chandler Collection; Figs. 18–5b, 18–6b, 18–9a,b,c; W.C. Still; Figs. 18–16b, 18–17b: Streetwieser and Owens, *Orbital and Electron Density Diagrams*, MacMillan Publishing Company, 1973; Fig. 18–24: Masterson, Slowinski, and Stanitski, *Chemical Principles*, 6th ed., Saunders College Publishing, 1986.

CHAPTER 19

Chapter opening photo p. 638, Fig. 19–25a,b: Charles Lieber/ Columbia University; Chapter opening photo p. 639, Figs. 19–1a,b,c, 19–14a,b,c, 19–16, 19–20c, 19–23, 19–30: AT & T Bell Laboratories; Fig. 19–19b: Columbia University; Fig. 19–31: Bruce Matheson; unnum. fig. p. 659, Fig. 19–26a,b: General Electric Company; unnum. fig. p. 663: Andrew Davidhazy.

CHAPTER 20

Chapter opening photo p. 670, Chapter opening photo p. 671, Figs. 20–1, 20–4a,b,c, unnum. fig. p. 676, Figs. 20–6a,b, 20–7: General Electric Plastics; Figs. 20–2, 20–3, 20–24b, 20–30c: General Electric Company; unnum. fig. p. 675 (top): Baseball Hall of Fame Library, Cooperstown, New York; unnum. fig. p. 675 (bottom), Figs. 20–30a,b, 20–31: C. Winters; Figs. 20–5, 20–12, 20–17a,b, unnum. fig. p. 692, Fig. 20–24a, unnum. fig. p. 701, Fig. 20–29: AT & T Bell Laboratories; Fig. 20–10a,b,c: Andrew Davidhazy; unnum. fig. p. 696: Particulate Minerology Unit, Avondale Research Center, U.S. Bureau of Mines; unnum. fig. p. 697: Prof. Carl Rettenmeyer, Museum of Natural History, University of Connecticut; unnum. fig. p. 698: Akzochemie; unnum. fig. p. 699: Ciba-Geigy.

CHAPTER 21

Chapter opening photos pp. 706, 707: Occidental Petroleum; unnum. fig. p. 709a,b,c, Figs. 21–3, 21–11, 21–16: C. Winters; Fig. 21–1, unnum. fig. p. 724: Marna G. Clarke; unnum. fig. p. 732 (bottom): Duracell Corporation; unnum. fig. p. 734a,b, unnum. fig. p. 744 (bottom), Fig. 21–17: General Electric Company; Fig. 21–9b: United Technologies; Fig. 21–10b: Beckman Industries.

CHAPTER 22

Chapter opening photo p. 756, unnum. fig. p. 765: General Electric Company; Chapter opening photo p. 757: BASF; Figs. 22–1a, 22–2, unnum. fig. p. 763a,b, Fig. 22–6: Marna G. Clarke; Fig. 22–1b, unnum. fig. p. 759, Figs. 22–14, 22–17a,b, 22–19, 22–20a: C. Winters; unnum. fig. p. 768: AT & T Bell Laboratories; Fig. 22–10: *Masterson, Slowinski, Stanitski*; unnum. fig. p. 785: J. Morganthaler; Fig. 22–20b: C. Steele.

CHAPTER 23

Chapter opening photo p. 792: Jean Lorre, Donald Lynne, and the Shroud of Turin Project; Chapter opening photo p. 793: Burndy Library; unnum. figs. pp. 796, 798a,b: C. Winters; Fig. 23–18: Nikolaas J. van der Merive, *American Scientist*, vol. 70, 1982, pp. 596–606.

CHAPTER 24

Chapter opening photo p. 830: Sandia National Laboratories; Chapter opening photo p. 831, Fig. 24–9: Nobel Foundation and the Library of the Royal Swedish Academy of Sciences; Fig. 24–2: David M. Phillips/Visuals Unlimited; Fig. 24–4: American Institute of Physics, Niels Bohr Library; Figs. 24–11, 24–13: C. Winters; Fig. 24–12: Thomas Eisner; unnum. fig. p. 851: Akzochemie.

CHAPTER 25

Chapter opening photo p. 858, Figs. 25–1, 25–4, 25–8, 25–12, 25–15: Du Pont/Conoco; Chapter opening photo p. 859, unnum. fig. p. 905 (Carothers): Du Pont, Hagley Library;

Fig. 25–2, unnum. fig. p. 873 (Bunsen): Columbia University, Chandler Collection; unnum. fig. p. 865, Fig. 25–3, unnum. figs. pp. 866, 867, 873 (Bunsen burner), 876, 880, Fig. 25–11, unnum. figs. pp. 883, 884, 887, 888, 900, Figs. 25–13, 25–14: C. Winter; Fig. 25–7: General Electric Company.

CHAPTER 26

Chapter opening photo p. 912: Merck and Company; Chapter opening photo p. 913: Nobel Foundation and the Library of the Royal Swedish Academy of Sciences; Fig. 26–3: David N. Harpp, McGill University; Fig. 26–12: David M. Phillips/Visuals Limited; Fig. 26–14: C. Winters; unnum. fig. p. 934: C. Steele; Fig. 26–26b: Drs. P.A. Dieppe, P.A. Bacon, A.N. Banji, I. Watt and Gower Medical Publishing Co., Ltd., London; Fig. 26–27: Burndy Library; Fig. 26–28: LWF Photos; Fig. 26–29b: General Electric Company.

Part One opening photo p. 1 and Part Two opening photo p. 471: Nobel Foundation and the Library of the Royal Swedish Academy of Sciences.